Fundamentals of Neutrino Physics and Astrophysics

Fundamentals of Neutrino Physics and Astrophysics

Carlo Giunti

Istituto Nazionale di Fisica Nucleare, Sezione di Torino and Dipartimento di Fisica Teorica, Universita di Torino, Italy

Chung W. Kim

Korea Institute for Advanced Study, Seoul, Korea and The Johns Hopkins University, Baltimore, MD, USA

OXFORD
UNIVERSITY PRESS

OXFORD
UNIVERSITY PRESS

Great Clarendon Street, Oxford OX2 6DP

Oxford University Press is a department of the University of Oxford.
It furthers the University's objective of excellence in research, scholarship,
and education by publishing worldwide in

Oxford New York

Auckland Cape Town Dar es Salaam Hong Kong Karachi
Kuala Lumpur Madrid Melbourne Mexico City Nairobi
New Delhi Shanghai Taipei Toronto

With offices in

Argentina Austria Brazil Chile Czech Republic France Greece
Guatemala Hungary Italy Japan Poland Portugal Singapore
South Korea Switzerland Thailand Turkey Ukraine Vietnam

Oxford is a registered trade mark of Oxford University Press
in the UK and in certain other countries

Published in the United States
by Oxford University Press Inc., New York

First published 2007

Reprinted 2007, 2011, 2012, 2014, 2015, 2016, 2017, 2018 (twice)

British Library Cataloguing in Publication Data
Data available

Library of Congress Cataloging in Publication Data
Data available

Printed and bound by
CPI Group (UK) Ltd, Croydon, CR0 4YY

ISBN 978–0–19–850871–7 (Hbk)

10

Dedicated to the late Young J. Kim, beloved wife of C.W.K. and dear friend of C.G., without whose unfailing encouragement and total sacrifice, this book would have never seen the light of day.

PREFACE

> An expert is a person who avoids the small errors while sweeping on to the grand fallacy.
> **Steven Weinberg**

Studying the properties and interactions of neutrinos has been one of the most exciting and vigorous activities in particle physics and astrophysics ever since Pauli first proposed their existence in 1930. In spite of their weakly interacting (or barely existing, as Lederman used to say) nature, we have so far accumulated an enormous amount of knowledge about neutrinos. From neutrino oscillation experiments, we learned a few years ago that neutrinos are massive and mixed. However, we still do not know the absolute values of their masses and some aspects of the mixing.

In this book we have tried to gather all the basic knowledge and tools that are necessary to understand and to infer the true nature of the neutrinos from the experimental data, using the theories that have been developed. We have also summarized the well-established facts concerning neutrinos and the important role played by neutrinos in the Sun and supernovae, and in shaping the Universe we live in. Special emphases are placed on the basic knowledge of how neutrinos interact, how they behave in matter as well as in vacuum, and on the formal aspects of the theory of neutrino oscillations. Salient features of the oscillation experiments in various settings and sources are given with careful analysis.

After a short history of the neutrino, leading up to the Standard Model (SM) of electroweak interactions and the discovery of neutrino oscillations in chapter 1, chapter 2 is devoted to a detailed discussion of the properties of spin 1/2 Dirac particles. This chapter can be used as an introduction to the quantum field theory of spin 1/2 fermions. We have tried to make chapter 2 as complete and self-contained as possible, especially for beginners, by including some details on the terminology used in quantum field theory and in gauge theory. The symmetry properties of spin 1/2 particles under charge conjugation, C, parity, P, and time reversal, T, as well as the space-time and Lorentz transformations are discussed. The realistic wave packet description of a particle is also presented.

In chapter 3, the ingredients necessary for understanding the Standard Model, such as gauge symmetry and the Higgs mechanism, are presented. The electroweak behavior of quarks and leptons is explained and summarized, together with some discussions of the gauge bosons involved.

One of the most remarkable discoveries in the past decade is the finding that neutrinos are massive and mixed. Chapter 4 is devoted to a detailed discussion of the three-generation mixing of quarks, which can be extended to the treatment of mixing of three Dirac neutrinos in a straightforward way (the mixing of three Majorana neutrinos implies the existence of two additional phases in the mixing matrix, as explained in chapter 6). A construction of the mixing matrix is presented with mathematical details. The possibility of CP violation due to mixing and how to describe and quantify it are explained.

Chapter 5 is devoted to neutrino interactions, with discussions of several important processes involving neutrinos, ranging from neutron decay to charged-current and neutral-current deep inelastic neutrino–nucleon scattering. All the discussions are based on the results obtained in chapter 3, where the SM was developed.

We present, in chapter 6, the description of the general, model-independent, theory of massive neutrinos. Depending on whether they are of Dirac or Majorana type, the mathematical description takes different forms. In addition, we discuss the famous see-saw mechanism, which explains the smallness of the neutrino masses in a natural way. Because of their speculative nature, we have intentionally omitted details concerning the possible neutrino mass-generating mechanisms that have been put forward by many authors.

Equipped with the mathematical and basic physics background in the previous chapters, we present, in chapter 7, the standard derivation of the neutrino oscillation probability. We have tried to make this chapter as complete as possible in the hope that the reader can find all the necessary information on the derivation of the standard oscillation formulas, together with how to use them in analyzing experimental data. The types of experiments are as numerous as they can be: experiments with energetic and low energy neutrinos, with reactor or accelerator neutrinos, solar, atmospheric, and even extragalactic (SN 1987A) neutrinos, and finally with short-baseline and long-baseline arrangements. We have included helpful discussions on all the possible cases. In addition, some consequences of C, CP, T and CPT violation for the oscillations, and special cases of oscillations with different oscillation parameters are discussed.

In chapter 8, in view of the important role that the oscillation plays in probing the neutrino properties, a derivation of the neutrino oscillation formulas is given with an emphasis on the more realistic relativistic wave packet treatment. Also presented in this chapter are answers to often raised questions concerning subtle and confusing issues in the oscillation formulas.

One of the most interesting findings in neutrino physics is the discovery that the properties of neutrinos change when they pass through dense matter. Due to the difference of the weak interaction potentials that different flavor neutrinos feel in matter, the effective mixing can be dramatically enhanced in the case of a very small mixing in vacuum. If the neutrinos pass through a region in which the effective mixing is maximal, it is possible to have large flavor transitions. In chapter 9, we explain this resonant effect, known as the MSW effect. We derive the necessary formulas and explained the methods for analyzing the data in the simplest case of two-neutrino mixing (three-neutrino mixing is discussed in chapter 13). Also presented is a geometrical description of the oscillations in vacuum as well as in matter, which, we hope, will help the reader to understand better this important phenomenon. The geometrical description shows that the neutrino oscillation is analogous to a classical magnetic moment precessing in an external magnetic field.

By far, the longest running neutrino experiment has been the Homestake solar neutrino experiment, started in the late 1960s by Ray Davis and collaborators after John Bahcall predicted a measurable rate on the basis of a solar model. In the early

1970s, Davis and collaborators discovered the so-called *solar neutrino problem*, which has remained unsolved until recent times. It consists of a solar neutrino detection rate which is substantially smaller than that predicted by standard solar models. In chapter 10, we start with a brief explanation of the standard solar models, which explain how the Sun shines and how neutrinos are produced. Then, we list and describe all the experiments that have so far been performed, as well as their results and their roles in the solar neutrino problem. We end this chapter by presenting the results of a global fit of all the solar neutrino data. The results of the combined analysis of the data of solar neutrino experiments and of the KamLAND very long-baseline reactor neutrino experiment is presented later in chapter 12.

Chapter 11 explains the generating mechanism and the flux of atmospheric neutrinos. A comprehensive survey of all the atmospheric neutrino experiments performed so far is presented. It is interesting to note that the first undisputed discovery of neutrino oscillations was made with atmospheric neutrinos which have been initially considered as unwanted background for other measurements. The confirmation of this discovery by the accelerator long-baseline K2K experiment is discussed later in chapter 12.

In chapter 12 we start with the introduction of the sensitivity of terrestrial reactor and accelerator neutrino experiments to the measurement of the oscillation parameters. Then, we present the main results of reactor and accelerator short-baseline, long-baseline and very long-baseline experiments, with specific reviews of the CHOOZ, Palo Verde, KamLAND, and K2K experiments, which are important for our present knowledge of the neutrino oscillation parameters. This chapter ends with a brief discussion of the future off-axis long-baseline experiments.

Deciphering the values of the squared-mass differences and the mixing angles from the data requires a careful analysis with use of the three-generation mixing matrix, which is often quite complex. Technical issues involved in the analysis and some useful approximations are discussed in chapter 13. Also discussed in this chapter are the results of a global analysis of all the existing oscillation data. The chapter ends with some comments on the absolute scale of neutrino masses.

The most important attempts to measure directly the values of the neutrino masses are the measurements of the end-point of the electron spectrum in nuclear β-decays, in particular that of tritium, and neutrinoless double-β-decays. In chapter 14, we review the current experimental upper bounds on the neutrino masses obtained with tritium β-decay experiments with the effects of neutrino mixings taken into account. Also given is a brief summary of neutrinoless double-β-decay theory and experiments. We emphasize the salience of neutrinoless double-β-decay experiments in determining whether the neutrino is a Dirac or a Majorana particle. Some useful comments on the results of the experiments are also given.

Chapter 15 is devoted to supernova neutrinos. In order to help the reader who is not familiar with the subject, a short introduction to supernova physics is presented, explaining the types of supernovae and their explosion mechanisms. The supernova which is of special interest to us, and so best studied, is SN1987A. The neutrino burst of SN1987A was detected by three experiments: Kamiokande-II, IMB, and Baksan. We present the data and compare them with theory. The limits on the neutrino masses are discussed, taking into account neutrino mixing, and the constraints on other neutrino properties are briefly summarized.

In chapter 16 we present a brief introduction to the main aspects of the standard cosmological model, which are necessary for understanding the relic neutrinos. These are the most abundant known, but not yet detected, relic particles in the Universe, next to the Cosmic Microwave Background Radiation (CMBR). After the introduction of the Standard Cosmological Model and the discussion of the dynamics of the expansion of the Universe, the thermodynamics of the early Universe and the decoupling of relic particles are explained. Finally, we present the main properties of the CMBR, which is one of the most important sources of cosmological information.

The final chapter 17 deals with the relic neutrinos. The decoupling of neutrinos, both light and possibly heavy ones, and the importance of neutrinos for nucleosynthesis in the early Universe are explained. Also discussed is the role of baryonic, cold, and hot dark matter in the formation of large-scale structures. A global fit of the cosmological data is presented to gain an insight into various limits on the neutrino masses, the number of neutrino species and the neutrino asymmetry in the Universe.

We wish to emphasize that this book is not a revised version of the book *Neutrinos in Physics and Astrophysics*, co-authored by one of the present authors more than 10 years ago. Here, by making the book as self-contained as possible, we have presented all the necessary details for the reader to follow and understand the subject matter. This is why we have chosen the title *Fundamentals of Neutrino Physics and Astrophysics*. For curious and studious readers who wish to find even more details, we have included as many references as possible, although no collection of references can be exhaustive. It is with our sincere regret that we have not discussed, in this book, some important topics, in particular, the theories of neutrino masses, the electromagnetic properties of neutrinos and the phenomenology of high-energy neutrinos from astrophysical sources. The interested readers can find introductions and reviews of these three topics, respectively, in Refs. [812, 74, 467, 781, 1076, 673, 810, 75, 763, 811, 815, 950, 813], Refs. [884, 52, 226, 158] and Refs. [487, 499, 584, 979, 785, 590, 583].

ACKNOWLEDGEMENTS

C.W.K. wishes to thank many colleagues at the Johns Hopkins University and the Korea Institute for Advanced Study; in particular, Gordon Feldman, Jon Bagger, E.J. Chun, P. Ko, K. Lee, C.B. Park, H. Park and P. Yi for valuable discussions, and for continued interest and encouragement. C.G. would like to thank: all the colleagues at the Torino Section of INFN and the Physics Department of Torino University, in particular A. Bottino and N. Fornengo, for encouragement and illuminating discussions; S.M. Bilenky for many enlightening discussions, long collaboration and for sharing his knowledge and experience; M. Laveder for stimulating collaboration, for cooperation in the creation and development of the *Neutrino Unbound* web pages at `http://www.nu.to.infn.it`; M.V. Garzelli and M. Laveder for reading the manuscript, finding several weak points, and suggesting corrections. Both of us are deeply indebted to the Korea Institute for Advanced Study for providing us with several opportunities to work together on this project. We should like to thank Sonke Adlung of the Oxford University Press for his patient and unfailing encouragement over the past few years.

CONTENTS

1

HISTORICAL INTRODUCTION

> I have done something very bad today by proposing a particle that cannot be detected; it is something no theorist should ever do.
>
> **Wolfgang Pauli**

The history of weak interactions dates back to 1896, when Becquerel discovered the radioactivity of uranium. Three years later, Rutherford discovered that there were two different by-products, α and β, γ being discovered later. In 1914, Chadwick demonstrated that the β-spectrum was continuous, in contrast to α- and γ-rays which were unique in energy. This surprising result was subsequently confirmed in 1927 by Ellis and Wooster. Meitner later demonstrated that the missing energy could not be ascribed to neutral γ-rays, which led to the idea that the missing energy could be explained by the existence of a new particle or, as N. Bohr suggested, perhaps energy conservation held only in a statistical sense.

In order to remedy this serious problem as well as the problem of spin statistics in β-decay, W. Pauli proposed, in an open letter to a physics conference at Tubingen on 4 December 1930, addressed to "Dear Radioactive Ladies and Gentlemen", that the existence of a neutral weakly interacting fermion emitted in β-decay could solve the problems (see Ref. [855]). He called this neutral fermion a *neutron*, with mass of the order of the electron. In June 1931, Pauli gave a talk at a meeting of the American Physical Society in Pasadena and reported for the first time on his idea. He did not have his talk printed, however, since he was still uncertain about his idea.

When J. Chadwick discovered in 1932 the neutron as we know it today [310], E. Fermi renamed the Pauli particle the *neutrino*. The first published reference to the neutrino is in the Proceedings of the Solvay Conference of October 1933. Fermi [430] and Perrin [865] independently concluded in 1933 that neutrinos could be massless!

The first milestone in the theory of weak interactions was established in 1934 when Fermi formulated a theory of β-decay [432, 431], now known as Fermi theory, in analogy with quantum electrodynamics (QED). In order to explain the observed change of one unit of the nuclear spin in some β-decays, G. Gamow and E. Teller in 1936 [505] extended the theory by introducing axial-vector currents in such a way that parity was still conserved, since parity violation at that time was unthinkable.

It was then realized that other couplings such as scalar, pseudoscalar, and tensor couplings could also participate in weak interactions. Due to this complication, for about two decades the real combination of the couplings was in a state of extreme confusion, in part, due to some erroneous experiments. A famous review article by E.J. Konopinski concluded in 1955 [693], just before the discovery of parity violation, that the correct form was a combination of scalar and tensor couplings, strongly biased by an impressive but wrong experiment with ^{6}He.

The horizon of the weak interactions was further extended by the discovery of the muon, μ, in 1937, by J.C. Street and E.C. Stevenson [989] and S.H. Neddermeyer and C.D. Anderson [825]. The observations of muon decay led B. Pontecorvo to propose in 1947 [879] the universality of the Fermi interactions of electrons and muons. This universality was further discussed by G. Puppi [885], O. Klein [682], J. Tiomno and J.A. Wheeler [1017] and T.D. Lee, M. Rosenbluth and C.N. Yang [726]. This may be the origin of the concept now known as *generation* or *family*.

Although the remarkable success of the Fermi theory left few in doubt of the neutrino's existence, none had yet been observed in interactions, partly because of the predicted strength of interactions by H. Bethe and R. Peierls, who claimed in 1934 that it might never be observed [220]. Urged, in particular by B. Pontecorvo in the early 1950s, F. Reines and C.L. Cowan searched for a way to measure inverse β-decay, in which an antineutrino can produce a positron. After considering several methods, including a nuclear explosion, they settled on using the large flux of antineutrinos from a nuclear reactor and 10 ton of equipment, including 1400 liters of liquid scintillators. This experiment was the first reactor-neutrino experiment. In June of 1956, Reines and Cowan sent a telegram informing Pauli of the discovery [899] (see Ref. [897]). Reines (Cowan passed away) was awarded the Nobel prize 40 years later!

First indicated in cosmic ray experiments and later confirmed by precise accelerator experiments, K^+ was found to decay into two different modes with opposite parity. This was the famous *θ-τ puzzle*; K^+, the one called θ, decays into two pions, whereas K^+, the one called τ, decays into three pions. The puzzle was that θ and τ have the same mass, spin, and charge, i.e. they are the same particle! This cannot happen if parity is conserved in weak interactions.

Some started to question the validity of parity conservation, but it was T.D. Lee and C.N. Yang who first noted in 1956 [728] that evidence for parity conservation in weak interactions was lacking, not just in K decays but in all observed weak interactions in the past. A number of tests to observe parity violation were suggested by Lee and Yang. Subsequently, parity violation was observed in the β-decay of polarized ^{60}Co [1073], $\pi^+ \to \mu^+ + \nu_\mu$ and $\mu^+ \to e^+ + \nu_e + \bar{\nu}_\mu$.

Once parity violation is allowed, the weak Lagrangian becomes even more complicated due to the appearance of parity violating couplings as well as the conserving ones. However, this seemingly confused situation was dramatically simplified in the form of the $V - A$ theory. The structure of the $V - A$ theory, formulated in 1958 by R.P. Feynman and M. Gell-Mann [434], E.C.G. Sudarshan and R.E. Marshak [993] and J.J. Sakurai [918] can easily be realized in the lepton sector by using the two-component theory of a massless neutrino, proposed in 1957 by L. Landau [711], T.D. Lee and C.N. Yang [727] and A. Salam [919]. (The idea was first developed by H. Weyl in 1929, but it was rejected by Pauli in 1933 on the grounds that it violates parity [854].) In this theory, neutrinos are left-handed and antineutrinos are right-handed, leading automatically to the $V - A$ couplings.

In 1958, Goldhaber, Grodzins and Sunyar [549] measured the polarization of a neutrino in the electron capture $e^- + {}^{152}\text{Eu} \rightarrow {}^{152}\text{Sm}^* + \nu_e$, with the subsequent decay ${}^{152}\text{Sm}^* \rightarrow {}^{152}\text{Sm} + \gamma$. They found that the measured polarization of the photon implies that the polarization of the ν_e was indeed in a direction opposite to its motion, within experimental uncertainties, in agreement with the two-component theory of a massless neutrino.

The concept of lepton number, L, was introduced in 1953 by E.J. Konopinski and H.M. Mahmoud to explain certain missing decay modes. The particles $e^-, \mu^-, \tau^-, \nu_e, \nu_\mu, \nu_\tau$ are assigned with $L = 1$, whereas their antiparticles have $L = -1$. In the $V - A$ theory and today's Standard Model of weak interactions, L is conserved. The Reines–Cowan experiment was consistent with lepton number conservation. On the other hand, R. Davis's attempts to observe $\bar{\nu}_e + {}^{37}\text{Cl} \rightarrow {}^{37}\text{Ar} + e^-$ turned out to produce only some limits on the cross-section for the process, because the process violates the lepton number. This effort, however, led Davis to the history-making Homestake solar neutrino experiment when he replaced $\bar{\nu}_e$ by ν_e from the Sun.

Although lepton number conservation allows the reaction $\mu \rightarrow e + \gamma$, its experimental limits were many orders of magnitude smaller than predicted. This suggested a new conservation law, one which assigns different lepton numbers to each lepton family, making $\mu \rightarrow e + \gamma$ forbidden. This assignment led to the prediction that $\nu_\mu + n \rightarrow p + e^-$ is forbidden. As suggested by Pontecorvo [882], if it is shown that ν_μ produced in $\pi^+ \rightarrow \mu^+ + \nu_\mu$ cannot induce e^-, then ν_μ and ν_e are indeed different particles. Encouraged by an estimate of the event rate by M. Schwartz [942], L.M. Lederman, M. Schwartz, J. Steinberger *et al.* [348] succeeded in 1962 at Brookhaven National Laboratory (BNL) in establishing the existence of the second neutrino ν_μ. This experiment, which utilized an enormous amount of iron shielding plates cut out of the battleship USS Missouri, marked the first serious accelerator neutrino experiment.

A crucial milestone in the theory of weak interactions is the formulation of the Glashow–Weinberg–Salam Standard Model (SM) by S. Weinberg [1051] and A. Salam [920] in 1967. The model is based on an $SU(2) \times U(1)$ gauge model proposed by S.L. Glashow in 1961 [543], which predicted the existence of weak neutral currents and the Z boson. The Standard Model incorporates the so-called *Higgs mechanism* into the Glashow model. The Higgs mechanism, which was discovered in 1964 by P.W. Higgs [611, 610, 612], F. Englert and R. Brout [412], and G.S. Guralnik, C.R. Hagen and T.W.B. Kibble [578, 666], allows the original massless gauge bosons that appear in the local gauge group model to acquire longitudinal degrees of freedom, finally making them massive as demanded in Nature. The renormalizability of the model was proved by G. 't Hooft and M.J.G. Veltman in 1971 [1002, 1001, 1003], elevating the model to an extremely viable one. The success of the SM was affirmed in 1973 by the discovery of neutral-current neutrino interactions in the Gargamelle experiment at CERN (1973) [600, 599, 601] and was subsequently confirmed at Fermilab [207].

The discovery in 1974 of the charm quark in the form of the J/ψ particle ($c\bar{c}$) at BNL (J) [124] and SLAC (ψ) [126] and the subsequent discovery of $W^\pm$ [108, 160] and Z [109, 135] at CERN firmly established the SM as the model for leptonic and hadronic weak and electromagnetic (electroweak) interactions. (The discovery of the charm quark was a triumph for the prediction of the existence of the charm quark as a result of the S.L. Glashow, J. Iliopoulos and L. Maiani (GIM) mechanism [544].) After the discovery of the third lepton, τ, by M. Perl in 1975 [862], the b and t quarks were discovered at Fermilab, respectively, in 1977 [608] and 1995 [22, 6], completing all the building blocks of the SM with three generations. Only the Higgs particle, which is necessary in the mass-generating Higgs mechanism in the SM, has not yet been found. The number of generations was fixed at three in 1989 by the impressive measurements by LEP experiments at CERN of the invisible width of the Z boson [362, 5, 29, 53].

In 1964 J.H. Christenson, J.W. Cronin, V.L. Fitch and R. Turlay [317] unexpectedly discovered a violation of CP symmetry in K^0-decay. The existence of this violation has been accommodated in the framework of the SM through the mixing of three generations of quarks by M. Kobayashi and T. Maskawa [685] in 1973, extending the theory of two-generation mixing developed by N. Cabibbo [292] in 1963.

No experiments that have been performed so far have detected conclusive deviations from the SM, except neutrino oscillation experiments, which have shown that neutrinos are massive and mixed. In the SM, this is not the case. This discovery has made the SM an effective theory of the yet unknown theory beyond the SM. The understanding of how the neutrinos would gain tiny masses and how they are mixed is an extremely challenging task that we have to face. The answer must be found in the theory beyond the SM. Thus, the neutrino is playing the role of a messenger of the new physics beyond the SM.

The concept of neutrino oscillations was first proposed in 1957 by Pontecorvo [880, 881], motivated by the $K^0 \leftrightarrows \bar{K}^0$ oscillation phenomenon (M. Gell-Mann and A. Pais [514], 1955), in which the strangeness quantum number is oscillating. The possible oscillations that he could find at that time were $\nu \leftrightarrows \bar{\nu}$ for Majorana neutrinos. Pontecorvo interpreted a rumor of Davis's successful observation of $\bar{\nu}+{}^{37}\mathrm{Cl} \to {}^{37}\mathrm{Ar}+e^-$ with reactor antineutrinos (which turned out to be false [353]) as a result of $\bar{\nu} \to \nu$ transitions and a subsequent $\nu + {}^{37}\mathrm{Cl} \to {}^{37}\mathrm{Ar} + e^-$ reaction. However, the $V - A$ theory of weak interactions implies that, in the ultrarelativistic limit applicable to neutrinos, reactor antineutrinos are right-handed. Even if they oscillate into right-handed neutrinos (helicity is conserved), these neutrinos cannot induce the process $\nu + {}^{37}\mathrm{Cl} \to {}^{37}\mathrm{Ar} + e^-$, which requires left-handed neutrinos. A more realistic case of oscillations became available with the assumption that ν_e and ν_μ are mixed states of two mass eigenstates, which was discussed by Z. Maki, M. Nakagawa and S. Sakata in 1967 [766]. However, there was only a vague hint of the present understanding of neutrino oscillations in their work. In 1967, Pontecorvo presented the first intuitive understanding of two-neutrino mixing and oscillations [883], which was later completed by V.N. Gribov and B. Pontecorvo in 1969 [567]. The theory of neutrino oscillations was finally developed in 1975–76 by S. Eliezer and A.R. Swift [404], H. Fritzsch and P. Minkowski [466], S.M. Bilenky and B. Pontecorvo [236, 239].

As in the case of many weak interaction experiments, neutrino oscillation experiments have had their own share of ups and downs in the early stages. The longest running experiments by far have been those of the solar neutrinos. We have recently witnessed some spectacular results finally confirming the oscillations of the solar and atmospheric neutrinos and the reactor and accelerator neutrinos in various settings.

The atmospheric neutrinos, first regarded as unwanted background for the experiments that had been designed to search for proton decay, have provided us with the first model-independent indication of oscillations of ν_μ's through the Super-Kamiokande experiment [476]. The atmospheric neutrino anomaly was discovered in the late 1980s in the Kamiokande and IMB experiments (see section 11.2). Nowadays, the high-precision measurements of the Super-Kamiokande experiment, confirmed by the measurements of the Soudan 2 and MACRO experiments (see section 11.2), give us precise information on the values of the atmospheric neutrino oscillation parameters, which are in good agreement with the independent results of the first accelerator long-baseline K2K experiment (see section 12.3.2).

In the past decade, there have been many spectacular successes in the pursuit of the solution of the solar neutrino problem, which, we believe, has finally been understood and solved to our satisfaction. This problem was discovered in the Homestake experiment [323] and confirmed by the observations of the Kamiokande, GALLEX/GNO, SAGE, Super-Kamiokande, and SNO experiments (see chapter 10). In particular, the results of the SNO experiment [43] have been instrumental in solving the solar neutrino problem in 2002. The depletion of the solar neutrinos is finally found to be due to the oscillations of ν_e into ν_μ and ν_τ inside the Sun by the Mikheev–Smirnov–Wolfenstein (MSW) resonance conversion effects [1065, 801, 802]. The solar model by the late John Bahcall and others [142, 137, 145, 152, 154] has become the Standard Solar Model which can now be used for the investigation of other solar properties. The reactor long-baseline KamLAND experiment has recently confirmed the values of the oscillation parameters obtained from a global analysis of the data of all solar neutrino experiments (see section 12.2.3).

The results of the atmospheric, solar, KamLAND and K2K neutrino experiments are nicely explained by neutrino oscillations in the framework of the simplest model of three-neutrino mixing, in which the three flavor neutrinos ν_e, ν_μ and ν_τ are unitary linear combinations of three massive neutrinos ν_1, ν_2, ν_3 (see chapter 13). As of this writing, we have a rather precise knowledge of the value of the neutrino squared-mass difference Δm^2_{21}, the absolute value of $\Delta m^2_{31} \simeq \Delta m^2_{32}$, and the values of two mixing angles, ϑ_{12} and ϑ_{23}. The value of the third mixing angle, ϑ_{13}, the values of the CP phases (one for Dirac neutrinos and three for Majorana neutrinos), the absolute scale of neutrino masses, and the sign of Δm^2_{31} remain unknown. On the value of ϑ_{13} we have only an upper bound, obtained from the absence of neutrino oscillations observed in the reactor long-baseline experiments CHOOZ and Palo Verde (see section 12.2.2). The absolute scale of neutrino masses is limited below a few eV by kinematical measurements of the electron spectrum in the recent tritium β-decay experiments Mainz and Troitzk (see section 14.1).

An open question of fundamental importance is the nature of neutrinos, which could be either of Dirac or Majorana type. Neutrinoless double-β-decay experiments (see section 14.3) are considered the most promising way to decide if neutrinos are Majorana particles. The current upper limit on the effective Majorana mass, obtained in the Heidelberg–Moscow ^{76}Ge experiment [680], is of the order of one eV, with an unfortunate uncertainty of a factor of about three due to the complexity of nuclear physics.

In recent years, we have learned that neutrinos play an important role in the early Universe in many subtle ways. For example, the number of neutrino species affects the primordial nucleosynthesis, which eventually decides the composition of elements in the Universe [857, 981]. From this argument alone one could conclude that the number of neutrino species is close to three, in agreement with the precise number determined through the measurement of the invisible width of the Z-boson by LEP experiments at CERN. Later in the evolution of the Universe, massive neutrinos influence the formation of the large-scale structures in the form of hot dark matter (see section 17.7.3). The recent observations (see chapters 16 and 17) of the Cosmic Microwave Background Radiation (WMAP, Boomerang, and others), the Large Scale Structures (2dFGRS and SDSS) and distant type Ia supernovae (High-z SN Search Team and Supernova Cosmology Project) and a precise determination of the Hubble constant by the Hubble Space Telescope Key Project made possible a better understanding of the evolution of the Universe. It turned out that only about 5% of the energy of the Universe is composed of ordinary matter (baryons). The other 95% is composed of invisible *dark matter* ($\sim$ 25%) and mysterious *dark energy* ($\sim$ 70%). Since the dark matter must be of the cold type (see section 17.7.2), hot dark matter in the form of massive neutrinos is severely constrained, leading to an upper bound on the sum of neutrino masses of the order of one eV (see section 17.8). This bound is remarkably close to the limits from totally independent experiments (tritium β-decay and neutrinoless double-β-decay experiments). The relic neutrinos, that decoupled from the rest of the primordial plasma when the Universe was about one second old, are the second most abundant particles in the Universe next to the photons, with a number density smaller only by a factor of 3/11 for each family. The direct detection of the relic neutrinos still remains, however, as one of the biggest scientific challenges in the twenty first century, because of their weak interactions.

2

QUANTIZED DIRAC FIELDS

To trace the unfamiliar to the familiar is to understand.
Algernon Blackwood, *The Damned*

In this chapter we review the main properties of quantized Dirac fields, which describe particles with spin 1/2 (see, for example, Refs. [943, 634, 821]). In the Standard Model, to be introduced in chapter 3, the fundamental fermions are quarks, charged leptons, and neutrinos, all of which have spin 1/2. In such a model, quarks and charged leptons are massive Dirac particles and neutrinos are massless Dirac particles. As discussed in chapters 10, 11, and 12, there is experimental evidence that neutrinos are massive. In this case, neutrinos can be either Dirac particles or Majorana particles, as discussed in chapter 6. In any case, the Dirac theory presented in this chapter represents the basis for the description of neutrinos, from which the Majorana theory can be derived (see chapter 6).

2.1 Dirac equation

The Dirac Lagrangian for a free fermion field $\psi(x)$ is[1]

$$\mathscr{L}(x) = \overline{\psi}(x)\,(i\overleftrightarrow{\partial\!\!\!/} - m)\,\psi(x)\,, \tag{2.1}$$

where $\psi(x)$ is a spinor field with four components, the adjoint field $\overline{\psi}(x)$ being given by

$$\overline{\psi}(x) \equiv \psi^{\dagger}(x)\,\gamma^{0}\,, \tag{2.2}$$

and

$$\overleftrightarrow{\partial}_{\mu} \equiv \frac{\overrightarrow{\partial}_{\mu} - \overleftarrow{\partial}_{\mu}}{2}\,, \tag{2.3}$$

where $\overrightarrow{\partial}_{\mu} \equiv \partial_{\mu}$ is the normal derivative operator which acts on the right and $\overleftarrow{\partial}_{\mu}$ is a derivative operator which acts on the left (i.e. $\overline{\psi}\,\overleftarrow{\partial}_{\mu} \equiv \partial_{\mu}\overline{\psi}$). For any four-vector

[1] Often the Dirac Lagrangian is written as $\mathscr{L}'(x) = \overline{\psi}(x)\,(i\partial\!\!\!/ - m)\,\psi(x)$, which differs from eqn (2.1) by the total derivative $\mathscr{L}' - \mathscr{L} = \frac{i}{2}\,\partial_{\mu}\left(\overline{\psi}\,\gamma^{\mu}\,\psi\right)$ that has no effect on the field equation derived from the Euler–Lagrange equation (C.9) (Gauss' theorem implies that the integral of a total derivative is a surface term, which is invariant under the variation in eqn (C.3)). However, it is better to write the Dirac Lagrangian in the form eqn (2.1) because it is explicitly real, as a Lagrangian should be.

A^μ

$$\not{A} \equiv \gamma^\mu A_\mu \,, \tag{2.4}$$

where γ^μ ($\mu = 0, 1, 2, 3$) is a set of four 4×4 matrices[2] called *Dirac γ matrices* that satisfy the anticommutation relations

$$\{\gamma^\mu, \gamma^\nu\} \equiv \gamma^\mu \gamma^\nu + \gamma^\nu \gamma^\mu = 2\, g^{\mu\nu} \tag{2.5}$$

and the condition

$$\gamma^0\, \gamma^{\mu\dagger}\, \gamma^0 = \gamma^\mu \,. \tag{2.6}$$

The γ matrices are constant in the sense that they do not transform under Lorentz transformations. In other words, the γ matrices have the same value in all inertial reference frames. Although they are not four-vectors, it is convenient to define formally the γ matrices with lowered indices as

$$\gamma_\mu \equiv g_{\mu\nu}\, \gamma^\nu \quad (\mu = 0, 1, 2, 3) \quad \Longleftrightarrow \quad \gamma_0 = \gamma^0 \,, \quad \gamma_k = -\gamma^k \quad (k = 1, 2, 3) \,. \tag{2.7}$$

For $\mu \neq \nu$ the relations in eqn (2.5) imply that the four γ matrices anticommute. For $\mu = \nu$, the relations in eqn (2.5) constrain the squares of the γ matrices:

$$(\gamma^0)^2 = \mathbf{1} \,, \qquad (\gamma^k)^2 = -\mathbf{1} \quad (k = 1, 2, 3) \,. \tag{2.8}$$

The additional constraint in eqn (2.6) implies that

$$(\gamma^0)^\dagger = \gamma^0 \,, \quad (\gamma^k)^\dagger = -\gamma^k \quad \Longleftrightarrow \quad (\gamma^\mu)^\dagger = \gamma_\mu \,. \tag{2.9}$$

Thus, γ^0 is Hermitian, whereas the γ^k matrices are anti-Hermitian.

Using the Euler–Lagrange procedure (see eqn (C.9)), the field equations are given by

$$\partial_\mu \frac{\partial \mathscr{L}}{\partial(\partial_\mu \overline{\psi})} - \frac{\partial \mathscr{L}}{\partial \overline{\psi}} = 0 \,, \tag{2.10}$$

from which one obtains the *Dirac equation*

$$(i\not{\partial} - m)\psi(x) = 0 \,. \tag{2.11}$$

The anticommutation relations in eqn (2.5) are necessary in order to guarantee the compatibility of the Dirac equation with the Klein–Gordon equation

$$(\Box + m^2)\psi(x) = 0 \,, \tag{2.12}$$

where $\Box \equiv \partial^\mu \partial_\mu$. The Klein–Gordon equation must be satisfied by any free field because it is equivalent to the relativistic energy–momentum dispersion relation

---

[2] The anticommutation relations in eqn (2.5) imply that the dimension N of the γ matrices is even. Indeed, let us consider the anticommutation relation with $\mu \neq \nu$, that is $\gamma^\mu \gamma^\nu = -\gamma^\nu \gamma^\mu = (-\mathbf{1})\, \gamma^\nu \gamma^\mu$. Taking the determinant, we obtain $\mathrm{Det}\gamma^\mu\, \mathrm{Det}\gamma^\nu = (-1)^N\, \mathrm{Det}\gamma^\nu\, \mathrm{Det}\gamma^\mu$. Since $\mathrm{Det}\gamma^\nu \neq \mathrm{Det}\gamma^\mu$, we have $(-1)^N = 1$, which implies that N is even.

The minimal dimension of the γ matrices is $N = 4$, because there are only three anticommuting matrices for $N = 2$, which are the Pauli matrices in eqn (A.29).

In practice the γ matrices are always represented in the minimal 4×4 dimensionality, because representations of higher dimensionality are equivalent but obviously more complicated.

in eqn (B.88). In fact, the Klein–Gordon equation has positive-energy plane-wave solutions $\psi(x) \propto e^{-ip\cdot x}$ with $p^2 = m^2$. The negative-energy plane-wave solutions $\psi(x) \propto e^{ip\cdot x}$ are interpreted as antiparticle states, as explained in section 2.9.

The Klein–Gordon equation (2.12) is obtained from the Dirac equation (2.11) by multiplying it on the left by $(i\partial\!\!\!/ + m)$ and using the identity $\partial\!\!\!/\partial\!\!\!/ = \partial^\mu\partial_\mu$. This identity is a particular case of the general identity

$$A\!\!\!/A\!\!\!/ = \gamma^\mu\gamma^\nu A_\mu A_\nu = \frac{1}{2}\left(\gamma^\mu\gamma^\nu + \gamma^\nu\gamma^\mu\right) A_\mu A_\nu = g^{\mu\nu} A_\mu A_\nu = A_\mu A^\mu\,, \tag{2.13}$$

which follows from the anticommutation relations in eqn (2.5).

The condition in eqn (2.6) is necessary in order to obtain, from the Dirac equation (2.11), a continuity equation with a quantum mechanical density

$$\varrho(x) = |\psi(x)|^2 = \psi^\dagger(x)\,\psi(x)\,. \tag{2.14}$$

To see this, we first take the Hermitian conjugate of the Dirac equation (2.11), taking into account eqn (2.8), and then multiply it on the right by $\gamma^0\psi$, leading to

$$-\,i\,\partial_\mu\overline{\psi}\,\gamma^0\,\gamma^\mu\,\gamma^0\,\psi - m\,\overline{\psi}\,\psi = 0\,. \tag{2.15}$$

Subtracting this equation from the Dirac equation (2.11) multiplied on the left by $\overline{\psi}$, we have

$$\overline{\psi}\,\gamma^\mu\partial_\mu\,\psi + \partial_\mu\overline{\psi}\,\gamma^0\,\gamma^\mu\,\gamma^0\,\psi = 0\,. \tag{2.16}$$

Imposing the condition in eqn (2.6) we obtain the continuity equation

$$\partial_\mu\, j^\mu = 0\,, \tag{2.17}$$

with the current

$$j^\mu(x) = \overline{\psi}(x)\,\gamma^\mu\,\psi(x)\,. \tag{2.18}$$

The temporal component of this current is the quantum mechanical density in eqn (2.14). In section 2.4 it will be shown that $j^\mu(x)$ is a well-behaved four-vector (see eqn (2.66) with $a = b$).

2.2 Representations of γ matrices

A specific choice of the four matrices γ^μ that satisfy the relations in eqns (2.5) and (2.6) is called a *representation of the Dirac matrices.* In his fundamental theorem on the representations of the Dirac matrices, Pauli proved that all representations are unitarily equivalent, i.e. any two sets of four matrices γ^μ and γ'^μ which fulfill the relations in eqns (2.5) and (2.6) are connected by the equivalence transformation

$$\gamma'^\mu = S\,\gamma^\mu\,S^{-1}\,, \tag{2.19}$$

where S is a unitary matrix ($S^\dagger = S^{-1}$). In oder to leave the Dirac equation invariant under a change of representation, a spinor field must transform as

$$\psi' = S\,\psi\,. \tag{2.20}$$

When an explicit expression of the γ matrices is needed, one can choose the most convenient one for the task under consideration. In any representation, at most one of the γ is diagonal, because of the anticommutation relation in eqn (2.5). From eqn (2.8) it follows that the eigenvalues of γ^0 are ± 1 and the eigenvalues of γ^k are $\pm i$.

The standard representation of the γ matrices is the Dirac representation

$$\gamma^0_{\rm D} = \begin{pmatrix} \mathbf{1} & 0 \\ 0 & -\mathbf{1} \end{pmatrix} , \qquad \vec{\gamma}_{\rm D} = \begin{pmatrix} 0 & \vec{\sigma} \\ -\vec{\sigma} & 0 \end{pmatrix} , \tag{2.21}$$

in which the matrices are written as 2×2 blocks and we put a subscript D to indicate the Dirac representation. The 2×2 matrices σ^k are the Pauli matrices given in eqn (A.29).

It is useful to define the chirality matrix

$$\gamma^5 \equiv \gamma_5 \equiv i\, \gamma^0\, \gamma^1\, \gamma^2\, \gamma^3 , \tag{2.22}$$

which has the following useful properties:

$$\{\gamma^5, \gamma^\mu\} = 0 \tag{2.23}$$

$$\left(\gamma^5\right)^2 = \mathbf{1} \tag{2.24}$$

$$\left(\gamma^5\right)^\dagger = \gamma^5 . \tag{2.25}$$

In the Dirac representation, the γ^5 matrix is given by

$$\gamma^5_{\rm D} = \begin{pmatrix} 0 & \mathbf{1} \\ \mathbf{1} & 0 \end{pmatrix} . \tag{2.26}$$

For a study of relativistic particles such as neutrinos, it is convenient to use the chiral representation in which the chirality matrix $\gamma^5_{\rm C} = \gamma^0_{\rm D}$ is diagonal and $\vec{\gamma}_{\rm C} = \vec{\gamma}_{\rm D}$. From eqn (2.22) it follows that $\gamma^0_{\rm C} = -\gamma^5_{\rm D}$. Thus, the 4×4 Dirac matrices in the chiral representation can be written in 2×2 block forms as

$$\gamma^0_{\rm C} = \begin{pmatrix} 0 & -\mathbf{1} \\ -\mathbf{1} & 0 \end{pmatrix} , \qquad \vec{\gamma}_{\rm C} = \begin{pmatrix} 0 & \vec{\sigma} \\ -\vec{\sigma} & 0 \end{pmatrix} , \qquad \gamma^5_{\rm C} = \begin{pmatrix} \mathbf{1} & 0 \\ 0 & -\mathbf{1} \end{pmatrix} , \tag{2.27}$$

or in the compact form

$$\gamma^\mu_{\rm C} = \begin{pmatrix} 0 & \bar{\sigma}^\mu \\ -\sigma^\mu & 0 \end{pmatrix} , \tag{2.28}$$

with the 2×2 matrices

$$\sigma^\mu = (\mathbf{1}\, , \vec{\sigma}) , \qquad \bar{\sigma}^\mu = (-\mathbf{1}\, , \vec{\sigma}) . \tag{2.29}$$

The unitary matrix $S_{\rm D \to C}$ which performs the equivalence transformation

$$\gamma^\mu_{\rm C} = S_{\rm D \to C}\, \gamma^\mu_{\rm D}\, S^{-1}_{\rm D \to C} \tag{2.30}$$

from the Dirac to the chiral representation is given by

$$S_{\rm D \to C} = \frac{1}{\sqrt{2}} \left(\mathbf{1} + \gamma^0_{\rm D}\, \gamma^5_{\rm D}\right) = \frac{1}{\sqrt{2}} \begin{pmatrix} \mathbf{1} & -\mathbf{1} \\ \mathbf{1} & \mathbf{1} \end{pmatrix} . \tag{2.31}$$

2.3 Products of γ matrices

Let us define 16 matrices Γ^a ($a = 1, 2, \ldots, 16$) obtained from products of γ matrices:

$$\Gamma^1 \equiv \mathbf{1} \qquad \text{(no } \gamma^\mu \text{ matrices)}, \tag{2.32}$$

$$\Gamma^2 - \Gamma^5 \equiv \gamma^\mu \qquad \text{(one } \gamma^\mu \text{ matrix)}, \tag{2.33}$$

$$\Gamma^6 - \Gamma^{11} \equiv \sigma^{\mu\nu} \equiv \frac{i}{2}\,[\gamma^\mu, \gamma^\nu] \qquad \text{(products of two } \gamma^\mu \text{ matrices)}, \tag{2.34}$$

$$\Gamma^{12} - \Gamma^{15} \equiv \gamma^\mu \gamma^5 \qquad \text{(products of three } \gamma^\mu \text{ matrices)}, \tag{2.35}$$

$$\Gamma^{16} \equiv \gamma^5 \qquad \text{(product of four } \gamma^\mu \text{ matrices)}. \tag{2.36}$$

In any product of γ matrices, all the pairs of identical matrices can be eliminated using the anticommutation relations in eqn (2.5). Since the number of γ matrices is four, any product of more than four γ matrices can be reduced to a product of four or fewer γ matrices.

The anticommutation relations in eqn (2.5) imply that products of γ matrices in which the matrices appear with different orders are equivalent up to a sign. Hence, the number of independent products of k γ matrices is given by the binomial coefficient

$$\binom{4}{k} = \frac{4!}{k!\,(4-k)!}\,. \tag{2.37}$$

A set of 16 Γ matrices represents all the irreducible products of γ matrices, as follows

1. The matrix $\Gamma^1 = \mathbf{1}$ is the only irreducible product with no γ matrices ($\binom{4}{0} = 1$).
2. The four matrices $\Gamma^2, \ldots, \Gamma^5 = \gamma^0, \gamma^1, \gamma^2, \gamma^3$ are trivially the four irreducible products of one γ matrix ($\binom{4}{1} = 4$).
3. The six[3] matrices $\Gamma^6, \ldots, \Gamma^{11} = \sigma^{01}, \sigma^{02}, \sigma^{03}, \sigma^{12}, \sigma^{23}, \sigma^{31}$, represent all the irreducible products of two γ matrices ($\binom{4}{2} = 6$). In fact, we have

$$\sigma^{\mu\nu} = i\gamma^\mu\gamma^\nu \qquad (\mu \neq \nu)\,. \tag{2.38}$$

4. The four matrices $\Gamma^{12}, \ldots, \Gamma^{15} = \gamma^0\gamma^5, \gamma^1\gamma^5, \gamma^2\gamma^5, \gamma^3\gamma^5$ represent all the irreducible products of three γ matrices ($\binom{4}{3} = 4$), i.e.

$$\gamma^0\gamma^5 = -i\gamma^1\gamma^2\gamma^3\,, \tag{2.39}$$

$$\gamma^1\gamma^5 = -i\gamma^0\gamma^2\gamma^3\,, \tag{2.40}$$

$$\gamma^2\gamma^5 = -i\gamma^0\gamma^3\gamma^1\,, \tag{2.41}$$

$$\gamma^3\gamma^5 = -i\gamma^0\gamma^1\gamma^2\,, \tag{2.42}$$

or, in compact form,

$$\gamma^\alpha\gamma^5 = \frac{i}{3!}\, g^{\alpha\beta}\epsilon_{\beta\mu\nu\rho}\gamma^\mu\gamma^\nu\gamma^\rho\,. \tag{2.43}$$

[3] By definition, the matrices $\sigma^{\mu\nu}$ are antisymmetric in the indices μ, ν. Thus, the number of independent matrices $\sigma^{\mu\nu}$ is $n(n-1)/2 = 6$ for $n = 4$.

TABLE 2.1. Order of the matrices Γ^a and the corresponding values of $s_a = \text{sign}(\Gamma^a)^2 = \frac{1}{4}\,\text{Tr}\big[(\Gamma^a)^2\big]$.

	a															
	1	2	3	4	5	6	7	8	9	10	11	12	13	14	15	16
Γ^a	$\mathbf{1}$	γ^0	γ^1	γ^2	γ^3	σ^{01}	σ^{02}	σ^{03}	σ^{12}	σ^{23}	σ^{31}	$\gamma^0\gamma^5$	$\gamma^1\gamma^5$	$\gamma^2\gamma^5$	$\gamma^3\gamma^5$	γ^5
s_a	1	1	-1	-1	-1	-1	-1	-1	1	1	1	-1	1	1	1	1

5. The matrix $\Gamma^{16} = \gamma^5$ is the only irreducible product of four γ matrices ($\binom{4}{4} = 1$).

Therefore, any product of γ matrices is proportional to one of the 16 Γ matrices. The coefficient of proportionality is equal to ± 1 or $\pm i$.

The Γ matrices enjoy the following useful properties:

A. As all products of γ matrices, the product of two Γ matrices is proportional to a Γ matrix. Moreover, if two Γ matrices are different, their product is different from unity:

$$\Gamma^a\,\Gamma^b \propto \Gamma^c \quad \text{with} \quad \Gamma^c \neq \mathbf{1} \quad \text{for} \quad a \neq b\,. \tag{2.44}$$

If Γ^a and Γ^b are different, their product contains an odd number of at least one γ matrix, which cannot be eliminated by using the anticommutation relations in eqn (2.5).

B. The square of all Γ matrices is equal to $\pm\mathbf{1}$:

$$(\Gamma^a)^2 = s_a\mathbf{1}\,, \quad \text{with} \quad s_a = \frac{1}{4}\,\text{Tr}\big[(\Gamma^a)^2\big] = \pm 1\,. \tag{2.45}$$

The values of s_a for the 16 matrices Γ^a are listed in Table 2.1.

C. For each Γ^a with $a > 1$ there is at least one Γ^b which anticommutes with Γ^a,

$$\Gamma^a\Gamma^b = -\Gamma^b\Gamma^a \quad \Longleftrightarrow \quad \{\Gamma^a\,,\,\Gamma^b\} = 0\,. \tag{2.46}$$

We have

$$\Gamma^a = \gamma^0 \quad (a = 2) \quad \Longrightarrow \quad \Gamma^b = \gamma^k, \gamma^5 \quad (b = 3, 4, 5, 16)\,, \tag{2.47}$$

$$\Gamma^a = \gamma^k \quad (a = 3, 4, 5) \quad \Longrightarrow \quad \Gamma^b = \gamma^0, \gamma^5 \quad (b = 1, 16)\,, \tag{2.48}$$

$$\Gamma^a = \sigma^{\mu\nu} \quad (a = 6 - 11) \quad \Longrightarrow \quad \Gamma^b = \sigma^{\mu\rho} \quad \rho \neq \nu\,, \tag{2.49}$$

$$\Gamma^a = \gamma^0\gamma^5 \quad (a = 12) \quad \Longrightarrow \quad \Gamma^b = \gamma^k\gamma^5 \quad (b = 13, 14, 15)\,, \tag{2.50}$$

$$\Gamma^a = \gamma^k\gamma^5 \quad (a = 13, 14, 15) \quad \Longrightarrow \quad \Gamma^b = \gamma^0\gamma^5 \quad (b = 12)\,, \tag{2.51}$$

$$\Gamma^a = \gamma^5 \quad (a = 16) \quad \Longrightarrow \quad \Gamma^b = \gamma^\mu \quad (b = 2, 3, 4, 5)\,. \tag{2.52}$$

D. The matrices Γ^a with $a > 1$ are traceless,

$$\mathrm{Tr}[\Gamma^a] = 0 \qquad \text{for} \qquad a > 1\,. \tag{2.53}$$

Using a matrix Γ^b which anticommutes with Γ^a, we have

$$\mathrm{Tr}[\Gamma^a] = s_b \mathrm{Tr}\left[\Gamma^a\left(\Gamma^b\right)^2\right] = -s_b \mathrm{Tr}\left[\Gamma^b \Gamma^a \Gamma^b\right] = -s_b \mathrm{Tr}\left[\left(\Gamma^b\right)^2 \Gamma^a\right] = -\mathrm{Tr}[\Gamma^a]\,, \tag{2.54}$$

where the second equality has been obtained by anticommuting Γ^a and Γ^b and the third equality has been obtained with a circular permutation of the argument of the trace.

E. From the properties in eqns (2.44), (2.45), and (2.53) it follows that

$$\mathrm{Tr}\left[\Gamma^a\,\Gamma^b\right] = 4\,s_a\,\delta_{ab}\,. \tag{2.55}$$

F. The Γ matrices are linearly independent[4], i.e. the relation

$$\sum_a c_a\,\Gamma^a = 0 \tag{2.56}$$

implies $c_a = 0$ for all $a = 1, \ldots, 16$. Taking the trace of eqn (2.56) and using the property in eqn (2.53), one finds that $c_1 = 0$. Similarly, taking the trace of

$$\Gamma^b \left(\sum_a c_a\,\Gamma^a\right) = 0 \tag{2.57}$$

and using the property in eqn (2.55) one finds that $c_b = 0$ for any $b = 2, \ldots, 16$.

G. From the previous properties it follows that any 4×4 matrix X can be written as a linear combination of the Γ^a matrices:

$$X = \sum_a x_a\,\Gamma^a\,, \quad \text{with} \quad x_a = \frac{s_a}{4}\,\mathrm{Tr}[X\,\Gamma^a]\,. \tag{2.58}$$

Therefore, the 16 matrices Γ^a form a basis in the vectorial space of 4×4 matrices.

2.4 Relativistic covariance

Under a Lorentz transformation in eqn (B.1), the Dirac field $\psi(x)$ transforms as

$$\psi(x) \to \psi'(x') = \mathcal{S}(\Lambda)\,\psi(x)\,, \tag{2.59}$$

with the 4×4 matrix $\mathcal{S}(\Lambda)$ such that

$$\mathcal{S}^{-1}(\Lambda)\,\gamma^\mu\,\mathcal{S}(\Lambda) = \Lambda^\mu{}_\nu\,\gamma^\nu\,. \tag{2.60}$$

In this way, the Dirac equation remains form-invariant under Lorentz transformations, according to the principle of relativistic covariance discussed in appendix B.

[4] This property implies that the minimal dimension of the Γ matrices is 4×4, in agreement with the minimal dimension of γ matrices discussed in footnote 2 on page 8. The reason is simply that since 4×4 matrices have 16 elements there are 16 independent 4×4 matrices, whereas the number of independent matrices with lower dimensionality is smaller.

The explicit form of $\mathcal{S}(\mathbf{1} + \varepsilon\omega)$ for an infinitesimal Lorentz transformation in eqn (B.22) is

$$\mathcal{S}(\mathbf{1} + \varepsilon\omega) = \mathbf{1} - \frac{i}{4}\,\varepsilon\,\omega_{\mu\nu}\,\sigma^{\mu\nu}\,. \tag{2.61}$$

Comparing this expression with eqn (B.63), we obtain the spin part of the generators of the Lorentz group in the Dirac spinor representation:

$$S^{\mu\nu} = -\frac{1}{2}\,\sigma^{\mu\nu}\,. \tag{2.62}$$

These generators satisfy the commutation relations in eqn (B.45) of the Lorentz group generators.

For a finite restricted Lorentz transformation $\Lambda = e^{\omega}$ in eqn (B.24), we have

$$\mathcal{S}(e^{\omega}) = \exp\left(-\frac{i}{4}\,\omega_{\mu\nu}\,\sigma^{\mu\nu}\right)\,. \tag{2.63}$$

Since the adjoint field $\overline{\psi}(x)$ transform as[5]

$$\overline{\psi}(x) \to \overline{\psi}'(x') = \overline{\psi}(x)\,\mathcal{S}^{-1}(\Lambda)\,, \tag{2.64}$$

the following five Hermitian covariant bilinears transform, respectively, as a scalar, vector, antisymmetric second-rank tensor, pseudovector, and pseudoscalar: for a restricted Lorentz transformation

$$S_{ab}(x) = \overline{\psi_a}(x)\,\psi_b(x) \to S'_{ab}(x') = \overline{\psi'_a}(x')\,\psi'_b(x') = \overline{\psi_a}(x)\,\psi_b(x) = S_{ab}(x)\,, \tag{2.65}$$

$$\begin{aligned} V^{\mu}_{ab}(x) = \overline{\psi_a}(x)\,\gamma^{\mu}\,\psi_b(x) \to V'^{\mu}_{ab}(x') &= \overline{\psi'_a}(x')\,\gamma^{\mu}\,\psi'_b(x') = \Lambda^{\mu}{}_{\nu}\,\overline{\psi_a}(x)\,\gamma^{\nu}\,\psi_b(x) \\ &= \Lambda^{\mu}{}_{\nu}\,V^{\nu}_{ab}(x)\,, \end{aligned} \tag{2.66}$$

$$\begin{aligned} T^{\mu\nu}_{ab}(x) = \overline{\psi_a}(x)\,\sigma^{\mu\nu}\,\psi_b(x) \to T'^{\mu\nu}_{ab}(x') &= \overline{\psi'_a}(x')\,\sigma^{\mu\nu}\,\psi'_b(x') \\ &= \Lambda^{\mu}{}_{\alpha}\,\Lambda^{\nu}{}_{\beta}\,\overline{\psi_a}(x)\,\sigma^{\alpha\beta}\,\psi_b(x) \\ &= \Lambda^{\mu}{}_{\alpha}\,\Lambda^{\nu}{}_{\beta}\,T^{\alpha\beta}_{ab}(x)\,, \end{aligned} \tag{2.67}$$

$$\begin{aligned} A^{\mu}_{ab}(x) = \overline{\psi_a}(x)\,\gamma^{\mu}\,\gamma^5\,\psi_b(x) \to A'^{\mu}_{ab}(x') &= \overline{\psi'_a}(x')\,\gamma^{\mu}\,\gamma^5\,\psi'_b(x') \\ &= \Lambda^{\mu}{}_{\nu}\,\overline{\psi_a}(x)\,\gamma^{\nu}\,\gamma^5\,\psi_b(x) = \Lambda^{\mu}{}_{\nu}\,A^{\nu}_{ab}(x)\,, \end{aligned} \tag{2.68}$$

$$\begin{aligned} P_{ab}(x) = \overline{\psi_a}(x)\,\gamma^5\,\psi_b(x) \to P'_{ab}(x') &= \overline{\psi'_a}(x')\,\gamma^5\,\psi'_b(x') = \overline{\psi_a}(x)\,\gamma^5\,\psi_b(x) \\ &= P_{ab}(x)\,, \end{aligned} \tag{2.69}$$

where we have considered the possibility of having different fields a and b. The transformations of the covariant bilinears in eqns (2.65)–(2.69) under space and

[5] Indeed, we have $\overline{\psi}(x) = \psi^{\dagger}(x)\,\gamma^0 \to \psi'^{\dagger}(x)\,\gamma^0 = \psi^{\dagger}(x)\,\mathcal{S}^{\dagger}(\Lambda)\,\gamma^0 = \overline{\psi}(x)\,\gamma^0\,\mathcal{S}^{\dagger}(\Lambda)\,\gamma^0$. Using the explicit form in eqn (2.61) of $\mathcal{S}(\Lambda)$ and the property in eqn (2.6), we obtain $\gamma^0\,\mathcal{S}^{\dagger}(\Lambda)\,\gamma^0 = \mathcal{S}^{-1}(\Lambda)$.

time inversions are discussed, respectively, in sections 2.11.2 and 2.11.4. Interactions between different fields ($a \neq b$) or self-interactions ($a = b$) must be expressed in the Lagrangian through the covariant bilinears in eqns (2.65)–(2.69), which allow us to write scalar Lagrangian terms. The bilinears in eqns (2.65)–(2.69) are also called *currents*.

2.4.1 Boosts

The spin part of the boost operator K^k in eqn (B.50) is

$$K^k_{\text{spin}} = -\frac{1}{2}\,\sigma^{0k} = -\frac{i}{2}\,\gamma^0\,\gamma^k = -\frac{i}{2}\,\alpha^k\,, \tag{2.70}$$

where we have used the Dirac notation

$$\alpha^k \equiv \gamma^0\,\gamma^k\,, \tag{2.71}$$

with

$$\left\{\alpha^k\,,\,\alpha^j\right\} = 2\,\delta^{kj}\,. \tag{2.72}$$

The explicit expressions of the matrices α^k in the Dirac and chiral representations of the γ matrices are, respectively,

$$\alpha^k_{\text{D}} = \begin{pmatrix} 0 & \sigma^k \\ \sigma^k & 0 \end{pmatrix}, \qquad \alpha^k_{\text{C}} = \begin{pmatrix} \sigma^k & 0 \\ 0 & -\sigma^k \end{pmatrix}. \tag{2.73}$$

The matrix $\mathcal{S}^k_{\text{boost}}(\varphi)$ for a boost with velocity $v = \tanh\varphi$ along the direction of the axis x^k can be written as

$$\mathcal{S}^k_{\text{boost}}(\varphi) = e^{-i\,\varphi\,K^k_{\text{spin}}} = e^{-\frac{1}{2}\,\varphi\,\alpha^k} = \cosh\frac{\varphi}{2} - \alpha^k\,\sinh\frac{\varphi}{2}\,, \tag{2.74}$$

where φ is the rapidity given in eqn (B.31), with

$$\cosh\frac{\varphi}{2} = \frac{\gamma+1}{2}\,, \qquad \sinh\frac{\varphi}{2} = \frac{\gamma-1}{2}\,, \qquad \gamma = \left(1-v^2\right)^{-1/2}. \tag{2.75}$$

Since $\vec{\alpha}$ is Hermitian, $\mathcal{S}^k_{\text{boost}}(\varphi)$ is also Hermitian,

$$\left[\mathcal{S}^k_{\text{boost}}(\varphi)\right]^\dagger = \mathcal{S}^k_{\text{boost}}(\varphi)\,, \tag{2.76}$$

but not unitary. This is consistent with the fact that under a boost $\psi^\dagger\psi$ is not a scalar, but transforms as the time-component of the four-vector $\overline{\psi}\gamma^\mu\psi$.

2.4.2 Rotations

The spin part of the angular momentum operators J^k in eqn (B.49) is

$$S^k = \frac{1}{4} \sum_{j,l} \epsilon^{kjl} \, \sigma^{jl} = \frac{1}{2} \Sigma^k \,, \tag{2.77}$$

with

$$\Sigma^k \equiv \frac{1}{2} \sum_{j,l} \epsilon^{kjl} \, \sigma^{jl} = \frac{i}{2} \sum_{j,l} \epsilon^{kjl} \, \gamma^j \, \gamma^l \qquad \Longrightarrow \qquad \vec{\Sigma} = (i\gamma^2\gamma^3, i\gamma^3\gamma^1, i\gamma^1\gamma^2) \,. \tag{2.78}$$

It is also possible to write Σ^k as

$$\Sigma^k = \gamma^0 \, \gamma^k \, \gamma^5 \,. \tag{2.79}$$

The matrices satisfy the commutation and anticommutation relations

$$[\Sigma^k, \Sigma^j] = 2\, i \sum_l \epsilon^{kjl} \, \Sigma^l \,, \qquad \{\Sigma^k, \Sigma^j\} = 2\, \delta^{kj} \,. \tag{2.80}$$

The Σ^k matrices have the same expression in the Dirac and chiral representations of the γ matrices:

$$\vec{\Sigma}_{\rm D} = \vec{\Sigma}_{\rm C} = \begin{pmatrix} \vec{\sigma} & 0 \\ 0 & \vec{\sigma} \end{pmatrix} \,, \tag{2.81}$$

from which it is clear that the $\vec{\Sigma}$ matrices represent a 4×4 generalization of the 2×2 Pauli matrices.

The matrix $\mathcal{S}^k_{\rm rot}(\theta)$ for a rotation through an angle θ around the axis x^k can be written as

$$\mathcal{S}^k_{\rm rot}(\theta) = e^{i\,\theta\, S^k} = e^{\frac{i}{2}\,\theta\,\Sigma^k} = \cos\frac{\theta}{2} + i\,\Sigma^k \sin\frac{\theta}{2} \,. \tag{2.82}$$

Since Σ^k is Hermitian, we have

$$\left[\mathcal{S}^k_{\rm rot}(\theta)\right]^\dagger = \left[\mathcal{S}^k_{\rm rot}(\theta)\right]^{-1} \,, \tag{2.83}$$

which implies that $\psi^\dagger\psi$ is a scalar under rotations.

A rotation of 2π changes the sign of ψ, because $\mathcal{S}^k_{\rm rot}(2\pi) = -1$. Therefore, the sign of ψ has no physical meaning. Indeed, all physical quantities depend on the covariant bilinears in eqns (2.65)–(2.69), which are invariant not only under a global change of sign of all fermion fields, but also under a global phase transformation of all fermion fields. This invariance corresponds, through Noether's theorem, to the conservation of the total fermion number, which is consistent with angular momentum conservation: the disappearance of a fermion without creation or annihilation of another fermion would imply an unrecoverable change of 1/2 of the total angular momentum.

2.4.3 Invariants

As discussed in section B.3 of appendix B, there are two Casimir operators of the Poincaré group, P^2 and W^2, which are used for the classification of its representations.

The first Casimir operator is $P^2 = P^\mu P_\mu$, where $P^\mu = i\partial^\mu$ is the momentum operator (eqn (B.86)). Therefore, $P^2 = -\Box$ and from the Klein–Gordon equation (2.12), satisfied by the Dirac field ψ, it is clear that the eigenvalue of P^2 is m^2 and the parameter m in the Dirac Lagrangian in eqn (2.1) can be interpreted as the mass of the particle.

The second Casimir operator is $W^2 = W^\mu W_\mu$, with the Pauli–Lubanski four-vector W_μ given in eqn (B.90), which gives the spin of the particle, according to eqn (B.95). From eqns (B.94) and (2.77), we have in the rest frame

$$W^0 = 0\,, \qquad \vec{W} = \frac{1}{2}\, m\, \vec{\Sigma}\,. \tag{2.84}$$

Consequently,

$$W^2 = -m^2\,\frac{1}{4}\,\vec{\Sigma}^2 = -m^2\,\frac{3}{4} = -m^2\,\frac{1}{2}\left(\frac{1}{2}+1\right), \tag{2.85}$$

which shows that a Dirac field describes particles with spin $s = 1/2$.

2.5 Helicity

From eqn (2.77), the helicity operator in eqn (B.96) is given by

$$\widehat{h} = \frac{\vec{S}\cdot\vec{P}}{s\,|\vec{P}|} = \frac{\vec{\Sigma}\cdot\vec{P}}{|\vec{P}|}\,. \tag{2.86}$$

Since the square of the helicity operator in eqn (2.86) is one, the eigenvalues of the helicity for a Dirac fermion are $h = \pm 1$.

As discussed at the end of section B.3 of appendix B, the mass m and the spin s distinguish (possibly together with other quantum numbers) different particles. The state of a particle is identified by the three components of the momentum $\vec{P}$ and by the helicity.

2.6 Gauge transformations

The Dirac Lagrangian in eqn (2.1) is invariant for global U(1) gauge transformations of the type

$$\psi(x) \to e^{i\theta}\,\psi(x)\,, \tag{2.87}$$

where θ is an arbitrary parameter. In this case, Noether's theorem implies that the electromagnetic current

$$j^\mu = q\,\overline{\psi}\,\gamma^\mu\,\psi \tag{2.88}$$

is conserved,

$$\partial_\mu j^\mu = 0 \,. \tag{2.89}$$

The associated conserved charge operator is

$$\mathsf{Q} = \int \mathrm{d}^3x \, j^0(x) = q \int \mathrm{d}^3x \, \psi^\dagger(x) \, \psi(x) \,, \tag{2.90}$$

where q is the electric charge of the particle (see eqn (2.248)).

2.7 Chirality

The matrix γ^5 is also called the *chirality* matrix. Since the chirality matrix is Hermitian (eqn (2.25)), it can be diagonalized with a unitary transformation $U\gamma^5U^\dagger = \gamma^5_{\text{diag}}$, with $U^\dagger = U^{-1}$. Since $(\gamma^5)^2 = \mathbf{1}$ (eqn (2.24)), the eigenvalues of γ^5 are ± 1. In fact, in the chiral representation in eqn (2.27), γ^5 is diagonal, equal to $\text{diag}(1, 1, -1, -1)$.

Let us denote by ψ_R and ψ_L the fields which are eigenfunctions of γ^5 with eigenvalues $+1$ and -1, respectively:

$$\gamma^5 \, \psi_R = + \psi_R \,, \tag{2.91}$$

$$\gamma^5 \, \psi_L = - \psi_L \,. \tag{2.92}$$

The chiral fields ψ_R and ψ_L are called, respectively, *right-handed* and *left-handed.*

It is always possible to split a generic spinor field ψ into its chiral right-handed and left-handed components:

$$\psi = \psi_R + \psi_L \,, \tag{2.93}$$

with

$$\psi_R = \frac{1+\gamma^5}{2} \, \psi \,, \tag{2.94}$$

$$\psi_L = \frac{1-\gamma^5}{2} \, \psi \,. \tag{2.95}$$

It is convenient to define the *chirality projection matrices*

$$P_R \equiv \frac{1+\gamma^5}{2} \,, \tag{2.96}$$

$$P_L \equiv \frac{1-\gamma^5}{2} \,, \tag{2.97}$$

which satisfy the properties

$$P_R + P_L = \mathbf{1} \,, \tag{2.98}$$

$$(P_R)^2 = P_R \,, \tag{2.99}$$

$$(P_L)^2 = P_L \,, \tag{2.100}$$

$$P_R\, P_L = P_L\, P_R = 0\,. \tag{2.101}$$

Let us consider the Dirac Lagrangian in eqn (2.1). Using the decomposition in eqn (2.93) of the spinor field ψ, we have

$$\mathscr{L} = (\overline{\psi_R} + \overline{\psi_L})\,(i\,\overleftrightarrow{\partial\!\!\!/} - m)\,(\psi_R + \psi_L)\,. \tag{2.102}$$

Since

$$\psi_R = P_R\,\psi\,, \qquad \psi_L = P_L\,\psi\,, \tag{2.103}$$

and

$$P_R^\dagger = P_R\,, \tag{2.104}$$

$$P_L^\dagger = P_L\,, \tag{2.105}$$

$$P_R\,\gamma^0 = \gamma^0\,P_L\,, \tag{2.106}$$

$$P_L\,\gamma^0 = \gamma^0\,P_R\,, \tag{2.107}$$

we obtain

$$\overline{\psi_R} = \overline{(P_R\psi)} = (P_R\psi)^\dagger\gamma^0 = \psi^\dagger\,P_R\gamma^0 = \psi^\dagger\gamma^0 P_L = \overline{\psi}P_L\,, \tag{2.108}$$

$$\overline{\psi_L} = \overline{\psi}P_R\,. \tag{2.109}$$

Therefore, the following four products in the Lagrangian in eqn (2.102) vanish identically:

$$i\overline{\psi_R}\overleftrightarrow{\partial\!\!\!/}\psi_L = i\overline{\psi}P_L\overleftrightarrow{\partial\!\!\!/}P_L\psi = i\overline{\psi}\overleftrightarrow{\partial\!\!\!/}P_R P_L\psi = 0\,, \tag{2.110}$$

$$i\overline{\psi_L}\overleftrightarrow{\partial\!\!\!/}\psi_R = 0\,, \tag{2.111}$$

$$m\overline{\psi_R}\psi_R = m\overline{\psi}P_L P_R\psi = 0\,, \tag{2.112}$$

$$m\overline{\psi_L}\psi_L = 0\,. \tag{2.113}$$

The Dirac Lagrangian in terms of the chiral fields ψ_R and ψ_L then becomes

$$\mathscr{L} = \overline{\psi_R}\,i\,\overleftrightarrow{\partial\!\!\!/}\,\psi_R + \overline{\psi_L}\,i\,\overleftrightarrow{\partial\!\!\!/}\,\psi_L - m\left(\overline{\psi_R}\,\psi_L + \overline{\psi_L}\,\psi_R\right)\,. \tag{2.114}$$

One can see that the chiral fields ψ_R and ψ_L have independent kinetic terms but they are coupled by the mass term. From the Lagrangian in eqn (2.114) one can find the field equations

$$i\,\partial\!\!\!/\,\psi_R = m\,\psi_L\,, \tag{2.115}$$

$$i\,\partial\!\!\!/\,\psi_L = m\,\psi_R\,, \tag{2.116}$$

which demonstrate that the space-time evolutions of the chiral fields ψ_R and ψ_L are related by the mass m.

The chiral fields ψ_R and ψ_L are also called *Weyl spinors*.

A Weyl spinor has only two independent components, as one can understand by noting that the decomposition in eqn (2.93) of a four-component spinor must split the four independent components equally into two groups, one for each chiral component. One can see this explicitly using a definite representation of the Dirac matrices. The most convenient one[6] is the chiral representation in eqn (2.27), in which

$$P_R = \begin{pmatrix} 1 & 0 \\ 0 & 0 \end{pmatrix} , \qquad P_L = \begin{pmatrix} 0 & 0 \\ 0 & 1 \end{pmatrix} . \tag{2.117}$$

Writing the four-component spinor ψ as

$$\psi = \begin{pmatrix} \chi_R \\ \chi_L \end{pmatrix} , \tag{2.118}$$

where χ_R and χ_L are two-component spinors, we have

$$\psi_R = \begin{pmatrix} \chi_R \\ 0 \end{pmatrix} , \qquad \psi_L = \begin{pmatrix} 0 \\ \chi_L \end{pmatrix} , \tag{2.119}$$

showing explicitly that ψ_R and ψ_L have only two independent components.

From eqns (2.28) and (2.114), the Dirac Lagrangian for the two-component fields χ_R and χ_L is

$$\mathscr{L} = i\chi_R^\dagger \sigma^\mu \partial_\mu \chi_R - i\chi_L^\dagger \bar{\sigma}^\mu \partial_\mu \chi_L + m \left(\chi_R^\dagger \chi_L + \chi_L^\dagger \chi_R \right) , \tag{2.120}$$

and the field equations are

$$i\,\sigma^\mu \partial_\mu \chi_R = -\, m\,\chi_L \,, \tag{2.121}$$

$$i\,\bar{\sigma}^\mu \partial_\mu \chi_L = m\,\chi_R \,. \tag{2.122}$$

Because

$$\partial_k = \frac{\partial}{\partial x^k} = \nabla^k \,, \tag{2.123}$$

these equations can be written in the explicit form

$$i \left(\partial_0 + \vec{\sigma} \cdot \vec{\nabla} \right) \chi_R = -\, m\,\chi_L \,, \tag{2.124}$$

$$i \left(\partial_0 - \vec{\sigma} \cdot \vec{\nabla} \right) \chi_L = -\, m\,\chi_R \,. \tag{2.125}$$

The two-component fields χ_R and χ_L are important from a relativistic point of view because in the chiral representation the explicit expressions for the matrices $\sigma^{\mu\nu}$ are

$$\sigma_{\rm C}^{0k} = i \begin{pmatrix} \sigma^k & 0 \\ 0 & -\sigma^k \end{pmatrix} , \qquad \sigma_{\rm C}^{kj} = \sum_\ell \epsilon^{kj\ell} \begin{pmatrix} \sigma^\ell & 0 \\ 0 & \sigma^\ell \end{pmatrix} . \tag{2.126}$$

From eqns (2.59) and (2.63) it follows that the spinor fields χ_R and χ_L transform independently under Lorentz transformations. This is a very important property

[6] In other representations one can show that ψ_L has only two independent components using the constraint $P_R\,\psi_L = 0$.

because it implies that two-component spinors are the simplest nontrivial representations of the Lorentz group and one must consider them as the fundamental quantities for the construction of the Lagrangian, which is a Lorentz scalar.

From the expressions in eqn (2.126) one can see that the spinor fields χ_R and χ_L transform in the same way under rotations, discussed in section 2.4.2. For a rotation through an angle θ around the axis x^k we have

$$\chi_{R,L} \to \left(\cos\frac{\theta}{2} + i\,\sigma^k \sin\frac{\theta}{2} \right) \chi_{R,L} \,. \tag{2.127}$$

On the other hand, the spinor fields χ_R and χ_L transform in different ways under Lorentz boosts, discussed in section 2.4.1. For a boost with velocity $v = \tanh\varphi$ along the direction of the axis x^k we have

$$\chi_R \to \left(\cosh\frac{\varphi}{2} - \sigma^k \sinh\frac{\varphi}{2} \right) \chi_R \,, \tag{2.128}$$

$$\chi_L \to \left(\cosh\frac{\varphi}{2} + \sigma^k \sinh\frac{\varphi}{2} \right) \chi_L \,. \tag{2.129}$$

Therefore, the spinor fields χ_R and χ_L belong to two different representations of the Lorentz group, which are traditionally called *dotted* and *undotted* (see, for example, Ref. [866]).

2.7.1 Massless field

From eqns (2.115) and (2.116) one can see that the space-time evolutions of the chiral fields ψ_R and ψ_L decouple for $m = 0$. In this case, we obtain the *Weyl equations*

$$i\,\partial\!\!\!/\,\psi_R = 0 \,, \tag{2.130}$$

$$i\,\partial\!\!\!/\,\psi_L = 0 \,. \tag{2.131}$$

Since the field equations of the chiral fields ψ_R and ψ_L are decoupled, the chiral fields are independent and it is possible that one of the two chiral fields is sufficient for the description of a massless fermion.

Let us consider a solution $\psi(x,p)$ of the massless Dirac equation

$$i\,\partial\!\!\!/\,\psi(x,p) = 0 \,, \tag{2.132}$$

which is also an eigenfunction of the four-momentum operator $P^\mu = i\partial^\mu$,

$$P^\mu\,\psi(x,p) = p^\mu\,\psi(x,p) \,, \tag{2.133}$$

with energy

$$p^0 = E = |\vec{p}| \,. \tag{2.134}$$

In this case, the massless Dirac equation (2.132) can be written as

$$\left(\gamma^0\,|\vec{p}| - \vec{\gamma}\cdot\vec{p} \right) \psi(x,p) = 0 \,. \tag{2.135}$$

Multiplying this equation on the left by $\gamma^5\gamma^0$ and using eqn (2.79), we obtain

$$\frac{\vec{\Sigma}\cdot\vec{p}}{|\vec{p}|}\,\psi(x,p) = \gamma^5\,\psi(x,p)\,. \tag{2.136}$$

Now we can see that the operator on the left-hand side is nothing but the helicity operator in eqn (2.86), with $P^\mu \to p^\mu$. Equation (2.136) shows that chirality coincides with helicity for the eigenfunctions of the four-momentum which are solutions of the massless Dirac equation. In particular, the eigenfunctions of the chirality matrix are eigenfunctions of the helicity operator with the same eigenvalue:

$$\frac{\vec{\Sigma}\cdot\vec{p}}{|\vec{p}|}\,\psi_R(x,p) = \psi_R(x,p)\,, \tag{2.137}$$

$$\frac{\vec{\Sigma}\cdot\vec{p}}{|\vec{p}|}\,\psi_L(x,p) = -\psi_L(x,p)\,. \tag{2.138}$$

Hence, a massless right-handed chiral field $\psi_R(x,p)$ with a definite four-momentum has positive helicity, and a massless left-handed chiral field $\psi_L(x,p)$ with a definite four-momentum has negative helicity.

2.8 Solution of the Dirac equation

The Dirac equation (2.11) can be solved using the Fourier expansion of the Dirac field

$$\psi(x) = \int \frac{\mathrm{d}^3p}{(2\pi)^3\,2E} \sum_{h=\pm1} \left[a^{(h)}(p)\,u^{(h)}(p)\,e^{-ip\cdot x} + b^{(h)\dagger}(p)\,v^{(h)}(p)\,e^{ip\cdot x} \right]\,, \tag{2.139}$$

where h is the helicity and

$$p^0 = E = \sqrt{\vec{p}^{\,2} + m^2}\,, \tag{2.140}$$

in order to satisfy the Klein–Gordon equation (2.12). The quantities $u^{(h)}(p)$, $v^{(h)}(p)$ are spinors and $a^{(h)}(p)$, $b^{(h)}(p)$ are numerical coefficients. We have adopted the notation

$$a^{(h)\dagger}(p) \equiv a^{(h)*}(p) \qquad \text{and} \qquad b^{(h)\dagger}(p) \equiv b^{(h)*}(p) \tag{2.141}$$

for later convenience in the discussion of the quantized Dirac field (in section 2.9).

The phase-space measure in the Fourier expansion of $\psi(x)$ in eqn (2.139) is invariant under restricted Lorentz transformations. This can be seen by writing it as

$$\frac{\mathrm{d}^3p}{(2\pi)^3\,2E} = \frac{\mathrm{d}^4p}{(2\pi)^4}\,2\pi\,\delta(p^2-m^2)\,\theta(p^0)\,, \tag{2.142}$$

where we have used the property in eqn (A.124) of the Dirac δ-function, which implies

$$\delta(p^2-m^2) = \delta({p^0}^2 - E^2) = \frac{\delta(p^0-E)+\delta(p^0+E)}{2E}\,. \tag{2.143}$$

Note that restricted Lorentz transformations do not change the sign of p^0.

From the Dirac equation in eqn (2.11) one can see that the spinors $u^{(h)}(p)$ and $v^{(h)}(p)$ satisfy the equations

$$(\not{p} - m)\, u^{(h)}(p) = 0\,, \tag{2.144}$$

$$(\not{p} + m)\, v^{(h)}(p) = 0\,. \tag{2.145}$$

For the adjoint spinors, using the property in eqn (2.6), we obtain

$$\overline{u^{(h)}}(p)\,(\not{p} - m) = 0\,, \tag{2.146}$$

$$\overline{v^{(h)}}(p)\,(\not{p} + m) = 0\,. \tag{2.147}$$

From eqns (2.144)–(2.147) it follows that

$$\overline{u^{(h)}}(p)\, v^{(h')}(p) = 0\,. \tag{2.148}$$

Let us now derive the helicity properties of $u^{(h)}(p)$ and $v^{(h)}(p)$. The field $\psi(x)$ can be written as the sum

$$\psi(x) = \sum_{h=\pm1} \psi^{(h)}(x)\,, \tag{2.149}$$

where

$$\psi^{(h)}(x) = \int \frac{\mathrm{d}^3p}{(2\pi)^3\, 2E} \left[a^{(h)}(p)\, u^{(h)}(p)\, e^{-ip\cdot x} + b^{(h)\dagger}(p)\, v^{(h)}(p)\, e^{ip\cdot x} \right] \tag{2.150}$$

is an eigenfield of the helicity operator in eqn (2.86) with eigenvalue h:

$$\widehat{h}\,\psi^{(h)}(x) = h\,\psi^{(h)}(x)\,. \tag{2.151}$$

Applying the helicity operator in eqn (2.86) to $\psi^{(h)}(x)$ we find

$$\begin{aligned} \widehat{h}\,\psi^{(h)}(x) &= \frac{\vec{\Sigma}\cdot\vec{P}}{|\vec{P}|}\,\psi^{(h)}(x) \\ &= \int \frac{\mathrm{d}^3p}{(2\pi)^3\, 2E} \left[a^{(h)}(p)\, \frac{\vec{\Sigma}\cdot\vec{p}}{|\vec{p}|}\, u^{(h)}(p)\, e^{-ip\cdot x} - b^{(h)\dagger}(p)\, \frac{\vec{\Sigma}\cdot\vec{p}}{|\vec{p}|}\, v^{(h)}(p)\, e^{ip\cdot x} \right]. \end{aligned} \tag{2.152}$$

In order to satisfy the eigenvalue equation (2.151), $u^{(h)}(p)$ and $v^{(h)}(p)$ must be eigenfunctions of the helicity operator in momentum space $\vec{\Sigma}\cdot\vec{p}/|\vec{p}|$ with opposite eigenvalues:

$$\frac{\vec{p}\cdot\vec{\Sigma}}{|\vec{p}|}\, u^{(h)}(p) = h\, u^{(h)}(p)\,, \tag{2.153}$$

$$\frac{\vec{p}\cdot\vec{\Sigma}}{|\vec{p}|}\, v^{(h)}(p) = -h\, v^{(h)}(p)\,. \tag{2.154}$$

The corresponding equations for the adjoint spinors are, with eqn (2.6),

$$\overline{u^{(h)}}(p)\,\frac{\vec{p}\cdot\vec{\Sigma}}{|\vec{p}|} = h\,\overline{u^{(h)}}(p)\,, \tag{2.155}$$

$$\overline{v^{(h)}}(p)\,\frac{\vec{p}\cdot\vec{\Sigma}}{|\vec{p}|} = -h\,\overline{v^{(h)}}(p)\,. \tag{2.156}$$

From eqns (2.153)–(2.156) it follows that $\overline{u^{(h)}}(p)u^{(h')}(p) \propto \delta_{hh'}$ and $\overline{v^{(h)}}(p)v^{(h')}(p) \propto \delta_{hh'}$. Here, we adopt the Lorentz-invariant normalization conditions

$$\overline{u^{(h)}}(p)\,u^{(h')}(p) = 2\,m\,\delta_{hh'}\,, \tag{2.157}$$

$$\overline{v^{(h)}}(p)\,v^{(h')}(p) = -2\,m\,\delta_{hh'}\,. \tag{2.158}$$

From these normalization conditions and the properties in eqns (2.144)–(2.147), one can derive the useful relations[7]

$$\overline{u^{(h)}}(p)\,\gamma^\mu\,u^{(h')}(p) = \overline{v^{(h)}}(p)\,\gamma^\mu\,v^{(h')}(p) = 2\,p^\mu\,\delta_{hh'}\,, \tag{2.159}$$

$$\overline{u^{(h)}}(p)\,\gamma^5\,u^{(h')}(p) = \overline{v^{(h)}}(p)\,\gamma^5\,v^{(h')}(p) = 0\,, \tag{2.160}$$

$$u^{(h)\dagger}(p)\,v^{(h')}(p_{\rm P}) = v^{(h)\dagger}(p)\,u^{(h')}(p_{\rm P}) = 0\,. \tag{2.161}$$

where $p_{\rm P} = (p^0, -\vec{p})$.

Using the relations in eqns (2.159) and (2.161), one can find that the coefficients $a^{(h)}(p)$ and $b^{(h)}(p)$ in eqn (2.139) are given by

$$a^{(h)}(p) = \int \mathrm{d}^3x\, u^{(h)\dagger}(p)\,\psi(x)\,e^{ip\cdot x}\,, \tag{2.162}$$

$$b^{(h)}(p) = \int \mathrm{d}^3x\, \psi^\dagger(x)\,v^{(h)}(p)\,e^{ip\cdot x}\,. \tag{2.163}$$

The normalization

$$\int \mathrm{d}^3x\,|\psi(x)|^2 = 1 \tag{2.164}$$

implies that the coefficients $a^{(h)}(p)$ and $b^{(h)}(p)$ are constrained by

$$\int \frac{\mathrm{d}^3p}{(2\pi)^3\,2E} \sum_{h=\pm1} \left[|a^{(h)}(p)|^2 + |b^{(h)}(p)|^2\right] = 1\,. \tag{2.165}$$

[7] For example,

$$\overline{u^{(h)}}(p)\,\gamma^\mu\,u^{(h')}(p) = \overline{u^{(h)}}(p)\,\frac{\gamma^\mu \not{p} + \not{p}\gamma^\mu}{2m}\,u^{(h')}(p) = \frac{p^\mu}{m}\,\overline{u^{(h)}}(p)\,u^{(h')}(p) = 2\,p^\mu\,\delta_{hh'}\,,$$

and

$$\begin{aligned}\overline{u^{(h)}}(p)\,\gamma^5\,u^{(h')}(p) &= \overline{u^{(h)}}(p)\,\gamma^5\,\frac{\not{p}}{m}\,u^{(h')}(p) = -\overline{u^{(h)}}(p)\,\frac{\not{p}}{m}\,\gamma^5\,u^{(h')}(p)\\ &= -\,\overline{u^{(h)}}(p)\,\gamma^5\,u^{(h')}(p) = 0\,.\end{aligned}$$

Since the four spinors $u^{(+)}(p)$, $u^{(-)}(p)$, $v^{(+)}(p)$, $v^{(-)}(p)$ are mutually orthogonal, they are linearly independent and form a basis of the vector space of four-dimensional spinors. The outer products

$$u^{(+)}(p)\,\overline{u^{(+)}}(p)\,, \quad u^{(-)}(p)\,\overline{u^{(-)}}(p)\,, \quad v^{(+)}(p)\,\overline{v^{(+)}}(p)\,, \quad v^{(-)}(p)\,\overline{v^{(-)}}(p) \tag{2.166}$$

form a basis of the vector space of 4×4 matrices. In particular, they satisfy the completeness relation

$$\sum_{h=\pm 1} \left[\frac{u^{(h)}(p)\,\overline{u^{(h)}}(p)}{2m} - \frac{v^{(h)}(p)\,\overline{v^{(h)}}(p)}{2m} \right] = \mathbf{1}\,. \tag{2.167}$$

The components of $\psi(x)$ proportional to $e^{-ip\cdot x}$ and $e^{ip\cdot x}$ are usually called, respectively, positive-energy and negative-energy components because ($P^0 = i\partial_0$)

$$P^0\,e^{-ip\cdot x} = E\,e^{-ip\cdot x}\,, \qquad P^0\,e^{ip\cdot x} = -E\,e^{ip\cdot x}\,. \tag{2.168}$$

It is useful to define the projection operators on the components with positive and negative energy:

$$\Lambda_\pm(p) \equiv \frac{m \pm \not{p}}{2m}\,, \tag{2.169}$$

with

$$\sum_{r=\pm} \Lambda_r(p) = \mathbf{1}\,, \qquad \Lambda_r(p)\,\Lambda_s(p) = \Lambda_r(p)\,\delta_{rs}\,, \tag{2.170}$$

and

$$\Lambda_+(p)\,u^{(h)}(p) = u^{(h)}(p)\,, \qquad \Lambda_+(p)\,v^{(h)}(p) = 0\,, \tag{2.171}$$

$$\Lambda_-(p)\,u^{(h)}(p) = 0\,, \qquad \Lambda_-(p)\,v^{(h)}(p) = v^{(h)}(p)\,. \tag{2.172}$$

From these equations and the completeness relation in eqn (2.167) one can derive the useful identities[8]

$$\Lambda_+(p) = \sum_{h=\pm 1} \frac{u^{(h)}(p)\,\overline{u^{(h)}}(p)}{2m}\,, \tag{2.173}$$

$$\Lambda_-(p) = -\sum_{h=\pm 1} \frac{v^{(h)}(p)\,\overline{v^{(h)}}(p)}{2m}\,. \tag{2.174}$$

From eqns (2.155) and (2.156), the projection operators on the u and v spinors with definite helicity are, respectively,

$$P_h^{(u)} = \frac{1}{2}\left(1 + h\,\frac{\vec{p}\cdot\vec{\Sigma}}{|\vec{p}|}\right)\,, \tag{2.175}$$

[8] For example, multiplying $\Lambda_+(p)\,u^{(h)}(p)$ on the right by $\overline{u^{(h)}}(p)$, summing over h and expressing $\sum_h u^{(h)}(p)\overline{u^{(h)}}(p)$ on the left-hand side as $2m + \sum_h v^{(h)}(p)\overline{v^{(h)}}(p)$, one obtains the identity in eqn (2.173).

$$P_h^{(v)} = \frac{1}{2}\left(1 - h\,\frac{\vec{p}\cdot\vec{\Sigma}}{|\vec{p}|}\right). \tag{2.176}$$

It is possible to write these projection operators in a unified covariant form. By using the property in eqn (2.144) and eqns (2.79), (A.64), we have

$$\begin{aligned}\frac{\vec{p}\cdot\vec{\Sigma}}{|\vec{p}|}\,u^{(h)}(p) &= \frac{\vec{p}\cdot\vec{\Sigma}}{|\vec{p}|}\,\frac{\not{p}}{m}\,u^{(h)}(p) = \frac{\gamma^5\,\gamma^0\,\vec{\gamma}\cdot\vec{p}}{|\vec{p}|}\,\frac{E\,\gamma^0 - \vec{\gamma}\cdot\vec{p}}{m}\,u^{(h)}(p)\\ &= \gamma^5\left(\frac{|\vec{p}|}{m}\,\gamma^0 - \frac{E}{m}\,\frac{\vec{\gamma}\cdot\vec{p}}{|\vec{p}|}\right)u^{(h)}(p) = h\,\gamma^5\,\not{s}_h\,u^{(h)}(p)\,,\end{aligned} \tag{2.177}$$

where s_h^μ is the polarization four-vector

$$s_h^\mu = h\left(\frac{|\vec{p}|}{m}\,,\,\frac{E}{m}\,\frac{\vec{p}}{|\vec{p}|}\right), \tag{2.178}$$

with

$$s_h^2 = -1\,, \qquad s_h\cdot p = 0\,. \tag{2.179}$$

In a similar way, one can obtain

$$\frac{\vec{p}\cdot\vec{\Sigma}}{|\vec{p}|}\,v^{(h)}(p) = -h\,\gamma^5\,\not{s}_h\,v^{(h)}(p)\,. \tag{2.180}$$

Hence, the helicity projection operators in eqn (2.175) and (2.176) can be written in a unified covariant form

$$P_h = \frac{1 + \gamma^5\,\not{s}_h}{2}\,. \tag{2.181}$$

The orthogonality of s_h and p guarantees that P_h commutes with $\Lambda_\pm$ because

$$\left[\gamma^5\,\not{s}_h\,,\,\not{p}\right] = \gamma^5\,\{\not{s}_h\,,\,\not{p}\} - \{\gamma^5\,,\,\not{p}\}\,\not{s}_h = 0\,. \tag{2.182}$$

Therefore, we can define the four projection operators on the components with definite energy and helicity as

$$\Lambda_\pm^h(p) \equiv \Lambda_\pm(p)\,P_h = P_h\,\Lambda_\pm(p) = \left(\frac{m \pm \not{p}}{2m}\right)\left(\frac{1 + \gamma^5\,\not{s}_h}{2}\right), \tag{2.183}$$

such that

$$\sum_{r=\pm}\sum_{h=\pm1}\Lambda_r^h(p) = \mathbf{1}\,, \qquad \Lambda_r^h(p)\,\Lambda_s^{h'}(p) = \Lambda_r^h(p)\,\delta_{rs}\,\delta_{hh'}\,, \tag{2.184}$$

and

$$\Lambda_+^h(p)\,u^{(h')}(p) = \delta_{hh'}\,u^{(h')}(p)\,, \qquad \Lambda_+^h(p)\,v^{(h')}(p) = 0\,, \tag{2.185}$$

$$\Lambda_-^h(p)\,u^{(h')}(p) = 0\,, \qquad \Lambda_-^h(p)\,v^{(h')}(p) = \delta_{hh'}\,v^{(h')}(p)\,. \tag{2.186}$$

From these equations and the completeness relation in eqn (2.167) one can derive the useful identities (using a method similar to that explained in footnote 8 on page 25)

$$\Lambda_+^h(p) = \frac{u^{(h)}(p)\,\overline{u^{(h)}}(p)}{2m}\,, \tag{2.187}$$

$$\Lambda_-^h(p) = -\frac{v^{(h)}(p)\,\overline{v^{(h)}}(p)}{2m}\,. \tag{2.188}$$

Let us finally derive some useful relations among the u and v spinors with different helicities. Since

$$(\not{p}_{\mathrm{P}} - m)\,\gamma^0\,u^{(-h)}(p_{\mathrm{P}}) = 0 \qquad \text{and} \qquad \frac{\vec{p}\cdot\vec{\Sigma}}{|\vec{p}|}\,\gamma^0\,u^{(-h)}(p_{\mathrm{P}}) = h\,\gamma^0\,u^{(-h)}(p_{\mathrm{P}})\,, \tag{2.189}$$

we have

$$\gamma^0\,u^{(-h)}(p_{\mathrm{P}}) = \eta(\vec{p},h)\,u^{(h)}(p)\,, \tag{2.190}$$

where $\eta(\vec{p},h)$ is a phase factor which depends on $\vec{p}$ and h. Changing the signs of $\vec{p}$ and h we obtain $\gamma^0\,u^{(h)}(p) = \eta(-\vec{p},-h)\,u^{(-h)}(p_{\mathrm{P}})$, which is compatible with eqn (2.190) only if

$$\eta(-\vec{p},-h) = \eta^*(\vec{p},h)\,. \tag{2.191}$$

A relation similar to eqn (2.190) holds for the v's. In the treatment of charge conjugation, to be discussed in section 2.11.1, it is convenient to choose the relative phase of the spinors $u^{(h)}(p)$ and $v^{(h)}(p)$ in order to satisfy the relation in eqn (2.354). In this case, we obtain

$$\gamma^0\,v^{(-h)}(p_{\mathrm{P}}) = -\eta^*(\vec{p},h)\,v^{(h)}(p)\,. \tag{2.192}$$

Furthermore, since

$$(\not{p} - m)\,\gamma^5\,v^{(-h)}(p) = 0 \qquad \text{and} \qquad \frac{\vec{p}\cdot\vec{\Sigma}}{|\vec{p}|}\,\gamma^5\,v^{(-h)}(p) = h\,\gamma^5\,v^{(-h)}(p)\,, \tag{2.193}$$

we have

$$\gamma^5\,v^{(-h)}(p) = \zeta(h)\,u^{(h)}(p)\,, \tag{2.194}$$

where $\zeta(h)$ is a phase factor which depends on h. A similar relation for $\gamma^5\,u^{(-h)}(p)$ is constrained by the relation in eqn (2.354) to be

$$\gamma^5\,u^{(-h)}(p) = -\zeta^*(h)\,v^{(h)}(p)\,. \tag{2.195}$$

This is compatible with eqn (2.194) only if

$$\zeta(-h) = -\zeta(h)\,. \tag{2.196}$$

2.8.1 Dirac representation

The explicit form of the free spinors $u^{(h)}(p)$ and $v^{(h)}(p)$ in the Dirac representation of the γ matrices (see eqn (2.21)) is

$$u_{\mathrm{D}}^{(h)}(p) = \begin{pmatrix} \sqrt{E+m}\,\chi^{(h)}(\vec{p}) \\ h\,\sqrt{E-m}\,\chi^{(h)}(\vec{p}) \end{pmatrix}, \tag{2.197}$$

$$v_{\mathrm{D}}^{(h)}(p) = \begin{pmatrix} -\sqrt{E-m}\,\chi^{(-h)}(\vec{p}) \\ h\,\sqrt{E+m}\,\chi^{(-h)}(\vec{p}) \end{pmatrix}, \tag{2.198}$$

where $\chi^{(h)}(\vec{p})$ are the orthonormal two-component helicity eigenstate spinors to be discussed in section 2.8.3. One can easily find that $\zeta(h)$ in eqns (2.194) and (2.195) is given by

$$\zeta(h) = -h\,. \tag{2.199}$$

In the nonrelativistic limit, $|\vec{p}| \ll m$, the spinors in eqns (2.197) and (2.198) are approximated by

$$u_{\mathrm{D}}^{(h)}(p) \simeq \sqrt{2m} \begin{pmatrix} \chi^{(h)}(\vec{p}) \\ h\,\frac{|\vec{p}|}{2m}\,\chi^{(h)}(\vec{p}) \end{pmatrix}, \tag{2.200}$$

$$v_{\mathrm{D}}^{(h)}(p) = \sqrt{2m} \begin{pmatrix} -\frac{|\vec{p}|}{2m}\,\chi^{(-h)}(\vec{p}) \\ h\,\chi^{(-h)}(\vec{p}) \end{pmatrix}. \tag{2.201}$$

Here, the two upper components of $u^{(h)}(p)$, called *large components*, are much larger than the two lower components, called *small components*. The opposite is true for $v^{(h)}(p)$. The Dirac representation is convenient to study nonrelativistic problems because of these properties.

2.8.2 Chiral representation

The explicit form of the free spinors $u^{(h)}(p)$ and $v^{(h)}(p)$ in the chiral representation of the γ matrices (see eqn (2.27)) is

$$u_{\mathrm{C}}^{(h)}(p) = \begin{pmatrix} -\sqrt{E+h\,|\vec{p}|}\,\chi^{(h)}(\vec{p}) \\ \sqrt{E-h\,|\vec{p}|}\,\chi^{(h)}(\vec{p}) \end{pmatrix}, \tag{2.202}$$

$$v_{\mathrm{C}}^{(h)}(p) = -\,h \begin{pmatrix} \sqrt{E-h\,|\vec{p}|}\,\chi^{(-h)}(\vec{p}) \\ \sqrt{E+h\,|\vec{p}|}\,\chi^{(-h)}(\vec{p}) \end{pmatrix}, \tag{2.203}$$

where $\chi^{(h)}(\vec{p})$ are the orthonormal two-component helicity eigenstate spinors discussed in section 2.8.3. One can see that $\zeta(h)$ in eqns (2.194) and (2.195) has the same expression in eqn (2.199) as in the Dirac representation,

$$\zeta(h) = -h \,. \tag{2.204}$$

In the relativistic limit $m \ll E$ the spinors in eqns (2.202) and (2.203) become

$$u_{\mathrm{C}}^{(+)}(p) \simeq -\sqrt{2E} \begin{pmatrix} \chi^{(+)}(\vec{p}) \\ -\frac{m}{2E}\,\chi^{(+)}(\vec{p}) \end{pmatrix} , \qquad u_{\mathrm{C}}^{(-)}(p) \simeq \sqrt{2E} \begin{pmatrix} -\frac{m}{2E}\,\chi^{(-)}(\vec{p}) \\ \chi^{(-)}(\vec{p}) \end{pmatrix} , \tag{2.205}$$

$$v_{\mathrm{C}}^{(+)}(p) \simeq -\sqrt{2E} \begin{pmatrix} \frac{m}{2E}\,\chi^{(-)}(\vec{p}) \\ \chi^{(-)}(\vec{p}) \end{pmatrix} , \qquad v_{\mathrm{C}}^{(-)}(p) \simeq \sqrt{2E} \begin{pmatrix} \chi^{(+)}(\vec{p}) \\ \frac{m}{2E}\,\chi^{(+)}(\vec{p}) \end{pmatrix} , \tag{2.206}$$

with two of the four components suppressed by the small ratio m/E. For this reason the chiral representation is convenient in the treatment of ultrarelativistic particles.

2.8.3 Two-component helicity eigenstate spinors

In sections 2.8.1 and 2.8.2 $\chi^{(h)}(\vec{p})$ are two-component helicity eigenstate spinors that satisfy the eigenvalue equation

$$\frac{\vec{p}\cdot\vec{\sigma}}{|\vec{p}|}\,\chi^{(h)}(\vec{p}) = h\,\chi^{(h)}(\vec{p}) \,. \tag{2.207}$$

Hence, $u^{(h)}(p)$ and $v^{(h)}(p)$ in eqns (2.197), (2.198) and (2.202), (2.203) satisfy the eigenvalue equations (2.153) and (2.154). The eigenvalue equation (2.207) guarantees that $\chi^{(h)}(\vec{p})$ and $\chi^{(-h)}(\vec{p})$ are orthogonal,

$$\left(\chi^{(h)}(\vec{p})\right)^{\dagger} \chi^{(-h)}(\vec{p}) = 0 \,. \tag{2.208}$$

They are also chosen to be normalized to one:

$$\left(\chi^{(h)}(\vec{p})\right)^{\dagger} \chi^{(h)}(\vec{p}) = 1 \,. \tag{2.209}$$

In the treatment of charge conjugation, to be discussed in section 2.11.1, it is convenient to choose the relative phase of $\chi^{(h)}(\vec{p})$ and $\chi^{(-h)}(\vec{p})$ in order to satisfy the relation in eqn (2.354). In this case, in both Dirac and chiral representations we obtain the relation

$$i\sigma^{2}\left(\chi^{(h)}(\vec{p})\right)^{*} = -h\,\chi^{(-h)}(\vec{p}) \,. \tag{2.210}$$

Since $\chi^{(-h)}(-\vec{p})$ is an eigenfunction of $\vec{p}\cdot\vec{\sigma}/|\vec{p}|$ with eigenvalue h, it is proportional to $\chi^{(h)}(\vec{p})$:

$$\chi^{(-h)}(-\vec{p}) = \eta(\vec{p},h)\,\chi^{(h)}(\vec{p}) \,, \tag{2.211}$$

where $\eta(\vec{p},h)$ is the same phase factor as in eqn (2.190). Changing the sign of both $\vec{p}$ and h in eqn (2.211) we obtain $\chi^{(h)}(\vec{p}) = \eta(-\vec{p},-h)\,\chi^{(-h)}(-\vec{p})$, which is compatible with eqn (2.211) only if

$$\eta(-\vec{p},-h) = \eta^*(\vec{p},h)\,, \tag{2.212}$$

in agreement with eqn (2.191). Furthermore, the compatibility of eqns (2.211) and (2.210) gives the constraint

$$\eta(\vec{p},-h) = -\eta^*(\vec{p},h)\,. \tag{2.213}$$

Let us finally remark that the two-component helicity eigenstate spinors satisfy the useful relation[9]

$$\left(\chi^{(h)}(\vec{p})\right)^\dagger \sigma^k\,\chi^{(h)}(\vec{p}) = h\,\frac{p^k}{|\vec{p}|}\,. \tag{2.214}$$

Using polar coordinates θ and ϕ, with $0 \leq \theta \leq \pi$ and $0 \leq \phi < 2\pi$, the three-momentum is written as $\vec{p} = |\vec{p}|\,(\sin\theta\cos\phi, \sin\theta\sin\phi, \cos\theta)$ and the two-component helicity eigenstate spinors are given by

$$\chi^{(+)}(\vec{p}) = \begin{pmatrix} \cos\frac{\theta}{2} \\ \sin\frac{\theta}{2}\,e^{i\phi} \end{pmatrix}, \qquad \chi^{(-)}(\vec{p}) = \begin{pmatrix} -\sin\frac{\theta}{2}\,e^{-i\phi} \\ \cos\frac{\theta}{2} \end{pmatrix}. \tag{2.215}$$

Since the transformation $\vec{p} \to -\vec{p}$ is equivalent to a transformation $\theta \to \pi - \theta$ and $\phi \to \phi \pm \pi$, with the plus sign if $0 \leq \phi < \pi$ and the minus sign if $\pi \leq \phi < 2\pi$, we have

$$\eta(\vec{p},h) = h\,e^{-ih\phi}\,. \tag{2.216}$$

It is often convenient to orient $\vec{p}$ along the z axis: $\vec{p} = (0,0,|\vec{p}|)$, which implies $\theta = 0$ and

$$\chi^{(+)}(\vec{p}) = \begin{pmatrix} 1 \\ 0 \end{pmatrix} \equiv \chi^{(+)}\,, \qquad \chi^{(-)}(\vec{p}) = \begin{pmatrix} 0 \\ 1 \end{pmatrix} \equiv \chi^{(-)}\,. \tag{2.217}$$

This is also a convenient form for the two-component spinors in the rest frame of the particle, where the helicity is undetermined. In the following, we will adopt such a definition.

2.8.4 Massless field

From eqns (2.205) and (2.206) it is clear that in the case of a massless fermion the four spinors $u^{(\pm)}(p)$ and $v^{(\pm)}(p)$ have only two nonvanishing components. In this subsection we discuss the implications of this property for the massless chiral fields ψ_R and ψ_L.

---

[9] From eqns (2.207) and (A.35) we have

$$\begin{aligned}\left(\chi^{(h)}(\vec{p})\right)^\dagger \sigma^k \chi^{(h)}(\vec{p}) &= \left(\chi^{(h)}(\vec{p})\right)^\dagger \frac{(\vec{p}\cdot\vec{\sigma})\sigma^k + \sigma^k(\vec{p}\cdot\vec{\sigma})}{2h|\vec{p}|}\chi^{(h)}(\vec{p}) \\ &= \frac{p^j}{2h|\vec{p}|}\left(\chi^{(h)}(\vec{p})\right)^\dagger \{\sigma^k,\sigma^j\}\chi^{(h)}(\vec{p}) = \frac{p^k}{h|\vec{p}|}\,.\end{aligned}$$

The Fourier expansions of the chiral fields ψ_R and ψ_L are given by

$$\psi_{R,L}(x) = \int \frac{\mathrm{d}^3p}{(2\pi)^3\,2E} \sum_{h=\pm 1} \left[a^{(h)}(p)\, u^{(h)}_{R,L}(p)\, e^{-ip\cdot x} + b^{(h)\dagger}(p)\, v^{(h)}_{R,L}(p)\, e^{ip\cdot x} \right] . \tag{2.218}$$

From eqns (2.205) and (2.206), for $m = 0$ we have

$$u_L^{(+)}(p) = u_R^{(-)}(p) = v_R^{(+)}(p) = v_L^{(-)}(p) = 0 \,, \tag{2.219}$$

and

$$u_R^{(+)}(p) = u^{(+)}(p) \,, \quad u_L^{(-)}(p) = u^{(-)}(p) \,, \quad v_L^{(+)}(p) = v^{(+)}(p) \,, \quad v_R^{(-)}(p) = v^{(-)}(p) \,. \tag{2.220}$$

Therefore, the Fourier expansions of the massless chiral fields ψ_R and ψ_L simplify to

$$\psi_R(x) = \int \frac{\mathrm{d}^3p}{(2\pi)^3\,2E} \left[a^{(+)}(p)\, u^{(+)}(p)\, e^{-ip\cdot x} + b^{(-)\dagger}(p)\, v^{(-)}(p)\, e^{ip\cdot x} \right] , \tag{2.221}$$

$$\psi_L(x) = \int \frac{\mathrm{d}^3p}{(2\pi)^3\,2E} \left[a^{(-)}(p)\, u^{(-)}(p)\, e^{-ip\cdot x} + b^{(+)\dagger}(p)\, v^{(+)}(p)\, e^{ip\cdot x} \right] . \tag{2.222}$$

One can see explicitly that the massless chiral fields $\psi_R(x)$ and $\psi_L(x)$ are independent, in agreement with the discussion in section 2.7.1. Moreover, the positive energy components of $\psi_R(x)$ and $\psi_L(x)$ have positive and negative helicity, respectively, in agreement with eqns (2.137) and (2.138).

2.9 Quantization

The quantization of the Dirac field can be implemented by imposing the canonical equal-time anticommutation relations in section C.2 of appendix C. Special care is necessary for the determination of the canonical conjugated momentum, because the Dirac Lagrangian in eqn (2.1) contains $\partial_0\psi$ and $\partial_0\psi^\dagger$, which are not independent. Separating the real and imaginary parts of each component ψ_α of ψ, and taking into account that $\mathfrak{Re}\pi_\alpha = \partial\mathscr{L}/\partial(\partial_0\mathfrak{Re}\psi_\alpha)$ and $\mathfrak{Im}\pi_\alpha = -\partial\mathscr{L}/\partial(\partial_0\mathfrak{Im}\psi_\alpha)$, one can find that

$$\pi = i\,\psi^\dagger \,. \tag{2.223}$$

Thus, the anticommutation relations in eqns (C.11) and (C.12) can be written as

$$\{\psi_\alpha(t,\vec{x}), \psi_\beta^\dagger(t,\vec{y})\} = \delta^3(\vec{x}-\vec{y})\,\delta_{\alpha\beta} \,, \tag{2.224}$$

$$\{\psi_\alpha(t,\vec{x}), \psi_\beta(t,\vec{y})\} = \{\psi_\alpha^\dagger(t,\vec{x}), \psi_\beta^\dagger(t,\vec{y})\} = 0 \,, \tag{2.225}$$

where α, β are Dirac indices. The anticommutation relations in eqn (2.224) and (2.225) are satisfied by writing the Fourier expansion of the quantized Dirac field $\psi(x)$ as in eqn (2.139), with coefficients $a^{(h)}(p)$ and $b^{(h)}(p)$ that satisfy the anticommutation relations

$$\{a^{(h)}(p), a^{(h')\dagger}(p')\} = \{b^{(h)}(p), b^{(h')\dagger}(p')\} = (2\pi)^3\, 2E\, \delta^3(\vec{p}-\vec{p}\,')\, \delta_{hh'}\,, \tag{2.226}$$

$$\{a^{(h)}(p), a^{(h')}(p')\} = \{a^{(h)\dagger}(p), a^{(h')\dagger}(p')\} = 0\,, \tag{2.227}$$

$$\{b^{(h)}(p), b^{(h')}(p')\} = \{b^{(h)\dagger}(p), b^{(h')\dagger}(p')\} = 0\,, \tag{2.228}$$

$$\{a^{(h)}(p), b^{(h')}(p')\} = \{a^{(h)\dagger}(p), b^{(h')\dagger}(p')\} = 0\,, \tag{2.229}$$

$$\{a^{(h)}(p), b^{(h')\dagger}(p')\} = \{a^{(h)\dagger}(p), b^{(h')}(p')\} = 0\,. \tag{2.230}$$

Since these are the canonical anticommutation relations for fermionic ladder operators, the coefficients $a^{(h)\dagger}(p)$ and $b^{(h)\dagger}(p)$ can be interpreted, respectively, as the particle and antiparticle creation operators which are necessary for constructing the Fock space of states starting from the ground state $|0\rangle$, such that

$$a^{(h)}(p)\,|0\rangle = 0\,, \qquad b^{(h)}(p)\,|0\rangle = 0\,. \tag{2.231}$$

Hence, the state $|0\rangle$ represents the vacuum and the states of a fermion $\mathfrak{f}$ and an antifermion $\bar{\mathfrak{f}}$ with definite momentum and helicity are given by

$$|\mathfrak{f}(p,h)\rangle = a^{(h)\dagger}(p)\,|0\rangle\,, \qquad |\bar{\mathfrak{f}}(p,h)\rangle = b^{(h)\dagger}(p)\,|0\rangle\,. \tag{2.232}$$

Since the vacuum is normalized to one,

$$\langle 0|0\rangle = 1\,, \tag{2.233}$$

we have

$$\langle \mathfrak{f}(p,h)|\mathfrak{f}(p',h')\rangle = \langle \bar{\mathfrak{f}}(p,h)|\bar{\mathfrak{f}}(p',h')\rangle = (2\pi)^3\, 2E\, \delta^3(\vec{p}-\vec{p}\,')\, \delta_{hh'}\,. \tag{2.234}$$

This expression can be written as

$$\langle \mathfrak{f}(p,h)|\mathfrak{f}(p',h')\rangle = \langle \bar{\mathfrak{f}}(p,h)|\bar{\mathfrak{f}}(p',h')\rangle = 2E\, \delta_{hh'} \int \mathrm{d}^3x\, e^{i(\vec{p}-\vec{p}\,')\cdot\vec{x}}\,. \tag{2.235}$$

Therefore, we have

$$\langle \mathfrak{f}(p,h)|\mathfrak{f}(p,h')\rangle = \langle \bar{\mathfrak{f}}(p,h)|\bar{\mathfrak{f}}(p,h')\rangle = 2E\, V\, \delta_{hh'}\,, \tag{2.236}$$

where $V = \int \mathrm{d}^3x$ is the total volume. Since the total volume is infinite, the one-particle states in eqn (2.232) are not properly normalized to describe real particles. In order to obtain well-normalized states one must construct wave packet superpositions of one-particle states in eqn (2.232) as described in section 2.12, or consider a finite normalization volume as described in section 2.13. The completeness in the Fock space of one-particle states is given by

$$\int \frac{\mathrm{d}^3p}{(2\pi)^3\, 2E} \sum_h \left[|\mathfrak{f}(p,h)\rangle\langle \mathfrak{f}(p,h)| + |\bar{\mathfrak{f}}(p,h)\rangle\langle \bar{\mathfrak{f}}(p,h)| \right] = 1\,. \tag{2.237}$$

The product $E\,\delta^3(\vec{p}-\vec{p}\,')$ in the anticommutation relations in eqns (2.226)–(2.230) and in the normalization of states in eqn (2.234) is invariant under restricted

Lorentz transformations. It is clearly invariant under rotations. In order to see that it is also invariant under boosts, let us consider, without loss of generality, a boost with velocity v in the direction of the z axis. In the boosted frame, the energy and the third component of the momentum are given by

$$\widetilde{E} = \gamma \left(E - v \, p^3 \right) , \qquad \widetilde{p}^3 = \gamma \left(p^3 - v \, E \right) . \tag{2.238}$$

Since

$$\frac{\mathrm{d}\widetilde{p}^3}{\mathrm{d}p^3} = \gamma \left(1 - v \, \frac{\mathrm{d}E}{\mathrm{d}p^3} \right) = \gamma \left(1 - v \, \frac{p^3}{E} \right) , \tag{2.239}$$

we obtain, using the property in eqn (A.124) of the Dirac δ-function,

$$\widetilde{E} \, \delta^3(\widetilde{\vec{p}} - \widetilde{\vec{p}}\,') = \gamma \left(E - v \, p^3 \right) \delta^3(\vec{p} - \vec{p}\,') \left| \frac{\mathrm{d}\widetilde{p}^3}{\mathrm{d}p^3} \right|^{-1} = E \, \delta^3(\vec{p} - \vec{p}\,') \, . \tag{2.240}$$

Hence, it follows that the anticommutation relations in eqns (2.226)–(2.230) and the normalization of states in eqn (2.234) are invariant under restricted Lorentz transformations.

From the general expression in eqn (C.44), the energy–momentum operator of a Dirac field is given by[10]

$$\mathsf{P}^\mu = \int \mathrm{d}^3x \, \psi^\dagger(x) \, i \overleftrightarrow{\partial^\mu} \, \psi(x) \, . \tag{2.241}$$

Using the Fourier expansion in eqn (2.139) of the Dirac field and the properties in eqns (2.159) and (2.161), we obtain

$$\mathsf{P}^\mu = \int \frac{\mathrm{d}^3p}{(2\pi)^3 \, 2E} \, p^\mu \sum_{h=\pm 1} \left(a^{(h)\dagger}(p) \, a^{(h)}(p) - b^{(h)}(p) \, b^{(h)\dagger}(p) \right) . \tag{2.242}$$

Unfortunately, the operator P^μ cannot be interpreted as the energy–momentum operator, because

$$\langle 0 | \mathsf{P}^\mu | 0 \rangle = - \int \mathrm{d}^3p \, p^\mu \, \delta^3(0) \tag{2.243}$$

would imply that the vacuum does not have zero energy and momentum (in fact, the energy and momentum of the vacuum would be infinite!). In order to get a well-behaved energy–momentum operator, we subtract from P^μ its vacuum expectation value $\langle 0 | \mathsf{P}^\mu | 0 \rangle$. In this way we obtain the normally ordered energy–momentum operator

$$: \mathsf{P}^\mu : \, = \int \frac{\mathrm{d}^3p}{(2\pi)^3 \, 2E} \, p^\mu \sum_{h=\pm 1} \left(a^{(h)\dagger}(p) \, a^{(h)}(p) + b^{(h)\dagger}(p) \, b^{(h)}(p) \right) . \tag{2.244}$$

Normal ordering is a general prescription for obtaining well-behaved operators with zero vacuum expectation value. In a normally ordered product of a creation

---

[10] The Dirac equation (2.11) implies that $\mathscr{L} = 0$.

operator $A^\dagger$ and a destruction operator B, the order of the operators is rearranged putting all the annihilation operators on the right, taking into account the anticommutation of fermion operators:

$$:A^\dagger B: \;= A^\dagger B\,, \qquad :B\,A^\dagger: \;= \pm A^\dagger B\,, \tag{2.245}$$

with the $+$ sign for bosonic operators and the $-$ sign for fermionic operators. All the other products remain unaltered.

Interpreting the normally ordered operator $:\mathsf{P}^\mu:$ as the energy–momentum operator of the Dirac field, we obtain that particles and antiparticles have the correct values of energy and momentum:

$$:\mathsf{P}^\mu:|\mathfrak{f}(p,h)\rangle = p^\mu\,|\mathfrak{f}(p,h)\rangle\,, \qquad :\mathsf{P}^\mu:|\bar{\mathfrak{f}}(p,h)\rangle = p^\mu\,|\bar{\mathfrak{f}}(p,h)\rangle\,. \tag{2.246}$$

The normally ordered electric charge operator in eqn (2.90) is given by

$$:\mathsf{Q}: \;= q\int \frac{\mathrm{d}^3p}{(2\pi)^3\,2E}\sum_{h=\pm1}\left(a^{(h)\dagger}(p)a^{(h)}(p) - b^{(h)\dagger}(p)b^{(h)}(p)\right). \tag{2.247}$$

Accordingly, particle and antiparticle have opposite electric charges:

$$:\mathsf{Q}:|\mathfrak{f}(p,h)\rangle = +q\,|\mathfrak{f}(p,h)\rangle\,, \qquad :\mathsf{Q}:|\bar{\mathfrak{f}}(p,h)\rangle = -q\,|\bar{\mathfrak{f}}(p,h)\rangle\,. \tag{2.248}$$

States with more than one fermion are totally antisymmetric under exchange of two fermions, as required by Fermi–Dirac statistics. For example, for a two-fermion state we have

$$\begin{aligned}|\mathfrak{f}_1(p_1,h_1),\mathfrak{f}_2(p_2,h_2)\rangle &= a_{\mathfrak{f}_1}^{(h_1)\dagger}(p_1)\,a_{\mathfrak{f}_2}^{(h_2)\dagger}(p_2)\,|0\rangle \\ &= -\,a_{\mathfrak{f}_2}^{(h_2)\dagger}(p_2)\,a_{\mathfrak{f}_1}^{(h_1)\dagger}(p_1)\,|0\rangle = -|\mathfrak{f}_2(p_2,h_2),\mathfrak{f}_1(p_1,h_1)\rangle\,,\end{aligned} \tag{2.249}$$

where the two fermions can be different (the creation and annihilation operators of different fermions anticommute). The Pauli principle follows: two identical fermions cannot exist in the same state.

An important quantity for the calculation of transition amplitudes is the Feynman propagator

$$G(x-x') = \langle 0|\mathrm{T}[\psi(x)\overline{\psi}(x')]|0\rangle = \lim_{\epsilon\to 0} i\int\frac{\mathrm{d}^4p}{(2\pi)^4}\,\frac{\not{p}+m}{p^2-m^2+i\,\epsilon}\,\mathrm{e}^{-ip\cdot(x-x')}\,, \tag{2.250}$$

where we have introduced Wick's chronological product $\mathrm{T}[A(x)B(y)]$ which is defined as

$$\mathrm{T}[A(x)B(y)] = \begin{cases} A(x)B(y) & \text{if} \quad x_0 > y_0\,, \\ \pm B(y)A(x) & \text{if} \quad y_0 > x_0\,, \end{cases} \tag{2.251}$$

with the $+$ sign for bosonic operators and the $-$ sign for fermionic operators.

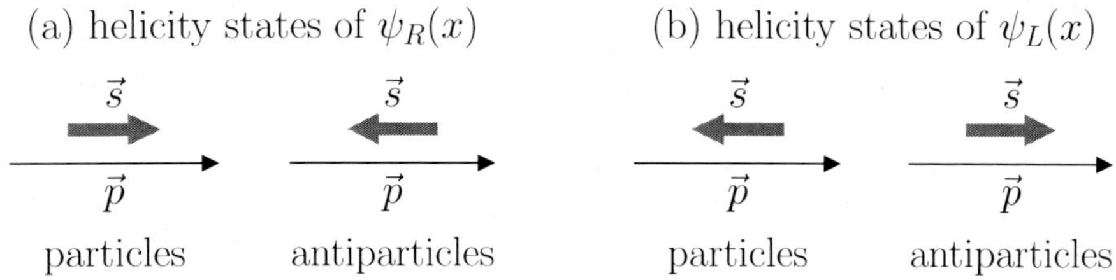

FIG. 2.1. Helicity states of the massless chiral fields $\psi_R(x)$ and $\psi_L(x)$.

2.9.1 Causality

In a local theory, causality demands that measurements at space-time points separated by a space-like interval do not interfere. This means that the corresponding operators must commute. Since all measurable quantities are represented by operators which are bilinear in the fermion fields, causality is satisfied if fermion fields at space-time points separated by a space-like interval commute or anticommute. The equal-time anticommutation relations in eqns (2.224) and (2.225) imply that fermion fields at space-like separated points must anticommute. Indeed, using the Fourier expansion in eqn (2.139) and the anticommutation relations in eqns (2.226)–(2.230) one can find that for any space-time points x and y

$$\{\psi_\alpha(x), \psi_\beta(y)\} = \{\psi^\dagger_\alpha(x), \psi^\dagger_\beta(y)\} = 0\,, \tag{2.252}$$

and

$$\{\psi_\alpha(x), \overline{\psi}_\beta(y)\} = \left(i\partial\!\!\!/^{(x)} + m\right)_{\alpha\beta} \Delta(x-y)\,, \tag{2.253}$$

where $\partial^{(x)}_\mu \equiv \partial/\partial x^\mu$ and

$$\Delta(x-y) = \int \frac{d^3p}{(2\pi)^3\, 2E} \left(e^{-ip\cdot(x-y)} - e^{ip\cdot(x-y)}\right) = -2i \int \frac{d^3p}{(2\pi)^3\, 2E} \sin[p\cdot(x-y)]\,. \tag{2.254}$$

Using eqn (2.143), this function of the space-time interval $x-y$ can also be written as

$$\Delta(x-y) = \int \frac{d^4p}{(2\pi)^3}\, \delta(p^2-m^2)\, \varepsilon(p^0)\, e^{-ip\cdot(x-y)}\,, \tag{2.255}$$

where $\varepsilon(p^0) = p^0/|p^0|$. This expression shows that $\Delta(x-y)$ is invariant under restricted Lorentz transformations, which maintain the sign of p^0. Since $\Delta(0, \vec{x}-\vec{y}) = 0$, the function $\Delta(x-y)$ vanishes for space-like $x-y$. Therefore,

$$\{\psi_\alpha(x), \overline{\psi}_\beta(y)\} = 0 \quad \text{for} \quad (x-y)^2 < 0\,, \tag{2.256}$$

as required by causality.

2.9.2 Massless field

As we have seen in sections 2.7.1 and 2.8.4, in the case of a massless field the chiral components $\psi_R(x)$ and $\psi_L(x)$ are independent and it is possible that only one of them is sufficient for the description of a fermion.

From the quantum versions of the Fourier expansions in eqns (2.221) and (2.222) of $\psi_R(x)$ and $\psi_L(x)$ one immediately see that $\psi_R(x)$ describes only particles with positive helicity and antiparticles with negative helicity, whereas $\psi_L(x)$ describes only particles with negative helicity and antiparticles with positive helicity, as summarized in Fig. 2.1.

2.10 Symmetry transformation of states

In order to understand how states transform under a symmetry transformation, let us consider a generic state $|\xi\rangle$ in the Hilbert space of a system. Under a symmetry transformation, the state $|\xi\rangle$ transforms into another state $|\xi'\rangle$ that belongs to the Hilbert space of the system under consideration. Thus, there must be an operator U which performs the transformation:

$$|\xi\rangle \to |\xi'\rangle = \mathsf{U}\,|\xi\rangle\,. \tag{2.257}$$

In order for the invariance under a transformation to be a symmetry of the system, the absolute value of the scalar product between any two states $|\xi_1\rangle$ and $|\xi_2\rangle$ of the system must be invariant:

$$|\langle\xi_1'|\xi_2'\rangle| = |\langle\xi_1|\xi_2\rangle|\,. \tag{2.258}$$

This means that

$$|\langle\xi_1|\mathsf{U}^\dagger\mathsf{U}|\xi_2\rangle| = |\langle\xi_1|\xi_2\rangle|\,. \tag{2.259}$$

Therefore, if the transformation is a symmetry of the system, U is a *unitary* operator:

$$\mathsf{U}^\dagger = \mathsf{U}^{-1}\,. \tag{2.260}$$

However, U can be either linear or antilinear. Given a state

$$|\xi\rangle = c_1\,|\xi_1\rangle + c_2\,|\xi_2\rangle\,, \tag{2.261}$$

with complex coefficients c_1, c_2, a linear operator satisfies

$$\mathsf{U}\,|\xi\rangle = c_1\,\mathsf{U}\,|\xi_1\rangle + c_2\,\mathsf{U}\,|\xi_2\rangle\,, \tag{2.262}$$

while an antilinear operator takes the complex conjugate of the coefficients:

$$\mathsf{U}\,|\xi\rangle = c_1^*\,\mathsf{U}\,|\xi_1\rangle + c_2^*\,\mathsf{U}\,|\xi_2\rangle\,. \tag{2.263}$$

Antilinear unitary operators are usually called *antiunitary*. Although this name is somewhat confusing, we use it in this book, whereas we call linear unitary operators simply *unitary*. As usual, we denote with the same letter U both unitary and antiunitary operators, their character being clear from the context.

For a unitary operator U, we have

$$\langle \xi_1' | \xi_2' \rangle = \langle \xi_1 | \mathsf{U}^\dagger \mathsf{U} | \xi_2 \rangle = \langle \xi_1 | \xi_2 \rangle . \tag{2.264}$$

This relation is consistent with the linearity in eqn (2.262), as one can see considering

$$|\xi'\rangle = \mathsf{U}\,|\xi\rangle = c_1\,\mathsf{U}\,|\xi_1\rangle + c_2\,\mathsf{U}\,|\xi_2\rangle = c_1\,|\xi_1'\rangle + c_2\,|\xi_2'\rangle . \tag{2.265}$$

We have

$$\begin{aligned}\langle \xi' | \xi' \rangle &= |c_1|^2\,\langle \xi_1' | \xi_1' \rangle + |c_2|^2\,\langle \xi_2' | \xi_2' \rangle + c_1^*\,c_2\,\langle \xi_1' | \xi_2' \rangle + c_1\,c_2^*\,\langle \xi_2' | \xi_1' \rangle \\ &= |c_1|^2\,\langle \xi_1 | \xi_1 \rangle + |c_2|^2\,\langle \xi_2 | \xi_2 \rangle + c_1^*\,c_2\,\langle \xi_1 | \xi_2 \rangle + c_1\,c_2^*\,\langle \xi_2 | \xi_1 \rangle ,\end{aligned} \tag{2.266}$$

which is consistent with

$$\langle \xi' | \xi' \rangle = \langle \xi | \xi \rangle = |c_1|^2\,\langle \xi_1 | \xi_1 \rangle + |c_2|^2\,\langle \xi_2 | \xi_2 \rangle + c_1^*\,c_2\,\langle \xi_1 | \xi_2 \rangle + c_1\,c_2^*\,\langle \xi_2 | \xi_1 \rangle . \tag{2.267}$$

On the other hand, for an antiunitary operator U we have

$$|\xi'\rangle = \mathsf{U}\,|\xi\rangle = c_1^*\,\mathsf{U}\,|\xi_1\rangle + c_2^*\,\mathsf{U}\,|\xi_2\rangle = c_1^*\,|\xi_1'\rangle + c_2^*\,|\xi_2'\rangle . \tag{2.268}$$

Hence,

$$\langle \xi' | \xi' \rangle = |c_1|^2\,\langle \xi_1' | \xi_1' \rangle + |c_2|^2\,\langle \xi_2' | \xi_2' \rangle + c_1\,c_2^*\,\langle \xi_1' | \xi_2' \rangle + c_1^*\,c_2\,\langle \xi_2' | \xi_1' \rangle . \tag{2.269}$$

This relation would not be consistent with eqn (2.267) if an antiunitary operator were to behave according to eqn (2.264). Hence, for an antiunitary operator U we have, in place of eqn (2.264),

$$\langle \xi_1' | \xi_2' \rangle = \langle \xi_1 | \mathsf{U}^\dagger \mathsf{U} | \xi_2 \rangle^* = \langle \xi_2 | \xi_1 \rangle . \tag{2.270}$$

The relation between the transformation of states and the transformation of operators can be found considering the expectation value $\langle \xi | \mathsf{O} | \xi \rangle$ of an operator O. Under the transformation in eqn (2.257) the operator transforms as

$$\mathsf{O} \to \mathsf{O}' . \tag{2.271}$$

If the transformation is a symmetry of the system we must have

$$|\langle \xi' | \mathsf{O}' | \xi' \rangle| = |\langle \xi | \mathsf{O} | \xi \rangle| . \tag{2.272}$$

From eqn (2.261) we can write $\langle \xi | \mathsf{O} | \xi \rangle$ as

$$\langle \xi | \mathsf{O} | \xi \rangle = |c_1|^2\,\langle \xi_1 | \mathsf{O} | \xi_1 \rangle + |c_2|^2\,\langle \xi_2 | \mathsf{O} | \xi_2 \rangle + c_1^*\,c_2\,\langle \xi_1 | \mathsf{O} | \xi_2 \rangle + c_1\,c_2^*\,\langle \xi_2 | \mathsf{O} | \xi_1 \rangle . \tag{2.273}$$

If the transformation is unitary, from eqn (2.265) we obtain

$$\begin{aligned}\langle \xi' | \mathsf{O}' | \xi' \rangle = {} & |c_1|^2 \langle \xi_1 | \mathsf{U}^\dagger \mathsf{O}' \mathsf{U} | \xi_1 \rangle + |c_2|^2 \langle \xi_2 | \mathsf{U}^\dagger \mathsf{O}' \mathsf{U} | \xi_2 \rangle \\ & + c_1^* c_2 \langle \xi_1 | \mathsf{U}^\dagger \mathsf{O}' \mathsf{U} | \xi_2 \rangle + c_1 c_2^* \langle \xi_2 | \mathsf{U}^\dagger \mathsf{O}' \mathsf{U} | \xi_1 \rangle .\end{aligned} \tag{2.274}$$

Comparing this with eqn (2.273) one can see that the equality in eqn (2.272) is satisfied for any value of c_1 and c_2 if

$$\mathsf{U}^\dagger\,\mathsf{O}'\,\mathsf{U} = \mathsf{O} \qquad \Longrightarrow \qquad \mathsf{O}' = \mathsf{U}\,\mathsf{O}\,\mathsf{U}^\dagger\,. \tag{2.275}$$

In a similar way, one can show that for an antiunitary transformation

$$\mathsf{U}^\dagger\mathsf{O}'^{*}\mathsf{U} = \mathsf{O} \qquad \Longrightarrow \qquad \mathsf{O}'^{*} = \mathsf{U}\,\mathsf{O}\,\mathsf{U}^\dagger\,. \tag{2.276}$$

Since quantized fields are operators, for a unitary transformation they must transform as

$$\psi'(x) = \mathsf{U}\,\psi(x)\,\mathsf{U}^\dagger\,, \tag{2.277}$$

whereas for an antiunitary transformation

$$\psi'^{*}(x) = \mathsf{U}\,\psi(x)\,\mathsf{U}^\dagger\,. \tag{2.278}$$

Most symmetry transformations are represented by unitary operators, but time reversal, to be discussed in section 2.11.4, and CPT, to be discussed in section 2.11.5, are represented by antiunitary operators.

Let us note that the transformations in eqns (2.277) and (2.278) are *active*, because they change the system, leaving the coordinates intact. In other words, eqns (2.277) and (2.278) give the transformation of a quantized field operator for a symmetry transformation of the system described by such a field, while the point of view of the observer, represented by the coordinate frame, remains unchanged. By contrast, for example the transformation in eqn (2.59) of the Dirac field under a Lorentz transformation in eqn (B.1) is *passive*, because the change of coordinates in the Lorentz transformation in eqn (B.1) represents a change of the point of view of the observer and the system represented by the Dirac field remains untouched.

2.10.1 Space-time translations

For an infinitesimal space-time translation

$$x^\mu \to x'^\mu = x^\mu + \varepsilon\, b^\mu\,, \tag{2.279}$$

with infinitesimal ε , the unitary transformation operator U can be written as

$$\mathsf{U} = 1 + i\,\varepsilon\, b_\mu\, \mathsf{P}^\mu\,, \tag{2.280}$$

with an Hermitian operator P^μ which is the generator of space-time translations. From eqn (2.277) we have

$$\psi'(x) = \psi(x) + i\,\varepsilon\, b_\mu\, [\mathsf{P}^\mu, \psi(x)]\,. \tag{2.281}$$

On the other hand, from eqns (B.77), (B.85), and (B.86) one can see that the transformation of the field $\psi(x)$ under the space-time translation in eqn (2.279) is given by

$$\psi'(x) = \psi(x) - i\,\varepsilon\, b_\mu\, \partial^\mu\, \psi(x)\,. \tag{2.282}$$

Comparing eqns (2.281) and (2.282), we obtain, in agreement with the general eqn (C.29),

$$[\mathsf{P}^\mu, \psi(x)] = -i\, \partial^\mu\, \psi(x)\,. \tag{2.283}$$

This commutation relation defines the Hermitian operator P^μ. One can check, by using the Fourier expansion of $\psi(x)$ in eqn (2.139) and the anticommutation relations in eqns (2.226)–(2.230), that eqn (2.283) is satisfied by the energy–momentum of the Dirac field in eqn (2.244) (here we omit the double colons because in this case the normal ordering is irrelevant).

Let us now consider the fermion and antifermion annihilation operators $a^{(h)}(p)$ and $b^{(h)}(p)$. Under the space-time translation

$$x^\mu \to x'^\mu = x^\mu + a^\mu\,, \tag{2.284}$$

from eqns (2.162), (2.163) and

$$\mathsf{U}\, \psi(x)\, \mathsf{U}^\dagger = \psi'(x) = \psi(x-a)\,, \tag{2.285}$$

we have

$$\begin{aligned} a'^{(h)}(p) = \mathsf{U}\, a^{(h)}(p)\, \mathsf{U}^\dagger &= \int \mathrm{d}^3x\, u^{(h)\dagger}(p)\, \mathsf{U}\, \psi(x)\, \mathsf{U}^\dagger\, e^{ip\cdot x} \\ &= \int \mathrm{d}^3x\, u^{(h)\dagger}(p)\, \psi(x-a)\, e^{ip\cdot x} = e^{ip\cdot a}\, a^{(h)}(p)\,, \end{aligned} \tag{2.286}$$

$$\begin{aligned} b'^{(h)}(p) = \mathsf{U}\, b^{(h)}(p)\, \mathsf{U}^\dagger &= \int \mathrm{d}^3x\, \mathsf{U}\, \psi^\dagger(x)\, \mathsf{U}^\dagger\, v^{(h)}(p)\, e^{ip\cdot x} \\ &= \int \mathrm{d}^3x\, \psi^\dagger(x-a)\, v^{(h)}(p)\, e^{ip\cdot x} = e^{ip\cdot a}\, b^{(h)}(p)\,. \end{aligned} \tag{2.287}$$

Hence, since the vacuum is invariant under space-time translations ($\mathsf{U}\,|0\rangle = |0\rangle$), the fermion and antifermion states in eqns (2.232) transform by a phase shift:

$$\mathsf{U}\,|\mathfrak{f}(p,h)\rangle = \mathsf{U}\, a^{(h)\dagger}(p)\, \mathsf{U}^\dagger\, \mathsf{U}\,|0\rangle = e^{-ip\cdot a}\,|\mathfrak{f}(p,h)\rangle\,, \qquad \mathsf{U}\,|\bar{\mathfrak{f}}(p,h)\rangle = e^{-ip\cdot a}\,|\bar{\mathfrak{f}}(p,h)\rangle\,. \tag{2.288}$$

Let us note that the phase shift is $e^{-ip\cdot a}$, as one may have expected from the space-time evolution of the wave functions

$$\langle 0|\psi(x)|\mathfrak{f}(p,h)\rangle = u^{(h)}(p)\, e^{-ip\cdot x}\,, \qquad \langle 0|\psi^C(x)|\bar{\mathfrak{f}}(p,h)\rangle = \xi_{\mathrm{C}}\, u^{(h)}(p)\, e^{-ip\cdot x}\,, \tag{2.289}$$

where $\psi^C(x)$ is the charge-conjugated field in eqn (2.356). Indeed, the transformation in eqn (2.285) is active, in agreement with the remark at the end of section 2.10.

2.10.2 Lorentz transformations

From eqns (2.59) and (2.277), there is a unitary operator U_Λ such that under a Lorentz transformation in eqn (B.1) the Dirac field operator $\psi(x)$ transforms as

$$\mathsf{U}_\Lambda\, \psi(x)\, \mathsf{U}_\Lambda^\dagger = \psi'(x) = \mathcal{S}(\Lambda)\, \psi(\Lambda^{-1}x)\,. \tag{2.290}$$

Let us derive the corresponding transformation for the fermion annihilation operators $a^{(h)}(p)$. Since the expression in eqn (2.162) is not written in a covariant form suitable to discuss Lorentz transformations, let us generalize it to

$$a^{(h)}(p) = \int_S \mathrm{d}S^\mu(x)\, \overline{u^{(h)}}(p)\, \gamma_\mu\, \psi(x)\, e^{ip\cdot x}\,, \tag{2.291}$$

where $\mathrm{d}S^\mu(x)$ is the surface element of an arbitrary three-dimensional space-like surface S, such that for any two points x_1 and x_2 on the surface the invariant squared distance $(x-y)^2$ is space-like (i.e. $(x-y)^2 < 0$). The unit four-vector $n^\mu(x)$ normal to S at the point x must be time-like (i.e. $n^\mu(x)n_\mu(x) = 1$). In general, $\mathrm{d}S^\mu(x)$ is given by

$$\mathrm{d}S^\mu(x) = \left(n^0(x)\mathrm{d}x^1\mathrm{d}x^2\mathrm{d}x^3, n^1(x)\mathrm{d}x^0\mathrm{d}x^2\mathrm{d}x^3, n^2(x)\mathrm{d}x^0\mathrm{d}x^1\mathrm{d}x^3, n^3(x)\mathrm{d}x^0\mathrm{d}x^1\mathrm{d}x^2\right). \tag{2.292}$$

A hyperplane $t =$ constant is a special case in which $n^\mu(x) = (1,0,0,0)$ for all x, which leads to $\mathrm{d}S^\mu(x) = \delta_{\mu 0}\mathrm{d}^3x$.

In order to prove that the expression in eqn (2.291) is equivalent to that in eqn (2.162), we use the Gauss theorem, which says that for any function $f_\mu(x)$

$$\int_S \mathrm{d}S^\mu(x)\, f_\mu(x) = \int_V \mathrm{d}^4x\, \partial^\mu\, f_\mu(x)\,, \tag{2.293}$$

where V is the volume enclosed by the hypersurface S. Let us introduce the functional derivative

$$\frac{\delta F(S)}{\delta S(x)} = \lim_{\Delta V(x)\to 0} \frac{F(S+\Delta S(x)) - F(S)}{\Delta V(x)}\,, \tag{2.294}$$

where $S+\Delta S(x)$ is a space-like hypersurface which differs from S by an infinitesimal deformation in the neighborhood of x and $\Delta V(x)$ is the infinitesimal four-dimensional volume between S and $S+\Delta S(x)$. For

$$F(S) = \int_S \mathrm{d}S^\mu(x)\, f_\mu(x)\,, \tag{2.295}$$

using Gauss theorem in eqn (2.293) we have

$$\frac{\delta F(S)}{\delta S(x)} = \lim_{\Delta V(x)\to 0} \frac{\int_{V+\Delta V(x)} \mathrm{d}^4x'\, \partial^\mu\, f_\mu(x') - \int_V \mathrm{d}^4x'\, \partial^\mu\, f_\mu(x')}{\Delta V(x)} = \partial^\mu\, f_\mu(x)\,. \tag{2.296}$$

Thus, $F(S)$ is invariant under a change of the space-like hypersurface of integration S if $\partial^\mu\, f_\mu(x) = 0$ for all space-time points x. This is what happens with eqn (2.291):

from the Dirac equation (2.11) and from eqn (2.146) we obtain

$$\partial^\mu \left(\overline{u^{(h)}}(p)\,\gamma_\mu\,\psi(x)\,e^{ip\cdot x} \right) = -i\,m\,\overline{u^{(h)}}(p)\,\psi(x)\,e^{ip\cdot x} + i\,\overline{u^{(h)}}(p)\,\not{p}\,\psi(x)\,e^{ip\cdot x} = 0\,. \tag{2.297}$$

Therefore, since eqn (2.291) is independent of the hypersurface of integration, as long as it remains space-like, it is equivalent to eqn (2.162).

Let us now use eqn (2.291) for the derivation of the transformation of the fermion annihilation operators $a^{(h)}(p)$ under the Lorentz transformation in eqn (2.290). We have

$$\mathsf{U}_\Lambda\,a^{(h)}(p)\,\mathsf{U}_\Lambda^\dagger = \int_S \mathrm{d}S^\mu(x)\,\overline{u^{(h)}}(p)\,\gamma_\mu\,\mathcal{S}(\Lambda)\,\psi(\Lambda^{-1}x)\,e^{ip\cdot x}\,. \tag{2.298}$$

Since

$$\widetilde{x} = \Lambda^{-1}\,x \quad \Longrightarrow \quad x = \Lambda\,\widetilde{x} \quad \Longrightarrow \quad \mathrm{d}S^\mu(x) = \Lambda^\mu{}_\nu\,\mathrm{d}\widetilde{S}^\nu(\widetilde{x})\,, \tag{2.299}$$

from eqn (2.60) we find that

$$\mathsf{U}_\Lambda\,a^{(h)}(p)\,\mathsf{U}_\Lambda^\dagger = \int_{\widetilde{S}} \mathrm{d}\widetilde{S}^\mu(\widetilde{x})\,\overline{u^{(h)}}(p)\,\mathcal{S}(\Lambda)\,\gamma_\mu\,\psi(\widetilde{x})\,e^{i\widetilde{p}\cdot\widetilde{x}}\,, \tag{2.300}$$

with

$$\widetilde{p} = \Lambda^{-1}\,p\,. \tag{2.301}$$

Let us now consider $\overline{u^{(h)}}(p)\,\mathcal{S}(\Lambda)$. From eqns (2.146) and (2.60) we obtain

$$\overline{u^{(h)}}(p)\,\mathcal{S}(\Lambda)\,(\not{\widetilde{p}} - m) = 0\,. \tag{2.302}$$

This means that $\overline{u^{(h)}}(p)\,\mathcal{S}(\Lambda)$ is a linear combination of $\overline{u^{(+)}}(\widetilde{p})$ and $\overline{u^{(-)}}(\widetilde{p})$:

$$\overline{u^{(h)}}(p)\,\mathcal{S}(\Lambda) = \sum_{\widetilde{h}} c_{h\widetilde{h}}(p,\Lambda)\,\overline{u^{(\widetilde{h})}}(\widetilde{p})\,, \tag{2.303}$$

with

$$c_{h\widetilde{h}}(p,\Lambda) = \frac{1}{2m}\,\overline{u^{(h)}}(p)\,\mathcal{S}(\Lambda)\,u^{(\widetilde{h})}(\Lambda^{-1}p)\,. \tag{2.304}$$

The coefficients $c_{h\widetilde{h}}(p,\Lambda)$ are obviously constrained by

$$\sum_{\widetilde{h}} |c_{h\widetilde{h}}(p,\Lambda)|^2 = 1\,. \tag{2.305}$$

In conclusion, we obtain

$$\mathsf{U}_\Lambda\,a^{(h)}(p)\,\mathsf{U}_\Lambda^\dagger = \sum_{\widetilde{h}} c_{h\widetilde{h}}(p,\Lambda) \int_{\widetilde{S}} \mathrm{d}\widetilde{S}^\mu(\widetilde{x})\,\overline{u^{(h)}}(\widetilde{p})\,\gamma_\mu\,\psi(\widetilde{x})\,e^{i\widetilde{p}\cdot\widetilde{x}}\,, \tag{2.306}$$

which implies

$$\mathsf{U}_\Lambda\,a^{(h)}(p)\,\mathsf{U}_\Lambda^\dagger = \sum_{\widetilde{h}} c_{h\widetilde{h}}(p,\Lambda)\,a^{(\widetilde{h})}(\Lambda^{-1}p)\,. \tag{2.307}$$

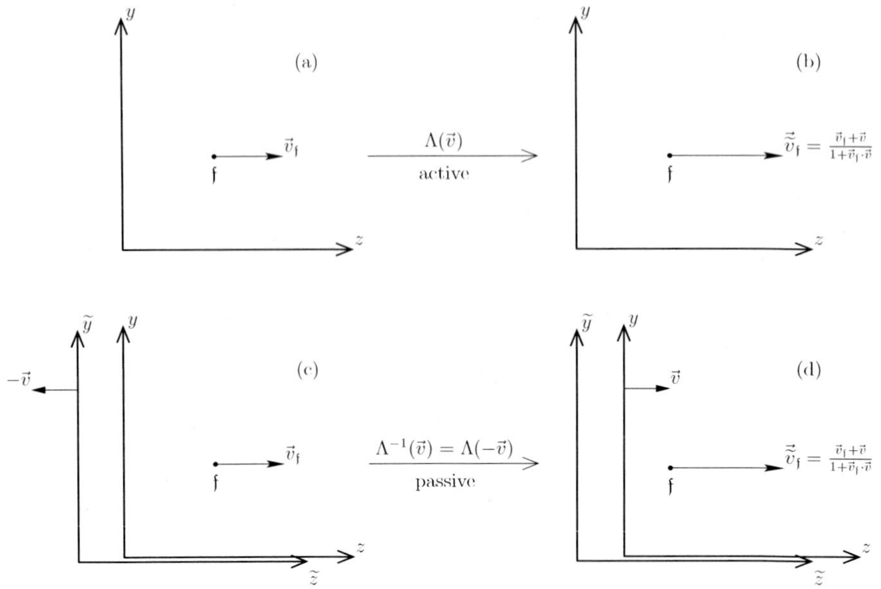

FIG. 2.2. Active boost corresponding to $\Lambda(\vec{v})$ in eqn (2.316), with velocity $\vec{v}$ in the direction of the particle velocity $\vec{v}_{\mathfrak{f}}$, which is oriented along the z axis.

In an analogous way, one can derive the transformation law of the antifermion annihilation operator $b^{(h)}(p)$:

$$\mathsf{U}_\Lambda \, b^{(h)}(p) \, \mathsf{U}_\Lambda^\dagger = \sum_{\widetilde{h}} d_{h\widetilde{h}}(p,\Lambda) \, b^{(\widetilde{h})}(\Lambda^{-1}p) \,, \tag{2.308}$$

with

$$d_{h\widetilde{h}}(p,\Lambda) = \frac{1}{2m} \, \overline{v^{(\widetilde{h})}}(\Lambda^{-1}p) \, \mathcal{S}^{-1}(\Lambda) \, v^{(h)}(p) \,, \tag{2.309}$$

such that

$$\sum_{\widetilde{h}} |d_{h\widetilde{h}}(p,\Lambda)|^2 = 1 \,. \tag{2.310}$$

Since the vacuum is invariant under Lorentz transformations ($\mathsf{U}_\Lambda |0\rangle = |0\rangle$), the fermion and antifermion states in eqns (2.232) transform as

$$\mathsf{U}_\Lambda |\mathfrak{f}(p,h)\rangle = \sum_{\widetilde{h}} c^*_{h\widetilde{h}}(p,\Lambda) \, |\mathfrak{f}(\Lambda^{-1}p,\widetilde{h})\rangle \,, \tag{2.311}$$

$$\mathsf{U}_\Lambda |\bar{\mathfrak{f}}(p,h)\rangle = \sum_{\widetilde{h}} d^*_{h\widetilde{h}}(p,\Lambda) \, |\bar{\mathfrak{f}}(\Lambda^{-1}p,\widetilde{h})\rangle \,. \tag{2.312}$$

One can understand that the transformation is active, i.e. a transformation of the system, by considering a boost with velocity $\vec{v}$ in the same direction as the particle velocity $\vec{v}_{\mathfrak{f}}$, which must increase the particle velocity to

$$\vec{\widetilde{v}}_{\mathfrak{f}} = \frac{\vec{v} + \vec{v}_{\mathfrak{f}}}{1 + \vec{v} \cdot \vec{v}_{\mathfrak{f}}} \, , \tag{2.313}$$

according to eqn (B.32). In fact, the boost has the effect of transforming a state with four-momentum p and velocity

$$\vec{v}_{\mathfrak{f}} = \frac{\vec{p}}{E} \tag{2.314}$$

into a state with four-momentum $\widetilde{p} = \Lambda^{-1} p$, which has the larger velocity

$$\vec{\widetilde{v}}_{\mathfrak{f}} = \frac{\vec{\widetilde{p}}}{\widetilde{E}} = \frac{\gamma \, (\vec{v} E + \vec{p})}{\gamma \, (E + \vec{v} \cdot \vec{p})} = \frac{\vec{v} + \vec{v}_{\mathfrak{f}}}{1 + \vec{v} \cdot \vec{v}_{\mathfrak{f}}} \, . \tag{2.315}$$

The active transformation from eqn (2.314) to eqn (2.315) is illustrated by panels (a) and (b) in Fig. 2.2, where we have considered $\vec{v}_{\mathfrak{f}}$ and $\vec{v}$ along the z axis. From eqn (B.28) we have

$$[\Lambda(\vec{v})]^{\mu}{}_{\nu} = \begin{pmatrix} \gamma & 0 & 0 & -\gamma v \\ 0 & 1 & 0 & 0 \\ 0 & 0 & 1 & 0 \\ -\gamma v & 0 & 0 & \gamma \end{pmatrix} , \tag{2.316}$$

$$[\Lambda^{-1}(\vec{v})]^{\mu}{}_{\nu} = [\Lambda(-\vec{v})]^{\mu}{}_{\nu} = \begin{pmatrix} \gamma & 0 & 0 & \gamma v \\ 0 & 1 & 0 & 0 \\ 0 & 0 & 1 & 0 \\ \gamma v & 0 & 0 & \gamma \end{pmatrix} , \tag{2.317}$$

where $v \equiv |\vec{v}|$. Comparing panels (a) and (b) in Fig. 2.2 with panels (c) and (d), one can see that the effect of the active transformation $\Lambda(\vec{v})$ is the same as that of a passive transformation $\Lambda^{-1}(\vec{v}) = \Lambda(-\vec{v})$ from the coordinate system x to the coordinate system $\widetilde{x} = \Lambda^{-1}(\vec{v}) \, x$.

By contrast, Fig. 2.3 shows that a passive transformation $\Lambda(\vec{v})$ from the coordinate system x to the coordinate system $x' = \Lambda(\vec{v}) \, x$ has the effect of decreasing the particle velocity from $\vec{v}_{\mathfrak{f}}$ in eqn (2.314) to

$$\vec{v}'_{\mathfrak{f}} = \frac{\vec{p}'}{\widetilde{E}} = \frac{\gamma \, (\vec{v} E - \vec{p})}{\gamma \, (E - \vec{v} \cdot \vec{p})} = \frac{\vec{v} - \vec{v}_{\mathfrak{f}}}{1 - \vec{v} \cdot \vec{v}_{\mathfrak{f}}} \, . \tag{2.318}$$

It is important to note that the linear combination of different helicities in eqns (2.311) and (2.312) implies that the helicity of a particle is not a Lorentz invariant. This fact should be expected on physical grounds. Let us consider for example an active boost in the opposite direction to the particle velocity. If the boost has a velocity which is larger than the original particle velocity, the motion of the boosted particle is inverted and the direction of the three-momentum of the particle is inverted. On the other hand, the spin is untouched and its vector points in the same direction before and after the boost. As a result, in this case the helicity is inverted by the boost and we expect that $c_{h\widetilde{h}} \propto d_{h\widetilde{h}} \propto \delta_{h(-\widetilde{h})}$.

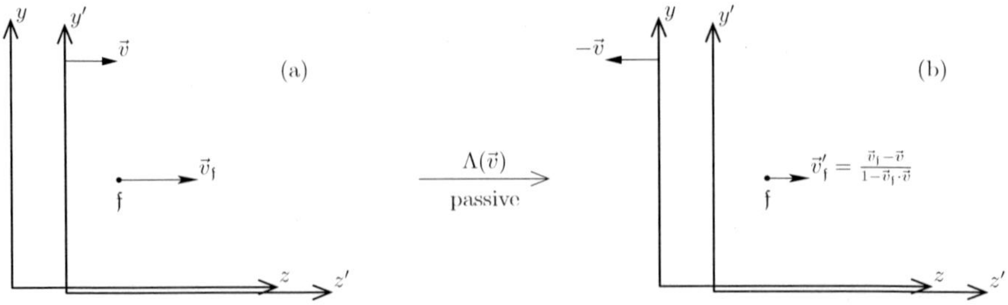

FIG. 2.3. Passive boost corresponding to $\Lambda(\vec{v})$ in eqn (2.316), with velocity $\vec{v}$ in the direction of the particle velocity $\vec{v}_\mathfrak{f}$ from the coordinate system x to the coordinate system x'. We have oriented the z and z' axes in the common direction of the velocities $\vec{v}_\mathfrak{f}$ and $\vec{v}$.

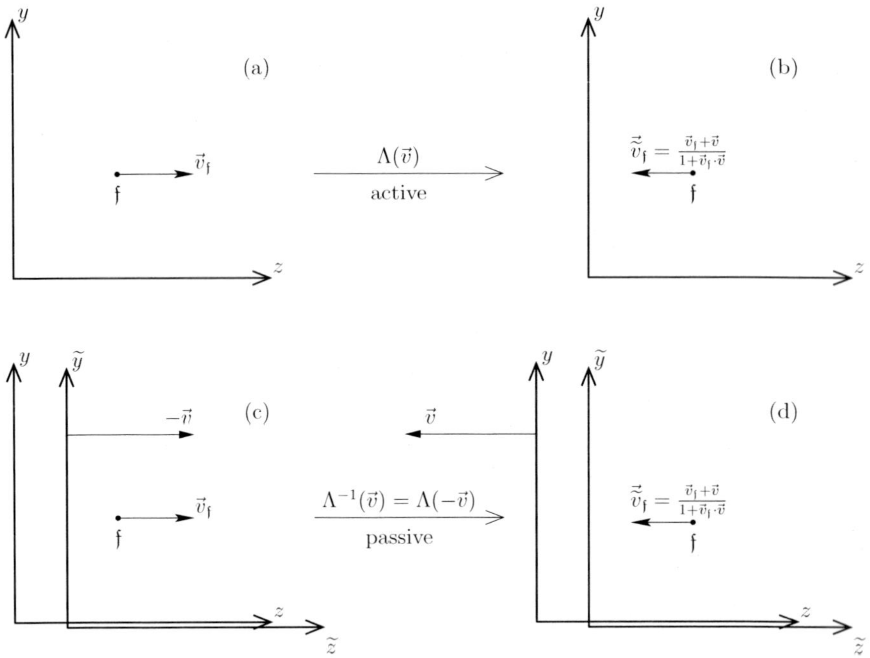

FIG. 2.4. Active boost corresponding to $\Lambda(\vec{v})$ in eqn (2.319), with velocity $\vec{v}$ in the direction opposite to the particle velocity $\vec{v}_\mathfrak{f}$ ($\vec{v} \cdot \vec{v}_\mathfrak{f} < 0$), which is oriented along the z axis, and such that $|\vec{v}| > |\vec{v}_\mathfrak{f}|$.

Let us calculate, for example, $c_{h\tilde{h}}(p, \Lambda)$. We consider an active boost corresponding to

$$[\Lambda(\vec{v})]^\mu{}_\nu = \begin{pmatrix} \gamma & 0 & 0 & \gamma v \\ 0 & 1 & 0 & 0 \\ 0 & 0 & 1 & 0 \\ \gamma v & 0 & 0 & \gamma \end{pmatrix}, \tag{2.319}$$

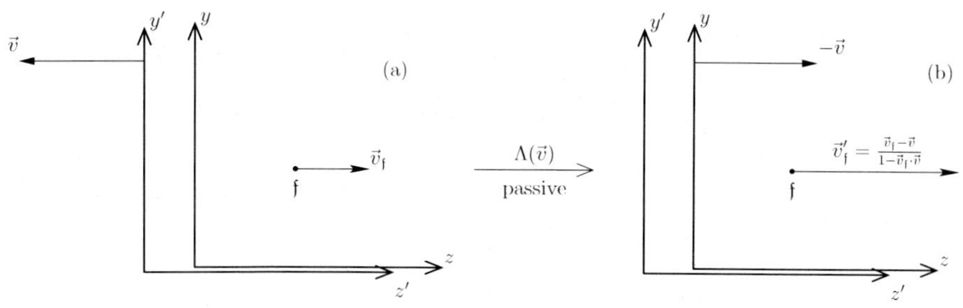

FIG. 2.5. Passive boost corresponding to $\Lambda(\vec{v})$ in eqn (2.319), with velocity $\vec{v}$ in the direction opposite of the particle velocity $\vec{v}_{\mathfrak{f}}$ ($\vec{v} \cdot \vec{v}_{\mathfrak{f}} < 0$), from the coordinate system x to the coordinate system x'. We have oriented the z and z' axes in the common direction of the velocities $\vec{v}_{\mathfrak{f}}$ and $\vec{v}$.

$$[\Lambda^{-1}(\vec{v})]^{\mu}{}_{\nu} = [\Lambda(-\vec{v})]^{\mu}{}_{\nu} = \begin{pmatrix} \gamma & 0 & 0 & -\gamma v \\ 0 & 1 & 0 & 0 \\ 0 & 0 & 1 & 0 \\ -\gamma v & 0 & 0 & \gamma \end{pmatrix}, \tag{2.320}$$

with velocity $\vec{v}$ in the opposite direction of the particle velocity $\vec{v}_{\mathfrak{f}}$ in eqn (2.314), which is oriented along the z axis, and such that

$$v \equiv |\vec{v}| > |\vec{v}_{\mathfrak{f}}| \equiv v_{\mathfrak{f}} . \tag{2.321}$$

As illustrated in Fig. 2.4, such an active boost inverts the particle velocity. One can also see that it is equivalent to a passive boost $\Lambda^{-1}(\vec{v}) = \Lambda(-\vec{v})$ from the coordinate system x to the coordinate system $\widetilde{x} = \Lambda^{-1}(\vec{v})\, x$, which moves with velocity $-\vec{v}$ along the z axis. The active boost corresponding to $\Lambda(\vec{v})$ must not be confused with the passive boost from the coordinate system x to the coordinate system $x' = \Lambda(\vec{v})\, x$, in which the particle velocity is larger than in the system x, as illustrated in Fig. 2.5.

Since the particle momentum $\vec{p}$ is directed along the z axis, from eqns (2.202) and (2.215) we have

$$u^{(h)}(p) = \sqrt{m} \begin{pmatrix} \left(\sinh \frac{\eta}{2} + h \cosh \frac{\eta}{2}\right) \chi^{(h)} \\ \left(\sinh \frac{\eta}{2} - h \cosh \frac{\eta}{2}\right) \chi^{(h)} \end{pmatrix}, \tag{2.322}$$

with $\chi^{(h)}$ given in eqn (2.217). Let us now calculate $u^{(\widetilde{h})}(\Lambda^{-1}p) = u^{(\widetilde{h})}(\widetilde{p})$. Since the active boost velocity $\vec{v}$ is in the opposite direction of the particle velocity $\vec{v}_{\mathfrak{f}}$, the corresponding rapidities subtract and the rapidity $\widetilde{\eta}$ corresponding to $\vec{\widetilde{p}}$ is given by

$$\widetilde{\eta} = |\eta - \varphi| = \varphi - \eta , \tag{2.323}$$

because

$$\tanh \varphi = v > v_{\mathfrak{f}} = \tanh \eta \tag{2.324}$$

implies that $\varphi > \eta$. Moreover, since $\vec{\widetilde{p}} = |\vec{\widetilde{p}}|(0\,,\,0\,,\,-1)$, the latitudinal polar angle $\widetilde{\theta}$ of $\vec{\widetilde{p}} = |\vec{\widetilde{p}}|\,(\sin\widetilde{\theta}\cos\widetilde{\phi}, \sin\widetilde{\theta}\sin\widetilde{\phi}, \cos\widetilde{\theta})$ is $\widetilde{\theta} = \pi$, and from eqn (2.215) we get

$$\chi^{(+)}(\vec{\widetilde{p}}) = \begin{pmatrix} 0 \\ e^{i\widetilde{\phi}} \end{pmatrix} = e^{i\widetilde{\phi}}\,\chi^{(-)}\,, \qquad \chi^{(-)}(\vec{\widetilde{p}}) = \begin{pmatrix} -e^{-i\widetilde{\phi}} \\ 0 \end{pmatrix} = -e^{-i\widetilde{\phi}}\,\chi^{(+)}\,. \tag{2.325}$$

We can summarize these relations with

$$\chi^{(\widetilde{h})}(\vec{\widetilde{p}}) = \widetilde{h}\,e^{i\widetilde{h}\widetilde{\phi}}\,\chi^{(-\widetilde{h})}\,, \tag{2.326}$$

which allows one to write $u^{(\widetilde{h})}(\widetilde{p})$ as

$$u^{(\widetilde{h})}(\widetilde{p}) = \widetilde{h}\,e^{i\widetilde{h}\widetilde{\phi}}\,\sqrt{m}\begin{pmatrix} \left(\sinh\frac{\widetilde{\eta}}{2} + \widetilde{h}\,\cosh\frac{\widetilde{\eta}}{2}\right)\chi^{(-\widetilde{h})} \\ \left(\sinh\frac{\widetilde{\eta}}{2} - \widetilde{h}\,\cosh\frac{\widetilde{\eta}}{2}\right)\chi^{(-\widetilde{h})} \end{pmatrix}\,. \tag{2.327}$$

From eqn (2.74), we have

$$\mathcal{S}^3_{\text{boost}}(-\varphi) = \cosh\frac{\varphi}{2} + \alpha^3\,\sinh\frac{\varphi}{2} = \begin{pmatrix} \cosh\frac{\varphi}{2} + \sigma^3\,\sinh\frac{\varphi}{2} & 0 \\ 0 & \cosh\frac{\varphi}{2} - \sigma^3\,\sinh\frac{\varphi}{2} \end{pmatrix}\,, \tag{2.328}$$

leading to

$$\begin{aligned}\mathcal{S}^3_{\text{boost}}(-\varphi)\,u^{(\widetilde{h})}(\widetilde{p}) &= -\,\widetilde{h}\,e^{i\widetilde{h}\widetilde{\phi}}\,\sqrt{m}\begin{pmatrix} \left(\sinh\frac{\eta}{2} - \widetilde{h}\,\cosh\frac{\eta}{2}\right)\chi^{(-\widetilde{h})} \\ \left(\sinh\frac{\eta}{2} + \widetilde{h}\,\cosh\frac{\eta}{2}\right)\chi^{(-\widetilde{h})} \end{pmatrix} \\ &= -\,\widetilde{h}\,e^{i\widetilde{h}\widetilde{\phi}}\,u^{(-\widetilde{h})}(p)\,,\end{aligned} \tag{2.329}$$

as one can see by comparison with eqn (2.322). Finally, using eqn (2.157), from eqn (2.304) we obtain

$$c_{h\widetilde{h}}(p, \Lambda(\vec{v})) = \frac{1}{2m}\,\overline{u^{(h)}}(p)\,\mathcal{S}^3_{\text{boost}}(-\varphi)\,u^{(\widetilde{h})}(\Lambda^{-1}p) = -\widetilde{h}\,e^{i\widetilde{h}\widetilde{\phi}}\,\delta_{h(-\widetilde{h})}\,, \tag{2.330}$$

as expected, which implies that in this case

$$\mathsf{U}_\Lambda\,a^{(h)}(p)\,\mathsf{U}_\Lambda^\dagger = -\widetilde{h}\,e^{i\widetilde{h}\widetilde{\phi}}\,a^{(-h)}(\Lambda^{-1}p)\,. \tag{2.331}$$

Hence, in this case the helicity of the particle is reversed, as illustrated in Fig. 2.6. It is straightforward to modify the above calculation to the case in which the boost does not reverse the direction of motion of the particle ($|\vec{v}| < |\vec{v}_{\mathrm{f}}|$), which yields $c_{h\widetilde{h}} = \delta_{h\widetilde{h}}$ and $\mathsf{U}_\Lambda\,a^{(h)}(p)\,\mathsf{U}_\Lambda^\dagger = a^{(h)}(\Lambda^{-1}p)$. Similar properties obviously hold for the antiparticle destruction operator $b^{(h)}(p)$. Let us emphasize, however, that a boost in a direction which is not parallel or antiparallel to the particle velocity produces a linear combination of the two helicities.

Let us now calculate $c_{h\widetilde{h}}(p, \Lambda)$ for a rotation. As an example, we consider a rotation by an angle ω in the y-z plane, with

$$[\Lambda(\omega)]^\mu{}_\nu = \begin{pmatrix} 1 & 0 & 0 & 0 \\ 0 & 1 & 0 & 0 \\ 0 & 0 & \cos\omega & \sin\omega \\ 0 & 0 & -\sin\omega & \cos\omega \end{pmatrix}\,, \tag{2.332}$$

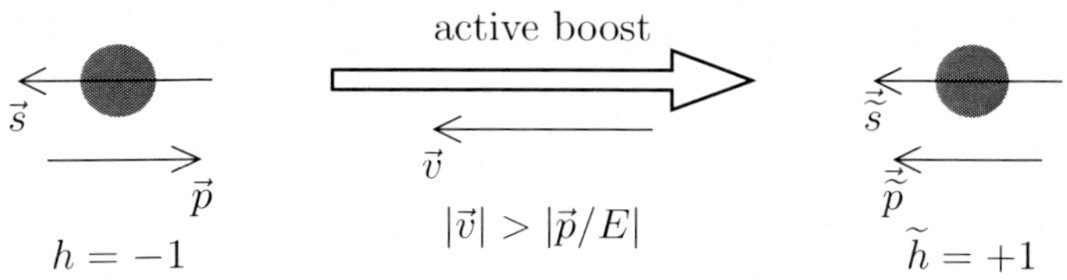

FIG. 2.6. Reversal of the helicity under an active boost with velocity $\vec{v}$ in the direction opposite to the particle velocity $\vec{p}/E$.

$$[\Lambda^{-1}(\omega)]^{\mu}{}_{\nu} = [\Lambda(-\omega)]^{\mu}{}_{\nu} = \begin{pmatrix} 1 & 0 & 0 & 0 \\ 0 & 1 & 0 & 0 \\ 0 & 0 & \cos\omega & -\sin\omega \\ 0 & 0 & \sin\omega & \cos\omega \end{pmatrix} . \tag{2.333}$$

Since

$$\widetilde{p}^{\mu} = [\Lambda^{-1}(\omega)]^{\mu}{}_{\nu}\, p^{\nu} = (E\,,\, 0\,,\, -\sin\omega|\vec{p}|\,,\, \cos\omega|\vec{p}|)\,, \tag{2.334}$$

we have

$$\widetilde{\eta} = \eta\,, \quad \widetilde{\theta} = \omega\,, \quad \widetilde{\phi} = \frac{3\pi}{2}\,, \tag{2.335}$$

and

$$u^{(\widetilde{h})}(\widetilde{p}) = \sqrt{m} \begin{pmatrix} \left(\sinh\frac{\eta}{2} + \widetilde{h}\cosh\frac{\eta}{2}\right)\chi^{(\widetilde{h})}(\widetilde{\vec{p}}) \\ \left(\sinh\frac{\eta}{2} - \widetilde{h}\cosh\frac{\eta}{2}\right)\chi^{(\widetilde{h})}(\widetilde{\vec{p}}) \end{pmatrix}, \tag{2.336}$$

with

$$\chi^{(+)}(\widetilde{\vec{p}}) = \begin{pmatrix} \cos\frac{\omega}{2} \\ -i\sin\frac{\omega}{2} \end{pmatrix}, \qquad \chi^{(-)}(\widetilde{\vec{p}}) = \begin{pmatrix} -i\sin\frac{\omega}{2} \\ \cos\frac{\omega}{2} \end{pmatrix}. \tag{2.337}$$

Since the rotation is around the x axis, from eqn (2.82) we have

$$\mathcal{S}^1_{\text{rot}}(\omega) = \cos\frac{\omega}{2} + i\,\Sigma^1\sin\frac{\omega}{2} = \begin{pmatrix} \cos\frac{\omega}{2} & i\sin\frac{\omega}{2} & 0 & 0 \\ i\sin\frac{\omega}{2} & \cos\frac{\omega}{2} & 0 & 0 \\ 0 & 0 & \cos\frac{\omega}{2} & i\sin\frac{\omega}{2} \\ 0 & 0 & i\sin\frac{\omega}{2} & \cos\frac{\omega}{2} \end{pmatrix}, \tag{2.338}$$

and

$$\mathcal{S}^1_{\text{rot}}(\omega)\,u^{(\widetilde{h})}(\widetilde{p}) = \sqrt{m} \begin{pmatrix} \left(\sinh\frac{\eta}{2} + \widetilde{h}\cosh\frac{\eta}{2}\right)\chi^{(\widetilde{h})} \\ \left(\sinh\frac{\eta}{2} - \widetilde{h}\cosh\frac{\eta}{2}\right)\chi^{(\widetilde{h})} \end{pmatrix} = u^{(\widetilde{h})}(p)\,, \tag{2.339}$$

as one can see by comparison with eqn (2.322). Finally, using eqn (2.157), from eqn (2.304) we obtain

$$c_{h\widetilde{h}}(p, \Lambda(\vec{v})) = \frac{1}{2m}\,\overline{u^{(h)}}(p)\,\mathcal{S}^1_{\text{rot}}(\omega)\,u^{(\widetilde{h})}(\Lambda^{-1}p) = \delta_{h\widetilde{h}}\,, \tag{2.340}$$

and from eqn (2.307)

$$\mathsf{U}_{\Lambda}\,a^{(h)}(p)\,\mathsf{U}^{\dagger}_{\Lambda} = a^{(h)}(\Lambda^{-1}p)\,. \tag{2.341}$$

Therefore, a rotation does not change the particle helicity. It is obvious that this property holds for any rotation and also for antiparticles.

In conclusion, a rotation does not change the particle helicity, whereas a boost produces a linear combination of the two different helicities. Only in the particular cases of a boost which is parallel or antiparallel to the particle velocity does the boosted particle conserve a definite helicity. Only if the boost is antiparallel to the particle velocity, and has a larger velocity, is the helicity reversed.

2.11 C, P, and T transformations

The Dirac Lagrangian in eqn (2.1) is invariant under the transformations of charge conjugation C, space inversion (parity) P, and time reversal T.

2.11.1 Charge conjugation

Under charge conjugation the spinor fields ψ and $\overline{\psi}$ transform as

$$\psi(x) \xrightarrow{\mathrm{C}} \psi^{\mathrm{C}}(x) = \xi_{\mathrm{C}}\, \mathcal{C}\, \overline{\psi}^{T}(x) = -\xi_{\mathrm{C}}\, \gamma^{0}\, \mathcal{C}\, \psi^{*}(x)\,, \tag{2.342}$$

$$\overline{\psi}(x) \xrightarrow{\mathrm{C}} \overline{\psi^{\mathrm{C}}}(x) = -\xi_{\mathrm{C}}^{*}\, \psi^{T}\, \mathcal{C}^{\dagger}(x)\,, \tag{2.343}$$

where $\mathcal{C}$ is the charge conjugation matrix, such that

$$\mathcal{C}\, \gamma_{\mu}^{T}\, \mathcal{C}^{-1} = -\gamma_{\mu}\,, \tag{2.344}$$

$$\mathcal{C}^{\dagger} = \mathcal{C}^{-1}\,, \tag{2.345}$$

$$\mathcal{C}^{T} = -\mathcal{C}\,. \tag{2.346}$$

Furthermore, we have

$$\mathcal{C}\, (\gamma^{5})^{T}\, \mathcal{C}^{-1} = \gamma^{5}\,, \tag{2.347}$$

$$\mathcal{C}\, (\sigma^{\mu\nu})^{T}\, \mathcal{C}^{-1} = -\sigma^{\mu\nu}\,. \tag{2.348}$$

In the Dirac representation in eqn (2.21) of the γ matrices

$$\mathcal{C}_{\mathrm{D}} = i\, \gamma_{\mathrm{D}}^{2}\, \gamma_{\mathrm{D}}^{0} = -i \begin{pmatrix} 0 & \sigma^{2} \\ \sigma^{2} & 0 \end{pmatrix}, \tag{2.349}$$

and in the chiral representation in eqn (2.27)

$$\mathcal{C}_{\mathrm{C}} = i\, \gamma_{\mathrm{C}}^{2}\, \gamma_{\mathrm{C}}^{0} = -i \begin{pmatrix} \sigma^{2} & 0 \\ 0 & -\sigma^{2} \end{pmatrix}. \tag{2.350}$$

Since two charge conjugation transformations,

$$\psi \xrightarrow{\mathrm{C}} \xi_{\mathrm{C}}\, \mathcal{C}\, \overline{\psi}^{T} \xrightarrow{\mathrm{C}} |\xi_{\mathrm{C}}|^{2}\, \psi\,, \tag{2.351}$$

must leave ψ invariant, the coefficient ξ_{C} is a phase with

$$|\xi_{\mathrm{C}}|^{2} = 1\,, \tag{2.352}$$

which represents the *intrinsic charge parity* of the field. A consequence of $|\xi_{\mathrm{C}}|^{2} = 1$ is that the expression for ψ in terms of ψ^{C} is equal to the expression in eqn (2.342) of ψ^C in terms of ψ:

$$\psi(x) = \xi_C \, \mathcal{C} \, \overline{\psi^C}^T(x) \,. \tag{2.353}$$

Since the invariance of the free Dirac Lagrangian in eqn (2.1) under charge conjugation does not put any other constraint on ξ_C, the intrinsic charge parities of free fields are arbitrary.

It is convenient to choose the relative phase of the spinors $u^{(h)}(p)$ and $v^{(h)}(p)$ in order to have

$$u^{(h)}(p) = \mathcal{C} \, \overline{v^{(h)}}^T(p) \,, \tag{2.354}$$

which is equivalent to

$$v^{(h)}(p) = \mathcal{C} \, \overline{u^{(h)}}^T(p) \,. \tag{2.355}$$

Then, the Fourier expansion of the charge-conjugated field $\psi^C(x)$ is given by

$$\psi^C(x) = \xi_C \int \frac{d^3p}{(2\pi)^3 \, 2E} \sum_{h=\pm 1} \left[b^{(h)}(p) \, u^{(h)}(p) \, e^{-ip\cdot x} + a^{(h)\dagger}(p) \, v^{(h)}(p) \, e^{ip\cdot x} \right] . \tag{2.356}$$

The transformation of the quantized field is implemented by

$$\mathsf{U}_C \, \psi(x) \, \mathsf{U}_C^\dagger = \psi^C(x) = \xi_C \, \mathcal{C} \, \overline{\psi}^T(x) \,, \tag{2.357}$$

where U_C is the charge conjugation unitary operator ($\mathsf{U}_C^\dagger = \mathsf{U}_C^{-1}$), such that $\mathsf{U}_C = \mathsf{U}_C^{-1}$, because $\mathsf{U}_C^2 = \mathbf{1}$. From eqns (2.139) and (2.356), one can see that

$$\mathsf{U}_C \, a^{(h)\dagger}(p) \, \mathsf{U}_C^\dagger = \xi_C^* \, b^{(h)\dagger}(p) \,, \tag{2.358}$$

$$\mathsf{U}_C \, b^{(h)\dagger}(p) \, \mathsf{U}_C^\dagger = \xi_C \, a^{(h)\dagger}(p) \,. \tag{2.359}$$

Therefore, the operation of charge conjugation converts particles into antiparticles. For the charge operator $:\mathsf{Q}:$ in eqn (2.247) we have

$$\mathsf{U}_C \; :\mathsf{Q}: \; \mathsf{U}_C^\dagger = - :\mathsf{Q}: \;, \tag{2.360}$$

leading to opposite charges for particles and antiparticles, in agreement with eqn (2.248).

The covariant bilinears in eqns (2.65)–(2.69) transform as

$$S_{ab} = \overline{\psi_a} \, \psi_b \xrightarrow{\;C\;} (S_{ab})^C = \xi_C^{a*} \, \xi_C^b \, \overline{\psi_b} \, \psi_a = \xi_C^{a*} \, \xi_C^b \, S_{ba} \,, \tag{2.361}$$

$$V_{ab}^\mu = \overline{\psi_a} \, \gamma^\mu \, \psi_b \xrightarrow{\;C\;} (V_{ab}^\mu)^C = -\xi_C^{a*} \, \xi_C^b \, \overline{\psi_b} \, \gamma^\mu \, \psi_a = -\xi_C^{a*} \, \xi_C^b \, V_{ba}^\mu \,, \tag{2.362}$$

$$T_{ab}^{\mu\nu} = \overline{\psi_a} \, \sigma^{\mu\nu} \, \psi_b \xrightarrow{\;C\;} (T_{ab}^{\mu\nu})^C = -\xi_C^{a*} \, \xi_C^b \, \overline{\psi_b} \, \sigma^{\mu\nu} \, \psi_a = -\xi_C^{a*} \, \xi_C^b \, T_{ba}^{\mu\nu} \,, \tag{2.363}$$

$$A_{ab}^\mu = \overline{\psi_a} \, \gamma^\mu \, \gamma^5 \, \psi_b \xrightarrow{\;C\;} (A_{ab}^\mu)^C = \xi_C^{a*} \, \xi_C^b \, \overline{\psi_b} \, \gamma^\mu \, \gamma^5 \, \psi_a = \xi_C^{a*} \, \xi_C^b \, A_{ba}^\mu \,, \tag{2.364}$$

$$P_{ab} = \overline{\psi_a} \, \gamma^5 \, \psi_b \xrightarrow{\;C\;} (P_{ab})^C = \xi_C^{a*} \, \xi_C^b \, \overline{\psi_b} \, \gamma^5 \, \psi_a = \xi_C^{a*} \, \xi_C^b \, P_{ba} \,. \tag{2.365}$$

The change of sign under charge conjugation of the vector current for $a = b$ is in agreement with the fact that particles and antiparticles have opposite charges, because for the electromagnetic current in eqn (2.88) we have

$$j^\mu = q\,\overline{\psi}\,\gamma^\mu\,\psi \xrightarrow{\mathrm{C}} (j^\mu)^{\mathrm{C}} = -q\,\overline{\psi}\,\gamma^\mu\,\psi = -j^\mu\,. \tag{2.366}$$

However, the electromagnetic interaction Lagrangian

$$\mathscr{L}_{\mathrm{I}}^{(\gamma)} = -e\,j^\mu\,A_\mu \tag{2.367}$$

is invariant under charge conjugation if

$$A_\mu \xrightarrow{\mathrm{C}} -A_\mu\,. \tag{2.368}$$

Since the Lagrangian of the free electromagnetic field,

$$\mathscr{L}^{(\gamma)} = -\frac{1}{4}\,F^{\mu\nu}\,F_{\mu\nu}\,, \tag{2.369}$$

with

$$F_{\mu\nu} = \partial_\mu A_\nu - \partial_\nu A_\mu\,, \tag{2.370}$$

is invariant under the transformation in eqn (2.368), the invariance under charge conjugation of the electromagnetic interaction Lagrangian fixes the transformation under charge conjugation of the electromagnetic field, to be given by eqn (2.368). In particular, one can see that the charge parity of the electromagnetic field is $\xi_{\mathrm{C}}^A = -1$.

The example of the electromagnetic interaction shows that the invariance of neutral-current interactions, which involve covariant bilinears with $a = b$, do not place any constraint on the intrinsic charge parity of the fermion fields, because $\xi_{\mathrm{C}}^{a*}\xi_{\mathrm{C}}^{b} = 1$ for $a = b$ on the right-hand sides of eqns (2.361)–(2.365). The intrinsic charge parities of the fermion fields can be constrained only through charge conjugation conserving charged-current interactions, which involve the covariant bilinears in eqns (2.361)–(2.365) with $a \neq b$. In this case, the charge parities of the fields ψ_a and ψ_b are related by the charge conjugation symmetry of the interaction Lagrangian. In any case, only the relative values of the charge parities of different fermion fields are determined by the symmetry under charge conjugation. In practice, it is convenient to define arbitrarily the charge parity of some fields and fix the charge parity of the other fields through the symmetry under charge conjugation of the interaction Lagrangian.

For example, let us consider the interaction Lagrangian

$$\mathscr{L}_{\mathrm{I}}^{(\Phi)} = -g\left(V_{ab}^\mu\,\Phi_\mu + V_{ba}^\mu\,\Phi_\mu^\dagger\right)\,, \tag{2.371}$$

where we have taken into account that $V_{ab}^{\mu\,\dagger} = V_{ba}^\mu$, with a vector field Φ_μ which transforms under charge conjugation as

$$\Phi_\mu \xrightarrow{\mathrm{C}} \xi_{\mathrm{C}}^{\Phi}\,\Phi_\mu^\dagger\,. \tag{2.372}$$

The transformation of the Lagrangian in eqn (2.371) is

$$\mathscr{L}_{\mathrm{I}}^{(\Phi)} \xrightarrow{\mathrm{C}} -g \left(-\xi_{\mathrm{C}}^{a*} \, \xi_{\mathrm{C}}^{b} \, \xi_{\mathrm{C}}^{\Phi} \, V_{ab}^{\mu\,\dagger} \, \Phi_{\mu}^{\dagger} - \xi_{\mathrm{C}}^{a} \, \xi_{\mathrm{C}}^{b\,*} \, \xi_{\mathrm{C}}^{\Phi\,*} \, V_{ab}^{\mu} \, \Phi_{\mu} \right) . \tag{2.373}$$

Invariance under charge conjugation is achieved with

$$\xi_{\mathrm{C}}^{a*} \, \xi_{\mathrm{C}}^{b} \, \xi_{\mathrm{C}}^{\Phi} = -1 \, . \tag{2.374}$$

This relation fixes the relative charge parities of ψ_a, ψ_b, and Φ_μ. For example, if we choose $\xi_{\mathrm{C}}^{\Phi} = -1$ and $\xi_{\mathrm{C}}^{a} = 1$, it follows that $\xi_{\mathrm{C}}^{b} = 1$.

Since the vector and axial currents transform with different signs, $V - A$ charged-current weak interactions violate the symmetry under charge conjugation in a maximal way. Let us consider, for example, the charged-current $V - A$ interaction Lagrangian

$$\mathscr{L}_{\mathrm{I}}^{(W)} = -\frac{g}{2\sqrt{2}} \left[(V_{ab}^{\mu} - A_{ab}^{\mu}) \, W_{\mu} + (V_{ba}^{\mu} - A_{ba}^{\mu}) \, W_{\mu}^{\dagger} \right] , \tag{2.375}$$

where we have taken into account $V_{ab}^{\mu\,\dagger} = V_{ba}^{\mu}$ and $A_{ab}^{\mu\,\dagger} = A_{ba}^{\mu}$. Since

$$W_{\mu} \xrightarrow{\mathrm{C}} \xi_{\mathrm{C}}^{W} \, W_{\mu}^{\dagger} , \tag{2.376}$$

we have

$$\mathscr{L}_{\mathrm{I}}^{(W)} \xrightarrow{\mathrm{C}} -\frac{g}{2\sqrt{2}} \left[-\xi_{\mathrm{C}}^{a*} \, \xi_{\mathrm{C}}^{b} \, \xi_{\mathrm{C}}^{W} \, (V_{ba}^{\mu} + A_{ba}^{\mu}) \, W_{\mu}^{\dagger} - \xi_{\mathrm{C}}^{a} \, \xi_{\mathrm{C}}^{b\,*} \, \xi_{\mathrm{C}}^{W\,*} \, (V_{ab}^{\mu} + A_{ab}^{\mu}) \, W_{\mu} \right] . \tag{2.377}$$

It is clear that there is no possible choice of the charge parities ξ_{C}^{a}, ξ_{C}^{b}, ξ_{C}^{W} which leaves invariant under charge conjugation the $V - A$ charged-current interaction Lagrangian in eqn (2.375). If the fields ψ_a, ψ_b, and W_μ do not take part in other interactions which are invariant under charge conjugation, their charge parities are arbitrary, because they are not fixed in any way by the free Lagrangians, which are invariant under charge conjugation but do not establish any connection between the charge parities of different fields. In this case, the charge parities of the fields ψ_a, ψ_b, and W_μ do not have a physical meaning and can be chosen arbitrarily (for example, $\xi_{\mathrm{C}}^{a} = \xi_{\mathrm{C}}^{b} = \xi_{\mathrm{C}}^{W} = 1$).

2.11.2 Parity

Under a parity transformation (space inversion)

$$x^{\mu} = (x^0, \vec{x}) \xrightarrow{\mathrm{P}} x_{\mathrm{P}}^{\mu} = (x^0, -\vec{x}) = x_{\mu} , \tag{2.378}$$

the spinor fields $\psi(x)$ and $\overline{\psi}(x)$ transform as

$$\psi(x) \xrightarrow{\mathrm{P}} \psi^{\mathrm{P}}(x_{\mathrm{P}}) = \xi_{\mathrm{P}} \, \gamma^0 \, \psi(x) , \tag{2.379}$$

$$\overline{\psi}(x) \xrightarrow{\mathrm{P}} \overline{\psi^{\mathrm{P}}}(x_{\mathrm{P}}) = \xi_{\mathrm{P}}^{*} \, \overline{\psi}(x) \, \gamma^0 . \tag{2.380}$$

The possible values of the *intrinsic parity* $\xi_{\rm P}$ are constrained by the fact that two successive parity transformations restore the system to the original state:

$$\psi(x) \xrightarrow{\rm P} \xi_{\rm P}\,\gamma^0\,\psi(x) \xrightarrow{\rm P} \xi_{\rm P}^2\,\psi(x)\,. \tag{2.381}$$

It then appears that $\xi_{\rm P}^2$ is constrained to be $+1$, but one must note that the sign of a fermion field has no physical meaning because it changes through a rotation of 2π, as discussed in section 2.4.2. Therefore $\xi_{\rm P}^2$ is constrained to be ± 1, which means that

$$\xi_{\rm P} = \pm 1\,, \pm i\,. \tag{2.382}$$

From eqns (2.190) and (2.192) we have

$$\gamma^0\,u^{(h)}(p_{\rm P}) = \eta(\vec{p},-h)\,u^{(-h)}(p)\,, \qquad \gamma^0\,v^{(h)}(p_{\rm P}) = -\eta^*(\vec{p},-h)\,v^{(-h)}(p)\,, \tag{2.383}$$

where $p_{\rm P} = (p^0, -\vec{p})$ and $\eta(\vec{p},-h)$ is a phase factor which depends on $\vec{p}$ and h (see eqn (2.216)). From eqn (2.139) and eqn (2.379), the Fourier expansion of the parity-transformed field $\psi^{\rm P}(x) = \xi_{\rm P}\,\gamma^0\,\psi(x_{\rm P})$ is given by

$$\psi^{\rm P}(x) = \xi_{\rm P} \int \frac{{\rm d}^3p}{(2\pi)^3\,2E} \sum_{h=\pm1} \Big[\eta(\vec{p},h)a^{(-h)}(p_{\rm P})u^{(h)}(p)e^{-ip\cdot x} - \eta^*(\vec{p},h)b^{(-h)\dagger}(p_{\rm P})v^{(h)}(p)e^{ip\cdot x}\Big]\,. \tag{2.384}$$

The transformation of the quantized field is implemented by

$$\mathsf{U}_{\rm P}\,\psi(x)\,\mathsf{U}_{\rm P}^\dagger = \psi^{\rm P}(x) = \xi_{\rm P}\,\gamma^0\,\psi(x_{\rm P})\,, \tag{2.385}$$

where $\mathsf{U}_{\rm P}$ is the unitary operator of a parity transformation ($\mathsf{U}_{\rm P}^\dagger = \mathsf{U}_{\rm P}^{-1}$). From the Fourier expansion in eqn (2.139) of the quantized Dirac field and eqn (2.384), the transformations of the particle and antiparticle creation operators are

$$\mathsf{U}_{\rm P}\,a^{(h)\dagger}(p)\,\mathsf{U}_{\rm P}^\dagger = \eta^*(\vec{p},h)\,\xi_{\rm P}^*\,a^{(-h)\dagger}(p_{\rm P})\,, \tag{2.386}$$

$$\mathsf{U}_{\rm P}\,b^{(h)\dagger}(p)\,\mathsf{U}_{\rm P}^\dagger = -\eta^*(\vec{p},h)\,\xi_{\rm P}\,b^{(-h)\dagger}(p_{\rm P})\,. \tag{2.387}$$

Therefore, a parity transformation reverses the helicity of a fermion, because the three-momentum changes sign, but the spin, which is an axial vector, as all angular momenta are, remains unchanged.

The covariant bilinears in eqns (2.65)–(2.69) transform as

$$S_{ab} = \overline{\psi_a}\,\psi_b \xrightarrow{\rm P} (S_{ab})^{\rm P} = \xi_{\rm P}^{a*}\,\xi_{\rm P}^b\,\overline{\psi_a}\,\psi_b = \xi_{\rm P}^{a*}\,\xi_{\rm P}^b\,S_{ab}\,, \tag{2.388}$$

$$V_{ab}^\mu = \overline{\psi_a}\,\gamma^\mu\,\psi_b \xrightarrow{\rm P} (V_{ab}^\mu)^{\rm P} = \xi_{\rm P}^{a*}\,\xi_{\rm P}^b\,\overline{\psi_a}\,\gamma_\mu\,\psi_b = \xi_{\rm P}^{a*}\,\xi_{\rm P}^b\,V_\mu^{ab}\,, \tag{2.389}$$

$$T_{ab}^{\mu\nu} = \overline{\psi_a}\,\sigma^{\mu\nu}\,\psi_b \xrightarrow{\rm P} (T_{ab}^{\mu\nu})^{\rm P} = \xi_{\rm P}^{a*}\,\xi_{\rm P}^b\,\overline{\psi_a}\,\sigma_{\mu\nu}\,\psi_b = \xi_{\rm P}^{a*}\,\xi_{\rm P}^b\,T_{\mu\nu}^{ab}\,, \tag{2.390}$$

$$A_{ab}^\mu = \overline{\psi_a}\,\gamma^\mu\,\gamma^5\,\psi_b \xrightarrow{\rm P} (A_{ab}^\mu)^{\rm P} = -\xi_{\rm P}^{a*}\,\xi_{\rm P}^b\,\overline{\psi_a}\,\gamma_\mu\,\gamma^5\,\psi_b = -\xi_{\rm P}^{a*}\,\xi_{\rm P}^b\,A_\mu^{ab}\,, \tag{2.391}$$

$$P_{ab} = \overline{\psi_a}\,\gamma^5\,\psi_b \xrightarrow{\rm P} (P_{ab})^{\rm P} = -\xi_{\rm P}^{a*}\,\xi_{\rm P}^b\,\overline{\psi_a}\,\gamma^5\,\psi_b = -\xi_{\rm P}^{a*}\,\xi_{\rm P}^b\,P_{ab}\,, \tag{2.392}$$

where the lowered Lorentz index embodies the change of sign of the spatial components.

As we have discussed in section 2.11.1 for the intrinsic charge parities, the intrinsic parities do not have a physical meaning per se. Only the relative values of the intrinsic parities of different particles which take part in parity conserving interactions have physical meaning.

For example, let us consider the interaction Lagrangian in eqn (2.371) with the vector field Φ_μ which transforms under a parity transformation as

$$\Phi_\mu \xrightarrow{\mathrm{P}} \xi_{\mathrm{P}}^{\Phi}\, \Phi^\mu\,. \tag{2.393}$$

The transformation of the interaction Lagrangian in eqn (2.371) is

$$\mathscr{L}_{\mathrm{I}}^{(\Phi)} \xrightarrow{\mathrm{P}} -g\left(\xi_{\mathrm{P}}^{a*}\,\xi_{\mathrm{P}}^{b}\,\xi_{\mathrm{P}}^{\Phi}\,V_\mu^{ab}\,\Phi^\mu + \xi_{\mathrm{P}}^{a}\,\xi_{\mathrm{P}}^{b\,*}\,\xi_{\mathrm{P}}^{\Phi\,*}\,V_\mu^{ba}\,\Phi^{\mu\dagger}\right). \tag{2.394}$$

Comparing with eqn (2.371), one can see that invariance under parity is achieved with

$$\xi_{\mathrm{P}}^{a*}\,\xi_{\mathrm{P}}^{b}\,\xi_{\mathrm{P}}^{\Phi} = 1\,. \tag{2.395}$$

This relation fixes the relative parities of ψ_a, ψ_b, and Φ_μ. For example, it is convenient to choose $\xi_{\mathrm{P}}^{a} = \xi_{\mathrm{P}}^{b} = \xi_{\mathrm{P}}^{\Phi} = 1$. However, since $|\xi_{\mathrm{P}}^{a}|^2 = 1$, if $a = b$ only the choice $\xi_{\mathrm{P}}^{\Phi} = 1$ is possible and the intrinsic parity ξ_{P}^{a} of the fermion field is undetermined. This is the case of the electromagnetic interaction Lagrangian $\mathscr{L}_{\mathrm{I}}^{(\gamma)}$ in eqn (2.367), with the further simplification that the electromagnetic field is real. One can immediately see that the invariance of $\mathscr{L}_{\mathrm{I}}^{(\gamma)}$ under space inversion implies that the intrinsic parity of the electromagnetic field is $\xi_{\mathrm{P}}^{A} = +1$, i.e.

$$A_\mu \xrightarrow{\mathrm{P}} A^\mu\,. \tag{2.396}$$

Charged-current weak interactions violate the parity symmetry, because of their $V - A$ structure and the different sign of the transformation under parity of the vector and axial current. Indeed, taking into account that the vector boson field W_μ transforms as

$$W_\mu \xrightarrow{\mathrm{P}} \xi_{\mathrm{P}}^{W}\, W^\mu\,, \tag{2.397}$$

for the charged-current $V - A$ interaction Lagrangian in eqn (2.375) we have the transformation

$$\mathscr{L}_{\mathrm{I}}^{(W)} \xrightarrow{\mathrm{P}} -\frac{g}{2\sqrt{2}}\left[-\xi_{\mathrm{P}}^{a*}\,\xi_{\mathrm{P}}^{b}\,\xi_{\mathrm{P}}^{W}\left(V_\mu^{ab} + A_\mu^{ab}\right)W^\mu - \xi_{\mathrm{P}}^{a}\,\xi_{\mathrm{P}}^{b\,*}\,\xi_{\mathrm{P}}^{W\,*}\left(V_\mu^{ba} + A_\mu^{ba}\right)W^{\mu\dagger}\right]. \tag{2.398}$$

Comparing with eqn (2.375), one can immediately see that there is no possible choice of the intrinsic parities ξ_{P}^{a}, ξ_{C}^{P}, ξ_{P}^{W} which would keep the $V - A$ charged-current interaction Lagrangian in eqn (2.375) invariant under a parity transformation. If the fields ψ_a, ψ_b, and W_μ do not take part in other interactions which are invariant under a space inversion, their intrinsic parities do not have a physical meaning, because they are not fixed in any way by the free Lagrangians, which are invariant under a space inversion but do not establish any connection between the intrinsic parities of different fields. In this case, the intrinsic parities of the fields ψ_a, ψ_b, and W_μ can be chosen arbitrarily (for example, $\xi_{\mathrm{P}}^a = \xi_{\mathrm{P}}^b = \xi_{\mathrm{P}}^W = 1$).

2.11.3 CP

The combined CP transformation is a parity transformation followed by charge conjugation:

$$\psi(x) \xrightarrow{\mathrm{CP}} \psi^{\mathrm{CP}}(x_{\mathrm{P}}) = \xi_{\mathrm{CP}}\, \gamma^0\, \mathcal{C}\, \overline{\psi}^T(x) = -\xi_{\mathrm{CP}}\, \mathcal{C}\, \psi^*(x)\,, \tag{2.399}$$

$$\overline{\psi}(x) \xrightarrow{\mathrm{CP}} \overline{\psi^{\mathrm{CP}}}(x_{\mathrm{P}}) = -\xi_{\mathrm{CP}}^*\, \psi^T(x)\, \mathcal{C}^\dagger\, \gamma^0\,. \tag{2.400}$$

The coefficient

$$\xi_{\mathrm{CP}} = \xi_{\mathrm{C}}\, \xi_{\mathrm{P}} \tag{2.401}$$

represents the *intrinsic* CP *parity*. Since $|\xi_{\mathrm{C}}|^2 = 1$ and $\xi_{\mathrm{P}} = \pm 1$ or $\pm i$, ξ_{CP} is a phase,

$$|\xi_{\mathrm{CP}}|^2 = 1\,, \tag{2.402}$$

which is usually called the CP *phase*. Indeed, performing two CP transformations we have

$$\psi \xrightarrow{\mathrm{CP}} -\xi_{\mathrm{CP}}\, \mathcal{C}\, \psi^* \xrightarrow{\mathrm{CP}} -|\xi_{\mathrm{CP}}|^2\, \psi\,. \tag{2.403}$$

As we have remarked after eqn (2.381), the sign of a fermion field has no physical meaning because it changes through a rotation of 2π (see section 2.4.2). Hence, the system is restored to its original state by two successive CP transformations if $|\xi_{\mathrm{CP}}|^2 = 1$.

The transformation of the quantized field $\psi(x)$ is given by

$$\mathsf{U}_{\mathrm{CP}} \psi(x) \mathsf{U}_{\mathrm{CP}}^\dagger = \psi^{\mathrm{CP}}(x) = \xi_{\mathrm{CP}}\, \gamma^0\, \mathcal{C}\, \overline{\psi}^T(x_{\mathrm{P}})\,, \tag{2.404}$$

with

$$\mathsf{U}_{\mathrm{CP}} = \mathsf{U}_{\mathrm{C}}\, \mathsf{U}_{\mathrm{P}}\,. \tag{2.405}$$

From eqns (2.358), (2.359), (2.386), and (2.387), the creation operators of fermions and antifermions transform as

$$\mathsf{U}_{\mathrm{CP}}\, a^{(h)\dagger}(p)\, \mathsf{U}_{\mathrm{CP}}^\dagger = \eta^*(\vec{p}, h)\, \xi_{\mathrm{CP}}^*\, b^{(-h)\dagger}(p_{\mathrm{P}})\,, \tag{2.406}$$

$$\mathsf{U}_{\mathrm{CP}}\, b^{(h)\dagger}(p)\, \mathsf{U}_{\mathrm{CP}}^\dagger = -\eta^*(\vec{p}, h)\, \xi_{\mathrm{CP}}\, a^{(-h)\dagger}(p_{\mathrm{P}})\,, \tag{2.407}$$

where $\eta(\vec{p}, h)$ is a phase factor which depends on $\vec{p}$ and h (see eqn (2.216)). Consequently, a CP transformation interchanges particles and antiparticles and reverses the three-momenta and helicities.

Under a CP transformation the covariant bilinears in eqns (2.65)–(2.69) transform as

$$S_{ab} = \overline{\psi_a}\, \psi_b \xrightarrow{\mathrm{CP}} (S_{ab})^{\mathrm{CP}} = \xi_{\mathrm{CP}}^{a*}\, \xi_{\mathrm{CP}}^b\, \overline{\psi_b}\, \psi_a = \xi_{\mathrm{CP}}^{a*}\, \xi_{\mathrm{CP}}^b\, S_{ba}\,, \tag{2.408}$$

$$V^{\mu}_{ab} = \overline{\psi_a}\,\gamma^{\mu}\,\psi_b \xrightarrow{\text{CP}} (V^{\mu}_{ab})^{\text{CP}} = -\xi^{a*}_{\text{CP}}\,\xi^{b}_{\text{CP}}\,\overline{\psi_b}\,\gamma_{\mu}\,\psi_a = -\xi^{a*}_{\text{CP}}\,\xi^{b}_{\text{CP}}\,V^{ba}_{\mu}\,, \tag{2.409}$$

$$T^{\mu\nu}_{ab} = \overline{\psi_a}\,\sigma^{\mu\nu}\,\psi_b \xrightarrow{\text{CP}} (T^{\mu\nu}_{ab})^{\text{CP}} = -\xi^{a*}_{\text{CP}}\,\xi^{b}_{\text{CP}}\,\overline{\psi_b}\,\sigma_{\mu\nu}\,\psi_a = -\xi^{a*}_{\text{CP}}\,\xi^{b}_{\text{CP}}\,T^{ba}_{\mu\nu}\,, \tag{2.410}$$

$$A^{\mu}_{ab} = \overline{\psi_a}\,\gamma^{\mu}\,\gamma^5\,\psi_b \xrightarrow{\text{CP}} (A^{\mu}_{ab})^{\text{CP}} = -\xi^{a*}_{\text{CP}}\,\xi^{b}_{\text{CP}}\,\overline{\psi_b}\,\gamma_{\mu}\,\gamma^5\,\psi_a = -\xi^{a*}_{\text{CP}}\,\xi^{b}_{\text{CP}}\,A^{ba}_{\mu}\,, \tag{2.411}$$

$$P_{ab} = \overline{\psi_a}\,\gamma^5\,\psi_b \xrightarrow{\text{CP}} (P_{ab})^{\text{CP}} = -\xi^{a*}_{\text{CP}}\,\xi^{b}_{\text{CP}}\,\overline{\psi_b}\,\gamma^5\,\psi_a = -\xi^{a*}_{\text{CP}}\,\xi^{b}_{\text{CP}}\,P_{ba}\,, \tag{2.412}$$

where we have lowered the Lorentz index when the spatial components change sign.

As we have discussed in section 2.11.1 for the intrinsic charge parities and in section 2.11.2 for the intrinsic parities, the intrinsic CP parities do not have a physical meaning per se, because the free Dirac Lagrangian is invariant for any value of the CP parity of a fermion field. Only the relative values of the intrinsic CP parities of different particles which take part in CP conserving interactions have physical meaning.

For example, we can consider the charged-current $V - A$ interaction Lagrangian in eqn (2.375), which is invariant under a CP transformation because the vector current in eqn (2.409) and the axial current in eqn (2.411) transform with the same sign. Indeed, if the vector boson field W_μ transforms as

$$W_{\mu} \xrightarrow{\text{CP}} \xi^{W}_{\text{CP}}\,W^{\mu\dagger}\,, \tag{2.413}$$

we have

$$\mathscr{L}^{(W)}_{\text{I}} \xrightarrow{\text{CP}} -\frac{g}{2\sqrt{2}}\Big[-\xi^{a*}_{\text{CP}}\,\xi^{b}_{\text{CP}}\,\xi^{W}_{\text{CP}}\,\big(V^{ab}_{\mu} - A^{ab}_{\mu}\big)\,W^{\mu\dagger}$$
$$-\xi^{a}_{\text{CP}}\,\xi^{b}_{\text{CP}}{}^{*}\,\xi^{W}_{\text{CP}}{}^{*}\,\big(V^{ba}_{\mu} - A^{ba}_{\mu}\big)\,W^{\mu}\Big]\,. \tag{2.414}$$

Comparing with eqn (2.375), one can see that $\mathscr{L}^{(W)}_{\text{I}}$ is invariant under a CP transformation if

$$\xi^{a*}_{\text{CP}}\,\xi^{b}_{\text{CP}}\,\xi^{W}_{\text{CP}} = -1\,. \tag{2.415}$$

This is the relation between the CP parities of ψ_a, ψ_b, and W_μ. For example, we can choose $\xi^{W}_{\text{CP}} = -1$ and $\xi^{a}_{\text{CP}} = \xi^{b}_{\text{CP}} = 1$.

2.11.4 Time reversal

Under a time reversal transformation

$$x^{\mu} = (x^0, \vec{x}) \xrightarrow{\text{T}} x^{\mu}_{\text{T}} = (-x^0, \vec{x}) = x_{\mu} \tag{2.416}$$

the spinor fields $\psi(x)$ and $\overline{\psi}(x)$ transform as

$$\psi(x) \xrightarrow{\text{T}} \psi^{\text{T}}(x_{\text{T}}) = \xi_{\text{T}}\,\gamma^0\,\gamma^5\,\mathcal{C}\,\overline{\psi}^{T}(x) = \xi_{\text{T}}\,\gamma^5\,\mathcal{C}\,\psi^{*}(x)\,, \tag{2.417}$$

$$\overline{\psi}(x) \xrightarrow{\text{T}} \overline{\psi^{\text{T}}}(x_{\text{T}}) = \xi^{*}_{\text{T}}\,\psi^{T}(x)\,\mathcal{C}^{\dagger}\,\gamma^5\,\gamma^0\,. \tag{2.418}$$

The possible values of the *intrinsic time reversal parity* ξ_{T} are constrained by the fact that two successive parity transformations restore the system to the original status:

$$\psi(x) \xrightarrow{\mathrm{T}} \xi_{\mathrm{T}}\,\gamma^5\,\mathcal{C}\,\psi^*(x) \xrightarrow{\mathrm{T}} -|\xi_{\mathrm{T}}|^2\,\psi(x)\,. \tag{2.419}$$

Since the sign of a fermion field has no physical meaning, as mentioned repeatedly, the system is restored to its original state by two successive time reversals if ξ_{T} is a phase,

$$|\xi_{\mathrm{T}}|^2 = 1\,. \tag{2.420}$$

From eqn (2.139) and eqn (2.417), the Fourier expansion of the time reversal-transformed field $\psi^{\mathrm{T}}(x) = \xi_{\mathrm{T}}\,\gamma^0\,\gamma^5\,\mathcal{C}\,\overline{\psi}^{T}(x_{\mathrm{T}})$ is given by

$$\psi^{\mathrm{T}}(x) = \xi_{\mathrm{T}} \int \frac{\mathrm{d}^3p}{(2\pi)^3\,2E} \sum_{h=\pm1} \Big[a^{(h)\dagger}(p_{\mathrm{P}})\,\gamma^0\,\gamma^5\,v^{(h)}(p_{\mathrm{P}})\,e^{-ip\cdot x} + b^{(h)}(p_{\mathrm{P}})\,\gamma^0\,\gamma^5\,u^{(h)}(p_{\mathrm{P}})\,e^{ip\cdot x} \Big]\,, \tag{2.421}$$

where we have used the relations in eqn (2.354). Furthermore, from eqns (2.190)–(2.196) we have

$$\gamma^0\,\gamma^5\,v^{(h)}(p_{\mathrm{P}}) = -\zeta(h)\,\eta(\vec{p},h)\,u^{(h)}(p)\,, \quad \gamma^0\,\gamma^5\,u^{(h)}(p_{\mathrm{P}}) = -\zeta^*(h)\,\eta^*(\vec{p},h)\,v^{(h)}(p)\,, \tag{2.422}$$

where $\zeta(h)$ and $\eta(\vec{p},h)$ are phases. Hence, we obtain

$$\psi^{\mathrm{T}}(x) = \xi_{\mathrm{T}} \int \frac{\mathrm{d}^3p}{(2\pi)^3\,2E} \sum_{h=\pm1} \Big[-\zeta(h)\,\eta(\vec{p},h)\,a^{(h)\dagger}(p_{\mathrm{P}})\,u^{(h)}(p)\,e^{-ip\cdot x} - \zeta^*(h)\,\eta^*(\vec{p},h)\,b^{(h)}(p_{\mathrm{P}})\,v^{(h)}(p)\,e^{ip\cdot x} \Big]\,. \tag{2.423}$$

Let us consider now the transformation of the quantized field $\psi(x)$. If we try to implement it with a unitary operator such that

$$\mathsf{U}_{\mathrm{T}}\,\psi(x)\,\mathsf{U}_{\mathrm{T}}^{\dagger} = \psi^{\mathrm{T}}(x)\,, \tag{2.424}$$

from eqn (2.139) and (2.423) it follows that

$$\mathsf{U}_{\mathrm{T}}\,a^{(h)\dagger}(p)\,\mathsf{U}_{\mathrm{T}}^{\dagger} = -\zeta^*(h)\,\eta^*(\vec{p},h)\,\xi_{\mathrm{T}}^*\,a^{(h)}(p_{\mathrm{P}})\,, \tag{2.425}$$

$$\mathsf{U}_{\mathrm{T}}\,b^{(h)\dagger}(p)\,\mathsf{U}_{\mathrm{T}}^{\dagger} = -\zeta^*(h)\,\eta^*(\vec{p},h)\,\xi_{\mathrm{T}}\,b^{(h)}(p_{\mathrm{P}})\,. \tag{2.426}$$

This transformation of creation operators into destruction operators leads to unacceptable results. For example, from the invariance of the vacuum under time reversal,

$$\mathsf{U}_{\mathrm{T}}\,|0\rangle = |0\rangle\,, \tag{2.427}$$

we obtain

$$\mathsf{U}_{\mathrm{T}}\,|\mathfrak{f}(p,h)\rangle = \mathsf{U}_{\mathrm{T}}\,a^{(h)\dagger}(p)\,\mathsf{U}_{\mathrm{T}}^{\dagger}\,\mathsf{U}_{\mathrm{T}}\,|0\rangle = -\zeta^*(h)\,\eta^*(\vec{p},h)\,\xi_{\mathrm{T}}^*\,a^{(h)}(p_{\mathrm{P}})\,|0\rangle = 0\,. \tag{2.428}$$

This is clearly an absurd result, because U_{T} cannot annihilate the one-fermion state $|\mathfrak{f}(p,h)\rangle$. The solution to such disastrous results is to implement time reversal with an antiunitary operator such that

$$\mathsf{U}_{\mathrm{T}}\,\psi(x)\,\mathsf{U}_{\mathrm{T}}^{\dagger} = \psi^{\mathrm{T}^*}(x)\,, \tag{2.429}$$

following eqn (2.276). From eqn (2.423), we have

$$\psi^{\mathrm{T}^*}(x) = \xi_{\mathrm{T}}^* \int \frac{\mathrm{d}^3p}{(2\pi)^3\,2E} \sum_{h=\pm1} \Big[-\zeta^*(h)\,\eta^*(\vec{p},h)\,a^{(h)}(p_{\mathrm{P}})\,u^{(h)^*}(p)\,e^{ip\cdot x} \\ -\zeta(h)\,\eta(\vec{p},h)\,b^{(h)\dagger}(p_{\mathrm{P}})\,v^{(h)^*}(p)\,e^{-ip\cdot x} \Big]\,, \tag{2.430}$$

and, from eqn (2.139) and the antilinearity of U_{T}, we have

$$\mathsf{U}_{\mathrm{T}}\,\psi(x)\,\mathsf{U}_{\mathrm{T}}^{\dagger} = \int \frac{\mathrm{d}^3p}{(2\pi)^3\,2E} \sum_{h=\pm1} \Big[\mathsf{U}_{\mathrm{T}} a^{(h)}(p) \mathsf{U}_{\mathrm{T}}^{\dagger}\,u^{(h)^*}(p)\,e^{ip\cdot x} \\ +\mathsf{U}_{\mathrm{T}} b^{(h)\dagger}(p) \mathsf{U}_{\mathrm{T}}^{\dagger}\,v^{(h)^*}(p)\,e^{-ip\cdot x} \Big]\,. \tag{2.431}$$

Comparing eqns (2.430) and (2.431), we obtain the transformation of creation operators

$$\mathsf{U}_{\mathrm{T}} a^{(h)\dagger}(p) \mathsf{U}_{\mathrm{T}}^{\dagger} = -\zeta(h)\,\eta(\vec{p},h)\,\xi_{\mathrm{T}}\,a^{(h)\dagger}(p_{\mathrm{P}})\,, \tag{2.432}$$

$$\mathsf{U}_{\mathrm{T}} b^{(h)\dagger}(p) \mathsf{U}_{\mathrm{T}}^{\dagger} = -\zeta(h)\,\eta(\vec{p},h)\,\xi_{\mathrm{T}}^*\,b^{(h)\dagger}(p_{\mathrm{P}})\,. \tag{2.433}$$

These are fine transformations. They show that time reversal changes a particle or antiparticle into itself, reversing the three-momentum and leaving intact the helicity, as expected on physical grounds (under time reversal three-momenta and angular momenta change sign and the scalar product of spin and three momentum remains invariant).

Under a time reversal transformation the covariant bilinears in eqns (2.65)–(2.69) transform as

$$S_{ab} = \overline{\psi_a}\,\psi_b \xrightarrow{\mathrm{T}} (S_{ab})^{\mathrm{T}} = -\xi_{\mathrm{T}}^{a*}\,\xi_{\mathrm{T}}^{b}\,\overline{\psi_b}\,\psi_a = -\xi_{\mathrm{T}}^{a*}\,\xi_{\mathrm{T}}^{b}\,S_{ba}\,, \tag{2.434}$$

$$V_{ab}^{\mu} = \overline{\psi_a}\,\gamma^\mu\,\psi_b \xrightarrow{\mathrm{T}} (V_{ab}^{\mu})^{\mathrm{T}} = -\xi_{\mathrm{T}}^{a*}\,\xi_{\mathrm{T}}^{b}\,\overline{\psi_b}\,\gamma_\mu\,\psi_a = -\xi_{\mathrm{T}}^{a*}\,\xi_{\mathrm{T}}^{b}\,V_{\mu}^{ba}\,, \tag{2.435}$$

$$T_{ab}^{\mu\nu} = \overline{\psi_a}\,\sigma^{\mu\nu}\,\psi_b \xrightarrow{\mathrm{T}} (T_{ab}^{\mu\nu})^{\mathrm{T}} = \xi_{\mathrm{T}}^{a*}\,\xi_{\mathrm{T}}^{b}\,\overline{\psi_b}\,\sigma_{\mu\nu}\,\psi_a = \xi_{\mathrm{T}}^{a*}\,\xi_{\mathrm{T}}^{b}\,T_{\mu\nu}^{ba}\,, \tag{2.436}$$

$$A_{ab}^{\mu} = \overline{\psi_a}\,\gamma^\mu\,\gamma^5\,\psi_b \xrightarrow{\mathrm{T}} (A_{ab}^{\mu})^{\mathrm{T}} = -\xi_{\mathrm{T}}^{a*}\,\xi_{\mathrm{T}}^{b}\,\overline{\psi_b}\,\gamma_\mu\,\gamma^5\,\psi_a = -\xi_{\mathrm{T}}^{a*}\,\xi_{\mathrm{T}}^{b}\,A_{\mu}^{ba}\,, \tag{2.437}$$

$$P_{ab} = \overline{\psi_a}\,\gamma^5\,\psi_b \xrightarrow{\mathrm{T}} (P_{ab})^{\mathrm{T}} = \xi_{\mathrm{T}}^{a*}\,\xi_{\mathrm{T}}^{b}\,\overline{\psi_b}\,\gamma^5\,\psi_a = \xi_{\mathrm{T}}^{a*}\,\xi_{\mathrm{T}}^{b}\,P_{ba}\,, \tag{2.438}$$

where the lowered Lorentz indices indicate a change of sign of the spatial components.

As the other parities discussed in previous subsections, the time reversal parity of a particle does not have a physical meaning per se, because the free Dirac equation is invariant under time reversal, which means that the time reversal parity is an arbitrary unphysical phase for a noninteracting particle, or a particle which interacts only through time reversal violating interactions. On the other hand, time reversal conserving interactions constraint the relative values of the time reversal parities of the participating particles.

For example, the charged-current $V - A$ interaction Lagrangian in eqn (2.375), is invariant under time reversal, because the vector current in eqn (2.435) and the axial current in eqn (2.437) transform with the same sign. Indeed, if the vector boson field W_μ transforms as

$$W_\mu \xrightarrow{\mathrm{T}} -\xi_{\mathrm{T}}^{W}\, W^{\mu\dagger} \,, \tag{2.439}$$

we have

$$\mathscr{L}_{\mathrm{I}}^{(W)} \xrightarrow{\mathrm{T}} -\frac{g}{2\sqrt{2}} \left[\xi_{\mathrm{T}}^{a*}\, \xi_{\mathrm{T}}^{b}\, \xi_{\mathrm{T}}^{W} \left(V_\mu^{ab} - A_\mu^{ab} \right) W^{\mu\dagger} + \xi_{\mathrm{T}}^{a}\, \xi_{\mathrm{T}}^{b\,*}\, \xi_{\mathrm{T}}^{W\,*} \left(V_\mu^{ba} - A_\mu^{ba} \right) W^{\mu} \right] . \tag{2.440}$$

Comparing with eqn (2.375), one can see that $\mathscr{L}_{\mathrm{I}}^{(W)}$ is invariant under time reversal if

$$\xi_{\mathrm{T}}^{a*}\, \xi_{\mathrm{T}}^{b}\, \xi_{\mathrm{T}}^{W} = 1 \,. \tag{2.441}$$

This relation bounds the relative time reversal parities of ψ_a, ψ_b, and W_μ. For example, one can choose $\xi_{\mathrm{T}}^{W} = \xi_{\mathrm{T}}^{a} = \xi_{\mathrm{T}}^{b} = 1$.

2.11.5 CPT

The combined CPT transformation is a time reversal followed by a CP transformation:

$$\psi(x) \xrightarrow{\mathrm{CPT}} \psi^{\mathrm{CPT}}(-x) = \xi_{\mathrm{CPT}}\, \gamma^5\, \psi(x) \,, \tag{2.442}$$

$$\overline{\psi}(x) \xrightarrow{\mathrm{CPT}} \overline{\psi^{\mathrm{CPT}}}(-x) = -\xi_{\mathrm{CPT}}^{*}\, \overline{\psi}(x)\, \gamma^5 \,. \tag{2.443}$$

The phase

$$\xi_{\mathrm{CPT}} = \xi_{\mathrm{T}}\, \xi_{\mathrm{CP}}^{*} \tag{2.444}$$

represents the *intrinsic CPT parity*. Performing two CPT transformations we obtain

$$\psi \xrightarrow{\mathrm{CPT}} \xi_{\mathrm{CPT}}\, \gamma^5\, \psi(x) \xrightarrow{\mathrm{CPT}} \xi_{\mathrm{CPT}}^{2}\, \psi(x) \,. \tag{2.445}$$

Since the sign of a fermion field has no physical meaning, the system is restored to its original state by two successive CPT transformations if

$$\xi_{\mathrm{CPT}} = \pm 1 \,, \pm i \,. \tag{2.446}$$

Therefore, although the time reversal parity ξ_{T} and the CP parity ξ_{CP} are arbitrary phases, their values are related by eqn (2.446):

$$\xi_{\mathrm{T}} = \pm \xi_{\mathrm{CP}} \,, \pm i \xi_{\mathrm{CP}} \,. \tag{2.447}$$

From eqns (2.139) and (2.442), the Fourier expansion of the CPT-transformed field $\psi^{\mathrm{CPT}}(x) = \xi_{\mathrm{CPT}}\, \gamma^5\, \psi(-x)$ is given by

$$\psi^{\mathrm{CPT}}(x) = \xi_{\mathrm{CPT}} \int \frac{\mathrm{d}^3 p}{(2\pi)^3\, 2E} \sum_{h=\pm 1} \left[h\, a^{(-h)}(p)\, v^{(h)}(p)\, e^{ip\cdot x} \right.$$

$$-h\, b^{(-h)\dagger}(p)\, u^{(h)}(p)\, e^{-ip\cdot x}\Big] , \tag{2.448}$$

where we have used

$$\gamma^5\, u^{(h)}(p) = -h\, v^{(-h)}(p) , \qquad \gamma^5\, v^{(h)}(p) = h\, u^{(-h)}(p) . \tag{2.449}$$

Since U_{T} is antiunitary and U_{C}, U_{P} are unitary, the operator

$$\mathsf{U}_{\mathrm{CPT}} = \mathsf{U}_{\mathrm{C}}\, \mathsf{U}_{\mathrm{P}}\, \mathsf{U}_{\mathrm{T}} \tag{2.450}$$

which transforms the quantized field $\psi(x)$ is antiunitary:

$$\mathsf{U}_{\mathrm{CPT}}\, \psi(x)\, \mathsf{U}_{\mathrm{CPT}}^{\dagger} = \psi^{\mathrm{CPT}^*}(x) . \tag{2.451}$$

From the Fourier expansion in eqn (2.139) of the quantized Dirac field and eqn (2.449), the CPT transformations of the creation operators are

$$\mathsf{U}_{\mathrm{CPT}}\, a^{(h)\dagger}(p)\, \mathsf{U}_{\mathrm{CPT}}^{\dagger} = -h\, \xi_{\mathrm{CPT}}\, b^{(-h)\dagger}(p) , \tag{2.452}$$

$$\mathsf{U}_{\mathrm{CPT}}\, b^{(h)\dagger}(p)\, \mathsf{U}_{\mathrm{CPT}}^{\dagger} = h\, \xi_{\mathrm{CPT}}^{*}\, a^{(-h)\dagger}(p) . \tag{2.453}$$

These relations can also be obtained from eqns (2.406), (2.407), (2.432), (2.433), and (2.444). They show that a CPT transformation changes a particle into its antiparticle with the same momentum and opposite helicity, as expected on physical grounds. Indeed, C changes particle into antiparticle and PT reverses $\vec{x}$ and t, maintaining three momenta invariant but changing the sign of angular momenta.

Under a CPT transformation the covariant bilinears in eqns (2.65)–(2.69) transform as

$$S_{ab} = \overline{\psi_a}\, \psi_b \xrightarrow{\mathrm{CPT}} (S_{ab})^{\mathrm{CPT}} = -\xi_{\mathrm{CPT}}^{a*}\, \xi_{\mathrm{CPT}}^{b}\, \overline{\psi_a}\, \psi_b = -\xi_{\mathrm{CPT}}^{a*}\, \xi_{\mathrm{CPT}}^{b}\, S_{ab} , \tag{2.454}$$

$$V_{ab}^{\mu} = \overline{\psi_a}\, \gamma^{\mu}\, \psi_b \xrightarrow{\mathrm{CPT}} (V_{ab}^{\mu})^{\mathrm{CPT}} = \xi_{\mathrm{CPT}}^{a*}\, \xi_{\mathrm{CPT}}^{b}\, \overline{\psi_a}\, \gamma_{\mu}\, \psi_b = \xi_{\mathrm{CPT}}^{a*}\, \xi_{\mathrm{CPT}}^{b}\, V_{ab}^{\mu} , \tag{2.455}$$

$$T_{ab}^{\mu\nu} = \overline{\psi_a}\, \sigma^{\mu\nu}\, \psi_b \xrightarrow{\mathrm{CPT}} (T_{ab}^{\mu\nu})^{\mathrm{CPT}} = -\xi_{\mathrm{CPT}}^{a*}\, \xi_{\mathrm{CPT}}^{b}\, \overline{\psi_a}\, \sigma_{\mu\nu}\, \psi_b = -\xi_{\mathrm{CPT}}^{a*}\, \xi_{\mathrm{CPT}}^{b}\, T_{ab}^{\mu\nu} , \tag{2.456}$$

$$A_{ab}^{\mu} = \overline{\psi_a}\, \gamma^{\mu}\, \gamma^5\, \psi_b \xrightarrow{\mathrm{CPT}} (A_{ab}^{\mu})^{\mathrm{CPT}} = \xi_{\mathrm{CPT}}^{a*}\, \xi_{\mathrm{CPT}}^{b}\, \overline{\psi_a}\, \gamma_{\mu}\, \gamma^5\, \psi_b = \xi_{\mathrm{CPT}}^{a*}\, \xi_{\mathrm{CPT}}^{b}\, A_{ab}^{\mu} , \tag{2.457}$$

$$P_{ab} = \overline{\psi_a}\, \gamma^5\, \psi_b \xrightarrow{\mathrm{CPT}} (P_{ab})^{\mathrm{CPT}} = -\xi_{\mathrm{CPT}}^{a*}\, \xi_{\mathrm{CPT}}^{b}\, \overline{\psi_a}\, \gamma^5\, \psi_b = -\xi_{\mathrm{CPT}}^{a*}\, \xi_{\mathrm{CPT}}^{b}\, P_{ab} . \tag{2.458}$$

Since all the covariant bilinears are left invariant by a CPT transformation, apart for a possible irrelevant phase (which is the same for the vector and axial currents), any possible interaction Lagrangian is invariant under CPT, in agreement with the CPT theorem, which says that CPT is a symmetry of any relativistic local field theory.

2.12 Wave packets

Since the states in eqn (2.232) are not properly normalized to one (see eqn (2.236)), they cannot be used to describe real particles. In reality, particles are always somehow localized and the uncertainty principle imposes a momentum spread. Therefore real particles and antiparticles with definite helicity are described by the wave packet states

$$|\mathfrak{f}_{\mathrm{wp}}(\langle\vec{p}\rangle, h)\rangle = \int \frac{\mathrm{d}^3 p}{(2\pi)^{3/2}\sqrt{2E}} \, \varphi(\vec{p}, \langle\vec{p}\rangle) \, |\mathfrak{f}(p, h)\rangle \,, \tag{2.459}$$

$$|\bar{\mathfrak{f}}_{\mathrm{wp}}(\langle\vec{p}\rangle, h)\rangle = \int \frac{\mathrm{d}^3 p}{(2\pi)^{3/2}\sqrt{2E}} \, \bar{\varphi}(\vec{p}, \langle\vec{p}\rangle) \, |\bar{\mathfrak{f}}(p, h)\rangle \,, \tag{2.460}$$

with normalized momentum distributions $\varphi(\vec{p}, \langle\vec{p}\rangle)$ and $\bar{\varphi}(\vec{p}, \langle\vec{p}\rangle)$ such that

$$\int \mathrm{d}^3 p \, |\varphi(\vec{p}, \langle\vec{p}\rangle)|^2 = \int \mathrm{d}^3 p \, |\bar{\varphi}(\vec{p}, \langle\vec{p}\rangle)|^2 = 1 \,, \tag{2.461}$$

where $\langle\vec{p}\rangle$ is the average momentum. In this way, the wave packet states $|\mathfrak{f}_{\mathrm{wp}}(\langle\vec{p}\rangle, h)\rangle$ and $|\bar{\mathfrak{f}}_{\mathrm{wp}}(\langle\vec{p}\rangle, h)\rangle$ are properly normalized to one,

$$\langle\mathfrak{f}_{\mathrm{wp}}(\langle\vec{p}\rangle, h)|\mathfrak{f}_{\mathrm{wp}}(\langle\vec{p}\rangle, h')\rangle = \langle\bar{\mathfrak{f}}_{\mathrm{wp}}(\langle\vec{p}\rangle, h)|\bar{\mathfrak{f}}_{\mathrm{wp}}(\langle\vec{p}\rangle, h')\rangle = \delta_{hh'} \,, \tag{2.462}$$

and describe one particle.

The wave function associated with the particle wave packet in eqn (2.459) is

$$\phi^{(h)}(x) = \langle 0|\psi(x)|\mathfrak{f}_{\mathrm{wp}}(\langle\vec{p}\rangle, h)\rangle = \int \frac{\mathrm{d}^3 p}{(2\pi)^{3/2}\sqrt{2E}} \, \varphi(\vec{p}, \langle\vec{p}\rangle) \, u^{(h)}(p) \, e^{-ip\cdot x} \,. \tag{2.463}$$

If the momentum distribution $\varphi(\vec{p}, \langle\vec{p}\rangle)$ is peaked around an average momentum $\langle\vec{p}\rangle$, it is possible to approximate it with the Gaussian

$$\varphi(\vec{p}, \langle\vec{p}\rangle) \simeq (2\pi)^{-3/4} \, |\mathrm{Det}\Lambda|^{1/4} \, \exp\left[-\frac{1}{4}\sum_{k,j} (\vec{p} - \langle\vec{p}\rangle)^k \, \Lambda^{kj} \, (\vec{p} - \langle\vec{p}\rangle)^j\right] \,, \tag{2.464}$$

with

$$\left.\frac{\partial \ln \varphi(\vec{p}, \langle\vec{p}\rangle)}{\partial \vec{p}}\right|_{\vec{p}=\langle\vec{p}\rangle} = 0 \,, \tag{2.465}$$

$$\left.\frac{\partial^2 \ln \varphi(\vec{p}, \langle\vec{p}\rangle)}{\partial p^k \partial p^j}\right|_{\vec{p}=\langle\vec{p}\rangle} = -\frac{1}{2}\Lambda^{kj} \,. \tag{2.466}$$

$\langle\vec{p}\rangle$ is the average momentum of the wave packet,

$$\langle\vec{p}\rangle = \int \mathrm{d}^3p\, \vec{p}\, |\varphi(\vec{p}, \langle\vec{p}\rangle)|^2 \,, \tag{2.467}$$

and Λ^{-1} is the covariance matrix of the momentum distribution

$$V_p^{kj} = \int \mathrm{d}^3p\, (\vec{p} - \langle\vec{p}\rangle)^k\, (\vec{p} - \langle\vec{p}\rangle)^j\, |\varphi(\vec{p}, \langle\vec{p}\rangle)|^2 = [\Lambda^{-1}]^{kj} \,. \tag{2.468}$$

Note that Λ is symmetric by definition and has positive eigenvalues because $f(\langle\vec{p}\rangle)$ must be a maximum of the momentum distribution. Since Λ is symmetric, it can be diagonalized with an orthogonal transformation, which corresponds to a rotation. Also $V_p = \Lambda^{-1}$ is symmetric, can be diagonalized by the same orthogonal transformation, and has positive eigenvalues. The square roots of the three eigenvalues of $V_p = \Lambda^{-1}$ give the widths of the three-momentum distribution. The square of the total momentum uncertainty δp is given by

$$(\delta p)^2 = \mathrm{Tr}[V_p^{kj}] = \mathrm{Tr}[\Lambda^{-1}] \,, \tag{2.469}$$

which is invariant under rotations.

Furthermore, it is possible to approximate the dependence of the energy $E = p^0$ on the momentum $\vec{p}$ given by the dispersion relation in eqn (2.140) as

$$E(\vec{p}) \simeq E(\langle\vec{p}\rangle) + \vec{v}\cdot(\vec{p} - \langle\vec{p}\rangle) + \frac{1}{2}\sum_{k,j} (\vec{p} - \langle\vec{p}\rangle)^k\, \Omega^{kj}\, (\vec{p} - \langle\vec{p}\rangle)^j \,, \tag{2.470}$$

with

$$E(\langle\vec{p}\rangle) = \sqrt{\langle\vec{p}\rangle^2 + m^2} \,, \tag{2.471}$$

$$\vec{v} = \left.\frac{\partial E(\vec{p})}{\partial \vec{p}}\right|_{\vec{p}=\langle\vec{p}\rangle} = \frac{\langle\vec{p}\rangle}{E(\langle\vec{p}\rangle)} \,, \tag{2.472}$$

$$\Omega^{kj} = \left.\frac{\partial^2 E(\vec{p})}{\partial p^k \partial p^j}\right|_{\vec{p}=\langle\vec{p}\rangle} = \frac{1}{E(\langle\vec{p}\rangle)} \left(\delta_{kj} - v^k\, v^j\right) . \tag{2.473}$$

It is clear that Ω is a symmetric matrix. The energy $E(\langle\vec{p}\rangle)$ corresponding to the average momentum $\langle\vec{p}\rangle$ does not coincide with the average energy $\langle E\rangle$, which, in the approximation in eqn (2.470), is given by

$$\begin{aligned}\langle E\rangle &\simeq E(\langle\vec{p}\rangle) + \frac{1}{2}\sum_{k,j} \Omega^{kj}\, V_p^{kj} = E(\langle\vec{p}\rangle) + \frac{1}{2}\sum_{k,j} \Omega^{kj}\, [\Lambda^{-1}]^{kj} \\ &= E(\langle\vec{p}\rangle) + \frac{\mathrm{Tr}[\Lambda^{-1}] - \sum_{k,j} v^k\, [\Lambda^{-1}]^{kj}\, v^j}{2\, E(\langle\vec{p}\rangle)} \,,\end{aligned} \tag{2.474}$$

where we have used the relations in eqns (2.468) and (2.473). It is interesting to evaluate also the energy uncertainty δE, whose square is given by

$$(\delta E)^2 = \int \mathrm{d}^3p\,(E(\vec{p}) - \langle E\rangle)^2\,|\varphi(\vec{p},\langle\vec{p}\rangle)|^2 = \sum_{k,j} v^k\,V_p^{kj}\,v^j = \sum_{k,j} v^k\,[\Lambda^{-1}]^{kj}\,v^j\,. \tag{2.475}$$

Working in the frame in which $V_p = \Lambda^{-1}$ is diagonal, one can see that $\delta E \leq \delta p$, because $|\vec{v}| \leq 1$ (with equality holding for massless particles). This is consistent with an estimation of the energy and momentum uncertainties through the dispersion relation in eqn (2.140), which gives $E\,\delta E \sim p\,\delta p$, leading to $\delta E \sim p\,\delta p/E \leq \delta p$. In particular, one can note that a localized particle at rest has a momentum uncertainty without energy uncertainty.

With the above approximations and

$$\frac{u^{(h)}(p)}{\sqrt{2E}} \simeq \frac{u^{(h)}(E(\langle\vec{p}\rangle),\langle\vec{p}\rangle)}{\sqrt{2E(\langle\vec{p}\rangle)}}\,, \tag{2.476}$$

the integral over d^3p in eqn (2.463) is Gaussian and can be calculated analytically, leading to

$$\phi^{(h)}(x) \simeq \left(\frac{2}{\pi}\right)^{3/4} \frac{|\mathrm{Det}\Lambda|^{1/4}}{|\mathrm{Det}(\Lambda + 2it\Omega)|^{1/2}} \frac{u^{(h)}(E(\langle\vec{p}\rangle),\langle\vec{p}\rangle)}{\sqrt{2E(\langle\vec{p}\rangle)}}$$
$$\times \exp\left[-iE(\langle\vec{p}\rangle)t + i\langle\vec{p}\rangle\cdot\vec{x} - \sum_{k,j}(\vec{x} - \vec{v}t)^k \left[(\Lambda + 2it\Omega)^{-1}\right]^{kj} (\vec{x} - \vec{v}t)^j\right]. \tag{2.477}$$

This expression shows that the wave packet in coordinate space moves with velocity $\vec{v}$, which is called the *group velocity*. Indeed, the average position is

$$\langle x(t)\rangle = \int \mathrm{d}^3x\,\vec{x}\,|\phi^{(h)}(x)|^2 = \vec{v}\,t\,. \tag{2.478}$$

The widths of the wave packet in coordinate space are given by the square roots of the three eigenvalues of the symmetric covariance matrix $V_x(t)$ with elements

$$V_x^{kj}(t) = \int \mathrm{d}^3x\,(\vec{x} - \langle\vec{x}(t)\rangle)^k\,(\vec{x} - \langle\vec{x}(t)\rangle)^j\,|\phi^{(h)}(x)|^2 = \frac{1}{4}\left[\Lambda + 4\,t^2\,\Omega\,\Lambda^{-1}\,\Omega\right]^{kj}. \tag{2.479}$$

Therefore, the value of $V_x^{kj}(t)$ is minimum for $t = 0$ and increases with time ($\Omega\,\Lambda^{-1}\,\Omega$ has positive eigenvalues, because Λ^{-1} has positive eigenvalues and all the elements of the matrix Ω in eqn (2.473) are positive), a phenomenon known as *spreading of the wave packet*, or *dispersion of the wave packet*. For $t = 0$ we have

$$V_x(t = 0) = \frac{1}{4}\,\Lambda\,, \tag{2.480}$$

which satisfies the minimal Heisenberg uncertainty relation

$$V_x(t = 0)\,V_p = \frac{1}{4}\,. \tag{2.481}$$

For antiparticles the wave packet in coordinate space is given by

$$\bar{\phi}^{(h)}(x) = \langle 0|\psi^{\mathrm{C}}(x)|\bar{\mathfrak{f}}_{\mathrm{WP}}(\langle\vec{p}\rangle, h)\rangle = \xi_{\mathrm{C}} \int \frac{\mathrm{d}^3 p}{(2\pi)^{3/2}\sqrt{2E}}\, \bar{\varphi}(\vec{p}, \langle\vec{p}\rangle)\, u^{(h)}(p)\, e^{-ip\cdot x}\,, \tag{2.482}$$

where $\psi^{\mathrm{C}}(x)$ is the charge-conjugated field in eqn (2.356). Adopting for $\bar{\varphi}(\vec{p}, \langle\vec{p}\rangle)$ an approximation analogous to that for $\varphi(\vec{p}, \langle\vec{p}\rangle)$ in eqn (2.464) and using the approximation in eqn (2.470), one obtains for $\bar{\phi}^{(h)}(x)$ an expression analogous to the right side of eqn (2.477), multiplied by the intrinsic charge parity ξ_{C}.

2.13 Finite normalization volume

Another way to get states which are properly normalized is to consider a finite normalization volume with periodic conditions at the boundaries. This is a method which is used by many authors, because it is simpler, albeit less realistic, than the wave packet approach. In practice one can do all calculations keeping the normalization volume finite through the intermediate steps and taking the limit to an infinite volume at the end.

It is convenient to consider a normalization cube with side L and volume $V = L^3$. The periodic conditions at the boundaries imply that the momentum $\vec{p}$ is quantized:

$$\vec{p} = \frac{2\pi}{L}\,\vec{n}\,, \tag{2.483}$$

with $\vec{n} = (n_1, n_2, n_3)$ and $n_i = 0, \pm1, \pm2, \ldots$. The discrete case can be obtained from the continuum through the replacements

$$\frac{\mathrm{d}^3 p}{(2\pi)^3} \to \frac{1}{V}\sum_{\vec{p}}\,, \qquad (2\pi)^3\,\delta^3(\vec{p}-\vec{p}\,') \to V\,\delta_{\vec{p}\vec{p}'}\,. \tag{2.484}$$

The Fourier expansion of the field is written as

$$\psi(x) = \frac{1}{V}\sum_{\vec{p}} \frac{1}{2E} \sum_{h=\pm1} \left[a^{(h)}(p)\, u^{(h)}(p)\, e^{-ip\cdot x} + b^{(h)\dagger}(p)\, v^{(h)}(p)\, e^{ip\cdot x} \right], \tag{2.485}$$

with the discrete anticommutation relations

$$\{a^{(h)}(p), a^{(h')\dagger}(p')\} = \{b^{(h)}(p), b^{(h')\dagger}(p')\} = 2EV\,\delta_{\vec{p}\vec{p}'}\,\delta_{hh'}\,. \tag{2.486}$$

Since in this approach the volume V is finite, one can define the one-particle states as

$$|\mathfrak{f}(p,h)\rangle = \frac{1}{2EV}\, a^{(h)\dagger}(p)\,|0\rangle\,, \qquad |\bar{\mathfrak{f}}(p,h)\rangle = \frac{1}{2EV}\, b^{(h)\dagger}(p)\,|0\rangle\,. \tag{2.487}$$

These states are properly normalized to one:

$$\langle \mathfrak{f}(p,h)|\mathfrak{f}(p,h')\rangle = \langle \bar{\mathfrak{f}}(p,h)|\bar{\mathfrak{f}}(p,h')\rangle = \delta_{hh'}\,. \tag{2.488}$$

2.14 Fierz transformations

Let us consider the five scalar interaction Lagrangians that can be constructed with four Dirac fields ψ_1, ψ_2, ψ_3, ψ_4:

$$\mathcal{L}^S(\psi_1,\psi_2,\psi_3,\psi_4) = \left(\overline{\psi_1}\,\psi_2\right)\left(\overline{\psi_3}\,\psi_4\right) , \tag{2.489}$$

$$\mathcal{L}^V(\psi_1,\psi_2,\psi_3,\psi_4) = \left(\overline{\psi_1}\,\gamma^\mu\,\psi_2\right)\left(\overline{\psi_3}\,\gamma_\mu\,\psi_4\right) , \tag{2.490}$$

$$\mathcal{L}^T(\psi_1,\psi_2,\psi_3,\psi_4) = \left(\overline{\psi_1}\,\sigma^{\mu\nu}\,\psi_2\right)\left(\overline{\psi_3}\,\sigma_{\mu\nu}\,\psi_4\right) , \tag{2.491}$$

$$\mathcal{L}^A(\psi_1,\psi_2,\psi_3,\psi_4) = \left(\overline{\psi_1}\,\gamma^\mu\,\gamma^5\,\psi_2\right)\left(\overline{\psi_3}\,\gamma_\mu\,\gamma^5\,\psi_4\right) , \tag{2.492}$$

$$\mathcal{L}^P(\psi_1,\psi_2,\psi_3,\psi_4) = \left(\overline{\psi_1}\,\gamma^5\,\psi_2\right)\left(\overline{\psi_3}\,\gamma^5\,\psi_4\right) . \tag{2.493}$$

These scalar interaction Lagrangians contain products of Γ matrices of type

$$\Gamma^a_{\alpha\beta}\,\Gamma^a_{\rho\eta} , \tag{2.494}$$

where $\alpha,\beta,\rho,\eta = 1,\ldots,4$ are Dirac indices. For example,

$$\begin{aligned}\mathcal{L}^V(\psi_1,\psi_2,\psi_3,\psi_4) &= \left(\overline{\psi_1}\,\gamma^0\,\psi_2\right)\left(\overline{\psi_3}\,\gamma^0\,\psi_4\right) - \sum_{k=1}^{3}\left(\overline{\psi_1}\,\gamma^k\,\psi_2\right)\left(\overline{\psi_3}\,\gamma^k\,\psi_4\right)\\ &= \sum_{\alpha,\beta,\rho,\eta}(\overline{\psi_1})_\alpha\,(\psi_2)_\beta\,(\overline{\psi_3})_\rho\,(\psi_4)_\eta\left(\Gamma^2_{\alpha\beta}\,\Gamma^2_{\rho\eta} - \Gamma^3_{\alpha\beta}\,\Gamma^3_{\rho\eta} - \Gamma^4_{\alpha\beta}\,\Gamma^4_{\rho\eta} - \Gamma^5_{\alpha\beta}\,\Gamma^5_{\rho\eta}\right) .\end{aligned} \tag{2.495}$$

From eqn (2.58), for fixed values of β and ρ, the 4×4 matrix $M^{a\beta\rho}$ with elements

$$M^{a\beta\rho}_{\alpha\eta} = \Gamma^a_{\alpha\beta}\,\Gamma^a_{\rho\eta} \tag{2.496}$$

can be written as the linear combination of Γ matrices

$$M^{a\beta\rho} = \frac{1}{4}\sum_b s_b\,\mathrm{Tr}\left[M^{a\beta\rho}\,\Gamma^b\right]\Gamma^b . \tag{2.497}$$

Since the traces in eqn (2.497) are given by

$$\mathrm{Tr}\left[M^{a\beta\rho}\,\Gamma^b\right] = \sum_{\alpha,\eta}\Gamma^a_{\alpha\beta}\,\Gamma^a_{\rho\eta}\,\Gamma^b_{\eta\alpha} = \left(\Gamma^a\,\Gamma^b\,\Gamma^a\right)_{\rho\beta} , \tag{2.498}$$

we obtain

$$\Gamma^a_{\alpha\beta}\,\Gamma^a_{\rho\eta} = \frac{1}{4}\sum_b s_b\,\left(\Gamma^a\,\Gamma^b\,\Gamma^a\right)_{\rho\beta}\Gamma^b_{\alpha\eta} . \tag{2.499}$$

The product $\Gamma^a\Gamma^b\Gamma^a$ is a 4×4 matrix which can also be written as a linear combination of Γ matrices using eqn (2.58):

$$\Gamma^a\,\Gamma^b\,\Gamma^a = \frac{1}{4}\sum_c s_c\,\mathrm{Tr}\left[\Gamma^a\,\Gamma^b\,\Gamma^a\,\Gamma^c\right]\Gamma^c . \tag{2.500}$$

TABLE 2.2. Values of the coefficients C_{XY} in eqn (2.506).

		Y				
		S	V	T	A	P
	S	$-1/4$	$-1/4$	$-1/8$	$1/4$	$-1/4$
	V	-1	$1/2$	0	$1/2$	1
X	T	-3	0	$1/2$	0	-3
	A	1	$1/2$	0	$1/2$	-1
	P	$-1/4$	$1/4$	$-1/8$	$-1/4$	$-1/4$

From the property in eqn (2.44) of Γ matrices it follows that each of the two products $\Gamma^a\Gamma^b$ and $\Gamma^a\Gamma^c$ is proportional to a Γ matrix,

$$\Gamma^a\Gamma^b \propto \Gamma^d \,, \qquad \Gamma^a\Gamma^c \propto \Gamma^e \,, \tag{2.501}$$

which implies that

$$\mathrm{Tr}\big[\Gamma^a\Gamma^b\Gamma^a\Gamma^c\big] \propto \mathrm{Tr}\big[\Gamma^d\Gamma^e\big] \propto \delta_{de} \,. \tag{2.502}$$

Since $\Gamma^d = \Gamma^e$ only if $\Gamma^b = \Gamma^c$, we obtain

$$\mathrm{Tr}\big[\Gamma^a\Gamma^b\Gamma^a\Gamma^c\big] = \mathrm{Tr}\big[(\Gamma^a\Gamma^b)^2\big]\,\delta_{bc} \,. \tag{2.503}$$

Taking into account that $s_b^2 = 1$, eqn (2.499) becomes

$$\Gamma^a_{\alpha\beta}\,\Gamma^a_{\rho\eta} = \frac{1}{16}\sum_b \mathrm{Tr}\big[(\Gamma^a\Gamma^b)^2\big]\,\Gamma^b_{\alpha\eta}\Gamma^b_{\rho\beta} \,. \tag{2.504}$$

This equation expresses the product $\Gamma^a_{\alpha\beta}\Gamma^a_{\rho\eta}$ as a linear combination of products of two Γ matrices in which the indices β and η have been exchanged. This allows us to express each of the five scalar interaction Lagrangians in eqns (2.489)–(2.493) as a linear combination of similar interaction Lagrangians in which the fields ψ_2 and ψ_4 have been exchanged. Indeed, from eqn (2.504) we have

$$\left(\overline{\psi}_1\Gamma^a\psi_2\right)\left(\overline{\psi}_3\Gamma^a\psi_4\right) = -\frac{1}{16}\sum_b \mathrm{Tr}\big[(\Gamma^a\Gamma^b)^2\big]\left(\overline{\psi}_1\Gamma^b\psi_4\right)\left(\overline{\psi}_3\Gamma^b\psi_2\right) , \tag{2.505}$$

where the minus sign takes into account the anticommutativity of the fermion fields. These are *Fierz transformations.*

Calculating the traces in eqn (2.505) one can find the relations

$$\mathcal{L}^X(\psi_1,\psi_2,\psi_3,\psi_4) = \sum_{Y=S,V,T,A,P} C_{XY}\,\mathcal{L}^Y(\psi_1,\psi_4,\psi_3,\psi_2) \,, \tag{2.506}$$

with the coefficients C_{XY} given in Table 2.2.

The five scalar interaction Lagrangians in eqns (2.489)–(2.493) are invariant under parity. If parity is not conserved, it is possible to have pseudoscalar interaction Lagrangians obtained from those in eqns (2.489)–(2.493) by multiplying one field by γ^5. The corresponding Fierz transformations can be obtained easily from eqn (2.506) by multiplying the same field by γ^5 on the left- and right-hand sides.

In particular, it is useful to consider the parity-violating interaction Lagrangian

$$\mathcal{L}^{V-A}(\psi_1, \psi_2, \psi_3, \psi_4) = \left(\overline{\psi_1}\, \gamma^\mu \left(\mathbf{1} - \gamma^5\right) \psi_2\right) \left(\overline{\psi_3}\, \gamma_\mu \left(\mathbf{1} - \gamma^5\right) \psi_4\right) , \qquad (2.507)$$

which finds application in the description of weak interactions at low energies (see sections 3.7). Since the VT and AT coefficients in Table 2.2 are zero, the VS and AS coefficients are opposite, the VP and AP coefficients are opposite, and the VV, AV, VA, AA coefficients are all equal to 1/2, we have

$$\mathcal{L}^{V-A}(\psi_1, \psi_2, \psi_3, \psi_4) = \mathcal{L}^{V-A}(\psi_1, \psi_4, \psi_3, \psi_2) . \qquad (2.508)$$

This is a useful result for the study of some weak interaction processes (see, for example, section 9.1).

3

THE STANDARD MODEL

> It doesn't matter how beautiful your theory is, it doesn't matter how smart you are. If it doesn't agree with experiment, it's wrong.
> **Richard P. Feynman**

The Standard Model (SM) [543, 1051, 920] describes the strong, electromagnetic, and weak interactions of elementary particles in the framework of quantum field theory. It is a gauge theory (see appendix D) based on the local symmetry group $\mathrm{SU}(3)_C \times \mathrm{SU}(2)_L \times \mathrm{U}(1)_Y$, where the subscripts C, L and Y denote color, left-handed chirality and weak hypercharge, respectively. The gauge group uniquely determines the interactions and the number of vector gauge bosons that correspond to the generators of the group. They are eight massless gluons, corresponding to the eight generators of $\mathrm{SU}(3)_C$, that mediate strong interactions; four gauge bosons, of which three are massive ($W^{\pm}$ and Z) and one is massless (γ, the photon), corresponding to the three generators of $\mathrm{SU}(2)_L$ and one generator of $\mathrm{U}(1)_Y$, responsible for electroweak interactions. In this chapter we present a brief review of the electroweak part of the SM, based on the symmetry group $\mathrm{SU}(2)_L \times \mathrm{U}(1)_Y$, which determines the interactions of neutrinos, among others. In the SM, electroweak interactions can be studied separately from strong interactions[11], because the symmetry under the color group $\mathrm{SU}(3)_C$ is unbroken and there is no mixing between the $\mathrm{SU}(3)_C$ and $\mathrm{SU}(2)_L \times \mathrm{U}(1)_Y$ sectors; on the other hand, as we will see in section 3.2, electromagnetic and weak interactions must be treated together because there can be a mixing between the neutral gauge bosons of $\mathrm{SU}(2)_L$ and $\mathrm{U}(1)_Y$ [543].

As in all gauge theories, the symmetry group of the SM fixes the interactions, i.e. the number and properties of the vector gauge bosons, with only three independent unknown parameters, the three coupling constants of the $\mathrm{SU}(3)_C$, $\mathrm{SU}(2)_L$ and $\mathrm{U}(1)_Y$ groups, all of which must be determined from experiments. On the other hand, the number and properties of scalar bosons and fermions are left unconstrained, except for the fact that they must transform in a definite way under the symmetry group, i.e. they must belong to the representations of the symmetry

[11] The theory of strong interactions based on the local symmetry group $\mathrm{SU}(3)_C$ is called *quantum chromodynamics* (QCD) (see section D.2 of appendix D). Its Lagrangian is relatively simple, but the dynamics is very complicated, because of the strong character of the interaction at low energies, which does not allow a perturbative approximation. The strong binding of quarks in nucleons and that of nucleons in nuclei are nonperturbative effects that have not been solved so far. The dynamics of QCD can be solved perturbatively only at high energies ($\gg 1$ GeV), where the effective coupling constant becomes small (*asymptotic freedom*). For an introduction see, for example, Refs. [720, 721].

group, and the fermion representations must lead to the cancellation of quantum anomalies [31, 201, 268, 572, 517]. In the SM, the number of scalar bosons and fermions and their arrangement in the representations of the symmetry group are chosen in a heuristic way. The scalar bosons are chosen in order to implement, in a minimal way, the Higgs mechanism for the generation of masses, whereas the number and properties of fermions are determined by experiments. It is remarkable that all the known elementary fermions can be accommodated in appropriate representations of the symmetry group of the SM with exact cancellation of quantum anomalies.

A puzzling feature of Nature is the existence of three generations of fermions with identical properties, except for different masses. This feature is unexplained in the SM.

The known elementary fermions are divided in two categories, quarks and leptons[12], according to the scheme

	1st generation	**2nd generation**	**3rd generation**
quarks:	u (up),	c (charm),	t (top),
	d (down),	s (strange),	b (bottom);
leptons:	ν_e (electron neutrino),	ν_μ (muon neutrino),	ν_τ (tau neutrino),
	e (electron),	μ (muon),	τ (tau).

They are distinguished by the fact that quarks participate in all the interactions (strong, electromagnetic, weak, and gravitational), whereas leptons participate in all the interactions except strong interactions. The masses and electric charges of quarks and leptons are given in Tables 3.1 and 3.2. The corresponding antiparticles have the same mass and opposite electric charge. All fundamental fermions have spin 1/2.

Quarks are elementary components of hadrons (see Tables 3.3 and 3.4) but do not exist as free particles. This means that their masses do not have the usual classical meaning and their values depend on how the masses are defined. The quark masses given in Table 3.1 are the so-called *current masses*, which are parameters in the QCD Lagrangian[13]. The top quark was discovered only in 1994 by the CDF and D0 experiments through the observation of the products of $p\bar{p}$ collisions produced by the Tevatron accelerator in Fermilab. Nevertheless, as can be seen from Table 3.1, the value of the top quark mass has the smallest relative uncertainty. This is due to the fact that, at the top mass scale, noncalculable nonperturbative effects of strong interactions are negligible, whereas they dominate in the case of the light quarks (u, d, s), whose mass uncertainty is very large.

[12] From the Greek $\lambda\epsilon\pi\tau\acute{o}\varsigma$ = thin. In each generation leptons are lighter than the quarks, which are constituents of the baryons, from the Greek $\beta\alpha\rho\acute{\upsilon}\varsigma$ = heavy (proton, neutron, etc.), and the mesons (pions, kaons, etc.).

[13] Technically speaking, the quark masses given in Table 3.1 are the current masses evaluated in the $\overline{\text{MS}}$ renormalization scheme at a renormalization scale of 2 GeV for light quarks (u, d, s) and a renormalization scale equal to the mass for the heavy quarks (c, b, t). For more explanations see [400] and references therein.

TABLE 3.1. Mass and charge of quarks [400]. The charge is given in units of the elementary charge e.

Flavor	Mass	Charge
u	1.5–4 MeV	2/3
d	4–8 MeV	−1/3
s	80–130 MeV	−1/3
c	1.15–1.35 GeV	2/3
b	4.1–4.4 GeV	−1/3
t	174.3 ± 5.1 GeV	2/3

TABLE 3.2. Mass, charge, lifetime, and magnetic moment of leptons [400]. The charge is given in units of the elementary charge e. The magnetic moment of the electron and the neutrinos is expressed in units of the Bohr magneton $\mu_{\mathrm{B}} \equiv e\hbar/2m_e$.

Flavor	Mass	Charge	Lifetime	Magn. Mom.
e	0.510 MeV	−1	$> 4.3 \times 10^{23}$ y	$1.001\ \mu_{\mathrm{B}}$
ν_e	< 15 eV	0	$> 7 \times 10^9 (m_{\nu_e}/\mathrm{eV})$	$< 1.8 \times 10^{-10}\ \mu_{\mathrm{B}}$
μ	105.65 MeV	−1	2.197×10^{-6} s	$1.001\ e/2m_\mu$
ν_μ	< 190 keV	0	$> 15.4 (m_{\nu_\mu}/\mathrm{eV})$ s	$< 7.4 \times 10^{-10}\ \mu_{\mathrm{B}}$
τ	1.777 GeV	−1	$(2.900 \pm 0.012) \times 10^{-13}$ s	$\simeq e/2m_\tau$
ν_τ	< 18.2 MeV	0	?	$< 5.4 \times 10^{-7}\ \mu_{\mathrm{B}}$

TABLE 3.3. Properties of the lightest baryons [400]. J is the spin, P is the parity, and I is the isospin.

Baryon	Quark content	Mass (GeV)	J^P	I	Lifetime (s)
p	uud	938.27	$1/2^+$	1/2	∞
n	udd	939.56			887
Λ^0	uds	1116	$1/2^+$	0	2.6×10^{-10}
Σ^+	uus	1189			8.0×10^{-11}
Σ^0	uds	1192	$1/2^+$	1	7×10^{-20}
Σ^-	dds	1197			1.5×10^{-10}
Δ^-, Δ^0 Δ^+, Δ^{++}	ddd, udd uud, uuu	1232	$3/2^+$	3/2	5×10^{-24}
Ξ^0	uss	1315	$1/2^+$	1/2	2.9×10^{-10}
Ξ^-	dss	1321			1.6×10^{-10}
Ω^-	sss	1672	$3/2^+$	0	8.2×10^{-11}

TABLE 3.4. Properties of the lightest mesons [400]. J is the spin, P is the parity, and I is the isospin.

Meson	Quark content	Mass (MeV)	J^P	I	Lifetime (s)
π^0 π^-, π^+	$(u\bar{u} - d\bar{d})/\sqrt{2}$ $\bar{u}d$, $u\bar{d}$	134.98 139.57	0^-	1	8×10^{-17} 2.6×10^{-8}
K^0, K^+ K^-, $\bar{K}^0$	$d\bar{s}$, $u\bar{s}$ $\bar{u}s$, $\bar{d}s$	$493.7\,(K^{\pm})$ $497.7\,(K^0)$	0^-	1/2 1/2	$1.2 \times 10^{-8}\,(K^{\pm})$ $8.9 \times 10^{-11}\,(K^0_S)$ $5.2 \times 10^{-8}\,(K^0_L)$
η	$(u\bar{u} + d\bar{d})/\sqrt{2}$	547	0^-	0	6×10^{-19}
ρ^0 ρ^-, ρ^+	$(u\bar{u} - d\bar{d})/\sqrt{2}$ $\bar{u}d$, $u\bar{d}$	768	1^-	1	4×10^{-24}
ω	$(u\bar{u} + d\bar{d})/\sqrt{2}$	782	1^-	0	8×10^{-23}
η'	$(u\bar{u} + d\bar{d})/\sqrt{2}$	957.78 ± 0.14	0^-	0	$(3.24 \pm 0.25) \times 10^{-21}$

The fermion sector of the SM depends on 13 independent parameters: six quark masses, three charged lepton masses (neutrinos are assumed to be massless in the SM), three quark mixing angles and one phase. The values of all these parameters must be determined from experimental measurements.

In addition to the three parameters in the gauge sector (the coupling constants), a very small QCD parameter related with the *strong CP problem* (see, for example, Ref. [721]) and the 13 parameters in the fermion sector (masses and mixing), the SM depends on two more parameters coming from the scalar Higgs sector, a Higgs mass and a quartic coupling constant, reaching a total of 19 independent parameters. It is clear that the large number of parameters and the unexplained existence of three generations, together with the fact that the SM does not include gravitational interactions, are unsatisfactory aspects of the SM. This justifies the widespread opinion that the SM is not the ultimate theory of the physics of elementary particles but must be a low-energy effective theory.

Since the SM is a renormalizable theory, even its quantum corrections are insensitive to the physics beyond the SM. Because of this reason, the SM is phenomenologically very successful and so far has been able to describe all the known phenomena, except for the indications in favor of neutrino oscillations that we will discuss in the following chapters. In particular, the SM interactions of neutrinos have been verified experimentally with high accuracy and are universally used for the analysis of the data of neutrino experiments. This is why this chapter is devoted to an important study of properties of neutrinos in the framework of the SM. Moreover, in chapters 15 and 16 we will see that the SM properties of neutrinos have important consequences for astrophysics and cosmology.

3.1 Electroweak Lagrangian

In this section, we present the electroweak part of the SM Lagrangian, which determines neutrino interactions in the SM. For this purpose, it is sufficient to consider only the $\mathrm{SU}(2)_L \times \mathrm{U}(1)_Y$ part of the SM symmetry group.

The symmetry group $\mathrm{SU}(2)_L$ is called *weak isospin*. Its subscript L indicates that the elements of the group act in a nontrivial way only on the left-handed chiral components of the fermion fields (the right-handed chiral components are singlets under weak isospin transformations). This group has three generators, for which we use the notation

$$I_a \qquad (a = 1, 2, 3) \,. \tag{3.1}$$

They satisfy the angular momentum commutation relations

$$[I_a, I_b] = i\varepsilon_{abc} I_c \,. \tag{3.2}$$

In eqn (3.2) ε_{abc} is the totally antisymmetric tensor with three indices having $\varepsilon_{123} = 1$. It is important to note that the nonabelian character of the weak isospin group embodied by the commutation relations in eqn (3.2) implies that for each representation of the group the scale of the generators is fixed. For example, in the two-dimensional representation the generators are $I_a = \tau_a/2$, where τ_1, τ_2, τ_3 are the three Pauli matrices[14]. A rescaling of the generators of the type $I_a \to c_a I_a$ with arbitrary constants c_a would spoil the commutation relations in eqn (3.2). This means that the action of the generators on each representation is unique.

The symmetry group $\mathrm{U}(1)_Y$ is called *hypercharge*. It is generated by the hypercharge operator Y, which is connected to I_3 and the charge operator Q by the Gell-Mann–Nishijima relation

$$Q = I_3 + \frac{Y}{2} \,. \tag{3.3}$$

This relation is necessary in order to fix the action of the hypercharge operator Y on the fermion fields, which is not constrained by the theory, because the $\mathrm{U}(1)_Y$ is abelian. Moreover, the Gell-Mann–Nishijima relation implies the unification of weak and electromagnetic interactions.

In order to have local gauge invariance, one must introduce three vector gauge boson fields A_a^μ $(a = 1, 2, 3)$ associated with the three generators I_a $(a = 1, 2, 3)$ of the group $\mathrm{SU}(2)_L$, and one vector gauge boson field B^μ associated with the generator Y of the group $\mathrm{U}(1)_Y$. The covariant derivative D_μ, which in gauge theories replaces the normal derivative ∂_μ in the Lagrangian, is

$$D_\mu = \partial_\mu + ig\,\underline{A}_\mu \cdot \underline{I} + ig'\,B_\mu\,\frac{Y}{2} \,, \tag{3.4}$$

where we have introduced the vector notation

$$\underline{A}^\mu \equiv (A_1^\mu, A_2^\mu, A_3^\mu) \,, \quad \underline{I} \equiv (I_1, I_2, I_3) \,, \tag{3.5}$$

[14] The Pauli matrices $\tau_a = \sigma_a$ are given in eqn (2.27). For the generators of the $\mathrm{SU}(2)_L$ group we use the notation $\tau_a/2$ in order to avoid confusion with the spin operators $\sigma_a/2$.

with the scalar product

$$\underline{A}^{\mu} \cdot \underline{I} \equiv \sum_{a=1}^{3} A_a^{\mu} I_a \,. \tag{3.6}$$

The covariant derivative in eqn (3.4) contains two independent coupling constants: g associated with the group $\mathrm{SU}(2)_L$ and g' associated with the group $\mathrm{U}(1)_Y$.

The next step in the construction of the electroweak $\mathrm{SU}(2)_L \times \mathrm{U}(1)_Y$ theory is to choose the representations for the fermion fields. Historically, this choice has been guided by the wisdom of previous experience, in particular the $V-A$ theory of weak interactions and the two-component theory of the neutrino. Here, we take the choice of the representations for the fermion fields as an assumption that leads to the correct phenomenology. The left-handed chiral components of the fermion fields are grouped into *weak isospin doublets*. For the sake of simplicity, let us consider only the first generation of leptons and quarks:

$$L_L = \begin{pmatrix} \nu_{eL} \\ e_L \end{pmatrix} , \qquad Q_L = \begin{pmatrix} u_L \\ d_L \end{pmatrix} . \tag{3.7}$$

The straightforward generalization for three generations is presented at the end of this section.

From the choice in eqn (3.7) of the weak isospin representation of the fermion fields, the generators of the $\mathrm{SU}(2)_L$ group are fixed to be $I_a = \tau_a/2$:

$$\underline{I}\, L_L = \frac{\underline{\tau}}{2}\, L_L \,, \qquad \underline{I}\, Q_L = \frac{\underline{\tau}}{2}\, Q_L \,, \tag{3.8}$$

where $\underline{\tau} = (\tau_1, \tau_2, \tau_3)$. On the other hand, the action of the hypercharge operator Y is fixed by the Gell-Mann–Nishijima relation in eqn (3.3):

$$Y\, L_L = -L_L \,, \qquad Y\, Q_L = \frac{1}{3}\, Q_L \,. \tag{3.9}$$

Hence, the left-handed lepton and quark doublets have, respectively, hypercharge $Y = -1$ and $Y = 1/3$.

We parameterize the elements g of the group of local $\mathrm{SU}(2)_L \times \mathrm{U}(1)_Y$ transformations with a set of 3+1 parameters $(\underline{\theta}(x), \eta(x))$, with $\underline{\theta}(x) = (\theta_1(x), \theta_2(x), \theta_3(x))$, which depend on space-time x:

$$g(\underline{\theta}(x), \eta(x)) \in \mathrm{SU}(2)_L \times \mathrm{U}(1)_Y \,, \tag{3.10}$$

with unitary representation of $g(\underline{\theta}(x), \eta(x))$ on the vector space of the fields

$$U(\underline{\theta}(x), \eta(x)) = e^{i\underline{\theta}(x) \cdot \underline{I} + i\eta(x)\frac{Y}{2}} = U(\underline{\theta}(x))\, U(\eta(x)) \,, \tag{3.11}$$

with

$$U(\underline{\theta}(x)) = e^{i\underline{\theta}(x) \cdot \underline{I}} \,, \qquad U(\eta(x)) = e^{i\eta(x)\frac{Y}{2}} \,. \tag{3.12}$$

The transformation of the left-handed fermion doublets under $g(\underline{\theta}(x), \eta(x))$ is given by

$$L_L \xrightarrow{g(\underline{\theta}(x),\eta(x))} L'_L = U(\underline{\theta}(x), \eta(x))\, L_L = U_L^l(\underline{\theta}(x), \eta(x))\, L_L \,, \tag{3.13}$$

$$Q_L \xrightarrow{g(\underline{\theta}(x),\eta(x))} Q'_L = U(\underline{\theta}(x),\eta(x))\, Q_L = U_L^q(\underline{\theta}(x),\eta(x))\, Q_L \,. \tag{3.14}$$

where

$$U_L^l(\underline{\theta}(x),\eta(x)) = e^{\frac{i}{2}\underline{\theta}(x)\cdot\underline{\tau}-\frac{i}{2}\eta(x)}\,, \qquad U_L^q(\underline{\theta}(x),\eta(x)) = e^{\frac{i}{2}\underline{\theta}(x)\cdot\underline{\tau}+\frac{i}{6}\eta(x)}\,. \tag{3.15}$$

In the SM, it is assumed that the neutrino fields have only left-handed components. This assumption follows from the two-component theory of Landau [711], Lee and Yang [727], and Salam [919], implying that neutrinos are massless. The right-handed components of the other fermions,

$$e_R\,, \quad u_R\,, \quad d_R\,, \tag{3.16}$$

are assumed to be singlets under the weak isospin group of transformations:

$$\underline{I}\, f_R = 0 \qquad (f = e, u, d)\,. \tag{3.17}$$

From the Gell-Mann–Nishijima relation in eqn (3.3), e_R, u_R, and d_R have, respectively, hypercharge $Y = -2, 4/3, -2/3$:

$$Y\, e_R = -2\, e_R\,, \qquad Y\, u_R = \frac{4}{3}\, u_R\,, \qquad Y\, d_R = -\frac{2}{3}\, d_R\,. \tag{3.18}$$

Then, the transformation of the right-handed components of the fermion fields under the transformation in eqn (3.10) is given by

$$f_R \xrightarrow{g(\underline{\theta}(x),\eta(x))} f'_R = U(\underline{\theta}(x),\eta(x))\, f_R = U_R^f(\eta(x))\, f_R \qquad (f = e, u, d)\,, \tag{3.19}$$

where

$$U_R^e(\eta(x)) = e^{-i\eta(x)}\,, \qquad U_R^u(\eta(x)) = e^{\frac{2}{3}i\eta(x)}\,, \qquad U_R^d(\eta(x)) = e^{-\frac{1}{3}i\eta(x)}\,. \tag{3.20}$$

The values of the weak isospin, hypercharge, and electric charge of the fermion doublets and singlets are listed in Table 3.5.

The electroweak SM Lagrangian is the most general renormalizable Lagrangian invariant under the local symmetry group $\mathrm{SU}(2)_L \times \mathrm{U}(1)_Y$ written in terms of the fermion fields, the gauge boson fields, and a Higgs doublet $\Phi(x)$ to be discussed in section 3.4: for the first generation of leptons and quarks, we have

$$\begin{aligned}
\mathscr{L} = {} & i\overline{L_L}\not{D}L_L + i\overline{Q_L}\not{D}Q_L + \sum_{f=e,u,d} i\overline{f_R}\not{D}f_R \\
& - \frac{1}{4}\,\underline{A}_{\mu\nu}\underline{A}^{\mu\nu} - \frac{1}{4}\,B_{\mu\nu}B^{\mu\nu} \\
& + (D_\mu\Phi)^\dagger(D^\mu\Phi) - \mu^2\,\Phi^\dagger\Phi - \lambda\,(\Phi^\dagger\Phi)^2 \\
& - y^e\left(\overline{L_L}\Phi e_R + \overline{e_R}\Phi^\dagger L_L\right) \\
& - y^d\left(\overline{Q_L}\Phi d_R + \overline{d_R}\Phi^\dagger Q_L\right) - y^u\left(\overline{Q_L}\widetilde{\Phi}u_R + \overline{u_R}\widetilde{\Phi}^\dagger Q_L\right).
\end{aligned} \tag{3.21}$$

TABLE 3.5. Eigenvalues of the weak isospin I, of its third component I_3, of the hypercharge Y, and of the charge $Q = I_3 + Y/2$ of the fermion doublets and singlets.

		I	I_3	Y	Q
lepton doublet	$L_L \equiv \begin{pmatrix} \nu_{eL} \\ e_L \end{pmatrix}$	1/2	1/2 −1/2	−1	0 −1
lepton singlet	e_R	0	0	−2	−1
quark doublet	$Q_L \equiv \begin{pmatrix} u_L \\ d_L \end{pmatrix}$	1/2	1/2 −1/2	1/3	2/3 −1/3
quark singlets	u_R d_R	0	0	4/3 −2/3	2/3 −1/3

The second line contains the kinetic terms and self-couplings of the gauge bosons, which will be discussed in section 3.6. The third line is the Lagrangian of the Higgs field which generates the spontaneous symmetry breaking to be discussed in section 3.4. The fourth and fifth lines describe the Higgs–fermion Yukawa couplings which generate lepton masses. It will be discussed in section 3.5 considering the more interesting case of three generations, which leads to quark mixing.

For the moment, let us consider only the first line in eqn (3.21). Since under the transformation in eqn (3.10) the lepton and quark fields transform according to eqns (3.13), (3.14), and (3.19), in order to satisfy gauge invariance, the covariant derivative must transform as

$$D_\mu \xrightarrow{g(\underline{\theta}(x),\eta(x))} D'_\mu = U(\underline{\theta}(x), \eta(x))\, D_\mu\, U^{-1}(\underline{\theta}(x), \eta(x))\,. \tag{3.22}$$

This means that the gauge boson fields transform as

$$\underline{A}_\mu \cdot \underline{I} \xrightarrow{g(\underline{\theta}(x),\eta(x))} \underline{A}'_\mu \cdot \underline{I} = U(\underline{\theta}(x)) \left[\underline{A}_\mu \cdot \underline{I} - \frac{i}{g}\, \partial_\mu \right] U^{-1}(\underline{\theta}(x))\,, \tag{3.23}$$

$$B_\mu \frac{Y}{2} \xrightarrow{g(\underline{\theta}(x),\eta(x))} B'_\mu \frac{Y}{2} = U(\eta(x)) \left[B_\mu \frac{Y}{2} - \frac{i}{g'}\, \partial_\mu \right] U^{-1}(\eta(x))\,. \tag{3.24}$$

The transformation of the B_μ field can be simplified to

$$B_\mu \xrightarrow{g(\underline{\theta}(x),\eta(x))} B'_\mu = B_\mu - \frac{1}{g'}\, \partial_\mu \eta(x)\,, \tag{3.25}$$

which is similar to the transformation under local $\mathrm{U}(1)_Q$ transformations of the electromagnetic field in QED (see, for example, Ref. [720]).

It is important to note that since the left-handed and right-handed components of the fermion fields transform in a different way under the transformations of the gauge group (see eqns (3.13), (3.14), and (3.19)), the presence of mass terms proportional to

$$\overline{f}f = \overline{f_L}f_R + \overline{f_R}f_L \qquad (f = e, u, d) \tag{3.26}$$

in the Lagrangian is forbidden by the gauge symmetry. As we will see in section 3.5, the generation of fermion masses in the SM is accomplished with spontaneous symmetry breaking through the Higgs mechanism.

3.2 Electroweak interactions

In this section we derive the interactions between the fermions and the physical gauge bosons.

Expanding the covariant derivatives in the first line of eqn (3.21) and omitting the kinetic terms, we obtain the interaction Lagrangian that describes the coupling of the fermions with the gauge bosons:

$$\begin{aligned}\mathscr{L}_{\mathrm{I}} = &-\frac{1}{2}\,\overline{L_L}\left(g\underline{\not{A}}\cdot\underline{\tau} - g'\not{B}\right)L_L - \frac{1}{2}\,\overline{Q_L}\left(g\underline{\not{A}}\cdot\underline{\tau} + \frac{1}{3}\,g'\not{B}\right)Q_L \\ &+ g'\overline{e_R}\not{B}e_R - \frac{2}{3}\,g'\overline{u_R}\not{B}u_R + \frac{1}{3}\,g'\overline{d_R}\not{B}d_R\,.\end{aligned} \tag{3.27}$$

In order to derive the explicit interaction terms for the fermions, let us first consider the leptons only:

$$\mathscr{L}_{\mathrm{I,L}} = -\frac{1}{2}\begin{pmatrix}\overline{\nu_{eL}} & \overline{e_L}\end{pmatrix}\begin{pmatrix} g\not{A}_3 - g'\not{B} & g(\not{A}_1 - i\not{A}_2) \\ g(\not{A}_1 + i\not{A}_2) & -g\not{A}_3 - g'\not{B}\end{pmatrix}\begin{pmatrix}\nu_{eL} \\ e_L\end{pmatrix} + g'\overline{e_R}\not{B}e_R\,. \tag{3.28}$$

Let us separate this interaction Lagrangian into a charged-current (CC) Lagrangian

$$\mathscr{L}_{\mathrm{I,L}}^{(\mathrm{CC})} = -\frac{g}{2}\left\{\overline{\nu_{eL}}(\not{A}_1 - i\not{A}_2)e_L + \overline{e_L}(\not{A}_1 + i\not{A}_2)\nu_{eL}\right\}, \tag{3.29}$$

which is given by the off-diagonal terms in eqn (3.28), and a neutral-current (NC) Lagrangian

$$\mathscr{L}_{\mathrm{I,L}}^{(\mathrm{NC})} = -\frac{1}{2}\left\{\overline{\nu_{eL}}(g\not{A}_3 - g'\not{B})\nu_{eL} - \overline{e_L}(g\not{A}_3 + g'\not{B})e_L - 2g'\overline{e_R}\not{B}e_R\right\}, \tag{3.30}$$

given by the diagonal terms in eqn (3.28). Now, let us consider first the charged-current Lagrangian in eqn (3.29). Defining a field W^μ that annihilates W^+ bosons and creates W^- bosons as

$$W^\mu \equiv \frac{A_1^\mu - iA_2^\mu}{\sqrt{2}}\,, \tag{3.31}$$

we obtain

$$\begin{aligned}\mathscr{L}_{\mathrm{I,L}}^{(\mathrm{CC})} &= -\frac{g}{\sqrt{2}}\left\{\overline{\nu_{eL}}\not{W}e_L + \overline{e_L}\not{W}^\dagger\nu_{eL}\right\} \\ &= -\frac{g}{2\sqrt{2}}\,\overline{\nu_e}\,\gamma^\mu\left(1-\gamma^5\right)e\,W_\mu + \mathrm{H.c.}\end{aligned}$$

$$= -\frac{g}{2\sqrt{2}}\, j^{\mu}_{W,\mathrm{L}} W_{\mu} + \text{H.c.}\,, \tag{3.32}$$

where $j^{\mu}_{W,\mathrm{L}}$ is the leptonic charged current

$$j^{\mu}_{W,\mathrm{L}} = \overline{\nu_e}\, \gamma^{\mu} \left(1 - \gamma^5\right) e = 2\, \overline{\nu_{eL}}\, \gamma^{\mu}\, e_L\,. \tag{3.33}$$

The interaction Lagrangian in eqn (3.32) generates the trilinear couplings represented by the diagrams

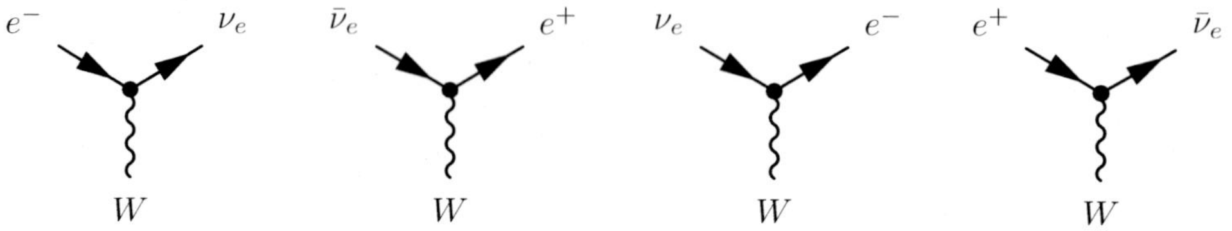

(3.34)

The first two diagrams are generated by the term $j^{\mu}_{W,\mathrm{L}} W_{\mu}$, and the last two by the Hermitian-conjugated term $j^{\mu\dagger}_{W,\mathrm{L}} W^{\dagger}_{\mu} = \overline{e}\gamma^{\mu}\left(1-\gamma^5\right)\nu_e\, W^{\dagger}_{\mu}$.

Next consider the neutral-current Lagrangian in eqn (3.30). The theory must include the electromagnetic interactions described by the quantum electrodynamic (QED) Lagrangian

$$\mathscr{L}^{(\gamma)}_{\mathrm{I,L}} = -e\, j^{\mu}_{\gamma,\mathrm{L}} A_{\mu}\,, \tag{3.35}$$

where e is the elementary electric charge, A^{μ} is the electromagnetic field, and $j^{\mu}_{\gamma,\mathrm{L}}$ is the leptonic electromagnetic current

$$j^{\mu}_{\gamma,\mathrm{L}} = -\overline{e}\, \gamma^{\mu}\, e\,. \tag{3.36}$$

The minus sign is due to the negative charge of the electron. The QED Lagrangian can be obtained as part of the neutral-current Lagrangian in eqn (3.30) expressing the electromagnetic field A^{μ} as an appropriate linear combination of A^{μ}_3 and B^{μ}. Let us write this linear combination and the orthogonal one, which defines the vector boson field Z^{μ}, performing a rotation in the plane of the A^{μ}_3, B^{μ} fields through an angle ϑ_{W}:

$$A^{\mu} = \sin\vartheta_{\mathrm{W}} A^{\mu}_3 + \cos\vartheta_{\mathrm{W}} B^{\mu}\,, \tag{3.37}$$

$$Z^{\mu} = \cos\vartheta_{\mathrm{W}} A^{\mu}_3 - \sin\vartheta_{\mathrm{W}} B^{\mu}\,. \tag{3.38}$$

The angle ϑ_{W} is called the *weak mixing angle* or *Weinberg angle* [1051], although it was introduced for the first time by Glashow in 1961 [543]. The weak mixing angle is chosen in order to obtain the QED Lagrangian for the coupling between the electromagnetic field and the fermion fields. Inserting the expressions in eqns (3.37) and (3.38) in the neutral-current Lagrangian in eqn (3.30), we obtain

$$\mathscr{L}^{(\mathrm{NC})}_{\mathrm{I,L}} = -\frac{1}{2}\Big\{\overline{\nu_{eL}}\left[\left(g\cos\vartheta_{\mathrm{W}} + g'\sin\vartheta_{\mathrm{W}}\right)\not{Z} + \left(g\sin\vartheta_{\mathrm{W}} - g'\cos\vartheta_{\mathrm{W}}\right)\not{A}\right]\nu_{eL}$$

$$- \overline{e_L}\left[\left(g\cos\vartheta_{\mathrm{W}} - g'\sin\vartheta_{\mathrm{W}}\right)\not{Z} + \left(g\sin\vartheta_{\mathrm{W}} + g'\cos\vartheta_{\mathrm{W}}\right)\not{A}\right]e_L$$

$$- 2g'\overline{e_R}\,[-\sin\vartheta_W \not{Z} + \cos\vartheta_W \not{A}]\, e_R \Big\} . \tag{3.39}$$

Since neutrinos are neutral particles, they do not have a coupling to the electromagnetic field. Setting the coefficient of the corresponding term in eqn (3.39) to zero, we obtain

$$g\sin\vartheta_W = g'\cos\vartheta_W \qquad \Longrightarrow \qquad \tan\vartheta_W = \frac{g'}{g} . \tag{3.40}$$

This is an important relation, which connects the coupling constants g and g' of the SM with the weak mixing angle ϑ_W.

Substituting eqn (3.40) into eqn (3.39), we obtain

$$\begin{aligned} \mathscr{L}_{\rm I,L}^{\rm (NC)} &= -\frac{g}{2\cos\vartheta_W}\left\{ \overline{\nu_{eL}} \not{Z} \nu_{eL} - (1 - 2\sin^2\vartheta_W)\, \overline{e_L} \not{Z} e_L + 2\sin^2\vartheta_W \overline{e_R} \not{Z} e_R \right\} \\ &\quad + g\sin\vartheta_W\, \overline{e} \not{A} e . \end{aligned} \tag{3.41}$$

Since the last term gives the coupling of the electron field with the electromagnetic field, it must coincide with the QED interaction Lagrangian in eqn (3.35) and thus we find

$$g\sin\vartheta_W = e . \tag{3.42}$$

Using the relation in eqn (3.40) we also have

$$g'\cos\vartheta_W = e . \tag{3.43}$$

These two relations are very important, because they give the relation between the coupling constants g and g' and the elementary electric charge e. The two relations in eqn (3.42) and (3.43) can be combined to give

$$g^2 + g'^2 = e^2 . \tag{3.44}$$

The neutral-current Lagrangian can be written as

$$\mathscr{L}_{\rm I,L}^{\rm (NC)} = \mathscr{L}_{\rm I,L}^{(Z)} + \mathscr{L}_{\rm I,L}^{(\gamma)} . \tag{3.45}$$

where $\mathscr{L}_{\rm I,L}^{(\gamma)}$ is the QED Lagrangian in eqn (3.35) and $\mathscr{L}_{\rm I,L}^{(Z)}$ is the *weak neutral-current* Lagrangian given by

$$\mathscr{L}_{\rm I,L}^{(Z)} = -\frac{g}{2\cos\vartheta_W}\, j^{\mu}_{Z,\rm L}\, Z_\mu , \tag{3.46}$$

with the leptonic weak neutral-current

$$j^{\mu}_{Z,\rm L} = 2\, g_L^{\nu}\, \overline{\nu_{eL}}\, \gamma^\mu\, \nu_{eL} + 2\, g_L^{l}\, \overline{e_L}\, \gamma^\mu\, e_L + 2\, g_R^{l}\, \overline{e_R}\, \gamma^\mu\, e_R . \tag{3.47}$$

Here, we have introduced the coefficients g_L^{ν}, g_L^{l}, and g_R^{l} (the superscript l indicates a charged lepton) whose values, obtained from eqn (3.41), are given in Table 3.6. In general, the values of the coefficients $g_L^{\mathfrak{f}}$ and $g_R^{\mathfrak{f}}$ for a fermion field $\mathfrak{f}$ are given by

$$g_L^{\mathfrak{f}} = I_3^{\mathfrak{f}} - q_{\mathfrak{f}} \sin^2\vartheta_W , \tag{3.48}$$

TABLE 3.6. Values of g_L, g_R, g_V, g_A for the fermion fields. The superscripts ν, l, U, D, indicate, respectively, a generic neutrino, charged lepton, up-type quark, and down-type quark. We have defined $s_W \equiv \sin\vartheta_{\mathrm{W}}$.

Fermions	g_L	g_R	g_V	g_A
ν_e, ν_μ, ν_τ	$g_L^\nu = \frac{1}{2}$	$g_R^\nu = 0$	$g_V^\nu = \frac{1}{2}$	$g_A^\nu = \frac{1}{2}$
e, μ, τ	$g_L^l = -\frac{1}{2} + s_W^2$	$g_R^l = s_W^2$	$g_V^l = -\frac{1}{2} + 2s_W^2$	$g_A^l = -\frac{1}{2}$
u, c, t	$g_L^U = \frac{1}{2} - \frac{2}{3}\,s_W^2$	$g_R^U = -\frac{2}{3}\,s_W^2$	$g_V^U = \frac{1}{2} - \frac{4}{3}\,s_W^2$	$g_A^U = \frac{1}{2}$
d, s, b	$g_L^D = -\frac{1}{2} + \frac{1}{3}\,s_W^2$	$g_R^D = \frac{1}{3}\,s_W^2$	$g_V^D = -\frac{1}{2} + \frac{2}{3}\,s_W^2$	$g_A^D = -\frac{1}{2}$

$$g_R^{\mathfrak{f}} = -q_{\mathfrak{f}} \sin^2\vartheta_{\mathrm{W}} \,, \tag{3.49}$$

where $I_3^{\mathfrak{f}}$ is the value of the third component of the weak isospin and $q_{\mathfrak{f}}$ is the electric charge of the fermion in units of the elementary electric charge e (see Table 3.5).

Because of the mixing of the gauge fields A_3^μ and B^μ in eqns (3.37) and (3.38), one can see that the weak neutral-current interactions of charged fermion fields involve not only their left-handed component, but also the right-handed one, with a strength proportional to the electric charge and to $\sin^2\vartheta_{\mathrm{W}}$.

The leptonic weak neutral current in eqn (3.47) can also be written as

$$j_{Z,\mathrm{L}}^\mu = \overline{\nu_e}\,\gamma^\mu \left(g_V^\nu - g_A^\nu\,\gamma^5\right) \nu_e + \overline{e}\,\gamma^\mu \left(g_V^l - g_A^l\,\gamma^5\right) e \,, \tag{3.50}$$

where we have introduced the widely used vector and axial couplings $g_V^{\nu,\mathrm{L}}$ and $g_A^{\nu,\mathrm{L}}$ for neutrinos and charged leptons, whose values are given in Table 3.6. In general, the values of $g_V^{\mathfrak{f}}$ and $g_A^{\mathfrak{f}}$ for a fermion field $\mathfrak{f}$ are given by

$$g_V^{\mathfrak{f}} = g_L^{\mathfrak{f}} + g_R^{\mathfrak{f}} = I_3^{\mathfrak{f}} - 2q_{\mathfrak{f}} \sin^2\vartheta_{\mathrm{W}} \,, \tag{3.51}$$

$$g_A^{\mathfrak{f}} = g_L^{\mathfrak{f}} - g_R^{\mathfrak{f}} = I_3^{\mathfrak{f}} \,. \tag{3.52}$$

The leptonic weak neutral-current interaction Lagrangian in eqn (3.46) describes the trilinear couplings represented by the diagrams

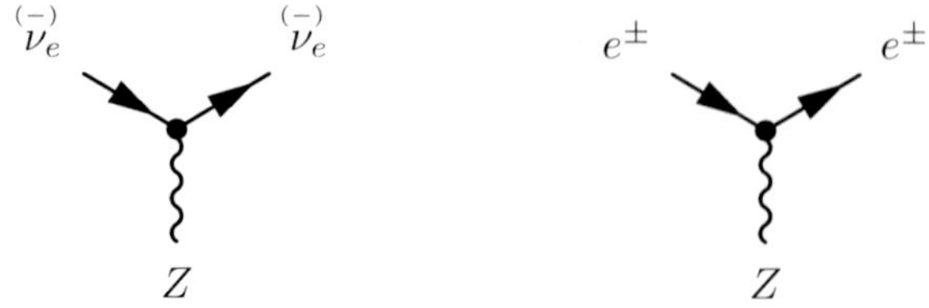

(3.53)

and the leptonic electromagnetic interaction Lagrangian in eqn (3.35) describes the trilinear couplings represented by the diagram

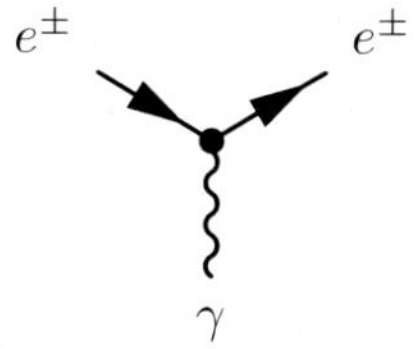

(3.54)

Coming back to the quark part of the interaction Lagrangian in eqn (3.27), we have

$$\begin{aligned}\mathscr{L}_{\mathrm{I,Q}} = & -\frac{1}{2}\begin{pmatrix}\overline{u_L} & \overline{d_L}\end{pmatrix}\begin{pmatrix} g\not{A}_3 + \frac{1}{3}g'\not{B} & g(\not{A}_1 - i\not{A}_2) \\ g(\not{A}_1 + i\not{A}_2) & -g\not{A}_3 + \frac{1}{3}g'\not{B}\end{pmatrix}\begin{pmatrix}u_L \\ d_L\end{pmatrix} \\ & -\frac{2}{3}g'\overline{u_R}\not{B}u_R + \frac{1}{3}g'\overline{d_R}\not{B}d_R\,. \end{aligned} \tag{3.55}$$

Following the same procedure as in the case of the leptons, we find the charged-current Lagrangian

$$\mathscr{L}_{\mathrm{I,Q}}^{(\mathrm{CC})} = -\frac{g}{2\sqrt{2}}\, j^{\mu}_{W,\mathrm{Q}} W_{\mu} + \mathrm{H.c.}\,, \tag{3.56}$$

where $j^{\mu}_{W,\mathrm{Q}}$ is the quark charged current

$$j^{\mu}_{W,\mathrm{Q}} = \overline{u}\,\gamma^{\mu}\left(1-\gamma^5\right)d = 2\,\overline{u_L}\,\gamma^{\mu}\,d_L\,. \tag{3.57}$$

The interaction Lagrangian in eqn (3.32) generates the trilinear couplings represented by the diagrams

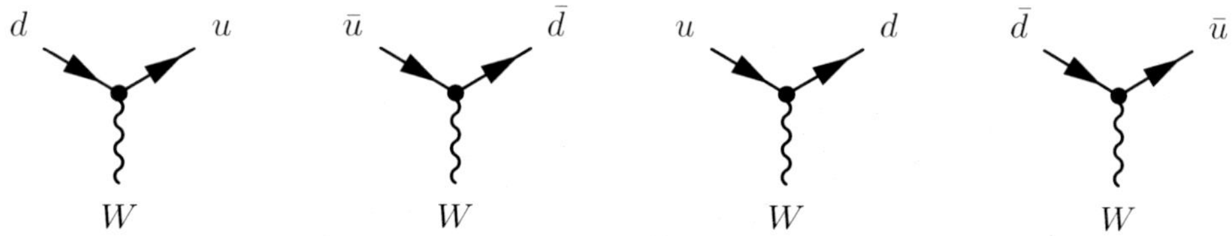

(3.58)

The first two diagrams are generated by the term $j^{\mu}_{W,\mathrm{Q}}W_{\mu}$ and the last two by the Hermitian-conjugated term $j^{\mu\dagger}_{W,\mathrm{Q}}W^{\dagger}_{\mu} = \overline{d}\gamma^{\mu}\left(1-\gamma^5\right)u\,W^{\dagger}_{\mu}$.

The quark neutral-current interaction Lagrangian can be separated into its weak and electromagnetic parts as in the case of the leptonic one in eqn (3.45),

$$\mathscr{L}_{\mathrm{I,Q}}^{(\mathrm{NC})} = \mathscr{L}_{\mathrm{I,Q}}^{(Z)} + \mathscr{L}_{\mathrm{I,Q}}^{(\gamma)}\,, \tag{3.59}$$

with

$$\mathscr{L}_{\mathrm{I,Q}}^{(\gamma)} = -\,e\,j^{\mu}_{\gamma,\mathrm{Q}}\,A_{\mu}\,, \tag{3.60}$$

$$\mathscr{L}_{\mathrm{I,Q}}^{(Z)} = -\frac{g}{2\cos\vartheta_{\mathrm{W}}}\,j^{\mu}_{Z,\mathrm{Q}}\,Z_{\mu}\,. \tag{3.61}$$

The quark electromagnetic current $j^{\mu}_{\gamma,\mathrm{Q}}$ is given by

$$j^{\mu}_{\gamma,\mathrm{Q}} = \frac{2}{3}\,\overline{u}\,\gamma^{\mu}\,u - \frac{1}{3}\,\overline{d}\,\gamma^{\mu}\,d\,, \tag{3.62}$$

and the quark weak neutral current $j^{\mu}_{Z,\mathrm{Q}}$ is given by

$$\begin{aligned} j^{\mu}_{Z,\mathrm{Q}} &= 2\,g_L^U\,\overline{u_L}\,\gamma^{\mu}\,u_L + 2\,g_R^U\,\overline{u_R}\,\gamma^{\mu}\,u_R + 2\,g_L^D\,\overline{d_L}\,\gamma^{\mu}\,d_L + 2\,g_R^D\,\overline{d_R}\,\gamma^{\mu}\,d_R \\ &= \overline{u}\,\gamma^{\mu}\left(g_V^U - g_A^U\gamma^5\right)u + \overline{d}\,\gamma^{\mu}\left(g_V^D - g_A^D\gamma^5\right)d\,. \end{aligned} \tag{3.63}$$

The coefficients $g_L^{U,D}$, $g_R^{U,D}$, $g_V^{U,D}$, and $g_A^{U,D}$, given by eqns (3.48), (3.49), (3.51), and (3.52), respectively, are listed in Table 3.6 (page 78).

The quark weak neutral-current interaction Lagrangian in eqn (3.61) describes the trilinear couplings represented by the diagrams

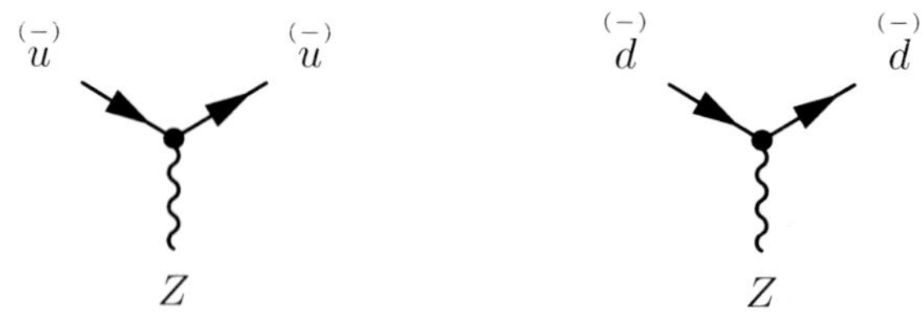

(3.64)

and the quark electromagnetic interaction Lagrangian in eqn (3.60) describes the trilinear couplings represented by the diagrams

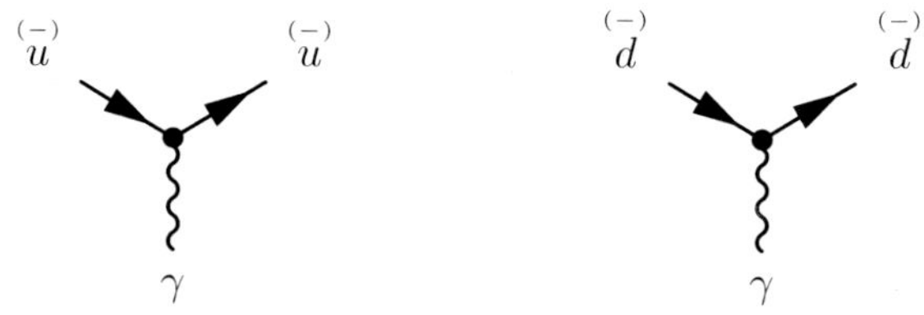

(3.65)

3.3 Three generations

In this section we generalize the expressions for the weak charged-current and neutral-current interaction Lagrangians in eqns (3.32), (3.46), (3.56), and (3.61) to the case of three generations of leptons and quarks which is realized in Nature. Let us define the three generations of left-handed weak isospin doublets

$$L'_{eL} \equiv \begin{pmatrix} \nu'_{eL} \\ e'_L \end{pmatrix}, \qquad L'_{\mu L} \equiv \begin{pmatrix} \nu'_{\mu L} \\ \mu'_L \end{pmatrix}, \qquad L'_{\tau L} \equiv \begin{pmatrix} \nu'_{\tau L} \\ \tau'_L \end{pmatrix}, \tag{3.66}$$

$$Q'_{1L} \equiv \begin{pmatrix} u'_L \\ d'_L \end{pmatrix}, \qquad Q'_{2L} \equiv \begin{pmatrix} c'_L \\ s'_L \end{pmatrix}, \qquad Q'_{3L} \equiv \begin{pmatrix} t'_L \\ b'_L \end{pmatrix}, \tag{3.67}$$

and singlets[15],

$$\ell'_{eR} \equiv e'_R \,, \qquad \ell'_{\mu R} \equiv \mu'_R \,, \qquad \ell'_{\tau R} \equiv \tau'_R \,, \tag{3.68}$$

$$q'^U_{uR} \equiv u'_R \,, \qquad q'^U_{cR} \equiv c'_R \,, \qquad q'^U_{tR} \equiv t'_R \,, \tag{3.69}$$

$$q'^D_{dR} \equiv d'_R \,, \qquad q'^D_{sR} \equiv s'_R \,, \qquad q'^D_{bR} \equiv= b'_R \,. \tag{3.70}$$

The primes on the fermion fields are necessary because, as we will see in section 3.5, the fields in eqns (3.66)–(3.70), in general, do not have definite masses, but are linear combinations of the fields with definite mass. The quantum numbers of the fields in eqns (3.66)–(3.70) are the same as the corresponding ones in the first generation, which are listed in Table 3.5.

The three-generation version of the electroweak SM Lagrangian in eqn (3.21) is

$$\begin{aligned}
\mathscr{L} = {} & i \sum_{\alpha=e,\mu,\tau} \overline{L'_{\alpha L}} \not{D} L'_{\alpha L} + i \sum_{\alpha=1,2,3} \overline{Q'_{\alpha L}} \not{D} Q'_{\alpha L} \\
& + i \sum_{\alpha=e,\mu,\tau} \overline{\ell'_{\alpha R}} \not{D} \ell'_{\alpha R} + i \sum_{\alpha=d,s,b} \overline{q'^D_{\alpha R}} \not{D} q'^D_{\alpha R} + i \sum_{\alpha=u,c,t} \overline{q'^U_{\alpha R}} \not{D} q'^U_{\alpha R} \\
& - \frac{1}{4} \underline{A}_{\mu\nu} \underline{A}^{\mu\nu} - \frac{1}{4} B_{\mu\nu} B^{\mu\nu} \\
& + (D_\rho \Phi)^\dagger (D^\rho \Phi) - \mu^2 \, \Phi^\dagger \Phi - \lambda \, (\Phi^\dagger \Phi)^2 \\
& - \sum_{\alpha,\beta=e,\mu,\tau} \left(Y'^\ell_{\alpha\beta} \, \overline{L'_{\alpha L}} \, \Phi \, \ell'_{\beta R} + Y'^{\ell *}_{\alpha\beta} \, \overline{\ell'_{\beta R}} \, \Phi^\dagger \, L'_{\alpha L} \right) \\
& - \sum_{\alpha=1,2,3} \sum_{\beta=d,s,b} \left(Y'^D_{\alpha\beta} \, \overline{Q'_{\alpha L}} \, \Phi \, q'^D_{\beta R} + Y'^{D *}_{\alpha\beta} \, \overline{q'^D_{\beta R}} \, \Phi^\dagger \, Q'_{\alpha L} \right) \\
& - \sum_{\alpha=1,2,3} \sum_{\beta=u,c,t} \left(Y'^U_{\alpha\beta} \, \overline{Q'_{\alpha L}} \, \widetilde{\Phi} \, q'^U_{\beta R} + Y'^{U *}_{\alpha\beta} \, \overline{q'^U_{\beta R}} \, \widetilde{\Phi}^\dagger \, Q'_{\alpha L} \right) .
\end{aligned} \tag{3.71}$$

The last three lines contain the Higgs–fermion Yukawa couplings that generate the fermion masses and quark mixing, as explained in section 3.5.

The electromagnetic interaction Lagrangian obtained from the first two lines of eqn (3.71) is

$$\mathscr{L}_{\mathrm{I}}^{(\gamma)} = -e \, j^\rho_\gamma \, A_\rho \,, \tag{3.72}$$

with the electromagnetic current

$$j^\rho_\gamma = j^\rho_{\gamma,\mathrm{L}} + j^\rho_{\gamma,\mathrm{Q}} \,. \tag{3.73}$$

Here $j^\rho_{\gamma,\mathrm{L}}$ and $j^\rho_{\gamma,\mathrm{Q}}$ are the leptonic and quark electromagnetic currents given by

$$j^\rho_{\gamma,\mathrm{L}} = - \sum_{\alpha=e,\mu,\tau} \overline{\ell'_\alpha} \, \gamma^\rho \, \ell'_\alpha \,, \tag{3.74}$$

[15] In the following we will use, for example, the notation $\ell'_{\alpha R}$ in compact formulas where a summation on the flavor index $\alpha = e, \mu, \tau$ is required. When the fields are needed explicitly, we write, for example, e'_R, μ'_R, τ'_R.

$$j^\rho_{\gamma,\mathrm{Q}} = \frac{2}{3} \sum_{\alpha=u,c,t} \overline{q'^U_\alpha} \, \gamma^\rho \, q'^U_\alpha - \frac{1}{3} \sum_{\alpha=d,s,b} \overline{q'^D_\alpha} \, \gamma^\rho \, q'^D_\alpha \,. \tag{3.75}$$

The charged-current weak interaction Lagrangian obtained from the first line of eqn (3.71) is

$$\mathscr{L}_{\mathrm{I}}^{(\mathrm{CC})} = -\frac{g}{2\sqrt{2}} \, j^\rho_W \, W_\rho + \mathrm{H.c.}\,, \tag{3.76}$$

where the fermion charged current j^ρ_W is the sum of the leptonic and quark charged-currents,

$$j^\rho_W = j^\rho_{W,\mathrm{L}} + j^\rho_{W,\mathrm{Q}} \,, \tag{3.77}$$

given by

$$j^\rho_{W,\mathrm{L}} = 2 \left(\overline{\nu'_{eL}} \, \gamma^\rho \, e'_L + \overline{\nu'_{\mu L}} \, \gamma^\rho \, \mu'_L + \overline{\nu'_{\tau L}} \, \gamma^\rho \, \tau'_L \right) , \tag{3.78}$$

$$j^\rho_{W,\mathrm{Q}} = 2 \left(\overline{u'_L} \, \gamma^\rho \, d'_L + \overline{c'_L} \, \gamma^\rho \, s'_L + \overline{t'_L} \, \gamma^\rho \, b'_L \right) . \tag{3.79}$$

The leptonic charged-current can be written in a compact form as

$$j^\rho_{W,\mathrm{L}} = 2 \sum_{\alpha=e,\mu,\tau} \overline{\nu'_{\alpha L}} \, \gamma^\rho \, \ell'_{\alpha L} \,, \tag{3.80}$$

with

$$\ell'_{eL} \equiv e'_L \,, \qquad \ell'_{\mu L} \equiv \mu'_L \,, \qquad \ell'_{\tau L} \equiv \tau'_L \,. \tag{3.81}$$

It is useful to define the weak isospin raising and lowering operators $I_\pm$ as

$$I_\pm = I_1 \pm i I_2 \,, \tag{3.82}$$

for which we have, from the commutation relations in eqn (3.2),

$$[I_3, I_\pm] = \pm I_\pm \quad \Longrightarrow \quad I_3 I_\pm = I_\pm \left(I_3 \pm 1 \right) . \tag{3.83}$$

Then, if $|i, i_3\rangle$ is an eigenstate of I and I_3 with eigenvalues i and i_3, $I_\pm |i, i_3\rangle$ is an eigenstate of I and I_3 with eigenvalues i and $i_3 \pm 1$:

$$I_3 \, |i, i_3\rangle = i_3 \, |i, i_3\rangle \quad \Longrightarrow \quad I_3 \, I_\pm \, |i, i_3\rangle = (i_3 \pm 1) \, I_\pm \, |i, i_3\rangle \,. \tag{3.84}$$

This is clear from the explicit matrix form of the operators $I_\pm$ in the doublet representation:

$$I_+ \xrightarrow{\text{doublet}} \frac{\tau_+}{2} = \frac{\tau_1 + i\tau_2}{2} = \begin{pmatrix} 0 & 1 \\ 0 & 0 \end{pmatrix} , \qquad I_- \xrightarrow{\text{doublet}} \frac{\tau_-}{2} = \frac{\tau_1 - i\tau_2}{2} = \begin{pmatrix} 0 & 0 \\ 1 & 0 \end{pmatrix} . \tag{3.85}$$

One can see that I_+ raises the lower component of a doublet, increasing the eigenvalue of I_3 by one unit, and I_- lowers the upper component, decreasing the eigenvalue of I_3 by one unit.

The fermion weak charged currents can be written in a compact form using the raising and lowering operators $I_\pm$:

$$j^\rho_{W,\mathrm{L}} = 2 \sum_{\alpha=e,\mu,\tau} \overline{L'_{\alpha L}} \, \gamma^\rho \, I_+ \, L'_{\alpha L} \,, \tag{3.86}$$

$$j^{\rho}_{W,Q} = 2 \sum_{\alpha=1,2,3} \overline{Q'_{\alpha L}} \, \gamma^{\rho} \, I_{+} \, Q'_{\alpha L} \,, \tag{3.87}$$

and the Hermitian conjugate currents are given by

$$j^{\mu\dagger}_{W,L} = 2 \sum_{\alpha=e,\mu,\tau} \overline{L'_{\alpha L}} \, \gamma^{\rho} \, I_{-} \, L'_{\alpha L} \,, \tag{3.88}$$

$$j^{\mu\dagger}_{W,Q} = 2 \sum_{\alpha=1,2,3} \overline{Q'_{\alpha L}} \, \gamma^{\rho} \, I_{-} \, Q'_{\alpha L} \,. \tag{3.89}$$

The neutral-current weak interaction Lagrangian obtained from the first two lines of eqn (3.71) is

$$\mathscr{L}^{(Z)}_{\mathrm{I}} = - \frac{g}{2 \cos \vartheta_{\mathrm{W}}} \, j^{\rho}_{Z} \, Z_{\rho} \,, \tag{3.90}$$

with the neutral current

$$j^{\rho}_{Z} = j^{\rho}_{Z,\mathrm{L}} + j^{\rho}_{Z,\mathrm{Q}} \,, \tag{3.91}$$

where $j^{\rho}_{Z,\mathrm{L}}$ and $j^{\rho}_{Z,\mathrm{Q}}$ are the leptonic and quark neutral currents given by

$$j^{\rho}_{Z,\mathrm{L}} = 2 \, g^{\nu}_{L} \sum_{\alpha=e,\mu,\tau} \overline{\nu'_{\alpha L}} \, \gamma^{\rho} \, \nu'_{\alpha L} + 2 \sum_{\alpha=e,\mu,\tau} \left(g^{l}_{L} \, \overline{\ell'_{\alpha L}} \, \gamma^{\rho} \, \ell'_{\alpha L} + g^{l}_{R} \, \overline{\ell'_{\alpha R}} \, \gamma^{\rho} \, \ell'_{\alpha R} \right) , \tag{3.92}$$

$$\begin{aligned} j^{\rho}_{Z,\mathrm{Q}} = {} & 2 \sum_{\alpha=u,c,t} \left(g^{U}_{L} \, \overline{q'^{U}_{\alpha L}} \, \gamma^{\rho} \, q'^{U}_{\alpha L} + g^{U}_{R} \, \overline{q'^{U}_{\alpha R}} \, \gamma^{\rho} \, q'^{U}_{\alpha R} \right) \\ & + 2 \sum_{\alpha=d,s,b} \left(g^{D}_{L} \, \overline{q'^{D}_{\alpha L}} \, \gamma^{\rho} \, q'^{D}_{\alpha L} + g^{D}_{R} \, \overline{q'^{D}_{\alpha R}} \, \gamma^{\rho} \, q'^{D}_{\alpha R} \right) , \end{aligned} \tag{3.93}$$

with

$$q'^{U}_{uL} \equiv u'_{L} \,, \qquad q'^{U}_{cL} \equiv c'_{L} \,, \qquad q'^{U}_{tL} \equiv t'_{L} \,, \tag{3.94}$$

$$q'^{D}_{dL} \equiv d'_{L} \,, \qquad q'^{D}_{sL} \equiv s'_{L} \,, \qquad q'^{D}_{bL} \equiv b'_{L} \,. \tag{3.95}$$

The values of the coefficients $g_{L}^{\nu,l,U,D}$ and $g_{R}^{\nu,l,U,D}$, given by eqns (3.48) and (3.49), are listed in Table 3.6 (page 78).

3.4 The Higgs mechanism

In the SM, the masses of the W and Z gauge bosons, as well as those of the fermions, are generated through the Higgs mechanism [611, 610, 612, 412, 578, 666] implemented by the Higgs doublet

$$\Phi(x) = \begin{pmatrix} \phi^{+}(x) \\ \phi^{0}(x) \end{pmatrix} , \tag{3.96}$$

where $\phi^{+}(x)$ is a charged complex scalar field and $\phi^{0}(x)$ is a neutral complex scalar field. The gauge quantum numbers of the Higgs fields are listed in Table 3.7. The

TABLE 3.7. Eigenvalues of the weak isospin I, of its third component I_3, of the hypercharge Y, and of the charge $Q = I_3 + \frac{Y}{2}$ of the Higgs doublet.

		I	I_3	Y	Q
Higgs doublet	$\Phi(x) \equiv \begin{pmatrix} \phi^+(x) \\ \phi^0(x) \end{pmatrix}$	$1/2$	$1/2$ $-1/2$	$+1$	1 0

transformation of the Higgs doublet under an element $g(\underline{\theta}(x), \eta(x))$ of the gauge group (see eqn (3.10)) is given by

$$\Phi \xrightarrow{g(\underline{\theta}(x),\eta(x))} \Phi' = U(\underline{\theta}(x), \eta(x))\,\Phi = e^{\frac{i}{2}\underline{\theta}(x)\cdot\underline{\tau}+\frac{i}{2}\eta(x)}\,\Phi\,. \tag{3.97}$$

Using this transformation and the transformations in eqn (3.22) of the covariant derivative, one can verify that the Higgs part of the SM Lagrangian,

$$\mathscr{L}_{\text{Higgs}} = (D_\mu\Phi)^\dagger(D^\mu\Phi) - \mu^2\,\Phi^\dagger\Phi - \lambda\,(\Phi^\dagger\Phi)^2\,, \tag{3.98}$$

is invariant under a gauge transformation $g(\underline{\theta}(x), \eta(x))$. In eqn (3.98), the coefficient λ of the quartic self-couplings of the Higgs fields must be positive, $\lambda > 0$, in order to have a potential

$$V(\Phi) = \mu^2\,\Phi^\dagger\Phi + \lambda\,(\Phi^\dagger\Phi)^2 \tag{3.99}$$

which is bounded from below. On the other hand, the squared mass-like coefficient μ^2 is assumed to be negative, $\mu^2 < 0$, in order to realize the spontaneous breaking of the symmetry

$$\text{SU}(2)_L \times \text{U}(1)_Y \to \text{U}(1)_Q\,, \tag{3.100}$$

where $\text{U}(1)_Q$ is the gauge symmetry group of electromagnetic interactions, associated with the conservation of the electric charge, which is well known to be unbroken.

Defining

$$v \equiv \sqrt{-\frac{\mu^2}{\lambda}}\,. \tag{3.101}$$

and neglecting an irrelevant constant term $v^4/4$, the Higgs potential in eqn (3.99) can be written as

$$V(\Phi) = \lambda\left(\Phi^\dagger\Phi - \frac{v^2}{2}\right)^2\,. \tag{3.102}$$

From this expression it is clear that the potential is minimum for

$$\Phi^\dagger\Phi = \frac{v^2}{2}\,. \tag{3.103}$$

In quantum field theory the minimum of the potential corresponds to the vacuum, which is the lowest energy state, and the quantized excitations of each field above the vacuum correspond to particle states. Fermion and vector boson fields, which carry nonzero spin, must have a zero value in the vacuum, in order to preserve the manifest invariance of Nature under spatial rotation. Also, charged scalar fields must have zero value in the vacuum, which is electrically neutral. On the other hand, neutral scalar fields, which do not have electric charge, can have a nonzero value in vacuum, which is called *vacuum expectation value* or *VEV*. From eqn (3.103) we can see that the Higgs fields have a nonzero VEV. In order to have an electrically neutral vacuum, the VEV $\langle\Phi\rangle$ of the Higgs fields must be due to ϕ^0:

$$\langle\Phi\rangle = \frac{1}{\sqrt{2}}\begin{pmatrix}0\\ v\end{pmatrix} . \tag{3.104}$$

The symmetry $SU(2)_L \times U(1)_Y$ is spontaneously broken by the VEV $\langle\Phi\rangle$:

$$I_1\,\langle\Phi\rangle = \frac{\tau_1}{2}\,\langle\Phi\rangle = \frac{1}{2\sqrt{2}}\begin{pmatrix}v\\ 0\end{pmatrix} \neq 0\,, \tag{3.105}$$

$$I_2\,\langle\Phi\rangle = \frac{\tau_2}{2}\,\langle\Phi\rangle = -\frac{i}{2\sqrt{2}}\begin{pmatrix}v\\ 0\end{pmatrix} \neq 0\,, \tag{3.106}$$

$$I_3\,\langle\Phi\rangle = \frac{\tau_3}{2}\,\langle\Phi\rangle = -\frac{1}{2\sqrt{2}}\begin{pmatrix}0\\ v\end{pmatrix} \neq 0\,, \tag{3.107}$$

$$Y\,\langle\Phi\rangle = \langle\Phi\rangle \neq 0\,, \tag{3.108}$$

but

$$Q\,\langle\Phi\rangle = \left(I_3 + \frac{Y}{2}\right)\langle\Phi\rangle = \frac{1}{\sqrt{2}}\begin{pmatrix}1 & 0\\ 0 & 0\end{pmatrix}\begin{pmatrix}0\\ v\end{pmatrix} = 0\,. \tag{3.109}$$

Therefore, the vacuum is invariant under gauge transformations that belong to the group $U(1)_Q$, of the type

$$e^{i\,\theta\,Q}\,\langle\Phi\rangle = \langle\Phi\rangle\,. \tag{3.110}$$

This invariance guarantees the existence of a massless gauge boson associated with the symmetry group $U(1)_Q$, which is identified with the photon.

It is important not to be misled by the words “spontaneous symmetry breaking”. The Lagrangian is perfectly symmetric. The symmetry is only broken by the vacuum, and consequently the physical states obtained by excitations of the fields above the vacuum do not manifest the symmetry of the Lagrangian. In fact, it is more proper to say that the symmetry becomes hidden.

In order to derive the physical properties of the particles resulting from the spontaneous breaking of the $SU(2)_L \times U(1)_Y$ symmetry to $U(1)_Q$, it is convenient to write the Higgs doublet as

$$\Phi(x) = \frac{1}{\sqrt{2}}\,\exp\left(\frac{i}{2\,v}\,\underline{\xi}(x)\cdot\underline{\tau}\right)\begin{pmatrix}0\\ v + H(x)\end{pmatrix} , \tag{3.111}$$

where $\underline{\xi}(x) = (\xi_1(x), \xi_2(x), \xi_3(x))$ and $H(x)$ are four real scalar fields (the presence of v in the argument of the exponential is necessary for dimensional reasons, because scalar fields have dimension of energy). The field $H(x)$ describes the physical Higgs boson, obtained by excitations of the neutral Higgs field above the vacuum. On the other hand, the fields $\underline{\xi}(x)$ are unphysical, because they can be rotated away by a gauge transformation of the type in eqn (3.97) with

$$\underline{\theta}(x) = -\frac{1}{v}\,\underline{\xi}(x)\,, \qquad \eta(x) = 0\,. \tag{3.112}$$

This transformation defines the so-called *unitary gauge*, in which the physical states of the theory appear explicitly. In the unitary gauge, the Higgs doublet reads

$$\Phi(x) = \frac{1}{\sqrt{2}}\begin{pmatrix} 0 \\ v + H(x) \end{pmatrix}, \tag{3.113}$$

and

$$\begin{aligned} D_\mu(x)\,\Phi(x) &= \left[\partial_\mu + \frac{i}{2} g\,\underline{A}_\mu(x)\cdot\underline{\tau} + \frac{i}{2} g'\,B_\mu(x)\right]\Phi(x) \\ &= \frac{1}{\sqrt{2}}\begin{pmatrix} i\,\frac{g}{\sqrt{2}}\,W_\mu(x)\,[v+H(x)] \\ \partial_\mu H(x) - \frac{i}{2}\,\frac{g}{\cos\vartheta_W}\,Z_\mu(x)\,[v+H(x)] \end{pmatrix}. \end{aligned} \tag{3.114}$$

In the unitary gauge, the Higgs Lagrangian in eqn (3.98) is given by

$$\begin{aligned} \mathscr{L}_{\text{Higgs}} = &\frac{1}{2}\,(\partial H)^2 + \frac{g^2}{4}\,(v+H)^2\,W^\dagger_\mu W^\mu + \frac{g^2}{8\cos^2\vartheta_W}\,(v+H)^2\,Z_\mu Z^\mu \\ &- \frac{\lambda}{4}\left(H^2 + 2v\,H\right)^2. \end{aligned} \tag{3.115}$$

Expanding the above, we obtain

$$\begin{aligned} \mathscr{L}_{\text{Higgs}} = &\frac{1}{2}\,(\partial H)^2 - \lambda v^2 H^2 - \lambda v H^3 - \frac{\lambda}{4}\,H^4 + \frac{g^2 v^2}{4}\,W^\dagger_\mu W^\mu + \frac{g^2 v^2}{8\cos^2\vartheta_W}\,Z_\mu Z^\mu \\ &+ \frac{g^2 v}{2}\,W^\dagger_\mu W^\mu H + \frac{g^2 v}{4\cos^2\vartheta_W}\,Z_\mu Z^\mu\,H \\ &+ \frac{g^2}{4}\,W^\dagger_\mu W^\mu H^2 + \frac{g^2}{8\cos^2\vartheta_W}\,Z_\mu Z^\mu H^2\,. \end{aligned} \tag{3.116}$$

The first term on the right-hand side is the kinetic term for the Higgs boson. The second term is the mass term for the Higgs boson, from which the mass of the Higgs boson is given by

$$m_H = \sqrt{2\,\lambda\,v^2} = \sqrt{-2\,\mu^2}\,. \tag{3.117}$$

Since μ^2 is a negative parameter specifically introduced in the SM, its value is not connected to other quantities already measured. Hence, the SM does not give a prediction for the value of the Higgs mass, which must be determined experimentally.

The third and fourth terms on the right-hand side of eqn (3.116) generate, respectively, trilinear and quadrilinear self-couplings of the Higgs field represented

by the diagrams

(3.118)

The fifth and sixth terms on the right-hand side of eqn (3.116) are of fundamental importance, because they are mass terms for the W and Z gauge bosons. From them, we see that the masses of the W and Z gauge bosons in the SM are given by

$$m_W = \frac{g\,v}{2} \qquad m_Z = \frac{g\,v}{2\,\cos\vartheta_{\mathrm{W}}} \,. \tag{3.119}$$

The widely used parameter ρ, defined by

$$\rho = \frac{m_W^2}{m_Z^2\,\cos^2\vartheta_{\mathrm{W}}} \,, \tag{3.120}$$

has the value

$$\rho = 1 \,, \tag{3.121}$$

in the SM[16]. The experimental test of eqn (3.121) is an important check of the value of the weak isospin of the Higgs. The reason is that, in principle, the Higgs sector of the SM could be extended including other Higgs multiplets, beside the standard doublet in eqn (3.96), all of which have vacuum expectation values which contribute to the generation of the masses of the W and Z gauge bosons through the Higgs mechanism. For an arbitrary number of such Higgs multiplets Φ_k, including the standard doublet, ρ is given by (see, for example, Ref. [829])

$$\rho = \frac{\sum_k \left[I^k\left(I^k+1\right) - \left(I_3^k\right)\right] v_k^2}{2\sum_k \left(I_3^k\right) v_k^2} \,. \tag{3.122}$$

Here I^k is the weak isospin of the Higgs multiplet Φ_k and I_3^k is the third component of the weak isospin of the component of Φ_k which has a vacuum expectation value v_k. Equation (3.122) implies that $\rho = 1$ for any number of Higgs doublets. The experimental value of ρ is [400]

$$\rho = 0.9998^{+0.0008}_{-0.0005} \,, \tag{3.123}$$

in perfect agreement with $\rho = 1$. Therefore, the experimental data leave open only the possibility that there might be other Higgs doublets, besides the standard

16 Equation (3.121) is valid only at the tree level. Radiative corrections lead to a departure from $\rho = 1$ which depends on the renormalization scheme, except in the on-shell scheme where $\rho = 1$ is assumed as the definition of the renormalized $\sin^2\vartheta_{\mathrm{W}}$ (see, for example, Ref. [400]).

one, which generate the masses of the W and Z gauge bosons through the Higgs mechanism.

The last four terms on the right-hand side of eqn (3.116) generate trilinear and quadrilinear couplings of the Higgs field with the gauge bosons, represented by the diagrams

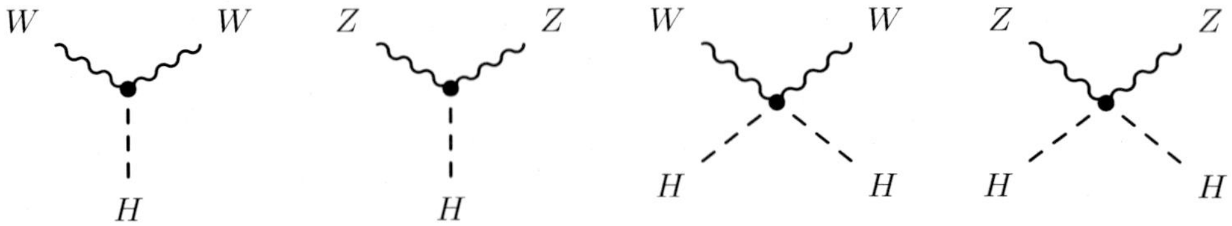

(3.124)

3.5 Fermion masses and mixing

In the SM, the mass of fermions arises as a result of the Higgs mechanism through the presence of Yukawa couplings of the fermion fields with the Higgs doublet.

Let us derive first the lepton masses. As shown in eqn (3.26), a fermion mass term must involve a coupling of left-handed and right-handed fields. So, it is clear that in the SM neutrinos are massless, because their fields do not have a right-handed component. Considering the charged leptons, the products $\overline{L'_{\alpha L}}\,\ell'_{\beta R}$, with $\alpha, \beta = e, \mu, \tau$, are isospin doublets with hypercharge $Y = -1$. Since the Higgs doublet has hypercharge $Y = +1$, the Higgs–lepton Yukawa Lagrangian

$$\mathscr{L}_{H,\mathrm{L}} = - \sum_{\alpha,\beta=e,\mu,\tau} Y'^{\ell}_{\alpha\beta}\, \overline{L'_{\alpha L}}\, \Phi\, \ell'_{\beta R} + \mathrm{H.c.} \tag{3.125}$$

is invariant under the transformations of the type in eqn (3.10) belonging to the $\mathrm{SU}(2)_L \times \mathrm{U}(1)_Y$ gauge group and appears in the fifth line of the SM electroweak Lagrangian in eqn (3.71). The matrix Y'^{ℓ} of Yukawa couplings is, in general, a complex 3×3 matrix. In the unitary gauge, the Higgs doublet has the expression given in eqn (3.113) and the Higgs–lepton Yukawa Lagrangian in eqn (3.125) becomes

$$\mathscr{L}_{H,\mathrm{L}} = - \left(\frac{v + H}{\sqrt{2}} \right) \sum_{\alpha,\beta=e,\mu,\tau} Y'^{\ell}_{\alpha\beta}\, \overline{\ell'_{\alpha L}}\, \ell'_{\beta R} + \mathrm{H.c.} \tag{3.126}$$

The term proportional to the VEV, v, of the Higgs doublet is a mass term for the charged fermion, whereas the term proportional to the Higgs boson field H gives trilinear couplings between the charged leptons and the Higgs boson. However, since the matrix Y'^{ℓ} is in general nondiagonal, the fields e', μ', and τ' do not have definite masses. In order to find the charged lepton fields with definite mass, it is necessary to diagonalize the matrix Y'^{ℓ}. In order to accomplish this task, let us define the arrays of charged lepton fields

$$\boldsymbol{\ell}'_L \equiv \begin{pmatrix} e'_L \\ \mu'_L \\ \tau'_L \end{pmatrix}, \qquad \boldsymbol{\ell}'_R \equiv \begin{pmatrix} e'_R \\ \mu'_R \\ \tau'_R \end{pmatrix}. \tag{3.127}$$

Using this notation, the Higgs–lepton Yukawa Lagrangian can be written in the matrix form

$$\mathscr{L}_{H,\mathrm{L}} = -\left(\frac{v+H}{\sqrt{2}}\right) \overline{\boldsymbol{\ell}'_L}\, Y'^{\ell}\, \boldsymbol{\ell}'_R + \mathrm{H.c.} \tag{3.128}$$

The matrix Y'^{ℓ} can be diagonalized through the biunitary transformation (see section 4.1)

$$V_L^{\ell\dagger}\, Y'^{\ell}\, V_R^{\ell} = Y^{\ell}, \qquad \text{with} \qquad Y^{\ell}_{\alpha\beta} = y^{\ell}_{\alpha}\, \delta_{\alpha\beta} \qquad (\alpha, \beta = e, \mu, \tau). \tag{3.129}$$

Here V_L^{ℓ} and V_R^{ℓ} are two appropriate 3×3 unitary matrices ($V_L^{\ell\dagger} = (V_L^{\ell})^{-1}$ and $V_R^{\ell\dagger} = (V_R^{\ell})^{-1}$).

The diagonalization in eqn (3.129) leads to

$$\mathscr{L}_{H,\mathrm{L}} = -\left(\frac{v+H}{\sqrt{2}}\right) \overline{\boldsymbol{\ell}_L}\, Y^{\ell}\, \boldsymbol{\ell}_R + \mathrm{H.c.}, \tag{3.130}$$

where

$$\boldsymbol{\ell}_L = V_L^{\ell\dagger}\, \boldsymbol{\ell}'_L \equiv \begin{pmatrix} e_L \\ \mu_L \\ \tau_L \end{pmatrix}, \qquad \boldsymbol{\ell}_R = V_R^{\ell\dagger}\, \boldsymbol{\ell}'_R \equiv \begin{pmatrix} e_R \\ \mu_R \\ \tau_R \end{pmatrix} \tag{3.131}$$

are the arrays containing the left-handed and right-handed components of the charged lepton fields with definite masses. Finally, the Higgs–lepton Yukawa Lagrangian can be written as

$$\mathscr{L}_{H,\mathrm{L}} = -\sum_{\alpha=e,\mu,\tau} \frac{y^{\ell}_{\alpha} v}{\sqrt{2}}\, \overline{\ell_\alpha}\, \ell_\alpha - \sum_{\alpha=e,\mu,\tau} \frac{y^{\ell}_{\alpha}}{\sqrt{2}}\, \overline{\ell_\alpha}\, \ell_\alpha\, H, \tag{3.132}$$

where

$$\ell_\alpha \equiv \ell_{\alpha L} + \ell_{\alpha R} \qquad (\alpha = e, \mu, \tau) \tag{3.133}$$

are the fields of the charged leptons with definite masses:

$$\ell_e \equiv e, \qquad \ell_\mu \equiv \mu, \qquad \ell_\tau \equiv \tau. \tag{3.134}$$

The first term on the right-hand side of eqn (3.132) is the mass term for the charged leptons, whose masses are given by

$$m_\alpha = \frac{y^{\ell}_{\alpha} v}{\sqrt{2}} \qquad (\alpha = e, \mu, \tau). \tag{3.135}$$

Since the coefficients y^{ℓ}_e, y^{ℓ}_μ, y^{ℓ}_τ are unknown parameters of the SM, the masses of the charged leptons cannot be predicted and must be obtained from experimental measurements.

An interesting property following from the second term on the right-hand sides of eqn (3.132) and eqn (3.135) is that the trilinear couplings between the charged leptons and the Higgs boson are proportional to the charged lepton masses. Indeed, that term can be written as

$$-\sum_{\alpha=e,\mu,\tau} \frac{m_\alpha}{v} \overline{\ell_\alpha} \, \ell_\alpha \, H \, . \tag{3.136}$$

On the other hand, neutrinos, being massless, do not couple to the Higgs boson. The trilinear coupling in eqn (3.136) of the charged lepton ℓ_α with the Higgs boson is represented by the diagram

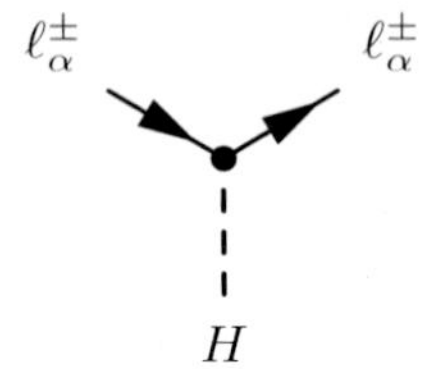

(3.137)

Let us now see what happens to the leptonic charged and neutral weak currents as a result of the transformations in eqn (3.131). Defining the array

$$\boldsymbol{\nu}'_L \equiv \begin{pmatrix} \nu'_{eL} \\ \nu'_{\mu L} \\ \nu'_{\tau L} \end{pmatrix} , \tag{3.138}$$

the leptonic charged weak current in eqn (3.86) can be written as

$$j^\rho_{W,L} = 2 \, \overline{\boldsymbol{\nu}'_L} \, \gamma^\rho \, \boldsymbol{\ell}'_L = 2 \, \overline{\boldsymbol{\nu}'_L} \, \gamma^\rho \, V^\ell_L \, \boldsymbol{\ell}_L \, . \tag{3.139}$$

Since we can freely transform the massless neutrino fields as

$$\boldsymbol{\nu}_L = V^{\ell\dagger}_L \, \boldsymbol{\nu}'_L \equiv \begin{pmatrix} \nu_{eL} \\ \nu_{\mu L} \\ \nu_{\tau L} \end{pmatrix} , \tag{3.140}$$

the leptonic charged weak current can be written as

$$j^\rho_{W,L} = 2 \, \overline{\boldsymbol{\nu}_L} \, \gamma^\rho \, \boldsymbol{\ell}_L = 2 \sum_{\alpha=e,\mu,\tau} \overline{\nu_{\alpha L}} \, \gamma^\rho \, \ell_{\alpha L} \, , \tag{3.141}$$

in terms of the massless neutrino fields ν_e, ν_μ, ν_τ and the charged lepton fields with definite mass in eqn (3.134). The neutrino fields ν_e, ν_μ, ν_τ are called *flavor neutrino fields* because each of them couples only with the corresponding charged lepton field in the charged weak current in eqn (3.141). The flavor neutrino fields have been defined in eqn (3.140) just to satisfy this property. In the SM the flavor neutrino fields are also mass eigenstates, because any linear combination of massless fields is a massless field. As we will see in chapter 6, in theories beyond the SM in which neutrinos are massive, flavor neutrino fields are, in general, not mass eigenstates, a phenomenon called *neutrino mixing*.

TABLE 3.8. Assignment of flavor lepton numbers.

	L_e	L_μ	L_τ		L_e	L_μ	L_τ
$(\nu_e\,,\,e^-)$	$+1$	0	0	$(\bar{\nu}_e\,,\,e^+)$	-1	0	0
$(\nu_\mu\,,\,\mu^-)$	0	$+1$	0	$(\bar{\nu}_\mu\,,\,\mu^+)$	0	-1	0
$(\nu_\tau\,,\,\tau^-)$	0	0	$+1$	$(\bar{\nu}_\tau\,,\,\tau^+)$	0	0	-1

With the leptonic charged weak current in eqn (3.141), the leptonic part of the charged-current weak interaction Lagrangian in eqn (3.76) describes trilinear couplings of the leptons with the W gauge boson represented by the diagrams

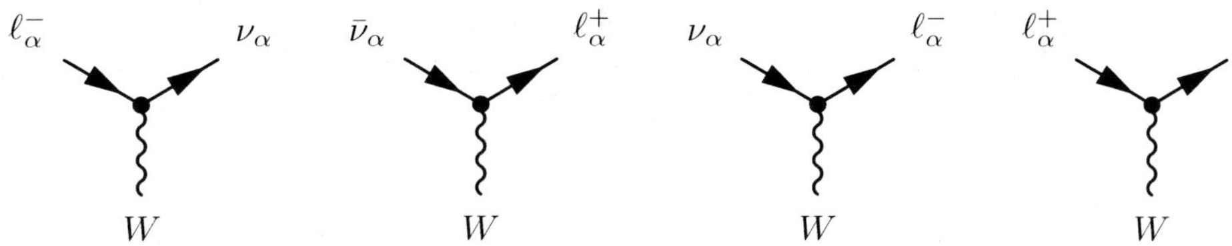

(3.142)

As can be seen from these diagrams, the leptonic current $j^\rho_{W,\mathrm{L}}$ connects each charged lepton with the corresponding flavor neutrino. Hence, the electron L_e, muon L_μ and tau L_τ flavor lepton numbers assigned as shown in Table 3.8 are conserved. A trivial consequence is that also the total lepton number

$$L = L_e + L_\mu + L_\tau \tag{3.143}$$

is also conserved. As we will see in the following chapters, the nonconservation of the lepton numbers L_e, L_μ, L_τ, L plays an important role in the physics of neutrinos beyond the SM.

The conservation of each flavor lepton number L_α is related, through Noether's theorem, to the invariance of the Lagrangian under the global U(1) gauge transformations

$$\nu_{\alpha L} \to e^{i\varphi_\alpha}\,\nu_{\alpha L}\,, \qquad \ell_{\alpha L} \to e^{i\varphi_\alpha}\,\ell_{\alpha L}\,, \qquad \ell_{\alpha R} \to e^{i\varphi_\alpha}\,\ell_{\alpha R}\,, \tag{3.144}$$

as explained in section C.7. From eqn (C.49), the associated conserved current $(\partial_\rho j^\rho_\alpha = 0)$ is

$$j^\rho_\alpha = \overline{\nu_{\alpha L}}\,\gamma^\rho\,\nu_{\alpha L} + \overline{\ell_\alpha}\,\gamma^\rho\,\ell_\alpha\,, \tag{3.145}$$

and the conserved charge $(\partial_0 \mathsf{L}_\alpha = 0)$ is

$$\mathsf{L}_\alpha = \int \mathrm{d}^3x\, j^0_\alpha(x)\,. \tag{3.146}$$

Using the Fourier expansion in eqn (2.139) for the massive Dirac charged lepton field $\ell_\alpha(x)$ and the Fourier expansion in eqn (2.222) for the massless chiral left-handed neutrino field $\nu_{\alpha L}(x)$, we obtain the normally ordered (see eqn (2.245)) flavor lepton number operator

$$:\!\mathsf{L}_\alpha\!: \;= \int \frac{\mathrm{d}^3p}{(2\pi)^3\,2E}\left[a^{(-)\dagger}_{\nu_\alpha}(p)\,a^{(-)}_{\nu_\alpha}(p) - b^{(+)\dagger}_{\nu_\alpha}(p)\,b^{(+)}_{\nu_\alpha}(p)\right]$$

$$+\int \frac{d^3p}{(2\pi)^3\, 2E} \sum_{h=\pm 1} \left[a_{\ell_\alpha}^{(h)\dagger}(p)\, a_{\ell_\alpha}^{(h)}(p) - b_{\ell_\alpha}^{(h)\dagger}(p)\, b_{\ell_\alpha}^{(h)}(p) \right] . \tag{3.147}$$

The neutrino contribution to the flavor lepton number operator $:\mathsf{L}_\alpha:$ agrees with the fact that the massless neutrino chiral field $\nu_{\alpha L}$ describes only neutrinos with negative helicity and antineutrinos with positive helicity, as we have explained in section 2.9.2.

Considering now the neutral weak current of the lepton fields in eqn (3.92), it can be written as

$$\begin{aligned} j^\rho_{Z,\mathrm{L}} =& 2\, g_L^\nu\, \overline{\boldsymbol{\nu}'_L}\, \gamma^\rho\, \boldsymbol{\nu}'_L + 2\, g_L^l\, \overline{\boldsymbol{\ell}'_L}\, \gamma^\rho \boldsymbol{\ell}'_L + 2\, g_R^l\, \overline{\boldsymbol{\ell}'_R}\, \gamma^\rho \boldsymbol{\ell}'_R \\ =& 2\, g_L^\nu\, \overline{\boldsymbol{\nu}_L}\, V_L^{\ell\dagger}\, \gamma^\rho\, V_L^\ell\, \boldsymbol{\nu}_L + 2\, g_L^l\, \overline{\boldsymbol{\ell}_L}\, V_L^{\ell\dagger}\, \gamma^\rho\, V_L^\ell\, \boldsymbol{\ell}_L + 2\, g_R^l\, \overline{\boldsymbol{\ell}_R}\, V_R^{\ell\dagger}\, \gamma^\rho\, V_R^\ell\, \boldsymbol{\ell}_R \\ =& 2\, g_L^\nu\, \overline{\boldsymbol{\nu}_L}\, \gamma^\rho\, \boldsymbol{\nu}_L + 2\, g_L^l\, \overline{\boldsymbol{\ell}_L}\, \gamma^\rho \boldsymbol{\ell}_L + 2\, g_R^l\, \overline{\boldsymbol{\ell}_R}\, \gamma^\rho \boldsymbol{\ell}_R\,, \end{aligned} \tag{3.148}$$

because the matrices V_L^ℓ and V_R^ℓ are unitary. Therefore, the expression of the neutral weak current in terms of the unprimed lepton fields with definite masses is the same as that in terms of the primed lepton fields. This phenomenon is called the GIM mechanism [544]. The GIM mechanism works also in the case of the electromagnetic current of leptons in eqn (3.92), which can be written as

$$j^\rho_{\gamma,\mathrm{L}} = - \sum_{\alpha=e,\mu,\tau} \overline{\ell_\alpha}\, \gamma^\rho\, \ell_\alpha\,, \tag{3.149}$$

in terms of the charged lepton fields in eqn (3.134) with definite masses. Thus, the trilinear couplings of the leptons with the Z gauge boson are represented by the diagrams

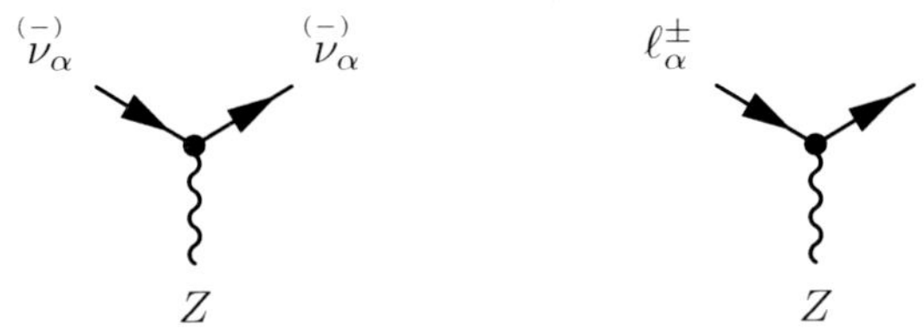

(3.150)

and the trilinear couplings of the charged lepton ℓ_α with the photon is represented by the diagram

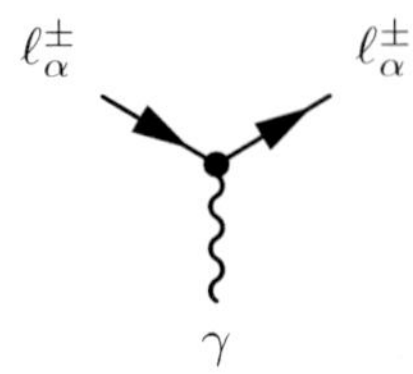

(3.151)

Let us consider now the Lagrangian mass terms of quarks. As for the leptons, we look at mass-like products of left-handed and right-handed fields. From the left-handed doublets in eqn (3.67) and the right-handed singlets in eqns (3.69) and

(3.70), it is possible to form two types of products:

$$\overline{Q'_{\alpha L}}\, q'^{D}_{\beta R} \quad \text{with} \quad \alpha = 1,2,3 \quad \text{and} \quad \beta = d,s,b\,, \tag{3.152}$$

and

$$\overline{Q'_{\alpha L}}\, q'^{U}_{\beta R} \quad \text{with} \quad \alpha = 1,2,3 \quad \text{and} \quad \beta = u,c,t\,. \tag{3.153}$$

The product in eqn (3.152) has hypercharge $Y = -1$ and can be coupled to the Higgs doublet with hypercharge $Y = +1$ in order to make a Yukawa Lagrangian term invariant under the $\mathrm{SU}(2)_L \times \mathrm{U}(1)_Y$ transformations in eqn (3.10),

$$- \sum_{\alpha=1,2,3} \sum_{\beta=d,s,b} Y'^{D}_{\alpha\beta}\, \overline{Q'_{\alpha L}}\, \Phi\, q'^{D}_{\beta R}\,, \tag{3.154}$$

where Y'^{D} is a complex 3×3 matrix of Yukawa couplings. This Yukawa Lagrangian term appears in the sixth line of the SM electroweak Lagrangian in eqn (3.71). It is analogous to the Higgs–lepton Yukawa Lagrangian in eqn (3.125) and gives rise to the masses of the down-like quarks d, s, b. Indeed, in the unitary gauge, where the Higgs doublet has the expression in eqn (3.113), the term in eqn (3.154) becomes

$$- \left(\frac{v+H}{\sqrt{2}} \right) \sum_{\alpha,\beta=d,s,b} Y'^{D}_{\alpha\beta}\, \overline{q'^{D}_{\alpha L}}\, q'^{D}_{\beta R}\,, \tag{3.155}$$

with $Y'^{D}_{d\beta} \equiv Y'^{D}_{1\beta}$, $Y'^{D}_{s\beta} \equiv Y'^{D}_{2\beta}$, $Y'^{D}_{b\beta} \equiv Y'^{D}_{3\beta}$. The terms proportional to v have the structure of mass terms for the d, s, b quarks.

The product in eqn (3.153) has hypercharge $Y = +1$ and in order to form an $\mathrm{SU}(2)_L \times \mathrm{U}(1)_Y$ invariant term it needs a Higgs doublet with hypercharge $Y = -1$. Such a Higgs doublet can be obtained from the Higgs doublet in eqn (3.96) with the transformation

$$\widetilde{\Phi} = i\tau_2\, \Phi^*\,. \tag{3.156}$$

Indeed, under a gauge transformation in eqn (3.10), taking into account the transformation in eqn (3.97) of Φ, we have

$$\begin{aligned} \widetilde{\Phi} \xrightarrow{g(\underline{\theta}(x),\eta(x))} i\tau_2\, e^{-i\underline{\theta}(x)\cdot\frac{\underline{\tau}^*}{2} - i\frac{\eta(x)}{2}}\, \Phi^* &= \left(\tau_2\, e^{-i\underline{\theta}(x)\cdot\frac{\underline{\tau}^*}{2} - i\frac{\eta(x)}{2}}\, \tau_2 \right) i\tau_2\, \Phi^* \\ &= e^{\frac{i}{2}\underline{\theta}(x)\cdot\underline{\tau} - \frac{i}{2}\eta(x)}\, \widetilde{\Phi}\,, \end{aligned} \tag{3.157}$$

where we have used the property $\tau_2 \underline{\tau}^* \tau_2 = -\underline{\tau}$. Therefore, $\widetilde{\Phi}$ transforms as a weak isospin doublet with hypercharge $Y = -1$. This allows us to write the gauge-invariant Yukawa term

$$- \sum_{\alpha=1,2,3} \sum_{\beta=u,c,t} Y'^{U}_{\alpha\beta}\, \overline{Q'_{\alpha L}}\, \widetilde{\Phi}\, q'^{U}_{\beta R}\,, \tag{3.158}$$

that appears in the last line of the SM electroweak Lagrangian in eqn (3.71). This term gives rise to the masses of up-like quarks u, c, t. Indeed, in the unitary gauge

$$\widetilde{\Phi} = \frac{1}{\sqrt{2}} \begin{pmatrix} v + H(x) \\ 0 \end{pmatrix}, \tag{3.159}$$

and the term in eqn (3.158) becomes

$$-\left(\frac{v+H}{\sqrt{2}}\right)\sum_{\alpha,\beta=u,c,t} Y'^{U}_{\alpha\beta}\,\overline{q'^{U}_{\alpha L}}\,q'^{U}_{\beta R}\,, \tag{3.160}$$

with $Y'^{U}_{u\beta} \equiv Y'^{U}_{1\beta}$, $Y'^{U}_{c\beta} \equiv Y'^{U}_{2\beta}$, $Y'^{U}_{t\beta} \equiv Y'^{U}_{3\beta}$. The terms proportional to v have the structure of mass terms for the u, c, t quarks.

Putting together the gauge-invariant Yukawa terms in eqns (3.154) and (3.158) and their Hermitian conjugates, we obtain the quark Yukawa Lagrangian

$$\mathscr{L}_{H,Q} = -\sum_{\alpha=1,2,3}\left[\sum_{\beta=d,s,b} Y'^{D}_{\alpha\beta}\,\overline{Q'_{\alpha L}}\,\Phi\,q'^{D}_{\beta R} + \sum_{\beta=u,c,t} Y'^{U}_{\alpha\beta}\,\overline{Q'_{\alpha L}}\,\widetilde{\Phi}\,q'^{U}_{\beta R}\right] + \text{H.c.} \tag{3.161}$$

In the unitary gauge, it becomes

$$\mathscr{L}_{H,Q} = -\left(\frac{v+H}{\sqrt{2}}\right)\left[\sum_{\alpha,\beta=d,s,b} Y'^{D}_{\alpha\beta}\overline{q'^{D}_{\alpha L}}\,q'^{D}_{\beta R} + \sum_{\alpha,\beta=u,c,t} Y'^{U}_{\alpha\beta}\overline{q'^{U}_{\alpha L}}\,q'^{U}_{\beta R}\right] + \text{H.c.} \tag{3.162}$$

The terms proportional to v are mass terms for the quarks. However, since the complex matrices Y'^{D} and Y'^{U} of Yukawa couplings are in general nondiagonal, the primed quark fields do not have definite masses. In order to find the massive fields, it is necessary to diagonalize the matrices Y'^{D} and Y'^{U}. Following the same procedure as in the case of the leptons, we define the arrays

$$\boldsymbol{q}'^{U}_{L} \equiv \begin{pmatrix} u'_L \\ c'_L \\ t'_L \end{pmatrix}, \qquad \boldsymbol{q}'^{U}_{R} \equiv \begin{pmatrix} u'_R \\ c'_R \\ t'_R \end{pmatrix}, \qquad \boldsymbol{q}'^{D}_{L} \equiv \begin{pmatrix} d'_L \\ s'_L \\ b'_L \end{pmatrix}, \qquad \boldsymbol{q}'^{D}_{R} \equiv \begin{pmatrix} d'_R \\ s'_R \\ b'_R \end{pmatrix}, \tag{3.163}$$

that allow us to write the Higgs–quark Yukawa Lagrangian in the matrix form

$$\mathscr{L}_{H,Q} = -\left(\frac{v+H}{\sqrt{2}}\right)\left[\overline{\boldsymbol{q}'^{D}_{L}}Y'^{D}\boldsymbol{q}'^{D}_{R} + \overline{\boldsymbol{q}'^{U}_{L}}Y'^{U}\boldsymbol{q}'^{U}_{R}\right] + \text{H.c.} \tag{3.164}$$

The matrices Y'^{D} and Y'^{U} can be diagonalized through the biunitary transformations (see section 4.1)

$$V^{D\dagger}_{L} Y'^{D} V^{D}_{R} = Y^{D}\,, \qquad \text{with} \qquad Y^{D}_{\alpha\beta} = y^{D}_{\alpha}\delta_{\alpha\beta} \qquad (\alpha,\beta = d,s,b)\,, \tag{3.165}$$

$$V^{U\dagger}_{L} Y'^{U} V^{U}_{R} = Y^{U}\,, \qquad \text{with} \qquad Y^{U}_{\alpha\beta} = y^{U}_{\alpha}\delta_{\alpha\beta} \qquad (\alpha,\beta = u,c,t)\,, \tag{3.166}$$

where V^{D}_{L}, V^{D}_{R}, V^{U}_{L}, and V^{U}_{R} are four appropriate 3×3 unitary matrices. Defining

$$\boldsymbol{q}^{U}_{L} = V^{U\dagger}_{L}\,\boldsymbol{q}'^{U}_{L} \equiv \begin{pmatrix} u_L \\ c_L \\ t_L \end{pmatrix}, \qquad \boldsymbol{q}^{U}_{R} = V^{U\dagger}_{R}\,\boldsymbol{q}'^{U}_{R} \equiv \begin{pmatrix} u_R \\ c_R \\ t_R \end{pmatrix}, \tag{3.167}$$

$$\boldsymbol{q}^{D}_{L} = V^{D\dagger}_{L}\,\boldsymbol{q}'^{D}_{L} \equiv \begin{pmatrix} d_L \\ s_L \\ b_L \end{pmatrix}, \qquad \boldsymbol{q}^{D}_{R} = V^{D\dagger}_{R}\,\boldsymbol{q}'^{D}_{R} \equiv \begin{pmatrix} d_R \\ s_R \\ b_R \end{pmatrix}. \tag{3.168}$$

we obtain

$$\begin{aligned}\mathscr{L}_{H,Q} &= -\left(\frac{v+H}{\sqrt{2}}\right)\left[\overline{\boldsymbol{q}_L^D}Y^D\boldsymbol{q}_R^D + \overline{\boldsymbol{q}_L^U}Y^U\boldsymbol{q}_R^U\right] + \text{H.c.}\\ &= -\sum_{\alpha=d,s,b}\frac{y_\alpha^D v}{\sqrt{2}}\overline{q_\alpha^D}\,q_\alpha^D - \sum_{\alpha=u,c,t}\frac{y_\alpha^U v}{\sqrt{2}}\overline{q_\alpha^U}\,q_\alpha^U\\ &\quad - \sum_{\alpha=d,s,b}\frac{y_\alpha^D}{\sqrt{2}}\overline{q_\alpha^D}\,q_\alpha^D\,H - \sum_{\alpha=u,c,t}\frac{y_\alpha^U}{\sqrt{2}}\overline{q_\alpha^U}\,q_\alpha^U\,H\,,\end{aligned} \tag{3.169}$$

where

$$q_\alpha^D \equiv q_{\alpha L}^D + q_{\alpha R}^D\,, \qquad q_\alpha^U \equiv q_{\alpha L}^U + q_{\alpha R}^U \tag{3.170}$$

are the fields of quarks with definite masses. The first two terms in the last line of eqn (3.169) are the mass terms for the quarks, whose masses are given by

$$m_\alpha = \frac{y_\alpha^D v}{\sqrt{2}} \qquad (\alpha = d,s,b)\,, \tag{3.171}$$

$$m_\alpha = \frac{y_\alpha^U v}{\sqrt{2}} \qquad (\alpha = u,c,t)\,. \tag{3.172}$$

As in the case of leptons, since the quantities y_d^D, y_s^D, y_b^D, y_u^U, y_c^U, y_t^U, are unknown parameters of the SM, the masses of the quarks cannot be predicted and must be obtained from experimental measurements.

Let us now discuss the practical effect of quark mixing, which is due to the mismatch between the unprimed quark fields with definite masses and the primed quark fields that appear in the weak charged current in eqn (3.79). Using the definitions in eqn (3.163), the quark weak charged current in eqn (3.79) can be written in the matrix form

$$j_{W,Q}^\rho = 2\,\overline{\boldsymbol{q}_L'^U}\,\gamma^\rho\,\boldsymbol{q}_L'^D\,. \tag{3.173}$$

Now we express the primed quark fields in terms of the unprimed ones using eqns (3.167) and (3.168):

$$j_{W,Q}^\rho = 2\,\overline{\boldsymbol{q}_L^U}\,V_L^{U\dagger}\,\gamma^\rho\,V_L^D\,\boldsymbol{q}_L^D = 2\,\overline{\boldsymbol{q}_L^U}\,\gamma^\rho\,V_L^{U\dagger}\,V_L^D\,\boldsymbol{q}_L^D\,. \tag{3.174}$$

Hence, the quark weak charged current does not depend separately on the matrices V_L^U and V_L^D, but only on their product

$$V = V_L^{U\dagger}\,V_L^D\,. \tag{3.175}$$

The matrix V is the quark mixing matrix, also called the Cabibbo–Kobayashi–Maskawa (CKM) matrix [292, 685], which embodies the physical effects of quark mixing. Indeed, the quark mixing matrix determines the weak charged-current interactions of quarks through the current

$$j_{W,Q}^\rho = 2\,\overline{\boldsymbol{q}_L^U}\,\gamma^\rho\,V\,\boldsymbol{q}_L^D\,. \tag{3.176}$$

The properties of the quark mixing matrix are discussed in chapter 4. The expression in eqn (3.176) is very important because it must be used in the calculation of weak interaction processes involving quarks, where the initial and final states describe particles with definite masses. Hence, the measurable quantities depend on the elements of the quark mixing matrix V. Indeed, with the quark charged weak current in eqn (3.176), the quark part of the charged-current weak interaction Lagrangian in eqn (3.76) describes trilinear couplings of the quarks with the W gauge boson represented by the diagrams

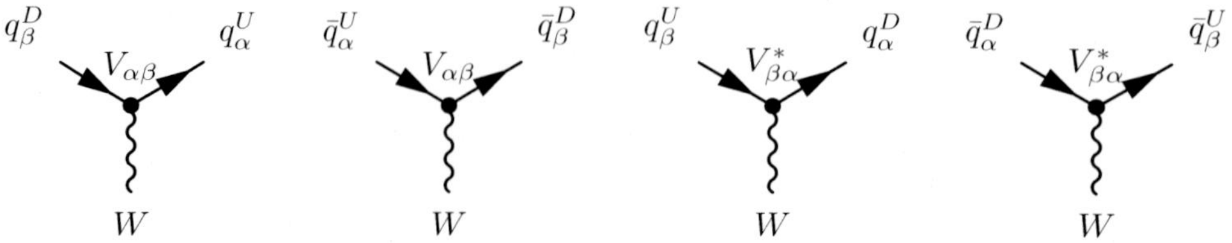

(3.177)

We have indicated in the vertex the mixing matrix contribution which emphasize the flavor-changing character of these interactions. There are no conserved flavor numbers for quarks. However, the quark charged current in eqn (3.79) conserves the baryon number, which is 1/3 for each quark and $-1/3$ for each antiquark.

Let us now consider the effect of quark mixing on the weak neutral current in eqn (3.93). Using the relations in eqns (3.167) and (3.168), the quark neutral current expressed in terms of the quark fields with definite masses, is given by

$$\begin{aligned} j^\rho_{Z,\mathrm{Q}} &= 2\, g^U_L\, \overline{\boldsymbol{q}'^U_L}\, \gamma^\rho\, \boldsymbol{q}'^U_L + 2\, g^U_R\, \overline{\boldsymbol{q}'^U_R}\, \gamma^\rho\, \boldsymbol{q}'^U_R \\ &\quad + 2\, g^D_L\, \overline{\boldsymbol{q}'^D_L}\, \gamma^\rho\, \boldsymbol{q}'^D_L + 2\, g^D_R\, \overline{\boldsymbol{q}'^D_R}\, \gamma^\rho\, \boldsymbol{q}'^D_R \\ &= 2\, g^U_L\, \overline{\boldsymbol{q}^U_L}\, V^{U\dagger}_L\, \gamma^\rho\, V^U_L\, \boldsymbol{q}^U_L + 2\, g^U_R\, \overline{\boldsymbol{q}^U_R}\, V^{U\dagger}_R\, \gamma^\rho\, V^U_R\, \boldsymbol{q}^U_R \\ &\quad + 2\, g^D_L\, \overline{\boldsymbol{q}^D_L}\, V^{D\dagger}_L\, \gamma^\rho\, V^D_L\, \boldsymbol{q}^D_L + 2\, g^D_R\, \overline{\boldsymbol{q}^D_R}\, V^{D\dagger}_R\, \gamma^\rho\, V^D_R\, \boldsymbol{q}^D_R \\ &= 2\, g^U_L\, \overline{\boldsymbol{q}^U_L}\, \gamma^\rho\, \boldsymbol{q}^U_L + 2\, g^U_R\, \overline{\boldsymbol{q}^U_R}\, \gamma^\rho\, \boldsymbol{q}^U_R \\ &\quad + 2\, g^D_L\, \overline{\boldsymbol{q}^D_L}\, \gamma^\rho\, \boldsymbol{q}^D_L + 2\, g^D_R\, \overline{\boldsymbol{q}^D_R}\, \gamma^\rho\, \boldsymbol{q}^D_R\,, \end{aligned} \tag{3.178}$$

because the matrices V^U_L, V^U_R, V^D_L, V^D_R are unitary. Hence, the neutral current has the same form if it is expressed in terms of the massive quark fields or in terms of the primed fields, which appear in the weak charged current. This is the GIM mechanism at work: the weak neutral current is invariant under mixing of the quark fields. The GIM mechanism operates also in the electromagnetic current of quarks, which can be written as

$$j^\rho_{\gamma,\mathrm{Q}} = \frac{2}{3} \sum_{\alpha=u,c,t} \overline{q^U_\alpha}\, \gamma^\rho\, q^U_\alpha - \frac{1}{3} \sum_{\alpha=d,s,b} \overline{q^D_\alpha}\, \gamma^\rho\, q^D_\alpha\,, \tag{3.179}$$

in terms of the massive quark fields in eqn (3.170). Therefore, in the SM there are no *flavor-changing neutral currents.* The trilinear couplings of the quarks with the

Z gauge boson are represented by the diagrams

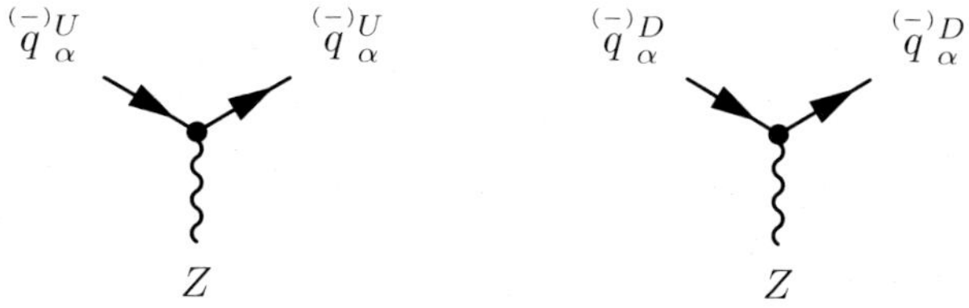

(3.180)

and the trilinear couplings of the quarks with the photon are represented by the diagrams

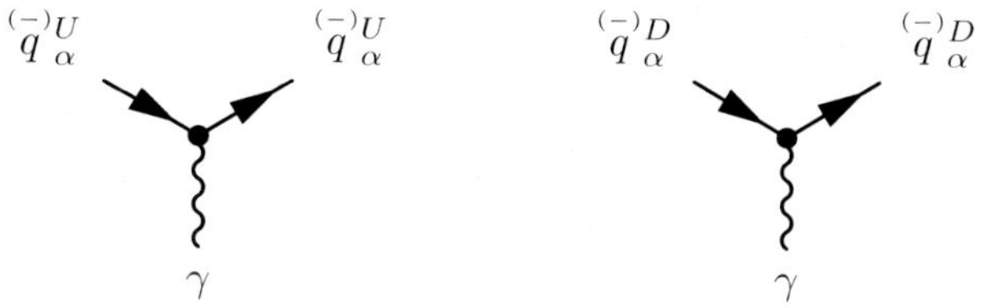

(3.181)

3.6 Gauge bosons

Let us consider the third line in the SM electroweak Lagrangian in eqn (3.71):

$$\mathscr{L}_{\text{gauge}} = -\frac{1}{4}\,\underline{A}_{\mu\nu}\underline{A}^{\mu\nu} - \frac{1}{4}\,B_{\mu\nu}B^{\mu\nu}\,. \tag{3.182}$$

These terms describe the kinetics and self-interactions of the gauge fields.

The expressions for $\underline{A}^{\mu\nu} \equiv (A_1^{\mu\nu}, A_2^{\mu\nu}, A_3^{\mu\nu})$ and $B^{\mu\nu}$ are:

$$A_a^{\mu\nu} = \partial^\mu A_a^\nu - \partial^\nu A_a^\mu - g \sum_{b,c=1}^{3} \varepsilon_{abc}\, A_b^\mu\, A_c^\nu \qquad (a = 1, 2, 3)\,, \tag{3.183}$$

$$B^{\mu\nu} = \partial^\mu B^\nu - \partial^\nu B^\mu\,. \tag{3.184}$$

The expression for $B^{\mu\nu}$ is a straightforward generalization of the electromagnetic tensor

$$F^{\mu\nu} = \partial^\mu A^\nu - \partial^\nu A^\mu\,, \tag{3.185}$$

where A^μ is the electromagnetic field in eqn (3.37). It is well known in QED that the kinetic term of the electromagnetic field is given by

$$\mathscr{L}_A = -\frac{1}{4}\,F_{\mu\nu}F^{\mu\nu}\,, \tag{3.186}$$

which is invariant under local $\mathrm{U}(1)_Q$ gauge transformations (see, for example, Ref. [720]). Similarly, the second term in eqn (3.182) is invariant under local $\mathrm{U}(1)_Y$ gauge transformations of the type in eqn (3.24).

On the other hand, the expression in eqn (3.183) for the tensor associated with the $\mathrm{SU}(2)_L$ gauge bosons has an additional term which is needed in order to achieve invariance of the first term in eqn (3.182) under local nonabelian $\mathrm{SU}(2)_L$ gauge transformations of the type in eqn (3.23). The derivation of the expression in eqn (3.183) is discussed in appendix D.

In order to obtain the physical consequences of the gauge Lagrangian in eqn (3.182) one must express it in terms of the physical gauge fields W^μ, Z^μ, and A^μ using eqns (3.31), (3.37), and (3.38). After a straightforward calculation, one obtains

$$\begin{aligned}
\mathcal{L}_{\text{gauge}} = & -\frac{1}{2}\,F^\dagger_{W\mu\nu}\,F^{\mu\nu}_W - \frac{1}{4}\,F_{Z\mu\nu}\,F^{\mu\nu}_Z - \frac{1}{4}\,F_{\mu\nu}\,F^{\mu\nu} \\
& + i\,g\,\cos\vartheta_{\mathrm{W}}\left[F^{\mu\nu}_W Z_\mu W^\dagger_\nu - F^\dagger_{W\mu\nu} Z^\mu W^\nu + F^{\mu\nu}_Z W^\dagger_\mu W_\nu\right] \\
& + i\,e\left[F^{\mu\nu}_W A_\mu W^\dagger_\nu - F^\dagger_{W\mu\nu} A^\mu W^\nu + F^{\mu\nu} W^\dagger_\mu W_\nu\right] \\
& + g^2\,\cos^2\vartheta_{\mathrm{W}}\left[(W_\mu Z^\mu)(W^\dagger_\nu Z^\nu) - (W^\mu W^\dagger_\mu)(Z^\nu Z_\nu)\right] \\
& + e^2\left[(W_\mu A^\mu)(W^\dagger_\nu A^\nu) - (W^\mu W^\dagger_\mu)(A^\nu A_\nu)\right] \\
& + e\,g\,\cos\vartheta_{\mathrm{W}}\left[(W_\mu Z^\mu)(W^\dagger_\nu A^\nu) + (W^\dagger_\mu Z^\mu)(W_\nu A^\nu) - 2\,(W^\mu W^\dagger_\mu)(Z_\nu A^\nu)\right] \\
& + \frac{1}{2}\,g^2\left[(W_\mu W^\mu)(W^\dagger_\nu W^{\dagger\nu}) - (W^\dagger_\mu W^\mu)^2\right],
\end{aligned} \tag{3.187}$$

where we recognize the last term in the first line as the kinetic term in eqn (3.186) of the electromagnetic field. The first two terms in the first line are, respectively, kinetic terms for the W and Z bosons, which are written in terms of the respective tensors

$$F^{\mu\nu}_W \equiv \partial^\mu W^\nu - \partial^\nu W^\mu\,, \tag{3.188}$$

$$F^{\mu\nu}_Z \equiv \partial^\mu Z^\nu - \partial^\nu Z^\mu\,, \tag{3.189}$$

which have the same structure as the electromagnetic tensor in eqn (3.185). The other terms represent trilinear and quadrilinear couplings of the gauge bosons which generate the vertices in eqn (E.24)–(E.29).

3.6.1 Electromagnetic field

From the first line in eqn (3.187) one can see that the free Lagrangian of the electromagnetic field A^μ is the classical one in eqn (3.186),

$$\mathscr{L}_A = -\frac{1}{4}\,F_{\mu\nu}F^{\mu\nu}\,. \tag{3.190}$$

The Euler–Lagrange procedure described in section C gives the classical field equation

$$\Box A^\nu - \partial^\nu(\partial_\mu A^\mu) = 0\,. \tag{3.191}$$

The free electromagnetic Lagrangian in eqn (3.190) is invariant under the local $\mathrm{U}(1)_Q$ gauge transformations

$$A^\mu \to A'_\mu = A^\mu - \frac{1}{e}\,\partial_\mu \varphi(x)\,. \tag{3.192}$$

Also the complete Lagrangian of the SM is invariant under the local $\mathrm{U}(1)_Q$ gauge transformations of the electromagnetic field, with the associated transformations of the fermion fields

$$f \to f' = e^{iQ\varphi(x)} f\,. \tag{3.193}$$

Therefore, one can choose an appropriate gauge for studying a physical problem. A convenient gauge is the *Lorentz gauge* in which the electromagnetic field satisfies the constraint

$$\partial_\mu A^\mu = 0\,. \tag{3.194}$$

Given a field A^μ it is always possible to make a gauge transformation in eqn (3.192) in such a way that the transformed field A'^μ satisfies the Lorentz constraint in eqn (3.194), $\partial_\mu A'^\mu = 0$. In fact, it is sufficient to choose a function $\varphi(x)$ such that $\Box\varphi(x) = -\partial_\mu A^\mu(x)$. The field $A'^\mu(x)$ is still defined up to gauge transformations with $\Box\varphi(x) = 0$.

The free electromagnetic field in the Lorentz gauge satisfies the d'Alembert equation

$$\Box A^\mu = 0\,. \tag{3.195}$$

The integral Fourier expansion of the electromagnetic field is (see, for example, Ref. [634])

$$A_\mu(x) = \int \frac{\mathrm{d}^3 p}{(2\pi)^3\, 2\omega} \sum_{\alpha=0}^{3} \left[a_\gamma^{(\alpha)}(p)\, \varepsilon_\mu^{(\alpha)}(p)\, e^{-ip\cdot x} + a_\gamma^{(\alpha)\dagger}(p)\, \varepsilon_\mu^{(\alpha)*}(p)\, e^{ip\cdot x} \right], \tag{3.196}$$

where

$$\omega = p^0 = |\vec{p}| \tag{3.197}$$

is the photon energy. The polarization four-vectors $\varepsilon_\mu^{(\alpha)}(p)$, with $\alpha = 0, 1, 2, 3$, form a set of four linearly independent four-vectors which satisfy the relations

$$\varepsilon^{(\alpha)}(p) \cdot \varepsilon^{(\beta)*}(p) = g^{\alpha\beta}\,, \tag{3.198}$$

$$\sum_{\alpha=0}^{3} \varepsilon_\mu^{(\alpha)}(p)\, \varepsilon_\nu^{(\alpha)*}(p)\, g^{\alpha\alpha} = g_{\mu\nu}\,. \tag{3.199}$$

It is convenient to choose $\varepsilon_\mu^{(0)}(p)$ time-like ($\varepsilon_\mu^{(0)}(p)\cdot p = \omega$), $\varepsilon_\mu^{(1)}(p)$ and $\varepsilon_\mu^{(2)}(p)$ transverse ($\varepsilon_\mu^{(1)}(p)\cdot p = \varepsilon_\mu^{(2)}(p)\cdot p = 0$) and $\varepsilon_\mu^{(3)}(p)$ longitudinal ($\varepsilon_\mu^{(3)}(p)\cdot p = -\omega$). The compatibility of these four polarizations with the two transverse polarizations of real photons is discussed, for example, in Ref. [634].

TABLE 3.9. Properties of vector bosons [400].

Boson	Mass (GeV)	Lifetime (s)
γ (photon)	$< 2 \times 10^{-25}$	∞
$W^{\pm}$	80.396 ± 0.061	$(3.195 \pm 0.077) \times 10^{-25}$
Z	91.187 ± 0.007	$(2.643 \pm 0.007) \times 10^{-25}$
g (8 gluons)	0	∞

The operators $a_\gamma^{(\alpha)}(p)$ satisfy the commutation relations

$$\left[a_\gamma^{(\alpha)}(p)\,,\,a_\gamma^{(\alpha')\dagger}(p')\right] = -g^{\alpha\alpha'}\,(2\pi)^3\,2\omega\,\delta^3(\vec{p}-\vec{p}')\,, \tag{3.200}$$

$$\left[a_\gamma^{(\alpha)}(p)\,,\,a_\gamma^{(\alpha')}(p')\right] = \left[a_\gamma^{(\alpha)\dagger}(p)\,,\,a_\gamma^{(\alpha')\dagger}(p')\right] = 0\,, \tag{3.201}$$

which lead to the photon propagator

$$G_{\mu\nu}(x-x') \equiv \langle 0|\mathrm{T}[A_\mu(x)\,A_\nu(x')]|0\rangle = \lim_{\epsilon\to 0} i \int \frac{\mathrm{d}^4p}{(2\pi)^4}\,\frac{-g_{\mu\nu}}{p^2+i\,\epsilon}\,e^{-ip\cdot(x-x')}\,. \tag{3.202}$$

3.6.2 W and Z bosons

In the SM, the vector bosons $W^{\pm}$ and Z acquire their masses m_W and m_Z, given in eqn (3.119) (with the experimental values in Table 3.9), through the Higgs mechanism that generates a spontaneous breaking of the $\mathrm{SU}(2)_L \times \mathrm{U}(1)_Y$ symmetry. This means that the Lagrangian is symmetric under $\mathrm{SU}(2)_L \times \mathrm{U}(1)_Y$ transformations, but the physical states are not. In other words, the symmetry of the Lagrangian is hidden in the real world. But the symmetry of the Lagrangian is essential in order to have a renormalizable theory. Therefore, explicit mass terms for the gauge boson fields are forbidden (as the fermion masses of the type in eqn (3.26)). For example, an explicit mass term for the gauge boson field B^μ of the type

$$\frac{1}{2}\,m_B^2 B_\mu B^\mu \tag{3.203}$$

is clearly not invariant under the gauge transformation in eqn (3.24).

From the first line of eqn (3.187) and the mass terms in eqn (3.116), the Lagrangians for the free W and Z boson fields in the unitary gauge are

$$\mathscr{L}^{(W)} = -\frac{1}{2}\,F_{W\mu\nu}^{\dagger}\,F_W^{\mu\nu} + m_W\,W_\mu^\dagger\,W^\mu\,, \tag{3.204}$$

$$\mathscr{L}^{(Z)} = -\frac{1}{4}\,F_{Z\mu\nu}\,F_Z^{\mu\nu} + \frac{1}{2}\,m_W\,Z_\mu\,Z^\mu\,. \tag{3.205}$$

These Lagrangians lead to Proca equations for the free W and Z boson fields:

$$\left(\Box + m_W^2\right) W^\mu - \partial^\mu \left(\partial_\nu W^\nu\right) = 0\,, \tag{3.206}$$

$$\left(\Box + m_Z^2\right) Z^\mu - \partial^\mu \left(\partial_\nu Z^\nu\right) = 0\,. \tag{3.207}$$

Taking the divergence of eqns (3.206) and (3.207) and taking into account that $m_W \neq 0$ and $m_Z \neq 0$, one obtains

$$\partial_\mu W^\mu = 0\,, \tag{3.208}$$

$$\partial_\mu Z^\mu = 0\,. \tag{3.209}$$

These constraints reduce the number of independent components of the W and Z boson fields from four to three and the Proca equations (3.206) and (3.207) are reduced to the Klein–Gordon equations

$$\left(\Box + m_W^2\right) W^\mu = 0\,, \tag{3.210}$$

$$\left(\Box + m_Z^2\right) Z^\mu = 0\,. \tag{3.211}$$

The integral Fourier expansions of the W and Z fields are

$$W_\mu(x) = \int \frac{\mathrm{d}^3p}{(2\pi)^3\, 2E_W} \sum_{r=1}^{3} \left[a_W^{(r)}(p)\, \varepsilon_{W\mu}^{(r)}(p)\, e^{-ip\cdot x} + b_W^{(r)\dagger}(p)\, \varepsilon_{W\mu}^{(r)*}(p)\, e^{ip\cdot x} \right], \tag{3.212}$$

with

$$E_W = p^0 = \sqrt{|\vec{p}|^2 + m_W^2}\,, \tag{3.213}$$

and

$$Z_\mu(x) = \int \frac{\mathrm{d}^3p}{(2\pi)^3\, 2E_Z} \sum_{r=1}^{3} \left[a_Z^{(r)}(p)\, \varepsilon_{Z\mu}^{(r)}(p)\, e^{-ip\cdot x} + a_Z^{(r)\dagger}(p)\, \varepsilon_{Z\mu}^{(r)*}(p)\, e^{ip\cdot x} \right], \tag{3.214}$$

with

$$E_Z = p^0 = \sqrt{|\vec{p}|^2 + m_Z^2}\,. \tag{3.215}$$

The polarization four-vectors $\varepsilon_W^{(r)}(p)$ and $\varepsilon_Z^{(r)}(p)$ form two sets of space-like orthonormal linearly independent four-vectors, such that

$$\varepsilon_W^{(r)}(p) \cdot \varepsilon_W^{(s)*}(p) = \varepsilon_Z^{(r)}(p) \cdot \varepsilon_Z^{(s)*}(p) = -\delta_{rs}\,. \tag{3.216}$$

From the conditions in eqns (3.208) and (3.209), the polarization four-vectors $\varepsilon_W^{(r)}(p)$ and $\varepsilon_Z^{(r)}(p)$ are transverse:

$$\varepsilon_W^{(r)}(p) \cdot p = \varepsilon_Z^{(r)}(p) \cdot p = 0\,. \tag{3.217}$$

Furthermore, from eqns (3.216) and (3.217) one can find the completeness relations

$$\sum_{r=1}^{3} \varepsilon_{W\mu}^{(r)}(p)\, \varepsilon_{W\nu}^{(r)*}(p) = -g_{\mu\nu} + \frac{p_\mu p_\nu}{m_W^2}\,, \qquad \sum_{r=1}^{3} \varepsilon_{Z\mu}^{(r)}(p)\, \varepsilon_{Z\nu}^{(r)*}(p) = -g_{\mu\nu} + \frac{p_\mu p_\nu}{m_Z^2}\,. \tag{3.218}$$

The operators $a_W^{(r)}(p)$, $b_W^{(r)}(p)$, and $a_Z^{(r)}(p)$ commute, except for the commutation relations

$$\left[a_W^{(\alpha)}(p)\,,\,a_W^{(\alpha')^\dagger}(p')\right] = \left[b_W^{(\alpha)}(p)\,,\,b_W^{(\alpha')^\dagger}(p')\right] = -g^{\alpha\alpha'}\,(2\pi)^3\,2E_W\,\delta^3(\vec{p}-\vec{p}')\,, \tag{3.219}$$

$$\left[a_Z^{(\alpha)}(p)\,,\,a_Z^{(\alpha')^\dagger}(p')\right] = -g^{\alpha\alpha'}\,(2\pi)^3\,2E_Z\,\delta^3(\vec{p}-\vec{p}')\,. \tag{3.220}$$

Using these commutation relations and the completeness in eqn (3.218), one can find that the propagators of the W and Z gauge boson fields are given by

$$G_{\mu\nu}^{(W)}(x-x') = \langle 0|\mathrm{T}[W_\mu(x)W_\nu^\dagger(x')]|0\rangle = \lim_{\epsilon\to 0} i\int\frac{\mathrm{d}^4p}{(2\pi)^4}\,\frac{-g_{\mu\nu}+\dfrac{p_\mu\,p_\nu}{m_W^2}}{p^2-m_W^2+i\,\epsilon}\,\mathrm{e}^{-ip\cdot(x-x')}\,, \tag{3.221}$$

$$G_{\mu\nu}^{(Z)}(x-x') = \langle 0|\mathrm{T}[Z_\mu(x)Z_\nu(x')]|0\rangle = \lim_{\epsilon\to 0} i\int\frac{\mathrm{d}^4p}{(2\pi)^4}\,\frac{-g_{\mu\nu}+\dfrac{p_\mu\,p_\nu}{m_Z^2}}{p^2-m_Z^2+i\,\epsilon}\,\mathrm{e}^{-ip\cdot(x-x')}\,. \tag{3.222}$$

3.7 Effective low-energy CC and NC Lagrangians

The energy involved in most phenomena under study in experimental high-energy physics is much smaller than the masses of the W and Z gauge bosons, which are of the order of 100 GeV (see Table 3.9 on page 100). For instance, the energy involved in a particle decay is given by its mass, and all particles, except those containing top quarks, have masses much smaller than 100 GeV. In these low-energy processes, the gauge boson propagators in momentum space can be approximated by

$$G_{\mu\nu}^{(W)}(p) \xrightarrow{|k|^2\ll m_W^2} i\,\frac{g_{\mu\nu}}{m_W^2}\,, \qquad G_{\mu\nu}^{(Z)}(p) \xrightarrow{|k|^2\ll m_Z^2} i\,\frac{g_{\mu\nu}}{m_Z^2}\,. \tag{3.223}$$

Therefore, the internal gauge boson lines in the Feynman diagrams representing low-energy processes can be contracted to a point. In the case of the CC processes, this contraction leads to an effective four-fermion interaction described by the current–current Lagrangian

$$\mathscr{L}_{\text{eff}}^{(\text{CC})} = -\frac{G_{\text{F}}}{\sqrt{2}}\,j_{W\mu}^\dagger j_W^\mu\,, \tag{3.224}$$

where G_{F} is the Fermi constant, whose value is given in eqn (A.151). As illustrated in Fig. 3.1 for a generic low-energy CC process, the contraction of the W-boson propagator leads to a four-fermion vertex with associated Feynman-rule quantity

$$-\,i\,\frac{g^2}{8m_W^2}\,V_{\mathfrak{f}_2\mathfrak{f}_1}\,V_{\mathfrak{f}_4\mathfrak{f}_3}\,\cdots\,\gamma^\mu\,(1-\gamma^5)\,\ldots\ldots\,\gamma_\mu\,(1-\gamma^5)\,\ldots\,, \tag{3.225}$$

where the dots must be filled with the appropriate spinors, according to the Feynman rules in section (E.1). The factors $V_{\mathfrak{f}_2\mathfrak{f}_1}$ and $V_{\mathfrak{f}_4\mathfrak{f}_3}$ are elements of the mixing matrix in the case of quarks.

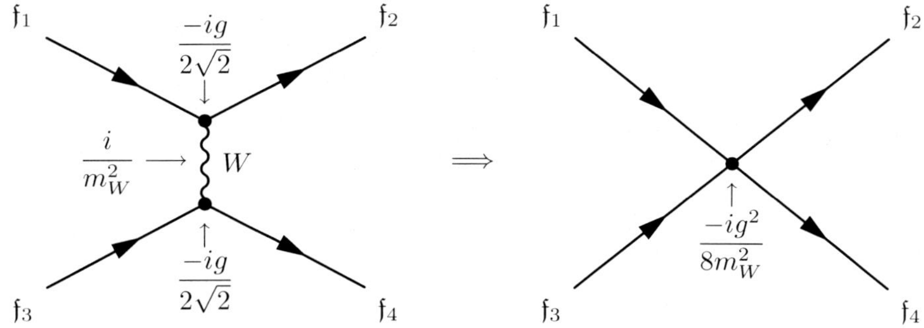

FIG. 3.1. Contraction of the W gauge boson propagator in a generic low-energy CC process.

For an interaction Lagrangian, the quantity associated with the corresponding vertex in the Feynman rules is given, apart from Dirac gamma factors, by the coupling constant with the sign of the interaction Lagrangian times a factor i. One can see it by comparing, for example, eqns (3.76) and (E.21). Taking into account the factor i, the effective Lagrangian in eqn (3.224) corresponds to a four-fermion vertex with associated Feynman-rule quantity

$$-i\,\frac{G_{\text{F}}}{\sqrt{2}}\,V_{\mathfrak{f}_2\mathfrak{f}_1}\,V_{\mathfrak{f}_4\mathfrak{f}_3}\,\ldots\,\gamma^{\mu}\left(1-\gamma^5\right)\,\ldots\,\ldots\,\gamma_{\mu}\left(1-\gamma^5\right)\,\ldots\,. \tag{3.226}$$

Comparing eqns (3.225) and (3.226), we obtain the important relation

$$\frac{G_{\text{F}}}{\sqrt{2}}=\frac{g^2}{8\,m_W^2}\,, \tag{3.227}$$

beween the coupling constant g in the SM, the W-boson mass m_W, and the well-measured Fermi constant G_{F} (see eqn (A.151)). Furthermore, from the expression in eqn (3.119) of m_W in terms of g and v, we obtain the value of the Higgs VEV:

$$v=\left(\sqrt{2}\,G_{\text{F}}\right)^{-1/2}=246\,\text{GeV}\,. \tag{3.228}$$

As an example of a process which is described remarkably well by the effective low-energy charged-current weak interaction Lagrangian in eqn (3.224), let us consider the decay of a muon,

$$\mu^-\to e^-+\bar{\nu}_e+\nu_\mu\,, \tag{3.229}$$

whose diagram is shown in Fig. 3.2. The Feynman rules in section E.1 lead to the amplitude

$$\mathcal{A}_{\mu^-\to e^-\bar{\nu}_e\nu_\mu}=-\frac{g^2}{8}\,\overline{u_e}(p_e)\gamma^{\rho}\left(1-\gamma^5\right)v_{\nu_e}(p_{\nu_e})\,G^{(W)}_{\rho\sigma}(q)\,\overline{u_{\nu_\mu}}(p_{\nu_\mu})\gamma^{\sigma}\left(1-\gamma^5\right)v_{\mu}(p_{\mu})\,, \tag{3.230}$$

with $q=p_\mu-p_{\nu_\mu}$. In the rest frame of the muon we have, from energy conservation, $p_\mu=(m_\mu,\underline{0})$ and $p_{\nu_\mu}=(E_{\nu_\mu},\vec{p}_{\nu_\mu})$, with $E_{\nu_\mu}=|\vec{p}_{\nu_\mu}|\leq m_\mu-m_e$. Hence, all

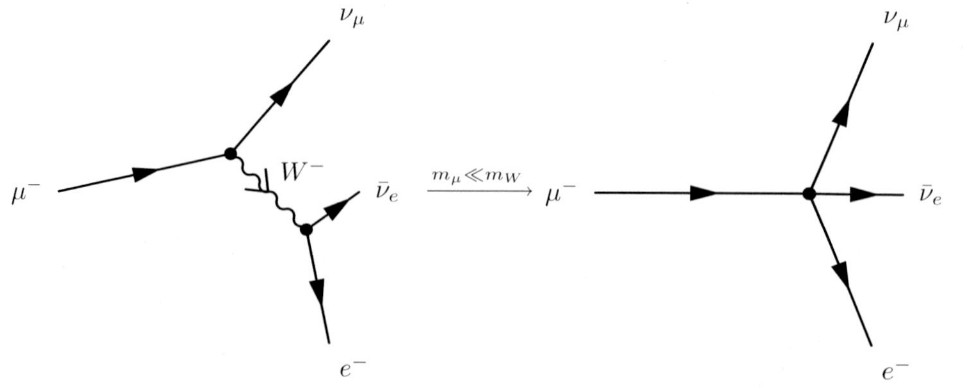

FIG. 3.2. Tree-level Feynman diagram of muon decay (left) and its low-energy approximation (right).

the components of the transferred four-momentum q are smaller than the muon mass and the approximation in eqn (3.223) is very accurate, yielding an effective amplitude which, taking into account eqn (3.227), coincides with that obtained from the effective Lagrangian in eqn (3.224),

$$\mathcal{A}^{(\text{eff})}_{\mu^- \to e^- \bar{\nu}_e \nu_\mu} = -i \, \frac{G_\text{F}}{\sqrt{2}} \, \overline{u_e}(p_e) \gamma^\rho \left(1 - \gamma^5\right) v_{\nu_e}(p_{\nu_e}) \, \overline{u_{\nu_\mu}}(p_{\nu_\mu}) \gamma_\rho \left(1 - \gamma^5\right) v_\mu(p_\mu) \, . \tag{3.231}$$

The resulting muon lifetime τ_μ is given by

$$\frac{1}{\tau_\mu} = \frac{G_\text{F}^2 \, m_\mu^5}{192 \, \pi^3} \left[1 - 8 \, \frac{m_e^2}{m_\mu^2} - 12 \left(\frac{m_e^2}{m_\mu^2} \right)^2 \ln \! \left(\frac{m_e^2}{m_\mu^2} \right) + 8 \left(\frac{m_e^2}{m_\mu^2} \right)^3 - \left(\frac{m_e^2}{m_\mu^2} \right)^4 \right] . \tag{3.232}$$

The correction due to the W boson propagator is $3m_\mu^2/5m_W^2 \simeq 1 \times 10^{-6}$, which is indeed small. The muon lifetime is very important, because it is the best process for the determination of G_F, whose value is given in eqn (A.151).

Let us consider now a low-energy weak neutral-current process. As illustrated in Fig. 3.3, the contraction of the Z-boson propagator leads to an effective four-fermion interaction with coupling constant

$$\frac{g^2}{4 \cos^2 \vartheta_\text{W} m_Z^2} = 2 \, \frac{G_\text{F}}{\sqrt{2}} \, \frac{m_W^2}{m_Z^2 \cos^2 \vartheta_\text{W}} \, , \tag{3.233}$$

where we used eqn (3.227) in order to express g^2 in terms of G_F and m_W^2. The last factor in eqn (3.233) is the ρ parameter introduced in eqn (3.119), which is equal to unity in the SM. Therefore, the neutral-current effective four-fermion coupling constant is given by

$$\frac{g^2}{4 \cos^2 \vartheta_\text{W} m_Z^2} = 2 \, \frac{G_\text{F}}{\sqrt{2}} \, . \tag{3.234}$$

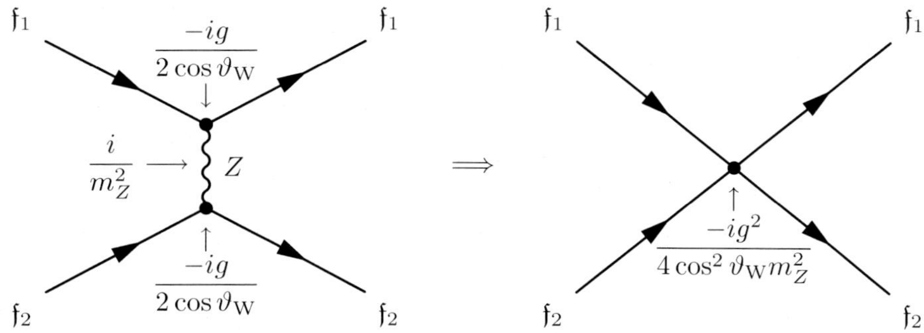

FIG. 3.3. Contraction of the Z gauge boson propagator in a generic low-energy NC process.

Thus, the effective Fermi-like NC Lagrangian appropriate for the description of low-energy NC weak processes is

$$\mathscr{L}_{\text{eff}}^{(\text{NC})} = -\frac{G_{\text{F}}}{\sqrt{2}}\, j_Z^{\mu} j_{Z\mu} \,, \tag{3.235}$$

where we halved the coupling constant in eqn (3.234) to avoid double counting. The Feynman-rule quantity associated with the corresponding vertex is

$$-2\,i\,\frac{G_{\text{F}}}{\sqrt{2}} \cdots \gamma^{\mu}\left(g_V^{f_1} - g_A^{f_1}\gamma^5\right) \cdots\cdots \gamma_{\mu}\left(g_V^{f_2} - g_A^{f_2}\gamma^5\right) \cdots \,, \tag{3.236}$$

where the dots must be filled with the appropriate spinors, according to the Feynman rules in section E.1. The coefficients g_V^f and g_A^f are given in Table 3.6 (page 78).

Let us finally note that eqns (3.42) and (3.227) allow us to express G_{F} in terms of the fine-structure constant $\alpha \equiv e^2/4\pi$, the weak mixing angle ϑ_{W}, and the W-boson mass m_W or the Z-boson mass $m_Z = m_W/\cos\vartheta_{\text{W}}$:

$$\frac{G_{\text{F}}}{\sqrt{2}} = \frac{\pi\,\alpha}{2\,\sin^2\vartheta_{\text{W}}\,m_W^2} = \frac{\pi\,\alpha}{2\,\sin^2\vartheta_{\text{W}}\,\cos^2\vartheta_{\text{W}}\,m_Z^2} \,. \tag{3.237}$$

Summarizing the results presented in this section, low-energy weak interaction processes are described, to an excellent approximation, by the CC and NC four-fermion Lagrangians in eqns (3.224) and (3.235). The amplitudes of these processes can be calculated at tree level with the Feynman rules in appendix E, contracting the gauge boson propagators and associating to the resulting four-fermion vertices the quantity in eqns (3.226) and (3.236) for CC and NC interactions, respectively.

4

THREE-GENERATION MIXING

Entia non sunt multiplicanda praeter necessitatem.*
William of Ockham (XIVth century)

---

* Entities should not be multiplied beyond necessity (*Ockham's razor*).

In this chapter we derive some important and useful properties of the Cabibbo–Kobayashi–Maskawa (CKM) [292, 685] mixing matrix of quarks. Besides its intrinsic importance, the understanding of quark mixing is useful in neutrino physics because the treatment of neutrino mixing is analogous to that of quark mixing in the case of Dirac neutrinos and follows similar methods in the case of Majorana neutrinos. For clarity, we will use, throughout the book, the notations V and U, respectively, for the quark and neutrino mixing matrices.

The CKM mixing matrix of quarks is a unitary 3×3 matrix given by eqn (3.175),

$$V = V_L^{U\dagger} V_L^D = \begin{pmatrix} V_{11} & V_{12} & V_{13} \\ V_{21} & V_{22} & V_{23} \\ V_{31} & V_{32} & V_{33} \end{pmatrix} \equiv \begin{pmatrix} V_{ud} & V_{us} & V_{ub} \\ V_{cd} & V_{cs} & V_{cb} \\ V_{td} & V_{ts} & V_{tb} \end{pmatrix}, \tag{4.1}$$

where V_L^D and V_L^U are, respectively, the unitary matrices that lead to the diagonalization of the mass matrices of down-like and up-like quarks in eqns (3.165) and (3.166). In the following we will use either the numerical indices or the quark flavor indices when they are more convenient for the clarity of discussions and for practical applications. Based on the experimental data so far, the 90% confidence intervals for the magnitude of the elements of the quark mixing matrix are [400]

$$|V| = \begin{pmatrix} 0.9739 - 0.9751 & 0.221 - 0.227 & 0.0029 - 0.0045 \\ 0.221 - 0.227 & 0.9730 - 0.9744 & 0.039 - 0.044 \\ 0.0048 - 0.014 & 0.037 - 0.043 & 0.9990 - 0.9992 \end{pmatrix}. \tag{4.2}$$

All properties of the quark mixing matrix V can be extended to the lepton mixing matrix U in the case of mixing of three Dirac neutrinos, to be discussed in section 6.1.2, with the replacements

$$V \to U = V_L^{\ell\dagger} V_L^{\nu} = \begin{pmatrix} U_{11} & U_{12} & U_{13} \\ U_{21} & U_{22} & U_{23} \\ U_{31} & U_{32} & U_{33} \end{pmatrix} \equiv \begin{pmatrix} U_{e1} & U_{e2} & U_{e3} \\ U_{\mu 1} & U_{\mu 2} & U_{\mu 3} \\ U_{\tau 1} & U_{\tau 2} & U_{\tau 3} \end{pmatrix}. \tag{4.3}$$

For historical reasons, the convention is that the quark mixing matrix connects weak isospin $I_3 = 1/2$ quarks on the left to $I_3 = -1/2$ quarks on the right, whereas the lepton mixing matrix connects $I_3 = -1/2$ charged leptons on the left to $I_3 = 1/2$ neutrinos on the right; obviously, one could switch to the same convention for quarks and leptons by transforming one of the two mixing matrix to its Hermitian conjugate.

We will denote the number of generations by N in order to present general properties valid for any number of generations. Later, however, we will consider explicitly only a realistic case $N = 3$ or a simple case $N = 2$.

4.1 Diagonalization of the mass matrix

A general $N\times N$ complex matrix M' can be diagonalized by a biunitary transformation

$$V_L^\dagger\, M'\, V_R = M\,, \qquad \text{with} \qquad M_{\alpha\beta} = m_\alpha\, \delta_{\alpha\beta}\,, \tag{4.4}$$

where V_L and V_R are two appropriate $N\times N$ unitary matrices ($V_L^\dagger = V_L^{-1}$ and $V_R^\dagger = V_R^{-1}$) and m_α are real and positive (see Ref. [235]).

In order to understand the fact that the biunitary diagonalization in eqn (4.4) is possible, let us first count the number of independent parameters. An arbitrary complex $N\times N$ matrix M' has $2N^2$ independent elements. Since each of the unitary matrices V_L and V_R has N^2 independent elements and M has N independent elements, the number of available parameters for the diagonalization is more than sufficient.

For the proof that the diagonalization in eqn (4.4) is possible, let us consider the product $M'M'^\dagger$, which is a Hermitian matrix which can be diagonalized through the unitary transformation ($V_L^\dagger = V_L^{-1}$)

$$V_L^\dagger\, M'\, M'^\dagger\, V_L = M^2\,, \qquad \text{where} \qquad M^2_{\alpha\beta} = m_\alpha^2\, \delta_{\alpha\beta}\,, \tag{4.5}$$

with real and positive eigenvalues

$$\begin{aligned} m_\alpha^2 &= \sum_\beta (V_L^\dagger\, M')_{\alpha\beta}\, (M'^\dagger\, V_L)_{\beta\alpha} = \sum_\beta (V_L^\dagger\, M')_{\alpha\beta}\, (V_L^\dagger\, M')^\dagger_{\beta\alpha} \\ &= \sum_\beta (V_L^\dagger\, M')_{\alpha\beta}\, (V_L^\dagger\, M')^*_{\alpha\beta} = \sum_\beta \left|(V_L^\dagger\, M')_{\alpha\beta}\right|^2 . \end{aligned} \tag{4.6}$$

Let us now write the matrix M' as

$$M' = V_L\, M\, V_R^\dagger\,, \qquad \text{with} \qquad M_{\alpha\beta} = \sqrt{m_\alpha^2}\, \delta_{\alpha\beta}\,, \tag{4.7}$$

and an appropriate matrix V_R. This matrix is given by

$$V_R = (M')^{-1}\, V_L\, M\,. \tag{4.8}$$

From eqns (4.5) and (4.8) one can see that, since $M^\dagger = M$, the matrix V_R is unitary:

$$V_R^\dagger V_R = M V_L^\dagger (M'^\dagger)^{-1} (M')^{-1} V_L M = M V_L^\dagger (M' M'^\dagger)^{-1} V_L M = M M^{-2} M = \mathbf{1}\,, \tag{4.9}$$

$$V_R V_R^\dagger = (M')^{-1} V_L M^2 V_L^\dagger (M'^\dagger)^{-1} = (M')^{-1} M' M'^\dagger (M'^\dagger)^{-1} = \mathbf{1}\,. \tag{4.10}$$

Using the unitarity of V_L and V_R, the biunitary diagonalization in eqn (4.4) can be realized by eqn (4.7), with $m_\alpha = \sqrt{m_\alpha^2}$.

4.2 Physical parameters in the mixing matrix

In general, a unitary $N \times N$ matrix depends on N^2 independent real parameters. These parameters can be divided into

$$\frac{N\,(N-1)}{2} \qquad \text{mixing angles}\,, \tag{4.11}$$

and

$$\frac{N\,(N+1)}{2} \qquad \text{phases}\,. \tag{4.12}$$

Hence, the quark mixing matrix with $N = 3$ can be written in terms of three mixing angles and six phases. However, not all the phases are physical observables, because the only physical effect of the quark mixing matrix occurs through its presence in the weak charged current in eqn (3.176) of quarks

$$j^\mu_{W,Q} = 2\,\overline{q^U_L}\,\gamma^\mu\,V\,q^D_L \tag{4.13}$$

(the quark mixing matrix has no effect on the quark weak neutral current, because of the GIM mechanism, see eqn (3.178)). Apart from the weak charged current, the Lagrangian is invariant under global phase transformations of the quark fields of the type

$$q^U_\alpha \to e^{i\psi^U_\alpha}\,q^U_\alpha\,, \qquad q^D_k \to e^{i\psi^D_k}\,q^D_k\,, \tag{4.14}$$

with $\alpha = u, c, t$ and $k = d, s, b$. Performing this transformation, the quark charged current in eqn (4.13) becomes

$$j^\mu_{W,Q} = 2 \sum_{\alpha=u,c,t} \sum_{k=d,s,b} \overline{q^U_{\alpha L}}\,\gamma^\mu\,e^{-i\psi^U_\alpha}\,V_{\alpha k}\,e^{i\psi^D_k}\,q^D_{kL}\,, \tag{4.15}$$

which can be written as

$$j^\mu_{W,Q} = 2\,\underbrace{e^{-i(\psi^U_c - \psi^D_s)}}_{1} \sum_{\alpha=u,c,t} \sum_{k=d,s,b} \overline{q^U_{\alpha L}}\,\gamma^\mu\,\underbrace{e^{-i(\psi^U_\alpha - \psi^U_c)}}_{N-1=2}\,V_{\alpha k}\,\underbrace{e^{i(\psi^D_k - \psi^D_s)}}_{N-1=2}\,q^D_{kL}\,, \tag{4.16}$$

where we have factorized an arbitrary phase $e^{-i(\psi^U_c - \psi^D_s)}$ and we have indicated the number of independent phases in each term. From this expression, it is clear that there are

$$1 + (N-1) + (N-1) = 2\,N - 1 = 5 \tag{4.17}$$

arbitrary phases of the quark fields that can be chosen to eliminate five of the six phases of the quark mixing matrix. The reason why $2N-1$ and not $2N$ phases of the mixing matrix can be eliminated is that a common rephasing of all the quark fields leaves the quark charged current invariant. Such an invariance is related, through Noether's theorem, to the conservation of baryon number.

Thus, the quark mixing matrix contains

$$\frac{N\,(N+1)}{2} - (2\,N-1) = \frac{(N-1)\,(N-2)}{2} = 1 \qquad \text{physical phase}\,, \tag{4.18}$$

and the total number of physical parameters (mixing angles and phases) in the quark mixing matrix is

$$\frac{N\,(N-1)}{2} + \frac{(N-1)\,(N-2)}{2} = (N-1)^2 = 4\,. \tag{4.19}$$

It is customary to express conveniently the quark mixing matrix only in terms of these four physical parameters, three mixing angles and one phase, which are measurable quantities.

Summarizing, for the quark mixing matrix we have

$$N=3 \qquad \Longrightarrow \qquad \begin{cases} \dfrac{N\,(N-1)}{2} = 3 & \text{mixing angles}\,, \\ \dfrac{(N-1)\,(N-2)}{2} = 1 & \text{physical phase}\,. \end{cases} \tag{4.20}$$

If the third generation is ignored,

$$N=2 \qquad \Longrightarrow \qquad \begin{cases} \dfrac{N\,(N-1)}{2} = 1 & \text{mixing angle}\,, \\ \dfrac{(N-1)\,(N-2)}{2} = 0 & \text{physical phase}\,. \end{cases} \tag{4.21}$$

Hence, in the case of two generations the mixing matrix depends only on one physical parameter, a mixing angle that is called the *Cabibbo angle* [292].

4.3 Parameterization of the mixing matrix

4.3.1 Two generations

A 2×2 unitary matrix can be written as

$$V = \begin{pmatrix} \cos\vartheta\, e^{i\omega_1} & \sin\vartheta\, e^{i(\omega_2+\eta)} \\ -\sin\vartheta\, e^{i(\omega_1-\eta)} & \cos\vartheta\, e^{i\omega_2} \end{pmatrix}, \tag{4.22}$$

in terms of one mixing angle ϑ and three phases ω_1, ω_2, η. The equivalent form

$$V = \begin{pmatrix} \omega_1 & 0 \\ 0 & \omega_2 \end{pmatrix} \begin{pmatrix} e^{i\eta} & 0 \\ 0 & 1 \end{pmatrix} \begin{pmatrix} \cos\vartheta & \sin\vartheta \\ -\sin\vartheta & \cos\vartheta \end{pmatrix} \begin{pmatrix} e^{-i\eta} & 0 \\ 0 & 1 \end{pmatrix} \tag{4.23}$$

shows explicitly that all three phases can be eliminated by rephasing the quark fields. Indeed, in the quark charged weak current in eqn (3.176),

$$j^{\mu}_{W,\mathrm{Q}} = 2\,\overline{q^U_L}\,\gamma^{\mu}\,V\,q^D_L\,, \tag{4.24}$$

for two generations we have

$$q_L^U = \begin{pmatrix} u_L \\ c_L \end{pmatrix} , \qquad q_L^D = \begin{pmatrix} d_L \\ s_L \end{pmatrix} . \tag{4.25}$$

The three phases ω_1, ω_2, η in V can be eliminated by the rephasing

$$u_L \to e^{i(\omega_1+\eta)} u_L , \qquad c_L \to e^{i\omega_2} c_L , \qquad d_L \to e^{i\eta} d_L . \tag{4.26}$$

Hence, it is convenient to write the mixing matrix V without the unphysical phases:

$$V = \begin{pmatrix} \cos\vartheta_{\rm C} & \sin\vartheta_{\rm C} \\ -\sin\vartheta_{\rm C} & \cos\vartheta_{\rm C} \end{pmatrix} , \tag{4.27}$$

where the mixing angle $\vartheta_{\rm C}$ is the Cabibbo angle [292]. The experimental value of the Cabibbo angle is [400]

$$\sin\vartheta_{\rm C} \simeq 0.2243 \pm 0.0016 . \tag{4.28}$$

Note, however, that the choice of parameterization of the two-generation quark mixing matrix is not unique: one could choose, for example,

$$V = \begin{pmatrix} \cos\vartheta'_{\rm C} & -\sin\vartheta'_{\rm C} \\ \sin\vartheta'_{\rm C} & \cos\vartheta'_{\rm C} \end{pmatrix} \quad \text{or} \quad V = \begin{pmatrix} \sin\vartheta''_{\rm C} & \cos\vartheta''_{\rm C} \\ -\cos\vartheta''_{\rm C} & \sin\vartheta''_{\rm C} \end{pmatrix} . \tag{4.29}$$

All the different parameterizations are related (for example, $\vartheta'_{\rm C} = -\vartheta_{\rm C}$, $\vartheta''_{\rm C} = \frac{\pi}{2} - \vartheta_{\rm C}$) and give the same physical results. Hence, one can choose any parameterization, but it is important to keep it consistently in a calculation and to perform the necessary transformation when the result is compared with other calculations.

The mixing angle $\vartheta_{\rm C}$ can be limited to the interval

$$0 \leq \vartheta_{\rm C} \leq \frac{\pi}{2} . \tag{4.30}$$

Let us consider, for example, a mixing matrix

$$\tilde{V} = \begin{pmatrix} \cos\tilde{\vartheta}_{\rm C} & \sin\tilde{\vartheta}_{\rm C} \\ -\sin\tilde{\vartheta}_{\rm C} & \cos\tilde{\vartheta}_{\rm C} \end{pmatrix} , \tag{4.31}$$

with

$$\frac{\pi}{2} \leq \tilde{\vartheta}_{\rm C} \leq \pi . \tag{4.32}$$

By taking

$$\vartheta_{\rm C} = \pi - \tilde{\vartheta}_{\rm C} , \tag{4.33}$$

with $\vartheta_{\rm C}$ in the interval in eqn (4.30), we can write $\tilde{V}$ as

$$\tilde{V} = \begin{pmatrix} -\cos\vartheta_{\rm C} & \sin\vartheta_{\rm C} \\ -\sin\vartheta_{\rm C} & -\cos\vartheta_{\rm C} \end{pmatrix} = \begin{pmatrix} -1 & 0 \\ 0 & 1 \end{pmatrix} \begin{pmatrix} \cos\vartheta_{\rm C} & \sin\vartheta_{\rm C} \\ -\sin\vartheta_{\rm C} & \cos\vartheta_{\rm C} \end{pmatrix} \begin{pmatrix} 1 & 0 \\ 0 & -1 \end{pmatrix} . \tag{4.34}$$

The two diagonal matrices on the right-hand side of eqn (4.34) can be eliminated by changing the signs of u_L and s_L. Therefore, the mixing matrix $\tilde{V}$ in eqn (4.31) with $\tilde{\vartheta}_{\rm C}$ in the interval in eqn (4.32) is physically equivalent to the mixing matrix V in eqn (4.27) with $\vartheta_{\rm C}$ in the interval in eqn (4.30).

4.3.2 Three generations

In the case of three generations, as one can imagine, the situation is more complicated because there are more possible parameterizations. A large degree of arbitrariness comes from the many possible choices of the phases of the elements of the mixing matrix, which correspond to only one physical phase (see eqn (4.20)). A convenient way to parameterize the quark mixing matrix is as follows (see [819, 932, 933, 537, 542]). Whenever possible we write formulas in a general form for N generations, keeping, however, the realistic case $N = 3$ in mind. Let us define the N^2 generators of unitary transformations as the matrices A^{ab} with elements

$$[A^{ab}]_{rs} = \delta_{ar}\,\delta_{bs}\,, \tag{4.35}$$

such that

$$A^{ab}\,A^{cd} = A^{ad}\,\delta_{bc}\,. \tag{4.36}$$

Considering the unitary and unimodular matrices

$$W^{ab}(\vartheta_{ab},\eta_{ab}) \equiv W^{ab}(\zeta_{ab}) = \exp\left(\zeta_{ab}\,A^{ab} - \zeta^*_{ab}\,A^{ba}\right) \quad \text{for} \quad a \neq b\,, \tag{4.37}$$

with

$$\zeta_{ab} = \vartheta_{ab}\,e^{i\eta_{ab}}\,, \tag{4.38}$$

it is easy to check that

$$\left(\zeta_{ab}\,A^{ab} - \zeta^*_{ab}\,A^{ba}\right)^{2k} = (-1)^k\,\vartheta_{ab}^{2k}\,\left(A^{aa} + A^{bb}\right) \quad \text{for} \quad k \geq 1\,, \tag{4.39}$$

$$\left(\zeta_{ab}\,A^{ab} - \zeta^*_{ab}\,A^{ba}\right)^{2k+1} = (-1)^k\,\vartheta_{ab}^{2k+1}\,\left(e^{i\eta_{ab}}\,A^{ab} - e^{-i\eta_{ab}}\,A^{ba}\right) \quad \text{for} \quad k \geq 0\,. \tag{4.40}$$

It then follows that

$$\begin{aligned}
W^{ab}(\vartheta_{ab},\eta_{ab}) &= \sum_{k=0}^{\infty} \frac{\left(\zeta_{ab}\,A^{ab} - \zeta^*_{ab}\,A^{ba}\right)^k}{k!} \\
&= \sum_{k=0}^{\infty} \frac{\left(\zeta_{ab}\,A^{ab} - \zeta^*_{ab}\,A^{ba}\right)^{2k}}{(2k)!} + \sum_{k=0}^{\infty} \frac{\left(\zeta_{ab}\,A^{ab} - \zeta^*_{ab}\,A^{ba}\right)^{2k+1}}{(2k+1)!} \\
&= \mathbf{1} - \left(A^{aa} + A^{bb}\right) + \left(A^{aa} + A^{bb}\right) \sum_{k=0}^{\infty} (-1)^k\,\frac{\vartheta_{ab}^{2k}}{(2k)!} \\
&\quad + \left(e^{i\eta_{ab}}\,A^{ab} - e^{-i\eta_{ab}}\,A^{ba}\right) \sum_{k=0}^{\infty} (-1)^k\,\frac{\vartheta_{ab}^{2k+1}}{(2k+1)!} \\
&= \mathbf{1} + (\cos\vartheta_{ab} - 1)\left(A^{aa} + A^{bb}\right) + \sin\vartheta_{ab}\left(e^{i\eta_{ab}}\,A^{ab} - e^{-i\eta_{ab}}\,A^{ba}\right).
\end{aligned} \tag{4.41}$$

Hence, the elements of $W^{ab}(\vartheta_{ab},\eta_{ab})$ are

$$\left[W^{ab}(\vartheta_{ab},\eta_{ab})\right]_{rs} = \delta_{rs} + (c_{ab} - 1)\,(\delta_{ra}\,\delta_{sa} + \delta_{rb}\,\delta_{sb})$$

$$+ s_{ab} \left(e^{i\eta_{ab}} \, \delta_{ra} \, \delta_{sb} - e^{-i\eta_{ab}} \, \delta_{rb} \, \delta_{sa} \right) , \tag{4.42}$$

with $c_{ab} \equiv \cos\vartheta_{ab}$ and $s_{ab} \equiv \sin\vartheta_{ab}$. For example, for $N = 3$, $a = 1$ and $b = 2$, we have

$$W^{12}(\vartheta_{12}, \eta_{12}) = \begin{pmatrix} \cos\vartheta_{12} & \sin\vartheta_{12} \, e^{i\eta_{12}} & 0 \\ -\sin\vartheta_{12} \, e^{-i\eta_{12}} & \cos\vartheta_{12} & 0 \\ 0 & 0 & 1 \end{pmatrix} . \tag{4.43}$$

One can see that the unitary and unimodular matrix $W^{ab}(\vartheta_{ab}, \eta_{ab})$ represents a complex rotation in the a-b plane, which reduces to a real rotation by the angle ϑ_{ab} if $\eta_{ab} = 0$.

Also the matrices A^{ab} with $a = b$ are useful, because they generate diagonal unitary matrices: an arbitrary diagonal unitary matrix can be written as

$$D(\boldsymbol{\omega}) = \mathrm{diag}\left(e^{i\omega_1}, \ldots, e^{i\omega_N}\right) = \exp\left(i \sum_{a=1}^{N} \omega_a \, A^{aa} \right) , \tag{4.44}$$

where $\boldsymbol{\omega}$ is the set of N phases $\boldsymbol{\omega} = (\omega_1, \ldots, \omega_N)$.

We parameterize an arbitrary $N \times N$ unitary matrix V as the product

$$V = D(\boldsymbol{\omega}) \left[\prod_{a<b} W^{ab}(\vartheta_{ab}, \eta_{ab}) \right] \qquad (a, b = 1, \ldots, N) , \tag{4.45}$$

through the N^2 generators in eqn (4.35). The angles ϑ_{ab} can be limited in the range

$$0 \leq \vartheta_{ab} \leq \frac{\pi}{2} , \tag{4.46}$$

whereas the phases ω_k and η_{ab} can have any value between 0 and 2π (excluded),

$$0 \leq \omega_k < 2\pi , \qquad 0 \leq \eta_{ab} < 2\pi . \tag{4.47}$$

The order of the product of the matrices W^{ab} in eqn (4.45) can be chosen in an arbitrary way. Different choices of order give different parameterizations. In agreement with eqns (4.11) and (4.12), there are $N(N-1)/2$ mixing angles ϑ_{ab} and $N(N+1)/2$ phases, divided into $N(N-1)/2$ phases η_{ab} and N phases ω_k.

The expression in eqn (4.45) for V can be written as

$$V = D(\boldsymbol{\omega} - \boldsymbol{\varphi}) \left[\prod_{a<b} D(\boldsymbol{\varphi}) W^{ab}(\vartheta_{ab}, \eta_{ab}) D^\dagger(\boldsymbol{\varphi}) \right] D(\boldsymbol{\varphi}) , \tag{4.48}$$

with a set of arbitrary phases $\boldsymbol{\varphi} = (\varphi_1, \ldots, \varphi_N)$. The matrices $D(\boldsymbol{\varphi})$ and $W^{ab}(\vartheta_{ab}, \eta_{ab})$ satisfy the useful identity

$$D(\boldsymbol{\varphi}) \, W^{ab}(\vartheta_{ab}, \eta_{ab}) D^\dagger(\boldsymbol{\varphi}) = W^{ab}(\vartheta_{ab}, \eta_{ab} + \varphi_a - \varphi_b) , \tag{4.49}$$

for any choice of the phases φ. Indeed, we have

$$\left[D(\boldsymbol{\varphi}) \, W^{ab}(\vartheta_{ab}, \eta_{ab}) D^\dagger(\boldsymbol{\varphi}) \right]_{rs} = \sum_{t,u} e^{i\varphi_r} \delta_{rt} \left[\delta_{tu} + (c_{ab} - 1) \left(\delta_{ta}\delta_{ua} + \delta_{tb}\delta_{ub} \right) \right.$$

$$
\begin{aligned}
&+s_{ab}\left(e^{i\eta_{ab}}\delta_{ta}\delta_{ub}-e^{-i\eta_{ab}}\delta_{tb}\delta_{ua}\right)\big]\,e^{-i\varphi_s}\delta_{us}\\
&=\delta_{rs}+(c_{ab}-1)\left(\delta_{ra}\delta_{sa}+\delta_{rb}\delta_{sb}\right)\\
&\quad+s_{ab}\left(e^{i(\eta_{ab}+\varphi_a-\varphi_b)}\delta_{ra}\delta_{sb}-e^{-i(\eta_{ab}+\varphi_a-\varphi_b)}\delta_{rb}\delta_{sa}\right)\\
&=\left[W^{ab}(\vartheta_{ab},\eta_{ab}+\varphi_a-\varphi_b)\right]_{rs}\,. \qquad (4.50)
\end{aligned}
$$

Using the identity in eqn (4.49), the expression in eqn (4.45) for the mixing matrix V can also be written with the diagonal phase matrix $D(\boldsymbol{\omega})$ on the right of the square brackets: choosing $\boldsymbol{\varphi}=\boldsymbol{\omega}$ we have

$$
V=\left[\prod_{a<b}W^{ab}(\vartheta_{ab},\eta_{ab}+\omega_a-\omega_b)\right]D(\boldsymbol{\omega}) \qquad (a,b=1,2,3)\,. \tag{4.51}
$$

The identity in eqn (4.49) allows one to write the matrix $W^{ab}(\vartheta_{ab},\eta_{ab})$ as[17]

$$
W^{ab}(\vartheta_{ab},\eta_{ab})=D^a(\eta_{ab})\,R^{ab}\,D^{a\dagger}(\eta_{ab})=D^{b\dagger}(\eta_{ab})\,R^{ab}\,D^b(\eta_{ab})\,, \tag{4.52}
$$

with

$$
[D^a(\eta)]_{rs}=\delta_{rs}+\left(e^{i\eta}-1\right)\delta_{ra}\,\delta_{sa}\,, \tag{4.53}
$$

$$
[R^{ab}]_{rs}=[R^{ab}(\vartheta_{ab})]_{rs}=\delta_{rs}+(\cos\vartheta_{ab}-1)\left(\delta_{ra}\,\delta_{sa}+\delta_{rb}\,\delta_{sb}\right) \tag{4.54}
$$

$$
+\sin\vartheta_{ab}\left(\delta_{ra}\,\delta_{sb}-\delta_{rb}\,\delta_{sa}\right)\,. \tag{4.55}
$$

The unitary matrix $D^a(\eta)$, such that $D^{a\dagger}(\eta)=D^a(-\eta)$, is diagonal with all diagonal elements equal to unity except for $[D^a(\eta)]_{aa}=e^{i\eta}$. The real orthogonal matrix R^{ab} operates a rotation of an angle ϑ_{ab} in the a-b plane ($R^{ab^T}=R^{ab^{-1}}$). For example, we have

$$
R^{12}=\begin{pmatrix}\cos\vartheta_{12} & \sin\vartheta_{12} & 0\\ -\sin\vartheta_{12} & \cos\vartheta_{12} & 0\\ 0 & 0 & 1\end{pmatrix}\,, \qquad D^1(\eta_{12})=\begin{pmatrix}e^{i\eta_{12}} & 0 & 0\\ 0 & 1 & 0\\ 0 & 0 & 1\end{pmatrix}\,. \tag{4.56}
$$

Using the identity in eqn (4.49), we obtain, from eqn (4.48)

$$
V=D(\boldsymbol{\omega}-\boldsymbol{\varphi})\left[\prod_{a<b}W^{ab}(\vartheta_{ab},\varphi_a+\eta_{ab}-\varphi_b)\right]D(\boldsymbol{\varphi})\,. \tag{4.57}
$$

The set of arbitrary phases $\boldsymbol{\varphi}=(\varphi_1,\ldots,\varphi_N)$ can be chosen in order to extract $N-1$ of the $N(N-1)/2$ phases η_{ab} from the product $\prod_{a<b}W^{ab}$. Only $N-1$ phases η_{ab} can be extracted because there are only $N-1$ independent differences $\varphi_a-\varphi_b$.

In the case of quarks $N=3$ and the unitary matrix V is parameterized in terms of three mixing angles ϑ_{12}, ϑ_{13}, ϑ_{23} and six phases ω_1, ω_2, ω_3, η_{12}, η_{13}, η_{23}. The

[17] Choosing $\varphi_a=-\eta_{ab}$, $\varphi_b=0$ and $\varphi_c=0$ for $c\neq a,b$.

set of arbitrary phases $\boldsymbol{\varphi} = (\varphi_1, \varphi_2, \varphi_3)$ can be chosen in order to extract two of the three phases η_{ab} from the product $\prod_{a<b} W^{ab}$. Let us extract η_{12} and η_{23} with the choice

$$\varphi_1 - \varphi_2 = -\eta_{12}\,, \qquad \varphi_2 - \varphi_3 = -\eta_{23}\,. \tag{4.58}$$

Then, η_{13} cannot be extracted because $\varphi_1 - \varphi_3 = -\eta_{12} - \eta_{23}$ is no longer arbitrary.

Making the choice in eqn (4.58), we have

$$\boldsymbol{\varphi} = (\varphi_2 - \eta_{12}, \varphi_2, \varphi_2 + \eta_{23})\,, \tag{4.59}$$

and the mixing matrix in eqn (4.57) can be written as

$$V = D(\boldsymbol{\omega} - \boldsymbol{\varphi})\, R^{23}\, W^{13}\, R^{12}\, D(\boldsymbol{\varphi})\,, \tag{4.60}$$

with

$$D(\boldsymbol{\omega} - \boldsymbol{\varphi}) = e^{-i\varphi_2}\, \mathrm{diag}\Big(e^{i(\omega_1+\eta_{12})}\,,\, e^{i\omega_2}\,,\, e^{i(\omega_3-\eta_{23})}\Big)\,, \tag{4.61}$$

$$W^{13} = W^{13}(\vartheta_{13}, \eta_{13}) = D^1(\eta_{13}) R^{13} D^{1\dagger}(\eta_{13})\,, \tag{4.62}$$

$$D(\boldsymbol{\varphi}) = e^{i\varphi_2}\, \mathrm{diag}(e^{-i\eta_{12}}\,,\, 1\,,\, e^{i\eta_{23}})\,. \tag{4.63}$$

In eqn (4.60) we have made a definite choice of ordering for the product of matrices.

It is clear that the factor $e^{-i\varphi_2}$ in $D(\boldsymbol{\omega} - \boldsymbol{\varphi})$ cancels with the factor $e^{i\varphi_2}$ in $D(\boldsymbol{\varphi})$, so that the value of φ_2 is irrelevant and can be taken for simplicity to be zero. Factorizing $e^{i\omega_2}$ out of $D(\boldsymbol{\omega} - \boldsymbol{\varphi})$, the unitary matrix in eqn (4.60) can be written as

$$V = e^{i\omega_2}\, D^L\, R^{23}\, W^{13}\, R^{12}\, D^R\,. \tag{4.64}$$

with

$$D^L \equiv \mathrm{diag}\Big(e^{i(\omega_1-\omega_2+\eta_{12})}\,,\, 1\,,\, e^{i(\omega_3-\omega_2-\eta_{23})}\Big)\,, \tag{4.65}$$

$$D^R \equiv \mathrm{diag}(e^{-i\eta_{12}}\,,\, 1\,,\, e^{i\eta_{23}})\,. \tag{4.66}$$

So we see that five phases have been factorized out of the matrix product $R^{23}W^{13}R^{12}$ in eqn (4.60): an overall phase ω_2, the two phases $\omega_1 - \omega_2 + \eta_{12}$ and $\omega_3 - \omega_2 - \eta_{23}$ on the left and the two phases $-\eta_{12}$ and η_{23} on the right.

Let us now insert the unitary matrix in eqn (4.64) in the expression in eqn (4.16) of the quark charged current. Defining

$$D^U \equiv \mathrm{diag}\Big(e^{i(\psi^U_u-\psi^U_c)}\,,\, 1\,,\, e^{i(\psi^U_t-\psi^U_c)}\Big)\,, \tag{4.67}$$

$$D^D \equiv \mathrm{diag}\Big(e^{i(\psi^D_d-\psi^D_s)}\,,\, 1\,,\, e^{i(\psi^D_b-\psi^D_s)}\Big)\,, \tag{4.68}$$

we can write the quark charged current in the matrix form

$$j^{\mu}_{W,Q} = 2\, e^{-i(\psi^U_c-\psi^D_s)}\, e^{i\omega_2}\, \overline{q^U_L}\, \gamma^\mu\, D^{U\dagger}\, D^L\, R^{23}\, W^{13}\, R^{12}\, D^R\, D^D\, q^D_L\,. \tag{4.69}$$

Now it is clear that the products $e^{-i(\psi^U_c-\psi^D_s)}e^{i\omega_2}$, $D^{U\dagger}D^L$ and $D^R D^D$ can be eliminated by the following choice of the arbitrary phases of the quark fields:

$$\psi^U_u = \psi^B + \omega_1 + \eta_{12}\,, \tag{4.70}$$

$$\psi_c^U = \psi^B + \omega_2 \,, \tag{4.71}$$

$$\psi_t^U = \psi^B + \omega_3 - \eta_{23} \,, \tag{4.72}$$

$$\psi_d^D = \psi^B + \eta_{12} \,, \tag{4.73}$$

$$\psi_s^D = \psi^B \,, \tag{4.74}$$

$$\psi_b^D = \psi^B - \eta_{23} \,, \tag{4.75}$$

with any value of the free phase ψ^B. The fact that a free common phase of the quark fields remains arbitrary is of great importance, because the invariance of the Lagrangian under a common phase rotation of the quark fields implies, through Noether's theorem, the conservation of the baryon number.

In conclusion, with the appropriate choice of the arbitrary phases of the quark fields, the quark charged current can be written as

$$j^\mu_{W,Q} = 2\,\overline{q_L^U}\,\gamma^\mu\,V\,q_L^D \,, \tag{4.76}$$

with the quark mixing matrix

$$V = R^{23}\,W^{13}\,R^{12} \,, \tag{4.77}$$

depending only on the three mixing angles ϑ_{12}, ϑ_{13}, ϑ_{23} and one physical phase η_{13}. Putting

$$\delta_{13} = -\eta_{13} \,, \tag{4.78}$$

the mixing matrix can be written explicitly as [312]

$$V = \begin{pmatrix} c_{12}c_{13} & s_{12}c_{13} & s_{13}e^{-i\delta_{13}} \\ -s_{12}c_{23} - c_{12}s_{23}s_{13}e^{i\delta_{13}} & c_{12}c_{23} - s_{12}s_{23}s_{13}e^{i\delta_{13}} & s_{23}c_{13} \\ s_{12}s_{23} - c_{12}c_{23}s_{13}e^{i\delta_{13}} & -c_{12}s_{23} - s_{12}c_{23}s_{13}e^{i\delta_{13}} & c_{23}c_{13} \end{pmatrix} , \tag{4.79}$$

with $c_{ab} \equiv \cos\vartheta_{ab}$ and $s_{ab} \equiv \sin\vartheta_{ab}$. This is the parameterization advocated by the Particle Data Group [400].

The experimental values of the sines of the mixing angles are [400]

$$s_{12} = 0.2243 \pm 0.0016 \,, \qquad s_{23} = 0.0413 \pm 0.0015 \,, \qquad s_{13} = 0.0037 \pm 0.0005 \,. \tag{4.80}$$

Since ϑ_{23} and ϑ_{13} are very small, the value of the mixing angle ϑ_{12} practically coincides with the value of the Cabibbo angle (see eqn (4.28)) and the two generation approximation in eqn (4.27) for the mixing matrix is a fairly good one for the study of processes that do not involve the b and t quarks. The experimental value of the phase δ_{13} is [400]

$$\delta_{13} = 1.05 \pm 0.24 = 60^\circ \pm 14^\circ \,. \tag{4.81}$$

A useful approximate parameterization of the quark mixing matrix that reflects the hierarchy $s_{13} \ll s_{23} \ll s_{12} \ll 1$ has been proposed by Wolfenstein [1068]. This parameterization can be obtained from eqn (4.79) by setting

$$s_{12} = \lambda \,, \qquad s_{23} = A\lambda^2 \,, \qquad s_{13}\, e^{-i\delta_{13}} = A\,\lambda^3\,(\rho - i\eta) \,. \tag{4.82}$$

Expanding each entry of the quark mixing matrix in powers of λ and keeping only the leading terms, one obtains the Wolfenstein parameterization

$$V = \begin{pmatrix} 1-\lambda^2/2 & \lambda & A\lambda^3(\rho - i\eta) \\ -\lambda & 1-\lambda^2/2 & A\lambda^2 \\ A\lambda^3(1-\rho-i\eta) & -A\lambda^2 & 1 \end{pmatrix} + \mathcal{O}(\lambda^4) \,, \tag{4.83}$$

in which only V_{ub} is exact. In order to achieve a high level of accuracy, experimental data are currently fitted using a generalization of the Wolfenstein parameterization in which terms of order λ^6 are neglected [277] (see Ref. [184]):

$$V = \begin{pmatrix} 1-\frac{\lambda^2}{2}-\frac{\lambda^4}{8} & \lambda + \mathcal{O}(\lambda^7) & A\lambda^3(\rho - i\eta) \\ -\lambda + \frac{A^2\lambda^5}{2}[1-2(\rho+i\eta)] & 1-\frac{\lambda^2}{2}-\frac{\lambda^4}{8}(1+4A^2) & A\lambda^2 + \mathcal{O}(\lambda^8) \\ A\lambda^3(1-\bar{\rho}-i\bar{\eta}) & -A\lambda^2 + \frac{A\lambda^4}{2}[1-2(\rho+i\eta)] & 1-\frac{A^2\lambda^4}{2} \end{pmatrix} \,, \tag{4.84}$$

with

$$\bar{\rho} = \rho\left(1 - \frac{\lambda^2}{2}\right) \,, \qquad \bar{\eta} = \eta\left(1 - \frac{\lambda^2}{2}\right) \,. \tag{4.85}$$

4.4 Degenerate masses

If two up-like quarks or two down-like quarks are degenerate in mass, the structure of the mixing matrix can be simplified. This case is obviously not realistic for quarks, because we know that the up-like quarks and down-like quarks have rather different masses (see Table 3.1). Yet, it may be a good approximation for experiments which are not sensitive to the quark mass differences (for example, in very high-energy experiments the d and s quarks could be considered as being approximately degenerate). Furthermore, this case is interesting for applications to the lepton sector, where the neutrino masses could be almost degenerate (see section 13.3.4).

Let us consider for example the case of degenerate d and s quarks. This means that the d and s quarks are not distinguishable, because all their properties are equal. Then, all observables are invariant under an arbitrary unitary rotation in the d-s field space,

$$q^D \to U^{12}\, q^D \,, \tag{4.86}$$

with unitary U^{12}, such that

$$U^{12} = \begin{pmatrix} U^{12}_{11} & U^{12}_{12} & 0 \\ U^{12}_{21} & U^{12}_{22} & 0 \\ 0 & 0 & 1 \end{pmatrix} \,. \tag{4.87}$$

Writing the mixing matrix in the form in eqn (4.57) it is convenient to extract the phases η_{13} and η_{23} choosing

$$\varphi_1 - \varphi_3 = -\eta_{13}\,, \qquad \varphi_2 - \varphi_3 = -\eta_{23}\,, \tag{4.88}$$

which leads to the following form for the mixing matrix:

$$V = D^L\, R^{23}\, R^{13}\, W^{12}\, D^R\,. \tag{4.89}$$

with

$$D^L = \mathrm{diag}\Big(e^{i(\omega_1+\eta_{13})}\,,\, e^{i(\omega_2+\eta_{23})}\,,\, e^{i\omega_3}\Big)\,, \tag{4.90}$$

$$D^R = \mathrm{diag}\big(e^{-i\eta_{13}}\,,\, e^{-i\eta_{23}}\,,\, 1\big)\,. \tag{4.91}$$

When this expression for the mixing matrix is inserted in the charged current in eqn (4.16), the phases ω_1, ω_2, ω_3, η_{13}, η_{23} can be eliminated by choosing the arbitrary phases of the quark fields as

$$\psi_u^U = \psi^B + \omega_1 + \eta_{13}\,, \tag{4.92}$$

$$\psi_c^U = \psi^B + \omega_2 + \eta_{23}\,, \tag{4.93}$$

$$\psi_t^U = \psi^B + \omega_3\,, \tag{4.94}$$

$$\psi_d^D = \psi^B + \eta_{13}\,, \tag{4.95}$$

$$\psi_s^D = \psi^B + \eta_{23}\,, \tag{4.96}$$

$$\psi_b^D = \psi^B\,, \tag{4.97}$$

with any value of the free phase ψ^B. Hence, we obtain the following expression for the mixing matrix in which the unphysical phases associated with the arbitrary rephasing of the quark fields have been eliminated:

$$V = R^{23}\, R^{13}\, W^{12}\,. \tag{4.98}$$

However, if d and s are degenerate in mass, we can still perform the arbitrary unitary transformation in eqn (4.86). The matrix W^{12} can be eliminated from the expression in eqn (4.98) for V by choosing $U^{12} = (W^{12})^\dagger$. Therefore, the physical mixing matrix can be written as

$$V = R^{23}\, R^{13}\,, \tag{4.99}$$

in terms of only two physical mixing angles ϑ_{23} and ϑ_{13}, which represent, respectively, rotations in the s-b and d-b planes of distinguishable down-like quarks. It is important to realize that not only has the third mixing angle been eliminated, but also the phase. Since the physical mixing matrix is real, in this case CP is conserved, as discussed in section 4.6.

It is clear that the case of mass degeneracy of any two up-like or down-like quarks can be treated as the example of d-s mass degeneracy leading to a real physical mixing matrix parameterized in terms of only two mixing angles, which respects CP symmetry.

If all the up-like or all the down-like quarks are degenerate in mass, all the mixing can be rotated away. Hence, mixing is not observable in experiments which are not sensitive to the mass differences of up-like or down-like quarks (in other words, experiments which cannot distinguish different up-like or down-like quarks).

4.5 Mixing matrix with one vanishing element

Another case in which mixing can be simplified is that in which the mixing matrix has at least one vanishing element. This can be easily seen by noticing that given a parameterization of the mixing matrix, one can obtain other parameterizations by arbitrary permutations of the rows and columns, which preserve the unitarity relations

$$V V^{\dagger} = \mathbf{1} \quad \Longleftrightarrow \quad \sum_{k=1}^{3} V_{\alpha k} V_{\beta k}^{*} = \delta_{\alpha\beta} \qquad (\alpha, \beta = 1, 2, 3) \,, \tag{4.100}$$

$$V^{\dagger} V = \mathbf{1} \quad \Longleftrightarrow \quad \sum_{\alpha=1}^{3} V_{\alpha k}^{*} V_{\alpha j} = \delta_{kj} \qquad (k, j = 1, 2, 3) \,. \tag{4.101}$$

With an appropriate permutation of the rows and columns of the mixing matrix in the parameterization in eqn (4.79) one can move $s_{13} e^{-i\delta_{13}}$ in place of the zero element. In this parameterization $s_{13} = 0$ and the phase δ_{13} disappears. Hence, one obtains a real mixing matrix which respects CP symmetry, as discussed in section 4.6.

For example, let us consider the case of $V_{22} = 0$. Switching the first and second rows of the matrix in eqn (4.79) and then switching the second and third column, we obtain

$$V = \begin{pmatrix} -s_{12}c_{23} - c_{12}s_{23}s_{13}e^{i\delta_{13}} & s_{23}c_{13} & c_{12}c_{23} - s_{12}s_{23}s_{13}e^{i\delta_{13}} \\ c_{12}c_{13} & s_{13}e^{-i\delta_{13}} & s_{12}c_{13} \\ s_{12}s_{23} - c_{12}c_{23}s_{13}e^{i\delta_{13}} & c_{23}c_{13} & -c_{12}s_{23} - s_{12}c_{23}s_{13}e^{i\delta_{13}} \end{pmatrix} . \tag{4.102}$$

Of course, after these permutations the angle ϑ_{ab} no longer represents a rotation in the a-b plane and could be renamed in an appropriate way. If $V_{22} = 0$, we have $s_{13} = 0$, which implies the real mixing matrix

$$V|_{V_{22}=0} = \begin{pmatrix} -s_{12}c_{23} & s_{23} & c_{12}c_{23} \\ c_{12} & 0 & s_{12} \\ s_{12}s_{23} & c_{23} & -c_{12}s_{23} \end{pmatrix} . \tag{4.103}$$

Given a specific parameterization in terms of three mixing angles, one can see easily that the phase disappears if any of the three angles is zero. In the parameterization in eqn (4.79) it is evident that if any of the three mixing angle is zero then there is one element of the mixing matrix which vanishes: if $\vartheta_{12} = 0$ we have

$V_{12} = 0$, if $\vartheta_{23} = 0$ we have $V_{23} = 0$, and if $\vartheta_{13} = 0$ we have $V_{13} = 0$. According to the reasoning above, it follows that in any of these cases the phase can be eliminated. This can be performed even without permutations of rows and columns. For example, if $\vartheta_{12} = 0$ the expression of the mixing matrix in eqn (4.79) becomes

$$V|_{\vartheta_{12}=0} = \begin{pmatrix} c_{13} & 0 & s_{13}e^{-i\delta_{13}} \\ -s_{23}s_{13}e^{i\delta_{13}} & c_{23} & s_{23}c_{13} \\ -c_{23}s_{13}e^{i\delta_{13}} & -s_{23} & c_{23}c_{13} \end{pmatrix}, \qquad (4.104)$$

which can be written as

$$V|_{\vartheta_{12}=0} = \begin{pmatrix} e^{-i\delta_{13}} & 0 & 0 \\ 0 & 1 & 0 \\ 0 & 0 & 1 \end{pmatrix} \begin{pmatrix} c_{13} & 0 & s_{13} \\ -s_{23}s_{13} & c_{23} & s_{23}c_{13} \\ -c_{23}s_{13} & -s_{23} & c_{23}c_{13} \end{pmatrix} \begin{pmatrix} e^{i\delta_{13}} & 0 & 0 \\ 0 & 1 & 0 \\ 0 & 0 & 1 \end{pmatrix}. \qquad (4.105)$$

In this form it is clear that the phase δ_{13} can be eliminated by rephasing the fields of the u and d quarks as $u \to e^{-i\delta_{13}}u$ and $d \to e^{-i\delta_{13}}d$.

In general, if one of the mixing angles is zero, one can take the phase which is associated with the sine of the vanishing mixing angle as the one which is not eliminated through the procedure described in section 4.3.2. In this way all the other phases are eliminated with the method described in section 4.3.2 and the surviving phase disappears because it is associated with the sine of the vanishing mixing angle.

The phase disappears also if a mixing angle has its maximal value of $\pi/2$. One can see it in the parameterization in eqn (4.79) by noting that $V_{11} = 0$ if $\vartheta_{12} = \pi/2$, $V_{12} = V_{23} = V_{33} = 0$ if $\vartheta_{13} = \pi/2$, and $V_{33} = 0$ if $\vartheta_{23} = \pi/2$.

In general, if the mixing angle ϑ_{ab} is equal to $\pi/2$ we have

$$\begin{aligned} D^{ab}(\eta)\, W^{ab}(\vartheta_{ab} = \pi/2, \eta_{ab}) &= W^{ab}(\vartheta_{ab} = \pi/2, \eta_{ab} + \eta) \\ &= W^{ab}(\vartheta_{ab} = \pi/2, \eta_{ab})\, D^{ab\dagger}(\eta)\,, \end{aligned} \qquad (4.106)$$

for an arbitrary phase η, where

$$[D^{ab}(\eta)]_{rs} = \delta_{rs} + \left(e^{i\eta} - 1\right) \delta_{ra}\, \delta_{sa} + \left(e^{-i\eta} - 1\right) \delta_{rb}\, \delta_{sb} \qquad (4.107)$$

is a diagonal unitary matrix with all diagonal entries equal to unity except $[D^{ab}(\eta)]_{aa} = e^{i\eta}$ and $[D^{ab}(\eta)]_{bb} = e^{-i\eta}$. Since $D^{ab}(-\eta) = D^{ab\dagger}(\eta)$, choosing $\eta = -\eta_{ab}$ one can write the matrix of complex rotation in the a-b plane as

$$W^{ab}(\vartheta_{ab} = \pi/2, \eta_{ab}) = D^{ab}(\eta_{ab})\, R^{ab}(\vartheta_{ab} = \pi/2) = R^{ab}(\vartheta_{ab} = \pi/2)\, D^{ab\dagger}(\eta_{ab})\,. \qquad (4.108)$$

One can parameterize the mixing matrix as a product of the type in eqn (4.65) with $W^{ab}(\vartheta_{ab} = \pi/2, \eta_{ab})$ on the extreme left or the extreme right. Using one of the relations in eqn (4.108) the phase η_{ab} can be moved to a diagonal matrix of type D^{ab} on the extreme left or the extreme right and eliminated by rephasing the appropriate quark fields. Since the other phases can be eliminated with the method described in section 4.3.2, the physical mixing matrix is real and respects the CP symmetry, as discussed in section 4.6.

4.6 CP violation

Experimentally, CP violation has been observed in the systems of neutral K [317] and B [123, 23] mesons. Experimental data are compatible with the hypothesis that CP violation is generated by the physical phase of the quark mixing matrix (see [400]). In order to understand how the presence of a nonzero physical phase in the quark mixing matrix implies the violation of CP symmetry in weak processes, let us derive how the quark charged current in eqn (4.76) behaves under a CP transformation. The arrays $\overline{q_L^U}$ and q_L^D that enter in the quark charged current in eqn (4.76) can be seen as transforming under CP as, by using eqns (2.399) and (2.400),

$$\overline{q_L^U} \xrightarrow{\text{CP}} -q_L^{U\,T}\, \mathcal{C}^{-1}\, \gamma^0\, D^\dagger(\vec{\xi}_U)\,, \tag{4.109}$$

$$q_L^D \xrightarrow{\text{CP}} D(\vec{\xi}_D)\, \gamma^0\, \mathcal{C}\, \overline{q_L^D}^{\,T}\,, \tag{4.110}$$

where $D(\vec{\xi}_U)$ and $D(\vec{\xi}_D)$ are diagonal matrices of the undetermined CP phases of the quark fields, with $\vec{\xi}_U = (\xi_u, \xi_c, \xi_t)$ and $\vec{\xi}_D = (\xi_d, \xi_s, \xi_b)$. Thus, the quark charged current transforms as

$$\begin{aligned}
j^\mu_{W,\mathrm{Q}} \xrightarrow{\text{CP}} & - 2\, q_L^{U\,T}\, \underbrace{\mathcal{C}^{-1}\, \overbrace{\gamma^0\, \gamma^\mu\, \gamma^0}^{\gamma^{\mu\dagger}}\, \mathcal{C}}_{-\gamma^{\mu *}}\, D^\dagger(\vec{\xi}_U)\, V\, D(\vec{\xi}_D)\, \overline{q_L^D}^{\,T} \\
&= 2\, q_L^{U\,T}\, \gamma^{\mu *}\, D^\dagger(\vec{\xi}_U)\, V\, D(\vec{\xi}_D)\, \overline{q_L^D}^{\,T} \\
&= -\, 2\, \overline{q_L^D}\, \gamma^{\mu\dagger}\, D(\vec{\xi}_D)\, V^T\, D^\dagger(\vec{\xi}_U)\, q_L^U \\
&= -\, 2\, \overline{q_L^D}\, D(\vec{\xi}_D)\, V^T\, D^\dagger(\vec{\xi}_U)\, \gamma_\mu\, q_L^U\,.
\end{aligned} \tag{4.111}$$

Note that in the last line we have lowered the Lorentz index by using the property in eqn (2.9). Under CP the W-boson field transforms as

$$W_\mu \xrightarrow{\text{CP}} e^{i\xi_W}\, W^{\mu\dagger}\,, \tag{4.112}$$

where ξ_W is an undetermined CP phase and the Lorentz index has been raised because the spatial components of W_μ change sign under space inversion. Putting together eqns (4.111) and (4.112), we have

$$j^\mu_{W,\mathrm{Q}}\, W_\mu \xrightarrow{\text{CP}} -2\, e^{i\xi_W}\, \overline{q_L^D}\, D(\vec{\xi}_D)\, V^T\, D^\dagger(\vec{\xi}_U)\, \gamma_\mu\, q_L^U\, W^{\mu\dagger}\,, \tag{4.113}$$

where the lowering and raising of the Lorentz index in $j^\mu_{W,\mathrm{Q}}$ and W_μ compensate each other, preserving the Lorentz invariance. The CP-transformed expression of $j^\mu_{W,\mathrm{Q}} W_\mu$ in eqn (4.113) has a structure similar to that of the Hermitian conjugate of $j^{\mu}_{W,Q} W_\mu$,

$$\left(j^{\mu}_{W,Q} W_\mu\right)^\dagger = 2\, \overline{q^D_L}\, V^\dagger \gamma^\mu q^U_L\, W^\dagger_\mu \,, \tag{4.114}$$

which is part of the CC quark Lagrangian in eqn (3.76) (the Lagrangian is a real scalar). Therefore, the Lagrangian is invariant under CP if

$$j^{\mu}_{W,Q} W_\mu \xrightarrow{\text{CP}} \left(j^{\mu}_{W,Q} W_\mu\right)^\dagger \,, \tag{4.115}$$

i.e. if

$$-\, e^{i\xi_W}\, D(\vec{\xi}_D)\, V^T\, D^\dagger(\vec{\xi}_U) = V^\dagger \,. \tag{4.116}$$

Choosing $\xi_W = \pi$, i.e.

$$W_\mu \xrightarrow{\text{CP}} -W^{\mu\dagger} \,, \tag{4.117}$$

and transposing eqn (4.115), we have the condition, with $D^\dagger = D^{-1}$,

$$D^{-1}(\vec{\xi}_U)\, V\, D(\vec{\xi}_D) = V^* \qquad \Longleftrightarrow \qquad e^{-i\xi_\alpha}\, V_{\alpha k}\, e^{i\xi_k} = V^*_{\alpha k} \,. \tag{4.118}$$

This condition is satisfied if the mixing matrix can be written in the form

$$V = D^{1/2}(\vec{\xi}_U)\, \mathcal{O}\, D^{-1/2}(\vec{\xi}_D) \qquad \Longleftrightarrow \qquad V_{\alpha k} = e^{i\xi_\alpha/2}\, \mathcal{O}_{\alpha k}\, e^{-i\xi_k/2} \,, \tag{4.119}$$

where $\mathcal{O}$ is a real orthogonal matrix ($\mathcal{O}^T = \mathcal{O}^{-1}$) and $\vec{\xi}_U = (\xi_u, \xi_c, \xi_t)$ and $\vec{\xi}_D = (\xi_d, \xi_s, \xi_b)$ are arbitrary phases. As shown in section 4.3, the physical phase in the mixing matrix V cannot be factorized into diagonal matrices lying on the sides of the mixing matrix. Hence, the condition in eqn (4.118) cannot be satisfied with any choice of the CP phases $\vec{\xi}_U$ and $\vec{\xi}_D$ if the physical phase in the mixing matrix is different from zero[18]. In other words, CP symmetry requires that the physical quark mixing matrix V is real:

$$V = V^* \qquad \Longleftrightarrow \qquad \text{CP symmetry} \,. \tag{4.120}$$

If the physical phase of the mixing matrix is nonzero, the symmetry of the Lagrangian under CP transformation is broken by the quark charged-current weak interaction Lagrangian. On the other hand, if the physical phase of the mixing matrix is equal to zero, CP is conserved and all the quark fields have the same CP phase, $\xi_u = \xi_c = \xi_t = \xi_d = \xi_s = \xi_b = \xi_q$, which is conventionally set to zero: $\xi_q = 0$.

[18] One can easily convince oneself of this fact using a specific parameterization of V (for example that given in eqn (4.79)) and writing explicitly the matrices in eqn (4.118). It is interesting to note that the proof presented here is rarely given in books (one exception is section 26.2 of Ref. [821]). Several books do not consider the CP phases of the quarks or assume that they are all equal. In some books it is even argued that if V is complex the interaction Lagrangian is not real, leading to violation of T and CP. This statement is clearly meaningless because the Lagrangian is Hermitian by construction, each non-Hermitian term being accompanied by its Hermitian conjugate.

It is interesting to ask what are the conditions on the mass matrices of up-type and down-type quarks for CP invariance. From eqn (3.162), the quark Lagrangian mass term is

$$\mathscr{L}_{q,\text{mass}} = -\overline{q_L'^U}\, M'^U\, q_R'^U - \overline{q_R'^U}\, M'^{U\dagger}\, q_L'^U - \overline{q_L'^D}\, M'^D\, q_R'^D - \overline{q_R'^D}\, M'^{D\dagger}\, q_L'^D\,, \quad (4.121)$$

where $M'^U = Y'^U v/\sqrt{2}$ and $M'^D = Y'^D v/\sqrt{2}$. The other parts of the Lagrangian are invariant under the CP transformation

$$q_L'^U \xrightarrow{\text{CP}} W_L\, \gamma^0\, \mathcal{C}\, \overline{q_L'^U}^T\,, \qquad q_R'^U \xrightarrow{\text{CP}} W_R^U\, \gamma^0\, \mathcal{C}\, \overline{q_R'^U}^T\,, \quad (4.122)$$

$$q_L'^D \xrightarrow{\text{CP}} W_L\, \gamma^0\, \mathcal{C}\, \overline{q_L'^D}^T\,, \qquad q_R'^D \xrightarrow{\text{CP}} W_R^D\, \gamma^0\, \mathcal{C}\, \overline{q_R'^D}^T\,, \quad (4.123)$$

where W_L, W_R^U, and W_R^D are unitary matrices that mix the quark fields. The mixing of the left-handed up and down quarks is constrained to be equal by the invariance of the charged-current weak interaction Lagrangian, whereas the right-handed up and down quarks can mix independently. The transformation of the quark mass term in eqn (4.121) is

$$\begin{aligned}\mathscr{L}_{q,\text{mass}} \xrightarrow{\text{CP}} & - \overline{q_R'^U}\, {W_R^U}^T\, {M'^U}^T\, {W_L}^*\, q_L'^U - \overline{q_L'^U}\, {W_L}^T\, {M'^U}^*\, {W_R^U}^*\, q_R'^U \\ & - \overline{q_R'^D}\, {W_R^D}^T\, {M'^D}^T\, {W_L}^*\, q_L'^D - \overline{q_L'^D}\, {W_L}^T\, {M'^D}^*\, {W_R^D}^*\, q_R'^D\,. \end{aligned} \quad (4.124)$$

Comparing with eqn (4.121), one can see that the Lagrangian mass term of quarks is invariant under the CP transformation in eqns (4.122) and (4.123) if there are unitary matrices W_L, W_R^U and W_R^D such that

$$W_L^\dagger\, M'^U\, W_R^U = {M'^U}^*\,, \quad (4.125)$$

$$W_L^\dagger\, M'^D\, W_R^D = {M'^D}^*\,. \quad (4.126)$$

Since these conditions can also be written as

$$W_L^*\, M'^U\, {W_R^U}^T = {M'^U}^*\,, \quad (4.127)$$

$$W_L^*\, M'^D\, {W_R^D}^T = {M'^D}^*\,, \quad (4.128)$$

the matrices W_L, W_R^U, and W_R^D must also be symmetric.

One of the two equations in eqns (4.125) and (4.126) can always be satisfied. However, if W_L is constrained by one of the two equations, in general, the other cannot be satisfied. For example, let us choose W_L and W_R^D in order to satisfy eqn (4.126). This can be done by using the diagonalization equation (3.165), with $M'^D = Y'^D v/\sqrt{2}$,

$${V_L^D}^\dagger\, M'^D V_R^D = M^D\,, \quad (4.129)$$

where $M^D_{\alpha\beta} = m^D_\alpha\, \delta^{\alpha\beta}$, with real and positive m^D_α. Using eqn (4.129), the condition in eqn (4.126) can be written as

$${V_L^D}^T\, W_L^\dagger\, V_L^D\, M^D\, {V_R^D}^\dagger\, W_R^D\, {V_R^D}^* = M^D\,. \quad (4.130)$$

If M^D does not have degenerate eigenvalues, the condition in eqn (4.130) is satisfied only if there is a diagonal phase matrix D^D such that

$${V_L^D}^T\, W_L^\dagger\, V_L^D = {D^D}^\dagger\,, \quad (4.131)$$

$$V_R^{D\dagger}\, W_R^D\, V_R^{D*} = D^D\,. \tag{4.132}$$

It then follows that

$$W_L = V_L^D\, D^D\, V_L^{DT}\,, \tag{4.133}$$

$$W_R^D = V_R^D\, D^D\, V_R^{DT}\,, \tag{4.134}$$

which are clearly unitary and symmetric. Let us now insert $W_L^\dagger$ in eqn (4.133) in the condition in eqn (4.125). From the diagonalization equation in eqn (3.166), with $M'^U = Y'^U v/\sqrt{2}$, we have

$$V_L^{U\dagger}\, M'^U V_R^U = M^U\,, \tag{4.135}$$

where $M^U_{\alpha\beta} = m^U_\alpha\, \delta^{\alpha\beta}$, with real and positive m^U_α, and using the definition of the quark mixing matrix in eqn (3.175),

$$V = V_L^{U\dagger}\, V_L^D\,, \tag{4.136}$$

we obtain

$$V^*\, D^{D\dagger}\, V^\dagger M^U\, V_R^{U\dagger}\, W_R^U\, V_R^{U*} = M^U\,. \tag{4.137}$$

If M^U does not have degenerate eigenvalues, the condition in eqn (4.137) is satisfied only if there is a diagonal phase matrix D^U such that

$$V^*\, D^{D\dagger}\, V^\dagger = D^{U\dagger}\,, \tag{4.138}$$

$$V_R^{U\dagger}\, W_R^U\, V_R^{U*} = D^U\,. \tag{4.139}$$

Equation (4.139) can always be satisfied by taking

$$W_R^U = V_R^U\, D^U\, V_R^{UT}\,, \tag{4.140}$$

which is obviously unitary and symmetric. On the other hand, in general the condition in eqn (4.138) is not satisfied, because V is fixed. CP becomes a symmetry only if the mixing matrix V is such that there are appropriate diagonal phase matrices D^U and D^D which satisfy the condition in eqn (4.138). We can write eqn (4.138) as

$$D^{U\dagger}\, V\, D^D = V^*\,, \tag{4.141}$$

which coincides with the condition in eqn (4.118), showing the consistency of the two methods that we have used in order to find the conditions for CP invariance (i.e. invariance of the charged-current weak interaction Lagrangian assuming diagonal mass terms and invariance of the mass terms assuming a diagonal charged-current weak interaction Lagrangian).

Let us now consider again the conditions in eqns (4.125) and (4.126) in order to find the constraints on the mass matrices M'^U and M'^D for CP invariance. A powerful, but mathematically complicated method, to find the general condition on the mass matrices for CP invariance is discussed in section 4.9. Now we consider a simple method which allows one to check CP invariance in some simple cases.

Eliminating W_L in eqns (4.125) and (4.126), we obtain

$$W_R^{D\dagger} \left[M'^{D\dagger} M'^{U} \right] W_R^U = \left[M'^{D\dagger} M'^{U} \right]^* . \tag{4.142}$$

This condition is satisfied if $M'^{D\dagger} M'^{U}$ can be written as

$$M'^{D\dagger} M'^{U} = P^\dagger R Q \,, \quad \text{with} \quad R = R^* \,, \quad P^\dagger = P^{-1} \,, \quad Q^\dagger = Q^{-1} \,, \tag{4.143}$$

i.e. real R and unitary P and Q. In this case, eqn (4.142) is solved by

$$W_R^D = P^\dagger P^* \,, \quad W_R^U = Q^\dagger Q^* \,, \tag{4.144}$$

which are unitary and symmetric.

Particularly simple cases in which the condition in eqn (4.143) is satisfied and CP is a symmetry are:

$M'^U = 0$ or $M'^D = 0$. This case is not realistic for quarks, which have nonzero masses. However it illustrates what happens in the Standard Models where neutrinos are massless. In their $\mathrm{SU}(2)_L$ doublets, neutrinos play the same role as up-like quarks, which have third component of weak isospin $I_3 = 1/2$, and charged leptons play the same role as down-like quarks. Since neutrino are massless, CP is a symmetry of the leptonic sector of the Standard Model for any value of the charged lepton mass matrix. This, however, is a trivial case because there is no mixing in the lepton sector, as we have seen in section 3.5.

Real $M'^{D\dagger} M'^{U}$. In this case W_R^U and W_R^D are equal to the identity matrix, up to an arbitrary phase. A particular simple case is

$$M'^U = M'^{U*} \quad \text{and} \quad M'^D = M'^{D*} \,. \tag{4.145}$$

One can easily check that this condition is consistent with the condition in eqn (4.120) of reality of the physical quark mixing matrix V. Indeed, if both M'^U and M'^D are real, they can be diagonalized as in eqns (3.165) and (3.166) with orthogonal matrices $V_L^U = \mathcal{O}_L^U$, $V_R^U = \mathcal{O}_R^U$, $V_L^D = \mathcal{O}_L^D$, $V_R^D = \mathcal{O}_R^D$, and the physical mixing matrix $V = V_L^{U\dagger} V_L^D = \mathcal{O}_L^{U\dagger} \mathcal{O}_L^D = \mathcal{O}$ is real.

4.7 Rephasing invariants

Since physical observables do not depend on the parameterization of the mixing matrix, it is possible to work only with quantities that are invariant under a reparameterization of the mixing matrix, often called *rephasing-invariants* (see, for example, the review in Ref. [394]) because they are invariant under the phase transformations in eqn (4.14) of the quark fields, which induces the phase transformations

$$V_{\alpha k} \to e^{-i\psi_\alpha^U} \, V_{\alpha k} \, e^{i\psi_k^D} \,. \tag{4.146}$$

of the elements of the mixing matrix.

Any observable which involves quarks is a polynomial function of products of elements of V and V^*, such that each product contains an equal number of elements of V and V^*, because an observable always comes from the squared modulus of an amplitude. Since the interacting quarks in each vertex are the same in the amplitude and its complex conjugate, in each product the set of indices $\{\alpha\}$ of the $V_{\alpha k}$'s is equal to that of the $V^*_{\alpha k}$'s, and the same property holds for the set of indices $\{k\}$. Hence, any observable is invariant under the rephasing in eqn (4.146).

Since we are interested in measurable quantities, it is useful to discuss which are the simplest products of elements of V and V^* which can represent the effects of quark mixing on the observables.

The simplest rephasing-invariant products of elements of V and V^* are the squared moduli of the matrix elements,

$$|V_{\alpha k}|^2 = V_{\alpha k} V^*_{\alpha k} \,. \tag{4.147}$$

The next simplest rephasing invariants are the quartic products

$${}^{\alpha k}\square_{\beta j} = V_{\alpha k} V_{\beta j} V^*_{\alpha j} V^*_{\beta k} \,, \tag{4.148}$$

with $\alpha \neq \beta$ and $k \neq j$, which have been named *plaques* in Ref. [246], *quartets* in Ref. [269], and *boxes* in Ref. [1045]. The quartic products in eqn (4.148) satisfy the conjugation relations

$${}^{\alpha k}\square_{\beta j} = {}^{\beta j}\square_{\alpha k} = {}^{\beta k}\square^*_{\alpha j} = {}^{\alpha j}\square^*_{\beta k} \,. \tag{4.149}$$

All the rephasing-invariant products of higher order can be expressed in terms of squared moduli and rephasing-invariant quartic products. For example, from the definition in eqn (4.148) it is easy to verify that

$$V_{\alpha k} V_{\beta j} V_{\gamma l} V^*_{\alpha j} V^*_{\beta l} V^*_{\gamma k} = \frac{{}^{\alpha k}\square_{\beta j} \, {}^{\beta k}\square_{\gamma l}}{|V_{\beta k}|^2} = \frac{{}^{\gamma l}\square_{\alpha k} \, {}^{\alpha l}\square_{\beta j}}{|V_{\alpha l}|^2} = \frac{{}^{\alpha k}\square_{\gamma j} \, {}^{\gamma l}\square_{\beta j}}{|V_{\gamma j}|^2} \,. \tag{4.150}$$

Since the rephasing-invariant quartic products are complex if V is complex, their imaginary parts

$${}^{\alpha k}\Im_{\beta j} \equiv \Im\mathfrak{m}\, {}^{\alpha k}\square_{\beta j} \equiv \Im\mathfrak{m}\left[V_{\alpha k} V_{\beta j} V^*_{\alpha j} V^*_{\beta k}\right] \tag{4.151}$$

may be useful for quantifying explicitly the complexity of V and CP violation in a rephasing-invariant way. From the relations in eqn (4.149) it follows that the ${}^{\alpha k}\Im_{\beta j}$'s are antisymmetric in the indices α, β and in the indices k, j, whereas the real parts of the rephasing-invariant quartic products,

$${}^{\alpha k}\Re_{\beta j} \equiv \Re\mathfrak{e}\, {}^{\alpha k}\square_{\beta j} \equiv \Re\mathfrak{e}\left[V_{\alpha k} V_{\beta j} V^*_{\alpha j} V^*_{\beta k}\right] \tag{4.152}$$

are symmetric in the indices α, β and in the indices k, j.

It appears that there are several imaginary parts in eqn (4.151), but the unitarity of the 3×3 quark mixing matrix implies that they are all equal, up to a sign. To see this, let us start with the unitarity relation $VV^\dagger = \mathbf{1}$,

$$\sum_{k=d,s,b} V_{\alpha k} V_{\beta k}^* = \delta_{\alpha\beta} \qquad (\alpha, \beta = u, c, t)\,, \tag{4.153}$$

and multiply it by $V_{\alpha j}^* V_{\beta j}$:

$$|V_{\alpha j}|^2 |V_{\beta j}|^2 + \sum_{k \neq j} V_{\alpha k} V_{\beta j} V_{\beta k}^* V_{\alpha j}^* = |V_{\alpha j}|^2 \, \delta_{\alpha\beta} \qquad (\alpha, \beta = u, c, t;\; j = d, s, b)\,. \tag{4.154}$$

Taking the imaginary part of this relation, we obtain

$$\sum_{k \neq j} \mathfrak{Im}\left[V_{\alpha k} \, V_{\beta j} \, V_{\beta k}^* \, V_{\alpha j}^*\right] = 0 \qquad (\alpha \neq \beta;\; \alpha, \beta = u, c, t;\; j = d, s, b)\,. \tag{4.155}$$

In a similar way, starting from the unitarity relation $V^\dagger V = \mathbf{1}$, one obtains the relations

$$\sum_{\alpha \neq \beta} \mathfrak{Im}\left[V_{\alpha k} \, V_{\beta j} \, V_{\beta k}^* \, V_{\alpha j}^*\right] = 0 \qquad (k {\neq} j;\; k, j = d, s, b;\; \beta = u, c, t)\,. \tag{4.156}$$

These relations show that all the imaginary parts in eqn (4.151) are equal up to a sign and CP violation can be quantified in a rephasing-invariant way in terms of the Jarlskog invariant [640, 639, 562, 395, 1075]

$$J = \mathfrak{Im}[V_{us} \, V_{cb} \, V_{ub}^* \, V_{cs}^*] = {}^{us}\Im_{cb}\,. \tag{4.157}$$

The sign of the other imaginary parts of quartic products is given by

$$\mathfrak{Im}\left[V_{\alpha k} \, V_{\beta j} \, V_{\alpha j}^* \, V_{\beta k}^*\right] = J \sum_{\gamma} \epsilon_{\alpha\beta\gamma} \sum_{l} \epsilon_{kjl}\,. \tag{4.158}$$

The experimental value of J is [400]

$$J = (2.88 \pm 0.33) \times 10^{-5}\,. \tag{4.159}$$

A convenient way to find the sign of a given ${}^{\alpha j}\Im_{\beta j}$ is through the graphical representation of a product of elements of V and V^* presented in Refs. [1074, 246, 394]: in a 3×3 matrix with empty entries represented by a dot $(\cdot)$, if $V_{\alpha j}$ appears in the product place a $\bigcirc$ in the αj entry, whereas if $V_{\alpha j}^*$ appears in the product place a $\times$ in the αj entry. For example, ${}^{us}\square_{cb}$ is represented by

$${}^{us}\square_{cb} = \begin{pmatrix} \cdot & \bigcirc & \times \\ \cdot & \times & \bigcirc \\ \cdot & \cdot & \cdot \end{pmatrix}\,. \tag{4.160}$$

Rephasing invariance implies that in this graphical representation each row and each column must have an equal number of $\bigcirc$'s and $\times$'s. A $\otimes$ in the αj entry represents $|V_{\alpha j}|^2$.

As another useful example, let us consider the unitarity relation in eqn (4.153) for $\alpha = u$ and $\beta = c$, which is represented by

$$\begin{pmatrix} \bigcirc & \cdot & \cdot \\ \times & \cdot & \cdot \\ \cdot & \cdot & \cdot \end{pmatrix} + \begin{pmatrix} \cdot & \bigcirc & \cdot \\ \cdot & \times & \cdot \\ \cdot & \cdot & \cdot \end{pmatrix} + \begin{pmatrix} \cdot & \cdot & \bigcirc \\ \cdot & \cdot & \times \\ \cdot & \cdot & \cdot \end{pmatrix} = 0 . \tag{4.161}$$

Multiplying it by $V_{ub}^* V_{cb}$ we obtain

$$\begin{pmatrix} \bigcirc & \cdot & \times \\ \times & \cdot & \bigcirc \\ \cdot & \cdot & \cdot \end{pmatrix} + \begin{pmatrix} \cdot & \bigcirc & \times \\ \cdot & \times & \bigcirc \\ \cdot & \cdot & \cdot \end{pmatrix} + \begin{pmatrix} \cdot & \cdot & \otimes \\ \cdot & \cdot & \otimes \\ \cdot & \cdot & \cdot \end{pmatrix} = 0 . \tag{4.162}$$

Taking the imaginary part, we get

$$\Im\mathfrak{m} \begin{pmatrix} \bigcirc & \cdot & \times \\ \times & \cdot & \bigcirc \\ \cdot & \cdot & \cdot \end{pmatrix} + \Im\mathfrak{m} \begin{pmatrix} \cdot & \bigcirc & \times \\ \cdot & \times & \bigcirc \\ \cdot & \cdot & \cdot \end{pmatrix} = 0 , \tag{4.163}$$

which represents the relation in eqn (4.155) with $\alpha = u$, $\beta = c$ and $j = b$:

$$^{ud}\Im_{cb} + {}^{us}\Im_{cb} = 0 \quad \Longrightarrow \quad {}^{ud}\Im_{cb} = -J . \tag{4.164}$$

In the parameterization in eqn (4.79), the Jarlskog invariant is given by

$$J = c_{12} s_{12} c_{23} s_{23} c_{13}^2 s_{13} \sin\delta_{13} = \frac{1}{8} \sin 2\vartheta_{12} \sin 2\vartheta_{23} \cos\vartheta_{13} \sin 2\vartheta_{13} \sin\delta_{13} . \tag{4.165}$$

The Jarlskog invariant is useful for quantifying CP violation in a parameterization-independent way. In particular, it is now possible to define a maximal CP violation as the case in which $|J|$ has its maximum possible value $|J|_{\max}$ [395], which can easily be calculated from eqn (4.165) to be

$$|J|_{\max} = \frac{1}{6\sqrt{3}} , \tag{4.166}$$

obtained for $\vartheta_{12} = \vartheta_{23} = \pi/4$, $s_{13} = 1/\sqrt{3}$ and $\sin\delta_{13} = \pm 1$. This is the case of trimaximal mixing with all the absolute values of the elements of the mixing matrix equal to $1/\sqrt{3}$,

$$V = \begin{pmatrix} \frac{1}{\sqrt{3}} & \frac{1}{\sqrt{3}} & \mp\frac{i}{\sqrt{3}} \\ -\frac{1}{2} \mp \frac{i}{2\sqrt{3}} & \frac{1}{2} \mp \frac{i}{2\sqrt{3}} & \frac{1}{\sqrt{3}} \\ \frac{1}{2} \mp \frac{i}{2\sqrt{3}} & -\frac{1}{2} \mp \frac{i}{2\sqrt{3}} & \frac{1}{\sqrt{3}} \end{pmatrix} = \frac{1}{\sqrt{3}} \begin{pmatrix} 1 & 1 & \mp i \\ -e^{\pm i\pi/6} & e^{\mp i\pi/6} & 1 \\ e^{\mp i\pi/6} & -e^{\pm i\pi/6} & 1 \end{pmatrix} . \tag{4.167}$$

The mixing angles and the phase can be expressed, in the parameterization in eqn (4.79), in terms of absolute values of the elements of the mixing matrix and J as

$$\tan\vartheta_{12} = \frac{|V_{us}|}{|V_{ud}|} , \tag{4.168}$$

$$\tan\vartheta_{23} = \frac{|V_{cb}|}{|V_{tb}|}\,, \tag{4.169}$$

$$\sin\vartheta_{13} = |V_{ub}|\,, \tag{4.170}$$

$$\sin\delta_{13} = \frac{8\,J}{c_{13}\sin 2\vartheta_{12}\sin 2\vartheta_{23}\sin 2\vartheta_{13}}\,. \tag{4.171}$$

These equations determine uniquely the mixing angles because they are bounded between 0 and $\pi/2$ (see eqn (4.30)). On the other hand, the phase δ_{13}, which lies between 0 and 2π, is not uniquely determined by eqn (4.171), because $\sin\delta_{13}$ is invariant under the transformation $\delta_{13} \to \pi - \delta_{13}$. Hence, apart from the discrete ambiguity $\delta_{13} \to \pi - \delta_{13}$, the mixing matrix can be parameterized in terms of the measurable rephasing invariants

$$|V_{us}|\,, \quad |V_{ub}|\,, \quad |V_{cb}|\,, \quad J\,, \tag{4.172}$$

because

$$|V_{ud}| = \sqrt{1 - |V_{us}|^2 - |V_{ub}|^2}\,, \tag{4.173}$$

$$|V_{tb}| = \sqrt{1 - |V_{cb}|^2 - |V_{ub}|^2}\,. \tag{4.174}$$

In order to determine δ_{13} uniquely it is necessary also to measure an observable depending on $\cos\delta_{13}$, which changes sign under the transformation $\delta_{13} \to \pi - \delta_{13}$.

A similar parameterization of the mixing matrix in terms of rephasing invariants has been proposed in Ref. [246, 394]. This parameterization uses three independent moduli of elements of the mixing matrix and the phase of an appropriate quartic product of the type in eqn (4.148). To have independent absolute values, three elements of the mixing matrix cannot belong to the same row or column, since in that case unitarity implies that the sum of the squares of the absolute values is equal to one. For example, a minimal set of rephasing invariant parameters is $\{V_{\alpha k}, \arg^{\alpha k}\square_{(\alpha+1)(k+1)}\}$ for $\alpha > k$, i.e.

$$|V_{us}|\,, \quad |V_{ub}|\,, \quad |V_{cb}|\,, \quad \arg^{us}\square_{cb}\,. \tag{4.175}$$

The mixing matrix can be constructed from the values of these quantities following the steps [246, 394]:

1. Use unitarity to determine $|V_{ud}|$ and $|V_{tb}|$ through eqns (4.173) and (4.174).
2. Choose the phases of V_{uk} and $V_{\alpha b}$ for $k = d, s, b$ and $\alpha = u, c, t$. Since there are $2N - 1 = 5$ such elements, these choices are allowed by the arbitrariness of the phases of the quark fields.
3. Since at this point all the elements in the first row and last column are determined, the phase of V_{cs} can be determined from $\arg^{us}\square_{cb}$:

$$\arg V_{cs} = \arg V_{us} + \arg V_{cb} - \arg V_{ub} - \arg^{us}\square_{cb}\,. \tag{4.176}$$

4. The values of $|V_{cs}|$ and V_{cd} are determined by combining the unitarity relations obtained from the orthogonality of the first and second rows and the normalization of the second row,

$$V_{cd}V_{ud}^{*} + V_{cs}V_{us}^{*} + V_{cb}V_{ub}^{*} = 0\,, \tag{4.177}$$

$$|V_{cd}|^2 + |V_{cs}|^2 + |V_{cb}|^2 = 1 \,. \tag{4.178}$$

Since the combination of these two equations leads to a quadratic equation for $|V_{cs}|$, there can be a two-fold ambiguity in the determination of $|V_{cs}|$ if both roots are positive.

5. The value of V_{ts} is determined by the unitarity equation obtained from the orthogonality of the second and third columns,

$$V_{us} V_{ub}^* + V_{cs} V_{cb}^* + V_{ts} V_{tb}^* = 0 \,. \tag{4.179}$$

6. The value of V_{td} is determined by the unitarity equation obtained from the orthogonality of the first and third columns,

$$V_{ud} V_{ub}^* + V_{cd} V_{cb}^* + V_{td} V_{tb}^* = 0 \,. \tag{4.180}$$

Hence, the mixing matrix is constructed from the rephasing invariants in eqn (4.175) up to a two-fold ambiguity generated in the calculation of the modulus of $|V_{cs}|$.

Another way of constructing the mixing matrix from the rephasing invariant parameters in eqn (4.175) is by using the parameterization in eqn (4.79) in which ϑ_{13} is determined directly from eqn (4.170), the angle ϑ_{12} is determined by eqn (4.168) through eqn (4.173), and the angle ϑ_{23} is determined by eqn (4.169) through eqn (4.174). The phase δ_{13} is determined by the relation $\arg {}^{us}\square_{cb} = \delta_{13} - \arg V_{cs}$, which leads to

$$c_{12} c_{23} \cos\delta_{13} = |V_{cs}| \cos(\arg[V_{cb}\, V_{us}\, V_{cs}^*\, V_{ub}^*]) + s_{12} s_{23} s_{13} \,, \tag{4.181}$$

$$c_{12} c_{23} \sin\delta_{13} = |V_{cs}| \sin(\arg[V_{cb}\, V_{us}\, V_{cs}^*\, V_{ub}^*]) \,, \tag{4.182}$$

with

$$|V_{cs}| = \sqrt{c_{12}^2 c_{23}^2 + s_{12}^2 s_{23}^2 s_{13}^2 - 2 c_{12} s_{12} c_{23} s_{23} s_{13} \cos\delta_{13}} \,. \tag{4.183}$$

Equation (4.182) is quadratic in $\cos\delta_{13}$,

$$c_{12}^2 c_{23}^2 \cos^2\delta_{13} - 2 s c_{12} s_{12} c_{23} s_{23} s_{13} \cos\delta_{13} - c^2 c_{12}^2 c_{23}^2 + s^2 s_{12}^2 s_{23}^2 s_{13}^2 = 0 \,, \tag{4.184}$$

with $c \equiv \cos(\arg {}^{us}\square_{cb})$ and $s \equiv \sin(\arg {}^{us}\square_{cb})$. If both roots are between -1 and $+1$, there is a two-fold ambiguity in the determination of $\cos\delta_{13}$. However, for a given value of $\cos\delta_{13}$, the value of $\sin\delta_{13}$, and hence the value of δ_{13}, is uniquely fixed by eqn (4.182).

4.8 Unitarity triangles

The unitarity relation $VV^\dagger = \mathbf{1}$ implies that the product of any row with the complex conjugate of another one must vanish:

$$\sum_{k=d,s,b} V_{\alpha k} V_{\beta k}^* = 0 \qquad (\alpha \neq \beta;\ \alpha, \beta = u, c, t) \,. \tag{4.185}$$

In other words, the sum of the three complex numbers $V_{\alpha d} V_{\beta d}^*$, $V_{\alpha s} V_{\beta s}^*$, $V_{\alpha b} V_{\beta b}^*$ vanishes. This relation can be represented in the complex plane by a triangle with sides $|V_{\alpha d}||V_{\beta d}|$, $|V_{\alpha s}||V_{\beta s}|$, $|V_{\alpha b}||V_{\beta b}|$, which is called the *unitarity triangle* [643]. Since $\alpha \neq \beta$, eqn (4.185) implies three unitarity triangles for ($\alpha = u$, $\beta = c$), ($\alpha = c$, $\beta = t$), and ($\alpha = t$, $\beta = u$).

Similarly, the unitarity relation $V^\dagger V = \mathbf{1}$ implies that the product of any column with the complex conjugate of another one must vanish:

$$\sum_{\alpha=u,c,t} V_{\alpha k}^* V_{\alpha j} = 0 \qquad (k \neq j;\, k, j = d, s, b)\,. \tag{4.186}$$

These relations determine three other unitarity triangles with sides $|V_{uk}||V_{uj}|$, $|V_{ck}||V_{cj}|$, and $|V_{tk}||V_{tj}|$, for ($k = d$, $j = s$), ($k = s$, $j = b$), and ($k = b$, $j = d$).

Therefore, in total there are six unitarity triangles in the complex plane. The most frequently used unitarity triangle is the so-called *bd* triangle obtained from eqn (4.186) with $k = b$ and $j = d$:

$$V_{ub}^* V_{ud} + V_{cb}^* V_{cd} + V_{tb}^* V_{td} = 0\,. \tag{4.187}$$

The *bd* unitarity triangle is shown in Fig. 4.1. Using the generalized Wolfenstein parameterization in eqn (4.84), we have

$$V_{ub}^* V_{ud} = A\lambda^3(\bar\rho + i\bar\eta) + \mathcal{O}(\lambda^7)\,, \tag{4.188}$$

$$V_{cb}^* V_{cd} = -A\lambda^3 + \mathcal{O}(\lambda^7)\,, \tag{4.189}$$

$$V_{tb}^* V_{td} = A\lambda^3(1 - \bar\rho - i\bar\eta) + \mathcal{O}(\lambda^7)\,. \tag{4.190}$$

This approximation gives the coordinates of the vertices shown in Fig. 4.1a. In practice, it is convenient to use the rescaled *bd* unitarity triangle in Fig. 4.1b, obtained by dividing eqn (4.187) by $A\lambda^3$, with

$$R_b = \frac{|V_{ub}^* V_{ud}|}{|V_{cb}^* V_{cd}|} = \sqrt{\bar\rho^2 + \bar\eta^2}\,, \qquad R_t = \frac{|V_{tb}^* V_{td}|}{|V_{cb}^* V_{cd}|} = \sqrt{(1-\bar\rho)^2 + \bar\eta^2}\,. \tag{4.191}$$

In this approximation,

$$\gamma = \arctan\frac{\eta}{\rho} = \delta_{13}\,. \tag{4.192}$$

From the experimental data [400]

$$\bar\rho = 0.20 \pm 0.09\,, \qquad \bar\eta = 0.33 \pm 0.05\,, \tag{4.193}$$

which imply the value of $\gamma = \delta_{13}$ in eqn (4.81) and

$$\beta = 23.4^\circ \pm 2^\circ\,. \tag{4.194}$$

It is interesting to note that all the unitarity triangles have the same area equal to a half of the Jarlskog invariant J defined in eqn (4.157). In fact, if we consider, for example, the *bd* unitarity triangle, and use a parameterization of the quark mixing matrix in which $V_{cb}^* V_{cd}$ is real, as done in Fig. 4.1, the area of the triangle is given by

$$2\mathcal{A} = -V_{cb}^* V_{cd}\, \Im\mathfrak{m}[V_{ub}^* V_{ud}] = -\Im\mathfrak{m}[V_{ud} V_{cb} V_{ub}^* V_{cd}^*] = -{}^{ud}\Im_{cb} = J\,, \tag{4.195}$$

where we have used the relation in eqn (4.164). This property of the unitarity triangles reflects the fact that if CP is conserved their area is obviously zero.

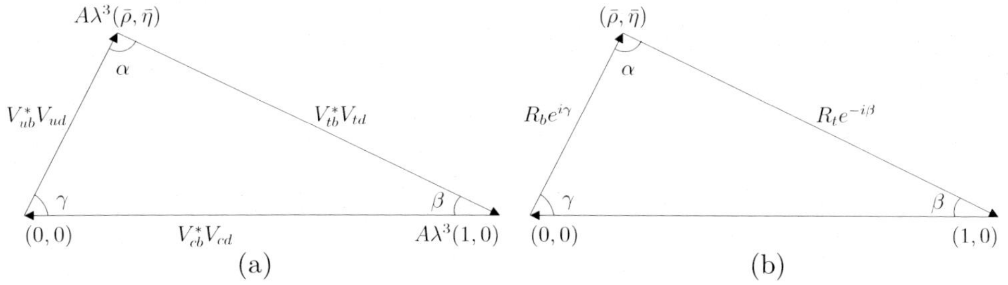

FIG. 4.1. (a) The bd unitarity triangle. The complex coordinates of the vertices are expressed in terms of the generalized Wolfenstein parameterization in eqn (4.84). (b) The rescaled bd unitarity triangle.

The area $\mathcal{A}$ of a triangle with sides a, b, c is given by

$$-16\,\mathcal{A}^2 = \lambda(a^2, b^2, c^2)\,, \tag{4.196}$$

where

$$\lambda(x, y, z) = x^2 + y^2 + z^2 - 2\,(xy + yz + zx) \tag{4.197}$$

is the Källen lambda function which appears in the kinematics of two-body decay. The area of the bd unitarity triangle can be calculated using eqn (4.196) with

$$a = |V_{ub}||V_{ud}|\,, \qquad b = |V_{cb}||V_{cd}|\,, \qquad c = |V_{tb}||V_{td}|\,. \tag{4.198}$$

Comparison with eqn (4.195) shows that the square of the Jarlskog invariant J can be expressed in terms of these three products of the moduli of elements of the mixing matrix [643]:

$$4\,J^2 = -\lambda(a^2, b^2, c^2)\,. \tag{4.199}$$

Similarly one can express the square of J in terms of the sides of any of the other five unitarity triangles, which lead in general to the relation in eqn (4.199) with

$$a = |V_{uk}||V_{uj}|\,, \quad b = |V_{ck}||V_{cj}|\,, \quad c = |V_{tk}||V_{tj}| \quad (k, \neq j,\ k = d, s, b,\ j = d, s, b) \tag{4.200}$$

for the three unitarity triangles obtained from eqn (4.186), and

$$a = |V_{\alpha d}||V_{\beta d}|\,, \quad b = |V_{\alpha s}||V_{\beta s}|\,, \quad c = |V_{\alpha b}||V_{\beta b}| \quad (\alpha \neq \beta,\ \alpha = u, c, t,\ \beta = u, c, t) \tag{4.201}$$

for the three unitarity triangles obtained from eqn (4.185).

The relation in eqn (4.199) shows that apart from a sign the Jarlskog invariant J is determined by the moduli of the elements of the mixing matrix. At the end of subsection 4.7 we saw that it is possible to parameterize the mixing matrix up to a two-fold ambiguity in terms of three independent moduli of the elements of the mixing matrix and the Jarlskog invariant J. Since J can be expressed in terms of moduli of the elements of the mixing matrix by eqn (4.199), it follows that it is possible to parameterize the mixing matrix in terms of four independent moduli of its matrix elements [925, 269, 589, 643, 1045].

In fact, four independent moduli of the matrix elements and the sign of J are sufficient for a unique parameterization of the mixing matrix. Let us consider, for example, the four moduli

$$|V_{us}|\,,\quad |V_{ub}|\,,\quad |V_{cs}|\,,\quad |V_{cb}|\,. \tag{4.202}$$

The orthogonality of the first two rows implies the unitarity relation

$$V_{ud}V_{cd}^* = -V_{us}V_{cs}^* - V_{ub}V_{cb}^*\,. \tag{4.203}$$

Taking the squared modulus of this relation, we obtain

$$|V_{ud}|^2|V_{cd}|^2 = |V_{us}|^2|V_{cs}|^2 + |V_{ub}|^2|V_{cb}|^2 + 2\,{}^{us}\mathfrak{R}_{cb}\,, \tag{4.204}$$

with ${}^{us}\mathfrak{R}_{cb} \equiv \mathfrak{Re}\,{}^{us}\square_{cb} \equiv \mathfrak{Re}[V_{us}V_{cs}^*V_{ub}^*V_{cb}]$ (see eqn (4.152)). Using the expressions of $|V_{ud}|^2$ and $|V_{cd}|^2$ in terms of the four moduli in eqn (4.202) obtained by the normalization of the first two rows,

$$|V_{ud}|^2 = 1 - |V_{us}|^2 - |V_{ub}|^2\,, \tag{4.205}$$

$$|V_{cd}|^2 = 1 - |V_{cs}|^2 - |V_{cb}|^2\,, \tag{4.206}$$

we find the relation

$${}^{us}\mathfrak{R}_{cb} = \frac{1}{2}\left(1 - |V_{us}|^2 - |V_{ub}|^2 - |V_{cs}|^2 - |V_{cb}|^2 + |V_{us}|^2|V_{cb}|^2 + |V_{ub}|^2|V_{cs}|^2\right)\,, \tag{4.207}$$

which gives the value of ${}^{us}\mathfrak{R}_{cb}$ in terms of the four moduli in eqn (4.202). Thus, J is determined up to a sign by

$$J^2 = |V_{us}|^2|V_{ub}|^2|V_{cs}|^2|V_{cb}|^2 - [{}^{us}\mathfrak{R}_{cb}]^2\,. \tag{4.208}$$

Given the sign of J, the standard parameterization in eqn (4.79) of the mixing matrix can be determined using eqns (4.168)–(4.171) and eqns (4.173) and (4.174). In this case there is no ambiguity in the determination of δ_{13}, because

$$\cos\delta_{13} = \frac{1}{c_{12}c_{23}}\left(s_{12}s_{23}s_{13} + \frac{{}^{us}\mathfrak{R}_{cb}}{|V_{us}||V_{cb}||V_{ub}|}\right)\,, \tag{4.209}$$

with ${}^{us}\mathfrak{R}_{cb}$ given by eqn (4.207).

If one does not want to use the standard parameterization in eqn (4.79), the mixing matrix can be constructed from the four moduli in eqn (4.202) and the sign of J following the steps below.

1. Use unitarity to determine $|V_{ud}|$ and $|V_{tb}|$ from eqns (4.173) and (4.174).
2. Choose the phases of V_{uk} and $V_{\alpha b}$ for $k = d, s, b$ and $\alpha = u, c, t$. Since there are $2N - 1 = 5$ such elements, these choices are allowed by the arbitrariness of the phases of the quark fields.
3. The phase of V_{cs} can be determined from eqn (4.176) with

$$\cos(\arg{}^{us}\square_{cb}) = \frac{{}^{us}\mathfrak{R}_{cb}}{|V_{us}||V_{cb}||V_{cs}||V_{ub}|}\,, \qquad \sin(\arg{}^{us}\square_{cb}) = \frac{J}{|V_{us}||V_{cb}||V_{cs}||V_{ub}|}\,. \tag{4.210}$$

4. The value of V_{cd} is determined by the unitarity equation obtained from the orthogonality of the first and second rows, given in eqn (4.177).
5. The value of V_{ts} is determined by the unitarity equation obtained from the orthogonality of the second and third columns, given in eqn (4.179).
6. The value of V_{td} is determined by the unitarity equation obtained from the orthogonality of the first and third columns, given in eqn (4.180).

This parameterization of the mixing matrix in terms of four independent moduli of its elements shows that it is possible to find CP violation measuring only CP conserving processes, which depend on the chosen moduli of elements of the mixing matrix.

4.9 Conditions for CP violation

In this section we discuss the powerful method for finding the general condition on the mass matrices M'^{U} and M'^{D} for CP violation presented in Ref. [639] (see also [640, 214, 571, 642, 641]).

As we have seen in sections 4.4 and 4.5, the physical mixing matrix is real and CP is conserved if two up-type quarks or two down-type quarks are degenerate in mass or if in a parameterization of the type in eqn (4.45) a mixing angle has its minimal value 0 or its maximal value $\pi/2$. Hence, in order to have CP violation, the mixing of quarks must satisfy the following 14 conditions [642]:

(A) No two up-type quarks or two down-type quarks are degenerate in mass, adding up to six conditions.
(B) No mixing angle is equal to 0 or $\pi/2$, adding up to six conditions.
(C) The physical phase is different from 0 or π, adding up to two conditions.

These 14 conditions are combined into a single condition [639]

$$\det C \neq 0 \,, \tag{4.211}$$

where C is the commutator

$$C = -i\,[M'^{U}\,M'^{U\dagger}\,,\,M'^{D}\,M'^{D\dagger}]\,. \tag{4.212}$$

One should not be surprised by the fact that the conditions for CP invariance constrain the products of mass matrices $M'^{U}\,M'^{U\dagger}$ and $M'^{D}\,M'^{D\dagger}$, because, as we have seen in subsection 4.1, these products determine, through the diagonalizing equation (4.5), the unitary matrices V_L^U and V_L^D, which determine the mixing matrix V in eqn (4.1). Therefore, the physical implications of mixing, such as the presence of CP violation, depend on the properties of the Hermitian matrices $M'^{U}\,M'^{U\dagger}$ and $M'^{D}\,M'^{D\dagger}$.

Using the diagonalizing equation (4.5) for $M'^U M'^{U\dagger}$ and $M'^D M'^{D\dagger}$,

$$M'^U M'^{U\dagger} = V_L^U M^2 V_L^{U\dagger} , \qquad M'^D M'^{D\dagger} = V_L^D M^2 V_L^{D\dagger} , \tag{4.213}$$

one can write the commutator C as

$$C = -i \det [M^{U^2} , V M^{D^2} V^\dagger] , \tag{4.214}$$

which shows that the condition in eqn (4.211) involves physical parameters: the quark masses contained in M^{U^2} and M^{D^2} and the mixing matrix V. Indeed, the determinant of C is given by [639]

$$\begin{aligned} \det C = -2\, J \left(m_c^2 - m_u^2\right) \left(m_t^2 - m_u^2\right) \left(m_t^2 - m_c^2\right) \\ \left(m_s^2 - m_d^2\right) \left(m_b^2 - m_d^2\right) \left(m_b^2 - m_s^2\right) . \end{aligned} \tag{4.215}$$

One can see that the quark masses appear in six factors. The nonvanishing of each of these six factors corresponds to one of the six conditions in item (A) above. The eight conditions in items (B) and (C) are equivalent to requiring the nonvanishing of J, which guarantees that the mixing matrix is not real. In fact, we have seen in section 4.7 that the quartic products in eqn (4.148) are the simplest rephasing invariant products of elements of the mixing matrix and all the rephasing-invariant products of higher order can be expressed in terms of squared moduli and rephasing-invariant quartic products. Since all the imaginary parts of the quartic products in eqn (4.148) are equal to J, up to a sign (see eqn (4.158)), the nonvanishing of J is a necessary and sufficient condition for the nonreality of V, which is equivalent to the eight conditions in items (B) and (C). For example, from eqn (4.165) one can see explicitly that in the parameterization in eqn (4.79) the nonvanishing of J is equivalent to the eight conditions in items (B) and (C).

5

NEUTRINO INTERACTIONS

> It is no good to try to stop knowledge from going forward. Ignorance is never better than knowledge.
>
> **Enrico Fermi**

Neutrino interactions are described, with an impressive accuracy, by the Standard Model. So far no deviations from the standard neutrino interactions have been found in experimental data.

The standard neutrino interactions are described by the leptonic charged current in eqn (3.141),

$$j^{\rho}_{W,\mathrm{L}} = 2 \sum_{\alpha=e,\mu,\tau} \overline{\nu_{\alpha L}} \, \gamma^{\rho} \, \ell_{\alpha L} = \sum_{\alpha=e,\mu,\tau} \overline{\nu_{\alpha}} \, \gamma^{\rho} \left(1 - \gamma^{5}\right) \ell_{\alpha} \, , \tag{5.1}$$

and the neutrino part of the leptonic neutral current in eqn (3.148),

$$j^{\rho}_{Z,\nu} = \sum_{\alpha=e,\mu,\tau} \overline{\nu_{\alpha L}} \, \gamma^{\rho} \, \nu_{\alpha L} = \frac{1}{2} \sum_{\alpha=e,\mu,\tau} \overline{\nu_{\alpha}} \, \gamma^{\rho} \left(1 - \gamma^{5}\right) \nu_{\alpha} \, , \tag{5.2}$$

which enter, respectively, into the leptonic charged-current weak interaction Lagrangian in eqn (3.76),

$$\mathscr{L}^{(\mathrm{CC})}_{\mathrm{I,L}} = -\frac{g}{2\sqrt{2}} \left(j^{\rho}_{W,\mathrm{L}} W_{\rho} + j^{\rho}_{W,\mathrm{L}}{}^{\dagger} W^{\dagger}_{\rho} \right) , \tag{5.3}$$

and the neutrino part of the leptonic neutral-current weak interaction Lagrangian in eqn (3.90),

$$\mathscr{L}^{(\mathrm{NC})}_{\mathrm{I},\nu} = -\frac{g}{2\cos\vartheta_{\mathrm{W}}} \, j^{\rho}_{Z,\nu} \, Z_{\rho} \, . \tag{5.4}$$

In this chapter we discuss neutrino–electron interactions in section 5.1, the most important hadron decays in section 5.2 and neutrino–nucleon scattering in section 5.3. Further information on neutrino interactions and weak interactions in general can be found in the books in Refs. [778, 328, 902, 227, 720] and in the reviews in Refs. [750, 672, 233, 249, 372, 774].

The introduction of neutrino masses, to be discussed in chapter 6, generates small kinematical effects in neutrino interaction processes, which will be neglected in this chapter. These small effects are searched for only in experiments aimed at the direct measurement of neutrino masses, which will be discussed in chapter 14.

5.1 Neutrino–electron interactions

Neutrino–electron interactions are the simplest interactions of neutrinos with components of matter in our world. At the lowest order in the weak interaction perturbation theory, neutrino–electron interactions involve only free leptons, whose interaction amplitude can be calculated exactly by using the Feynman rules of the Standard Model listed in appendix E. In the following we discuss neutrino–electron elastic scattering in subsection 5.1.1, the neutrino–electron quasielastic scattering process $\nu_\mu + e^- \to \nu_e + \mu^-$ and the neutrino–antineutrino pair production process $e^+ + e^- \to \nu + \bar{\nu}$ in subsection 5.1.3.

5.1.1 Neutrino–electron elastic scattering

Low-energy neutrinos and antineutrinos with flavor $\alpha = e, \mu, \tau$ interact with electrons through the elastic scattering process

$$\overset{(-)}{\nu}_\alpha + e^- \to \overset{(-)}{\nu}_\alpha + e^- . \tag{5.5}$$

This process is used, for example, in water Cherenkov solar neutrino detectors (see section 10.6). The elastic scattering process does not have a threshold, since the final state is the same as the initial state. The only effect of an elastic scattering process is a redistribution of the total energy and momentum between the two participating particles.

Figure 5.1 shows the two tree-level Feynman diagrams which contribute to the elastic scattering

$$\nu_e + e^- \to \nu_e + e^- . \tag{5.6}$$

In the case of the elastic scattering

$$\bar{\nu}_e + e^- \to \bar{\nu}_e + e^- , \tag{5.7}$$

the charged-current t-channel diagram in Fig. 5.1a is replaced by the s-channel diagram in Fig. 5.2a. The diagram in Fig. 5.2b also contributes to this process.

At tree-level, the elastic scattering process[19]

$$\overset{(-)}{\nu}_{\mu,\tau} + e^- \to \overset{(-)}{\nu}_{\mu,\tau} + e^- \tag{5.8}$$

receives contribution only from the neutral-current diagram in Fig. 5.3a, which is analogous to those in Fig. 5.1b and Fig. 5.2b.

For low neutrino energies, where the effects of the W and Z propagators can be neglected, the above processes are described by the effective charged-current and neutral-current Lagrangians in eqns (3.224) and (3.235). For example, the effective

[19] The observation in 1973 of the process $\bar{\nu}_\mu + e^- \to \bar{\nu}_\mu + e^-$ in the Gargamelle experiment [600], together with the observation of the processes $\overset{(-)}{\nu}_\mu + N \to \overset{(-)}{\nu}_\mu + X$ (see section 5.3.4) in the Gargamelle experiment at CERN [599, 601] and an experiment at Fermilab [207], led to the experimental confirmation of the existence of the neutral-current interactions predicted by the SM.

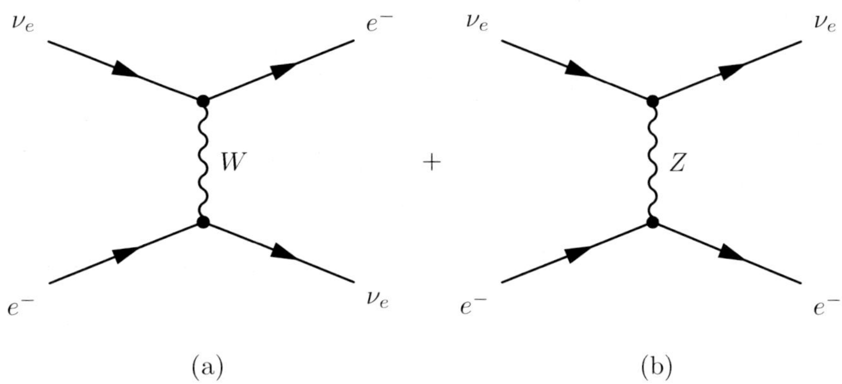

FIG. 5.1. The two tree-level Feynman diagrams for the elastic scattering process $\nu_e + e^- \to \nu_e + e^-$: charged current (a) and neutral current (b).

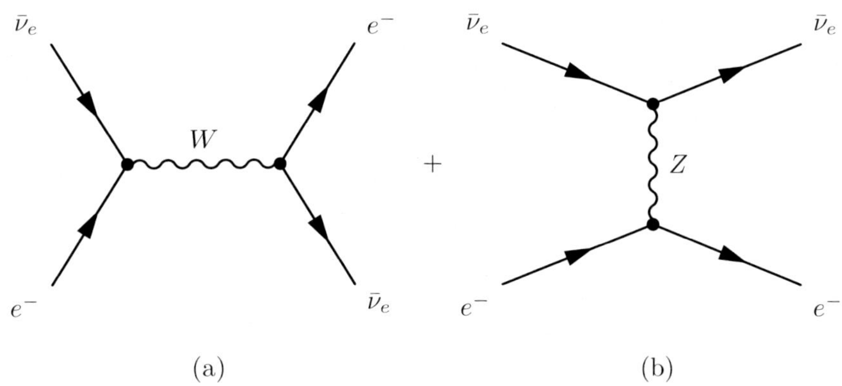

FIG. 5.2. The two tree-level Feynman diagrams for the elastic scattering process $\bar{\nu}_e + e^- \to \bar{\nu}_e + e^-$: charged current (a) and neutral current (b).

low-energy Lagrangian for the elastic scattering processes in eqns (5.6) and (5.7) is given by

$$\mathscr{L}_{\text{eff}}(\overset{(-)}{\nu}_e e^- \to \overset{(-)}{\nu}_e e^-) = -\frac{G_F}{\sqrt{2}} \left\{ \left[\overline{\nu_e} \, \gamma^\rho \left(1 - \gamma^5\right) e \right] \left[\overline{e} \, \gamma_\rho \left(1 - \gamma^5\right) \nu_e \right] \right.$$
$$\left. + \left[\overline{\nu_e} \, \gamma^\rho \left(1 - \gamma^5\right) \nu_e \right] \left[\overline{e} \, \gamma_\rho \left(g_V^l - g_A^l \gamma^5\right) e \right] \right\} , \quad (5.9)$$

with the coefficients g_V^l and g_A^l given in Table 3.6 (page 78). The first term on the right-hand side is the charged-current contribution. The second term is the neutral-current contribution. The charged-current contribution can be rearranged with the Fierz transformation in eqn (2.508), leading to an expression which has the same form as the neutral-current contribution. This allows us to write the effective

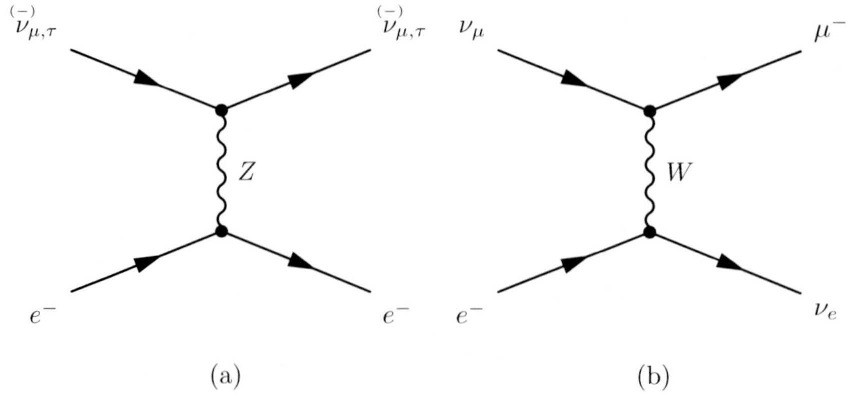

FIG. 5.3. (a) Tree-level Feynman diagram for the elastic scattering process $\overset{(-)}{\nu}_{\mu,\tau} + e^- \to \overset{(-)}{\nu}_{\mu,\tau} + e^-$. (b) Tree-level Feynman diagram for the charged-current process $\nu_\mu + e^- \to \nu_e + \mu^-$.

low-energy Lagrangian in eqn (5.9) in the compact form

$$\mathscr{L}_{\text{eff}}(\overset{(-)}{\nu}_e e^- \to \overset{(-)}{\nu}_e e^-) = -\frac{G_F}{\sqrt{2}} \left[\overline{\nu_e}\, \gamma^\rho \left(1 - \gamma^5\right) \nu_e\right] \left[\overline{e}\, \gamma_\rho \left(\left(1 + g_V^l\right) - \left(1 + g_A^l\right) \gamma^5\right) e\right] . \tag{5.10}$$

On the other hand, the effective Lagrangian for the process in eqn (5.8) contains only a neutral-current term:

$$\mathscr{L}_{\text{eff}}(\overset{(-)}{\nu}_\alpha e^- \to \overset{(-)}{\nu}_\alpha e^-) = -\frac{G_F}{\sqrt{2}} \left[\overline{\nu_\alpha}\, \gamma^\rho \left(1 - \gamma^5\right) \nu_\alpha\right] \left[\overline{e}\, \gamma_\rho \left(g_V^l - g_A^l \gamma^5\right) e\right] \quad (\alpha = \mu, \tau) . \tag{5.11}$$

Since the above processes and effective Lagrangians have similar structures, they share some common important properties. The cross-sections are proportional to G_F^2. Since a cross-section has dimension $(\text{length})^2 \sim (\text{energy})^{-2}$ and G_F^2 has dimension $(\text{energy})^{-4}$, in order to write a cross-section, a factor with dimension $(\text{energy})^2$ is needed. Let us consider the general process

$$\nu_i + e_i^- \to \nu_f + e_f^- . \tag{5.12}$$

In the center-of-mass frame the only quantity with dimension of squared energy which depends only on the initial state is the total squared energy $s = \left(E_{\nu_i} + E_{e_i}\right)^2$. This is one of the three relativistic invariant Mandelstam variables[20]:

$$s = \left(p_{\nu_i} + p_{e_i}\right)^2 = \left(p_{\nu_f} + p_{e_f}\right)^2 , \tag{5.13}$$

[20] In eqn (5.14) we introduced the usual notation q for the four-momentum transfer, $q = p_{\nu_i} - p_{\nu_f} = p_{e_f} - p_{e_i}$, and $Q^2 \equiv -q^2$. The reason for this definition is that q^2 is negative. This can be easily seen by calculating the Lorentz-invariant q^2 in the laboratory frame, where e_i is at rest: $q^2 = -2m_e T_e$, where $T_e \equiv T_{e_f}$ is the kinetic energy of the recoil electron.

TABLE 5.1. Total neutrino–electron elastic scattering cross-sections for $\sqrt{s} \gg m_e$. The numerical values are in units of $10^{-46}\,\text{cm}^2$.

Process	Total cross-section
$\nu_e + e^-$	$(G_F^2\, s/4\,\pi)\left[(1+2\,\sin^2\vartheta_W)^2 + \frac{4}{3}\,\sin^4\vartheta_W\right] \simeq 93\,s/\text{MeV}^2$
$\bar{\nu}_e + e^-$	$(G_F^2\, s/4\,\pi)\left[\frac{1}{3}\,(1+2\,\sin^2\vartheta_W)^2 + 4\,\sin^4\vartheta_W\right] \simeq 39\,s/\text{MeV}^2$
$\nu_{\mu,\tau} + e^-$	$(G_F^2\, s/4\,\pi)\left[(1-2\,\sin^2\vartheta_W)^2 + \frac{4}{3}\,\sin^4\vartheta_W\right] \simeq 15\,s/\text{MeV}^2$
$\bar{\nu}_{\mu,\tau} + e^-$	$(G_F^2\, s/4\,\pi)\left[\frac{1}{3}\,(1-2\,\sin^2\vartheta_W)^2 + 4\,\sin^4\vartheta_W\right] \simeq 13\,s/\text{MeV}^2$

$$t = (p_{\nu_i} - p_{\nu_f})^2 = (p_{e_f} - p_{e_i})^2 = q^2 \equiv -Q^2\,, \tag{5.14}$$

$$u = (p_{\nu_i} - p_{e_f})^2 = (p_{\nu_f} - p_{e_i})^2\,, \tag{5.15}$$

where we used the energy–momentum conservation $p_{\nu_i} + p_{e_i} = p_{\nu_f} + p_{e_f}$. Hence, in any frame the neutrino–electron cross-section is given by

$$\sigma \propto G_F^2\, s\,. \tag{5.16}$$

The values of these total cross-sections for $\sqrt{s} \gg m_e$ are given in Table 5.1. In the laboratory frame, where the electron is initially at rest, neglecting the neutrino mass, we have

$$s = 2\, m_e\, E_\nu\,, \tag{5.17}$$

where $E_\nu \equiv E_{\nu_i}$ is the energy of the incoming neutrino. From the values in Table 5.1 one can see that for $\sqrt{s} \gg m_e$ the approximate ratios of the four cross-sections are

$$\sigma_{\nu_e} : \sigma_{\bar{\nu}_e} : \sigma_{\nu_{\mu,\tau}} : \sigma_{\bar{\nu}_{\mu,\tau}} \simeq 1 : 0.42 : 0.16 : 0.14\,. \tag{5.18}$$

Hence, the ν_e-e^- cross-section is about 2.4 times larger than the $\bar{\nu}_e$-e^- cross-section, about 6.2 times larger than the $\nu_{\mu,\tau}$-e^- cross-section, and about 7.1 times larger than the $\bar{\nu}_{\mu,\tau}$-e^- cross-section. These ratios will become useful when we discuss solar neutrino detection in water Cherenkov experiments (see section 10.6).

For the differential cross-section, one can find, after lengthy calculations, the result

$$\frac{d\sigma}{dQ^2} = \frac{G_F^2}{\pi}\left[g_1^2 + g_2^2\left(1 - \frac{Q^2}{2\,p_{\nu_i}\cdot p_{e_i}}\right)^2 - g_1\, g_2\, m_e^2\,\frac{Q^2}{2\,(p_{\nu_i}\cdot p_{e_i})^2}\right]. \tag{5.19}$$

The quantities g_1 and g_2 depend on the flavor of the neutrino: for ν_e and $\bar{\nu}_e$ we have

$$g_1^{(\nu_e)} = g_2^{(\bar{\nu}_e)} = 1 + \frac{g_V^l + g_A^l}{2} = 1 + g_L^l = \frac{1}{2} + \sin^2\vartheta_W \simeq 0.73\,, \tag{5.20}$$

$$g_2^{(\nu_e)} = g_1^{(\bar{\nu}_e)} = \frac{g_V^l - g_A^l}{2} = g_R^l = \sin^2 \vartheta_{\mathrm{W}} \simeq 0.23\,, \tag{5.21}$$

whereas for $\nu_{\mu,\tau}$ and $\bar{\nu}_{\mu,\tau}$ we have

$$g_1^{(\nu_{\mu,\tau})} = g_2^{(\bar{\nu}_{\mu,\tau})} = \frac{g_V^l + g_A^l}{2} = g_L^l = -\frac{1}{2} + \sin^2 \vartheta_{\mathrm{W}} \simeq -0.27\,, \tag{5.22}$$

$$g_2^{(\nu_{\mu,\tau})} = g_1^{(\bar{\nu}_{\mu,\tau})} = \frac{g_V^l - g_A^l}{2} = g_R^l = \sin^2 \vartheta_{\mathrm{W}} \simeq 0.23\,. \tag{5.23}$$

In the laboratory frame, where $\vec{p}_{e_{\mathrm{i}}} = 0$, we have

$$Q^2 = 2\, m_e\, T_e\,, \tag{5.24}$$

where $T_e \equiv T_{e_{\mathrm{f}}}$ is the kinetic energy of the recoil electron. The differential cross-section as a function of T_e in the laboratory frame is given by

$$\frac{\mathrm{d}\sigma}{\mathrm{d}T_e}(E_\nu, T_e) = \frac{\sigma_0}{m_e} \left[g_1^2 + g_2^2 \left(1 - \frac{T_e}{E_\nu}\right)^2 - g_1\, g_2\, \frac{m_e\, T_e}{E_\nu^2} \right], \tag{5.25}$$

with

$$\sigma_0 = \frac{2\, G_{\mathrm{F}}^2\, m_e^2}{\pi} \simeq 88.06 \times 10^{-46}\, \mathrm{cm}^2\,. \tag{5.26}$$

In the laboratory frame we obtain, from energy–momentum conservation,

$$T_e = \frac{2\, m_e\, E_\nu^2 \cos^2\theta}{(m_e + E_\nu)^2 - E_\nu^2 \cos^2\theta}\,, \tag{5.27}$$

where θ is the electron scattering angle depicted in Fig. 5.4. Since

$$\mathrm{d}T_e = \frac{4\, m_e\, E_\nu^2\, (m_e + E_\nu)^2}{\left[(m_e + E_\nu)^2 - E_\nu^2 \cos^2\theta\right]^2} \cos\theta\, \mathrm{d}\cos\theta\,, \tag{5.28}$$

the differential cross-section as a function of the electron scattering angle in the laboratory frame is given by

$$\frac{\mathrm{d}\sigma}{\mathrm{d}\cos\theta} = \sigma_0\, \frac{4\, E_\nu^2\, (m_e + E_\nu)^2 \cos\theta}{\left[(m_e + E_\nu)^2 - E_\nu^2 \cos^2\theta\right]^2} \left[g_1^2 + g_2^2 \left(1 - \frac{2\, m_e\, E_\nu \cos^2\theta}{(m_e + E_\nu)^2 - E_\nu^2 \cos^2\theta}\right)^2 \right.$$
$$\left. - g_1\, g_2\, \frac{2\, m_e^2 \cos^2\theta}{(m_e + E_\nu)^2 - E_\nu^2 \cos^2\theta} \right]. \tag{5.29}$$

Since $\cos\theta \leq 1$, from eqn (5.27) one can deduce that for a given neutrino energy E_ν there is a maximum kinetic energy of the recoil electron,

$$T_e^{\max}(E_\nu) = \frac{2\, E_\nu^2}{m_e + 2\, E_\nu}\,, \tag{5.30}$$

which corresponds to its motion in the forward direction ($\cos\theta = 1$). From eqn (5.27) one can also see that there is a minimum neutrino energy which can

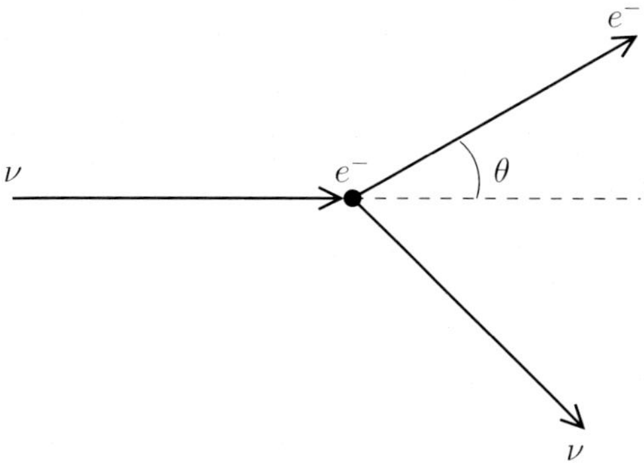

FIG. 5.4. Neutrino–electron elastic scattering in the laboratory frame.

produce a given kinetic energy T_e of the recoil electron, given by

$$E_\nu^{\min}(T_e) = \frac{T_e}{2}\left(1+\sqrt{1+\frac{2\,m_e}{T_e}}\right) = \frac{T_e+|\vec{p}_e|}{2} \simeq \begin{cases} \sqrt{m_e T_e/2} & \text{for} \quad T_e \ll m_e\,, \\ T_e + m_e/2 & \text{for} \quad T_e \gg m_e\,. \end{cases} \tag{5.31}$$

In practice, it is not possible to measure neutrino–electron elastic scattering without a threshold T_e^{th} for the kinetic energy of the recoil electron above the background. For example, in the Super-Kamiokande solar neutrino experiment we have $T_e^{\text{th}} \simeq 4.5\,\text{MeV}$ [625]. Therefore, one can measure a total cross-section which is a function of the neutrino energy and the kinetic energy threshold of the recoil electron:

$$\begin{aligned}\sigma(E_\nu, T_e^{\text{th}}) = \frac{\sigma_0}{m_e}\Bigg[&\left(g_1^2+g_2^2\right)\left(T_e^{\max}-T_e^{\text{th}}\right) \\ &-\left(g_2^2+g_1 g_2\,\frac{m_e}{2\,E_\nu}\right)\left(\frac{{T_e^{\max}}^2-{T_e^{\text{th}}}^2}{E_\nu}\right)+\frac{1}{3}\,g_2^2\left(\frac{{T_e^{\max}}^3-{T_e^{\text{th}}}^3}{E_\nu^2}\right)\Bigg]\,,\end{aligned} \tag{5.32}$$

with $T_e^{\max} = T_e^{\max}(E_\nu)$, as given by eqn (5.30). The values of these neutrino–electron cross-sections for $T_e^{\text{th}} = 0$ and $T_e^{\text{th}} = 4.50\,\text{MeV}$ are plotted in Fig. 5.5 as functions of the neutrino energy E_ν. The high-energy part of the cross-sections with $T_e^{\text{th}} = 0$ corresponds to the values given in Table 5.1,

$$\sigma(E_\nu, T_e^{\text{th}}=0) \simeq \sigma_0\,\frac{E_\nu}{m_e}\left(g_1^2+\frac{1}{3}\,g_2^2\right)\,, \qquad \text{for} \qquad E_\nu \gg m_e\,, \tag{5.33}$$

which are proportional to E_ν in the laboratory frame.

Precise measurements of leptonic cross-sections such as those listed in Table 5.1 can provide the value of $\sin^2\vartheta_{\text{W}}$. In particular, the ratio of the ν_μ-e^- and $\bar{\nu}_\mu$-e^- cross-section measured in the same experiment has been used to extract the value of $\sin^2\vartheta_{\text{W}}$ by canceling out many systematic errors. The CHARM-II collaboration

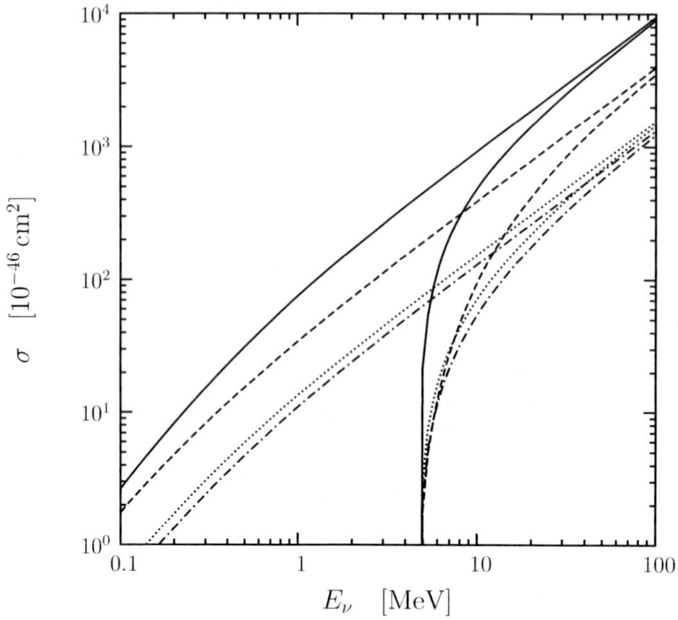

FIG. 5.5. Neutrino–electron cross-sections in eqn (5.32) as functions of the neutrino energy E_ν. Solid line: $\nu_e + e^- \to \nu_e + e^-$. Dashed line: $\bar{\nu}_e + e^- \to \bar{\nu}_e + e^-$. Dotted line: $\nu_{\mu,\tau} + e^- \to \nu_{\mu,\tau} + e^-$. Dash-dotted line: $\bar{\nu}_{\mu,\tau} + e^- \to \bar{\nu}_{\mu,\tau} + e^-$. For each scattering process the upper curve is the cross-section without a threshold for the kinetic energy of the recoil electron, whereas the lower curve is obtained with $T_e^{\text{th}} = 4.50\,\text{MeV}$, which corresponds to $E_\nu^{\text{th}} = 4.74\,\text{MeV}$, according to eqn (5.31).

obtained [1038]

$$\sin^2 \vartheta_{\text{W}} = 0.2324 \pm 0.0058 \pm 0.0059\,, \tag{5.34}$$

which is compatible with the *Review of Particle Physics* value in eqn (A.171), obtained from measurements at e^+e^- collider experiments (see section 5.1.3).

5.1.2 Neutrino–electron quasielastic scattering

Muon neutrinos with energy above the μ production threshold can interact with electrons through the quasielastic charged-current process

$$\nu_\mu + e^- \to \nu_e + \mu^-\,. \tag{5.35}$$

This process is sometimes called *inverse muon decay*.

In general, the threshold for a scattering process of type

$$\nu + A \to \sum_X X\,. \tag{5.36}$$

TABLE 5.2. Threshold neutrino energy in eqn (5.37) for some charged-current reactions used for neutrino detection.

Reaction	Masses	E_ν^{th}
$\nu_e + {}^{71}\text{Ga} \to {}^{71}\text{Ge} + e^-$	$m({}^{71}\text{Ga}) = 66050.093\,\text{MeV}$ $m({}^{71}\text{Ge}) = 66049.814\,\text{MeV}$	0.23 MeV
$\nu_e + {}^{37}\text{Cl} \to {}^{37}\text{Ar} + e^-$	$m({}^{37}\text{Cl}) = 34424.829\,\text{MeV}$ $m({}^{37}\text{Ar}) = 34425.132\,\text{MeV}$	0.82 MeV
$\bar{\nu}_e + p \to n + e^+$	$m_p = 938.272\,\text{MeV}$ $m_n = 939.565\,\text{MeV}$	1.81 MeV
$\nu + d \to p + n + \nu$	$m_d = 1875.613\,\text{MeV}$	2.23 MeV
$\nu_\mu + n \to p + \mu^-$	$m_\mu = 105.658\,\text{MeV}$	110.16 MeV
$\nu_\tau + n \to p + \tau^-$	$m_\tau = 1777.03\,\text{MeV}$	3.45 GeV
$\nu_\mu + e^- \to \mu^- + \nu_e$	$m_e = 0.511\,\text{MeV}$	10.92 GeV

with the target particle A at rest can be calculated by taking into account that the squared center-of-mass energy $s = 2E_\nu m_A + m_A^2$ (neglecting the neutrino mass) must be bigger than $(\sum_X m_X)^2$. This constraint leads to the energy threshold

$$E_\nu^{\text{th}} = \frac{(\sum_X m_X)^2}{2m_A} - \frac{m_A}{2} \,. \tag{5.37}$$

Some examples are given in Table 5.2, from which one can see that the energy threshold for the process in eqn (5.35) is quite high: $T_e^{\text{th}} = 10.92\,\text{GeV}$.

Figure 5.3b shows the tree-level Feynman diagram which contributes to the charged-current process in eqn (5.35). The effective Lagrangian for this process is

$$\mathscr{L}_{\text{eff}}^{(\nu_\mu + e^- \to \mu^- + \nu_e)} = -\frac{G_{\text{F}}}{\sqrt{2}} \left[\overline{\mu}\,\gamma^\rho \left(1 - \gamma^5\right) \nu_\mu\right] \left[\overline{\nu_e}\,\gamma_\rho \left(1 - \gamma^5\right) e\right] . \tag{5.38}$$

The differential cross-section is given by

$$\frac{\mathrm{d}\sigma}{\mathrm{d}\cos\theta} = \frac{G_{\text{F}}^2\, s}{2\,\pi} \left(1 - \frac{m_\mu^2}{s}\right) , \tag{5.39}$$

where θ is the scattering angle of the outgoing muon (analogous to that in Fig. 5.4), and $s = \left(p_{\nu_\mu} + p_e\right)^2$. Taking into account the W propagator in Fig. 5.3b would add a factor $m_W^2 / \left(q^2 - m_W^2\right)$, with $q^2 = t = \left(p_{\nu_\mu} - p_\mu\right)^2$. The total cross-section is given by

$$\sigma = \frac{G_{\text{F}}^2\, s}{\pi} \left(1 - \frac{m_\mu^2}{s}\right) , \tag{5.40}$$

which is in good agreement with the experimental data [1039].

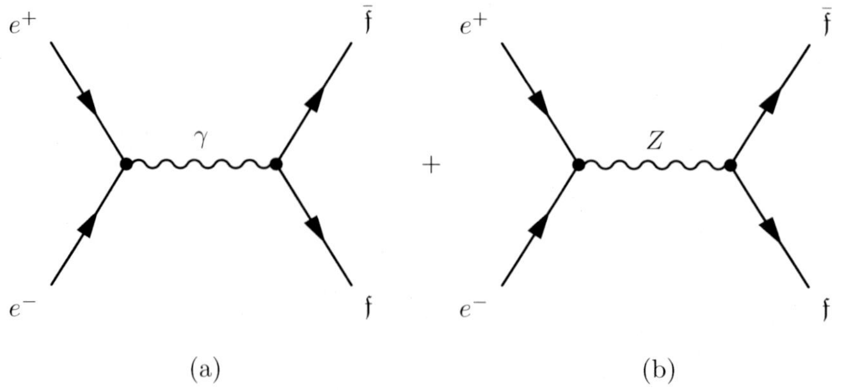

FIG. 5.6. The two tree-level s-channel Feynman diagrams for $e^+ + e^- \to \mathfrak{f} + \bar{\mathfrak{f}}$ through photon exchange (a) and Z-boson exchange (b).

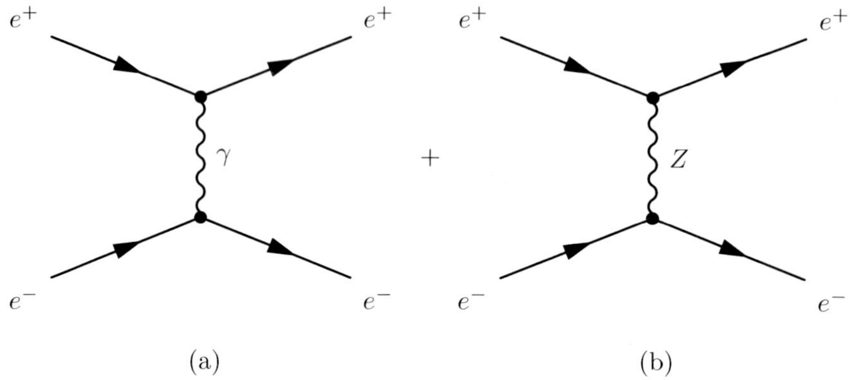

FIG. 5.7. The two tree-level t-channel Feynman diagrams for $e^+ + e^- \to e^+ + e^-$ through photon exchange (a) and Z-boson exchange (b).

5.1.3 $e^+ + e^- \to \nu + \bar{\nu}$ and the number of neutrino species

In this section, we discuss the process $e^+ + e^- \to \nu + \bar{\nu}$ and its application to the determination of the number of neutrino species in e^+-e^- collider experiments. First, we will consider a generic process $e^+ + e^- \to \mathfrak{f} + \bar{\mathfrak{f}}$, where $\mathfrak{f}$ is either a neutrino or a charged lepton or a quark. Figure 5.6 shows the two tree-level s-channel Feynman diagrams which contribute to this process through exchange of a photon or a Z-boson. In the case of $e^+ + e^- \to e^+ + e^-$ scattering (called *Bhabha scattering*), one must add the contributions of the two tree-level t-channel Feynman diagrams in Fig. 5.7 (see Ref. [233]).

For simplicity, we will present only the cross-section for $e^+ + e^- \to \mathfrak{f} + \bar{\mathfrak{f}}$, with $\mathfrak{f} \neq e$, which includes, in particular, the neutrino–antineutrino production process

$e^+ + e^- \to \nu + \bar{\nu}$. Working in the center-of-mass frame, which is the laboratory frame in this case, we call θ the scattering angle between the directions of the initial e^- and the final fermion $\mathfrak{f}$. The differential cross-section (neglecting fermion masses) is given by (see Refs. [720, 167, 931])

$$\frac{\mathrm{d}\sigma}{\mathrm{d}\cos\theta} = \frac{\pi}{2s} N_{\mathrm{c}}^{\mathfrak{f}} \Big\{ \alpha^2 \, q_{\mathfrak{f}}^2 \left(1 + \cos^2\theta\right) \\ + 16|\chi(s)|^2 \left[\left((g_V^e)^2 + (g_A^e)^2\right)\left((g_V^{\mathfrak{f}})^2 + (g_A^{\mathfrak{f}})^2\right)\left(1 + \cos^2\theta\right) + 8 g_V^e g_A^e g_V^{\mathfrak{f}} g_A^{\mathfrak{f}} \cos\theta \right] \\ - 8\, \alpha\, q_{\mathfrak{f}}\, \mathfrak{Re}[\chi(s)] \left[g_V^e g_V^{\mathfrak{f}} \left(1 + \cos^2\theta\right) + 2 g_A^e g_A^{\mathfrak{f}} \cos\theta \right] \Big\} , \tag{5.41}$$

where $N_{\mathrm{c}}^{\mathfrak{f}}$ is a color factor ($N_{\mathrm{c}}^{\mathfrak{f}} = 3$ for quarks and $N_{\mathrm{c}}^{\mathfrak{f}} = 1$ for leptons), $q_{\mathfrak{f}}$ is the charge of the fermion $\mathfrak{f}$ (in units of the elementary charge e) and

$$\chi(s) = \frac{G_{\mathrm{F}}\, m_Z^2}{8\pi\sqrt{2}} \, \frac{s}{s - m_Z^2 + i s \Gamma_Z / m_Z} . \tag{5.42}$$

The term $\chi(s)$ is due to the Z-boson propagator, modified according to the Breit-Wigner prescription in order to take into account the decay width Γ_Z of the Z (see ref. [248]). In the center-of-mass frame we have $\sqrt{s} = 2E_{\mathrm{cm}}$, where E_{cm} is the energy of the colliding electrons and positrons. The first line in eqn (5.41) is the electromagnetic contribution (γ exchange in Fig. 5.6a), the second line is the weak neutral-current contribution (Z exchange in Fig. 5.6b) and the third line is the interference of the two contributions.

Figure 5.8 shows the total hadronic cross-section as a function of $E_{\mathrm{cm}} = \sqrt{s}/2$. One can see that there is a maximum at $s = 0$, due to the photon propagator. With increasing energy, the photon propagator suppresses the cross-section, until the Z-resonance is reached, around $E_{\mathrm{cm}} \simeq m_Z$. Near the peak of the Z-resonance, the contribution of the γ-exchange diagram in Fig. 5.6a can be neglected. In this case, the total cross-section for $e^+ + e^- \to \mathfrak{f} + \bar{\mathfrak{f}}$, with $\mathfrak{f} \neq e$, can be written as

$$\sigma_{e^+e^- \to \mathfrak{f}\bar{\mathfrak{f}}}(s) = \frac{12\pi\, s}{m_Z^2} \, \frac{\Gamma_{Z\to e^+e^-}\, \Gamma_{Z\to\mathfrak{f}\bar{\mathfrak{f}}}}{\left(s - m_Z^2\right)^2 + s^2\Gamma_Z^2/m_Z^2} , \tag{5.43}$$

where $\Gamma_{Z\to e^+e^-}$ and $\Gamma_{Z\to\mathfrak{f}\bar{\mathfrak{f}}}$ are, respectively, the widths of the decays of the Z into e^+e^- and $\mathfrak{f}\bar{\mathfrak{f}}$ pairs. The partial Z-decay width into a $\mathfrak{f}\bar{\mathfrak{f}}$ pair is given by (neglecting the fermion mass)

$$\Gamma_{Z\to\mathfrak{f}\bar{\mathfrak{f}}} = \frac{G_{\mathrm{F}}\, m_Z^3}{6\sqrt{2}\pi} N_{\mathrm{c}}^{\mathfrak{f}} \left[(g_V^{\mathfrak{f}})^2 + (g_A^{\mathfrak{f}})^2 \right] . \tag{5.44}$$

The total decay width of the Z-boson is given by

$$\Gamma_Z = \sum_{\ell=e,\mu,\tau} \Gamma_{Z\to\ell\bar{\ell}} + \sum_{q\neq t} \Gamma_{Z\to q\bar{q}} + \Gamma_{\mathrm{inv}} , \tag{5.45}$$

where Γ_{inv} is the so-called *invisible width*, which represents the Z-decays into neutrinos (and maybe other invisible particles, which we will not consider). The invisible

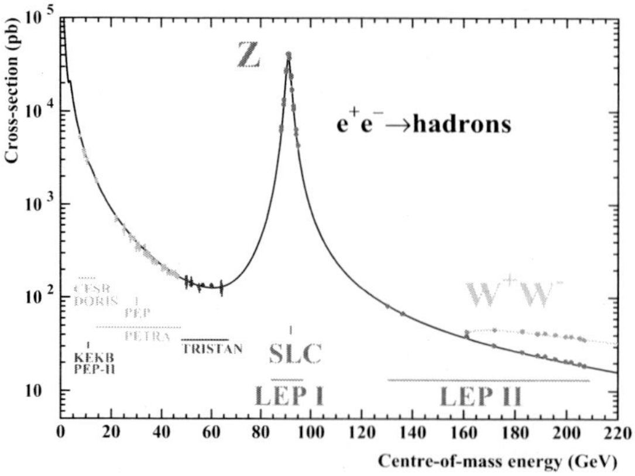

FIG. 5.8. The hadronic e^+-e^- cross-section as a function of the center-of-mass energy. The solid line is the prediction of the SM and the points are the experimental measurements [931]. Also indicated are the energy ranges of various e^+-e^- accelerators.

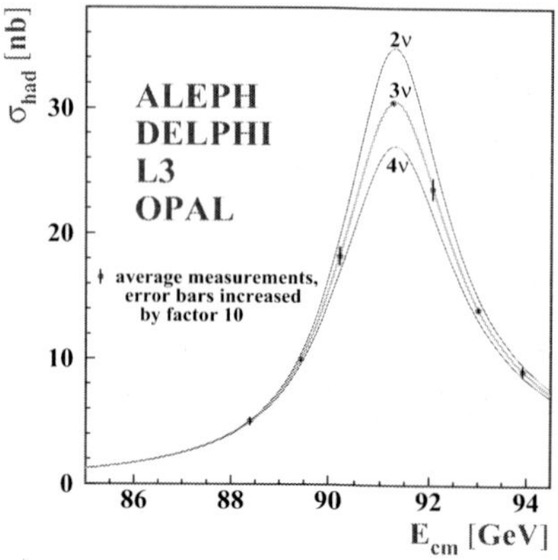

FIG. 5.9. The hadron production cross-section around the Z resonance measured by the four LEP experiments as a function of the center-of-mass energy $E_{\text{cm}} = \sqrt{s}/2$ [931]. The curves indicate the predicted cross–section for two, three, and four neutrino species with SM couplings and negligible mass.

width can be written as

$$\Gamma_{\text{inv}} = N_\nu \, \Gamma_{Z\to\nu\bar{\nu}} \,, \tag{5.46}$$

where N_ν is the number of neutrino species. More precisely, N_ν is the number of light active neutrinos, i.e. the number of neutrinos with mass much smaller than $m_Z/2$, which couple to the Z with the SM neutral current.

Figure 5.9 shows that the Z resonance is quite sensitive to the number of neutrino species. The experimental data shown in Fig. 5.9 have been obtained by the four LEP (Large Electron Positron) experiments ALEPH, DELPHI, L3, and OPAL at CERN. The small error bars are due to the high statistics accumulated by these experiments from about 1989 to 1995: about 17 million Z-decays. One can see, from Fig. 5.9 that the best fit of the Z resonance corresponds to $N_\nu \simeq 3$ and other integer values of N_ν are excluded.

A precise determination of N_ν from eqn (5.46) requires an accurate measurement of Γ_{inv} and a detailed calculation of $\Gamma_{Z\to\nu\bar{\nu}}$, taking into account radiative corrections which are not included in eqn (5.44). However, on the experimental side, the ratio $\Gamma_{\text{inv}}/\Gamma_{Z\to\ell\bar{\ell}}$, where $\Gamma_{Z\to\ell\bar{\ell}}$ is defined as the partial decay width into a massless charged lepton, is determined with higher precision than Γ_{inv}. On the theoretical side, the ratio $\Gamma_{Z\to\nu\bar{\nu}}/\Gamma_{Z\to\ell\bar{\ell}}$ is less dependent on the uncertainties of the SM. Therefore, the number of neutrino species is determined with higher accuracy from

$$N_\nu = \frac{\Gamma_{\text{inv}}}{\Gamma_{Z\to\ell\bar{\ell}}} \left(\frac{\Gamma_{Z\to\ell\bar{\ell}}}{\Gamma_{Z\to\nu\bar{\nu}}} \right)_{\text{SM}} , \tag{5.47}$$

where $\Gamma_{\text{inv}}/\Gamma_{Z\to\ell\bar{\ell}}$ is measured and $(\Gamma_{Z\to\nu\bar{\nu}}/\Gamma_{Z\to\ell\bar{\ell}})_{\text{SM}} = 1.99125 \pm 0.00083$ is calculated. The LEP experiments measured $\Gamma_{\text{inv}}/\Gamma_{Z\to\ell\bar{\ell}} = 5.943 \pm 0.016$, leading to [931]

$$N_\nu = 2.9840 \pm 0.0082\,. \tag{5.48}$$

This value is in good agreement with the observed three generations of fundamental fermions. It implies that there are no further generations with SM properties and a neutrino lighter than $m_Z/2$.

From eqns (5.41)–(5.44) one can see that the e^+-e^- cross-section is also sensitive to the value of the weak mixing angle ϑ_{W}, through the quantities g_V^{f} and g_A^{f} (see Table 3.6 on page 78). The analysis of LEP data, of data collected by the SLD (SLAC Large Detector) detector at the e^+-e^- Stanford Linear Collider (SLC) and the top mass measurement $m_t = 177.9 \pm 4.4\,\text{GeV}$ by the CDF (Collider Detector at Fermilab) and D0 experiments at the Tevatron p-$\bar{p}$ collider gave the value of $\sin^2\vartheta_{\text{W}}$ in eqn (A.171) [421, 400].

5.2 Hadron decays

5.2.1 Charged pion decay

The most common process to produce muon neutrinos and antineutrinos is through the decay of charged pions, which occur mainly through the two-body decay processes

$$\pi^\pm \to \ell^\pm + \overset{(-)}{\nu_\ell} \qquad (\ell = e, \mu)\,. \tag{5.49}$$

These processes are sometimes called $\pi_{\ell 2}$ decay, where ℓ gives the flavor of the final leptons and the number 2 represents a two-body decay. Charged pions are produced copiously in hadron interactions.

Let us consider the π^- decay process. The corresponding tree-level Feynman diagram is shown in Fig. 5.10a, where $\bar{u}\,d$ is the valence quark composition of a π^-. Considering these valence quarks, the Feynman rules in appendix E and the effective low-energy four-fermion CC vertex in eqn (3.226) lead to the amplitude

$$\mathcal{A}_{\pi^-\to\ell^-\bar{\nu}_\ell} = -i\,\frac{G_{\text{F}}}{\sqrt{2}}\,V_{ud}\,\overline{v_u}(p_u)\,\gamma^\rho\left(1-\gamma^5\right)u_d(p_d)\,\overline{u_\nu}(p_\nu)\,\gamma_\rho\left(1-\gamma^5\right)v_\ell(p_\ell)\,. \tag{5.50}$$

This amplitude, however, would be correct if the $\bar{u}$ and d quarks were free. Since they are bound as valence quarks in the pion, the quark current must be replaced with the appropriate hadronic transition matrix element which takes into account the incalculable effect of strong interactions:

$$\overline{v_u}(p_u)\,\gamma^\rho\left(1-\gamma^5\right)u_d(p_d) \to \frac{1}{m_\pi}\,\langle 0|h_W^\rho(0)|\pi^-(p_\pi)\rangle\,, \tag{5.51}$$

where

$$h_W^\rho(x) = \overline{u}(x)\gamma^\mu\left(1-\gamma^5\right)d(x) \tag{5.52}$$

is the relevant part of the quark charged current in eqn (3.176), stripped of the factor V_{ud}. The factor $1/m_\pi$ serves to keep the dimensions right (the left-hand side has

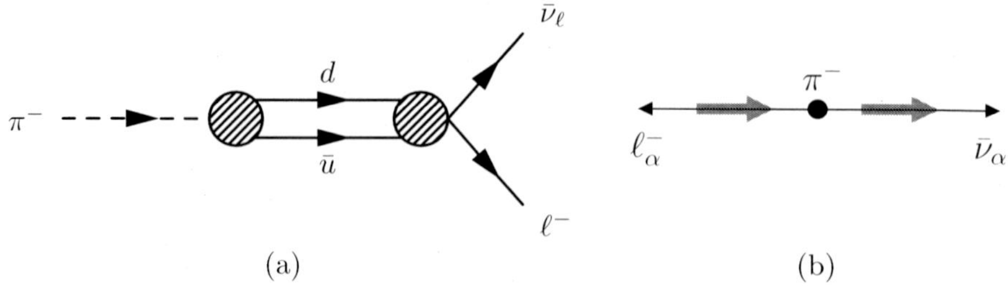

FIG. 5.10. (a) Tree-level Feynman diagram of the π^- decay process in eqn (5.49), with the π^- represented as a bound state of the valence quarks $\bar{u}\,d$. The blobs represent strong interaction binding. (b) Configuration of the final state in the pion rest frame in the limit $m_\ell \to 0$. The thin and thick arrows represent, respectively, momentum and spin.

dimension of energy, the current h_W^ρ has dimension of E^3, and the one-pion state has dimension of E^{-1}, with the normalization $\langle \pi^-(p_\pi)|\pi^-(p'_\pi)\rangle = (2\pi)^3 2E_\pi \delta^3(\vec{p}_\pi - \vec{p}\,'_\pi)$ as the one-fermion state in eqn (2.234)). Since the leptonic current is of the $V-A$ type, the hadronic matrix element in eqn (5.51) must be a linear combination of a vector and an axial-vector, constructed from the available kinematical quantities. Since the pion is a pseudoscalar particle ($J^P = 0^-$), the four-momentum p_π is the only available four-vector. Therefore, the hadronic matrix element can be written as[21]

$$\langle 0|h_W^\rho(0)|\pi^-(p_\pi)\rangle = i\, f_\pi\, p_\pi^\rho\,, \tag{5.53}$$

where the coefficient f_π has dimension of energy. In principle, f_π may depend on p_π^2, but since $p_\pi^2 = m_\pi^2$, it is a constant, called the *pion decay constant*. The pion-decay amplitude is given by

$$\mathcal{A}_{\pi^-\to\ell^-\bar{\nu}_\ell} = \frac{G_F}{\sqrt{2}}\, V_{ud}\, f_\pi\, \overline{u_\nu}(p_\nu)\, \frac{\not{p}_\pi}{m_\pi}\left(1-\gamma^5\right) v_\ell(p_\ell)\,. \tag{5.54}$$

Using the energy–momentum conservation relation $p_\pi = p_\ell + p_\nu$ and the Dirac equations of the charged lepton and the massless antineutrino, the decay amplitude

[21] Note that only the axial part of the quark current, $a_W^\rho = \overline{u}\gamma^\mu\gamma^5 d$, contributes to the hadronic matrix element. Since the pion is a pseudoscalar particle, under space inversion $\mathsf{U}_\mathrm{P}\,|\pi^-(p_\pi)\rangle = -|\pi^-(p_{\pi\mathrm{P}})\rangle$, where U_P is the unitary operator of the parity transformation (see section 2.11.2), $p_\pi^\rho = (p_\pi^0, \vec{p}_\pi)$ and $p_{\pi\mathrm{P}}^\rho = (p_\pi^0, -\vec{p}_\pi)$. Under parity, the vector part of the quark current, $v_W^\rho = \overline{u}\gamma^\mu d$ transforms as $\mathsf{U}_\mathrm{P} v_W^\rho \mathsf{U}_\mathrm{P}^\dagger = v_{W\mathrm{P}}^\rho$, where $v_W^\rho = (v^0, \vec{v})$ and $v_{W\mathrm{P}}^\rho = (v^0, -\vec{v})$. Taking into account that the vacuum is parity-invariant, for the hadronic vector matrix element $\langle 0|v_W^\rho(0)|\pi^-(p_\pi)\rangle = i\, f_v\, p_\pi^\rho$ we have $\langle 0|v_W^\rho|\pi^-(p_\pi)\rangle = \langle 0|\mathsf{U}_\mathrm{P}^\dagger \mathsf{U}_\mathrm{P} v_W^\rho \mathsf{U}_\mathrm{P}^\dagger \mathsf{U}_\mathrm{P}|\pi^-(p_\pi)\rangle = -\langle 0|v_{W\mathrm{P}}^\rho|\pi^-(p_{\pi\mathrm{P}})\rangle = -i\, f_v\, p_\pi^\rho$. Thus, $f_v = 0$, i.e. the hadronic vector matrix element vanishes. On the other hand, parity does not constrain the hadronic axial matrix element, because under parity $\mathsf{U}_\mathrm{P} a_W^\rho \mathsf{U}_\mathrm{P}^\dagger = -a_{W\mathrm{P}}^\rho$, where $a_W^\rho = (a^0, \vec{a})$ and $a_{W\mathrm{P}}^\rho = (a^0, -\vec{a})$.

can be simplified to

$$\mathcal{A}_{\pi^- \to \ell^- \bar{\nu}_\ell} = -\frac{G_F}{\sqrt{2}} V_{ud} f_\pi \frac{m_\ell}{m_\pi} \overline{u_\nu}(p_\nu) \left(1 + \gamma^5\right) v_\ell(p_\ell) . \tag{5.55}$$

Hence, the matrix element is proportional to the mass of the final charged lepton and vanishes in the limit $m_\ell \to 0$. This is due to the helicities of the final leptons produced by the $V - A$ CC weak interaction: as illustrated in Fig. 5.10b, in the pion rest frame the charged lepton and antineutrino are emitted in opposite directions. The final massless antineutrino is right-handed, i.e. it has positive helicity (see section 2.7.1). In the limit $m_\ell \to 0$, the final charged lepton is left-handed. In this case, the total angular momentum of the final state is equal to unity and the decay of the pseudoscalar pion is forbidden by angular momentum conservation. In other words, the decay of the pion is allowed only if the charged lepton is massive and the $V - A$ CC weak interaction can generate it in a right-handed state, with positive helicity, such that the final state has zero angular momentum, as the initial state.

From the amplitude in eqn (5.55) one can derive the decay rate

$$\Gamma_{\pi^\pm \to \ell^\pm \overset{(-)}{\nu_\ell}} = \frac{G_F^2}{8\pi} {f_\pi}^2 |V_{ud}|^2 m_\pi m_\ell^2 \left(1 - \frac{m_\ell^2}{m_\pi^2}\right)^2 , \tag{5.56}$$

which is the same for the CP-conjugated π^+ and π^- decay processes in eqn (5.49). From the charged pion lifetime in eqn (A.162) and the branching ratio $\Gamma_{\pi^\pm \to \ell^\pm \overset{(-)}{\nu_\ell}}/\Gamma_{\pi^\pm} = 0.999\,877\,0 \pm 0.000\,000\,4$ [400], after taking into account radiative corrections, one obtains [775, 1000]

$$f_\pi = 130.7 \pm 0.1 \pm 0.36\,\text{MeV} , \tag{5.57}$$

where the first uncertainty comes from the experimental uncertainty on $|V_{ud}|$ and the second uncertainty comes from the uncertainty of the calculation of radiative corrections.

In the decay rate $\Gamma_{\pi^\pm \to \ell^\pm \overset{(-)}{\nu_\ell}}$, the factor m_ℓ^2, due to the angular momentum problem discussed above, suppresses the decay rate for small values of the final charged lepton mass. On the other hand, the factor $\left(1 - m_\ell^2/m_\pi^2\right)^2$, due to the phase space of the decay, is maximal for $m_\ell \to 0$. However, the first factor dominates in the ratio between the decay rates into an electron and a muon:

$$R_{e/\mu} = \frac{\Gamma_{\pi^\pm \to e^\pm \overset{(-)}{\nu_e}}}{\Gamma_{\pi^\pm \to \mu^\pm \overset{(-)}{\nu_\mu}}} = \frac{m_e^2 \left(1 - m_e^2/m_\pi^2\right)^2}{m_\mu^2 \left(1 - m_\mu^2/m_\pi^2\right)^2} \simeq 1.28 \times 10^{-4} , \tag{5.58}$$

where we used the best-fit values of the electron, muon, and pion masses in section A.5. The more precise theoretical prediction including radiative corrections is

[775]

$$R^{\rm the}_{e/\mu} = (1.2352 \pm 0.0005) \times 10^{-4}\,, \tag{5.59}$$

in good agreement with the measured value [400]

$$R^{\rm exp}_{e/\mu} = (1.230 \pm 0.004) \times 10^{-4}\,. \tag{5.60}$$

Let us finally mention that high-energy hadron interactions produce, besides pions and other particles, charged kaons, whose decay can generate neutrinos. Since kaons are pseudoscalar particles, just like pions, the $K_{\ell 2}$ decay of charged kaons,

$$K^{\pm} \to \ell^{\pm} + \overset{(-)}{\nu_\ell} \qquad (\ell = e, \mu)\,, \tag{5.61}$$

can be treated as the decay of charged pions. In analogy with eqn (5.53), the hadronic matrix element of the decay of a $K^- = \bar{u}\,s$ can be written as

$$\langle 0 | h^{\rho}_{W}(0) | K^-(p_K) \rangle = i\, f_K\, p^{\rho}_{K}\,, \tag{5.62}$$

where $h^{\rho}_{W}(x) = \overline{u}(x)\gamma^{\mu}\left(1 - \gamma^5\right) s(x)$ is the relevant part of the quark charged current in eqn (3.176), stripped of the factor V_{us}. The coefficient f_K is the *kaon decay constant*, with experimental value [400]

$$f_K = 159.8 \pm 1.4 \pm 0.44\,\text{MeV}\,, \tag{5.63}$$

where the first uncertainty comes from the experimental uncertainty on $|V_{us}|$ and the second comes from that of the calculation of radiative corrections. The tree-level decay rate of $K^{\pm}$ can be obtained from the charged pion tree-level decay rate in eqn (5.56) with the replacements $f_{\pi}{}^2|V_{ud}|^2 \to f_K{}^2|V_{us}|^2$ and $m_\pi \to m_K$.

5.2.2 Neutron decay, CVC, and PCAC

The neutron decay process

$$n \to p + e^- + \bar{\nu}_e \tag{5.64}$$

is the basic β-decay process, which was considered first by Fermi in his theory of β-decay [432, 431]. The tree-level Feynman diagram of neutron decay is shown in Fig. 5.11, where the neutron and proton are represented in terms of their valence quarks, $n = udd$ and $p = uud$. Considering the neutron and proton valence quarks, neutron decay is a $d \to u$ transition. In this case, the Feynman rules in appendix E and the effective low-energy four-fermion CC vertex in eqn (3.226) lead to the amplitude

$$\mathcal{A}_{n \to p + e^- \bar{\nu}_e} = -i\,\frac{G_{\rm F}}{\sqrt{2}}\,V_{ud}\,\overline{u_u}(p_u)\,\gamma^{\rho}\left(1 - \gamma^5\right) u_d(p_d)\,\overline{u_e}(p_e)\,\gamma_{\rho}\left(1 - \gamma^5\right) v_{\nu_e}(p_{\nu_e})\,. \tag{5.65}$$

However, the valence quarks are not free, but bound in the nucleons. Thus, as in the case of pion decay discussed in section 5.2.1, the quark current must be replaced

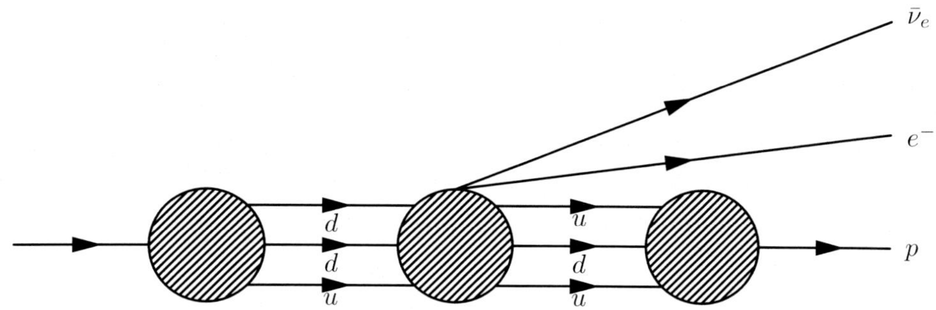

FIG. 5.11. Tree-level Feynman diagram of the neutron decay process in eqn (5.64), with the neutron and proton represented as a bound states of valence quarks. The blobs represent strong interaction binding.

by the appropriate hadronic transition matrix element which takes into account the incalculable effect of strong interactions:

$$\overline{u_u}(p_u)\,\gamma^\rho\left(1-\gamma^5\right)u_d(p_d)\to\langle p(p_p)|h_W^\rho(0)|n(p_n)\rangle\,, \tag{5.66}$$

with $h_W^\rho(x)$ given in eqn (5.52).

In general, the hadronic matrix element must be a linear combination of a vector and an axial-vector, which can only be constructed from the available kinematical quantities. These are the neutron and proton four-momenta p_n and p_p (we consider unpolarized particles). It is convenient to split the vector and axial parts of the hadronic current:

$$h_W^\rho(x)=v_W^\rho(x)-a_W^\rho(x)\,, \tag{5.67}$$

with

$$v_W^\rho(x)=\overline{u}(x)\,\gamma^\mu\,d(x)\,, \tag{5.68}$$

$$a_W^\rho(x)=\overline{u}(x)\,\gamma^\mu\,\gamma^5\,d(x)\,. \tag{5.69}$$

Since the initial neutron and the final proton are free fermions, in momentum space they are described, respectively, by spinors $u_n(p_n)$ and $u_p(p_p)$ which obey the corresponding Dirac equations in momentum space of the type in eqn (2.144). Taking into account the fact that the only vector bilinear of spinors is of the type in eqn (2.66), the most general hadronic vector matrix element can be written as

$$\begin{aligned}\langle p(p_p)|v_W^\rho(0)|n(p_n)\rangle=\overline{u_p}(p_p)\left[f_1(q^2)\,\gamma^\rho+f_2(q^2)\left(p_n^\rho+p_p^\rho\right)\right.\\ \left.+f_3(q^2)\left(p_n^\rho-p_p^\rho\right)\right]u_n(p_n)\,,\end{aligned} \tag{5.70}$$

where $f_1(q^2)$, $f_2(q^2)$, and $f_3(q^2)$ are *form factors*, which depend on the only scalar quantity available which can be constructed from p_n and p_p:

$$p_n\cdot p_p=\frac{1}{2}\left[m_n^2+m_p^2-(p_p-p_n)^2\right]\equiv\frac{1}{2}\left[m_n^2+m_p^2-q^2\right]\,, \tag{5.71}$$

where $q\equiv p_p-p_n$. Since the neutron and proton masses are constants, the form factors depend only on q^2.

Using the Dirac equation for the u-spinors (eqn (2.144)), we have

$$\overline{u_p}(p_p) i\,\sigma^{\rho\eta}\,q_\eta u_n(p_n) = \overline{u_p}(p_p)\left[p_p^\rho + p_n^\rho - (m_n + m_p)\,\gamma^\rho\right] u_n(p_n)\,. \tag{5.72}$$

Using this relation and approximating $m_n \simeq m_p \simeq m_N$, where m_N is the nucleon mass, the vector matrix element in eqn (5.70) can be written in the usual form

$$\langle p(p_p)|v_W^\rho(0)|n(p_n)\rangle = \overline{u_p}(p_p)\left[\gamma^\rho\,F_1(Q^2) + \frac{i\,\sigma^{\rho\eta}\,q_\eta}{2\,m_N}\,F_2(Q^2) + \frac{q^\rho}{m_N}\,F_3(Q^2)\right] u_n(p_n)\,. \tag{5.73}$$

In a similar way, the most general axial matrix element can be written as

$$\langle p(p_p)|a_W^\rho(0)|n(p_n)\rangle = \overline{u_p}(p_p)\left[\gamma^\rho\gamma^5 G_A(Q^2) + \frac{q^\rho}{m_N}\,\gamma^5 G_P(Q^2) + \frac{p_p^\rho + p_n^\rho}{m_N}\,\gamma^5 G_3(Q^2)\right] u_n(p_n)\,. \tag{5.74}$$

The form factors have been written, as usual, as functions of $Q^2 \equiv -q^2$, although in neutron decay $q^2 > 0$. The reason is that in neutron decay q^2 is small: from kinematics one can find

$$m_e^2 \le q^2 \le (m_n - m_p)^2\,, \tag{5.75}$$

with $m_e \simeq 0.51\,\text{MeV}$ and $m_n - m_p \simeq 1.29\,\text{MeV}$. Hence, the dependence on Q^2 of the form factors is practically irrelevant in neutron decay. On the other hand, the same form factors enter in the quasielastic charged-current reactions discussed in section 5.3.1, where Q^2 can be large and the dependence on Q^2 of the form factors is important.

It can be shown that the invariance under time reversal of strong interactions implies that the six form factors in eqns (5.73) and (5.74) are real (see Ref. [227]). Moreover, the invariance of strong interactions under isospin transformations implies that

$$F_3(Q^2) = 0\,, \qquad G_3(Q^2) = 0\,. \tag{5.76}$$

These are the form factors of the so-called *second-class currents* [1054], whose absence has been verified by experiments.

Isospin is the approximate symmetry obtained neglecting the small mass difference between the up and down quarks. Under this approximation, the Lagrangian of the quark fields, including strong interactions, is invariant under the global $\text{SU}(2)_\text{I}$ gauge transformations

$$Q(x) \to \mathsf{U}(\underline{\theta})\,Q(x)\,\mathsf{U}^\dagger(\underline{\theta}) = \exp\left(i\,\underline{\theta}\cdot\underline{T}\right) Q(x)\,, \tag{5.77}$$

where $Q(x)$ is the quark doublet,

$$Q(x) = \begin{pmatrix} u(x) \\ d(x) \end{pmatrix}, \tag{5.78}$$

and $\mathsf{U}(\underline{\theta})$ is the unitary isospin operator, which depends on a set of three real numbers $\underline{\theta} = (\theta_1, \theta_2, \theta_3)$. The set of matrices $\underline{T} = (T_1, T_2, T_3)$ are a two-dimensional representation of the generators of the $\mathrm{SU}(2)_{\mathrm{I}}$ isospin group. Since the generators must satisfy angular momentum commutation relations of the type in eqn (3.2), we have $T_a = \tau_a/2$, where τ_1, τ_2, τ_3, are the three Pauli matrices, as in chapter 3 (see footnote 14 on page 71).

Defining the isospin raising and lowering matrices

$$T_\pm = T_1 \pm i\, T_2\,, \tag{5.79}$$

the vector and axial currents in eqns (5.68) and (5.69) can be written as

$$v_W^\rho(x) = \overline{Q}(x)\, \gamma^\mu\, T_+\, Q(x)\,, \tag{5.80}$$

$$a_W^\rho(x) = \overline{Q}(x)\, \gamma^\mu\, \gamma^5\, T_+\, Q(x)\,. \tag{5.81}$$

Let us consider an isospin rotation of an angle π around the second axis:

$$Q(x) \to \mathsf{U}_\pi\, Q(x)\, \mathsf{U}_\pi^\dagger = e^{i\pi T_2}\, Q(x) = i\, \tau_2\, Q(x) = \begin{pmatrix} d(x) \\ -u(x) \end{pmatrix}, \tag{5.82}$$

where $\mathsf{U}_\pi \equiv \mathsf{U}(0,\pi,0)$. Invariance under such a transformation is called *charge symmetry*. Under this transformation

$$v_W^\rho(x) \to \mathsf{U}_\pi\, v_W^\rho(x)\, \mathsf{U}_\pi^\dagger = -v^{\rho\dagger}(x)\,, \tag{5.83}$$

$$a_W^\rho(x) \to \mathsf{U}_\pi\, a_W^\rho(x)\, \mathsf{U}_\pi^\dagger = -a^{\rho\dagger}(x)\,. \tag{5.84}$$

Since under such transformation $u \to d$ and $d \to -u$, we have $p = uud \to -ddu = -n$ and $n = udd \to duu = p$. Thus, the proton and neutron states transform as

$$\mathsf{U}_\pi\, |p(p_p)\rangle = -|n(p_p)\rangle\,, \qquad \mathsf{U}_\pi\, |n(p_n)\rangle = |p(p_n)\rangle\,. \tag{5.85}$$

Considering the vector matrix element in eqn (5.73), we have

$$\begin{aligned}\langle p(p_p)|v_W^\rho(0)|n(p_n)\rangle &= \langle p(p_p)|\mathsf{U}_\pi^\dagger \mathsf{U}_\pi v_W^\rho(0)\mathsf{U}_\pi^\dagger \mathsf{U}_\pi|n(p_n)\rangle = \langle n(p_p)|v^{\rho\dagger}(0)|p(p_n)\rangle \\ &= \langle p(p_n)|v_W^\rho(0)|n(p_p)\rangle^*\,. \end{aligned}\tag{5.86}$$

Substituting the expression on the right-hand side of eqn (5.73), one can find that the equality in eqn (5.86) implies that $F_3(Q^2) = 0$. In an analogous way, for the axial matrix element in eqn (5.74), one can obtain the equality

$$\langle p(p_p)|a_W^\rho(0)|n(p_n)\rangle = \langle p(p_n)|a_W^\rho(0)|n(p_p)\rangle^*\,, \tag{5.87}$$

which implies $G_3(Q^2) = 0$. Therefore, the charged-current vector and axial matrix element are given by

$$\langle p(p_p)|v_W^\rho(0)|n(p_n)\rangle = \overline{u_p}(p_p)\left[\gamma^\rho\, F_1(Q^2) + \frac{i}{2\,m_N}\, \sigma^{\rho\eta}\, q_\eta\, F_2(Q^2)\right] u_n(p_n)\,, \tag{5.88}$$

$$\langle p(p_p)|a_W^\rho(0)|n(p_n)\rangle = \overline{u_p}(p_p)\left[\gamma^\rho\, \gamma^5\, G_A(Q^2) + \frac{q^\rho}{m_N}\, \gamma^5\, G_P(Q^2)\right] u_n(p_n)\,. \tag{5.89}$$

The form factors $F_1(Q^2)$, $F_2(Q^2)$, $G_A(Q^2)$, and $G_P(Q^2)$ are called, respectively, *Dirac*, *Pauli*, *axial*, and *pseudoscalar* weak charged-current form factors of the nucleon.

A further improvement of the expression for the vector matrix element follows from the isospin symmetry, which allows us to express the form factors $F_1(Q^2)$ and $F_2(Q^2)$ in terms of the electromagnetic form factors of the nucleons, whose values for $q^2 = 0$ are known. Noether's theorem (see section C.7) implies that the isovector currents

$$v_a^\rho(x) = \overline{Q}(x)\,\gamma^\rho\,T_a\,Q(x) \tag{5.90}$$

are conserved:

$$\partial_\rho v_a^\rho(x) = 0\,. \tag{5.91}$$

This property, which was formulated as an hypothesis in the 1950s [519, 434], is called the *conserved vector current* (CVC) hypothesis. It is now understood that it is a consequence of the isospin invariance of the strong interaction QCD Lagrangian. As explained in appendix C, the associated charges

$$\mathsf{T}_a = \int \mathrm{d}^3x\, v_a^0(x) = \int \mathrm{d}^3x\, Q^\dagger(x)\,T_a\,Q(x) \tag{5.92}$$

form a representation of the generators of the $\mathrm{SU}(2)_\mathrm{I}$ isospin group. They are Hermitian operators ($\mathsf{T}_a^\dagger = \mathsf{T}_a$), which satisfy the group commutation relations (see eqn (C.57)),

$$[\mathsf{T}_a\,,\,\mathsf{T}_b] = i\,\epsilon_{abc}\,\mathsf{T}_c\,, \tag{5.93}$$

and generate the isospin transformations in eqn (5.77) through

$$[\mathsf{T}_a\,,\,Q(x)] = -T_a\,Q(x)\,. \tag{5.94}$$

The unitary isospin operator in eqn (5.77) is given by $\mathsf{U}(\underline{\theta}) = \exp(-i\,\underline{\theta}\cdot\underline{\mathsf{T}})$, with $\underline{\mathsf{T}} = (\mathsf{T}_1, \mathsf{T}_2, \mathsf{T}_3)$.

Using the relations in eqn (5.94) and the expression of the conserved current in eqn (5.90), one can derive the useful commutation relations

$$[\mathsf{T}_a\,,\,v_b^\rho(x)] = i\,\epsilon_{abc}\,v_c^\rho(x)\,, \tag{5.95}$$

which imply

$$[\mathsf{T}_\pm\,,\,v_3^\rho(x)] = \mp\,v_\pm^\rho(x)\,, \tag{5.96}$$

where

$$\mathsf{T}_\pm = \mathsf{T}_1 \pm i\,\mathsf{T}_2\,, \qquad v_\pm^\rho = v_1^\rho \pm i\,v_2^\rho\,. \tag{5.97}$$

Now, note that v_+^ρ is the vector current v_W^ρ in eqn (5.80), which enters in the vector matrix element in eqn (5.88). Using eqn (5.96), we can write the vector matrix element as

$$\langle p(p_p)|v_W^\rho(0)|n(p_n)\rangle = \langle p(p_p)|\,[v_3^\rho(0)\,,\,\mathsf{T}_\pm]\,|n(p_n)\rangle\,. \tag{5.98}$$

Since $\mathsf{T}_\pm$ are the isospin raising and lowering operators and $\mathsf{T}_-^\dagger = \mathsf{T}_+$, we have

$$\mathsf{T}_+\,|n(p_n)\rangle = |p(p_n)\rangle\,, \tag{5.99}$$

$$\mathsf{T}_{-}\,|p(p_p)\rangle = |n(p_p)\rangle \qquad \Longrightarrow \qquad \langle p(p_p)|\,\mathsf{T}_{+} = \langle n(p_p)|\,, \tag{5.100}$$

which imply

$$\langle p(p_p)|v_W^\rho(0)|n(p_n)\rangle = \langle p(p_p)|v_3^\rho(0)|p(p_n)\rangle - \langle n(p_p)|v_3^\rho(0)|n(p_n)\rangle\,. \tag{5.101}$$

This relation is useful, because the proton and nucleon matrix elements of $v_3^\rho(0)$ are the vectorial parts of the corresponding electromagnetic matrix elements. From eqn (3.179), the electromagnetic current of quarks,

$$j_{\gamma,\mathrm{Q}}^\rho = \frac{2}{3}\sum_{\alpha=u,c,t} \overline{q_\alpha^U}\,\gamma^\mu\,q_\alpha^U - \frac{1}{3}\sum_{\alpha=d,s,b} \overline{q_\alpha^D}\,\gamma^\mu\,q_\alpha^D\,, \tag{5.102}$$

can be written as

$$j_{\gamma,\mathrm{Q}}^\rho = v_3^\rho + v_0^\rho\,, \tag{5.103}$$

where

$$v_0^\rho = \frac{1}{6}\,\overline{Q}\,\gamma^\rho\,Q + \frac{2}{3}\sum_{q=c,t} \overline{q}\,\gamma^\mu\,q - \frac{1}{3}\sum_{q=s,b} \overline{q}\,\gamma^\mu\,q \tag{5.104}$$

is the isoscalar part of the current. The difference of the proton and neutron electromagnetic matrix elements is equal to the difference of the proton and neutron matrix elements of v_3^ρ in eqn (5.101):

$$\begin{aligned}\langle p(p_p)|j_{\gamma,\mathrm{Q}}^\rho(0)|p(p_n)\rangle &- \langle n(p_p)|j_{\gamma,\mathrm{Q}}^\rho(0)|n(p_n)\rangle \\ &= \langle p(p_p)|v_3^\rho(0)|p(p_n)\rangle - \langle n(p_p)|v_3^\rho(0)|n(p_n)\rangle\,.\end{aligned} \tag{5.105}$$

The contributions of the isoscalar current cancel out, because

$$\langle p(p_p)|v_0^\rho|p(p_n)\rangle = \langle p(p_p)|v_0^\rho \mathsf{T}_{+}|n(p_n)\rangle = \langle p(p_p)|\mathsf{T}_{+}v_0^\rho|n(p_n)\rangle = \langle n(p_p)|v_0^\rho|n(p_n)\rangle\,, \tag{5.106}$$

since $[\mathsf{T}_a\,,\,v_0^\rho(x)] = 0$. Thus, eqns (5.101) and (5.105) imply the following relation between the weak charged-current vector matrix element and the electromagnetic matrix elements of the nucleons:

$$\langle p(p_p)|v_W^\rho(0)|n(p_n)\rangle = \langle p(p_p)|j_{\gamma,\mathrm{Q}}^\rho(0)|p(p_n)\rangle - \langle n(p_p)|j_{\gamma,\mathrm{Q}}^\rho(0)|n(p_n)\rangle\,. \tag{5.107}$$

In analogy with eqn (5.88), the nucleon matrix element of the electromagnetic current can be written as

$$\langle N(p_2)|j_{\gamma,\mathrm{Q}}^\rho(0)|N(p_1)\rangle = \overline{u_N}(p_2)\left[\gamma^\rho\,F_1^N(Q^2) + \frac{i}{2\,m_N}\,\sigma^{\rho\eta}\,q_\eta\,F_2^N(Q^2)\right]u_N(p_1)\,, \tag{5.108}$$

where $N = p, n$ and $q = p_2 - p_1$. It can be shown (see Ref. [227]) that an additional term proportional to $q^\rho\overline{u_N}(p_2)u_N(p_1)$ vanishes, because the electromagnetic current is conserved, $\partial_\rho j_\gamma^\rho = 0$. From eqns (5.88), (5.107), and (5.108), we finally obtain the relations between the charged-current vector form factors and the electromagnetic form factors

$$F_1(Q^2) = F_1^p(Q^2) - F_1^n(Q^2)\,, \tag{5.109}$$

$$F_2(Q^2) = F_2^p(Q^2) - F_2^n(Q^2) . \tag{5.110}$$

The electromagnetic form factors $F_1^N(Q^2)$ and $F_2^N(Q^2)$ are called, respectively, *Dirac* and *Pauli* form factors of the nucleon N. Since for $q^2 = 0$ the Dirac form factors reduce to the electric charges of the nucleons[22] and the Pauli form factors reduce to the anomalous magnetic moments of the nucleons, we have

$$F_1^p(0) = 1 , \qquad F_1^n(0) = 0 , \tag{5.111}$$

$$F_2^p(0) = \frac{\mu_p}{\mu_{\rm N}} - 1 , \qquad F_2^n(0) = \frac{\mu_n}{\mu_{\rm N}} , \tag{5.112}$$

where $\mu_{\rm N} \equiv e\hbar/2m_p$ is the nuclear magneton and μ_p and μ_n are, respectively, the proton and neutron magnetic moments, which have the values in eqns (A.168) and (A.159). It is often convenient to use the Sachs electric and magnetic form factors

$$G_E^N(Q^2) = F_1^N(Q^2) - \frac{Q^2}{4m_N^2} F_2^N(Q^2) , \tag{5.113}$$

$$G_M^N(Q^2) = F_1^N(Q^2) + F_2^N(Q^2) , \tag{5.114}$$

with

$$G_E^p(0) = 1 , \qquad G_E^n(0) = 0 , \tag{5.115}$$

$$G_M^p(0) = \mu_p/\mu_{\rm N} , \qquad G_M^n(0) = \mu_n/\mu_{\rm N} . \tag{5.116}$$

The electron scattering data are usually fitted by assuming the same Q^2 dependence for the electromagnetic form factors of the nucleons[23]:

$$G_E^p(Q^2) = G_D(Q^2) , \qquad G_E^n(Q^2) = 0 , \tag{5.117}$$

$$G_M^p(Q^2) = \frac{\mu_p}{\mu_{\rm N}} G_D(Q^2) , \qquad G_M^n(Q^2) = \frac{\mu_n}{\mu_{\rm N}} G_D(Q^2) , \tag{5.118}$$

where $G_D(Q^2)$ is the *dipole* function

$$G_D(Q^2) = \left(1 + \frac{Q^2}{M_V^2} \right)^{-2} , \tag{5.119}$$

with $M_V \simeq 0.84\,\text{GeV}$.

The electric and magnetic form factors characterize the nucleon matrix element of the electromagnetic current in the so-called *Breit frame*, which is the reference frame in which $q^0 = 0$ and $|\vec{q}| = \sqrt{Q^2}$. In order to determine the Breit frame

[22] For $q = 0$, we have, from eqn (2.159), $\langle N(p)|j^0_{\gamma,\mathsf{Q}}(0)|N(p)\rangle = 2EF_1^N(0)$. Since, considering a finite total volume V and the normalization of states in eqn (2.236) and using eqn (C.32), we have $\langle N(p)|j^0_{\gamma,\mathsf{Q}}(0)|N(p)\rangle = \frac{1}{V}\int \mathrm{d}^3x \, \langle N(p)|e^{-ix\cdot\mathsf{P}} j^0_{\gamma,\mathsf{Q}}(x) e^{ix\cdot\mathsf{P}}|N(p)\rangle = \frac{1}{V}\langle N(p)|\mathsf{Q}|N(p)\rangle = 2Eq_N$, where Q is the electric charge operator (see eqns (2.90) and (2.247)) and q_N is the electric charge of the nucleon N. Therefore, $F_1^N(0) = q_N$.

[23] A more accurate fit of the data is discussed in Ref. [273].

with respect to the laboratory frame, let us denote by $q_{\rm L}^{\alpha}$ the coordinates of the four-momentum transfer in the laboratory frame. Performing a boost with velocity v along the direction of $\vec{q}_{\rm L}$, in the new reference frame we have $q^0 = \gamma\left(q_{\rm L}^0 - v\,|\vec{q}_{\rm L}|\right)$ and $|\vec{q}| = \gamma\left(-v\,q_{\rm L}^0 + |\vec{q}_{\rm L}|\right)$, with $\gamma = \left(1 - v^2\right)^{-1/2}$. We obtain the Breit frame by choosing $v = q_{\rm L}^0/|\vec{q}_{\rm L}|$. In the Breit frame, the three-momenta of the initial and final nucleons are opposite:

$$\vec{p}_2 = -\vec{p}_1 = \frac{\vec{q}}{2}\,. \tag{5.120}$$

Using the Gordon identity in eqn (A.90), one can find that, in the Breit frame, we have

$$\langle N(p_2)|j^0_{\gamma,{\rm Q}}(0)|N(p_1)\rangle = \overline{u_N}(p_2)\,\gamma^0\,u_N(p_1)\,G_E^N(Q^2)\,, \tag{5.121}$$

$$\langle N(p_2)|\vec{j}_{\gamma,{\rm Q}}(0)|N(p_1)\rangle = \overline{u_N}(p_2)\,\vec{\gamma}\,u_N(p_1)\,G_M^N(Q^2)\,. \tag{5.122}$$

Furthermore, since in the Breit frame the form factor $G_i^N(Q^2)$ is a function of $\sqrt{Q^2} = |\vec{q}|$, it can be interpreted as the Fourier transform of a spherically symmetric charge distribution $\rho_i^N(r)$:

$$G_i^N(Q^2) = \int {\rm d}^3x\,\rho_i^N(r)\,e^{-i\vec{q}\cdot\vec{x}} = 4\pi\int {\rm d}r\,r^2\,\rho_i^N(r)\,\frac{\sin(\sqrt{Q^2}\,r)}{\sqrt{Q^2}\,r}\,. \tag{5.123}$$

For $\sqrt{Q^2}\,r \ll 1$, we have

$$G_i^N(Q^2) \simeq 1 - \frac{Q^2}{6}\,\langle(r_i^N)^2\rangle\,, \tag{5.124}$$

where $\langle(r_i^N)^2\rangle = \int {\rm d}^3x\,r^2\,\rho_i^N(r)$ is the average squared radius. The quantity $\langle(r_i^N)^2\rangle^{1/2}$ is the associated *charge radius*. For example, from the dipole approximation of the electric form factor of the proton in eqn (5.117), the electric charge radius of the proton is given by $\langle(r_E^p)^2\rangle^{1/2} \simeq 12/M_V^2 \sim 10^{-13}\,{\rm cm}$, which is the same as the usual value of the proton radius of the order of one fermi.

Coming back to the weak charged-current form factors, for $Q^2 = 0$ we have, from eqns (5.109)–(5.112),

$$F_1(0) = 1\,, \tag{5.125}$$

$$F_2(0) = \frac{\mu_p - \mu_n}{\mu_{\rm N}} - 1 \simeq 3.706\,. \tag{5.126}$$

Thus, the charged-current vector matrix element for zero four-momentum transfer is given by

$$\langle p(p)|v_W^\rho(0)|n(p)\rangle = \overline{u_p}(p)\,\gamma^\rho\,u_n(p)\,, \tag{5.127}$$

as if the charged-current vector transition from a neutron to a proton were reduced to a transition of a valence d quark of the neutron to a valence u quark of the proton. This result is due to the conservation[24] of the quark vector charged-current

[24] Since $v_W^\rho = v_+^\rho = v_1^\rho + i v_2^\rho$, from $\partial_\rho v_a^\rho(x) = 0$ (eqn (5.91)) it follows that $\partial_\rho v_W^\rho(x) = 0$.

v_W^ρ. The corresponding vectorial weak charge, which is equal to unity, must be conserved in all strong interaction vertices. Thus, it remains unaffected by strong interactions. For the same reason, the electric charge of a nucleon is equal to the sum of the electric charges of its valence quarks because the electromagnetic current is conserved.

We now discuss neutron decay in the rest frame of the neutron, where all the components of q^ρ are small. In this case, it is possible to approximate the hadronic matrix element with

$$\langle p(p_p)|h_W^\rho(0)|n(p_n)\rangle \simeq \overline{u_p}(p_p)\,\gamma^\rho\left(g_V - g_A\,\gamma^5\right)u_n(p_n)\,, \tag{5.128}$$

with the usual definitions

$$g_V \equiv F_1(0) \simeq 1\,, \qquad g_A \equiv G_A(0)\,. \tag{5.129}$$

In the nonrelativistic limit, the transitions generated by g_V and g_A are called, respectively, *Fermi* and *Gamow–Teller* transitions. Using the Dirac representation of the γ matrices (in eqn (2.21)), one can see, from eqn (2.200), that, in Fermi transitions, the components $\overline{u_p}(p_p)\gamma^k u_n(p_n)$ are suppressed and $\overline{u_p^{(h_p)}}(p_p)\gamma^0 u_n^{(h_n)}(p_n) \simeq 2m_N \chi^{(h_p)\dagger}\chi^{(h_n)} = 2m_N\delta_{h_p h_n}$ (in the rest frame of the neutron, $\chi^{(\pm)}$ are defined by eqn (2.217), with the z axis oriented in the direction of the three-momentum of the final proton). On the other hand, in Gamow–Teller transitions, $\overline{u_p}(p_p)\gamma^0\gamma^5 u_n(p_n)$ is suppressed and $\overline{u_p^{(h_p)}}(p_p)\gamma^k\gamma^5 u_n^{(h_n)}(p_n) \simeq 2m_N \chi^{(h_p)\dagger}\sigma^k\chi^{(h_n)}$. Therefore, in Fermi transitions the spin of the nucleon remains unchanged, whereas in Gamow–Teller transitions the spin of the nucleon can change (through transitions generated by σ^1 and σ^2) or remain unchanged (through transitions generated by σ^3). In nuclear physics, the two-component spinors $\chi^{(\pm)}$ are replaced by appropriate wavefunctions of the nucleons. In this case, the Fermi and Gamow–Teller contributions to the matrix element are usually indicated, respectively, with $\langle \mathbf{1}\rangle$ and $\langle\sigma^k\rangle$.

The value of g_A is different from unity, since the axial current a_W^ρ is not conserved. In fact, the quark mass terms in the Lagrangian are not invariant under chiral $\mathrm{SU}(2)_\mathrm{I}$ transformations. Moreover, it can be shown that charged pions would not decay if the axial current were conserved[25]. This fact motivated Gell-Mann and Levy [513] to introduce the *partially conserved axial current* (PCAC) hypothesis

$$\partial_\rho a_W^\rho(x) = f_\pi\, m_\pi^2\, \pi^-(x)\,, \tag{5.130}$$

where $\pi^-(x)$ is the negative pion field, which satisfies the Klein–Gordon equation

$$\left(\Box + m_\pi^2\right)\pi^-(x) = j_{\pi^-}(x)\,, \tag{5.131}$$

and $j_{\pi^-}(x)$ is the source of the pion field. Since, from eqn (C.32), we have

$$\langle p(p_p)|\pi^-(x)|n(p_n)\rangle = \langle p(p_p)|e^{ix\cdot\mathsf{P}}\pi^-(0)e^{-ix\cdot\mathsf{P}}|n(p_n)\rangle$$

[25] As shown in footnote 21 on page 148, only the axial current contributes to charged pion decay. From eqn (5.53), taking into account that $a_W^\rho(x) = e^{ix\cdot\mathsf{P}}a_W^\rho(0)e^{-ix\cdot\mathsf{P}}$ (see eqn (C.32)), where P^ρ is the four-momentum operator which generates space-time translations, we have $\langle 0|a_W^\rho(x)|\pi^-(p_\pi)\rangle = i\,f_\pi\,p_\pi^\rho\,e^{-ip_\pi\cdot x}$. Hence $\langle 0|\partial_\rho a_W^\rho(x)|\pi^-(p_\pi)\rangle = f_\pi\,m_\pi^2\,e^{-ip_\pi\cdot x}$. If the axial current were conserved ($\partial_\rho a_W^\rho(x) = 0$), f_π would have to vanish and charged pions would not decay.

$$= e^{ix\cdot q}\langle p(p_p)|\pi^-(0)|n(p_n)\rangle\,, \tag{5.132}$$

equation (5.131) implies that

$$\left(-q^2+m_\pi^2\right)\langle p(p_p)|\pi^-(x)|n(p_n)\rangle = \langle p(p_p)|j_{\pi^-}(x)|n(p_n)\rangle\,. \tag{5.133}$$

In general, the matrix element on the right-hand side can be written as

$$\langle p(p_p)|j_{\pi^-}(x)|n(p_n)\rangle = i\,\sqrt{2}\,g_{\pi N}(Q^2)\,\overline{u_p}(p_p)\,\gamma^5\,u_n(p_n)\,e^{ix\cdot q}\,, \tag{5.134}$$

where $g_{\pi N}(Q^2)$ is a real function of q^2, which is called the *pion–nucleon form factor*. By definition, $g_{\pi N}\equiv g_{\pi N}(m_\pi^2)$ is the pion–nucleon interaction coupling constant. Its value, determined in pion–nucleon scattering experiments, is [215]

$$g_{\pi N} = 13.10\pm 0.35\,. \tag{5.135}$$

Using the expression in eqn (5.89) for the axial matrix element, eqns (5.130), (5.133), and (5.134) lead to the relation

$$\left(-q^2+m_\pi^2\right)\left[2\,m_N\,G_A(Q^2)+\frac{q^2}{m_N}\,G_P(Q^2)\right] = \sqrt{2}\,g_{\pi N}(Q^2)\,f_\pi\,m_\pi^2\,. \tag{5.136}$$

For $q^2=0$, assuming $g_{\pi N}(0)\simeq g_{\pi N}$, we obtain the Goldberger–Treiman relation [548]

$$g_A \simeq \frac{f_\pi\,g_{\pi N}}{\sqrt{2}\,m_N}\,. \tag{5.137}$$

From the values of f_π and $g_{\pi N}$ in eqns (5.57) and (5.135), one can find $g_A\simeq 1.3$. Furthermore, eqn (5.136) allows one to determine $G_P(Q^2)$ in terms of $G_A(Q^2)$ and $g_{\pi N}(Q^2)$. In particular, we have

$$G_P(0)\simeq \frac{2\,m_N^2}{m_\pi^2}\,g_A\,. \tag{5.138}$$

Using the hadronic matrix element in eqn (5.128), the differential decay rate of the neutron in its rest frame is given by

$$\frac{\mathrm{d}\Gamma_n}{\mathrm{d}E_e} = \frac{G_{\mathrm{F}}^2\,|V_{ud}|^2}{2\pi^3}\,(E_0-E_e)^2\,E_e\,|\vec{p}_e|\,\left(g_V^2+3g_A^2\right)\,, \tag{5.139}$$

where $E_0 = m_n - m_p = 1.293\,331\,7\pm 0.000\,000\,5\,\mathrm{MeV}$ [400] is the maximal energy of the emitted electron and the recoil energy of the proton has been neglected.

Integrating over the electron energy, we obtain the neutron lifetime ($\tau_n = \Gamma_n^{-1}$)

$$\tau_n = \frac{2\pi^3}{G_F^2 \, |V_{ud}|^2 \, m_e^5 \, f} \left(g_V^2 + 3 g_A^2 \right)^{-1} , \tag{5.140}$$

where f is the dimensionless phase-space integral

$$f = \int_{m_e}^{E_0} \mathrm{d}E_e \, \frac{(E_0 - E_e)^2 \, E_e \, |\vec{p}_e|}{m_e^5} \, . \tag{5.141}$$

Taking into account Coulomb, radiative, and other corrections, the value of the phase-space integral is [1063]

$$f = 1.71465 \pm 0.00015 \, . \tag{5.142}$$

The experimental value of τ_n is given in eqn (A.158). Although the value of g_A can be derived from the measurement of the neutron lifetime, assuming $g_V = 1$, measurements of the decays of polarized neutrons (see Ref. [24]) yield the most precise value of the ratio g_A/g_V [400]:

$$\frac{g_A}{g_V} = 1.2695 \pm 0.0029 \, . \tag{5.143}$$

Since $g_V \simeq 1$, this value is in approximate agreement with the Goldberger–Treiman relation in eqn (5.137).

5.3 Neutrino–nucleon scattering

5.3.1 Quasielastic charged-current reactions

The quasielastic charged-current interactions of neutrinos and antineutrinos with nucleons are

$$\nu_\ell + n \to p + \ell^- \, , \tag{5.144}$$

$$\bar{\nu}_\ell + p \to n + \ell^+ \, , \tag{5.145}$$

with $\ell = e, \mu, \tau$, In practice, only electron and muon neutrino and antineutrino beams are available in the laboratory. Tau neutrino beams are generated in astrophysical environments and by high-energy cosmic ray interactions with the Earth's atmosphere.

The process in eqn (5.145) with $\ell = e$, sometimes called *inverse neutron decay*, has been used in the historical experiment of Cowan and Reines [899] in which neutrinos have been observed for the first time. Such a process is currently used in detectors of electron antineutrinos produced in reactors (for example, CHOOZ [100], Palo Verde [255], KamLAND [103]; see section 12.2). Here, however, we will consider the high-energy case, in which the mass difference between the neutron and the proton is neglected. The low-energy case appropriate for reactor $\bar{\nu}_e$ detection is discussed in section 12.2.

For $\ell = e$, the ν_ℓ-n scattering process in eqn (5.144) can be obtained by crossing from the neutron decay process in eqn (5.64). This means that the hadronic vertex can be treated with the same method employed in the analysis of neutron decay in section 5.2.2. The amplitude of the ν_ℓ-n scattering process in eqn (5.144) is given by

$$\begin{aligned}\mathcal{A}_{\nu_\ell n \to p \ell^-} = & -i\,\frac{G_{\mathrm{F}}}{\sqrt{2}}\,V_{ud}\,\overline{u_\ell}(p_\ell)\,\gamma_\rho\left(1-\gamma^5\right)u_{\nu_\ell}(p_\nu) \\ & \times \left\{\overline{u_p}(p_p)\left[\gamma^\rho\,F_1(Q^2) + \frac{i}{2\,m_N}\,\sigma^{\rho\eta}\,q_\eta\,F_2(Q^2)\right.\right. \\ & \left.\left. -\,\gamma^\rho\,\gamma^5\,G_A(Q^2) - \frac{q^\rho}{m_N}\,\gamma^5\,G_P(Q^2)\right]u_n(p_n)\right\}, \qquad (5.146)\end{aligned}$$

where $q = p_\nu - p_\ell = p_p - p_n$ is the four-momentum transferred from the neutrino to the nucleon, and $Q^2 \equiv -q^2$.

The differential cross-sections for the ν_ℓ and $\bar{\nu}_\ell$ scattering processes in eqns (5.144) and (5.145) in the laboratory frame are given by (see Ref. [750])

$$\frac{\mathrm{d}\sigma_{\mathrm{CC}}^{\nu_\ell n, \bar{\nu}_\ell p}}{\mathrm{d}Q^2} = \frac{G_{\mathrm{F}}^2\,|V_{ud}|^2\,m_N^4}{8\pi\,(p_\nu \cdot p_{N_{\mathrm{i}}})^2}\left[A(Q^2) \pm B(Q^2)\,\frac{s-u}{m_N^2} + C(Q^2)\,\frac{(s-u)^2}{m_N^4}\right], \qquad (5.147)$$

with the Mandelstam variables

$$s = (p_\nu + p_N)^2\,, \qquad (5.148)$$

$$t = (p_\nu - p_\ell)^2 = q^2 \equiv -Q^2\,, \qquad (5.149)$$

$$u = (p_\ell - p_N)^2\,, \qquad (5.150)$$

where p_N is the four-momentum of the initial nucleon. In eqn (5.147), the plus and minus signs refer, respectively, to ν_ℓ-n and $\bar{\nu}_\ell$-p scatterings. The functions A, B, and C of Q^2 are given by

$$\begin{aligned}A = & \frac{m_\ell^2 + Q^2}{m_N^2}\left\{\left(1+\frac{Q^2}{4m_N^2}\right)G_A^2 - \left(1-\frac{Q^2}{4m_N^2}\right)\left(F_1^2 - \frac{Q^2}{4m_N^2}\,F_2^2\right) + \frac{Q^2}{m_N^2}\,F_1\,F_2\right. \\ & \left. -\frac{m_\ell^2}{4m_N^2}\left[(F_1+F_2)^2 + (G_A + 2\,G_P)^2 - \frac{1}{4}\left(1+\frac{Q^2}{4m_N^2}\right)G_P^2\right]\right\}, \qquad (5.151)\end{aligned}$$

$$B = \frac{Q^2}{m_N^2}\,G_A\,(F_1+F_2)\,, \qquad (5.152)$$

$$C = \frac{1}{4}\left(G_A^2 + F_1^2 + \frac{Q^2}{4m_N^2}\,F_2^2\right). \qquad (5.153)$$

Since $m_e^2/m_N^2 \simeq 2.5 \times 10^{-7}$ and $m_\mu^2/m_N^2 \simeq 1.3 \times 10^{-2}$, the term in A proportional to m_ℓ^2/m_N^2 can be neglected for $\overset{(-)}{\nu}_e$ and $\overset{(-)}{\nu}_\mu$ scattering. In this approximation, which we will adopt in the following, the cross-sections do not depend on the pseudoscalar form factor $G_P(Q^2)$.

Since $F_1(Q^2)$ and $F_2(Q^2)$ are known from eqns (5.109) and (5.110) and measurements of the electromagnetic form factors of the nucleon, the measurements of the differential cross-sections in eqn (5.147) give information on the Q^2-dependence of the axial form factor $G_A(Q^2)$. In principle, it is convenient to obtain directly the value of $G_A(Q^2)$ from measurements of the difference

$$\frac{\mathrm{d}\sigma_{\mathrm{CC}}^{\nu n}}{\mathrm{d}Q^2} - \frac{\mathrm{d}\sigma_{\mathrm{CC}}^{\bar{\nu} p}}{\mathrm{d}Q^2} = \frac{G_{\mathrm{F}}^2 \, |V_{ud}|^2 \, (s-u) \, Q^2}{4\pi \, (p_\nu \cdot p_{N_{\mathrm{i}}})^2} \, G_A \, (F_1 + F_2) \, . \tag{5.154}$$

However, in practice it is difficult to have beams of neutrinos and antineutrinos. Hence, most experiments have obtained information on the axial form factor through the scattering of ν_μ's or $\bar{\nu}_\mu$'s produced in accelerators with protons or nuclei (see Ref. [215]).

The neutrino scattering data can be fitted with F_1 and F_2 expressed in terms of the dipole electromagnetic form factors in eqns (5.117)–(5.119) (see eqns (5.109), (5.110), (5.113), and (5.114)) and the dipole axial form factor

$$G_A(Q^2) = \frac{g_A}{(1 + Q^2/m_A^2)^2} \, , \tag{5.155}$$

where m_A is the so-called *axial mass*, with experimental value [215]

$$m_A = 1.026 \pm 0.021 \, \mathrm{GeV} \, . \tag{5.156}$$

In the near future, the MINERνA experiment [391] will measure $G_A(Q^2)$ with high accuracy.

In the laboratory frame, where the initial nucleon is at rest, we have[26]

$$s = m_N^2 + 2 \, m_N \, E_\nu \, , \tag{5.157}$$

$$t = q^2 \equiv -Q^2 = m_\ell^2 - 2 \, E_\nu \, (E_\ell - p_\ell \cos\theta) \, , \tag{5.158}$$

$$u = m_N^2 - 2 \, m_N \, E_\nu + 2 \, E_\nu \, (E_\ell - p_\ell \cos\theta) \, , \tag{5.159}$$

where θ is the scattering angle of the outgoing lepton (analogous to that in Fig. 5.4). The angular differential cross-sections for the quasielastic ν_ℓ and $\bar{\nu}_\ell$ scattering processes in eqns (5.144) and (5.145) in the laboratory frame are given by

$$\frac{\mathrm{d}\sigma_{\mathrm{CC}}^{\nu n, \bar{\nu} p}}{\mathrm{d}\cos\theta} = -\frac{G_{\mathrm{F}}^2 \, |V_{ud}|^2 \, m_N^2}{4\pi} \, \frac{p_\ell}{E_\nu} \left[A(Q^2) \pm B(Q^2) \, \frac{s-u}{m_N^2} + C(Q^2) \, \frac{(s-u)^2}{m_N^4} \right] , \tag{5.160}$$

with $s - u = 4 \, m_N \, E_\nu - 2 \, E_\nu \, (E_\ell - p_\ell \cos\theta)$.

[26] The expression for u can be straightforwardly obtained from those of s and t by using the relation in eqn (E.43).

For low neutrino energies in the laboratory frame, $E_\nu \ll m_N$, we have $Q^2 \ll m_N^2$ and $s - u \simeq 4\, m_N\, E_\nu$. Taking into account that also $m_\ell \ll m_N$, for $\ell = e, \mu$, we obtain

$$A(Q^2) = \frac{2\, E_\nu\, (E_\ell - p_\ell \cos\theta)}{m_N^2}\, (g_A^2 - g_V^2) + \mathcal{O}\left(\frac{E_\nu^3}{m_N^3}\right) , \tag{5.161}$$

$$B(Q^2)\, \frac{s-u}{m_N^2} = \mathcal{O}\left(\frac{E_\nu^3}{m_N^3}\right) , \tag{5.162}$$

$$C(Q^2)\, \frac{(s-u)^2}{m_N^4} = \frac{4\, E_\nu^2}{m_N^2}\, (g_A^2 + g_V^2) + \mathcal{O}\left(\frac{E_\nu^3}{m_N^3}\right) . \tag{5.163}$$

Hence, in this approximation the contribution of $B(Q^2)$ is negligible and the cross-sections of the ν_ℓ-n and $\bar{\nu}_\ell$-p charged-current scattering processes are the same. Integrating over the scattering angle, neglecting the charged lepton mass and taking $g_V \simeq 1$, we obtain the total low-energy cross-section

$$\sigma_{\text{CC}}^{\nu n, \bar{\nu} p} \simeq \frac{G_\text{F}^2\, |V_{ud}|^2}{\pi}\, (1 + 3\, g_A^2)\, E_\nu^2 \simeq 1.601 \times 10^{-44}\, (1 + 3\, g_A^2) \left(\frac{E_\nu}{\text{MeV}}\right)^2 \text{cm}^2 . \tag{5.164}$$

5.3.2 Elastic neutral-current reactions

Neutrinos and antineutrinos interact elastically with nucleons through the neutral-current processes

$$\overset{(-)}{\nu}_\ell + N \to \overset{(-)}{\nu}_\ell + N , \tag{5.165}$$

where $N = p, n$.

In this case, in analogy with the charged-current case, the nucleon interaction is described by the hadronic matrix element

$$\langle N(p_f) | j_{Z,\text{Q}}^\rho(0) | N(p_i) \rangle , \tag{5.166}$$

where $j_{Z,\text{Q}}^\rho$ is the quark neutral current in eqn (3.178),

$$\begin{aligned} j_{Z,\text{Q}}^\rho = {} & 2 \sum_{\alpha=u,c,t} \left(g_L^U\, \overline{q_{\alpha L}^U}\, \gamma^\mu\, q_{\alpha L}^U + g_R^U\, \overline{q_{\alpha R}^U}\, \gamma^\mu\, q_{\alpha R}^U \right) \\ & + 2 \sum_{\alpha=d,s,b} \left(g_L^D\, \overline{q_{\alpha L}^D}\, \gamma^\mu\, q_{\alpha L}^D + g_R^D\, \overline{q_{\alpha R}^D}\, \gamma^\mu\, q_{\alpha R}^D \right) . \end{aligned} \tag{5.167}$$

From the values of the coefficients $g_L^{U,D}$ and $g_R^{U,D}$, which are summarized in Table 3.6 (page 78), the neutral current $j_{Z,\text{Q}}^\rho$ can be written as

$$j_{Z,\text{Q}}^\rho = v_3^\rho - a_3^\rho - 2\, \sin^2\vartheta_\text{W}\, j_{\gamma,\text{Q}}^\rho - \frac{1}{2}\, (v_s^\rho - a_s^\rho) , \tag{5.168}$$

with the isovector current v_3^ρ given in eqn (5.90), the axial currents

$$a_a^\rho(x) = \overline{Q}(x)\, \gamma^\rho\, \gamma^5\, T_a\, Q(x) , \tag{5.169}$$

and the electromagnetic current $j^{\rho}_{\gamma,\mathrm{Q}}$ given in eqn (5.103). The isoscalar currents v^{ρ}_{s} and a^{ρ}_{s} are given by

$$v^{\rho}_{s} = \sum_{q=s,c,b,t} \overline{q}\,\gamma^{\mu}\,q\,, \qquad a^{\rho}_{s} = \sum_{q=s,c,b,t} \overline{q}\,\gamma^{\mu}\,\gamma^{5}\,q\,. \tag{5.170}$$

They represent, respectively, the vector and axial contributions of the strange and heavier quarks.

Let us separate the vector and axial parts of the hadronic current:

$$j^{\rho}_{Z,\mathrm{Q}}(x) = v^{\rho}_{Z}(x) - a^{\rho}_{Z}(x)\,, \tag{5.171}$$

with

$$v^{\rho}_{Z}(x) = v^{\rho}_{3}(x) - 2\,\sin^{2}\vartheta_{\mathrm{W}}\,j^{\rho}_{\gamma,\mathrm{Q}}(x) - \frac{1}{2}\,v^{\rho}_{s}\,, \tag{5.172}$$

$$a^{\rho}_{Z}(x) = a^{\rho}_{3}(x) - \frac{1}{2}\,a^{\rho}_{s}\,. \tag{5.173}$$

In analogy with the charged-current case (see eqns (5.88) and (5.89)), the neutral-current vector and axial matrix element can be written as

$$\langle N(p_{\mathrm{f}})|v^{\rho}_{Z}(0)|N(p_{\mathrm{i}})\rangle = \overline{u_{N}}(p_{\mathrm{f}})\left[\gamma^{\rho}\,F^{ZN}_{1}(Q^{2}) + \frac{i}{2\,m_{N}}\,\sigma^{\rho\eta}\,q_{\eta}\,F^{ZN}_{2}(Q^{2})\right]u_{N}(p_{\mathrm{i}})\,, \tag{5.174}$$

$$\langle N(p_{\mathrm{f}})|a^{\rho}_{Z}(0)|N(p_{\mathrm{i}})\rangle = \overline{u_{N}}(p_{\mathrm{f}})\left[\gamma^{\rho}\,\gamma^{5}\,G^{ZN}_{A}(Q^{2}) + \frac{q^{\rho}}{m_{N}}\,\gamma^{5}\,G^{ZN}_{P}(Q^{2})\right]u_{N}(p_{\mathrm{i}})\,, \tag{5.175}$$

with the neutral-current Dirac, Pauli, axial, and pseudoscalar form factors $F^{ZN}_{1}(Q^{2})$, $F^{ZN}_{2}(Q^{2})$, $G^{ZN}_{A}(Q^{2})$, and $G^{ZN}_{P}(Q^{2})$, respectively.

The one-nucleon matrix element of the isovector current v^{ρ}_{3} can be expressed in terms of the proton and neutron matrix elements of the electromagnetic current $j^{\rho}_{\gamma,\mathrm{Q}}$ in eqn (5.103). This can be done by means of the charge-symmetry transformation in eqn (5.82). Since

$$\mathsf{U}_{\pi}\,v^{\rho}_{3}\,\mathsf{U}^{\dagger}_{\pi} = -v^{\rho}_{3}\,, \qquad \mathsf{U}_{\pi}\,v^{\rho}_{0}\,\mathsf{U}^{\dagger}_{\pi} = v^{\rho}_{0}\,, \tag{5.176}$$

one can obtain, from eqn (5.85),

$$\langle p(p_{f})|v^{\rho}_{3}|p(p_{i})\rangle = -\langle n(p_{f})|v^{\rho}_{3}|n(p_{i})\rangle = \frac{1}{2}\left[\langle p(p_{f})|j^{\rho}_{\gamma,\mathrm{Q}}|p(p_{i})\rangle - \langle n(p_{f})|j^{\rho}_{\gamma,\mathrm{Q}}|n(p_{i})\rangle\right]. \tag{5.177}$$

Using this relation, we have, from the expression of the nucleon electromagnetic matrix element in eqn (5.108),

$$F^{ZN}_{i} = \pm\frac{1}{2}\,(F^{p}_{i} - F^{n}_{i}) - 2\,\sin^{2}\vartheta_{\mathrm{W}}\,F^{N}_{i} - \frac{1}{2}\,F^{sN}_{i} \qquad (i = 1, 2;\ N = p, n)\,, \tag{5.178}$$

with the plus sign for $N = p$ and the minus sign for $N = n$. The form factors $F_1^{sN}(Q^2)$ and $F_2^{sN}(Q^2)$, which are defined by

$$\langle N(p_\mathrm{f})|v_s^\rho(0)|N(p_\mathrm{i})\rangle = \overline{u_N}(p_\mathrm{f}) \left[\gamma^\rho F_1^{sN}(Q^2) + \frac{i}{2\,m_N}\,\sigma^{\rho\eta}\,q_\eta\,F_2^{sN}(Q^2)\right] u_N(p_\mathrm{i})\,, \tag{5.179}$$

are usually called *strange vector form factors*, since it is believed that the strange quarks give the dominant contribution (see Ref. [406]). Hence, the vector hadronic neutral-current matrix element is determined by the nucleon electromagnetic form factors, whose values are reasonably well known (see eqns (5.111)–(5.119)), and the strange vector form factors.

Let us now consider the axial currents $a_a^\rho(x)$. Since under the charge-symmetry transformation in eqn (5.82) we have

$$\mathsf{U}_\pi\,a_3^\rho\,\mathsf{U}_\pi^\dagger = -a_3^\rho\,, \tag{5.180}$$

using eqn (5.85), one can find that

$$\langle p(p_f)|a_3^\rho|p(p_i)\rangle = -\langle n(p_f)|a_3^\rho|n(p_i)\rangle\,. \tag{5.181}$$

Furthermore, it is possible to express these matrix elements in terms of the charged-current axial matrix element in eqn (5.89). Using eqn (5.94), one can obtain commutation relations which are analogous to those in eqns (5.95) and (5.96) for the vector currents:

$$[\mathsf{T}_a\,,\,a_b^\rho(x)] = i\,\epsilon_{abc}\,a_c^\rho(x)\,,\qquad [\mathsf{T}_\pm\,,\,a_3^\rho(x)] = \mp\,a_\pm^\rho(x)\,,\qquad \text{with}\quad a_\pm^\rho = a_1^\rho \pm i\,a_2^\rho\,. \tag{5.182}$$

Since the axial charged current a_W^ρ in eqn (5.69) coincides with a_+^ρ, we have

$$\begin{aligned}\langle p(p_p)|a_W^\rho(0)|n(p_n)\rangle &= \langle p(p_p)|\,[a_3^\rho(0)\,,\,\mathsf{T}_\pm]\,|n(p_n)\rangle \\ &= \langle p(p_p)|a_3^\rho(0)|p(p_n)\rangle - \langle n(p_p)|a_3^\rho(0)|n(p_n)\rangle\,.\end{aligned} \tag{5.183}$$

Finally, we obtain, from eqns (5.181) and (5.183),

$$\langle p(p_p)|a_3^\rho(0)|p(p_n)\rangle = -\langle n(p_p)|a_3^\rho(0)|n(p_n)\rangle = \frac{1}{2}\,\langle p(p_p)|a_W^\rho(0)|n(p_n)\rangle\,. \tag{5.184}$$

Thus, the neutral-current axial and pseudoscalar form factors in eqn (5.175) are related to the corresponding charged-current form factors in eqn (5.89) by

$$G_i^{ZN}(Q^2) = \pm\frac{1}{2}\,G_i(Q^2) - \frac{1}{2}\,G_i^{sN}(Q^2) \qquad (i = A, P;\; N = p, n)\,, \tag{5.185}$$

with the plus sign for $N = p$ and the minus sign for $N = n$. The *strange axial and pseudoscalar form factors* $G_A^{sN}(Q^2)$ and $G_P^{sN}(Q^2)$ are defined by

$$\langle N(p_\mathrm{f})|a_s^\rho(0)|N(p_\mathrm{i})\rangle = \overline{u_N}(p_\mathrm{f}) \left[\gamma^\rho\,G_A^{sN}(Q^2) + \frac{q^\rho}{m_N}\,\gamma^5\,G_P^{sN}(Q^2)\right] u_N(p_\mathrm{i})\,. \tag{5.186}$$

The kinematics in the neutral-current elastic scattering processes in eqn (5.165) is the same as in the charged-current quasielastic scattering processes in eqns (5.144) and (5.145), with the only exception that the final lepton, being a neutrino, is practically massless. This implies that the differential cross-sections of the neutral-current elastic scattering processes in eqn (5.165) have the same form as the charged-current cross-sections in eqn (5.147), without $|V_{ud}|^2$:

$$\frac{d\sigma_{\rm NC}^{\nu N, \bar{\nu} N}}{dQ^2} = \frac{G_{\rm F}^2 \, m_N^4}{8\pi \, (p_\nu \cdot p_{N_{\rm i}})^2} \left[A_N(Q^2) \pm B_N(Q^2) \, \frac{s-u}{m_N^2} + C_N(Q^2) \, \frac{(s-u)^2}{m_N^4} \right] , \tag{5.187}$$

with the plus and minus signs referring, respectively, to ν_ℓ-N and $\bar{\nu}_\ell$-N scattering. The functions A^{ZN}, B^{ZN}, and C^{ZN} of Q^2 have the same form as those in eqns (5.151)–(5.153), with the charged-current form factors replaced with the corresponding neutral-current form factors in eqns (5.178) and (5.185) and $m_\ell^2 \to 0$:

$$A_N = \frac{Q^2}{m_N^2} \left\{ \left(1 + \frac{Q^2}{4m_N^2} \right) (G_A^{ZN})^2 - \left(1 - \frac{Q^2}{4m_N^2} \right) \left[(F_1^{ZN})^2 - \frac{Q^2}{4m_N^2} \, (F_2^{ZN})^2 \right] + \frac{Q^2}{m_N^2} \, F_1^{ZN} \, F_2^{ZN} \right\} , \tag{5.188}$$

$$B_N = \frac{Q^2}{m_N^2} \, G_A^{ZN} \left(F_1^{ZN} + F_2^{ZN} \right) , \tag{5.189}$$

$$C_N = \frac{1}{4} \left[(G_A^{ZN})^2 + (F_1^{ZN})^2 + \frac{Q^2}{4m_N^2} \, (F_2^{ZN})^2 \right] . \tag{5.190}$$

Thus, the neutral-current pseudoscalar form factors $G_P^{ZN}(Q^2)$ do not contribute to the cross-sections.

In the same approximations as eqn (5.164), the total elastic neutral-current cross-sections of neutrinos and antineutrinos on protons and neutrons are given by

$$\sigma_{\rm NC}^{\overset{(-)}{\nu} p} \simeq \frac{G_{\rm F}^2}{4\pi} \left[\left(1 - 4 \, \sin^2 \vartheta_{\rm W} \right)^2 + 3 \, g_A^2 \right] E_\nu^2 \, , \tag{5.191}$$

$$\sigma_{\rm NC}^{\overset{(-)}{\nu} n} \simeq \frac{G_{\rm F}^2}{4\pi} \left[1 + 3 \, g_A^2 \right] E_\nu^2 \, . \tag{5.192}$$

One can see, from eqns (5.178) and (5.185), that the nucleon form factors entering in the cross-section of the neutral-current elastic scattering processes in eqn (5.187) are functions of the electromagnetic form factors, the axial form factor and the strange vector form factors. Since the values of the electromagnetic form factors are reasonably well known (see eqns (5.111)–(5.119)) and the axial form factor can be determined through the measurement of the charged-current quasielastic scattering processes (see eqns (5.155) and (5.156)), the measurements of the neutral-current elastic scattering processes in eqn (5.165) give information on the strange form factors of the nucleon [657, 407, 508, 55, 57, 57] (see the review in Ref. [56]).

Assuming the absence of the strange form factors of the nucleon, the elastic scattering of neutrinos and antineutrinos on protons has been used for the measurement of the weak mixing angle, which appears in the relation in eqn (5.178) which connects the values of the nucleon weak and electromagnetic form factors. The average of experimental results gives [400]

$$\sin^2\vartheta_W = 0.203 \pm 0.033\,. \tag{5.193}$$

5.3.3 Charged-current deep inelastic scattering

At high energies, $E_\nu \gg m_N$ in the laboratory frame, charged-current neutrino–nucleon interactions are dominated by the inclusive deep inelastic scattering (DIS) processes

$$\nu_\ell + N \to \ell^- + X\,, \qquad \bar{\nu}_\ell + N \to \ell^+ + X\,, \tag{5.194}$$

where $N = p, n$ and X denotes any set of final hadrons. In the following, we denote with p_ν, p_ℓ, p_N, and p_X the four-momenta of the neutrino, the charged lepton, the nucleon and the sum of the four-momenta of the final hadrons, respectively, and we neglect the mass of the final lepton.

The diagram of the process $\nu_\ell(p_\nu)+N(p_N) \to \ell^-(p_\ell)+X(p_X)$ at lowest order in the weak interaction perturbation expansion is shown in Fig. 5.12a. It is customary to define the four-momentum transfer q as

$$q \equiv p_\nu - p_\ell = p_X - p_N\,. \tag{5.195}$$

The process is described by three kinematic variables. One of them is the Lorentz-invariant squared center-of-mass energy

$$s = (p_\nu + p_N)^2 = m_N^2 + 2\,p_\nu \cdot p_N\,. \tag{5.196}$$

The other two can be chosen among the Lorentz invariants

$$Q^2 \equiv -q^2 = 2\,p_\nu \cdot p_\ell \geq 0\,, \qquad x \equiv \frac{Q^2}{2\,p_N \cdot q}\,, \qquad y \equiv \frac{p_N \cdot q}{p_N \cdot p_\nu}\,. \tag{5.197}$$

These four Lorentz-invariant kinematical variables are related by

$$x\,y = \frac{Q^2}{s - m_N^2}\,. \tag{5.198}$$

The DIS region of the kinematic variables is defined by

$$Q^2 \gg m_N^2\,, \qquad p_N \cdot q \gg m_N^2\,. \tag{5.199}$$

Since the values of the variables x and y are confined in the intervals

$$0 < x \leq 1\,, \qquad 0 < y \leq 1\,, \tag{5.200}$$

we also have

$$s \gg m_N^2\,, \qquad x\,y \simeq Q^2/s\,. \tag{5.201}$$

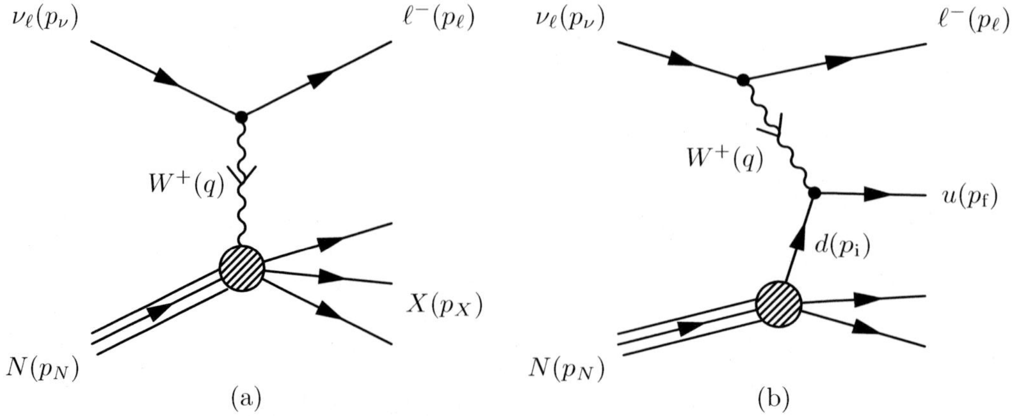

FIG. 5.12. (a) Diagram of the $\nu_\ell(p_\nu) + N(p_N) \to \ell^-(p_\ell) + X(p_X)$ charged-current DIS process at lowest order in the weak interaction perturbation expansion. (b) Diagram of the same process in the quark–parton model, with elementary $W^+(q) + d(p_\mathrm{i}) \to u(p_\mathrm{f})$ transition (see eqn (5.215)).

The neutrino and antineutrino DIS differential cross-sections are given by (see Refs. [902, 227, 720])

$$\frac{\mathrm{d}^2\sigma_{\mathrm{CC}}^{\overset{(-)}{\nu} N}}{\mathrm{d}x\,\mathrm{d}y} = \sigma_{\mathrm{CC}}^0 \left[x\,y^2\,F_1^{W^\pm N} + (1-y)\,F_2^{W^\pm N} \pm x\,y\left(1-\frac{y}{2}\right) F_3^{W^\pm N} \right] , \tag{5.202}$$

with

$$\sigma_{\mathrm{CC}}^0 = \frac{G_\mathrm{F}^2}{2\pi}\,s\left(1+\frac{Q^2}{m_W^2}\right)^{-2} . \tag{5.203}$$

The plus and minus signs in eqn (5.202) refer, respectively, to ν_ℓ and $\bar{\nu}_\ell$ scattering, in which the vector boson absorbed by the nucleon are, respectively, a W^+ and a W^-, as indicated by the superscripts of the *structure functions* $F_i^{W^\pm N}$. These are real functions of two independent kinematical variables which depend on the four-momentum transfer q. It is common to choose $F_i^{W^\pm N} = F_i^{W^\pm N}(x, Q^2)$. Isospin symmetry implies that (see Ref. [227])

$$F_i^{W^+ p} = F_i^{W^- n} , \qquad F_i^{W^+ n} = F_i^{W^- p} . \tag{5.204}$$

Many experiments use so-called *isoscalar targets*, composed of nuclei with an equal number of protons and neutrons. In this case, the average neutrino and antineutrino cross-sections are given by

$$\begin{aligned} \frac{\mathrm{d}^2\sigma_{\mathrm{CC}}^{\overset{(-)}{\nu}}}{\mathrm{d}x\,\mathrm{d}y} &= \frac{1}{2}\left(\frac{\mathrm{d}^2\sigma_{\mathrm{CC}}^{\overset{(-)}{\nu} p}}{\mathrm{d}x\,\mathrm{d}y} + \frac{\mathrm{d}^2\sigma_{\mathrm{CC}}^{\overset{(-)}{\nu} n}}{\mathrm{d}x\,\mathrm{d}y} \right) \\ &= \sigma_{\mathrm{CC}}^0 \left[x\,y^2\,F_1^{W^\pm} + (1-y)\,F_2^{W^\pm} \pm x\,y\left(1-\frac{y}{2}\right) F_3^{W^\pm} \right] , \end{aligned} \tag{5.205}$$

with the same structure functions for neutrinos and antineutrinos:

$$F_i^{W^+} = \frac{1}{2}\left(F_i^{W^+p} + F_i^{W^+n}\right) = \frac{1}{2}\left(F_i^{W^-n} + F_i^{W^-p}\right) = F_i^{W^-} . \tag{5.206}$$

In the laboratory frame, where the initial nucleon is at rest,

$$x = \frac{Q^2}{2\,m_N\,(E_\nu - E_\ell)}\,, \qquad y = 1 - \frac{E_\ell}{E_\nu}\,, \qquad Q^2 = 4\,E_\nu\,E_\ell\,\sin^2\frac{\theta}{2}\,, \tag{5.207}$$

where θ is the scattering angle of the outgoing lepton (analogous to that in Fig. 5.4). For $Q^2 \ll m_W^2$ we have, in the laboratory frame,

$$\sigma_{\text{CC}}^0 \simeq \frac{G_{\text{F}}^2}{\pi}\,m_N\,E_\nu \simeq 1.58 \times 10^{-38}\left(\frac{E_\nu}{\text{GeV}}\right)\text{cm}^2 . \tag{5.208}$$

In the laboratory frame, the DIS conditions in eqn (5.199) imply that the neutrino energy, the energy and scattering angle of the final charged lepton, and the energy transfer from the neutrino to the hadrons must be large.

The current interpretation of the DIS processes is based on the *quark–parton model* of hadrons[27]. According to this model, a nucleon is a composite system of three valence quarks and a sea of quark–antiquark pairs of all flavors. In the DIS processes, the intermediate virtual gauge boson (W in the neutrino charged-current processes in eqn (5.194), Z in the neutrino neutral-current processes discussed in section (5.3.4), and γ in charged lepton–nucleon DIS processes) which connects the lepton vertex to the hadronic vertex interacts directly with the quark constituents of the nucleon, as illustrated by the tree-level diagram in Fig. 5.12b.

In order to introduce quantitatively the quark–parton model, it is convenient to work in the Breit frame, which was introduced in section 5.2.2 in the discussion of the electromagnetic form factors of the nucleon (page 156). In the Breit frame, the intermediate virtual gauge boson does not carry any energy, i.e. $q^0 = 0$ and $\sqrt{Q^2} = |\vec{q}|$. In the DIS reactions, we have

$$\vec{q} = -2\,x\,\vec{p}_N\,. \tag{5.209}$$

Thus, $\vec{q}$ and the three-momentum $\vec{p}_N$ of the initial nucleon are collinear and directed in opposite directions. Moreover, in the DIS processes, where the conditions in eqn (5.199) hold, we have

$$|\vec{p}_N|^2 \gg m_N^2\,. \tag{5.210}$$

The basic assumptions of the quark–parton model are:

(A) The nucleon is a composite system made of elementary quarks.
(B) The interactions among constituent quarks can be neglected in the DIS region defined in eqn (5.199) (*asymptotic freedom*).
(C) In the Breit frame (where the inequality in eqn (5.210) is satisfied for the DIS processes) the constituent quarks have three-momenta in the same direction of

---

[27] The parton model of hadrons was introduced by Feynman in 1969 [435] (see Ref. [436]). It was later understood that the partons are quarks.

the nucleon. In other words, the transverse momenta of constituent quarks can be neglected.

(D) In the Breit frame, the masses of the constituent quarks can be neglected in comparison with their energy.

Let us consider the elementary process of interaction of the vector boson with a constituent quark. Since (from assumption (B)) the initial and final quarks are free particles, we have

$$p_{\mathrm{i}} + q = p_{\mathrm{f}} \,, \tag{5.211}$$

where p_{i} and p_{f} are the initial and final quark four-momenta, respectively. In the Breit frame where $q^0 = 0$, we have

$$p_{\mathrm{i}}^0 = p_{\mathrm{f}}^0 \,, \tag{5.212}$$

$$\vec{p}_{\mathrm{i}} + \vec{q} = \vec{p}_{\mathrm{f}} \,. \tag{5.213}$$

Since (from assumption (D)) the quark masses can be neglected, we obtain, from eqn (5.212), that $|\vec{p}_{\mathrm{i}}| = |\vec{p}_{\mathrm{f}}|$. In the Breit frame, the three-momentum of the nucleon is collinear with the three-momentum of the vector boson. From the assumption (C), the three-momentum of the initial quark is also collinear with the three-momentum of the vector boson and from eqn (5.213) it follows that also the three-momentum of the final quark is collinear: $\vec{p}_{\mathrm{f}}$, $\vec{p}_{\mathrm{i}}$, $\vec{p}_N$, and $\vec{q}$ are all parallel. Then, there are only two possibilities: $\vec{p}_{\mathrm{f}} = \pm\vec{p}_{\mathrm{i}}$. On the basis of the momentum conservation law (5.213), since $\vec{q} \neq 0$, only $\vec{p}_{\mathrm{f}} = -\vec{p}_{\mathrm{i}}$ can be realized. Substituting this in eqn (5.213) and using eqn (5.209), we obtain[28]

$$\vec{p}_{\mathrm{i}} = -\frac{\vec{q}}{2} = x\,\vec{p}_N \,. \tag{5.214}$$

Since the quark mass has been neglected, we have also $p_{\mathrm{i}}^0 = x\,p_N^0$, which implies $p_{\mathrm{i}} = x\,p_N$. Therefore, in the quark–parton model the four-momentum conservation law in eqn (5.211) implies that a virtual vector boson with four-momentum q interacts with those quarks which have, in the Breit frame, a four-momentum p_{i} equal to a fraction x of the nucleon four-momentum p_N.

The elementary W-quark and W-antiquark processes which contribute to CC neutrino–nucleon DIS are

$$W^+ + q_{\mathrm{i}} \to q_{\mathrm{f}} \qquad (q_{\mathrm{i}} = d, s, b\,;\; q_{\mathrm{f}} = u, c, t)\,, \tag{5.215}$$

$$W^+ + \bar{q}_{\mathrm{i}} \to \bar{q}_{\mathrm{f}} \qquad (\bar{q}_{\mathrm{i}} = \bar{u}, \bar{c}, \bar{t}\,;\; \bar{q}_{\mathrm{f}} = \bar{d}, \bar{s}, \bar{b})\,. \tag{5.216}$$

The process in eqn (5.215) for $q_{\mathrm{i}} = d$ and $q_{\mathrm{f}} = u$ is illustrated in Fig. 5.12b. Since different quarks generate different final states, the total ν-N cross-section is given by the sum of the cross-sections for each elementary process in eqns (5.215) and (5.216). For each value of x, the cross-section of the elementary process with initial

[28] This can also be seen as follows: let us write $\vec{p}_{\mathrm{i}}$ as $\vec{p}_{\mathrm{i}} = \xi\,\vec{p}_N$. Since the quark mass is negligible, we have also $p_{\mathrm{i}}^0 = \xi\,p_N^0$, which implies $p_{\mathrm{i}} = \xi\,p_N$. Then the four-momentum conservation law in eqn (5.211) can be written as $\xi\,p_N + q = p_{\mathrm{f}}$. Taking the square and neglecting the quark masses, we obtain $\xi = -q^2/2(p_N \cdot q) = x$.

(anti)quark $\overset{(-)}{q}_{\mathrm{i}}$ is proportional to the probability density $f^N_{\overset{(-)}{q}_{\mathrm{i}}}(x)$ of finding within the nucleon a $\overset{(-)}{q}_{\mathrm{i}}$ with a four-momentum $p_{\mathrm{i}} = x\, p_N$. The probability densities $f^N_{\overset{(-)}{q}_{\mathrm{i}}}(x)$ are called *parton distribution functions* (PDF) of the nucleon. The structure functions $F_i^{W^\pm N}$ in the differential cross-section in eqn (5.202) are given by the sum of the contributions of each elementary process in eqns (5.215) and (5.216), which can be calculated to be (see Ref. [227])

$$F^{W^+N}_{i,q_{\mathrm{f}}q_{\mathrm{i}}}(x) = \xi_i\, |V_{q_{\mathrm{f}}q_{\mathrm{i}}}|^2\, f^N_{q_{\mathrm{i}}}(x) \qquad (N = p, n\,;\ i = 1,2,3\,;\ q_{\mathrm{i}} = d,s,b\,;\ q_{\mathrm{f}} = u,c,t)\,, \tag{5.217}$$

$$F^{W^+N}_{i,\bar{q}_{\mathrm{f}}\bar{q}_{\mathrm{i}}}(x) = \bar{\xi}_i\, |V_{q_{\mathrm{i}}q_{\mathrm{f}}}|^2\, f^N_{\bar{q}_{\mathrm{i}}}(x) \qquad (N = p, n\,;\ i = 1,2,3\,;\ \bar{q}_{\mathrm{i}} = \bar{u},\bar{c},\bar{t}\,;\ \bar{q}_{\mathrm{f}} = \bar{d},\bar{s},\bar{b})\,, \tag{5.218}$$

with

$$\xi_1 = \bar{\xi}_1 = 1\,, \qquad \xi_2 = \bar{\xi}_2 = 2\,x\,, \qquad \xi_3 = -\bar{\xi}_3 = 2\,. \tag{5.219}$$

In the following we will consider, for simplicity, the DIS processes above the charm threshold, but below the bottom threshold (the treatment of heavy quark thresholds is reviewed in Refs. [276, 720]). We will also neglect the small mixing of the first two quark generations with the third generation (see eqn (4.2)). In this approximation, the 2×2 mixing matrix of the first two generations is unitary and the sum over the final quarks and antiquarks of $F^{W^+N}_{i,q_{\mathrm{f}}q_{\mathrm{i}}}(x)$ and $F^{W^+N}_{i,\bar{q}_{\mathrm{f}}\bar{q}_{\mathrm{i}}}(x)$ leads to the disappearance of the mixing matrix elements:

$$F^{W^+N}_{i,q_{\mathrm{i}}}(x) = \sum_{q_{\mathrm{f}}=u,c} F^{W^+N}_{i,q_{\mathrm{f}}q_{\mathrm{i}}}(x) = \xi_i\, f^N_{q_{\mathrm{i}}}(x) \qquad (N = p, n\,;\ i = 1,2,3\,;\ q_{\mathrm{i}} = d,s)\,, \tag{5.220}$$

$$F^{W^+N}_{i,\bar{q}_{\mathrm{i}}}(x) = \sum_{\bar{q}_{\mathrm{f}}=\bar{d},\bar{s}} F^{W^+N}_{i,\bar{q}_{\mathrm{f}}\bar{q}_{\mathrm{i}}}(x) = \bar{\xi}_i\, f^N_{\bar{q}_{\mathrm{i}}}(x) \qquad (N = p, n\,;\ i = 1,2,3\,;\ \bar{q}_{\mathrm{i}} = \bar{u},\bar{c})\,. \tag{5.221}$$

Finally, summing over the initial quarks and antiquarks, one obtains

$$F_i^{W^+N}(x) = \xi_i \sum_{q=d,s} f^N_q(x) + \bar{\xi}_i \sum_{\bar{q}=\bar{u},\bar{c}} f^N_{\bar{q}}(x) \qquad (N = p, n\,;\ i = 1,2,3)\,. \tag{5.222}$$

These are the expressions of the structure functions of the nucleons in CC neutrino–nucleon DIS processes in the quark–parton model.

For CC antineutrino–nucleon DIS, the elementary W-quark and W-antiquark processes are

$$W^- + q_{\mathrm{i}} \to q_{\mathrm{f}} \qquad (q_{\mathrm{i}} = u,c,t\,;\ q_{\mathrm{f}} = d,s,b)\,, \tag{5.223}$$

$$W^- + \bar{q}_{\mathrm{i}} \to \bar{q}_{\mathrm{f}} \qquad (\bar{q}_{\mathrm{i}} = \bar{d},\bar{s},\bar{b}\,;\ \bar{q}_{\mathrm{f}} = \bar{u},\bar{c},\bar{t})\,. \tag{5.224}$$

The structure functions $F_i^{W^-N}$ are obtained from the sum of the contributions of each elementary process, which are given by

$$F^{W^-N}_{i,q_{\mathrm{f}}q_{\mathrm{i}}}(x) = \xi_i\, |V_{q_{\mathrm{f}}q_{\mathrm{i}}}|^2\, f^N_{q_{\mathrm{i}}}(x) \qquad (N = p, n\,;\ i = 1,2,3\,;\ q_{\mathrm{i}} = u,c,t\,;\ q_{\mathrm{f}} = d,s,b)\,, \tag{5.225}$$

$$F_{i,\bar{q}_{\rm f}\bar{q}_{\rm i}}^{W^- N}(x) = \bar{\xi}_i \, |V_{q_{\rm i} q_{\rm f}}|^2 \, f_{\bar{q}_{\rm i}}^N(x) \qquad (N = p, n\,;\; i = 1,2,3\,;\; \bar{q}_{\rm i} = \bar{d}, \bar{s}, \bar{b}\,;\; \bar{q}_{\rm f} = \bar{u}, \bar{c}, \bar{t})\,. \tag{5.226}$$

Under the same approximations which led to eqn (5.222), we obtain

$$F_i^{W^- N}(x) = \xi_i \sum_{q=u,c} f_q^N(x) + \bar{\xi}_i \sum_{\bar{q}=\bar{d},\bar{s}} f_{\bar{q}}^N(x) \qquad (N = p, n\,;\; i = 1,2,3)\,. \tag{5.227}$$

Equations (5.222) and (5.227) show that in the quark–parton model at the tree-level the nucleon structure functions depend only on the kinematic variable x. This property, called *scaling*, was predicted by Bjorken in 1969 [245], before the formulation of the quark–parton model. It is possible to show that higher order QCD radiative corrections imply that the nucleon structure functions have a weak logarithmic Q^2-dependence (see Refs. [73, 721, 176]), which can be neglected in a first approximation.

Equations (5.222) and (5.227) imply the Callan–Gross relation [295]

$$F_2^{W^\pm N}(x) = 2\,x\,F_1^{W^\pm N}(x)\,. \tag{5.228}$$

From the expressions of the structure functions in eqn (5.222) and (5.227), one can write the differential cross-sections in eqn (5.202) as

$$\frac{\mathrm{d}^2\sigma_{\rm CC}^{\nu N}}{\mathrm{d}x\,\mathrm{d}y} = 2\,x\,\sigma_{\rm CC}^0 \left[\sum_{q=d,s} f_q^N(x) + (1-y)^2 \sum_{\bar{q}=\bar{u},\bar{c}} f_{\bar{q}}^N(x) \right], \tag{5.229}$$

$$\frac{\mathrm{d}^2\sigma_{\rm CC}^{\bar{\nu} N}}{\mathrm{d}x\,\mathrm{d}y} = 2\,x\,\sigma_{\rm CC}^0 \left[\sum_{\bar{q}=\bar{d},\bar{s}} f_{\bar{q}}^N(x) + (1-y)^2 \sum_{q=u,c} f_q^N(x) \right]. \tag{5.230}$$

It is interesting to note that in the limit $y = 1$ the cross-section of neutrinos depends only on the quark PDFs and the cross-section of antineutrinos depends only on the antiquark PDFs. This is due to angular-momentum conservation, taking into account that we have neglected the quark and charged lepton masses. For example, in the case of scattering of left-handed neutrinos, in the center-of-mass frame the initial state in the elementary interaction process with a right-handed antiquark has a projection of the total angular momentum along the direction of the neutrino momentum equal to -1, as illustrated in Fig. 5.13a. Since in the center-of-mass frame $\vec{p}_N = -\vec{p}_\nu$ and the DIS conditions in eqn (5.199) imply that $E_\nu \gg m_N$, we have $E_N \simeq E_\nu$ and

$$y \simeq \frac{E_\nu\,q^0 + \vec{p}_\nu \cdot \vec{q}}{E_\nu^2 + |\vec{p}_\nu|^2} = 1 - \frac{E_\nu\,E_\ell\,(1+\cos\theta_{\rm cms})}{E_\nu^2 + |\vec{p}_\nu|^2}\,, \tag{5.231}$$

where $\theta_{\rm cms}$ is the scattering angle of the final lepton with respect to the direction of the neutrino momentum in the center-of-mass frame. Thus, $y = 1$ implies that $\cos\theta_{\rm cms} = -1$. In this case, the final lepton is emitted backwards with respect to the direction of the neutrino momentum and the final antiquark is emitted along

FIG. 5.13. (a) Configuration of the initial state in the process $\nu_\ell + d \to \ell^- + u$ in the center-of-mass frame. The thin and thick arrows represent, respectively, momentum and spin. (b) Configuration of the final state in the same process with the charged lepton emitted backwards ($y = 1$).

the direction of the neutrino momentum, as illustrated in Fig. 5.13b. Since the final charged lepton is left-handed and the final antiquark is right-handed, the final state has a projection of the total angular momentum along the direction of the neutrino momentum equal to $+1$, which is different from the initial -1 and hence forbidden. On the other hand, one can easily see with the same reasoning that the elementary interaction process of a neutrino with a left-handed quark has a vanishing total angular momentum in both the initial and final state in the center-of-mass frame, leading to the absence of a $(1-y)$-dependence of the quark contribution in eqn (5.229). Hence, the contribution to the neutrino and antineutrino CC DIS differential cross-sections of a quark or antiquark q is given by

$$\left(\frac{\mathrm{d}^2\sigma_{\mathrm{CC}}^{\overset{(-)}{\nu} N}}{\mathrm{d}x\,\mathrm{d}y} \right)_q = 2\,x\,\sigma_{\mathrm{CC}}^0\, g(y)\, f_q^N(x)\,, \tag{5.232}$$

where $g(y) = 1$ if the helicities of the initial (anti)neutrino and (anti)quark are equal and $g(y) = (1-y)^2$ if the helicities are opposite. In the case of ν-N scattering, we have $q = d, s, \bar{u}, \bar{c}$, whereas in the case of $\bar{\nu}$-N scattering, we have $q = u, c, \bar{d}, \bar{s}$.

By integrating eqns (5.229) and (5.230) over $\mathrm{d}y$ in the range in eqn (5.200), we obtain

$$\frac{\mathrm{d}\sigma_{\mathrm{CC}}^{\nu N}}{\mathrm{d}x} = 2\,x\,\sigma_{\mathrm{CC}}^0 \left[\sum_{q=d,s} f_q^N(x) + \frac{1}{3} \sum_{\bar{q}=\bar{u},\bar{c}} f_{\bar{q}}^N(x) \right], \tag{5.233}$$

$$\frac{\mathrm{d}\sigma_{\mathrm{CC}}^{\bar{\nu} N}}{\mathrm{d}x} = 2\,x\,\sigma_{\mathrm{CC}}^0 \left[\sum_{\bar{q}=\bar{d},\bar{s}} f_{\bar{q}}^N(x) + \frac{1}{3} \sum_{q=u,c} f_q^N(x) \right]. \tag{5.234}$$

Further integration over $\mathrm{d}x$ leads to the cross-sections

$$\sigma_{\mathrm{CC}}^{\nu N} = 2\,\sigma_{\mathrm{CC}}^0 \left[\sum_{q=d,s} \langle x \rangle_q^N + \frac{1}{3} \sum_{\bar{q}=\bar{u},\bar{c}} \langle x \rangle_{\bar{q}}^N \right], \tag{5.235}$$

$$\sigma_{\mathrm{CC}}^{\bar{\nu} N} = 2\,\sigma_{\mathrm{CC}}^0 \left[\sum_{\bar{q}=\bar{d},\bar{s}} \langle x \rangle_{\bar{q}}^N + \frac{1}{3} \sum_{q=u,c} \langle x \rangle_q^N \right], \tag{5.236}$$

where $\langle x \rangle_{q,\bar{q}}^N = \int \mathrm{d}x\, x\, f_{q,\bar{q}}^N(x)$ are the average fractional momenta carried by quarks and antiquarks. Note that the fractional momenta carried by quarks and antiquarks are constrained by the momentum sum rule

$$\int \mathrm{d}x\, x \left\{ \sum_{q=u,d,s,c} \left[f_q^N(x) + f_{\bar{q}}^N(x) \right] + f_g^N(x) \right\} = 1\,, \tag{5.237}$$

where $f_g^N(x)$ is the gluon distribution function of the nucleon N.

In the standard notation, the quark, antiquark, and gluon distributions in the proton are denoted by

$$q(x) = f_q^p(x)\,, \qquad \bar{q}(x) \equiv f_{\bar{q}}^p(x)\,, \qquad g(x) \equiv f_g^p(x)\,. \tag{5.238}$$

Isospin symmetry implies that the up and down quark and antiquark distributions in the neutron are given by

$$f_u^n(x) = d(x)\,, \qquad f_{\bar{u}}^n(x) = \bar{d}(x)\,, \qquad f_d^n(x) = u(x)\,, \qquad f_{\bar{d}}^n(x) = \bar{u}(x)\,, \tag{5.239}$$

whereas the strange and charm quark and antiquark distributions and the gluon distributions are the same,

$$f_s^n(x) = s(x)\,, \quad f_{\bar{s}}^n(x) = \bar{s}(x)\,, \quad f_c^n(x) = c(x)\,, \quad f_{\bar{c}}^n(x) = \bar{c}(x)\,, \quad f_g^n(x) = g(x)\,. \tag{5.240}$$

The quark and antiquark distributions must satisfy the following sum rules, in order to reproduce the quantum numbers of the nucleons:

No Strangeness and Charm:

$$\int_0^1 [s(x) - \bar{s}(x)]\, \mathrm{d}x = \int_0^1 [c(x) - \bar{c}(x)]\, \mathrm{d}x = 0\,; \tag{5.241}$$

Unit Electric Charge:

$$\int_0^1 \left[\frac{2}{3} \left(u(x) - \bar{u}(x) \right) - \frac{1}{3} \left(d(x) - \bar{d}(x) \right) \right] \mathrm{d}x = 1\,; \tag{5.242}$$

Unit Baryon Number:

$$\frac{1}{3} \sum_{q=u,d} \int_0^1 \left(q(x) - \bar{q}(x) \right) \mathrm{d}x = 1\,. \tag{5.243}$$

Equations (5.242) and (5.243) imply the conditions

$$\int_0^1 \left(u(x) - \bar{u}(x) \right) \mathrm{d}x = 2\,, \qquad \int_0^1 \left(d(x) - \bar{d}(x) \right) \mathrm{d}x = 1\,. \tag{5.244}$$

Defining the valence quark distributions $u_v(x) = u(x) - \bar{u}(x)$ and $d_v(x) = d(x) - \bar{d}(x)$, the sum rules in eqn (5.244) reflect the fact that the proton is composed of two valence u quarks and one valence d quark. Since in QCD quarks and antiquarks are created together by gluons, it is also plausible that

$$s(x) = \bar{s}(x) , \qquad c(x) = \bar{c}(x) . \tag{5.245}$$

For the structure functions, using the expressions in eqns (5.222) and (5.227), one can derive the Bjorken sum rule [244]

$$\int_0^1 \left[F_1^{W^+ n}(x) - F_1^{W^+ p}(x) \right] \mathrm{d}x = \int_0^1 \left[F_1^{W^- p}(x) - F_1^{W^- n}(x) \right] \mathrm{d}x = 1 \tag{5.246}$$

and the Gross–Llewellyn Smith sum rule [573]

$$\int_0^1 \left[F_3^{W^+ p}(x) + F_3^{W^- p}(x) \right] \mathrm{d}x = \int_0^1 \left[F_3^{W^+ n}(x) + F_3^{W^- n}(x) \right] \mathrm{d}x = 6 . \tag{5.247}$$

Furthermore, from the Callan–Gross relations in eqn (5.228), one can obtain the Adler sum rule [30]

$$\int_0^1 \left[F_2^{W^+ n}(x) - F_2^{W^+ p}(x) \right] \frac{\mathrm{d}x}{x} = \int_0^1 \left[F_2^{W^- p}(x) - F_2^{W^- n}(x) \right] \frac{\mathrm{d}x}{x} = 2 . \tag{5.248}$$

For scattering on an isoscalar target, the neutrino and antineutrino cross-sections $\sigma_{\mathrm{CC}}^{\overset{(-)}{\nu}} = \left(\sigma_{\mathrm{CC}}^{\overset{(-)}{\nu} p} + \sigma_{\mathrm{CC}}^{\overset{(-)}{\nu} n} \right) /2$ are given by

$$\sigma_{\mathrm{CC}}^{\nu} = \sigma_{\mathrm{CC}}^{0} \left[\langle x \rangle_u + \langle x \rangle_d + 2 \langle x \rangle_s + \frac{1}{3} \left(\langle x \rangle_{\bar{u}} + \langle x \rangle_{\bar{d}} + 2 \langle x \rangle_{\bar{c}} \right) \right] , \tag{5.249}$$

$$\sigma_{\mathrm{CC}}^{\bar{\nu}} = \sigma_{\mathrm{CC}}^{0} \left[\langle x \rangle_{\bar{u}} + \langle x \rangle_{\bar{d}} + 2 \langle x \rangle_{\bar{s}} + \frac{1}{3} \left(\langle x \rangle_{u} + \langle x \rangle_{d} + 2 \langle x \rangle_{c} \right) \right] , \tag{5.250}$$

where $\langle x \rangle_{q,\bar{q}} = \langle x \rangle_{q,\bar{q}}^{p}$ are the average fractional momenta of the proton carried by quarks and antiquarks. Assuming the equalities in eqn (5.245), one can find that the difference of the neutrino and antineutrino cross-sections on an isoscalar target have the simple expression

$$\sigma_{\mathrm{CC}}^{\nu} - \sigma_{\mathrm{CC}}^{\bar{\nu}} = \frac{2}{3} \sigma_{\mathrm{CC}}^{0} \left(\langle x \rangle_u + \langle x \rangle_d - \langle x \rangle_{\bar{u}} - \langle x \rangle_{\bar{d}} \right) , \tag{5.251}$$

which will be useful in the derivation of the Paschos–Wolfenstein relation in eqn (5.266).

5.3.4 Neutral-current deep inelastic scattering

High-energy neutrinos and antineutrinos interact with nucleons also through the neutral-current DIS reactions

$$\overset{(-)}{\nu_\ell} + N \to \overset{(-)}{\nu_\ell} + X , \tag{5.252}$$

where $N = p, n$ and X denotes any set of final hadrons. The observation in 1973 of these neutral-current processes, with $\ell = \mu$, in the Gargamelle experiment at CERN

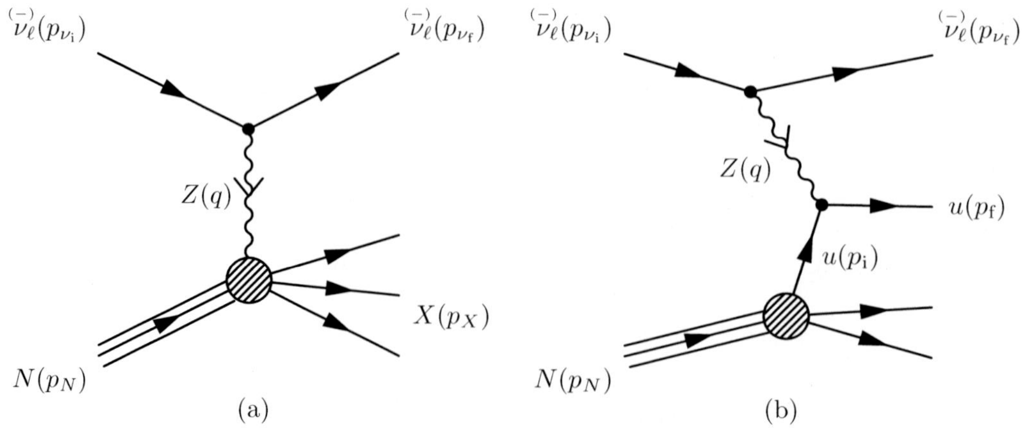

FIG. 5.14. (a) Diagram of the $\overset{(-)}{\nu}_\ell(p_{\nu_\mathrm{i}}) + N(p_N) \to \overset{(-)}{\nu}_\ell(p_{\nu_\mathrm{f}}) + X(p_X)$ neutral-current DIS process at lowest order in the weak interaction perturbation expansion. (b) Diagram of the same process in the quark–parton model, with elementary $Z(q) + u(p_\mathrm{i}) \to u(p_\mathrm{f})$ transition (see eqn (5.255)).

[599, 601] and an experiment at Fermilab [207], together with the observation of $\bar{\nu}_\mu + e^- \to \bar{\nu}_\mu + e^-$ reactions in the Gargamelle experiment [600], led to the experimental confirmation of the existence of the neutral-current interactions predicted by the SM.

The kinematical variables are the same as in the charged-current DIS processes discussed in section 5.3.3, with the obvious replacements $p_\nu \to p_{\nu_\mathrm{i}}$, $p_\ell \to p_{\nu_\mathrm{f}}$, where p_{ν_i} and p_{ν_f} are, respectively, the four-momenta of the initial and final neutrinos. The NC DIS processes are mediated by exchange of a Z-boson, as shown in Fig. 5.14a, with differential cross-sections given by

$$\frac{\mathrm{d}^2\sigma_{\mathrm{NC}}^{\overset{(-)}{\nu} N}}{\mathrm{d}x\,\mathrm{d}y} = \sigma_{\mathrm{NC}}^0 \left[x\,y^2\,F_1^{ZN} + (1-y)\,F_2^{ZN} \pm x\,y\left(1-\frac{y}{2}\right) F_3^{ZN} \right], \tag{5.253}$$

where

$$\sigma_{\mathrm{NC}}^0 = \frac{G_\mathrm{F}^2}{2\pi}\,s\left(1+\frac{Q^2}{m_Z^2}\right)^{-2}. \tag{5.254}$$

The plus and minus signs in eqn (5.253) refer, respectively, to ν_ℓ and $\bar{\nu}_\ell$ scattering. For $Q^2 \ll m_Z^2$, we have $\sigma_{\mathrm{NC}}^0 \simeq \sigma_{\mathrm{CC}}^0$ (see eqn (5.203)).

In the quark–parton model, the elementary Z-quark and Z-antiquark processes which contribute to NC neutrino–nucleon DIS are

$$Z + q \to q\,, \qquad Z + \bar{q} \to \bar{q}\,. \tag{5.255}$$

The elementary Z-u process is illustrated in Fig. 5.14b. The quark and antiquark contributions to the structure functions are given by

$$F_{1,\mathrm{Q}}^{ZN} = \frac{1}{2}\left[(g_V^q)^2 + (g_A^q)^2\right] f_q^N(x)\,, \qquad F_{1,\bar{q}}^{ZN} = \frac{1}{2}\left[(g_V^q)^2 + (g_A^q)^2\right] f_{\bar{q}}^N(x)\,, \tag{5.256}$$

$$F_{2,\mathrm{Q}}^{ZN} = 2\,x\,F_{1,\mathrm{Q}}^{ZN}\,, \qquad F_{2,\bar{q}}^{ZN} = 2\,x\,F_{1,\bar{q}}^{ZN}\,, \tag{5.257}$$

$$F_{3,\mathrm{Q}}^{ZN} = 2\,g_V^q\,g_A^q\,f_q^N(x)\,, \qquad F_{3,\bar{q}}^{ZN} = -2\,g_V^q\,g_A^q\,f_{\bar{q}}^N(x)\,, \tag{5.258}$$

with the neutral-current vector and axial couplings g_V^q and g_A^q listed in Table 3.6 (page 78). In the following, for simplicity, we will consider the NC DIS processes above the charm threshold, but below the bottom threshold, as in the discussion of the CC DIS processes in section 5.3.3 (the treatment of heavy quark thresholds is reviewed in Refs. [276, 720]). Hence, we consider $q = u, d, s, c$ in eqn (5.255). Expressing the couplings g_V^q and g_A^q in terms of the corresponding g_L^q and g_R^q (see eqns (3.51) and (3.52))[29], one can write the neutrino differential cross-section as

$$\begin{aligned}\frac{\mathrm{d}^2\sigma_{\mathrm{NC}}^{\nu N}}{\mathrm{d}x\,\mathrm{d}y} = 2\,x\,\sigma_{\mathrm{NC}}^0\Big\{ &\left[(g_L^U)^2 + (1-y)^2\,(g_R^U)^2\right] \sum_{q=u,c} f_q^N(x) \\ &+ \left[(g_L^D)^2 + (1-y)^2\,(g_R^D)^2\right] \sum_{q=d,s} f_q^N(x) \\ &+ \left[(g_R^U)^2 + (1-y)^2\,(g_L^U)^2\right] \sum_{\bar{q}=\bar{u},\bar{c}} f_{\bar{q}}^N(x) \\ &+ \left[(g_R^D)^2 + (1-y)^2\,(g_L^D)^2\right] \sum_{\bar{q}=\bar{d},\bar{s}} f_{\bar{q}}^N(x)\Big\}\,. \end{aligned} \tag{5.259}$$

Note that the quark and antiquark contributions follow a rule similar to that in eqn (5.232) for the CC processes. The antineutrino cross-section can be obtained from the neutrino cross-section through the exchange $L \leftrightarrows R$:

$$\frac{\mathrm{d}^2\sigma_{\mathrm{NC}}^{\bar{\nu} N}}{\mathrm{d}x\,\mathrm{d}y} = \left.\frac{\mathrm{d}^2\sigma_{\mathrm{NC}}^{\nu N}}{\mathrm{d}x\,\mathrm{d}y}\right|_{L\leftrightarrows R}\,. \tag{5.260}$$

By integrating eqn (5.259) over $\mathrm{d}y$ and $\mathrm{d}x$ in the ranges in eqn (5.200), we obtain

$$\begin{aligned}\sigma_{\mathrm{NC}}^{\nu N} = 2\,\sigma_{\mathrm{NC}}^0\Big\{ &\left[(g_L^U)^2 + \frac{1}{3}\,(g_R^U)^2\right] \sum_{q=u,c} \langle x\rangle_q^N + \left[(g_L^D)^2 + \frac{1}{3}\,(g_R^D)^2\right] \sum_{q=d,s} \langle x\rangle_q^N \\ &+ \left[(g_R^U)^2 + \frac{1}{3}\,(g_L^U)^2\right] \sum_{\bar{q}=\bar{u},\bar{c}} \langle x\rangle_{\bar{q}}^N + \left[(g_R^D)^2 + \frac{1}{3}\,(g_L^D)^2\right] \sum_{\bar{q}=\bar{d},\bar{s}} \langle x\rangle_{\bar{q}}^N \Big\}\,, \end{aligned} \tag{5.261}$$

and $\sigma_{\mathrm{NC}}^{\bar{\nu}N} = \sigma_{\mathrm{NC}}^{\nu N}\big|_{L\leftrightarrows R}$. For scattering on an isoscalar target, the neutrino and antineutrino cross-sections $\sigma_{\mathrm{NC}}^{\overset{(-)}{\nu}} = \left(\sigma_{\mathrm{NC}}^{\overset{(-)}{\nu}p} + \sigma_{\mathrm{NC}}^{\overset{(-)}{\nu}n}\right)/2$ are given by

$$\sigma_{\mathrm{NC}}^{\nu} = \sigma_{\mathrm{NC}}^0\Big\{ \left[(g_L^U)^2 + \frac{1}{3}\,(g_R^U)^2\right] \left[\langle x\rangle_u + \langle x\rangle_d + 2\,\langle x\rangle_c\right]$$

[29] In the literature [672, 233, 421] it is common to use the notation $\epsilon_L^q \equiv g_L^q$ and $\epsilon_R^q \equiv g_R^q$, or something similar.

$$
\begin{aligned}
&+\left[(g_L^D)^2 + \frac{1}{3}\,(g_R^D)^2\right] \left[\langle x\rangle_u + \langle x\rangle_d + 2\,\langle x\rangle_s\right] \\
&+\left[(g_R^U)^2 + \frac{1}{3}\,(g_L^U)^2\right] \left[\langle x\rangle_{\bar{u}} + \langle x\rangle_{\bar{d}} + 2\,\langle x\rangle_{\bar{c}}\right] \\
&+\left[(g_R^D)^2 + \frac{1}{3}\,(g_L^D)^2\right] \left[\langle x\rangle_{\bar{u}} + \langle x\rangle_{\bar{d}} + 2\,\langle x\rangle_{\bar{s}}\right]\Big\}\,,
\end{aligned} \tag{5.262}
$$

and $\sigma_{\rm NC}^{\bar{\nu}} = \sigma_{\rm NC}^{\nu}|_{L\leftrightarrows R}$. The equalities in eqn (5.245) imply that the difference of the neutrino and antineutrino cross-sections on an isoscalar target is given by

$$
\sigma_{\rm NC}^{\nu} - \sigma_{\rm NC}^{\bar{\nu}} = \frac{2}{3}\,\sigma_{\rm NC}^{0}\,(g_L^2 - g_R^2)\,\left[\langle x\rangle_u + \langle x\rangle_d - \langle x\rangle_{\bar{u}} + \langle x\rangle_{\bar{d}}\right], \tag{5.263}
$$

where we adopted the common notation [421]

$$
g_L^2 \equiv (g_L^U)^2 + (g_L^D)^2 = \frac{1}{2}\left(1 - \sin^2\vartheta_{\rm W}\right) + \frac{5}{9}\,\sin^4\vartheta_{\rm W}\,, \tag{5.264}
$$

$$
g_R^2 \equiv (g_R^U)^2 + (g_R^D)^2 = \frac{5}{9}\,\sin^4\vartheta_{\rm W}\,. \tag{5.265}
$$

Here we have used the SM values of $g_L^{U,D}$ and $g_R^{U,D}$ listed in Table 3.6 (page 78). From eqns (5.251) and (5.263), one can immediately derive the Paschos–Wolfenstein relation [852]

$$
\frac{\sigma_{\rm NC}^{\nu} - \sigma_{\rm NC}^{\bar{\nu}}}{\sigma_{\rm CC}^{\nu} - \sigma_{\rm CC}^{\bar{\nu}}} = g_L^2 - g_R^2 = \frac{1}{2}\left(1 - \sin^2\vartheta_{\rm W}\right)\,. \tag{5.266}
$$

This relation is useful in practice, since it allows the measurement of the value of the weak mixing angle by measuring the NC and CC neutrino DIS cross-sections on an isoscalar target. If these measurements are made in the same experiment, it is expected that most of the systematic uncertainties cancel in the ratio of the cross-section differences, leading to the possibility of an accurate measurement of $\sin^2\vartheta_{\rm W}$. In fact, such a measurement has been done in the NuTeV experiment [1087], with the result

$$
\sin^2\vartheta_{\rm W} = 0.2277 \pm 0.0013 \pm 0.0009\,. \tag{5.267}
$$

Let us finally remark that the neutrino and antineutrino CC and NC DIS processes give information on the nucleon PDFs which are complementary to those obtained in charged lepton DIS on nucleons,

$$
\ell^{\pm} + N \to \ell^{\pm} + X\,, \tag{5.268}
$$

which is due to electromagnetic interactions. Since the scattering is mediated by a virtual photon, this process is insensitive to the charge of the lepton. Neglecting the charged lepton mass, the differential cross-section for unpolarized charged leptons is given by

$$\frac{\mathrm{d}^2\sigma_{\mathrm{em}}}{\mathrm{d}x\,\mathrm{d}y} = \frac{4\pi\,\alpha^2\,s}{Q^4}\left[x\,y^2\,F_1^{\gamma N} + (1-y)\,F_2^{\gamma N}\right] . \tag{5.269}$$

In the quark–parton model, the structure function $F_1^{\gamma N}$ is given by

$$F_1^{\gamma N}(x) = \frac{1}{2}\sum_{q=u,d,s,c} e_q^2\left(f_q^N(x) + f_{\bar{q}}^N(x)\right) , \tag{5.270}$$

where e_q is the electric charge of the quark q ($e_q = 2/3$ for $q = u, c, t$ and $e_q = -1/3$ for $q = d, s, b$). The structure function $F_2^{\gamma N}$ is related to $F_1^{\gamma N}$ by Callan–Gross relations analogous to that in eqn (5.228):

$$F_2^{\gamma N}(x) = 2\,x\,F_1^{\gamma N}(x) . \tag{5.271}$$

6

MASSIVE NEUTRINOS

> Although it is perhaps still not possible to decide experimentally between this new theory and what is the simple extension of the Dirac equations to neutral particles, one must consider that the first one introduces, in this still little explored field, a smaller number of hypothetical entities. ... The advantage of this procedure over the elementary interpretation of the Dirac equations is that there is no reason to presume the existence of antineutrons or antineutrinos.
> **Ettore Majorana** [765]

Neutrino mass is by far the most important subject of study in neutrino physics. Ever since the proposal by Pauli, the mass of neutrinos has been the topic of intense experimental and theoretical investigation. At the time of the Pauli proposal, the neutrino mass was postulated to be of the order of the electron mass and even massless. Finally, we know now that neutrinos have mass, although only two small values of squared-mass differences are known.

In this chapter we discuss, in detail, how to describe a massive neutrino in the case of a Dirac or a Majorana mass, as well as in the most general Dirac–Majorana case. We discuss also the realistic case of three-generation mixing.

The origin of the small neutrino mass is still a mystery. It is commonly believed that neutrino masses are a low-energy manifestation of physics beyond the Standard Model and their smallness is due to a suppression generated by a new high-energy scale, perhaps related to the unification of forces. This is achieved, for example, with the celebrated *see-saw mechanism* (introduced in section 6.4.6). Numerous models and hypotheses on the origin of neutrino masses and possible implementations of the see-saw mechanism have been presented, especially in the past few years. Although this subject is very important, we do not treat it here, since a proper discussion would require a lengthy introduction of theoretical ideas which is beyond the scope of this book. The interested reader can find detailed information in Refs. [812, 74, 467, 781, 1076, 673, 810, 75, 763, 811, 815, 950, 813].

6.1 Dirac masses

A Dirac neutrino mass can be generated with the same Higgs mechanism that gives masses to quarks and charged leptons in the SM (see section 3.4). The only extension of the SM that is needed is the introduction of right-handed components $\nu_{\alpha R}$ of the neutrino fields ($\alpha = e, \mu, \tau$). Such a model is sometimes called the *minimally extended Standard Model*, in which the asymmetry in the SM between the lepton and quark sectors due to the absence of right-handed neutrino fields is eliminated.

Let us recall, however, that the right-handed neutrino fields are fundamentally different from the other elementary fermion fields because they are invariant under the symmetries of the SM: they are singlets of $\mathrm{SU}(3)_C \times \mathrm{SU}(2)_L$ and have hypercharge $Y = 0$. The right-handed neutrino fields are called *sterile* [883] because they do not participate in weak interactions (as well as strong and electromagnetic interactions, as all neutrino fields); their only interaction is gravitational. On the other hand, the normal left-handed neutrino fields that participate in weak interactions are usually called *active*.

It is also important to note that the presence of sterile right-handed neutrino fields is totally irrelevant for the cancellation of quantum anomalies [31, 201], which constrain the properties of the other elementary fermion fields [268, 572, 517, 516, 606, 805, 915, 551, 459, 692]. As a consequence, the number of sterile right-handed neutrino fields is not constrained by the theory, and the introduction of three right-handed neutrino fields, one for each generation, is not even the minimal extension of the SM, because the presence of only one right-handed neutrino field cannot be excluded.

Let us emphasize that right-handedness is not an essential quality of the new chiral fields, because we could work as well with the left-handed chiral fields $\widetilde{\nu}_{\alpha L} = \nu^C_{\alpha R}$. The essential characteristic of these fields is that they are singlets under the SM symmetries, and hence sterile. We follow the usual convention of calling the new sterile neutrino fields *right-handed*, in order to distinguish them from the active neutrino fields of the SM, which participate in weak interactions through their left-handed chiral components.

In the minimally extended Standard Model with three right-handed neutrino fields, the SM Higgs–lepton Yukawa Lagrangian in eqn (3.125) is extended by adding a lepton term with the same structure as the second term on the right-hand side of eqn (3.161), which generates the masses of up-type quarks:

$$\mathscr{L}_{H,\mathrm{L}} = - \sum_{\alpha,\beta=e,\mu,\tau} Y'^{\ell}_{\alpha\beta} \, \overline{L_{\alpha L}} \, \Phi \, \ell'_{\beta R} - \sum_{\alpha,\beta=e,\mu,\tau} Y'^{\nu}_{\alpha\beta} \, \overline{L_{\alpha L}} \, \widetilde{\Phi} \, \nu'_{\beta R} + \mathrm{H.c.}\,, \tag{6.1}$$

where Y'^{ν} is a new matrix of Yukawa couplings. In the unitary gauge the Higgs–lepton Yukawa Lagrangian can be written in matrix form as

$$\mathscr{L}_{H,\mathrm{L}} = - \left(\frac{v + H}{\sqrt{2}} \right) \left[\overline{\boldsymbol{\ell}'_L} \, Y'^{\ell} \, \boldsymbol{\ell}'_R + \overline{\boldsymbol{\nu}'_L} \, Y'^{\nu} \, \boldsymbol{\nu}'_R \right] + \mathrm{H.c.}\,, \tag{6.2}$$

with the chiral charged lepton arrays in eqn (3.127), the left-handed neutrino array in eqn (3.138), and the new right-handed neutrino array

$$\boldsymbol{\nu}'_R \equiv \begin{pmatrix} \nu'_{eR} \\ \nu'_{\mu R} \\ \nu'_{\tau R} \end{pmatrix} . \tag{6.3}$$

The matrix Y'^{ℓ} of charged lepton Yukawa couplings can be diagonalized as in eqn (3.129), and the matrix Y'^{ν} of neutrino Yukawa couplings can be diagonalized in a similar way:

$$V_L^{\nu\dagger}\, Y'^{\nu}\, V_R^{\nu} = Y^{\nu}\,, \qquad \text{with} \qquad Y^{\nu}_{kj} = y^{\nu}_k\, \delta_{kj} \qquad (k,j=1,2,3)\,, \tag{6.4}$$

with real and positive y^{ν}_k. Here V_L^{ν} and V_R^{ν} are two appropriate 3×3 unitary matrices ($V_L^{\nu\dagger} = (V_L^{\nu})^{-1}$ and $V_R^{\nu\dagger} = (V_R^{\nu})^{-1}$).

Defining the massive chiral charged lepton arrays in eqn (3.131) and the chiral massive neutrino arrays

$$\boldsymbol{n}_L = V_L^{\nu\dagger}\, \boldsymbol{\nu}'_L \equiv \begin{pmatrix} \nu_{1L} \\ \nu_{2L} \\ \nu_{3L} \end{pmatrix}, \qquad \boldsymbol{n}_R = V_R^{\nu\dagger}\, \boldsymbol{\nu}'_R \equiv \begin{pmatrix} \nu_{1R} \\ \nu_{2R} \\ \nu_{3R} \end{pmatrix}, \tag{6.5}$$

the diagonalized Higgs–lepton Yukawa Lagrangian reads

$$\begin{aligned} \mathscr{L}_{H,\mathrm{L}} &= -\left(\frac{v+H}{\sqrt{2}}\right) \left[\overline{\boldsymbol{\ell}_L}\, Y^{\ell}\, \boldsymbol{\ell}_R + \overline{\boldsymbol{n}_L}\, Y^{\nu}\, \boldsymbol{n}_R\right] + \text{H.c.} \\ &= -\left(\frac{v+H}{\sqrt{2}}\right) \left[\sum_{\alpha=e,\mu,\tau} y^{\ell}_{\alpha}\, \overline{\ell_{\alpha L}}\, \ell_{\alpha R} + \sum_{k=1}^{3} y^{\nu}_k\, \overline{\nu_{kL}}\, \nu_{kR}\right] + \text{H.c.} \end{aligned} \tag{6.6}$$

Using the Dirac charged lepton fields in eqn (3.133) and the Dirac neutrino fields

$$\nu_k = \nu_{kL} + \nu_{kR} \qquad (k = 1,2,3)\,, \tag{6.7}$$

we finally obtain

$$\mathscr{L}_{H,\mathrm{L}} = -\sum_{\alpha=e,\mu,\tau} \frac{y^{\ell}_{\alpha}\, v}{\sqrt{2}}\, \overline{\ell_{\alpha}}\, \ell_{\alpha} - \sum_{k=1}^{3} \frac{y^{\nu}_k\, v}{\sqrt{2}}\, \overline{\nu_k}\, \nu_k - \sum_{\alpha=e,\mu,\tau} \frac{y^{\ell}_{\alpha}}{\sqrt{2}}\, \overline{\ell_{\alpha}}\, \ell_{\alpha}\, H - \sum_{k=1}^{3} \frac{y^{\nu}_k}{\sqrt{2}}\, \overline{\nu_k}\, \nu_k\, H\,. \tag{6.8}$$

Therefore, the neutrino masses are given by

$$m_k = \frac{y^{\nu}_k\, v}{\sqrt{2}} \qquad (k = 1,2,3)\,, \tag{6.9}$$

and massive Dirac neutrinos couple to the Higgs field through the last term in eqn (6.8).

Note that the neutrino masses that we have obtained with this mechanism are proportional to the Higgs VEV v, as the masses of charged leptons and quarks. However, it is known that the masses of neutrinos are much smaller than those of charged leptons and quarks. In the mechanism that we have just described, however, there is no explanation of the very small values of the eigenvalues y^{ν}_k of the Higgs-neutrino Yukawa coupling matrix that are needed.

In fact, the SM Higgs mechanism discussed in section 3.4 leaves completely open the question of the value of the Yukawa couplings with the Higgs of all particles. Therefore, the origin of the values of quark and lepton masses is a mystery in the framework of the SM. Among other problems, this mystery leads us to believe that the SM must be considered as an effective theory obtained from the low-energy limit of a more complete theory, in which the masses of quarks and leptons can be derived from first principles. Seen from this point of view, the values of quark and lepton masses represent a useful handle that may allow us to understand the physics beyond the SM. Indeed, the comparison of the predicted values with the observed quark and lepton spectra is one of the strong constraints that enable us to select among the new theories. In this context, the smallness of neutrino masses is an additional clue of the physics beyond the SM and a very strong constraint that the new theories must satisfy.

The mixing of three Dirac neutrinos is similar to the SM mixing of quarks, discussed in section 3.5 and chapter 4. From eqns (3.131) and (6.5), the leptonic weak charged current in eqn (3.80) can be written as

$$j^{\rho}_{W,L} = 2\,\overline{\boldsymbol{\nu}'_L}\,\gamma^{\rho}\,\boldsymbol{\ell}'_L = 2\,\overline{\boldsymbol{n}_L}\,V^{\nu\dagger}_L\,\gamma^{\rho}\,V^{\ell}_L\,\boldsymbol{\ell}_L = 2\,\overline{\boldsymbol{n}_L}\,V^{\nu\dagger}_L\,V^{\ell}_L\,\gamma^{\rho}\,\boldsymbol{\ell}_L\,. \tag{6.10}$$

This current depends on the product

$$U = V^{\ell\dagger}_L\,V^{\nu}_L\,, \tag{6.11}$$

which is the mixing matrix in the lepton sector, analogous to the CKM mixing matrix of quarks. In terms of the mixing matrix, the leptonic weak charged current reads

$$j^{\rho}_{W,L} = 2\,\overline{\boldsymbol{n}_L}\,U^{\dagger}\,\gamma^{\rho}\,\boldsymbol{\ell}_L\,. \tag{6.12}$$

It is customary to define left-handed flavor neutrino fields as

$$\boldsymbol{\nu}_L = U\,\boldsymbol{n}_L = V^{\ell\dagger}_L\,\boldsymbol{\nu}'_L\,, \qquad \text{with} \qquad \boldsymbol{\nu}_L = \begin{pmatrix} \nu_{eL} \\ \nu_{\mu L} \\ \nu_{\tau L} \end{pmatrix}, \tag{6.13}$$

which allow us to write the leptonic weak charged current as in the SM eqn (3.141):

$$j^{\rho}_{W,L} = 2\,\overline{\boldsymbol{\nu}_L}\,\gamma^{\rho}\,\boldsymbol{\ell}_L = 2\sum_{\alpha=e,\mu,\tau}\overline{\nu_{\alpha L}}\,\gamma^{\rho}\,\ell_{\alpha L}\,. \tag{6.14}$$

We wish to emphasize that the flavor neutrino fields must be treated with caution. In practice they are useful only for calculations in which the effects of neutrino masses are neglected, i.e. in the SM limit. On the other hand, if neutrino masses are taken into account the flavor neutrino fields do not have a definite mass and are not independent, being coupled by the mass terms. In that case it is much more convenient to work with the independent massive neutrino fields.

6.1.1 Lepton numbers

The definition of left-handed flavor neutrino fields is useful for defining flavor lepton numbers as in the SM (see Table 3.8). The flavor lepton numbers are conserved in weak interactions, but in general they are violated by the neutrino part of the Higgs–lepton Yukawa Lagrangian in eqn (6.6), which can be written as

$$\mathscr{L}_{H,L} = -\left(\frac{v+H}{\sqrt{2}}\right)\left[\overline{\boldsymbol{\ell}_L}\,Y^{\ell}\,\boldsymbol{\ell}_R + \overline{\boldsymbol{\nu}_L}\,U\,Y^{\nu}\,\boldsymbol{n}_R\right] + \text{H.c.}$$

$$= -\left(\frac{v+H}{\sqrt{2}}\right) \sum_{\alpha=e,\mu,\tau} \left[y^{\ell}_{\alpha} \, \overline{\ell_{\alpha L}} \, \ell_{\alpha R} + \overline{\nu_{\alpha L}} \sum_{k=1}^{3} U_{\alpha k} \, y^{\nu}_{k} \, \nu_{kR} \right] + \text{H.c.} \tag{6.15}$$

Indeed, the weak charged-current in eqn (6.14) is invariant under the global U(1) gauge transformations

$$\ell_{\alpha L} \to e^{i\varphi_\alpha} \, \ell_{\alpha L} \,, \qquad \nu_{\alpha L} \to e^{i\varphi_\alpha} \, \nu_{\alpha L} \qquad (\alpha = e, \mu, \tau) \,, \tag{6.16}$$

with different phases for each flavor. The charged lepton terms in eqn (6.15) are invariant if the right-handed chiral components of the charged lepton fields transform with the corresponding phases:

$$\ell_{\alpha R} \to e^{i\varphi_\alpha} \, \ell_{\alpha R} \qquad (\alpha = e, \mu, \tau) \,. \tag{6.17}$$

But it is not possible to find any transformation of the right-handed neutrino fields that leaves simultaneously invariant the neutrino part of the Higgs–lepton Yukawa Lagrangian in eqn (6.15) and the kinetic part of the neutrino Lagrangian

$$\begin{aligned} \mathscr{L}^{(\nu)}_{\text{kinetic}} &= \sum_{k=1}^{3} \overline{\nu_k} i \overleftrightarrow{\partial\!\!\!/} \nu_k = \sum_{k=1}^{3} \left(\overline{\nu_{kL}} i \overleftrightarrow{\partial\!\!\!/} \nu_{kL} + \overline{\nu_{kR}} i \overleftrightarrow{\partial\!\!\!/} \nu_{kR} \right) \\ &= \sum_{\alpha=e,\mu,\tau} \overline{\nu_{\alpha L}} i \overleftrightarrow{\partial\!\!\!/} \nu_{\alpha L} + \sum_{k=1}^{3} \overline{\nu_{kR}} i \overleftrightarrow{\partial\!\!\!/} \nu_{kR} \,. \end{aligned} \tag{6.18}$$

In fact, the only way to leave the neutrino part of the Higgs–lepton Yukawa Lagrangian in eqn (6.15) invariant under the transformation in eqn (6.16) is to let the combination $\sum_{k=1}^{3} U_{\alpha k} \, y^{\nu}_{k} \, \nu_{kR}$ transform as $e^{i\varphi_\alpha} \sum_{k=1}^{3} U_{\alpha k} \, y^{\nu}_{k} \, \nu_{kR}$. But, in general, this transformation does not leave invariant the kinetic part in eqn (6.18) of the neutrino Lagrangian, because $\sum_{k=1}^{3} U_{\alpha k} \, y^{\nu}_{k} \, \nu_{kR}$ is not a unitary combination of the right-handed neutrino fields.

In chapter 7 we will see that the nonconservation of the flavor lepton numbers leads to neutrino oscillations, which is one of the most interesting and extensively studied phenomena in neutrino physics. The only exceptions are the cases in which all the Yukawa couplings y^{ν}_{k} are equal, i.e. the three neutrino masses given by eqn (6.9) are equal, or the mixing matrix U is unity, i.e. there is no mixing. In these cases the flavor lepton numbers are conserved and, as we will see in chapter 7, neutrinos do not oscillate.

For massive Dirac neutrinos, even if the flavor lepton numbers are not conserved, the total lepton number in eqn (3.143) is conserved. In fact, the Lagrangian is invariant under the global U(1) gauge transformations

$$\nu_{kL} \to e^{i\varphi} \, \nu_{kL} \,, \qquad \nu_{kR} \to e^{i\varphi} \, \nu_{kR} \qquad (k = 1, 2, 3) \,, \tag{6.19}$$

$$\ell_{\alpha L} \to e^{i\varphi} \, \ell_{\alpha L} \,, \qquad \ell_{\alpha R} \to e^{i\varphi} \, \ell_{\alpha R} \qquad (\alpha = e, \mu, \tau) \,, \tag{6.20}$$

with the same phase φ for the independent chiral neutrino and charged lepton fields. Noether's theorem discussed in section C.7 implies that there is a conserved current. From eqn (C.49) and the kinetic part of the neutrino and charged lepton Lagrangian, this conserved current ($\partial_\rho j^\rho = 0$) is

$$j^\rho = \sum_{k=1}^{3} \overline{\nu_k}\, \gamma^\rho\, \nu_k + \sum_{\alpha=e,\mu,\tau} \overline{\ell_\alpha}\, \gamma^\rho\, \ell_\alpha \,, \tag{6.21}$$

and the conserved charge ($\partial_0 \mathsf{L} = 0$) is

$$\mathsf{L} = \int \mathrm{d}^3x\, j^0(x) = \int \mathrm{d}^3x \left[\sum_{k=1}^{3} \nu_k^\dagger(x)\, \nu_k(x) + \sum_{\alpha=e,\mu,\tau} \ell_\alpha^\dagger(x)\, \ell_\alpha(x) \right] . \tag{6.22}$$

The Dirac massive neutrino fields in eqn (6.7) can be quantized in the standard way for Dirac fields, as explained in chapter 2. Using the Fourier expansion in eqn (2.139) of the Dirac neutrino and charged lepton Lagrangian, the conserved charge is the lepton number

$$\begin{aligned} :\mathsf{L}: \; = & \sum_{k=1}^{3} \int \frac{\mathrm{d}^3p}{(2\pi)^3\, 2E} \sum_{h=\pm 1} \left[a_{\nu_k}^{(h)\dagger}(p)\, a_{\nu_k}^{(h)}(p) - b_{\nu_k}^{(h)\dagger}(p)\, b_{\nu_k}^{(h)}(p) \right] \\ & + \sum_{\alpha=e,\mu,\tau} \int \frac{\mathrm{d}^3p}{(2\pi)^3\, 2E} \sum_{h=\pm 1} \left[a_{\ell_\alpha}^{(h)\dagger}(p)\, a_{\ell_\alpha}^{(h)}(p) - b_{\ell_\alpha}^{(h)\dagger}(p)\, b_{\ell_\alpha}^{(h)}(p) \right] , \end{aligned} \tag{6.23}$$

where the double colon denotes normal ordering[30] of the creation and annihilation operators. As for other charges, in quantum field theory the normal ordering is necessary in order to assign to the vacuum a zero charge. Equation (6.23) shows that neutrinos and negatively charged leptons have $L = +1$, whereas antineutrinos and positively charged leptons have $L = -1$, as in the SM (see Table 3.8). Therefore, the lepton quantum numbers are different for neutrinos and antineutrinos (charged leptons and antileptons are also distinguished by the lepton number, but in this case this fact is less important, because they are also distinguished by the electric charge).

The conserved charge in eqn (6.23), which distinguishes neutrinos from antineutrinos, is important for the classification of the states describing physical systems with one or more neutrinos and antineutrinos. Hence, we see that the Dirac character of massive neutrinos, which implies that neutrinos and antineutrinos are different particles, is closely related to the invariance of the total Lagrangian under the global U(1) gauge transformations in eqns (6.19) and (6.20).

The right-handed components ν_{kR} of the massive Dirac neutrino fields do not enter in the weak charged current in eqn (6.12). Hence, they represent sterile degrees of freedom, that do not participate in weak interactions. It is possible to define superpositions of right-handed neutrino fields in analogy to the definition in eqn (6.13) of left-handed flavor neutrino fields. However, such a definition would

[30] In a normally ordered product of creation and annihilation operators, the order of the operators is rearranged putting all the destruction operators on the right, taking into account the anticommutation of fermion operators.

be useless, because of the sterility of right-handed neutrino fields. Since the mixing in eqn (6.13) of the left-handed neutrino fields is independent of the right-handed neutrino fields, the active and sterile degrees of freedom remain decoupled in the presence of Dirac mixing and oscillations between active and sterile states is not possible.

In practice, one does not consider the possible existence of additional sterile Dirac neutrino fields, i.e. fields of additional Dirac neutral fermions that are singlets of the $\mathrm{SU}(2)_L \times \mathrm{U}(1)_Y$ electroweak symmetry group. These fields do not participate in weak interaction with both their left and right components, but can couple with the ordinary neutrinos through the mass term, generating a complicated mixing between active and sterile degrees of freedom. Since at present there is no indication of the existence of such additional sterile Dirac neutrino fields, the principle of simplicity[31] forces us to ignore them.

6.1.2 Mixing

As in the case of quarks discussed in section 4.2, the mixing matrix of Dirac neutrinos depends on four physical parameters: three mixing angles and one CP-violating phase. As in the case of quarks, five of the six phases in the unitary mixing matrix are unphysical because they can be eliminated by rephasing the neutrino and charged lepton fields. This is possible because, apart from the weak CC part, the Lagrangian is invariant under the global phase transformations

$$\nu_{kL} \to e^{i\varphi_k}\, \nu_{kL}\,, \qquad \nu_{kR} \to e^{i\varphi_k}\, \nu_{kR} \qquad (k = 1, 2, 3)\,, \tag{6.24}$$

$$\ell_{\alpha L} \to e^{i\varphi_\alpha}\, \ell_{\alpha L}\,, \qquad \ell_{\alpha R} \to e^{i\varphi_\alpha}\, \ell_{\alpha R} \qquad (\alpha = e, \mu, \tau)\,, \tag{6.25}$$

with three independent phases φ_k for the massive neutrino fields and three independent phases φ_α for the massive charged lepton fields.

The Dirac three-neutrino mixing matrix has properties analogous to the quark mixing matrix, discussed in chapter 4. In particular, CP violation can be quantified in a rephasing-invariant way in terms of the Jarlskog invariant [640, 639, 562, 395, 1075]

$$J = \mathfrak{Im}\big[U_{\mu 3}\, U_{e2}\, U_{\mu 2}^{*}\, U_{e3}^{*}\big] \,. \tag{6.26}$$

The CP and T asymmetries in neutrino oscillations depend on J (see section 13.1.1 for a general discussion of CP and T violations in neutrino oscillations and section 13.1.1 for a discussion of CP and T violations in neutrino oscillations in the case of three-neutrino mixing).

As discussed in section 4.7 for the quark case, in the case of three-neutrino mixing all the imaginary parts of the rephasing invariant quartic products $U_{\alpha k}^{*}\, U_{\beta k}\, U_{\alpha j}\, U_{\beta j}^{*}$ are equal up to a sign,

$$\mathfrak{Im}\big[U_{\alpha k}^{*}\, U_{\beta k}\, U_{\alpha j}\, U_{\beta j}^{*}\big] = s_{\alpha\beta;kj}\, J\,, \tag{6.27}$$

[31] This principle is also known as *Ockham's razor* (see the quote at the beginning of chapter 4 on page 106).

because of the unitarity relations analogous to eqns (4.155) and (4.156)

$$\sum_{k \neq j} \Im\mathfrak{m}\left[U_{\alpha k}^{*} U_{\beta k} U_{\alpha j} U_{\beta j}^{*}\right]=0 \qquad (\alpha \neq \beta)\,, \tag{6.28}$$

$$\sum_{\alpha \neq \beta} \Im\mathfrak{m}\left[U_{\alpha k}^{*} U_{\beta k} U_{\alpha j} U_{\beta j}^{*}\right]=0 \qquad (k \neq j)\,. \tag{6.29}$$

The coefficients $s_{\alpha\beta;kj} = \pm 1$ are antisymmetric in the indices α, β and in the indices k, j:

$$s_{\alpha\beta;kj} = -s_{\beta\alpha;kj} = s_{\beta\alpha;jk}\,. \tag{6.30}$$

Their values are given in Table 13.1 below (page 454).

A convenient parameterization of the Dirac neutrino mixing matrix is the same as given in eqn (4.79):

$$U=\begin{pmatrix} c_{12}c_{13} & s_{12}c_{13} & s_{13}e^{-i\delta_{13}} \\ -s_{12}c_{23}-c_{12}s_{23}s_{13}e^{i\delta_{13}} & c_{12}c_{23}-s_{12}s_{23}s_{13}e^{i\delta_{13}} & s_{23}c_{13} \\ s_{12}s_{23}-c_{12}c_{23}s_{13}e^{i\delta_{13}} & -c_{12}s_{23}-s_{12}c_{23}s_{13}e^{i\delta_{13}} & c_{23}c_{13} \end{pmatrix}, \tag{6.31}$$

where $c_{ab} \equiv \cos\vartheta_{ab}$ and $s_{ab} \equiv \sin\vartheta_{ab}$. The three mixing angles ϑ_{12}, ϑ_{13}, ϑ_{23} take values in the ranges $0 \leq \vartheta_{ab} \leq \pi/2$ and δ_{13} is the CP-violating phase with a value in the range $0 \leq \delta_{13} < 2\pi$. In this parameterization the Jarlskog invariant is given by

$$J = c_{12}s_{12}c_{23}s_{23}c_{13}^{2}s_{13}\sin\delta_{13} = \frac{1}{8}\sin 2\vartheta_{12}\sin 2\vartheta_{23}\cos\vartheta_{13}\sin 2\vartheta_{13}\sin\delta_{13}\,. \tag{6.32}$$

As in the SM (see eqn (3.148)), the unitarity of the charged lepton and neutrino mixing matrices V_L^{ℓ}, V_R^{ℓ}, and V_L^{ν} implies the GIM mechanism: the expression of the neutral weak current in terms of the lepton fields with definite masses is the same as that in terms of the primed lepton fields, as shown by

$$\begin{aligned} j_{Z,\mathrm{L}}^{\rho} &= 2\,g_L^{\nu}\,\overline{\boldsymbol{\nu}_L'}\,\gamma^{\rho}\,\boldsymbol{\nu}_L' + 2\,g_L^{l}\,\overline{\boldsymbol{\ell}_L'}\,\gamma^{\rho}\boldsymbol{\ell}_L' + 2\,g_R^{l}\,\overline{\boldsymbol{\ell}_R'}\,\gamma^{\rho}\boldsymbol{\ell}_R' \\ &= 2\,g_L^{\nu}\,\overline{\boldsymbol{n}_L}\,V_L^{\nu\dagger}\,\gamma^{\rho}\,V_L^{\nu}\,\boldsymbol{n}_L + 2\,g_L^{l}\,\overline{\boldsymbol{\ell}_L}\,V_L^{\ell\dagger}\,\gamma^{\rho}\,V_L^{\ell}\,\boldsymbol{\ell}_L + 2\,g_R^{l}\,\overline{\boldsymbol{\ell}_R}\,V_R^{\ell\dagger}\,\gamma^{\rho}\,V_R^{\ell}\,\boldsymbol{\ell}_R \\ &= 2\,g_L^{\nu}\,\overline{\boldsymbol{n}_L}\,\gamma^{\rho}\,\boldsymbol{n}_L + 2\,g_L^{l}\,\overline{\boldsymbol{\ell}_L}\,\gamma^{\rho}\boldsymbol{\ell}_L + 2\,g_R^{l}\,\overline{\boldsymbol{\ell}_R}\,\gamma^{\rho}\boldsymbol{\ell}_R\,. \end{aligned} \tag{6.33}$$

An interesting question is why the mixing is always applied, as we have done, to the neutrinos, whereas the charged leptons are treated as particles with definite mass. The reason is that the only characteristic that distinguishes the three charged leptons is their mass and the flavor of a charged lepton is identified by measuring its mass. The mass determines its kinematical properties and its decay modes, which can be measured directly through long-range electromagnetic interactions. Hence, charged leptons with a definite flavor are, by definition, particles with definite mass. On the other hand, neutrinos can be detected only indirectly by identifying the charged particles produced in weak interactions and the flavor of a neutrino created or destroyed in a charged-current weak interaction process is, by definition, the flavor of the associated charged lepton. Therefore, flavor neutrinos are not required to have a definite mass and the mixing implies that they are superpositions of neutrinos with definite masses.

6.2 Majorana neutrinos

As we have seen in chapter 3, chiral fermion fields are the building blocks of the SM as well as blocks of modern gauge theories, because chiral spinors are the smallest irreducible representations of the Lorentz group, from which all representations can be constructed. In particular, a massless fermion can be described by a chiral field, as shown by Landau [711], Lee and Yang [727], and Salam [919] with their two-component theory of massless neutrinos, which we are going to describe.

As shown in section 2.7, the Dirac equation

$$(i\gamma^\mu \partial_\mu - m)\, \psi = 0 \tag{6.34}$$

for a fermion field

$$\psi = \psi_L + \psi_R\,, \tag{6.35}$$

is equivalent to the equations

$$i\gamma^\mu \partial_\mu \psi_L = m\, \psi_R\,, \tag{6.36}$$

$$i\gamma^\mu \partial_\mu \psi_R = m\, \psi_L\,, \tag{6.37}$$

for the chiral fields ψ_L and ψ_R, whose space-time evolutions are coupled by the mass m.

If a fermion is massless, the two equations in eqns (6.36) and (6.37) are decoupled:

$$i\gamma^\mu \partial_\mu \psi_L = 0\,, \tag{6.38}$$

$$i\gamma^\mu \partial_\mu \psi_R = 0\,. \tag{6.39}$$

Hence, a massless fermion can be described by a single chiral field (left-handed or right-handed), which has only two independent components (see section 2.7). The equations in eqns (6.38) and (6.39) are called the *Weyl equations* and the spinors ψ_L and ψ_R are called *Weyl spinors.*

It is interesting that the possibility of describing a physical particle with a Weyl spinor was rejected by Pauli in 1933 because it leads to the violation of parity. In fact, space inversion transforms ψ_L into ψ_R and vice versa (see section 2.11.2), implying that parity conservation requires the simultaneous existence of both chiral components. However, the discovery of parity violation in 1956–57 invalidated Pauli's reasoning, renewing the possibility to describe massless particles with Weyl spinor fields. In particular, since there was no indication of the existence of a neutrino mass and it was likely that the neutrino[32] participates in weak interactions through its left-handed chiral component, Landau [711], Lee and Yang [727], and Salam [919] proposed to describe the neutrino with a left-handed Weyl spinor ν_L. This is the so-called *two-component theory of massless neutrinos*, which has been incorporated in the SM, where neutrinos are massless and described by left-handed Weyl spinors.

[32] At that time only one type of neutrino was known.

If a two-component spinor is sufficient for the description of a massless fermion, it is natural to ask if a four-component spinor is necessary for the description of a massive particle. The answer, which may be surprising at first sight after our discussion of the system of coupled equations in eqns (6.36) and (6.37), is *no*, as discovered by E. Majorana in 1937 [765]. The trick lies in the assumption that ψ_R and ψ_L are not independent. Obviously, the relation connecting ψ_R and ψ_L must be compatible with eqns (6.36) and (6.37). This means that the two equations must be two ways of writing the same equation for one independent field, say ψ_L. In order to obtain eqn (6.36) from eqn (6.37), we take the Hermitian conjugate of eqn (6.37) and we multiply it on the right with γ^0. By using the property in eqn (2.6) of Dirac matrices, we obtain

$$-\, i\partial_\mu \overline{\psi_R}\, \gamma^\mu = m\, \overline{\psi_L}\,. \tag{6.40}$$

Now, in order to obtain the same structure as eqn (6.36), we take the transpose of eqn (6.40) and multiply on the left with the charge conjugation matrix $\mathcal{C}$. Using the defining property in eqn (2.344) of $\mathcal{C}$, we finally obtain

$$i\gamma^\mu \partial_\mu\, \mathcal{C}\, \overline{\psi_R}^T = m\, \mathcal{C}\, \overline{\psi_L}^T\,. \tag{6.41}$$

This equation has the same structure as eqn (6.36) and we can consider them as identical if we set

$$\psi_R = \xi\, \mathcal{C}\, \overline{\psi_L}^T\,, \tag{6.42}$$

where ξ is an arbitrary phase factor ($|\xi|^2 = 1$). This is the Majorana relation between ψ_R and ψ_L, which makes sense, because $\mathcal{C}\, \overline{\psi_L}^T$ is right-handed: using eqn (2.347) we have $P_L\, \mathcal{C} = \mathcal{C}\, P_L^T$, which leads to

$$P_L \left(\mathcal{C}\, \overline{\psi_L}^T \right) = \mathcal{C}\, \left(\overline{\psi_L}\, P_L \right)^T = \mathcal{C} \left[\left(P_R\, \psi_L \right)^\dagger\, \gamma^0 \right]^T = 0\,. \tag{6.43}$$

From eqns (6.36) and (6.42) we obtain the Majorana equation for the chiral field ψ_L:

$$i\gamma^\mu \partial_\mu \psi_L = m\, \xi\, \mathcal{C}\, \overline{\psi_L}^T\,. \tag{6.44}$$

We can eliminate the phase factor ξ by rephasing the field ψ_L as

$$\psi_L \to \xi^{1/2}\, \psi_L\,. \tag{6.45}$$

Then the Majorana equation for the rephased chiral field ψ_L reads

$$i\gamma^\mu \partial_\mu \psi_L = m\, \mathcal{C}\, \overline{\psi_L}^T\,. \tag{6.46}$$

The Majorana condition for the field ψ in eqn (6.35),

$$\psi = \psi_L + \psi_R = \psi_L + \mathcal{C}\, \overline{\psi_L}^T\,, \tag{6.47}$$

is

$$\psi = \mathcal{C}\, \overline{\psi}^T \tag{6.48}$$

From the discussion in section 2.11.1, one can see that $\mathcal{C}\, \overline{\psi_L}^T$ is identical, apart from a possible phase factor, to the charge conjugated field ψ_L^C. However, as discussed at the end of section 2.11.1, charged-current $V - A$ weak interactions maximally violate the symmetry under charge conjugation. Since neutrinos interact only through weak interactions, the charge parity of the neutrino fields has no physical meaning and we can choose it arbitrarily. Choosing for simplicity the charge parity of the neutrino field ψ_L to be equal to unity, we have

$$\psi_L^C = \mathcal{C}\,\overline{\psi_L}^T . \tag{6.49}$$

Using this convention, the Majorana field ψ in eqn (6.47) can be written as

$$\psi = \psi_L + \psi_L^C , \tag{6.50}$$

and the Majorana condition in eqn (6.48) can also be written as

$$\psi = \psi^C . \tag{6.51}$$

The Majorana condition in eqn (6.51) implies the equality of particle and antiparticle. Hence, only neutral fermions can be described by a Majorana field. This can also be seen by considering the Dirac equation for a fermion with charge q coupled to the electromagnetic field A_μ:

$$(i\gamma^\mu \partial_\mu - q\,\gamma^\mu A_\mu - m)\,\psi = 0 \qquad \text{(particle)}, \tag{6.52}$$

$$(i\gamma^\mu \partial_\mu + q\,\gamma^\mu A_\mu - m)\,\psi^C = 0 \qquad \text{(antiparticle)}. \tag{6.53}$$

If $q \neq 0$, ψ and ψ^C obey different equations and the Majorana equality in eqn (6.51) cannot be imposed. On the other hand, for a neutral fermion we have $q = 0$ and thus the two equations in eqns (6.52) and (6.53) become the same, allowing the imposition of the Majorana equality in eqn (6.51).

In the Majorana case, the electromagnetic current $j^\mu = q\overline{\psi}\gamma^\mu\psi$ vanishes identically:

$$\overline{\psi}\gamma^\mu\psi = \overline{\psi^C}\gamma^\mu\psi^C = -\psi^T\mathcal{C}^\dagger\gamma^\mu\mathcal{C}\overline{\psi}^T = \overline{\psi}\mathcal{C}\gamma^{\mu T}\mathcal{C}^\dagger\psi = -\overline{\psi}\gamma^\mu\psi = 0 . \tag{6.54}$$

One can obtain the same result $j^\mu = -j^\mu = 0$ from the transformation in eqn (2.366) of the electromagnetic current j^μ under charge conjugation.

Among known elementary fermions only the neutrinos are neutral and they can be Majorana particles. As noted already by Majorana [765], since a Majorana spinor has only two independent components, the Majorana theory is simpler and more economical than the Dirac theory. Hence, the Majorana nature of massive neutrinos may be more natural than the Dirac nature. In fact, neutrinos are Majorana particles in most theories beyond the SM.

The Dirac and Majorana descriptions of a neutrino have different phenomenological consequences only if the neutrino is massive. In the massless Dirac theory, the independent left-handed and right-handed chiral components of the neutrino field obey the decoupled Weyl equations (6.38) and (6.39) (with $\psi_L \to \nu_L$ and $\psi_R \to \nu_R$). In the massless Majorana theory, the same Weyl equations (6.38) and (6.39) hold, with the left-handed and right-handed chiral fields related by eqn (6.42).

However, only the left-handed chiral component of the neutrino field interacts. If the neutrino is massless, since the left-handed chiral component of the neutrino field obeys the Weyl equation (6.38) in both the Dirac and Majorana descriptions and the right-handed chiral component is irrelevant for neutrino interactions, the Dirac and Majorana theories are physically equivalent.

From these considerations, it is clear that in practice one can distinguish a Dirac from a Majorana neutrino only by measuring some effect due to the neutrino mass, since otherwise the massless theory applies in an effective way. Moreover, the mass effect must not be of kinematical nature, because the kinematical effects of Dirac and Majorana masses are the same. For example, the Dirac or Majorana nature of neutrinos cannot be revealed through neutrino oscillations (see chapter 7). The most promising way to find if neutrinos are Majorana particles is the search for neutrinoless double-β-decay (see section 14.3).

In the next subsection 6.2.1 we introduce the Lagrangian mass term of a Majorana neutrino field. In subsection 6.2.2 we compare the numbers of degrees of freedom of a Dirac and a Majorana neurino. In subsection 6.2.3 we discuss the quantization of a Majorana neutrino field. In the following subsections 6.2.4, 6.2.5, 6.2.6, and 6.2.7 we discuss, respectively, the violation of lepton number conservation, the number of degrees of freedom, the two-component formalism, CP symmetry, and the concept of an effective Majorana mass in the case of a Majorana neutrino field. The mixing of three Majorana neutrinos is introduced in the next section 6.3.

6.2.1 Majorana mass term

In order to understand the theory of Majorana neutrinos, let us consider first a single neutrino type ν. A Majorana mass is generated by a Lagrangian mass term with a chiral fermion field alone. Since neutrinos are left-handed, we use the left-handed chiral field ν_L.

Is it possible to write a mass term using ν_L alone? In order to answer this question let us consider first a Dirac mass term for a Dirac neutrino field $\nu = \nu_L + \nu_R$,

$$\mathscr{L}^{\mathrm{D}}_{\mathrm{mass}} = -m\,\overline{\nu}\,\nu = -m\,(\overline{\nu_R}\,\nu_L + \overline{\nu_L}\,\nu_R) = -m\,\overline{\nu_R}\,\nu_L + \mathrm{H.c.} \tag{6.55}$$

Note that only the $\overline{\nu_R}\nu_L$ and $\overline{\nu_L}\nu_R$ couplings survive, since we have, from $P_L\nu_L = \nu_L$ and $P_R\nu_L = 0$,

$$\overline{\nu_L}\,\nu_L = \nu_L^\dagger\,\gamma^0\,\nu_L = \nu_L^\dagger\,\gamma^0\,P_L\,\nu_L = \nu_L^\dagger\,P_R\,\gamma^0\,\nu_L = (P_R\,\nu_L)^\dagger\,\gamma^0\,\nu_L = 0\,, \tag{6.56}$$

and similarly $\overline{\nu_R}\,\nu_R = 0$. The Dirac mass term in eqn (6.55) is a Lorentz scalar, as all the terms in the Lagrangian must be, because under a Lorentz transformation the chiral fields $\nu_L(x)$ and $\nu_R(x)$ transform as the Dirac field $\nu(x)$ (see eqn (2.59)), and the adjoint chiral fields $\overline{\nu_L}(x)$ and $\overline{\nu_R}(x)$ transform as $\overline{\nu}(x)$ (see eqn (2.64)):

$$\nu_L(x) \to \nu_L'(x') = \mathcal{S}\,\nu_L(x)\,, \qquad \nu_R(x) \to \nu_R'(x') = \mathcal{S}\,\nu_R(x)\,, \tag{6.57}$$

$$\overline{\nu_L}(x) \to \overline{\nu_L'}(x') = \overline{\nu_L}(x)\,\mathcal{S}^{-1}\,, \qquad \overline{\nu_R}(x) \to \overline{\nu_R'}(x') = \overline{\nu_R}(x)\,\mathcal{S}^{-1}\,, \tag{6.58}$$

with $\mathcal{S}$ given in eqn (2.61). In the above, we have used the commutation relation in eqn (A.53) to find that

$$\nu'_L(x') = P_L\,\nu'(x') = P_L\,\mathcal{S}\,\nu(x) = \mathcal{S}\,P_L\,\nu(x) = \mathcal{S}\,\nu_L(x)\,, \tag{6.59}$$

and analogous derivations for ν_R, $\overline{\nu_L}(x)$ and $\overline{\nu_R}(x)$.

In order to write a Majorana mass term using ν_L alone, we must find a right-handed function of ν_L which transforms as ν_L under Lorentz transformations and can be substituted in place of ν_R in eqn (6.55). This function of ν_L is precisely the charge conjugated field

$$\nu_L^C = \mathcal{C}\,\overline{\nu_L}^T\,, \tag{6.60}$$

where we have chosen the arbitrary intrinsic charge parity to be equal to unity (as already noted before eqn (6.60), the intrinsic charge parity of a neutrino has no physical meaning because charged-current $V-A$ weak interactions violate maximally the symmetry under charge conjugation). Since ν_L^C is right-handed, as we proved in eqn (6.43), the coupling $\overline{\nu_L^C}\,\nu_L$ does not vanish. Furthermore, under a Lorentz transformation the charge conjugated field $\nu_L^C(x)$ transforms as

$$\nu_L^C(x) = \mathcal{C}\left(\nu_L^\dagger(x)\,\gamma^0\right)^T \to \mathcal{C}\left(\nu_L'^\dagger(x')\,\gamma^0\right)^T = \mathcal{C}\left(\mathcal{S}^{-1}\right)^T\mathcal{C}^{-1}\,\nu_L^C(x)\,. \tag{6.61}$$

Using the relation in eqn (2.348), we obtain $\mathcal{C}\left(\mathcal{S}^{-1}\right)^T\mathcal{C}^{-1} = \mathcal{S}$. Hence, $\nu_L^C(x)$ transforms as $\nu_L(x)$, whereas $\overline{\nu_L^C}(x)$ transforms as $\overline{\nu_L}(x)$:

$$\nu_L^C(x) \to \mathcal{S}\,\nu_L^C(x)\,, \qquad \overline{\nu_L^C}(x) \to \overline{\nu_L^C}(x)\,\mathcal{S}^{-1}\,. \tag{6.62}$$

Therefore, ν_L^C has the correct properties to be used in place of ν_R in eqn (6.55), leading to the Majorana mass term

$$\mathscr{L}^{\mathrm{M}}_{\mathrm{mass}} = -\frac{1}{2}\,m\,\overline{\nu_L^C}\,\nu_L + \text{H.c.} \tag{6.63}$$

The full Majorana Lagrangian consisting of the kinetic terms for ν_L and ν_L^C and the Majorana mass term in eqn (6.63) is given by

$$\mathscr{L}^{\mathrm{M}} = \frac{1}{2}\left[\overline{\nu_L}\,i\overleftrightarrow{\partial\!\!\!/}\,\nu_L + \overline{\nu_L^C}\,i\overleftrightarrow{\partial\!\!\!/}\,\nu_L^C - m\left(\overline{\nu_L^C}\,\nu_L + \overline{\nu_L}\,\nu_L^C\right)\right]. \tag{6.64}$$

In eqns (6.63) and (6.64) we introduced an overall factor 1/2 in order to avoid double counting due to the fact that ν_L^C and $\overline{\nu_L}$ are not independent ($\nu_L^C = \mathcal{C}\overline{\nu_L}^T$). For the derivation of the field equation using the Euler–Lagrange equation (see eqn (C.9))

$$\partial_\mu \frac{\partial \mathscr{L}^{\mathrm{M}}}{\partial(\partial_\mu \overline{\nu_L})} - \frac{\partial \mathscr{L}^{\mathrm{M}}}{\partial \overline{\nu_L}} = 0\,, \tag{6.65}$$

it is useful to express ν_L^C in terms of ν_L using eqn (6.60) and

$$\overline{\nu_L^C} = \left(\mathcal{C}\,\overline{\nu_L}^T\right)^\dagger\gamma^0 = \nu_L^T\,(\gamma^0)^T\,\mathcal{C}^\dagger\,\gamma^0 = -\nu_L^T\,\mathcal{C}^\dagger\,, \tag{6.66}$$

where we have used the properties in eqn (2.344) of the charge conjugation matrix $\mathcal{C}$. Then, the Majorana Lagrangian in eqn (6.64) can also be written as

$$\mathscr{L}^{\mathrm{M}} = \frac{1}{2} \left[\overline{\nu_L}\, i \overleftrightarrow{\partial\!\!\!/}\, \nu_L + \nu_L^T\, i \overleftrightarrow{\partial\!\!\!/}^T\, \overline{\nu_L}^T - m \left(-\nu_L^T\, \mathcal{C}^\dagger\, \nu_L + \overline{\nu_L}\, \mathcal{C}\, \overline{\nu_L}^T \right) \right] . \tag{6.67}$$

From eqn (6.65) and

$$\frac{\partial \mathscr{L}^{\mathrm{M}}}{\partial(\partial_\mu \overline{\nu_L})} = -\frac{1}{2}\, i\gamma^\mu\, \nu_L \,, \tag{6.68}$$

$$\frac{\partial \mathscr{L}^{\mathrm{M}}}{\partial \overline{\nu_L}} = \frac{1}{2}\, i\gamma^\mu\, \nu_L - m\, \mathcal{C}\, \overline{\nu_L}^T \,, \tag{6.69}$$

one can get the Majorana field equation (see eqn (6.46))

$$i \partial\!\!\!/\, \nu_L = m\, \mathcal{C}\, \overline{\nu_L}^T \,. \tag{6.70}$$

We note that in the derivatives of $\mathscr{L}^{\mathrm{M}}$ with respect to $\partial_\mu \overline{\nu_L}$ and $\overline{\nu_L}$, we took into account the anticommuting character of the field by adding a minus sign when $\overline{\nu_L}$ lies on the right[33]. For example,

$$\frac{\partial}{\partial(\partial_\mu \overline{\nu_L})}\, \nu_L^T\, i \overleftrightarrow{\partial\!\!\!/}^T\, \overline{\nu_L}^T = -\frac{1}{2}\, i\gamma^\mu\, \nu_L \,. \tag{6.71}$$

One can convince oneself that this procedure is correct by considering the variation of $\nu_L^T\, i \overleftrightarrow{\partial\!\!\!/}^T\, \overline{\nu_L}^T$ under a variation $\delta_{\mathrm{v}}(\partial_\mu \overline{\nu_L})$ of the type in eqn (C.5):

$$\delta_{\mathrm{v}} \left(\nu_L^T\, i \overleftrightarrow{\partial\!\!\!/}^T\, \overline{\nu_L}^T \right) = \frac{1}{2}\, \nu_L^T\, i\gamma^{\mu T}\, \delta_{\mathrm{v}}(\partial_\mu \overline{\nu_L}^T) = \delta_{\mathrm{v}}(\partial_\mu \overline{\nu_L}) \left(-\frac{1}{2}\, i\gamma^\mu\, \nu_L \right) , \tag{6.72}$$

where the minus sign is due to the anticommutation of the fields under transposition.

It is convenient to define the Majorana field

$$\nu = \nu_L + \nu_L^C \,, \tag{6.73}$$

which satisfies the Majorana condition

$$\nu^C = \nu \,. \tag{6.74}$$

The Majorana Lagrangian in eqn (6.64) can be written in terms of ν as

$$\mathscr{L}^{\mathrm{M}} = \frac{1}{2}\, \overline{\nu} \left(i \overleftrightarrow{\partial\!\!\!/} - m \right) \nu \,. \tag{6.75}$$

Note the factor 1/2, which distinguishes the Majorana Lagrangian from the Dirac Lagrangian in eqn (2.1).

---

[33] This problem does not appear in the derivation of the Dirac equation in eqn (2.11) from the Dirac Lagrangian in eqn (2.1) because $\overline{\psi}(x)$ lies always on the left.

Let us finally emphasize that the anticommutation property of fermion fields is essential in order to have a Majorana Lagrangian, because, for commuting fields, $\nu_L^T\, \mathcal{C}^\dagger\, \nu_L = \nu_L^T\, (\mathcal{C}^\dagger)^T\, \nu_L = -\nu_L^T\, \mathcal{C}^\dagger\, \nu_L$, and the Majorana mass term vanishes identically. Also the kinetic term would vanish. On the other hand, for anticommuting fields the second part of the kinetic term in eqn (6.67) can be written as

$$\nu_L^T\, i\overleftrightarrow{\partial\!\!\!/}^T\, \overline{\nu_L}^T = \overline{\nu_L}\, i\overleftrightarrow{\partial\!\!\!/}\, \nu_L\,. \tag{6.76}$$

In order to prove this equality, we use the anticommutation property in eqn (2.253) of fermion fields, which implies

$$\{\nu_{L\alpha}(x), \overline{\nu}_{L\beta}(y)\} = \left(\frac{1-\gamma^5}{2} i\partial\!\!\!/^{(x)}\right)_{\alpha\beta} \Delta(x-y)\,, \tag{6.77}$$

where α and β are Dirac indices, $\partial_\mu^{(x)} \equiv \partial/\partial x^\mu$, and $\Delta(x-y)$ is given by eqn (2.254). Then, we have

$$\begin{aligned}
\nu_L^T(x)\, i\overleftrightarrow{\partial\!\!\!/}^T\, \overline{\nu_L}^T(x) &= \frac{1}{2} \lim_{y\to x} \partial_\mu^{(x)} \left[\nu_L^T(y)\, i\gamma^{\mu T}\, \overline{\nu_L}^T(x) - \nu_L^T(x)\, i\gamma^{\mu T}\, \overline{\nu_L}^T(y)\right] \\
&= \overline{\nu_L}(x)\, i\overleftrightarrow{\partial\!\!\!/}\, \nu_L(x) + \delta\,,
\end{aligned} \tag{6.78}$$

with

$$\begin{aligned}
\delta &= -\frac{1}{4} \lim_{y\to x} \partial_\mu^{(x)} \left\{\mathrm{Tr}\left[\left(1-\gamma^5\right) \partial\!\!\!/^{(y)} \gamma^\mu\right] \Delta(y-x) - \mathrm{Tr}\left[\left(1-\gamma^5\right) \partial\!\!\!/^{(x)} \gamma^\mu\right] \Delta(x-y)\right\} \\
&= -\lim_{y\to x} \partial_\mu^{(x)} \left[\partial^\mu_{(y)} \Delta(y-x) - \partial^\mu_{(x)} \Delta(x-y)\right]\,.
\end{aligned} \tag{6.79}$$

where $\partial^\mu_{(x)} \equiv \partial/\partial x_\mu$. Since $\partial^\mu_{(y)}\Delta(y-x) = -\partial^\mu_{(x)}\Delta(y-x)$ and $\Delta(y-x) = -\Delta(x-y)$, we obtain $\delta = 0$, leading to the equality in eqn (6.76). Hence, the Majorana Lagrangian in eqn (6.67) is commonly written as

$$\mathscr{L}^{\mathrm{M}} = \overline{\nu_L}\, i\overleftrightarrow{\partial\!\!\!/}\, \nu_L - \frac{m}{2} \left(-\nu_L^T\, \mathcal{C}^\dagger\, \nu_L + \overline{\nu_L}\, \mathcal{C}\, \overline{\nu_L}^T\right)\,. \tag{6.80}$$

This expression for the Majorana Lagrangian is more convenient than that in eqn (6.67), because the kinetic term has the same form as that of a massless neutrino in the SM. The Majorana mass term represents a physical effect beyond the SM.

6.2.2 Degrees of freedom

The invariance of the Lagrangian under CPT and Lorentz transformations allows us to understand the difference in the number of degrees of freedom of a Dirac and a Majorana neutrino with a given momentum. Considering a Dirac or a Majorana neutrino with momentum $\vec{p}$ and helicity h, we will apply CPT and Lorentz transformations in order to find which are the other states with the same momentum $\vec{p}$.

The existence of these states is guaranteed by the CPT and Lorentz symmetries. The number of states with the same momentum is the number of degrees of freedom for each value of the momentum.

As shown in section 2.11.5, a CPT operation transforms a particle into its antiparticle with the same momentum and opposite helicity. This is expected on physical grounds. First, under a space inversion (P, parity), we have

$$t \xrightarrow{\text{P}} t\,, \qquad \vec{x} \xrightarrow{\text{P}} -\vec{x}\,, \qquad \vec{p} \xrightarrow{\text{P}} -\vec{p}\,, \qquad \vec{L} = \vec{x} \times \vec{p} \xrightarrow{\text{P}} \vec{L}\,. \tag{6.81}$$

Thus, the spin $\vec{s}$, which is an angular momentum, and the helicity h behave as

$$\vec{s} \xrightarrow{\text{P}} \vec{s}\,, \qquad h = \frac{\vec{s}\cdot\vec{p}}{|\vec{p}|} \xrightarrow{\text{P}} -h\,. \tag{6.82}$$

Under a time reversal (T), we have

$$t \xrightarrow{\text{T}} -t\,, \qquad \vec{x} \xrightarrow{\text{T}} \vec{x}\,, \qquad \vec{p} \xrightarrow{\text{T}} -\vec{p}\,, \qquad \vec{L} = \vec{x} \times \vec{p} \xrightarrow{\text{T}} -\vec{L}\,. \tag{6.83}$$

Hence, for the spin and the helicity we obtain

$$\vec{s} \xrightarrow{\text{T}} -\vec{s}\,, \qquad h = \frac{\vec{s}\cdot\vec{p}}{|\vec{p}|} \xrightarrow{\text{T}} h\,. \tag{6.84}$$

Combining, under a space-time inversion (PT), we have

$$t \xrightarrow{\text{PT}} -t\,, \qquad \vec{x} \xrightarrow{\text{PT}} -\vec{x}\,, \qquad \vec{p} \xrightarrow{\text{PT}} \vec{p}\,, \qquad \vec{s} \xrightarrow{\text{PT}} -\vec{s}\,, \qquad h = \frac{\vec{s}\cdot\vec{p}}{|\vec{p}|} \xrightarrow{\text{PT}} -h\,. \tag{6.85}$$

Finally under a CPT transformation, a Dirac neutrino with momentum $\vec{p}$ and helicity h is transformed into a Dirac antineutrino with momentum $\vec{p}$ and helicity $-h$:

$$\nu(\vec{p},h) \xrightarrow{\text{CPT}} \bar{\nu}(\vec{p},-h)\,. \tag{6.86}$$

This transformation is shown by the upper horizontal line in Fig. 6.1a. The upper vertical line on the right in Fig. 6.1a shows an appropriate Lorentz boost that inverts the momentum $\vec{p}$ and the helicity h

$$\bar{\nu}(\vec{p},-h) \xrightarrow{\text{boost}} \bar{\nu}(-\vec{p},h)\,, \tag{6.87}$$

as explained in section 2.10.2. The lower horizontal line in Fig. 6.1a shows a CPT transformation

$$\bar{\nu}(-\vec{p},h) \xrightarrow{\text{CPT}} \nu(-\vec{p},-h)\,. \tag{6.88}$$

Finally, the upper vertical line on the left in Fig. 6.1a shows that with an appropriate Lorentz boost that inverts the momentum $-\vec{p}$ and the helicity $-h$ we get the original state $\nu(\vec{p},h)$:

$$\nu(-\vec{p},-h) \xrightarrow{\text{boost}} \nu(\vec{p},h)\,. \tag{6.89}$$

The direction of the momentum of $\bar{\nu}(-\vec{p},h)$ and $\nu(-\vec{p},-h)$ can be reversed with a 180° rotation, which leaves the helicity invariant,

$$\bar{\nu}(-\vec{p},h) \xrightarrow{180^\circ\ \text{rotation}} \bar{\nu}(\vec{p},h)\,, \qquad \nu(-\vec{p},-h) \xrightarrow{180^\circ\ \text{rotation}} \nu(\vec{p},-h)\,, \tag{6.90}$$

as shown by the lower vertical lines in Fig. 6.1a. Hence, from CPT and Lorentz invariance we find that there are four possible states of a Dirac neutrino for each

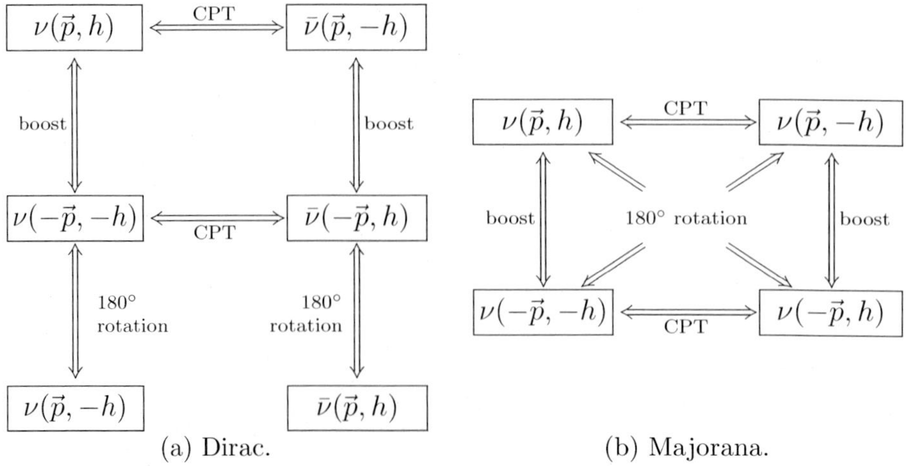

FIG. 6.1. Dirac (a) and Majorana (b) degrees of freedom.

value of the momentum $\vec{p}$:

$$\nu(\vec{p},h)\,, \qquad \nu(\vec{p},-h)\,, \qquad \bar{\nu}(\vec{p},h)\,, \qquad \bar{\nu}(\vec{p},-h)\,. \tag{6.91}$$

These are the states of a Dirac neutrino and antineutrino with positive and negative helicity.

Let us consider now the Majorana case. Since a Majorana neutrino is self-charge conjugated, a CPT transformation changes only the helicity:

$$\nu(\vec{p},h) \xrightarrow{\text{CPT}} \nu(\vec{p},-h)\,, \tag{6.92}$$

as shown by the upper horizontal line in Fig. 6.1b. An appropriate Lorentz boost can invert the momentum $\vec{p}$ and the helicity h,

$$\nu(\vec{p},-h) \xrightarrow{\text{boost}} \nu(-\vec{p},h)\,, \tag{6.93}$$

as shown by the vertical line on the right in Fig. 6.1b. However, a diagonal line in Fig. 6.1b shows that a 180° rotation transforms the state $\nu(-\vec{p},h)$ to the original state $\nu(\vec{p},h)$,

$$\nu(-\vec{p},h) \xrightarrow{180^\circ\ \text{rotation}} \nu(\vec{p},h)\,. \tag{6.94}$$

The lower horizontal line in Fig. 6.1b shows that a CPT operation transforms $\nu(-\vec{p},h)$ into $\nu(-\vec{p},-h)$,

$$\nu(-\vec{p},h) \xrightarrow{\text{CPT}} \nu(-\vec{p},-h)\,, \tag{6.95}$$

which can be transformed either into the original state $\nu(\vec{p},h)$ by an appropriate Lorentz boost,

$$\nu(-\vec{p},-h) \xrightarrow{\text{boost}} \nu(\vec{p},h)\,, \tag{6.96}$$

as shown by the vertical line on the left in Fig. 6.1b, or into $\nu(\vec{p}, -h)$ with a 180° rotation,

$$\nu(-\vec{p}, -h) \xrightarrow{180^\circ \text{ rotation}} \nu(\vec{p}, -h)\,, \tag{6.97}$$

as shown by a diagonal line in Fig. 6.1b. In both cases we do not get anything new. Therefore, from CPT and Lorentz invariance we have only two possible states of a Majorana neutrino for each value of the momentum $\vec{p}$:

$$\nu(\vec{p}, h)\,, \qquad \nu(\vec{p}, -h)\,. \tag{6.98}$$

This means that a Majorana neutrino has half the degrees of freedom of a Dirac neutrino. This is obviously due to the Majorana constraint in eqn (6.74), which implies that in the Majorana case particle and antiparticle are identical, halving the allowed degrees of freedom.

6.2.3 Quantized Majorana field

A Majorana field has all the properties of a Dirac field, discussed in chapter 2, with the additional constraint in eqn (6.74). Because of the properties in eqn (2.354) of the spinors $u^{(h)}(p)$ and $v^{(h)}(p)$, the Majorana condition in eqn (6.74) implies the constraint $b^{(h)}(p) = a^{(h)}(p)$ for the coefficients of the Fourier expansion in eqn (2.139) of a Majorana field. From this constraint and eqn (2.139), the Fourier expansion of a Majorana neutrino field $\nu(x)$ is

$$\nu(x) = \int \frac{\mathrm{d}^3 p}{(2\pi)^3\, 2E} \sum_{h=\pm 1} \left[a^{(h)}(p)\, u^{(h)}(p)\, e^{-ip\cdot x} + a^{(h)\dagger}(p)\, v^{(h)}(p)\, e^{ip\cdot x} \right]. \tag{6.99}$$

When the Majorana field is quantized, the coefficients $a^{(h)}(p)$ become operators which satisfy the canonical anticommutation relations in eqns (2.226)–(2.230),

$$\{a^{(h)}(p), a^{(h')\dagger}(p')\} = (2\pi)^3\, 2E\, \delta^3(\vec{p} - \vec{p}\,')\, \delta_{hh'}\,, \tag{6.100}$$

$$\{a^{(h)}(p), a^{(h')}(p')\} = \{a^{(h)\dagger}(p), a^{(h')\dagger}(p')\} = 0\,. \tag{6.101}$$

Since there is only one type of operator, there is no distinction between particles and antiparticles. In other words, a Majorana neutrino is the same as a Majorana antineutrino, as should be clear from the constraint $b^{(h)}(p) = a^{(h)}(p)$.

In practice, however, it is customary to speak of neutrinos and antineutrinos in the Majorana case also, the reason being that detectable neutrinos are always ultrarelativistic (see the introduction of chapter 7). For one lepton generation, the leptonic charged-current weak interaction Lagrangian is

$$\mathscr{L}_{\mathrm{I,L}}^{\mathrm{CC}} = -\frac{g}{\sqrt{2}} \left(\overline{\nu_L}\, \gamma^\mu\, \ell_L\, W_\mu + \overline{\ell_L}\, \gamma^\mu\, \nu_L\, W_\mu^\dagger \right), \tag{6.102}$$

where ℓ is the charged lepton field. The Hermitian-conjugated leptonic charged current

$$j_{W,\mathrm{L}}^{\mu\dagger} = 2\, \overline{\ell_L}\, \gamma^\mu\, \nu_L\,, \tag{6.103}$$

which creates antineutrinos in the Dirac case, mainly creates ultrarelativistic Majorana neutrinos with positive helicity. In fact, using the formulas in section 2.9, we have

$$\langle \nu(p,h) | \nu_L(x) | 0 \rangle = v_L^{(h)}(p)\, e^{ip\cdot x} \,, \tag{6.104}$$

which gives, the approximations in eqn (2.206),

$$\langle \nu(p,+) | \nu_L(x) | 0 \rangle \simeq -\sqrt{2E} \begin{pmatrix} 0 \\ \chi^{(-)}(\vec{p}) \end{pmatrix} e^{ip\cdot x} \,, \tag{6.105}$$

$$\langle \nu(p,-) | \nu_L(x) | 0 \rangle \simeq -\sqrt{2E} \begin{pmatrix} 0 \\ \frac{m}{2E}\, \chi^{(+)}(\vec{p}) \end{pmatrix} e^{ip\cdot x} \,. \tag{6.106}$$

Hence, the creation by $j^{\mu\dagger}_{W,\mathrm{L}}$ of ultrarelativistic Majorana neutrinos with negative helicity is suppressed by the ratio m/E with respect to that of ultrarelativistic Majorana neutrinos with positive helicity.

On the other hand, the leptonic charged-current

$$j^{\mu}_{W,\mathrm{L}} = 2\, \overline{\nu_L}\, \gamma^\mu\, \ell_L \,, \tag{6.107}$$

which creates neutrinos in the Dirac case, mainly creates ultrarelativistic Majorana neutrinos with negative helicity:

$$\langle \nu(p,h) | \overline{\nu_L}(x) | 0 \rangle = \overline{u_L^{(h)}}(p)\, e^{ip\cdot x} \,. \tag{6.108}$$

Using the approximations in eqn (2.205) we have

$$\langle \nu(p,+) | \overline{\nu_L}(x) | 0 \rangle \simeq \sqrt{2E} \begin{pmatrix} \frac{m}{2E}\, \chi^{(+)\dagger}(\vec{p}) & 0 \end{pmatrix} e^{ip\cdot x} \,, \tag{6.109}$$

$$\langle \nu(p,-) | \overline{\nu_L}(x) | 0 \rangle \simeq -\sqrt{2E} \begin{pmatrix} \chi^{(-)\dagger}(\vec{p}) & 0 \end{pmatrix} e^{ip\cdot x} \,. \tag{6.110}$$

Hence, the creation by $j^{\mu}_{W,\mathrm{L}}$ of ultrarelativistic Majorana neutrinos with positive helicity is suppressed by the ratio m/E with respect to that of ultrarelativistic Majorana neutrinos with negative helicity.

Therefore, an ultrarelativistic Majorana neutrino with positive helicity interacts as a Dirac antineutrino with positive helicity, for which we have

$$\langle \bar{\nu}(p,h) | \nu_L(x) | 0 \rangle = v_L^{(h)}(p)\, e^{ip\cdot x} \,, \tag{6.111}$$

and an ultrarelativistic Majorana neutrino with negative helicity interacts as a Dirac neutrino with negative helicity, for which eqn (6.108) holds. For this reason it is customary to call Majorana neutrinos with negative helicity *neutrinos*,

$$|\nu(p)\rangle \equiv |\nu(p,-)\rangle \,, \tag{6.112}$$

and Majorana neutrinos with positive helicity *antineutrinos*,

$$|\bar{\nu}(p)\rangle \equiv |\nu(p,+)\rangle \,. \tag{6.113}$$

From the general expression in eqn (C.44), the energy–momentum operator of a Majorana field is given by

$$\mathsf{P}^\mu = \frac{1}{2} \int \mathrm{d}^3x\, \nu^\dagger(x)\, i\overleftrightarrow{\partial^\mu}\, \nu(x) \,. \tag{6.114}$$

The factor 1/2 stems from the coefficient 1/2 in the Majorana Lagrangian in eqn (6.75). This factor is crucial for getting the correct energy–momentum operator, as one can see by imposing the Majorana constraint $b^{(h)}(p) = a^{(h)}(p)$ in the expression in eqn (2.244) for the normally ordered energy–momentum operator of a Dirac field: it is clear that it would give twice the correct energy–momentum operator. On the other hand, from eqn (6.114) and the Fourier expansion in eqn (6.99) of the quantized Majorana field, we obtain the correct normally ordered energy–momentum operator

$$:\mathsf{P}^\mu: = \int \frac{\mathrm{d}^3p}{(2\pi)^3\, 2E}\, p^\mu \sum_h a^{(h)\dagger}(p)\, a^{(h)}(p)\,, \tag{6.115}$$

which gives

$$:\mathsf{P}^\mu: |\nu(p,h)\rangle = p^\mu\, |\nu(p,h)\rangle\,, \tag{6.116}$$

with

$$p^0 = E = \sqrt{\vec{p}^{\,2} + m^2}\,. \tag{6.117}$$

Let us emphasize that the essential difference between Dirac and Majorana neutrinos is that in the case of Dirac neutrinos particle and antiparticle are distinct, whereas in the case of Majorana neutrinos they coincide. However, from the kinematical point of view Dirac and Majorana mass are equivalent, because they satisfy the same energy–momentum dispersion relation in eqn (6.117) (which coincides with eqn (2.140)).

6.2.4 Lepton number

In section 6.1.1 we have seen that if massive neutrinos are Dirac particles, the total lepton number in eqn (3.143), which is associated, through Noether's theorem, with the global U(1) gauge transformations in eqns (6.19) and (6.20), is conserved. In the case of massive Majorana neutrinos, this is no longer true, because the Majorana mass term in eqn (6.63) is not invariant under the global U(1) gauge transformation

$$\nu_L \to e^{i\varphi}\, \nu_L\,. \tag{6.118}$$

This becomes clear by writing the Majorana mass term in eqn (6.63) as

$$\mathscr{L}^{\mathrm{M}}_{\mathrm{mass}} = \frac{1}{2}\, m \left(\nu_L^T\, \mathcal{C}^\dagger\, \nu_L + \nu_L^\dagger\, \mathcal{C}\, \nu_L^* \right)\,, \tag{6.119}$$

which then changes to

$$\mathscr{L}^{\mathrm{M}}_{\mathrm{mass}} \to \frac{1}{2}\, m \left(e^{2i\varphi}\, \nu_L^T\, \mathcal{C}^\dagger\, \nu_L + e^{-2i\varphi}\, \nu_L^\dagger\, \mathcal{C}\, \nu_L^* \right) \tag{6.120}$$

under the transformation in eqn (6.118). One may ask if it is possible to consider, instead of the transformation in eqn (6.118) of the left-handed chiral field, a transformation of the Majorana field in eqn (6.73)

$$\nu \to e^{i\varphi}\, \nu\,, \tag{6.121}$$

which leaves invariant the Majorana Lagrangian in eqn (6.75). The answer, however, is that such a transformation is not possible because it is incompatible with the

Majorana constraint in eqn (6.74), since

$$\nu^C \to e^{-i\varphi} \, \nu^C \,. \tag{6.122}$$

One can also see that the transformation in eqn (6.121) is incompatible with the definition in eqn (6.73) of the Majorana field, because if $\nu_L \to e^{i\varphi} \, \nu_L$ we have $\nu_L^C \to e^{-i\varphi} \, \nu_L^C$.

The absence of a conserved lepton number is also evident from the fact that Dirac neutrinos have $L = +1$ and Dirac antineutrinos have $L = -1$. Since in the Majorana case neutrinos and antineutrinos are the same object, it is clear that there cannot be a conserved lepton number. Indeed, the neutrino part of the Dirac lepton number operator in eqn (6.23) vanishes under the Majorana constraint $b_{\nu_k}^{(h)}(p) = a_{\nu_k}^{(h)}(p)$.

However, since neutrino masses are very small and, apart from the mass term of Majorana neutrinos, the Lagrangian is invariant under the global U(1) gauge transformations in eqns (6.19) and (6.20), it is possible to assign to charged leptons and neutrinos an effective total lepton number which is conserved in all processes that are not sensitive to the Majorana masses of neutrinos. In these processes, the neutrinos can be considered to be massless. The corresponding lepton number operator is obtained from the invariance of the effective Lagrangian

$$\mathscr{L}_{\text{eff}} = \overline{\nu_L^{\text{eff}}} \, i \overleftrightarrow{\partial\!\!\!/} \, \nu_L^{\text{eff}} + \overline{\ell} \left(i \overleftrightarrow{\partial\!\!\!/} - m_\ell \right) \ell - \frac{g}{\sqrt{2}} \left(\overline{\nu_L^{\text{eff}}} \, \gamma^\mu \, \ell_L \, W_\mu + \overline{\ell_L} \, \gamma^\mu \, \nu_L^{\text{eff}} \, W_\mu^\dagger \right) \tag{6.123}$$

under the global U(1) gauge transformation

$$\nu_L^{\text{eff}} \to e^{i\varphi} \, \nu_L^{\text{eff}} \,, \qquad \ell_L \to e^{i\varphi} \, \ell_L \,, \qquad \ell_R \to e^{i\varphi} \, \ell_R \,, \tag{6.124}$$

where ν_L^{eff} is an effective massless Weyl neutrino field. From eqn (C.49), the associated effective conserved current $(\partial_\mu j_{\text{eff}}^\mu)$ is

$$j_{\text{eff}}^\mu = \overline{\nu_L^{\text{eff}}} \, \gamma^\mu \, \nu_L^{\text{eff}} + \overline{\ell_\alpha} \, \gamma^\mu \, \ell_\alpha \,, \tag{6.125}$$

and the effective conserved charge $(\partial_0 \mathsf{L}_{\text{eff}} = 0)$ is

$$\mathsf{L}_{\text{eff}} = \int \mathrm{d}^3x \, j_{\text{eff}}^0(x) \,. \tag{6.126}$$

In this case, the effective massless neutrino chiral field ν_L^{eff} is quantized according to the discussion in sections 2.8.4 and 2.9.2:

$$\nu_L^{\text{eff}}(x) = \int \frac{\mathrm{d}^3p}{(2\pi)^3 \, 2E} \left[a^{(-)}(p) \, u^{(-)}(p) \, e^{-ip \cdot x} + a^{(+)\dagger}(p) \, v^{(+)}(p) \, e^{ip \cdot x} \right] , \tag{6.127}$$

where we relabeled $b^{(+)}(p) \to a^{(+)}(p)$, in agreement with the convention in eqn (6.113) in order to show the connection with the massive Majorana field in eqn (6.99). Using the relativistic expressions in eqn (2.9) of the u and v spinors, one can see that in the massless limit the left-handed part of the massive Majorana field in eqn (6.99) reduces to $\nu_L^{\text{eff}}(x)$.

Using eqn (6.127) and the Fourier expansion in eqn (2.139) for the massive Dirac charged lepton field $\ell(x)$, we obtain the normally ordered (see eqn (2.245)) effective lepton number operator

$$
\begin{aligned}
:\mathsf{L}_{\text{eff}}: = & \int \frac{\mathrm{d}^3 p}{(2\pi)^3\, 2E} \left[a^{(-)\dagger}(p)\, a^{(-)}_{\nu_\alpha}(p) - a^{(+)\dagger}(p)\, a^{(+)}_{\nu_\alpha}(p) \right] \\
& + \int \frac{\mathrm{d}^3 p}{(2\pi)^3\, 2E} \sum_{h=\pm 1} \left[a^{(h)\dagger}_{\ell_\alpha}(p)\, a^{(h)}_{\ell_\alpha}(p) - b^{(h)\dagger}_{\ell_\alpha}(p)\, b^{(h)}_{\ell_\alpha}(p) \right] .
\end{aligned} \tag{6.128}
$$

Hence, we see that neutrinos with negative helicity have $L_{\text{eff}} = +1$ and neutrinos with positive helicity have $L_{\text{eff}} = -1$, in agreement with the convention in eqn (6.113) of calling an antineutrino a neutrino with positive helicity. These assignments of the effective lepton number lead to its conservation in all weak interaction processes which are not sensitive to the Majorana neutrino mass, because in such processes a charged lepton with $L_{\text{eff}} = +1$ can produce only a negative helicity neutrino and a charged antilepton with $L_{\text{eff}} = -1$ can produce only a positive helicity neutrino.

If the Majorana mass term in eqn (6.119) is considered as a perturbation of the massless Lagrangian, it generates transitions with

$$
\Delta L_{\text{eff}} = \pm 2 . \tag{6.129}
$$

The most promising process which could allow the discovery of such transitions is neutrinoless double-β-decay (see section 14.3).

6.2.5 Two-component formalism

The fact that a Majorana neutrino has half the degrees of freedom of a Dirac neutrino is explicitly clear in the two-component formalism presented in section 2.7. In the chiral representation in eqn (2.27) of the γ matrices we have

$$
[\nu_L(x)]_{\mathrm{C}} = \begin{pmatrix} 0 \\ \varphi(x) \end{pmatrix} , \qquad [\overline{\nu_L}(x)]_{\mathrm{C}} = \begin{pmatrix} -\varphi^\dagger(x) & 0 \end{pmatrix} , \tag{6.130}
$$

and the Majorana Lagrangian in eqn (6.67) can be written as

$$
\mathscr{L}^{\mathrm{M}} = -\varphi^\dagger\, i\bar{\sigma}^\mu \overleftrightarrow{\partial}_\mu \varphi - \frac{m}{2} \left(\varphi^T\, i\sigma^2\, \varphi - \varphi^\dagger\, i\sigma^2\, \varphi^* \right) . \tag{6.131}
$$

The corresponding field equation follows from the Euler–Lagrange equation (see eqn (C.9))

$$
\partial_\mu \frac{\partial \mathscr{L}^{\mathrm{M}}}{\partial(\partial_\mu \varphi^\dagger)} - \frac{\partial \mathscr{L}^{\mathrm{M}}}{\partial \varphi^\dagger} = 0 . \tag{6.132}
$$

We have

$$
\frac{\partial \mathscr{L}^{\mathrm{M}}}{\partial(\partial_\mu \varphi^\dagger)} = \frac{i}{2}\, \bar{\sigma}^\mu\, \varphi , \tag{6.133}
$$

$$\frac{\partial \mathscr{L}^{\mathrm{M}}}{\partial \varphi^{\dagger}} = -\frac{i}{2}\,\bar{\sigma}^{\mu}\,\partial_{\mu}\varphi + m\,i\sigma^{2}\,\varphi^{*}\,, \tag{6.134}$$

where we have used the anticommuting character of the field in the derivatives with respect to $\varphi^{\dagger}$ when $\varphi^{\dagger}$ lies on the right (as $\varphi^{*} = (\varphi^{\dagger})^{T}$). The resulting Majorana field equation for the two-component field $\varphi(x)$ is

$$\bar{\sigma}^{\mu}\partial_{\mu}\,\varphi - m\,\sigma^{2}\,\varphi^{*} = 0\,, \tag{6.135}$$

which can also be written as

$$\left(\partial_{0} - \vec{\sigma}\cdot\vec{\nabla}\right)\varphi + m\,\sigma^{2}\,\varphi^{*} = 0\,. \tag{6.136}$$

From eqn (6.99) and the explicit expressions in eqn (2.202) of the $u^{(h)}(p)$ and $v^{(h)}(p)$ spinors in the chiral representation, it follows that the Fourier expansion of the two-component field $\varphi(x)$ is

$$\varphi(x) = \int \frac{\mathrm{d}^{3}p}{(2\pi)^{3}\,2E} \sum_{h=\pm 1} \Big[\sqrt{E - h\,|\vec{p}|}\,a^{(h)}(p)\,\chi^{(h)}(\vec{p})\,e^{-ip\cdot x}$$
$$-h\,\sqrt{E + h\,|\vec{p}|}\,a^{(h)\dagger}(p)\,\chi^{(-h)}(\vec{p})\,e^{ip\cdot x}\Big]\,. \tag{6.137}$$

Using the properties in eqns (2.207) and (2.210) of the two-component helicity eigenstate spinors $\chi^{(h)}(\vec{p})$, one can verify that this expression satisfies the field equation (6.136).

Since the derivation of the energy–momentum operator is rather tricky, we discuss it in some detail. From the general expression in eqn (C.44), the energy–momentum operator is given by

$$\mathsf{P}^{\mu} = \int \mathrm{d}^{3}x\,\varphi^{\dagger}(x)\,i\overleftrightarrow{\partial^{\mu}}\,\varphi(x)\,. \tag{6.138}$$

Using the Fourier expansion in eqn (6.137) and the properties of the two-component helicity eigenstate spinors discussed in section 2.8.3, one can obtain

$$\mathsf{P}^{\mu} = \int \frac{\mathrm{d}^{3}p}{(2\pi)^{3}4E^{2}} \sum_{h=\pm 1} \Big\{p^{\mu}\Big[(E - h|\vec{p}|)\,a^{(h)\dagger}(p)a^{(h)}(p) - (E + h|\vec{p}|)\,a^{(h)}(p)a^{(h)\dagger}(p)\Big]$$
$$+\,hm\left(\frac{p_{\mathrm{P}}^{\mu} - p^{\mu}}{2}\right)\Big[\eta(\vec{p},h)a^{(h)\dagger}(p)a^{(h)\dagger}(p_{\mathrm{P}})e^{2iEt}$$
$$+\eta^{*}(\vec{p},h)a^{(h)}(p)a^{(h)}(p_{\mathrm{P}})e^{-2iEt}\Big]\Big\}\,, \tag{6.139}$$

where $\eta(\vec{p},h)$ is the phase in eqn (2.211) and $p_{\mathrm{P}}^{\mu} = (E, -\vec{p})$. For $\mu = 0$, we have $p_{\mathrm{P}}^{0} - p^{0} = 0$. For $\mu = k$, we have $p_{\mathrm{P}}^{k} - p^{k} = -2p^{k}$. However, changing $\vec{p} \to -\vec{p}$ in the integrand we obtain

$$\int \frac{\mathrm{d}^{3}p}{E^{2}}p^{k}\eta(\vec{p},h)a^{(h)\dagger}(p)a^{(h)\dagger}(p_{\mathrm{P}})e^{2iEt} = -\int \frac{\mathrm{d}^{3}p}{E^{2}}p^{k}\eta(-\vec{p},h)a^{(h)\dagger}(p_{\mathrm{P}})a^{(h)\dagger}(p)e^{2iEt}$$

$$= -\int \frac{d^3p}{E^2} p^k \eta(\vec{p}, h) a^{(h)\dagger}(p) a^{(h)\dagger}(p_P) e^{2iEt} . \tag{6.140}$$

In the last equality we used the anticommuting property of $a^{(h)\dagger}(p)$ and $a^{(h)\dagger}(p_P)$, and the properties in eqn (2.212) and (2.213) of the phase $\eta(\vec{p}, h)$, which imply $\eta(-\vec{p}, h) = -\eta(\vec{p}, h)$. Hence, the term proportional to $a^{(h)\dagger}(p_P)\, a^{(h)\dagger}(p)$ vanishes. With an analogous calculation one can see that also the term proportional to $a^{(h)\dagger}(p)\, a^{(h)\dagger}(p_P)$ vanishes. Finally, taking into account the anticommutation relation in eqn (6.100), from the two terms in the first line of eqn (6.139) we obtain the correct normal ordered energy–momentum operator in eqn (6.115),

$$:\mathsf{P}^\mu: = \int \frac{d^3p}{(2\pi)^3\, 2E} p^\mu \sum_h a^{(h)\dagger}(p)\, a^{(h)}(p) . \tag{6.141}$$

The Fourier expansion in eqn (6.137) assumes a very simple form in the case of ultrarelativistic neutrinos, which is of practical interest in most experiments. In this case $|\vec{p}| \gg m$ and the energy can be approximated by

$$E \simeq |\vec{p}| + \frac{m^2}{2|\vec{p}|} . \tag{6.142}$$

The effective two-component field operator can be written as

$$\varphi_{\text{eff}}(x) = \int_{|\vec{p}|\gg m} \frac{d^3p}{(2\pi)^3 \sqrt{2|\vec{p}|}} \left[a^{(-)}(p)\, \chi^{(-)}(\vec{p})\, e^{-ip\cdot x} - a^{(+)\dagger}(p)\, \chi^{(-)}(\vec{p})\, e^{ip\cdot x} \right] , \tag{6.143}$$

neglecting terms of order $m^2/|\vec{p}|^2$, which are safely negligible in practice. Under this approximation the effective neutrino field destroys negative helicity neutrinos and creates positive helicity neutrinos, as a massless chiral left-handed field (see sections 2.8.4 and 2.9.2).

6.2.6 CP symmetry

Since neutrinos interact only through weak interactions which violate maximally the charge-conjugation and parity symmetries (see sections 2.11.1 and 2.11.2), their properties under these transformations do not lead to physical constraints. On the other hand, weak interactions do not necessarily violate the symmetry under a CP transformation (see sections 2.11.3). A violation of the CP symmetry in weak interactions may be a consequence of the fundamental structure of weak interactions, but is now presumed to depend on the value of the fermion mass matrices and the resulting mixing matrix, as discussed in section 4.6 for the case of quarks, which can be extended by analogy to the mixing of Dirac neutrinos.

In the simplest, albeit unrealistic, case of one Majorana neutrino CP cannot be violated, as we see below.

Let us first consider the Majorana Lagrangian in eqn (6.64). Under a CP transformation the chiral field ν_L transforms as

$$\mathsf{U}_{\rm CP}\nu_L(x)\mathsf{U}_{\rm CP}^{-1} = \xi_\nu^{\rm CP}\,\gamma^0\,\nu_L^C(x_{\rm P})\,, \tag{6.144}$$

where $\xi_\nu^{\rm CP}$ is the CP phase and $x_{\rm P}^\mu = (x^0, -\vec{x})$. Recall that the theory is CP-symmetric if there are values of the phase $\xi_\nu^{\rm CP}$ such that the Lagrangian transforms as

$$\mathsf{U}_{\rm CP}\mathscr{L}(x)\mathsf{U}_{\rm CP}^{-1} = \mathscr{L}(x_{\rm P})\,. \tag{6.145}$$

From eqn (6.144), the CP transformation of

$$\nu_L^C(x) = \mathcal{C}\,\overline{\nu_L}^T(x) = \mathcal{C}\,(\gamma^0)^T\,\nu_L^*(x) \tag{6.146}$$

is given by

$$\mathsf{U}_{\rm CP}\nu_L^C(x)\mathsf{U}_{\rm CP}^{-1} = -\xi_\nu^{\rm CP\,*}\,\mathcal{C}\,(\gamma^0)^T\,\mathcal{C}^*\,\nu_L(x_{\rm P}) = -\xi_\nu^{\rm CP\,*}\,\gamma^0\,\nu_L(x_{\rm P})\,. \tag{6.147}$$

Here we have used the properties in eqns (2.344)–(2.346) of the charge conjugation matrix $\mathcal{C}$, which imply $\mathcal{C}^* = -\mathcal{C}^{-1}$. Therefore, $\overline{\nu_L}(x)$ and $\overline{\nu_L^C}(x)$ transform as

$$\mathsf{U}_{\rm CP}\overline{\nu_L}(x)\mathsf{U}_{\rm CP}^{-1} = \xi_\nu^{\rm CP\,*}\,\overline{\nu_L^C}(x_{\rm P})\,\gamma^0\,, \tag{6.148}$$

$$\mathsf{U}_{\rm CP}\overline{\nu_L^C}(x)\mathsf{U}_{\rm CP}^{-1} = -\xi_\nu^{\rm CP}\,\overline{\nu_L}(x_{\rm P})\,\gamma^0\,. \tag{6.149}$$

The kinetic term

$$\mathscr{L}_{\rm kin}^{\rm M}(x) = \frac{1}{2}\left(\overline{\nu_L}(x)\,i\overleftrightarrow{\partial\!\!\!/}\,\nu_L(x) + \overline{\nu_L^C}(x)\,i\overleftrightarrow{\partial\!\!\!/}\,\nu_L^C(x)\right)\,. \tag{6.150}$$

is CP invariant for any value of the phase $\xi_\nu^{\rm CP}$:

$$\begin{aligned}\mathsf{U}_{\rm CP}\mathscr{L}_{\rm kin}^{\rm M}(x)\mathsf{U}_{\rm CP}^{-1} &= \frac{1}{2}\left(\overline{\nu_L^C}(x_{\rm P})\,\gamma^0\,i\overleftrightarrow{\partial\!\!\!/}\,\gamma^0\,\nu_L^C(x_{\rm P}) + \overline{\nu_L}(x_{\rm P})\,\gamma^0\,i\overleftrightarrow{\partial\!\!\!/}\,\gamma^0\,\nu_L(x_{\rm P})\right)\\ &= \frac{1}{2}\left(\overline{\nu_L^C}(x_{\rm P})\,i\overleftrightarrow{\partial\!\!\!/}^{\rm P}\,\nu_L^C(x_{\rm P}) + \overline{\nu_L}(x_{\rm P})\,i\overleftrightarrow{\partial\!\!\!/}^{\rm P}\,\nu_L(x_{\rm P})\right) = \mathscr{L}_{\rm kin}^{\rm M}(x_{\rm P})\,,\end{aligned} \tag{6.151}$$

where $\partial_\mu^{\rm P} = \partial/\partial x_{\rm P}^\mu$. On the other hand, the Majorana mass term in eqn (6.63),

$$\mathscr{L}_{\rm mass}^{\rm M}(x) = -\frac{1}{2}\,m\left[\overline{\nu_L^C}(x)\,\nu_L(x) + \overline{\nu_L}(x)\,\nu_L^C(x)\right]\,, \tag{6.152}$$

transforms as

$$\mathsf{U}_{\rm CP}\mathscr{L}_{\rm mass}^{\rm M}(x)\mathsf{U}_{\rm CP}^{-1} = -\frac{1}{2}\,m\left[-(\xi_\nu^{\rm CP})^2\,\overline{\nu_L}(x_{\rm P})\,\nu_L^C(x_{\rm P}) - (\xi_\nu^{\rm CP\,*})^2\,\overline{\nu_L^C}(x_{\rm P})\,\nu_L(x_{\rm P})\right]\,. \tag{6.153}$$

Thus, we obtain, from the criterion in eqn (6.145),

$$\xi_\nu^{\rm CP} = \pm i\,. \tag{6.154}$$

Note that the free Majorana case is different from the free Dirac case, in which the CP phase $\xi_\nu^{\rm CP}$ is arbitrary, as explained in section 2.11.3.

The purely imaginary CP parity of a Majorana neutrino does not pose any problem for the invariance under CP of the charged-current weak interaction Lagrangian in eqn (6.102). From the CP transformation

$$\mathsf{U}_{\text{CP}}\,\ell\,\mathsf{U}_{\text{CP}}^{-1} = \xi_\ell^{\text{CP}}\,\gamma^0\,\mathcal{C}\,\overline{\ell}^T \tag{6.155}$$

of the Dirac charged lepton field ℓ and the CP transformation in eqn (4.117) of the gauge boson field W_μ,

$$\mathsf{U}_{\text{CP}}\,W_\mu\,\mathsf{U}_{\text{CP}}^{-1} = -W^{\mu\dagger}\,, \tag{6.156}$$

we obtain the following CP transformation for the charged-current weak interaction Lagrangian in eqn (6.102):

$$\mathsf{U}_{\text{CP}}\,\mathscr{L}_{\text{I,L}}^{\text{CC}}\,\mathsf{U}_{\text{CP}}^{-1} = -\frac{g}{\sqrt{2}}\left(\xi_\nu^{\text{CP}*}\,\xi_\ell^{\text{CP}}\,\overline{\ell_L}\,\gamma^{\mu\dagger}\,\nu_L\,W^{\mu\dagger} - \xi_\nu^{\text{CP}}\,\xi_\ell^{\text{CP}*}\overline{\nu_L}\,\gamma^{\mu\dagger}\,\ell_L\,W^\mu\right). \tag{6.157}$$

Since

$$\gamma^{\mu\dagger} = \left(\gamma^{0\dagger}, \vec{\gamma}^\dagger\right) = \left(\gamma^0, -\vec{\gamma}\right) = \gamma_\mu\,, \tag{6.158}$$

we obtain

$$\mathsf{U}_{\text{CP}}\,\mathscr{L}_{\text{I,L}}^{\text{CC}}\,\mathsf{U}_{\text{CP}}^{-1} = -\frac{g}{\sqrt{2}}\left(\xi_\nu^{\text{CP}*}\,\xi_\ell^{\text{CP}}\,\overline{\ell_L}\,\gamma^{\mu}\,\nu_L\,W_\mu^{\dagger} - \xi_\nu^{\text{CP}}\,\xi_\ell^{\text{CP}*}\overline{\nu_L}\,\gamma^{\mu}\,\ell_L\,W_\mu\right). \tag{6.159}$$

Comparing with eqn (6.102), one can see that the charged-current weak interaction Lagrangian is invariant under a CP transformation if

$$\xi_\nu^{\text{CP}*}\,\xi_\ell^{\text{CP}} = 1\,. \tag{6.160}$$

Therefore, the purely imaginary CP parity of a Majorana neutrino implies a purely imaginary CP parity of the charged lepton:

$$\xi_\nu^{\text{CP}} = \pm i \quad \Longrightarrow \quad \xi_\ell^{\text{CP}} = \pm i\,. \tag{6.161}$$

Such a choice of the CP parities lead to CP conservation.

6.2.7 Effective Majorana mass

The Majorana mass term in eqn (6.63) involves only the neutrino left-handed chiral field ν_L, which is present in the SM (one for each lepton generation). Hence, one should ask if it is possible that SM neutrinos have Majorana masses. The answer is negative, because, as one can see from Table 3.5, a neutrino left-handed chiral field ν_L has third component I_3 of the weak isospin equal to 1/2 and hypercharge Y equal to -1. It then follows that

$$\overline{\nu_L^C}\,\nu_L = -\nu_L^T\,\mathcal{C}^\dagger\,\nu_L \tag{6.162}$$

has $I_3 = 1$ and $Y = -2$. Since the SM does not contain any weak isospin triplet with $Y = 2$, it is not possible to have a renormalizable Lagrangian term which can generate a Majorana neutrino mass. In fact, from power counting of divergent diagrams, Lagrangian terms which contain products of fields with energy dimension larger than four are not renormalizable (see, for example, Ref. [634]). Let us recall that fermion fields with spin 1/2 have dimension $[E]^{3/2}$, as one can easily deduce from the Dirac Lagrangian in eqn (2.1), using the fact that the action $\int \mathrm{d}^4x\, \mathscr{L}(x)$ is dimensionless (hence $\mathscr{L}(x) \sim [E]^4$). On the other hand, boson fields have dimension of energy.

Considering for simplicity only one generation, the lowest dimensional term which could generate a Majorana neutrino mass that one can construct with the SM fields in Tables 3.5 and 3.7, respecting the SM symmetries, is the lepton number violating term[34]

$$\mathscr{L}_5 = \frac{g}{\mathcal{M}}\, (L_L^T\, \tau_2\, \Phi)\, \mathcal{C}^\dagger\, (\Phi^T\, \tau_2\, L_L) + \text{H.c.}\,, \tag{6.163}$$

where g is a dimensionless coupling coefficient and $\mathcal{M}$ is a constant with dimension of mass. In eqn (6.163) L_L is the one-generation SM lepton doublet in eqn (3.7) and Φ is the Higgs doublet in eqn (3.96). Sometimes $\mathscr{L}_5$ is also written as

$$\mathscr{L}_5 = \frac{1}{2}\,\frac{g}{\mathcal{M}}\, (L_L^T\, \mathcal{C}^\dagger\, \tau_2\, \vec{\tau}\, L_L) \cdot (\Phi^T\, \tau_2\, \vec{\tau}\, \Phi) + \text{H.c.}\,, \tag{6.164}$$

which is equivalent to the expression in eqn (6.163).

As a consequence of the electroweak symmetry breaking (see section 3.4)

$$\Phi = \begin{pmatrix} \phi^+ \\ \phi^0 \end{pmatrix} \xrightarrow[\text{Breaking}]{\text{Symmetry}} \frac{1}{\sqrt{2}} \begin{pmatrix} 0 \\ v + H \end{pmatrix}, \tag{6.165}$$

$\mathscr{L}_5$ generates the Majorana mass term for ν_L

$$\mathscr{L}^{\text{M}}_{\text{mass}} = \frac{1}{2}\,\frac{g\,v^2}{\mathcal{M}}\, \nu_L^T\, \mathcal{C}^\dagger\, \nu_L + \text{H.c.}\,, \tag{6.166}$$

which, by comparison with eqn (6.119), corresponds to the Majorana mass

$$m = \frac{g\,v^2}{\mathcal{M}}\,. \tag{6.167}$$

The Lagrangian term $\mathscr{L}_5$ is not acceptable in the framework of the SM because it contains a product of fields with energy dimension five, which is not renormalizable. It is important, however, to realize that the SM cannot be considered as the final theory of everything, but only as an effective low-energy theory which is the low-energy product of the symmetry breaking of a high-energy unified theory (see Refs. [812, 714]). Hence, it is plausible that there are effective low-energy Lagrangian terms which respect the symmetries of the SM, but are nonrenormalizable [1053, 1062, 1055, 1056] (see the review in Ref. [331]). These effective Lagrangian terms must respect the symmetries of the SM above the electroweak scale because they are generated by the high-energy theory which must include the gauge symmetries of the SM in order to be effectively reduced to the SM at low energies. This approach is analogous to that adopted in the study of weak interaction processes at energies well below the electroweak scale through the effective

[34] As already explained in footnote 14 on page 71, τ_a are the Pauli matrices.

nonrenormalizable Fermi theory of weak interactions (see section 3.7), which is a low-energy manifestation of the SM.

A Lagrangian term which contains a product of fields with energy dimension d is called a *dimension-d operator*, since the fields are operators in quantum field theory.

Since any Lagrangian term must have a total dimension of $[E]^4$, the coupling constant of a dimension-d operator with $d > 4$ is proportional to $\mathcal{M}^{4-d}$, where $\mathcal{M}$ is a heavy mass characteristic of the symmetry-breaking scale of the high-energy unified theory. This is analogous to the relation in eqn (3.227), in which the dimensional Fermi constant in the effective low-energy dimension-six operator which constitutes the Fermi Lagrangian is proportional to m_W^{-2}.

The proportionality to $\mathcal{M}^{4-d}$ of the coupling constant of dimension-d operators represents in practice a strong suppressing factor that limits the observability of the low-energy effects of the new physics beyond the SM. It is clear that the difficulty of observing the effects of the effective low-energy nonrenormalizable operators increase rapidly with their dimensionality.

The effective low-energy nonrenormalizable operator with lowest dimensionality, compatible with the symmetries of the SM, is the dimension-five effective operator $\mathscr{L}_5$ in eqn (6.163), which generates the Majorana neutrino mass in eqn (6.167), which is proportional to $\mathcal{M}^{-1}$. From these considerations, it is expected that the study of neutrino masses provides the most accessible low-energy window on the new physics beyond the SM.

It is clear that a similar dimension-five effective operator cannot be written for quarks, because it would generate a quark Majorana mass term that is forbidden, because quarks are charged particles. In fact, a replacement of L_L with Q_L in eqn (6.163) would produce a Lagrangian term that is not invariant under a $\mathrm{U}(1)_Y$ gauge transformation, because $Y(Q_L) = 1/3$ and $Y(\Phi) = 1$. The effective operator of lowest dimension that involve quark fields are the dimension-six lepton and baryon number violating operators

$$\begin{aligned} &\left(q_R^{D^T} q_R^U\right)\left(Q_L^T L_L\right), \qquad &&\left(Q_L^T Q_L\right)\left(q_R^{U^T} \ell_R\right), \qquad &&\left(Q_L^T Q_L\right)\left(Q_L^T L_L\right), \\ &\left(q_R^{D^T} q_R^U\right)\left(q_R^{U^T} \ell_R\right), \qquad &&\left(q_R^{U^T} q_R^U\right)\left(q_R^{D^T} \ell_R\right). && \end{aligned} \tag{6.168}$$

The Lagrangian terms containing these operators are suppressed by coefficients proportional to $\mathcal{M}^{-2}$. However, they are very interesting and potentially observable, because they generate proton decay (for example, $p \to e^+ \pi^0$).

Turning back to the dimension-five operator $\mathscr{L}_5$, it is important to note that the generated neutrino Majorana mass in eqn (6.167) is proportional to the ratio $v^2/\mathcal{M}$. Since v is the scale of the electroweak symmetry breaking, it sets the scale of the Dirac fermion masses generated through the Higgs mechanism. Hence, eqn (6.167) can be written as

$$m \propto \frac{m_{\mathrm{D}}^2}{\mathcal{M}}, \tag{6.169}$$

where m_{D} is a typical Dirac mass, which could be of the same order as the charged lepton mass or a quark mass of the same generation. The relation in eqn (6.169) has the same structure as that obtained with the *see-saw mechanism*, that will be discussed in section 6.4.6. It is so-called because the heavier the mass $\mathcal{M}$, the lighter is the neutrino mass m.

The relation in eqn (6.169) can explain the observed smallness of neutrino masses. For example, if we take $m_{\mathrm{D}} \sim v \sim 10^2\,\mathrm{GeV}$ and $\mathcal{M} \sim 10^{15}\,\mathrm{GeV}$, which is a plausible grand unification scale [532, 79, 79, 408], we get $m \sim 10^{-2}\,\mathrm{eV}$, which is a plausible scale for the neutrino mass, according to the experimental data (see section 13.3.4).

6.3 Mixing of three Majorana neutrinos

Let us consider now three generations of massive Majorana neutrinos. From the array of left-handed flavor neutrino fields in eqn (3.138),

$$\boldsymbol{\nu}'_L \equiv \begin{pmatrix} \nu'_{eL} \\ \nu'_{\mu L} \\ \nu'_{\tau L} \end{pmatrix}, \tag{6.170}$$

we can construct the Majorana mass term

$$\mathscr{L}^{\mathrm{M}}_{\mathrm{mass}} = \frac{1}{2}\,\boldsymbol{\nu}'^{T}_L\,\mathcal{C}^\dagger\,M^L\,\boldsymbol{\nu}'_L + \mathrm{H.c.} = \frac{1}{2}\sum_{\alpha,\beta=e,\mu,\tau} \nu'^{T}_{\alpha L}\,\mathcal{C}^\dagger\,M^L_{\alpha\beta}\,\nu'_{\beta L} + \mathrm{H.c.} \tag{6.171}$$

In general, the matrix M^L is a complex symmetric matrix. In fact, we have

$$\sum_{\alpha,\beta} \nu'^{T}_{\alpha L}\,\mathcal{C}^\dagger\,M^L_{\alpha\beta}\,\nu'_{\beta L} = -\sum_{\alpha,\beta} \nu'^{T}_{\beta L}\,M^L_{\alpha\beta}\,(\mathcal{C}^\dagger)^T\,\nu'_{\alpha L}\,, \tag{6.172}$$

where the last expression has been obtained taking the transpose and changing sign because of the anticommutation property of fermion fields. Since $\mathcal{C}^T = -\mathcal{C}$ (see eqn (2.346)), we obtain

$$\sum_{\alpha,\beta} \nu'^{T}_{\alpha L}\,\mathcal{C}^\dagger\,M^L_{\alpha\beta}\,\nu'_{\beta L} = \sum_{\alpha,\beta} \nu'^{T}_{\beta L}\,\mathcal{C}^\dagger\,M^L_{\alpha\beta}\,\nu'_{\alpha L} = \sum_{\alpha,\beta} \nu'^{T}_{\alpha L}\,\mathcal{C}^\dagger\,M^L_{\beta\alpha}\,\nu'_{\beta L}\,, \tag{6.173}$$

where in the last expression we have just renamed $\alpha \leftrightarrows \beta$. Comparing the left and right sides of eqn (6.173), one can see that M^L is symmetric:

$$M^L_{\alpha\beta} = M^L_{\beta\alpha}\,. \tag{6.174}$$

As in the case of Dirac neutrinos discussed in section 6.1, the fields of massive neutrinos are obtained diagonalizing the Majorana mass term in eqn (6.171). As shown in section 6.7.1, the symmetric mass matrix M^L can be diagonalized with the transformation

$$(V^\nu_L)^T\,M^L\,V^\nu_L = M\,, \qquad \text{with} \qquad M_{kj} = m_k\,\delta_{kj} \qquad (k,j = 1,2,3)\,, \tag{6.175}$$

with a unitary matrix V^ν_L and with real and positive masses m_k. The diagonalization is achieved by expressing the left-handed flavor fields as unitary linear combinations of the left-handed components of fields with definite mass:

$$\boldsymbol{\nu}'_L = V_L^{\nu}\, \boldsymbol{n}_L \qquad \text{with} \qquad \boldsymbol{n}_L = \begin{pmatrix} \nu_{1L} \\ \nu_{2L} \\ \nu_{3L} \end{pmatrix} . \tag{6.176}$$

Using eqn (6.175), the Majorana mass term in eqn (6.171) can be written in the diagonal form

$$\mathscr{L}^{\text{M}}_{\text{mass}} = \frac{1}{2}\, \boldsymbol{n}_L^T\, \mathcal{C}^\dagger\, M\, \boldsymbol{n}_L + \text{H.c.} = \frac{1}{2} \sum_{k=1}^{3} m_k\, \nu_{kL}^T\, \mathcal{C}^\dagger\, \nu_{kL} + \text{H.c.}\,, \tag{6.177}$$

or as

$$\mathscr{L}^{\text{M}}_{\text{mass}} = -\frac{1}{2}\, \overline{\boldsymbol{n}_L^C}\, M\, \boldsymbol{n}_L + \text{H.c.} = -\frac{1}{2} \sum_{k=1}^{3} m_k\, \overline{\nu_{kL}^C}\, \nu_{kL} + \text{H.c.} \tag{6.178}$$

The Majorana fields of massive neutrinos

$$\nu_k = \nu_{kL} + \nu_{kL}^C \tag{6.179}$$

satisfy the Majorana condition

$$\nu_k^C = \nu_k \tag{6.180}$$

and allow one to write the three-generation Majorana Lagrangian as

$$\mathscr{L}^{\text{M}} = \frac{1}{2} \sum_{k=1}^{3} \overline{\nu_k} \left(i \overleftrightarrow{\partial\!\!\!/} - m_k \right) \nu_k = \frac{1}{2}\, \overline{\boldsymbol{n}} \left(i \overleftrightarrow{\partial\!\!\!/} - M \right) \boldsymbol{n}\,, \tag{6.181}$$

with the column matrix of massive Majorana neutrino fields

$$\boldsymbol{n} = \begin{pmatrix} \nu_1 \\ \nu_2 \\ \nu_3 \end{pmatrix} . \tag{6.182}$$

The Majorana mass term in eqn (6.177) is not invariant under the global U(1) gauge transformations

$$\nu_{kL} \to e^{i\varphi}\, \nu_{kL} \qquad (k = 1, 2, 3)\,, \tag{6.183}$$

with the same phase φ for all massive neutrinos. As discussed in section 6.2.4, this fact implies a violation of the conservation of the total lepton number, leading to interesting phenomena such as the neutrinoless double-β-decay discussed in section 14.3.

6.3.1 Weak interactions

Let us now examine the effects of the mixing relations in eqn (6.176) on the leptonic weak charged current in eqn (3.80). As in the case of Dirac neutrinos (see eqns (6.11) and (6.12)), the leptonic weak charged current can be written as

$$j^{\rho}_{\mathrm{W,L}} = 2\,\overline{\boldsymbol{n}_L}\,U^{\dagger}\,\gamma^{\rho}\,\boldsymbol{\ell}_L\,, \tag{6.184}$$

with the mixing matrix

$$U = V_L^{\ell\dagger}\,V_L^{\nu}\,. \tag{6.185}$$

The left-handed flavor neutrino fields are conventionally defined as in eqn (6.13),

$$\boldsymbol{\nu}_L = U\,\boldsymbol{n}_L = V_L^{\ell\dagger}\,\boldsymbol{\nu}'_L\,, \qquad \text{with} \qquad \boldsymbol{\nu}_L = \begin{pmatrix} \nu_{eL} \\ \nu_{\mu L} \\ \nu_{\tau L} \end{pmatrix}, \tag{6.186}$$

in order to write the leptonic weak charged current as in the SM eqn (3.141),

$$j^{\rho}_{\mathrm{W,L}} = 2\,\overline{\boldsymbol{\nu}_L}\,\gamma^{\rho}\,\boldsymbol{\ell}_L = 2\sum_{\alpha=e,\mu,\tau}\overline{\nu_{\alpha L}}\,\gamma^{\rho}\,\ell_{\alpha L}\,. \tag{6.187}$$

However, there is an important difference with respect to the mixing of Dirac neutrinos: the physical CP-violating phases in the Majorana mixing matrix are three instead of one. This is due to the fact that the Majorana mass term in eqn (6.177) is not invariant under the global U(1) gauge transformations

$$\nu_{kL} \to e^{i\varphi_k}\,\nu_{kL} \qquad (k = 1, 2, 3)\,. \tag{6.188}$$

In sections 4.2 and 4.3 it was explained that five of the six phases of the unitary 3×3 CKM mixing matrix of quarks are not physical because they can be eliminated by a suitable phase transformation of the quark fields, which leaves the Lagrangian invariant. Similar arguments apply to the mixing of three Dirac neutrinos discussed in section 6.1, because, apart from the weak CC part, the Lagrangian is invariant under the global phase transformations in eqns (6.24) and (6.25) of the neutrino and charged lepton fields. Since in the Majorana case the mass term is not invariant under the phase transformations in eqn (6.188), the left-handed massive neutrino fields cannot be rephased in order to eliminate the two phases that can be factorized on the right of the mixing matrix, as explained in section 4.3. Therefore, the unitary 3×3 mixing matrix of Majorana neutrinos depends on three mixing angles and three physical CP-violating phases. This mixing matrix can be written as a product of a unitary matrix U^{D}, with three mixing angles and one phase similar to the mixing matrix in the Dirac case, and a diagonal unitary matrix D^{M} with two independent phases:

$$U = U^{\mathrm{D}}\,D^{\mathrm{M}}\,. \tag{6.189}$$

The phase in U^{D} is usually called *Dirac phase*, and the two phases in D^{M} are usually called *Majorana phases*. The Dirac three-neutrino mixing matrix U^{D} has properties analogous to the quark mixing matrix, discussed in chapter 4. In particular, CP violation due to the Dirac phase can be quantified in a rephasing-invariant way in terms of the Jarlskog invariant [640, 639, 562, 395, 1075]

$$J = \Im\mathfrak{m}\left[U_{\mu 3}\, U_{e2}\, U_{\mu 2}^{*}\, U_{e3}^{*}\right] = \Im\mathfrak{m}\left[U_{\mu 3}^{\mathrm{D}}\, U_{e2}^{\mathrm{D}}\, U_{\mu 2}^{\mathrm{D}*}\, U_{e3}^{\mathrm{D}*}\right] , \tag{6.190}$$

where the second equality is due to the rephasing invariance of J, which implies that the Majorana phases do not contribute. All CP and T asymmetries due to the Dirac phase depend on J (see section 13.1.1 for a discussion of CP and T violations in neutrino oscillations).

A convenient parameterization for U^{D} is equal to that in eqn (6.31) for Dirac neutrinos:

$$U^{\mathrm{D}} = \begin{pmatrix} c_{12}c_{13} & s_{12}c_{13} & s_{13}e^{-i\delta_{13}} \\ -s_{12}c_{23} - c_{12}s_{23}s_{13}e^{i\delta_{13}} & c_{12}c_{23} - s_{12}s_{23}s_{13}e^{i\delta_{13}} & s_{23}c_{13} \\ s_{12}s_{23} - c_{12}c_{23}s_{13}e^{i\delta_{13}} & -c_{12}s_{23} - s_{12}c_{23}s_{13}e^{i\delta_{13}} & c_{23}c_{13} \end{pmatrix} , \tag{6.191}$$

where $c_{ab} \equiv \cos\vartheta_{ab}$ and $s_{ab} \equiv \sin\vartheta_{ab}$. ϑ_{12}, ϑ_{13}, ϑ_{23} are the three mixing angles ($0 \leq \vartheta_{ab} \leq \pi/2$) and δ_{13} is the Dirac CP-violating phase ($0 \leq \delta_{13} < 2\pi$).

The diagonal unitary matrix D^{M} can be written as

$$D^{\mathrm{M}} = \mathrm{diag}\left(e^{i\lambda_1}\, , \, e^{i\lambda_2}\, , \, e^{i\lambda_3}\right) , \quad \text{with} \quad \lambda_1 = 0 \, . \tag{6.192}$$

The phases λ_2 and λ_3 are the two physical Majorana CP-violating phases. Since all measurable quantities depend only on the differences of the three Majorana phases λ_1, λ_2, λ_3, the choice $\lambda_1 = 0$ is a matter of convention and other choices are equivalent from the physical point of view. In fact, rephasing all the charged lepton fields by $e^{i\varphi}$, we have $e^{i\lambda_k} \to e^{i(\lambda_k - \varphi)}$, whereas $e^{i(\lambda_k - \lambda_j)}$ remains constant.

In the parameterization in eqn (6.191) of U^{D} the Jarlskog invariant is given by eqn (6.32), i.e. by the same expression as in the Dirac case.

6.3.2 CP invariance

We now discuss the conditions on the elements of the mixing matrix for CP invariance. We have seen in subsection 6.2.6 that, in order to satisfy the CP invariance of the Majorana mass term, a left-handed massive neutrino field ν_{kL} must transform under CP as

$$\mathsf{U}_{\mathrm{CP}} \nu_{kL}(x) \mathsf{U}_{\mathrm{CP}}^{-1} = \eta_k \, i \, \gamma^0 \, \mathcal{C} \, \overline{\nu_{kL}}^T(x_{\mathrm{P}}) \, , \tag{6.193}$$

with $\eta_k = \pm 1$. First, we investigate the conditions for CP invariance of the leptonic charged-current weak interaction Lagrangian

$$\mathscr{L}_{\mathrm{I,L}}^{(\mathrm{CC})} = -\frac{g}{\sqrt{2}} \sum_{\alpha=e,\mu,\tau} \sum_{k=1}^{3} \left(U_{\alpha k}^{*} \, \overline{\nu_{kL}} \, \gamma^{\rho} \, \ell_{\alpha L} \, W_{\rho} + U_{\alpha k} \, \overline{\ell_{\alpha L}} \, \gamma^{\rho} \, \nu_{kL} \, W_{\rho}^{\dagger} \right) . \tag{6.194}$$

Under a CP transformation, the Dirac charged lepton field ℓ transforms as

$$\mathsf{U}_{\mathrm{CP}} \, \ell_{\alpha} \, \mathsf{U}_{\mathrm{CP}}^{-1} = \xi_{\ell_\alpha}^{\mathrm{CP}} \, \gamma^0 \, \mathcal{C} \, \overline{\ell_{\alpha}}^T \, . \tag{6.195}$$

Using the transformation of W_μ in eqn (4.117),

$$\mathsf{U}_{\mathrm{CP}}\, W_\mu\, \mathsf{U}_{\mathrm{CP}}^{-1} = -W^{\mu\dagger}\,, \tag{6.196}$$

and using the relation in eqn (6.158), we obtain

$$\mathsf{U}_{\mathrm{CP}}\, \mathscr{L}_{\mathrm{I,L}}^{(\mathrm{CC})}\, \mathsf{U}_{\mathrm{CP}}^{-1} = -\frac{g}{\sqrt{2}} \sum_{\alpha=e,\mu,\tau} \sum_{k=1}^{3} \Big(- U_{\alpha k}^{*}\, \eta_k\, i\, \xi_{\ell_\alpha}^{\mathrm{CP}}\, \overline{\ell_{\alpha L}}\, \gamma^\rho\, \nu_{kL}\, W_\rho^\dagger$$
$$+ U_{\alpha k}\, \eta_k\, i\, \xi_{\ell_\alpha}^{\mathrm{CP}\,*}\, \overline{\nu_{kL}}\, \gamma^\rho\, \ell_{\alpha L}\, W_\rho \Big)\,. \tag{6.197}$$

Comparing with eqn (6.194), one can see that CP invariance is realized if

$$U_{\alpha k}\, \eta_k\, i\, \xi_{\ell_\alpha}^{\mathrm{CP}\,*} = U_{\alpha k}^{*}\,. \tag{6.198}$$

As we have shown in eqn (6.189), the mixing matrix can always be written as a product of a Dirac unitary matrix U^{D} and the diagonal matrix of phases in eqn (6.192),

$$U_{\alpha k} = U_{\alpha k}^{\mathrm{D}}\, e^{i\lambda_k}\,. \tag{6.199}$$

Hence, the condition in eqn (6.198) can be written as

$$U_{\alpha k}^{\mathrm{D}}\, e^{2i\lambda_k}\, \eta_k\, i\, \xi_{\ell_\alpha}^{\mathrm{CP}\,*} = U_{\alpha k}^{\mathrm{D}\,*}\,. \tag{6.200}$$

The Dirac unitary matrix U^{D} can be parameterized in terms of three mixing angles and one phase, which cannot be eliminated by rephasing the charged lepton and neutrino fields in the charged-current interaction Lagrangian. Therefore, the phase in U^{D} cannot be eliminated by an appropriate choice of the charged lepton CP phases $\xi_{\ell_\alpha}^{\mathrm{CP}}$ and of the η_k's. This means that a necessary condition for CP invariance is the reality of U^{D}, as in the case of Dirac neutrinos:

$$U^{\mathrm{D}} = U^{\mathrm{D}\,*} = \mathcal{O}\,, \tag{6.201}$$

where $\mathcal{O}$ is an orthogonal matrix ($\mathcal{O}^T = \mathcal{O}^{-1}$). In the parameterization in eqn (6.191), CP is conserved if

$$\delta_{13} = 0\,,\ \pi\,. \tag{6.202}$$

However, the reality of U^{D} is not sufficient for CP invariance, because of the presence of the Majorana phases λ_k. If the condition in eqn (6.201) is satisfied, eqn (6.200) reduces to

$$\eta_k = -i\, \xi_{\ell_\alpha}^{\mathrm{CP}}\, e^{-2i\lambda_k}\,, \tag{6.203}$$

with $\eta_k = \pm 1$. Choosing the CP phases $\xi_{\ell_\alpha}^{\mathrm{CP}}$ of the charged leptons all equal to i, we find

$$\eta_k = e^{-2i\lambda_k}\,. \tag{6.204}$$

It turns out that there are four possibilities for $0 \le \lambda_k < 2\pi$, according to Table 6.1. Since the diagonal matrix in eqn (6.192) has the factors $e^{i\lambda_k}$ on the right of the

TABLE 6.1. Correspondence among the CP parity $\eta_k i$ of ν_k, the Majorana phase λ_k, and the coefficient $e^{i\lambda_k}$ in the diagonal matrix in eqn (6.192).

η_k	λ_k	$e^{i\lambda_k}$
$+1$	$0,\ \pi$	± 1
-1	$\pi/2,\ 3\pi/2$	$\pm i$

mixing matrix, the Lagrangian is symmetric under a CP transformation if all the elements of each column of the mixing matrix are either real or purely imaginary. If one column has real elements and another has imaginary elements, the corresponding massive neutrinos have opposite values of η_k, i.e. opposite CP parities. Obviously, the value of η_k and the associated CP parity of a massive neutrino have no physical meaning per se, because it depends on the arbitrary choice of the charged lepton CP phases $\xi^{\text{CP}}_{\ell_\alpha}$ (for example, switching to $\xi^{\text{CP}}_{\ell_\alpha} = -i$ corresponds to $\eta_k \to -\eta_k$). Only the differences of the CP parities of different massive neutrinos is physically meaningful. The arbitrariness in the value of η_k is reflected in the arbitrariness of the value of the phase λ_k discussed after eqn (6.192), whereas the physical meaning of the relative CP phases is reflected in the physical meaning of the differences of the Majorana phases in the mixing matrix. Indeed, from eqn (6.203) we have

$$\frac{\eta_k}{\eta_j} = e^{-2i(\lambda_k - \lambda_j)}\,, \tag{6.205}$$

which does not depend on the arbitrary CP phases $\xi^{\text{CP}}_{\ell_\alpha}$ of the charged leptons and is invariant under rephasing of the charged lepton fields, as discussed after eqn (6.192).

Let us now follow a different approach to CP invariance, analogous to that in the second part of section 4.6: we consider the charged-current weak interaction Lagrangian in eqn (3.76) written in terms of the primed lepton fields in eqn (3.66) and we find which are the conditions on the mass matrices of charged leptons and neutrinos in order to have CP invariance.

From eqns (3.128) and (6.171) the Lagrangian mass term of leptons is

$$\mathscr{L}_{l,\text{mass}} = -\overline{\boldsymbol{\ell}'_L}\, M'^{\ell}\, \boldsymbol{\ell}'_R - \overline{\boldsymbol{\ell}'_R}\, M'^{\ell\dagger}\, \boldsymbol{\ell}'_L + \frac{1}{2}\, \boldsymbol{\nu}'^T_L\, \mathcal{C}^\dagger\, M^L\, \boldsymbol{\nu}'_L + \frac{1}{2}\, \nu'^\dagger_L\, M^{L*}\, \mathcal{C}\, \nu'^*_L\,, \tag{6.206}$$

where we have taken into account that $M^{L^T} = M^L$ (see eqn (6.174)). Let us consider the general CP transformations of the lepton fields

$$\boldsymbol{\ell}'_R \xrightarrow{\text{CP}} W_R\, \gamma^0\, \mathcal{C}\, \overline{\boldsymbol{\ell}'_R}^T\,, \tag{6.207}$$

$$\boldsymbol{\ell}'_L \xrightarrow{\text{CP}} W_L\, \gamma^0\, \mathcal{C}\, \overline{\boldsymbol{\ell}'_L}^T\,, \tag{6.208}$$

$$\boldsymbol{\nu}'_L \xrightarrow{\text{CP}} W_L\, \gamma^0\, \mathcal{C}\, \overline{\boldsymbol{\nu}'_L}^T\,, \tag{6.209}$$

where W_R and W_L are unitary matrices that mix the fields. The mixing of the left-handed charged lepton and neutrino fields are equal, in order to keep invariant the charged-current weak interaction Lagrangian. In this way, the Lagrangian is trivially invariant under the CP transformations in eqns (6.207)–(6.209), apart from the mass term in eqn (6.206) (and the charged lepton interaction term with the Higgs boson in eqn (3.128) whose invariance conditions are the same as those for the charged lepton mass term). Under the CP transformations in eqns (6.207)–(6.209) the mass term in eqn (6.206) transforms as

$$\begin{aligned}\mathscr{L}_{l,\mathrm{mass}} \xrightarrow{\mathrm{CP}} & -\overline{\boldsymbol{\ell}'_R}\, W_R^T\, M'^{\ell^T}\, W_L^*\, \boldsymbol{\ell}'_L - \overline{\boldsymbol{\ell}'_L}\, W_L^T\, M'^{\ell^*}\, W_R^*\, \boldsymbol{\ell}'_R \\ & - \frac{1}{2}\, \nu_L'^{\dagger}\, W_L^T\, M^L\, W_L\, \mathcal{C}\, \nu_L'^{*} - \frac{1}{2}\, \boldsymbol{\nu}_L'^T\, \mathcal{C}^{\dagger}\, W_L^{\dagger}\, M^{L^*}\, W_L^*\, \boldsymbol{\nu}'_L\,. \end{aligned} \tag{6.210}$$

The mass term in eqn (6.206) is invariant under a CP transformation if it is possible to find appropriate unitary matrices W_R and W_L such that

$$W_L^{\dagger}\, M'^{\ell}\, W_R = M'^{\ell^*}\,, \tag{6.211}$$

$$W_L^T\, M^L\, W_L = -M^{L^*}\,. \tag{6.212}$$

Since these two conditions can be written as

$$W_L^*\, M'^{\ell}\, W_R^T = M'^{\ell^*}\,, \tag{6.213}$$

$$W_L\, M^L\, W_L^T = -M^{L^*}\,, \tag{6.214}$$

the matrices W_R and W_L must also be symmetric.

The condition in eqn (6.211) can always be satisfied with appropriate matrices W_R and W_L. Using the same method employed in section 4.6 for quarks we find

$$W_L = V_L^{\ell}\, D^{\ell} V_L^{\ell^T}\,, \tag{6.215}$$

$$W_R = V_R^{\ell}\, D^{\ell} V_R^{\ell^T}\,, \tag{6.216}$$

where V_L^{ℓ} and V_R^{ℓ} are the unitary matrices that diagonalize the charged lepton mass matrix through eqn (3.129), and D^{ℓ} is an arbitrary diagonal matrix of phases.

Substituting W_L given by eqn (6.215) in the condition in eqn (6.212) and using the diagonalization equation (6.175) for M^L, we obtain the condition

$$U^{\dagger}\, D^{\ell}\, U^*\, M\, U^{\dagger}\, D^{\ell}\, U^* = -M\,, \tag{6.217}$$

with the mixing matrix U given in eqn (6.185). This condition is satisfied if

$$U^{\dagger}\, D^{\ell}\, U^* = \eta\, i\,, \tag{6.218}$$

where η is a diagonal matrix with eigenvalues equal to ± 1. CP is a symmetry if it is possible to find appropriate matrices D^{ℓ} and η such that the condition in eqn (6.218) is satisfied.

Let us note that the condition in eqn (6.218) can also be written as

$$D^{\ell^{\dagger}}\, U\, \eta\, i = U^*\,, \tag{6.219}$$

which coincides with the condition in eqn (6.198) for $D^{\ell}_{\alpha\beta} = \xi^{\mathrm{CP}}_{\ell_\alpha}\delta_{\alpha\beta}$ and $\eta_{kj} = \eta_k \delta_{kj}$. Hence, we see that the two methods that we have used for finding the conditions for CP invariance are equivalent (invariance of the charged-current weak interaction Lagrangian assuming diagonal mass terms and invariance of the mass terms assuming a diagonal charged-current weak interaction Lagrangian).

The condition for CP invariance on the lepton mass matrices M'^{ℓ} and M^L can be obtained from eqns (6.211) and (6.212) by eliminating W_L:

$$W_R^T \left[M'^{\ell^T} M^L M'^{\ell} \right] W_R = - \left[M'^{\ell^T} M^L M'^{\ell} \right]^* . \tag{6.220}$$

This condition is satisfied if $M'^{\ell^T} M^L M'^{\ell}$ can be written as

$$M'^{\ell^T} M^L M'^{\ell} = P^T R P , \quad \text{with} \quad R = R^T = R^* , \quad P^\dagger = P^{-1} . \tag{6.221}$$

In this case, the unitary and symmetric matrix W_R is given by

$$W_R = i \, P^\dagger P^* . \tag{6.222}$$

Particular simple cases in which the condition in eqn (6.221) for CP invariance is satisfied are

$\boldsymbol{M^L = 0}$**.** This is the trivial case of the SM, in which there is no mixing in the lepton sector.

M'^{ℓ}**.** This is an unrealistic case, because we know that charged leptons are massive. Let us only note, for the sake of curiosity, that also this case is trivial, because it would imply that there is no mixing in the lepton sector.

Real $M'^{\ell^T} M^L M'^{\ell}$**.** In this case, $P = \pm\mathbf{1}$ and $W_R = \pm i\mathbf{1}$. A particular simple case is

$$M'^{\ell} = M'^{\ell^*} \quad \text{and} \quad M^L = M^{L^*} . \tag{6.223}$$

In this case, M'^{ℓ} can be diagonalized with orthogonal matrices $V_L^{\ell} = \mathrm{O}_L^{\ell}$, $V_R^{\ell} = \mathrm{O}_R^{\ell}$. However, as explained in section 6.7.1, the matrix V_L^{ν} is given by a real orthogonal matrix multiplied by a diagonal matrix with elements equal to ± 1 or $\pm i$. Hence, the columns of the mixing matrix U in eqn (6.185) are either real or imaginary. This is consistent with the result in eqn (6.204).

6.3.3 Effective three-neutrino mixing

As we have discussed in section 6.2.7, it is natural to expect that the SM is an effective low-energy theory obtained through the symmetry breaking of a unified theory which describes the physics at very high energies. In this case, the high-energy theory can manifest itself at low energies through nonrenormalizable effective Lagrangian terms with product of fields of dimension larger than four, whose effects are suppressed by the dimensional coupling constants inversely proportional to the appropriate power of the high-energy mass scale. In the case of one neutrino the effective Lagrangian term with lowest dimension is the dimension-five lepton number violating term in eqn (6.163). This term can be straightforwardly generalized to the case of three neutrinos:

$$\mathscr{L}_5 = \frac{1}{\mathcal{M}} \sum_{\alpha\beta} g_{\alpha\beta} \, (L'^T_{\alpha L} \, \tau_2 \, \Phi) \, \mathcal{C}^\dagger \, (\Phi^T \, \tau_2 \, L'_{\beta L}) + \text{H.c.}\,, \tag{6.224}$$

where g is a symmetric 3×3 matrix of coupling constants. The electroweak symmetry breaking VEV of the Higgs field (see section 3.4) leads to the neutrino Majorana mass term

$$\mathscr{L}^{\text{M}}_{\text{mass}} = \frac{1}{2} \frac{v^2}{\mathcal{M}} \sum_{\alpha\beta} g_{\alpha\beta} \, \nu'^T_{\alpha L} \, \mathcal{C}^\dagger \, \nu'_{\beta L} + \text{H.c.} \tag{6.225}$$

In this case, the Majorana mass matrix in eqn (6.171) has elements

$$M^L_{\alpha\beta} = \frac{v^2}{\mathcal{M}} \, g_{\alpha\beta} \,. \tag{6.226}$$

Hence, the scale of the Majorana neutrino masses is set by the small ratio $v^2/\mathcal{M}$, providing a natural explanation of the strong suppression of neutrino masses with respect to the electroweak scale.

6.4 One-generation Dirac–Majorana mass term

From the discussion in the previous sections of this chapter it should be clear that, considering for simplicity only one generation, the chiral fields ν_L and ν_R are the building blocks of the neutrino Lagrangian. We know that the chiral field ν_L exists, because it is present in the SM and enters in the charged-current weak interaction Lagrangian. We do not know if the chiral field ν_R exists, but it is allowed by the symmetries of the SM. If only ν_L exists, the neutrino Lagrangian can contain only the Majorana mass term

$$\mathscr{L}^{L}_{\text{mass}} = \frac{1}{2} \, m_L \, \nu_L^T \, \mathcal{C}^\dagger \, \nu_L + \text{H.c.}\,, \tag{6.227}$$

and the neutrino is a Majorana particle. If ν_R also exists, the neutrino Lagrangian can contain the Dirac mass term

$$\mathscr{L}^{\text{D}}_{\text{mass}} = -m_{\text{D}} \, \overline{\nu_R} \, \nu_L + \text{H.c.}\,, \tag{6.228}$$

which would imply that the neutrino is a Dirac particle. However, in addition to the Dirac mass term in eqn (6.228), the neutrino Lagrangian can also contain the Majorana mass term in eqn (6.227) for ν_L and the Majorana mass term

$$\mathscr{L}^{R}_{\text{mass}} = \frac{1}{2} \, m_R \, \nu_R^T \, \mathcal{C}^\dagger \, \nu_R + \text{H.c.} \tag{6.229}$$

for ν_R. Therefore, in general it is possible to have the Dirac–Majorana neutrino mass term

$$\mathscr{L}^{\text{D+M}}_{\text{mass}} = \mathscr{L}^{\text{D}}_{\text{mass}} + \mathscr{L}^{L}_{\text{mass}} + \mathscr{L}^{R}_{\text{mass}} \,. \tag{6.230}$$

This is a surprise, because, among all the known elementary particles, all of which have both chiral left-handed and right-handed chiral field components, only neutrinos can have the Majorana mass terms $\mathscr{L}^{L}_{\text{mass}}$ and $\mathscr{L}^{R}_{\text{mass}}$. This possibility implies that neutrinos are very special particles that can generate new physics through the lepton number violating Majorana mass terms. The Majorana mass term in eqn (6.227) for ν_L is not allowed by the symmetries of the SM because it is not invariant under $\mathrm{SU}(2)_L \times \mathrm{U}(1)_Y$ transformations (it can, however, be generated by new physics beyond the SM, as discussed in section 6.2.7). On the other hand, the Majorana mass term in eqn (6.229) for ν_R is allowed by the symmetries of the SM, because ν_R is a singlet of $\mathrm{SU}(3)_C \times \mathrm{SU}(2)_L \times \mathrm{U}(1)_Y$. Therefore, the Dirac–Majorana mass term in eqn (6.230) with $m_L = 0$ is allowed in the framework of the SM with the only addition of the right-handed chiral field ν_R.

We note that the Majorana mass m_R in eqn (6.229) can be chosen to be real and positive by an appropriate rephasing of the chiral field ν_R. Once the phase of ν_R is fixed, the Dirac mass m_D in eqn (6.228) can be chosen to be real and positive by an appropriate rephasing of the chiral field ν_L. However, once the phases of ν_L and ν_R are fixed, there is no additional freedom to cancel a possible complex phase of the Majorana mass m_L in eqn (6.227). Therefore, in the following we will consider real and positive m_R and m_D and complex m_L. Obviously, other choices, such as real and positive m_R and m_L and complex m_D, are equivalent from the physical point of view.

In order to understand the implications of the Dirac–Majorana mass term in eqn (6.230) it is useful to define the column matrix of left-handed chiral fields

$$N_L = \begin{pmatrix} \nu_L \\ \nu_R^C \end{pmatrix} = \begin{pmatrix} \nu_L \\ \mathcal{C}\, \overline{\nu_R}^T \end{pmatrix} . \tag{6.231}$$

Using eqn (6.66), we can write the Dirac–Majorana mass term in eqn (6.230) as

$$\mathscr{L}^{\mathrm{D+M}}_{\mathrm{mass}} = \frac{1}{2}\, N_L^T\, \mathcal{C}^\dagger\, M\, N_L + \mathrm{H.c.}\,, \tag{6.232}$$

with the symmetric mass matrix

$$M = \begin{pmatrix} m_L & m_\mathrm{D} \\ m_\mathrm{D} & m_R \end{pmatrix} . \tag{6.233}$$

From the expression in eqn (6.232) it is clear that the chiral fields ν_L and ν_R do not have a definite mass, because of the off-diagonal Dirac mass. In order to find the fields of massive neutrinos it is necessary to diagonalize the mass matrix in eqn (6.233). This can be done with a unitary transformation of the chiral fields,

$$N_L = U\, n_L\,, \tag{6.234}$$

where

$$n_L = \begin{pmatrix} \nu_{1L} \\ \nu_{2L} \end{pmatrix} \tag{6.235}$$

is the column matrix of chiral left-handed massive neutrino fields. The unitary matrix U must be such that

$$U^T\, M\, U = \begin{pmatrix} m_1 & 0 \\ 0 & m_2 \end{pmatrix} , \tag{6.236}$$

with real $m_k \geq 0$. This can always be done, as we have shown in section 6.3 for the symmetric Majorana mass matrix M^L.

With the transformation in eqn (6.234), the Dirac–Majorana mass term in eqn (6.232) can be written as

$$\mathscr{L}_{\text{mass}}^{\text{D+M}} = \frac{1}{2} \sum_{k=1,2} m_k \, \nu_{kL}^T \, \mathcal{C}^\dagger \, \nu_{kL} + \text{H.c.} = -\frac{1}{2} \sum_{k=1,2} m_k \, \overline{\nu_k} \, \nu_k \,, \tag{6.237}$$

where we have defined the Majorana massive neutrino field

$$\nu_k = \nu_{kL} + \nu_{kL}^C = \nu_{kL} + \mathcal{C} \, \overline{\nu_{kL}}^T \,. \tag{6.238}$$

Hence, we have obtained the important result that a Dirac–Majorana mass term implies that massive neutrinos are Majorana particles. This should not be a surprise, because the Dirac–Majorana mass term in eqn (6.232) has the structure of a Majorana mass term for the two chiral fields ν_L and ν_R^C.

The values of the masses m_k and the elements of the mixing matrix can be calculated with the method presented in section 6.7.1. Remembering that we consider real and positive m_R and m_{D} and complex m_L, the masses m_1 and m_2 are the two positive eigenvalues of the matrix

$$\mathfrak{M} = \begin{pmatrix} \mathfrak{Re}[m_L] & m_{\text{D}} & -\mathfrak{Im}[m_L] & 0 \\ m_{\text{D}} & m_R & 0 & 0 \\ -\mathfrak{Im}[m_L] & 0 & -\mathfrak{Re}[m_L] & -m_{\text{D}} \\ 0 & 0 & -m_{\text{D}} & -m_R \end{pmatrix} . \tag{6.239}$$

We obtain

$$m_{2,1}^2 = \frac{1}{2} \Bigg[|m_L|^2 + m_R^2 + 2\,m_{\text{D}}^2 \pm \Bigg((\mathfrak{Re}[m_L] + m_R)^2 \left[(\mathfrak{Re}[m_L] - m_R)^2 + 4\,m_{\text{D}}^2 \right] \\ + (\mathfrak{Im}[m_L])^4 + 2(\mathfrak{Im}[m_L])^2 \left((\mathfrak{Re}[m_L])^2 - m_R^2 + 2\,m_{\text{D}}^2 \right) \Bigg)^{1/2} \Bigg] . \tag{6.240}$$

It is easy to check that the expression on the right-hand side is always positive.

Let us write the mixing matrix U as

$$U = \begin{pmatrix} \cos\vartheta & \sin\vartheta \\ -\sin\vartheta & \cos\vartheta \end{pmatrix} \begin{pmatrix} e^{i\lambda} & 0 \\ 0 & 1 \end{pmatrix} = \begin{pmatrix} \cos\vartheta \, e^{i\lambda} & \sin\vartheta \\ -\sin\vartheta \, e^{i\lambda} & \cos\vartheta \end{pmatrix} , \tag{6.241}$$

with $0 \leq \vartheta \leq \pi/2$ and $0 \leq \lambda < 2\pi$. The value of the mixing angle ϑ is determined by the second eigenvalue equation of the type in eqn (6.397),

$$\begin{pmatrix} \mathfrak{Re}[m_L] & m_{\text{D}} & -\mathfrak{Im}[m_L] & 0 \\ m_{\text{D}} & m_R & 0 & 0 \\ -\mathfrak{Im}[m_L] & 0 & -\mathfrak{Re}[m_L] & -m_{\text{D}} \\ 0 & 0 & -m_{\text{D}} & -m_R \end{pmatrix} \begin{pmatrix} \sin\vartheta \\ \cos\vartheta \\ 0 \\ 0 \end{pmatrix} = m_2 \begin{pmatrix} \sin\vartheta \\ \cos\vartheta \\ 0 \\ 0 \end{pmatrix} , \tag{6.242}$$

which gives

$$\tan 2\vartheta = \frac{2 m_{\text{D}}}{m_R - \mathfrak{Re}[m_L]} \,. \tag{6.243}$$

One can then determine the value of the phase λ using the first eigenvalue equation

$$\begin{pmatrix} \mathfrak{Re}[m_L] & m_{\rm D} & -\mathfrak{Im}[m_L] & 0 \\ m_{\rm D} & m_R & 0 & 0 \\ -\mathfrak{Im}[m_L] & 0 & -\mathfrak{Re}[m_L] & -m_{\rm D} \\ 0 & 0 & -m_{\rm D} & -m_R \end{pmatrix} \begin{pmatrix} \cos\vartheta\cos\lambda \\ -\sin\vartheta\cos\lambda \\ \cos\vartheta\sin\lambda \\ -\sin\vartheta\sin\lambda \end{pmatrix} = m_1 \begin{pmatrix} \cos\vartheta\cos\lambda \\ -\sin\vartheta\cos\lambda \\ \cos\vartheta\sin\lambda \\ -\sin\vartheta\sin\lambda \end{pmatrix} . \tag{6.244}$$

We obtain

$$\tan 2\lambda = -\frac{2\,\mathfrak{Im}[m_L]}{\mathfrak{Re}[m_L] + m_R - \sqrt{\left(\mathfrak{Re}[m_L] - m_R\right)^2 + 4m_{\rm D}^2}} . \tag{6.245}$$

Since $0 \leq 2\lambda < 4\pi$, eqn (6.245) gives four values for λ. The values which are really possible must be selected by checking that the mass eigenvalues obtained through eqn (6.236) are positive. If CP is conserved, the mass matrix is real and $\tan 2\lambda$ vanishes. In this case, the four possibilities for $e^{i\lambda}$ are ± 1 and $\pm i$. The selection of the possible value is discussed in subsection 6.4.2.

Summarizing, in the case of one generation with both left-handed and right-handed chiral neutrino fields ν_L and ν_R, the diagonalization of the most general Dirac–Majorana mass term implies that there are two massive Majorana neutrino fields ν_1 and ν_2. It is usual to refer to ν_L and ν_R^C as the left-handed fields in the flavor basis and refer to ν_1 and ν_2 as the fields in the mass basis. The two flavor fields ν_L and ν_R are, respectively, active and sterile, because ν_L participates in weak interactions, whereas ν_R is a singlet of the gauge symmetries of the SM. The mixing between ν_L and ν_R^C implies that in general oscillations between active and sterile neutrinos are possible (see chapter 7). The oscillations between the active ν_L and the sterile ν_R^C depend on the squared-mass difference

$$\Delta m^2 = \Big[\left(\mathfrak{Re}[m_L] + m_R\right)^2 \left[\left(\mathfrak{Re}[m_L] - m_R\right)^2 + 4\,m_{\rm D}^2 \right] + (\mathfrak{Im}[m_L])^4 + 2(\mathfrak{Im}[m_L])^2 \left((\mathfrak{Re}[m_L])^2 - m_R^2 + 2\,m_{\rm D}^2 \right) \Big]^{1/2} . \tag{6.246}$$

6.4.1 Weak interactions

The mixing relations in eqn (6.234),

$$\nu_L = U_{11}\,\nu_{1L} + U_{12}\,\nu_{2L} , \tag{6.247}$$

$$\nu_R^C = U_{21}\,\nu_{1L} + U_{22}\,\nu_{2L} , \tag{6.248}$$

imply that the active neutrino field ν_L and the sterile field ν_R are linear combinations of the same massive neutrino fields ν_{1L}, ν_{2L}. In this case, oscillations between active and sterile states are possible. Indeed, the one-generation charged-current weak interaction Lagrangian in eqn (6.102) written in terms of massive fields reads

$$\mathscr{L}_{\mathrm{I,L}}^{\mathrm{CC}} = -\frac{g}{\sqrt{2}} \sum_{k=1,2} \left(U_{1k}^{*} \, \overline{\nu_{kL}} \, \gamma^{\mu} \, \ell_L \, W_{\mu} + U_{1k} \, \overline{\ell_L} \, \gamma^{\mu} \, \nu_{kL} \, W_{\mu}^{\dagger} \right) . \tag{6.249}$$

Hence, charged-current weak interactions create a superposition of the two massive neutrinos. Since their phases evolve in space-time in a different way, depending on the masses, the superposition can evolve into a linear combination of active and sterile neutrinos with an oscillatory probability to detect the active component. The sterile component cannot be detected and manifests itself through the disappearance of active neutrinos which corresponds to a survival probability of active neutrinos smaller than one.

For neutral-current weak interactions we have the interesting effect that the GIM mechanism does not work [933]. Indeed, let us consider the neutrino part of the one-generation neutral-current weak interaction Lagrangian in eqn (3.41)

$$\mathscr{L}_{\mathrm{I},\nu}^{\mathrm{NC}} = -\frac{g}{2\cos\vartheta_{\mathrm{W}}} \, \overline{\nu_L} \, \gamma^{\mu} \, \nu_L \, Z_{\mu} . \tag{6.250}$$

Under the mixing in eqn (6.247) we obtain

$$\mathscr{L}_{\mathrm{I},\nu}^{\mathrm{NC}} = -\frac{g}{2\cos\vartheta_{\mathrm{W}}} \sum_{k,j=1,2} U_{1k}^{*} \, U_{1j} \, \overline{\nu_{kL}} \, \gamma^{\mu} \, \nu_{jL} \, Z_{\mu} . \tag{6.251}$$

Hence, there can be neutral-current transitions among different massive neutrinos.

6.4.2 CP invariance

It is useful to consider the simplest case of a real mass matrix, which implies CP invariance, as we will see in the following. This means that, besides the real and positive m_R and m_{D} we consider a real m_L, which can be either positive or negative.

Since in this case the mass matrix in eqn (6.233) is real and symmetric, it can be diagonalized through the transformation in eqn (6.236) with the unitary matrix

$$U = \mathcal{O} \, \rho , \tag{6.252}$$

where $\mathcal{O}$ is an orthogonal 2×2 matrix that can be written as

$$\mathcal{O} = \begin{pmatrix} \cos\vartheta & \sin\vartheta \\ -\sin\vartheta & \cos\vartheta \end{pmatrix} , \tag{6.253}$$

and ρ is a diagonal matrix of phases

$$\rho = \begin{pmatrix} \rho_1 & 0 \\ 0 & \rho_2 \end{pmatrix} , \tag{6.254}$$

with $\rho_k^2 = \pm 1$. The orthogonal matrix $\mathcal{O}$ must be chosen in order to diagonalize the mass matrix in eqn (6.233):

$$\mathcal{O}^T \, M \, \mathcal{O} = \begin{pmatrix} m_1' & 0 \\ 0 & m_2' \end{pmatrix} , \tag{6.255}$$

where m'_1 and m'_2 are the eigenvalues of the mass matrix

$$m'_{2,1} = \frac{1}{2}\left[m_L + m_R \pm \sqrt{(m_L - m_R)^2 + 4\,m_{\rm D}^2}\right]. \tag{6.256}$$

This is accomplished with

$$\tan 2\vartheta = \frac{2m_{\rm D}}{m_R - m_L}. \tag{6.257}$$

One must note that m'_1 is negative if $m_L m_R < m_{\rm D}^2$. The role of the matrix ρ is to change the sign of the first mass eigenvalue if $m'_1 < 0$ through

$$U^T M U = \rho^T \mathcal{O}^T M \mathcal{O} \rho = \begin{pmatrix} \rho_1 & 0 \\ 0 & \rho_2 \end{pmatrix} \begin{pmatrix} m'_1 & 0 \\ 0 & m'_2 \end{pmatrix} \begin{pmatrix} \rho_1 & 0 \\ 0 & \rho_2 \end{pmatrix} = \begin{pmatrix} \rho_1^2 m'_1 & 0 \\ 0 & \rho_2^2 m'_2 \end{pmatrix}. \tag{6.258}$$

Hence, the two real and positive neutrino masses are given by

$$m_k = \rho_k^2\, m'_k\,, \tag{6.259}$$

with $\rho_2^2 = 1$, whereas $\rho_1^2 = 1$ if $m'_1 > 0$ and $\rho_1^2 = -1$ if $m'_1 < 0$. It is easy to check that the values of the masses agree with the general expression in eqn (6.240) in the case $\mathfrak{Im}[m_L] = 0$.

Hence, m'_2 is always positive and we have

$$m_2 = \frac{1}{2}\left[m_L + m_R + \sqrt{(m_L - m_R)^2 + 4\,m_{\rm D}^2}\right]. \tag{6.260}$$

If $m_L m_R \geq m_{\rm D}^2$, we have $m'_1 \geq 0$ and $\rho_1^2 = 1$, leading to

$$m_1 = \frac{1}{2}\left[m_L + m_R - \sqrt{(m_L - m_R)^2 + 4\,m_{\rm D}^2}\right]. \tag{6.261}$$

Choosing $\rho_1 = 1$ and $\rho_2 = 1$ we obtain the mixing matrix

$$U = \begin{pmatrix} \cos\vartheta & \sin\vartheta \\ -\sin\vartheta & \cos\vartheta \end{pmatrix}. \tag{6.262}$$

If $m_L m_R < m_{\rm D}^2$, we have $m'_1 < 0$ and $\rho_1^2 = -1$, which implies

$$m_1 = \frac{1}{2}\left[\sqrt{(m_L - m_R)^2 + 4\,m_{\rm D}^2} - (m_L + m_R)\right]. \tag{6.263}$$

Choosing $\rho_1 = i$ and $\rho_2 = 1$, the mixing matrix is given by

$$U = \begin{pmatrix} i\cos\vartheta & \sin\vartheta \\ -i\sin\vartheta & \cos\vartheta \end{pmatrix}. \tag{6.264}$$

The oscillations between the active ν_L and the sterile ν_R^C depend on the mixing angle given by eqn (6.257) and on the squared-mass difference

$$\Delta m^2 = m_2^2 - m_1^2 = (m_L + m_R)\sqrt{(m_L - m_R)^2 + 4\,m_{\rm D}^2}. \tag{6.265}$$

Note that Δm^2 may be negative. This happens if $m_L < -m_R$, which is possible, because m_L can be negative.

In spite of the possible imaginary elements in the mixing matrix U, the Lagrangian is invariant under CP transformations, and the phase

$$\xi_k^{\mathrm{CP}} = i\,\rho_k^2 \tag{6.266}$$

is the CP parity of ν_k, as we are going to show. The CP transformation of the column matrix N_L of flavor left-handed fields in eqn (6.231) is determined by the request of CP invariance of the Dirac–Majorana mass term in eqn (6.232),

$$\mathscr{L}_{\text{mass}}^{\text{D+M}} = \frac{1}{2}\left(N_L^T\,\mathcal{C}^\dagger\,M\,N_L - \overline{N_L}\,M\,\mathcal{C}\,\overline{N_L}^T\right), \tag{6.267}$$

where we have taken into account the symmetry and reality of M, which imply $M^\dagger = M$. Under a CP transformation, N_L and $\overline{N_L}$ transform as

$$\mathsf{U}_{\mathrm{CP}}\,N_L\,\mathsf{U}_{\mathrm{CP}}^{-1} = \Xi_{\mathrm{CP}}^{(N_L)}\,\gamma^0\,\mathcal{C}\,\overline{N_L}^T, \qquad \mathsf{U}_{\mathrm{CP}}\,\overline{N_L}\,\mathsf{U}_{\mathrm{CP}}^{-1} = -N_L^T\,\mathcal{C}^\dagger\,\gamma^0\,(\Xi_{\mathrm{CP}}^{(N_L)})^*, \tag{6.268}$$

with a diagonal matrix of phases $\Xi_{\mathrm{CP}}^{(N_L)}$. For the mass term in eqn (6.267) we have

$$\mathsf{U}_{\mathrm{CP}}\,\mathscr{L}_{\text{mass}}^{\text{D+M}}\,\mathsf{U}_{\mathrm{CP}}^{-1} = \overline{N_L}\,\Xi_{\mathrm{CP}}^{(N_L)}\,M\,\Xi_{\mathrm{CP}}^{(N_L)}\,\mathcal{C}\,\overline{N_L}^T - N_L^T\,\mathcal{C}^\dagger\,(\Xi_{\mathrm{CP}}^{(N_L)})^*\,M\,(\Xi_{\mathrm{CP}}^{(N_L)})^*\,N_L\,. \tag{6.269}$$

In order to have CP invariance the diagonal matrix of phases $\Xi_{\mathrm{CP}}^{(N_L)}$ must be such that

$$\Xi_{\mathrm{CP}}^{(N_L)}\,M\,\Xi_{\mathrm{CP}}^{(N_L)} = -M\,. \tag{6.270}$$

Let us take

$$\Xi_{\mathrm{CP}}^{(N_L)} = \begin{pmatrix} i & 0 \\ 0 & i \end{pmatrix} = i\,\mathbf{1}, \tag{6.271}$$

which implies that

$$\mathsf{U}_{\mathrm{CP}}\,N_L\,\mathsf{U}_{\mathrm{CP}}^{-1} = i\,\gamma^0\,\mathcal{C}\,\overline{N_L}^T. \tag{6.272}$$

Considering now the CP transformation of the column matrix in eqn (6.235) of chiral left-handed massive neutrino fields, we have

$$\begin{aligned}\mathsf{U}_{\mathrm{CP}}\,n_L\,\mathsf{U}_{\mathrm{CP}}^{-1} &= U^\dagger\,\mathsf{U}_{\mathrm{CP}}\,N_L\,\mathsf{U}_{\mathrm{CP}}^{-1} = i\,U^\dagger\,\gamma^0\,\mathcal{C}\,\overline{N_L}^T \\ &= i\,U^\dagger\,U^*\,\gamma^0\,\mathcal{C}\,\overline{n_L}^T = \Xi_{\mathrm{CP}}^{(n_L)}\,\gamma^0\,\mathcal{C}\,\overline{n_L}^T.\end{aligned} \tag{6.273}$$

The diagonal matrix of CP phases of n_L is

$$\Xi_{\mathrm{CP}}^{(n_L)} = i\,U^\dagger\,U^* = i\,U^\dagger\,U^* = i\left(U^T\,U\right)^* = i\left(\rho\,\mathcal{O}^T\,\mathcal{O}\,\rho\right)^* = i\,\rho^2, \tag{6.274}$$

where we have used the expression in eqn (6.252) for U. Hence, the CP parities of the massive Majorana fields are determined by the phases ρ_k according to eqn (6.266),

$$\xi_k^{\mathrm{CP}} = (\Xi_{\mathrm{CP}}^{(n_L)})_{kk} = i\,\rho_k^2\,. \tag{6.275}$$

Indeed, the transformation of the column matrix

$$n = n_L + n_L^C = n_L + \mathcal{C}\,\overline{n_L}^T \tag{6.276}$$

of massive neutrino fields is

$$\mathsf{U}_{\mathrm{CP}}\, n\, \mathsf{U}_{\mathrm{CP}}^{-1} = \Xi_{\mathrm{CP}}^{(n_L)}\, \gamma^0\, \mathcal{C}\, \overline{n_L}^T - \Xi_{\mathrm{CP}}^{(n_L)^*}\, \gamma^0\, n_L = i\, \rho^2\, \gamma^0\, n\,. \tag{6.277}$$

This result is consistent with the result in section 6.2.6, where we have shown that a Majorana neutrino has imaginary CP parity. Here we have two massive Majorana neutrinos with imaginary CP parities whose sign is determined by ρ_k^2.

The CP transformation

$$\mathsf{U}_{\mathrm{CP}}\, \nu_L\, \mathsf{U}_{\mathrm{CP}}^{-1} = i\, \gamma^0\, \mathcal{C}\, \overline{\nu_L}^T \tag{6.278}$$

implicit in eqn (6.272) does not pose any problem for the CP invariance of weak interactions. For one lepton generation the charged-current weak interaction Lagrangian is given in eqn (6.102),

$$\mathscr{L}_{\mathrm{I,L}}^{\mathrm{CC}} = -\frac{g}{\sqrt{2}} \left(\overline{\nu_L}\, \gamma^\mu\, \ell_L\, W_\mu + \overline{\ell_L}\, \gamma^\mu\, \nu_L\, W_\mu^\dagger \right) . \tag{6.279}$$

Under a CP transformation the Dirac charged lepton field ℓ transforms as given in eqn (6.155),

$$\mathsf{U}_{\mathrm{CP}}\, \ell\, \mathsf{U}_{\mathrm{CP}}^{-1} = \xi_\ell^{\mathrm{CP}}\, \gamma^0\, \mathcal{C}\, \overline{\ell}^T , \tag{6.280}$$

and the gauge boson field W_μ transforms as given in eqn (4.117),

$$\mathsf{U}_{\mathrm{CP}}\, W_\mu\, \mathsf{U}_{\mathrm{CP}}^{-1} = -W^{\mu\dagger} . \tag{6.281}$$

Taking into account the relation in eqn (6.158), we obtain

$$\mathsf{U}_{\mathrm{CP}}\, \mathscr{L}_{\mathrm{I,L}}^{\mathrm{CC}}\, \mathsf{U}_{\mathrm{CP}}^{-1} = -\frac{g}{\sqrt{2}} \left(i\, \xi_\ell^{\mathrm{CP}}\, \overline{\ell_L}\, \gamma^\mu\, \nu_L\, W_\mu^\dagger - i\, \xi_\ell^{\mathrm{CP}*}\, \overline{\nu_L}\, \gamma^\mu\, \ell_L\, W_\mu \right) , \tag{6.282}$$

which coincides with $\mathscr{L}_{\mathrm{I,L}}^{\mathrm{CC}}$ in eqn (6.102) if

$$\xi_\ell^{\mathrm{CP}} = -i\,. \tag{6.283}$$

Therefore, CP invariance of the one-generation Dirac–Majorana mass term implies that the charged lepton has an imaginary CP parity (this result is analogous to that obtained in section 6.2.6 for the simplest case of one massive Majorana neutrino).

6.4.3 Maximal mixing

From the expression in eqn (6.257) for the mixing angle, one can see that the mixing is maximal, i.e. $\vartheta = \pi/4$, for

$$m_L = m_R\,. \tag{6.284}$$

In this case, the eigenvalues in eqn (6.256) of the mass matrix reduce to

$$m'_{2,1} = m_L \pm m_{\mathrm{D}}\,. \tag{6.285}$$

If $m_{\mathrm{D}} < m_L$, m'_1 is positive, $\rho_1^2 = 1$ and the mixing is of the type in eqn (6.262), with $\vartheta = \pi/4$. On the other hand, if $m_{\mathrm{D}} > m_L$, m'_1 is negative, $\rho_1^2 = -1$ and the mixing is of the type in eqn (6.264), with $\vartheta = \pi/4$, leading to

$$\nu_{1L} = \frac{-i}{\sqrt{2}} \left(\nu_L - \nu_R^C \right) , \tag{6.286}$$

$$\nu_{2L} = \frac{1}{\sqrt{2}} \left(\nu_L + \nu_R^C \right) . \tag{6.287}$$

In this case, the two massive Majorana neutrino fields are given by

$$\nu_1 = \nu_{1L} + \nu_{1L}^C = \frac{-i}{\sqrt{2}} \left[(\nu_L + \nu_R) - \left(\nu_L^C + \nu_R^C \right) \right] , \tag{6.288}$$

$$\nu_2 = \nu_{2L} + \nu_{2L}^C = \frac{1}{\sqrt{2}} \left[(\nu_L + \nu_R) + \left(\nu_L^C + \nu_R^C \right) \right] , \tag{6.289}$$

with respective masses

$$m_1 = m_{\rm D} - m_L , \tag{6.290}$$

$$m_2 = m_{\rm D} + m_L , \tag{6.291}$$

and the squared-mass difference

$$\Delta m^2 = m_2^2 - m_1^2 = 4\, m_L\, m_{\rm D} , \tag{6.292}$$

which is important for neutrino oscillations (see chapter 7).

6.4.4 Dirac limit

It is interesting to ask what happens if

$$m_L = m_R = 0 . \tag{6.293}$$

In this case, since the Dirac–Majorana mass term in eqn (6.230) reduces to the Dirac mass term in eqn (6.228), we must obtain a massive Dirac neutrino. Indeed, from eqn (6.285) we have

$$m'_{2,1} = \pm m_{\rm D} . \tag{6.294}$$

Therefore,

$$\rho_1^2 = -1 \qquad m_1 = m_{\rm D} , \tag{6.295}$$

$$\rho_2^2 = +1 \qquad m_2 = m_{\rm D} . \tag{6.296}$$

In other words, the two massive Majorana neutrinos have the same mass and opposite CP parities. In this case, the two Majorana fields ν_1 and ν_2 can be combined to give one Dirac field ν:

$$\nu = \frac{1}{\sqrt{2}} \left(i\nu_1 + \nu_2 \right) = \nu_L + \nu_R . \tag{6.297}$$

Similarly, a Dirac field ν can always be split in two Majorana fields by writing it as

$$\nu = \frac{1}{2} \left[\left(\nu - \nu^C \right) + \left(\nu + \nu^C \right) \right] = \frac{i}{\sqrt{2}} \left(-i \frac{\nu - \nu^C}{\sqrt{2}} \right) + \frac{1}{\sqrt{2}} \left(\frac{\nu + \nu^C}{\sqrt{2}} \right) = \frac{i\nu_1 + \nu_2}{\sqrt{2}} , \tag{6.298}$$

with the two Majorana fields ($\nu_1 = \nu_1^C$, $\nu_2 = \nu_2^C$) given by

$$\nu_1 = \frac{-i}{\sqrt{2}} \left(\nu - \nu^C \right) , \tag{6.299}$$

$$\nu_2 = \frac{1}{\sqrt{2}} \left(\nu + \nu^C \right) . \tag{6.300}$$

Therefore, in general, *a Dirac field is equivalent to two Majorana fields with the same mass and opposite CP parities.*

One can ask which is the CP parity of the Dirac field. Since $\rho_1^2 = -1$ and $\rho_2^2 = 1$, we have $\xi_1^{\mathrm{CP}} = -i$ and $\xi_2^{\mathrm{CP}} = i$, which imply that

$$\nu_1 \xrightarrow{\mathrm{CP}} -i\,\gamma^0\,\nu_1 , \tag{6.301}$$

$$\nu_2 \xrightarrow{\mathrm{CP}} i\,\gamma^0\,\nu_2 . \tag{6.302}$$

Using eqns (6.298)–(6.300), for the Dirac field ν we obtain

$$\nu = \frac{1}{\sqrt{2}} \left(i\nu_1 + \nu_2 \right) \xrightarrow{\mathrm{CP}} i\,\gamma^0 \, \frac{1}{\sqrt{2}} \left(-i\nu_1 + \nu_2 \right) = i\,\gamma^0\,\nu^C . \tag{6.303}$$

Therefore, the Dirac field ν has the definite CP parity $\xi_\nu^{\mathrm{CP}} = i$.

6.4.5 Pseudo-Dirac neutrinos

Another interesting case is that in which

$$|m_L| \, , \, m_R \ll m_{\mathrm{D}} . \tag{6.304}$$

In this case, from eqn (6.256) we have

$$m'_{2,1} \simeq \frac{m_L + m_R}{2} \pm m_{\mathrm{D}} . \tag{6.305}$$

Since m'_1 is negative, we have $\rho_1^2 = -1$ and

$$m_{2,1} \simeq m_{\mathrm{D}} \pm \frac{m_L + m_R}{2} . \tag{6.306}$$

The two massive Majorana neutrinos have opposite CP parities and are almost degenerate in mass, the mass splitting being given by $m_L + m_R \ll m_{2,1}$. These two almost degenerate Majorana neutrinos are usually called *pseudo-Dirac* neutrinos because it is very difficult to distinguish them from a Dirac neutrino, which corresponds to a pair of degenerate Majorana neutrinos, as we have seen in the previous subsection 6.4.4. The best way to reveal pseudo-Dirac neutrinos are active–sterile neutrino oscillations due to the small squared-mass difference

$$\Delta m^2 \simeq m_{\mathrm{D}} \left(m_L + m_R \right) . \tag{6.307}$$

The oscillations occur with practically maximal mixing:

$$\tan 2\vartheta = \frac{2 m_{\mathrm{D}}}{m_R - m_L} \gg 1 \quad \Longrightarrow \quad \vartheta \simeq \pi/4 . \tag{6.308}$$

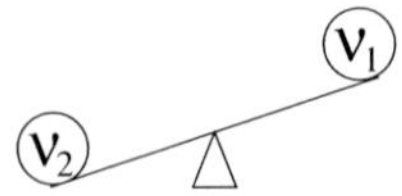

FIG. 6.2. Illustration of the see-saw mechanism: the heavier the mass of ν_2 is, the lighter the mass of ν_1 is.

6.4.6 See-saw mechanism

A very interesting case is

$$m_{\mathrm{D}} \ll m_R \,, \qquad m_L = 0 \,. \tag{6.309}$$

From eqn (6.256) we obtain

$$m_1' \simeq -\frac{m_{\mathrm{D}}^2}{m_R} \,, \qquad m_2' \simeq m_R \,. \tag{6.310}$$

Since m_1' is negative, we have $\rho_1^2 = -1$ and

$$m_1 \simeq \frac{m_{\mathrm{D}}^2}{m_R} \,, \tag{6.311}$$

$$m_2 \simeq m_R \,. \tag{6.312}$$

Therefore, ν_2 is as heavy as m_R and ν_1 is very light, because its mass is suppressed with respect to m_{D} by the small ratio m_{D}/m_R. This is the famous *see-saw mechanism* [807, 1079, 512, 814], which acts as the see-saw depicted in Fig. 6.2: the heavy mass $m_2 \simeq m_R$ of ν_2 is responsible for the lightness of ν_1.

The mixing angle is very small:

$$\tan 2\vartheta = 2\,\frac{m_{\mathrm{D}}}{m_R} \ll 1 \,, \tag{6.313}$$

which implies that ν_1 is composed mainly of active ν_L and ν_2 is composed mainly of sterile ν_R,

$$\nu_{1L} \simeq -i\,\nu_L \,, \qquad \nu_{2L} \simeq \nu_R^C \,. \tag{6.314}$$

The see-saw mechanism is very important, because it provides a very plausible explanation of the smallness of neutrino masses with respect to the masses of the other fermions in the SM, i.e. charged leptons and quarks. The assumption $m_L = 0$ is natural, since a Majorana mass term for the left-handed chiral field ν_L is forbidden by the symmetries and renormalizability of the SM (see section 6.2.7). The Dirac mass m_{D}, which can be generated through the Higgs mechanism of the SM, is expected to be of the order of the charged lepton mass of the same generation or of the order of the up-like quark mass of the same generation. In any case, the order of magnitude of m_{D} cannot be much larger than the electroweak scale, which is of the order of 10^2 GeV. The reason is that a Dirac mass term is forbidden by the unbroken symmetries of the SM. It can arise only as a consequence of symmetry breaking, as for the other particles in the SM. Hence, $m_{\rm D}$ is proportional to the symmetry-breaking scale. This fact is often summarized by saying that $m_{\rm D}$ is *protected* by the symmetries of the SM. On the other hand, since the Majorana mass term in eqn (6.211) is a singlet of the SM symmetries, the Majorana mass m_R of the right-handed chiral neutrino field ν_R is not protected by the SM symmetries. It is plausible that the Majorana mass m_R is generated by new physics beyond the SM and the right-handed chiral neutrino field ν_R belongs to a nontrivial multiplet of the symmetries of the high-energy theory. In this case, the mass m_R is protected by the symmetries of the high-energy theory and its order of magnitude corresponds to the breaking scale of these symmetries, which may be at the grand unification scale of the order of 10^{14}–10^{16} GeV [532, 79, 79, 408]. Hence the see-saw expression in eqn (6.298) may give a light-neutrino mass m_1 which is suppressed with respect to the Dirac mass $m_{\rm D}$ by the small ratio $m_{\rm D}/m_R \sim 10^{-14}$–$10^{-12}$.

We also note that the expression for the light-neutrino Majorana mass m_1 given by the see-saw mechanism has the same structure as that of the Majorana mass in eqn (6.169), obtained with the effective Lagrangian term $\mathscr{L}_5$ in eqn (6.163), if the mass m_R is identified with the high-energy mass scale $\mathcal{M}$. This is not a coincidence, because the see-saw mechanism corresponds to a particular case of the general effective dimension-five operator $\mathscr{L}_5$. This can be seen through the following alternative derivation of the see-saw mechanism [74].

Let us consider the Dirac–Majorana neutrino mass term in eqn (6.230) with $m_L = 0$:

$$\mathscr{L}^{\rm D+M} = -m_{\rm D}\left(\overline{\nu_R}\,\nu_L + \overline{\nu_L}\,\nu_R\right) + \frac{1}{2}\,m_R\left(\nu_R^T\,\mathcal{C}^\dagger\,\nu_R + \nu_R^\dagger\,\mathcal{C}\,\nu_R^*\right) . \qquad (6.315)$$

Above the electroweak symmetry-breaking scale, the symmetries of the SM require that $\mathscr{L}^{\rm D+M}$ is written as

$$\mathscr{L}^{\rm D+M} = -y^\nu\left(\overline{\nu_R}\,\widetilde{\Phi}^\dagger\,L_L + \overline{L_L}\,\widetilde{\Phi}\,\nu_R\right) + \frac{1}{2}\,m_R\left(\nu_R^T\,\mathcal{C}^\dagger\,\nu_R + \nu_R^\dagger\,\mathcal{C}\,\nu_R^*\right) , \qquad (6.316)$$

where y^ν is a Yukawa coupling and $\widetilde{\Phi}$ is the transformed Higgs doublet in eqn (3.156). Below the electroweak symmetry-breaking scale (see section 3.4), the Dirac part of $\mathscr{L}^{\rm D+M}$ generates the Dirac neutrino mass

$$m_{\rm D} = \frac{y^\nu\,v}{\sqrt{2}} , \qquad (6.317)$$

whose expression is analogous to those of the Dirac masses of the charged leptons (eqn (3.135)) and quarks (eqns (3.171) and (3.172)), generated through the same Higgs mechanism.

If the mass m_R is very heavy, at SM energies the right-handed chiral field can be integrated away by considering it in the static limit in which the kinetic term in the equation of motion is neglected:

$$0 \simeq \frac{\partial \mathscr{L}^{\rm D+M}}{\partial \nu_R} = m_R\,\nu_R^T\,\mathcal{C}^\dagger - y^\nu\,\overline{L_L}\,\widetilde{\Phi} . \qquad (6.318)$$

Solving for ν_R, we obtain its value in the static approximation:

$$\nu_R \simeq -\frac{y^\nu}{m_R} \, \widetilde{\Phi}^T \, \mathcal{C} \, \overline{L_L}^T \,. \tag{6.319}$$

Substituting this expression in the Dirac–Majorana mass term in eqn (6.316), we get the dimension-five operator

$$\mathscr{L}_5^{\text{D+M}} \simeq -\frac{1}{2} \frac{(y^\nu)^2}{m_R} \left[(L_L^T \, \tau_2 \, \Phi) \, \mathcal{C}^\dagger \, (\Phi^T \, \tau_2 \, L_L) - (\overline{L_L} \, \tau_2 \, \Phi^*) \, \mathcal{C} \, (\Phi^\dagger \, \tau_2 \, \overline{L_L}^T) \right] \,, \tag{6.320}$$

which coincides with the Lagrangian in eqn (6.163) if

$$g = -\frac{(y^\nu)^2}{2} \,, \qquad \mathcal{M} = m_R \,. \tag{6.321}$$

Below the electroweak symmetry-breaking scale (see section 3.4), $\mathscr{L}_5^{\text{D+M}}$ generates the effective Majorana mass term for the light left-handed chiral neutrino field ν_L

$$\mathscr{L}_5^{\text{D+M}} \simeq -\frac{1}{2} \frac{m_{\text{D}}^2}{m_R} \left(\nu_L^T \, \mathcal{C}^\dagger \, \nu_L + \nu_L^\dagger \, \mathcal{C} \, \nu_L^* \right) \,, \tag{6.322}$$

where we have taken into account of the definition in eqn (6.317) of the Dirac mass m_{D}. Comparing with eqn (6.119), one can see that the sign of the mass in eqn (6.322) is wrong. To remedy this, we define the massive field ν_{1L} as in eqn (6.314),

$$\nu_{1L} \simeq -i \, \nu_L \,, \tag{6.323}$$

which allows one to write $\mathscr{L}_5^{\text{D+M}}$ as

$$\mathscr{L}_5^{\text{D+M}} \simeq \frac{1}{2} \frac{m_{\text{D}}^2}{m_R} \left(\nu_{1L}^T \, \mathcal{C}^\dagger \, \nu_{1L} + \nu_{1L}^\dagger \, \mathcal{C} \, \nu_{1L}^* \right) \,. \tag{6.324}$$

This is a correct Majorana mass term for ν_{1L}, whose mass m_1 is given by the see-saw formula in eqn (6.311),

$$m_1 \simeq \frac{m_{\text{D}}^2}{m_R} \,. \tag{6.325}$$

This procedure shows that the see-saw mechanism is a particular case of the effective Lagrangian approach discussed in section 6.2.7. It is obtained when the dimension-five operator in eqn (6.163) is generated only by the presence of a right-handed chiral neutrino field (in the case of one generation) having a Majorana mass of the order of the high-energy scale $\mathcal{M}$. In general, however, there can be other mechanisms that contribute to the dimension-five effective Lagrangian term in eqn (6.163).

It is also possible to investigate the viability of the see-saw mechanism when the left-handed Majorana mass is small but nonzero. For example, in the so-called *type-II see-saw mechanism* [814] (see Refs. [810, 813]), the mass m_L is generated by the VEV of a Higgs triplet, which is induced by the breaking of a $\text{SU}(2)_L \times$

$\text{SU}(2)_R \times \text{U}(1)_{B-L}$ local gauge symmetry to $\text{SU}(2)_L \times \text{U}(1)_Y$ at a high-energy scale v_R, where v_R is the VEV of another Higgs triplet, and by the symmetry breaking $\text{SU}(2)_L \times \text{U}(1)_Y \to \text{U}(1)_Q$ at the SM electroweak scale v. In this so-called *left-right symmetric model*, we have $m_L \propto v^2/v_R$.

Without considering a specific model, let us consider the general possibility

$$m_L \ll m_{\text{D}} \ll m_R \,, \tag{6.326}$$

with

$$m_L = g \, \frac{m_{\text{D}}^2}{\mathcal{M}} \,, \tag{6.327}$$

where g is a numerical coefficient and $\mathcal{M}$ is a high-energy scale of new physics beyond the SM. Since the Majorana mass m_L violates the conservation of the total lepton number L, if the new physics beyond the SM conserves L (or $B - L$), $\mathcal{M}$ is the scale of L-symmetry (or $B - L$-symmetry) breaking. In this so-called *mixed see-saw* (see Ref. [813]), the masses of ν_1 and ν_2 are given by

$$m_1 \simeq \left| g \, \frac{m_{\text{D}}^2}{\mathcal{M}} - \frac{m_{\text{D}}^2}{m_R} \right| \,, \qquad m_2 \simeq m_R \,. \tag{6.328}$$

Conventionally, one speaks of a *type-II see-saw* when $|m_L| \gg m_{\text{D}}^2/m_R$ whereas the normal see-saw, or *type-I see-saw*, refers to the case in which $|m_L| \ll m_{\text{D}}^2/m_R$. In any case, the light neutrino mass m_1 is strongly suppressed with respect to the Dirac mass m_{D} by the small ratios $m_{\text{D}}/\mathcal{M}$ and m_{D}/m_R. If $m_R \sim \mathcal{M}$, we obtain the normal see-saw type relation

$$m_1 \sim \frac{m_{\text{D}}^2}{\mathcal{M}} \,. \tag{6.329}$$

From eqn (6.257), one can see that the mixing angle is extremely small, as in the normal see-saw case with $m_L = 0$ (see eqn (6.314)). The only difference is that $\nu_1 \simeq \nu_L$ if $m_L m_R > m_{\text{D}}^2$, whereas $\nu_1 \simeq -i \, \nu_L$ if $m_L m_R < m_{\text{D}}^2$.

In conclusion, considering one generation, we have seen that the see-saw mechanism can provide a natural explanation of the smallness of the light neutrino either if $m_L = 0$ or with a small $m_L \neq 0$ generated by new physics beyond the SM.

6.5 Three-generation Dirac–Majorana mixing

It is possible that in addition to the three known active left-handed neutrino fields ν'_{eL}, $\nu'_{\mu L}$, $\nu'_{\tau L}$, there are N_{s} sterile right-handed neutrino fields ν_{sR}, with $s = s_1, \ldots, s_{N_{\text{s}}}$. The notation of the sterile right-handed neutrino fields does not need a prime, because these fields do not take part in weak interactions (on the other hand, as explained in section 6.5.1, the active left-handed neutrino fields need to be redefined in order to diagonalize the leptonic weak charged current). With all these fields, the most general mass term is the Dirac–Majorana mass term

$$\mathscr{L}^{\text{D+M}}_{\text{mass}} = \mathscr{L}^{L}_{\text{mass}} + \mathscr{L}^{R}_{\text{mass}} + \mathscr{L}^{\text{D}}_{\text{mass}} \,, \tag{6.330}$$

with the Majorana mass terms

$$\mathscr{L}^{L}_{\text{mass}} = \frac{1}{2} \sum_{\alpha,\beta=e,\mu,\tau} \nu'^{T}_{\alpha L}\, \mathcal{C}^{\dagger}\, M^{L}_{\alpha\beta}\, \nu'_{\beta L} + \text{H.c.}\,, \tag{6.331}$$

$$\mathscr{L}^{R}_{\text{mass}} = \frac{1}{2} \sum_{s,s'=s_1,\ldots,s_{N_\text{s}}} \nu^{T}_{sR}\, \mathcal{C}^{\dagger}\, M^{R}_{ss'}\, \nu_{s'R} + \text{H.c.}\,, \tag{6.332}$$

and the Dirac mass term

$$\mathscr{L}^{\text{D}}_{\text{mass}} = - \sum_{s=s_1,\ldots,s_{N_\text{s}}} \sum_{\alpha=e,\mu,\tau} \overline{\nu_{sR}}\, M^{\text{D}}_{s\alpha}\, \nu'_{\alpha L} + \text{H.c.} \tag{6.333}$$

All the three mass matrices M^L, M^R, and M^{D} are complex. The Majorana mass matrices M^L and M^R are symmetric, as we have shown in eqn (6.172). The left-handed Majorana mass matrix M^L is a 3×3 square matrix, the right-handed Majorana mass matrix M^R is a $N_\text{s} \times N_\text{s}$ square matrix, and the Dirac mass matrix M^{D} is a $N_\text{s} \times 3$ rectangular matrix.

The neutrino fields with definite masses are obtained through the diagonalization of the Dirac–Majorana mass term in eqn (6.330). It is convenient to define the column matrix of $N = 3 + N_\text{s}$ left-handed fields

$$\boldsymbol{N}'_L \equiv \begin{pmatrix} \boldsymbol{\nu}'_L \\ \boldsymbol{\nu}^C_R \end{pmatrix} , \tag{6.334}$$

with the column matrix of left-handed active neutrinos $\boldsymbol{\nu}'_L$ defined in eqn (6.170) and the column matrix of charge-conjugated right-handed sterile neutrinos

$$\boldsymbol{\nu}^C_R \equiv \begin{pmatrix} \nu^C_{s_1 R} \\ \vdots \\ \nu^C_{s_{N_\text{s}} R} \end{pmatrix} . \tag{6.335}$$

In this way, the Dirac–Majorana mass term in eqn (6.330) can be written in the compact form

$$\mathscr{L}^{\text{D+M}}_{\text{mass}} = \frac{1}{2} \boldsymbol{N}'^{T}_L\, \mathcal{C}^{\dagger}\, M^{\text{D+M}}\, \boldsymbol{N}'_L + \text{H.c.}\,, \tag{6.336}$$

with the $N \times N$ symmetric mass matrix

$$M^{\text{D+M}} \equiv \begin{pmatrix} M^L & M^{\text{D}^T} \\ M^{\text{D}} & M^R \end{pmatrix} . \tag{6.337}$$

Now the diagonalization of the Dirac–Majorana mass term is formally written in a simple form analogous to the diagonalization of the Majorana mass term discussed in section 6.3, because the Dirac–Majorana mass term in eqn (6.336) has the structure of a Majorana mass term, as one can see by comparing it with eqn (6.171). We write the left-handed flavor fields as unitary linear combinations of the left-handed components of N fields with definite mass,

$$\boldsymbol{N}'_L = V_L^\nu \, \boldsymbol{n}_L \,, \qquad \text{with} \qquad \boldsymbol{n}_L = \begin{pmatrix} \nu_{1L} \\ \vdots \\ \nu_{NL} \end{pmatrix} . \tag{6.338}$$

The unitary matrix V_L^ν is chosen in order to diagonalize the symmetric mass matrix $M^{\text{D+M}}$ (see section 6.7.1):

$$(V_L^\nu)^T \, M^{\text{D+M}} \, V_L^\nu = M \,, \qquad \text{where} \qquad M_{kj} = m_k \, \delta_{kj} \qquad (k, j = 1, \ldots, N) \,, \tag{6.339}$$

with real and positive masses m_k. In this way, the Dirac–Majorana mass term in eqn (6.336) can be written in terms of the massive left-handed fields as

$$\mathscr{L}^{\text{D+M}}_{\text{mass}} = \frac{1}{2} \, \boldsymbol{n}_L^T \, \mathcal{C}^\dagger \, M \, \boldsymbol{n}_L + \text{H.c.} = \frac{1}{2} \sum_{k=1}^{N} m_k \, \nu_{kL}^T \, \mathcal{C}^\dagger \, \nu_{kL} + \text{H.c.} \,, \tag{6.340}$$

or, using the charge-conjugated fields, as

$$\mathscr{L}^{\text{D+M}}_{\text{mass}} = -\frac{1}{2} \, \overline{\boldsymbol{n}_L^C} \, M \, \boldsymbol{n}_L + \text{H.c.} = -\frac{1}{2} \sum_{k=1}^{N} m_k \, \overline{\nu_{kL}^C} \, \nu_{kL} + \text{H.c.} \tag{6.341}$$

We can define the column matrix of massive Majorana neutrino fields

$$\boldsymbol{n} = \begin{pmatrix} \nu_1 \\ \vdots \\ \nu_N \end{pmatrix} , \tag{6.342}$$

where

$$\nu_k = \nu_{kL} + \nu_{kL}^C \tag{6.343}$$

are Majorana fields which satisfy the constraint

$$\nu_k^C = \nu_k \,. \tag{6.344}$$

Using these Majorana fields, we can write the free neutrino Lagrangian as

$$\mathscr{L}^{\text{D+M}} = \frac{1}{2} \sum_{k=1}^{N} \overline{\nu_k} \left(i \overleftrightarrow{\partial\!\!\!/} - m_k \right) \nu_k = \overline{\boldsymbol{n}} \left(i \overleftrightarrow{\partial\!\!\!/} - M \right) \boldsymbol{n} \,. \tag{6.345}$$

Let us emphasize that a general result of the diagonalization of a Dirac–Majorana mass term is that massive neutrinos are Majorana particles.

6.5.1 Weak interactions

The mixing of active and sterile neutrino fields in eqn (6.338) has important consequences for weak interactions.

Let us first examine the leptonic part of the charged-current weak interaction Lagrangian in eqn (3.76). The leptonic charged-current is given by eqn (3.80), which can be written as

$$j^{\rho}_{W,L} = 2\,\overline{\boldsymbol{\nu}'_L}\,\gamma^{\rho}\,\boldsymbol{\ell}'_L\,, \tag{6.346}$$

with the column matrix $\boldsymbol{\nu}'_L$ of active left-handed neutrino fields in eqn (6.170) and the column matrix $\boldsymbol{\ell}'_L$ of the left-handed components of the charged lepton fields in eqn (3.127). From eqns (3.131) and (6.338), the leptonic charged-current written in terms of massive fields is

$$j^{\rho}_{W,L} = 2\,\overline{\boldsymbol{n}_L}\,U^{\dagger}\,\gamma^{\rho}\,\boldsymbol{\ell}_L\,, \tag{6.347}$$

where U is the mixing matrix with components

$$U_{\alpha k} = \sum_{\beta=e,\mu,\tau} (V_L^{\ell\,\dagger})_{\alpha\beta}\,(V_L^{\nu})_{\beta k}\,. \tag{6.348}$$

The matrix V_L^{ℓ} that diagonalizes the charged lepton mass matrix is a 3×3 unitary matrix, whereas the matrix V_L^{ν} that diagonalizes the Dirac–Majorana mass matrix is an $N \times N$ unitary matrix. It follows that the mixing matrix U given by the product in eqn (6.348) is a rectangular $3 \times N$ matrix. It is not unitary, because although

$$U\,U^{\dagger} = \mathbf{1}\,, \tag{6.349}$$

we have

$$U^{\dagger}\,U \neq \mathbf{1}\,. \tag{6.350}$$

Although the expression in eqn (6.347) of the leptonic charged-current is formally equal to that in eqn (6.12) for the Dirac case and to that in eqn (6.184) for the mixing of three Majorana neutrinos, the Dirac–Majorana case under consideration is very different, because the mixing matrix is a $3 \times N$ matrix which connects the three active flavor neutrinos to N massive neutrinos. Keeping in mind this warning, it is conventional to define the left-handed flavor neutrino fields as in eqns (6.13) and (6.186),

$$\boldsymbol{\nu}_L = U\,\boldsymbol{n}_L = V_L^{\ell\,\dagger}\,\boldsymbol{\nu}'_L\,, \qquad \text{with} \qquad \boldsymbol{\nu}_L = \begin{pmatrix} \nu_{eL} \\ \nu_{\mu L} \\ \nu_{\tau L} \end{pmatrix}. \tag{6.351}$$

In this way, the leptonic weak charged current can be written as in the SM eqn (3.141):

$$j^{\rho}_{W,L} = 2\,\overline{\boldsymbol{\nu}_L}\,\gamma^{\rho}\,\boldsymbol{\ell}_L = 2\sum_{\alpha=e,\mu,\tau} \overline{\nu_{\alpha L}}\,\gamma^{\rho}\,\ell_{\alpha L}\,. \tag{6.352}$$

Since the parameterization of the rectangular mixing matrix U is rather complicated, it is discussed in separately in section 6.7.3. Here, we remark only that the rectangular $3\times N$ mixing matrix U can be parameterized in terms of $3+3\,N_{\mathrm{s}}$ mixing angles and $3+3\,N_{\rm s}$ phases, which are divided into $1+2\,N_{\rm s}$ Dirac phases and $2+N_{\rm s}$ Majorana phases.

The mixing of active and sterile neutrinos is explicitly given by

$$\nu_{\alpha L} = \sum_{k=1}^{N} U_{\alpha k}\,\nu_{kL} \qquad (\alpha = e, \mu, \tau)\,, \tag{6.353}$$

$$\nu_{sR}^{C} = \sum_{k=1}^{N} (V_L^{\nu})_{sk}\,\nu_{kL} \qquad (s = s_1, \ldots, s_{N_{\rm s}})\,. \tag{6.354}$$

It is possible to write these mixing relations in the compact matrix form

$$\boldsymbol{N}_L = \mathcal{U}\,\boldsymbol{n}_L\,, \tag{6.355}$$

where we have defined

$$\boldsymbol{N}_L \equiv \begin{pmatrix} \boldsymbol{\nu}_L \\ \boldsymbol{\nu}_R^C \end{pmatrix}, \qquad \mathcal{U} \equiv \begin{pmatrix} U \\ V_L^{\nu}|_{N_{\rm s}\times N} \end{pmatrix} = \begin{pmatrix} V_L^{\ell\dagger}\, V_L^{\nu}|_{3\times N} \\ V_L^{\nu}|_{N_{\rm s}\times N} \end{pmatrix}. \tag{6.356}$$

Here $V_L^{\nu}|_{3\times N}$ and $V_L^{\nu}|_{N_{\rm s}\times N}$ are, respectively, the $3\times N$ and $N_{\rm s}\times N$ matrices obtained by taking the first three rows and the remaining $N_{\rm s}$ rows of V_L^{ν}. Note that the $N\times N$ matrix $\mathcal{U}$ is unitary.

As discussed in section 6.4.1 for the one-generation case, since active and sterile neutrino fields are linear combinations of the same massive neutrino fields, oscillations between active and sterile states are possible. This can also be seen by noting that the nonunitarity of the mixing matrix U implies that the total probability of active flavors is not conserved.

Let us now consider the neutrino part of the neutral-current weak interaction Lagrangian. From eqns (3.92) and (6.351), the neutrino neutral current is given by

$$j_{Z,\nu}^{\rho} = \overline{\boldsymbol{\nu}_L'}\,\gamma^{\rho}\,\boldsymbol{\nu}_L' = \overline{\boldsymbol{\nu}_L}\,\gamma^{\rho}\,\boldsymbol{\nu}_L\,, \tag{6.357}$$

since V_L^{ℓ} is unitary. However, the nonunitarity of U (eqn (6.350)) implies that the GIM mechanism does not work. In fact, the neutrino neutral current is not diagonal in the massive fields:

$$j_{Z,\nu}^{\rho} = \overline{\boldsymbol{n}_L}\,\gamma^{\rho}\,U^{\dagger}\,U\,\boldsymbol{n}_L\,. \tag{6.358}$$

This means that in the general Dirac–Majorana case it is possible to have neutral-current transitions among different massive neutrinos.

6.5.2 See-saw mechanism

The see-saw mechanism presented in section 6.4.6 for the one-generation case is generalized to three generations by considering the Dirac–Majorana mass matrix in eqn (6.337) with $M^L = 0$,

$$M^{\rm D+M} = \begin{pmatrix} 0 & M^{{\rm D}^T} \\ M^{\rm D} & M^R \end{pmatrix}, \tag{6.359}$$

and a right-handed Majorana mass matrix M^R with elements which are much larger than the elements of the Dirac mass matrix $M^{\rm D}$. The physical justifications of these assumptions are the same as those in the one-generation case: M^L is forbidden by the SM symmetries and M^R is generated by new physics beyond the SM. It is clear, however, that the three-generation see-saw mechanism is more complicated than the one-generation case due to the matrix character of the entries in eqn (6.359).

If all the eigenvalues of M^R are much larger than all the elements of M^{D}, the mass matrix can be diagonalized by blocks [654, 934], up to corrections of the order $(M^R)^{-1}M^{\text{D}}$:

$$W^T \, M^{\text{D+M}} \, W \simeq \begin{pmatrix} M_{\text{light}} & 0 \\ 0 & M_{\text{heavy}} \end{pmatrix} , \tag{6.360}$$

with

$$W \simeq \begin{pmatrix} 1 - \frac{1}{2}\, M^{\text{D}\dagger}(M^R M^{R\dagger})^{-1} M^{\text{D}} & [(M^R)^{-1} M^{\text{D}}]^\dagger \\ -(M^R)^{-1} M^{\text{D}} & 1 - \frac{1}{2}\,(M^R)^{-1} M^{\text{D}} M^{\text{D}\dagger}(M^{R\dagger})^{-1} \end{pmatrix} . \tag{6.361}$$

The light 3×3 mass matrix M_{light} and the heavy $N_{\text{s}} \times N_{\text{s}}$ mass matrix M_{heavy} are given by

$$M_{\text{light}} \simeq -M^{\text{D}^T} \, (M^R)^{-1} \, M^{\text{D}} \,, \qquad M_{\text{heavy}} \simeq M^R \,. \tag{6.362}$$

The heavy masses are given by the eigenvalues of M^R, whereas the light masses are given by the eigenvalues of M_{light}, whose elements are suppressed with respect to the elements of the Dirac mass matrix M^{D} by the small matrix factor $M^{\text{D}^T}(M^R)^{-1}$. However, the values of the light neutrino masses and their relative sizes can vary over wide ranges, depending on the specific values of the elements of M^{D} and M^R. The following two simple possibilities are often considered in the literature (see Refs. [251, 250]):

Quadratic See-Saw. This is the case with

$$M^R = \mathcal{M}\, I \,, \tag{6.363}$$

where I is the $N_{\text{s}} \times N_{\text{s}}$ identity matrix and $\mathcal{M}$ is the high-energy scale of new physics beyond the SM in which the total lepton number is violated. In this case, we have

$$M_{\text{light}} \simeq - \frac{M^{\text{D}^T} M^{\text{D}}}{\mathcal{M}} \,. \tag{6.364}$$

The light neutrino masses are given by

$$m_k = \frac{(m_k^{\text{D}})^2}{\mathcal{M}} \qquad (k = 1, 2, 3) \,, \tag{6.365}$$

where $(m_k^{\text{D}})^2$ are the three eigenvalues of the 3×3 matrix $M^{\text{D}^T} M^{\text{D}}$. Since the Dirac mass matrix M^{D} is generated by the SM Higgs mechanism, m_k^{D} are expected to be of the order of the charged lepton masses or of the order of the up-like quark masses. The large energy scale in the denominator of the right-hand side of eqn (6.365) suppresses the light neutrino masses with respect to the masses of the charged leptons and quarks as in the one-generation see-saw mechanism (see eqn (6.311)).

The case under consideration is called the *quadratic see-saw*, because the light neutrino masses m_k scale as the squares of the masses m_k^{D}:

$$m_1 : m_2 : m_3 = (m_1^{\mathrm{D}})^2 : (m_2^{\mathrm{D}})^2 : (m_3^{\mathrm{D}})^2 \,. \tag{6.366}$$

Linear See-Saw. If $N_{\mathrm{s}} = 3$, it is possible to have

$$M^R = \frac{\mathcal{M}}{\mathcal{M}^{\mathrm{D}}} \, M^{\mathrm{D}} \,, \tag{6.367}$$

where $\mathcal{M}^{\mathrm{D}}$ is the scale of the elements of M^{D} and $\mathcal{M} \gg \mathcal{M}^{\mathrm{D}}$ is the high-energy scale of new physics beyond the SM in which the total lepton number is violated. Since the Dirac mass matrix M^{D} is generated by the SM Higgs mechanism, it is expected that $\mathcal{M}^{\mathrm{D}}$ is the energy scale of the SM electroweak symmetry breaking, i.e. $\mathcal{M}^{\mathrm{D}} \sim 10^2$ GeV. From eqn (6.362), we obtain

$$M_{\mathrm{light}} \simeq -\frac{\mathcal{M}^{\mathrm{D}}}{\mathcal{M}} \, M^{\mathrm{D}} \,. \tag{6.368}$$

The light neutrino masses are given by

$$m_k = \frac{\mathcal{M}^{\mathrm{D}}}{\mathcal{M}} \, m_k^{\mathrm{D}} \qquad (k = 1, 2, 3) \,, \tag{6.369}$$

where m_k^{D} are the eigenvalues of M^{D}. Hence, the light neutrino masses are suppressed by the small ratio $\mathcal{M}^{\mathrm{D}}/\mathcal{M}$ with respect to the masses m_k^{D}, which are expected to be of the order of the charged lepton masses or of the order of the up-like quark masses.

The name *linear see-saw* follows from the proportionality of the light neutrino masses m_k to the masses m_k^{D}:

$$m_1 : m_2 : m_3 = m_1^{\mathrm{D}} : m_2^{\mathrm{D}} : m_3^{\mathrm{D}} \,. \tag{6.370}$$

6.6 Special cases

In this section we discuss briefly two special cases which are interesting for neutrino oscillations studies (see sections 13.1.3 and 13.1.4): three-neutrino trimaximal mixing in subsection 6.6.1 and three-neutrino bilarge mixing in subsection 6.6.2.

6.6.1 Three-neutrino trimaximal mixing

Trimaximal mixing means that the absolute values of all the elements of the mixing matrix are equal. The unitarity relations

$$\sum_k |U_{\alpha k}|^2 = 1 \,, \quad \sum_\alpha |U_{\alpha k}|^2 = 1 \tag{6.371}$$

imply that the absolute values of all the elements of the mixing matrix must be equal to $1/\sqrt{3}$:

$$|U_{\alpha k}| = \frac{1}{\sqrt{3}} \,. \tag{6.372}$$

However, as discussed in Refs. [831, 1066], it is not possible to construct a real mixing matrix with this requirement. This means that CP violation is necessary in order to have trimaximal mixing. Using the parameterization in eqn (6.191) of the Dirac part of the mixing matrix, trimaximal mixing is obtained for $\vartheta_{12} = \vartheta_{23} = \pi/4$, $s_{13} = 1/\sqrt{3}$ and $\sin\delta_{13} = \pm 1$:

$$U^{\mathrm{D}} = \begin{pmatrix} \frac{1}{\sqrt{3}} & \frac{1}{\sqrt{3}} & \mp\frac{i}{\sqrt{3}} \\ -\frac{1}{2} \mp \frac{i}{2\sqrt{3}} & \frac{1}{2} \mp \frac{i}{2\sqrt{3}} & \frac{1}{\sqrt{3}} \\ \frac{1}{2} \mp \frac{i}{2\sqrt{3}} & -\frac{1}{2} \mp \frac{i}{2\sqrt{3}} & \frac{1}{\sqrt{3}} \end{pmatrix} = \frac{1}{\sqrt{3}} \begin{pmatrix} 1 & 1 & \mp i \\ e^{\mp i5\pi/6} & e^{\mp i\pi/6} & 1 \\ e^{\mp i\pi/6} & e^{\mp i5\pi/6} & 1 \end{pmatrix}. \tag{6.373}$$

In particular, the Jarlskog invariant in eqn (6.190) has its maximum absolute value,

$$J = \pm\frac{1}{6\sqrt{3}} \quad \Longrightarrow \quad |J| = \frac{1}{6\sqrt{3}} = |J|_{\max} \tag{6.374}$$

(see eqn (4.165)). Therefore, according to the definition proposed in Ref. [395], in trimaximal mixing there is maximal CP violation.

By rephasing appropriately the charged lepton and neutrino fields, the trimaximal mixing matrix in eqn (6.373) can be transformed to a more convenient form with elements given by $1/\sqrt{3}$ times one of the three cubic roots of one (1, $e^{\pm i2\pi/3}$). Extracting a phase $\pm i$ from the μ, τ, and ν_3 fields (i.e. multiplying the second and third rows by $\mp i$ and the third column by $\pm i$) we obtain

$$U^{\mathrm{D}} = \frac{1}{\sqrt{3}} \begin{pmatrix} 1 & 1 & 1 \\ e^{\pm i2\pi/3} & e^{\mp i2\pi/3} & 1 \\ e^{\mp i2\pi/3} & e^{\pm i2\pi/3} & 1 \end{pmatrix}. \tag{6.375}$$

If ν_3 is a Majorana particle, the phase $\pm i$ can be extracted from the Majorana phase $e^{i\lambda_3}$, which changes its value accordingly. Furthermore, by extracting a phase $e^{\pm i2\pi/3}$ from the e, μ, τ, and ν_3 fields and a phase $e^{\mp i2\pi/3}$ from the ν_2 field (or a phase $e^{\pm i2\pi/3}$ from $e^{i\lambda_3}$ and a phase $e^{\mp i2\pi/3}$ from $e^{i\lambda_2}$ if ν_3 and ν_2 are Majorana particles), we obtain

$$U^{\mathrm{D}} = \frac{1}{\sqrt{3}} \begin{pmatrix} e^{\mp i2\pi/3} & e^{\pm i2\pi/3} & 1 \\ 1 & 1 & 1 \\ e^{\pm i2\pi/3} & e^{\mp i2\pi/3} & 1 \end{pmatrix}. \tag{6.376}$$

One can easily see that the two physically equivalent mixing matrices in eqns (6.375) and (6.376) can be obtained by a cyclic permutation of the rows of the mixing matrix, which is equivalent to the cyclic permutation of the flavors $e, \mu, \tau \to \mu, \tau, e$. It is clear that by similar reasoning one can show that the following cyclic permutation of the flavors $\mu, \tau, e \to \tau, e, \mu$ yields the physically equivalent mixing matrix

$$U^{\mathrm{D}} = \frac{1}{\sqrt{3}} \begin{pmatrix} e^{\pm i2\pi/3} & e^{\mp i2\pi/3} & 1 \\ e^{\mp i2\pi/3} & e^{\pm i2\pi/3} & 1 \\ 1 & 1 & 1 \end{pmatrix}, \tag{6.377}$$

and it is also possible to obtain physically equivalent mixing matrices by cyclic permutations of the columns, which is equivalent to cyclic permutations of the massive neutrinos $\nu_1, \nu_2, \nu_3 \leftrightarrows \nu_2, \nu_3, \nu_1 \leftrightarrows \nu_3, \nu_1, \nu_2$. Using these properties one can easily see that the exchange of two columns or the exchange of two rows, which is equivalent to the operation of Hermitian conjugation, inverts all the phases in the trimaximal mixing matrix. Therefore, it exchanges the two inequivalent mixing matrices with opposite Jarlskog invariant in eqn (6.374).

Summarizing, a trimaximal Dirac mixing matrix is physically invariant under the permutations $e, \mu, \tau \leftrightarrows \mu, \tau, e \leftrightarrows \tau, e, \mu$ and $\nu_1, \nu_2, \nu_3 \leftrightarrows \nu_2, \nu_3, \nu_1 \leftrightarrows \nu_3, \nu_1, \nu_2$, whereas an exchange $e \leftrightarrows \mu$ or $\mu \leftrightarrows \tau$ or $\tau \leftrightarrows e$ or $\nu_1 \leftrightarrows \nu_2$ or $\nu_2 \leftrightarrows \nu_3$ or $\nu_3 \leftrightarrows \nu_1$ transforms the two inequivalent trimaximal Dirac mixing matrices with opposite Jarlskog invariant one into the other. These properties have important consequences for the symmetries of neutrino oscillation probabilities, which will be discussed in section 13.1.3.

6.6.2 Three-neutrino bilarge mixing

We have three-neutrino bilarge mixing if all the elements of the mixing matrix are large, except one. In this case, the small element of the mixing matrix can be neglected in practice and it is possible to express the mixing matrix in terms of only two mixing angles.

In general, in three-neutrino mixing the Jarlskog invariant is zero if any of the elements of the mixing matrix is zero. This follows obviously from the relation in eqn (6.27). Therefore, in the case of three-neutrino bilarge mixing the Jarlskog invariant is very small and it is very difficult to observe Dirac-type CP violations in the lepton sector.

As a useful example, we consider the case in which $U_{e3} = 0$, which is a realistic approximation of the true mixing matrix, as we will see in chapter 13. In this case, in the parameterization in eqn (6.191) we have $\vartheta_{13} = 0$ and the Dirac part of the mixing matrix reduces to

$$U^{\mathrm{D}} = \begin{pmatrix} c_{12} & s_{12} & 0 \\ -s_{12}c_{23} & c_{12}c_{23} & s_{23} \\ s_{12}s_{23} & -c_{12}s_{23} & c_{23} \end{pmatrix} . \tag{6.378}$$

Since this matrix is real, it is clear that it cannot induce any CP violation effect. In section 13.1.4 we will discuss the implications of the mixing matrix in eqn (6.378) for neutrino oscillations in vacuum.

6.7 Majorana mass matrix

In this section we discuss the properties of a symmetric $N \times N$ Majorana mass matrix M^{M}. In subsection 6.7.1 we discuss the diagonalization of the mass matrix. In subsection 6.7.2 we discuss the parameterization of the mixing matrix in the case of three-neutrino mixing introduced in section 6.3. In subsection 6.7.3 we discuss the parameterization of the mixing matrix in the general case of N-neutrino mixing with $N > 3$, which can be applied to the case of Dirac–Majorana mixing discussed in section 6.5.

6.7.1 Diagonalization

A symmetric complex $N \times N$ Majorana mass matrix M^{M} can be diagonalized with the transformation

$$(V_L^\nu)^T \, M^{\mathrm{M}} \, V_L^\nu = M \,, \quad \text{with} \quad M_{kj} = m_k \, \delta_{kj} \quad (k,j = 1,\ldots,N) \,, \tag{6.379}$$

where V_L^ν is a unitary matrix and m_k are real and positive masses (see Ref. [235]). The diagonalization equation (6.379) corresponds to eqn (6.175) with $M^{\mathrm{M}} = M^L$ in the case of mixing of three Majorana neutrinos, and to eqn (6.339) with $M^{\mathrm{M}} = M^{\mathrm{D+M}}$ in the case of Dirac–Majorana mixing.

Let us first note that the number of available parameters for the diagonalization is sufficient: the $N \times N$ symmetric complex matrix M^{M} is determined by $N(N+1)$ independent real parameters, which is the same amount as the N^2 independent real parameters in the $N \times N$ unitary matrix V_L^ν plus the N independent real elements of the diagonal $N \times N$ matrix M.

The proof of the diagonalization in eqn (6.7.1) makes use of the fact that, as shown in subsection 4.1, an arbitrary complex matrix can be diagonalized by the biunitary transformation

$$V^\dagger \, M^{\mathrm{M}} \, W = M \,, \qquad \text{with} \qquad M_{kj} = m_k \, \delta_{kj} \qquad (k,j = 1,\ldots,N) \,, \tag{6.380}$$

where V and W are unitary matrices. Therefore,

$$M^{\mathrm{M}} = V \, M \, W^\dagger \,, \tag{6.381}$$

from which we have

$$M^{\mathrm{M}} \, (M^{\mathrm{M}})^\dagger = V \, M \, W^\dagger \, W M^\dagger \, V^\dagger = V \, M^2 \, V^\dagger \,, \tag{6.382}$$

and

$$(M^{\mathrm{M}})^T \, ((M^{\mathrm{M}})^T)^\dagger = (W^\dagger)^T \, M \, V^T \, (V^T)^\dagger \, M^\dagger \, W^T = (W^\dagger)^T \, M^2 \, W^T \,. \tag{6.383}$$

Since M^{M} is symmetric, the two expressions in eqns (6.382) and (6.383) are equal:

$$V \, M^2 \, V^\dagger = (W^\dagger)^T \, M^2 \, W^T \,. \tag{6.384}$$

Multiplying on the left by W^T and on the right by V, we see that $W^T \, V$ commutes with M^2,

$$W^T \, V \, M^2 = M^2 \, W^T \, V \,. \tag{6.385}$$

Since $W^T \, V$ is unitary, it must be a diagonal matrix of phases:

$$W^T \, V = D \qquad \text{with} \qquad D_{kj} = e^{i\varphi_k} \, \delta_{kj} \,, \tag{6.386}$$

with real numbers φ_k. Using eqns (6.381) and (6.386), M^{M} can be written as

$$M^{\mathrm{M}} = V \, M \, W^\dagger = (W^T)^\dagger \, W^T \, V \, M \, W^\dagger = (W^\dagger)^T \, D \, M \, W^\dagger \,. \tag{6.387}$$

Since D is diagonal, also $D^{1/2}$ is diagonal and commutes with the diagonal matrix M, leading to

$$M^{\mathrm{M}} = (W^{\dagger})^{T}\, D^{1/2}\, M\, D^{1/2}\, W^{\dagger} = (D^{1/2}\, W^{\dagger})^{T}\, M\, D^{1/2}\, W^{\dagger} \,. \tag{6.388}$$

It is clear now that the diagonalization in eqn (6.379) is achieved taking

$$V_L^{\nu\dagger} = D^{1/2}\, W^{\dagger} \qquad \Longrightarrow \qquad V_L^{\nu} = W\, D^{-1/2} \,. \tag{6.389}$$

Let us note that if M^{M} is real,

$$M^{\mathrm{M}} = M^{\mathrm{M}*} \,, \tag{6.390}$$

the diagonalization in eqn (6.380) is performed with real orthogonal matrices

$$V = V^{*} \,, \quad V^{T} = V^{-1} \,, \qquad W = W^{*} \,, \quad W^{T} = W^{-1} \,. \tag{6.391}$$

In this case, the diagonal matrix D is real with elements equal to ± 1. From eqn (6.389) it follows that in this case the matrix V_L^{ν} is given by a real orthogonal matrix W multiplied by a diagonal matrix $D^{-1/2}$ with elements equal to ± 1 or $\pm i$. Hence, the columns of the matrix V_L^{ν} are either real or imaginary.

In order to find the masses m_k, let us multiply eqn (6.379) on the left by $V_L^{\nu *}$ to get

$$M^{\mathrm{M}} V_L^{\nu} = V_L^{\nu *} M \,. \tag{6.392}$$

Let us define the vectors $v^{(j)}$ with components

$$v_k^{(j)} \equiv [V_L^{\nu}]_{kj} \,. \tag{6.393}$$

In other words, the vectors $v^{(j)}$ are the columns of the unitary matrix V_L^{ν}. The unitarity relation $V_L^{\nu\dagger} V_L^{\nu} = \mathbf{1}$ implies the orthonormality relation

$$\sum_k v_k^{(i)*}\, v_k^{(j)} = \delta_{ij} \,. \tag{6.394}$$

Using the vectors $v^{(j)}$, eqn (6.392) can be written as

$$M^{\mathrm{M}} v^{(j)} = m_j v^{(j)*} \,. \tag{6.395}$$

Let us separate the real and imaginary parts of eqn (6.395):

$$\left(\mathfrak{Re}\left[M^{\mathrm{M}}\right] + i\,\mathfrak{Im}\left[M^{\mathrm{M}}\right]\right) \left(\mathfrak{Re}\left[v^{(j)}\right] + i\,\mathfrak{Im}\left[v^{(j)}\right]\right) = m_j \left(\mathfrak{Re}\left[v^{(j)}\right] - i\,\mathfrak{Im}\left[v^{(j)}\right]\right) , \tag{6.396}$$

which can be written as

$$\begin{pmatrix} \mathfrak{Re}\left[M^{\mathrm{M}}\right] & -\mathfrak{Im}\left[M^{\mathrm{M}}\right] \\ -\mathfrak{Im}\left[M^{\mathrm{M}}\right] & -\mathfrak{Re}\left[M^{\mathrm{M}}\right] \end{pmatrix} \begin{pmatrix} \mathfrak{Re}\left[v^{(j)}\right] \\ \mathfrak{Im}\left[v^{(j)}\right] \end{pmatrix} = m_j \begin{pmatrix} \mathfrak{Re}\left[v^{(j)}\right] \\ \mathfrak{Im}\left[v^{(j)}\right] \end{pmatrix} . \tag{6.397}$$

This is a real eigenvalue equation for the symmetric matrix

$$\mathfrak{M} = \begin{pmatrix} \mathfrak{Re}\left[M^{\mathrm{M}}\right] & -\mathfrak{Im}\left[M^{\mathrm{M}}\right] \\ -\mathfrak{Im}\left[M^{\mathrm{M}}\right] & -\mathfrak{Re}\left[M^{\mathrm{M}}\right] \end{pmatrix} , \tag{6.398}$$

which can be solved numerically. The eigenvalues of $\mathfrak{M}$ are the masses and the eigenvectors give the real and imaginary parts of the columns of the unitary diagonalizing matrix V_L^{ν}.

Let us note that, if the dimension of M^{M} is $N \times N$, the eigenvalue equation (6.397) has $2N$ solutions. However, if m_j is the eigenvalue of eqn (6.397), also $-m_j$ is an eigenvalue: using the relations obtained with eqn (6.397) it is straightforward to obtain

$$\begin{pmatrix} \mathfrak{Re}\left[M^{\mathrm{M}}\right] & -\mathfrak{Im}\left[M^{\mathrm{M}}\right] \\ -\mathfrak{Im}\left[M^{\mathrm{M}}\right] & -\mathfrak{Re}\left[M^{\mathrm{M}}\right] \end{pmatrix} \begin{pmatrix} -\mathfrak{Im}\left[v^{(j)}\right] \\ \mathfrak{Re}\left[v^{(j)}\right] \end{pmatrix} = -m_j \begin{pmatrix} -\mathfrak{Im}\left[v^{(j)}\right] \\ \mathfrak{Re}\left[v^{(j)}\right] \end{pmatrix} . \qquad (6.399)$$

Therefore, there are correctly only N positive mass eigenvalues whose corresponding eigenvectors give the real and imaginary parts of the columns of the matrix V_L^{ν}.

6.7.2 Three-neutrino mixing

In this section we consider the case of mixing of three Majorana neutrinos introduced in section 6.3, for which $N = 3$. The leptonic mixing matrix U is given by eqn (6.185),

$$U = V_L^{\ell\dagger} \, V_L^{\nu} \,, \qquad (6.400)$$

where V_L^{ℓ} is the unitary matrix that participates in the diagonalization of the charged lepton mass matrix in eqn (3.129).

In general, a unitary $N \times N$ matrix depends on N^2 independent real parameters, which can be divided as in eqns (4.11) and (4.12): $N(N-1)/2$ angles and $N(N+1)/2$ phases. However, as in the case of CKM quark mixing discussed in section 4.2, not all the phases have physical meaning. The mixing matrix enters only in the leptonic charged-current weak interaction Lagrangian in eqn (6.194)

$$\mathscr{L}_{\mathrm{I,L}}^{(\mathrm{CC})} = -\frac{g}{\sqrt{2}} \sum_{\alpha=e,\mu,\tau} \sum_{k=1}^{3} \overline{\ell_{\alpha L}} \gamma^{\rho} U_{\alpha k} \nu_{kL} W_{\rho}^{\dagger} + \mathrm{H.c.} = -\frac{g}{\sqrt{2}} \overline{\boldsymbol{\ell}_L} \gamma^{\rho} U \boldsymbol{n}_L W_{\rho}^{\dagger} + \mathrm{H.c.} \,, \qquad (6.401)$$

with the array $\boldsymbol{\ell}_L$ of left-handed charged lepton fields in eqn (3.131) and the array $\boldsymbol{n}_L$ of left-handed massive Majorana fields in eqn (6.176). Since the other terms in the Lagrangian are invariant under a global phase shift of the charged lepton fields, $N = 3$ phases of the mixing matrix can be eliminated. The neutrino fields cannot be rephased, because the Majorana mass term is not invariant under such transformation, as shown in eqn (6.120). Therefore, the number of physical parameters in the mixing matrix is

$$N^2 - N = N\,(N-1) = 6 \,, \qquad (6.402)$$

which can be equally divided into

$$\frac{N\,(N-1)}{2} = 3 \qquad \text{mixing angles} \,, \qquad (6.403)$$

$$\frac{N\,(N-1)}{2} = 3 \qquad \text{physical phases} \,. \qquad (6.404)$$

Since in the case of Dirac neutrinos $N - 1 = 2$ of the phases in eqn (6.404) can be further eliminated by appropriate rephasing of the neutrino fields, in analogy to the case of quark mixing discussed in section 4.2, in general, the mixing matrix U can be written as in eqn (6.189),

$$U = U^{\mathrm{D}}\, D^{\mathrm{M}}\,, \tag{6.405}$$

where U^{D} is a Dirac-like mixing matrix with

$$\frac{(N-1)\,(N-2)}{2} = 1 \qquad \text{Dirac phase} \tag{6.406}$$

(see eqn (4.18)), and D^{M} is a diagonal matrix with

$$N - 1 = 2 \qquad \text{Majorana phases}\,. \tag{6.407}$$

For the parameterization of the 3×3 leptonic mixing matrix we can follow a method similar to that described in section 4.3.2 for the parameterization of the 3×3 CKM mixing matrix of quarks. We can start with the parameterization of a 3×3 unitary mixing matrix in eqn (4.64), which can be written as

$$U = D^L\, R^{23}\, W^{13}\, R^{12}\, D^R\,, \tag{6.408}$$

with the new definitions

$$D^L \equiv \operatorname{diag}\!\left(e^{i\omega_1}\,,\, e^{i(\omega_2-\eta_{12})}\,,\, e^{i(\omega_3-\eta_{12}-\eta_{23})}\right)\,, \tag{6.409}$$

$$D^R \equiv \operatorname{diag}\!\left(1\,,\, e^{i\eta_{12}}\,,\, e^{i(\eta_{12}+\eta_{23})}\right)\,. \tag{6.410}$$

Five phases have been factorized out of the matrix product $R^{23}W^{13}R^{12}$: the three phases ω_1, $\omega_2 - \eta_{12}$ and $\omega_3 - \eta_{12} - \eta_{23}$ on the left and the two phases η_{12} and $\eta_{12} + \eta_{23}$ on the right.

Considering the leptonic charged-current weak interaction Lagrangian in eqn (6.401), we make a change of phase of the charged lepton fields

$$\ell_{\alpha L} \to e^{i\psi_\alpha}\, \ell_{\alpha L}\,, \tag{6.411}$$

with

$$\psi_e = \omega_1\,, \tag{6.412}$$

$$\psi_\mu = \omega_2 - \eta_{12}\,, \tag{6.413}$$

$$\psi_\tau = \omega_3 - \eta_{12} - \eta_{23}\,, \tag{6.414}$$

which eliminate the three phases in D^L. Hence, we can write the leptonic mixing matrix in terms of the physical parameters as in eqn (6.189),

$$U = U^{\mathrm{D}}\, D^{\mathrm{M}}\,, \tag{6.415}$$

with the Dirac-like mixing matrix in eqn (6.191), which is identical to the quark mixing matrix in eqn (4.79), and the diagonal matrix with the two Majorana phases

$$D^{\mathrm{M}} = D^L\,. \tag{6.416}$$

Calling $\lambda_2 = \eta_{12}$ and $\lambda_3 = \eta_{12} + \eta_{23}$, D^{M} has the form in eqn (6.192),

$$D^{\mathrm{M}} = \operatorname{diag}\!\left(e^{i\lambda_1}\,,\, e^{i\lambda_2}\,,\, e^{i\lambda_3}\right)\,, \quad \text{with} \quad \lambda_1 = 0\,. \tag{6.417}$$

We have written explicitly all the phases λ_k because one could choose another physically equivalent parameterization of D^{M} changing the global phase of the charged lepton fields by $e^{i\psi}$, which entails $e^{i\lambda_k} \to e^{i(\lambda_k - \psi)}$, leaving $e^{i(\lambda_k - \lambda_j)}$ constant. Thus, only the relative differences $\lambda_k - \lambda_j$ have physical meaning.

6.7.3 N-neutrino mixing

In this section we consider the general case of a symmetric $N \times N$ Majorana neutrino mass matrix with $N > 3$, as in the case of Dirac–Majorana mixing discussed in section 6.5 (see eqn (6.337)).

As explained in section 6.5.1, the mixing matrix U in the leptonic charged-current weak interaction Lagrangian

$$\mathscr{L}_{\text{I,L}}^{(\text{CC})} = -\frac{g}{\sqrt{2}} \sum_{k=1}^{N} \sum_{\alpha=e,\mu,\tau} \overline{\ell_{\alpha L}} \gamma^{\rho} U_{\alpha k} \nu_{kL} + \text{H.c.} = -\frac{g}{\sqrt{2}} \overline{\boldsymbol{\ell}_L} \gamma^{\rho} U \, \boldsymbol{n}_L + \text{H.c.} \quad (6.418)$$

is the rectangular $3 \times N$ matrix

$$U = V_L^{\ell\dagger} \, V_L^{\nu} \,, \quad (6.419)$$

where V_L^{ℓ} is the unitary matrix that participates in the diagonalization of the charged lepton mass matrix in eqn (3.129). The mixing matrix U is not unitary, because

$$U^{\dagger} U \neq \mathbf{1} \,, \quad (6.420)$$

although

$$U U^{\dagger} = \mathbf{1} \,. \quad (6.421)$$

In order to discuss the parameterization of U in a general way, let us write N as $N = N_A + N_{\text{s}}$, where $N_A = 3$ is the number of active flavors and N_{s} is the number of sterile neutrino fields. Thus, U is a $N_A \times N$ rectangular matrix which satisfies eqn (6.421).

A general complex $N_A \times N$ rectangular matrix depends on $2 N_A N$ real parameters. Since in eqn (6.421) $\mathbf{1}$ is the $N_A \times N_A$ identity matrix, eqn (6.421) imposes N_A^2 constraints, which lower the number of independent real parameters to

$$N_A (N_A + 2 N_{\text{s}}) \,. \quad (6.422)$$

In order to understand how many of these real parameters are mixing angles and how many are phases, let us consider a unitary $N \times N$ matrix V, which depends on $N(N-1)/2$ angles and $N(N+1)/2$ phases. We can construct V following the method described in section 4.3.2. From eqn (4.45), we can write V as

$$V = D(\boldsymbol{\omega}) \left[\prod_{a<b} W^{ab}(\vartheta_{ab}, \eta_{ab}) \right] \qquad (a,b = 1,\ldots,N) \,, \quad (6.423)$$

where $D(\boldsymbol{\omega})$ is the diagonal unitary matrix in eqn (4.44), which depends on the set of N phases $\boldsymbol{\omega} = (\omega_1, \ldots, \omega_N)$, and $W^{ab}(\vartheta_{ab}, \eta_{ab})$ is the matrix in eqn (4.41) of a complex rotation in the a-b plane depending on the angle ϑ_{ab} and the phase

η_{ab}. We can now construct the $N_A \times N$ rectangular matrix U by taking the first N_A rows of V. It is rather obvious that in so doing the phases ω_a with $a > N_A$ are truncated away. With a little more thinking one can see that also the complex rotations $W^{ab}(\vartheta_{ab}, \eta_{ab})$ with $a > N_A$ (which implies $b > a > N_A$) are truncated away. Therefore, we can write U as

$$U = [D(\boldsymbol{\omega})]_{N_A \times N_A} \left[\prod_{a=1}^{N_A} \prod_{b=a+1}^{N} W^{ab}(\vartheta_{ab}, \eta_{ab}) \right]_{N_A \times N} , \qquad (6.424)$$

where the square brackets with subscript $N_A \times N_B$ indicate that the matrix is truncated to the first N_A rows and N_B columns.

Equation (6.424) gives us a simple practical way for the construction of the rectangular $N_A \times N$ mixing matrix U. It allows also an easy determination of the numbers of mixing angles and phases in U. The combinations of the indices ab allowed in eqn (6.424) are

$$\left.\begin{array}{ccccc} 12 & & & & \\ 13 & 23 & & & \\ \vdots & \vdots & \ddots & & \\ 1N_A & 2N_A & \cdots & (N_A-1)N_A & \end{array}\right\} \frac{N_A\,(N_A-1)}{2}$$

$$\left.\begin{array}{ccccc} 1(N_A+1) & 2(N_A+1) & \cdots & (N_A-1)(N_A+1) & N_A(N_A+1) \\ \vdots & \vdots & & \vdots & \vdots \\ 1N & 2N & \cdots & (N_A-1)N & N_A N \end{array}\right\} N_A\,N_{\mathrm{s}} \,. \qquad (6.425)$$

Therefore, there are

$$\frac{N_A\,(N_A-1)}{2} + N_A\,N_{\mathrm{s}} = 3 + 3\,N_{\mathrm{s}} \qquad \text{mixing angles}\,, \qquad (6.426)$$

$$\frac{N_A\,(N_A+1)}{2} + N_A\,N_{\mathrm{s}} = 6 + 3\,N_{\mathrm{s}} \qquad \text{phases}\,, \qquad (6.427)$$

counting both the phases η_{ab} and the phases ω_a. The sum of the numbers of mixing angles and phases coincides correctly with the number of real parameters in eqn (6.422).

Not all the phases counted in eqn (6.427) are physical, because N_A phases can be eliminated by appropriate rephasing of the charged lepton fields in the charged-current weak interaction Lagrangian in eqn (6.418). Thus, there are

$$\frac{N_A\,(N_A-1)}{2} + N_A\,N_{\mathrm{s}} = 3 + 3\,N_{\mathrm{s}} \qquad \text{physical phases}\,, \qquad (6.428)$$

whose number is the same as that of mixing angles in eqn (6.426), as in the case of absence of sterile neutrino fields (see eqns (6.403) and (6.404)). This result is rather obvious from the expression in eqn (6.424) of U, which shows clearly that the phases ω_a can be eliminated by appropriate rephasing of the charged lepton fields, leaving one physical phase η_{ab} for each mixing angle ϑ_{ab}. Thus, a possible parameterization of the mixing in terms of physical parameters is obtained from eqn (6.424) by eliminating the diagonal unitary matrix $[D(\boldsymbol{\omega})]_{N_A \times N_A}$. However, it may be more convenient to adopt a parameterization which reduces to that presented in section 6.7.2 in the three-neutrino mixing case (i.e. $N_\mathrm{s} = 0$). To this end, we can write the mixing matrix U in a form as close as possible to that in eqn (6.408):

$$U = D^L \left[R^{23}\, W^{13}\, R^{12}\, D^R \prod_{a=1}^{N_A} \prod_{b=N_A+1}^{N} W^{ab}(\vartheta_{ab}, \eta_{ab}) \right]_{N_A \times N} , \tag{6.429}$$

where D^L is the diagonal $N_A \times N_A$ matrix of phases defined in eqn (6.409), whereas D^R is the diagonal $N \times N$ matrix of phases

$$D^R \equiv \operatorname{diag}\Big(1\,,\, e^{i\eta_{12}}\,,\, e^{i(\eta_{12}+\eta_{23})}\,,\, 1\,,\, \ldots\,,\, 1\Big)\,. \tag{6.430}$$

The matrix D^L on the left can be eliminated by rephasing the charged lepton fields as in eqns (6.411)–(6.414). The matrix D^R can be moved to the right, leading to

$$U = \left[R^{23}\, W^{13}\, R^{12} \prod_{a=1}^{N_A} \prod_{b=N_A+1}^{N} W^{ab}(\vartheta_{ab}, \eta'_{ab}) \right]_{N_A \times N} D^{\mathrm{M}} , \tag{6.431}$$

where

$$D^{\mathrm{M}} = \operatorname{diag}\big(e^{i\lambda_1}\,,\, e^{i\lambda_2}\,,\, \ldots\,,\, e^{i\lambda_N}\big)\,, \tag{6.432}$$

and

$$\lambda_1 = 0\,, \tag{6.433}$$

$$\lambda_2 = \eta_{12}\,, \tag{6.434}$$

$$\lambda_3 = \eta_{12} + \eta_{23}\,, \tag{6.435}$$

$$\eta'_{ab} = \eta_{ab} + \lambda_a - \lambda_b\,, \quad \text{for} \quad a = 1, \ldots, N_A \quad \text{and} \quad b = N_A + 1, \ldots, N\,. \tag{6.436}$$

The N_s phases λ_b for $b = N_A + 1, \ldots, N$ can be chosen in order to cancel N_s of the phases η'_{ab}. In this way, the physical phases are separated into

$$N - 1 = 2 + N_\mathrm{s} \qquad \text{Majorana phases } \lambda_a \ (a = 2, \ldots, N)\,, \tag{6.437}$$

which can be factorized in the diagonal matrix D^{M} on the right of U in eqn (6.431), and

$$\frac{(N_A - 1)\,(N_A - 2)}{2} + (N_A - 1)\, N_\mathrm{s} = 1 + 2\, N_\mathrm{s} \qquad \text{Dirac phases}\,. \tag{6.438}$$

For example, we can choose $\eta'_{1b} = 0$ for $b = 4, \ldots, N$. In this case, the Dirac phases are $\eta_{13}, \eta'_{24}, \ldots, \eta'_{2N}, \eta'_{34}, \ldots, \eta'_{3N}$.

7

NEUTRINO OSCILLATIONS IN VACUUM

> The ability to perceive or think differently is more important than the knowledge gained.
> **David Bohm**

Neutrino oscillation is a quantum mechanical phenomenon proposed in the late 1950s by Pontecorvo [880, 881] in analogy with K^0-$\bar{K}^0$ oscillations. The oscillations are generated by the interference of different massive neutrinos, which are produced and detected coherently because of their very small mass differences.

Since in the late 1950s only one *active* neutrino was known, the electron neutrino, in order to discuss neutrino oscillations, Pontecorvo invented the concept of a *sterile* neutrino [883], which is a neutral fermion which does not take part in weak interactions. The muon neutrino was discovered in 1962 in the Brookhaven experiment of Lederman, Schwartz, Steinberger, *et al.* [348], who followed up a proposal made by Pontecorvo in 1959 [882]. Since then, it became clear that oscillations between different active neutrino flavors are possible if neutrinos are massive and mixed. In 1962 Maki, Nakagawa, and Sakata [766] considered for the first time a model with the mixing of different neutrino flavors. In 1967 Pontecorvo [883] predicted the Solar Neutrino Problem as a consequence of $\nu_e \to \nu_\mu$ (or $\nu_e \to \nu_{\text{sterile}}$) transitions even before the first measurement of the solar electron neutrino flux in the Homestake experiment [323], and in 1969 Gribov and Pontecorvo discussed solar neutrino oscillations due to neutrino mixing [567].

However, in the above and other papers, the probability of neutrino oscillations was not calculated in a rigorous way, but simply estimated on the basis of the analogy with kaon oscillations. As a result, the phase of the oscillations was correct within a factor of two.

The standard theory of neutrino oscillations in the plane-wave approximation was developed in 1975–76 by Eliezer and Swift [404], Fritzsch and Minkowski [466], Bilenky and Pontecorvo [236, 239], and elegantly reviewed by Bilenky and Pontecorvo in Ref. [237].

In this chapter, we review the standard plane-wave derivation of the neutrino oscillation probability in section 7.1. In the following sections, we discuss the main phenomenological aspects of neutrino oscillations in vacuum.

An important feature necessary for the derivation of a simple and general expression for the probability of neutrino oscillations is the fact that neutrinos in oscillation experiments are ultrarelativistic, since neutrino masses are smaller than about one eV (see chapter 14) and only neutrinos with energy larger than about 100 keV can be detected. Neutrinos are detected in:

1. Charged-current or neutral-current weak scattering processes which have an energy threshold given in eqn (5.37). As shown in Table 5.2 (page 143), current experimental thresholds are always larger than some fraction of an MeV. The lowest threshold reached so far is that of the first reaction in Table 5.2, $\nu_e + {}^{71}\text{Ga} \to {}^{71}\text{Ge} + e^-$, which is used in gallium solar neutrino experiments (see section 10.5).
2. The elastic scattering process $\nu + e^- \to \nu + e^-$, whose cross-section is proportional to the neutrino energy ($\sigma(E_\nu) \sim \sigma_0 E_\nu / m_e$, with $\sigma_0 \sim 10^{-44}\,\text{cm}^2$). An energy threshold of some MeV's is needed in order to have a signal above the background. For example, $E_\nu^{\text{th}} \simeq 5\,\text{MeV}$ in the Super-Kamiokande [472] solar neutrino experiment.

Neutrinos ν_α with flavor $\alpha = e, \mu, \tau$ are produced[35] in charged-current (CC) weak interaction processes from a charged lepton ℓ_α^- (i.e. $\ell_\alpha^- \to \nu_\alpha$ transitions) or together with a charged antilepton ℓ_α^+ (i.e. creation of a $\ell_\alpha^+\,\nu_\alpha$ pair). These processes are generated by the charged-current leptonic interaction Lagrangian (see eqns (3.76) and (3.77))

$$\mathscr{L}_{\text{I,L}}^{(\text{CC})} = -\frac{g}{2\sqrt{2}} \left(j^\rho_{W,\text{L}} W_\rho + j^{\rho\dagger}_{W,\text{L}} W^\dagger_\rho \right) , \tag{7.1}$$

where $j^\rho_{W,\text{L}}$ is the leptonic charged current

$$j^\rho_{W,\text{L}} = 2 \sum_{\alpha=e,\mu,\tau} \overline{\nu_{\alpha L}}\, \gamma^\rho\, \ell_{\alpha L} = 2 \sum_{\alpha=e,\mu,\tau} \sum_k U^*_{\alpha k}\, \overline{\nu_{kL}}\, \gamma^\rho\, \ell_{\alpha L} , \tag{7.2}$$

which is valid both in the case of Dirac (see eqns (6.14) and (6.187)) or Majorana massive neutrinos (see eqn (6.184)). In the case of Dirac neutrinos, the Fourier expansion of the field operator $\overline{\nu_k}$ is (see eqn (2.139))

$$\overline{\nu_{kL}}(x) = \int \frac{\text{d}^3p}{(2\pi)^3\, 2E} \sum_{h=\pm 1} \left[a^{(h)\dagger}_{\nu_k}(p)\, \overline{u^{(h)}_{\nu_k L}}(p)\, e^{-ip\cdot x} + b^{(h)}_{\nu_k}(p)\, \overline{v^{(h)}_{\nu_k L}}(p)\, e^{ip\cdot x} \right] . \tag{7.3}$$

Hence, the leptonic charged current in eqn (7.2) contains creation operators $a^{(h)\dagger}_{\nu_k}(p)$ of neutrinos with mass m_k and, through the charged lepton field ℓ_α, it contains destruction operators of ℓ_α^- and creation operators of ℓ_α^+, which lead to the generation of $\ell_\alpha^- \to \nu_k$ transitions or $\ell_\alpha^+\,\nu_k$ pair creation.

In the case of Majorana neutrinos, we have $b^{(h)}_{\nu_k}(p) = a^{(h)}_{\nu_k}(p)$ (see eqn (6.99)). Hence, creation operators $a^{(h)\dagger}_{\nu_k}(p)$ of neutrinos with mass m_k are present both in

[35] Neutrinos can also be produced in the neutral-current (NC) weak interaction process $Z \to \nu\bar{\nu}$, generated by the interaction Lagrangian in eqn (3.90). The Z can be either real (Z-decay) or virtual (for example in $e^- e^+ \to \nu\bar{\nu}$). In this case, the neutrinos do not have a definite flavor. However, the produced neutrinos are active and can oscillate into sterile neutrinos if light sterile neutrinos exist. These oscillations could be observed by measuring the disappearance of active neutrinos. The theory of active–sterile oscillations can be obtained by straightforward modifications to the theory of flavor oscillations discussed in this chapter.

$j^{\rho}_{W,L}$ and $j^{\rho\dagger}_{W,L}$. However, $j^{\rho\dagger}_{W,L}$ does not contribute to $\ell_\alpha^- \to \nu_k$ transitions or $\ell_\alpha^+ \, \nu_k$ pair creation because it contains the charged lepton field in the adjoint form $\overline{\ell_\alpha}$, which does not contain destruction operators of ℓ_α^- and creation operators of ℓ_α^+ (it contains creation operators of ℓ_α^- and destruction operators of ℓ_α^+).

The leptonic charged current in eqn (7.2) generates a superposition of massive neutrinos if the energies and momenta of the particles which participate in the neutrino production process are not measured with a degree of accuracy which would allow the determination, through energy–momentum conservation, of the massive neutrino which is emitted. This is characteristic of neutrino oscillation experiments, in which a flavor neutrino ν_α is a superposition of massive neutrinos ν_k with weights proportional to $U_{\alpha k}^*$.

7.1 Standard Derivation of the Neutrino Oscillation Probability

In the standard theory of neutrino oscillations [404, 466, 236, 239, 237] a neutrino with flavor α and momentum $\vec{p}$, created in a charged-current weak interaction process from a charged lepton ℓ_α^- or together with a charged antilepton ℓ_α^+, is described by the flavor state

$$|\nu_\alpha\rangle = \sum_k U_{\alpha k}^* \, |\nu_k\rangle \qquad (\alpha = e, \mu, \tau) \, . \tag{7.4}$$

The presence of the weight $U_{\alpha k}^*$ of the $|\nu_k\rangle$ in the flavor state $|\nu_\alpha\rangle$ is due to the decomposition in eqn (7.2) of the leptonic charged current $j^{\rho}_{W,L}$ in terms of the massive neutrino contributions, which contain the creation operators of massive neutrinos. As we will see in subsection 8.1.1, additional coefficients due to the effect of the difference of the neutrino masses on the interaction process are negligible for neutrino oscillations.

For simplicity, we consider a finite normalization volume V, according to the method discussed in section 2.13, in order to have orthonormal massive neutrino states:

$$\langle \nu_k | \nu_j \rangle = \delta_{kj} \, . \tag{7.5}$$

The unitarity of the mixing matrix implies that also the flavor states are orthonormal:

$$\langle \nu_\alpha | \nu_\beta \rangle = \delta_{\alpha\beta} \, . \tag{7.6}$$

In eqn (7.4) we have not limited the number of massive neutrinos. Since it is known that the number of active flavor neutrinos is three, corresponding to ν_e, ν_μ, ν_τ, the number of massive neutrinos must be equal to or greater than three. If the number of massive neutrinos is greater than three, the additional neutrinos in the flavor basis are sterile, i.e. they do not participate in weak interactions (since neutrinos are electrically neutral, sterile neutrinos interact with ordinary matter only through gravitational interactions or exotic interactions beyond those in the SM). Transitions of active flavor neutrinos into sterile ones can be observed only through the disappearance of active neutrinos.

The massive neutrino states $|\nu_k\rangle$ are eigenstates of the Hamiltonian,

$$\mathscr{H}|\nu_k\rangle = E_k|\nu_k\rangle\,, \tag{7.7}$$

with energy eigenvalues

$$E_k = \sqrt{\vec{p}^2 + m_k^2}\,. \tag{7.8}$$

The Schrödinger equation

$$i\,\frac{\mathrm{d}}{\mathrm{d}t}\,|\nu_k(t)\rangle = \mathscr{H}\,|\nu_k(t)\rangle \tag{7.9}$$

implies that the massive neutrino states evolve in time as plane waves:

$$|\nu_k(t)\rangle = e^{-iE_kt}\,|\nu_k\rangle\,. \tag{7.10}$$

Let us consider now a flavor state $|\nu_\alpha(t)\rangle$ which describes a neutrino created with a definite flavor α at time $t = 0$. From eqns (7.4) and (7.10), the time evolution of this state is given by

$$|\nu_\alpha(t)\rangle = \sum_k U_{\alpha k}^*\,e^{-iE_kt}\,|\nu_k\rangle\,, \tag{7.11}$$

such that

$$|\nu_\alpha(t=0)\rangle = |\nu_\alpha\rangle\,. \tag{7.12}$$

Using the unitarity relation

$$U^\dagger\,U = \mathbf{1} \quad \Longleftrightarrow \quad \sum_\alpha U_{\alpha k}^*\,U_{\alpha j} = \delta_{jk}\,, \tag{7.13}$$

the massive states can be expressed in terms of flavor states inverting eqn (7.4):

$$|\nu_k\rangle = \sum_\alpha U_{\alpha k}\,|\nu_\alpha\rangle\,. \tag{7.14}$$

Substituting eqn (7.14) into eqn (7.11), we obtain

$$|\nu_\alpha(t)\rangle = \sum_{\beta=e,\mu,\tau}\left(\sum_k U_{\alpha k}^*\,e^{-iE_kt}\,U_{\beta k}\right)|\nu_\beta\rangle\,. \tag{7.15}$$

Hence, the superposition of massive neutrino states $|\nu_\alpha(t)\rangle$, which is the pure flavor state given in eqn (7.4) at $t = 0$, becomes a superposition of different flavor states at $t > 0$ (if the mixing matrix U is not diagonal, i.e. neutrinos are mixed). The coefficient of $|\nu_\beta\rangle$,

$$A_{\nu_\alpha \to \nu_\beta}(t) \equiv \langle \nu_\beta | \nu_\alpha(t) \rangle = \sum_k U_{\alpha k}^* \, U_{\beta k} \, e^{-iE_k t} \,, \tag{7.16}$$

is the amplitude of $\nu_\alpha \to \nu_\beta$ transitions as a function of time. The transition probability is, then, given by

$$P_{\nu_\alpha \to \nu_\beta}(t) = \left| A_{\nu_\alpha \to \nu_\beta}(t) \right|^2 = \sum_{k,j} U_{\alpha k}^* \, U_{\beta k} \, U_{\alpha j} \, U_{\beta j}^* \, e^{-i(E_k - E_j)t} \,. \tag{7.17}$$

For ultrarelativistic neutrinos, the dispersion relation in eqn (7.8) can be approximated by

$$E_k \simeq E + \frac{m_k^2}{2E} \,. \tag{7.18}$$

In this case,

$$E_k - E_j \simeq \frac{\Delta m_{kj}^2}{2E} \,, \tag{7.19}$$

where Δm_{kj}^2 is the squared-mass difference

$$\Delta m_{kj}^2 \equiv m_k^2 - m_j^2 \,, \tag{7.20}$$

and

$$E = |\vec{p}| \tag{7.21}$$

is the neutrino energy, neglecting the mass contribution. Therefore, the transition probability in eqn (7.17) can be approximated by

$$P_{\nu_\alpha \to \nu_\beta}(t) = \sum_{k,j} U_{\alpha k}^* U_{\beta k} U_{\alpha j} U_{\beta j}^* \exp\left(-i \frac{\Delta m_{kj}^2 t}{2E} \right) \,. \tag{7.22}$$

The final step in the standard derivation of the neutrino oscillation probability is based on the fact that, in neutrino oscillation experiments, the propagation time t is not measured. What is known is the distance L between the source and the detector. Since ultrarelativistic neutrinos propagate almost at the speed of light, it is possible to approximate $t = L$, leading to

$$P_{\nu_\alpha \to \nu_\beta}(L, E) = \sum_{k,j} U_{\alpha k}^* \, U_{\beta k} \, U_{\alpha j} \, U_{\beta j}^* \, \exp\left(-i \, \frac{\Delta m_{kj}^2 L}{2E} \right) \,. \tag{7.23}$$

This expression shows that the source–detector distance L and the neutrino energy E are the quantities depending on the experiment which determine the phases of neutrino oscillations

$$\Phi_{kj} = -\frac{\Delta m_{kj}^2 \, L}{2 \, E} \,. \tag{7.24}$$

Of course, the phases are determined also by the squared-mass differences Δm_{kj}^2, which are physical constants. The amplitude of the oscillations is specified only by the elements of the mixing matrix U, which are constants of nature. Therefore measurements of neutrino oscillations allow one to shed some light on the values of the squared-mass differences Δm^2_{kj} and the elements of the mixing matrix U.

Although positive measurements of neutrino oscillations imply massive neutrinos, they yield precise information only on the values of the squared-mass differences Δm^2_{kj}, but not on the absolute values of neutrino masses, except that obviously m_k^2 or m_j^2 must be larger than $|\Delta m^2_{kj}|$.

The oscillation probability in eqn (7.23) depends on the elements of the mixing matrix U through the quartic products

$$U_{\alpha k}^* \, U_{\beta k} \, U_{\alpha j} \, U_{\beta j}^* \,, \tag{7.25}$$

which do not depend on the specific parameterization of the mixing matrix and on the choice of phases. In fact, the quartic products in eqn (7.25) are invariant under the rephasing transformation

$$U_{\alpha k} \to e^{i\psi_\alpha} \, U_{\alpha k} \, e^{i\phi_k} \,. \tag{7.26}$$

Hence, the quartic products eqn (7.25) do not depend on the phases that can be factorized on the left or on the right of the mixing matrix. This corresponds to a rephasing of the charged lepton and neutrino fields.

As we have discussed in section 6.3, in the case of Majorana neutrinos the three-neutrino mixing matrix contains, in addition to the Dirac phase analogous to that in the CKM mixing matrix of quarks, two Majorana phases which appear in a diagonal matrix at the right of the mixing matrix. In other words, as shown in eqn (6.189), we can write

$$U_{\alpha k} = U_{\alpha k}^{\mathrm{D}} \, e^{i\lambda_k} \,. \tag{7.27}$$

The rephasing invariants in eqn (7.25) are free from the Majorana phases. This means that the Majorana phases cannot be measured in neutrino oscillation experiments (see section 9.2 for the proof that this statement holds also in the case of neutrino oscillations in matter). This statement holds in general for any number of generations: neutrino oscillations are independent of the Majorana phases, which are always factorized in a diagonal matrix on the right of the mixing matrix. In particular, CP and T violations in neutrino oscillations, discussed in sections 7.3.2 and 7.3.3, depend only on the Dirac phases.

It is clear that transitions among different flavors manifest for $L > 0$, because the unitarity relation

$$U \, U^\dagger = \mathbf{1} \quad \Longleftrightarrow \quad \sum_k U_{\alpha k} \, U_{\beta k}^* = \delta_{\alpha\beta} \tag{7.28}$$

implies that

$$P_{\nu_\alpha \to \nu_\beta}(L = 0, E) = \delta_{\alpha\beta} \,. \tag{7.29}$$

Sometimes it is convenient to write the probability in eqn (7.23) as

$$P_{\nu_\alpha \to \nu_\beta}(L, E) = \sum_k |U_{\alpha k}|^2 \, |U_{\beta k}|^2 + 2\mathfrak{Re} \sum_{k>j} U_{\alpha k}^* \, U_{\beta k} \, U_{\alpha j} \, U_{\beta j}^* \, \exp\left(-2\pi i \, \frac{L}{L_{kj}^{\mathrm{osc}}} \right) \,, \tag{7.30}$$

in which we have separated a constant term from the oscillating term and we have defined the oscillation lengths

$$L_{kj}^{\text{osc}} = \frac{4\pi E}{\Delta m_{kj}^2} \,. \tag{7.31}$$

The oscillation length L_{kj}^{osc} is the distance at which the phase generated by Δm_{kj}^2 becomes equal to 2π.

The oscillating term in eqn (7.30) is produced by the interference of the different massive neutrino components of the state in eqn (7.11). Therefore, its existence depends on the coherence of the massive neutrino components. If, for some reason, different massive neutrinos are produced or detected in an incoherent way, as discussed in section 8.2.2, the probability of $\nu_\alpha \to \nu_\beta$ reduces to the constant term in eqn (7.30), as can also be seen by calculating the incoherent transition probability

$$P_{\nu_\alpha \to \nu_\beta}^{\text{incoherent}} = \sum_k \left| \langle \nu_\beta | \nu_k \rangle \, e^{-iE_k t} \, \langle \nu_k | \nu_\alpha \rangle \right|^2 = \sum_k |U_{\alpha k}|^2 \, |U_{\beta k}|^2 \,. \tag{7.32}$$

As discussed in section 7.6, the incoherent average of the oscillation probability in eqn (7.30) over the energy resolution of the detector or over the uncertainty of the distance L can also lead to an effectively constant measurable probability which has the same value as the incoherent transition probability in eqn (7.32):

$$\langle P_{\nu_\alpha \to \nu_\beta} \rangle = \sum_k |U_{\alpha k}|^2 \, |U_{\beta k}|^2 \,. \tag{7.33}$$

This result follows simply from the fact that the average of the exponential functions in eqn (7.30) is zero.

The oscillation probability in eqn (7.23) satisfies the two rules of the conservation of probability which are consequences of the unitary evolution of the states:

1. The sum of the probabilities of transition from a flavor neutrino ν_α to all flavor neutrinos ν_β (including $\alpha = \beta$) is equal to unity:

$$\sum_\beta P_{\nu_\alpha \to \nu_\beta}(L, E) = 1 \,. \tag{7.34}$$

2. The sum of the probabilities of transition from any flavor neutrino ν_α (including $\alpha = \beta$) to a flavor neutrino ν_β is equal to unity:

$$\sum_\alpha P_{\nu_\alpha \to \nu_\beta}(L, E) = 1 \,. \tag{7.35}$$

These sum rules can be derived from eqn (7.23) by using the unitarity relations in eqns (7.13) and (7.28).

Another useful way to write the oscillation probability in eqn (7.23) is to separate the real and imaginary parts of $U_{\beta k} \, U_{\alpha k}^* \, U_{\beta j}^* \, U_{\alpha j}$. From the square of the unitarity relation in eqn (7.28), we obtain

$$\sum_k |U_{\alpha k}|^2 \, |U_{\beta k}|^2 = \delta_{\alpha\beta} - 2 \sum_{k>j} \mathfrak{Re}\big[U_{\alpha k}^* \, U_{\beta k} \, U_{\alpha j} \, U_{\beta j}^*\big] \,, \tag{7.36}$$

which allows one to write the oscillation probability as

$$\begin{aligned} P_{\nu_\alpha \to \nu_\beta}(L, E) = \delta_{\alpha\beta} &- 2 \sum_{k>j} \mathfrak{Re}\big[U_{\alpha k}^* \, U_{\beta k} \, U_{\alpha j} \, U_{\beta j}^*\big] \left[1 - \cos\left(\frac{\Delta m_{kj}^2 L}{2E}\right)\right] \\ &+ 2 \sum_{k>j} \mathfrak{Im}\big[U_{\alpha k}^* \, U_{\beta k} \, U_{\alpha j} \, U_{\beta j}^*\big] \sin\left(\frac{\Delta m_{kj}^2 L}{2E}\right) \,, \end{aligned} \tag{7.37}$$

or in the form

$$\begin{aligned} P_{\nu_\alpha \to \nu_\beta}(L, E) = \delta_{\alpha\beta} &- 4 \sum_{k>j} \mathfrak{Re}\big[U_{\alpha k}^* \, U_{\beta k} \, U_{\alpha j} \, U_{\beta j}^*\big] \sin^2\left(\frac{\Delta m_{kj}^2 L}{4E}\right) \\ &+ 2 \sum_{k>j} \mathfrak{Im}\big[U_{\alpha k}^* \, U_{\beta k} \, U_{\alpha j} \, U_{\beta j}^*\big] \sin\left(\frac{\Delta m_{kj}^2 L}{2E}\right) \,. \end{aligned} \tag{7.38}$$

The oscillation probabilities of the channels with $\alpha \neq \beta$ are usually called *transition probabilities*, whereas the oscillation probabilities of the channels with $\alpha = \beta$ are usually called *survival probabilities*. Since, in the case of the survival probabilities, the quartic products in eqn (7.25) are real and equal to $|U_{\alpha k}|^2 |U_{\alpha j}|^2$, the survival probabilities can be written in the simple form

$$P_{\nu_\alpha \to \nu_\alpha}(L, E) = 1 - 4 \sum_{k>j} |U_{\alpha k}|^2 \, |U_{\alpha j}|^2 \sin^2\left(\frac{\Delta m_{kj}^2 L}{4E}\right) \,. \tag{7.39}$$

It is interesting to see for which values of $|U_{\alpha k}|^2$ and $|U_{\beta k}|^2$ with $\alpha \neq \beta$, the average transition probability in eqn (7.33) has its maximum. Since the values of $|U_{\alpha k}|^2$ and $|U_{\beta k}|^2$ are subject to the unitarity constraints

$$\sum_k |U_{\alpha k}|^2 = 1 \qquad \text{and} \qquad \sum_k |U_{\beta k}|^2 = 1 \,, \tag{7.40}$$

we can use the method of the Lagrange multipliers and calculate the stationary point of

$$f(|U_{\alpha k}|^2, |U_{\beta k}|^2) = \sum_k |U_{\alpha k}|^2 \, |U_{\beta k}|^2 - a\left(1 - \sum_k |U_{\alpha k}|^2\right) - b\left(1 - \sum_k |U_{\beta k}|^2\right) \,, \tag{7.41}$$

where a and b are the Lagrange multipliers. The stationary point is given by

$$0 = \frac{\mathrm{d}f}{\mathrm{d}|U_{\alpha k}|^2} = |U_{\beta k}|^2 + a \,, \qquad 0 = \frac{\mathrm{d}f}{\mathrm{d}|U_{\beta k}|^2} = |U_{\alpha k}|^2 + b \,. \tag{7.42}$$

Hence, for fixed $\alpha \neq \beta$ and different values of k, all $|U_{\alpha k}|^2$ are the same and all $|U_{\beta k}|^2$ are the same too. From the constraints in eqn (7.40) we find the solution

$$|U_{\alpha k}|^2 = |U_{\beta k}|^2 = \frac{1}{N} \quad \text{for all} \quad k \quad \Longrightarrow \quad \langle P_{\nu_\alpha \to \nu_\beta} \rangle_{\text{max}} = \frac{1}{N} \qquad (\alpha \neq \beta) \,, \quad (7.43)$$

where N is the number of massive neutrinos. Similarly, one can show that in the case of the survival probabilities

$$|U_{\alpha k}|^2 = \frac{1}{N} \quad \text{for all} \quad k \quad \Longrightarrow \quad \langle P_{\nu_\alpha \to \nu_\alpha} \rangle_{\text{min}} = \frac{1}{N} \,. \qquad (7.44)$$

Therefore, the case in which all the elements of the $N \times N$ mixing matrix have the same absolute value, usually called *N-maximal mixing*, corresponds to minimal average survival probability equal to $1/N$ and maximal average transition probability equal to $1/N$ in each possible channel, in such a way that the sum of the average survival probability and all the average transition probabilities add up correctly to unity.

Let us finally summarize and comment on the three main assumptions adopted in the standard derivation of the neutrino oscillation probability.

(A1) Neutrinos produced or detected in CC weak interaction processes are described by the flavor states in eqn (7.4). As we will see in subsection 8.1.1 the states which describe flavor neutrinos can be derived in the framework of quantum field theory. We will show that these flavor states reduce to the standard ones in eqn (7.4) in the case of experiments which are not sensitive to the differences of the contributions of the different neutrino masses to the production and detection processes. This is the case for all neutrino oscillation experiments.

(A2) Flavor neutrinos have a definite momentum $\vec{p}$, i.e. all the massive neutrino components have the same momentum. This is sometimes called the *equal momentum assumption*. It may have been motivated by the fact that all the components propagate in the same direction from source to detector. However, there is no justification for this assumption and we will see in section 8.4.3 that it is, in fact, unrealistic. Luckily, however, the equal momentum assumption is irrelevant in the derivation of the oscillation probability, as will be seen in section 8.1.2.

(A3) The propagation time t is equal to the distance L traveled by the neutrino between production and detection. It is sometimes called the *light-ray approximation*. This assumption is unjustified in a plane-wave treatment of oscillations, because plane waves extend with the same amplitude over the whole space-time. However, in quantum theory, localized particles are described by wave packets. As will be discussed in section 8.2, in the case of neutrino oscillation experiments, neutrinos are described by wave packets [660, 533, 528, 225, 523, 526] that are localized in the production process at the production time and propagate between the production and the detection processes with a group velocity close to the velocity of light, justifying the approximation $t = L$.

From eqn (2.472), the group velocity $\vec{v}_k$ of an ultrarelativistic neutrino with mass m_k, average wave packet momentum $\vec{p}_k$ and average wave packet energy

E_k is given by

$$\vec{v}_k = \frac{\vec{p}_k}{E_k} \simeq \frac{\vec{p}_k}{|\vec{p}_k|}\left(1 - \frac{m_k^2}{2E^2}\right) . \tag{7.45}$$

Since, in oscillation experiments, all massive neutrinos are produced and detected in the same localized processes, their group velocities have the same direction. However, since the absolute value of the group velocities of different massive neutrinos are different, there is a separation of the wave packets which increases with the distance from the production process. Nevertheless, different massive neutrino wave packets can contribute coherently to the detection process if they overlap in a sufficient way.

If the massive neutrinos are ultrarelativistic and contribute coherently to the detection process, their wave packets overlap with the detection process for an interval of time $[t - \Delta t\,,\, t + \Delta t]$, with

$$t = \frac{L}{\overline{v}} \simeq L\left(1 + \frac{\overline{m^2}}{2E^2}\right) \qquad \text{and} \qquad \Delta t \sim \sigma_x\,, \tag{7.46}$$

where $\overline{v} = 1 - \overline{m^2}/2E^2$ is the average group velocity, $\overline{m^2}$ is the average of the squared neutrino masses, and σ_x is given by the spatial uncertainties of the production and detection processes summed in quadrature [528] (the spatial uncertainty of the production process determines the size of the massive neutrino wave packets). The correction $L\overline{m^2}/2E^2$ to $t = L$ in eqn (7.46) can be neglected, because it gives corrections to the oscillation phases which are of higher order in the very small ratios m_k^2/E^2. The corrections due to $\Delta t \sim \sigma_x$ are also negligible, because in all real experiments σ_x is much smaller than the oscillation length $L_{kj}^{\text{osc}} = 4\pi E/\Delta m_{kj}^2$; otherwise oscillations cannot be observed [660, 533, 225, 526]. One can summarize these arguments by saying that the approximation $t = L$ is correct because the phase of the oscillations is practically constant over the interval of time in which the massive neutrino wave packets overlap in the detection process.

7.2 Antineutrino case

As we have discussed in the introduction to this chapter, flavor neutrinos are produced in weak interaction processes through the action of the leptonic charged-current in eqn (7.2), which contains the creation operators of massive neutrinos. Antineutrinos $\bar{\nu}_\alpha$ with flavor $\alpha = e, \mu, \tau$ are similarly produced in charged-current (CC) weak interaction processes from a charged antilepton ℓ_α^+ (i.e. $\ell_\alpha^+ \to \bar{\nu}_\alpha$ transitions) or together with a charged lepton ℓ_α^- (i.e. creation of a $\ell_\alpha^-\,\bar{\nu}_\alpha$ pair), through the action of the Hermitian conjugate of the leptonic charged-current in eqn (7.2),

$$j_{W,L}^{\rho\dagger} = 2 \sum_{\alpha=e,\mu,\tau} \overline{\ell_{\alpha L}}\,\gamma^\rho\,\nu_{\alpha L} = 2 \sum_{\alpha=e,\mu,\tau}\sum_k U_{\alpha k}\,\overline{\ell_{\alpha L}}\,\gamma^\rho\,\nu_{kL}\,, \tag{7.47}$$

which is included in the charged-current leptonic interaction Lagrangian in eqn (7.1). In fact, the Hermitian-conjugated leptonic charged-current in eqn (7.47) contains, through the adjoint charged lepton field $\overline{\ell_\alpha}$, destruction operators of ℓ_α^+ for the generation of $\ell_\alpha^+ \to \bar{\nu}_\alpha$ transitions, and creation operators of ℓ_α^- for the creation of $\ell_\alpha^- \, \bar{\nu}_\alpha$ pairs.

In the case of Dirac neutrinos, the Fourier expansion of the field operator ν_{kL} (see eqn (2.139)),

$$\nu_{kL}(x) = \int \frac{\mathrm{d}^3 p}{(2\pi)^3 \, 2E} \sum_{h=\pm 1} \left[a_{\nu_k}^{(h)}(p) \, u_{\nu_k L}^{(h)}(p) \, e^{-ip \cdot x} + b_{\nu_k}^{(h)\dagger}(p) \, v_{\nu_k L}^{(h)}(p) \, e^{ip \cdot x} \right] , \tag{7.48}$$

contains creation operators $b_{\nu_k}^{(h)}(p)$ of antineutrinos with mass m_k. In neutrino oscillation experiments, the energies and momenta of the particles involved in the neutrino production process are not measured with a degree of accuracy which would allow one to determine, through energy–momentum conservation, which massive neutrino is emitted. In this case, the Hermitian-conjugated leptonic charged-current in eqn (7.47) produces flavor antineutrinos $\bar{\nu}_\alpha$ which are superpositions of massive antineutrinos $\bar{\nu}_k$ with weights proportional to $U_{\alpha k}$. Neglecting additional coefficients due to the effect of the difference of the neutrino masses in the interaction process (see subsection 8.1.1), these flavor antineutrinos are described by the standard flavor antineutrino states

$$|\bar{\nu}_\alpha\rangle = \sum_k U_{\alpha k} \, |\bar{\nu}_k\rangle \qquad (\alpha = e, \mu, \tau) . \tag{7.49}$$

It is to be noted that the coefficients of the massive neutrino components of flavor antineutrinos are simply related to the corresponding coefficients of the massive neutrino components of flavor antineutrinos by complex conjugation.

In the case of Majorana neutrinos, there is no difference between neutrinos and antineutrinos. However, as we have explained in section 6.2.3, it is customary, by convention, to call Majorana neutrinos with negative helicity *neutrinos* and Majorana neutrinos with positive helicity *antineutrinos*. Hence, the standard flavor neutrino states in eqn (7.4) describe Dirac or Majorana neutrinos with negative helicity and the states in eqn (7.49) describe Dirac antineutrinos with positive helicity or Majorana neutrinos with positive helicity.

Let us now consider the probability of $\bar{\nu}_\alpha \to \bar{\nu}_\beta$ oscillations. Since the kinematical properties of massive antineutrinos are identical to those of neutrinos, the derivation of the probability of $\bar{\nu}_\alpha \to \bar{\nu}_\beta$ oscillations follows the same lines as that of $\nu_\alpha \to \nu_\beta$ oscillations presented in the previous section 7.1, the only difference being that in this case we start with the flavor antineutrino states in eqn (7.49) in which the elements of the mixing matrix are complex conjugated with respect to the flavor neutrino states in eqn (7.4). Applying the same transformation to the neutrino oscillation probability in eqn (7.23), we obtain the antineutrino oscillation probability

$$P_{\bar{\nu}_\alpha \to \bar{\nu}_\beta}(L, E) = \sum_{k,j} U_{\alpha k} \, U_{\beta k}^* \, U_{\alpha j}^* \, U_{\beta j} \, \exp\left(-i \, \frac{\Delta m_{kj}^2 L}{2E} \right) . \tag{7.50}$$

In particular, we note that the oscillation length of antineutrinos is the same as that of neutrinos, given in eqn (7.31), since it depends only on the same kinematical properties of massive neutrinos and antineutrinos. Writing the antineutrino oscillation probability as

$$P_{\bar{\nu}_\alpha \to \bar{\nu}_\beta}(L,E) = \delta_{\alpha\beta} - 4 \sum_{k>j} \mathfrak{Re}\big[U_{\alpha k}^* \, U_{\beta k} \, U_{\alpha j} \, U_{\beta j}^*\big] \sin^2\left(\frac{\Delta m_{kj}^2 L}{4E}\right)$$
$$- 2 \sum_{k>j} \mathfrak{Im}\big[U_{\alpha k}^* \, U_{\beta k} \, U_{\alpha j} \, U_{\beta j}^*\big] \sin\left(\frac{\Delta m_{kj}^2 L}{2E}\right) , \qquad (7.51)$$

one can see that it differs from the corresponding neutrino oscillation probability in eqn (7.38) only in the sign of the terms depending on the imaginary parts of the quartic products of the elements of the mixing matrix.

7.3 CPT, CP, and T transformations

Physical neutrinos and antineutrinos are related by a CP transformation which interchanges neutrinos with antineutrinos and reverses the helicity[36] (see subsection 2.11.3):

$$\nu_\alpha \stackrel{\text{CP}}{\longleftrightarrow} \bar{\nu}_\alpha \, . \qquad (7.52)$$

Moreover, a T transformation interchanges the initial and final states. Therefore, as schematized in Fig. 7.1, a CP transformation interchanges the $\nu_\alpha \to \nu_\beta$ and $\bar{\nu}_\alpha \to \bar{\nu}_\beta$ channels,

$$\nu_\alpha \to \nu_\beta \stackrel{\text{CP}}{\longleftrightarrow} \bar{\nu}_\alpha \to \bar{\nu}_\beta \, . \qquad (7.53)$$

A T transformation interchanges the $\nu_\alpha \to \nu_\beta$ and $\nu_\beta \to \nu_\alpha$ channels,

$$\nu_\alpha \to \nu_\beta \stackrel{\text{T}}{\longleftrightarrow} \nu_\beta \to \nu_\alpha \, , \qquad (7.54)$$

or the $\bar{\nu}_\alpha \to \bar{\nu}_\beta$ and $\bar{\nu}_\beta \to \bar{\nu}_\alpha$ channels,

$$\bar{\nu}_\alpha \to \bar{\nu}_\beta \stackrel{\text{T}}{\longleftrightarrow} \bar{\nu}_\beta \to \bar{\nu}_\alpha \, . \qquad (7.55)$$

Finally, CPT interchanges the $\nu_\alpha \to \nu_\beta$ and $\bar{\nu}_\beta \to \bar{\nu}_\alpha$ channels,

$$\nu_\alpha \to \nu_\beta \stackrel{\text{CPT}}{\longleftrightarrow} \bar{\nu}_\beta \to \bar{\nu}_\alpha \, . \qquad (7.56)$$

[36] As discussed in the previous section 7.2, in the case of Majorana neutrinos, where the C transformation coincides with the identity, it is conventional to call *neutrinos* the states with negative helicity and *antineutrinos* the states with positive helicity.

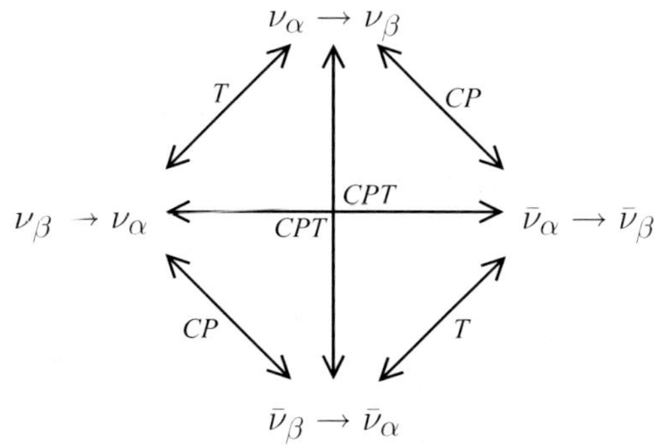

FIG. 7.1. Scheme of the CPT, CP, and T transformations that relate different flavor transition channels.

7.3.1 CPT

The CPT transformation is a symmetry of any local quantum field theory (see Refs. [988]), such as the SM, presented in chapter 3, and its extensions with the inclusion of neutrino masses, discussed in chapter 6. Since the theory of neutrino oscillations which we consider in this book is formulated in the framework of a local quantum field theory, CPT is a symmetry of the oscillation probabilities. Hence, we have

$$P_{\nu_\alpha \to \nu_\beta} = P_{\bar{\nu}_\beta \to \bar{\nu}_\alpha} \,. \tag{7.57}$$

In fact, we have, from eqn (7.50),

$$P_{\bar{\nu}_\beta \to \bar{\nu}_\alpha}(L, E) = \sum_{k,j} U_{\beta k} \, U_{\alpha k}^{*} \, U_{\beta j}^{*} \, U_{\alpha j} \, \exp\left(-i \, \frac{\Delta m_{kj}^2 L}{2E} \right) , \tag{7.58}$$

which coincides with $P_{\nu_\alpha \to \nu_\beta}(L, E)$ in eqn (7.23). The equality in eqn (7.57) is also easily verified by noting that the probability in eqn (7.58) is invariant under a CP transformation ($\bar{\nu}_\beta \to \bar{\nu}_\alpha \Rightarrow \nu_\beta \to \nu_\alpha$) which changes $U \leftrightarrows U^*$, followed by a T transformation ($\nu_\beta \to \nu_\alpha \Rightarrow \nu_\alpha \to \nu_\beta$) which changes $\alpha \leftrightarrows \beta$.

A special case of the equality in eqn (7.57) is the equality of the survival probabilities of neutrinos and antineutrinos,

$$P_{\nu_\alpha \to \nu_\alpha} = P_{\bar{\nu}_\alpha \to \bar{\nu}_\alpha} \,, \tag{7.59}$$

which has important phenomenological implications.

It is possible, however, that the description of nature through local Quantum Field Theories is approximate. In this case, there may be small violations of the CPT symmetry. Neutrino oscillations experiments could reveal such violations by measuring a nonzero value of the CPT asymmetry

$$A_{\alpha\beta}^{\mathrm{CPT}} = P_{\nu_\alpha \to \nu_\beta} - P_{\bar{\nu}_\beta \to \bar{\nu}_\alpha} \,. \tag{7.60}$$

7.3.2 CP

As shown in eqns (7.52), (7.53) and Fig. 7.1, the CP transformation interchanges neutrinos with negative helicity and antineutrinos with positive helicity, transforming the $\nu_\alpha \to \nu_\beta$ channel into the $\bar{\nu}_\alpha \to \bar{\nu}_\beta$ channel. As we have discussed in section 6.1.2, in the case of three-neutrino mixing, the mixing matrix is, in general, complex and leads to violations of the CP symmetry. Such violations can be revealed in neutrino oscillation experiments by measuring the CP asymmetry

$$A^{\mathrm{CP}}_{\alpha\beta} = P_{\nu_\alpha \to \nu_\beta} - P_{\bar{\nu}_\alpha \to \bar{\nu}_\beta} \,. \tag{7.61}$$

The CPT symmetry implies that the CP asymmetry is antisymmetric in the flavor indices α and β,

$$A^{\mathrm{CP}}_{\alpha\beta} = -A^{\mathrm{CP}}_{\beta\alpha} \,, \tag{7.62}$$

because the CPT relation in eqn (7.57), $P_{\nu_\alpha \to \nu_\beta} = P_{\bar{\nu}_\beta \to \bar{\nu}_\alpha}$, also implies $P_{\bar{\nu}_\alpha \to \bar{\nu}_\beta} = P_{\nu_\beta \to \nu_\alpha}$. Hence, it is clear that a CP asymmetry can be measured only in transitions between different flavors.

As discussed in section 7.2, the oscillation probabilities of neutrinos and antineutrinos are related by complex conjugation of the elements of the mixing matrix. Confronting eqns (7.38) and (7.51), one can see that they differ only in the sign of the terms depending on the imaginary parts of the quartic products of the elements of the mixing matrix. Thus, only these terms contribute to the CP asymmetry, leading to

$$A^{\mathrm{CP}}_{\alpha\beta}(L,E) = 4 \sum_{k>j} \mathfrak{Im}\left[U^*_{\alpha k}\, U_{\beta k}\, U_{\alpha j}\, U^*_{\beta j}\right] \sin\left(\frac{\Delta m^2_{kj} L}{2E}\right) \,. \tag{7.63}$$

This expression confirms that a CP asymmetry can be measured only in the transitions between different flavors, since, for $\alpha = \beta$, the imaginary parts in eqn (7.63) vanish.

Since the quartic products of elements of the mixing matrix in eqn (7.63) depend only on the Dirac phases in the mixing matrix, as we have discussed in section 6.2.1, CP violation in neutrino oscillations depend only on the Dirac phases in the mixing matrix.

A characteristic of the CP asymmetry in eqn (7.63), which is important for practical applications, is that its average over the distance L or the energy E vanishes. Therefore, in order to be able to measure CP violations in neutrino oscillations, experiments must be sensitive to the oscillatory behavior of the neutrino and antineutrino transition probabilities. This means that the distance L and energy E must be such that at least one of the phases $\Delta m^2_{kj} L/2E$ is of order one. If all the phases are much smaller than unity, the transition probabilities are too small to be measurable, whereas if all the phases are much larger than unity, the average of the CP asymmetry in eqn (7.63) over the uncertainty in L and E washes out all the sine functions. However, as we will see in section 7.7, the phase of order one should not be generated by the largest squared-mass difference, because in that case the effective CP asymmetries are identically zero.

7.3.3 T

If CPT is a symmetry of nature, the violation of CP symmetry implies the violation of T symmetry, which interchanges the initial and final states in neutrino oscillations (see eqns (7.54) and (7.55) and Fig. 7.1), as in other processes. In neutrino oscillation experiments it is possible to observe T violations by measuring the T asymmetries of neutrinos and antineutrinos

$$A^{\rm T}_{\alpha\beta} = P_{\nu_\alpha\to\nu_\beta} - P_{\nu_\beta\to\nu_\alpha}\,, \tag{7.64}$$

$$\bar{A}^{\rm T}_{\alpha\beta} = P_{\bar{\nu}_\alpha\to\bar{\nu}_\beta} - P_{\bar{\nu}_\beta\to\bar{\nu}_\alpha}\,. \tag{7.65}$$

The CPT relation in eqn (7.57) implies that

$$A^{\rm T}_{\alpha\beta} = -\bar{A}^{\rm T}_{\alpha\beta} = A^{\rm CP}_{\alpha\beta}\,. \tag{7.66}$$

In this case, measuring a CP asymmetry is equivalent to measuring a T asymmetry. As noted for the case of CP violation, T violations in neutrino oscillations depend only on the Dirac phases of the mixing matrix. In order to measure T violations, one should carry out neutrino oscillation experiments which are sensitive to the oscillatory behavior of the flavor transition probabilities.

7.4 Two-neutrino mixing

Two-neutrino mixing is an approximation in which only two massive neutrinos out of three are considered. That is, one neglects the coupling of the flavor neutrinos with the third massive neutrino that exists in nature. This approximation is very useful in practice for two reasons:

1. The oscillation formulas in the case of two-neutrino mixing are much simpler and depend on fewer parameters than in the case of three-neutrino mixing.
2. Since many experiments are not sensitive to the influence of three-neutrino mixing, the data can be analyzed by using an effective model with two-neutrino mixing.

In the case of two-neutrino mixing, we consider two flavor neutrinos ν_α and ν_β, which can be pure flavor neutrinos ($\alpha,\beta = e,\mu$ or $\alpha,\beta = e,\tau$ or $\alpha,\beta=\mu,\tau$) or linear combinations of pure flavor neutrinos (for example, $\nu_\alpha = \nu_e$ and $\nu_\beta = c_\mu\nu_\mu + c_\tau\nu_\tau$, with $c_\mu^2 + c_\tau^2 = 1$ in electron neutrino disappearance experiments or $\nu_e \to \nu_{\mu,\tau}$ experiments where ν_μ and ν_τ are not distinguished). The two flavor neutrino states are linear superpositions of the two massive neutrinos ν_1 and ν_2 with coefficients given by the elements of the two-neutrino effective mixing matrix

$$U = \begin{pmatrix} \cos\vartheta & \sin\vartheta \\ -\sin\vartheta & \cos\vartheta \end{pmatrix}, \tag{7.67}$$

where ϑ is the mixing angle, with a value in the interval $0 \le \vartheta \le \pi/2$. In the case of two-neutrino mixing, there is only one squared-mass difference:

$$\Delta m^2 \equiv \Delta m^2_{21} \equiv m_2^2 - m_1^2\,. \tag{7.68}$$

For convenience, we define ν_1 as the lightest of the two massive neutrinos, so that Δm^2 is positive[37].

From eqn (7.23), it is straightforward to derive the expression for the probability of $\nu_\alpha \to \nu_\beta$ transitions with $\alpha \neq \beta$:

$$P_{\nu_\alpha \to \nu_\beta}(L,E) = \frac{1}{2} \sin^2 2\vartheta \left[1 - \cos\left(\frac{\Delta m^2 L}{2E} \right) \right] \qquad (\alpha \neq \beta)\,, \tag{7.69}$$

or, equivalently,

$$P_{\nu_\alpha \to \nu_\beta}(L,E) = \sin^2 2\vartheta \, \sin^2\left(\frac{\Delta m^2 L}{4E} \right) \qquad (\alpha \neq \beta)\,. \tag{7.70}$$

In the case $\alpha = \beta$, the *survival probability* $P_{\nu_\alpha \to \nu_\alpha}(L,E)$ is easily obtained by unitarity from the transition probability in eqn (7.70):

$$P_{\nu_\alpha \to \nu_\alpha}(L,E) = 1 - P_{\nu_\alpha \to \nu_\beta}(L,E) = 1 - \sin^2 2\vartheta \, \sin^2\left(\frac{\Delta m^2 L}{4E} \right)\,. \tag{7.71}$$

From eqn (7.31), the oscillation length in this case is given by

$$L^{\text{osc}} = \frac{4\pi E}{\Delta m^2}\,, \tag{7.72}$$

and the average transition probability in eqn (7.33), which is the same as the incoherent transition probability in eqn (7.32), becomes

$$\langle P_{\nu_\alpha \to \nu_\beta} \rangle = \frac{1}{2} \sin^2 2\vartheta \qquad (\alpha \neq \beta)\,. \tag{7.73}$$

This expression can simply be derived from eqn (7.69) by noting that the average of the cosine function is zero.

One can see from eqn (7.70) that the mixing angle dependence of the transition probability is expressed by $\sin^2 2\vartheta$, which is symmetric under the exchange $\vartheta \leftrightarrows \pi/2 - \vartheta$. Since the allowed range of ϑ is $0 \leq \vartheta \leq \pi/2$, there is a degeneracy of the transition probability for ϑ and $\pi/2 - \vartheta$. However, it is important to keep in mind that the two possibilities correspond to two physically different mixings: if

---

[37] Some authors prefer to constrain the mixing angle in the interval $0 \leq \vartheta \leq \pi/4$, considering positive and negative Δm^2. The usefulness of this convention stems from the consideration that $0 \leq \vartheta \leq \pi/4$ corresponds to $0 \leq \sin^2 2\vartheta \leq 1$. Since the flavor transition probability in eqn (7.69) depends on the mixing angle through $\sin^2 2\vartheta$, considering $0 \leq \vartheta \leq \pi/4$ covers all the possibilities, leaving a complete uncertainty on the sign of Δm^2. We did not adopt such a convention because it may give the false impression that there is a discontinuity for $\vartheta = \pi/4$, related to the change of sign of Δm^2. This inconvenience becomes evident [359] in the case of neutrino oscillations in matter (see section 9.3), where the mixing-angle dependence of the flavor transition probability is not limited to $\sin^2 2\vartheta$. Hence, our convention is more appropriate for a clear unified treatment of neutrino oscillations in vacuum and in matter. In fact, it is adopted by most authors in the latest analyses of solar neutrino data, in which the mixing is quantified through $\tan^2\vartheta$ (see section 10.10).

$\vartheta < \pi/4$ the electron neutrino is composed more of ν_1 than ν_2 and vice versa for ν_μ; if $\vartheta > \pi/4$ the electron neutrino is composed more of ν_2 than ν_1 and vice versa for ν_μ. The degeneracy is resolved in neutrino oscillations in matter, as will be explained in section 9.3.

For the analyses of the data of reactor oscillation experiments, in which $E \sim 1\,\text{MeV}$, as well as those of accelerator oscillation experiments, in which $E \sim 1\,\text{GeV}$, it is convenient to write the transition probability in eqn (7.70) as

$$\begin{aligned} P_{\nu_\alpha \to \nu_\beta}(L, E) &= \sin^2 2\vartheta \, \sin^2\left(1.27 \, \frac{\Delta m^2[\text{eV}^2] \, L[\text{m}]}{E[\text{MeV}]}\right) \\ &= \sin^2 2\vartheta \, \sin^2\left(1.27 \, \frac{\Delta m^2[\text{eV}^2] \, L[\text{km}]}{E[\text{GeV}]}\right) . \end{aligned} \tag{7.74}$$

and the oscillation length as

$$L^{\text{osc}} = 2.47 \, \frac{E\,[\text{MeV}]}{\Delta m^2\,[\text{eV}^2]} \, \text{m} = 2.47 \, \frac{E\,[\text{GeV}]}{\Delta m^2\,[\text{eV}^2]} \, \text{km} . \tag{7.75}$$

The behavior of the transition probability in eqn (7.70) for $\sin^2 2\vartheta = 1$ as a function of $L/E\,[\text{km/GeV}]\,\Delta m^2\,[\text{eV}^2]$ is shown by the dashed line in Fig. 7.2. For fixed values of the squared-mass difference Δm^2 and of the energy E, the axis represents the distance L. The oscillation length in eqn (7.75) corresponds to the location of the first dip of the transition probability at $L/E\,[\text{km/GeV}]\,\Delta m^2\,[\text{eV}^2] = 2.47$, where the phase in the cosine function in eqn (7.69) is equal to 2π and the phase in the sine function in eqn (7.70) is equal to π. The transition probability is very small for $L \ll L^{\text{osc}}$ and oscillates very rapidly for $L \gg L^{\text{osc}}$ in the logarithmic scale of L.

From the absence of any phase in the two-neutrino effective mixing matrix in eqn (7.67), it is clear that there are no CP or T violations and the transition probabilities of neutrinos and antineutrinos as well as the probabilities of direct and inverted channels are all equal:

$$P_{\nu_\alpha \to \nu_\beta}(L, E) = P_{\nu_\beta \to \nu_\alpha}(L, E) = P_{\bar{\nu}_\alpha \to \bar{\nu}_\beta}(L, E) = P_{\bar{\nu}_\beta \to \bar{\nu}_\alpha}(L, E) . \tag{7.76}$$

7.5 Types of neutrino oscillation experiments

The discussion in the previous section helps us to understand the usual classification of the different types of neutrino oscillation experiments.

Neutrino oscillation experiments are divided into:

Appearance experiments. These experiments measure transitions between different neutrino flavors. If the final flavor to be searched for in the detector is not present in the initial beam, the background can be very small. In this case, an experiment can be sensitive to rather small values of the mixing angle.

Disappearance experiments. These experiments measure the survival probability of a neutrino flavor by counting the number of interactions in the detector

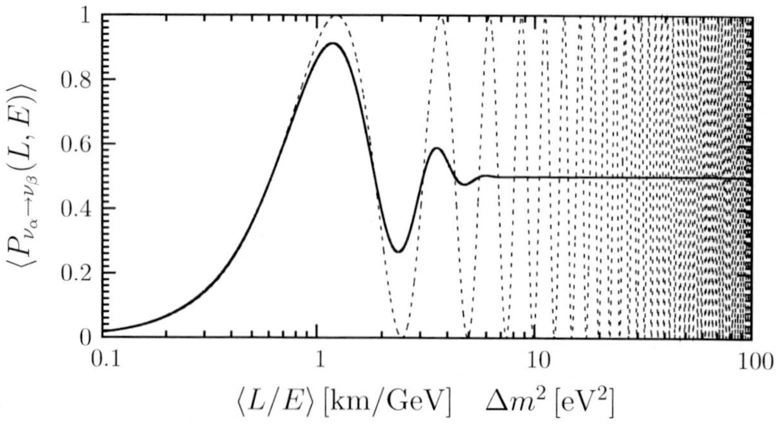

FIG. 7.2. Probability of $\nu_\alpha \to \nu_\beta$ transitions for $\sin^2 2\vartheta = 1$ as a function of $\langle L/E \rangle$ [km/GeV] Δm^2 [eV^2]. The average ratio $\langle L/E \rangle$ can also be expressed in units [m/MeV]. Solid line: transition probability averaged over a Gaussian L/E distribution with $\sigma_{L/E} = 0.2 \langle L/E \rangle$ (see eqn (7.93)). Dashed line: unaveraged transition probability (see eqn (7.70)), with $L/E = \langle L/E \rangle$.

and comparing it with the expected one. Since, even in the absence of oscillations, the number of detected events has statistical fluctuations, it is very difficult to reveal a small disappearance. Therefore, in this type of experiment, it is hard to measure small values of the mixing angle.

In the simplest case of two-neutrino mixing, an important characteristic of neutrino oscillations is that the transitions to different flavors cannot be measured if

$$\frac{\Delta m^2 L}{2E} \ll 1 \,. \tag{7.77}$$

On the other hand, for

$$\frac{\Delta m^2 L}{2E} \gg 1 \tag{7.78}$$

only the average transition probability in eqn (7.73) is observable, yielding information only on $\sin^2 2\vartheta$.

Since the value of Δm^2 is fixed by nature, different experiments can be designed in order to be sensitive to different values of Δm^2, by choosing appropriate values of the ratio L/E. The so-called *sensitivity* to Δm^2 of an experiment is the value of Δm^2 for which

$$\frac{\Delta m^2 L}{2E} \sim 1 \,. \tag{7.79}$$

Different types of neutrino oscillation experiments are traditionally classified depending on the average value of the ratio L/E for an experiment, which determines its sensitivity to Δm^2 through eqn (7.79) (see Table 7.1):

Short BaseLine experiments (SBL). They are divided as follows:

TABLE 7.1. Types of neutrino oscillation experiments with their typical source–detector distance, energy, and sensitivity to Δm^2, which is given by $E[\text{MeV}]/L[\text{m}] = E[\text{GeV}]/L[\text{km}]$ (see eqn (7.79)).

Type of experiment	L	E	Δm^2 sensitivity
Reactor SBL	$\sim 10\,\text{m}$	$\sim 1\,\text{MeV}$	$\sim 0.1\,\text{eV}^2$
Accelerator SBL (Pion DIF)	$\sim 1\,\text{km}$	$\gtrsim 1\,\text{GeV}$	$\gtrsim 1\,\text{eV}^2$
Accelerator SBL (Muon DAR)	$\sim 10\,\text{m}$	$\sim 10\,\text{MeV}$	$\sim 1\,\text{eV}^2$
Accelerator SBL (Beam Dump)	$\sim 1\,\text{km}$	$\sim 10^2\,\text{GeV}$	$\sim 10^2\,\text{eV}^2$
Reactor LBL	$\sim 1\,\text{km}$	$\sim 1\,\text{MeV}$	$\sim 10^{-3}\,\text{eV}^2$
Accelerator LBL	$\sim 10^3\,\text{km}$	$\gtrsim 1\,\text{GeV}$	$\gtrsim 10^{-3}\,\text{eV}^2$
ATM	20–$10^4\,\text{km}$	0.5–$10^2\,\text{GeV}$	$\sim 10^{-4}\,\text{eV}^2$
Reactor VLB	$\sim 10^2\,\text{km}$	$\sim 1\,\text{MeV}$	$\sim 10^{-5}\,\text{eV}^2$
Accelerator VLB	$\sim 10^4\,\text{km}$	$\gtrsim 1\,\text{GeV}$	$\gtrsim 10^{-4}\,\text{eV}^2$
SOL	$\sim 10^{11}\,\text{km}$	0.2–$15\,\text{MeV}$	$\sim 10^{-12}\,\text{eV}^2$

Reactor SBL. These are experiments that utilize large isotropic fluxes of electron antineutrinos produced in nuclear reactors by β^--decays of heavy nuclei (mainly fission fragments of ^{235}U, ^{238}U, ^{239}Pu, ^{241}Pu). A typical energy of reactor ν_e's is of the order of a few MeV and the source–detector distance in the reactor SBL experiments is several tens of meters. The range of L/E covered by the reactor SBL experiments and their sensitivity to Δm^2 are

$$\frac{L}{E} \lesssim 10\,\text{m/MeV} \quad \Longrightarrow \quad \Delta m^2 \gtrsim 0.1\,\text{eV}^2\,. \tag{7.80}$$

Since the antineutrino energy is too low to produce μ's or τ's, only the survival probability of $\bar{\nu}_e$'s can be measured by detecting, in a liquid scintillator, the inverse β-decay reaction

$$\bar{\nu}_e + p \to n + e^+\,, \tag{7.81}$$

with a threshold $E_{\text{th}} = 1.8\,\text{MeV}$. Experiments of this type which have been performed in the past are: ILL [710], Gosgen [1082], Rovno [32], Krasnoyarsk [1037], Bugey [363], Savannah River [563].

Accelerator SBL. These are experiments with beams of neutrinos produced by decay of pions, kaons, and muons created by a proton beam hitting a target. They can be divided as:

Pion Decay In Flight (DIF). These are the experiments with a neutrino beam composed mainly of muon neutrinos produced by the decay of pions and kaons initially produced by a proton beam hitting a target. The pions and kaons are allowed to decay in a decay tunnel of length of the order of 100 m. The beam is composed of ν_μ's or $\bar{\nu}_\mu$'s, depending on the polarity of the horn which focalizes the pions and kaons. In the case of a ν_μ beam produced by $\pi^+, K^+ \to \mu^+ + \nu_\mu$, there are about 1% of $\bar{\nu}_\mu$'s and about 1% of ν_e's, which are mainly due to $\mu^+ \to e^+ + \nu_e + \bar{\nu}_\mu$ decays[38]. The typical energy of the neutrinos is of the order of a few GeV, but can be much larger, depending on the energy of the proton beam. The typical source–detector distance in the accelerator SBL experiments is of the order of 1 km. The range of L/E covered by these experiments and their sensitivity to Δm^2 are

$$\frac{L}{E} \lesssim 1\,\text{km/GeV} \quad \Longrightarrow \quad \Delta m^2 \gtrsim 1\,\text{eV}^2\,. \tag{7.82}$$

In the case of $\nu_\mu \to \nu_\tau$ and $\bar{\nu}_\mu \to \bar{\nu}_\tau$ transitions, the energy must be about one order of magnitude larger in order to have sufficient τ productions (the τ production threshold is about 3.5 GeV; see Table 5.2), leading to $L/E \lesssim 0.1\,\text{km/GeV}$ and a sensitivity to $\Delta m^2 \gtrsim 10\,\text{eV}^2$. Experiments of this type which have been performed in the past are: BEBC [89] ($\nu_\mu \to \nu_e$), FNAL-E531 [1030] ($\nu_\mu \to \nu_\tau$), CDHSW [396] ($\overset{(-)}{\nu}_\mu \to \overset{(-)}{\nu}_\mu$), CCFR ($\overset{(-)}{\nu}_\mu \to \overset{(-)}{\nu}_\mu$ [983], $\overset{(-)}{\nu}_\mu \to \overset{(-)}{\nu}_\tau$ [789], $\overset{(-)}{\nu}_\mu \to \overset{(-)}{\nu}_e$ [910], $\overset{(-)}{\nu}_e \to \overset{(-)}{\nu}_\tau$ [823]), CHARM [212] ($\nu_\mu \to \nu_\mu$, $\nu_\mu \to \nu_e$, $\nu_\mu \to \nu_\tau$), BNL-E776 [265] ($\overset{(-)}{\nu}_\mu \to \overset{(-)}{\nu}_e$), CHORUS [422] ($\nu_\mu \to \nu_\tau$, $\nu_e \to \nu_\tau$), NOMAD [116] ($\nu_\mu \to \nu_\tau$, $\nu_e \to \nu_\tau$, $\nu_\mu \to \nu_e$), LSND [121] ($\nu_\mu \to \nu_e$), NuTeV [130] ($\overset{(-)}{\nu}_\mu \to \overset{(-)}{\nu}_e$).

Muon Decay At Rest (DAR). These are lower energy experiments with a beam composed of muon antineutrinos coming from the decay

$$\mu^+ \to e^+ + \nu_e + \bar{\nu}_\mu \tag{7.83}$$

of the μ^+ produced in the pion decay

$$\pi^+ \to \mu^+ + \nu_\mu \tag{7.84}$$

(the π^- are mostly absorbed by nuclei) and stopped in the target. These $\bar{\nu}_\mu$'s have energy of the order of several tens of MeV and can be used for measuring $\bar{\nu}_\mu \to \bar{\nu}_e$ transitions, because $\bar{\nu}_e$ are not present in the final products of π^+ and μ^+-decay. These experiments have a typical source–detector distance of the order of several tens of meters, with a range of L/E and a sensitivity to Δm^2 given by

$$\frac{L}{E} \lesssim 1\,\text{m/MeV} \quad \Longrightarrow \quad \Delta m^2 \gtrsim 1\,\text{eV}^2\,. \tag{7.85}$$

Experiments of this type that have been performed in the past are the $\bar{\nu}_\mu \to \bar{\nu}_e$ experiments LAMPF-0645 [460], LSND [37], KARMEN [105].

[38] Since pions and muons are ultrarelativistic and have about the same time for decaying in the decay tunnel, the ratio of the numbers of muon and pion decays is approximately equal to the ratio of the pion and muon lifetimes, which is about 1% (see eqns (A.155) and (A.162)). The small branching ratio of the $\pi^+ \to e^+ + \nu_e$ decay channel increases the electron neutrino fraction only by about 0.01% (see eqn (5.60)). Kaons, which are typically about 10% of pions, have a branching ratio of about 5% for $K^+ \to \pi^0 + e^+ + \nu_e$ decays, which give an additional contribution of about 0.5% to the ν_e contamination.

Beam Dump. Also called *prompt* neutrino experiments. In these experiments a proton beam with very high energy, of the order of some hundreds of GeV, is completely stopped in a thick target, called the *beam dump*, where the proton nucleon interactions generate heavy hadrons. The charmed heavy hadrons decay promptly with practically equal branching ratios into electrons and muons, emitting equal fluxes of electron and muon neutrinos with energies of the order of 10^2 GeV. A detector at a distance of the order of 1 km can measure the ratio of the electron and muon neutrino fluxes, whose deviations from unity would signal the presence of oscillations. The typical range of L/E and sensitivity to Δm^2 are given by

$$\frac{L}{E} \lesssim 10^{-2}\,\text{m/MeV} \quad \Longrightarrow \quad \Delta m^2 \gtrsim 10^2\,\text{eV}^2\,. \tag{7.86}$$

Experiments of this type that have been performed in the past are: BEBC [465, 559], CHARM [388], CDHSW [208].

Long BaseLine experiments (LBL). These are experiments which have sources similar to SBL experiments, but the source–detector distance is about two or three orders of magnitude larger. LBL experiments are classified as follows:

Reactor LBL. These are reactor neutrino experiments in which the source–detector distance is of the order of 1 km. The range of L/E covered by these experiments and their sensitivity to Δm^2 are

$$\frac{L}{E} \lesssim 10^{3}\,\text{m/MeV} \quad \Longrightarrow \quad \Delta m^2 \gtrsim 10^{-3}\,\text{eV}^2\,. \tag{7.87}$$

Experiments of this type which have been performed in the past are: CHOOZ [100] and Palo Verde [255]. Several future experiments are under study or in preparation: Double CHOOZ [104], KASKA [994], and others (see the reviews in Refs. [87, 557]).

Accelerator LBL. These are neutrino experiments with a muon neutrino or antineutrino beam produced by the decay in flight of pions and kaons created by shooting a proton beam to a target. The source–detector distance is about 10^2–10^3 km, leading to a range of L/E and a sensitivity to Δm^2 given by

$$\frac{L}{E} \lesssim 10^{3}\,\text{km/GeV} \quad \Longrightarrow \quad \Delta m^2 \gtrsim 10^{-3}\,\text{eV}^2\,. \tag{7.88}$$

The only experiment of this type which has produced results is K2K, which is still running. The results obtained so far have been published in Ref. [48, 46, 66] for the $\nu_\mu \to \nu_\mu$ channel and in Ref. [47] for the $\nu_\mu \to \nu_e$ channel. The MINOS experiment [131] ($\nu_\mu \to \nu_\mu$, $\nu_\mu \to \nu_e$) started in 2005. Two future experiments are under preparation: ICARUS [101] ($\nu_\mu \to \nu_\tau$, $\nu_\mu \to \nu_e$), OPERA [577] ($\nu_\mu \to \nu_\tau$). Another experiment called T2K [633] ($\nu_\mu \to \nu_\mu$, $\nu_\mu \to \nu_e$) with a neutrino superbeam, i.e. a conventional accelerator neutrino beam with much larger intensity, is being planned at Tokai, in Japan.

ATMospheric neutrino experiments (ATM). Primary cosmic rays interact with the upper layers of the atmosphere producing a large flux of pions and kaons which decay in the atmosphere into muons and neutrinos. Many muons further decay into electrons and neutrinos before hitting the ground. Atmospheric neutrino experiments detect these neutrinos. The energy of detectable atmospheric neutrinos cover a very wide range, from about 500 MeV to about 100 GeV. The source–detector distance ranges from about 20 km for neutrinos coming from above, to about 1.3×10^4 km for neutrinos coming from below, initially produced on the other side of the Earth. Therefore, in atmospheric neutrino experiments, typical values of L/E and the associated sensitivity to Δm^2 are

$$\frac{L}{E} \lesssim 10^4\,\mathrm{km/GeV} \quad \Longrightarrow \quad \Delta m^2 \gtrsim 10^{-4}\,\mathrm{eV}^2\,. \tag{7.89}$$

Atmospheric neutrino experiments which have been performed in the past are: Kamiokande [474], IMB [199], NUSEX [34], Frejus [352], Super-Kamiokande [110], MACRO [84], Soudan-2 [922]. The Super-Kamiokande and Soudan-2 experiments are still operating. Also the MINOS detector, which started in 2005, is sensitive to atmospheric neutrinos [745].

Very Long-Baseline experiments (VLB). These are experiments with a source–detector distance larger than LBL experiments by one or two orders of magnitude. They are:

Reactor VLB. These experiments measure the combined neutrino flux of many reactors at a distance of the order of 100 km, with a range of L/E and a sensitivity to Δm^2 given by

$$\frac{L}{E} \lesssim 10^5\,\mathrm{m/MeV} \quad \Longrightarrow \quad \Delta m^2 \gtrsim 10^{-5}\,\mathrm{eV}^2\,. \tag{7.90}$$

Only one experiment, KamLAND [398], is still in operation. In the future, the Borexino [1008] experiment may be able to perform as a reactor VLB experiment.

Accelerator VLB. These are accelerator neutrino experiments with a source–detector distance of the order of several thousands of km, comparable with the diameter of the Earth. They cover the range of L/E and the sensitivity to Δm^2 given by

$$\frac{L}{E} \lesssim 10^4\,\mathrm{km/GeV} \quad \Longrightarrow \quad \Delta m^2 \gtrsim 10^{-4}\,\mathrm{eV}^2\,. \tag{7.91}$$

These experiments are under study; new and more intense neutrino beams are needed in order to observe a sufficient number of events at such large distances. Candidate types of beam are: Super-Beam [552], Beta-Beam [1091, 796, 59, 576], and Neutrino Factory [511, 58, 99, 59, 576].

SOLar neutrino experiments (SOL). These are experiments which detect the neutrinos generated in the core of the Sun by the thermonuclear reactions that power the Sun. The Sun–Earth distance is about 1.5×10^{11} m. Since the energy of detectable solar neutrinos is in the range 0.2–15 MeV, the range of L/E and the sensitivity to Δm^2 in solar neutrino experiments are

$$\frac{L}{E} \lesssim 10^{12}\,\mathrm{m/MeV} \quad \Longrightarrow \quad \Delta m^2 \gtrsim 10^{-12}\,\mathrm{eV}^2 \,. \tag{7.92}$$

Hence, solar neutrino experiments are sensitive to extremely small values of Δm^2, much smaller than the sensitivity of the other experiment discussed above. Several solar neutrino experiments have been performed in the past: Homestake [323], Kamiokande [475], GALLEX [588], SAGE [19], GNO [76], SNO [45]. The SNO experiment is still in operation. In the future, the Borexino [1008] experiment, which is under construction, will start data taking and the KamLAND experiments may be improved to detect solar neutrinos [399]. Several other experiments are under study (see Ref. [938]).

The types of neutrino oscillation experiments with their typical source–detector distance, energy, and sensitivity to Δm^2 are summarized in Table 7.1.

7.6 Averaged transition probability

In practice, it is impossible to measure the oscillation probabilities for precise values of the neutrino propagation distance L and the neutrino energy E, because in any experiment both the source and the detection processes have some spatial uncertainty, i.e. the source has an energy spectrum and the energy resolution of the detector is finite. Therefore, in practice it is always necessary to average the oscillation probability over the appropriate distributions of the distance L and the energy E.

Considering the simplest case of two-neutrino mixing, the transition probability measured in practice is obtained by averaging the cosine function in eqn (7.69) over the appropriate distribution $\phi(L/E)$ of L/E:

$$\langle P_{\nu_\alpha \to \nu_\beta}(L,E) \rangle = \frac{1}{2} \sin^2 2\vartheta \left[1 - \left\langle \cos\left(\frac{\Delta m^2 L}{2E} \right) \right\rangle \right] \qquad (\alpha \neq \beta) \,, \tag{7.93}$$

with

$$\left\langle \cos\left(\frac{\Delta m^2 L}{2E} \right) \right\rangle = \int \cos\left(\frac{\Delta m^2 L}{2E} \right) \phi\left(\frac{L}{E} \right) \mathrm{d}\frac{L}{E} \,. \tag{7.94}$$

In order to illustrate the effect of these averages, let us consider the simplest case of a Gaussian L/E distribution with average $\langle L/E \rangle$ and standard deviation $\sigma_{L/E}$:

$$\phi\left(\frac{L}{E} \right) = \frac{1}{\sqrt{2\pi\sigma_{L/E}^2}} \exp\left[\frac{(L/E - \langle L/E \rangle)^2}{2\,\sigma_{L/E}^2} \right] \,. \tag{7.95}$$

In this case, the average of the cosine in eqn (7.93) can be calculated analytically, yielding

$$\left\langle \cos\left(\frac{\Delta m^2 L}{2E} \right) \right\rangle = \cos\left(\frac{\Delta m^2}{2} \left\langle \frac{L}{E} \right\rangle \right) \exp\left[-\frac{1}{2} \left(\frac{\Delta m^2}{2} \sigma_{L/E} \right)^2 \right] \,. \tag{7.96}$$

The solid line in Fig. 7.2 shows the corresponding averaged transition probability as a function of $\langle L/E \rangle\,[\mathrm{km/GeV}]\,\Delta m^2\,[\mathrm{eV}^2]$ for $\sin^2 2\vartheta = 1$ and $\sigma_{L/E} = 0.2\,\langle L/E \rangle$. It is reasonable to assume an uncertainty $\sigma_{L/E}$ proportional to $\langle L/E \rangle$. Since the uncertainties of L and E are independent, we have

$$\left(\frac{\sigma_{L/E}}{\langle L/E \rangle}\right)^2 = \left(\frac{\sigma_L}{\langle L \rangle}\right)^2 + \left(\frac{\sigma_E}{\langle E \rangle}\right)^2 , \tag{7.97}$$

where $\langle L \rangle$ is the average distance, σ_L is the distance uncertainty, $\langle E \rangle$ is the average energy, and σ_E is the energy uncertainty.

For fixed values of the squared-mass difference Δm^2 and of the average energy $\langle E \rangle$, the horizontal axis in Fig. 7.2 is approximately proportional to the average distance $\langle L \rangle$ and the solid curve represents the behavior of the transition probability in eqn (7.70) as a function of $\langle L \rangle$. One can see that for distances $\langle L \rangle \lesssim L^{\mathrm{osc}}$, which corresponds to $\langle L/E \rangle\,[\mathrm{km/GeV}]\,\Delta m^2\,[\mathrm{eV}^2] = 2.47$, the averaged transition probability oscillates as the unaveraged transition probability given by the dashed line, with amplitude somewhat suppressed, the suppression being dependent on the width $\sigma_{L/E}$ of the energy distribution (a larger width gives more suppression). For distances $\langle L \rangle \gg L^{\mathrm{osc}}$, the oscillations are completely suppressed and one can measure only the average transition probability in eqn (7.73).

If an oscillation experiment does not observe any oscillation, the data imply an upper limit on the averaged transition probability:

$$\langle P_{\nu_\alpha \to \nu_\beta}(L, E) \rangle \leq P^{\max}_{\nu_\alpha \to \nu_\beta} , \tag{7.98}$$

which implies an upper limit for $\sin^2 2\vartheta$ as a function of Δm^2: considering the averaged transition probability in eqn (7.93) we have

$$\sin^2 2\vartheta \leq \frac{2\,P^{\max}_{\nu_\alpha \to \nu_\beta}}{1 - \left\langle \cos\left(\dfrac{\Delta m^2 L}{2E}\right) \right\rangle} . \tag{7.99}$$

The solid line in Fig. 7.3a shows this upper limit as a function of $\Delta m^2\,[\mathrm{eV}^2]\,\langle L/E \rangle\,[\mathrm{km/GeV}]$ for $P^{\max}_{\nu_\alpha \to \nu_\beta} = 0.1$, and the Gaussian energy distribution in eqn (7.95) with $\sigma_{L/E} = 0.2\,\langle L/E \rangle$ (the dashed line shows the upper limit corresponding to the unaveraged transition probability, i.e. $\sigma_{L/E} = 0$). The solid line in Fig. 7.3a is usually called the *exclusion curve*, because it separates the allowed region from the indicated excluded region. From Fig. 7.3a one can see that there is no limit on $\sin^2 2\vartheta$ for

$$\frac{\Delta m^2}{2} \left\langle \frac{L}{E} \right\rangle \ll \pi \iff \Delta m^2\,[\mathrm{eV}^2] \left\langle \frac{L}{E} \right\rangle \frac{[\mathrm{km}]}{[\mathrm{GeV}]} \ll 1 . \tag{7.100}$$

The reason is that in this case the cosine function in eqn (7.99) is practically constant and equal to unity in the range of L/E relevant for the integration. This is true when the variance of L/E is always much smaller than the square of the

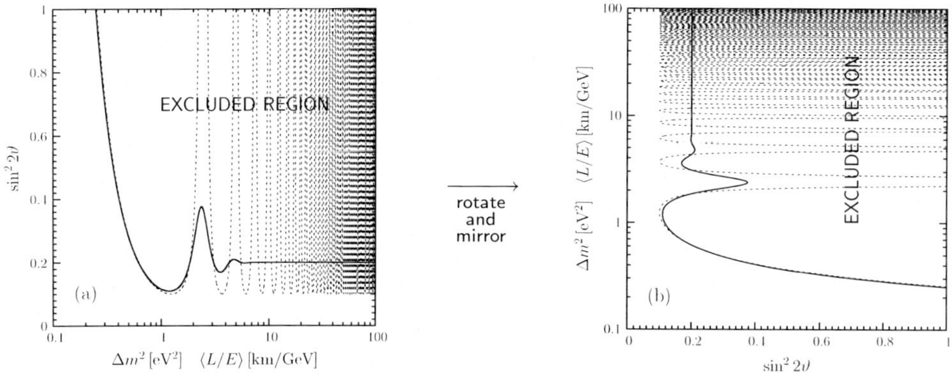

FIG. 7.3. Upper limit in eqn (7.99) with a Gaussian L/E distribution as a function of $\Delta m^2\,[\mathrm{eV}^2]\,\langle L/E\rangle\,[\mathrm{km/GeV}]$ for $P^{\mathrm{max}}_{\nu_\alpha\to\nu_\beta} = 0.1$. Solid line: $\sigma_{L/E} = 0.2\,\langle L/E\rangle$. Dashed line: $\sigma_{L/E} = 0$, which corresponds to $P_{\nu_\alpha\to\nu_\beta}(L,E) \leq P^{\mathrm{max}}_{\nu_\alpha\to\nu_\beta}$ with $L/E = \langle L/E\rangle$, where $P_{\nu_\alpha\to\nu_\beta}(L,E)$ is the unaveraged transition probability in eqn (7.70).

average value,

$$\left\langle\left(\frac{L}{E}-\left\langle\frac{L}{E}\right\rangle\right)^2\right\rangle = \left\langle\left(\frac{L}{E}\right)^2\right\rangle - \left(\left\langle\frac{L}{E}\right\rangle\right)^2 \ll \left(\left\langle\frac{L}{E}\right\rangle\right)^2, \tag{7.101}$$

which is the case in real experiments.

The most stringent bound on $\sin^2 2\vartheta$ is obtained for

$$\frac{\Delta m^2}{2}\left\langle\frac{L}{E}\right\rangle \simeq \pi \quad \Longleftrightarrow \quad \Delta m^2\,[\mathrm{eV}^2]\left\langle\frac{L}{E}\right\rangle\frac{[\mathrm{km}]}{[\mathrm{GeV}]} \simeq 1.24\,. \tag{7.102}$$

In this case, the cosine function in eqn (7.99) is approximately equal to -1 over the relevant range of L/E.

For much larger values of Δm^2, the limit on $\sin^2 2\vartheta$ does not depend on Δm^2:

$$\sin^2 2\vartheta \leq 2\,P^{\mathrm{max}}_{\nu_\alpha\to\nu_\beta}\,, \tag{7.103}$$

for

$$\frac{\Delta m^2}{2}\left\langle\frac{L}{E}\right\rangle \gg \pi \quad \Longleftrightarrow \quad \Delta m^2\,[\mathrm{eV}^2]\left\langle\frac{L}{E}\right\rangle\frac{[\mathrm{km}]}{[\mathrm{GeV}]} \gg 1\,. \tag{7.104}$$

In this case, the argument of the cosine function in eqn (7.99) is very large and rapidly oscillating in the relevant range of L/E, leading to a vanishing average.

Let us examine more closely the inequality in eqn (7.99) for small values of $\Delta m^2\,\langle L/E\rangle$. From eqn (7.101), if $\Delta m^2\,\langle L/E\rangle \ll 2\pi$, the argument of the cosine function in eqn (7.99) becomes small in the relevant range of L/E and the cosine function can be approximated by

$$\cos\left(\frac{\Delta m^2 L}{2E}\right) \simeq 1 - \frac{1}{2}\left(\frac{\Delta m^2 L}{2E}\right)^2 . \tag{7.105}$$

This approximation allows one to write the bound in eqn (7.99) as

$$\sin^2 2\vartheta \lesssim \frac{2\, P^{\max}_{\nu_\alpha\to\nu_\beta}}{\frac{1}{2}\left(\frac{\Delta m^2}{2}\right)^2 \left\langle\left(\frac{L}{E}\right)^2\right\rangle} . \tag{7.106}$$

Because of eqn (7.101), we can approximate $\langle (L/E)^2 \rangle$ with $(\langle L/E \rangle)^2$ and obtain

$$\sin^2 2\vartheta \lesssim \frac{2\, P^{\max}_{\nu_\alpha\to\nu_\beta}}{\frac{1}{2}\left(\frac{\Delta m^2}{2}\left\langle\frac{L}{E}\right\rangle\right)^2} \iff \sin^2 2\vartheta \lesssim \frac{0.62\, P^{\max}_{\nu_\alpha\to\nu_\beta}}{\left(\Delta m^2\,[\mathrm{eV}^2]\left\langle\frac{L}{E}\right\rangle\frac{[\mathrm{km}]}{[\mathrm{GeV}]}\right)^2} . \tag{7.107}$$

The right-hand side of this inequality is a quadratic function of Δm^2, which appears as the straight-line part of the exclusion curve for small $\Delta m^2\,[\mathrm{eV}^2]\,\langle L/E \rangle\,[\mathrm{km/GeV}]$ in Fig. 7.3a, due to the logarithmic scale. The intercept of the exclusion curve with the $\sin^2 2\vartheta = 1$ axis occurs at

$$\frac{\Delta m^2}{2}\left\langle\frac{L}{E}\right\rangle \simeq 2\sqrt{P^{\max}_{\nu_\alpha\to\nu_\beta}} \iff \Delta m^2\,[\mathrm{eV}^2]\left\langle\frac{L}{E}\right\rangle\frac{[\mathrm{km}]}{[\mathrm{GeV}]} \simeq 0.79\sqrt{P^{\max}_{\nu_\alpha\to\nu_\beta}} . \tag{7.108}$$

For example, one can see in Fig. 7.3a, which has been drawn for $\sqrt{P^{\max}_{\nu_\alpha\to\nu_\beta}} = 0.32$, that the intercept of the exclusion curve with the $\sin^2 2\vartheta = 1$ axis occurs at $\Delta m^2\,[\mathrm{eV}^2]\,\langle L/E \rangle\,[\mathrm{km/GeV}] \simeq 0.25$.

The experimental bound on $\sin^2 2\vartheta$ as a function of Δm^2 is usually presented by rotating and mirroring Fig. 7.3a, as shown in Fig. 7.3b, with $\sin^2 2\vartheta$ on the horizontal axis and Δm^2, for a fixed value of $\langle L/E \rangle$, on the vertical axis. In these figures which are called the *exclusion plots*, the excluded region of the oscillation parameters $\sin^2 2\vartheta$ and Δm^2 always lies on the right of the exclusion curve. Figure 7.4a shows an example of an exclusion curve obtained for $P^{\max}_{\nu_\alpha\to\nu_\beta} = 0.1$ and a Gaussian L/E distribution with $\langle L/E \rangle = 1\,\mathrm{km/GeV}$ and $\sigma_{L/E} = 0.2\,\mathrm{km/GeV}$.

In some disappearance experiments, the initial neutrino flux is not well known. In such a case, if the shape of the initial energy spectrum is known, information on transitions in different flavors can be obtained by measuring the distortions of the spectrum. Obviously, in this case one cannot obtain any information on oscillations if $\Delta m^2 \langle L/E \rangle \gg 2\pi$, because, as shown in Fig. 7.2 and eqn (7.103), the averaged transition probability is independent of energy. A similar situation is realized in experiments in which a *near* detector is placed upstream of the neutrino beam in order to measure the neutrino flux, whereas disappearance of the original neutrino flavor is measured at a *far* detector. No information on oscillations can be obtained if $\Delta m^2 \langle L/E \rangle_{\mathrm{near}} \gtrsim 2\pi$, because in this case the near detector already measures the averaged oscillated neutrino flux.

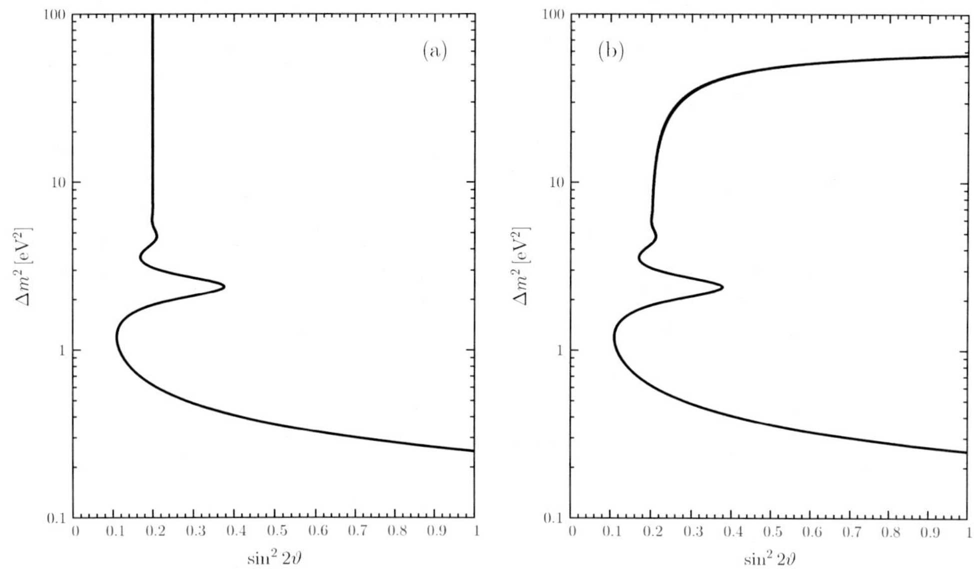

FIG. 7.4. Examples of exclusion plots in the plane of the two-neutrino oscillation parameters $\sin^2 2\vartheta$ and Δm^2. (a) Example of an exclusion plot obtained for $P^{\text{max}}_{\nu_\alpha\to\nu_\beta} = 0.1$ and a Gaussian L/E distribution with $\langle L/E\rangle = 1\,\text{km/GeV}$ and $\sigma_{L/E} = 0.2\,\text{km/GeV}$. (b) Example of an exclusion plot that could be obtained in a disappearance experiment with two detectors using eqn (7.111) with $R = 0.9$ and a Gaussian L/E distribution with $\langle L/E\rangle_{\text{near}} = 0.01\,\text{km/GeV}$, $\sigma^{\text{near}}_{L/E} = 0.002\,\text{km/GeV}$ and $\langle L/E\rangle_{\text{far}} = 1\,\text{km/GeV}$, $\sigma^{\text{far}}_{L/E} = 0.2\,\text{km/GeV}$.

As an example, let us consider a disappearance experiment with two detectors which measure

$$\frac{\langle P_{\nu_\alpha\to\nu_\alpha}(L,E)\rangle_{\text{far}}}{\langle P_{\nu_\alpha\to\nu_\alpha}(L,E)\rangle_{\text{near}}} \geq R\,, \tag{7.109}$$

where $0 \leq R < 1$. The inequality in eqn (7.109) means that the experiment did not find any indication of flavor transitions, which then yields an exclusion curve in the $\sin^2 2\vartheta$-Δm^2 plane. Since the averaged survival probability is given by

$$\langle P_{\nu_\alpha\to\nu_\alpha}(L,E)\rangle = 1 - \frac{1}{2}\sin^2 2\vartheta \left[1 - \left\langle \cos\left(\frac{\Delta m^2 L}{2E}\right)\right\rangle\right], \tag{7.110}$$

the inequality in eqn (7.109) can be written as

$$\sin^2 2\vartheta \leq 2\left[1 - (1-R)^{-1}\left(\left\langle \cos\left(\frac{\Delta m^2 L}{2E}\right)\right\rangle_{\text{far}} - R\left\langle \cos\left(\frac{\Delta m^2 L}{2E}\right)\right\rangle_{\text{near}}\right)\right]^{-1}. \tag{7.111}$$

Figure 7.4b shows the exclusion curve obtained with eqn (7.111) for $R = 0.9$ and a Gaussian L/E distribution with $\langle L/E\rangle_{\text{near}} = 0.01\,\text{km/GeV}$, $\sigma^{\text{near}}_{L/E} = 0.002\,\text{km/GeV}$ and $\langle L/E\rangle_{\text{far}} = 1\,\text{km/GeV}$, $\sigma^{\text{far}}_{L/E} = 0.2\,\text{km/GeV}$. One can see that there is no limit on $\sin^2 2\vartheta$ for $\Delta m^2 \gtrsim 60\,\mathrm{eV}^2$, which corresponds to $\Delta m^2 \langle L/E \rangle_{\mathrm{near}} \gtrsim 3$. For such large values of Δm^2, oscillations occur between the source and the near detector, preventing a measurement of the initial neutrino flux.

If an experiment detects a positive signal, the flavor transition probability is limited in a range which corresponds to an allowed band in the $\sin^2 2\vartheta$-Δm^2 plane, as shown in the example in Fig. 7.5a, obtained for $0.05 \leq \langle P_{\nu_\alpha \to \nu_\beta} \rangle \leq 0.15$ and a Gaussian L/E distribution with $\langle L/E \rangle = 1\,\mathrm{km/GeV}$ and $\sigma_{L/E} = 0.15\,\mathrm{km/GeV}$. However, in most experiments the position and energy of each event is measured, with some uncertainty. Therefore, there is not only one global L/E distribution, but typically the data are binned and each bin has its own L/E distribution. The allowed bands of the different bins are combined in a statistical analysis that typically yields one or more allowed regions in the $\sin^2 2\vartheta$-Δm^2 plane. The effect of combining the results of different L/E distributions is illustrated by Fig. 7.5b, where we have plotted two curves obtained with Gaussian L/E distributions, corresponding to $\langle P_{\nu_\alpha \to \nu_\beta} \rangle = 0.11$, $\langle L/E \rangle = 2\,\mathrm{km/GeV}$, $\sigma_{L/E} = 0.25\,\mathrm{km/GeV}$ and $\langle P_{\nu_\alpha \to \nu_\beta} \rangle = 0.09$, $\langle L/E \rangle = 0.4\,\mathrm{km/GeV}$, $\sigma_{L/E} = 0.05\,\mathrm{km/GeV}$. One can see that the two curves overlap at several points in the $\sin^2 2\vartheta$-Δm^2 plane. These points are compatible with both $\langle P_{\nu_\alpha \to \nu_\beta} \rangle$'s. If the uncertainties of the two $\langle P_{\nu_\alpha \to \nu_\beta} \rangle$'s are taken into account, a combined fit would yield allowed regions around the points of overlap of the two curves. It is clear that adding more information, i.e. more measurements of $P_{\nu_\alpha \to \nu_\beta}$ at different values of $\langle L/E \rangle$, results in fewer points of overlap of the corresponding curves and a smaller number of allowed regions in the $\sin^2 2\vartheta$-Δm^2 plane. The final experimental goal is, of course, to select only one, possibly small, allowed region in the $\sin^2 2\vartheta$-Δm^2 plane.

So far we have considered averaged oscillations only in the simplest case of two-neutrino mixing. In a general case of mixing of any number of neutrinos, the average of the oscillation probability in eqn (7.37) over an appropriate distribution $\phi(L/E)$ of L/E is given by

$$\begin{aligned}\langle P_{\nu_\alpha \to \nu_\beta}(L,E) \rangle = \delta_{\alpha\beta} &- 2 \sum_{k>j} \mathfrak{Re}\left[U_{\alpha k}^* \, U_{\beta k} \, U_{\alpha j} \, U_{\beta j}^*\right] \left[1 - \left\langle \cos\left(\frac{\Delta m_{kj}^2 L}{2E}\right) \right\rangle\right] \\ &+ 2 \sum_{k>j} \mathfrak{Im}\left[U_{\alpha k}^* \, U_{\beta k} \, U_{\alpha j} \, U_{\beta j}^*\right] \left\langle \sin\left(\frac{\Delta m_{kj}^2 L}{2E}\right) \right\rangle . \end{aligned} \tag{7.112}$$

In the case of a Gaussian L/E distribution, the averaged cosine is given by eqn (7.96) and the averaged sine is given by

$$\left\langle \sin\left(\frac{\Delta m_{kj}^2 L}{2E}\right) \right\rangle = \sin\left(\frac{\Delta m^2}{2} \left\langle \frac{L}{E} \right\rangle\right) \exp\left[-\frac{1}{2}\left(\frac{\Delta m^2}{2}\sigma_{L/E}\right)^2\right] . \tag{7.113}$$

It is clear that, in general, the expression in eqn (7.112) is a complicated function of several parameters (squared mass differences, mixing angles, and CP phases).

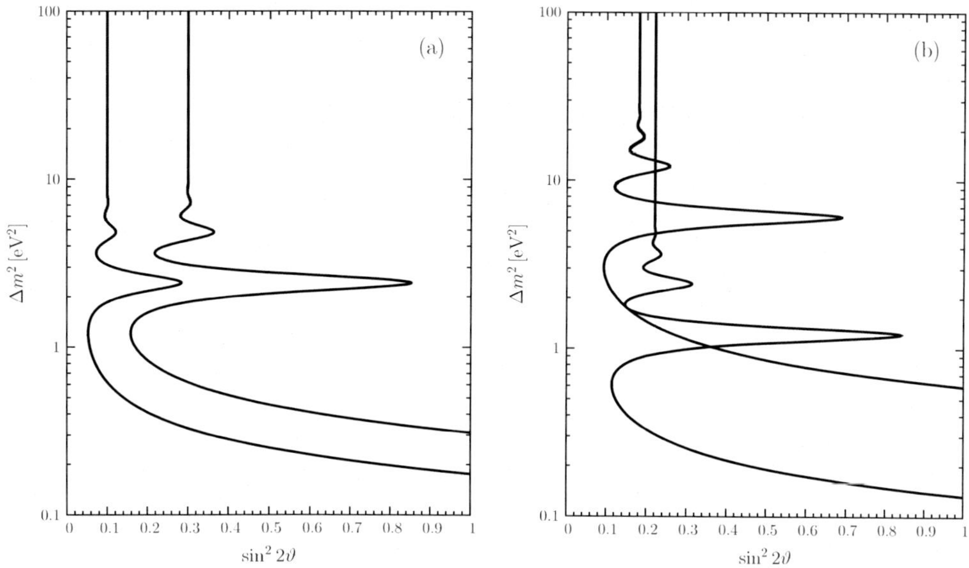

FIG. 7.5. Examples of plots in the plane of the two-neutrino oscillation parameters $\sin^2 2\vartheta$ and Δm^2. (a) Example of an allowed region, between the two curves, obtained for $0.05 \leq P_{\nu_\alpha \to \nu_\beta} \leq 0.15$ and a Gaussian L/E distribution with $\langle L/E \rangle = 1\,\text{km/GeV}$ and $\sigma_{L/E} = 0.15\,\text{km/GeV}$. (b) Example of two best-fit curves obtained with a Gaussian L/E distribution; the curve which reaches lower values of Δm^2 corresponds to $P_{\nu_\alpha \to \nu_\beta} = 0.11$, $\langle L/E \rangle = 2\,\text{km/GeV}$, and $\sigma_{L/E} = 0.25\,\text{km/GeV}$; the other curve, which reaches lower values of $\sin^2 2\vartheta$ for large Δm^2, corresponds to $P_{\nu_\alpha \to \nu_\beta} = 0.09$, $\langle L/E \rangle = 0.4\,\text{km/GeV}$, and $\sigma_{L/E} = 0.05\,\text{km/GeV}$.

7.7 Large Δm^2 dominance

In this section we consider the case in which one scale of neutrino squared-mass differences is much larger than the others. This means that the massive neutrinos can be divided in two groups, which we call A and B, such that all the squared-mass differences between two neutrinos belonging to different groups are much larger than all the squared-mass differences between two neutrinos belonging to one of the two groups, as illustrated schematically in Fig. 7.6.

This type of schemes is realized, for example, in the case of three-neutrino mixing with a hierarchy of Δm^2's (see Fig. 13.1 on page 453).

Let us denote with N_A and N_B the numbers of massive neutrinos belonging, respectively, to A and B, with the total number of massive neutrinos equal to $N = N_A + N_B$. We assign the numbers of the massive neutrinos in such a way that $\nu_1, \ldots, \nu_{N_A} \in A$ and $\nu_{N_A+1}, \ldots, \nu_N \in B$. Then, there is dominance of the largest Δm^2 if

$$|\Delta m^2_{N1}| \gg |\Delta m^2_{kj}| \quad \text{for} \quad k, j \leq N_A \quad \text{or} \quad k, j > N_A \,. \tag{7.114}$$

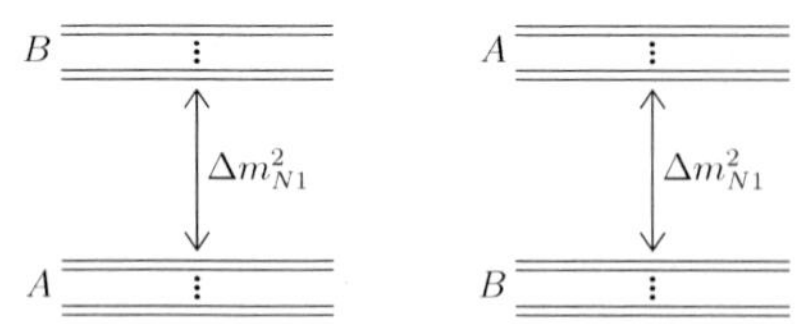

FIG. 7.6. Schematic illustration of the two possible types of neutrino mass spectra with a dominant scale of squared-mass difference. The dots indicate possible massive neutrinos with squared-mass differences much smaller than $|\Delta m^2_{N1}|$.

In this case

$$|\Delta m^2_{kj}| \simeq |\Delta m^2_{N1}| \quad \text{for} \quad (k \leq N_A \text{ and } j > N_A) \quad \text{or} \quad (k > N_A \text{ and } j \leq N_A) \,. \tag{7.115}$$

Hence, Δm^2_{N1} is the dominant squared-mass difference.

Let us consider an experiment which is sensitive to oscillations generated by the dominant squared-mass difference Δm^2_{N1}. This means that

$$\frac{|\Delta m^2_{N1}|}{2} \left\langle \frac{L}{E} \right\rangle \sim \pi \,, \tag{7.116}$$

and

$$\frac{|\Delta m^2_{kj}|}{2} \left\langle \frac{L}{E} \right\rangle \ll \pi \quad \text{for} \quad k,j \leq N_A \quad \text{or} \quad k,j > N_A \,. \tag{7.117}$$

In this case, the oscillation probability in eqn (7.23),

$$P_{\nu_\alpha \to \nu_\beta}(L,E) = \left| \sum_k U^*_{\alpha k} \, U_{\beta k} \, \exp\left(-i \, \frac{\Delta m^2_{k1} L}{2E} \right) \right|^2 \,, \tag{7.118}$$

can be separated into the contributions with $k \leq N_A$ and those with $k > N_A$,

$$\begin{aligned} P_{\nu_\alpha \to \nu_\beta}(L,E) = \Bigg| & \sum_{k \leq N_A} U^*_{\alpha k} \, U_{\beta k} \, \exp\left(-i \, \frac{\Delta m^2_{k1} L}{2E} \right) \\ & + \sum_{k > N_A} U^*_{\alpha k} \, U_{\beta k} \, \exp\left(-i \, \frac{\Delta m^2_{k1} L}{2E} \right) \Bigg|^2 \,. \end{aligned} \tag{7.119}$$

Because of eqns (7.116) and (7.117), we can neglect the phases of the first group and we can approximate all the phases of the second group with $\Delta m^2_{N1} L/2E$, leading to the effective probability

$$P^{\text{eff}}_{\nu_\alpha \to \nu_\beta}(L,E) = \left| \sum_{k \leq N_A} U^*_{\alpha k} \, U_{\beta k} + \exp\left(-i \, \frac{\Delta m^2_{N1} L}{2E} \right) \sum_{k > N_A} U^*_{\alpha k} \, U_{\beta k} \right|^2 \,. \tag{7.120}$$

Using the unitarity relation in eqn (7.28), the above probability can be written as

$$P^{\text{eff}}_{\nu_\alpha\to\nu_\beta}(L,E) = \left|\delta_{\alpha\beta} - \left[1 - \exp\left(-i\,\frac{\Delta m^2_{N1} L}{2E}\right)\right] \widetilde{\sum_k} U^*_{\alpha k}\, U_{\beta k}\right|^2 , \qquad (7.121)$$

where the symbol $\widetilde{\sum}_k$ indicates a sum over the index k that can be performed either over the range $1, \ldots, N_A$ or over the range $N_A + 1, \ldots, N$:

$$\widetilde{\sum_k} U^*_{\alpha k}\, U_{\beta k} = \sum_{k \leq N_A} U^*_{\alpha k}\, U_{\beta k} \qquad \text{or} \qquad \widetilde{\sum_k} U^*_{\alpha k}\, U_{\beta k} = \sum_{k > N_A} U^*_{\alpha k}\, U_{\beta k} . \qquad (7.122)$$

The squared modulus in eqn (7.121) is easily evaluated, yielding

$$P^{\text{eff}}_{\nu_\alpha\to\nu_\beta}(L,E) = \delta_{\alpha\beta} - 4\left[\delta_{\alpha\beta}\widetilde{\sum_k}|U^*_{\alpha k}|^2 - \left|\widetilde{\sum_k} U^*_{\alpha k}\, U_{\beta k}\right|^2\right] \sin^2\left(\frac{\Delta m^2_{N1} L}{4E}\right) . \qquad (7.123)$$

For the transition probabilities we have

$$P^{\text{eff}}_{\nu_\alpha\to\nu_\beta}(L,E) = 4\left|\widetilde{\sum_k} U^*_{\alpha k}\, U_{\beta k}\right|^2 \sin^2\left(\frac{\Delta m^2_{N1} L}{4E}\right) \qquad (\alpha \neq \beta) . \qquad (7.124)$$

Comparing with eqn (7.70), we see that the transition probability in eqn (7.124) corresponds to an effective two-neutrino-like transition probability,

$$P^{\text{eff}}_{\nu_\alpha\to\nu_\beta}(L,E) = \sin^2 2\vartheta^{\text{eff}}_{\alpha\beta} \sin^2\left(\frac{\Delta m^2_{N1} L}{4E}\right) \qquad (\alpha \neq \beta) , \qquad (7.125)$$

with effective squared-mass difference Δm^2_{N1} and effective mixing angle $\vartheta^{\text{eff}}_{\alpha\beta}$ given by

$$\sin^2 2\vartheta^{\text{eff}}_{\alpha\beta} = 4\left|\widetilde{\sum_k} U^*_{\alpha k}\, U_{\beta k}\right|^2 . \qquad (7.126)$$

This definition of an effective mixing angle is consistent, because using the Cauchy–Schwartz inequality on the two equivalent expressions in eqn (7.122) we have

$$\left|\widetilde{\sum_k} U^*_{\alpha k}\, U_{\beta k}\right|^2 \leq \left(\sum_{k \leq N_A} |U_{\alpha k}|^2\right)\left(\sum_{k \leq N_A} |U_{\beta k}|^2\right) \qquad (7.127)$$

and

$$\begin{aligned} \left|\widetilde{\sum_k} U^*_{\alpha k}\, U_{\beta k}\right|^2 &\leq \left(\sum_{k > N_A} |U_{\alpha k}|^2\right)\left(\sum_{k > N_A} |U_{\beta k}|^2\right) \\ &= \left(1 - \sum_{k \leq N_A} |U_{\alpha k}|^2\right)\left(1 - \sum_{k \leq N_A} |U_{\beta k}|^2\right) . \end{aligned} \qquad (7.128)$$

Hence, the maximum of $\left|\widetilde{\sum}_k U_{\alpha k}^* U_{\beta k}\right|^2$ is 1/4, reached for $\sum_{k \leq N_A} |U_{\alpha k}|^2 = \sum_{k \leq N_A} |U_{\beta k}|^2 = 1/2$. The definition in eqn (7.126) corresponds to two possible values of the effective mixing angle symmetric with respect to $\pi/4$, which are given by

$$\sin \vartheta_{\alpha\beta}^{\text{eff}} = \frac{1}{\sqrt{2}} \left(1 \pm \sqrt{1 - 4 \left| \widetilde{\sum_k} U_{\alpha k}^* U_{\beta k} \right|^2} \right)^{1/2} . \tag{7.129}$$

The effective transition probability in eqn (7.125) is very useful for the analysis of experimental data of appearance experiments which satisfy the conditions in eqns (7.116) and (7.117), since it implies that the two effective parameters are sufficient for the description of the data.

It is interesting to note that, although the effective mixing angle in eqn (7.129) may depend on the Dirac phases in the mixing matrix, it is invariant under CP or T transformation, which correspond to a charge conjugation of the elements of the mixing matrix. Hence, the effective mixing angle can depend only on the cosines of the Dirac phases in the mixing matrix. Its invariance under CP and T transformation implies that

$$P_{\nu_\alpha \to \nu_\beta}^{\text{eff}} = P_{\bar{\nu}_\alpha \to \bar{\nu}_\beta}^{\text{eff}} = P_{\bar{\nu}_\beta \to \bar{\nu}_\alpha}^{\text{eff}} = P_{\nu_\beta \to \nu_\alpha}^{\text{eff}} , \tag{7.130}$$

i.e. all the CP and T asymmetries are zero:

$$A_{\alpha\beta}^{\text{T}} = -\bar{A}_{\alpha\beta}^{\text{T}} = A_{\alpha\beta}^{\text{CP}} = 0 . \tag{7.131}$$

Therefore, an experiment which is sensitive only to the largest squared-mass difference cannot probe CP violations.

From eqn (7.121), the survival probabilities in the case of large Δm^2 dominance are given by

$$P_{\nu_\alpha \to \nu_\alpha}^{\text{eff}}(L, E) = 1 - 4 \left(\widetilde{\sum_k} |U_{\alpha k}|^2 \right) \left(1 - \widetilde{\sum_k} |U_{\alpha k}|^2 \right) \sin^2 \left(\frac{\Delta m_{N1}^2 L}{4E} \right) . \tag{7.132}$$

Comparing with eqn (7.71), one can see that the survival probability in eqn (7.132) corresponds to an effective two-neutrino-like survival probability,

$$P_{\nu_\alpha \to \nu_\alpha}^{\text{eff}}(L, E) = 1 - \sin^2 2\vartheta_{\alpha\alpha}^{\text{eff}} \sin^2 \left(\frac{\Delta m_{N1}^2 L}{4E} \right) , \tag{7.133}$$

with an effective squared-mass difference Δm_{N1}^2 and an effective mixing angle $\vartheta_{\alpha\alpha}^{\text{eff}}$ given by

$$\sin^2 2\vartheta_{\alpha\alpha}^{\text{eff}} = 4 \left(\widetilde{\sum_k} |U_{\alpha k}|^2 \right) \left(1 - \widetilde{\sum_k} |U_{\alpha k}|^2 \right) . \tag{7.134}$$

This definition of an effective mixing angle makes sense, since the upper bound of the right-hand side is one, which is reached for $\widetilde{\sum}_k |U_{\alpha k}|^2 = \frac{1}{2}$. Equation (7.134) corresponds to two possible values of the effective mixing angle symmetric with respect to $\pi/4$, which are given by

$$\sin \vartheta_{\alpha\alpha}^{\text{eff}} = \widetilde{\sum_k |U_{\alpha k}|^2} \quad \text{or} \quad \sin \vartheta_{\alpha\alpha}^{\text{eff}} = 1 - \widetilde{\sum_k |U_{\alpha k}|^2} \,. \tag{7.135}$$

The effective survival probability in eqn (7.125), which depends only on two parameters, is very useful for the analysis of experimental data of disappearance experiments in which the conditions in eqns (7.116) and (7.117) are satisfied.

The effective two-neutrino-like oscillation probabilities averaged over the appropriate distributions of the distance L and the energy E have the same properties as the averaged oscillations in two-neutrino mixing discussed in section 7.6.

Let us emphasize that from a theoretical point of view the large Δm^2 dominance is a realistic possibility. The simplest and most plausible case in which it is realized is through a hierarchy of squared-mass differences, which could be due to a hierarchy of neutrino masses. If there is a hierarchy of squared-mass differences, there is only one squared-mass which is much larger than the others, leading to a considerable simplification of the expressions of the effective mixing angles in eqns (7.126) and (7.134):

$$\sin^2 2\vartheta_{\alpha\beta}^{\text{eff}} = 4\,|U_{\alpha N}|^2\,|U_{\beta N}|^2 \quad (\alpha \neq \beta)\,, \qquad \sin^2 2\vartheta_{\alpha\alpha}^{\text{eff}} = 4\,|U_{\alpha N}|^2\left(1 - |U_{\alpha N}|^2\right) . \tag{7.136}$$

We will see in chapter 13 that the results of neutrino oscillation experiments show that the neutrino spectrum has a hierarchy of squared-mass differences and the effective mixing angles in eqn (7.136) can be applied with success to the analysis of the data of experiments which are sensitive only to the largest Δm^2. The relations in eqn (7.131) imply that these experiments cannot measure CP or T violations.

7.8 Active small Δm^2

In this section we discuss a case in which there is a scale of neutrino squared-mass differences which is much smaller than the largest scale of squared-mass differences. We derive the effective oscillation probabilities in experiments which are sensitive to such a small Δm^2. In these experiments the small Δm^2 under consideration is *active*, because it generates the measurable oscillations which depend on L and E. It is possible that in addition to the largest squared-mass difference there are also squared-mass differences much smaller than the active one. Therefore, we consider a general case in which the massive neutrinos can be divided in three groups that we call A_1, A_2, and B, with the following properties. All the squared-mass differences between a neutrino belonging to A_1 and a neutrino belonging to A_2 are much larger than all the squared-mass differences between two neutrinos belonging either to A_1 or to A_2 or to B. Moreover, all the squared-mass differences between a neutrino belonging to A_1 and a neutrino belonging to A_2 are much smaller than all the squared-mass differences between a neutrino belonging to A_1 or A_2 and a neutrino belonging to B. These properties are illustrated schematically in Fig. 7.7. A more general case in which there can also be squared-mass differences between two

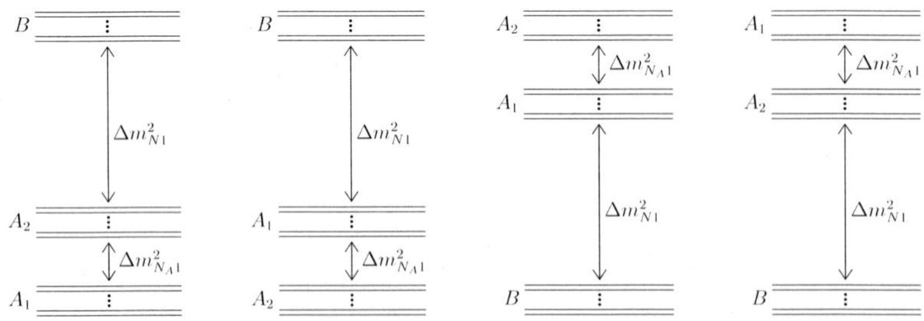

FIG. 7.7. Schematic illustration of the four possible types of neutrino mass spectra considered in section 7.8. The dots indicate possible massive neutrinos with squared-mass differences much smaller than $|\Delta m^2_{N_A 1}|$.

neutrinos belonging to B which are much larger than the squared-mass differences between a neutrino belonging to A_1 and a neutrino belonging to A_2 can also be treated, but we avoid it because the formalism becomes too cumbersome.

The general case discussed in this section is useful in practice, for example, in the study of neutrino oscillations in vacuum due to Δm^2_{21} in a three-neutrino mixing scheme with a hierarchy of Δm^2's, as explained in section 13.1.2.

Let us denote by N_{A_1}, N_{A_2}, and N_B the numbers of massive neutrinos belonging, respectively, to A_1, A_2, and B, with the total number of massive neutrinos equal to $N = N_{A_1} + N_{A_2} + N_B$. It is also convenient to call $A = A_1 \cup A_2$ and denote with N_A the number of massive neutrinos belonging to A ($N_A = N_{A_1} + N_{A_2}$). Let us assign the numbers of the massive neutrinos in such a way that $\nu_1, \ldots, \nu_{N_{A_1}} \in A_1$, $\nu_{N_{A_1}+1}, \ldots, \nu_{N_A} \in A_2$ and $\nu_{N_A+1}, \ldots, \nu_N \in B$. We consider a neutrino mass spectrum such that

$$|\Delta m^2_{N_A 1}| \gg |\Delta m^2_{kj}| \quad \text{for} \quad \left\{ \begin{array}{c} k, j \leq N_{A_1} \\ \text{or} \\ N_{A_1} < k, j \leq N_A \\ \text{or} \\ k, j > N_A \,, \end{array} \right. \tag{7.137}$$

$$|\Delta m^2_{N_A 1}| \ll |\Delta m^2_{kj}| \quad \text{for} \quad \left\{ \begin{array}{c} k \leq N_A \quad \text{and} \quad j > N_A \\ \text{or} \\ k > N_A \quad \text{and} \quad j \leq N_A \,. \end{array} \right. \tag{7.138}$$

We are interested in the oscillations generated by $\Delta m^2_{N_A 1}$. Because of the inequalities in eqn (7.137), we have

$$|\Delta m^2_{kj}| \simeq |\Delta m^2_{N_A 1}|$$
$$\text{for} \quad (k \leq N_{A_1} \text{ and } N_{A_1} < j \leq N_A) \quad \text{or} \quad (N_{A_1} < k \leq N_A \text{ and } j \leq N_{A_1}) \,. \tag{7.139}$$

In an experiment which is sensitive to the oscillations generated by $\Delta m^2_{N_A 1}$ we have

$$\frac{|\Delta m^2_{N_A 1}|}{2} \left\langle \frac{L}{E} \right\rangle \sim \pi \,, \tag{7.140}$$

$$\frac{|\Delta m^2_{kj}|}{2}\left\langle\frac{L}{E}\right\rangle \ll \pi \quad \text{for} \quad \begin{cases} k,j \leq N_{A_1} \\ \text{or} \\ N_{A_1} < k,j \leq N_A \\ \text{or} \\ k,j > N_A\,, \end{cases} \tag{7.141}$$

$$\frac{|\Delta m^2_{kj}|}{2}\left\langle\frac{L}{E}\right\rangle \gg \pi \quad \text{for} \quad \begin{cases} k \leq N_A \quad \text{and} \quad j > N_A \\ \text{or} \\ k > N_A \quad \text{and} \quad j \leq N_A\,. \end{cases} \tag{7.142}$$

The inequalities in eqns (7.141) and (7.142) imply that

$$\left\langle\cos\left(\frac{\Delta m^2_{kj}L}{2E}\right)\right\rangle \simeq 1\,, \quad \left\langle\sin\left(\frac{\Delta m^2_{kj}L}{2E}\right)\right\rangle \ll 1 \quad \text{for} \quad \begin{cases} k,j \leq N_{A_1} \\ \text{or} \\ N_{A_1} < k,j \leq N_A \\ \text{or} \\ k,j > N_A\,, \end{cases} \tag{7.143}$$

$$\left\langle\cos\left(\frac{\Delta m^2_{kj}L}{2E}\right)\right\rangle \ll 1\,, \quad \left\langle\sin\left(\frac{\Delta m^2_{kj}L}{2E}\right)\right\rangle \ll 1 \quad \text{for} \quad \begin{cases} k \leq N_A \,\text{and}\, j > N_A \\ \text{or} \\ k > N_A \,\text{and}\, j \leq N_A\,. \end{cases} \tag{7.144}$$

Using these approximations, the effective probability obtained from eqn (7.37) is

$$\begin{aligned} P^{\text{eff}}_{\nu_\alpha\to\nu_\beta}(L,E) = \delta_{\alpha\beta} \\ & - 2\left(\sum_{j\leq N_{A_1}}\sum_{k=N_{A_1}+1}^{N_A} \mathfrak{Re}\left[U^*_{\alpha k}\,U_{\beta k}\,U_{\alpha j}\,U^*_{\beta j}\right]\right)\left[1-\cos\left(\frac{\Delta m^2_{N_A 1}L}{2E}\right)\right] \\ & + 2\left(\sum_{j\leq N_{A_1}}\sum_{k=N_{A_1}+1}^{N_A} \mathfrak{Im}\left[U^*_{\alpha k}\,U_{\beta k}\,U_{\alpha j}\,U^*_{\beta j}\right]\right)\sin\left(\frac{\Delta m^2_{N_A 1}L}{2E}\right) \\ & - 2\sum_{j\leq N_A}\sum_{k>N_A} \mathfrak{Re}\left[U^*_{\alpha k}\,U_{\beta k}\,U_{\alpha j}\,U^*_{\beta j}\right]\,. \end{aligned} \tag{7.145}$$

Let us consider first the survival probabilities. From eqn (7.145) for $\alpha = \beta$ we get

$$\begin{aligned} P^{\text{eff}}_{\nu_\alpha\to\nu_\alpha}(L,E) = 1 - 4 & \left(\sum_{j\leq N_{A_1}} |U_{\alpha j}|^2\right)\left(\sum_{k=N_{A_1}+1}^{N_A} |U_{\alpha k}|^2\right)\sin^2\left(\frac{\Delta m^2_{N_A 1}L}{4E}\right) \\ & - 2\left(\sum_{k=1}^{N_A}|U_{\alpha k}|^2\right)\left(\sum_{k>N_A}|U_{\alpha k}|^2\right)\,, \end{aligned} \tag{7.146}$$

which can be written as

$$P^{\text{eff}}_{\nu_\alpha\to\nu_\alpha}(L,E) = 1 - 4\left(\sum_{j\leq N_{A_1}} |U_{\alpha j}|^2\right)\left(\sum_{k=N_{A_1}+1}^{N_A} |U_{\alpha k}|^2\right)\sin^2\left(\frac{\Delta m^2_{N_A 1} L}{4E}\right)$$
$$- 2\left(\sum_{k>N_A} |U_{\alpha k}|^2\right)\left(1 - \sum_{k>N_A} |U_{\alpha k}|^2\right), \tag{7.147}$$

Let us separate the contributions due to the mixing of ν_α with the massive neutrinos ν_k belonging to the group B from those belonging to the group A. This task can be accomplished by writing the survival probability in eqn (7.147) as

$$P^{\text{eff}}_{\nu_\alpha\to\nu_\alpha}(L,E) = \left(1 - \sum_{k>N_A} |U_{\alpha k}|^2\right)^2 P^{(N_A,1)}_{\nu_\alpha\to\nu_\alpha}(L,E) + \left(\sum_{k>N_A} |U_{\alpha k}|^2\right)^2, \tag{7.148}$$

with

$$P^{(N_A,1)}_{\nu_\alpha\to\nu_\alpha}(L,E) = 1 - 4\,\frac{\left(\sum_{k\leq N_{A_1}} |U_{\alpha k}|^2\right)\left(\sum_{k=N_{A_1}+1}^{N_A} |U_{\alpha k}|^2\right)}{\left(\sum_{k=1}^{N_A} |U_{\alpha k}|^2\right)^2}\sin^2\left(\frac{\Delta m^2_{N_A 1} L}{4E}\right). \tag{7.149}$$

This is a two-neutrino-like survival probability that can be written as

$$P^{(N_A,1)}_{\nu_\alpha\to\nu_\alpha}(L,E) = 1 - \sin^2 2\vartheta^{\text{eff}}_{\alpha\alpha}\,\sin^2\left(\frac{\Delta m^2_{N_A 1} L}{4E}\right), \tag{7.150}$$

with effective squared-mass difference $\Delta m^2_{N_A 1}$ and effective mixing angle $\vartheta^{\text{eff}}_{\alpha\alpha}$ given by

$$\sin^2 2\vartheta^{\text{eff}}_{\alpha\alpha} = 4\,\frac{\left(\sum_{k\leq N_{A_1}} |U_{\alpha k}|^2\right)\left(\sum_{k=N_{A_1}+1}^{N_A} |U_{\alpha k}|^2\right)}{\left(\sum_{k=1}^{N_A} |U_{\alpha k}|^2\right)^2}. \tag{7.151}$$

This definition of an effective mixing angle is consistent because

$$\frac{\left(\sum_{k\leq N_{A_1}} |U_{\alpha k}|^2\right)\left(\sum_{k=N_{A_1}+1}^{N_A} |U_{\alpha k}|^2\right)}{\left(\sum_{k=1}^{N_A} |U_{\alpha k}|^2\right)^2} = \frac{\sum_{k\leq N_{A_1}} |U_{\alpha k}|^2}{\sum_{k=1}^{N_A} |U_{\alpha k}|^2}\left(1 - \frac{\sum_{k\leq N_{A_1}} |U_{\alpha k}|^2}{\sum_{k=1}^{N_A} |U_{\alpha k}|^2}\right) \tag{7.152}$$

reaches its maximum value 1/4 for $\sum_{k\leq N_{A_1}} |U_{\alpha k}|^2 / \sum_{k=1}^{N_A} |U_{\alpha k}|^2 = 1/2$. There are two possible values of the effective mixing angles symmetric with respect to $\pi/4$, given by

$$\sin\vartheta^{\text{eff}}_{\alpha\alpha} = \sqrt{\frac{\sum_{k\leq N_{A_1}} |U_{\alpha k}|^2}{\sum_{k=1}^{N_A} |U_{\alpha k}|^2}} \quad \text{or} \quad \sin\vartheta^{\text{eff}}_{\alpha\alpha} = \sqrt{\frac{\sum_{k=N_{A_1}+1}^{N_A} |U_{\alpha k}|^2}{\sum_{k=1}^{N_A} |U_{\alpha k}|^2}}. \tag{7.153}$$

Equation (7.148) is rather meaningful, because it shows that the effective survival probability $P^{\text{eff}}_{\nu_\alpha\to\nu_\alpha}(L,E)$ deviates from being two-neutrino-like because of the mixing of ν_α with the massive neutrinos belonging to the group B, which contributes with a constant term and with a suppression factor of the two-neutrino-like survival probability $P^{(N_A,1)}_{\nu_\alpha\to\nu_\alpha}(L,E)$. Both effects depend only on the sum of the squared absolute values of the elements of the mixing matrix that connect ν_α with the massive neutrinos belonging to the group B.

The effective survival probability $P^{\text{eff}}_{\nu_\alpha\to\nu_\alpha}(L,E)$ represents a great simplification with respect to the general form of the survival probability, which depends on many parameters, whereas eqn (7.148) depends on only three parameters: $\Delta m^2_{N_A 1}$, $\sin^2 2\vartheta^{\text{eff}}_{\alpha\alpha}$ and $\sum_{k>N_A} |U_{\alpha k}|^2$. This is very useful in the analysis of experimental data.

In the limit of zero or negligible mixing of ν_α with the massive neutrinos belonging to the group B, the present case becomes coincident with the case of dominance of one scale of squared-mass differences discussed in the previous section 7.7. Indeed, the effective survival probability in eqn (7.148) is reduced to the two-neutrino-like survival probability in eqn (7.150), which is analogous to in eqn (7.133) with the appropriate change of the mixing parameters. The definition in eqn (7.151) of effective mixing angle becomes analogous to that in eqn (7.134) because in this case $\sum_{k=1}^{N_A} |U_{\alpha k}|^2 = 1$.

Unfortunately, the calculation of the transition probabilities is more cumbersome. Taking into account that $\alpha \neq \beta$, the unitary relation in eqn (7.28) implies that

$$\sum_{j\leq N_A}\sum_{k>N_A} U^*_{\alpha k}\,U_{\beta k}\,U_{\alpha j}\,U^*_{\beta j} = -\left|\widetilde{\sum_k} U^*_{\alpha k}\,U_{\beta k}\right|^2 , \tag{7.154}$$

where the symbol $\widetilde{\sum}_k$ has the same meaning as in eqn (7.122). Moreover, we have

$$\begin{aligned}\sum_{j\leq N_{A_1}}\sum_{k=N_{A_1}+1}^{N_A} U^*_{\alpha k}\,U_{\beta k}\,U_{\alpha j}\,U^*_{\beta j} &= -\left|\sum_{k\leq N_{A_1}} U^*_{\alpha k}\,U_{\beta k}\right|^2 - \sum_{j\leq N_{A_1}}\sum_{k>N_A} U^*_{\alpha k}\,U_{\beta k}\,U_{\alpha j}\,U^*_{\beta j} \\ &= -\left|\sum_{k=N_{A_1}+1}^{N_A} U^*_{\alpha k}\,U_{\beta k}\right|^2 - \sum_{j=N_A+1}^{N}\sum_{k=N_{A_1}+1}^{N_A} U^*_{\alpha k}\,U_{\beta k}\,U_{\alpha j}\,U^*_{\beta j} \,.\end{aligned} \tag{7.155}$$

Choosing the first equality in eqn (7.155) and the second equality in eqn (7.122), the effective transition probabilities can be written as

$$\begin{aligned}P^{\text{eff}}_{\nu_\alpha\to\nu_\beta}(L,E) = 4&\left|\sum_{k\leq N_{A_1}} U^*_{\alpha k}\,U_{\beta k}\right|^2 \sin^2\left(\frac{\Delta m^2_{N_A 1} L}{4E}\right) \\ &+ 4\left(\sum_{j\leq N_{A_1}}\sum_{k>N_A} \mathfrak{Re}\big[U^*_{\alpha k}\,U_{\beta k}\,U_{\alpha j}\,U^*_{\beta j}\big]\right)\sin^2\left(\frac{\Delta m^2_{N_A 1} L}{4E}\right)\end{aligned}$$

$$+2\left(\sum_{j\leq N_{A_1}}\sum_{k>N_A}\Im\mathfrak{m}\left[U_{\alpha k}^* U_{\beta k} U_{\alpha j} U_{\beta j}^*\right]\right)\sin\left(\frac{\Delta m_{N_A 1}^2 L}{2E}\right)$$

$$+2\left|\sum_{k>N_A} U_{\alpha k}^* U_{\beta k}\right|^2 . \tag{7.156}$$

The last term, which is constant, depends on the mixing of ν_α and ν_β with the massive neutrinos belonging to the group B. The first term, which is oscillating, depends on the mixing of ν_α and ν_β with the massive neutrinos belonging to the group A. The other two oscillating terms depend on the mixing of ν_α and ν_β with both the massive neutrinos belonging to the groups A and B. Clearly, it is not possible to separate completely the contributions due to the mixing of ν_α and ν_β with the massive neutrinos belonging to the groups A and B and write the contribution due to the mixing of ν_α and ν_β with the massive neutrinos belonging to the groups A in terms of an effective two-neutrino-like transition probability, because a two-neutrino-like transition probability cannot account for the effect of the phases in the mixing matrix, which in general are present in eqn (7.156) if the mixing of ν_α and ν_β with the massive neutrinos belonging to the group B is not zero. Only if such mixing is zero is there no effect of the phases, because the present case becomes coincident with the case of dominance of one scale of squared-mass differences discussed in the previous section 7.7. Indeed, in this limit only the first term on the right-hand side of eqn (7.156) survives and one can see that it is analogous to eqn (7.124) with the appropriate change of the mixing parameters.

Nevertheless, one must realize that the effective transition probability $P^{\text{eff}}_{\nu_\alpha\to\nu_\beta}(L,E)$ represents a great simplification with respect to the general form of the transition probability, which depends on many parameters. Indeed, eqn (7.148) depends on only five parameters: $\Delta m^2_{N_A 1}$, $\left|\sum_{k\leq N_{A_1}} U_{\alpha k}^* U_{\beta k}\right|^2$, $\sum_{j\leq N_{A_1}}\sum_{k>N_A}\Re\mathfrak{e}\left[U_{\alpha k}^* U_{\beta k} U_{\alpha j} U_{\beta j}^*\right]$, $\sum_{j\leq N_{A_1}}\sum_{k>N_A}\Im\mathfrak{m}\left[U_{\alpha k}^* U_{\beta k} U_{\alpha j} U_{\beta j}^*\right]$, and $\left|\sum_{k>N_A} U_{\alpha k}^* U_{\beta k}\right|^2$. This is very useful in practice, because it simplifies considerably the analysis of experimental data.

8

THEORY OF NEUTRINO OSCILLATIONS IN VACUUM

> You cannot teach a man anything; you can only help him discover it in himself.
> **Galileo Galilei**

In section 7.1, we have seen that the standard theory of neutrino oscillations formulated in Refs. [404, 466, 236, 239, 237] is based on the three main assumptions listed at the end of that section. In this chapter, we present two versions of the theory of neutrino oscillations which depend on fewer assumptions.

In section 8.1, we present a derivation of the probability of neutrino oscillations in the plane-wave approximation, in which only the physically motivated assumption (A3) on page 253, of equal propagation time and distance, is needed. In particular, we will derive the neutrino flavor states from first principles in the framework of quantum field theory and we will show that no assumption is needed on the values of the energies and momenta of massive neutrinos.

However, the plane-wave approximation is not totally consistent, because energy–momentum conservation implies that the creation of massive neutrinos with definite energies and momenta is possible only if all the particles involved in the production process have definite energies and momenta. The problem is that in this case energy–momentum conservation cannot hold simultaneously for different massive neutrinos and the production of a superposition of different massive neutrinos is forbidden. In order to overcome this problem, it is necessary to treat neutrinos and the other particles participating in the production and detection processes as wave packets, as discussed in section 8.2. We will see that the wave-packet treatment of neutrino oscillations is free from the three standard assumptions listed at the end of section 7.1, including the assumption (A3) of equal propagation time and distance.

From an historical point of view, in 1976, Nussinov [831], for the first time, considered the wave packet nature of propagating neutrinos and inferred the existence of a coherence length, beyond which the interference of different massive neutrinos is not observable. This is due to the different group velocities of different massive neutrinos, leading to a separation of their wave packets. In 1981, Kayser [660] presented the first detailed discussion of the problems in the quantum mechanics of neutrino oscillations, pointing out the necessity of a wave-packet treatment. In 1996, Kiers, Nussinov, and Weiss [667] first pointed out the importance of the detection process for the coherence of neutrino oscillations and discussed some implications for the wave packet approach.

Detailed wave packet models of neutrino oscillations have been developed in the framework of quantum mechanics [533, 534, 528, 525] and in the framework of quantum field theory [531, 536, 668, 301, 224, 523] (see also Ref. [670] and the reviews in Refs. [1090, 225]).

In section 8.3, we discuss the problem of the determination of the size of neutrino wave packets. We also estimate the size of neutrino wave packets in some cases of practical interest. Finally, in section 8.4, we answer some interesting questions that are often asked and have been debated in the literature.

8.1 Plane-wave approximation

In the standard approach described in chapter 7, the massive neutrino components of a flavor state are described by plane waves. In this section we adopt this so-called *plane wave approximation.* In subsection 8.1.1 we present a derivation of the flavor neutrino states. In subsection 8.1.2, using these flavor states, we derive the probability of neutrino oscillations, taking into account the general possibility that different massive neutrinos may have different momenta as well as different energies [1064, 533, 540, 522, 526]. In subsection 8.1.3, we show that the corrections to the standard oscillation phase due to violations of the ultrarelativistic time = distance approximation and the one-dimensional propagation approximation, adopted in subsection 8.1.2, are negligible. Furthermore, in subsection 8.1.4 we discuss the invariance of the neutrino oscillation probability under Lorentz transformations and in section 8.1.5 we discuss the compatibility of the flavor neutrino states with the calculation of neutrino production and detection rates.

8.1.1 Flavor neutrino states

In general, in quantum field theory the asymptotic final state $|f\rangle$ resulting from an interaction process with asymptotic initial state $|i\rangle$ is given by

$$|f\rangle = \mathsf{S}\,|i\rangle\,, \tag{8.1}$$

where S is the S-matrix operator. If the final state is a superposition of orthogonal and normalized states $|f_k\rangle$, it can be written as

$$|f\rangle = \sum_k \mathcal{A}_k\,|f_k\rangle\,, \tag{8.2}$$

with the coefficients

$$\mathcal{A}_k = \langle f_k|f\rangle = \langle f_k|\mathsf{S}|i\rangle\,, \tag{8.3}$$

which are just the amplitudes of production of the corresponding states $|f_k\rangle$.

Let us emphasize that in this section we assume that one-particle states are normalized and have definite energy and momentum. Therefore, the fermion one-particle states are not those derived in eqn (2.232) from the quantization of the corresponding fermion field, which are not normalized. The one-particle states considered in this section correspond to wave packets of the type in eqn (2.459) with a very narrow momentum distribution, whose effects are neglected in the plane-wave approximation (for simplicity, in this section we consider only neutrinos with negative helicity), or states obtained with a finite normalization volume, as discussed in section 2.13.

The state of a flavor neutrino ν_α is defined as the state which describes a neutrino produced in a charged-current weak interaction process together with a charged lepton ℓ_α^+ or from a charged lepton ℓ_α^- (where $\ell_\alpha^\pm = e^\pm, \mu^\pm, \tau^\pm$ for $\alpha = e, \mu, \tau$, respectively), or the state which describes a neutrino detected in a charged-current weak interaction process with a charged lepton ℓ_α^- in the final state. In fact, the neutrino flavor can only be measured through the identification of the charged lepton associated with the neutrino in a charged-current weak interaction process.

Let us first consider a neutrino produced in the generic decay process

$$P_\mathrm{I} \to P_\mathrm{F} + \ell_\alpha^+ + \nu_\alpha \,, \tag{8.4}$$

where $\mathrm{P_I}$ is the decaying particle and $\mathrm{P_F}$ represents any number of final particles. For example: in the pion decay process

$$\pi^+ \to \mu^+ + \nu_\mu \,, \tag{8.5}$$

we have $P_\mathrm{I} = \pi^+$, P_F is absent and $\alpha = \mu$; in a nuclear β^+-decay process

$$\mathrm{N}(A, Z) \to \mathrm{N}(A, Z-1) + e^+ + \nu_e \,, \tag{8.6}$$

we have $P_\mathrm{I} = \mathrm{N}(A, Z)$, $P_\mathrm{F} = \mathrm{N}(A, Z-1)$ and $\alpha = e$; in the μ^+-decay process

$$\mu^+ \to \bar{\nu}_\mu + e^+ + \nu_e \,, \tag{8.7}$$

considering the final ν_e, we have $P_\mathrm{I} = \mu^+$, $P_\mathrm{F} = \bar{\nu}_\mu$ and $\alpha = e$. The following method can easily be modified in the case of a ν_α produced in the generic scattering process $\ell_\alpha^- + \mathrm{P_I} \to \mathrm{P_F} + \nu_\alpha$ by replacing the ℓ_α^+ in the final state with a ℓ_α^- in the initial state.

The final state resulting from the decay of the initial particle $\mathrm{P_I}$ is given by

$$|f\rangle = \mathsf{S}\,|\mathrm{P_I}\rangle \,, \tag{8.8}$$

where S is the S-matrix operator. Since the final state $|f\rangle$ contains all the decay channels of $\mathrm{P_I}$, it can be written as

$$|f\rangle = \sum_k \mathcal{A}^\mathrm{P}_{\alpha k}\,|\nu_k, \ell_\alpha^+, \mathrm{P_F}\rangle + \ldots \,, \tag{8.9}$$

where we have singled out the decay channel in eqn (8.4) and we have taken into account that the flavor neutrino ν_α is a coherent superposition of massive neutrinos ν_k. Since the states of the other decay channels represented by dots in eqn (8.9) are orthogonal to $|\nu_k, \ell_\alpha^+, \mathrm{P_F}\rangle$ and the different states $|\nu_k, \ell_\alpha^+, \mathrm{P_F}\rangle$ are orthogonal and normalized, the coefficients $\mathcal{A}^\mathrm{P}_{\alpha k}$ are the amplitudes of production of the corresponding state in the decay channel in eqn (8.4):

$$\mathcal{A}^{\mathrm{P}}_{\alpha k} = \langle \nu_k, \ell^+_\alpha, \mathrm{P_F} | f \rangle = \langle \nu_k, \ell^+_\alpha, \mathrm{P_F} | \mathsf{S} | \mathrm{P_I} \rangle \,. \tag{8.10}$$

Projecting the final state in eqn (8.9) over $|\ell^+_\alpha, \mathrm{P_F}\rangle$ and normalizing, we obtain the flavor neutrino state [535, 229, 54, 538, 539]

$$|\nu^{\mathrm{P}}_\alpha\rangle = \left(\sum_i |\mathcal{A}^{\mathrm{P}}_{\alpha i}|^2 \right)^{-1/2} \sum_k \mathcal{A}^{\mathrm{P}}_{\alpha k} \, |\nu_k\rangle \,. \tag{8.11}$$

Therefore, a flavor neutrino state is a coherent superposition of massive neutrino states $|\nu_k\rangle$ and the coefficient $\mathcal{A}^{\mathrm{P}}_{\alpha k}$ of the kth massive neutrino component is given by the amplitude of production of ν_k. Since, in general, the amplitudes $\mathcal{A}^{\mathrm{P}}_{\alpha k}$ depend on the production process, a flavor neutrino state depends on the production process. In the following, we will call a flavor neutrino state of the type in eqn (8.11) a *production flavor neutrino state.*

Let us now consider the detection of a flavor neutrino ν_α through the generic charged-current weak interaction process

$$\nu_\alpha + \mathrm{D_I} \to \mathrm{D_F} + \ell^-_\alpha \,, \tag{8.12}$$

where $\mathrm{D_I}$ is the target particle and $\mathrm{D_F}$ represents one or more final particles. In general, since the incoming neutrino state in the detection process is a superposition of massive neutrino states, it may not have a definite flavor. Therefore, we must consider the generic process

$$\nu + \mathrm{D_I} \,, \tag{8.13}$$

with a generic incoming neutrino state $|\nu\rangle$. In this case, the final state of the scattering process is given by

$$|f\rangle = \mathsf{S} \, |\nu, \mathrm{D_I}\rangle \,, \tag{8.14}$$

This final state contains all the possible scattering channels:

$$|f\rangle = |\mathrm{D_F}, \ell^-_\alpha\rangle + \ldots \,, \tag{8.15}$$

where we have singled out the scattering channel in eqn (8.12). We want to find the component

$$|\nu_\alpha, \mathrm{D_I}\rangle = \sum_k \mathcal{A}^{\mathrm{D}}_{\alpha k} |\nu_k, \mathrm{D_I}\rangle \tag{8.16}$$

of the initial state $|\nu, \mathrm{D_I}\rangle$ which corresponds to the flavor α, i.e. the component which generates only the scattering channel in eqn (8.12). This means that

$|\mathrm{D_F}, \ell_\alpha^-\rangle = \mathsf{S}\,|\nu_\alpha, \mathrm{D_I}\rangle$. Using the unitarity of the mixing matrix, we obtain

$$|\nu_\alpha, \mathrm{D_I}\rangle = \mathsf{S}^\dagger\,|\mathrm{D_F}, \ell_\alpha^-\rangle\,. \tag{8.17}$$

From eqns (8.16) and (8.17), the coefficients $\mathcal{A}^{\mathrm{D}}_{\alpha k}$ are the complex conjugate of the amplitude of detection of ν_k in the detection process in eqn (8.12):

$$\mathcal{A}^{\mathrm{D}}_{\alpha k} = \langle \nu_k, \mathrm{D_I}|\mathsf{S}^\dagger\,|\mathrm{D_F}, \ell_\alpha^-\rangle\,. \tag{8.18}$$

Projecting $|\nu_\alpha, \mathrm{D_I}\rangle$ over $|\mathrm{D_I}\rangle$ and normalizing, we finally obtain the flavor neutrino state in the detection process in eqn (8.12):

$$|\nu_\alpha^{\mathrm{D}}\rangle = \left(\sum_i |\mathcal{A}^{\mathrm{D}}_{\alpha i}|^2\right)^{-1/2} \sum_k \mathcal{A}^{\mathrm{D}}_{\alpha k}\,|\nu_k\rangle\,. \tag{8.19}$$

In the following, we will call a flavor neutrino state of this type a *detection flavor neutrino state.*

Although the expressions in eqns (8.11) and (8.19) for the production and detection flavor neutrino states have the same structure, these states have different meanings. A production flavor neutrino state describes the neutrino created in a charged-current interaction process, which propagates out of a source. Hence, it describes the initial state of a propagating neutrino. A detection flavor neutrino state does not describe a propagating neutrino. It describes the component of the state of a propagating neutrino which can generate a charged lepton with appropriate flavor through a charged-current weak interaction with an appropriate target particle. In other words, the scalar product

$$A_\alpha = \langle \nu_\alpha^{\mathrm{D}}|\nu\rangle \tag{8.20}$$

is the probability amplitude to find a ν_α by observing the scattering channel in eqn (8.12) with the scattering process in eqn (8.13).

In order to understand the connection of the production and detection flavor neutrino states with the standard flavor neutrino states in eqn (7.4), it is useful to express the S-matrix operator as

$$\mathsf{S} = 1 - i\int \mathrm{d}^4x\,\mathscr{H}_{\mathrm{CC}}(x)\,, \tag{8.21}$$

where we have considered only the first order perturbative contribution of the effective low-energy charged-current weak interaction Hamiltonian

$$\mathscr{H}_{\mathrm{CC}}(x) = \frac{G_{\mathrm{F}}}{\sqrt{2}}\,j_\rho^\dagger(x)\,j^\rho(x)\,, \tag{8.22}$$

where G_{F} is the Fermi constant. The weak charged current $j^\rho(x)$ is given by

$$j^\rho(x) = \sum_{\alpha=e,\mu,\tau} \overline{\nu_\alpha}(x)\,\gamma^\rho\left(1-\gamma^5\right)\ell_\alpha(x) + h^\rho(x)$$

$$= \sum_{\alpha=e,\mu,\tau} \sum_{k} U_{\alpha k}^{*} \, \overline{\nu_{k}}(x) \, \gamma^{\rho} \left(1 - \gamma^{5}\right) \ell_{\alpha}(x) + h^{\rho}(x) \,, \tag{8.23}$$

where $h^{\rho}(x)$ is the hadronic weak charged current. The production and detection amplitudes $\mathcal{A}_{\alpha k}^{\mathrm{P}}$ and $\mathcal{A}_{\alpha k}^{\mathrm{D}}$ can be written as

$$\mathcal{A}_{\alpha k}^{\mathrm{P}} = U_{\alpha k}^{*} \, \mathcal{M}_{\alpha k}^{\mathrm{P}} \,, \qquad \mathcal{A}_{\alpha k}^{\mathrm{D}} = U_{\alpha k}^{*} \, \mathcal{M}_{\alpha k}^{\mathrm{D}} \,, \tag{8.24}$$

with the interaction matrix elements

$$\mathcal{M}_{\alpha k}^{\mathrm{P}} = -i \, \frac{G_{\mathrm{F}}}{\sqrt{2}} \int \mathrm{d}^{4}x \, \langle \nu_{k}, \ell_{\alpha}^{+} | \, \overline{\nu_{k}}(x) \, \gamma^{\rho} \left(1 - \gamma^{5}\right) \ell_{\alpha}(x) \, |0\rangle \, J_{\rho}^{\mathrm{P_I} \to \mathrm{P_F}}(x) \,, \tag{8.25}$$

$$\mathcal{M}_{\alpha k}^{\mathrm{D}} = i \, \frac{G_{\mathrm{F}}}{\sqrt{2}} \int \mathrm{d}^{4}x \, \langle \nu_{k} | \, \overline{\nu_{k}}(x) \, \gamma^{\rho} \left(1 - \gamma^{5}\right) \ell_{\alpha}(x) \, |\ell_{\alpha}^{-}\rangle \, J_{\rho}^{\mathrm{D_I} \to \mathrm{D_F} \, *}(x) \,. \tag{8.26}$$

Here, $J_{\rho}^{\mathrm{P_I} \to \mathrm{P_F}}(x)$ and $J_{\rho}^{\mathrm{D_I} \to \mathrm{D_F}}(x)$ are, respectively, the matrix elements of the $\mathrm{P_I} \to \mathrm{P_F}$ and $\mathrm{D_I} \to \mathrm{D_F}$ transitions.

Using eqn (8.24), the production and detection flavor neutrino states can be written as

$$|\nu_{\alpha}^{\mathrm{P}}\rangle = \sum_{k} \frac{\mathcal{M}_{\alpha k}^{\mathrm{P}}}{\sqrt{\sum_{j} |U_{\alpha j}|^{2} \, |\mathcal{M}_{\alpha j}^{\mathrm{P}}|^{2}}} \, U_{\alpha k}^{*} \, |\nu_{k}\rangle \,, \tag{8.27}$$

$$|\nu_{\alpha}^{\mathrm{D}}\rangle = \sum_{k} \frac{\mathcal{M}_{\alpha k}^{\mathrm{D}}}{\sqrt{\sum_{j} |U_{\alpha j}|^{2} \, |\mathcal{M}_{\alpha j}^{\mathrm{D}}|^{2}}} \, U_{\alpha k}^{*} \, |\nu_{k}\rangle \,. \tag{8.28}$$

These states have a structure which is similar to the standard flavor states in eqn (7.4), with the relative contribution of the massive neutrino ν_{k} proportional to $U_{\alpha k}^{*}$. The additional factors are due to the dependence of the production and detection processes on the neutrino masses.

In experiments which are not sensitive to the dependence of $\mathcal{M}_{\alpha k}^{\mathrm{P}}$ or $\mathcal{M}_{\alpha k}^{\mathrm{D}}$ on the difference of the neutrino masses it is possible to approximate

$$\mathcal{M}_{\alpha k}^{\mathrm{P}} \simeq \mathcal{M}_{\alpha}^{\mathrm{P}} \,, \qquad \mathcal{M}_{\alpha k}^{\mathrm{D}} \simeq \mathcal{M}_{\alpha}^{\mathrm{D}} \,. \tag{8.29}$$

In this case, since

$$\sum_{k} |U_{\alpha k}|^{2} = 1 \,, \tag{8.30}$$

we obtain, up to an irrelevant phase, the standard flavor neutrino states in eqn (7.4), which do not depend on the production or detection process. Hence, the standard flavor neutrino states are approximations of the production and detection flavor neutrino states in experiments which are not sensitive to the dependence of the neutrino interaction rate on the difference of the neutrino masses.

In the following section 8.1.2, we will derive the neutrino oscillation probability starting from the flavor neutrino states in eqns (8.11) and (8.19). We will show that, with the appropriate approximations, the oscillation probability reduces to the standard one in eqn (7.23), which was derived from the approximate standard flavor neutrino states in eqn (7.4). In section 8.1.5, we show that the expressions in eqns (8.11) and (8.19) for the flavor neutrino states are important in order to be able to describe, in a consistent framework, neutrino oscillations and neutrino production and detection.

8.1.2 Neutrino oscillations

Let us consider a neutrino oscillation experiment in which $\nu_\alpha \to \nu_\beta$ transitions are studied with a production process of the type in eqn (8.4) and a detection process of the type in eqn (8.12). In this case, the produced flavor neutrino ν_α is described by the production flavor state $|\nu_\alpha^{\text{P}}\rangle$ in eqn (8.11). If the neutrino production and detection processes are separated by a space-time interval $(\vec{L}, T)$, the neutrino propagates freely between production and detection, evolving into the state

$$|\nu(\vec{L}, T)\rangle = e^{-i\mathsf{P}^0 T + i\vec{\mathsf{P}}\cdot\vec{L}}\,|\nu_\alpha^{\text{P}}\rangle\,, \tag{8.31}$$

where P^0 and $\vec{\mathsf{P}}$ are, respectively, the energy and momentum operators. This is the incoming neutrino state in the detection process. The amplitude of the measurable $\nu_\alpha \to \nu_\beta$ transitions is given by the scalar product in eqn (8.20):

$$A_{\nu_\alpha\to\nu_\beta}(\vec{L}, T) = \langle\nu_\beta^{\text{D}}|\nu(\vec{L}, T)\rangle = \langle\nu_\beta^{\text{D}}|e^{-i\mathsf{P}^0 T + i\vec{\mathsf{P}}\cdot\vec{L}}|\nu_\alpha^{\text{P}}\rangle\,, \tag{8.32}$$

with the detection flavor state $|\nu_\beta^{\text{D}}\rangle$ in eqn (8.19).

Since the massive neutrinos have definite kinematical properties (energy and momentum), we have, in the plane wave approximation,

$$\mathsf{P}^\mu\,|\nu_k\rangle = p_k^\mu\,|\nu_k\rangle\,, \tag{8.33}$$

with

$$p_k^0 = E_k = \sqrt{|\vec{p}_k|^2 + m_k^2}\,. \tag{8.34}$$

Using the normalization $\langle\nu_k|\nu_j\rangle = \delta_{kj}$, we obtain the flavor transition amplitude

$$A_{\nu_\alpha\to\nu_\beta}(\vec{L}, T) = \left(\sum_i |\mathcal{A}_{\alpha i}^{\text{P}}|^2\right)^{-1/2} \left(\sum_i |\mathcal{A}_{\beta i}^{\text{D}}|^2\right)^{-1/2} \sum_k \mathcal{A}_{\alpha k}^{\text{P}}\,\mathcal{A}_{\beta k}^{\text{D}*}\,e^{-iE_k T + i\vec{p}_k\cdot\vec{L}}\,, \tag{8.35}$$

Notice that the consideration of the space-time interval between neutrino production and detection allows one to take into account both the differences in energy and momentum of massive neutrinos [1064, 533, 540, 522, 526].

Let us consider the simplest case in which all massive neutrino momenta $\vec{p}_k$ are aligned along $\vec{L}$. This assumption is reasonable, because all massive neutrino components are created in the same microscopic production process and detected in the same microscopic detection process, which are separated by the large macroscopic space interval $\vec{L}$. In subsection 8.1.3, we will show that possible deviations from this assumption do not lead to any observable effect.

In this *one-dimensional approximation*, the transition amplitude depends on $L \equiv |\vec{L}|$ and T:

$$A_{\nu_\alpha \to \nu_\beta}(L,T) = \left(\sum_i |\mathcal{A}^{\mathrm{P}}_{\alpha i}|^2 \right)^{-1/2} \left(\sum_i |\mathcal{A}^{\mathrm{D}}_{\beta i}|^2 \right)^{-1/2} \sum_k \mathcal{A}^{\mathrm{P}}_{\alpha k} \, \mathcal{A}^{\mathrm{D}*}_{\beta k} \, e^{-iE_k T + i p_k L} \,, \tag{8.36}$$

where $p_k \equiv |\vec{p}_k|$. It is important to note that the consideration of the space-time interval between neutrino production and detection allows one to take into account both the differences in energy and momentum of massive neutrinos.

In oscillation experiments in which the neutrino propagation time T is not measured, it is possible to approximate $T = L$, as we have discussed in item (A3) on page 253 (the effects of possible deviations from $T = L$ are shown to be negligible in subsection 8.1.3 below). In this case, the phase in eqn (8.36) becomes

$$-E_k T + p_k L = -(E_k - p_k)\, L = -\frac{E_k^2 - p_k^2}{E_k + p_k}\, L = -\frac{m_k^2}{E_k + p_k}\, L \simeq -\frac{m_k^2}{2E}\, L \,, \tag{8.37}$$

where E is the neutrino energy neglecting mass contributions. It is important to note that eqn (8.37) shows that the phases of massive neutrinos relevant for the oscillations are independent of the values of the energies and momenta of different massive neutrinos [1064, 533, 540, 522, 526], because of the relativistic dispersion relation in eqn (8.34). In particular, eqn (8.37) shows that the equal momentum assumption (A2) on page 253, which was adopted in the standard derivation of the neutrino oscillation probability (see section 7.1), is not necessary in the above derivation.

The probability of $\nu_\alpha \to \nu_\beta$ transitions in space is given by

$$P_{\nu_\alpha \to \nu_\beta}(L) = \left(\sum_i |\mathcal{A}^{\mathrm{P}}_{\alpha i}|^2 \right)^{-1} \left(\sum_i |\mathcal{A}^{\mathrm{D}}_{\beta i}|^2 \right)^{-1} \times \sum_{k,j} \mathcal{A}^{\mathrm{P}}_{\alpha k} \, \mathcal{A}^{\mathrm{D}*}_{\beta k} \, \mathcal{A}^{\mathrm{P}*}_{\alpha j} \, \mathcal{A}^{\mathrm{D}}_{\beta j} \, \exp\left(-i \frac{\Delta m^2_{kj} L}{2E} \right) . \tag{8.38}$$

Using the decomposition in eqn (8.24), the oscillation probability can now be written as

$$P_{\nu_\alpha \to \nu_\beta}(L,E) = \sum_{k,j} \left(\frac{\mathcal{M}^{\mathrm{P}}_{\alpha k} \, \mathcal{M}^{\mathrm{P}*}_{\alpha j}}{\sum_i |U_{\alpha i}|^2 |\mathcal{M}^{\mathrm{P}}_{\alpha i}|^2} \right) \left(\frac{\mathcal{M}^{\mathrm{D}*}_{\beta k} \, \mathcal{M}^{\mathrm{D}}_{\beta j}}{\sum_i |U_{\beta i}|^2 |\mathcal{M}^{\mathrm{D}}_{\beta i}|^2} \right) \times U^*_{\alpha k} \, U_{\beta k} \, U_{\alpha j} \, U^*_{\beta j} \, \exp\left(-i \frac{\Delta m^2_{kj} L}{2E} \right) . \tag{8.39}$$

This probability has the same structure as the standard oscillation probability in eqn (7.23), with additional factors that take into account the effect of the neutrino masses in the production and detection processes. It is clear from eqn (8.39) that these effects have an influence on the amplitude of the oscillations, but not on the phase, which coincides with the standard one in eqn (7.24).

Since neutrinos in oscillation experiments are ultrarelativistic and the experiments are not sensitive to the dependence of neutrino interactions on the neutrino masses, the dependence of $\mathcal{M}^{\mathrm{P}}_{\alpha k}$ and $\mathcal{M}^{\mathrm{D}}_{\beta k}$ on the neutrino masses can be neglected, leading to the approximation in eqn (8.29). In this case, the transition probability in eqn (8.39) reduces to the standard one in eqn (7.23), which can be obtained starting from the standard flavor states in eqn (7.4). As shown in section 8.1.1, the standard flavor states are obtained from the production and detection flavor states under the same approximations. Therefore, the standard flavor states are appropriate for the description of neutrino oscillation experiments in the plane wave approximation, as long as the dependence of the production and detection probabilities on the neutrino masses is negligible. These considerations justify the assumption (A1) on page 253, which was adopted in the standard derivation of the neutrino oscillation probability.

8.1.3 Universality of the oscillation phase

In the previous subsection 8.1.2, we have derived the probability of neutrino oscillations under the assumptions that $T = L$ and all massive neutrino momenta $\vec{p}_k$ are aligned along $\vec{L}$. In this subsection we will show that possible deviations from these assumptions do not affect in a significant way the oscillation phases measured in neutrino oscillation experiments, which are correctly given by the standard expression in eqn (7.24).

The momentum of a massive neutrino created in a production process depends on the characteristics of the interaction, on the nature of the other particles taking part in the process and on the neutrino mass. Let us call $\vec{p}$ and $E = |\vec{p}|$, respectively, the momentum and energy of a massless neutrino. The first order contribution of the neutrino mass to $\vec{p}_k$ and E_k must be proportional to m_k^2, because the energy–momentum dispersion relation in eqn (8.34) depends on m_k^2. Therefore, in general, the momentum $\vec{p}_k$ can be written to first order in m_k^2 as

$$\vec{p}_k \simeq \vec{p} - \vec{\xi}\,\frac{m_k^2}{2E}\,, \tag{8.40}$$

where

$$\frac{\vec{\xi}}{2E} = -\left.\frac{\partial \vec{p}_k}{\partial m_k^2}\right|_{m_k=0}\,. \tag{8.41}$$

The value of the vector $\vec{\xi}$ depends on the production process. However, as we will see in the following, the measurable oscillation phases are independent of $\vec{\xi}$. Therefore, they are universal, i.e. independent of the specific nature of the neutrino production process, as well as detection.

In the approximation in eqn (8.40), the dispersion relation in eqn (8.34) implies that the energy of ν_k is given by

$$E_k \simeq E + \left(1 - \frac{\vec{p}\cdot\vec{\xi}}{E}\right)\frac{m_k^2}{2E}\,. \tag{8.42}$$

Note that eqn (8.40) implies that the directions of propagation of different massive neutrinos are slightly different if $\vec{\xi}$ is not collinear with $\vec{p}$. How this is possible can be illustrated in the case of the pion decay production process in eqn (8.5) as follows. In the rest frame of the pion, the energy of ν_k can be calculated from energy–momentum conservation to be given by

$$E_k = \frac{m_\pi}{2}\left(1 - \frac{m_\mu^2}{m_\pi^2}\right) + \frac{m_k^2}{2\,m_\pi}\,, \tag{8.43}$$

where m_π and m_μ are the masses of the pion and muon, respectively. Since, in the rest frame of the pion, the neutrino and the muon are emitted back-to-back, all the massive neutrinos are emitted in the same direction. Thus, $\vec{\xi}$ is collinear with $\vec{p}$, leading to

$$E_k = E + (1-\xi)\,\frac{m_k^2}{2E}\,, \qquad p_k \equiv |\vec{p}_k| \simeq E - \xi\,\frac{m_k^2}{2E}\,, \tag{8.44}$$

with

$$E = \frac{m_\pi}{2}\left(1 - \frac{m_\mu^2}{m_\pi^2}\right) \simeq 30\,\text{MeV}\,, \qquad \xi \equiv |\vec{\xi}| = \frac{1}{2}\left(1 + \frac{m_\mu^2}{m_\pi^2}\right) \simeq 0.8\,. \tag{8.45}$$

Note that the different values of the momenta and energies of different massive neutrinos imply different corresponding values for the momentum and energy of the outgoing muon. Two different massive neutrinos can be produced coherently in the same decay process only if the outgoing muon has energy and momentum uncertainties which are larger, respectively, of the difference of the energies and momenta of the two massive neutrinos. These uncertainties must come from corresponding uncertainties for the pion. Hence, in a rigorous treatment, the pion and the muon must be described by wave packets [531, 536, 1090, 824, 301, 225, 224, 523]. This implies that also the massive neutrinos must be described by wave packets [831, 660, 533, 667, 668, 528, 1090, 523, 526]. The spatial extension of the massive neutrino wave packets explains how it is possible that the different massive neutrinos can be detected in the same interaction process even if their space-time trajectories are different.

Let us now consider a boosted reference frame. If the frame is boosted in the direction of the emitted massive neutrinos, obviously they are still collinear. Instead, if the frame is boosted in another direction, collinearity is lost. For example, let us consider a frame boosted with velocity V in a direction which is orthogonal to the neutrino direction in the pion rest frame. In this frame, we have, in an obvious notation,

$$p'_{k\|} = p_{k\|} = p_k\,, \qquad p'_{k\perp} = -\frac{V}{\sqrt{1-V^2}}\,E_k\,. \tag{8.46}$$

Hence, the angle of propagation of ν_k in the boosted frame, with respect to its direction in the rest frame, is given by

$$\tan\theta'_k = \frac{p'_{k\perp}}{p'_{k\|}} \simeq -\frac{V}{\sqrt{1-V^2}}\left(1 + \frac{m_k^2}{2E^2}\right)\,. \tag{8.47}$$

Since this angle depends on the neutrino mass, in the boosted frame different massive neutrinos are seen to propagate in slightly different directions. In fact, in this case, $\vec{p}'$ and $\vec{\xi}'$ are not collinear, since

$$p'_{\parallel} = p_{\parallel} = E\,, \quad p'_{\perp} = -\frac{V}{\sqrt{1-V^2}}\,E\,, \quad \xi'_{\parallel} = \xi_{\parallel} = \xi\,, \quad \xi'_{\perp} = \frac{V}{\sqrt{1-V^2}}\,(1-\xi)\,. \tag{8.48}$$

Since many experiments measure the oscillations of neutrinos produced by pion decay in flight (e.g. atmospheric and accelerator neutrino oscillation experiments), in which the neutrino direction is not constrained to be collinear with the pion direction, in order to be realistic, in this section we consider the general case of noncollinear $\vec{p}$ and $\vec{\xi}$.

Let us now calculate the phases

$$\phi_k = -E_k T + \vec{p}_k \cdot \vec{L} \tag{8.49}$$

in the transition amplitude in eqn (8.35). We consider a deviation of the time T between neutrino production and detection from the light-ray approximation $T = L$ of the type

$$T = \frac{L}{|\langle\vec{v}\rangle|}\,(1+\varepsilon_T)\,. \tag{8.50}$$

Here, $\langle\vec{v}\rangle$ is the average velocity

$$\langle\vec{v}\rangle = \frac{1}{N}\sum_{k=1}^{N}\frac{\vec{p}_k}{E_k} \simeq \frac{\vec{p}}{E} - \left[\frac{\vec{p}}{E}\left(1-\frac{\vec{p}\cdot\vec{\xi}}{E}\right)+\vec{\xi}\right]\frac{\overline{m^2}}{2E^2}\,, \tag{8.51}$$

where N is the number of massive neutrinos and

$$\overline{m^2} = \frac{1}{N}\sum_{k=1}^{N} m_k^2 \tag{8.52}$$

is the average of the squared neutrino masses. The absolute value of the average velocity is given by the usual ultrarelativistic expression

$$|\langle\vec{v}\rangle| \simeq 1 - \frac{\overline{m^2}}{2E^2}\,. \tag{8.53}$$

In eqn (8.50), we consider $\varepsilon_T \ll 1$, because neutrinos are ultrarelativistic and the deviation from $T = L/|\langle\vec{v}\rangle|$ cannot be larger than the size of the neutrino wave packets, which must be much smaller than L in order to observe the oscillations [660, 533]. To first order in the small ratio $\overline{m^2}/E^2$, eqn (8.50) becomes

$$T \simeq L\,(1+\varepsilon_T)\left(1+\frac{\overline{m^2}}{2E^2}\right)\,. \tag{8.54}$$

Now, we consider also deviations from the assumption that the momenta of all massive neutrinos are aligned along $\vec{L}$. As discussed above, the expression for the neutrino momenta in eqn (8.40) already implies that, in general, the momenta of different massive neutrinos are not collinear. Moreover, we can consider a deviation from the collinearity of $\vec{L}$ and the average momentum

$$\langle\vec{p}\rangle = \frac{1}{N}\sum_{k=1}^{N}\vec{p}_k \simeq \vec{p} - \vec{\xi}\,\frac{\overline{m^2}}{2E}\,, \tag{8.55}$$

which we can write as

$$\frac{\langle\vec{p}\rangle}{|\langle\vec{p}\rangle|} = \frac{\vec{L}}{L} + \vec{\varepsilon}_L\,, \tag{8.56}$$

with $|\vec{\varepsilon}_L| \ll 1$, for the same reasons as $\varepsilon_T \ll 1$. At zeroth order in the small ratio $\overline{m^2}/E^2$, we have

$$\vec{p} \simeq E\left(\frac{\vec{L}}{L} + \vec{\varepsilon}_L\right)\,, \tag{8.57}$$

which implies, from eqn (8.42),

$$E_k \simeq E + \left(1 - \frac{\vec{L}\cdot\vec{\xi}}{L} - \vec{\varepsilon}_L\cdot\vec{\xi}\right)\frac{m_k^2}{2E}\,. \tag{8.58}$$

From eqns (8.40), (8.54), and (8.58), for the difference $\Delta\phi_{kj} = \phi_k - \phi_j$ of the phases in eqn (8.49) we obtain, at first order in the small quantities ε_T and $|\vec{\varepsilon}_L|$,

$$\Delta\phi_{kj} \simeq -\frac{\Delta m_{kj}^2 L}{2E} + \varepsilon_{kj}\,, \tag{8.59}$$

with the contribution

$$\varepsilon_{kj} = \left[\vec{\varepsilon}_L\cdot\vec{\xi} - \varepsilon_T\left(1 - \frac{\vec{L}\cdot\vec{\xi}}{L}\right)\right]\frac{\Delta m_{kj}^2 L}{2E}\,, \tag{8.60}$$

in addition to the standard oscillation phase in eqn (7.24). However, since $\varepsilon_T \ll 1$ and $|\vec{\varepsilon}_L| \ll 1$, the contribution ε_{kj} is nonnegligible only for $\Delta m_{kj}^2 L/2E \gg 1$. But in this case oscillations are not measurable, since they are washed out by the average over the energy resolution of the detector (see section 7.6). In the case of $\Delta m_{kj}^2 L/2E \sim 1$, in which oscillations are measurable, ε_{kj} is extremely small and can be safely neglected ($e^{i\varepsilon_{kj}} \simeq 1$), leading to the validity of the standard expression in eqn (7.24) for the oscillation phases.

Note the irrelevance for $\Delta\phi_{kj}$ of the lack of collinearity of the trajectories of different massive neutrinos if $\varepsilon_T = |\vec{\varepsilon}_L| = 0$. It is due to the fact that the deviation from collinearity of ν_k and ν_j is proportional to $\Delta m_{kj}^2/2E^2$ (see the example in eqn (8.47)). Thus, it induces in the phase difference $\Delta\phi_{kj}$ effects of higher order in $\Delta m_{kj}^2/2E^2$, which are completely negligible.

In conclusion, in this section we have shown that possible small deviations from the light-ray approximation $T = L$ and from the collinearity of the massive neutrino momenta $\vec{p}_k$ and $\vec{L}$ are irrelevant for the measurable oscillation phases, which are universally independent of the specific characteristics of the process of neutrino production, as well as detection. Hence, the time = distance assumption (A3) on page 253, which was adopted in the standard derivation of the neutrino oscillation probability, is correct.

8.1.4 Lorentz invariance

An important aspect of the derivation of the neutrino oscillation probability in subsection 8.1.2 is that it is manifestly Lorentz covariant.

This is important because flavor, which is the quantum number that distinguishes different types of quarks and leptons, is a Lorentz-invariant quantity. For example, an electron is seen as an electron by any observer, never as a muon. Therefore, the probability of flavor neutrino oscillations must be Lorentz invariant [540, 527].

The flavor transition amplitude in eqn (8.35) is manifestly Lorentz invariant. In fact, the phases are given by scalar products of four-vectors and the amplitudes $\mathcal{A}^{\mathrm{P}}_{\alpha k}$ and $\mathcal{A}^{\mathrm{D}}_{\beta k}$ are Lorentz invariant[39]. Also the transition probabilities in eqns (7.23) and (8.38) are Lorentz invariant, because L transforms as E [527]. Indeed, the approximation $T = L$ implies that L transforms as T, which transforms as E, because they are both the time components of four-vectors.

For example, let us consider an inertial system $\mathcal{O}$ and another inertial system $\mathcal{O}'$ moving with respect to $\mathcal{O}$ with velocity v along the direction of neutrino propagation. The Lorentz transformations of space and time intervals are

$$L' = \gamma \, (L - v \, T) \, , \tag{8.61}$$

$$T' = \gamma \, (T - v \, L) \, , \tag{8.62}$$

with $\gamma \equiv \left(1 - v^2\right)^{-1/2}$. If we have $L = T$ in the system $\mathcal{O}$, from eqns (8.61) and (8.62) in the system $\mathcal{O}'$ we have

$$L' = T' = \gamma \, (1 - v) \, L \, . \tag{8.63}$$

On the other hand, the Lorentz transformations of momentum and energy are

$$p' = \gamma \, (p - v \, E) \, , \tag{8.64}$$

$$E' = \gamma \, (E - v \, p) \, . \tag{8.65}$$

In the massless limit the relation $E = p$ in the system $\mathcal{O}$ implies that in the system $\mathcal{O}'$

$$E' = p' = \gamma \, (1 - v) \, E \, . \tag{8.66}$$

From eqns (8.63) and (8.66), it is clear that the ratio L/E is Lorentz invariant and the phases in the transition probabilities in eqns (7.23) and (8.38) are Lorentz invariant.

In general, L is not the instantaneous source–detector distance, which would transform according to the well-known contraction of distances law $\Delta x' = \gamma \left(1 - v^2\right) \Delta x = \Delta x / \gamma$. This is easily understood by considering an inertial system in which the detector is moving. Since the detector moves after the propagating neutrino has left the source, the spatial distance traveled by the neutrino is

[39] Considering, for example, the amplitude $\mathcal{A}_{\alpha k}$ in eqn (8.10), we have $\mathcal{A}_{\alpha k} = \langle \nu_k, \ell_\alpha^+, P_{\mathrm{F}} | \mathsf{S} | P_{\mathrm{I}} \rangle = \langle \nu_k, \ell_\alpha^+, P_{\mathrm{F}} | \mathsf{U}_\Lambda^\dagger \, \mathsf{U}_\Lambda \, \mathsf{S} \, \mathsf{U}_\Lambda^\dagger \, \mathsf{U}_\Lambda \, | P_{\mathrm{I}} \rangle = \langle \nu'_k, \ell'^+_\alpha, P'_{\mathrm{F}} | \mathsf{S} | P'_{\mathrm{I}} \rangle = \mathcal{A}'_{\alpha k}$, since the S-matrix operator is a Lorentz scalar.

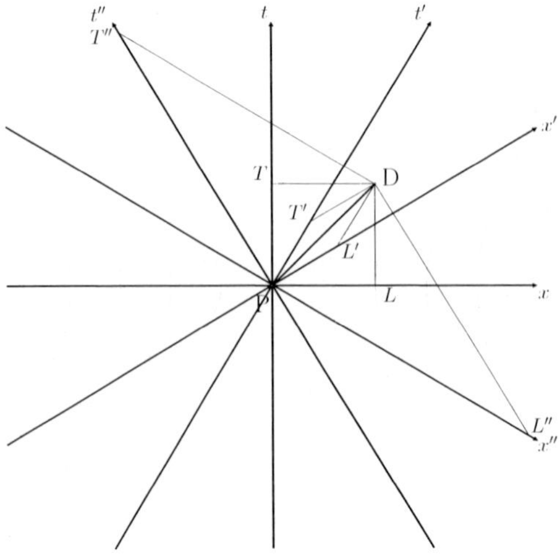

FIG. 8.1. Minkowski diagram showing the distance and time between the space–time events of neutrino production P and detection D measured in three coordinate systems: $\mathcal{O}$ with axes x, t; $\mathcal{O}'$ with axes x', t' moving with respect to $\mathcal{O}$ with velocity $v = 3/5$ in the x direction along the direction of neutrino propagation; $\mathcal{O}''$ with axes x'', t'' moving with respect to $\mathcal{O}$ with velocity $v = -3/5$ in the x direction. The x' (x'') axis is inclined with respect to the x axis by an angle $\arctan 3/5$ ($-\arctan 3/5$); the t' (t'') axis is inclined with respect to the t axis by an angle $-\arctan 3/5$ ($\arctan 3/5$). The hyperbolas $t^2 - x^2 = t'^2 - x'^2 = t''^2 - x''^2 =$ constant fix the scale on the axes. One can see that $L' = L/2$ and $L'' = 2L$, in perfect agreement with eqn (8.63).

different from the instantaneous source–detector distance at the time of neutrino production. The spatial distance traveled by the neutrino coincides with the instantaneous source–detector distance at the time of neutrino detection only if the source is stationary. However, even in this case, since the instantaneous source–detector distance is time dependent, one cannot apply to L the contraction of distances law (which would contradict eqn (8.63)).

The correct transformation law of the propagation distance in eqn (8.63) is illustrated by the Minkowski diagram in Fig. 8.1, in which one can clearly see that in a system in which source and detector are in motion the spatial distance between the two space-time events of neutrino production and detection is different from the instantaneous source–detector distance. The two distances coincide only in the system $\mathcal{O}$, in which source and detector are at rest.

One can ask what happens if, instead of considering the source and detector at rest in the system $\mathcal{O}$, we consider a case in which source and detector are in relative motion. Since the oscillation probability is measured by the detector, the velocity of the source with respect to the detector does not matter[40]. The system $\mathcal{O}$ in which the distance L covered by the neutrino coincides with the instantaneous distance between source and detector at the time of neutrino emission is always the rest system of the detector. In Fig. 8.1 the world-line of the detector is constrained to be the vertical line passing through the space-time event D of neutrino detection, but the world-line of the source can be any time-like line passing through the space-time event P of neutrino production.

Let us emphasize that the Lorentz invariance of the phases in the transition probabilities in eqns (7.23) and (8.38) follows from the fact that L is the distance traveled by the neutrino in the time T [527]. One could ask[41] what happens to the relation in eqn (8.37) in a frame in which the massive neutrinos are not ultrarelativistic, where it is not possible to justify the approximations $T \simeq L$ and $E_k + p_k \simeq 2E$. The answer is that, since the phase $-E_k T + p_k L$ is Lorentz invariant, it can be evaluated in any frame. In eqn (8.37), we have chosen to evaluate it in a frame where the massive neutrinos are ultrarelativistic. Since, as we have just shown above, the result in the right-hand side of eqn (8.37) is Lorentz invariant, it applies in any frame, including all the frames in which the massive neutrinos are not ultrarelativistic.

Let us finally remark that the probability of flavor neutrino oscillations must be invariant not only under Lorentz transformation, but also under the larger group of Poincaré transformations, which include also space-time translations. The invariance of the transition amplitudes in eqns (8.32) and (8.36) and the transition probabilities in eqns (7.23) and (8.38) under space-time translations is obvious, because E_k, $\vec{p}_k$, T, $\vec{L}$ are invariant and states transform as in eqn (2.288), leading to the invariance of the amplitudes $\mathcal{A}^{\mathrm{P}}_{\alpha k}$, $\mathcal{A}^{\mathrm{D}}_{\beta k}$.

8.1.5 Production and detection rates

In order to measure $\nu_\alpha \to \nu_\beta$ oscillations it is necessary to calculate the neutrino production rate Γ_α of ν_α in the source and the detection cross-section σ_β of ν_β. The number of transition events as a function of the distance L traveled by the neutrino between production and detection is given by

$$N_{\alpha\beta}(L) \propto \Gamma_\alpha \, P_{\nu_\alpha \to \nu_\beta}(L) \, \sigma_\beta \,, \tag{8.67}$$

with a constant of proportionality which depends on the size and composition of the source and detector and on the running time of the experiment. From the measurement of $N_{\alpha\beta}(L)$ and the knowledge of Γ_α and σ_β, the experimentalist infers the value of $P_{\nu_\alpha \to \nu_\beta}(L)$, which gives information on the mixing parameters (elements of the mixing matrix and squared-mass differences) through eqn (7.23).

As discussed in section 14.1.1 below for β-decay, decay rates and cross-sections are given by the incoherent sum over the different channels corresponding to different massive neutrinos [958, 790, 687, 959, 960]. The reason is that massive neutrinos

[40] Of course, the velocity of the source must be taken into account in the calculation of the neutrino energy. For example, if the source is a decaying pion, the neutrino energy depends on the velocity of the pion.

[41] We are grateful to Prof. D.V. Ahluwalia-Khalilova for asking this question.

are the physical particles which propagate in space-time with definite kinematical properties. However, in subsection 8.1.2 we derived the oscillation probability $P_{\nu_\alpha \to \nu_\beta}(L)$ starting from the description of neutrinos through flavor states, which are coherent superpositions of massive neutrino states. Then, it is natural to ask if the derivation of the oscillation probability is consistent with the calculation of decay rates and cross-sections. In the following, we show that the description of neutrinos through the flavor states in eqns (8.11) and (8.19) leads to the correct expression for the decay rates and cross-sections. Therefore, all quantities in eqn (8.67) can be calculated in a consistent way [538, 539].

In order to be definite, let us consider again the generic decay process in eqn (8.4) in which a flavor neutrino ν_α is produced. Any other process of neutrino production or detection can be treated in an analogous way.

Using the flavor state in eqn (8.11) and taking into account eqn (8.10), the amplitude of the general decay process in eqn (8.4) is given by [538, 539]

$$\begin{aligned} \mathcal{A}^{\mathrm{P}}_{\alpha} = \langle \nu^{\mathrm{P}}_{\alpha}, \ell^{+}_{\alpha}, \mathrm{P_F} | \, \mathsf{S} \, | \mathrm{P_I} \rangle &= \left(\sum_i |\mathcal{A}^{\mathrm{P}}_{\alpha i}|^2 \right)^{-1/2} \sum_k \mathcal{A}^{\mathrm{P}*}_{\alpha k} \, \langle \nu_k, \ell^{+}_{\alpha}, \mathrm{P_F} | \mathsf{S} | \mathrm{P_I} \rangle \\ &= \sqrt{\sum_i |\mathcal{A}^{\mathrm{P}}_{\alpha i}|^2} \, . \end{aligned} \tag{8.68}$$

Therefore, the decay probability is correctly given by an incoherent sum of the probabilities of production of different massive neutrinos,

$$|\mathcal{A}^{\mathrm{P}}_{\alpha}|^2 = \sum_i |\mathcal{A}^{\mathrm{P}}_{\alpha i}|^2 \, . \tag{8.69}$$

In other words, the coherent character of the flavor state in eqn (8.11) is irrelevant for the decay probability, which can be obtained either using the flavor neutrino state in eqn (8.11) or an incoherent mixture of massive neutrino states. The decay rate is then obtained by integrating each massive neutrino contribution to the decay probability over its phase space.

Using the expression in eqn (8.24) for the amplitude $\mathcal{A}^{\mathrm{P}}_{\alpha k}$, the decay probability in eqn (8.69) can be written as

$$|\mathcal{A}^{\mathrm{P}}_{\alpha}|^2 = \sum_k |U_{\alpha k}|^2 \, |\mathcal{M}^{\mathrm{P}}_{\alpha k}|^2 \, , \tag{8.70}$$

which is an incoherent sum of the probabilities of production of the different massive neutrinos weighted by $|U_{\alpha k}|^2$ [958, 790, 687, 959, 960].

Therefore, the flavor neutrino state in eqn (8.11) leads to the correct decay rate for the general decay process in eqn (8.4). It is clear that this proof can easily be generalized to any charged-current weak interaction process in which flavor neutrinos are produced or detected.

If an experiment is not sensitive to the dependence of $\mathcal{M}^{\mathrm{P}}_{\alpha k}$ on the different neutrino masses, it is possible to use the approximation in eqn (8.29). In this case, using eqn (8.30), we obtain

$$|\mathcal{A}_{\alpha}^{\mathrm{P}}|^2 = |\mathcal{M}_{\alpha}^{\mathrm{P}}|^2 \,. \tag{8.71}$$

If the scale of neutrino masses is negligible in comparison with the experimental resolution, the dependence of $\mathcal{M}_{\alpha}^{\mathrm{P}}$ on the neutrino masses is negligible and the decay probability in eqn (8.71) reduces to the standard decay probability for massless neutrinos.

The decay probability in eqn (8.71) can also be obtained starting from the standard flavor states in eqn (7.4), which are obtained from eqn (8.11) through the approximation in eqn (8.29). Indeed, in this case the decay amplitude is given by [524]

$$\mathcal{A}_{\alpha}^{\mathrm{P}} = \langle \nu_\alpha, \ell_\alpha^+, \mathrm{P_F} | \, \mathsf{S} \, | \mathrm{P_I} \rangle = \sum_k U_{\alpha k} \, \mathcal{A}_{\alpha k}^{\mathrm{P}} = \sum_k |U_{\alpha k}|^2 \, \mathcal{M}_{\alpha}^{\mathrm{P}} = \mathcal{M}_{\alpha}^{\mathrm{P}} \,. \tag{8.72}$$

Let us remark, however, that in the case of an experiment which is sensitive to the dependence of $\mathcal{M}_{\alpha k}^{\mathrm{P}}$ on the different neutrino masses, a derivation of the decay amplitude starting from the standard flavor states would lead to an incorrect result. This is due to the fact that in this case the approximation in eqn (8.29) is not valid and one must take into account the dependence of $\mathcal{M}_{\alpha k}^{\mathrm{P}}$ on the different neutrino masses in the definition of the flavor states.

8.2 Wave-packet treatment

So far, we have considered massive neutrinos as particles described by plane waves with definite energy and momentum. However, we have seen that the approximation $T = L$ requires a wave packet description. The reason is clear: since plane waves cover all space-time in a periodic way they cannot describe the localized events of neutrino production and detection. As discussed in introductory books on optics (see [263, 644]) and quantum mechanics (see [936, 259]), real localized particles are described by superpositions of plane waves known as *wave packets*.

Moreover, different massive neutrinos can be produced and detected coherently only if the energies and momenta in the production and detection processes have sufficiently large uncertainties [660, 667]. The uncertainty of the production process implies that the massive neutrinos propagating between production and detection have a momentum distribution [523], i.e. they are described by wave packets.

The propagation of a massless particle between localized production and detection processes is illustrated schematically in the space-time diagram in Fig. 8.2a. The interesting case of propagation of a superposition of two neutrinos with definite masses, one massless (ν_1) and one massive but ultrarelativistic (ν_2) is illustrated schematically in Fig. 8.2b. One can note that in these diagrams both the production and detection processes occupy a finite region in space-time, called the *coherence region*, in which the propagating particles are produced or detected coherently. Indeed, the uncertainty principle implies that any interaction process I has a space

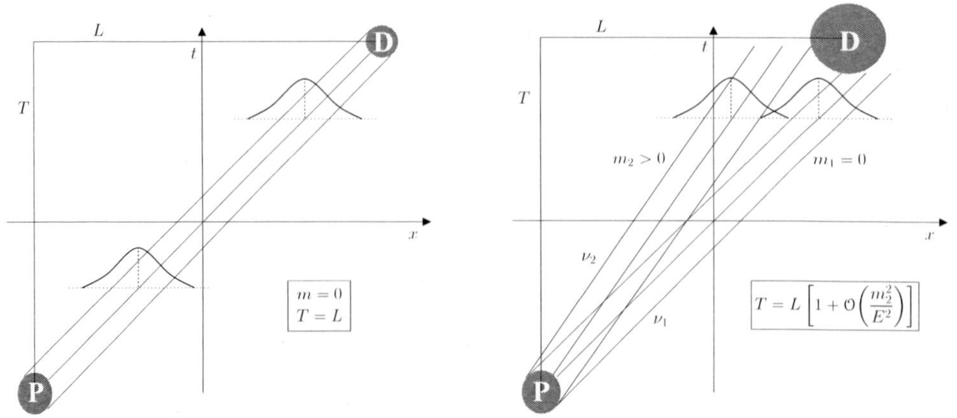

(a) Propagation of the wave packet of a massless particle.

(b) Propagation of the wave packets of a superposition of a massless and a massive ultrarelativistic particle.

FIG. 8.2. Space-time diagram representing schematically the propagation of the wave packets of one particle (a) or a superposition of two particles (b) from a production process P to a detection process D.

uncertainty σ_x^{I} related to the momentum uncertainty σ_p^{I} by

$$\sigma_x^{\mathrm{I}} \, \sigma_p^{\mathrm{I}} \sim \frac{1}{2} \,. \tag{8.73}$$

A process with vanishing space uncertainty would have an infinite momentum uncertainty. Vice versa, a process with definite momentum would be completely delocalized in space. The momentum uncertainty can be estimated as the quadratic sum of the uncertainties of the momenta of the localized particles taking part in the process:

$$(\sigma_p^{\mathrm{I}})^2 \sim \sum_i (\sigma_p^i)^2 \,. \tag{8.74}$$

The sum is over the initial particles and the final particles which are localized through interaction with the environment and their momentum uncertainties σ_p^i are related to the size σ_x^i of their wave packets by the uncertainty relation analogous to eqn (8.73),

$$\sigma_x^i \, \sigma_p^i \sim \frac{1}{2} \,. \tag{8.75}$$

Therefore, the space uncertainty of the process is given by

$$(\sigma_x^{\mathrm{I}})^{-2} \sim \sum_i (\sigma_x^i)^{-2} \,. \tag{8.76}$$

It is clear that the particle with larger momentum uncertainty and associated smaller space uncertainty gives the dominant contribution.

The coherence time σ_t^{I} of an interaction process I is the time over which the wave packets of the interacting particles overlap. If the process is the decay of a particle in vacuum, the localization of such particle and its decay products is very poor and the coherence time σ_t^{I} is of the order of the particle lifetime. On the other hand, if the decay occurs in a medium where the decay particle and its products are well localized or if the production process is a scattering process, the coherence time can be estimated by

$$(\sigma_t^{\mathrm{I}})^{-2} \sim \sum_i \left(\frac{\sigma_x^i}{v_i} \right)^{-2} , \tag{8.77}$$

where v_i is the velocity of the particle i, because σ_t^{I} must be dominated by the particle with smaller ratio σ_x^i/v_i, which is the first to leave the interaction region. Therefore, in general $\sigma_t^{\mathrm{I}} \gtrsim \sigma_p^{\mathrm{I}}$, in agreement with the physical expectation that the coherence region of a process must be causally connected.

As illustrated in Fig. 8.2, one can estimate the size of the wave packet of a massive neutrino created in a production process P as the coherence time σ_t^{P} of the production process,

$$\sigma_x^{\nu} \sim \sigma_t^{\mathrm{P}} . \tag{8.78}$$

Let us emphasize that there is a profound difference between the behavior of final neutrinos and other particles in the production process. The initial particles have wave packets which are determined by their creation process or by previous interactions. The initial particles and the final particles which interact with the environment contribute to the coherence time σ_t^{P} through their contribution to the momentum uncertainty in eqn (8.74). An initial decaying particle contributes directly to the coherence time σ_t^{P} with its lifetime. On the other hand, neutrinos are stable and leave the production process without interacting with the environment. Therefore, they do not contribute to the determination of the coherence time σ_t^{P} and the size of their wave packets is determined by σ_t^{P}.

Considering now the detection process D, if there is only one particle propagating between the production and detection processes, as shown in Fig. 8.2a, the coherence size of the detection process is determined by eqn (8.74), with the sum over all the participating particles which interact with the environment and the propagating particle, which is described by a wave packet. In the case of neutrino mixing, the neutrino propagating between the production and detection processes is in general a superposition of massive neutrino wave packets which propagate with different phase velocity, as illustrated in Fig. 8.2b. In this case, in the detection process, the wave packets of different massive neutrinos are separated by a distance $\Delta x = \Delta v\, T$, where Δv is the velocity difference. If the source–detector distance is very large, the separation of the massive neutrino wave packets at detection may be larger than their size, leading to the lack of overlap [831]. In this case, the effective coherence size of the neutrino wave function at the detection process is

$$\sigma_{x,\mathrm{eff}}^{\nu} \sim \sqrt{(\sigma_x^{\nu})^2 + (\Delta x)^2} \sim \sqrt{(\sigma_t^{\mathrm{P}})^2 + (\Delta v\, T)^2} . \tag{8.79}$$

However, eqn (8.76) shows that the particle with smaller space uncertainty gives the dominant contribution to the coherence size of the detection process. Therefore, if the effective coherence size in eqn (8.79) of the neutrino wave function is dominated by the separation of the wave packets ($\Delta v\, T \gg \sigma_t^{\rm P}$) and there is another particle participating in the detection process which has much smaller space uncertainty, the different massive neutrinos cannot be detected coherently. In this case, there cannot be any interference between the different massive neutrino contributions to the detection process and the probability of transitions between different flavors does not oscillate as a function of the source–detector distance. In other words, the incoherent transition probability in eqn (7.32), which coincides with the average oscillation probability in eqn (7.33), is measured. On the other hand, if all the other particles participating in the detection process have space uncertainties which are larger than effective coherence size in eqn (8.79) of the neutrino wave function, the different massive neutrinos are detected coherently [667], leading to the interference of their contributions to the detection process which manifests itself as oscillations of the probability of flavor transitions, according to eqn (7.23).

Hence, a wave-packet treatment of massive neutrinos is important in order to understand the coherence properties of neutrino oscillations.

8.2.1 Flavor neutrino states

In this subsection, we derive the states which describe flavor neutrinos in the wave packet approach by following a method which is similar to that presented in subsection 8.1.1.

For definiteness, we consider again neutrino production in the generic decay process in eqn (8.4). The final state which is analogous to the final state in eqn (8.9) and takes into account the momentum distributions of the massive neutrinos is

$$|f\rangle = \sum_k \int \frac{\mathrm{d}^3 p}{(2\pi)^3\, 2E_k} \sum_h \mathcal{A}^{\rm P}_{\alpha k}(\vec{p}, h)\, |\nu_k(E_k, \vec{p}, h), \ell^+_\alpha, P_{\rm F}\rangle + \ldots \,, \tag{8.80}$$

where $\vec{p}$ is the neutrino momentum, $E_k = \sqrt{\vec{p}^{\,2} + m_k^2}$ is the associated energy of the kth massive neutrino, and h is the neutrino helicity. The coefficient $\mathcal{A}^{\rm P}_{\alpha k}(\vec{p}, h)$ is given by the amplitude of production of $|\nu_k(E_k, \vec{p}, h)\rangle$:

$$\mathcal{A}^{\rm P}_{\alpha k}(\vec{p}, h) = \langle \nu_k(E_k, \vec{p}, h), \ell^+_\alpha, P_{\rm F} | f\rangle = \langle \nu_k(E_k, \vec{p}, h), \ell^+_\alpha, P_{\rm F} |\, \mathsf{S}\, | P_{\rm I}\rangle \,. \tag{8.81}$$

Projecting the final state in eqn (8.80) on $|\ell^+_\alpha, P_{\rm F}\rangle$, we obtain the normalized wave packet flavor neutrino state

$$|\nu^{\rm P}_\alpha\rangle = N^{\rm P}_\alpha \sum_k \int \frac{\mathrm{d}^3 p}{(2\pi)^3\, 2E_k} \sum_h \mathcal{A}^{\rm P}_{\alpha k}(\vec{p}, h)\, |\nu_k(E_k, \vec{p}, h)\rangle \,, \tag{8.82}$$

with the normalization factor

$$N^{\rm P}_\alpha = \left(\sum_k \int \frac{\mathrm{d}^3 p}{(2\pi)^3\, 2E_k} \sum_h |\mathcal{A}^{\rm P}_{\alpha k}(\vec{p}, h)|^2 \right)^{-1/2} \,. \tag{8.83}$$

The flavor neutrino state $|\nu^{\rm P}_\alpha\rangle$ in eqn (8.82) is a wave packet state, because it consists of a superposition of massive neutrino states with momentum distributions given by the production amplitudes in eqn (8.81). Obviously, in order to have momentum distributions which are not delta-functions, at least one of the particles $P_{\rm I}$, $P_{\rm F}$, ℓ_α^+ must be described by a wave packet with a momentum distribution. The properties of the massive neutrino wave packets can be derived with the method presented in section 2.12.

It is important to note that if all the particles $P_{\rm I}$, $P_{\rm F}$, ℓ_α^+ are described by plane waves, the production process is not localized and energy–momentum conservation forbids the coherent production of different massive neutrinos. Therefore, in order to have neutrino oscillations, at least one of the particles $P_{\rm I}$, $P_{\rm F}$, ℓ_α^+ must be described by a localized wave packet with sufficient energy–momentum uncertainty. If the total energy–momentum uncertainty is so small that different massive neutrinos cannot be produced coherently, the state in eqn (8.82) becomes effectively an incoherent mixture of massive neutrino wave packets, because the different energy–momentum conservations contained in the amplitudes in eqn (8.81) cannot be satisfied simultaneously.

In the discussion of the plane-wave approximation in section 8.1 we swept this problem under the rug. However, there are no implications for the neutrino production and detection rates discussed in subsection 8.1.5 which, as we have seen in eqn (8.69), are given by an incoherent sum of the probabilities of production or detection of the different massive neutrinos. On the other hand, the oscillation probability in eqn (8.38) has, strictly speaking, no physical meaning, because the exact energy–momentum conservation delta-functions contained in the amplitudes in eqn (8.10) for different massive neutrinos are mutually exclusive. However, these delta-functions have been factorized out of the standard oscillation probability in eqn (7.23), which acquires a physical meaning as the approximation of the oscillation probability in the limit of negligible wave packet effects, as we will see in subsection 8.2.2.

8.2.2 Neutrino oscillations

In experiments in which $\nu_\alpha \to \nu_\beta$ oscillations are studied, the produced (P) and detected (D) neutrinos are described by the wave packet states

$$|\nu_\alpha^{\rm I}\rangle = N_\alpha^{\rm I} \sum_k \int \frac{{\rm d}^3 p}{(2\pi)^3\, 2E_k} \sum_h \mathcal{A}_{\alpha k}^{\rm I}(\vec{p},h)\, |\nu_k(\vec{p},h)\rangle\,, \tag{8.84}$$

with the normalization factors

$$N_\alpha^{\rm I} = \left(\sum_k \int \frac{{\rm d}^3 p}{(2\pi)^3\, 2E_k} \sum_h |\mathcal{A}_{\alpha k}^{\rm I}(\vec{p},h)|^2 \right)^{-1/2}, \tag{8.85}$$

for I = P, D. The amplitude of $\nu_\alpha \to \nu_\beta$ transitions is given by eqn (8.32). Using the flavor states in eqn (8.84), we obtain

$$A_{\nu_\alpha \to \nu_\beta}(\vec{L},T) = N_\alpha^{\rm P}\, N_\beta^{\rm D} \sum_k \int \frac{{\rm d}^3 p}{(2\pi)^3\, 2E_k} \sum_h \mathcal{A}_{\alpha k}^{\rm P}(\vec{p},h)\, \mathcal{A}_{\beta k}^{{\rm D}*}(\vec{p},h)\, e^{-iE_k T + i\vec{p}\cdot\vec{L}}\,. \tag{8.86}$$

The derivation of the neutrino oscillation probability from the explicit values of the production and detection amplitudes in eqn (8.86) has been discussed in Ref. [523]. Since it is rather complicated, here we consider the following approximation which leads to a similar result for the oscillation probability:

$$\mathcal{A}^{\mathrm{P}}_{\alpha k}(\vec{p},h)\,\mathcal{A}^{\mathrm{D}*}_{\beta k}(\vec{p},h) \propto U^*_{\alpha k}\,U_{\beta k}\,\tilde{\mathcal{M}}^{\mathrm{P}}_{\alpha k}(\vec{p},h)\,\tilde{\mathcal{M}}^{\mathrm{D}*}_{\beta k}(\vec{p},h)\,\exp\left[-\frac{(\vec{p}-\vec{p}_k)^2}{4\sigma_p^2}\right], \tag{8.87}$$

where $\vec{p}_k$ is the average momentum of the kth massive neutrino contribution to the transition amplitude. The exponential in eqn (8.87) describes energy–momentum conservation within the momentum uncertainty σ_p, which can be estimated as [536, 528, 523, 526]

$$\sigma_p^{-2} \sim (\sigma_p^{\mathrm{P}})^{-2} + (\sigma_p^{\mathrm{D}})^{-2}\,, \tag{8.88}$$

where σ_p^{P} and σ_p^{D} are, respectively, the momentum uncertainties in the production and detection processes. In eqn (8.88), the inverses of the momentum uncertainties in the production and detection processes are summed in quadrature, because the smaller momentum uncertainty must dominate in the determination of the total effective momentum uncertainty. On the other hand, the total effective space uncertainty σ_x, related to the momentum uncertainty σ_p by the minimal Heisenberg relation

$$\sigma_x\,\sigma_p = \frac{1}{2}\,, \tag{8.89}$$

is dominated by the largest of the space uncertainties of the production and detection processes,

$$\sigma_x \sim (\sigma_x^{\mathrm{P}})^2 + (\sigma_x^{\mathrm{D}})^2\,. \tag{8.90}$$

In eqn (8.87), the value of $\vec{p}_k$ is determined by both the production and detection processes. One can understand this fact with the following reasoning. Let us first consider a massless neutrino propagating between the production and detection processes. Its wave packet, with average momentum $\vec{p}$ and corresponding energy $\tilde{E} = |\vec{p}|$, is determined by the production process. Since the neutrino and the target particle are the initial states in the detection process, their wave packets determine the kinematics of the interaction and the wave packets of the final particles. Thus, if the detection matrix element is a smooth function of the energies and momenta of the interacting particles, the average momenta of the final particles are determined by energy–momentum conservation. For example, in the detection process in eqn (8.12) we have $\tilde{p}^\rho + \tilde{p}^\rho_{D_{\mathrm{I}}} = \tilde{p}^\rho_{D_{\mathrm{F}}} + \tilde{p}^\rho_{\ell^-_\beta}$, in an obvious notation. Let us consider now a propagating superposition of massive neutrinos which are produced and detected coherently. According to the argument presented at the beginning of section 8.1.3, the average momentum of each propagating ν_k is given by $\vec{p}^{\mathrm{P}}_k \simeq \vec{p} - \vec{\xi}_{\mathrm{P}} m_k^2/2\tilde{E}$, where $\vec{\xi}_{\mathrm{P}}$ depends on the production process. In this case, in the detection process it is not possible to satisfy simultaneously the different energy–momentum conservation equations $\tilde{p}^{\mathrm{P}\rho}_k + \tilde{p}^\rho_{D_{\mathrm{I}}} = \tilde{p}^\rho_{D_{\mathrm{F}}} + \tilde{p}^\rho_{\ell^-_\beta}$ for different values of the massive neutrino index k and the same values of $\tilde{p}^\rho_{D_{\mathrm{I}}}$, $\tilde{p}^\rho_{D_{\mathrm{F}}}$ and $\tilde{p}^\rho_{\ell^-_\beta}$. Therefore, the detection process selects values of $\vec{p}_k$ which are different from $\vec{p}_k^{\rm P}$. The selection is done in order to minimize the violation of energy–momentum conservation within the energy–momentum uncertainties $\sigma_p^{\rm P}$ and $\sigma_p^{\rm D}$ of the production and detection processes [523].

In the approximation in eqn (8.87), we choose a Gaussian approximation for the momentum distribution in order to be able to perform the integral over d^3p analytically. Furthermore, in order to simplify the derivation of the oscillation probability for real experimental setups as much as possible, we consider ultrarelativistic neutrinos and we adopt the following assumptions, which in practice are always satisfied: (a) $\sigma_p \ll \langle p_k \rangle$, with $\langle p_k \rangle = |\vec{p}_k|$; (b) $\tilde{\mathcal{M}}_{\alpha k}^{\rm P}(\vec{p}, h)\, \tilde{\mathcal{M}}_{\beta k}^{\rm D*}(\vec{p}, h)$ is a smooth function of $\vec{p}$; and (c) the experiment is not sensitive to the dependence of $\tilde{\mathcal{M}}_{\alpha k}^{\rm P}(\vec{p}, h)\, \tilde{\mathcal{M}}_{\beta k}^{\rm D*}(\vec{p}, h)$ on the neutrino masses. Under these assumptions, we can approximate

$$\frac{\tilde{\mathcal{M}}_{\alpha k}^{\rm P}(\vec{p}, h)\, \tilde{\mathcal{M}}_{\beta k}^{\rm D*}(\vec{p}, h)}{E_k} \simeq \frac{\tilde{\mathcal{M}}_{\alpha}^{\rm P}(\vec{\bar{p}}, h)\, \tilde{\mathcal{M}}_{\beta}^{\rm D*}(\vec{\bar{p}}, h)}{E}, \tag{8.91}$$

where $\vec{\bar{p}}$ is the average neutrino momentum neglecting mass effects and $E = |\vec{\bar{p}}|$ is the corresponding energy. The approximation in eqn. (8.91) is important, because it allows us to factorize the contribution of the production and detection matrix elements out of the sum over the massive neutrino index k in eqn. (8.86). In this way, the production and detection matrix elements become overall factors which are irrelevant for the flavor transition probability.

The energies $E_k(\vec{p})$ in the phases in eqn (8.86) can be approximated by

$$E_k(\vec{p}) \simeq \tilde{E}_k + \vec{v}_k \left(\vec{p} - \vec{\bar{p}}_k \right), \tag{8.92}$$

with

$$\tilde{E}_k = E_k(\vec{\bar{p}}) = \sqrt{\vec{\bar{p}}_k^{\,2} + m_k^2}\,, \tag{8.93}$$

$$\vec{v}_k = \left. \frac{\partial E_k(\vec{p})}{\partial \vec{p}} \right|_{\vec{p}=\vec{\bar{p}}_k} = \frac{\vec{\bar{p}}_k}{\tilde{E}_k}. \tag{8.94}$$

The vector $\vec{v}_k$ has dimensions of velocity, but it cannot be interpreted as the velocity of the wave packet of ν_k, because the average momentum $\vec{\bar{p}}_k$ is determined by both the production and detection processes, as discussed after eqn (8.90). On the other hand, from causality reasons it is clear that the velocity of the wave packet of ν_k which propagates between source and detector is determined only by the production process. Its value can be obtained from a calculation of the wave packet flavor state in eqn (8.82) for a specific production process [523].

With the above approximations, the flavor transition amplitude in eqn (8.86) can be written as

$$A_{\nu_\alpha \to \nu_\beta}(\vec{L}, T) \propto \sum_k U_{\alpha k}^* \, U_{\beta k} \, e^{-i\tilde{E}_k T + i\vec{\bar{p}}_k \cdot \vec{L}} \int \mathrm{d}^3p \, e^{i\left(\vec{p}-\vec{\bar{p}}_k\right) \cdot \left(\vec{L}-\vec{v}_k T\right) - \frac{(\vec{p}-\vec{\bar{p}}_k)^2}{4\sigma_p^2}}, \tag{8.95}$$

where we have omitted all the overall factors which are irrelevant for the normalized transition probability (including the production and detection matrix elements in eqn (8.91)). The integral over d^3p in eqn (8.95) is Gaussian, leading to

$$A_{\nu_\alpha \to \nu_\beta}(\vec{L}, T) \propto \sum_k U_{\alpha k}^* U_{\beta k} \exp\left[-i\tilde{E}_k T + i\vec{\tilde{p}}_k \cdot \vec{L} - \frac{(\vec{L} - \vec{v}_k T)^2}{4\sigma_x^2}\right] . \tag{8.96}$$

In order to obtain the oscillation probability as a function of the known distance $\vec{L}$ traveled by the neutrino between production and detection, the probability

$$P_{\nu_\alpha \to \nu_\beta}(\vec{L}, T) = |A_{\nu_\alpha \to \nu_\beta}(\vec{L}, T)|^2 \tag{8.97}$$

must be averaged over the unknown time T [533]:

$$P_{\nu_\alpha \to \nu_\beta}(\vec{L}) \propto \int \mathrm{d}T \, P_{\nu_\alpha \to \nu_\beta}(\vec{L}, T) . \tag{8.98}$$

Since the integral over T is Gaussian, we obtain

$$\begin{aligned} P_{\nu_\alpha \to \nu_\beta}(\vec{L}) \propto \sum_{k,j} U_{\alpha k}^* U_{\beta k} U_{\alpha j} U_{\beta j}^* \exp\left\{-i\left[\left(\tilde{E}_k - \tilde{E}_j\right)\frac{\vec{v}_k + \vec{v}_j}{v_k^2 + v_j^2} - \left(\vec{\tilde{p}}_k - \vec{\tilde{p}}_j\right)\right] \cdot \vec{L}\right\} \\ \times \exp\left\{-\frac{L^2}{2\sigma_x^2} + \frac{(\vec{v}_k \cdot \vec{L})^2 + (\vec{v}_j \cdot \vec{L})^2}{2\sigma_x^2\left(v_k^2 + v_j^2\right)} - \frac{\left[(\vec{v}_k - \vec{v}_j)\cdot\vec{L}\right]^2}{4\sigma_x^2\left(v_k^2 + v_j^2\right)} - \frac{\left(\tilde{E}_k - \tilde{E}_j\right)^2}{4\sigma_p^2\left(v_k^2 + v_j^2\right)}\right\} . \end{aligned} \tag{8.99}$$

In each term of the sum, the first exponential generates the oscillations, whereas the second exponential is a damping term, which sets the conditions for the observability of the oscillations.

In order to simplify the expression in eqn (8.99), let us expand, as in section 8.1.3, the average momentum $\vec{\tilde{p}}_k$ in powers of m_k^2 up to the first order:

$$\vec{\tilde{p}}_k \simeq \vec{p} - \vec{\xi}\,\frac{m_k^2}{2E} . \tag{8.100}$$

From now on, in this section, we simplify the notation denoting with $\vec{p}$ and $E = |\vec{p}|$ the average momentum and the corresponding energy of a massless neutrino. From eqns (8.93) and (8.94), the energy $\tilde{E}_k$ and the velocity $\vec{v}_k$ are given by

$$\tilde{E}_k \simeq E + \left(1 - \frac{\vec{p}\cdot\vec{\xi}}{E}\right)\frac{m_k^2}{2E} , \tag{8.101}$$

$$\vec{v}_k \simeq \frac{\vec{p}}{E} - \left[\frac{\vec{p}}{E}\left(1 - \frac{\vec{p}\cdot\vec{\xi}}{E}\right) + \vec{\xi}\right]\frac{m_k^2}{2E^2} . \tag{8.102}$$

Note that, although $\vec{v}_k$ is not a physical velocity, for its absolute value we have the usual ultrarelativistic expression

$$v_k \equiv |\vec{v}_k| \simeq 1 - \frac{m_k^2}{2E^2} \,. \tag{8.103}$$

The average of the average momenta in eqn (8.100) is given by

$$\langle \vec{p} \rangle = \frac{1}{N} \sum_{k=1}^{N} \vec{p}_k \simeq \vec{p} - \vec{\xi}\, \frac{\overline{m^2}}{2E} \,, \tag{8.104}$$

where $\overline{m^2}$ is the average squared neutrino mass in eqn (8.52). We consider the possible deviation from the collinearity of $\langle \vec{p} \rangle$ and $\vec{L}$ as in eqn (8.56):

$$\frac{\langle \vec{p} \rangle}{|\langle \vec{p} \rangle|} = \frac{\vec{L}}{L} + \vec{\varepsilon}_L \,. \tag{8.105}$$

Since the absolute values of both sides of this equation are equal to unity, we have

$$\vec{L} \cdot \vec{\varepsilon}_L = -\frac{1}{2} \, |\vec{\varepsilon}_L|^2 \, L \,, \tag{8.106}$$

which implies that

$$|\vec{\varepsilon}_L| \leq 2 \,. \tag{8.107}$$

At zeroth order in the expansion in powers of the neutrino masses, we have

$$\vec{p} \simeq E \left(\frac{\vec{L}}{L} + \vec{\varepsilon}_L \right) , \qquad \vec{v}_k \simeq \frac{\vec{L}}{L} + \vec{\varepsilon}_L \,. \tag{8.108}$$

Then, we obtain, from eqn (8.106),

$$-\frac{L^2}{2\sigma_x^2} + \frac{(\vec{v}_k \cdot \vec{L})^2 + (\vec{v}_j \cdot \vec{L})^2}{2\sigma_x^2 \left(v_k^2 + v_j^2 \right)} \simeq \frac{2L(\vec{\varepsilon}_L \cdot \vec{L}) + (\vec{\varepsilon}_L \cdot \vec{L})^2}{2\sigma_x^2} = -|\vec{\varepsilon}_L|^2 \left(1 - \frac{1}{4} \, |\vec{\varepsilon}_L|^2 \right) \frac{L^2}{2\sigma_x^2} \,. \tag{8.109}$$

In real experiments, we always have $\sigma_x \ll L$, because the production and detection processes are localized in small regions with respect to the source-detector distance. From the inequality in eqn (8.107), the transition probability in eqn (8.99) is not suppressed only if

$$|\vec{\varepsilon}_L|^2 \ll \sigma_x^2 / L^2 \ll 1 \,. \tag{8.110}$$

This constraint on $|\vec{\varepsilon}_L|^2$ is obvious from a physical point of view: if the deviation in eqn (8.105) from the collinearity of $\langle \vec{p} \rangle$ and $\vec{L}$ is not extremely small, the detector is out of the trajectories of the massive neutrino wave packets and the detection probability is suppressed. In fact, σ_x^2 is approximately proportional to the largest between the areas of coherent neutrino emission and detection (see eqn (8.90)) and the geometrical divergence of the neutrino trajectory is proportional to L^2.

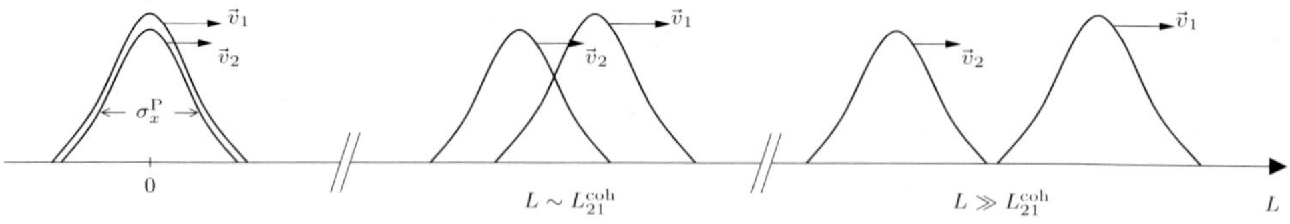

FIG. 8.3. Schematic illustration of the separation of two wave packets with different group velocities, produced coherently at $L = 0$ with widths $\sigma_x^{\rm P}$ determined by the coherence size of the production process. The coherence size of the detection process is assumed to be negligible.

Since $|\vec{\varepsilon}_L|^2 \ll 1$, the effect of $\vec{\varepsilon}_L$ is negligible in the damping terms in the expression in eqn (8.99) for the oscillation probability. The effect of $\vec{\varepsilon}_L$ could be relevant only for the oscillation phase, which is given, at first order in $|\vec{\varepsilon}_L|$, by

$$-\left[\left(\tilde{E}_k - \tilde{E}_j\right)\frac{\vec{v}_k + \vec{v}_j}{v_k^2 + v_j^2} - \left(\tilde{\vec{p}}_k - \tilde{\vec{p}}_j\right)\right] \cdot \vec{L} \simeq -\frac{\Delta m_{kj}^2 L}{2E}\left(1 - \vec{\varepsilon}_L \cdot \vec{\xi}\right) . \tag{8.111}$$

However, as discussed in section 8.1.3, such effect of $\vec{\varepsilon}_L$ is negligible when the oscillation phase is measurable, i.e. $\Delta m_{kj}^2 L/2E \sim 1$. In this case, the oscillation probability at the lowest order in the small quantities m_k^2/E^2 is given by

$$P_{\nu_\alpha \to \nu_\beta}(\vec{L}) = \sum_{k,j} U_{\alpha k}^* U_{\alpha j} U_{\beta k} U_{\beta j}^* \exp\left[-2\pi i \frac{L}{L_{kj}^{\rm osc}} - \left(\frac{L}{L_{kj}^{\rm coh}}\right)^2 \right.$$
$$\left. -2\pi^2 \left(1 - \frac{\vec{L}\cdot\vec{\xi}}{L}\right)^2 \left(\frac{\sigma_x}{L_{kj}^{\rm osc}}\right)^2\right] , \tag{8.112}$$

with the standard oscillation lengths in eqn (7.31) and the coherence lengths

$$L_{kj}^{\rm coh} = \frac{4\sqrt{2}E^2}{|\Delta m_{kj}^2|}\,\sigma_x . \tag{8.113}$$

As remarked at the end of subsection 8.2.1, in the limit of negligible wave packet effects, i.e. for $L \ll L_{kj}^{\rm coh}$ and $\sigma_x \ll L_{kj}^{\rm osc}$, the oscillation probability in the wave packet approach reduces to the standard one in eqn (7.23), obtained in the plane wave approximation.

Let us discuss the physical meaning of the *localization term*

$$\exp\left[-2\pi^2 \left(1 - \frac{\vec{L}\cdot\vec{\xi}}{L}\right)^2 \left(\frac{\sigma_x}{L_{kj}^{\rm osc}}\right)^2\right] , \tag{8.114}$$

and the *coherence term*

$$\exp\left[-\left(\frac{L}{L_{kj}^{\rm coh}}\right)^2\right] , \tag{8.115}$$

which appear in eqn (8.112) in addition to the standard oscillation phase in eqn (7.24) [533, 528, 225, 523, 526, 531, 536, 301, 224, 538, 539].

The localization term in eqn (8.115) suppresses the oscillations due to Δm^2_{kj} if $\sigma_x \gtrsim L^{\text{osc}}_{kj}$. This means that in order to measure the interference of the massive neutrino components ν_k and ν_j the production and detection processes must be localized in space-time regions much smaller than the oscillation length L^{osc}_{kj}. In practice this requirement is satisfied in all neutrino oscillation experiments.

The localization term allows one to distinguish neutrino oscillation experiments from experiments on the measurement of neutrino masses. As first shown in Ref. [660], neutrino oscillations are suppressed in experiments which are able to measure, through energy–momentum conservation, the mass of the neutrino. Indeed, from the energy–momentum dispersion relation in eqn (8.34) the uncertainty of the mass determination is

$$\delta {m_k}^2 = \sqrt{\left(2\,\tilde{E}_k\,\delta\tilde{E}_k\right)^2 + \left(2\,|\vec{\tilde{p}}_k|\,\delta|\vec{\tilde{p}}_k|\right)^2} \simeq 2\sqrt{2}\,E\,\sigma_p\,, \tag{8.116}$$

where the approximation holds for ultrarelativistic neutrinos. If $\delta {m_k}^2 < |\Delta m^2_{kj}|$, the mass of ν_k is measured with an accuracy better than the difference Δm^2_{kj}. In this case the neutrino ν_j is not produced or detected and the interference of ν_k and ν_j which would generate oscillations does not occur. The localization term in the oscillation probability in eqn (8.112) automatically suppresses the interference of ν_k and ν_j, because

$$-\,2\pi^2\left(\frac{\sigma_x}{L^{\text{osc}}_{kj}}\right)^2 = -\left(\frac{\Delta m^2_{kj}}{4\sqrt{2}E\sigma_p}\right)^2 \simeq -\frac{1}{4}\left(\frac{\Delta m^2_{kj}}{\delta {m_k}^2}\right)^2 . \tag{8.117}$$

It is important to note, however, that the validity of this interpretation of the localization term hinges on the validity of the flavor states in eqn (8.84) not only for the description of neutrino oscillations but also for the description of neutrino production and detection, which we will prove in the next subsection 8.2.3. The suppression of oscillations due to the localization term reflects the effective loss of coherence of the flavor states discussed after eqn (8.83).

The localization term is the only one in the oscillation probability in eqn (8.112) which depends on the direction of $\vec{L}$. If the condition

$$\sigma_x \ll L^{\text{osc}}_{kj}\,, \tag{8.118}$$

which is necessary for unsuppressed interference of ν_k and ν_j, is satisfied, as usual in neutrino oscillation experiments, the localization term can be neglected, leading to the transition probability

$$P_{\nu_\alpha\to\nu_\beta}(L) = \sum_{k,j} U^*_{\alpha k} U_{\alpha j} U_{\beta k} U^*_{\beta j} \exp\left[-2\pi i \frac{L}{L^{\text{osc}}_{kj}} - \left(\frac{L}{L^{\text{coh}}_{kj}}\right)^2\right], \tag{8.119}$$

which is a function of the distance L, depending on the oscillation lengths in eqn (7.31) and the coherence lengths in eqn (8.113).

In eqn (8.119), each k, j term contains, in addition to the standard oscillation phase in eqn (7.24), the coherence term in eqn (8.115), which suppresses the interference of the massive neutrinos ν_k and ν_j for distances larger than the corresponding coherence length, i.e. for $L \gg L_{kj}^{\mathrm{coh}}$. This suppression is due to the separation of the different massive neutrino wave packets, which propagate with different velocities, as illustrated in Figs. 8.2b and 8.3. When the wave packets of ν_k and ν_j are so much separated that they cannot both overlap with the detection process, the massive neutrinos ν_k and ν_j cannot be absorbed coherently [831, 667]. In this case, only one of the two massive neutrinos contributes to the detection process and the interference effect which produces the oscillations is absent. However, in general, the transition probability does not vanish. For example, if $L \gg L_{kj}^{\mathrm{coh}}$ for all k and j, the transition probability reduces to the incoherent transition probability in eqn (7.32). It is possible also to have intermediate cases, in which $L \gg L_{kj}^{\mathrm{coh}}$ for some values of k and j and $L \ll L_{kj}^{\mathrm{coh}}$ for other values of k and j. For example, if $L \ll L_{21}^{\mathrm{coh}}$ and $L \gg L_{kj}^{\mathrm{coh}}$ for all $k > 2$, the effective transition probability is

$$P_{\nu_\alpha \to \nu_\beta}(L) = \sum_k |U_{\alpha k}|^2 \, |U_{\beta k}|^2 + 2\, \mathfrak{Re}\left[U_{\alpha 2}^* U_{\beta 2} U_{\alpha 1} U_{\beta 1}^* \exp\left(-i \frac{\Delta m_{21}^2 \, L}{2E} \right) \right] . \tag{8.120}$$

In this case, it is possible to observe only the unsuppressed oscillations generated by Δm_{21}^2.

8.2.3 Production and detection rates

As explained in subsection 8.1.5, in order to have a theoretical interpretation of neutrino oscillation experiments, it is necessary to describe in a consistent framework neutrino oscillations and neutrino production and detection. In this subsection, we show that the wave packet flavor states in eqn (8.84) are suitable for the calculation of neutrino production and detection rates. These rates must result from the incoherent sum of the rates of the different channels corresponding to different massive neutrinos, which are the physical particles with definite kinematical properties.

For definiteness, we will consider a decay process of the type in eqn (8.4). Any other type of neutrino production or detection processes can be treated in a similar way.

Using the flavor state in eqn (8.82), the amplitude of the general decay process in eqn (8.4) becomes

$$\begin{aligned} \mathcal{A}_\alpha &= \langle \nu_\alpha, \ell_\alpha^+, P_{\mathrm{F}} | \, \mathsf{S} \, | P_{\mathrm{I}} \rangle \\ &= N_\alpha \sum_k \int \frac{\mathrm{d}^3 p}{(2\pi)^3 \, 2E_k} \sum_h \mathcal{A}_{\alpha k}^*(\vec{p}, h) \, \langle \nu_k(E_k, \vec{p}, h), \ell_\alpha^+, P_{\mathrm{F}} | \, \mathsf{S} \, | P_{\mathrm{I}} \rangle \\ &= \sqrt{ \sum_k \int \frac{\mathrm{d}^3 p}{(2\pi)^3 \, 2E_k} \sum_h |\mathcal{A}_{\alpha k}(\vec{p}, h)|^2 } \, . \end{aligned} \tag{8.121}$$

Hence, the decay probability is given by

$$|\mathcal{A}_\alpha|^2 = \sum_k \int \frac{d^3p}{(2\pi)^3\, 2E_k} \sum_h |\mathcal{A}_{\alpha k}(\vec{p}, h)|^2 \,, \tag{8.122}$$

which, correctly, is an incoherent sum of the probabilities of production of the different massive neutrinos. As in the plane-wave approximation discussed in subsection 8.1.5, the coherent character of the flavor states is irrelevant for the calculation of neutrino production and detection rates.

Using the first-order perturbative expansion of the S-matrix operator in eqn (8.21), the amplitudes $\mathcal{A}_{\alpha k}(\vec{p}, h)$ can be written as

$$\mathcal{A}_{\alpha k}(\vec{p}, h) = U_{\alpha k}^* \, \mathcal{M}_{\alpha k}(\vec{p}, h) \,, \tag{8.123}$$

where $\mathcal{M}_{\alpha k}(\vec{p}, h)$ are the matrix elements

$$\mathcal{M}_{\alpha k}(\vec{p}, h) = -i \frac{G_F}{\sqrt{2}} \int d^4x \langle \nu_k(E_k, \vec{p}, h), \ell_\alpha^+, P_F | \overline{\nu_k}(x) \gamma^\rho (1 - \gamma^5) \ell_\alpha(x) J_\rho^{P_I \to P_F}(x) | P_I \rangle \,. \tag{8.124}$$

Using the expression in eqn (8.123) in eqn (8.122), it becomes clear that the decay probability is given by an incoherent sum of the probabilities of production of the different massive neutrinos weighted by $|U_{\alpha k}|^2$,

$$|\mathcal{A}_\alpha|^2 = \sum_k |U_{\alpha k}|^2 \int \frac{d^3p}{(2\pi)^3\, 2E_k} \sum_h |\mathcal{M}_{\alpha k}(\vec{p}, h)|^2 \,. \tag{8.125}$$

Therefore, a description of the flavor neutrino created in the process in eqn (8.4) with the process-dependent coherent state in eqn (8.82) leads to the correct neutrino production and detection rates [958, 790, 687, 959, 960], for which the coherent character of the superposition of massive neutrinos is irrelevant.

If the experiment is not sensitive to the dependence of $\mathcal{M}_{\alpha k}(\vec{p}, h)$ on the neutrino masses, it is possible to approximate

$$\mathcal{M}_{\alpha k}(\vec{p}, h) \simeq \mathcal{M}_\alpha(\vec{p}, h) \,. \tag{8.126}$$

In this case we obtain

$$|\mathcal{A}_\alpha|^2 = \int \frac{d^3p}{(2\pi)^3\, 2E_k} \sum_h |\mathcal{M}_\alpha(\vec{p}, h)|^2 \,, \tag{8.127}$$

which coincides with the decay probability for massless neutrinos in the wave-packet treatment if the common scale of neutrino masses is negligible in comparison with the experimental resolution.

8.3 Size of neutrino wave packets

As discussed in the introduction of section 8.2, the size of the neutrino wave packet is determined by the coherence size of the production process. This is a general property of the wave packets of all particles which are created in some process, because *no source of waves vibrates indefinitely* (quote from section 11.11 of Ref. [644]). For example [644],

> In light sources, the radiating atoms emit wave trains of finite length. Usually, because of collisions or damping arising from other causes, these packets are very short. According to the theorem mentioned above[42], the consequence is that the spectrum lines will not be very narrow but will have an appreciable width $\Delta\lambda$. A measurement of this width will yield the effective *lifetime* of the electromagnetic oscillators in the atoms and the average length of the wave packets. A low-pressure discharge through the vapor of mercury containing the single isotope ^{198}Hg yields very sharp spectral lines, of width about 0.005 Å. Taking the wavelength of one of the brightest lines, 5461 Å, we may estimate that there are roughly 10^6 waves in a packet and that the packets themselves are some 50 cm long.

The broadening of optical lines due to the finite lifetime of atomic transitions is known as the *natural linewidth.*

It is interesting to note that the broadening of optical lines was well known to experimental physicists in the nineteenth century and explained by classical models before the advent of quantum theory (see Ref. [271] and references therein). In 1895, Michelson listed among the hypotheses formulated before that time to account for the line broadening [798]

> The exponential diminution in amplitude of the vibrations due to communication of energy to the surrounding medium or to other causes.

As explained in Ref. [271],

> We consider an emitting atom which we shall proceed to remove to infinity and reduce the *temperature* to the point where, classically at least, no translational motion exists. Now from the classical picture of a vibrating electron or the simple picture of a pair of energy levels between which our radiation transition takes place, we should expect these conditions to yield a spectral line of a single frequency. We, of course, do not obtain this result, but, rather, we obtain the familiar natural line shape which is attributable to Michelson's Cause 3.

A classical derivation of the natural linewidth can be found in section 17.7 of Ref. [636]. Let us emphasize that the natural linewidth of atomic lines has been *observed experimentally* (see section 21.4 of Ref. [1059])!

Another important cause of line broadening known in 1895 was [798]

> The change in wavelength due to the Doppler effect of the component of the velocity of the vibrating atom in the line of sight.

---

[42] "The largest the number N of waves in the group, the smaller the spread $\Delta\lambda$, and in fact theory shows that $\Delta\lambda/\lambda_0$ is approximately equal to $1/N$". (section 11.11 of Ref. [644]).

The Doppler line broadening due to the thermal motion of atoms in a medium was calculated by Rayleigh in 1889 [753]. This important effect, which must be aways taken into account in calculating the spectral shape of monochromatic beams, does not concern us here, because it does not generate a coherent broadening. It is simply due to the different motion of different atoms, whose radiation is incoherent.

To these (and other) causes of line broadening Michelson added in 1895 [798]

> The limitation of the number of regular vibrations by more or less abrupt changes of phase amplitude or plane of vibration caused by collisions.

This important coherent effect has been given several names, among which *collision broadening*, *pressure broadening*, and *interruption broadening*. It has been studied in depth by many authors (see, for example, Refs.[271, 755]). Even more important, it has been *observed experimentally* (see section 21.5 of Ref. [1059])!

In quantum theory, the fact that *no source of waves vibrates indefinitely* implies that all particles are produced as wave packets, whose size is determined by the finite lifetime of the parent particle, or by its finite mean free path if the production process occurs in a medium. Since also no wave detector vibrates indefinitely, it is clear that all particles are also detected as wave packets.

Usually, at least in high-energy physics, the wave packet character of particles is not important and they can be well approximated by plane waves. The phenomenon of neutrino oscillations is an exception, as shown in 1981 by Kayser [660], because the localization of the source and detector requires a wave-packet treatment.

In the following subsections, we estimate the size of the massive neutrino wave packets for some specific production processes. We will write the production processes in terms of flavor neutrinos, but one must always keep in mind that a flavor neutrino is a superposition of massive neutrinos and the massive neutrinos have a wave packet character, because they are the ones that propagate in space-time with definite kinematical properties.

Depending on the physical situation, the size of the massive neutrino wave packets depends on one of the following two physical effects:

Natural Linewidth. If a neutrino is produced in the decay of a particle X with lifetime τ_X, the collapse of the particle wave function due to the decay interrupts the coherent emission of the neutrino wave train. In the rest frame of the decaying particle

$$\sigma_x^\nu \sim \tau_X \,. \tag{8.128}$$

If the particle decays in flight, its lifetime and the corresponding size of the neutrino wave packets are dilated by the relativistic γ factor (given in eqn (B.29)),

$$\sigma_x^\nu \sim \gamma\,\tau_X \,, \tag{8.129}$$

with

$$\gamma = \frac{E_X}{m_X} \,, \tag{8.130}$$

where E_X and m_X are, respectively, the energy and mass of the decaying particle.

Collision Broadening. If a neutrino is created in a decay process or in a scattering process occurring in a medium, the coherent emission of the neutrino wave train is interrupted when one of the other particles taking part to the process interacts with the medium. Hence, the minimum among the mean times between two collisions of the particles participating in the production process gives the size of the massive neutrino wave packets.

If we denote by ℓ_X the mean free path of the particle X participating in the production process and by v_X its velocity, the average time between two collisions of the particle X with the medium is given by

$$\tau_X = \frac{\ell_X}{v_X}, \tag{8.131}$$

and the size of the massive neutrino wave packets due to collision broadening can be estimated as

$$\sigma_x^\nu \sim \operatorname*{Min}_X [\tau_X] . \tag{8.132}$$

Usually, the velocity v_X can be easily evaluated. For the initial particles in thermal equilibrium with the medium, v_X is given by the mean thermal velocity. For the particles emitted in the production process, v_X is given by the kinematics of the process.

On the other hand, the determination of the mean free path ℓ_X is not trivial, because collisions between particles are due to long-range electromagnetic interactions. Hence, particles interact all the time and it is difficult to establish when an interaction has been so strong as to be called a *collision* and to be considered as being effective for the interruption of the coherent emission of the neutrino wave train. In practice, the mean free path is evaluated following the heuristic way often used in optics: for example, from section IV.A of Ref. [271],

> A collision has been undergone by the emitter when the phase of the emitted radiation, η, has changed by unity. This is an admittedly arbitrary collision definition, but one which has proved quite satisfactory for certain physical situations and, further, one which forms one of the bases for the simple form of the interruption theory.

In his 1976 pioneering paper [831], Nussinov wrote:

> The effective time τ_e for the atom (or nucleus) to emit an uninterrupted train of waves is the time during which the random phase change of the emitter due to collisions, i.e. changes of potential energy, with neighboring atoms does not exceed $\approx \pi/2$.

8.3.1 Pion decay

Let us consider a muon neutrino which is created in the decay of a free pion,

$$\pi^+ \to \mu^+ + \nu_\mu . \tag{8.133}$$

In the rest frame of the pion, the natural linewidth relation in eqn (8.128) gives $\sigma_x^\nu \sim \tau_\pi \sim 10\,\text{m}$ for the size of the massive neutrino wave packets, where $\tau_\pi = 2.6 \times 10^{-8}\,\text{s}$ is the lifetime of the pion. This value of σ_x^ν corresponds to $\sigma_p^\nu \sim 10^{-8}\,\text{eV}$.

If a free pion is decaying in flight, eqn (8.129) applies, with $\gamma = E_\pi/m_\pi$. For example, for a pion with 1 GeV energy we have $\gamma \simeq 7$, which implies $\sigma_x^\nu \sim 70\,\mathrm{m}$.

If a high-energy pion decays in a decay tunnel, as in many oscillation experiments with accelerator neutrinos, the size σ_x^ν of the neutrino wave packets is of the order of the minimum between the length of the decay tunnel and $\gamma\tau_\pi$. For example, for a pion with 10 GeV energy, we have $\gamma\tau_\pi \sim 700\,\mathrm{m}$. If the decay tunnel is as long as 100 m, it is much shorter than $\gamma\tau_\pi$ and σ_x^ν is of the order of the length of the decay tunnel.

If the pion decays in a medium, both the pion and the outgoing antimuon interact with the surrounding nuclei. The time allowed for a coherent emission of the decay products is of the order of the time elapsed before the pion or the antimuon has interacted with the material. In this case $\sigma_x^\nu \sim \mathrm{Min}[\ell_\pi/v_\pi, \ell_\mu/v_\mu]$, where ℓ_π and ℓ_μ are the pion and antimuon mean free paths and v_π and v_μ are the pion and antimuon velocities. The pion and antimuon mean free paths are of the order of

$$\ell \sim \mathrm{Max}\left[\frac{1}{\pi b^2 N}, \frac{1}{N^{1/3}}\right], \tag{8.134}$$

where N is the nuclear number density and b is the impact parameter for Coulomb scattering of the pion or the antimuon in the electric field of the nucleus. For a nucleus with atomic number Z the impact parameter b is given by

$$b \approx \frac{4\pi\,\alpha\,Z}{T}, \tag{8.135}$$

where α is the fine-structure constant and T is the kinetic energy of the pion or of the antimuon. The term $1/N^{1/3}$ in eqn (8.134) is the order of magnitude of the internuclear distance and we have taken into account the fact that b cannot be larger than the internuclear distance and the mean free path cannot be smaller than the internuclear distance. In other words, eqn (8.135) is not applicable when T is too small and what happens physically is that the pion or antimuon trajectory is strongly deflected as soon as it passes close to a nucleus, interrupting the coherence of the decay process.

Let us consider, for example, a pion decaying in graphite, which has atomic number $Z = 6$, atomic mass about 12, a density of about $2\,\mathrm{g/cm^3}$, a nuclear number density $N \approx 10^{23}\mathrm{cm}^{-3}$, and an internuclear distance $1/N^{1/3} \approx 2\times10^{-8}\,\mathrm{cm}$. In this case, we have

$$\ell \sim \mathrm{Max}\left[3\times10^{-2}\left(\frac{T}{\mathrm{MeV}}\right)^2\mathrm{cm}\,, 2\times10^{-8}\,\mathrm{cm}\right]. \tag{8.136}$$

A pion decaying *at rest* in reality does not decay precisely at rest, because it has a kinetic energy of the order of the thermal energy of the medium. At 300 K the kinetic energy of the pion is $T_\pi \approx 4\times10^{-2}\,\mathrm{eV}$, corresponding to $\ell_\pi \sim 5\times10^{-17}\,\mathrm{cm}$, which is much smaller than the internuclear distance. Hence ℓ_π is given by the internuclear distance and is of the order of $2\times10^{-8}\,\mathrm{cm}$. Since the pion velocity is of the order of 2×10^{-5}, from the pion interaction with the medium we have $\sigma_x^\nu \sim 10^{-3}\mathrm{cm}$.

Let us consider now the outgoing antimuon. The antimuon kinetic energy is $(m_\pi - m_\mu)^2/2m_\pi \simeq 4\,\text{MeV}$, which corresponds to an antimuon velocity $v_\mu \simeq 0.3$. From eqn (8.136) we have $\ell_\mu \sim 5 \times 10^{-1}\,\text{cm}$ and $\sigma_x^\nu \sim 1\,\text{cm}$.

We see that the collisional broadening of the neutrino line due to the pion collisions with the medium is dominant over that due to the antimuon collisions with the medium and the natural linewidth. Hence for a neutrino produced by pion decay at rest in graphite, we estimate $\sigma_x^\nu \sim 10^{-3}\text{cm}$ and $\sigma_p^\nu \sim 10^{-2}\text{eV}$.

8.3.2 Muon decay

Let us now consider an electron neutrino ν_e or a muon antineutrino $\bar{\nu}_\mu$ produced in the antimuon decay process

$$\mu^+ \to e^+ + \nu_e + \bar{\nu}_\mu \,. \tag{8.137}$$

In the rest frame of the antimuon the collapse of the antimuon wave function due to the decay gives $\sigma_x^\nu \sim \tau_\mu \sim 10^5\,\text{cm}$, where $\tau_\mu = 2.2 \times 10^{-6}\,\text{s}$ is the lifetime of the muon. This natural linewidth of the neutrino line corresponds to $\sigma_p^\nu \sim 10^{-10}\,\text{eV}$.

If the muon decays in a medium, both the antimuon and the outgoing positron interact with the surrounding nuclei and, as in the case discussed in the previous subsection, $\sigma_x^\nu \sim \text{Min}[\ell_\mu/v_\mu, \ell_e/v_e]$, where ℓ_μ and ℓ_e are the antimuon and positron mean free paths and v_μ and v_e are the antimuon and positron velocities. The antimuon and positron mean free paths are given by eqns (8.134) and (8.135), which lead to eqn (8.136) in graphite.

An antimuon decaying *at rest*, e.g. in the LAMPF-0645 [460], LSND [37], and KARMEN [105] experiments, has a kinetic energy of the order of the thermal energy of the medium: $T_\mu \approx 4 \times 10^{-2}\,\text{eV}$. As in the case of the pion in the previous subsection, according to eqn (8.136) a kinetic energy of the order of the thermal energy corresponds to a mean free path which is much smaller than the internuclear distance. Hence ℓ_μ is given by the internuclear distance: $\ell_\mu \sim 2 \times 10^{-8}\,\text{cm}$. Since the antimuon velocity is of the order of 2×10^{-5}, from the antimuon interaction with the medium we have $\sigma_x^\nu \sim 10^{-3}\text{cm}$.

Let us consider now the outgoing positron, which is annihilated as soon as it meets an electron in the medium. Hence, its mean free path is of the order of the internuclear distance: $\ell_e \sim 2 \times 10^{-8}\,\text{cm}$. Since the positron is relativistic, $v_e \simeq 1$ and we have $\sigma_x^\nu \sim \ell_e \sim 2 \times 10^{-8}\,\text{cm}$.

In the case of antimuon decay the collisional broadening of the neutrino line due to the annihilation of the outgoing positron with the electrons in the medium is dominant with respect to that due to the antimuon collisions with the medium and to the natural linewidth. Hence for a neutrino produced by antimuon decay at rest in graphite we estimate $\sigma_x^\nu \sim 2 \times 10^{-8}\text{cm}$ and $\sigma_p^\nu \sim 2 \times 10^3\text{eV}$.

8.4 Questions

In this section we answer some interesting questions that are often asked and have been debated in the literature.

8.4.1 Do charged leptons oscillate?

Flavor oscillations of charged leptons are not possible, because the flavor of a charged lepton is defined by its mass, which is the only property that distinguishes between different charged leptons. In practice, all detectors reveal particles through electromagnetic interactions. Since all charged leptons have the same electromagnetic interactions, as well as other interactions, they are distinguished either through kinematics or through their decay products. Both these characteristics depend directly on the charged lepton mass. In other words, flavor is measured through mass. That is why charged leptons are defined as mass eigenstates and, as such, cannot undergo flavor oscillations, i.e. periodic changes of flavor during propagation.

A different type of oscillations of charged leptons is conceivable in principle [843], but very difficult to realize in practice. Imagine an experiment in which a massive neutrino interacts with a nucleon through a charged-current interaction. Since a massive neutrino is a superposition of ν_e, ν_μ, and ν_τ, in appropriate conditions it could generate a superposition of e, μ, and τ. If this superposition propagates freely in space-time and interacts with another nucleon through a charged-current interaction, in appropriate conditions a massive neutrino can be produced. The phase differences of e, μ, and τ at detection imply that the the final massive neutrino may be different from the initial one and the probability of massive neutrino conversion oscillates as a function of the source-detector distance. However, it must be clear that this type of oscillations do not involve any conversion of the flavor of charged leptons, which is a constant, as explained in the previous paragraph. The new type of oscillations was called *charged-lepton oscillations* in Ref. [843] because of the following analogy with neutrino oscillations. In neutrino oscillations, massive neutrinos propagate unchanged between production and detection, with a difference of mass (flavor) of the charged leptons involved in the production and detection processes. In charged-lepton oscillations, massive charged leptons propagate unchanged between production and detection, with a difference of mass of the neutrinos involved in the production and detection processes. Of course, an experiment of charged-lepton oscillations is extremely difficult, because it would have to overcome, at least, the following problems: (1) The initial and final neutrinos in the production and detection processes must be massive neutrinos of known type; this means that the neutrinos masses must be measured. (2) The propagating charged leptons must be ultrarelativistic, in order to be produced and detected coherently (if τ is not ultrarelativistic, only e and μ contribute to the phase). (3) The energy of the propagating charged leptons must be extremely high, in order to have a measurable oscillation length; for example, the oscillation length due to the interference of the e and μ components of the wave function is $4\pi E/(m_\mu^2 - m_e^2) \simeq 4\pi E/m_\mu^2 \simeq 2\times 10^{-11}\,(E/\text{GeV})\,\text{cm}$, where E is the charged-lepton energy.

Let us finally consider the claim presented in Refs. [926, 975, 978] that the probability of detecting a charged lepton oscillates in space-time. This claim has been refuted in Ref. [384]. A similar effect, called *Lambda oscillations*, which has been claimed to exist [976, 1061] for the Λ's produced together with a neutral kaon, as in the process $\pi^- + p \to \Lambda + K^0$, has been refuted in Refs. [756, 283]. Considering the pion decay process $\pi^- \to \mu^- + \bar{\nu}_\mu$, the authors of Refs. [926, 975, 978, 974] argued that, since the final state of the muon and antineutrino is entangled, if the probability of detecting the antineutrino oscillates as a function of distance, also the probability of detecting the muon must oscillate. However, it is well known that the probability of detecting the antineutrino, irrespective of its flavor, does not oscillate. This property is usually called *conservation of probability* or *unitarity* and is represented mathematically by the general relation $\sum_\beta P_{\bar{\nu}_\mu \to \bar{\nu}_\beta} = 1$. Hence, this argument in fact proves that the probability of detecting a charged lepton does *not* oscillate in space-time!

8.4.2 Is the standard oscillation phase wrong by a factor of 2?

It has been claimed that the standard phase of neutrino oscillations in eqn (7.24) is wrong by a factor of two [437] or there is an ambiguity by a factor of two in the oscillation phase [575, 741, 729, 360]. A similar discrepancy by a factor of two in the phase of $K^0 - \bar{K}^0$ oscillations has been claimed in Refs. [977, 976, 1061] and an ambiguity by a factor of two has been claimed in Ref. [740]. These claims, which have been refuted in Refs. [826, 659, 756, 662, 661, 540, 525, 284], stem from the following fallacious reasoning.

Different massive neutrinos propagate with different velocities

$$v_k = \frac{p_k}{E_k}, \tag{8.138}$$

where E_k and p_k are, respectively, the energy and momentum of the neutrino with mass m_k, related by the relativistic dispersion relation in eqn (8.34). According to the fallacious reasoning, the phases of the different massive neutrinos wave functions after a propagation distance L should be, due to the different times of propagation of different massive neutrinos,

$$\tilde{\Phi}_k = p_k\, L - E_k\, t_k\,. \tag{8.139}$$

The propagation times are given by

$$t_k = \frac{L}{v_k} = \frac{E_k}{p_k}\, L\,, \tag{8.140}$$

which lead, in the relativistic approximation, to the phase difference

$$\Delta\tilde{\Phi}_{kj} \equiv \tilde{\Phi}_k - \tilde{\Phi}_j = -\frac{\Delta m^2_{kj}\, L}{E}\,. \tag{8.141}$$

This phase difference is twice as large as the standard one in eqn (7.24).

Let us note that in eqn (8.139) we have considered the possibility of different energies and momenta for different massive neutrino wave functions, as we have done throughout this chapter (see the following item). The authors of Refs. [740, 575, 741] claimed that a correct way to obtain the standard oscillation phase is to assume the same energy for the different massive neutrino wave functions. Apart from the fact that this is an unphysical assumption (see the following item), it is not true that the disagreement of a factor of two disappears assuming the same energy for the different massive neutrino wave functions, as clearly shown by the above derivation of eqn (8.141), in which the energies of the different massive neutrino wave functions could have been taken to be equal [729, 525]. Indeed, even if the different massive neutrino wave functions have the same energy, the time contribution $-Et_k + Et_j$ to the phase difference $\Delta\tilde{\Phi}_{kj}$ does not disappear if $t_k \neq t_j$. This contribution has been missed in Refs. [740, 575, 741].

8.4.3 What are the energies and momenta of massive neutrinos?

In sections 8.1.2 and 8.2.2, we have seen that, because of eqn (8.37), in both the plane-wave approximation and the wave-packet treatment the phases of massive neutrinos relevant for the oscillations are independent of the values of the energies and momenta of different massive neutrinos as long as the relativistic dispersion relation in eqn (8.34) is satisfied and massive neutrinos are ultrarelativistic.

However, one can still ask if it is physically realistic to assume equal momenta for the massive neutrinos, as done in the standard derivation of the oscillation probability presented in section 7.1, or even equal energies, as claimed in Refs. [740, 575, 986, 739, 742].

It is rather easy to understand that, in general, massive neutrinos have neither equal momenta nor equal energies [540, 522, 526, 526]. The reason is that in general the energy and momentum of a particle created in some process depend on the interaction, on the properties of the other particles participating in the process and on the particle mass. Since different massive neutrinos have different masses, it is obvious that, in general, their energies and momenta are different. In the ultrarelativistic approximation, the production-dependent vector $\vec{\xi}$ in eqns (8.40) and (8.42) is, in general, different from the special values zero, which would give equal momenta, and $\vec{p}/E$, which would give equal energies.

Moreover, it is easy to show that special relativity implies that in a general reference frame different massive neutrinos have different energies and momenta [522]. Indeed, from the Lorentz transformations in eqn (B.28) the transformations of the energy difference $\Delta E_{kj} = E_k - E_j$ and momentum difference $\Delta\vec{p}_{kj} = \vec{p}_k - \vec{p}_j$ from an inertial system $\mathcal{O}$ to another inertial system $\mathcal{O}'$ moving with respect to $\mathcal{O}$ with velocity $\vec{v}$ are given by

$$\Delta E'_{kj} = \gamma\left[\Delta E_{kj} - \vec{v}\cdot\Delta\vec{p}_{kj}\right] , \tag{8.142}$$

$$\Delta\vec{p}\,'_{kj} = \Delta\vec{p}_{kj} - \left[(1-\gamma)\,\frac{\vec{v}\cdot\Delta\vec{p}_{kj}}{v^2} + \gamma\,\Delta E_{kj}\right]\vec{v}\,. \tag{8.143}$$

It is clear that since ΔE_{kj} and $\Delta\vec{p}_{kj}$ cannot both be zero, even if one of them is zero in $\mathcal{O}$, in the boosted frame $\mathcal{O}'$ we have $\Delta E'_{kj} \neq 0$ and $\Delta\vec{p}\,'_{kj} \neq 0$. Therefore, even if there is a peculiar inertial frame in which different massive neutrinos have either the same momentum or the same energy, this property is not satisfied in any other inertial frame.

In any case, the possible existence of a reference frame in which different massive neutrinos have either the same momentum or the same energy is irrelevant in practice, because neutrinos are produced by sources in which the production process has a distribution of velocities, sometimes very wide (for example, pion or muon decay production of atmospheric neutrinos), and different sources may produce neutrinos with the same production process at very different velocities (for example pion decay production of neutrinos in different accelerator experiments). If the equal momentum or the equal energy assumption is valid for a specific velocity of a specific production process, it does not hold for any other velocity.

Let us finally emphasize that this discussion on the exact determination of the energies and momenta of massive neutrinos is only of academic interest. It is irrelevant for neutrino oscillations, which are independent of the energies and momenta of massive neutrinos, as we have shown in sections 8.1.2 and 8.1.3.

8.4.4 Are wave packets necessary?

As we have discussed several times in this chapter, a wave-packet treatment of neutrino oscillations is necessary in order to derive the oscillation probability in a consistent quantum framework, which must take into account the localization of the production and detection processes and the associated momentum uncertainties. In particular, the wave-packet treatment is necessary in order to allow the time average in eqn (8.98) of the space and time dependent probability in eqn (8.97), which leads to the probability in eqn (8.99) of neutrino oscillations as a function of the source–detector distance $\vec{L}$.

However, in practical applications, the wave-packet treatment can be avoided, because the oscillation probability in eqn (8.112) obtained with the wave-packet treatment cannot be distinguished from the plane-wave oscillation probability in eqn (7.112) averaged over the Gaussian L/E distribution in eqn (7.95). Indeed, in this case the averaged plane-wave oscillation probability in eqn (7.112) can be written as

$$\langle P_{\nu_\alpha \to \nu_\beta}(L, E) \rangle = \sum_{k,j} U_{\alpha k}^* \, U_{\beta k} \, U_{\alpha j} \, U_{\beta j}^* \, \exp\left[-i \, \frac{\Delta m^2}{2} \left\langle \frac{L}{E} \right\rangle - \frac{1}{2} \left(\frac{\Delta m^2}{2} \, \sigma_{L/E} \right)^2 \right] . \tag{8.144}$$

If the uncertainty $\sigma_{L/E}$ is proportional to the average $\langle L/E \rangle$, as in the example of Fig. 7.2 on page 262 (see eqn (7.97)), the behavior of the averaged probability in eqn (8.144) as a function of $\langle L/E \rangle$ is identical to the behavior of the wave packet probability in eqn (8.112) as a function of L/E (the localization term can be neglected in real oscillation experiments) [667, 526]. This means that in practice one can take into account the quantum space and momentum uncertainties by adding them to the classical uncertainties of L and E.

As emphasized in Ref. [667], if one does not have control of both the production and detection processes, one cannot know if oscillations are suppressed because of incoherent averaging over the energy and location of different microscopic processes or because of decoherence due to the separation of wave packets. However, having good control of both the production and detection processes, it may be possible to reduce the causes of incoherent broadening of the energy spectrum and the uncertainty of the spatial location of the production and detection processes in such a way to prove experimentally that oscillations can be suppressed because of wave packet effects (in practice using an approach similar to the one adopted for the measurement of the natural linewidth and collision broadening of atomic lines discussed in the introduction of section 8.3).

The fact that coherent and incoherent stationary beams are indistinguishable without a theoretical analysis is well known in optics (see section 7.5.8 of Ref. [263]), in neutron interferometry [329, 652], and in general stationary particle beams [216].

Eventually, the existence of massive neutrino wave packets may be revealed experimentally by measuring the flavor transition probability in eqn (8.97) as a function of both the distance and time interval between the events of neutrino production and detection.

9

NEUTRINO OSCILLATIONS IN MATTER

> All truths are easy to understand once they are discovered; the point is to discover them.
> **Galileo Galilei**

In 1978, L. Wolfenstein [1065] discovered that neutrinos propagating in matter are subject to a potential due to the coherent forward elastic scattering with the particles in the medium (electrons and nucleons). This potential, which is equivalent to an index of refraction, modifies the mixing of neutrinos. In the case of two-neutrino mixing, the mixing angle in vacuum is replaced by an effective mixing angle in matter, which, for suitable matter densities, can become large, even if the mixing angle in vacuum is very small. Wolfenstein [1065] and other authors in the early 1980s [171] studied neutrino propagation in a medium with constant matter density[43], or the possibility of detecting relic neutrinos through their coherent interactions with matter [731, 715]. Unfortunately, in early papers the sign of the matter potential was incorrect, leading to an enhancement of the transitions of electron antineutrinos instead of neutrinos. The correct sign for the matter potential was introduced in Ref. [715].

In 1985 S.P. Mikheev and A.Yu. Smirnov [801, 802] discovered that it is possible to have resonant flavor transitions when neutrinos propagate in a medium with varying density, and there is a region along the neutrino path in which the effective mixing angle passes through the maximal mixing value of $\pi/4$ (see the clear discussion in Ref. [222]). This so-called *MSW mechanism* quickly become very famous, because it could explain the flavor conversion of solar neutrinos during their propagation out of the Sun, even in the case of a small vacuum mixing angle. Today we know that the vacuum mixing angle relevant for solar neutrino oscillations is large but not maximal and that the flavor transitions of solar neutrinos occur through the MSW effect (see chapter 10).

Before we proceed to the calculation of neutrino potentials in matter, let us remark that neutrinos in matter are affected not only by coherent forward elastic scattering, but also by incoherent scatterings with the particles in the medium. However, the amount of these incoherent scatterings is extremely small in most

[43] It is interesting to note that Wolfenstein [1065] considered not only the effects of matter potentials for the standard two-neutrino oscillations, but also the possibility of matter potentials generated by flavor-changing neutral currents, which can generate flavor transitions in matter even in the case of massless neutrinos, as studied later in Ref. [1031, 580]. He also considered the matter effects on the oscillations of long-baseline accelerator neutrinos propagating in the Earth, a topic which is presently under intense study (see Refs. [58, 59]).

situations and can be safely neglected. We can estimate easily the order of magnitude of a mean free path due to incoherent scatterings as follows. From dimensional arguments, the cross-section of neutrino weak interactions with a charged lepton or hadron in the center-of-mass frame is given by

$$\sigma_{\mathrm{cm}} \sim G_{\mathrm{F}}\, s\,, \tag{9.1}$$

where s is the Lorentz invariant Mandelstam variable which represents the square of the total energy in the center-of-mass frame. In the laboratory frame, where the target particle is at rest, s is given by $2EM$, where E is the neutrino energy and M is the mass of the target particle (we neglected the neutrino mass), yielding

$$\sigma_{\mathrm{lab}} \sim G_{\mathrm{F}}\, E\, M \sim 10^{-38}\, \mathrm{cm}^2\, \frac{E\, M}{\mathrm{GeV}^2}\,. \tag{9.2}$$

The mean free path of the neutrino in a medium with number density N of target particles is given by

$$\ell \sim \frac{1}{N\,\sigma} \sim \frac{10^{38}\,\mathrm{cm}}{(N\,\mathrm{cm}^3)\left(E\,M/\mathrm{GeV}^2\right)}\,. \tag{9.3}$$

In normal matter, the main target particles are nucleons with mass $M \sim 1\mathrm{GeV}$ and number density $N \sim N_{\mathrm{A}}/\mathrm{cm}^3 \sim 10^{24}/\mathrm{cm}^3$, yielding

$$\ell_{\mathrm{matter}} \sim \frac{10^{14}\,\mathrm{cm}}{(E/\mathrm{GeV})}\,. \tag{9.4}$$

One can see that the Earth, which has a diameter of about 10^9 cm, is opaque to neutrinos only if their energy exceeds about 10^5 GeV, which is extremely high. Solar neutrinos have energies of the order of 1 MeV, which implies a mean free path in normal matter of the order of 10^{17} cm, which is about 0.1 light years! On the other hand, if the nucleon density is of the order of $10^{12} N_{\mathrm{A}}/\mathrm{cm}^3$, neutrinos with energy of the order of 1 MeV have a mean free path of only 1 km. Such extremely high matter densities can be reached in neutron stars and supernova cores, which have diameters of the order of 1 km and are thus opaque to low-energy neutrinos. Only in these extreme cases, is it necessary to take into account neutrino incoherent scattering.

This chapter's discussion includes the calculation of the effective neutrino potentials in matter in section 9.1, the general evolution of neutrino flavors in matter in section 9.2, and the MSW effect for two generations in section 9.3. In sections 9.4 and 9.5 we present, respectively, solutions of the evolution equation of neutrino flavors in the case of a medium whose density can be approximated by a series of slabs with constant density and in the case of a medium which varies periodically along the neutrino path. Finally, section 9.6 presents an instructive geometrical representation of neutrino oscillations.

9.1 Effective potentials in matter

When active flavor neutrinos propagate in matter, their evolution equation is affected by effective potentials due to the coherent interactions with the medium

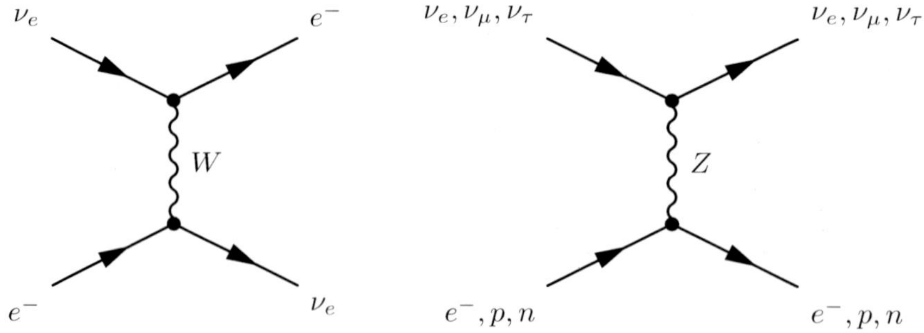

FIG. 9.1. Feynman diagrams of the coherent forward elastic scattering processes that generate the CC potential V_{CC} through W exchange and the NC potential V_{NC} through Z exchange.

through coherent forward elastic weak CC and NC scatterings [1065, 731, 715, 222, 828, 844, 530, 418, 845, 747, 387, 716, 464]. The Feynman diagrams of CC and NC scattering are shown in Fig. 9.1.

Let us first calculate the CC potential V_{CC} for an electron neutrino propagating in a homogeneous and isotropic gas of unpolarized electrons [530, 716]. From the effective low-energy charged-current weak interaction Lagrangian in eqn (3.224), the effective CC Hamiltonian corresponding to the left diagram in Fig. 9.1 is

$$\mathscr{H}_{\text{eff}}^{(\text{CC})}(x) = \frac{G_{\text{F}}}{\sqrt{2}} \left[\overline{\nu_e}(x)\gamma^\rho \left(1-\gamma^5\right) e(x)\right] \left[\overline{e}(x)\gamma_\rho \left(1-\gamma^5\right) \nu_e(x)\right] . \tag{9.5}$$

In order to separate the neutrino and electron contributions, we apply to eqn (9.5) the Fierz transformation in eqn (2.508):

$$\mathscr{H}_{\text{eff}}^{(\text{CC})}(x) = \frac{G_{\text{F}}}{\sqrt{2}} \left[\overline{\nu_e}(x)\gamma^\rho \left(1-\gamma^5\right) \nu_e(x)\right] \left[\overline{e}(x)\gamma_\rho \left(1-\gamma^5\right) e(x)\right] . \tag{9.6}$$

The average of the effective Hamiltonian over the electron background in the rest frame of the medium is given by

$$\begin{aligned}\overline{\mathscr{H}_{\text{eff}}^{(\text{CC})}}(x) = {} & \frac{G_{\text{F}}}{\sqrt{2}} \overline{\nu_e}(x)\gamma^\rho \left(1-\gamma^5\right) \nu_e(x) \int \mathrm{d}^3p_e \, f(E_e, T) \\ & \times \frac{1}{2} \sum_{h_e=\pm1} \langle e^-(p_e, h_e)|\overline{e}(x)\gamma_\rho \left(1-\gamma^5\right) e(x)|e^-(p_e, h_e)\rangle .\end{aligned} \tag{9.7}$$

The electron states correspond to the electrons in the left diagram in Fig. 9.1. Their four-momenta and helicities before and after the scattering are identical, because the interaction must leave the medium unchanged in order to contribute coherently to the neutrino potential. For simplicity, for the electron background we consider a finite normalization volume V according to the method discussed in section 2.13, with the one-electron states $|e^-(p_e, h_e)\rangle$ defined according to eqn (2.487):

$$|e^-(p_e, h_e)\rangle = \frac{1}{2E_e V} \, a_e^{(h_e)\dagger}(p_e) \, |0\rangle \, . \tag{9.8}$$

The function $f(E_e, T)$ is the statistical distribution of the electron energy E_e, which depends on the temperature T of the electron background and is normalized by

$$\int \mathrm{d}^3p \, f(E_e, T) = N_e \, V \, , \tag{9.9}$$

where N_e is the electron density of the medium and $N_e \, V$ is the total number of electrons.

The average over helicities of the electron matrix element is given by

$$\begin{aligned}
\frac{1}{2} \sum_{h_e=\pm 1} & \langle e^-(p_e, h_e)|\overline{e}(x)\gamma_\rho \left(1 - \gamma^5\right) e(x)|e^-(p_e, h_e)\rangle \\
&= \frac{1}{4E_e V} \sum_{h_e=\pm 1} \overline{u_e^{(h_e)}}(p_e)\gamma_\rho \left(1 - \gamma^5\right) u_e^{(h_e)}(p_e) \\
&= \frac{1}{4E_e V} \, \mathrm{Tr}\left[\left(\sum_{h_e=\pm 1} u_e^{(h_e)}(p_e)\overline{u_e^{(h_e)}}(p_e)\right) \gamma_\rho \left(1 - \gamma^5\right)\right] \\
&= \frac{1}{4E_e V} \, \mathrm{Tr}\big[(\not{p}_e + m_e) \, \gamma_\rho \left(1 - \gamma^5\right)\big] = \frac{p_{e\rho}}{E_e V} \, .
\end{aligned} \tag{9.10}$$

Hence, we obtain

$$\overline{\mathscr{H}_{\mathrm{eff}}^{(\mathrm{CC})}}(x) = \frac{G_{\mathrm{F}}}{\sqrt{2}} \frac{1}{V} \int \mathrm{d}^3p_e \, f(E_e, T) \, \overline{\nu_e}(x) \frac{\not{p}_e}{E_e} \left(1 - \gamma^5\right) \nu_e(x) \, . \tag{9.11}$$

The integral over d^3p_e becomes

$$\int \mathrm{d}^3p_e \, f(E_e, T) \, \frac{\not{p}_e}{E_e} = \int \mathrm{d}^3p_e \, f(E_e, T) \left(\gamma^0 - \frac{\vec{p}_e \cdot \vec{\gamma}}{E_e}\right) = N_e \, V \, \gamma^0 \, , \tag{9.12}$$

where the integral of the $\vec{p}_e/E_e$ term vanishes because the integrand is odd under $\vec{p}_e \to -\vec{p}_e$. Finally, the normalization volume cancels, leading to

$$\overline{\mathscr{H}_{\mathrm{eff}}^{(\mathrm{CC})}}(x) = V_{\mathrm{CC}} \, \overline{\nu_{eL}}(x)\gamma^0 \nu_{eL}(x) \, , \tag{9.13}$$

with the charged-current potential given by

$$V_{\mathrm{CC}} = \sqrt{2} \, G_{\mathrm{F}} \, N_e \, . \tag{9.14}$$

The neutral-current potential of neutrinos propagating in a medium with density $N_{\mathfrak{f}}$ of fermions $\mathfrak{f}$ can be calculated in a similar way starting from the effective low-energy neutral-current weak interaction Lagrangian in eqn (3.235). The effective

NC Hamiltonian corresponding to the right diagram in Fig. 9.1 is

$$\mathscr{H}_{\text{eff}}^{(\text{NC})}(x) = \frac{G_{\text{F}}}{\sqrt{2}} \sum_{\alpha=e,\mu,\tau} \left[\overline{\nu_\alpha}(x)\gamma^\rho \left(1-\gamma^5\right)\nu_\alpha(x)\right] \sum_{\mathfrak{f}} \left[\bar{\mathfrak{f}}(x)\gamma_\rho \left(g_V^{\mathfrak{f}} - g_A^{\mathfrak{f}}\gamma^5\right)\mathfrak{f}(x)\right] . \tag{9.15}$$

Comparing with the effective CC Hamiltonian which generates the potential in eqn (9.14) one can see that the neutral-current potential of any flavor neutrino ν_α due to coherent interaction with fermions $\mathfrak{f}$ is

$$V_{\text{NC}}^{\mathfrak{f}} = \sqrt{2}\, G_{\text{F}}\, N_{\mathfrak{f}}\, g_V^{\mathfrak{f}} . \tag{9.16}$$

From Table 3.6 on page 78, we have for electrons

$$g_V^e = -\frac{1}{2} + 2\sin^2\vartheta_{\text{W}} . \tag{9.17}$$

Since $p = uud$ and $n = udd$, we have for protons

$$g_V^p = 2\, g_V^u + g_V^d = \frac{1}{2} - 2\sin^2\vartheta_{\text{W}} , \tag{9.18}$$

and for neutrons

$$g_V^n = g_V^u + 2\, g_V^d = -\frac{1}{2} . \tag{9.19}$$

In astrophysical environments with low temperature and density, matter is composed of neutrons, protons, and electrons. Since electrical neutrality implies an equal number density of protons and electrons, the neutral-current potentials of protons and electrons cancel each other and only neutrons contribute, yielding

$$V_{\text{NC}} = -\frac{1}{2}\sqrt{2}\, G_{\text{F}}\, N_n . \tag{9.20}$$

Summarizing, the effective neutrino potential Hamiltonian in astrophysical environments with low temperature and density is

$$\overline{\mathscr{H}_{\text{eff}}}(x) = \sum_{\alpha=e,\mu,\tau} V_\alpha\, \overline{\nu_{\alpha L}}(x)\gamma^0 \nu_{\alpha L}(x) , \tag{9.21}$$

with the potentials

$$V_\alpha = V_{\text{CC}}\,\delta_{\alpha e} + V_{\text{NC}} = \sqrt{2}\, G_{\text{F}} \left(N_e\,\delta_{\alpha e} - \frac{1}{2} N_n\right) . \tag{9.22}$$

These potentials are very small, because

$$\sqrt{2}\, G_{\text{F}} \simeq 7.63 \times 10^{-14}\, \frac{\text{eV}\,\text{cm}^3}{N_{\text{A}}} , \tag{9.23}$$

where N_{A} is Avogadro's number in eqn (A.144).

In order to understand the physical meaning of the potentials V_α, let us calculate the potential energy of a neutrino of flavor α propagating through the medium, which is given by

$$\mathscr{V}_\alpha^{(h)} = \langle \nu_\alpha^{\mathrm{wp}}(\langle\vec{p}\rangle, h)| \int \mathrm{d}^3x \, \overline{\mathscr{H}_{\mathrm{eff}}}(x) |\nu_\alpha^{\mathrm{wp}}(\langle\vec{p}\rangle, h)\rangle \,. \tag{9.24}$$

Here, we have considered a neutrino wave packet of the type in eqn (2.459),

$$|\nu_\alpha^{\mathrm{wp}}(\langle\vec{p}\rangle, h)\rangle = \int \frac{\mathrm{d}^3p}{(2\pi)^{3/2}\sqrt{2E}} \varphi(\vec{p}, \langle\vec{p}\rangle) \, |\nu_\alpha(p, h)\rangle \,, \tag{9.25}$$

which describes a neutrino with helicity h and average momentum $\langle\vec{p}\rangle$, as discussed in section 2.12. Expanding the wave packets in eqn (9.24) and calculating the matrix element, we obtain, in the case of Dirac neutrinos,

$$\mathscr{V}_{\mathrm{D}\alpha}^{(h)} = \frac{1}{2} V_\alpha \int \frac{\mathrm{d}^3p}{2E} |\varphi(\vec{p}, \langle\vec{p}\rangle)|^2 \, \overline{u_{\nu_\alpha}^{(h)}}(p) \, \gamma^0 \left(1-\gamma^5\right) u_{\nu_\alpha}^{(h)}(p) \,. \tag{9.26}$$

Using eqn (A.94), we have

$$\begin{aligned} \overline{u_{\nu_\alpha}^{(h)}}(p) \, \gamma^0 \left(1-\gamma^5\right) u_{\nu_\alpha}^{(h)}(p) &= \mathrm{Tr}\left[u_{\nu_\alpha}^{(h)}(p) \, \overline{u_{\nu_\alpha}^{(h)}}(p) \, \gamma^0 \left(1-\gamma^5\right)\right] \\ &= \mathrm{Tr}\left[(\not{p} + m_{\nu_\alpha}) \left(\frac{1+\gamma^5 \not{s}_h}{2}\right) \gamma^0 \left(1-\gamma^5\right)\right] \\ &= 2\,(E - h\,|\vec{p}|) \simeq \begin{cases} 4\,E & \text{for} \quad h=-1 \\ m_{\nu_\alpha}^2/E & \text{for} \quad h=+1 \,, \end{cases} \end{aligned} \tag{9.27}$$

where the last approximation holds for ultrarelativistic neutrinos and m_{ν_α} is the mass of ν_α if mixing is neglected. We consider this unrealistic mass only to show the order of magnitude of the strong suppression of the potential energy of ultrarelativistic right-handed neutrinos:

$$\mathscr{V}_{\mathrm{D}\alpha}^{(+)} \simeq V_\alpha \frac{m_{\nu_\alpha}^2}{4} \left\langle \frac{1}{E^2} \right\rangle \,, \tag{9.28}$$

where $\langle 1/E^2 \rangle$ is the average of $1/E^2$ over the momentum distribution of the wave packet, which is approximately equal to $1/\langle E^2 \rangle$ for sharp wave packets.

On the other hand, taking into account the normalization in eqn (2.461) of the momentum distribution of the wave packet, the potential energy of ultrarelativistic left-handed neutrinos is given by

$$\mathscr{V}_{\mathrm{D}\alpha}^{(-)} \simeq V_\alpha \,. \tag{9.29}$$

Thus, V_α is the potential energy of ultrarelativistic left-handed neutrinos of flavor α propagating through the medium.

These results for $\mathscr{V}_\alpha^{(+)}$ and $\mathscr{V}_\alpha^{(-)}$ can also be obtained with the finite normalization volume approach described in section 2.13. In this case, the potential energy is given by

$$\mathscr{V}_\alpha^{(h)} = \langle \nu_\alpha(p,h)| \int_V \mathrm{d}^3x \, \overline{\mathscr{H}_{\mathrm{eff}}}(x) |\nu_\alpha(p,h)\rangle \,. \tag{9.30}$$

In the case of Dirac neutrinos, we have

$$\mathscr{V}_{\mathrm{D}\alpha}^{(h)} = \frac{V_\alpha}{4E} \, \overline{u_{\nu_\alpha}^{(h)}}(p) \, \gamma^0 \left(1-\gamma^5\right) u_{\nu_\alpha}^{(h)}(p) \,. \tag{9.31}$$

Using eqn (9.27), one obtains the results in eqns (9.28) and (9.29).

Let us now consider Dirac antineutrinos. Considering an antineutrino wave packet of the type in eqn (2.460),

$$|\bar\nu_\alpha^{\mathrm{wp}}(\langle\vec p\rangle, h)\rangle = \int \frac{\mathrm{d}^3p}{(2\pi)^{3/2}\sqrt{2E}} \, \bar\varphi(\vec p, \langle\vec p\rangle) \, |\bar\nu_\alpha(p,h)\rangle \,. \tag{9.32}$$

we have

$$\bar{\mathscr{V}}_{\mathrm{D}\alpha}^{(h)} = -\frac{1}{2} V_\alpha \int \frac{\mathrm{d}^3p}{2\,E} \, |\bar\varphi(\vec p, \langle\vec p\rangle)|^2 \, \overline{v_{\nu_\alpha}^{(h)}}(p) \, \gamma^0 \left(1-\gamma^5\right) v_{\nu_\alpha}^{(h)}(p) \,, \tag{9.33}$$

and

$$\overline{v_{\nu_\alpha}^{(h)}}(p) \, \gamma^0 \left(1-\gamma^5\right) v_{\nu_\alpha}^{(h)}(p) = 2\left(E + h\,|\vec p|\right) \simeq \begin{cases} m_{\nu_\alpha}^2/E & \text{for} \quad h=-1 \\ 4\,E & \text{for} \quad h=+1 \,, \end{cases} \tag{9.34}$$

which leads to

$$\bar{\mathscr{V}}_{\mathrm{D}\alpha}^{(+)} \simeq -V_\alpha \,, \tag{9.35}$$

and

$$\bar{\mathscr{V}}_{\mathrm{D}\alpha}^{(-)} \simeq -V_\alpha \, \frac{m_{\nu_\alpha}^2}{4} \left\langle \frac{1}{E^2} \right\rangle \,. \tag{9.36}$$

Hence, the potential energy of ultrarelativistic right-handed Dirac antineutrinos is opposite in sign to that of left-handed neutrinos and the potential energy of ultrarelativistic left-handed Dirac antineutrinos is strongly suppressed by the ratio $(\text{mass}/\text{energy})^2$.

These results for $\mathscr{V}_{\mathrm{D}\alpha}^{(\pm)}$ and $\bar{\mathscr{V}}_{\mathrm{D}\alpha}^{(\pm)}$ are clear from a physical point of view: in the ultrarelativistic limit, neutrinos can be considered massless in interactions and eqn (2.222) tells us that, in this approximation, only Dirac neutrinos with negative helicity and Dirac antineutrinos with positive helicity take part in weak interactions.

On the other hand, in the case of Majorana neutrinos we have

$$\begin{aligned} \mathscr{V}_{\mathrm{M}\alpha}^{(h)} = \frac{1}{2} V_\alpha \int \frac{\mathrm{d}^3p}{2\,E} \, |\varphi(\vec p, \langle\vec p\rangle)|^2 \Big[& \overline{u_{\nu_\alpha}^{(h)}}(p) \, \gamma^0 \left(1-\gamma^5\right) u_{\nu_\alpha}^{(h)}(p) \\ & - \overline{v_{\nu_\alpha}^{(h)}}(p) \, \gamma^0 \left(1-\gamma^5\right) v_{\nu_\alpha}^{(h)}(p) \Big] \,. \end{aligned} \tag{9.37}$$

From eqns (9.27) and (9.34) it follows that

$$\mathscr{V}_{\mathrm{M}\alpha}^{(\pm)} \simeq \mp V_{\alpha} \,. \tag{9.38}$$

Hence, the potential energy of ultrarelativistic left-handed Dirac and Majorana neutrinos are the same, whereas the potential energy of ultrarelativistic right-handed Majorana neutrinos is the same as that of ultrarelativistic right-handed Dirac antineutrinos. These equalities are a particular case of the general correspondence discussed in section 6.2.3.

Let us finally remark that the coherent interactions with matter conserve the neutrino helicity, because using the properties in eqns (2.153), (2.155), and (A.62) we have

$$\begin{aligned}
\overline{u_{\nu_\alpha}^{(h)}}(p)\,\gamma^0\left(1-\gamma^5\right)u_{\nu_\alpha}^{(h')}(p) &= \left(\overline{u_{\nu_\alpha}^{(h)}}(p)\,\frac{\vec{p}\cdot\vec{\Sigma}}{h\,|\vec{p}|}\right)\gamma^0\left(1-\gamma^5\right)\left(\frac{\vec{p}\cdot\vec{\Sigma}}{h'\,|\vec{p}|}\,u_{\nu_\alpha}^{(h')}(p)\right) \\
&= h\,h'\,\overline{u_{\nu_\alpha}^{(h)}}(p)\,\gamma^0\left(1-\gamma^5\right)u_{\nu_\alpha}^{(h')}(p)\,,
\end{aligned} \tag{9.39}$$

which implies

$$\overline{u_{\nu_\alpha}^{(h)}}(p)\,\gamma^0\left(1-\gamma^5\right)u_{\nu_\alpha}^{(h')}(p) \propto \delta_{hh'} \,. \tag{9.40}$$

9.2 Evolution of neutrino flavors

Let us consider[44] an ultrarelativistic left-handed neutrino with flavor α ($\alpha = e, \mu, \tau$) and momentum $\vec{p}$. This neutrino is described by the flavor state in eqn (7.4),

$$|\nu_\alpha\rangle = \sum_k U_{\alpha k}^{*}\,|\nu_k\rangle \,, \tag{9.41}$$

with the normalizations in eqns (7.5) and (7.6). The massive neutrino state $|\nu_k\rangle$ with momentum $\vec{p}$ is an eigenstate of the vacuum Hamiltonian $\mathscr{H}_0$:

$$\mathscr{H}_0\,|\nu_k\rangle = E_k\,|\nu_k\rangle \,, \qquad \text{with} \qquad E_k = \sqrt{\vec{p}^{\,2} + m_k^2} \,. \tag{9.42}$$

The total Hamiltonian in matter is

$$\mathscr{H} = \mathscr{H}_0 + \mathscr{H}_I \,, \qquad \text{with} \qquad \mathscr{H}_I\,|\nu_\alpha\rangle = V_\alpha\,|\nu_\alpha\rangle \,, \tag{9.43}$$

where V_α, given in eqn (9.22), is the effective potential felt by the ultrarelativistic left-handed flavor neutrino (see eqn (9.29)).

[44] A more complicated wave-packet treatment is necessary for the derivation of neutrino oscillations in matter by taking into account different energies and momenta of the different massive neutrino components [534].

In the Schrödinger picture, a neutrino state with initial flavor α obeys the evolution equation

$$i \frac{\mathrm{d}}{\mathrm{d}t} |\nu_\alpha(t)\rangle = \mathscr{H} |\nu_\alpha(t)\rangle \,, \qquad \text{with} \qquad |\nu_\alpha(0)\rangle = |\nu_\alpha\rangle \,. \tag{9.44}$$

The amplitude of $\nu_\alpha \to \nu_\beta$ transitions after a time t is given by

$$\psi_{\alpha\beta}(t) = \langle \nu_\beta | \nu_\alpha(t) \rangle \,, \qquad \text{with} \qquad \psi_{\alpha\beta}(0) = \delta_{\alpha\beta} \,. \tag{9.45}$$

Thus, the probability that a neutrino born at $t = 0$ with flavor α is found to have flavor β after a time t is

$$P_{\nu_\alpha \to \nu_\beta}(t) = |\psi_{\alpha\beta}(t)|^2 \,. \tag{9.46}$$

From eqns (9.41)–(9.45), the time evolution equation of the flavor transition amplitudes is

$$i \frac{\mathrm{d}}{\mathrm{d}t} \psi_{\alpha\beta}(t) = \sum_\eta \left(\sum_k U_{\beta k} \, E_k \, U_{\eta k}^* + \delta_{\beta\eta} \, V_\beta \right) \psi_{\alpha\eta}(t) \,. \tag{9.47}$$

It is possible to show that[45]

$$\sum_\eta \psi_{\alpha\eta}(t) \, \psi_{\beta\eta}^*(t) = \delta_{\alpha\beta} \,. \tag{9.48}$$

For $\alpha = \beta$, one obtains the equation of conservation of probability

$$\sum_\beta P_{\nu_\alpha \to \nu_\beta}(t) = \sum_\beta |\psi_{\alpha\beta}(t)|^2 = 1 \,. \tag{9.49}$$

For ultrarelativistic neutrinos, we have

$$E_k \simeq E + \frac{m_k^2}{2E} \,, \qquad p \simeq E \,, \qquad t \simeq x \,, \tag{9.50}$$

where x is the distance from the source. With these approximations, eqn (9.47) becomes the evolution equation in space

$$\begin{aligned} i \frac{\mathrm{d}}{\mathrm{d}x} \psi_{\alpha\beta}(x) = {} & \left(p + \frac{m_1^2}{2E} + V_{\mathrm{NC}} \right) \psi_{\alpha\beta}(x) \\ & + \sum_\eta \left(\sum_k U_{\beta k} \frac{\Delta m_{k1}^2}{2E} U_{\eta k}^* + \delta_{\beta e} \, \delta_{\eta e} \, V_{\mathrm{CC}} \right) \psi_{\alpha\eta}(x) \,. \end{aligned} \tag{9.51}$$

In this expression, we separated out the term $\left(p + m_1^2/2E + V_{\mathrm{NC}} \right) \psi_{\alpha\beta}(x)$, which is irrelevant for the flavor transitions, since it generates a phase common to all flavors.

[45] The initial conditions in eqn (9.45) imply that eqn (9.48) is trivially satisfied at $t = 0$ and, from eqn (9.47), one can find $\frac{\mathrm{d}}{\mathrm{d}t} \psi_{\alpha\eta}(t) \, \psi_{\beta\eta}^*(t) = 0$.

It can be eliminated by the phase shift

$$\psi_{\alpha\beta}(x) \to \psi_{\alpha\beta}(x)\, e^{-i\left(p+m_1^2/2E\right)x - i\int_0^x V_{\mathrm{NC}}(x')\,\mathrm{d}x'}\,, \tag{9.52}$$

which does not have any effect on the probability of $\nu_\alpha \to \nu_\beta$ transitions in eqn (9.46) (with $t = x$). Therefore, the relevant evolution equation for the flavor transition amplitudes is

$$i\,\frac{\mathrm{d}}{\mathrm{d}x}\,\psi_{\alpha\beta}(x) = \sum_\eta \left(\sum_k U_{\beta k}\,\frac{\Delta m_{k1}^2}{2E}\,U_{\eta k}^* + \delta_{\beta e}\,\delta_{\eta e}\,V_{\mathrm{CC}} \right) \psi_{\alpha\eta}(x)\,, \tag{9.53}$$

which shows that neutrino oscillations in matter, as neutrino oscillations in vacuum, depend on the differences of the squared neutrino masses, not on the absolute value of neutrino masses. Equation (9.53) can be written in matrix form as

$$i\,\frac{\mathrm{d}}{\mathrm{d}x}\,\Psi_\alpha = \mathcal{H}_{\mathrm{F}}\,\Psi_\alpha\,. \tag{9.54}$$

This equation has the structure of a Schrödinger equation with the effective Hamiltonian matrix $\mathcal{H}_{\mathrm{F}}$ in the flavor basis given by

$$\mathcal{H}_{\mathrm{F}} = \frac{1}{2E}\left(U\,\mathbb{M}^2\,U^\dagger + \mathbb{A} \right)\,. \tag{9.55}$$

In the case of three-neutrino mixing, we have

$$\Psi_\alpha = \begin{pmatrix} \psi_{\alpha e} \\ \psi_{\alpha\mu} \\ \psi_{\alpha\tau} \end{pmatrix}, \qquad \mathbb{M}^2 = \begin{pmatrix} 0 & 0 & 0 \\ 0 & \Delta m_{21}^2 & 0 \\ 0 & 0 & \Delta m_{31}^2 \end{pmatrix}, \qquad \mathbb{A} = \begin{pmatrix} A_{\mathrm{CC}} & 0 & 0 \\ 0 & 0 & 0 \\ 0 & 0 & 0 \end{pmatrix}, \tag{9.56}$$

where

$$A_{\mathrm{CC}} \equiv 2\,E\,V_{\mathrm{CC}} = 2\,\sqrt{2}\,E\,G_{\mathrm{F}}\,N_e\,. \tag{9.57}$$

One can easily check that $\mathcal{H}_{\mathrm{F}}$ is Hermitian.

Based on the evolution equation (9.54), one can prove that the Majorana phases in the mixing matrix do not have any effect on neutrino oscillations in vacuum [234, 376] as well as in matter [713]. In fact, the diagonal matrix of Majorana phases D^{M} on the right of the mixing matrix in eqn (6.189) cancels in the product $U\mathbb{M}^2U^\dagger$. Therefore, the Dirac or Majorana nature of neutrinos cannot be distinguished in neutrino oscillations.

9.3 The MSW effect

Since the case of three neutrino mixing is too complicated for an introductory discussion, in this section we consider the simplest case of two neutrino mixing between ν_e, ν_μ and ν_1, ν_2. The case of ν_e-ν_τ mixing is identical, since ν_μ and ν_τ have the same matter potential (see eqn (9.22)). In the case of ν_e-ν_s mixing, where

ν_s is a sterile neutrino, one can use the following formulas with the replacement $V_{\mathrm{CC}} \to V_{\mathrm{CC}} + V_{\mathrm{NC}}$, which corresponds to the change $N_e \to N_e - N_n/2$.

We assume the initial neutrino to be an electron neutrino, i.e. $\alpha = e$ in the equations in section 9.2. This is the most interesting case, because of its application to solar neutrinos, which are created and propagate in the dense interior of the Sun.

Neglecting an irrelevant common phase

$$\exp\left[-i\Delta m^2 x/4E - \frac{i}{2}\int_0^x V_{\mathrm{CC}}(x')\,\mathrm{d}x'\right], \tag{9.58}$$

the evolution equation (9.54) can be written as

$$i\frac{\mathrm{d}}{\mathrm{d}x}\begin{pmatrix}\psi_{ee}\\ \psi_{e\mu}\end{pmatrix} = \frac{1}{4E}\begin{pmatrix}-\Delta m^2\cos 2\vartheta + A_{\mathrm{CC}} & \Delta m^2 \sin 2\vartheta \\ \Delta m^2 \sin 2\vartheta & \Delta m^2 \cos 2\vartheta - A_{\mathrm{CC}}\end{pmatrix}\begin{pmatrix}\psi_{ee}\\ \psi_{e\mu}\end{pmatrix}, \tag{9.59}$$

where $\Delta m^2 \equiv m_2^2 - m_1^2$ and ϑ is the mixing angle, defined by

$$\nu_e = \cos\vartheta\,\nu_1 + \sin\vartheta\,\nu_2\,, \qquad \nu_\mu = -\sin\vartheta\,\nu_1 + \cos\vartheta\,\nu_2\,. \tag{9.60}$$

For an initial ν_e, the initial condition for the evolution equation (9.59) is

$$\Psi_e(0) = \begin{pmatrix}\psi_{ee}(0)\\ \psi_{e\mu}(0)\end{pmatrix} = \begin{pmatrix}1\\ 0\end{pmatrix}, \tag{9.61}$$

and the probabilities of $\nu_e \to \nu_\mu$ transitions and ν_e survival are, respectively,

$$P_{\nu_e\to\nu_\mu}(x) = |\psi_{e\mu}(x)|^2\,, \qquad P_{\nu_e\to\nu_e}(x) = |\psi_{ee}(x)|^2 = 1 - P_{\nu_e\to\nu_\mu}(x)\,. \tag{9.62}$$

In practice, the evolution equation of the flavor transition amplitudes can always be solved numerically with sufficient degree of precision given enough computational power. In the following we will discuss an approximate analytical solution of the evolution equation (9.59) in the case of a matter density profile which is sufficiently smooth. This solution is useful in order to understand the qualitative physical aspects of the problem. It is also very useful in practice for the analysis of experimental data when the space of mixing parameters must be scanned to find the allowed regions. In these cases, the numerical solution of the evolution equation is impractical, since it is too slow.

The evolution equation (9.59) has the structure of a Schrödinger equation with the effective Hamiltonian matrix in the flavor basis

$$\mathcal{H}_{\mathrm{F}} = \frac{1}{4E}\begin{pmatrix}-\Delta m^2\cos 2\vartheta + A_{\mathrm{CC}} & \Delta m^2 \sin 2\vartheta \\ \Delta m^2 \sin 2\vartheta & \Delta m^2 \cos 2\vartheta - A_{\mathrm{CC}}\end{pmatrix}. \tag{9.63}$$

This matrix can be diagonalized by the orthogonal transformation

$$U_{\mathrm{M}}^T\,\mathcal{H}_{\mathrm{F}}\,U_{\mathrm{M}} = \mathcal{H}_{\mathrm{M}}\,, \tag{9.64}$$

where

$$\mathcal{H}_{\mathrm{M}} = \frac{1}{4E}\,\mathrm{diag}(-\Delta m_{\mathrm{M}}^2, \Delta m_{\mathrm{M}}^2) \tag{9.65}$$

is the effective Hamiltonian matrix in the mass basis in matter. The unitary matrix

$$U_{\mathrm{M}} = \begin{pmatrix} \cos\vartheta_{\mathrm{M}} & \sin\vartheta_{\mathrm{M}} \\ -\sin\vartheta_{\mathrm{M}} & \cos\vartheta_{\mathrm{M}} \end{pmatrix} \tag{9.66}$$

is the effective mixing matrix in matter, and

$$\Delta m^2_{\mathrm{M}} = \sqrt{\left(\Delta m^2 \cos 2\vartheta - A_{\mathrm{CC}}\right)^2 + \left(\Delta m^2 \sin 2\vartheta\right)^2} \tag{9.67}$$

is the effective squared-mass difference. The effective mixing angle in matter ϑ_{M} is given by

$$\tan 2\vartheta_{\mathrm{M}} = \frac{\tan 2\vartheta}{1 - \dfrac{A_{\mathrm{CC}}}{\Delta m^2 \cos 2\vartheta}} . \tag{9.68}$$

The interesting new phenomenon, discovered by Mikheev and Smirnov in 1985 [801, 802] (see also the lucid explanation in Ref. [222]), is that there is a resonance when A_{CC} becomes equal to

$$A^{\mathrm{R}}_{\mathrm{CC}} = \Delta m^2 \cos 2\vartheta \,, \tag{9.69}$$

which corresponds to the electron number density

$$N_e^{\mathrm{R}} = \frac{\Delta m^2 \cos 2\vartheta}{2\sqrt{2} E G_{\mathrm{F}}} . \tag{9.70}$$

At the resonance the effective mixing angle is equal to $\pi/4$, i.e. the mixing is maximal, leading to the possibility of total transitions between the two flavors if the resonance region is wide enough. This mechanism is called the *MSW effect*, named after Mikheev, Smirnov, and Wolfenstein. Note that the effective squared-mass difference in matter in eqn (9.67) has its minimum value at the resonance, where

$$\Delta m^2_{\mathrm{M}}\big|_{\mathrm{R}} = \Delta m^2 \sin 2\vartheta \,. \tag{9.71}$$

Since, in normal matter, A_{CC} is positive, a resonance can exist only if $\vartheta < \pi/4$, because for $\vartheta > \pi/4$ we have $\cos 2\vartheta < 0$. Therefore, the behavior of neutrino oscillations in matter is different from that of neutrino oscillations in vacuum, whose probability is symmetric under the exchange $\vartheta \to \pi/2 - \vartheta$ (see eqn (7.70)). For antineutrinos the potential is reversed, as explained in section 9.1, and there can be a resonance only if $\vartheta > \pi/4$.

Since

$$\cos 2\vartheta_{\mathrm{M}} = \frac{\Delta m^2 \cos 2\vartheta - A_{\mathrm{CC}}}{\Delta m^2_{\mathrm{M}}} , \qquad \sin 2\vartheta_{\mathrm{M}} = \frac{\Delta m^2 \sin 2\vartheta}{\Delta m^2_{\mathrm{M}}} , \tag{9.72}$$

the evolution equation (9.59) can be written, in terms of the mixing angle in matter, as

$$i \frac{\mathrm{d}}{\mathrm{d}x} \begin{pmatrix} \psi_{ee} \\ \psi_{e\mu} \end{pmatrix} = \frac{1}{4E} \begin{pmatrix} -\Delta m^2_{\mathrm{M}} \cos 2\vartheta_{\mathrm{M}} & \Delta m^2_{\mathrm{M}} \sin 2\vartheta_{\mathrm{M}} \\ \Delta m^2_{\mathrm{M}} \sin 2\vartheta_{\mathrm{M}} & \Delta m^2_{\mathrm{M}} \cos 2\vartheta_{\mathrm{M}} \end{pmatrix} \begin{pmatrix} \psi_{ee} \\ \psi_{e\mu} \end{pmatrix} . \tag{9.73}$$

Performing the transformation

$$\Psi_e = U_{\rm M}\, \Phi_e\,, \quad \text{with} \quad \Psi_e = \begin{pmatrix} \psi_{ee} \\ \psi_{e\mu} \end{pmatrix}, \quad \Phi_e = \begin{pmatrix} \phi_{e1} \\ \phi_{e2} \end{pmatrix}, \tag{9.74}$$

which diagonalizes the effective Hamiltonian matrix in eqn (9.63), the evolution equation (9.59) becomes

$$i \frac{\mathrm{d}}{\mathrm{d}x} \begin{pmatrix} \phi_{e1} \\ \phi_{e2} \end{pmatrix} = \frac{1}{4E} \begin{pmatrix} -\Delta m^2_{\rm M} & -4Ei\mathrm{d}\vartheta_{\rm M}/\mathrm{d}x \\ 4Ei\mathrm{d}\vartheta_{\rm M}/\mathrm{d}x & \Delta m^2_{\rm M} \end{pmatrix} \begin{pmatrix} \phi_{e1} \\ \phi_{e2} \end{pmatrix}. \tag{9.75}$$

The off-diagonal terms proportional to $\mathrm{d}\vartheta_{\rm M}/\mathrm{d}x$ are due to the derivative $\mathrm{d}\Psi_e/\mathrm{d}x = (\mathrm{d}U_{\rm M}/\mathrm{d}x)\Phi_e + U_{\rm M}(\mathrm{d}\Phi_e/\mathrm{d}x)$. From eqn (9.61), the initial conditions of the differential equation (9.75) are

$$\begin{pmatrix} \phi_{e1}(0) \\ \phi_{e2}(0) \end{pmatrix} = \begin{pmatrix} \cos\vartheta^{(\rm i)}_{\rm M} & -\sin\vartheta^{(\rm i)}_{\rm M} \\ \sin\vartheta^{(\rm i)}_{\rm M} & \cos\vartheta^{(\rm i)}_{\rm M} \end{pmatrix} \begin{pmatrix} 1 \\ 0 \end{pmatrix} = \begin{pmatrix} \cos\vartheta^{(\rm i)}_{\rm M} \\ \sin\vartheta^{(\rm i)}_{\rm M} \end{pmatrix}, \tag{9.76}$$

where $\vartheta^{(\rm i)}_{\rm M}$ is the effective mixing angle in matter at the point of neutrino production. The quantities ψ_r, with $r = 1, 2$, can be thought of as the amplitudes of the effective massive neutrino $\nu^{\rm M}_r$ in matter (although such a probability is not measurable, since only flavor neutrinos can be detected).

If the matter density is constant, $\mathrm{d}\vartheta_{\rm M}/\mathrm{d}x = 0$ and the evolutions of the amplitudes of the effective massive neutrinos in matter are decoupled, leading to the transition probability

$$P_{\nu_e\to\nu_\mu}(x) = \sin^2 2\vartheta_{\rm M}\, \sin^2\left(\frac{\Delta m^2_{\rm M} x}{4E}\right), \tag{9.77}$$

which has the same structure as the two-neutrino transition probability in vacuum in eqn (7.70), with the mixing angle and the squared-mass difference replaced by their effective values in matter. The oscillation length in matter is

$$L^{\rm osc}_{\rm M} = \frac{4\pi E}{\Delta m^2_{\rm M}}. \tag{9.78}$$

If the matter density is not constant, it is necessary to take into account the effect of $\mathrm{d}\vartheta_{\rm M}/\mathrm{d}x$, which is given by

$$\frac{\mathrm{d}\vartheta_{\rm M}}{\mathrm{d}x} = \frac{1}{2} \frac{\sin 2\vartheta_{\rm M}}{\Delta m^2_{\rm M}} \frac{\mathrm{d}A_{\rm CC}}{\mathrm{d}x}. \tag{9.79}$$

The effect of the off-diagonal terms proportional to $\mathrm{d}\vartheta_{\rm M}/\mathrm{d}x$ in the evolution equation (9.75) is to generate transitions between $\nu^{\rm M}_1$ and $\nu^{\rm M}_2$. Such transitions, however, are negligible if the off-diagonal terms are much smaller than the difference between the diagonal terms. In order to quantify the amount of such transitions, it is useful to introduce the so-called *adiabaticity parameter*

$$\gamma = \frac{\Delta m^2_{\rm M}}{4E|\mathrm{d}\vartheta_{\rm M}/\mathrm{d}x|} = \frac{(\Delta m^2_{\rm M})^2}{2E\sin 2\vartheta_{\rm M}|\mathrm{d}A_{\rm CC}/\mathrm{d}x|}. \tag{9.80}$$

If $\gamma \gg 1$ in all points of the neutrino trajectory, the evolution is adiabatic, which means that the transitions between $\nu^{\rm M}_1$ and $\nu^{\rm M}_2$ are negligible. This is the simplest case, because it implies that each amplitude ψ_r, with $r = 1, 2$, evolves independently and the effect of the evolution is a simple phase factor:

$$\phi_{e1}(x) = \exp\left(i \int_0^x \frac{\Delta m_{\rm M}^2(x')}{4E}\, {\rm d}x'\right) \phi_{e1}(0)\,, \tag{9.81}$$

$$\phi_{e2}(x) = \exp\left(-i \int_0^x \frac{\Delta m_{\rm M}^2(x')}{4E}\, {\rm d}x'\right) \phi_{e2}(0)\,. \tag{9.82}$$

Taking into account the initial conditions in eqn (9.76), the ν_e survival probability in eqn (9.62) is given by

$$\begin{aligned} P_{\nu_e\to\nu_e}^{\rm adiabatic}(x) = {} & \frac{1}{2} + \frac{1}{2}\cos 2\vartheta_{\rm M}^{(\rm i)} \cos 2\vartheta_{\rm M}^{(\rm f)} \\ & + \frac{1}{2}\sin 2\vartheta_{\rm M}^{(\rm i)} \sin 2\vartheta_{\rm M}^{(\rm f)} \cos\left(\int_0^x \frac{\Delta m_{\rm M}^2(x')}{2E}\, {\rm d}x'\right), \end{aligned} \tag{9.83}$$

where $\vartheta_{\rm M}^{(\rm f)}$ is the effective mixing angle at the detection point. This survival probability is composed of a constant term which depends on the mixing angle and an oscillating term whose phase depends on the effective squared-mass difference integrated over the neutrino path.

In practice, the adiabatic evolution can be realized in the case of neutrinos produced in stars, such as solar neutrinos (see section 10.8) or supernova neutrinos (see chapter 15), because stars are composed of plasma, with a smooth density variation. In such cases, neutrinos are detected very far away, practically in vacuum[46], and the final effective mixing angle $\vartheta_{\rm M}^{(\rm f)}$ is equal to the vacuum mixing angle ϑ. Moreover, since the distance between the neutrino source and the detector is huge, the phase of the cosine in eqn (9.83) is very large and has a variation much larger than 2π in the energy resolution interval of the detector. In this case, the cosine averages to zero and the measurable average survival probability is given by

$$\overline{P}_{\nu_e\to\nu_e}^{\rm adiabatic} = \frac{1}{2} + \frac{1}{2}\cos 2\vartheta_{\rm M}^{(\rm i)} \cos 2\vartheta\,, \tag{9.84}$$

which is independent of the source–detector distance.

The behavior of $\vartheta_{\rm M}$ as a function of the electron number density N_e is illustrated in Fig. 9.2a, for $\Delta m^2 = 7\times10^{-6}\,{\rm eV}^2$, $\sin^2 2\vartheta = 10^{-3}$, and $E = 1\,{\rm MeV}$. One can see that for $N_e \ll N_e^{\rm R}$ the effective mixing angle is practically equal to the mixing angle in vacuum, $\vartheta_{\rm M} \simeq \vartheta$, for $N_e \simeq N_e^{\rm R}$ the effective mixing angle varies very rapidly with the electron number density, passing through 45° at $N_e = N_e^{\rm R}$ and increasing rapidly to 90° for $N_e > N_e^{\rm R}$.

[46] What is important is the density of the medium in which the neutrino propagates before reaching the detector. The density of the detector is irrelevant, because the size of the detector is too small to have significant flavor transitions.

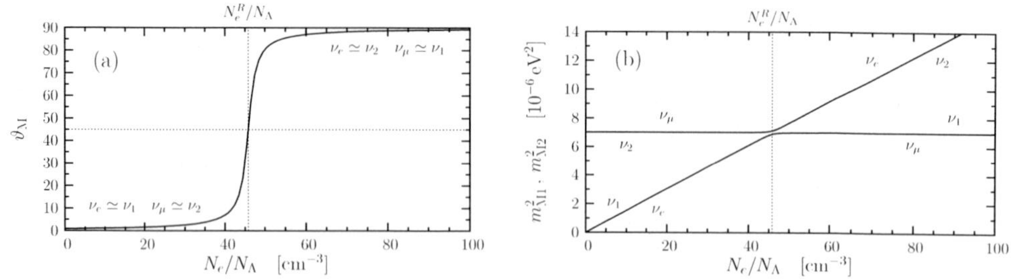

FIG. 9.2. Effective mixing angle $\vartheta_{\rm M}$ (a) and the effective squared-masses $m^2_{\rm M1}$, $m^2_{\rm M2}$ (b) in matter as functions of the electron number density N_e divided by the Avogadro's number $N_{\rm A}$, for $m_1 = 0$, $\Delta m^2 = 7 \times 10^{-6}\,{\rm eV}^2$, $\sin^2 2\vartheta = 10^{-3}$ and $E = 1\,{\rm MeV}$. $N_e^{\rm R} \equiv \Delta m^2 \cos 2\vartheta / 2\sqrt{2} E G_{\rm F}$ is the electron number density at the resonance, where $\vartheta_{\rm M} = 45°$.

Figure 9.2b shows the corresponding behavior of the effective squared-masses $m^2_{\rm M1}$ and $m^2_{\rm M2}$, which are given by

$$m^2_{{\rm M}2,1} = \frac{1}{2}\left(m_1^2 + m_2^2 + A_{\rm CC} \pm \Delta m^2_{\rm M}\right) , \tag{9.85}$$

where we took into account the common phases in eqns (9.51) and (9.58), except for the neutral-current contribution $2EV_{\rm NC}$. From Fig. 9.2b one can see that the effective squared-mass difference $\Delta m^2_{\rm M}$ reaches its minimum value at the resonance. This figure is useful in order to understand how the presence of a resonance can induce an almost complete $\nu_e \to \nu_\mu$ conversion of solar neutrinos. If the mixing parameters are such that $N_e \gg N_e^{\rm R}$ at the center of the Sun, the effective mixing angle is practically 90° and electron neutrinos are produced as almost pure ν_2. As the neutrino propagates out, it crosses the resonance at $N_e = N_e^{\rm R}$, where the energy gap between ν_1 and ν_2 is minimum. If the resonance is crossed adiabatically, the neutrino remains ν_2 and exits the Sun as $\nu_2 = \sin\vartheta\,\nu_e + \cos\vartheta\,\nu_\mu$, which is almost equal to ν_μ if the mixing angle is small, leading to almost complete $\nu_e \to \nu_\mu$ conversion. This is the case in which the MSW effect is most effective and striking, since a large conversion is achieved in spite of a small mixing angle.

We now discuss the case of nonadiabatic evolution, in which the off-diagonal terms proportional to ${\rm d}\vartheta_{\rm M}/{\rm d}x$ in the evolution equation (9.75) generate transitions between $\nu_1^{\rm M}$ and $\nu_2^{\rm M}$. The transitions are sizable in the regions where the adiabaticity parameter γ in eqn (9.80) is of the order of or less than unity, and reach their maximal size at the minimum of γ [803, 804, 463, 744]. In order to find such a minimum, called the point of *maximum violation of adiabaticity* (MVA), we find, from the expression of $\cos 2\vartheta_{\rm M}$ in eqn (9.72),

$$\frac{{\rm d}A_{\rm CC}}{{\rm d}x} = -\frac{\Delta m^2_{\rm M}}{\sin^2 2\vartheta_{\rm M}} \frac{{\rm d}\cos 2\vartheta_{\rm M}}{{\rm d}x} , \tag{9.86}$$

from which γ can be written as

$$\gamma = \frac{\Delta m^2 \sin 2\vartheta}{2E|\mathrm{d}\cos 2\vartheta_{\mathrm{M}}/\mathrm{d}x|}\,. \tag{9.87}$$

For the MVA point x_{MVA}, we obtain the condition [744]

$$\left.\frac{\mathrm{d}^2 \cos 2\vartheta_{\mathrm{M}}}{\mathrm{d}x^2}\right|_{x=x_{\mathrm{MVA}}} = 0\,. \tag{9.88}$$

Hence, the MVA point is, in general, different from the resonance point x_{R}, which, from eqns (9.69) and (9.72), is given by

$$\cos 2\vartheta_{\mathrm{M}}|_{x=x_{\mathrm{R}}} = 0\,. \tag{9.89}$$

However, calculating the second derivative in eqn (9.88), we obtain

$$\left[3\cos 2\vartheta_{\mathrm{M}}\,\sin 2\vartheta_{\mathrm{M}}\left(\frac{\mathrm{d}A_{\mathrm{CC}}}{\mathrm{d}x}\right)^2 + \Delta m^2\,\sin 2\vartheta\,\frac{\mathrm{d}^2 A_{\mathrm{CC}}}{\mathrm{d}x^2}\right]_{x=x_{\mathrm{MVA}}} = 0\,. \tag{9.90}$$

For a linear density, $\mathrm{d}^2 A_{\mathrm{CC}}/\mathrm{d}x^2 = 0$. In this case, the MVA point is equal to the resonance point. If the mixing is small, the second term in eqn (9.90) can be neglected and the MVA point is well approximated by the resonance point. In practice, as one can see from the expression in eqn (9.87), for relatively large values of Δm^2, the adiabaticity parameter is small only for small values of the mixing angle. Hence, in practice, the nonadiabatic transitions between ν_1^{M} and ν_2^{M} can be calculated with good accuracy by approximating the MVA point with the resonance point (exceptions for solar neutrinos are discussed in Refs. [463, 744]). In the following we will adopt this approximation. Note, however, that the resonance may exist only if $\vartheta \leq \pi/4$, whereas the MVA point is defined for any value of ϑ in its defining interval $[0, \pi/2]$. Thus, in the following we will consider the most interesting case of $\vartheta \leq \pi/4$.

We consider neutrinos created at a high density, above the resonance, propagating in a medium with monotonously decreasing density, and detected practically in vacuum (see footnote 46 on page 335), where $\vartheta_{\mathrm{M}}^{(\mathrm{f})} \simeq \vartheta$. This may be the case of solar neutrinos or supernova neutrinos. If the resonance is not crossed adiabatically, $\nu_1^{\mathrm{M}} \leftrightarrows \nu_2^{\mathrm{M}}$ transitions occur in an interval around the resonance. Neglecting the width of the resonance region, we can write the amplitudes of ν_1^{M} and ν_2^{M} at any point x after the resonance crossing as

$$\begin{aligned}\phi_{e1}(x) = {} & \left[\cos\vartheta_{\mathrm{M}}^{(\mathrm{i})}\exp\left(i\int_0^{x_{\mathrm{R}}}\frac{\Delta m_{\mathrm{M}}^2(x')}{4E}\mathrm{d}x'\right)\mathcal{A}_{11}^{\mathrm{R}}\right.\\ & \left.+\sin\vartheta_{\mathrm{M}}^{(\mathrm{i})}\exp\left(-i\int_0^{x_{\mathrm{R}}}\frac{\Delta m_{\mathrm{M}}^2(x')}{4E}\mathrm{d}x'\right)\mathcal{A}_{21}^{\mathrm{R}}\right]\exp\left(i\int_{x_{\mathrm{R}}}^{x}\frac{\Delta m_{\mathrm{M}}^2(x')}{4E}\mathrm{d}x'\right),\end{aligned} \tag{9.91}$$

$$\phi_{e2}(x) = \left[\cos\vartheta_{\mathrm{M}}^{(\mathrm{i})}\exp\left(i\int_0^{x_{\mathrm{R}}}\frac{\Delta m_{\mathrm{M}}^2(x')}{4E}\mathrm{d}x'\right)\mathcal{A}_{12}^{\mathrm{R}}\right.$$

$$+ \sin \vartheta_{\mathrm{M}}^{(\mathrm{i})} \exp\left(-i \int_0^{x_\mathrm{R}} \frac{\Delta m_\mathrm{M}^2(x')}{4E} \mathrm{d}x'\right) \mathcal{A}_{22}^\mathrm{R}\Bigg] \exp\left(-i \int_{x_\mathrm{R}}^{x} \frac{\Delta m_\mathrm{M}^2(x')}{4E} \mathrm{d}x'\right) , \tag{9.92}$$

where $\mathcal{A}_{kj}^\mathrm{R}$ is the amplitude of $\nu_k^\mathrm{M} \to \nu_j^\mathrm{M}$ transitions in the resonance. As we have already noted in connection with eqn (9.83), if the source–detector distance is very large and Δm_M^2 is not too small, all the phases in eqns (9.91) and (9.92) are very large and rapidly oscillating as functions of the neutrino energy. In this case, the average of the transition probability over the energy resolution of the detector washes out all interference terms and the measurable quantity is the averaged survival probability

$$\begin{aligned} \overline{P}_{\nu_e \to \nu_e} = {} & \cos^2 \vartheta \cos^2 \vartheta_\mathrm{M}^{(\mathrm{i})} \, |\mathcal{A}_{11}^\mathrm{R}|^2 + \cos^2 \vartheta \sin^2 \vartheta_\mathrm{M}^{(\mathrm{i})} \, |\mathcal{A}_{21}^\mathrm{R}|^2 \\ & + \sin^2 \vartheta \cos^2 \vartheta_\mathrm{M}^{(\mathrm{i})} \, |\mathcal{A}_{12}^\mathrm{R}|^2 + \sin^2 \vartheta \sin^2 \vartheta_\mathrm{M}^{(\mathrm{i})} \, |\mathcal{A}_{22}^\mathrm{R}|^2 , \end{aligned} \tag{9.93}$$

which is independent of the source–detector distance. From the conservation of probability, we have

$$|\mathcal{A}_{11}^\mathrm{R}|^2 = |\mathcal{A}_{22}^\mathrm{R}|^2 = 1 - P_\mathrm{c} , \qquad |\mathcal{A}_{12}^\mathrm{R}|^2 = |\mathcal{A}_{21}^\mathrm{R}|^2 = P_\mathrm{c} , \tag{9.94}$$

where P_c is the $\nu_1^\mathrm{M} \leftrightarrows \nu_2^\mathrm{M}$ crossing probability at the resonance. Thus, we obtain the so-called Parke formula [851] for the averaged ν_e survival probability:

$$\overline{P}_{\nu_e \to \nu_e} = \frac{1}{2} + \left(\frac{1}{2} - P_\mathrm{c}\right) \cos 2\vartheta_\mathrm{M}^{(\mathrm{i})} \cos 2\vartheta . \tag{9.95}$$

This formula has been widely used for the analysis of the solar neutrino data.

The main problem in the application of the Parke formula in eqn (9.95) is the calculation of the crossing probability. This probability must involve the energy gap $\Delta m_\mathrm{M}^2/2E$ between ν_1^M and ν_2^M and the off-diagonal terms proportional to $\mathrm{d}\vartheta_\mathrm{M}/\mathrm{d}x$ in eqn (9.75), which cause the $\nu_1^\mathrm{M} \leftrightarrows \nu_2^\mathrm{M}$ transitions. For the electron density profiles[47] in Table 9.1, the crossing probability can be written as [869, 697, 870, 708]

$$P_\mathrm{c} = \frac{\exp\left(-\frac{\pi}{2}\gamma_\mathrm{R} F\right) - \exp\left(-\frac{\pi}{2}\gamma_\mathrm{R} \frac{F}{\sin^2 \vartheta}\right)}{1 - \exp\left(-\frac{\pi}{2}\gamma_\mathrm{R} \frac{F}{\sin^2 \vartheta}\right)} , \tag{9.96}$$

where γ_R is the adiabaticity parameter at the resonance,

$$\gamma_\mathrm{R} = \left.\frac{\Delta m_\mathrm{M}^2/2E}{2|\mathrm{d}\vartheta_\mathrm{M}/\mathrm{d}x|}\right|_\mathrm{R} = \frac{\Delta m^2 \sin^2 2\vartheta}{2E \cos 2\vartheta \, |\mathrm{d} \ln N_e/\mathrm{d}x|_\mathrm{R}} . \tag{9.97}$$

The value of the parameter F depends on the electron density profile. The results of its calculations by approximate analytic solution of the evolution equation for three useful density profiles are given in Table 9.1. Especially useful is the case of an exponential density profile, which is a good approximation for the solar neutrinos (see section 10.8).

[47] The values of P_c for other density profiles have been calculated in Refs. [827, 708].

TABLE 9.1. Values of the parameter F for various electron density profiles.

Density profile	F	Refs.
$N_e \propto x$	1	[851, 604, 875] [868, 708, 157]
$N_e \propto 1/x$	$\left(1-\tan^2\vartheta\right)^2 / \left(1+\tan^2\vartheta\right)$	[708]
$N_e \propto e^{-x}$	$1-\tan^2\vartheta$	[869, 697, 870, 875, 1020] [655, 632, 708, 157]

Since $F \sim 1$, one can see, from eqn (9.96), that if $\gamma_{\rm R}$ is large,

$$P_{\rm c}^{(\gamma_{\rm R} \gg 1)} \ll 1\,, \tag{9.98}$$

and the resonance is crossed adiabatically. In this case, eqn (9.95) reduces to the adiabatic survival probability in eqn (9.84).

On the other hand, if $\gamma_{\rm R}$ is very small, we have the so-called *extreme nonadiabatic limit*, for which

$$P_{\rm c}^{(\gamma_{\rm R} \ll 1)} \simeq \cos^2\vartheta\,. \tag{9.99}$$

Considering a neutrino produced in a very dense environment, where $\cos 2\vartheta_{\rm M}^{(\rm i)} \simeq -1$, the averaged ν_e survival probability in eqn (9.95) becomes

$$\overline{P}_{\nu_e\to\nu_e}^{(\gamma_{\rm R} \ll 1)} \simeq 1 - \frac{1}{2}\sin^2 2\vartheta\,. \tag{9.100}$$

This is the average survival probability of electron neutrino in vacuum (see eqn (7.73)). The physical reason for this equality can be understood as follows. Since $\cos 2\vartheta_{\rm M}^{(\rm i)} \simeq -1$, an electron neutrino is created practically as a $\nu_2^{\rm M}$. The $\nu_e \simeq \nu_2^{\rm M}$ propagates undisturbed towards the resonance, where the density variation is very rapid. In this case, there is no time to generate flavor transitions and the neutrino emerges from the resonance as a ν_e. After the resonance, the effective mixing is practically equal to the mixing in vacuum and the oscillations from the resonance to the detector are oscillations in vacuum.

The most important application of the MSW effect is in solar neutrino flavor transitions, to be discussed in section 10.8.

9.4 Slab approximation

Sometimes neutrinos propagate through a medium whose density can be approximated by a series of slabs with constant density. In such cases, according to eqn (9.75), the effective massive neutrinos propagate as plane waves in regions of constant density, with phases $\exp\left(\pm i\Delta m_{\rm M}^2 \Delta x/4E\right)$, where Δx is the width of the slab. At the boundaries of slabs the wave functions of flavor neutrinos

$$\Psi_e = \begin{pmatrix} \psi_{ee} \\ \psi_{e\mu} \end{pmatrix} \tag{9.101}$$

are joined according to the scheme

$$\begin{aligned}\Psi_e(x_n) = & \left[U_{\rm M}\,\mathscr{U}_{\rm M}(x_n - x_{n-1})\,U_{\rm M}^\dagger\right]_{(n)} \left[U_{\rm M}\,\mathscr{U}_{\rm M}(x_{n-1} - x_{n-2})\,U_{\rm M}^\dagger\right]_{(n-1)} \\ & \ldots \left[U_{\rm M}\,\mathscr{U}_{\rm M}(x_2 - x_1)\,U_{\rm M}^\dagger\right]_{(2)} \left[U_{\rm M}\,\mathscr{U}_{\rm M}(x_1 - x_0)\,U_{\rm M}^\dagger\right]_{(1)} \Psi_e(x_0)\,, \end{aligned} \tag{9.102}$$

where x_0 is the coordinate of the starting point, x_1, x_2, ..., x_n are the boundaries of n slabs, and

$$\mathscr{U}_{\rm M}(\Delta x) = \mathrm{diag}\big(\exp\left(i\Delta m^2_{\rm M}\Delta x/4E\right)\,,\, \exp\left(-i\Delta m^2_{\rm M}\Delta x/4E\right)\big) \tag{9.103}$$

is the unitary evolution operator in each slab diagonal basis. The notation $[\ldots]_{(i)}$ indicates that all the matter-dependent quantities in the square brackets must be evaluated with the matter density in the ith slab, which extends from x_{i-1} to x_i.

The unitary evolution operator in the flavor basis in each slab with constant density

$$\mathscr{U}(x) = U_{\rm M}\,\mathscr{U}_{\rm M}(x)\,U_{\rm M}^\dagger \tag{9.104}$$

can also be determined as follows. Let us write the effective Hamiltonian matrix in eqn (9.63) in terms of the Pauli matrices:

$$\mathcal{H}_{\rm F} = \frac{1}{4E}\left[\Delta m^2 \sin 2\vartheta\,\sigma^1 - \left(\Delta m^2 \cos 2\vartheta - A_{\rm CC}\right)\sigma^3\right]. \tag{9.105}$$

Defining the unit vector

$$\begin{aligned}\vec{n} &= \frac{1}{\Delta m^2_{\rm M}}\left(\Delta m^2 \sin 2\vartheta\,,\, 0\,,\, -\left(\Delta m^2 \cos 2\vartheta - A_{\rm CC}\right)\right) \\ &= \left(\sin 2\vartheta_{\rm M}\,,\, 0\,,\, -\cos 2\vartheta_{\rm M}\right)\,,\end{aligned} \tag{9.106}$$

we have

$$\mathcal{H}_{\rm F} = \frac{\Delta m^2_{\rm M}}{4E}\,\vec{n}\cdot\vec{\sigma}\,. \tag{9.107}$$

Then, the unitary evolution operator in the flavor basis in each slab with constant density is given by

$$\mathscr{U}(x) = \exp\left(-i\mathcal{H}_{\rm F}x\right) = \cos\varphi(x) - i\left(\vec{n}\cdot\vec{\sigma}\right)\sin\varphi(x)\,, \tag{9.108}$$

with the phase

$$\varphi(x) = \frac{\Delta m^2_{\rm M}\,x}{4E}\,. \tag{9.109}$$

This expression is useful in the study of the parametric resonance, discussed in the following section.

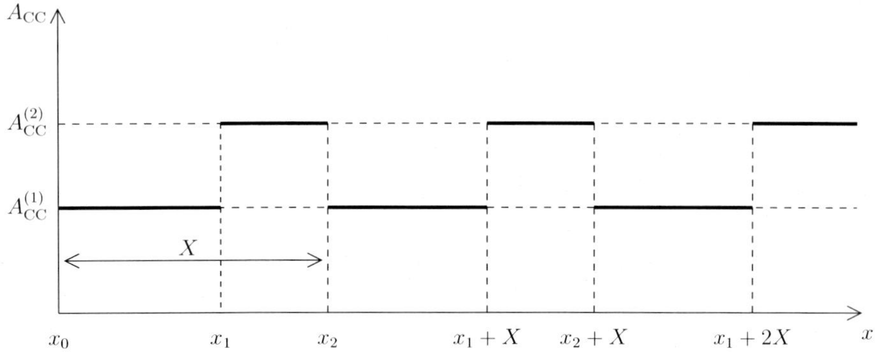

FIG. 9.3. *Castle wall* density profile.

9.5 Parametric resonance

The MSW effect can occur when neutrinos propagate through a medium with density profile which is monotonically varying and smooth. In such a case there may be a resonance region where the effective mixing angle is close to maximal, leading to large flavor transitions. However, it is also possible to have large flavor transitions in another case, the so-called *parametric resonance* [51, 698, 748, 871, 50, 315, 316]. In general, a parametric resonance can occur in dynamical systems whose parameters vary periodically in time. In the case of neutrino oscillations a parametric resonance can occur if the matter potential varies periodically along the neutrino path. In this case, the origin of the resonance is different from the MSW effect, because, in general, the mixing angle does not need to become large along the neutrino path. The amplification of the transition probability is due to the modification of the phase of oscillations.

In this section, we discuss the parametric resonance in the case of the so-called *castle wall* density profile depicted in Fig. 9.3, which is a periodic step function with

$$A_{CC}(x) = \begin{cases} A_{CC}^{(1)} & \text{for} \quad x_0 \leq x < x_1 \,, \\ A_{CC}^{(2)} & \text{for} \quad x_1 \leq x < x_2 \,, \\ A_{CC}(x + X) \,, & \end{cases} \tag{9.110}$$

with the period $X = x_2 - x_0 = X_1 + X_2$, where $X_1 = x_1 - x_0$ and $X_2 = x_2 - x_1$. Using the expression in eqn (9.108) for the unitary evolution operator in the flavor basis in each of the two periodic slabs with constant density, we have

$$\mathscr{U}_k = c_k - i\,(\vec{n}_k \cdot \vec{\sigma})\,s_k \qquad (k = 1, 2)\,, \tag{9.111}$$

with $c_k \equiv \cos\varphi_k$ and $s_k \equiv \sin\varphi_k$. Here $\varphi_k = \Delta m^2_{\mathrm{M}k} X_k / 4E$ is the phase in the kth slab, with effective squared-mass difference $\Delta m^2_{\mathrm{M}k}$. The evolution operator for one period is given by [50]

$$\mathscr{U}_X = \mathscr{U}_2 \mathscr{U}_1 = R - i\vec{\sigma} \cdot \vec{I}\,, \tag{9.112}$$

with

$$R = c_1 c_2 - s_1 s_2 \left(\vec{n}_1 \cdot \vec{n}_2 \right) \,, \tag{9.113}$$

$$\vec{I} = s_1 c_2 \vec{n}_1 + c_1 s_2 \vec{n}_2 - s_1 s_2 \vec{n}_1 \times \vec{n}_2 \,. \tag{9.114}$$

Since $R^2 + |\vec{I}|^2 = 1$, as required by unitarity ($\mathscr{U}_X \mathscr{U}_X^\dagger = \mathscr{U}_X^\dagger \mathscr{U}_X = \mathbf{1}$), the evolution operator for one period can also be written as

$$\mathscr{U}_X = \exp\left[-i \left(\vec{\sigma} \cdot \vec{N} \right) \xi \right] \,, \tag{9.115}$$

with

$$\xi = \arccos R = \arcsin |\vec{I}| \,, \tag{9.116}$$

$$\vec{N} = \vec{I} / |\vec{I}| \,. \tag{9.117}$$

Furthermore, we have

$$\vec{n}_1 \cdot \vec{n}_2 = \cos 2(\vartheta_{\mathrm{M}1} - \vartheta_{\mathrm{M}2}) \,, \tag{9.118}$$

$$\vec{n}_1 \times \vec{n}_2 = (0 \,, \sin 2(\vartheta_{\mathrm{M}1} - \vartheta_{\mathrm{M}2}) \,, 0) \,, \tag{9.119}$$

and the components of the vector $\vec{I}$ are given by

$$I_1 = s_1 c_2 \sin 2\vartheta_{\mathrm{M}1} + c_1 s_2 \sin 2\vartheta_{\mathrm{M}2} \,, \tag{9.120}$$

$$I_2 = - s_1 s_2 \sin 2(\vartheta_{\mathrm{M}1} - \vartheta_{\mathrm{M}2}) \,, \tag{9.121}$$

$$I_3 = - s_1 c_2 \cos 2\vartheta_{\mathrm{M}1} - c_1 s_2 \cos 2\vartheta_{\mathrm{M}2} \,. \tag{9.122}$$

Considering now a castle wall potential with p periods, the evolution operator is simply given by the pth power of $\mathscr{U}_X$:

$$\mathscr{U}_{pX} = \exp\left[-i \left(\vec{\sigma} \cdot \vec{N} \right) p\xi \right] \,. \tag{9.123}$$

Taking into account the initial condition in eqn (9.61), we have

$$\Psi_e(pX) = \mathscr{U}_{pX} \, \Psi_e(0) = \begin{pmatrix} [\mathscr{U}_{pX}]_{11} \\ [\mathscr{U}_{pX}]_{21} \end{pmatrix} \,. \tag{9.124}$$

Therefore, the probability of $\nu_e \to \nu_\mu$ transitions is given by

$$P_{\nu_e \to \nu_\mu}(pX) = |[\mathscr{U}_{pX}]_{21}|^2 = \left[(N_1)^2 + (N_2)^2 \right] \sin^2 p\xi = \left(1 - \frac{(I_3)^2}{|\vec{I}|^2} \right) \sin^2 p\xi \,. \tag{9.125}$$

Therefore, the amplitude of the oscillations can be large if the third component of $\vec{I}$ is small. Parametric resonance occurs for a vanishing I_3, which, according to eqn (9.122), corresponds to the condition

$$s_1 c_2 \cos 2\vartheta_{\mathrm{M}1} + c_1 s_2 \cos 2\vartheta_{\mathrm{M}2} = 0 \,. \tag{9.126}$$

It is to be noted that both the amplitude and the phase of the oscillations are different from those in either of the two slabs with constant density in the castle

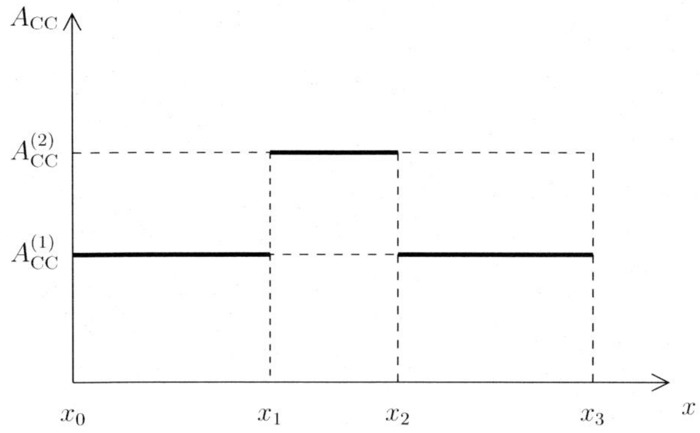

FIG. 9.4. *One and half period* castle wall density profile.

wall. In particular, the amplitude can be maximal even if the effective mixing angles in both slabs are small.

Let us finally consider the *one and half period* castle wall potential depicted in Fig. 9.4. This is, for example, a good approximation of the Earth's density (see section 10.9). In this case, the evolution operator is given by [50]

$$\mathscr{U} = \mathscr{U}_1 \mathscr{U}_2 \mathscr{U}_1 = \mathscr{U}_1 \mathscr{U}_X = Z - i\vec{\sigma} \cdot \vec{W} \,, \tag{9.127}$$

with

$$Z = 2c_1 R - c_2 \,, \tag{9.128}$$

$$\vec{W} = 2s_1 R \vec{n}_1 + s_2 \vec{n}_2 \,. \tag{9.129}$$

Using eqn (9.113), one can easily check that $Z^2 + |\vec{W}|^2 = 1$, as required by unitarity ($\mathscr{U}\mathscr{U}^\dagger = \mathscr{U}^\dagger \mathscr{U} = \mathbf{1}$). Since $W_2 = 0$, the probability of $\nu_e \to \nu_\mu$ transitions is given by

$$P_{\nu_e \to \nu_\mu} = (W_1)^2 = \left[2s_1 \sin 2\vartheta_{\rm M1} \left(c_1 c_2 - s_1 s_2 \cos 2(\vartheta_{\rm M1} - \vartheta_{\rm M2})\right) + s_2 \sin 2\vartheta_{\rm M2}\right]^2 \,. \tag{9.130}$$

This probability can be large, depending on the values of the quantities involved (for detailed discussions of the maxima of the transition probability see Ref. [50]).

9.6 Geometrical representation

In this section we describe a geometrical representation of neutrino oscillations developed in Refs. [671, 669]. It is convenient to introduce the density matrix formalism, which is useful for the description of oscillations of neutrino ensembles with more than one initial flavor, with possible loss of coherence (for example, the case of neutrinos in the early Universe; see Ref. [379]).

A neutrino beam can be described by the Hermitian *density matrix operator*[48]

$$\widehat{\rho}(x) = \sum_{\alpha} |\nu_\alpha(x)\rangle \, W_\alpha \, \langle\nu_\alpha(x)| \,, \tag{9.131}$$

where W_α is the initial statistical weight of flavor α, i.e. the probability of the flavor α at $x = t = 0$. The initial statistical weights are real numbers between zero and one, such that

$$\sum_{\alpha} W_\alpha = 1 \,. \tag{9.132}$$

The density matrix operator is suitable for the description of a neutrino beam composed of an initial incoherent mixture of different neutrino flavors, which are generated in different processes. The special case of one initial flavor β is obtained by setting $W_\alpha = \delta_{\alpha\beta}$. The density matrix is normalized by

$$\mathrm{Tr}[\widehat{\rho}(x)] \equiv \sum_{a} \langle\nu_a|\widehat{\rho}(x)|\nu_a\rangle = 1 \,, \tag{9.133}$$

where $\{|\nu_a\rangle\}$ is a complete set of states. The probability of detection of a neutrino with flavor β at a distance x is given by

$$P_\beta(x) = \mathrm{Tr}[\widehat{\rho}(x)|\nu_\beta\rangle\langle\nu_\beta|] = \langle\nu_\beta|\widehat{\rho}(x)|\nu_\beta\rangle = \rho^{\mathrm{F}}_{\beta\beta}(x) \,, \tag{9.134}$$

where $|\nu_\beta\rangle\langle\nu_\beta|$ is the β-flavor operator and $\rho^{\mathrm{F}}(x)$ is the density matrix in the flavor basis, with elements given by

$$\rho^{\mathrm{F}}_{\eta\xi}(x) = \langle\nu_\eta|\widehat{\rho}(x)|\nu_\xi\rangle = \sum_{\alpha} W_\alpha \, \psi_{\alpha\eta}(x) \, \psi^*_{\alpha\xi}(x) \,. \tag{9.135}$$

One can easily verify that the density matrix is Hermitian and its trace is equal to unity (from the conservation of probability in eqn (9.49)). Moreover, by using eqn (9.48), one can find that

$$\mathrm{Tr}\left[\widehat{\rho}^2(x)\right] = \sum_{\eta,\xi} \langle\nu_\eta|\widehat{\rho}(x)|\nu_\xi\rangle\langle\nu_\xi|\widehat{\rho}(x)|\nu_\eta\rangle = \sum_{\alpha} W_\alpha^2 \,. \tag{9.136}$$

Hence, the trace of the square of the density matrix is equal to unity only if there is only one initial flavor. In this case, the beam is coherent. On the other hand, if the beam is initially composed of more than one flavor, it is incoherent and $\mathrm{Tr}\left[\widehat{\rho}^2(x)\right]$ is less than unity. Note, however, that the trace of $\widehat{\rho}^2(x)$ depends only on the initial conditions. This means that there is no loss of coherence during the evolution of the neutrino beam[49].

[48] Also called the *density operator* or *statistical operator*.

[49] The loss of coherence due to inelastic collisions with the medium can be described by adding a damping term to the evolution equation of the density matrix [985]. A loss of coherence can also be due to the separation of the massive neutrino wave packets [534].

The evolution equation of the density matrix in the flavor basis, obtained from the evolution (9.53) of the flavor transition amplitudes, is

$$i\,\frac{\mathrm{d}\rho^{\mathrm{F}}}{\mathrm{d}x} = \left[\mathcal{H}_{\mathrm{F}}\,,\,\rho^{\mathrm{F}}\right]\,, \tag{9.137}$$

with the initial condition, from eqn (9.45),

$$\rho^{\mathrm{F}}_{\eta\xi}(0) = W_{\eta}\,\delta_{\eta\xi}\,. \tag{9.138}$$

Since the effective Hamiltonian matrix $\mathcal{H}_{\mathrm{F}}$ is Hermitian, it can be diagonalized by the unitary transformation

$$U_{\mathrm{M}}^{\dagger}\,\mathcal{H}_{\mathrm{F}}\,U_{\mathrm{M}} = \mathcal{H}_{\mathrm{M}}\,, \qquad \text{with} \qquad (\mathcal{H}_{\mathrm{M}})_{kj} = (\mathcal{H}_{\mathrm{M}})_{kk}\,\delta_{kj}\,. \tag{9.139}$$

U_{M} is the unitary effective mixing matrix in matter. For the density matrix in the effective mass basis in matter,

$$\rho^{\mathrm{M}} = U_{\mathrm{M}}^{\dagger}\,\rho^{\mathrm{F}}\,U_{\mathrm{M}}\,, \tag{9.140}$$

we find that the evolution equation

$$i\,\frac{\mathrm{d}\rho^{\mathrm{M}}}{\mathrm{d}x} = \left[\mathcal{H}_{\mathrm{M}}\,,\,\rho^{\mathrm{M}}\right] - i\left[U_{\mathrm{M}}^{\dagger}\,\frac{\mathrm{d}U_{\mathrm{M}}}{\mathrm{d}x}\,,\,\rho^{\mathrm{M}}\right]. \tag{9.141}$$

Considering, for simplicity, the adiabatic case in which the second term in the right-hand side of eqn (9.141) is negligible, we obtain the decoupled evolution equations

$$i\,\frac{\mathrm{d}\rho^{\mathrm{M}}_{kj}}{\mathrm{d}x} = \left[(\mathcal{H}_{\mathrm{M}})_{kk} - (\mathcal{H}_{\mathrm{M}})_{jj}\right]\rho^{\mathrm{M}}_{kj}\,, \tag{9.142}$$

for the components of ρ^{M}. In this case, the diagonal elements of ρ^{M} remain constant and, for $k \neq j$,

$$\rho^{\mathrm{M}}_{kj}(x) = \rho^{\mathrm{M}}_{kj}(0)\,\exp\!\left(-i\int_0^x \left[(\mathcal{H}_{\mathrm{M}}(x'))_{kk} - (\mathcal{H}_{\mathrm{M}}(x'))_{jj}\right]\mathrm{d}x'\right). \tag{9.143}$$

Let us now consider the simplest case of two-neutrino mixing[50] in the ν_e-ν_μ sector discussed in section 9.3, with the effective Hamiltonian matrix $\mathcal{H}_{\mathrm{F}}$ in eqn (9.63). Since $\mathcal{H}_{\mathrm{F}}$ is traceless and the trace of ρ^{F} is equal to unity, by using the decomposition in eqn (A.42), we can write these two 2×2 matrices as

$$\mathcal{H}_{\mathrm{F}} = -\frac{1}{2}\,\vec{\sigma}_{\mathrm{F}}\cdot\vec{B}\,, \qquad \rho^{\mathrm{F}} = \frac{1}{2}\,\mathbf{1} + \frac{1}{2}\,\vec{\sigma}_{\mathrm{F}}\cdot\vec{S}\,, \tag{9.144}$$

with the vectors

$$\vec{\sigma}_{\mathrm{F}} = \sum_{a=1}^{3}\sigma^{a}\,\vec{e}^{\,\mathrm{F}}_{a}\,, \qquad \vec{B} = \sum_{a=1}^{3} B^{a}_{\mathrm{F}}\,\vec{e}^{\,\mathrm{F}}_{a}\,, \qquad \vec{S} = \sum_{a=1}^{3} S^{a}_{\mathrm{F}}\,\vec{e}^{\,\mathrm{F}}_{a}\,, \tag{9.145}$$

where $\vec{e}^{\,\mathrm{F}}_{1}$, $\vec{e}^{\,\mathrm{F}}_{2}$, and $\vec{e}^{\,\mathrm{F}}_{3}$ are three orthonormal vectors which form the flavor basis ($\vec{e}^{\,\mathrm{F}}_{a}\cdot\vec{e}^{\,\mathrm{F}}_{b} = \delta_{ab}$). The components of the vectors $\vec{B}$ and $\vec{S}$ in the flavor basis are

$$B^{1}_{\mathrm{F}} = -\,\frac{\Delta m^{2}\,\sin 2\vartheta}{2E} = -\,\frac{\Delta m^{2}_{\mathrm{M}}\,\sin 2\vartheta_{\mathrm{M}}}{2E}\,, \tag{9.146}$$

[50] The present treatment can be generalized to three-neutrino mixing by using, instead of the Pauli SU(2) matrices, the Gell-Mann SU(3) matrices [669, 670].

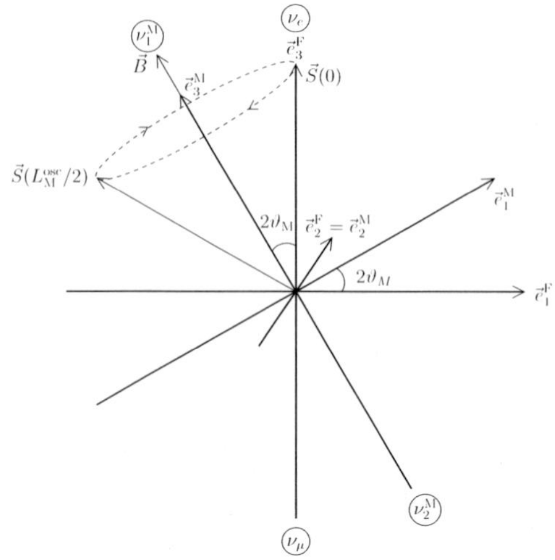

FIG. 9.5. Precession of $\vec{S}$ around $\vec{B}$ in a medium with constant $N_e < N_e^{\mathrm{R}}$ in the case of an initially pure ν_e beam and $\vartheta < \pi/4$.

$$B_{\mathrm{F}}^2 = 0\,, \tag{9.147}$$

$$B_{\mathrm{F}}^3 = \frac{\Delta m^2 \cos 2\vartheta - A_{\mathrm{CC}}}{2E} = \frac{\Delta m_{\mathrm{M}}^2 \cos 2\vartheta_{\mathrm{M}}}{2E} \tag{9.148}$$

and

$$S_{\mathrm{F}}^1 = 2\,\mathfrak{Re}\rho_{e\mu}^{\mathrm{F}}\,, \quad S_{\mathrm{F}}^2 = -2\,\mathfrak{Im}\rho_{e\mu}^{\mathrm{F}}\,, \quad S_{\mathrm{F}}^3 = \rho_{ee}^{\mathrm{F}} - \rho_{\mu\mu}^{\mathrm{F}}\,. \tag{9.149}$$

From the initial condition for the density matrix in eqn (9.138), we obtain

$$S_{\mathrm{F}}^1(0) = 0\,, \quad S_{\mathrm{F}}^2(0) = 0\,, \quad S_{\mathrm{F}}^3(0) = W_e - W_\mu\,. \tag{9.150}$$

Hence, if the initial state is a pure electron neutrino ($W_e = 1$ and $W_\mu = 0$), the vector $\vec{S}(0)$ has unit length and is aligned along $\vec{e}_3^{\,\mathrm{F}}$, as shown in Fig. 9.5. If the initial state is a pure muon neutrino ($W_e = 0$ and $W_\mu = 1$), the vector $\vec{S}(0)$ also has unit length, but it is aligned along $-\vec{e}_3^{\,\mathrm{F}}$. If the initial state is an incoherent mixture of ν_e and ν_μ, the length of $\vec{S}(0)$ is shorter than unity and it is directed along $\pm\vec{e}_3^{\,\mathrm{F}}$, depending on the relative amount of ν_e and ν_μ. In general, a length of $\vec{S}$ smaller than unity indicates an incoherent mixture.

From eqns (9.134), (9.144), and (9.145), the probabilities of detection of a ν_e or a ν_μ at a distance x are given by

$$P_e(x) = \rho_{ee}^{\mathrm{F}}(x) = \frac{1}{2}\left(1 + S_{\mathrm{F}}^3\right) = \frac{1}{2}\left(1 + \vec{S}\cdot\vec{e}_3^{\,\mathrm{F}}\right)\,, \tag{9.151}$$

$$P_\mu(x) = \rho_{\mu\mu}^{\mathrm{F}}(x) = \frac{1}{2}\left(1 - S_{\mathrm{F}}^3\right) = \frac{1}{2}\left(1 - \vec{S}\cdot\vec{e}_3^{\,\mathrm{F}}\right)\,. \tag{9.152}$$

The evolution equation of the vector $\vec{S}$ is

$$\frac{d\vec{S}}{dx} = \vec{S} \times \vec{B} \,. \tag{9.153}$$

This equation is analogous to that of a magnetic moment $\vec{S}$, with a gyromagnetic ratio $g = 1$, precessing around a magnetic field $\vec{B}$. The precession frequency is

$$\omega = -|\vec{B}| = -\frac{\Delta m^2_{\mathrm{M}}}{2E} \,, \tag{9.154}$$

where Δm^2_{M} is the effective squared-mass difference in eqn (9.67) (which reduces to Δm^2 in vacuum). The negative sign of ω means that $\vec{S}$ is precessing in the anticlockwise sense when viewed along the direction of $\vec{B}$, as shown in Fig. 9.5 for a medium with constant density (or in vacuum), in which $\vec{B}$ is constant. In this case, $\vec{S}$ describes the surface of a cone with the axis along $\vec{B}$ and opening angle $2\vartheta_{\mathrm{M}}$. In fact, $\vec{S} \cdot \vec{B}/|\vec{S}||\vec{B}|$ is a constant of motion and from eqns (9.72), (9.150) and eqns (9.146)–(9.148), we have $\vec{S}(0) \cdot \vec{B}/|\vec{S}(0)||\vec{B}| = \cos 2\vartheta_M$.

The precession of $\vec{S}$ around $\vec{B}$ becomes clear by rotating the reference frame from the flavor basis to the mass basis $\vec{e}^{\,\mathrm{M}}_1$, $\vec{e}^{\,\mathrm{M}}_2$, and $\vec{e}^{\,\mathrm{M}}_3$, with

$$\vec{\sigma}_{\mathrm{M}} = \sum_{a=1}^{3} \sigma^a \, \vec{e}^{\,\mathrm{M}}_a \,, \qquad \vec{B} = \sum_{a=1}^{3} B^a_{\mathrm{M}} \, \vec{e}^{\,\mathrm{M}}_a \,, \qquad \vec{S} = \sum_{a=1}^{3} S^a_{\mathrm{M}} \, \vec{e}^{\,\mathrm{M}}_a \,. \tag{9.155}$$

The mass basis is obtained by writing the diagonal 2×2 effective Hamiltonian in eqn (9.139) as

$$\mathcal{H}_{\mathrm{M}} = -\frac{1}{2} \, \vec{\sigma}_{\mathrm{M}} \cdot \vec{B} \,. \tag{9.156}$$

From eqns (9.139) and (9.144), we obtain

$$\vec{\sigma}_{\mathrm{M}} = U^{\dagger}_{\mathrm{M}} \, \vec{\sigma}_{\mathrm{F}} \, U_{\mathrm{M}} = \mathcal{H}_{\mathrm{M}} \,, \tag{9.157}$$

which yields

$$\vec{e}^{\,\mathrm{M}}_1 = \cos 2\vartheta_{\mathrm{M}} \, \vec{e}^{\,\mathrm{F}}_1 + \sin 2\vartheta_{\mathrm{M}} \, \vec{e}^{\,\mathrm{F}}_3 \,, \tag{9.158}$$

$$\vec{e}^{\,\mathrm{M}}_2 = \vec{e}^{\,\mathrm{F}}_2 \,, \tag{9.159}$$

$$\vec{e}^{\,\mathrm{M}}_3 = -\sin 2\vartheta_{\mathrm{M}} \, \vec{e}^{\,\mathrm{F}}_1 + \cos 2\vartheta_{\mathrm{M}} \, \vec{e}^{\,\mathrm{F}}_3 \,. \tag{9.160}$$

Hence, the mass basis is rotated by an angle $2\vartheta_{\mathrm{M}}$ in the 1-3 plane with respect to the flavor basis. This means that $\vec{e}^{\,\mathrm{M}}_3$ lies along $\vec{B}$, as shown in Fig. 9.5. In fact, we have

$$B^{\mathrm{M}}_1 = \cos 2\vartheta_{\mathrm{M}} \, B^{\mathrm{F}}_1 + \sin 2\vartheta_{\mathrm{M}} \, B^{\mathrm{F}}_3 = 0 \,, \tag{9.161}$$

$$B^{\mathrm{M}}_2 = B^{\mathrm{F}}_2 = 0 \,, \tag{9.162}$$

$$B^{\mathrm{M}}_3 = -\sin 2\vartheta_{\mathrm{M}} \, B^{\mathrm{F}}_1 + \cos 2\vartheta_{\mathrm{M}} \, B^{\mathrm{F}}_3 = \Delta m^2_{\mathrm{M}}/2E \,. \tag{9.163}$$

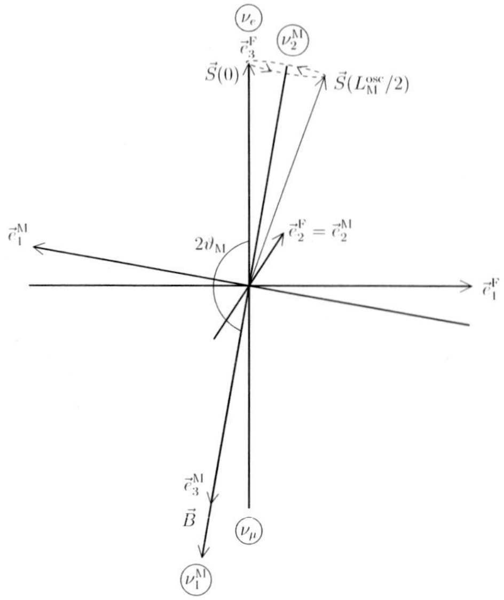

FIG. 9.6. Precession of $\vec{S}$ around $\vec{B}$ in a medium with constant $N_e \gg N_e^{\rm R}$ in the case of an initially pure ν_e beam and $\vartheta < \pi/4$.

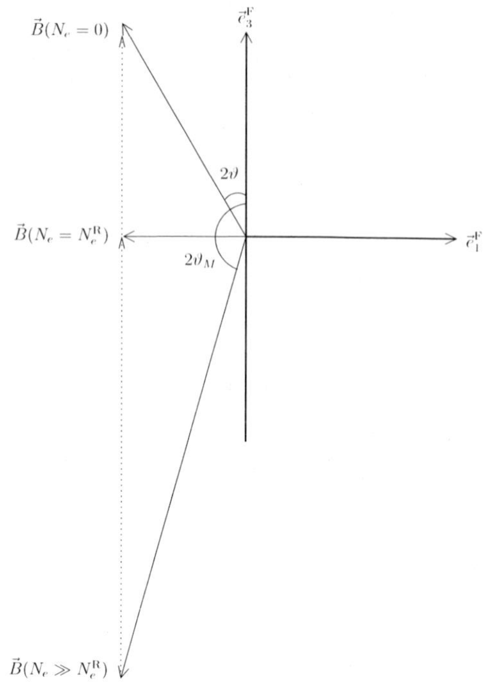

FIG. 9.7. Change of $\vec{B}$ from a matter density much larger than the resonant density ($N_e \gg N_e^{\rm R}$) to the resonance ($N_e = N_e^{\rm R}$) and finally to the vacuum ($N_e = 0$).

In this case, the evolution equation (9.153) becomes

$$\frac{\mathrm{d}S_{\rm M}^1}{\mathrm{d}x} = -\omega\, S_{\rm M}^2\,, \qquad \frac{\mathrm{d}S_{\rm M}^2}{\mathrm{d}x} = \omega\, S_{\rm M}^1\,, \qquad \frac{\mathrm{d}S_{\rm M}^3}{\mathrm{d}x} = 0\,. \tag{9.164}$$

Thus, $S_{\rm M}^3$ is constant, whereas $S_{\rm M}^1$ and $S_{\rm M}^2$ obey the harmonic oscillator equations

$$\frac{\mathrm{d}^2 S_{\rm M}^{1,2}}{\mathrm{d}x^2} + \omega^2\, S_{\rm M}^{1,2} = 0\,. \tag{9.165}$$

For the initial conditions in eqn (9.150), the solution is

$$S_{\rm M}^1(x) = \sin 2\vartheta_{\rm M}\,(W_e - W_\mu)\cos(\omega\, x)\;, \tag{9.166}$$

$$S_{\rm M}^2(x) = \sin 2\vartheta_{\rm M}\,(W_e - W_\mu)\sin(\omega\, x)\;, \tag{9.167}$$

$$S_{\rm M}^3(x) = \cos 2\vartheta_{\rm M}\,(W_e - W_\mu)\;, \tag{9.168}$$

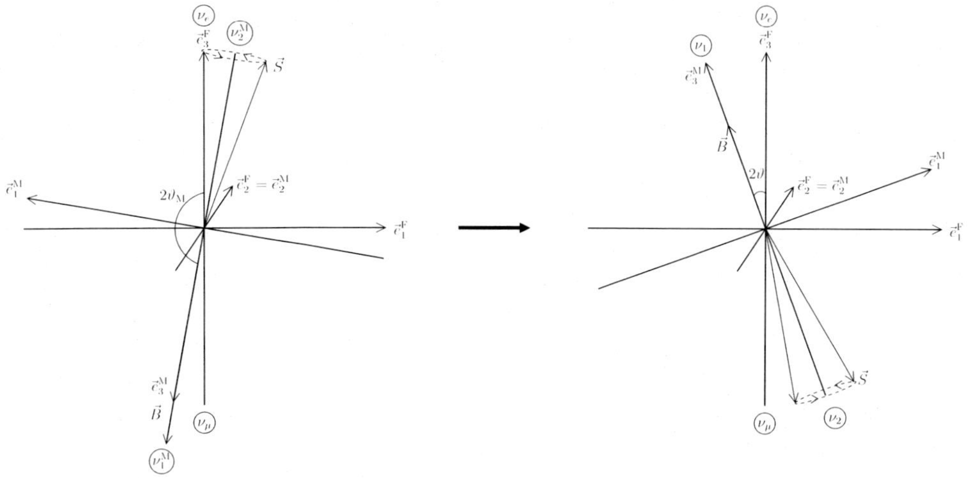

FIG. 9.8. Change of $\vec{B}$ and rotation of $\vec{S}$ for adiabatic solar neutrino transitions.

which confirms that $\vec{S}$ describes the surface of a cone with the axis along $\vec{B}$ and opening angle $2\vartheta_{\rm M}$. For the third component of $\vec{S}$ in the flavor basis we obtain

$$S_{\rm F}^3(x) = \sin 2\vartheta_{\rm M}\, S_{\rm M}^1(x) + \cos 2\vartheta_{\rm M}\, S_{\rm M}^3(x) = (W_e - W_\mu)\left[1 - 2\,\sin^2 2\vartheta_{\rm M}\,\sin^2\frac{\omega x}{2}\right]. \tag{9.169}$$

Hence, the probabilities of ν_e and ν_μ detection are, respectively, from eqns (9.151) and (9.152),

$$P_e(x) = \frac{1}{2} + (W_e - W_\mu)\left[\frac{1}{2} - \sin^2 2\vartheta_{\rm M}\,\sin^2\frac{\omega x}{2}\right], \tag{9.170}$$

$$P_\mu(x) = \frac{1}{2} + (W_\mu - W_e)\left[\frac{1}{2} - \sin^2 2\vartheta_{\rm M}\,\sin^2\frac{\omega x}{2}\right]. \tag{9.171}$$

One can easily check that in the case of an initial pure ν_e beam ($W_e = 1$ and $W_\mu = 0$), the probability of $\nu_e \to \nu_\mu$ transitions coincides with that in eqn (9.77).

In Fig. 9.5 we considered the case of an initially pure ν_e beam in a medium with constant $N_e < N_e^{\rm R}$. We also assumed $\vartheta < \pi/4$, which implies $\vartheta_{\rm M} < \pi/4$ for a density which is smaller than the resonance density. On the other hand, for a density which is larger than the resonance density, we have $\vartheta_{\rm M} > \pi/4$. Figure 9.6 illustrates the case of an initially pure ν_e beam in a medium with constant $N_e \gg N_e^{\rm R}$, which implies $\vartheta_{\rm M} \simeq \pi/2$ (in the figure $\vartheta_{\rm M}$ is taken to be significantly less than $\pi/2$ in order to have a visible cone). In this case, ν_e almost coincides with $\nu_2^{\rm M}$ and $\vec{S}$ describes the surface of a narrow cone around the negative $\vec{e}_3^{\,\rm M}$ axis.

Let us now consider the more interesting case of a variable matter density. For simplicity we will consider only the case of solar neutrinos which are created as ν_e in the core of the Sun with an electron density much larger than the resonance electron density. Since well above the resonance the effective mixing angle $\vartheta_{\rm M}$ is almost constant and close to $\pi/2$ (see Fig. 9.2a), initially $\vec{S}$ behaves as shown in

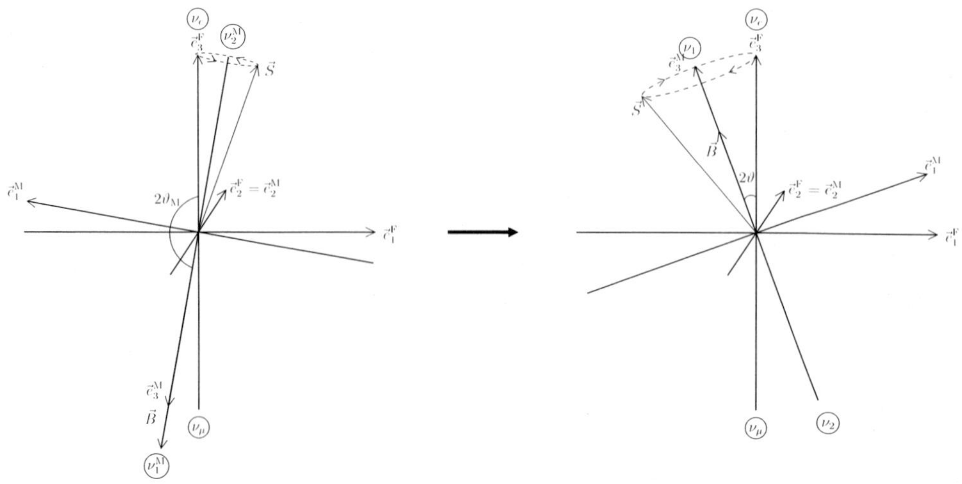

FIG. 9.9. Change of $\vec{B}$ and rotation of $\vec{S}$ for completely nonadiabatic solar neutrino transitions.

Fig. 9.6, describing the surface of a narrow cone around the negative $\vec{e}_3^{\,M}$ axis. When the neutrino passes through the resonance region, the vector $\vec{B}$ changes from a value similar to that in Fig. 9.6 to a value similar to that in Fig. 9.5. Note that only the component B_F^3 changes in matter (see eqns (9.146)–(9.148)). Thus, in going from a region where the matter density is much larger than the resonant density to the resonance and finally to the vacuum, the vector $\vec{B}$ changes as illustrated in Fig. 9.7.

If the resonance is crossed adiabatically, the speed of rotation of $\vec{S}$ around $\vec{B}$ is much faster than the change of $\vec{B}$. In this case, $\vec{S}$ is dragged by $\vec{B}$ and the transition from the core of the Sun to the vacuum is of the type depicted in Fig. 9.8. One can see that the neutrino is created as a ν_e, which is almost a ν_2^M in the core of the Sun. As the neutrino propagates out of the Sun, the cone swept by $\vec{S}$ is rotated upside-down. The neutrino emerging from the Sun almost as a ν_2 is almost a ν_μ if the mixing angle is small. Thus, there is a large probability of $\nu_e \to \nu_\mu$ conversion due to the propagation in the Sun, which is nicely illustrated in the geometrical representation in Fig. 9.8.

The other extreme case is the one of a completely nonadiabatic transition, in which the change of $\vec{B}$ in the resonance region is much faster than the speed of rotation of $\vec{S}$ around $\vec{B}$. In this case, $\vec{S}$ is practically frozen during the resonance crossing and it is left behind by $\vec{B}$, as shown in Fig. 9.9. The neutrino is created as a ν_e, which is almost a ν_2^M in the core of the Sun. It remains a ν_e through the resonance and oscillates in the vacuum from the Sun to the Earth. If the vacuum mixing angle is small, the neutrino emerges from the Sun almost as a ν_1. Consequently, the stationarity of $\vec{S}$ during resonance crossing in the nonadiabatic case represents large $\nu_2^M \to \nu_1^M$ transitions in the resonance.

Other details and examples of the geometrical representation of neutrino oscillations are discussed in Refs. [671, 669, 670]. Finally, it is to be mentioned that another graphic description of neutrino oscillations has been developed in Refs. [803, 804].

10

SOLAR NEUTRINOS

> Only neutrinos, with their extremely small interaction cross-sections, can enable us to see into the interior of a star, and thus verify directly the hypothesis of nuclear energy generation in stars.
> **John N. Bahcall** [136]

The experimental and theoretical study of solar neutrinos[51] is one of the main areas of research in neutrino physics. The Sun is a very powerful source of electron neutrinos with energy of the order of 1 MeV, produced in the thermonuclear fusion reactions in the solar core[52]. Since neutrino interactions with matter is extremely weak, practically all the neutrinos produced in the core of the Sun pass undisturbed through the solar interior and flow in space. The solar neutrino flux on the Earth is about $6 \times 10^{10}\,\mathrm{cm}^{-2}\,\mathrm{s}^{-1}$. In spite of this extremely large flux, the detection of solar neutrinos is difficult and requires large detectors because of the small neutrino interaction cross-section. These detectors must be placed underground in order to be shielded by rock from cosmic rays whose interactions in the detector would largely outnumber and dominate solar neutrino interactions.

Solar neutrinos were detected for the first time in 1970 in the Homestake experiment, which has monitored the solar flux for the next 24 years. In the late 1980s the Kamiokande experiment obtained the first real-time neutrino image of the Sun. From 1990 the GALLEX/GNO and SAGE experiments have measured the low-energy neutrinos produced in the fundamental *pp*. Starting in the late 1990s, the Super-Kamiokande and SNO experiments have provided important high-precision data on the high-energy part of the solar neutrino flux.

The detection of solar neutrinos is considered today mainly an activity of high-energy physicists interested in the properties of neutrinos. However, in our opinion it is important to be aware that the first success of solar neutrino experiments is the detection of solar neutrinos and the proof that the theory of thermonuclear energy generation in stars is correct. The second success of solar neutrino experiments

---

[51] This chapter is dedicated to the late John N. Bahcall, who was a pioneer and the champion of solar neutrino physics.

[52] In general, nuclear fusion produces electron neutrinos, because heavy stable nuclei contain a higher fraction of neutrons with respect to light stable nuclei. In this way, the binding nuclear force acting among nucleons can dominate over the electric repulsion among protons. Thus, in nuclear fusion processes, protons must be transformed into neutrons through the weak process $p \to n + e^+ + \nu_e$, producing electron neutrinos. On the other hand, for the same reason, the fission of heavy nuclei produces electron antineutrinos in the transformation $n \to p + e^- + \bar{\nu}_e$ of neutrons into protons.

is the discovery (Homestake), confirmation (Kamiokande, GALLEX/GNO, SAGE, Super-Kamiokande), and solution (SNO) of the *solar neutrino problem* (SNP) in favor of neutrino oscillations.

The SNP is a deficit of observed solar ν_e's with respect to the *standard solar model* (SSM) prediction. This chapter presents a pedagogical introduction to the main ingredients of the SSM in sections 10.1 and 10.2. In section 10.3 we discuss the model-independent constraints on solar neutrino fluxes. In sections 10.4–10.6 we present brief summaries of the main aspects and results of solar neutrino experiments. In section 10.7 we briefly discuss vacuum oscillations of solar neutrinos and in sections 10.8 and 10.9 we discuss, respectively, matter effects in the Sun and in the Earth. Finally, section 10.10 presents the information on the neutrino oscillation parameters obtained in global fits of solar neutrino data.

10.1 Thermonuclear energy production

The Sun shines because energy is produced in its core by thermonuclear reactions. The theory of thermonuclear energy production in the hot and dense core of stars has been developed since the late 1920s [357], after the crucial discovery by Gamow in 1928 and independently by Condon and Gurney in 1929 of the tunnel effect [503, 330], which allows the penetration of the Coulomb barrier between ions (see Refs. [909, 148]). The modern theory of stellar nucleosynthesis was developed by Bethe [221] and others [282, 296] (see Ref. [457]).

Thermonuclear reactions release energy because the total mass of a nucleus is less than the total mass of the constituent nucleons,

$$m(A,Z) = Z\,m_p + (A-Z)\,m_n - B(A,Z)\,, \tag{10.1}$$

where A and Z are, respectively, the atomic mass and number of the nucleus, $m_p = 938.272\,\text{MeV}$ and $m_n = 939.565\,\text{MeV}$ are, respectively, the mass of the proton and the neutron, and $B(A,Z)$ is the binding energy. For example, the binding energy of deuterium (d or ^{2}H) is $B(2,1) = 2.22\,\text{MeV}$, and the binding energy of ^{4}He is $B(4,2) = 28.296\,\text{MeV}$.

The Sun is powered by the two groups of thermonuclear reactions known as the *pp* chain and the CNO cycle, which are depicted, respectively, in Figs. 10.1 and 10.2. The result of both the *pp* chain and the CNO cycle is the conversion of four protons and two electrons into a ^{4}He nucleus plus two electron neutrinos:

$$4\,p + 2\,e^- \to {}^4\text{He} + 2\,\nu_e + Q\,, \tag{10.2}$$

where the energy release (usually called the Q-value of the process) is given by ($m_e = 0.511\,\text{MeV}$ is the mass of the electron)

$$Q = 4\,m_p + 2\,m_e - m_{^4\text{He}} = B(4,2) + 2\,m_e - 2\,(m_n - m_p) = 26.731\,\text{MeV}\,. \tag{10.3}$$

This energy is released in the form of photons or kinetic energy of the neutrinos (the kinetic energy of the ^{4}He nucleus is negligible because of its large mass).

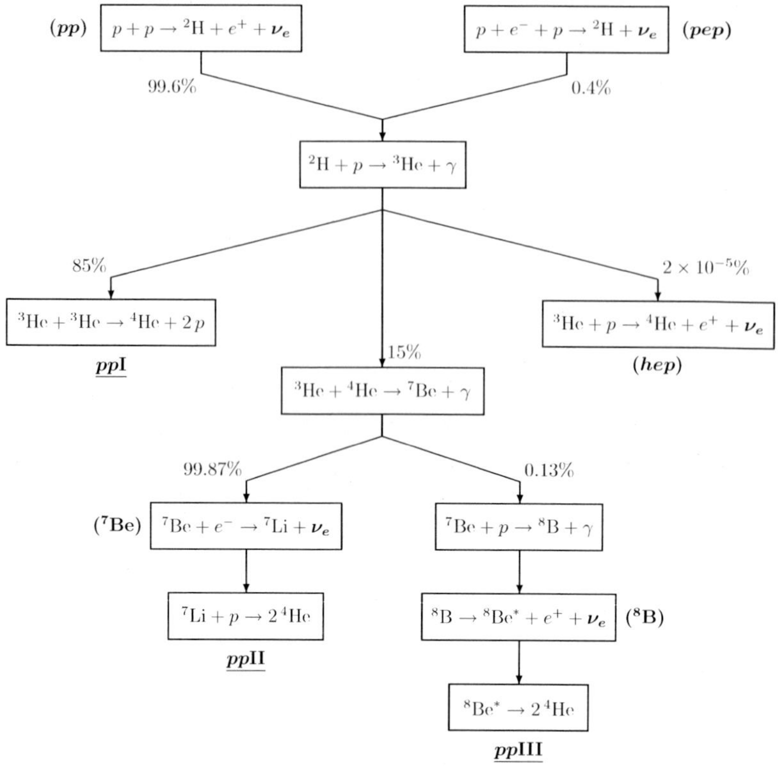

FIG. 10.1. The *pp* chain of stellar thermonuclear reactions. The produced neutrinos are typed in boldface characters. The traditional names of the neutrino-producing reactions and the corresponding neutrino fluxes are given in parentheses. The underlined labels indicate the three main branches of the *pp* chain.

The electron neutrinos produced in the process in eqn (10.2) in the core of the Sun can be detected on the Earth. They provide a unique direct probe of the interior of the Sun.

In order to understand the basic principles of energy generation in the core of the Sun, let us consider a generic nuclear reaction

$$\text{A} + \text{B} \to \text{anything} \tag{10.4}$$

occurring in a stellar gas where there are N_{A} and N_{B} particles per unit volume of type A and B, respectively. The cross-section σ of the process in eqn (10.4) depends only on the relative velocity v and one can consider either A or B as a projectile and the other as a target. Let us consider A as the projectile, with velocity v, and B as the target at rest. The rate of reaction for each projectile particle A is given by the cross-section σ times the number density N_{B} of targets. Since the flux of projectiles is $N_{\text{A}}v$, the rate of the reaction in eqn (10.4) per unit volume is given

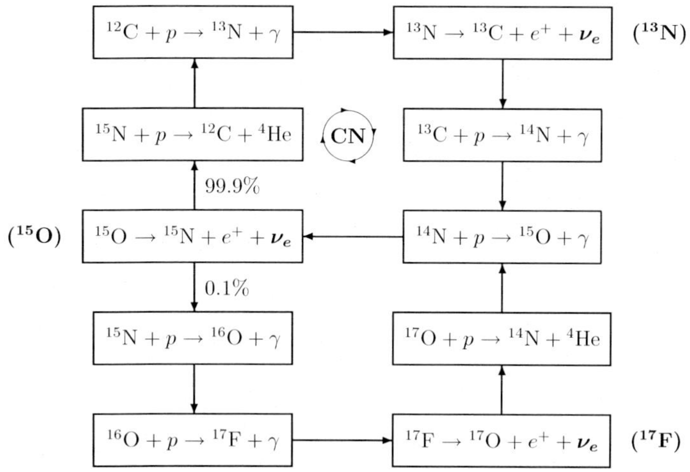

FIG. 10.2. The CNO cycle of stellar thermonuclear reactions. The produced neutrinos are typed in boldface characters. The traditional names of the neutrino-producing reactions and the corresponding neutrino fluxes are given in parentheses.

by

$$\mathcal{R}_{\mathrm{AB}} = \frac{N_{\mathrm{A}}\, N_{\mathrm{B}}\, \langle \sigma v \rangle_{\mathrm{AB}}}{1 + \delta_{\mathrm{AB}}} \,, \tag{10.5}$$

where $\langle \sigma v \rangle_{\mathrm{AB}}$ is the average of the product σv over the thermal distribution of velocities in the hot gas and the Kronecker delta prevents double counting of identical particles. Using a Maxwell–Boltzmann velocity distribution, one obtains (see Refs. [909])

$$\langle \sigma v \rangle_{\mathrm{AB}} = \sqrt{\frac{8}{\pi \mu \, (k_B T)^3}} \int_0^\infty \mathrm{d}E \, \sigma(E) \, E \, \exp\left(- \frac{E}{k_B T} \right) \,, \tag{10.6}$$

where T is the temperature, $k_B = 8.617 \times 10^{-5}\,\mathrm{eV\,K^{-1}}$ is the Boltzmann constant, $E = \frac{1}{2}\,\mu\,v^2$ is the center-of-mass kinetic energy, and $\mu = m_{\mathrm{A}} m_{\mathrm{B}} / (m_{\mathrm{A}} + m_{\mathrm{B}})$ is the reduced mass of the system of interacting particles. The energy distribution is proportional to E for $E \ll k_B T$, reaches a peak at $E = k_B T$, and decreases proportionally to $\exp(-E/k_B T)$ for $E \gg k_B T$.

In the formation of a star, the gas is initially heated by the contraction of the proto-stellar cloud under the force of gravity until it reaches a temperature sufficient for the onset of thermonuclear reactions. The central temperature of the Sun is about $1.5 \times 10^7\,\mathrm{K}$ and the gas is in the form of plasma, with ionized atoms that carry positive charges. In particular, protons (^{1}H nuclei) are stripped of electrons. In order to ignite the nuclear reaction, the positive A and B ions must overcome the repulsive Coulomb force, whose potential energy is given by

$$V_{\mathrm{C}}(r) = \frac{Z_{\mathrm{A}}\, Z_{\mathrm{B}}\, \alpha}{r} \,, \tag{10.7}$$

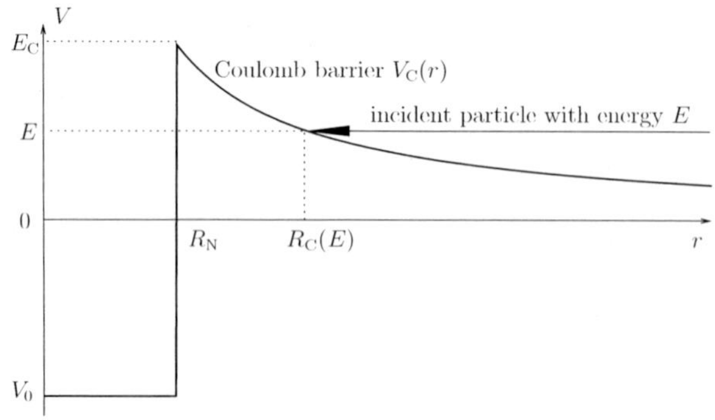

FIG. 10.3. Schematic representation of the Coulomb barrier. The repulsive Coulomb potential dominates for $r > R_N$. If the incident particle has energy E smaller than the height E_C of the Coulomb barrier, it can interact with the target nucleus only if it penetrates the Coulomb barrier and reaches distances smaller than the nuclear radius R_N, where the attractive nuclear potential dominates. According to classical physics, the incident particle cannot penetrate the Coulomb barrier and would reach the closest distance to the nucleus at the classical turning point R_C. The penetration of the Coulomb barrier through the tunnel effect is allowed in quantum mechanics.

where $\alpha = e^2/4\pi$ is the fine-structure constant, Z_A and Z_B are the atomic numbers of the A and B nuclei, and r is the distance between them. The repulsive Coulomb force dominates nuclear interactions for distances larger than the sum $R_N = R_A + R_B$ of the two nuclear radii. At smaller distances nuclear forces dominate, and the nuclear reaction can occur. The average radius R of a nucleus with atomic mass number A is given by

$$R \simeq R_0 \, A^{1/3} \,, \tag{10.8}$$

with $R_0 = 1.3 \times 10^{-13}$ cm.

The potential in eqn (10.7) is called *Coulomb barrier*. As illustrated in Fig. 10.3, classical physics forbids the penetration of the Coulomb barrier for an incident particle with energy E smaller than the height $E_C = Z_A Z_B \alpha / R_N$ of the Coulomb barrier. The particle cannot get closer than the classical turning point

$$R_C = \frac{Z_A \, Z_B \, \alpha}{E} \,. \tag{10.9}$$

For example, the height of the Coulomb barrier for the interaction of two protons is $E_C \simeq 550$ keV. According to classical physics, the pp interaction should be strongly suppressed in the interior of the Sun, where $T \simeq 1.5 \times 10^7$ K and the average energy of protons is $k_B T \simeq 1.3$ keV.

This obstacle was removed in 1928 with the discovery of the quantum tunnel effect [503, 330], which allows the penetration of the Coulomb barrier for an incident particle with energy $E < E_C$. The probability of penetrating the Coulomb barrier is given by

$$P_{\mathrm{C}} = \frac{|\psi(R_{\mathrm{N}})|^2}{|\psi(R_{\mathrm{C}})|^2}, \tag{10.10}$$

where $\psi(r)$ is the wavefunction of the incident particle that can be calculated by solving the Schrödinger equation for the Coulomb potential in eqn (10.7). The probability P_{C} is unity for $E \geq E_{\mathrm{C}}$ and decreases rapidly for $E < E_{\mathrm{C}}$. For low-energy particles, with $E \ll E_{\mathrm{C}}$, the tunneling probability in eqn (10.10) can be approximated by a simple expression known as the *Gamow factor*

$$P_{\mathrm{C}} \simeq e^{-2\pi\eta}, \tag{10.11}$$

with the *Sommerfeld parameter*

$$\eta = \frac{Z_{\mathrm{A}}\, Z_{\mathrm{B}}\, \alpha}{v} = \frac{Z_{\mathrm{A}}\, Z_{\mathrm{B}}\, \alpha}{\sqrt{2}} \sqrt{\frac{\mu}{E}}, \tag{10.12}$$

where v is the relative velocity between the two interacting particles, μ is the reduced mass, and E is the center-of-mass energy.

The total cross-section of the reaction in eqn (10.4) is given by the nuclear cross-section times the Gamow factor in eqn (10.11) and can be written as

$$\sigma(E) = \frac{1}{E}\, e^{-2\pi\eta}\, S(E), \tag{10.13}$$

where the function $S(E)$ contains all nuclear effects and is known as the *astrophysical S-factor*. For nonresonant reactions, the astrophysical S-factor is a smooth and slowly varying function of energy E. Hence, $S(E)$ is useful for extrapolating the value of a cross-section out of the range of energies in which it is measured. An important branch of research in solar physics focuses on the determination of the astrophysical S-factors of the nuclear reactions of the *pp* chain and the CNO cycle in the core of the Sun.

Inserting the expression in eqn (10.13) for the cross-section in eqn (10.6), we obtain the reaction rate per particle pair

$$\langle \sigma v \rangle_{\mathrm{AB}} = \sqrt{\frac{8}{\pi\mu\, (k_B T)^3}} \int_0^\infty \mathrm{d}E\, S(E)\, \exp\left(-\frac{E}{k_B T} - \sqrt{\frac{E_{\mathrm{G}}}{E}} \right), \tag{10.14}$$

where E_{G} is the Gamow energy

$$E_{\mathrm{G}} = 2\,\mu\, (\pi\, \alpha\, Z_{\mathrm{A}}\, Z_{\mathrm{B}})^2\,. \tag{10.15}$$

Usually, the Gamow energy E_{G} is much larger than the thermal energy $k_B T$.

Since the energy dependence of the astrophysical S-factor $S(E)$ is weak, the energy dependence of the integrand in eqn (10.14) is determined by the exponential term, which is a product of $\exp\left(-\frac{E}{k_B T}\right)$, due to the Maxwell–Boltzmann energy distribution, and $\exp\left(-\sqrt{\frac{E_{\mathrm{G}}}{E}}\right)$, due to the probability of tunneling through the

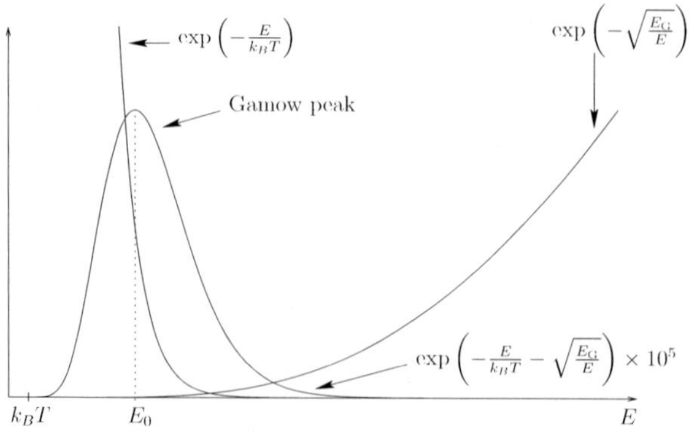

FIG. 10.4. Schematic representation of the Gamow peak, given by the product of the tail of the Maxwell–Boltzmann energy distribution $\propto \exp(-E/k_BT)$ and the probability of penetration of the Coulomb barrier $\propto \exp\left(-\sqrt{E_G/E}\right)$. For illustration purposes the Gamow peak has been enhanced by a factor $\sim 10^5$. The energy E_0 of the Gamow peak is generally much larger than the thermal energy k_BT.

Coulomb barrier. The product results in the *Gamow peak* of the integrand around the energy

$$E_0 = \left(\frac{\sqrt{E_G}\,k_BT}{2}\right)^{2/3} = \left[\frac{\mu}{2}\,(\pi\,\alpha\,Z_A\,Z_B\,k_BT)^2\right]^{1/3}, \tag{10.16}$$

as illustrated in Fig. 10.4. Usually, since $E_G \gg k_BT$, the energy E_0 of the Gamow peak is much larger than k_BT. Nuclear reactions in a stellar core with temperature T occur in the relatively narrow energy interval $E_0 - \Delta_0 \lesssim E \lesssim E_0 + \Delta_0$, where Δ_0 is the width of the Gamow peak. Approximating the exponential term in the integrand of eqn (10.14) with a Gaussian function,

$$\exp\left(-\frac{E}{k_BT}-\sqrt{\frac{E_G}{E}}\right) \simeq I_0 \exp\left[-\left(\frac{E-E_0}{\Delta_0/2}\right)^2\right], \tag{10.17}$$

where I_0 is the peak value of the integrand

$$I_0 = \exp\left(-3\,\frac{E_0}{k_BT}\right), \tag{10.18}$$

one obtains

$$\Delta_0 \simeq 4\sqrt{\frac{E_0\,k_BT}{3}}. \tag{10.19}$$

Since $k_BT \ll E_0$, the width Δ_0 of the Gamow peak is much smaller than its central energy E_0. Therefore, the smoothly varying astrophysical S-factor can be approximated with its value at E_0, leading to

$$\langle\sigma v\rangle_{AB} \simeq \sqrt{\frac{8}{\pi\mu\,(k_BT)^3}}\,S(E_0)\int_0^\infty dE\,\exp\left(-\frac{E}{k_BT}-\sqrt{\frac{E_G}{E}}\right)$$

$$\simeq \sqrt{\frac{8}{\pi\mu\,(k_B T)^3}}\, S(E_0)\, I_0\, \sqrt{\pi}\, \frac{\Delta_0}{2}\,, \tag{10.20}$$

where the integration has been performed using the Gaussian approximation in eqn (10.17). Using the expressions in eqns (10.18) and (10.19) in eqn (10.20), we obtain the final approximate expression for the reaction rate per particle pair:

$$\langle \sigma v \rangle_{\mathrm{AB}} \simeq 4 \sqrt{\frac{2}{3\mu}}\, S(E_0)\, \frac{\sqrt{E_0}}{k_B T}\, \exp\!\left(-3\, \frac{E_0}{k_B T} \right). \tag{10.21}$$

This reaction rate depends strongly on the exponential I_0 in eqn (10.18), which is due to the Coulomb barrier. This is very important for the life of a star, since it leads to the existence of different epochs in stellar evolution, called *stages of nuclear burning*: hydrogen burning, helium burning, and heavy-ion burning.

From eqn (10.16) one can see that the energy E_0 of the Gamow peak increases with the atomic numbers of the interacting nuclei. This implies that the exponential I_0 and the reaction rate in eqn (10.21) decrease rapidly as the atomic numbers of the interacting nuclei are increased, keeping the temperature constant. Therefore, if the astrophysical S-factors do not differ much, reactions involving light nuclei proceed much faster than reactions involving heavy nuclei. On the other hand, the exponential I_0 increases with the temperature T. Hence, reactions involving heavy nuclei can proceed with sufficient speed to heat a star when the temperature is sufficiently high.

When a star starts to form, the contraction of the proto-stellar cloud under the force of gravity heats the gas until the temperature is high enough to allow a significant penetration of the Coulomb barrier between the lightest nuclei, i.e. protons (hydrogen nuclei), igniting the weak interaction reaction

$$p + p \to d + e^+ + \nu_e\,. \tag{10.22}$$

This is the basic reaction of the *pp* chain (see Fig. 10.1 on page 354), which heats the star during the hydrogen burning stage. This is the longest epoch in the life of a star (for the Sun it is estimated to be about 10 Gy long, of which about 4.5 Gy have passed), during which the star lies in the main sequence of the Hertzsprung–Russel diagram (luminosity versus surface temperature). For a brief description of the main stages of stellar evolution, see chapter 15.

10.2 Standard solar models

A standard solar model (SSM) is a "solar model that is constructed with the best available physics and input data" [145] and is "required to fit the observed luminosity and radius of the Sun at the present epoch, as well as the observed heavy-element-to-hydrogen ratio at the surface of the Sun" (see Table 10.1).

In recent years, several experiments have tested the SSMs by measuring p-mode oscillations of the Sun (were p stands for *pressure*). The science that studies these oscillations is called *helioseismology*. It has become of primary importance for

TABLE 10.1. Fundamental characteristics of the Sun and Sun–Earth system [400]. One astronomical unit is the mean Sun–Earth distance. The solar constant $K_\odot$ is the mean solar photon flux on the Earth.

Solar luminosity	$\mathscr{L}_\odot$	=	$(3.846 \pm 0.008) \times 10^{26}$ W
		=	$(2.400 \pm 0.005) \times 10^{39}$ MeV s^{-1}
Solar radius	$\mathcal{R}_\odot$	=	6.961×10^{10} cm
Solar mass	$\mathcal{M}_\odot$	=	$(1.98844 \pm 0.00030) \times 10^{33}$ g
Astronomical unit	1 au	=	$(149.597870660 \pm 0.000000020) \times 10^{6}$ km
Solar constant	$K_\odot$	$\equiv$	$\mathscr{L}_\odot/4\pi(1\text{au})^2 \simeq 8.534 \times 10^{11}$ MeV cm^{-2} s^{-1}
Year	1 yr	=	3.15569252×10^{7} s

TABLE 10.2. Sources of solar neutrinos [137, 141, 147, 138]. For each reaction r, $\langle E\rangle_r$ is the average neutrino energy, $E_r^{\max}$ is the maximum neutrino energy, and α_r is the average thermal energy released together with a neutrino from the source r [139], which enters in the luminosity constraint in eqn (10.23).

Source r	Reaction	$\langle E\rangle_r$ (MeV)	$E_r^{\max}$ (MeV)	α_r (MeV)
pp	$p + p \to d + e^+ + \nu_e$	0.2668	0.423 ± 0.03	13.0987
pep	$p + e^- + p \to d + \nu_e$	1.445	1.445	11.9193
hep	$^3\text{He} + p \to {}^4\text{He} + e^+ + \nu_e$	9.628	18.778	3.7370
^{7}Be	$e^- + {}^7\text{Be} \to {}^7\text{Li} + \nu_e$	0.3855 0.8631	0.3855 0.8631	12.6008
^{8}B	$^8\text{B} \to {}^8\text{Be}^* + e^+ + \nu_e$	6.735 ± 0.036	~ 15	6.6305
^{13}N	$^{13}\text{N} \to {}^{13}\text{C} + e^+ + \nu_e$	0.7063	1.1982 ± 0.0003	3.4577
^{15}O	$^{15}\text{O} \to {}^{15}\text{N} + e^+ + \nu_e$	0.9964	1.7317 ± 0.0005	21.5706
^{17}F	$^{17}\text{F} \to {}^{17}\text{O} + e^+ + \nu_e$	0.9977	1.7364 ± 0.0003	2.363

understanding the physics of the Sun, since the helioseismological observations give detailed information on the sound speed and matter density in the interior of the Sun. It is important to note that the large amount of available helioseismological data is not used in the construction of SSMs. Therefore, the beautiful agreement of SSM predictions with helioseismological data is indeed an impressive success (see Ref. [145]).

Many SSMs have been constructed by several groups and updated in the course of time as the understanding of the physics of the Sun, the quality and quantity of input data, and computational power have improved. The SSMs that have played the major role in neutrino physics are those developed by Bahcall and collaborators

TABLE 10.3. BP00 SSM [145] neutrino fluxes, average neutrino cross-sections [137, 141, 147], and BP00 SSM predictions for the neutrino capture rates [145] in the chlorine (Cl) Homestake experiment and in the gallium (Ga) GALLEX, SAGE and GNO experiments.

Source r	Flux Φ_r $[\text{cm}^{-2}\,\text{s}^{-1}]$	$\langle\sigma_{\text{Cl}}\rangle_r$ $[10^{-44}\,\text{cm}^2]$	$R^{(r)}_{^{37}\text{Cl}}$ [SNU]	$\langle\sigma_{\text{Ga}}\rangle_r$ $[10^{-44}\,\text{cm}^2]$	$R^{(r)}_{^{71}\text{Ga}}$ [SNU]
pp	$5.95 \times 10^{10}\,(1 \pm 0.01)$	–	–	0.117 ± 0.003	69.7
pep	$1.40 \times 10^{8}\,(1 \pm 0.015)$	0.16	0.22	$2.04\,^{+0.35}_{-0.14}$	2.8
hep	9.3×10^{3}	390	0.04	$714\,^{+228}_{-114}$	0.1
^{7}Be	$4.77 \times 10^{9}\,(1 \pm 0.10)$	0.024	1.15	$0.717\,^{+0.050}_{-0.0.021}$	34.2
^{8}B	$5.05 \times 10^{6}\,(1\,^{+0.20}_{-0.16})$	114 ± 11	5.76	$240\,^{+77}_{-36}$	12.1
^{13}N	$5.48 \times 10^{8}\,(1\,^{+0.21}_{-0.17})$	0.017	0.09	$0.604\,^{+0.036}_{-0.018}$	3.4
^{15}O	$4.80 \times 10^{8}\,(1\,^{+0.25}_{-0.19})$	0.068 ± 0.001	0.33	$1.137\,^{+0.136}_{-0.057}$	5.5
^{17}F	$5.63 \times 10^{6}\,(1 \pm 0.25)$	0.069	0.0	$1.139\,^{+0.137}_{-0.057}$	0.1
Total	6.54×10^{10}		$7.6\,^{+1.3}_{-1.1}$		$128\,^{+9}_{-7}$

TABLE 10.4. Solar neutrino fluxes and radiochemical rates in the BP00 [145], BP04 [152], BSB05(GS98) [154], and BSB05(AGS05) [154] SSMs. Fluxes are given in units of $\text{cm}^{-2}\,\text{s}^{-1}$.

	BP00	BP04	BSB05(GS98)	BSB05(AGS05)
$\Phi_{pp}\,/\,10^{10}$	$5.95\,(1 \pm 0.01)$	$5.94\,(1 \pm 0.01)$	$5.99\,(1 \pm 0.009)$	$6.06\,(1 \pm 0.007)$
$\Phi_{pep}\,/\,10^{8}$	$1.40\,(1 \pm 0.015)$	$1.40\,(1 \pm 0.02)$	$1.42\,(1 \pm 0.015)$	$1.45\,(1 \pm 0.011)$
$\Phi_{hep}\,/\,10^{3}$	9.3	$7.88\,(1 \pm 0.16)$	$7.93\,(1 \pm 0.155)$	$8.25\,(1 \pm 0.155)$
$\Phi_{^7\text{Be}}\,/\,10^{9}$	$4.77\,(1 \pm 0.10)$	$4.86\,(1 \pm 0.12)$	$4.84\,(1 \pm 0.105)$	$4.34\,(1 \pm 0.093)$
$\Phi_{^8\text{B}}\,/\,10^{6}$	$5.05\,(1\,^{+0.20}_{-0.16})$	$5.79\,(1 \pm 0.23)$	$5.69\,(1\,^{+0.173}_{-0.147})$	$4.51\,(1\,^{+0.127}_{-0.113})$
$\Phi_{^{13}\text{N}}\,/\,10^{8}$	$5.48\,(1\,^{+0.21}_{-0.17})$	$5.71\,(1\,^{+0.37}_{-0.35})$	$3.05\,(1\,^{+0.366}_{-0.268})$	$2.00\,(1\,^{+0.145}_{-0.127})$
$\Phi_{^{15}\text{O}}\,/\,10^{8}$	$4.80\,(1\,^{+0.25}_{-0.19})$	$5.03\,(1\,^{+0.43}_{-0.39})$	$2.31\,(1\,^{+0.374}_{-0.272})$	$1.44\,(1\,^{+0.165}_{-0.142})$
$\Phi_{^{17}\text{F}}\,/\,10^{6}$	$5.63\,(1 \pm 0.25)$	$5.91\,(1\,^{+0.44}_{-0.44})$	$5.83\,(1\,^{+0.724}_{-0.420})$	$3.25\,(1\,^{+0.166}_{-0.142})$
$R_{^{37}\text{Cl}}$ [SNU]	$7.6\,^{+1.3}_{-1.1}$	$8.5\,^{+1.8}_{-1.8}$	8.12	6.58
$R_{^{71}\text{Ga}}$ [SNU]	$128\,^{+9}_{-7}$	$131\,^{+12}_{-10}$	126.08	118.88

in an impressive series of papers starting in 1962 [142] (see Refs. [137, 145, 152, 154] and references therein).

In order to study the physics of solar neutrinos, it is convenient to treat separately the neutrino fluxes produced by the individual thermonuclear reactions of the *pp* chain and the CNO cycle, which are shown in Figs. 10.1 and 10.2, where the produced neutrinos are typed in boldface characters and the neutrino-producing reactions are associated with the corresponding traditional names. These

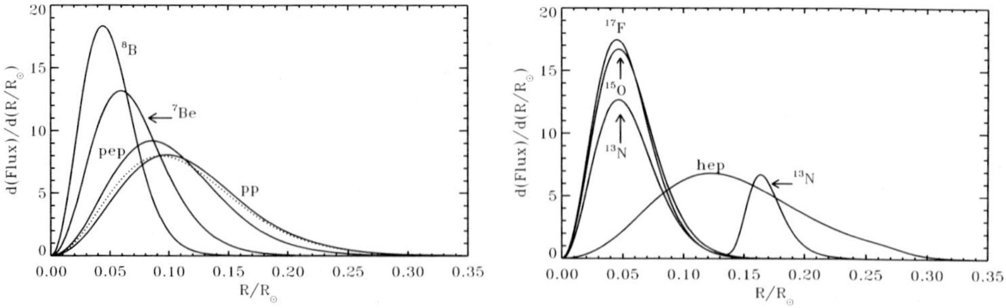

FIG. 10.5. Distribution of the neutrino production as a function of radius for each of the solar neutrino fluxes in the BSB(GS98) SSM [154]. The dotted line close to the distribution for the *pp* flux represents the distribution of the production of the solar luminosity.

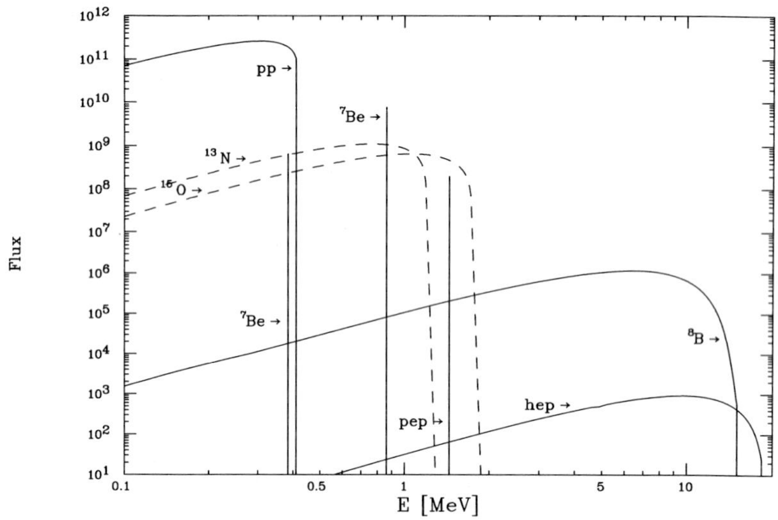

FIG. 10.6. Energy spectra of neutrino fluxes from the *pp* and CNO chains, as predicted by the SSM. For continuous sources, the differential flux is in $\mathrm{cm}^{-2}\,\mathrm{s}^{-1}\,\mathrm{MeV}^{-1}$. For the lines, the flux is in $\mathrm{cm}^{-2}\,\mathrm{s}^{-1}$. Figure taken from Ref. [307].

sources of solar neutrinos are listed in Table 10.2, together with the corresponding average and maximum neutrino energy. Table 10.3 presents the values of the neutrino fluxes according to the BP00 SSM [145], together with the corresponding average cross-sections and rates in the radiochemical chlorine and gallium detectors (see, respectively, sections 10.4 and 10.5). The rate of radiochemical detectors is conventionally measured in *Solar Neutrino Units* (SNU), such that $1\,\mathrm{SNU} \equiv 10^{-36}\,\mathrm{events\,atom}^{-1}\,\mathrm{s}^{-1}$.

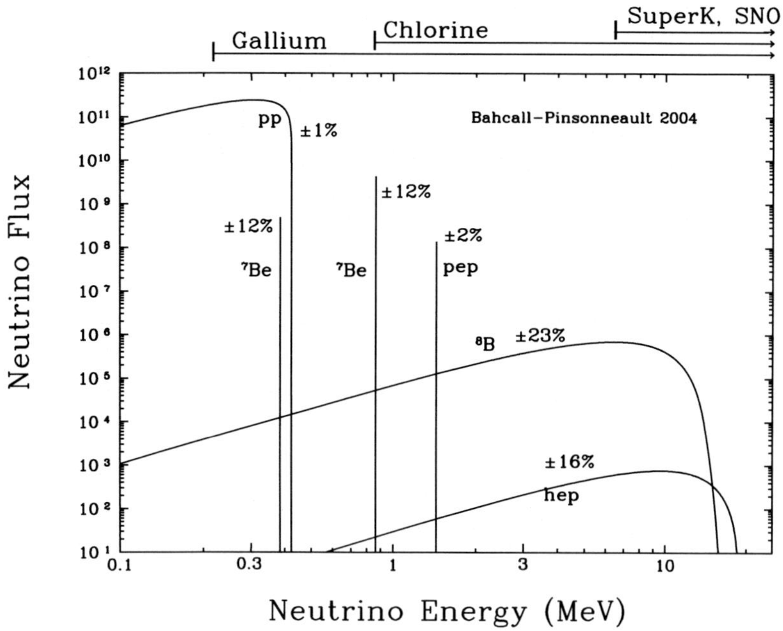

FIG. 10.7. Energy spectra of neutrino fluxes from the *pp* chain, as predicted by the BP04 SSM [152]. On the top, the energy ranges probed by different solar neutrino experiments are indicated. For continuous sources, the differential flux is in $\mathrm{cm}^{-2}\,\mathrm{s}^{-1}\,\mathrm{MeV}^{-1}$. For the lines, the flux is in $\mathrm{cm}^{-2}\,\mathrm{s}^{-1}$. The percentages indicate the uncertainties in the values of the fluxes. Figure taken from Ref. [151].

Table 10.4 compares the neutrino fluxes and radiochemical rates of the BP00 [145], BP04 [152], BSB05(GS98) [154] and BSB05(AGS05) [154] SSMs. The difference between the BSB05(GS98) and BSB05(AGS05) models is the adopted input abundance of heavy elements: BSB05(GS98) assumes the old heavy element abundances in Ref. [566], whereas BSB05(AGS05) uses the new heavy element abundances in Ref. [112], which lead to a disagreement between the predictions of the model and helioseismological observations [96]. Figure 10.5 shows the distribution of the neutrino production as a function of radius for each of the solar neutrino fluxes in the BSB(GS98) SSM [154]. Figure 10.6 shows the energy spectra of the neutrino fluxes from the *pp* chain and the CNO cycle in SSMs [307]. The precise energy spectra of neutrino fluxes of the *pp* chain in the BP00 SSM [145] are shown in Fig. 10.7, together with the energy ranges probed by different solar neutrino experiments.

TABLE 10.5. Mass excess $\Delta M_{\mathcal{I}}$ and binding energy $B(A,Z) = Z\,\Delta M_{^1\mathrm{H}} + (A-Z)\,\Delta M_n - \Delta M_{\mathcal{I}}$ of the isotopes $\mathcal{I}$ that participate in the *pp* chain and the CNO cycle [125]. The mass excess of an isotope with mass $M_{\mathcal{I}}$ and mass number Z is $\Delta M_{\mathcal{I}} = M_{\mathcal{I}} - Z\,\mathrm{amu}$, where one atomic mass unit (amu) is, by definition, one twelfth of the mass of ^{12}C (see eqn (A.143)). The difference between nuclear and atomic mass excesses due to the electron binding energy is negligible (a few eV's).

Isotope $\mathcal{I}$	$\Delta M_{\mathcal{I}}$ [MeV]	$B(A,Z)$ [MeV]
n	8.0713171 ± 0.0000005	
^{1}H	7.2889705 ± 0.0000001	
^{2}H	13.1357216 ± 0.0000003	2.2245660 ± 0.0000006
^{3}He	14.931215 ± 0.000002	7.718043 ± 0.000002
^{4}He	2.4249156 ± 0.0000001	28.295660 ± 0.000001
^{7}Li	14.90814 ± 0.00008	39.24404 ± 0.00008
^{7}Be	15.7700 ± 0.0001	37.5998 ± 0.0001
^{8}Be	4.94167 ± 0.00004	56.49948 ± 0.00004
^{8}B	22.922 ± 0.001	37.737 ± 0.001
^{12}C	0	92.161726 ± 0.000003
^{13}C	3.1250113 ± 0.0000009	97.108031 ± 0.000004
^{13}N	5.3455 ± 0.0003	94.1052 ± 0.0003
^{14}N	2.8634170 ± 0.0000006	104.658596 ± 0.000004
^{15}N	0.1014380 ± 0.0000007	115.491892 ± 0.000004
^{15}O	2.8556 ± 0.0005	111.9554 ± 0.0005
^{16}O	-4.7370014 ± 0.0000001	127.619302 ± 0.000004
^{17}O	-0.8088 ± 0.0001	131.7624 ± 0.0001
^{17}F	1.9517 ± 0.0003	128.2196 ± 0.0003

10.3 Model-independent constraints on solar neutrino fluxes

Thermonuclear reactions in the core of the Sun produce thermal energy and neutrinos. Neutrinos escape the Sun in about two seconds, whereas thermal energy takes more than 10^4 years to reach the surface, and is liberated into free space as radiation. In spite of the huge difference between the time scales of neutrino and thermal energy emission, since the Sun is in a stable state, the solar luminosity is connected with the neutrino flux through the so-called *luminosity constraint*, which can be written as

$$\sum_r \alpha_r\, \Phi_r = K_\odot \qquad (r = pp, pep, hep, {}^7\mathrm{Be}, {}^8\mathrm{B}, {}^{13}\mathrm{N}, {}^{15}\mathrm{O}, {}^{17}\mathrm{F})\,, \tag{10.23}$$

where $K_\odot$ is the solar constant in Table 10.1, ϕ_r is the flux of the source r, and α_r is the average thermal energy released together with a neutrino from the source r.

The values of α_r listed in Table 10.2 are independent of the details of the solar model [139]. They depend on the differences among nuclear masses, with small corrections due to the average thermal energy of the fusing particles.

Let us consider first the CNO cycle. In order to calculate the value of average thermal energy α_r released together with a neutrino from the source r, one must take into account that the reactions of the CNO cycle proceed with different speeds. In particular, since the $^{14}\mathrm{N}(p,\gamma)^{15}\mathrm{O}$ reaction[53] is the slowest process in the CN cycle (see Fig. 10.2), the rate of emission of $^{13}\mathrm{N}$ neutrinos is faster than the rate of completion of CN cycles, leading to the model-independent inequality

$$\Phi_{^{15}\mathrm{O}} \leq \Phi_{^{13}\mathrm{N}} \,. \tag{10.24}$$

Hence, the average thermal energy $\alpha_{^{13}\mathrm{N}}$ released together with a $^{13}\mathrm{N}$ is given by the thermal energy produced by the $^{12}\mathrm{C}(p,\gamma)^{13}\mathrm{N}$ and $^{13}\mathrm{N} \to {}^{13}\mathrm{C} + e^+ + \nu_e$ reactions minus the average $^{13}\mathrm{N}$ neutrino energy:

$$\alpha_{^{13}\mathrm{N}} = M_{^{12}\mathrm{C}} + M_{^{1}\mathrm{H}} - M_{^{13}\mathrm{C}} - \langle E \rangle_{^{13}\mathrm{N}} \,. \tag{10.25}$$

Since the masses of the constituent nucleons cancel, the calculation of eqn (10.25) can easily be performed replacing each isotopic mass $M_{\mathcal{I}}$ with the corresponding mass excess $\Delta M_{\mathcal{I}}$ given in Table 10.5. The same consideration applies to the following equations which are similar to eqn (10.25), leading to the values of α_r given in Table 10.2.

The average thermal energy $\alpha_{^{15}\mathrm{O}}$ released together with an $^{15}\mathrm{O}$ derives from the reactions $^{13}\mathrm{C}(p,\gamma)^{14}\mathrm{N}$, $^{14}\mathrm{N}(p,\gamma)^{15}\mathrm{O}$, $^{15}\mathrm{O} \to {}^{15}\mathrm{N} + e^+ + \nu_e$ and $^{15}\mathrm{N}(p,\alpha)^{12}\mathrm{C}$:

$$\alpha_{^{15}\mathrm{O}} = M_{^{13}\mathrm{C}} + 3\, M_{^{1}\mathrm{H}} - M_{^{12}\mathrm{C}} - M_{^{4}\mathrm{He}} - \langle E \rangle_{^{15}\mathrm{O}} \,. \tag{10.26}$$

The $^{17}\mathrm{F}$ neutrino flux is practically negligible, having an energy spectrum similar to the $^{15}\mathrm{O}$ neutrino flux, but being about 100 times smaller. Nevertheless, one can calculate the corresponding average thermal energy $\alpha_{^{17}\mathrm{F}}$ from the reactions $^{16}\mathrm{O}(p,\gamma)^{17}\mathrm{F}$ and $^{17}\mathrm{F} \to {}^{17}\mathrm{O} + e^+ + \nu_e$:

$$\alpha_{^{17}\mathrm{F}} = M_{^{16}\mathrm{O}} + M_{^{1}\mathrm{H}} - M_{^{17}\mathrm{O}} - \langle E \rangle_{^{17}\mathrm{F}} \,. \tag{10.27}$$

Neglecting the small fraction of energy carried away by neutrinos, the luminosity constraint gives the approximate value of the total solar neutrino flux $\Phi = \sum_r \Phi_r$:

$$\Phi \simeq \frac{2\, K_{\odot}}{Q} = 6.4 \times 10^{10}\, \mathrm{cm}^{-2}\, \mathrm{s}^{-1} \,. \tag{10.28}$$

The real total flux of solar neutrinos must have a value slightly larger than that in eqn (10.28), since more reactions are needed in order to produce the observed luminosity if some fraction of the energy is carried away by neutrinos. Comparing the total neutrino flux in the BP00 SSM reported in Table 10.3 with the minimum value in eqn (10.28), one can see that about 2% of the thermonuclear energy produced in the Sun is carried away by neutrinos.

---

[53] A reaction $a + X \to Y + b$, in which a particle a strikes a nucleus X producing a nucleus Y and a new particle b, is often symbolized by $X(a,b)Y$.

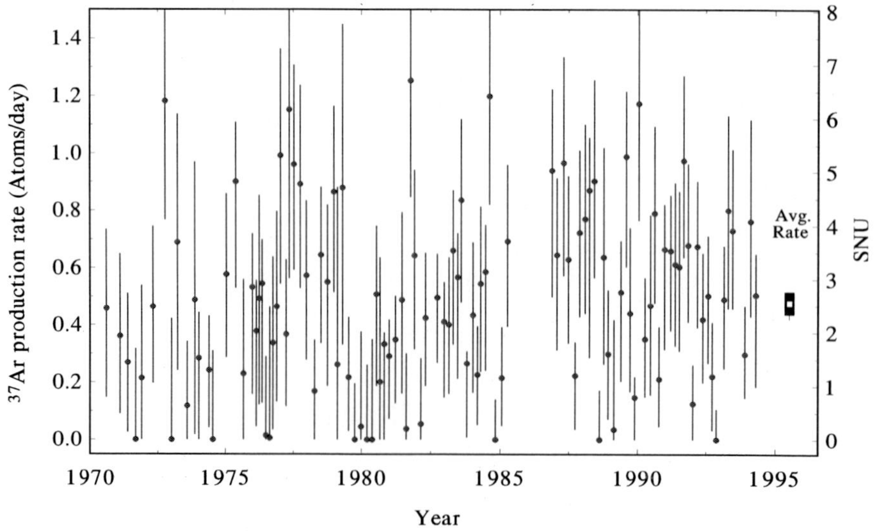

FIG. 10.8. Results of the 108 individual solar neutrino observations made with the Homestake chlorine experiment from March 1970 to February 1994 [323]. The uncertainty shown for individual measurements is statistical only and is significantly non-Gaussian for results near zero. The uncertainty shown for the cumulative result is the combination of the statistical and systematic uncertainties in quadrature.

Since the ^{3}He nuclei necessary for the formation of ^{7}Be and ^{8}B and for the *hep* reaction are created by the *pp* or *pep* reactions, there is another model-independent constraint for the solar neutrino fluxes of the *pp* chain (see [144]):

$$\Phi_{^7\mathrm{Be}} + \Phi_{^8\mathrm{B}} + \Phi_{hep} \leq \Phi_{pp} + \Phi_{pep} \,. \tag{10.29}$$

10.4 Homestake experiment

The pioneering Homestake experiment [323] is a radiochemical experiment which detects solar neutrinos through the Pontecorvo–Alvarez inverse β-decay Cl-Ar reaction [878, 78]

$$\nu_e + {}^{37}\mathrm{Cl} \to {}^{37}\mathrm{Ar} + e^- \,, \tag{10.30}$$

with neutrino energy threshold $E_\nu^{\mathrm{th}} = 0.814\,\mathrm{MeV}$. Therefore, this experiment can detect only intermediate and high-energy neutrinos. Since the detection cross-section increases with the neutrino energy, the main contribution to the Homestake event rate comes from high-energy ^{8}B neutrinos, as shown in Table 10.3.

The Homestake Solar Neutrino Observatory is located 1478 m below the surface, in the Homestake Gold Mine at Lead, in South Dakota, USA, at latitude 44°20′ N and longitude 103°50′ W. The depth of the detector is 4200 ± 100 meters of water equivalent[54] (mwe), such that the intensity of cosmic ray muons is about 4 per square meter per day, with an average energy of about 300 GeV. This cosmic ray flux generates an ^{37}Ar production background of $0.047 \pm 0.013\,\text{atoms}\,\text{day}^{-1}$. The detector consists of a single horizontal steel tank 6.1 m in diameter and 14.6 m long (a volume of 6×10^5 liters), containing 2.16×10^{30} atoms of ^{37}Cl (133 ton) in the form of 615 ton of tetrachloroethylene (C_2Cl_4).

The problem in such an experiment is to collect and count the few atoms of ^{37}Ar produced by solar neutrinos through the Cl-Ar reaction. In the Homestake experiment the Argon in the tank is extracted through chemical methods and the radioactive ^{37}Ar is counted using miniature proportional counters which detect the Auger electron produced in the electron-capture of the ^{37}Ar nuclei with a lifetime of about 35 days[55]. This is the reason why such experiments are called *radiochemical*.

About 0.5 atoms of ^{37}Ar are produced each day by solar neutrinos (the average measured rate is $0.478 \pm 0.030 \pm 0.029\,\text{atoms}\,\text{day}^{-1}$) and about 16 atoms are measured in each extraction about every two months. The number of extracted ^{37}Ar atoms is smaller than the number of atoms produced by solar neutrinos (about 30), because some of them decay before extraction and because the extraction efficiency is about 90%. The choice of a time interval of about two months between extractions is due to the fact that the number of ^{37}Ar in the tank produced by solar neutrinos grows only for about two months. After this time interval the number of ^{37}Ar in the tank remains approximately constant, because the production rate is equal to the decay rate.

The Homestake experiment was proposed in 1964 [354] and built in the period 1965–1967, following the measurement in 1959 of a relatively high value of the cross-section of the process $^3\text{He} + {}^4\text{He} \to {}^7\text{Be} + \gamma$ (more than a thousand times larger than was previously believed), which leads to the *pp*I and *pp*II branches of the *pp* chain. In these branches high-energy ^{7}Be and ^{8}B are produced, which can be detected with the Cl-Ar reaction in eqn (10.30). Furthermore, Bahcall found that the cross-section of the Cl-Ar reaction is greatly enhanced above 5.8 MeV (about 20 times larger than previous calculations). This is due to a superallowed transition of the ground state of the ^{37}Cl nucleus to its isobaric analog ^{37}Ar state [136]. Hence, a chlorine experiment is particularly sensitive to the high-energy part of the solar ^{8}B flux. This was rightly considered as a great chance to test the theory of stellar thermonuclear energy generation.

The first data of the Homestake experiment indicated in 1968 [356] that the solar neutrino flux was less than 3 SNU, well below the rate predicted by the SSM of Ref. [149]. The Homestake collaboration performed several tests (see [355, 323])

---

[54] One meter of water equivalent of a material is the thickness of that material which provides a shielding equivalent to one meter of water. It can easily be converted into *slant depth*, which is usually measured in units of $\text{g}\,\text{cm}^{-2}$, by multiplying for the density of water, which is $1\,\text{g}\,\text{cm}^{-3}$.

[55] The electron-capture process $e^- + {}^{37}\text{Ar} \to {}^{37}\text{Cl} + \nu_e$, which is the inverse of the neutrino capture process in eqn (10.30), occurs through the capture of an inner-shell electron. The orbit left empty by the captured electron is immediately filled by an higher-level electron. The released energy is carried away by the Auger electron, first observed by Pierre V. Auger in the 1920s, which is ejected from a high energy level.

to make sure that the detector was operating properly. In particular, artificially produced Argon was introduced in the tank and fully recovered.

In 1970 a new technique (pulse rise time) that improved the signal versus background discrimination was introduced for the counting of the radioactive ^{37}Ar atoms. Figure 10.8 shows the ^{37}Ar rates measured in 108 extractions from March 1970 to February 1994. The average solar neutrino rate is [323]

$$R^{\text{exp}}_{^{37}\text{Cl}} = 2.56 \pm 0.16 \pm 0.16\,\text{SNU} = 2.56 \pm 0.23\,\text{SNU}\,, \tag{10.31}$$

with the statistical and systematic uncertainties shown separately after the first equality and added in quadrature after the second equality.

Because of the extremely small production rate of ^{37}Ar, the uncertainties associated with the results of the individual extractions shown in Fig. 10.8 are large and dominated by statistical fluctuations. However, the accumulation of data over 25 years of observation has resulted in a statistical uncertainty for the average rate of about 6%, which is about equal to the estimated systematic uncertainty.

The solar neutrino rate measured in the Homestake experiment is about one-third of that predicted by the SSM, with a discrepancy of more than 3σ (see Tables 10.4, 10.3, and 10.6).

10.5 Gallium experiments

There are three gallium solar neutrino experiments: GALLEX/GNO (section 10.5.1) and SAGE (section 10.5.2). These experiments detect solar neutrinos through the reaction [709]

$$\nu_e + {}^{71}\text{Ga} \to {}^{71}\text{Ge} + e^-\,, \tag{10.32}$$

which has the low neutrino energy threshold $E_\nu^{\text{th}} = 0.233\,\text{MeV}$. This makes the detection of solar neutrinos from all sources possible (see Fig. 10.7).

The detection cross-section was calculated by Bahcall in Ref. [147]. The event rate expected from the BP00 SSM is given in Table 10.3. Since the contributions to the total predicted event rate from *pp*, ^{7}Be and ^{8}B neutrinos are, respectively, 54%, 27% and 9%, gallium experiments are very important for the measurement of the neutrino flux produced in the basic *pp* reaction of the *pp* chain, which is closely related to the luminosity of the Sun.

In gallium experiments, the atoms of ^{71}Ge produced by solar neutrinos are extracted with chemical methods from a large detector mass containing about 30–50 ton of ^{71}Ga and counted in small proportional counters by observing their decay back to ^{71}Ga. In each exposure, lasting about 30 days, a few tens of atoms of ^{71}Ge must be extracted from about 10^{29} atoms of ^{71}Ga.

10.5.1 GALLEX/GNO

The GALLium EXperiment (GALLEX) [91, 90, 92, 93, 95, 585, 588] was located in the Laboratori Nazionali del Gran Sasso (LNGS), in Italy, having an overhead

TABLE 10.6. Neutrino capture reaction, neutrino energy threshold, and ratio of the measured rate ($R^{\rm exp}$) and the best-fit BP04 SSM prediction ($R_{\rm bf}^{\rm BP04}$) [152] (see Table 10.4) for the solar neutrino experiments. The last column gives the number of standard deviations n_σ of the discrepancy between the measured and predicted rated, calculated as $n_\sigma = \left(R_{\rm bf}^{\rm BP04} - R_{\rm bf}^{\rm exp}\right) / \left(\sqrt{(\delta R^{\rm BP04})^2 + (\delta R^{\rm exp})^2}\right)$. In the ES and NC reactions $\alpha = e, \mu, \tau$.

Experiment	Reaction	$E_\nu^{\rm th}$ (MeV)	$R^{\rm exp}/R_{\rm bf}^{\rm BP04}$	n_σ
GALLEX/GNO [77]	$\nu_e + {}^{71}{\rm Ga} \to {}^{71}{\rm Ge} + e^-$	0.233	0.529 ± 0.042	5.4
SAGE [19]			0.540 ± 0.040	5.0
Homestake [323]	$\nu_e + {}^{37}{\rm Cl} \to {}^{37}{\rm Ar} + e^-$	0.814	0.301 ± 0.027	3.3
Kamiokande [475]	$\nu_\alpha + e^- \to \nu_\alpha + e^-$	6.7	0.484 ± 0.066	2.2
SK [625]		4.7	0.406 ± 0.014	2.6
SNO – D_2O [43]	$\nu_e + d \to p + p + e^-$	6.9	0.304 ± 0.019	3.0
	$\nu_\alpha + d \to p + n + \nu_\alpha$	2.224	0.879 ± 0.111	0.5
	$\nu_\alpha + e^- \to \nu_\alpha + e^-$	5.7	0.413 ± 0.047	2.5
SNO – NaCl [40]	$\nu_e + d \to p + p + e^-$	6.9	0.290 ± 0.017	3.1
	$\nu_\alpha + d \to p + n + \nu_\alpha$	2.224	0.853 ± 0.075	0.6
	$\nu_\alpha + e^- \to \nu_\alpha + e^-$	5.7	0.406 ± 0.046	2.5

shielding of 3300 mwe. The detector consisted of 101 ton of a liquid gallium chloride ($GaCl_3$-HCl) solution containing 30.3 ton of gallium. GALLEX operated from May 1991 to January 1997 and was followed by the Gallium Neutrino Observatory (GNO), which operated from May 1998 to April 2003 with the same detector and an improved extraction equipment [76, 77].

The functionality of GALLEX was successfully tested by exposing the detector to an intense artificial ^{51}Cr neutrino source[56] [94, 586]. There have been two exposures: the first from June to October 1994 and the second from October 1995 to February 1996. The combined ratio of the measured and predicted neutrino capture rates in the two source experiments is

$$R_{^{71}{\rm Ga}}^{\rm GALLEX}(^{51}{\rm Cr}) = 0.93 \pm 0.08\,. \tag{10.33}$$

Moreover, the ^{71}Ge recovery efficiency has been tested by adding to the gallium solution known amounts of ^{71}As, which decays through electron capture and positron

[56] ^{51}Cr decays through electron capture ($e^- + {}^{51}{\rm Cr} \to {}^{51}{\rm V} + \nu_e$), with a Q-value of 752.73 ± 0.24 keV [1] and a half-life of 27.7025 ± 0.0024 days [125].

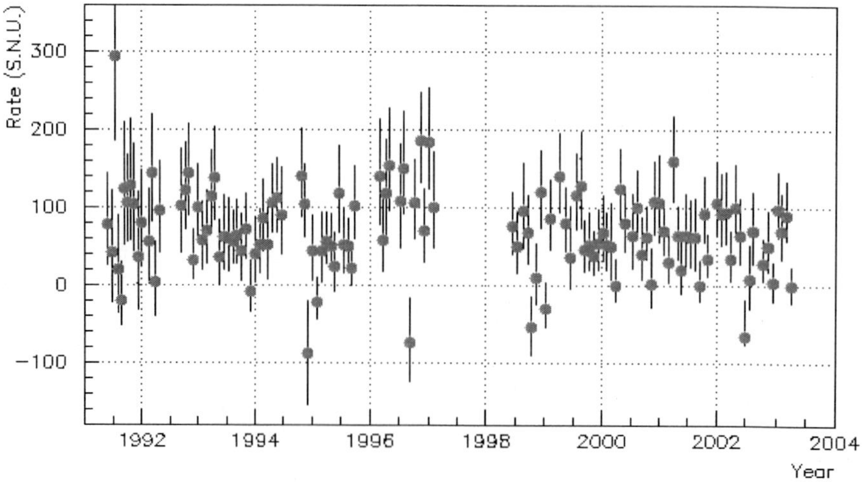

FIG. 10.9. Results of the 123 extraction runs of GALLEX/GNO from May 1991 to April 2003 [77]. The error bars show the statistical uncertainty.

emission into ^{71}Ge with a half-life of 65.28 ± 0.15 days [125]. The measured recovery rate was 0.998 ± 0.008 [587]. These results led the GALLEX collaboration to conclude that the detector was working properly.

The GALLEX/GNO experiment measured the solar neutrino flux from May 1991 to April 2003 with a total of 123 extraction runs [77]. Figure 10.9 shows the single-run results. The resulting average solar neutrino capture rate is [77]

$$R_{^{71}\mathrm{Ga}}^{\mathrm{GALLEX/GNO}} = 69.3 \pm 4.1 \pm 3.6\,\mathrm{SNU} = 69.3 \pm 5.5\,\mathrm{SNU}\,, \tag{10.34}$$

where the statistical and systematic uncertainties have been added in quadrature. This rate is about half of that predicted by the SSM, with a discrepancy of more than 5σ (see Tables 10.4, 10.3, and 10.6).

10.5.2 SAGE

The Soviet–American Gallium Experiment (SAGE) [12, 14, 18, 17, 19, 21] is located in the Baksan Neutrino Observatory (BNO) of the Russian Academy of Sciences in the northern Caucasus mountains. BNO is situated 3.5 km from the entrance of a horizontal adit excavated into the side of Mount Andyrchi. It is at a depth of about 2000 m from the top of the mountain, with an overhead shielding of about 4700 mwe, which reduces the cosmic muon flux to $(3 \pm 0.15) \times 10^{-9}\,\mathrm{muons\,cm^{-2}\,s^{-1}}$. In order to reduce the neutron and gamma background from the rock, the BNO laboratory is entirely lined with 60 cm of low-radioactivity concrete, with an outer 6 mm steel shell.

The experiment began data-taking in 1990. From December 1994 to May 1995, the functionality of the SAGE experiment was successfully tested with an intense artificial ^{51}Cr neutrino source [15, 16]. The ratio of the measured and predicted

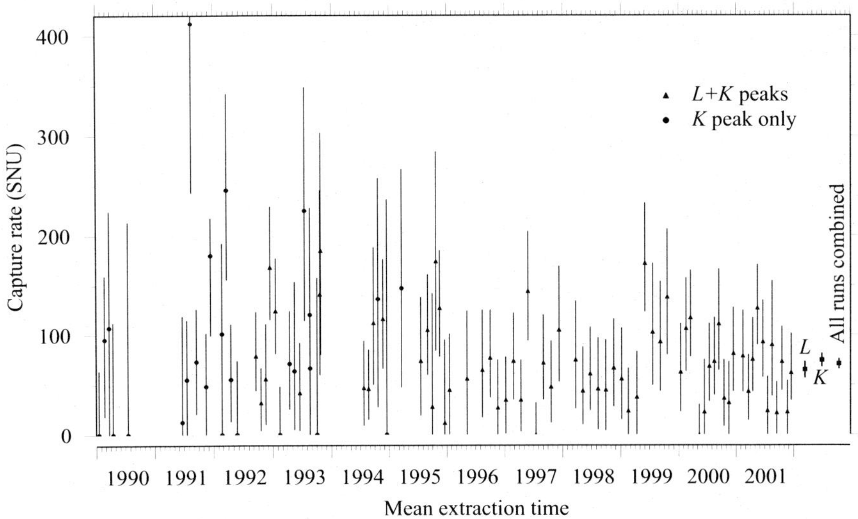

FIG. 10.10. Results of the 92 extraction runs of SAGE from January 1990 to December 2001 [19]. The error bars show the statistical uncertainty.

neutrino capture rates has been found to be [16]

$$R^{\text{SAGE}}_{^{71}\text{Ga}}(^{51}\text{Cr}) = 0.95^{+0.11+0.06+0.035}_{-0.10-0.05-0.027} = 0.95^{+0.13}_{-0.12}\,, \tag{10.35}$$

where, after the first equality, the first uncertainty is statistical, the second is systematic, and the third is theoretical. The three uncertainties have been added in quadrature after the second equality. As the value of $R(^{51}\text{Cr})$ is consistent with one, the SAGE collaboration concluded that the experiment was working properly.

More recently, the SAGE collaboration performed another test of the experiment from April to September 2004 with an artificial ^{37}Ar source[57] [20]. The ratio of the measured and predicted neutrino capture rates is [20]

$$R^{\text{SAGE}}_{^{71}\text{Ga}}(^{37}\text{Ar}) = 0.79^{+0.09}_{-0.10}\,. \tag{10.36}$$

Since this result is almost 2.5σ smaller than unity and the weighted average of the ratios of measured and predicted neutrino capture rates in the GALLEX/GNO and SAGE source experiments is 0.88 ± 0.05, the SAGE collaboration suggested that the calculated detection cross-section may have been slightly overestimated [20].

The capture rates of all SAGE extractions as a function of time are shown in Fig. 10.10. The average neutrino capture rate from January 1990 to December 2001 is [19]

$$R^{\text{SAGE}}_{^{71}\text{Ga}} = 70.8^{+5.3+3.7}_{-5.2-3.2}\,\text{SNU} = 70.8^{+6.5}_{-6.1}\,\text{SNU}\,, \tag{10.37}$$

with the statistical and systematic uncertainties added in quadrature after the second equality. This rate is in agreement with that measured in GALLEX/GNO

[57] ^{37}Ar decays through electron capture ($e^- + {}^{37}\text{Ar} \to {}^{37}\text{Cl} + \nu_e$), with a Q-value of $813.5 \pm 0.3\,\text{keV}$ [1] and a half-life of 35.04 ± 0.04 days [125].

(eqn (10.34)). It is about half of that predicted by the SSM, with a discrepancy of about 5σ (see Tables 10.4, 10.3, and 10.6).

10.6 Water Cherenkov detectors

Water Cherenkov detectors allow the detection of neutrinos in real time by observing the tracks of the ultrarelativistic charged leptons produced by neutrino interactions. In general, when a charged particle passes with velocity $v > 1/n$ through a medium with index of refraction n, the particle emits Cerenkov light in a cone around the direction of motion. The half-opening angle θ of the cone is given by $\cos\theta = 1/nv$ and the spectrum is given by

$$\frac{\mathrm{d}N}{\mathrm{d}\lambda\,\mathrm{d}x} = 2\pi\alpha\left[1 - \left(\frac{1}{nv}\right)^2\right]\lambda^{-2}, \tag{10.38}$$

where N is the number of photons, λ is the wavelength, and x is a coordinate along the track.

Water has an index of refraction $n \simeq 1.33$ [400], leading to $\theta \simeq 41°$ for relativistic particles. For every cm of track length about 340 photons are produced in the wavelength range between 300 and 600 nm, which is appropriate for detection by photomultiplier tube (PMT). Through observation of these photons, with a precise determination of the arrival time at each PMT, it is possible to determine the neutrino interaction point, the direction of the track of the produced charged lepton and its energy.

In underground water Cherenkov detectors a large mass of water is surrounded by an array of PMTs. In order to have a good reconstruction of the neutrino events, it is important to have a substantial coverage of the enclosing surface, of the order of 20%.

In the following subsections we describe the setup and the results of the three solar neutrino water Cherenkov experiments: Kamiokande, Super-Kamiokande, and SNO.

10.6.1 Kamiokande

The Kamiokande detector was originally built for the search of nucleon decay. In fact, the name is a contraction of *Kamioka Nucleon Decay Experiment.* It was located in the Kamioka mine in Japan, about 1000 m underground, with an overburden of 2600 mwe.

The first phase of the Kamiokande experiment, called Kamiokande-I, started in 1983. The detector had a water mass of 3000 ton in a cylindrical steel tank of 15.6 m diameter and 16 m height. An array of 1000 very large PMTs, with a diameter of 50 cm, covered 20% of the inner surface (one PMT every square meter). The detector had an inner fiducial volume of about 1 kton of water equipped to observe events with energy of the order of 1 GeV, which was appropriate for searching nucleon decays.

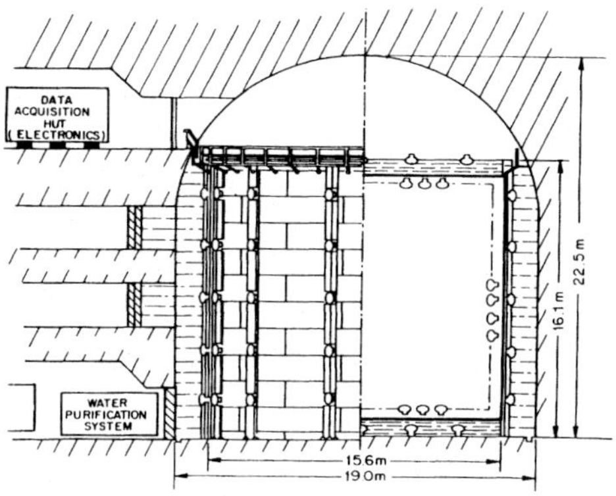

FIG. 10.11. The Kamiokande-II detector [619].

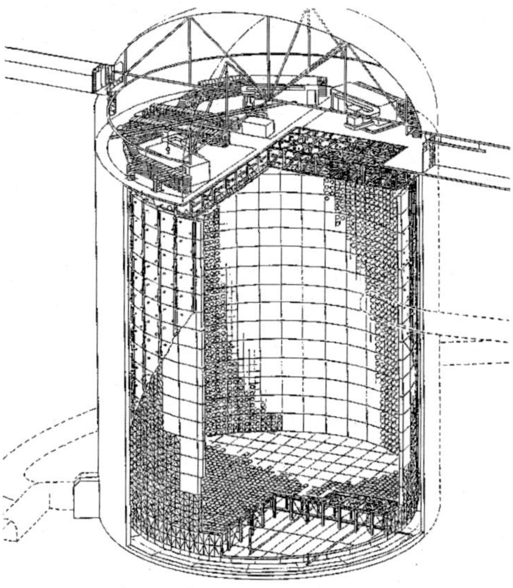

FIG. 10.12. The Super-Kamiokande detector [111].

In 1986 the detector was upgraded to Kamiokande-II in order to be able to observe ^{8}B solar neutrinos [616, 617, 618, 620, 619], which generate events with an energy of the order of 10 MeV. In order to lower the threshold to such low energies, the background had to be reduced. For this purpose, the detector was enlarged, dividing it into an inner detector and an outer detector which acted as an active shield, which was able to reject background events through anticoincidence. In order to achieve this configuration, the bottom layer of PMTs was lifted by 1.4 m and the space between the tank and the cave wall on the sides and on the top of the tank was filled with water, providing a water shield of about 1.7 m on the sides and 1 m on the top. In this configuration, the inner volume measured 14.4 m in diameter and 13.1 m in height, contained 2142 ton of purified water, and was equipped with 948 PMTs giving a 20% surface coverage. The outer volume, containing 1500 ton of water, was equipped with 123 PMTs, one PMT every $9\,\mathrm{m}^2$. The fiducial volume was defined in different ways for different measurements. For solar neutrino detection it was defined as the volume of $680\,\mathrm{m}^3$ lying 2.0 m inside of the bottom and side inner detector and 3.14 m below the top of the inner detector. The Kamiokande-II is shown schematically in Fig. 10.11. A detailed description of the Kamiokande experiment can be found in Ref. [695].

From 1987, the Kamiokande-II experiment observed ^{8}B solar neutrinos with a recoil electron energy threshold $E_e^{\mathrm{th}} = 9.3\,\mathrm{MeV}$. In 1990 about 100 dead PMTs were replaced, starting the final phase of the experiment, called Kamiokande-III, which lasted until 1996, when the Super-Kamiokande experiment started. The energy threshold was lowered to 7.5 MeV during Kamiokande-II and further lowered to 7.0 MeV during Kamiokande-III [475]. Using eqn (5.31), one can find that the three neutrino energy thresholds corresponding to $E_e^{\mathrm{th}} = 9.3\,, 7.5\,, 7.0\,\mathrm{MeV}$ are, respectively, $E_\nu^{\mathrm{th}} = 9.0\,, 7.2\,, 6.7\,\mathrm{MeV}$.

Kamiokande measured the solar neutrino flux through the elastic scattering reaction

$$\nu_\alpha + e^- \to \nu_\alpha + e^- \,, \tag{10.39}$$

which is mainly sensitive to electron neutrinos, whose cross-section is about six times larger than those of muon and tau neutrinos (see eqn (5.18)). Since the recoil electron has a sharp forward peak, one can distinguish solar neutrino events from the isotropic background by measuring the directional correlation of the recoil electron and the Sun.

The average ^{8}B neutrino flux measured in the Kamiokande experiment from January 1987 to February 1995 (2079 days) is [475]

$$\Phi_{^8\mathrm{B}}^{\mathrm{Kam}} = (2.80 \pm 0.19 \pm 0.33) \times 10^6\,\mathrm{cm}^{-2}\,\mathrm{s}^{-1} = (2.80 \pm 0.38) \times 10^6\,\mathrm{cm}^{-2}\,\mathrm{s}^{-1}\,, \tag{10.40}$$

where we have added the statistical and systematic uncertainties in quadrature. In eqn (10.40) it is assumed that the ^{8}B solar ν_e's do not oscillate into other flavors on their way from the production region in the core of the Sun to the detector on the Earth. In this case, the flux of ^{8}B solar ν_e's measured by Kamiokande is about half of the SSM flux, with a discrepancy of more than 2σ (see Tables 10.4, 10.3, and 10.6).

The Kamiokande measurement of solar neutrinos, which covers an entire cycle of the solar sunspot activity, does not show a correlation of the solar neutrino flux with the sunspot activity [475].

10.6.2 Super-Kamiokande

Super-Kamiokande (SK) is a 50 kton water Cherenkov detector located in the Kamioka mine, about 500 m from the cavity which was previously occupied by the Kamiokande detector and now contains the KamLAND experiment (see section 12.2.3). The SK detector, shown schematically in Fig. 10.12, consists of two concentric, optically separated, water Cherenkov detectors contained in a cylindrical steel tank with a diameter of 39.3 m and a height of 42 m. The inner detector has a diameter of 33.8 m and a height of 36.2 m. Its surface is equipped with 11146 large PMTs with a diameter of 50 cm, giving a coverage of 40%. The surface of the outer detector is instrumented with 1885 PMTs with a diameter of 20 cm.

The first phase of the SK experiment, called SK-I, started in April 1996 and ended in July 2001. On 12 November 2001, during the refill of the detector after a maintenance drain, an accidental explosion of a phototube triggered the explosion of about half of the phototubes. In the second phase, called SK-II, from January 2003 to February 2004, the inner detector was equipped with the surviving 5182 PMTs, having a coverage of 19%. This was enough to run the second phase of the K2K experiment (see section 12.3.2). In the following we will present only the main results of SK-I, which have been published by the SK collaboration [478, 480, 482, 472, 471, 473, 507, 1081, 968, 746, 625].

The SK-I experiment collected 1496 days of solar neutrino data with a recoil electron energy threshold $E_e^{\mathrm{th}} = 6.5\,\mathrm{MeV}$ for the first 280 days and $E_e^{\mathrm{th}} = 5.0\,\mathrm{MeV}$

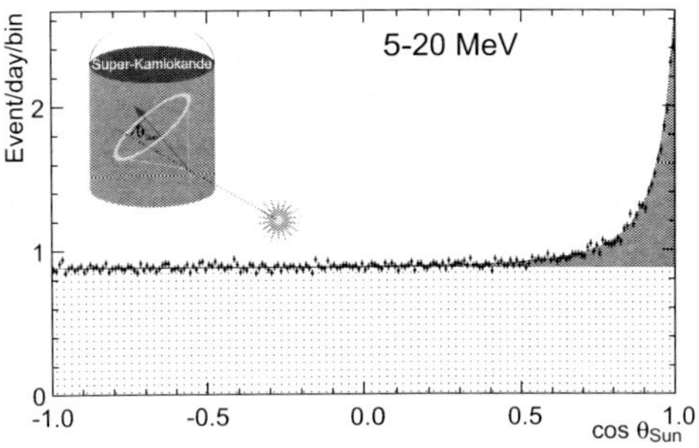

FIG. 10.13. Angular distribution of the SK solar neutrino event candidates [625]. $\theta_{\rm sun}$ is the angle between the reconstructed recoil electron direction and the direction of the Sun, which is the incoming direction of solar neutrinos. The shaded area indicates the elastic scattering peak. The dotted area is the contribution from background events.

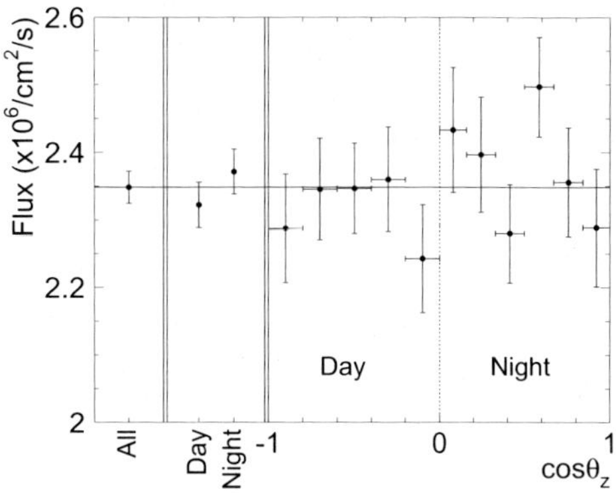

FIG. 10.14. The solar zenith angle dependence of the solar neutrino flux measured in SK (error bars show statistical error) [625]. The width of the night-time bins was chosen to separate solar neutrinos that pass through the Earth's dense core (the right-most night bin) from those that pass through the mantle. The horizontal line shows the flux for all data.

for the remaining 1216 days. The corresponding neutrino energy thresholds are, from eqn (5.31), $E_\nu^{\rm th} = 6.2\,{\rm MeV}$ and $E_\nu^{\rm th} = 4.7\,{\rm MeV}$.

Figure 10.13 shows the angular distribution of solar neutrino event candidates with respect to the direction of the Sun. One can see the forward peak in the direction of the Sun due to the elastic scattering[58] of solar neutrinos. The width of the peak is due to the multiple scattering of the recoil electron and to the angular resolution of the detector. The dotted area is the contribution from background events. A fit of the distribution in Fig. 10.13 with an isotropic background yielded $22404 \pm 226^{+784}_{-717}$ solar neutrino events. In the absence of neutrino oscillations[59], the measured interaction rate corresponds to [625]

$$\Phi_{^8{\rm B}}^{\rm SK} = (2.35 \pm 0.02 \pm 0.08) \times 10^6\,{\rm cm}^{-2}\,{\rm s}^{-1} = (2.35 \pm 0.08) \times 10^6\,{\rm cm}^{-2}\,{\rm s}^{-1}\,, \quad (10.41)$$

where we have added the statistical and systematic uncertainties in quadrature. This flux is smaller than half of the SSM flux, with a discrepancy of more than 2σ (see Tables 10.4, 10.3, and 10.6).

Figure 10.14 shows the solar zenith angle dependence of the solar neutrino flux measured in SK. The day and night fluxes could be different because of the Earth

[58] From eqn (5.29) one can see that, for $E_\nu \gg m_e$ the differential cross-section is strongly peaked in the forward direction.

[59] In the extraction from the data of the ^{8}B flux in eqn (10.41) it is assumed that all solar neutrinos are ν_e.

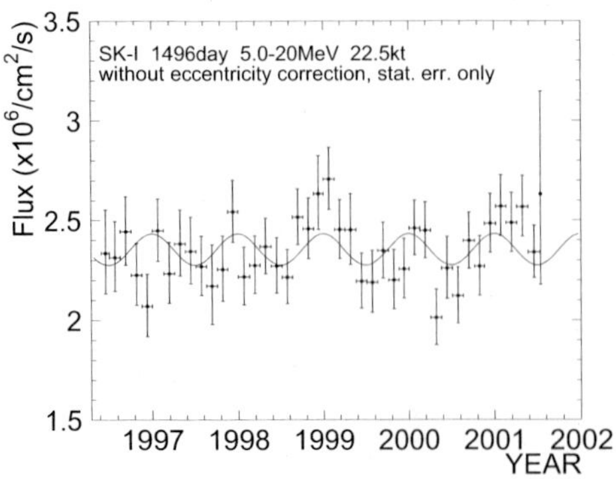

FIG. 10.15. Solar neutrino flux measured in SK as a function of time [625]. The binning of the horizontal axis is 1.5 months. The solid line is the prediction based on the eccentricity of the Earth's orbit.

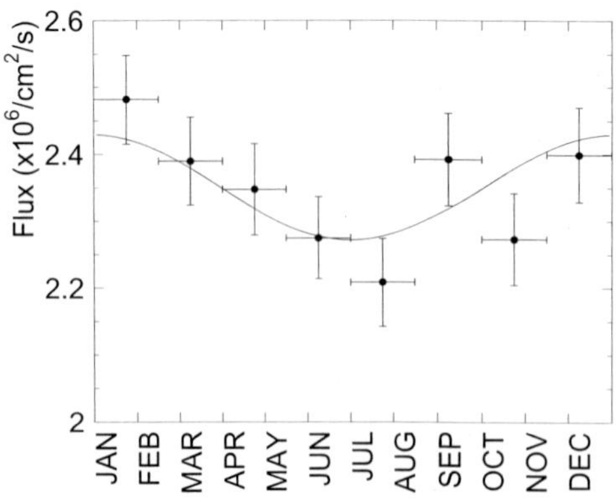

FIG. 10.16. The seasonal variation of the solar neutrino flux measured in SK [625]. The solid line is the prediction based on the eccentricity of the Earth's orbit.

matter effect for neutrinos which pass through the Earth during the night (see section 10.9). The measured day and night fluxes are [625]

$$\Phi_{^8\text{B}}^{\text{SK,day}} = (2.32 \pm 0.03^{+0.08}_{-0.07}) \times 10^6 \, \text{cm}^{-2} \, \text{s}^{-1} \,, \tag{10.42}$$

$$\Phi_{^8\text{B}}^{\text{SK,night}} = (2.37 \pm 0.03^{+0.08}_{-0.08}) \times 10^6 \, \text{cm}^{-2} \, \text{s}^{-1} \,, \tag{10.43}$$

with an asymmetry

$$\mathcal{A}_{\text{day-night}}^{\text{SK}} \equiv \frac{\Phi_{^8\text{B}}^{\text{SK,day}} - \Phi_{^8\text{B}}^{\text{SK,night}}}{\frac{1}{2}\left(\Phi_{^8\text{B}}^{\text{SK,day}} + \Phi_{^8\text{B}}^{\text{SK,night}}\right)} = -0.021 \pm 0.020^{+0.013}_{-0.012} \,. \tag{10.44}$$

Hence, the data are compatible with a null asymmetry.

Figure 10.15 shows the solar neutrino flux measured in SK as a function of time. The seasonal variation of the flux is displayed in Fig. 10.16. One can see that the temporal dependence of the solar neutrino flux is in agreement with the seasonal variation due to the eccentricity of the Earth's orbit, which causes a modulation of the flux proportional to $1/L^2$, where L is the Sun–Earth distance. Since the eccentricity is about 1.7%, the predicted flux variation is about 7%. From the data, there is no indication of an additional seasonal variation of the flux which could be due to vacuum oscillations (see section 10.7).

The energy spectrum of the recoil electrons is depicted in Fig. 10.17, where it is compared with the Monte Carlo prediction based on the BP04 SSM. One can see that there is no significant distortion of the shape of the measured spectrum with respect to the prediction. Therefore, the possible energy dependence of the rate suppression is severely limited.

The SK collaboration searched for a possible signal of *hep* neutrinos, whose flux is expected to be about three orders of magnitudes smaller than the ^{8}B neutrino

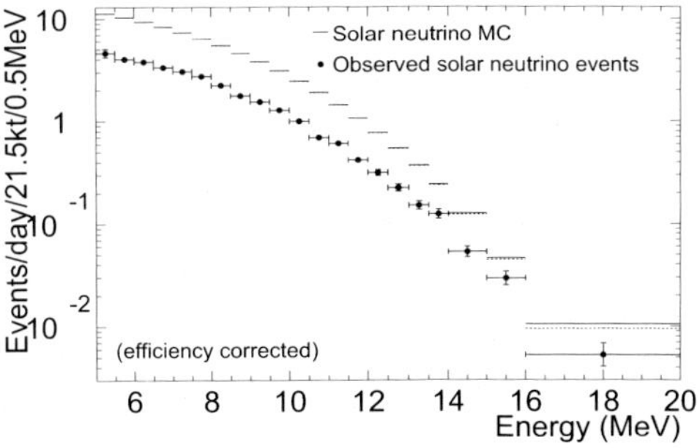

FIG. 10.17. Energy spectrum of the SK solar neutrino signal [625]. The horizontal axis gives the total energy of the recoil electrons. The error bars represent the statistical and uncorrelated systematic uncertainties added in quadrature. The BP04 SSM flux values are used for the ^{8}B and *hep* Monte Carlo (MC) fluxes. The dashed lines shows the contribution of only ^{8}B.

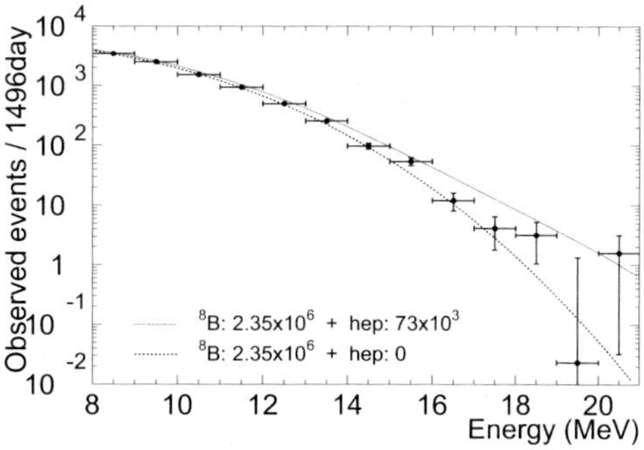

FIG. 10.18. Energy spectrum of the SK recoil electrons in the high-energy region [625]. The points show data with statistical error bars. The curves show the expected spectra for $\Phi_{hep}{}^{\mathrm{SK}} = 73 \times 10^3\,\mathrm{cm}^{-2}\,\mathrm{s}^{-1}$ (solid line) and $\Phi_{hep}{}^{\mathrm{SK}} = 0$ (dashed line). The ^{8}B flux is assumed to be equal to the best-fit value (eqn (10.41)), without spectral distortions with respect to the SSM prediction.

flux. The contribution of *hep* neutrinos can be investigated through a study of high-energy events, because the *hep* neutrino has an end-point of about 18.8 MeV, which is higher than the end-point of about 16 MeV of the ^{8}B neutrino spectrum (see Figs. 10.6 and 10.7). Since no significant excess of high-energy events was observed (see Fig. 10.18), the flux of *hep* neutrinos has been limited by

$$\Phi_{hep}{}^{\mathrm{SK}} < 73 \times 10^3\,\mathrm{cm}^{-2}\,\mathrm{s}^{-1} \quad (90\%\ \mathrm{CL})\,, \tag{10.45}$$

which is compatible with the SSM.

10.6.3 SNO

The Sudbury Neutrino Observatory (SNO) is a real-time heavy-water Cherenkov detector located in the Creighton mine (INCO Ltd.), near Sudbury (Ontario, Canada) [258]. It consists of one kiloton of 99.92% isotopically pure D_2O contained inside a spherical 12 m diameter acrylic vessel. Cherenkov light is detected by 9456 20-cm photomultiplier tubes (PMTs) mounted on a spherical stainless steel geodesic structure with a diameter of 18 m. The acrylic vessel and the geodesic sphere are immersed in ultrapure water (H_2O) which provides shielding against radioactive background from the geodesic structure and the cavity rock. Schematic views of the SNO laboratory and detector are shown in Fig. 10.19. The detector is at a depth of 2092 m, which correspond to 6010 meters of water equivalent. The background of cosmic ray muons is about 65 per day.

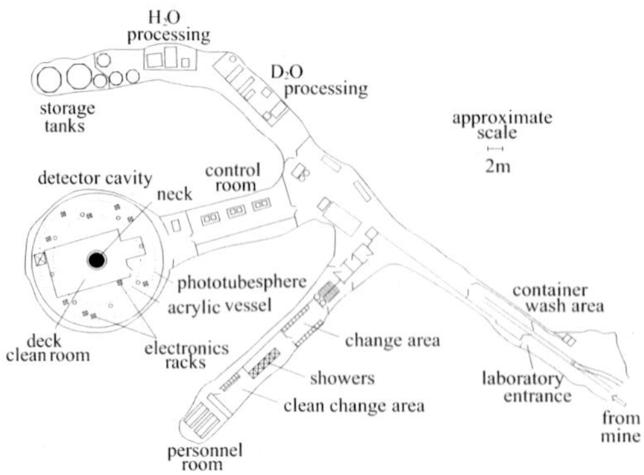

(a) Layout of the SNO laboratory.

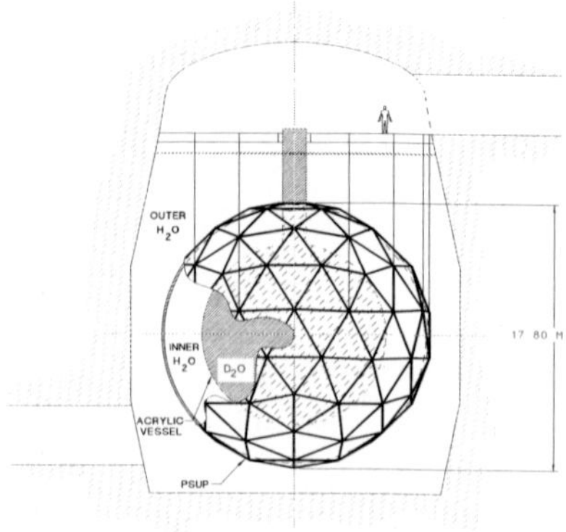

(b) The SNO detector.

FIG. 10.19. The Sudbury Neutrino Observatory [258].

The SNO experiment detects solar neutrinos through the three reactions [42, 43, 44, 45, 38, 40, 39]

$$\text{CC:} \qquad \nu_e + d \to p + p + e^- \,, \tag{10.46}$$

$$\text{NC:} \qquad \nu_\alpha + d \to p + n + \nu_\alpha \,, \tag{10.47}$$

$$\text{ES:} \qquad \nu_\alpha + e^- \to \nu_\alpha + e^- \,, \tag{10.48}$$

with $\alpha = e, \mu, \tau$.

The charged-current (CC) reaction on deuterium in eqn (10.46) is used to find the energy spectrum of electron neutrinos above the energy threshold $E_\nu^{\text{th,CC}} = 2m_p + m_e - m_d = 1.442\,\text{MeV}$ by measuring the kinetic energy T_e of the final electron: $E_\nu = T_e + E_\nu^{\text{th,CC}}$. However, since at low energies there are high backgrounds, the SNO experiment has a threshold of 5.5 MeV for T_e, which corresponds to a neutrino energy threshold of 6.9 MeV. Therefore, the SNO CC reaction is sensitive only to ^{8}B solar neutrinos.

The neutral-current (NC) reaction on deuterium in eqn (10.47) is extremely important for checking the neutrino oscillation solution of the SNP, because it is equally sensitive to all active neutrinos [313]. Hence, the NC reaction can measure the total flux of active neutrinos coming from the Sun. It has a neutrino energy threshold $E_\nu^{\text{th,NC}} = m_p + m_n - m_d = 2.224\,\text{MeV}$ and it is also sensitive only to ^{8}B solar neutrinos (see Table 10.2).

The elastic scattering (ES) reaction on electrons in eqn (10.48) is the same as that used in H_2O water Cherenkov detectors. The electron kinetic energy threshold is the same as in the CC process, $T_e^{\text{th}} = 5.5\,\text{MeV}$. Since the corresponding neutrino energy threshold is $E_\nu^{\text{th,ES}} = 5.7\,\text{MeV}$, the SNO ES reaction is also sensitive only to ^{8}B solar neutrinos.

The SNO experiment is divided in three phases:

D_2O Phase. In this phase the final neutron in the NC process in eqn (10.47) was detected through the reaction with deuterium

$$n + d \to {}^3\mathrm{H} + \gamma\,(6.25\,\mathrm{MeV})\,. \tag{10.49}$$

This phase operated for 306.4 live days, from 2 November 1999 to 28 May 2001 [42, 43].

NaCl Phase. In this so-called *salt phase*, about 2 ton of NaCl have been added to the heavy water in order to detect the final neutron in the NC process in eqn (10.47) with the reaction

$$n + {}^{35}\mathrm{Cl} \to {}^{36}\mathrm{Cl} + \text{several } \gamma\text{'s}\,(8.57\,\mathrm{MeV})\,. \tag{10.50}$$

This reaction improved SNO's detection of solar neutrinos in the following three ways. First, the neutron capture efficiency increased (the thermal neutron capture cross-section of ^{35}Cl is 44 b, whereas that of deuterium is 0.5 mb), allowing a significant statistical improvement of the measurement of the NC process. Second, the higher total γ-ray energy allowed a precise measurement well above the low-energy radioactive backgrounds. Third, the isotropy of the hit distribution of the PMTs from the multiple γ-rays emitted after neutron capture on ^{35}Cl is significantly different from that produced by the Cherenkov light emitted by a single relativistic electron. Hence, the signals produced in the CC and NC reactions can be separated with good accuracy. This phase operated for 391.4 live days, between 26 July 2001 and 28 August 2003 [45, 40].

Third Phase. The SNO collaboration hung in a grid within the heavy water three hundred ^{3}He proportional counter tubes. Helium-3 has a very large cross-section for the capture of thermal neutrons, which produces an energetic proton–triton pair resulting in an electrical pulse in the counter wire. In this phase, which started in January 2005 and is currently operating, the measurement of the NC process will be further improved.

The results from the first D_2O phase [42, 43, 44] confirmed the deficit of solar electron neutrinos observed previously. The NC measurement of the total flux of active neutrinos demonstrated that about two electron solar neutrinos out of three change their flavor to ν_μ or ν_τ on their way from the center of the Sun to the Earth. These important results have been confirmed by the more precise salt phase data [45, 40]. Hence, the results of the SNO experiment have proved that the SNP is due to neutrino flavor transitions.

In the SNO salt phase 2176 ± 78 CC events, 2010 ± 85 NC events, and 279 ± 26 ES events have been observed, corresponding, respectively, to the equivalent fluxes of ^{8}B electron neutrinos [40]

$$\Phi_{\mathrm{CC}}^{\mathrm{SNO}} = \left(1.68 \pm 0.06^{+0.08}_{-0.09}\right) \times 10^6\,\mathrm{cm}^{-2}\,\mathrm{s}^{-1}\,, \tag{10.51}$$

$$\Phi_{\mathrm{NC}}^{\mathrm{SNO}} = \left(4.94 \pm 0.21^{+0.38}_{-0.34}\right) \times 10^6\,\mathrm{cm}^{-2}\,\mathrm{s}^{-1}\,, \tag{10.52}$$

$$\Phi_{\mathrm{ES}}^{\mathrm{SNO}} = (2.35 \pm 0.22 \pm 0.15) \times 10^6\,\mathrm{cm}^{-2}\,\mathrm{s}^{-1}\,. \tag{10.53}$$

These are the fluxes of ^{8}B electron neutrinos, with the energy spectrum of ^{8}B decay, which would be needed to fit the SNO data. However, one can immediately see that

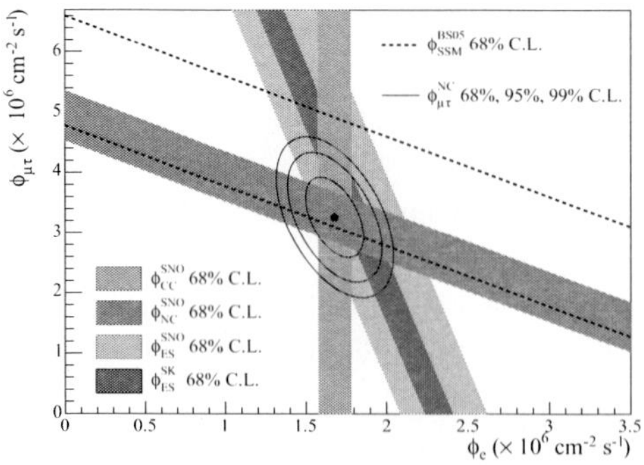

FIG. 10.20. Flux of ν_μ and ν_τ ($\phi_{\mu\tau} = \Phi^{\text{SNO}}_{\nu_\mu,\nu_\tau}$) versus flux of ν_e ($\phi_e = \Phi^{\text{SNO}}_{\nu_e}$) [40]. CC, NC and, ES flux measurements are indicated by the filled bands. The total ^{8}B solar neutrino flux predicted by the BS05 $\equiv$ BSB05(GS98) SSM [153] (see Table 10.4) is represented by the band enclosed by the dashed lines, and that measured with the NC channel is shown as the solid band parallel to the model prediction. The narrow band parallel to the SNO ES result corresponds to the Super-Kamiokande result in Ref. [473].

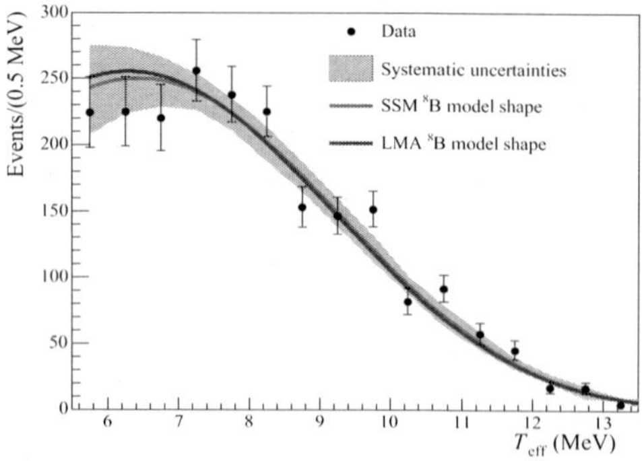

FIG. 10.21. Spectrum of SNO CC events as a function of $T_{\text{eff}} \equiv T_e$ [153], compared to that predicted with the best-fit LMA parameters (see section 10.10) and that predicted by the SSM. Only statistical uncertainties are shown in the data spectrum. The shadowed band represents the 1σ uncertainty determined from detector systematic uncertainties.

they are not compatible with each other. This can also be seen from the ratio [40]

$$\frac{\Phi^{\text{SNO}}_{\text{CC}}}{\Phi^{\text{SNO}}_{\text{NC}}} = 0.340 \pm 0.023^{+0.029}_{-0.031}\,, \tag{10.54}$$

which deviates from unity by 17 standard deviations! Such a large discrepancy implies that SNO data exclude that ^{8}B solar electron neutrinos arrive on the Earth unchanged. About two thirds of the ^{8}B solar electron neutrino flux are converted into ν_μ or ν_τ on their way to the Earth. Assuming that the energy spectrum of the electron neutrino flux on the Earth is not distorted with respect to the ^{8}B decay spectrum, the CC flux in eqn (10.51) corresponds to the electron neutrino flux:

$$\Phi^{\text{SNO}}_{\text{CC}} = \Phi^{\text{SNO}}_{\nu_e}\,. \tag{10.55}$$

The flux $\Phi^{\text{SNO}}_{\nu_\mu,\nu_\tau}$ of μ and τ neutrinos on the Earth can be determined by subtracting $\Phi^{\text{SNO}}_{\nu_e}$ from the NC and ES fluxes, which are given by

$$\Phi^{\text{SNO}}_{\text{NC}} = \Phi^{\text{SNO}}_{\nu_e} + \Phi^{\text{SNO}}_{\nu_\mu,\nu_\tau}\,, \tag{10.56}$$

TABLE 10.7. Day and night fluxes and asymmetries measured in SNO [153].

Type of signal	$\Phi^{\rm SNO,day}$ [cm^{-2} s^{-1}]	$\Phi^{\rm SNO,night}$ [cm^{-2} s^{-1}]	$\mathcal{A}^{\rm SNO}_{\rm day-night}$
CC	$1.73 \pm 0.09 \pm 0.10$	$1.64 \pm 0.08 \pm 0.09$	$-0.056 \pm 0.074 \pm 0.053$
NC	$4.81 \pm 0.31 \pm 0.39$	$5.02 \pm 0.29 \pm 0.41$	$0.042 \pm 0.086 \pm 0.072$
ES	$2.17 \pm 0.34 \pm 0.14$	$2.52 \pm 0.32 \pm 0.16$	$0.146 \pm 0.198 \pm 0.033$

$$\Phi^{\rm SNO}_{\rm ES} = \Phi^{\rm SNO}_{\nu_e} + 0.1553\, \Phi^{\rm SNO}_{\nu_\mu,\nu_\tau} \,. \tag{10.57}$$

The resulting values for the flux of μ and τ neutrinos on the Earth are

$$\Phi^{\rm SNO,NC}_{\nu_\mu,\nu_\tau} = \left(3.26 \pm 0.25^{+0.40}_{-0.35}\right) \times 10^6\,{\rm cm}^{-2}\,{\rm s}^{-1}\,, \tag{10.58}$$

$$\Phi^{\rm SNO,ES}_{\nu_\mu,\nu_\tau} = \left(4.36 \pm 1.52^{+0.90}_{-0.87}\right) \times 10^6\,{\rm cm}^{-2}\,{\rm s}^{-1}\,, \tag{10.59}$$

which are in good agreement with each other. Figure 10.20 shows $\Phi^{\rm SNO}_{\nu_\mu,\nu_\tau}$ versus $\Phi^{\rm SNO}_{\nu_e}$ and illustrates the agreement between $\Phi^{\rm SNO}_{\rm NC}$ and the total ^{8}B solar neutrino flux predicted by the BSB05(GS98) SSM [153]. From Table 10.6 one can see that the SNO NC data are also compatible with the BP04 SSM prediction, whose difference from the BSB05(GS98) SSM is compatible with the theoretical uncertainties.

Table 10.7 gives the day and night fluxes measured in the SNO salt phase [153], together with the resulting asymmetries, defined similarly to that in eqn (10.44). One can see that there is no significant day–night asymmetry, within the uncertainties.

The energy spectrum of SNO CC events is depicted in Fig. 10.21. A statistical analysis [153] show that the data are compatible with the SSM prediction and with the slightly distorted LMA solution of the SNP (see section 10.10), both shown in the figure.

10.7 Vacuum oscillations

In 1967 Pontecorvo [883] predicted that solar electron neutrinos may be converted into ν_μ's (or sterile neutrinos) because of oscillations on their way from the Sun to the Earth. He considered only the possibility of an averaged constant suppression of the solar electron neutrino flux. In 1968 the first data of the Homestake experiment [356] indicated that the solar neutrino flux on the Earth was significantly smaller than that predicted by the SSM [149]. This discrepancy was called the *solar neutrino problem* (SNP). This confirmation of Pontecorvo's brilliant idea prompted Gribov and Pontecorvo [567] to elaborate the theory of neutrino oscillations considering the general case of two-neutrino mixing. Also other authors discussed the possibility of neutrino oscillations as the most simple and attractive solution of the SNP (see Ref. [237]).

In the simplest case of two-neutrino mixing, the survival probability of solar electron neutrinos due to vacuum oscillations is given by

$$P_{\nu_e \to \nu_e}(E, L(t)) = 1 - \sin^2 2\vartheta \, \sin^2 \left(\frac{\Delta m^2 \, L(t)}{4 \, E} \right) , \tag{10.60}$$

where $L(t)$ is the Sun–Earth distance, which depends on the time t of the year due to the eccentricity of the Earth's orbit:

$$L(t) = L_0 \left[1 - \epsilon \, \cos \left(2\pi \, \frac{t}{T} \right) \right] . \tag{10.61}$$

Here $L_0 = 1\,\text{au}$ (see Table 10.1), $\epsilon = 0.0167$ is the eccentricity of the Earth's orbit, $T = 1\,\text{yr}$ (see Table 10.1) and $t = 0$ at the perihelion (in early January). Hence, a fingerprint of vacuum oscillations would be the observation of a seasonal variation of the rates measured in solar neutrino experiments (see Refs. [567, 237, 427, 776, 358] and references therein).

Since no significant seasonal variation was observed in solar neutrino data and the measured energy spectra do not show a distortion compatible with eqn (10.60), the vacuum oscillation solution of the SNP is currently disfavored. However, the data can be explained by neutrino oscillations in matter, as explained in the following sections.

10.8 Resonant flavor transitions in the Sun

Solar neutrinos are produced in the solar core, where the matter density is about $150\,\text{g}\,\text{cm}^{-3}$ and the electron density is about $100\,N_{\text{A}}\,\text{cm}^{-3}$, corresponding to the initial charged-current potential $V_{\text{CC}}^{(\text{i})} \sim 10^{-11}\,\text{eV}$. For solar neutrinos with energy of the order of $10\,\text{MeV}$, we have $A_{\text{CC}}^{(\text{i})} \sim 10^{-4}\,\text{eV}^2$, which means that the density is larger than the resonance density if $\Delta m^2 \cos 2\vartheta \lesssim 10^{-4}\,\text{eV}^2$. In this case, solar neutrinos can undergo MSW resonant transitions on their way towards the surface of the Sun. Depending on the values of the electron density profile, the neutrino energy and the mixing parameters Δm^2 and ϑ, the resonance can be crossed adiabatically or nonadiabatically, leading to different amounts of flavor transitions.

Figure 10.22a shows the SSM electron density profile in the Sun [145], which, in the interval $0.1 \lesssim r/R_\odot \lesssim 0.9$, is well approximated by the exponential

$$N_e(r) = N_e(0) \exp\left(-\frac{r}{r_0} \right) , \quad \text{with} \quad N_e(0) = 245\,N_{\text{A}}/\text{cm}^3 \quad \text{and} \quad r_0 = \frac{R_\odot}{10.54} , \tag{10.62}$$

where r is the distance from the center of the Sun and $R_\odot$ is the solar radius. Hence, solar neutrino data can be analyzed with a good approximation by using the Parke formula in eqn (9.95) with the crossing probability in eqn (9.96) and the parameter F in the third row of Table 9.1. In order to calculate the adiabaticity parameter at the resonance, the logarithmic derivative $|\text{d} \ln N_e / \text{d}x|_{\text{R}}$ is needed. The authors of Ref. [744] suggested the practical prescription, verified with numerical

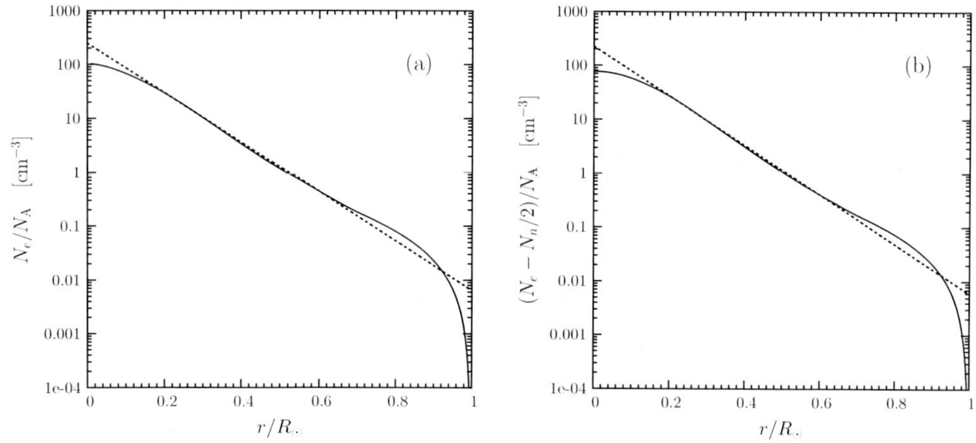

FIG. 10.22. (a) SSM electron density profile in the Sun as a function of the ratio $r/R_\odot$ [145] (solid curve). The straight dotted line represents the approximation in eqn (10.62). (b) The solid curve represents the density $N_e - N_n/2$, which replaces N_e in the case of $\nu_e \to \nu_s$ transitions. The straight line represents the approximation $223 \exp(-r/r_0)$, with r_0 given in eqn (10.62).

solutions of the differential evolution equation, to calculate it numerically from the SSM electron density profile for $r \leq 0.904 R_\odot$ and take the constant value $18.9/R_\odot$ for $r > 0.904 R_\odot$, where the exponential approximation in eqn (10.62) breaks down. Figure 10.22a shows the density $N_e - N_n/2$ as a function of $r/R_\odot$ [145]. This density must be used in place of N_e for the calculation of $\nu_e \to \nu_s$ transitions, where ν_s is a sterile neutrino. The figure shows that, in the interval $0.1 \lesssim r/R_\odot \lesssim 0.9$, the density $N_e - N_n/2$ is also well approximated by an exponential with the same slope as N_e.

Figure 10.23a shows the conventional acronyms for the regions in the $\tan^2 \vartheta$-Δm^2 plane obtained from the analysis of solar neutrino data. The Small Mixing Angle (SMA) region is the one where the mixing angle is very small and the resonant enhancement of flavor transitions due to the MSW effect is more efficient. However, as explained later, there is currently very strong evidence in favor of the large mixing angle (LMA) region, in which both the mixing angle and Δm^2 are large. Other regions with large mixing are: the low Δm^2 (LOW) region, the quasi-vacuum-oscillations (QVO) region, and the vacuum oscillations region (VAC). In the SMA, LMA, and LOW regions, vacuum oscillations due to the Sun–Earth distance are not observable because Δm^2 is too large and interference effects are washed out by the average over the energy resolution of the detector (in these cases the Parke formula in eqn (9.95) applies). In the QVO region, both matter effects and vacuum oscillations are important [462, 453, 463, 744]. In the VAC region matter effects are negligible and the vacuum oscillations discussed in section 10.7 are dominant.

In the following, we will illustrate the main features of the matter effects in the SMA, LMA, and LOW regions using the formalism derived in section 9.3.

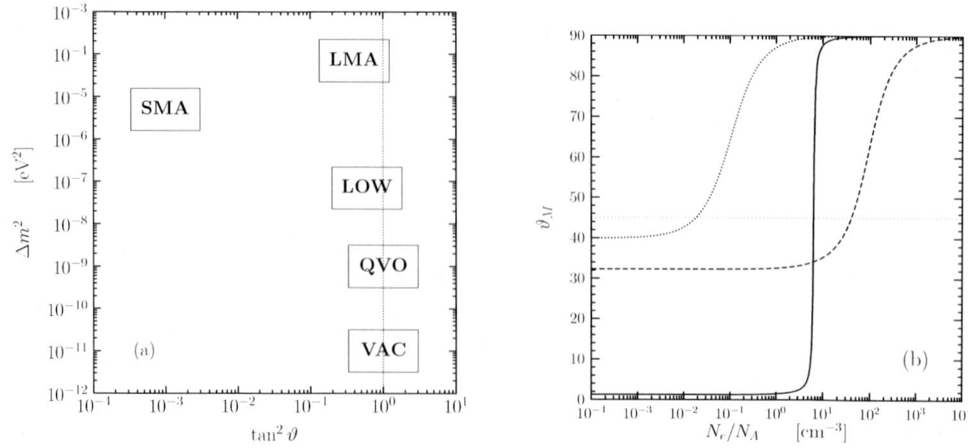

FIG. 10.23. (a) The conventional names for regions in the $\tan^2 \vartheta$-Δm^2 plane obtained from the analysis of solar neutrino data. The vertical dotted line correspond to maximal mixing. (b) Effective mixing angle in matter as a function of $N_e/N_{\rm A}$ for $E = 5\,\text{MeV}$. The horizontal dotted line correspond to maximal mixing (resonance). Solid line: $\Delta m^2 = 5 \times 10^{-6}\,\text{eV}^2$, $\tan^2 \vartheta = 5 \times 10^{-4}$ (typical SMA). Dashed line: $\Delta m^2 = 7 \times 10^{-5}\,\text{eV}^2$, $\tan^2 \vartheta = 0.4$ (typical LMA). Dotted line: $\Delta m^2 = 8 \times 10^{-8}\,\text{eV}^2$, $\tan^2 \vartheta = 0.7$ (typical LOW).

Figure 10.23b shows the effective mixing angle in matter as a function of $N_e/N_{\rm A}$ for three typical SMA, LMA, and LOW solutions of the SNP. One can see that when the mixing angle is very small, as in the SMA case, the size of the resonance region, where the value of the effective mixing angle changes from its vacuum value to $\simeq 90^\circ$, is narrow. On the other hand, if the mixing angle is large as in the LMA and LOW solutions, the transition of the effective mixing angle from its vacuum value to $\simeq 90^\circ$ occurs over some orders of magnitude of $N_e/N_{\rm A}$.

In Fig. 10.24 we have plotted the effective squared masses in matter

$$m^2 = \frac{1}{2}\left[\Delta m^2 + A_{\rm CC} \pm \Delta m^2_{\rm M}\right] , \tag{10.63}$$

where we assumed that $m_1 = 0$ and we neglected the neutral-current contribution, which does not affect the effective squared-mass difference $\Delta m^2_{\rm M}$ given in eqn (9.67). From Fig. 10.24a one can see that the mass gap at the resonance is rather small in the case of a typical SMA solution of the SNP. This means that the crossing of the resonance can likely be nonadiabatic, with large $\nu_1^{\rm M} \leftrightarrows \nu_2^{\rm M}$ transitions. This is also seen in Fig. 10.25a, in which the solid line shows the value of the resonance adiabaticity parameter $\gamma_{\rm R}$ in eqn (9.97) for the same typical SMA solution. One can see that $\gamma_{\rm R}$ is smaller than unity in the energy range of solar neutrinos, signaling nonadiabatic transitions. The solid line in Fig. 10.25b shows that the crossing probability $P_{\rm c}$ is large for $E \gtrsim 1\,\text{MeV}$. On the other hand, from Figs. 10.24b and c, one can see that the mass gap at the resonance is quite wide in the case of typical LMA and LOW solutions. Figures 10.25b and 10.25c show that the resonance crossing is

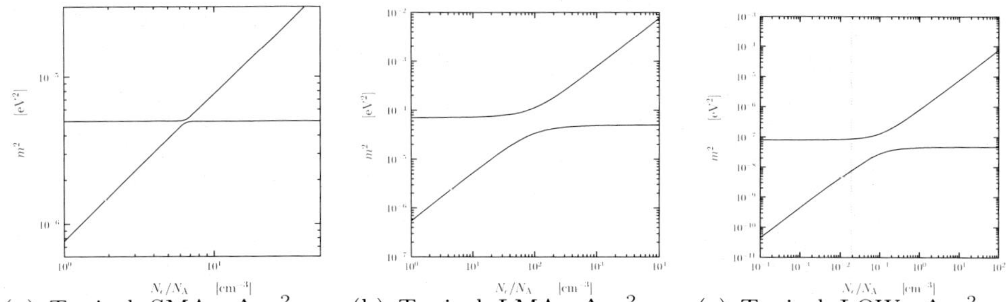

(a) Typical SMA: $\Delta m^2 = 5 \times 10^{-6}\,\text{eV}^2$, $\tan^2\vartheta = 5 \times 10^{-4}$.

(b) Typical LMA: $\Delta m^2 = 7 \times 10^{-5}\,\text{eV}^2$, $\tan^2\vartheta = 0.4$.

(c) Typical LOW: $\Delta m^2 = 8 \times 10^{-8}\,\text{eV}^2$, $\tan^2\vartheta = 0.7$.

FIG. 10.24. Effective squared masses in matter in eqn (10.63) as functions of N_e/N_A for $E = 5\,\text{MeV}$. The dotted vertical lines show the location of the resonance (eqn (9.69)), where the effective squared-mass difference Δm^2_M in eqn (9.67) is minimal (in b and c, the location of the resonance appears off-center because of the logarithmic scale).

always adiabatic in the case of the LMA solution, whereas in the LOW case it is adiabatic for $E \lesssim 10\,\text{MeV}$.

Figure 10.25c shows the survival probability of solar electron neutrinos as a function of energy for the same typical SMA, LMA, and LOW solutions of the SNP discussed above. The plotted survival probability is given by the Parke formula in eqn (9.95) modified in order to take into account the fact that for sufficiently low-energy neutrinos the matter potential in the center of the Sun is below the resonance value:

$$\overline{P}_{\nu_e \to \nu_e} = \frac{1}{2} + \left(\frac{1}{2} - P_\text{c}\, \theta(A^0_\text{CC} - \Delta m^2 \cos 2\vartheta) \right) \cos 2\vartheta^{(\text{i})}_\text{M} \cos 2\vartheta \,, \qquad (10.64)$$

where A^0_CC is the value of A_CC in the center of the Sun and $\theta(x) = \frac{1}{2}(1 + x/|x|)$. This is a rather crude but reasonably effective approximation [697]. For simplicity, we considered only neutrinos produced at the center of the Sun, whereas in a realistic calculation the distribution of neutrino production in the core of the Sun must be taken into account. From Fig. 10.25c one can see that the solar electron neutrino survival probability is quite different in the typical SMA, LMA, and LOW solutions.

In the case of the SMA solution, low-energy neutrinos are created below the resonance and their survival probability is practically equal to unity. For $0.3\,\text{MeV} \lesssim E \lesssim 1\,\text{MeV}$, neutrinos are created above the resonance and cross the resonance region adiabatically, leading to large flavor conversion. For $E \gtrsim 1\,\text{MeV}$, the resonance crossing is nonadiabatic and the survival probability increases with energy.

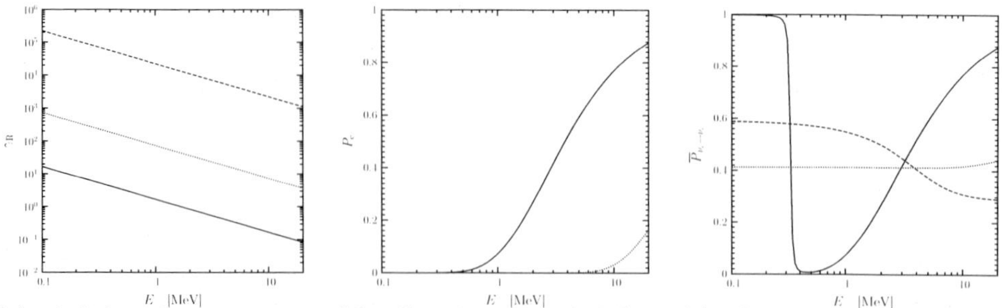

(a) Adiabaticity parameter at the resonance (eqn (9.97)). (b) Crossing probability (eqn (9.96)). (c) Survival probability (eqn (9.95)).

FIG. 10.25. Solid line: $\Delta m^2 = 5 \times 10^{-6}\,\text{eV}^2$, $\tan^2\vartheta = 5 \times 10^{-4}$ (typical SMA). Dashed line: $\Delta m^2 = 7 \times 10^{-5}\,\text{eV}^2$, $\tan^2\vartheta = 0.4$ (typical LMA). Dotted line: $\Delta m^2 = 8 \times 10^{-8}\,\text{eV}^2$, $\tan^2\vartheta = 0.7$ (typical LOW).

In the case of the LMA solution, there is no resonance for $E \lesssim 2\,\text{MeV}$, where the survival probability is equal to its average value in vacuum (see section 7.4),

$$\overline{P}_{\nu_e\to\nu_e} \simeq \overline{P}^{\text{vac}}_{\nu_e\to\nu_e} = 1 - \frac{1}{2}\sin^2 2\vartheta\,. \tag{10.65}$$

For $E \gtrsim 2\,\text{MeV}$, the resonance is crossed adiabatically ($P_c \ll 1$ in eqn (10.64)), leading to

$$\overline{P}_{\nu_e\to\nu_e} \simeq \frac{1}{2} + \frac{1}{2}\cos 2\vartheta^{(i)}_{\text{M}}\cos 2\vartheta\,. \tag{10.66}$$

For $E \gg 2\,\text{MeV}$, the central density of the Sun is much larger than the resonance density. In this case, $\vartheta^{(i)}_{\text{M}} \simeq \pi/2$ and the survival probability is independent of energy,

$$\overline{P}_{\nu_e\to\nu_e} \simeq \sin^2\vartheta\,. \tag{10.67}$$

These characteristics of the LMA solution are relevant for the analysis of the present-day solar neutrino data [151], which strongly support the LMA solution (see section 10.10).

In the case of the LOW solution, there is a resonance for all solar neutrino energies and the resonance is always crossed adiabatically, leading to a flat survival probability, except for a small increase for $E \gtrsim 10\,\text{MeV}$.

The different behavior of the survival probability as a function of energy depending on the values of the mixing parameters implies that different solutions can be distinguished by measuring the solar neutrino flux on the Earth at different energies, either with a detector which is energy-sensitive (as real-time water Cherenkov detectors) or with energy-insensitive detectors with different energy thresholds (as radiochemical detectors).

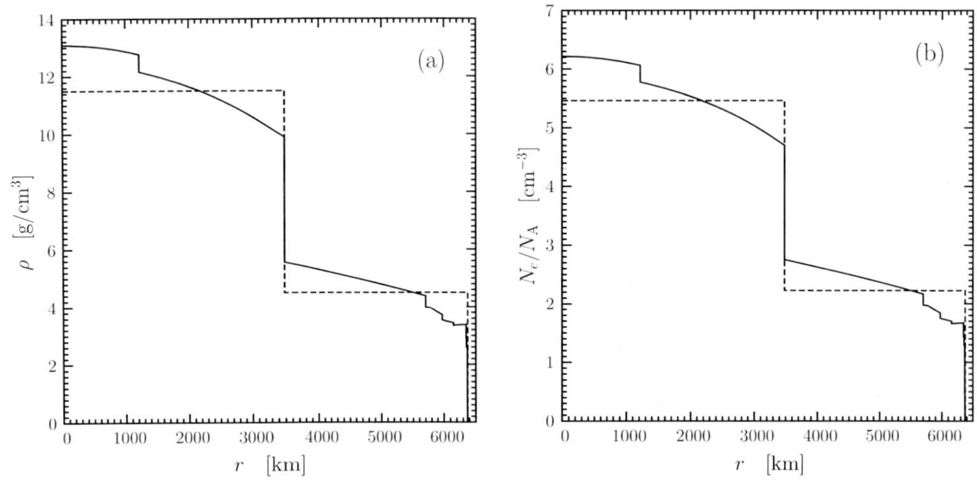

FIG. 10.26. Solid lines: (a) Earth's density ρ as a function of the radius r [86]; (b) electron number density $N_e = N_A\,(\rho/\mathrm{g})\,\langle Z/A\rangle$, with $\langle Z/A\rangle = 0.475$ for $r \leq 3480\,\mathrm{km}$ (core) and $\langle Z/A\rangle = 0.495$ for $r > 3480\,\mathrm{km}$ (mantle). Dashed lines: Two-steps approximation $\rho_{\mathrm{core}} = 11.5\,\mathrm{g\,cm^{-3}}$ and $\rho_{\mathrm{mantle}} = 4.5\,\mathrm{g\,cm^{-3}}$.

10.9 Regeneration of solar ν_e's in the Earth

Solar neutrino detectors are located in underground laboratories at latitudes around 35°-45° in the northern hemisphere. Therefore, during the day solar neutrinos come from above, crossing the atmosphere and the thin part of the Earth crust above the detector. The matter effects generated during these passages are negligible. On the other hand, during the night, solar neutrinos come from below, passing through the Earth. In this case, matter effects in the Earth may not be negligible and may cause the so-called ν_e *regeneration in the Earth*. This mechanism can generate a day–night asymmetry of solar neutrino data if

$$A_{\mathrm{CC}} \sim \Delta m^2 \cos 2\vartheta\,, \tag{10.68}$$

according to the resonance criterion in eqn (9.69). Since the electron number density in the Earth core is about $6\,N_A\,\mathrm{cm}^{-3}$, there can be an Earth matter effect for solar neutrinos of energy of the order of 1 MeV if

$$\Delta m^2 \cos 2\vartheta \sim 10^{-6}\,\mathrm{eV}^2\,. \tag{10.69}$$

The probability of solar electron neutrino survival after passing through the Earth can be calculated as follows. For values of Δm^2 as large as those in eqn (10.69), solar neutrinos arrive at the Earth as an effectively incoherent sum of the two mass eigenstates ν_1 and ν_2 [373]. In this case, the survival probability after the Earth crossing is given by

$$P^{\mathrm{Sun+Earth}}_{\nu_e\to\nu_e} = P^{\mathrm{Sun}}_{\nu_e\to\nu_1}\,P^{\mathrm{Earth}}_{\nu_1\to\nu_e} + P^{\mathrm{Sun}}_{\nu_e\to\nu_2}\,P^{\mathrm{Earth}}_{\nu_2\to\nu_e}\,, \tag{10.70}$$

where $P^{\text{Sun}}_{\nu_e\to\nu_k}$ is the probability of $\nu_e \to \nu_k$ transitions of solar neutrinos arriving on the Earth and $P^{\text{Earth}}_{\nu_k\to\nu_e}$ is the probability of $\nu_k \to \nu_e$ transitions during the Earth crossing. Unitarity implies that

$$P^{\text{Sun}}_{\nu_e\to\nu_1} = 1 - P^{\text{Sun}}_{\nu_e\to\nu_2}\,, \qquad P^{\text{Earth}}_{\nu_1\to\nu_e} = 1 - P^{\text{Earth}}_{\nu_2\to\nu_e}\,. \tag{10.71}$$

Furthermore, since the survival probability of solar electron neutrinos arriving on the Earth (given by the Parke formula in eqn (10.64)) is given by

$$P^{\text{Sun}}_{\nu_e\to\nu_e} = P^{\text{Sun}}_{\nu_e\to\nu_1}\,\cos^2\vartheta + P^{\text{Sun}}_{\nu_e\to\nu_2}\,\sin^2\vartheta\,, \tag{10.72}$$

we have

$$P^{\text{Sun}}_{\nu_e\to\nu_2} = \frac{\cos^2\vartheta - P^{\text{Sun}}_{\nu_e\to\nu_e}}{\cos 2\vartheta}\,. \tag{10.73}$$

Then, the survival probability after the Earth crossing can be written as [804, 159]

$$P^{\text{Sun+Earth}}_{\nu_e\to\nu_e} = P^{\text{Sun}}_{\nu_e\to\nu_e} + \frac{\left(1 - 2P^{\text{Sun}}_{\nu_e\to\nu_e}\right)\left(P^{\text{Earth}}_{\nu_2\to\nu_e} - \sin^2\vartheta\right)}{\cos 2\vartheta}\,. \tag{10.74}$$

One can see that, if the matter effects in the Earth are negligible, $P^{\text{Earth}}_{\nu_2\to\nu_e} = \sin^2\vartheta$ and $P^{\text{Sun+Earth}}_{\nu_e\to\nu_e}$ is reduced to $P^{\text{Sun}}_{\nu_e\to\nu_e}$. On the other hand, if matter effects in the Earth generate additional $\nu_2 \to \nu_e$ transitions, there is a regeneration of ν_e's in the Earth if $P^{\text{Sun}}_{\nu_e\to\nu_e}$ is small and $\vartheta < \pi/4$.

Since the Earth's density profile, shown in Fig. 10.26, is not a smooth function, the probability $P^{\text{Earth}}_{\nu_2\to\nu_e}$ must be calculated numerically. A good approximation is obtained by approximating the Earth's density profile with a step function and using the solution of the evolution equation discussed in section 9.4. The simplest approximation takes into account only the mantle and the core of the Earth, as shown by the dashed lines in Fig. 10.26, approximating its density with the *one and half period castle wall* profile discussed at the end of section 9.5 (see Refs. [871, 50, 315, 316]). The probability $P^{\text{Earth}}_{\nu_2\to\nu_e}$ can be calculated using the evolution operator in eqn (9.127) taking into account that the initial state is $|\nu_2\rangle = \sin\vartheta|\nu_e\rangle + \cos\vartheta|\nu_\mu\rangle$:

$$\begin{pmatrix}\psi_{2e}\\ \psi_{2\mu}\end{pmatrix} = \mathscr{U}\begin{pmatrix}\sin\vartheta\\ \cos\vartheta\end{pmatrix} = \begin{pmatrix}[\mathscr{U}]_{11}\sin\vartheta + [\mathscr{U}]_{12}\cos\vartheta\\ [\mathscr{U}]_{21}\sin\vartheta + [\mathscr{U}]_{22}\cos\vartheta\end{pmatrix}\,. \tag{10.75}$$

Hence, we have

$$P^{\text{Earth(Mantle+Core)}}_{\nu_2\to\nu_e} = |\psi_{2e}|^2 = \sin^2\vartheta + (W_1)^2\,\cos 2\vartheta + W_1\,W_3\,\sin 2\vartheta\,, \tag{10.76}$$

where we have taken into account the unitarity relation $Z^2 + |\vec{W}|^2 = 1$ for the quantities in eqns (9.128) and (9.129). Note that the first term in eqn (10.76), i.e. $\sin^2\vartheta$, is just the probability to find ν_e in ν_2. The other terms express the effect of oscillations in the Earth.

Using the expression in eqn (10.76), one can calculate rather easily the probability $P^{\text{Earth}}_{\nu_2\to\nu_e}$ for any neutrino path which crosses the mantle and the core of the Earth. For the paths which cross only the mantle, one can use a constant density approximation with the evolution operator in eqn (9.108), which leads to

$$P^{\text{Earth(Mantle)}}_{\nu_2\to\nu_e} = \sin^2\vartheta + \sin^2\phi\,\sin 2\vartheta_{\text{M}}\left(\sin 2\vartheta_{\text{M}}\,\cos 2\vartheta + \cos 2\vartheta_{\text{M}}\,\sin 2\vartheta\right)\,, \tag{10.77}$$

with the phase ϕ given in eqn (9.109) and the effective mixing angle ϑ_{M} given in eqn (9.72).

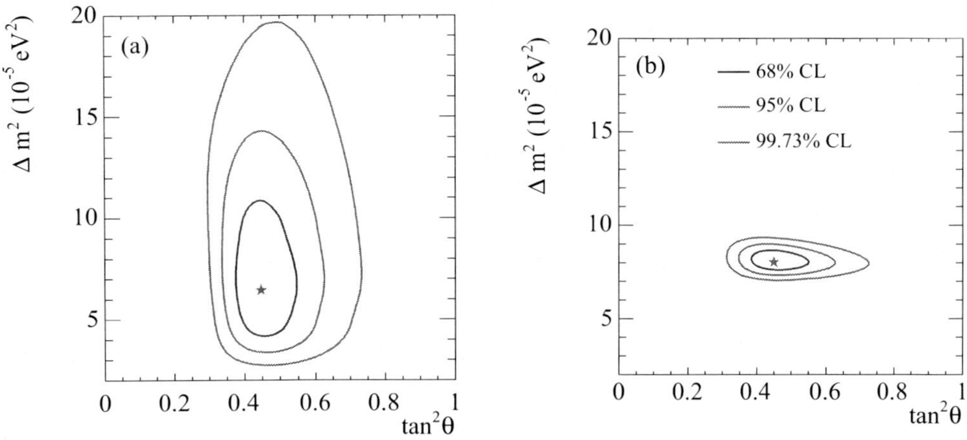

FIG. 10.27. (a) Allowed region in the $\tan^2\vartheta$-Δm^2 plane (LMA) obtained from a global analysis of solar neutrino data [40]. (b) Restricted region obtained by including KamLAND 766 ton-year data [103].

10.10 Global fit of solar neutrino data

A global fit of all solar neutrino data leads to the conclusion that the LMA solution of the SNP is correct. This is shown in Fig. 10.27a, taken from Ref. [40], which depicts the allowed region in the $\tan^2\vartheta$-Δm^2 plane. The solar neutrino data used to calculate this region are: the day and night spectra of SNO's D_2O phase, the day and night CC spectra and ES and NC fluxes of SNO's salt phase, the rates measured in the Homestake [323], SAGE [19], GALLEX/GNO [588, 76], and Super-Kamiokande-I zenith spectra [968]. The ^{8}B flux was left free in the fit, since it is determined by the SNO NC measurement (see Fig. 10.20). The best-fit oscillation parameters with 1σ uncertainties are [40]

$$\Delta m^2 = \left(6.5^{+4.4}_{-2.3}\right) \times 10^{-5}\,\mathrm{eV}^2\,, \qquad \tan^2\vartheta = 0.45^{+0.09}_{-0.08}\,. \tag{10.78}$$

Hence, the mixing angle is quite large, but not maximal ($\tan^2\vartheta = 1$ is excluded at about 6σ).

Figure 10.27b shows that the determination of the allowed region in the $\tan^2\vartheta$-Δm^2 plane can be improved by including KamLAND's data (see section 12.2.3), which leads to a significant restriction of the allowed range of Δm^2. The resulting best-fit oscillation parameters are [40]

$$\Delta m^2 = \left(8.0^{+0.6}_{-0.4}\right) \times 10^{-5}\,\mathrm{eV}^2\,, \qquad \tan^2\vartheta = 0.45^{+0.09}_{-0.07}\,. \tag{10.79}$$

Thus, we do have now a rather precise determination of Δm^2 and ϑ.

11

ATMOSPHERIC NEUTRINOS

> How wonderful that we have met with a paradox. Now we have some hope of making progress.
>
> **Niels Bohr**

Atmospheric neutrinos are created by the interactions of primary cosmic rays with the nuclei in the atmosphere. The primary cosmic rays are mainly composed of protons, with a small component of heavier nuclei. The interactions of these primary cosmic rays with the nuclei in the atmosphere generate secondary cosmic rays, which include all the hadrons and their decay products, with an energy spectrum which has a peak in the GeV range and extends to higher energy with an approximate power law. In particular, many secondary pions are produced. As illustrated in Fig. 11.1, these pions decay mainly into muons and muon neutrinos (see eqns (5.58)–(5.60)):

$$\pi^+ \to \mu^+ + \nu_\mu \,, \qquad \pi^- \to \mu^- + \bar{\nu}_\mu \,. \tag{11.1}$$

At high energies, also kaons contribute to the production of neutrinos and muons. The muons which decay before hitting the ground generate electrons, electron neutrinos, and muon neutrinos through the processes

$$\mu^+ \to e^+ + \nu_e + \bar{\nu}_\mu \,, \qquad \mu^- \to e^- + \bar{\nu}_e + \nu_\mu \,. \tag{11.2}$$

The neutrinos generated in these reactions are called *atmospheric neutrinos*. Those with energies from about 100 MeV to about 100 GeV can be detected in underground laboratories through scattering on nuclei. The detectors must be underground in order to provide good shielding from the flux of secondary cosmic-ray muons, whose frequent interactions would otherwise overwhelm the detector.

From the multiplicities of neutrinos in the processes in eqns (11.1) and (11.2) it is clear that at low energies ($E \lesssim 1\,\text{GeV}$), for which most muons decay before hitting the ground, the neutrino fluxes satisfy the following ratios

$$\frac{\phi_{\nu_\mu} + \phi_{\bar{\nu}_\mu}}{\phi_{\nu_e} + \phi_{\bar{\nu}_e}} \simeq 2 \,, \qquad \frac{\phi_{\nu_\mu}}{\phi_{\bar{\nu}_\mu}} \simeq 1 \,, \qquad \frac{\phi_{\nu_e}}{\phi_{\bar{\nu}_e}} \simeq \frac{\phi_{\mu^+}}{\phi_{\mu^-}} \,. \tag{11.3}$$

At energies higher than 1 GeV the fraction of muons which hit the ground before decaying increases, leading to an increase of the flavor ratio $\phi_{\nu_\mu} + \phi_{\bar{\nu}_\mu} / \phi_{\nu_e} + \phi_{\bar{\nu}_e}$.

The basic principles of detection of atmospheric neutrinos were already known in the 1960s [564, 777, 1084]. In 1965 two groups reported the observation of atmospheric neutrinos by detectors in the Kolar Gold Field Mine in South India [26, 27]

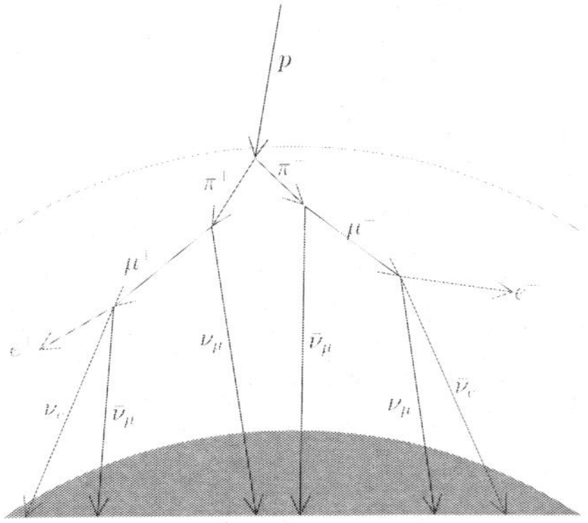

FIG. 11.1. Schematic view of neutrino production by cosmic-ray proton interactions in the atmosphere, with generation of pions and muons.

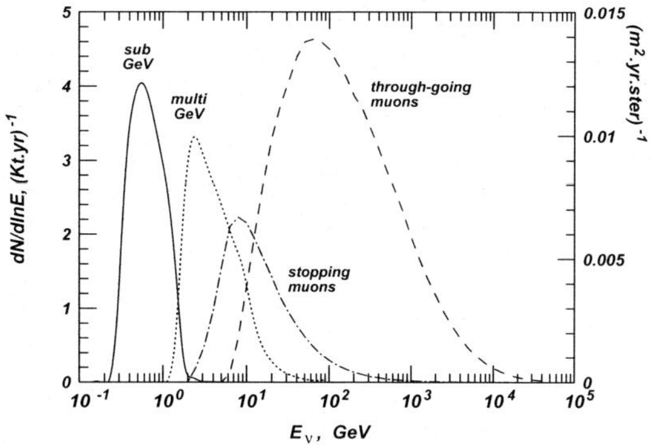

FIG. 11.2. Distributions of neutrino energies that give rise to four classes of events [489]. Sub-GeV and multi-GeV refer to contained events in Kamiokande and Super-Kamiokande. Stopping and through-going muons refer to neutrino-induced muons produced outside the detector.

and in the East Rand Proprietary Gold Mine in South Africa [900]. Both experiments were located very deep underground, with overburdens of about 8000 mwe[60]. The detectors were made of scintillator, which recorded the tracks of muons. Deep underground the residual secondary cosmic-ray muon flux is strongly peaked in the downward-going vertical direction. On the other hand, the atmospheric neutrino flux is almost isotropic and can generate upward-going and horizontal muons by interacting with the rock surrounding the detector. Nowadays detectors can distinguish upward-going muons from the downward-going secondary cosmic-ray muons, but at that time it was only possible to reveal the atmospheric neutrino flux by measuring horizontal muons. The events reported by the Indian and South-African experiments were of horizontal type, with a very low probability of being generated by cosmic-ray muons. In the following years, the observation with scintillator detectors of muons generated in the rock by atmospheric neutrinos continued in India [704, 705], in South Africa [901, 341], in Utah (USA) [211], and in the Baksan Laboratory in Russia [260, 261, 65].

In the second half of the 1980s, atmospheric neutrinos began to be observed by the large underground Kamiokande [614, 840, 621, 474, 603] and IMB [581, 243, 306, 199, 200, 322] water Cherenkov experiments, which have been built for the search of nucleon decay. These detectors could observe events generated by atmospheric neutrino interactions in the detector, as well as upward-going muons generated by atmospheric neutrino interactions in the rock below the detector. Initially, the interactions of atmospheric neutrinos in the detectors were mainly considered as a

[60] For an explanation of mwe units, see footnote 54 on page 367.

background for the nucleon decay searches, although the possibility of atmospheric neutrino oscillations have been studied before [237, 174, 175, 458, 133, 351, 493] (neutrino oscillations have been suggested in Refs. [174, 175] as an explanation of the low fluxes of neutrino-induced muons observed in the Indian [704] and South African [341] experiment on the basis of the analysis in Ref. [1043], although these observations had large uncertainties). To take into account such background, the rates of atmospheric neutrino interactions in the detectors were calculated rather precisely [495, 349, 493, 490, 822]. However, it turned out that the Kamiokande [614, 840, 621, 474, 603] and IMB [581, 306, 200] experiments observed a number of atmospheric muon neutrino interactions which was significantly smaller than the predicted one. This deficit was commonly called for many years the *atmospheric neutrino anomaly*. It was rather controversial, because the *fine-grained* iron tracking detectors NUSEX [185, 186, 33, 34] and Frejus [209, 210, 352] did not see any anomaly. Moreover, the IMB experiment found an anomaly only for the contained events with energy below 1.5 GeV [581, 306, 200], whereas the upward-going muon data [199] and the partially contained events with energy above 0.95 GeV [322] did not show any deviation from expectations.

The breakthrough for the solution of the atmospheric neutrino anomaly in favor of neutrino oscillations came from the observation of an up-down asymmetry of high-energy events generated by atmospheric muon neutrinos in the Super-Kamiokande water Cherenkov detector [476]. This detector started operations in 1996 as a follow-up of the Kamiokande experiment in the Kamioka mine in Japan. Until 2001, it has provided high-statistics data on the atmospheric neutrino flux [477, 479, 476, 481, 486, 483, 470, 110, 111], which allowed the Super-Kamiokande collaboration to determine that the oscillations are from muon to tau neutrinos [470] and to infer the values of the oscillation parameters with unprecedented accuracy [111]. These results are in agreement with those obtained with the fine-grained iron tracking Soudan 2 [67, 68, 922, 69] and MACRO [41, 81, 82, 83, 84] detectors. Furthermore, the values of the atmospheric neutrino oscillation parameters are in agreement with the observed disappearance of muon neutrinos in the accelerator long-baseline K2K experiment [48, 46, 47, 66].

Figure 11.2 shows the distributions of neutrino energies that generate four classes of events in atmospheric neutrino detectors (Sub-GeV and multi-GeV contained events in Kamiokande and Super-Kamiokande and stopping and through-going muons produced outside the detector). One can see that the range of neutrino energies in atmospheric experiments is huge, covering about five orders of magnitude. The range of pathlengths of atmospheric neutrinos is also very wide, going from about 20 km for vertical downward-going neutrinos to about 1.3×10^4 km for vertical upward-going neutrinos. These characteristics imply that the ratio L/E, which determines the sensitivity of an experiment to Δm^2 (see eqn (7.74) and the discussion in section 7.5), covers a wide range, from about $10^{-4}\,\mathrm{eV}^2$ to about $10\,\mathrm{eV}^2$ (in the estimate of the upper limit one must consider pathlengths larger than about 100 km, since only upward-going stopping and through-going muons can be distinguished from secondary cosmic-ray muons). Hence, atmospheric neutrino experiments are a powerful tool for the investigation of neutrino oscillations due to a Δm^2 in such a wide range.

In this chapter, we describe the basic principles of the calculation of the flux of atmospheric neutrinos in section 11.1. In section 11.2, we review the setup and the results of atmospheric neutrino experiments.

Further information on the general principles of the calculation of neutrino fluxes and on atmospheric neutrino experiments can be found, respectively, in Refs. [487, 489] and Refs. [695, 1023, 653, 645, 785].

11.1 Flux of atmospheric neutrinos

Figure 11.3 shows the impressive energy spectrum of primary cosmic rays, from 200 MeV up to about 10^{20} eV (see Ref. [85]). The observed energy and isotropy of cosmic rays imply that they come from outside the solar system (apart from a small component coming from solar flares). Only below about 100 MeV does the solar wind dominate the cosmic ray flux. However, the solar wind decelerates the galactic cosmic rays with energies below about 10 GeV, causing a so-called *modulation* of the cosmic ray flux, which is an anticorrelation of the flux with the eleven-year cycle of solar activity.

Above a few GeV, the cosmic ray spectrum is approximately proportional to E^{-3}. More precisely, it is possible to describe it approximately with a series of power laws: at about $10^{15.5}$ eV (the *knee*) the spectrum steepens from $E^{-2.7}$ to $E^{-3.0}$; at about $10^{17.7}$ eV (the *dip*) the spectrum steepens further to $E^{-3.3}$; at about $10^{18.5}$ eV (the *ankle*) the spectrum flattens to $E^{-2.7}$. Cosmic rays have been observed up to energies of a few 10^{20} eV. However, it is believed that the cosmic ray spectrum should be suppressed above about 10^{20} eV because of the so-called Greisen–Zatsepin–Kuzmin (GZK) cutoff [565, 1085] due to the interaction of the cosmic rays with the cosmic microwave background. The investigation of the GZK cutoff is presently a very active field of theoretical and experimental research (see Refs. [963, 85, 647, 500]). However, in the following we will be interested in cosmic rays of much lower energies, from about 1 GeV to about 10^4 GeV, whose flux is sufficiently high to generate an observable flux of atmospheric neutrinos. The differential flux of primary cosmic-ray nucleons for energies between a few GeV to some hundreds TeV is given approximately by

$$\phi_N(E) \simeq \phi_N(E = 1\,\text{GeV}) \left(\frac{E}{\text{GeV}}\right)^{-(1+\gamma)}, \tag{11.4}$$

with the integral spectral index $\gamma = 1.7$ and

$$\phi_N(E = 1\,\text{GeV}) \simeq 1.8 \times 10^4 \, \frac{\text{nucleons}}{\text{m}^2\ \text{sr s GeV}}. \tag{11.5}$$

Integrating eqn (11.4) from 1 GeV to infinity, we find the integral flux of nucleons with energy bigger than 1 GeV:

$$\phi_N(E > 1\,\text{GeV}) \simeq \phi_N(E = 1\,\text{GeV})\,\gamma^{-1} \simeq 1 \times 10^4 \, \frac{\text{nucleons}}{\text{m}^2\ \text{sr s GeV}}. \tag{11.6}$$

About 79% of primary cosmic-ray nucleons are protons and about 15% are bound in helium nuclei.

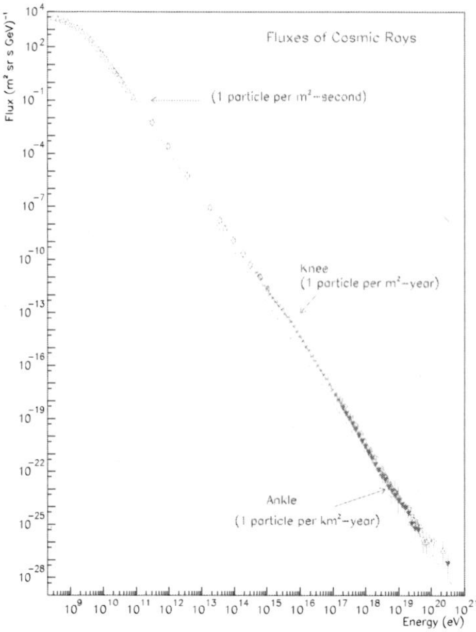

FIG. 11.3. Compilation of measurements of the differential energy spectrum of cosmic rays. The dotted line shows an E^{-3} power law for comparison. Approximate integral fluxes (per steradian) are also shown. The energy in the abscissa is the kinetic energy. Figure taken from Ref. [85].

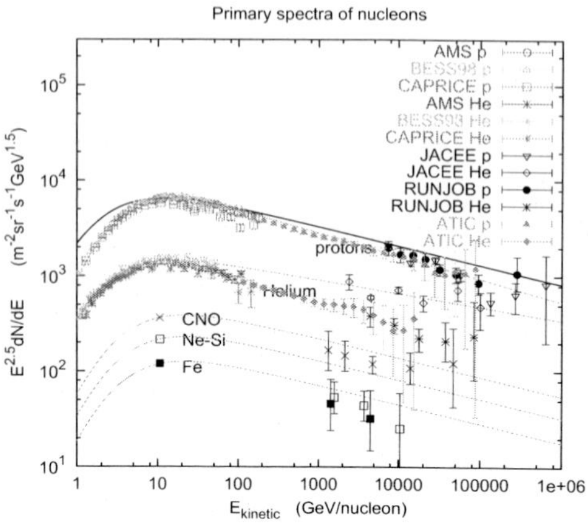

FIG. 11.4. Summary of measurements of the energy of primary cosmic-ray nuclei. The heavy black line corresponds to the power-law approximation in eqn (11.4). The lighter lines show extrapolations of fits [492] to measurements of protons, helium and three heavier groups below 100 GeV/nucleon. Figure taken from Ref. [498].

A summary of the measurements of the energy of the main primary cosmic-ray nuclei is shown in Fig. 11.4. One can see that there are agreements and discrepancies among the measurements of different experiments (for a more detailed discussion, see Refs. [498, 497]) Hence, there is some uncertainty on the energy spectrum of primary cosmic rays. The approximated power-law flux of nucleons in eqn (11.4) is shown by the heavy black line in Fig. 11.4.

Figure 11.5 shows the elemental abundances in the cosmic-ray flux at energies below about 10^5 GeV, compared with the abundances in the solar system. One can see that the cosmic-ray abundances of several nuclei, such as lithium, beryllium, and boron, are orders of magnitude larger than in the solar system. This is the consequence of a process called *spallation*: when they propagate in the Galaxy, cosmic ray nuclei interact with the interstellar medium, breaking apart. The products of these fragmentations are called *secondary* nuclei, to be distinguished from the so-called *primary* nuclei, which are the ones generated by a cosmic-ray source. The cosmic-ray abundance of primary nuclei such as carbon, oxygen, and iron, which

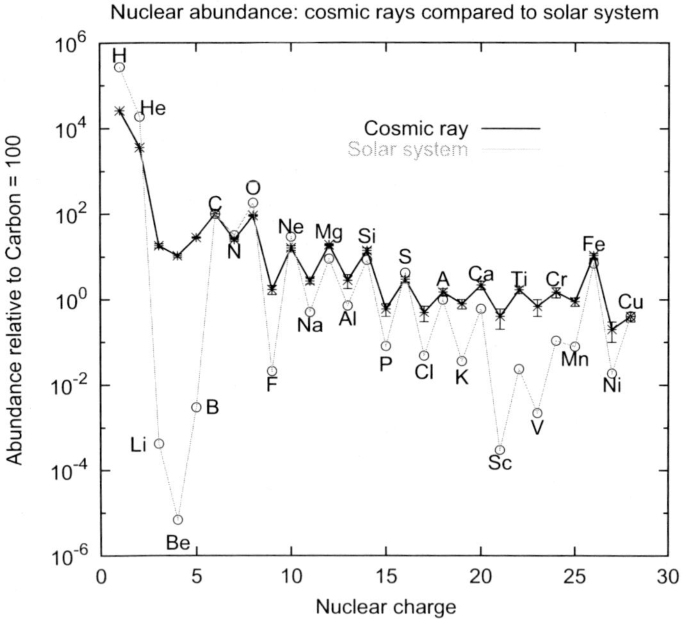

FIG. 11.5. Comparison of solar system [1089] and cosmic-ray elemental abundances. Cosmic-ray nuclear abundances are from Ref. [411], whereas those of protons and helium are from Refs. [60, 61, 923]. Figure taken from Ref. [500].

are abundant end-products of stellar nucleosynthesis, is similar to that in the solar system. Also protons and electrons are primaries. On the other hand, antiprotons and positrons are in large part secondaries. The amounts of secondary nuclei give information on the time of propagation of cosmic rays in the galaxy and on the composition of the interstellar medium. Usually, it is clear from the context if the adjective *secondary* refers to particles created by spallation in the Galaxy or by interaction of cosmic rays in the atmosphere. Otherwise, it is common to use the more specific adjectives *galactic secondary* or *atmospheric secondary*.

In the following subsection 11.1.1 we describe an approximate analytical calculation of the atmospheric neutrino fluxes, which is a simplified version of that presented in Ref. [736]. Such an analytical calculation can be used to obtain an order-of-magnitude estimate of the atmospheric neutrino fluxes. However, its main interest is pedagogical, since it gives an idea of the basic physical problems involved in the calculations of the atmospheric neutrino fluxes. In subsection 11.1.2 we summarize the main result of the most refined calculations, which are done with Monte Carlo numerical methods. Finally, in subsection 11.1.3 we discuss the up-down asymmetry, whose observation in the Super-Kamiokande experiment led in 1998 to a model-independent proof of atmospheric neutrino oscillations.

11.1.1 An approximate analytical calculation

In a one-dimensional approximation, the evolution in the atmosphere of the flux ϕ_j of a cosmic ray of type j is given by the cascade equation (see Refs. [487, 736, 553, 498])

$$\frac{\mathrm{d}\phi_j(E,X)}{\mathrm{d}X} = -\frac{\phi_j(E,X)}{\lambda_j(E)} - \frac{\phi_j(E,X)}{d_j(E,X)} + \sum_k S_{k\to j}(E,X)\,, \tag{11.7}$$

where X is the so-called *slant depth*, or *atmospheric depth*, which is usually measured in units of $\mathrm{g\,cm^{-2}}$. For simplicity, we will consider a downward-going particle in the vertical direction[61]. In this case, if h is the height of the position of the particle, the slant depth as a function of h is given by

$$X(h) = \int_h^\infty \rho(h')\,\mathrm{d}h'\,. \tag{11.8}$$

Here, $\rho(h)$ is the density of the atmosphere as a function of the height h. Most cosmic ray interactions occur between 10 and 40 km of height, in the stratosphere, which is the layer of the atmosphere between the tropopause (from $h \simeq 7\,\mathrm{km}$ at the poles to $h \simeq 17\,\mathrm{km}$ at the equator) and the stratopause ($h \simeq 50\,\mathrm{km}$). In a first approximation, the density in the stratosphere is given by

$$\rho(h) \simeq \rho_0\, e^{-h/h_0}\,, \tag{11.9}$$

with $\rho_0 \simeq 2.03 \times 10^{-3}\,\mathrm{g\,cm^{-3}}$ and $h_0 \simeq 6.4\,\mathrm{km}$. This is the so-called *barometric formula*, or *isothermal atmosphere*, or *exponential atmosphere*. In such an approximation, we have the simple relation between slant depth and height

$$X(h) \simeq \rho_0\, h_0\, e^{-h/h_0} \simeq h_0\, \rho(h)\,, \tag{11.10}$$

with $\rho_0 h_0 \simeq 1300\,\mathrm{g\,cm^{-2}}$.

The first term on the right-hand side of eqn (11.7) describes the disappearance of the particle j due to the interactions with the atmosphere with an interaction length $\lambda_j(E)$ in g cm^{-2}. Strictly speaking, the interaction length depends on the slant depth, since it is given by $\lambda_j(E,X) = \rho(h(X))/\sum_A \sigma_{jA}(E) n_A(h(X))$, where σ_{jA} is the inelastic cross-section of the interaction with nuclei with mass number A (the isospin symmetry of hadronic interactions implies that only the mass number is relevant) and n_A is the number density of these nuclei. However, since the atmospheric composition is approximately constant up to a height of about 100 km, it is possible to approximate $\rho(h) = \sum_A A n_A(h)/N_\mathrm{A} \simeq \bar{A} \sum_A n_A(h)/N_\mathrm{A}$, where N_A is

[61] The case of a particle which is downward-going along a straight trajectory forming an angle θ_z with the vertical direction (θ_z is the *zenith angle*) can be treated taking into account that the height h corresponding to a distance ℓ from the ground along the trajectory is given by $h(\ell,\theta_z) = \sqrt{(\ell\cos\theta_z + R_\oplus)^2 + (\ell\sin\theta_z)^2} - R_\oplus$, where $R_\oplus$ is the radius of the Earth. Since the atmosphere is very thin in comparison with the radius of the Earth, one can approximate $h(\ell,\theta_z) \simeq \ell\cos\theta_z + \ell^2\sin^2\theta_z/2R_\oplus$.

the Avogadro's number and $\bar{A} \simeq 14.5$ is the average mass number of air. Approximating $\sum_A \sigma_{jA}(E) n_A(h(X)) \simeq \bar{\sigma}_j(E) \sum_A n_A(h(X))$, where $\bar{\sigma}_j(E)$ is the average cross-section, we obtain

$$\lambda_j(E) \simeq \frac{\bar{A}}{N_{\mathrm{A}}\, \bar{\sigma}_j(E)} \,. \tag{11.11}$$

The second term on the right-hand side of eqn (11.7) describes the decay of the particle, with a decay length $d_j(E, X)$ in g cm^{-2}, which is given by (in the ultrarelativistic approximation)

$$d_j(E, X) = \rho(h(X))\, \frac{E}{m_j}\, \tau_j \,, \tag{11.12}$$

where τ_j is the lifetime and E/m_j is the Lorentz time-dilation factor (m_j is the mass of the particle j).

The third term on the right-hand side of eqn (11.7) describes the generation of secondary cosmic rays of type j due to the interactions of their parent cosmic rays with the atmosphere. For a hadron j which is produced by the interaction of a parent hadron k with the atmosphere, the source term is given by

$$S_{k \to j}(E, X) = \int_E^\infty \mathrm{d}E_k\, \frac{\phi_k(E_k, X)}{\lambda_k(E_k)}\, \frac{\mathrm{d}n_{k \to j}(E, E_k)}{\mathrm{d}E} \,. \tag{11.13}$$

The energy distribution $\mathrm{d}n_{k \to j}(E, E_k)/\mathrm{d}E$ of the hadron j in the $k \to j$ production process is given by

$$\frac{\mathrm{d}n_{k \to j}(E, E_k)}{\mathrm{d}E} = \frac{1}{\sigma_k(E_k)}\, \frac{\mathrm{d}\sigma_{k \to j}(E, E_k)}{\mathrm{d}E} \,. \tag{11.14}$$

Here, $\sigma_k(E_k)$ is the inclusive cross-section for interactions with air of a hadron of type k with energy E_k and $\mathrm{d}\sigma_{k \to j}(E, E_k)/\mathrm{d}E$ is the differential exclusive cross-section of production of a hadron of type j with energy E. In the evaluation of the source term in eqn (11.13), it is convenient to make the following approximations (see Refs. [487, 736, 553, 498]):

(i) the interaction length λ_k is independent of energy;

(ii) the dimensionless distribution $\mathrm{d}n_{k \to j}/\mathrm{d}x_{\mathrm{F}}$, where $x_{\mathrm{F}} = E/E_k$ is the Feynman x-variable, depends only on x_{F} (Feynman scaling). With these approximations, the source term in eqn (11.13) can be written as

$$S_{k \to j}(E, X) = \frac{1}{\lambda_k} \int_0^1 \frac{\mathrm{d}x_{\mathrm{F}}}{x_{\mathrm{F}}}\, \phi_k\!\left(\frac{E}{x_{\mathrm{F}}}, X\right) \frac{\mathrm{d}n_{k \to j}}{\mathrm{d}x_{\mathrm{F}}} \,. \tag{11.15}$$

In this case, the cascade equation (11.7) can be solved by factorizing the dependences on energy and slant-depth of the fluxes:

$$\phi_j(E, X) = \phi_j(E)\, \phi_j(X) \,, \quad \text{with} \quad \phi_j(E) \propto E^{-(1+\gamma)} \,, \tag{11.16}$$

where the energy dependence comes from that of the primary nucleon flux in eqn (11.4). Then, the source term becomes

$$S_{k \to j}(E, X) = \frac{\phi_k(E, X)}{\lambda_k}\, Z_{kj} \,, \tag{11.17}$$

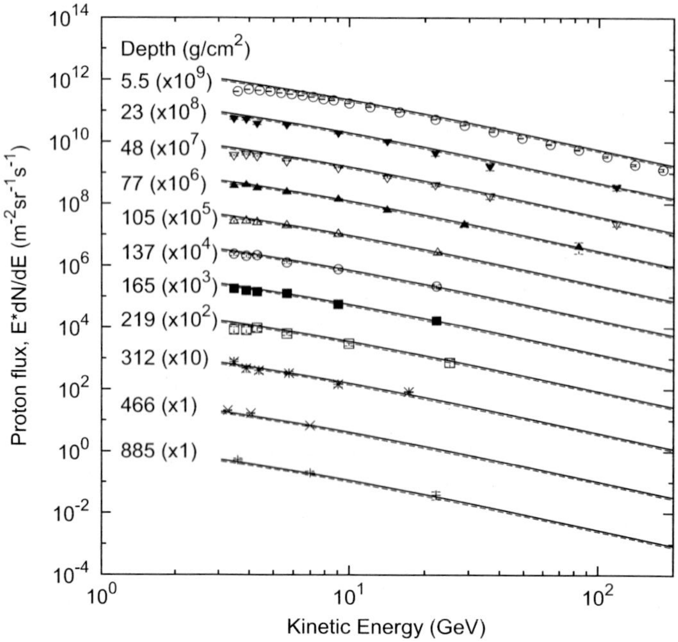

FIG. 11.6. The approximate energy spectrum of protons at various slant depths given by eqn (11.21), after subtraction of the neutron flux [487]. The data are from measurements with the CAPRICE detector [809]. Figure taken from Ref. [498].

with the so-called Z-factors given by[62]

$$Z_{kj} = \int_0^1 \mathrm{d}x_\mathrm{F}\, x_\mathrm{F}^{\gamma} \frac{\mathrm{d}n_{k\to j}}{\mathrm{d}x_\mathrm{F}} \,. \tag{11.18}$$

As a result, the approximations leading to eqn (11.18) allow us to write the general cascade equation (11.7) for hadrons in the simple form

$$\frac{\mathrm{d}\phi_j(E,X)}{\mathrm{d}X} = -\frac{\phi_j(E,X)}{\lambda_j(E)} - \frac{\phi_j(E,X)}{d_j(E,X)} + \sum_k \frac{\phi_k(E,X)}{\lambda_k} Z_{kj} \,. \tag{11.19}$$

For nucleons, which are stable (neutrons can be considered stable, since their lifetime is much longer than their transit time through the atmosphere), the cascade equation reduces to (see Refs. [487, 736, 553, 498])

$$\frac{\mathrm{d}\phi_N(E,X)}{\mathrm{d}X} = -\frac{\phi_N(E,X)}{\lambda_N} + \frac{\phi_N(E,X)}{\lambda_N} Z_{NN} \,, \tag{11.20}$$

[62] Since Feynman scaling is not exact, an accurate calculation must consider the Z-factors as energy dependent [553, 488].

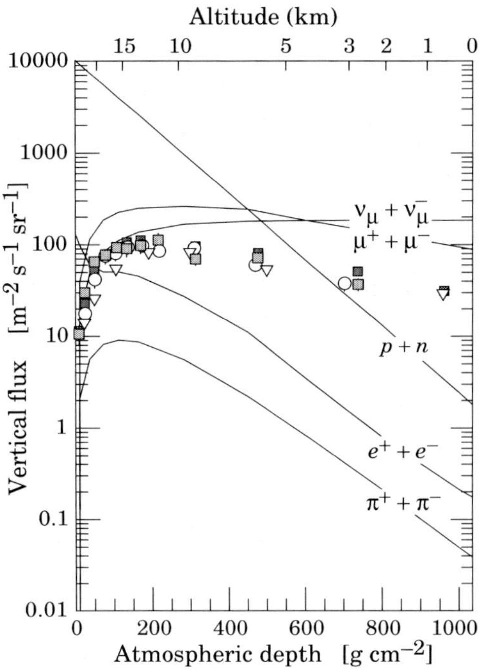

FIG. 11.7. Integral fluxes of cosmic rays with $E > 1\,\mathrm{GeV}$ as a function of the atmospheric depth. The nucleon flux corresponds to eqn (11.23). The data points show measurements of the integral flux of negative muons [203, 204, 333, 257]. Figure taken from Ref. [491].

under the approximation that nucleons are only generated by other nucleons. The solution with the boundary condition in eqn (11.4) at $X = 0$ is given by

$$\phi_N(E, X) \simeq \phi_N(E)\, e^{-X/\Lambda_N}\,, \tag{11.21}$$

where $\phi_N(E)$ is given by eqn (11.4) and Λ_N is the attenuation length

$$\Lambda_N = \frac{\lambda_N}{1 - Z_{NN}}\,. \tag{11.22}$$

For $\gamma = 1.7$, we have $Z_{NN} \simeq 0.30$ [736]. Since $\lambda_N \simeq 86\,\mathrm{g\,cm^{-2}}$ [487], we obtain $\Lambda_N \simeq 123\,\mathrm{g\,cm^{-2}}$, which, from eqn (11.10), corresponds to an altitude of about 15 km. From Fig. 11.6 one can see that eqn (11.21) gives a reasonable approximation of the energy spectrum of protons measured at different slant depths. Since the attenuation factor in eqn (11.21) does not depend on energy, the integral flux of nucleons with energy bigger than 1 GeV is given by

$$\phi_N(E > 1\,\mathrm{GeV}, X) \simeq \phi_N(E > 1\,\mathrm{GeV}, X = 0)\, e^{-X/\Lambda_N}\,, \tag{11.23}$$

with $\phi_N(E > 1\,\mathrm{GeV}, X = 0)$ given by eqn (11.6). Figure 11.7 shows the behavior of $\phi_N(E > 1\,\mathrm{GeV}, X)$ as a function of the atmospheric depth X.

The cascade equation for pions is (see Refs. [487, 736, 553])

$$\frac{\mathrm{d}\phi_\pi(E,X)}{\mathrm{d}X} = -\frac{\phi_\pi(E,X)}{\lambda_\pi} - \frac{\phi_\pi(E,X)}{d_\pi(E,X)} + \frac{\phi_N(E,X)}{\lambda_N} Z_{N\pi} + \frac{\phi_\pi(E,X)}{\lambda_\pi} Z_{\pi\pi} \,, \quad (11.24)$$

where we have made the approximation that pions are only generated by nucleons and pions and, for simplicity, we did not distinguish the charge of the pion. For $\gamma = 1.7$, we have $Z_{N\pi} \simeq 0.08$ and $Z_{\pi\pi} \simeq 0.27$ [736]. The value of the pion interaction length is $\lambda_\pi \simeq 116\,\mathrm{g\,cm^{-2}}$ [487]. There is a critical energy, given by the equality $\lambda_\pi = d_\pi$, below which the decay is more important than re-interactions. Since the decay length depends on the slant depth X, it is convenient to define the critical energy at the depth of cascade maximum [487], i.e. for $X = \lambda_\pi$. In general, with such a definition and the approximation in eqn (11.10), the critical energy for a decaying particle j is given by

$$\epsilon_j = \frac{h_0\, m_j}{\tau_j} \,. \quad (11.25)$$

For pions, we have $\epsilon_\pi \simeq 115\,\mathrm{GeV}$. Below the critical energy, the interactions of pions in the cascade equation (11.24) can be neglected, including the regeneration terms, leading to the low-energy approximate solution [736, 553]

$$\phi_\pi(E,X) = \frac{Z_{N\pi}}{\lambda_N}\, d_\pi(E,X)\, \phi_N(E,X) \qquad (E \ll \epsilon_\pi) \,, \quad (11.26)$$

with $Z_{N\pi} \simeq 0.08$ for $\gamma = 1.7$ [736]. On the other hand, in the high-energy regime, the decay term in eqn (11.24) can be neglected and the solution is

$$\phi_\pi(E,X) = \frac{Z_{N\pi}}{1 - Z_{NN}}\, \frac{\Lambda_\pi}{\Lambda_\pi - \Lambda_N} \left(e^{-X/\Lambda_\pi} - e^{-X/\Lambda_N} \right) \phi_N(E) \qquad (E \gg \epsilon_\pi) \,, \quad (11.27)$$

with the pion attenuation length $\Lambda_\pi \simeq 159\,\mathrm{g\,cm^{-2}}$, which is defined analogously to the nucleon attenuation length in eqn (11.22): $\Lambda_\pi = \lambda_\pi/(1 - Z_{\pi\pi})$. A similar treatment applies to kaons (see Refs. [736, 553]), which have the critical energy $\epsilon_K \simeq 850\,\mathrm{GeV}$ [498]. Note that for energies below the critical energy, the spectral index of mesons is smaller by one unit of the spectral index of nucleons ($\phi_\pi(E,X) \propto E^{-\gamma}$), because the decay length is proportional to the energy (see eqn (11.12)), whereas for energies above the critical energy the spectral index of mesons is the same as that of nucleons ($\phi_\pi(E,X) \propto E^{-(1+\gamma)}$). In other words, the pion spectrum and the spectrum of the neutrinos generated by pion decay are flatter than the primary nucleon spectrum below the critical energy and steepens around the critical energy until it reaches the same slope as the primary nucleon spectrum well above the critical energy. The same is true for kaons. Below the pion critical energy, pions are the main source of neutrinos. However, since $\epsilon_K \gg \epsilon_\pi$, the kaon spectrum remains flatter above the pion critical energy and rapidly becomes the dominant source of neutrinos.

In the following we will consider, for simplicity, only the low-energy case. We will discuss only the calculation of the neutrino fluxes from pion decay (and the subsequent muon decay) below the pion critical energy, using the pion flux in eqn (11.26).

The cascade equation for the flux $\phi^{(\pi)}_{\nu_\mu}(E,X)$ of atmospheric muon neutrinos and antineutrinos produced by pion decay (we neglect the small fraction of pion decays into electron neutrinos and antineutrinos) is simply

$$\frac{\mathrm{d}\phi^{(\pi)}_{\nu_\mu}(E,X)}{\mathrm{d}X} = S_{\pi\to\nu_\mu}(E,X)\,, \tag{11.28}$$

since neutrinos are stable and practically do not interact with the atmosphere. The source term $S_{\pi\to\nu_\mu}(E,X)$ of production of ν_μ from pion decay is given by [736]

$$S_{\pi\to\nu_\mu}(E,X) = \int_E^\infty \mathrm{d}E_\pi\,\frac{\phi_\pi(E_\pi,X)}{d_\pi(E_\pi,X)}\,F_{\pi\to\nu_\mu}(E,E_\pi)\,, \tag{11.29}$$

where $F_{\pi\to\nu_\mu}(E,E_\pi)$ is the energy spectrum of ν_μ from the decay. It can be shown that in the ultrarelativistic limit the inclusive spectrum of a particle B produced in the decay of a particle A has the scaling form [736]

$$F_{A\to B}(E_B,E_A) = \frac{1}{E_A}\,F_{A\to B}\left(\frac{E_B}{E_A}\right). \tag{11.30}$$

Defining, in analogy with eqn (11.18), the Z-factors

$$Z_{AB} = \int_0^1 \mathrm{d}x\,x^\gamma\,F_{A\to B}(x) \tag{11.31}$$

and taking into account that from the low-energy pion flux in eqn (11.26) we have $\phi_\pi/d_\pi \propto \phi_N(E,X) \propto E^{-(1+\gamma)}$, the source term in eqn (11.29) becomes

$$S_{\pi\to\nu_\mu}(E,X) = \frac{Z_{N\pi}}{\lambda_N}\,\phi_N(E,X)\,Z_{\pi\nu_\mu}\,. \tag{11.32}$$

Since [736]

$$F_{\pi\to\nu_\mu}(x) = \frac{1}{1-r_\pi}\,\left[1-\theta(1-(1-r_\pi))\right]\,, \tag{11.33}$$

where $r_\pi \equiv (m_\mu/m_\pi)^2 \simeq 0.57$, we obtain

$$Z_{\pi\nu_\mu} = \frac{1}{1-r_\pi}\int_0^{1-r_\pi}\mathrm{d}x\,x^\gamma = \frac{(1-r_\pi)^\gamma}{\gamma+1}\,. \tag{11.34}$$

For $\gamma = 1.7$, we have $Z_{\pi\nu_\mu} \simeq 0.087$.

Using the source term in eqn (11.32), the cascade equation (11.28) can be integrated analytically taking into account that $\phi_N(E,X) \propto e^{-X/\Lambda_{NN}}$. The result is

$$\phi^{(\pi)}_{\nu_\mu}(E,X) = \frac{Z_{N\pi}\,Z_{\pi\nu_\mu}}{1-Z_{NN}}\,\phi_N(E)\left(1-e^{-X/\Lambda_N}\right). \tag{11.35}$$

Since the slant depth of the whole atmosphere is much larger than Λ_N, the flux of muon neutrinos from pion decay (below the critical energy ϵ_π) measurable in an underground detector is given by

$$\phi^{(\pi)}_{\nu_\mu}(E) = \frac{Z_{N\pi}\,Z_{\pi\nu_\mu}}{1-Z_{NN}}\,\phi_N(E)\,. \tag{11.36}$$

This is a remarkably simple expression, once the values of the Z-factors are known. Note that the neutrino energy spectrum has the same spectral index as the spectrum of the primary nucleons. The integral flux of muon neutrinos from pion decay above 1 GeV is easily obtained by using eqn (11.6).

Let us now consider the muon flux produced by pion decays. The evolution of the muon flux with slant depth is complicated by the decay of muons and by their energy loss due to interactions with air. Treating the energy loss as a continuous process, it can be described approximately by

$$\frac{\mathrm{d}E}{\mathrm{d}X} = -a - b\,E\,, \tag{11.37}$$

where $a \simeq 2\,\mathrm{MeV\,cm^2\,g^{-1}}$ and $b \simeq 3.9 \times 10^{-6}\,\mathrm{cm^2\,g^{-1}}$ for muons in the GeV–TeV energy range [181, 493]. If a muon with energy E_0 is created at the slant depth X_0, its energy $E(X - X_0, E_0)$ at the slant depth X is obtained by integrating eqn (11.37), with the result

$$E(X - X_0, E_0) = \left(\frac{a}{b} + E_0\right) e^{-b(X - X_0)} - \frac{a}{b} \simeq E_0 - a\,(X - X_0)\,. \tag{11.38}$$

The last approximation is due to the smallness of $b\,(X - X_0)$ in the atmosphere: since $X - X_0 \lesssim 1.3 \times 10^3\,\mathrm{g\,cm^{-2}}$, we have $b\,(X - X_0) \lesssim 5 \times 10^{-3}$.

Considering now the effect of muon decay, the probability $P_\mu(E, X, X_0)$ that a muon created with energy E_0 at the slant depth X_0 reaches the slant depth X with an energy $E(X - X_0, E_0)$ is given by

$$P_\mu(E, X, X_0) = \exp\left(-\frac{m_\mu}{\tau_\mu} \int_{X_0}^{X} \frac{\mathrm{d}X'}{E(E_0, X' - X_0)\,\rho(h(X'))}\right)\,. \tag{11.39}$$

With the approximations in eqns (11.10) and (11.38), the integral in eqn (11.39) is easily calculated, leading to

$$P_\mu(E, X, X_0) = \left(\frac{E + a\,(X - X_0)}{E}\,\frac{X}{X_0}\right)^{-\epsilon_\mu/(E + aX)}\,, \tag{11.40}$$

where $\epsilon_\mu = h_0 m_\mu/\tau_\mu$ is the muon critical energy, according to eqn (11.25).

Finally, the muon flux at slant depth X generated by pion decay is given by

$$\phi_\mu(E, X) = \int_0^X \mathrm{d}X_0\, S_{\pi\to\mu}(E_0, X_0)\, P_\mu(E, X, X_0)\,, \tag{11.41}$$

where the source term $S_{\pi\to\mu}(E_0, X_0)$ has the same expression as that in eqn (11.29), with ν_μ replaced by μ. Using eqns (11.30) and (11.31), we find that

$$S_{\pi\to\mu}(E, X) = \frac{Z_{N\pi}}{\lambda_N}\,\phi_N(E, X)\, Z_{\pi\mu}\,. \tag{11.42}$$

The expression of $F_{\pi\to\mu}(x)$ is [736]

$$F_{\pi\to\mu}(x) = \frac{1}{1 - r_\pi}\,\theta(x - r_\pi)\,. \tag{11.43}$$

Thus, we obtain

$$Z_{\pi\mu} = \frac{1}{1 - r_\pi} \int_{r_\pi}^{1} \mathrm{d}x\, x^\gamma = \frac{1 - r_\pi^{\gamma+1}}{(\gamma + 1)\,(1 - r_\pi)}\,. \tag{11.44}$$

For $\gamma = 1.7$, we have $Z_{\pi\mu} \simeq 0.675$.

The last quantity to calculate is the neutrino flux from muon decays. The cascade equation is similar to eqn (11.28), with the difference that now we consider electron and muon neutrinos (for simplicity, we do not distinguish here between neutrinos and antineutrinos)[63]:

$$\frac{\mathrm{d}\phi_{\nu_\alpha}^{(\mu)}(E,X)}{\mathrm{d}X} = \int_E^\infty \mathrm{d}E_\mu \, \frac{\phi_\mu(E_\mu,X)}{d_\mu(E_\mu,X)} \, F_{\mu\to\nu_\alpha}(E,E_\mu) \qquad (\alpha = e,\mu)\,, \tag{11.45}$$

with the muon flux in eqn (11.41). The quantity on the right-hand side is the source term and $F_{\mu\to\nu_\alpha}(E,E_\mu)$ is the energy spectrum of ν_α from muon decay. Using the property in eqn (11.30), the neutrino flux from muon decays obtained by integration of eqn (11.45) is given by

$$\phi_{\nu_\alpha}^{(\mu)}(E,X) = \int_E^\infty \frac{\mathrm{d}E_\mu}{E_\mu} \, F_{\mu\to\nu_\alpha}\left(\frac{E}{E_\mu}\right) \int_0^X \mathrm{d}X' \, \frac{\phi_\mu(E_\mu,X')}{d_\mu(E_\mu,X')} \qquad (\alpha = e,\mu)\,, \tag{11.46}$$

where we have exchanged, for convenience, the order of the integrations over slant depth and energy. The neutrino spectra from muon decay are [736]

$$F_{\mu\to\nu_e}\left(\frac{E}{E_\mu}\right) = 2 - 6\left(\frac{E}{E_\mu}\right)^2 + 4\left(\frac{E}{E_\mu}\right)^3\,, \tag{11.47}$$

$$F_{\mu\to\nu_\mu}\left(\frac{E}{E_\mu}\right) = \frac{5}{3} - 3\left(\frac{E}{E_\mu}\right)^2 + \frac{4}{3}\left(\frac{E}{E_\mu}\right)^3\,. \tag{11.48}$$

In order to obtain the fluxes of electron and muon neutrinos from muon decay measurable in an underground detector one must integrate numerically eqn (11.46) up to the slant depth corresponding to the ground level. The muons which do not decay in the atmosphere rapidly lose energy after hitting the ground. When they decay, they generate low-energy neutrinos which are not observable with present-day underground neutrino detectors.

The final result for the fluxes of atmospheric neutrinos and antineutrinos in the approximate analytical approximation presented above is given by eqn (11.46) for the electron flavor and by the sum of the fluxes in eqns (11.35) and (11.46) for the muon flavor. It is possible to refine the above analytical calculations by improving some approximations and by taking into account sub-leading effects [487, 736, 553, 853, 496, 724], or introducing semi-analytical methods [274, 445]. The resulting estimates of muon fluxes typically provide acceptable fits with the data (see, for example, Fig. 11.8).

Among other applications, analytical calculations of the atmospheric neutrino fluxes allow the estimate of the most probable height of production of neutrinos in the atmosphere [494], which is interesting for oscillation studies, since the probability of flavor transitions depends on the distance between neutrino production

---

[63] In a calculation of the neutrino and antineutrino fluxes from muon decays, it is necessary to take into account the polarization of the muons, which modifies the neutrino spectra [179]. The analytical treatment of muon polarization is discussed in Ref. [736].

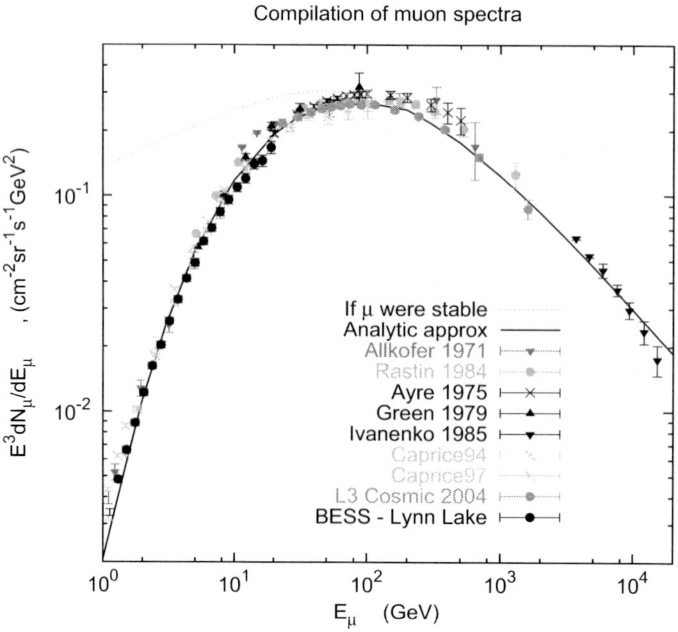

FIG. 11.8. Summary of measurements of the vertical muon intensity at the ground. The solid line shows an analytic calculation [496]. The dotted line shows the spectrum in the absence of decay and energy loss, or equivalently the muon production spectrum integrated over the atmosphere. Figure taken from Ref. [498].

and detection. The distance L covered by a neutrino produced at a height h and detected underground at a depth d is given by

$$L = \sqrt{(R_\oplus + h)^2 - (R_\oplus - d)^2 \sin^2\theta_z} + (R_\oplus - d)\cos\theta_z\,, \tag{11.49}$$

where $R_\oplus$ is the radius of the Earth and θ_z is the zenith angle of the neutrino trajectory. The most probable height of production of muon neutrinos from pion decay can be estimated in a completely analytical way as follows [494]. From eqns (11.21), (11.28), and (11.32), we can write the differential equation of neutrino generation as

$$\frac{\mathrm{d}\phi^{(\pi)}_{\nu_\mu}(E,X)}{\mathrm{d}X} = \frac{Z_{N\pi}\,Z_{\pi\nu_\mu}}{\lambda_N}\,\phi_N(E)\,e^{-X/\Lambda_N}\,. \tag{11.50}$$

Using the approximation in eqn (11.10), the differential equation of neutrino generation as a function of the height h is given by

$$\frac{\mathrm{d}\phi^{(\pi)}_{\nu_\mu}(E,h)}{\mathrm{d}h} = \rho_0\,\frac{Z_{N\pi}\,Z_{\pi\nu_\mu}}{\lambda_N}\,\phi_N(E)\,\exp\!\left(-\frac{h}{h_0} - -\frac{\rho_0\,h_0}{\Lambda_N}\,e^{-h/h_0}\right)\,. \tag{11.51}$$

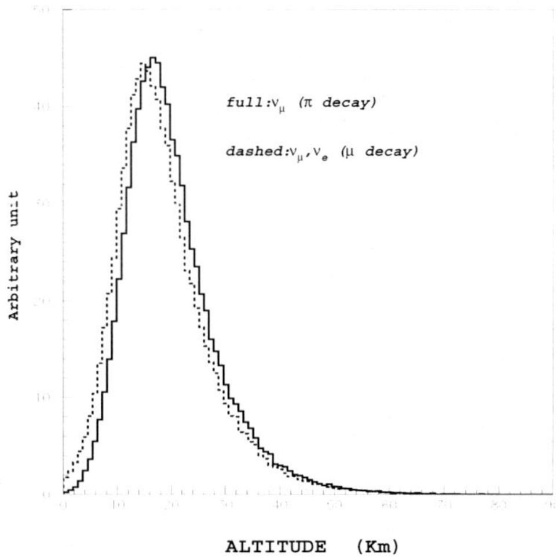

FIG. 11.9. Distributions of production altitudes of neutrinos from pion and muon decays [428].

Taking the derivative of this equation with respect to h and equating it to zero gives the most probable production height

$$h_{\text{most probable}} = h_0 \ln\left(\frac{\rho_0 h_0}{\Lambda_N}\right) \simeq 15\,\text{km}\,, \tag{11.52}$$

which is independent of the neutrino energy. This estimate is in good agreement with the result of more precise calculations [494]. The most probable height of production of neutrinos from muon decay is more complicated to calculate, because of the integral in the expression of the muon flux in eqn (11.41). Numerical calculations yield a value which is about 2 km lower than the most probable height of production of neutrinos from pion decay, because of the longer muon lifetime.

Figure 11.9 shows the distributions of production altitudes of neutrinos from pion and muon decays obtained in Ref. [428]. One can see that the distribution of altitudes of production of neutrinos from muon decay is slightly shifted towards lower altitudes with respect to the distribution of altitudes of production of neutrinos from pion decay. As a consequence, the mean pathlength of electron neutrinos, which are produced only in muon decays, is slightly shorter than the mean pathlength of muon neutrinos. This is illustrated in Fig. 11.10, which shows the relative difference of the mean pathlengths of electron and muon neutrinos as a function of zenith angle. As one can see from the figure, in practice the difference of the mean pathlengths of electron and muon neutrinos is noticeable only in the case of downward-going neutrinos, since for upward-going neutrinos the difference is a tiny fraction of the total pathlength from source to detector. Figure 11.11 shows the value of the mean pathlength distance in the atmosphere of muon neutrinos and antineutrinos in the range 300–500 MeV as a function of the zenith angle, calculated in Ref. [178]. One can see the geometrical enhancement of the mean pathlength

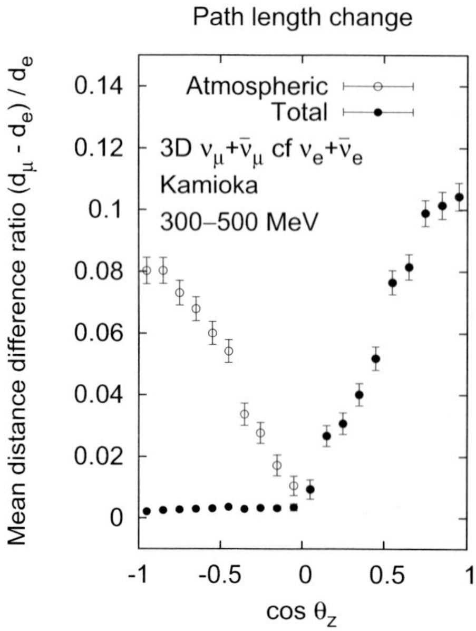

FIG. 11.10. Relative difference of the mean pathlengths of electron and muon neutrinos as a function of zenith angle θ_z [178]. Full circles compare the total distances traveled (Earth plus atmosphere) and open circles compare the distance traveled in the atmosphere only.

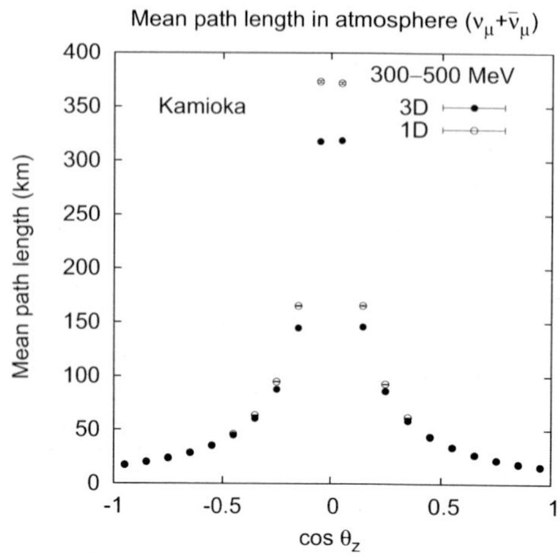

FIG. 11.11. Mean pathlength distance in the atmosphere of muon neutrinos and antineutrinos in the range 300–500 MeV as a function of the zenith angle θ_z, comparing three-dimensional (3D) and one-dimensional (1D) calculations [178].

near the horizon. Finally, Fig. 11.12 show the integral probabilities of neutrino production height at near-vertical and near-horizontal directions as functions of the neutrino energy for $\nu_\mu + \bar{\nu}_\mu$ and $\nu_e + \bar{\nu}_e$. One can see that the integral probabilities above about 50% are almost independent of the neutrino energy, especially near the vertical direction.

11.1.2 Monte Carlo calculations

The most refined calculations of atmospheric neutrino fluxes are carried out, nowadays, with three-dimensional (3D) Monte Carlo numerical methods [187, 738, 737, 876, 624, 188, 749, 1057, 428, 178, 797], which are improved versions of the previous one-dimensional (1D) Monte Carlo calculations [177, 623, 36]. Calculations done before 1996 are compared in Ref. [488]. Recent reviews and comparison of various calculations can be found in Refs. [489, 1057]. In the Monte Carlo method, primary cosmic rays are generated numerically at the top of the atmosphere. Their evolution and the evolution of the secondary shower generated in the atmosphere are calculated numerically until the detector is reached.

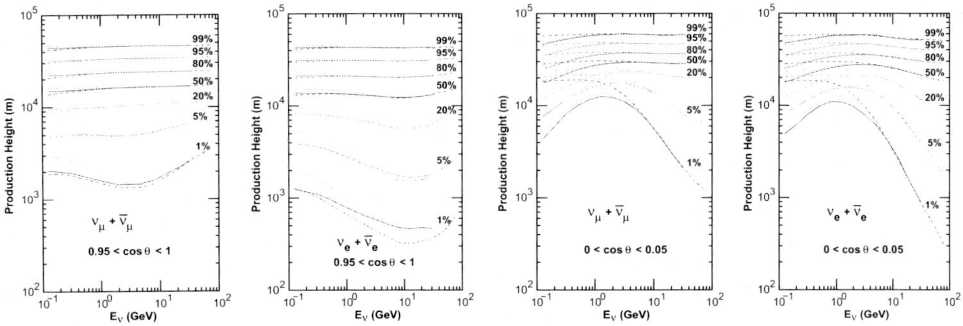

FIG. 11.12. 1%, 5%, 20%, 50%, 80%, 95%, and 99% integral probabilities of neutrino production height at near-vertical and near-horizontal directions as functions of the neutrino energy for $\nu_\mu + \bar{\nu}_\mu$ and $\nu_e + \bar{\nu}_e$ [797]. The solid (dashed) lines correspond to the 3D (1D) Monte Carlo calculation in Ref. [797] for Kamioka. The dotted lines correspond to the 3D calculation in Ref. [797] for Soudan. Here $\theta = \pi - \theta_z$ is the nadir angle.

In 1D calculations, it is assumed that each secondary particle has the same direction as its parent primary. In such an approximation, the transverse momenta of the secondaries and the bending of their trajectories in the geomagnetic field are neglected. Hence, it is sufficient to generate at the top of the atmosphere primaries with directions which pass through the detector. Three-dimensional calculations are much more computer-time consuming, because primaries must be generated at the top of the atmosphere with random directions and very few of the secondaries produced in the atmosphere hit the detector. As a consequence, the efficiency of a full 3D calculations is extremely small, about $A/R_\oplus^2 \sim 10^{-10}$ where $A \sim 10^3\,\mathrm{m}^2$ is the area of the detector and $R_\oplus$ is the radius of the Earth. Thus, all calculations must resort to some approximation which allows an increase of the efficiency. The two approximations used in the existing 3D Monte Carlo calculations are:

(i) consider a very large detector, with size comparable with the Earth's radius;

(ii) assume the absence or a dipole symmetry of the geomagnetic field (in this way, a shower can be shifted along the symmetry direction to find if a neutrino hits the detector).

Outside of the atmosphere, the geomagnetic field acts as a shield for cosmic rays with low energy. For the evaluation of the effect of the geomagnetic field on the trajectory of a charged particle, the important quantity is the magnetic rigidity R of the particle, which is given

$$R = \frac{p}{|q|}, \tag{11.53}$$

where p is the momentum and q is the charge. Since the gyroradius r of the trajectory of the particle is given by

$$r = \frac{R}{B}, \tag{11.54}$$

where B is the magnetic field, a big rigidity implies a large gyroradius, i.e. a small bending of the particle trajectory.

The rigidity of a nucleus with mass number A and atomic number Z carrying an energy E_N per nucleon is $R \simeq Ap_N/Ze$, with $p_N = \sqrt{E_N^2 - m_N^2}$. Therefore, the rigidities of different nuclei carrying the same energy per nucleon are different. This effect implies that the nuclear composition of cosmic rays must be taken into account in a precise calculation.

The shielding of cosmic rays with low energy outside of the atmosphere is taken into account by applying a rigidity cut-off, which is calculated with the so-called *backtracing* technique. In this method, when a cosmic ray primary is generated at the top of the atmosphere its trajectory is traced back in time to check if it reaches a distance of some tenths of Earth radii, where the geomagnetic field is negligible and the cosmic ray flux is isotropic. The events with backtraced trajectories which spiral back and either remain in the vicinity of the Earth or hit the Earth are discarded.

Three-dimensional calculations can take into account the bending of the trajectories of cosmic rays in the atmosphere due to the geomagnetic field. This effect is most important for muons and the neutrinos generated from their decays, because muons have the longest decay length. The bending angle θ of the trajectory of an ultrarelativistic muon with a decay length which is shorter than the length of its potential trajectory in the atmosphere is given by

$$\theta \sim \frac{E_\mu}{m_\mu} \frac{\tau_\mu}{r_\mu} = \frac{E_\mu}{m_\mu} \frac{\tau_\mu}{R_\mu} B \simeq \frac{e\,\tau_\mu}{m_\mu} B \,, \tag{11.55}$$

where E_μ/m_μ is the Lorentz time-dilation factor, r_μ is the gyroradius of the muon trajectory, R_μ is the muon rigidity, and we have used the ultrarelativistic approximation $p_\mu \simeq E_\mu$. Hence, the bending of muon trajectories is independent of energy when the potential trajectory in the atmosphere is longer than the decay length. As a consequence, muon bending has some effect even on the flux of neutrinos with energy in the multi-GeV range, especially for trajectories close to the horizon, which have the longest path in the atmosphere.

Basically, in the Monte Carlo approach, for each type of primary cosmic-ray nucleus $\mathcal{N}$, the flux of generated atmospheric ν_α's is given by a convolution of the spectrum $\phi_\mathcal{N}$ of $\mathcal{N}$ outside of the Earth's magnetosphere, a filter function $F_\mathcal{N}$ which represents the effect of the geomagnetic field outside of the Earth's atmosphere and the yield $Y_{\mathcal{N}\to\nu_\alpha}$ of ν_α's for each $\mathcal{N}$ due to the interactions and decays in the atmosphere of the primary nucleus and its secondaries. Summing up all nuclear types, the flux ϕ_{ν_α} of ν_α's is given by

$$\phi_{\nu_\alpha} = \sum_{\mathcal{N}} \phi_\mathcal{N} \otimes F_\mathcal{N} \otimes Y_{\mathcal{N}\to\nu_\alpha} \,. \tag{11.56}$$

Monte Carlo calculations allow the treatment of the primary cosmic ray flux with a better approximation than the simple power law in eqn (11.4). The primary spectra of different nuclei with mass number A can be better fitted with the

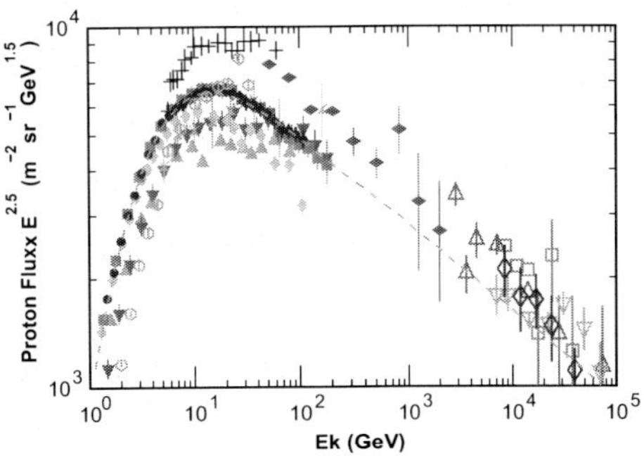

FIG. 11.13. Measurements of the primary cosmic ray proton flux as a function of kinetic energy (see Ref. [489] for the experimental references). The dashed line shows the spectrum given by eqn (11.57) with the parameters in Table 11.1. Figure taken from Ref. [489].

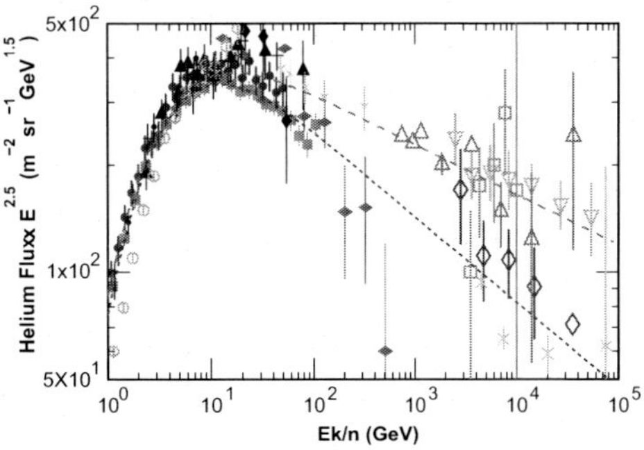

FIG. 11.14. Measurements of the primary cosmic ray helium flux as a function of kinetic energy per nucleon (see Ref. [489] for the experimental references). The dashed lines show the high (long-dash) and low (short-dash) spectra given by eqn (11.57) with the corresponding parameters in Table 11.1. Figure taken from Ref. [489].

expression [492]

$$\phi_A(E_k) \simeq K_A \left[E_k + b_A \exp\left(-c_A \sqrt{E_k} \right) \right]^{-(1+\gamma_A)} , \tag{11.57}$$

where E_k is the kinetic energy. The values of the parameters γ_A, K_A, b_A, and b_A, which have been obtained from a fit of the experimental data [489], are listed in Table 11.1. The resulting proton and helium spectra are shown in Figs. 11.13 and 11.14, together with the experimental data. However, it must be emphasized that, as one can see from Figs. 11.4, 11.13, and 11.14, there are uncertainties in the measurements of the primary cosmic ray spectra (since the helium data of different experiments have some incompatibilities, two fits are given in Table 11.1). Hence, different groups use different fits of the primary cosmic ray spectra in their Monte Carlo calculations, leading to somewhat different results for the atmospheric neutrino fluxes. The other most important source of differences between the different calculations is the treatment of the hadronic interactions. Since measurements of nucleon–nucleus and nucleus–nucleus cross-sections are limited, models are needed in order to extrapolate them over the needed wide range of energies and over the phase space of final states.

In the following part of this subsection we will present some selected results of recent Monte Carlo calculations, which are, at least, indicative of the state-of-the-art in the field.

Figure 11.15 shows the electron ($\nu_e + \bar{\nu}_e$) and muon ($\nu_\mu + \bar{\nu}_\mu$) neutrino fluxes as functions of energy obtained with the 3D and 1D Monte Carlo calculations in Ref. [178]. The large difference between the fluxes of low-energy neutrinos at

TABLE 11.1. Parameters of the fit in eqn (11.57) of the primary cosmic ray fluxes [489].

Mass number	γ_A	K_A	b_A	c_A
$A = 1$ (H)	1.74 ± 0.01	$(14.9 \pm 0.6) \times 10^3$	2.15	0.21
$A = 4$ (He, high)	1.64 ± 0.01	$(0.60 \pm 0.03) \times 10^3$	1.25	0.14
$A = 4$ (He, low)	1.74 ± 0.03	$(0.75 \pm 0.10) \times 10^3$	1.50	0.30
$A = 14$ (CNO)	1.60 ± 0.07	33.2 ± 5	0.97	0.01
$A = 25$ (Mg–Si)	1.79 ± 0.08	34.2 ± 6	2.14	0.01
$A = 56$ (Fe)	1.68 ± 0.01	4.45 ± 0.50	3.07	0.41

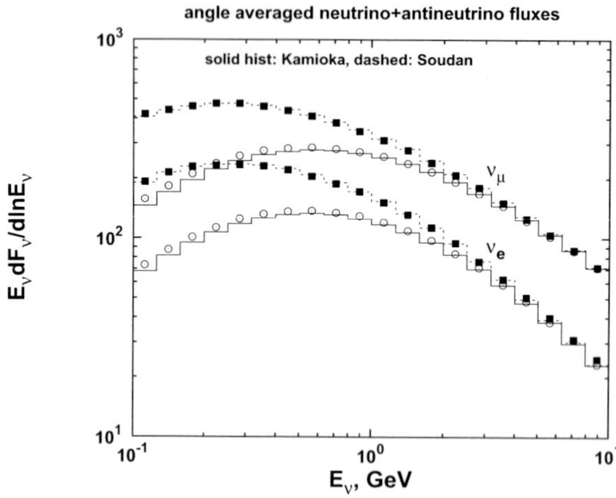

FIG. 11.15. Electron ($\nu_e + \bar{\nu}_e$) and muon ($\nu_\mu + \bar{\nu}_\mu$) neutrino fluxes as functions of energy obtained with the 3D (points) and 1D (lines) calculations in Ref. [178]. Empty points and solid lines give the fluxes at Kamioka. Filled points and dotted lines give the fluxes at Soudan.

Kamioka and Soudan is due to the difference in the corresponding geomagnetic cutoffs. Kamioka is near the geomagnetic equator, where the cutoff for downward-going primaries near the vertical direction is large, around 20 GeV. On the other hand, Soudan is near the geomagnetic pole, where the cutoff for downward-going primaries near the vertical direction is only a few GeV. Thus, more low-energy primaries are allowed by the geomagnetic field to reach the top of the atmosphere and generate neutrinos which are detected at Soudan than at Kamioka. From Fig. 11.15 one can also see that the low-energy 3D fluxes at Kamioka are slightly larger than the corresponding 1D fluxes (about 3% [178]). This is a small improvement of 3D calculations due to their better treatment of geomagnetic effects.

Figure 11.16a shows the separate fluxes of ν_e, $\bar{\nu}_e$, ν_μ, and $\bar{\nu}_\mu$ as functions of the neutrino energy obtained in four different Monte Carlo calculations. Figure 11.16a

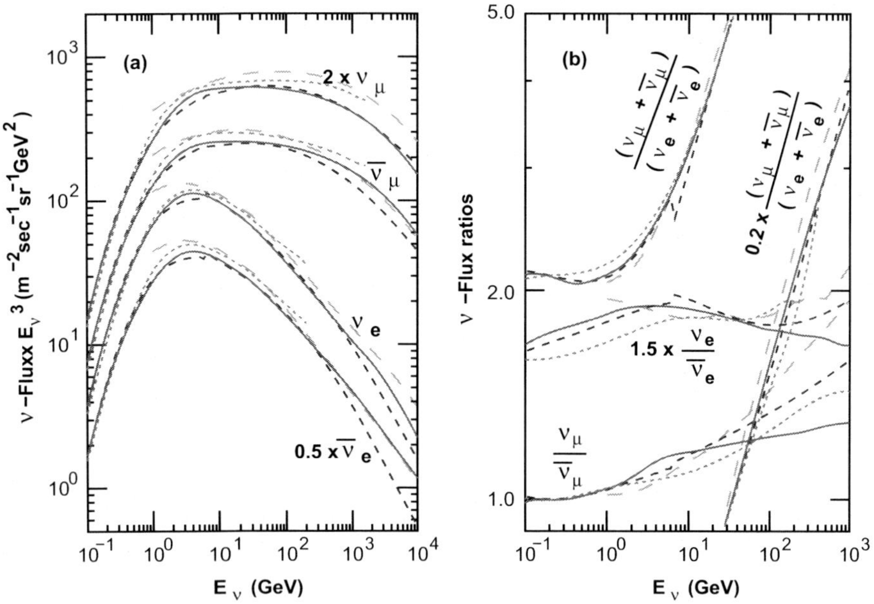

FIG. 11.16. (a) Atmospheric neutrino fluxes as functions of energy. (b) Flux ratios $(\nu_\mu + \bar{\nu}_\mu)/(\nu_e + \bar{\nu}_e)$, $\nu_\mu/\bar{\nu}_\mu$, and $\nu_e/\bar{\nu}_e$ as functions of energy. Solid, dotted, short-dashed, and long-dashed lines represent, respectively, the results of the Monte Carlo calculations in Ref. [797] (3D), Ref. [623] (1D), Ref. [187] (3D), and Ref. [36] (1D). Figures taken from Ref. [797].

shows the values of the ratios in eqn (11.3) as functions of the neutrino energy obtained in the same four different Monte Carlo calculations. One can see that the values of $(\phi_{\nu_\mu} + \phi_{\bar{\nu}_\mu})/(\phi_{\nu_e} + \phi_{\bar{\nu}_e})$ and $\phi_{\nu_\mu}/\phi_{\bar{\nu}_\mu}$ in eqn (11.3) are confirmed by detailed calculations for low energies, for which most parent muons have time to decay before hitting the ground.

Three-dimensional calculations improve the estimate of the pathlength distribution of the trajectories near the horizon of low-energy neutrinos. This is due to the fact that in 1D calculations a horizontal neutrino can only be produced by a primary which grazes the atmosphere and interacts at a large altitude. On the other hand, in 3D calculations horizontal neutrinos can also be produced by primaries which have directions closer to the vertical and interact at lower altitudes. Such a difference in 3D and 1D calculation is significant only for low-energy neutrinos, since the trajectory of a high-energy neutrino is practically collinear with that of its parent primary. This effect can be seen in Fig. 11.11 and, in more detail, in Fig. 11.17, which shows the distributions of the pathlengths from production to detection of low-energy electron and muon neutrinos near the vertical and near the horizon [178]. One can see that the results of 3D and 1D calculations are similar for the vertical direction, whereas the distributions obtained with the 3D calculation peaks at a lower pathlength with respect to the corresponding distributions

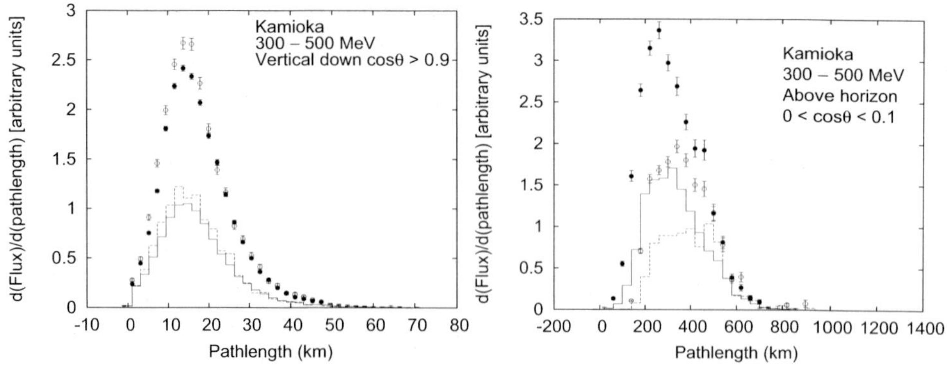

(a) (b)

FIG. 11.17. Distributions of neutrino pathlength from production to detection for vertically downward-going neutrinos (a) and near-horizontal neutrinos (b) [178]. Full (open) circles are muon neutrinos from the 3D (1D) calculation. Full (dashed) histograms are electron neutrinos from the 3D (1D) calculation. Here $\theta = \pi - \theta_z$ is the nadir angle.

obtained with the 1D calculation. The 3D reduction of pathlengths is about 15% near the horizon for neutrinos with energies in the range 300–500 MeV and rapidly decreases towards the vertical direction (it is about 2% for $0.2 < |\cos\theta_z| < 0.5$, where θ_z is the zenith angle) [178]. As one can see in Fig. 11.12, the 3D pathlength reduction at the horizon decreases at higher energies: it is about 10% at 1 GeV [797] and decreases to an average of about 4% for neutrinos above 1 GeV [178].

In Fig. 11.17 one can also see that near the horizon the 3D fluxes are much larger than the corresponding 1D fluxes and near the vertical the 3D fluxes are slightly smaller than the corresponding 1D fluxes [187]. This is a geometrical effect [738] which is significant for low-energy neutrinos, as one can see from Fig. 11.18, which shows the zenith angle dependence of the atmospheric neutrino fluxes at Kamioka in three energy bins. For high-energy neutrinos, there is a broad peak around the horizon, which is mainly due to the longer pathlength in the atmosphere available to muons, which allow a larger number of muons to decay before hitting the ground.

An important quantity which affects the analysis of atmospheric neutrino data is the uncertainty of our knowledge of the neutrino fluxes generated by cosmic rays in the atmosphere. In the analysis of atmospheric neutrino data in terms of neutrino oscillations this uncertainty determines the precision of the extraction of the mixing parameters (the squared-mass difference Δm^2 and the mixing angle ϑ in the simplest case of two-neutrino mixing). The uncertainties come mainly from the lack of a precise knowledge of the primary cosmic ray spectrum and of the hadronic interactions of cosmic rays in the atmosphere. A possibly conservative estimate of the uncertainty of the cosmic ray spectrum is obtained by considering the range of measurements, which is about 20% below 100 GeV and about 30% above [489]. A similar estimate of the uncertainty of hadronic interactions is given by the range of model predictions, which is about 20–25% [489]. Summing these

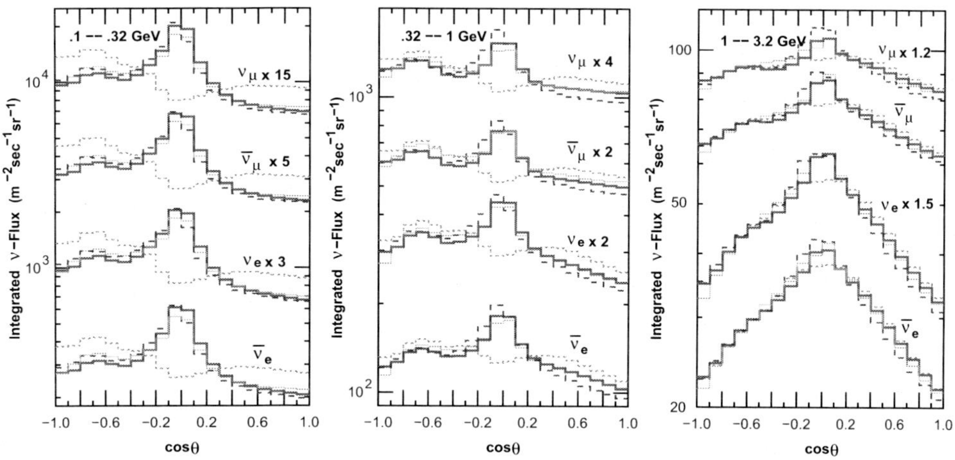

FIG. 11.18. Zenith angle dependence of the atmospheric neutrino fluxes at Kamioka in three energy bins [797]. The thick solid lines and the dotted lines correspond, respectively, to the 3D and 1D calculations in Ref. [797]. The dashed lines and the thin solid lines correspond, respectively, to the 3D calculations in Ref. [187] and Ref. [624]. Here $\theta = \pi - \theta_z$ is the nadir angle.

uncertainties in quadrature, as if they were independent statistical uncertainties, the total uncertainty of the absolute values of the neutrino fluxes can be estimated to be around 30% for low-energy neutrinos[64], which are relevant for contained events in atmospheric neutrino detectors, and around 35% for higher neutrino energies. Since these uncertainties are quite large, the absolute fluxes of atmospheric neutrinos cannot be used to obtain reliable information on neutrino physics from the analysis of the data. In fact, all analyses use ratios of fluxes, whose estimated uncertainties are much smaller. Commonly used ratios are $(\phi_{\nu_\mu} + \phi_{\bar{\nu}_\mu})/(\phi_{\nu_e} + \phi_{\nu_e})$ and $\phi_{\text{up}}/\phi_{\text{down}}$ for contained events and $\phi_{\text{stopping}}/\phi_{\text{through-going}}$ and $\phi_{\text{vertical}}/\phi_{\text{horizontal}}$ for neutrino-induced upward-going muons.

The flavor ratio $(\phi_{\nu_\mu} + \phi_{\bar{\nu}_\mu})/(\phi_{\nu_e} + \phi_{\nu_e})$, which has been used as an indicator of neutrino oscillations in the analysis of experimental data, has an estimated uncertainty of less than 5% at energies smaller than about 5 GeV [653, 188, 111]. At higher energies, the contribution of kaon decays to the neutrino production in the atmosphere becomes more important, as shown in Fig. 11.19. Contrary to pions, which decay mainly into muons and muon neutrinos according to eqn (11.1), the kaon branching ratio for such a decay is 0.6343 ± 0.0017 [400]. The other important semileptonic decay channels of $K^{\mp}$ are $K^{\mp} \to \pi^0 + e^{\mp} + \overset{(-)}{\nu}_e$, with branching ratio 0.0487 ± 0.0006, and $K^{\mp} \to \pi^0 + \mu^{\mp} + \overset{(-)}{\nu}_\mu$, with branching ratio 0.0327 ± 0.0006 [400]. Moreover, the K_L^0 has the semileptonic decay channels $K_L^0 \to \pi^{\pm} + e^{\mp} + \overset{(-)}{\nu}_e$

64 Less conservatively, the authors of Ref. [797] estimate an uncertainty of about 10% for neutrino energies below 10 GeV.

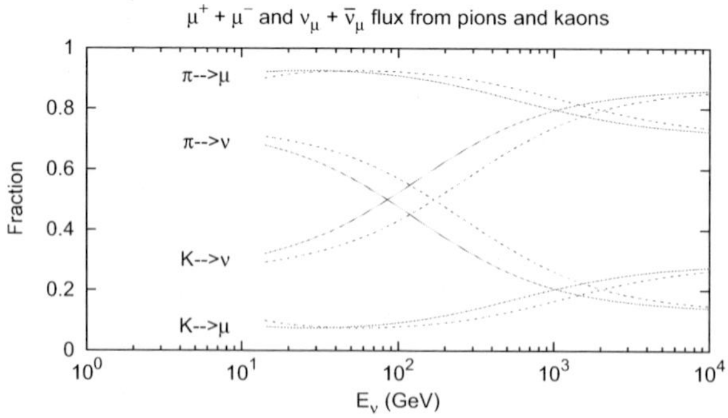

FIG. 11.19. Fractional contribution of pions and kaons to the flux of muons and neutrinos [489]. Solid lines correspond to the vertical direction and dashed lines to the directions with a zenith angle of 60°.

with branching ratio 0.3881 ± 0.0027, and $K_L^0 \to \pi^\pm + \mu^\mp + \overset{(-)}{\nu}_\mu$ with branching ratio 0.2719 ± 0.0025 [400]. Since the kaon production cross-section has larger uncertainties than the pion production cross-section, the uncertainty in the flavor ratio increases at high energies. The comparison of different calculations indicate an uncertainty of about 10% at 100 GeV [111].

The flavor ratio cannot be measured directly in real experiments, because what is seen are not the neutrinos, but the Cherenkov light produced by the charged leptons generated by neutrino interactions. Electron and muon neutrinos have different cross-sections and the detection efficiencies and selection criteria are different for events generated by electron and muon neutrinos. Thus, experimental data are reported in terms of e-like and μ-like events and the ratio used to reveal an anomaly is the *ratio-of-ratios*

$$R_{\mu/e} \equiv \frac{(N_{\mu\text{-like}}/N_{e\text{-like}})_{\text{data}}}{(N_{\mu\text{-like}}/N_{e\text{-like}})_{\text{MC}}}, \tag{11.58}$$

where $N_{\mu\text{-like}}$ and $N_{e\text{-like}}$ are, respectively, the numbers of μ-like and e-like events. The numerator in eqn (11.58) is the measured ratio and the denominator is that calculated with Monte Carlo. A measurement of a ratio-of-ratios different from unity is evidence of an anomaly.

In the following subsection we discuss the up-down asymmetry, which have allowed the Super-Kamiokande collaboration to find, in 1998, a model-independent proof of the oscillations of atmospheric neutrinos.

11.1.3 The up-down asymmetry

At low energies, the directional flux of atmospheric neutrinos depends on the effect of the geomagnetic field on the parent cosmic ray primaries and secondaries. This effect disappears at high energies, above a few GeV. Hence, the production of high-energy neutrinos in the atmosphere is practically uniform around the globe. This implies that, in the absence of oscillations, at any location on the surface of the Earth or below, the fluxes of each neutrino flavor arriving from opposite directions

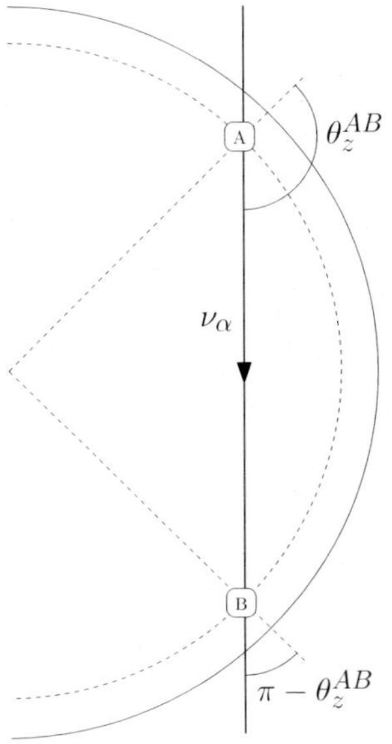

FIG. 11.20. Illustration of the equality in eqn (11.59).

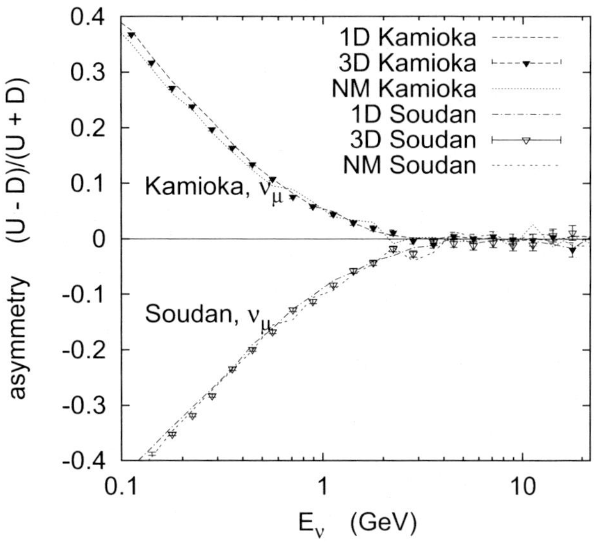

FIG. 11.21. Up-down asymmetry of the muon neutrino flux as a function of neutrino energy for the Kamioka and Soudan sites obtained with different Monte Carlo calculations in Ref. [178] (NM denotes a 3D Monte Carlo calculation with bending of particles within the atmosphere turned off).

are the same, i.e. $\phi_{\nu_\alpha}(\theta_z) = \phi_{\nu_\alpha}(\pi - \theta_z)$. This fact can be understood on the basis of the following elementary geometrical argument illustrated in Fig. 11.20. Let us consider two underground detectors A and B at the same depth. As shown in the figure, neglecting the extremely small probability of neutrino interactions in the Earth, a neutrino passing through the detector A with a direction having zenith angle θ_z^{AB} later passes through the detector B with a direction having zenith angle $\pi - \theta_z^{AB}$. Hence, we have

$$\phi_{\nu_\alpha}^{(A)}(\theta_z^{AB}) = \phi_{\nu_\alpha}^{(B)}(\pi - \theta_z^{AB}) \,. \tag{11.59}$$

Since the production of high-energy neutrinos in the atmosphere is uniform around the globe, we also have

$$\phi_{\nu_\alpha}^{(A)}(\theta_z^{AB}) = \phi_{\nu_\alpha}^{(B)}(\theta_z^{AB}) \,. \tag{11.60}$$

Finally, since the location of the detector B is arbitrary, the zenith angle θ_z^{AB} can assume any value. Then, for a generic zenith angle θ_z we obtain, from eqn (11.59) and (11.60)

$$\phi_{\nu_\alpha}^{(A)}(\theta_z) = \phi_{\nu_\alpha}^{(A)}(\pi - \theta_z) \,, \tag{11.61}$$

for any underground detector A.

This elementary geometrical property of the neutrino fluxes implies that flavor oscillations of atmospheric neutrinos can be revealed in a model-independent way by measuring a difference of the upward-going and downward-going fluxes of high-energy neutrinos with a definite flavor [351, 447]. In fact, downward-going neutrinos are produced in the atmosphere above the detector and travel a distance of some tens of km, whereas upward-going neutrinos come from the other side of the globe, with traveling distances of several thousands km. The larger distances covered by upward-going neutrinos may allow them to oscillate into other flavors, leading to a suppression of the upward-going flux with respect to the downward going one. We will see in section 11.2.3 that irrefutable evidence of atmospheric neutrino oscillations was obtained in 1998 by the Super-Kamiokande experiment just measuring a nonzero asymmetry of the upward-going and downward-going fluxes of high-energy muon neutrinos.

In the argument above we did not discuss the problem of the determination of the minimum neutrino energy above which the up-down symmetry holds. The relation between the upward-going and downward-going neutrino fluxes are usually quantified through the asymmetry

$$A_{\alpha}^{\text{up-down}} = \left(\frac{U-D}{U+D}\right)_{\alpha} \tag{11.62}$$

or the up-down ratio

$$R_{\alpha}^{\text{up-down}} = (U/D)_{\alpha} \,, \tag{11.63}$$

where U and D are, respectively, the neutrino fluxes neutrinos integrated in the ranges $0.2 < \cos\theta_z < 1$ and $-1 < \cos\theta_z < -0.2$. The subscript α denotes ν_α fluxes in theoretical calculations and α-like events in experimental measurements, with $\alpha = e, \mu$.

Figure 11.21 shows the value of the asymmetry $A_{\mu}^{\text{up-down}}$ as a function of the neutrino energy, obtained with different Monte Carlo calculations in Ref. [178]. In Fig. 11.21, one can see that the up-down muon asymmetry $A_{\mu}^{\text{up-down}}$ vanishes for $E_\nu \gtrsim 2\,\text{GeV}$. The different signs of the deviations from $A_{\mu}^{\text{up-down}} = 0$ at lower energies at the locations of Kamioka and Soudan are due to the different geomagnetic cutoffs. Since Kamioka is near the geomagnetic equator, where the cutoff for downward-going primaries near the vertical direction is large, the low-energy upward-going flux is larger than the downward-going one. On the other hand, Soudan is near the geomagnetic pole, where the cutoff for downward-going primaries near the vertical direction is small. In this case, the low-energy downward-going flux is larger than the upward-going one, which is generated all around the globe, where the cutoff is larger on average.

11.2 Atmospheric neutrino experiments

In this section we describe the setup and main results of the atmospheric neutrino experiments which have been carried out since the 1980s and have been important for the study of atmospheric neutrino oscillations. For simplicity, we omit the description of the NUSEX [185, 186, 33, 34] and Frejus [209, 210, 352] experiments, which did not find any indication of an anomaly in their data[65].

In atmospheric neutrino experiments, neutrino fluxes of different flavors are measured by detection of the charged leptons produced in the neutrino–nucleon collisions

$$\nu_\ell + N \to \ell^- + X\,, \qquad \bar{\nu}_\ell + N \to \ell^+ + X \qquad (\ell = e, \mu, \tau)\,. \tag{11.64}$$

In the experiments carried out so far the charge of the lepton could not be distinguished, because the detectors were not magnetized. It is also extremely difficult to reveal tau neutrinos, because the produced tau leptons decay immediately into leptons and hadrons, without leaving a clear track.

For a discussion of the data, it is necessary to know that for atmospheric neutrino detection three classes of events are defined:

Contained Events. These are events in which the neutrinos interact with matter inside the detector and the trajectories of all the particles which are generated are contained in the detector.

Stopping Muons. These are tracks of muons which enter the detector and stop inside. They are generated by neutrino interactions in the rock outside the detector.

Through-Going Muons. These are tracks of muons which enter the detector and exit without stopping. They are also generated by neutrino interactions in the rock outside the detector.

In the Kamiokande and Super-Kamiokande experiments, which have an inner *fiducial volume* detector surrounded by an outer detector, two subclasses of contained events have been defined:

Fully Contained (FC) Events. These are events totally contained in the inner detector.

Partially Contained (PC) Events. These are contained events in which the neutrinos interact with the material inside the inner detector and some of the produced particles exit the inner detector, stopping in the outer detector.

Furthermore, in the Kamiokande and Super-Kamiokande experiments, for historical reasons, two subclasses of FC events have been defined:

Sub-GeV Events. These are FC events with total visual energy $E_{\text{vis}} < 1.33\,\text{GeV}$.

Multi-GeV Events. These are FC events with total visual energy $E_{\text{vis}} > 1.33\,\text{GeV}$.

As shown in Fig. 11.2, the energy ranges of neutrinos which generate sub-GeV, multi-GeV, stopping and through-going events are, respectively, ~ 0.2–$20\,\text{GeV}$, ~ 10–$500\,\text{GeV}$, ~ 20–$1000\,\text{GeV}$, and ~ 50–$20\,000\,\text{GeV}$.

---

[65] The value of the ratio-of-ratios in eqn (11.58) obtained in the Frejus experiment is $R_{\mu/e} = 1.00 \pm 0.15 \pm 0.05$ [352] and that found in the NUSEX experiment is $R_{\mu/e} = 0.96^{+0.32}_{-0.28}$ [33].

11.2.1 Kamiokande

The Kamiokande experiment has been described in section 10.6.1 in the context of solar neutrino detection. The whole inner detector, with a mass of 2142 ton of purified water, was used as the fiducial volume for the detection of atmospheric neutrinos [614, 840, 621, 474, 603]. The outer volume, used as shield and anti-counter, contained 1500 ton of water. For the calculation of geomagnetic effects it is useful to know that the Kamioka mine is located at 36°25′36″ N, 137°18′43″ E.

The first indication that there is an atmospheric neutrino anomaly came in 1988 [614] from the measurement in Kamiokande of a number of contained sub-GeV μ-like events which was a fraction 0.59 ± 0.07 of that predicted by a Monte Carlo calculation, whereas the number of e-like events was in agreement with the Monte Carlo prediction.

In terms of the ratio-of-ratios in eqn (11.58) the final results of the Kamiokande experiment for sub-GeV and multi-GeV events are:

$$R_{\mu/e}^{\text{sub-GeV}} = 0.60^{+0.07}_{-0.06} \pm 0.05 \qquad [621]\,, \tag{11.65}$$

$$R_{\mu/e}^{\text{multi-GeV}} = 0.57^{+0.08}_{-0.07} \pm 0.07 \qquad [474]\,. \tag{11.66}$$

These anomalies are well explained by ν_μ disappearance due to oscillations. Furthermore, as shown in Fig. 11.22, the Kamiokande collaboration found a zenith angle dependence of upward through-going muons which supported the interpretation of the contained events anomaly in terms of neutrino oscillations [603]. The Kamiokande data could not allow discrimination between $\nu_\mu \leftrightarrows \nu_e$ and $\nu_\mu \to \nu_\tau$ transitions. However, in 1997 the results of the CHOOZ [97] long-baseline reactor experiment excluded the $\nu_\mu \leftrightarrows \nu_e$ channel. Figure 11.23 shows the regions in the $\sin^2 2\vartheta$-Δm^2 plane for $\nu_\mu \to \nu_\tau$ oscillations which are allowed by the different sets of Kamiokande data.

11.2.2 IMB

The IMB (Irvine–Michigan–Brookhaven) [581, 243, 306, 199, 200, 322] detector was a $24\,\text{m} \times 18\,\text{m} \times 19\,\text{m}$ tank filled with about 8000 ton of water. The fiducial mass was of 3300 ton. It was located deep underground in the Morton Thiokol salt mine near Cleveland, Ohio (USA), at a depth of 610 m, with an overburden of 1570 mwe. The IMB detector took data from 1982 to 1991. In a first phase called IMB-1 it was equipped with 2048 photomultiplier tubes (PMTs) with a diameter of 12.5 cm. The light collection was improved in the second IMB-2 phase. In the third and final IMB-3 phase, the 2048 PMTs were replaced by larger ones with a diameter of 20 cm.

The IMB experiment found an anomaly compatible with the Kamiokande data only in the contained events with energy below 1.5 GeV [581, 306, 200]. On the other hand, no anomaly was found in the upward-going muon data [199] and in the partially contained events with energy above 0.95 GeV [322].

The anomaly found in the low-energy contained events is quantified through the ratio-of-ratios in eqn (11.58) as [663]

$$R_{\mu/e}^{\text{multi-GeV}} = 0.54 \pm 0.05 \pm 0.11\,, \tag{11.67}$$

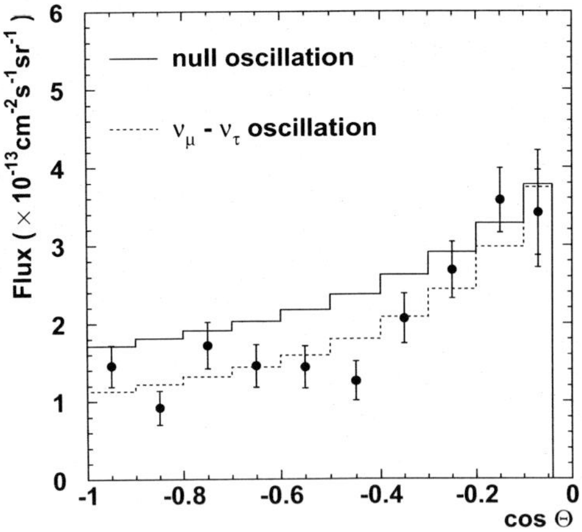

FIG. 11.22. Zenith angle distribution of the upward through-going muon flux observed in Kamiokande [603]. The inner (outer) error bars indicate statistical (uncorrelated experimental systematic + statistical added in quadrature) errors. The solid and dashed histograms show, respectively, the expected upward through-going muon flux without oscillations and with $\nu_\mu \to \nu_\tau$ oscillations ($\sin^2 2\vartheta = 1.00$ and $\Delta m^2 = 3.2 \times 10^{-3}\,\mathrm{eV}^2$). Here $\Theta = \pi - \theta_z$ is the nadir angle.

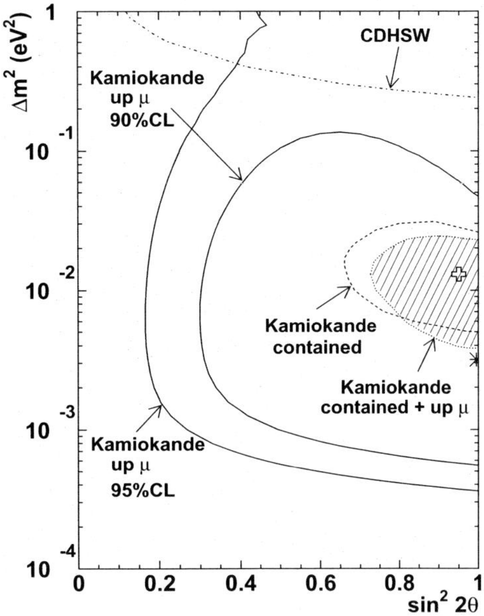

FIG. 11.23. Allowed regions at 90% CL in the $\sin^2 2\vartheta$-Δm^2 plane for $\nu_\mu \to \nu_\tau$ oscillations obtained with different sets of Kamiokande data [603]. The exclusion curve of the CDHSW experiment [396] is also shown. The asterisk indicates the best-fit point for upward–going muons at $\sin^2 2\vartheta = 1.00$ and $\Delta m^2 = 3.2 \times 10^{-3}\,\mathrm{eV}^2$. The cross indicates the best-fit point for contained events plus upward-going muons at $\sin^2 2\vartheta = 0.95$ and $\Delta m^2 = 1.3 \times 10^{-2}\,\mathrm{eV}^2$.

which is in agreement with the values in eqn (11.65) and (11.66) obtained in the Kamiokande experiment.

11.2.3 Super-Kamiokande

The Super-Kamiokande (SK) experiment has been described in section 10.6.2 in the context of solar neutrino detection. The fiducial volume for atmospheric neutrino detection [477, 479, 476, 481, 486, 483, 470, 110, 111] is 22.5 ktons.

In 1998 the SK collaboration presented the result [476]

$$A_\mu^{\text{up-down}} = -0.296 \pm 0.048 \pm 0.01\,, \tag{11.68}$$

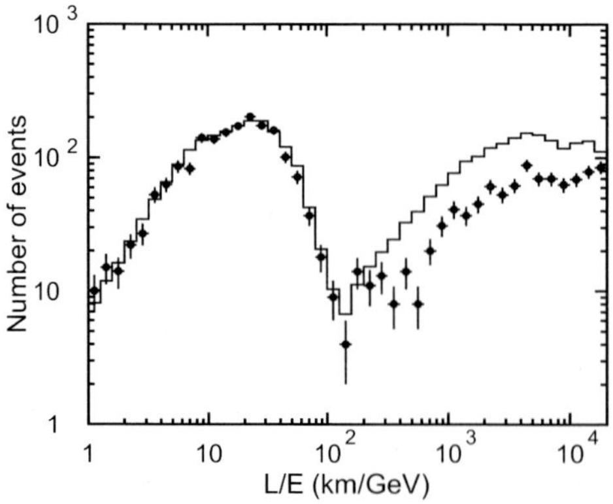

FIG. 11.24. Number of μ-like events in Super-Kamiokande as a function of L/E (points) [110]. The histogram shows the Monte Carlo prediction without oscillations.

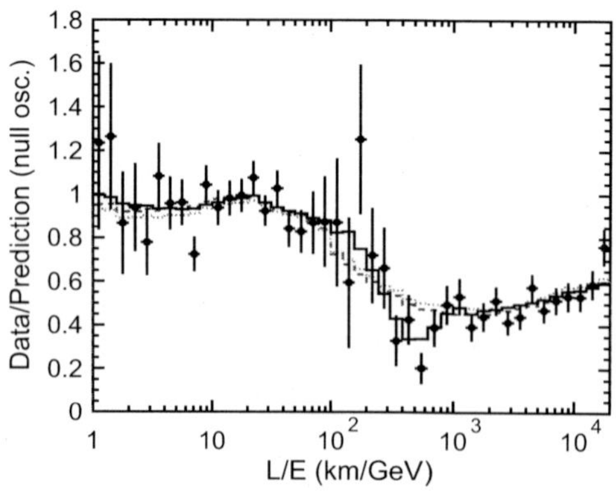

FIG. 11.25. Ratio of the data and the Monte Carlo prediction for the number of μ-like events in Super-Kamiokande as a function of L/E (points) [110]. The solid histogram shows the best-fit expectation for $\nu_\mu \to \nu_\tau$ oscillations ($\sin^2 2\vartheta = 1$ and $\Delta m^2 = 2.4 \times 10^{-3}\,\text{eV}^2$). The error bars are statistical only. Also shown are the best-fit expectation for neutrino decay [173, 172] (dashed line) and neutrino decoherence [574, 743] (dotted line).

for the up-down asymmetry in eqn (11.62) of multi-GeV and PC μ-like events, whereas the corresponding electron up-down asymmetry was consistent with zero ($A_e^{\text{up-down}} = -0.036 \pm 0.067 \pm 0.02$). This was considered as a 6σ model-independent evidence that part of the upward-going atmospheric muon neutrino flux disappears.

The most likely explanation of the up-down muon asymmetry is neutrino oscillations, since the source–detector distance covered by upward-going neutrinos is much larger than that covered by downward-going neutrinos. This explanation is confirmed by the distribution of the observed number of μ-like events as a function of L/E shown in Figs. 11.24 and 11.25. The ratio L/E determines the phase of the oscillation probability, as explained in chapter 7. From Fig. 11.24 one can see that the number of events with $L/E \lesssim 10^2\,\text{km}\,\text{GeV}^{-1}$ agrees with the Monte Carlo calculation. This means that for the corresponding neutrinos the phase of the oscillations is too small, i.e. they did not have enough time to oscillate. On the other hand, the number of events with $L/E \gtrsim 10^2\,\text{km}\,\text{GeV}^{-1}$ shows a deficit with respect to the Monte Carlo calculation, because the oscillation phase of the corresponding neutrinos was large enough to produce a transition. Figure 11.25 shows that other explanations are disfavored, since the ratio of the data and the Monte Carlo prediction shows the dip at $L/E \simeq 500\,\text{km}\,\text{GeV}^{-1}$ which corresponds to the first maximum of the oscillation probability for $\sin^2 2\vartheta = 1$ and $\Delta m^2 = 2.4 \times 10^{-3}\,\text{eV}^2$.

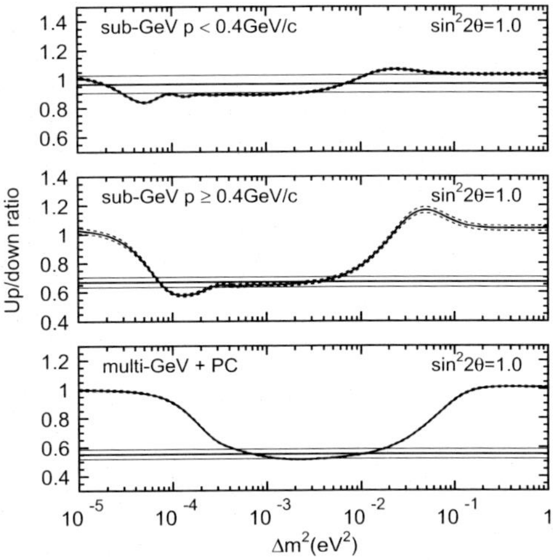

FIG. 11.26. Expected up-down ratio $R_\mu^{\text{up-down}}$ in Super-Kamiokande in the case of $\nu_\mu \to \nu_\tau$ oscillations as a function of Δm^2, for $\sin^2 2\vartheta = 1$ [111]. The ratio for the data together with the $\pm 1\sigma$ statistical error are shown by the horizontal lines. The systematic errors are shown by the band in the expectation.

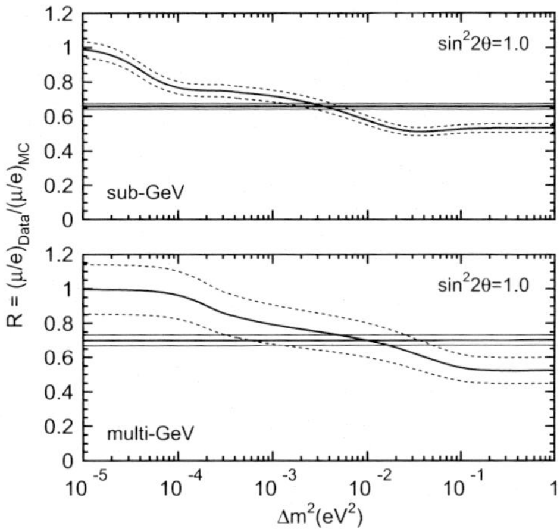

FIG. 11.27. Expected value of the ratio of ratios $R_{\mu/e}$ in Super-Kamiokande in the case of $\nu_\mu \to \nu_\tau$ oscillations as a function of Δm^2, for $\sin^2 2\vartheta = 1$ [111]. The values for the data together with $\pm 1\sigma$ statistical errors are shown by the horizontal lines. The systematic errors are shown by the band in the expectation.

A crucial problem is the determination of the flavor channel of the atmospheric neutrino oscillations. The results of the CHOOZ [97, 98, 100] and Palo Verde [254, 253, 255] long-baseline reactor experiment disfavor the $\nu_\mu \leftrightarrows \nu_e$ channel for $\Delta m^2 \gtrsim 10^{-3}\,\text{eV}^2$. The data of SK favor the $\nu_\mu \to \nu_\tau$ channel over the $\nu_\mu \leftrightarrows \nu_e$ and $\nu_\mu \to \nu_s$ channels, where ν_s is a sterile neutrino (see chapter 6). The oscillations $\nu_\mu \leftrightarrows \nu_e$ are disfavored [111] by the absence of any anomaly in the e-like data. The $\nu_\mu \to \nu_s$ channel is disfavored [470] by the observed absence of matter effects for neutrino oscillations through the Earth[66] and by the up-down symmetry of a sample of events with a considerable neutral-current fraction[67].

Recently [111], the SK collaboration presented the final results of SK-I. The measured values of the up-down ratio $R_\alpha^{\text{up-down}}$ in eqn (11.63) for $\alpha = e, \mu$ are listed in Table 11.2. One can see that e-like events are up-down symmetric at all energies. At low energies also μ-like events are up-down symmetric, since the

[66] In the case of $\nu_\mu \to \nu_\tau$ transitions there are no matter effects, since ν_μ and ν_τ have the same interactions with matter. On the other hand, in the case of $\nu_\mu \to \nu_s$ transitions there are matter effects, because sterile neutrinos do not interact with matter.

[67] In the case of $\nu_\mu \to \nu_s$ transitions neutral-current events should show the same up-down asymmetry as μ-like events. On the other hand, $\nu_\mu \to \nu_\tau$ transitions do not affect the amount of neutral-current events.

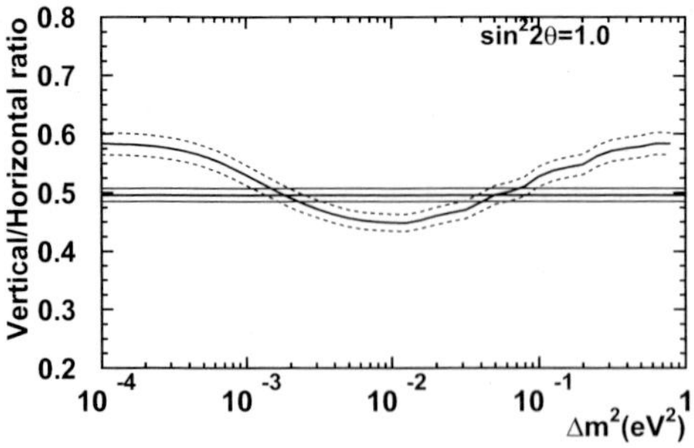

FIG. 11.28. Expected ratio V/H of upward through-going muons in the case of $\nu_\mu \to \nu_\tau$ oscillations as a function of Δm^2, for $\sin^2 2\vartheta = 1$ [111]. The ratio for the data together with $\pm 1\sigma$ combined statistical and systematic error is also shown by the horizontal lines. The systematic error is shown by the band in the expectation.

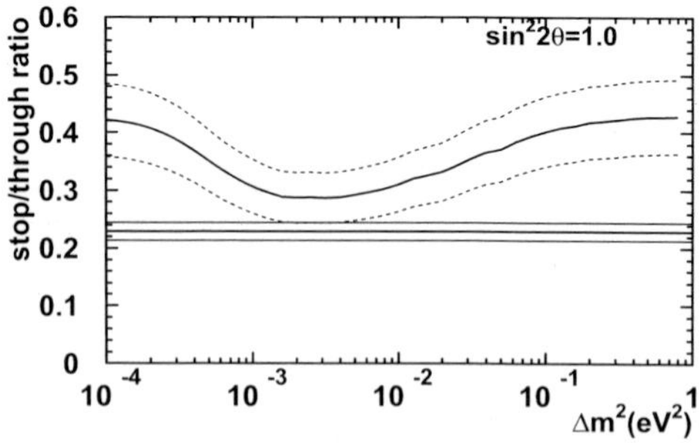

FIG. 11.29. Expected ratio $R_{\text{stop/through}}$ in the case of $\nu_\mu \to \nu_\tau$ oscillations as a function of Δm^2, for $\sin^2 2\vartheta = 1$ [111]. The ratio for the data together with $\pm 1\sigma$ combined statistical and systematic error is also shown by the horizontal lines. The systematic error is shown by the band in the expectation.

TABLE 11.2. Values of the up-down ratio $R_\alpha^{\text{up-down}}$ in eqn (11.63) for $\alpha = e, \mu$ measured in the Super-Kamiokande experiment [111]. p is the charged lepton momentum.

Flavor	sub-GeV ($p < 400$ MeV)	sub-GeV ($p > 400$ MeV)	multi-GeV + PC
e	$1.133^{+0.062}_{-0.059} \pm 0.009$	$1.082^{+0.063}_{-0.060} \pm 0.024$	$0.961^{+0.086}_{-0.079} \pm 0.016$
μ	$0.964^{+0.062}_{-0.058} \pm 0.008$	$0.670^{+0.035}_{-0.034} \pm 0.012$	$0.551^{+0.035}_{-0.033} \pm 0.004$

angular correlation between the neutrino and the produced charged lepton is very poor, washing out any zenith-angle dependence. The up-down ratio of multi-GeV and PC μ-like events confirms the result in eqn (11.68) with a statistical significance of 12σ. Figure 11.26 shows a comparison of the measured values of $R_\mu^{\text{up-down}}$ with that expected in the case of $\nu_\mu \to \nu_\tau$ oscillations as a function of Δm^2, for $\sin^2 2\vartheta = 1$. One can see that neutrino oscillations explain the data for Δm^2 between about $10^{-3.5}$ and $10^{-2}\,\text{eV}^2$.

The results [111] of the SK experiment for the ratio-of-ratios in eqn (11.58) are, for sub-GeV and multi-GeV plus PC events,

$$R_{\mu/e}^{\text{sub-GeV}} = 0.658 \pm 0.016 \pm 0.035\,, \tag{11.69}$$

$$R_{\mu/e}^{\text{multi-GeV}} = 0.702^{+0.032}_{-0.030} \pm 0.101\,. \tag{11.70}$$

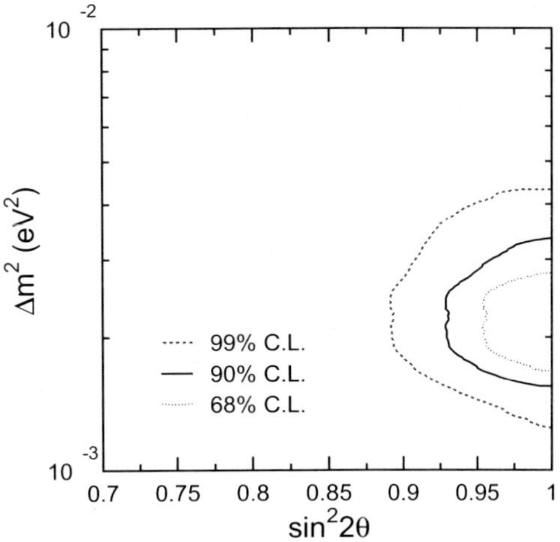

FIG. 11.30. Allowed region in the $\sin^2 2\vartheta$-Δm^2 plane for $\nu_\mu \to \nu_\tau$ oscillations obtained in the Super-Kamiokande experiment [111].

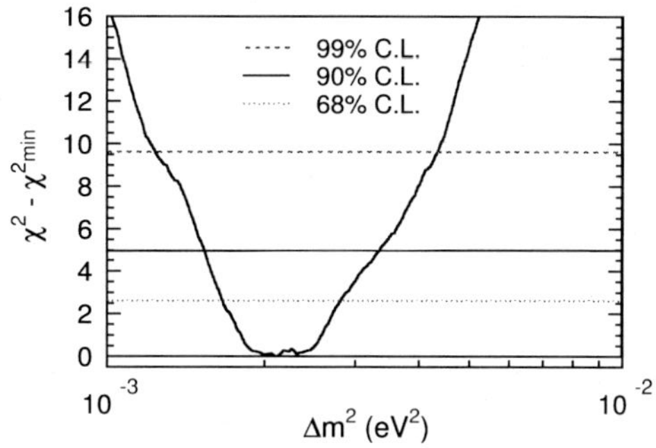

FIG. 11.31. Super-Kamiokande $\chi^2 - \chi^2_{\rm min}$ as a function of Δm^2 in the case of $\nu_\mu \to \nu_\tau$ oscillations [111].

These values confirm those measured in Kamiokande (eqns (11.65) and (11.66)) and IMB (eqn (11.67)). From Fig. 11.27 one can see that these values of the ratio of ratios are consistently explained by $\nu_\mu \to \nu_\tau$ oscillations with maximal mixing ($\sin^2 2\vartheta = 1$) and Δm^2 in the range between 10^{-3} and $10^{-2}\,\mathrm{eV}^2$.

Super-Kamiokande also measured the flux of upward-going neutrino-induced muons. In order to reduce the uncertainty due to the calculation of the neutrino fluxes, the SK collaboration considered the vertical/horizontal ratio V/H of through-going muons, where V and H are, respectively, the numbers of events with $0.5 < \cos\theta_z < 1$ and $0 < \cos\theta_z < 0.5$. The measured value and the value expected in the absence of oscillations are

$$(V/H)_{\rm data} = 0.497 \pm 0.022 \pm 0.003\,, \qquad (V/H)_{\rm MC} = 0.586 \pm 0.019\,, \qquad (11.71)$$

which show a 3σ discrepancy. Figure 11.28 shows the expected value of V/H in the case of $\nu_\mu \to \nu_\tau$ oscillations as a function of Δm^2, for $\sin^2 2\vartheta = 1$. One can see that neutrino oscillations can reproduce the observed ratio if $1 \times 10^{-3} \lesssim \Delta m^2 \lesssim 3 \times 10^{-3}\,\mathrm{eV}^2$ or $5 \times 10^{-2} \lesssim \Delta m^2 \lesssim 1 \times 10^{-1}\,\mathrm{eV}^2$.

Another quantity with reduced uncertainty due to the calculation of the neutrino fluxes is the ratio $R_{\rm stop/through}$ of stopping and through-going muons. The measured value and the value expected in the absence of oscillations are

$$R^{\rm data}_{\rm stop/through} = 0.229 \pm 0.015 \pm 0.003\,, \qquad R^{\rm MC}_{\rm stop/through} = 0.430 \pm 0.065\,, \quad (11.72)$$

with a discrepancy of more than 3σ. Figure 11.29 shows the expected value of $R_{\rm stop/through}$ in the case of $\nu_\mu \to \nu_\tau$ oscillations as a function of Δm^2, for $\sin^2 2\vartheta = 1$, compared with the observed ratio. One can see that neutrino oscillations are

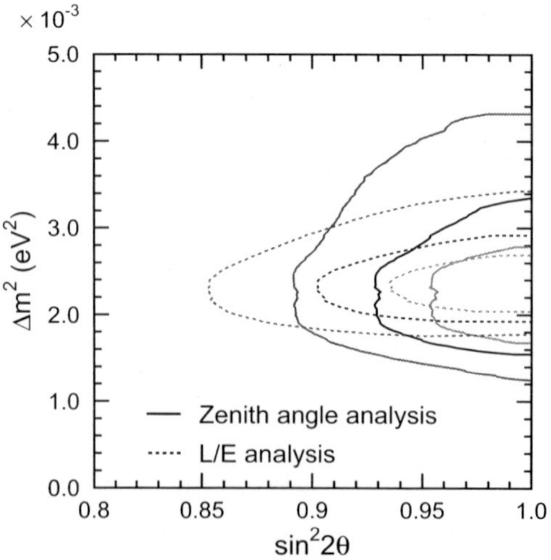

FIG. 11.32. Comparison of the 68%, 90%, and 99% CL allowed regions in Fig. 11.30 with the corresponding ones obtained with the L/E analysis of Super-Kamiokande data in Ref. [110]. Figure taken from Ref. [111].

consistent with the data if Δm^2 is in the range between about 2×10^{-3} and $4 \times 10^{-3}\,\text{eV}^2$.

Super-Kamiokande has measured the atmospheric neutrino fluxes with an accuracy which is by far better than those of other experiments (between 1996 and 2001, SK-I observed more than 1.5×10^4 atmospheric neutrino events with energy between 100 MeV and 10 TeV). Thus, SK data determine our present knowledge of the oscillations of atmospheric neutrinos. Figure 11.30 shows the regions in the $\sin^2 2\vartheta$-Δm^2 plane for $\nu_\mu \to \nu_\tau$ oscillations allowed at 68%, 90%, and 99% CL. The best-fit is at

$$\sin^2 2\vartheta = 1.00\,, \qquad \Delta m^2 = 2.1 \times 10^{-3}\,\text{eV}^2\,. \tag{11.73}$$

The 90% CL ranges of the oscillation parameters are

$$\sin^2 2\vartheta > 0.92\,, \qquad 1.5 \times 10^{-3} < \Delta m^2 < 3.4 \times 10^{-3}\,\text{eV}^2\,. \tag{11.74}$$

Figure 11.31 shows the value of $\chi^2 - \chi^2_{\text{min}}$ as a function of Δm^2. One can see that the χ^2 is rather flat between $2.0 \times 10^{-3}\,\text{eV}^2$ and $2.5 \times 10^{-3}\,\text{eV}^2$. Thus, in practice, the best-fit value of Δm^2 is not determined within this range. On the other hand, since the value of $\chi^2 - \chi^2_{\text{min}}$ increases when $\sin^2 2\vartheta$ is decreased from the best-fit value, the data show a preference for maximal mixing.

In Fig. 11.30 and in eqns (11.73) and (11.74) the mixing is quantified through $\sin^2 2\vartheta$, because in $\nu_\mu \to \nu_\tau$ oscillations there are no matter effects, since ν_μ and ν_τ have the same interactions with matter. Hence, the transition probability depends on the mixing angle through $\sin^2 2\vartheta$, according to eqn (7.70). Notice that in our convention Δm^2 is defined to be positive (see section 7.4) and each value of $\sin^2 2\vartheta$ corresponds to two values of ϑ related by $\vartheta \leftrightarrows \pi/2 - \vartheta$.

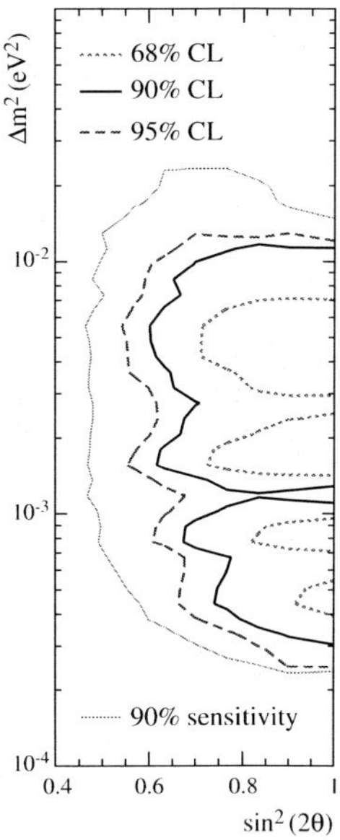

FIG. 11.33. Allowed region in the $\sin^2 2\vartheta$-Δm^2 plane for $\nu_\mu \to \nu_\tau$ oscillations obtained in the Soudan 2 experiment [69].

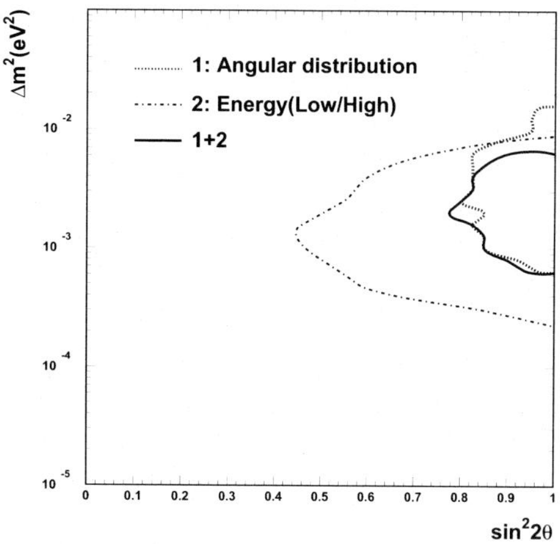

FIG. 11.34. Regions in the $\sin^2 2\vartheta$-Δm^2 plane for $\nu_\mu \to \nu_\tau$ oscillations allowed at 90% CL by different data samples of the MACRO experiment [84].

In Ref. [110], the SK collaboration obtained the allowed region in the $\sin^2 2\vartheta$-Δm^2 plane corresponding to the L/E analysis in Figs. 11.24 and 11.25. This allowed region is compared in Fig. 11.32 with that in Fig. 11.30. In spite of the fact that the L/E analysis does not use the full set of atmospheric data, but only the subset of the data which allow the determination of L/E, the constraint on the value of Δm^2 is more stringent than that obtained from the analysis of full data set. The reason is that the location of the dip in Fig. 11.25, corresponding to the first maximum of the oscillation probability, is very sensitive to the value of Δm^2. On the other hand, from Fig. 11.32 one can see that the analysis of the full data set provides a better determination of the value of $\sin^2 2\vartheta$ due to a higher statistics.

11.2.4 Soudan 2

The Soudan 2 experiment was located 710 m underground in the Soudan underground Mine State Park, in Minnesota (USA), with an overburden of 2070 mwe. The main detector was an iron tracking calorimeter which operates as a time projection chamber. The total mass was of 963 ton, with a fiducial mass of 770 ton. The main detector was surrounded by an active shield of aluminum proportional tubes. The experiment took data from 1989 to 2001 [67, 68, 922, 69].

The Soudan 2 experiment measured the ratio-of-ratios [922]

$$R_{\mu/e}^{\text{sub-GeV}} = 0.69 \pm 0.10 \pm 0.06\,, \tag{11.75}$$

for neutrinos with energy smaller than about 1 GeV. This value, which is in agreement with those measured in Kamiokande, IMB and SK, indicates a disappearance of muon neutrinos. The zenith-angle distribution of contained events [922] and upward-going muons [69] in Soudan 2 show a disappearance of muon neutrinos coming from below, in agreement with the SK measurements.

Figure 11.33 shows the region in the $\sin^2 2\vartheta$-Δm^2 plane for $\nu_\mu \to \nu_\tau$ oscillations allowed by the data of the Soudan 2 experiment [69]. Comparing with the SK allowed region in Fig. 11.30, one can see that the results of the two experiments are compatible and the Soudan 2 allowed region is much larger than that of the SK experiment. This is due to the small statistics collected in the Soudan 2 experiment, compared with that obtained in the SK experiment. Therefore, the results of the Soudan 2 experiment are considered as a confirmation of the results of Kamiokande, IMB and SK. However, it must be emphasized that such confirmation is very important, since it has been obtained through a detector which is completely different from the water Cherenkov detectors Kamiokande, IMB, and SK.

11.2.5 MACRO

MACRO (Monopole Astrophysics and Cosmic Ray Observatory) was a large-area scintillator detector which could distinguish upward-going muons by measuring their time of flight [41, 81, 82, 83, 84]. It was located in the Laboratori Nazionali del Gran Sasso (LNGS), in Italy, with an overburden of 3700 mwe. It operated from 1989 to 1995 with part of the apparatus. After its construction was completed in 1995, MACRO took data with the whole apparatus until 2000. It had dimensions of $76.6 \times 12 \times 9.3\,\mathrm{m}^3$.

The zenith-angle distribution of upward through-going muons measured by MACRO show a deviation from the Monte Carlo expectation without neutrino oscillations. The best fit is obtained with $\sin^2 2\vartheta \simeq 1.0$ and $\Delta m^2 \simeq 2.5 \times 10^{-3}\,\mathrm{eV}^2$ in the case of $\nu_\mu \to \nu_\tau$ oscillations [84].

In order to reduce the uncertainty due to the calculation of the neutrino fluxes and optimize the oscillation signal, the MACRO collaboration considered the ratio $N_{\mathrm{low}}/N_{\mathrm{high}}$, where N_{low} and N_{high} are, respectively, the number of events with reconstructed neutrino energy lower than 30 GeV and higher than 130 GeV. The measured value of $N_{\mathrm{low}}/N_{\mathrm{high}}$ is 0.85 ± 0.16, whereas the Monte Carlo expectation without neutrino oscillations is 1.50 ± 0.25, with a discrepancy of about 2σ.

Figure 11.34 shows the regions in the $\sin^2 2\vartheta$-Δm^2 plane for $\nu_\mu \to \nu_\tau$ oscillations allowed by the angular distribution of MACRO events, by the observed ratio $N_{\mathrm{low}}/N_{\mathrm{high}}$ and by their combination. Comparing with the SK allowed region in Fig. 11.30, one can see that the results of MACRO and SK are compatible and the MACRO allowed region includes the SK allowed region.

Furthermore, MACRO confirmed the SK preference of $\nu_\mu \to \nu_\tau$ oscillations with respect to transitions into sterile neutrinos, since the matter effects in $\nu_\mu \to \nu_s$ oscillations are disfavored by the zenith-angle distribution of upward-going muons observed in MACRO [83].

11.2.6 Connection with long-baseline experiments

Long-baseline (LBL) experiments are sensitive to the range of Δm^2 probed by atmospheric neutrino experiments (see section 7.5). Thus, LBL experiments can be used to confirm and improve the results of atmospheric neutrino experiments. Three LBL experiments have been performed so far: the reactor $\bar{\nu}_e$-disappearance experiments CHOOZ and Palo Verde (see section 12.2.2) and the accelerator ν_μ-disappearance experiment K2K (see section 12.3.2).

The CHOOZ [97, 98, 100] and Palo Verde [254, 253, 255] experiments did not measure any disappearance of reactor $\bar{\nu}_e$'s, as previous SBL reactor experiments (see Fig. 12.5). However, the LBL character of the CHOOZ and Palo Verde experiments led to the extension of the exclusion curve for oscillations of $\bar{\nu}_e$ into other states to values of Δm^2 as low as about $10^{-3}\,\text{eV}^2$, excluding the $\nu_\mu \leftrightarrows \nu_e$ oscillation solution of the atmospheric neutrino anomaly (see Figs. 12.6 and 12.7).

The K2K experiment, on the other hand, observed a disappearance of ν_μ's which is compatible with the results of atmospheric neutrino experiments. This can be seen by comparing the K2K allowed region in Fig. 12.17 and the SK allowed region in Fig. 11.30.

Let us emphasize that the K2K observation of neutrino oscillations with the same values of the mixing parameters as those found in atmospheric neutrino experiments is extremely important, because it shows without doubt that the disappearance of atmospheric muon neutrinos is due to neutrino oscillations, eliminating other mechanisms. Of course, other mechanisms are still possible as subleading effects, but the important point is that atmospheric neutrino experiments and K2K have proved that neutrinos are massive, that there is a squared-mass difference of the order of $2 - 3 \times 10^{-3}\,\text{eV}^2$ and that the mixing relevant for the observed oscillations is maximal or close to maximal.

12

TERRESTRIAL NEUTRINO OSCILLATION EXPERIMENTS

> Through all one sees the spirit of catch as catch can, trial and error, progress by making almost all possible mistakes, the great point being only to make them as quickly as possible and to learn from them.
>
> **John A. Wheeler**, *At Home in the Universe*

Terrestrial neutrino oscillation experiments started in the late 70s using neutrinos produced in reactors and accelerators. For a long time, terrestrial neutrino oscillation experiments did not find neutrino oscillations. Their results led to exclusion curves which constrained the allowed region in the parameter space of neutrino oscillations. Recently, the KamLAND reactor experiment [398, 103] and the K2K accelerator experiment [48, 46, 47, 66] succeeded in finding important evidences of neutrino oscillations. Both are long-baseline (LBL) experiments, which have been carried out in order to check the evidence of neutrino oscillations found in solar and atmospheric neutrino experiments.

In this chapter we discuss mainly the setup and results of KamLAND and K2K (the controversial results of the LSND experiment are briefly discussed in section 12.3.1). The other terrestrial neutrino oscillation experiments are only mentioned and their exclusion curves shown in figures, with the exception of the CHOOZ [97, 98, 100] and Palo Verde [254, 253, 255] experiments, which are discussed in some detail, since they have been important to exclude the $\nu_\mu \leftrightarrows \nu_e$ channel as an explanation of the atmospheric neutrino anomaly found in Kamiokande and IMB (see section 11.2) and for constraining the value of the element U_{e3} of the neutrino mixing matrix in the case of three-neutrino mixing (see section 13.3.1).

In a discussion of the capability of an experiment to measure the oscillation parameters, it is useful to consider, as a measure of merit, the *sensitivity* of the experiment, which is defined and discussed in section 12.1. Generally, the sensitivity of a neutrino oscillation experiment depends on many factors. The most important are the source–detector distance, the neutrino energy, the power of the source, the mass of the detector, the detection cross-section and the background. In the case of atmospheric neutrinos and neutrinos from astrophysical sources (Sun, supernovae, etc.), the values of the first three factors are fixed by nature and the experimenter can improve its measurements only by increasing the mass of the detector, by choosing a better detection material and by decreasing the background. On the other hand, in terrestrial neutrino experiments there is some freedom to vary the values of all the five above-mentioned factors. Hence, in the discussions of the capabilities of different terrestrial experiments it is useful to understand the dependence of the sensitivity of an experiment on the five factors above. This is explained in section 12.1. In the following section 12.2 we review reactor neutrino oscillation experiments (see also Refs. [205, 800]), with special attention to CHOOZ, Palo Verde and KamLAND. Finally, in section 12.3 we discuss accelerator neutrino experiments, focusing on the K2K experiment.

12.1 Sensitivity

The sensitivity of an experiment is a measure of merit which characterizes the potential of the experiment for the measurement of the oscillation parameters. In the plane of the two-neutrino oscillation parameters $\sin^2 2\vartheta$ and Δm^2, the sensitivity of an experiment is expressed by its *sensitivity curve*, which is the exclusion curve that the experiment would yield in the absence of a signal[68]. From the discussion in section 7.6, it is clear that the sensitivity curve depends on the value of the upper limit $(P^{\text{max}}_{\nu_\alpha \to \nu_\beta})_{\text{s}}$ on the averaged transition probability that an experiment can obtain in the absence of a signal. It is also useful to define separately the sensitivities to $\sin^2 2\vartheta$ and Δm^2.

The sensitivity to $\sin^2 2\vartheta$, denoted by $(\sin^2 2\vartheta)_{\text{s}}$, is defined as the limit on $\sin^2 2\vartheta$ obtained in the absence of a signal for large Δm^2. From eqn (7.103), we have

$$(\sin^2 2\vartheta)_{\text{s}} = 2\,(P^{\text{max}}_{\nu_\alpha \to \nu_\beta})_{\text{s}}\,, \tag{12.1}$$

which is independent of Δm^2. Since the sensitivity to $\sin^2 2\vartheta$ is proportional to $(P^{\text{max}}_{\nu_\alpha \to \nu_\beta})_{\text{s}}$, experiments which aim to explore small values of the mixing angle must reach small values of $(P^{\text{max}}_{\nu_\alpha \to \nu_\beta})_{\text{s}}$.

The sensitivity to Δm^2, denoted by $(\Delta m^2)_{\text{s}}$, is defined as the value of Δm^2 at which the exclusion curve intercepts the $\sin^2 2\vartheta = 1$ boundary. From eqn (7.108), we obtain

$$(\Delta m^2)_{\text{s}} \simeq 4\,\sqrt{(P^{\text{max}}_{\nu_\alpha \to \nu_\beta})_{\text{s}}}\,\frac{E}{L}\,, \tag{12.2}$$

where E is the average energy and we neglect the small sizes of the source and the detector compared with the source–detector distance L. Thus, improvements in the value of $(P^{\text{max}}_{\nu_\alpha \to \nu_\beta})_{\text{s}}$ have a weak effect on $(\Delta m^2)_{\text{s}}$. In order to improve effectively the sensitivity to small values of Δm^2 it is necessary to increase the ratio L/E.

In the case of ν_α disappearance experiments, the value of $(P^{\text{max}}_{\nu_\alpha \to \nu_\beta})_{\text{s}}$, with $\beta \neq \alpha$, is determined by the statistical fluctuation of the number of events. In fact, we have

$$P_{\nu_\alpha \to \nu_\beta} = 1 - P_{\nu_\alpha \to \nu_\alpha} = 1 - \frac{N^{\text{obs}}_{\nu_\alpha}}{N^{\text{no-osc}}_{\nu_\alpha}}\,, \tag{12.3}$$

where $N^{\text{obs}}_{\nu_\alpha}$ and $N^{\text{no-osc}}_{\nu_\alpha}$ are, respectively, the number of observed ν_α events and the number of expected ν_α events in the absence of oscillations. Then, neglecting

68 See, however, the discussion in Ref. [541].

the uncertainty of $N_{\nu_\alpha}^{\text{no-osc}}$, the uncertainty of $P_{\nu_\alpha \to \nu_\beta}$ is given by

$$\Delta P_{\nu_\alpha \to \nu_\beta} \simeq \frac{\Delta N_{\nu_\alpha}^{\text{obs}}}{N_{\nu_\alpha}^{\text{no-osc}}} \,. \tag{12.4}$$

The observed number of events $N_{\nu_\alpha}^{\text{obs}}$ has a Poisson distribution with variance $(\Delta N_{\nu_\alpha}^{\text{obs}})^2 = N_{\nu_\alpha}^{\text{obs}}$. Since the sensitivity is determined in the absence of a signal, if the background is negligible, we have $N_{\nu_\alpha}^{\text{obs}} \simeq N_{\nu_\alpha}^{\text{no-osc}}$. Thus, we obtain

$$(P_{\nu_\alpha \to \nu_\beta}^{\text{max}})_{\text{s}} \simeq \Delta P_{\nu_\alpha \to \nu_\beta}\big|_{\text{no-signal}} \simeq \left(N_{\nu_\alpha}^{\text{no-osc}}\right)^{-1/2} \,. \tag{12.5}$$

From eqn (12.1), it follows that the sensitivity to $\sin^2 2\vartheta$ can improve only as the square root of the number of events.

In order to understand the behavior of $(\sin^2 2\vartheta)_{\text{s}}$ and $(\Delta m^2)_{\text{s}}$ as functions of the main quantities which characterize a neutrino oscillation experiment, we note that the number of events is given by

$$N_{\nu_\alpha}^{\text{no-osc}} \propto \frac{N_{\nu_\alpha}^{\text{src}} \, \sigma_{\nu_\alpha}(E) \, \eta \, M_{\text{det}}}{L^2} \,, \tag{12.6}$$

where $N_{\nu_\alpha}^{\text{src}}$ is the number of ν_α's emitted by the source, $\sigma_{\nu_\alpha}(E)$ is the average detection cross-section, η is the detection efficiency, and M_{det} is the mass of the detector. Then, we have

$$(P_{\nu_\alpha \to \nu_\beta}^{\text{max}})_{\text{s}} \propto \frac{L}{\sqrt{N_{\nu_\alpha}^{\text{src}} \, \sigma_{\nu_\alpha}(E) \, \eta \, M_{\text{det}}}} \,. \tag{12.7}$$

It is clear that in order to reach a small value of $(P_{\nu_\alpha \to \nu_\beta}^{\text{max}})_{\text{s}}$ and hence of $(\sin^2 2\vartheta)_{\text{s}}$ a small source–detector distance would be convenient. However, such a choice is undesirable for the sensitivity to Δm^2, which is given by

$$(\Delta m^2)_{\text{s}} \propto E \, L^{-1/2} \left(N_{\nu_\alpha}^{\text{src}} \, \sigma_{\nu_\alpha}(E) \, \eta \, M_{\text{det}}\right)^{-1/4} \,. \tag{12.8}$$

Thus, lowering the energy and increasing the source–detector distance are the main factors which determine an improvement of the sensitivity to Δm^2.

Considering the neutrino energy, the cross-section $\sigma_{\nu_\alpha}(E)$ increases with neutrino energy, but it cannot increase faster than E^2, since dimensionally at most we have $\sigma_{\nu_\alpha}(E) \propto G_{\text{F}} E^2$. Therefore, eqn (12.8) implies that $(\Delta m^2)_{\text{s}}$ can be lowered by decreasing the energy E. However, in practice there are limits below which the energy cannot be decreased. One limit is due to the detection threshold, below which $\sigma_{\nu_\alpha}(E) = 0$ (the above dimensional argument is valid well above threshold). Another limit applies to accelerator experiments (see section 12.3), where the neutrino beam is produced by pions generated by accelerated protons hitting a target. In this case, the intensity of the neutrino beam increases with the proton energy and hence with the energy of the neutrino beam itself. Thus, $N_{\nu_\alpha}^{\text{src}}$ depends on the neutrino energy, complicating the energy dependence of $(\Delta m^2)_{\text{s}}$ in eqn (12.8). In this case, the experimental design must be studied in order to optimize the energy

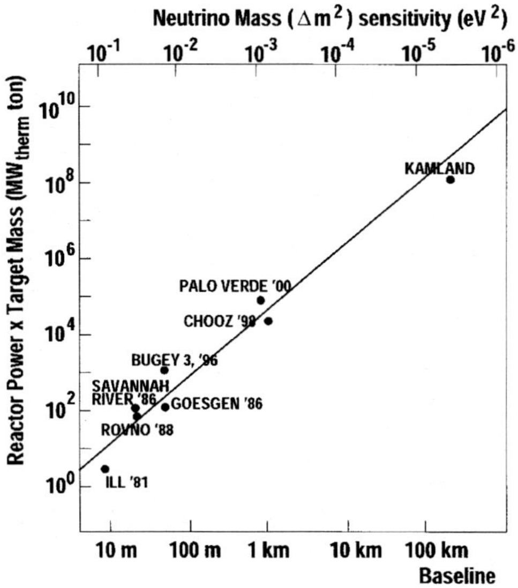

FIG. 12.1. Sensitivity to Δm^2 as a function of the product of the total reactor power and the detector fiducial mass [205].

of the beam for the desired sensitivity to Δm^2 within the existing practical limitations. Thus, accelerator experiments typically have neutrino energies from 100 MeV to 100 GeV. The smallest neutrino energy, of the order of a few MeV, is reached in reactor experiments, discussed in section 12.2. Typically, these experiments allow the exploration of very small values of Δm^2.

If the neutrino energy cannot be lowered, the best way to increase the sensitivity to Δm^2 is to increase the source–detector distance. The resulting loss in sensitivity to $\sin^2 2\vartheta$ can be avoided or alleviated by increasing the power of the source and increasing the mass of the detector. For example, $(\sin^2 2\vartheta)_{\rm s}$ remains constant if $N^{\rm src}_{\nu_\alpha}\,\sigma_{\nu_\alpha}(E)\,\eta\,M_{\rm det} \propto L^2$. Such an increase of $N^{\rm src}_{\nu_\alpha}\,\sigma_{\nu_\alpha}(E)\,\eta\,M_{\rm det}$ is also advantageous for the sensitivity to Δm^2, since it implies that $(\Delta m^2)_{\rm s} \propto E/L$. This is illustrated, for the case of reactor experiments, in Fig. 12.1 (taken from Ref. [205]), where the bottom abscissa is the source–detector length L and the ordinate is the product of the thermal power of the reactor, which is proportional to $N^{\rm src}_{\nu_\alpha}$, and the mass $M_{\rm det}$ of the detector (all reactor neutrino detectors have the same E and $\sigma_{\nu_\alpha}(E)$, with similar efficiencies η). The diagonal line represents the function $N^{\rm src}_{\nu_\alpha}\,M_{\rm det} \propto L^2$. The top abscissa gives the value of $(\Delta m^2)_{\rm s} \propto L^{-1}$ for experiments lying on the diagonal line. One can see that all experiments lie near the diagonal line, in order to reach different sensitivities to Δm^2, keeping the sensitivity $\sin^2 2\vartheta$ at a significant level. Obviously, the experiments above the diagonal line have better $\sin^2 2\vartheta$ sensitivity and the sensitivity to Δm^2 increases towards the right.

Let us consider now a $\nu_\alpha \to \nu_\beta$ appearance experiment, with $\beta \neq \alpha$. The transition probability is given by

$$P_{\nu_\alpha \to \nu_\beta} \propto \frac{N^{\text{obs}}_{\nu_\beta} - N^{\text{bck}}_{\nu_\beta}}{\phi_{\nu_\alpha} \, T_{\text{obs}} \, \sigma_{\nu_\beta}(E) \, \eta \, M_{\text{det}}} \,, \tag{12.9}$$

where $N^{\text{bck}}_{\nu_\beta}$ is the estimated number of background events, ϕ_{ν_α} is the flux of ν_α through the detector, and T_{obs} is the observation time. Neglecting the uncertainties of ϕ_{ν_α}, $\sigma_{\nu_\beta}(E)$, and $N^{\text{bck}}_{\nu_\beta}$, the uncertainty of the transition probability is given by

$$\Delta P_{\nu_\alpha \to \nu_\beta} \propto \frac{\Delta N^{\text{obs}}_{\nu_\beta}}{\phi_{\nu_\alpha} \, T_{\text{obs}} \, \sigma_{\nu_\beta}(E) \, \eta \, M_{\text{det}}} \,, \tag{12.10}$$

where $\Delta N^{\text{obs}}_{\nu_\beta}$ is the statistical fluctuation of the number of observed events. Since the sensitivity is defined in the absence of a signal, the number of observed events is due to the background and, from the Poisson distribution of the background, we have $\Delta N^{\text{obs}}_{\nu_\beta} = \sqrt{N^{\text{bck}}_{\nu_\beta}}$, leading to

$$(P^{\max}_{\nu_\alpha \to \nu_\beta})_{\text{s}} \simeq \Delta P_{\nu_\alpha \to \nu_\beta}\big|_{\text{no-signal}} \propto \frac{\sqrt{N^{\text{bck}}_{\nu_\beta}}}{\phi_{\nu_\alpha} \, T_{\text{obs}} \, \sigma_{\nu_\beta}(E) \, \eta \, M_{\text{det}}} \,. \tag{12.11}$$

It is clear that decreasing the background is crucial in order to improve the sensitivity. Unfortunately, in practice there is always a form of background that cannot be eliminated, which is the so-called beam-related background. For example, in accelerator experiments measuring $\nu_\mu \to \nu_e$ transitions with a beam of ν_μ's produced by $\pi^+ \to \mu^+ + \nu_\mu$ decays, there is an unavoidable beam-related background of ν_e's mainly due to the $\mu^+ \to e^+ + \nu_e + \bar{\nu}_\mu$ decays. Another form of beam-related background are the ν_α-induced events which are misidentified as ν_β events. From these two examples, it is clear that the beam-related background is approximately proportional to the intensity of the beam. Considering $N^{\text{bck}}_{\nu_\beta} \propto \phi_{\nu_\alpha} \, T_{\text{obs}} \, \sigma_{\nu_\beta}(E) \, \eta \, M_{\text{det}}$, we have

$$(P^{\max}_{\nu_\alpha \to \nu_\beta})_{\text{s}} \propto \frac{L}{\sqrt{N^{\text{src}}_{\nu_\alpha} \, \sigma_{\nu_\beta}(E) \, \eta \, M_{\text{det}}}} \,, \tag{12.12}$$

where $N^{\text{src}}_{\nu_\alpha} \propto \phi_{\nu_\alpha} \, T_{\text{obs}} \, L^2$ is the number of ν_α's emitted by the source. This expression has the same structure as that in eqn (12.7) obtained for disappearance experiments. Therefore, once all sources of background above the unavoidable beam-related background have been eliminated, the sensitivity of appearance experiments follows the same rules as that of disappearance experiments, discussed after eqn (12.7).

12.2 Reactor experiments

Fission reactors are copious sources of electron antineutrinos produced in the β-decays of neutron-rich nuclei. The power of a reactor is mainly due to the fission of four isotopes: ^{235}U ($\sim$ 56%), ^{238}U ($\sim$ 8%), ^{239}Pu ($\sim$ 30%), and ^{241}Pu ($\sim$ 6%). The electron antineutrinos are produced by the chain of β-decays of the fission products. Since, on average, each fission produces about 200 MeV, with release of about six $\bar{\nu}_e$'s, the electron antineutrino yield is about $2 \times 10^{20}\,\mathrm{s}^{-1}$ for each $\mathrm{GW_{th}}$ of thermal power. Typical modern light-water nuclear power plants have several reactor cores, each with a thermal power of the order of $3\,\mathrm{GW_{th}}$. Thus, the total rate of antineutrino production of a typical nuclear power plant is very high. Unfortunately, however, since the antineutrino flux is isotropic, it decreases rapidly with distance. This is obviously a problem for neutrino oscillation experiments, which require an appropriate source–detector distance in order to reveal the oscillations. This drawback is partially compensated by the low-energy of reactor antineutrinos, of the order of a few MeV, which implies a relatively short oscillation length. Such a low energy also implies that only $\bar{\nu}_e$ disappearance can be investigated in reactor experiments, since the energy is not sufficient to produce, in a detector, muons or taus through charged-current interactions. The neutral-current interactions of $\bar{\nu}_\mu$'s and $\bar{\nu}_\tau$'s produced by oscillations are too weak to be distinguished from the background. The disappearance character of reactor neutrino oscillation experiments implies that it is very difficult to measure a small mixing, which gives a small disappearance, because of the statistical fluctuations of the signal.

Reactor electron antineutrinos are detected through the *inverse neutron decay process*

$$\bar{\nu}_e + p \to n + e^+ \,, \tag{12.13}$$

which was already used by Cowan and Reines in their first detection of electron antineutrinos produced in the Savannah River power plant [899]. This reaction liberates a total visible energy $E_e + m_e$ where E_e is the energy of the positron, which annihilates immediately with a surrounding electron. This energy can be seen in scintillator detectors [898, 334]. The antineutrino events are distinguished from the background by the coincidence of the prompt positron signal with the delayed signal produced by the nuclear capture of the neutron. Neglecting the small recoil energy of the neutron, the neutrino and positron energies are related by

$$E_\nu = E_e + T_n + m_n - m_p \simeq E_e + 1.293\,\mathrm{MeV}\,, \tag{12.14}$$

where T_n is the negligibly small recoil kinetic energy of the neutron. From eqn (5.37), the neutrino energy threshold is given by

$$E_\nu^{\mathrm{th}} = \frac{(m_n + m_e)^2 - m_p^2}{2\,m_p} \simeq 1.806\,\mathrm{MeV}\,, \tag{12.15}$$

which is slightly larger than the naive $m_n - m_p + m_e \simeq 1.804\,\mathrm{MeV}$. The cross-section is given by[69]

$$\sigma_{\mathrm{CC}}^{\bar{\nu}_e p} = \frac{G_{\mathrm{F}}^2\,|V_{ud}|^2}{\pi}\left(g_V^2 + 3\,g_A^2\right) E_e\,p_e\,, \tag{12.16}$$

[69] The energy-dependent radiative corrections have been calculated in Refs. [1041, 429]. The corrections of the order E_ν/m_N and the angular distribution of the positrons are discussed in Ref. [1042].

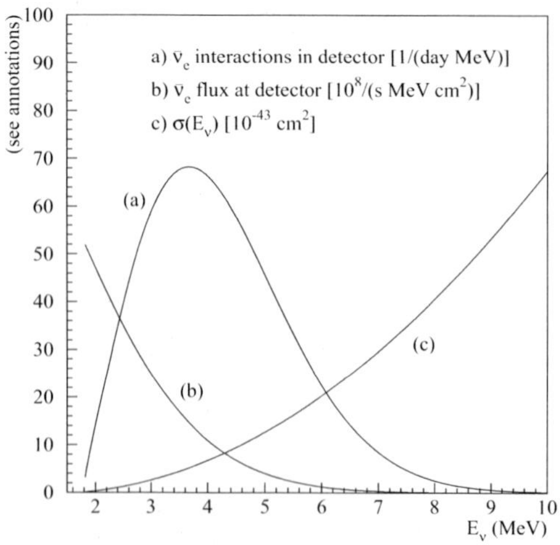

FIG. 12.2. Reactor $\bar{\nu}_e$ flux, inverse neutron decay cross-section, and $\bar{\nu}_e$ interaction spectrum at a detector based on such a reaction: (a) and (b) refer to a 12 ton fiducial mass detector located 0.8 km from a 12 $\mathrm{GW_{th}}$ power reactor [205].

where $g_V \simeq 1$ and $g_A \simeq 1.27$ (see eqn (5.143)). The cross-section can be conveniently written in terms of the neutron lifetime in eqn (5.140) as

$$\sigma_{\mathrm{CC}}^{\bar{\nu}_e p} = \frac{2\pi^2}{\tau_n\, m_e^5\, f}\, E_e\, p_e \simeq 9.56 \times 10^{-44} \left(\frac{E_e\, p_e}{\mathrm{MeV}^2}\right) \left(\frac{\tau_n}{886\,\mathrm{s}}\right)^{-1} \mathrm{cm}^2\,, \tag{12.17}$$

where f is the phase space integral in eqns (5.141) and (5.142). This form has the advantage of expressing the cross-section in terms of the well-measured quantities m_e and τ_n (see eqns (A.150) and (A.158)), eliminating the need to know the values of $|V_{ud}|$, g_V, and g_A.

The threshold of about 1.8 MeV implies that only about 25% of the antineutrinos produced in a reactor can be detected, since the others are below threshold. The response of a detector to a reactor $\bar{\nu}_e$ flux is illustrated by the (a) curve in Fig. 12.2, which is proportional to the product of the antineutrino spectrum represented by the (b) curve and the detection cross-section represented by the (c) curve. One can see that the peak of the antineutrino detection is at the energy of about 3.6 MeV.

The calculation of a reactor antineutrino spectrum is a difficult task, since the decay of each isotope produces a different neutrino spectrum. The qualitative features of the existing calculations are reviewed in Refs. [205, 800]. Here we only mention the following interesting points:

- The threshold of about 1.8 MeV implies that only antineutrinos produced by β-decays with a relatively large Q-value are detected. Since these β-decays are relatively fast, the intensity of the $\bar{\nu}_e$ flux is closely correlated in time with the thermal power of the reactor.

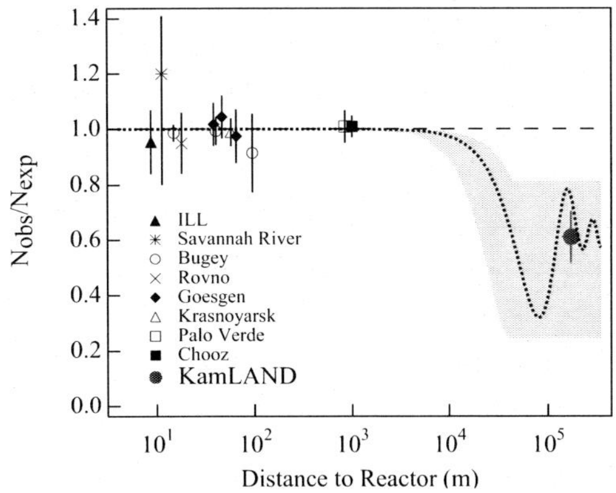

FIG. 12.3. The ratio of measured to expected $\bar{\nu}_e$ flux of different reactor neutrino experiments as a function of their source–detector distance. The shaded region indicates the range of flux predictions corresponding to the 95% CL LMA region found in a global analysis of the solar neutrino data [450]. The dotted curve corresponds to the best-fit values $\Delta m^2_{\rm sol} = 5.5 \times 10^{-5}\,{\rm eV}^2$ and $\sin^2 2\vartheta_{\rm sol} = 0.83$. Figure from Ref. [398].

- Reactors require shutdowns for refueling every 12–24 months. The temporal flux modulation due to these shutdowns can be used to check the background subtraction methods.
- The intensity of the $\bar{\nu}_e$ flux is proportional to the thermal power of the reactor, which is monitored with accuracy better than 1% by the power plant operators. Also the initial isotope composition of the nuclear fuel is known with high accuracy.
- Several SBL reactor neutrino oscillation experiments (see section 12.2.1) measured antineutrino fluxes and energy spectra in agreement with the calculations. Since these measurements have been done at several different source–detector distances, it is safe to assume that they represent direct determinations of the antineutrino spectrum produced in a reactor. These measurements have been used to reduce the uncertainty of the calculated antineutrino spectrum for LBL experiments.

Figure 12.3 shows the ratio of measured to expected $\bar{\nu}_e$ flux of different reactor neutrino experiments as a function of their source–detector distance L. The experiments with $L \sim$ 10–100 m are short-baseline (SBL) experiments, the two experiments (CHOOZ and Palo Verde) with $L \sim 1\,{\rm km}$ are long-baseline (LBL) experiments, and the experiment (KamLAND) with $L \sim 200\,{\rm km}$ is the only very-long-baseline (VLBL) experiment. These types of reactor experiments are discussed, respectively, in the following subsections 12.2.1, 12.2.2, and 12.2.3. From Fig. 12.3, one can see that SBL and LBL reactor experiments, which did not have a sufficient source–detector distance, were not sensitive to the disappearance of $\bar{\nu}_e$ due to the small solar Δm^2. As explained in section 12.1, the sensitivity to small values

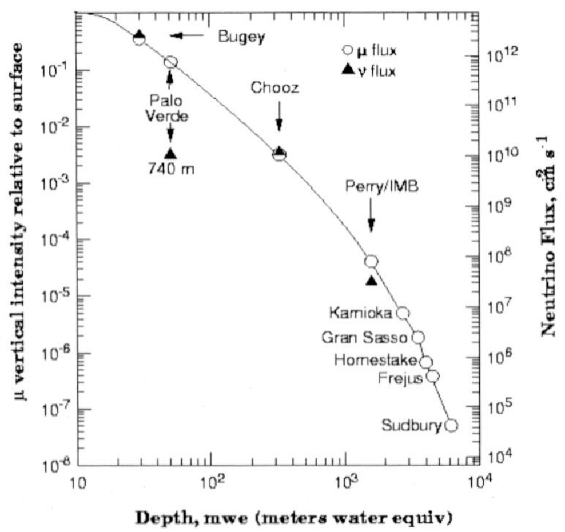

FIG. 12.4. Cosmic muon flux and reactor neutrino flux in different underground experimental sites [100].

of Δm^2 improves by increasing the source–detector distance. In fact, the VLBL experiment KamLAND was successfully carried out at a very large source detector distance, in order to reach the sensitivity to small values of Δm^2 required for the observation of $\bar{\nu}_e$ disappearance due to the solar Δm^2.

In reactor neutrino experiments, it is important to reduce the background below the neutrino event rate. This means that reactor neutrino experiments must be shielded from the cosmic ray background. However, since the neutrino flux from reactors is very intense, in SBL and LBL experiments it is typically necessary only to have shielding from the hadronic component of cosmic rays and to have a moderate reduction of the muon component. In this case, an underground location with a shallow depth is sufficient. This is fortunate, because it could be very difficult to find an appropriate deep underground location within 100 m or 1 km from a nuclear power plant. Figure 12.4 shows the cosmic ray muon flux at some underground experimental sites. For the Bugey, Palo Verde, and CHOOZ experiments, the reactor neutrino flux is shown to be similar to the muon flux (the Perry experiment [982], named in Fig. 12.4, is an unrealized project). The VLBL KamLAND experiment is located in the Kamioka mine, where the low cosmic muon flux allows a measurement of the flux of neutrinos from reactors at a distance of more than 100 km, which is of the order of 10^4 times smaller than the flux at CHOOZ.

12.2.1 Short-baseline reactor experiments

SBL reactor experiments have source–detector distances between 10 and 100 m. Several experiments of this type have been performed from the late 1970s to the 1990s: ILL [710], Gosgen [1082], Rovno [32], Krasnoyarsk [1037], Bugey [363], and Savannah River [563]. The masses of the detectors were of the order of a few hundred kg. As shown in Fig. 12.3, none of these experiments observed a $\bar{\nu}_e$ disappearance,

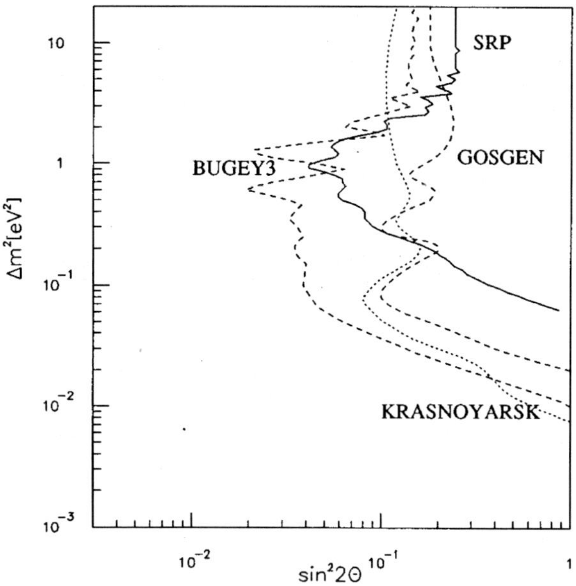

FIG. 12.5. Exclusion curves of the most sensitive reactor experiments: Gosgen [1082], Krasnoyarsk [1037], Bugey [363], Savannah River (SRP) [563]. Figure taken from Ref. [563].

because their source–detector distance was not sufficient to reach the sensitivity to the small values of Δm^2 required for the observation of $\bar{\nu}_e$ disappearance due to the solar Δm^2.

The negative results of SBL reactor neutrino oscillation experiments led to exclusion curves in the $\sin^2 2\vartheta$-Δm^2 plane which extend to values of Δm^2 as small as about $10^{-2}\,\text{eV}^2$, since the neutrino energy is relatively small. The most stringent exclusion curves are shown in Fig. 12.5.

12.2.2 Long-baseline reactor experiments: CHOOZ and Palo Verde

LBL reactor experiment have source–detector distances of the order of 1 km, which allow the extension of the Δm^2 sensitivity down to about $10^{-3}\,\text{eV}^2$. This is a very interesting range, because it covers the region of Δm^2 corresponding to atmospheric neutrino oscillations (see section 11.2). In fact, the CHOOZ [97, 98, 100] and Palo Verde [254, 253, 255] experiments have been carried out in the 1990s in order to check the $\nu_\mu \leftrightarrows \nu_e$ oscillation explanation of the atmospheric neutrino anomaly observed in the Kamiokande [614, 840, 621, 474, 603] and IMB [581, 306, 200] experiments. With a detector mass of a few ton, the CHOOZ and Palo Verde experiments were sensitive to values of $\sin^2 2\vartheta$ as small as about 10^{-1}.

The CHOOZ experiment [97, 98, 100] was located near the CHOOZ power plant of Électricité de France in the Ardennes region of France, which is composed of two

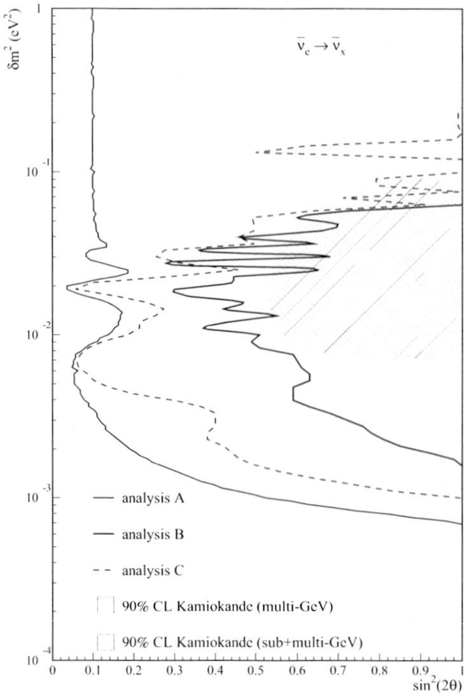

FIG. 12.6. Exclusion plot contours at 90% CL obtained in the CHOOZ experiment with three different analyses of the data [100]. Analysis A uses both shape and normalization of the background-subtracted positron spectrum; analysis B uses the baseline difference between the two reactors; analysis C uses the spectrum shape. The Kamiokande allowed region for atmospheric $\nu_\mu \leftrightarrows \nu_e$ oscillations [474] is also shown for comparison.

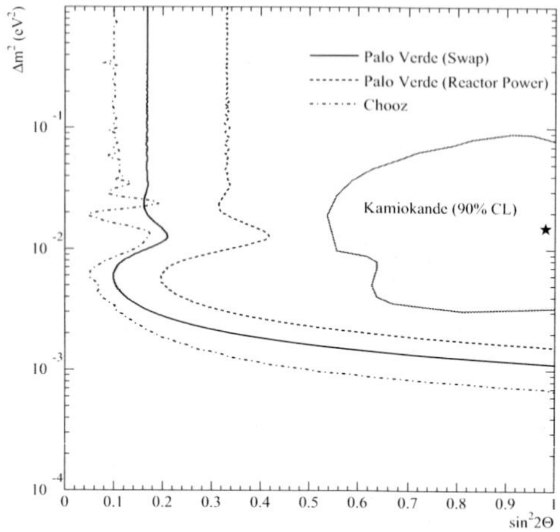

FIG. 12.7. Region of the Δm^2-$\sin^2 2\vartheta$ plane excluded at the 90% CL by the *reactor power* analysis (dashed curve) and *swap* analysis (solid curve) of the data of the Palo Verde experiment [255]. Also shown are the Kamiokande allowed region and best fit (star) [474] and the region excluded by the CHOOZ experiment [98].

reactors with a total thermal power of $8.5\,\mathrm{GW_{th}}$. The detector was built in a pre-existing underground cavity under about 100 m of rock, providing an overburden of about 300 mwe. The distances from the two reactors were 1115 m and 998 m. The detector had an active mass of 5 ton composed of liquid scintillators loaded with gadolinium, which has a high thermal neutron capture cross-section with the release of photons with a total energy of 8 MeV. This allowed the tagging of the inverse neutron decay events through the delayed coincidence between the prompt e^+ signal and the delayed neutron-capture signal. Since the two reactors started operation after the CHOOZ detector started data taking, the background was measured with high accuracy and the response of the detector to the increase of the reactor power was monitored. The CHOOZ experiment took data from April 1997 to July 1998, finding the ratio of measured to expected $\bar{\nu}_e$ events [100]

$$R = 1.01 \pm 0.028 \pm 0.027\,. \tag{12.18}$$

Also the spectral shape of the events was in agreement with the one expected in the case of no oscillations. Since no disappearance of $\bar{\nu}_e$ was observed, the CHOOZ collaboration obtained, with three different analyses of the data, the three exclusion curves shown in Fig. 12.6, together with the Kamiokande allowed region for atmospheric $\nu_\mu \leftrightarrows \nu_e$ oscillations [474]. The solid curve on the left in Fig. 12.6 implies that, at 90% CL,

$$\sin^2 2\vartheta \lesssim 0.1 \qquad \text{for} \qquad \Delta m^2 \gtrsim 2 \times 10^{-3}\,\text{eV}\,, \tag{12.19}$$

$$\Delta m^2 < 7 \times 10^{-4}\,\text{eV} \qquad \text{for} \qquad \sin^2 2\vartheta = 1\,. \tag{12.20}$$

These results are based on an analysis which assumes the knowledge of the absolute value of reactor neutrino flux.

From Fig. 12.6, one can see that the $\nu_\mu \leftrightarrows \nu_e$ explanation of the atmospheric neutrino anomaly found in Kamiokande and IMB is excluded by the CHOOZ result. The CHOOZ result is also important for our knowledge of the value of the element U_{e3} of the neutrino mixing matrix in the case of three-neutrino mixing, as it will be discussed in section 13.3.1.

The Palo Verde detector [254, 253, 255] was built near the Palo Verde Nuclear Generating Station in the Arizona desert, which is the largest nuclear plant in America. It has three reactors with a total thermal power of 11.6 GW_{th}. The detector was located 890 m from two reactors and 750 m from the third. Since no pre-existing underground location was available, the detector was located in an underground bunker 12 m deep, with an overburden of only 32 mwe. The detector had a fiducial mass of 12 ton which was made, as the CHOOZ detector, of gadolinium-loaded liquid scintillators. The data taking of the Palo Verde experiment from October 1998 to July 2000 led to the ratio of measured to expected $\bar{\nu}_e$ events [255]

$$R = 1.01 \pm 0.024 \pm 0.053\,, \tag{12.21}$$

with no spectral distortion. The exclusion curves obtained by the Palo Verde collaboration with two different analyses of the data are shown in Fig. 12.7, together with the CHOOZ exclusion curve from Ref. [98] and the Kamiokande allowed region for atmospheric $\nu_\mu \leftrightarrows \nu_e$ oscillations [474]. One can see that the results of the Palo Verde experiment confirmed those of the CHOOZ experiment.

There is an intense activity to develop LBL reactor neutrino experiments which can improve the sensitivities of the CHOOZ and Palo Verde experiments, with the hope to measure the $\bar{\nu}_e$ disappearance due to the atmospheric Δm^2 (see Refs. [87, 718, 557]). Such an observation would provide crucial information on the value of the element U_{e3} of the neutrino mixing matrix, as discussed in section 13.3.1. Among the projects, the first one to be realized will be Double-CHOOZ [104, 218], with a near and a far detector at distances, respectively, of about 100 m and 1.05 km from the CHOOZ power plant (data taking is scheduled to begin in 2008).

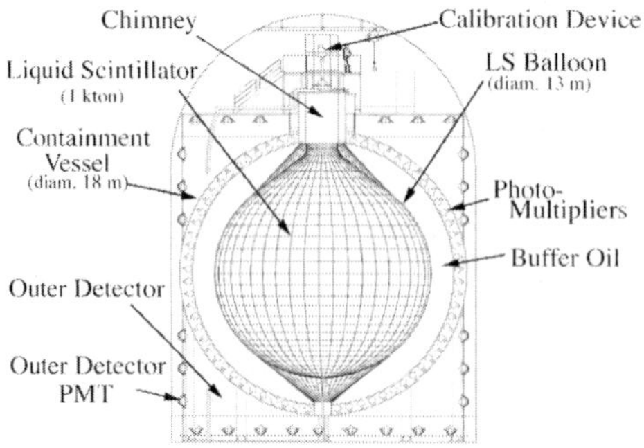

FIG. 12.8. Schematic diagram of the KamLAND detector [398].

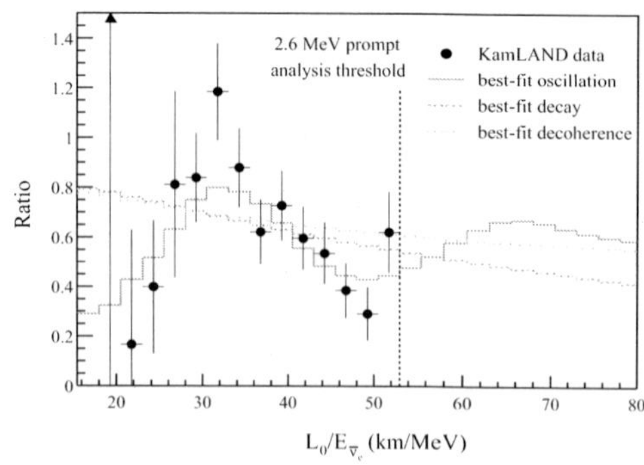

FIG. 12.9. Ratio of the observed $\bar{\nu}_e$ spectrum to the expectation for no-oscillation versus L_0/E [103], with $L_0 = 180\,\text{km}$. The curves show the expectation for the best-fit oscillation, best-fit decay [173, 172], and best-fit decoherence [574, 743] models.

12.2.3 Very long-baseline reactor experiments: KamLAND

The KamLAND (Kamioka Liquid scintillator AntiNeutrino Detector) experiment [398, 103] has been designed to detect electron antineutrinos produced by 53 nuclear power reactors in Japan, with a small contribution from reactors in South Korea ($\sim$ 2%) and in the rest of the world ($\lesssim$ 1%), at distances varying from 80 km to 800 km. About 80% of the detected neutrinos come from reactors at distances between 140 km and 215 km, with an average distance of about 180 km.

KamLAND is located in the Kamioka mine, in the cavity that was previously occupied by the Kamiokande experiment (see section 10.6.1). It consists of 1200 m^3 of liquid scintillator, confined in a spherical balloon with a diameter of 13 m. The luminescence from the scintillator is picked up by 1879 PMTs. The active scintillator volume is surrounded by a 2.5 m thick layer of ultrapure mineral oil that shields it from external neutron and gamma radiation. The 3000 ton of liquid scintillator, buffer oil, and photomultipliers are contained and supported by a stainless-steel sphere with a diameter of 18 m, as shown schematically in Fig. 12.8. The volume between the sphere and the cylindrical cavity in the rock is filled with water that allows the detection of the background of cosmic-ray muons through their Cherenkov light, which is picked up by the old Kamiokande photomultipliers.

In KamLAND electron neutrinos are detected through the inverse neutron decay reaction in eqn (12.13) with a threshold of 1.8 MeV (see eqn (12.15)). However, since at low energy there is a geoneutrino background (see the end of this subsection), the threshold for the oscillation analysis is at 2.6 MeV. The $\bar{\nu}_e$ energy is estimated from eqn (12.14).

The KamLAND experiment has been constructed in order to check the LMA solution of the solar neutrino problem (see chapter 10). As shown in Fig. 12.3, the

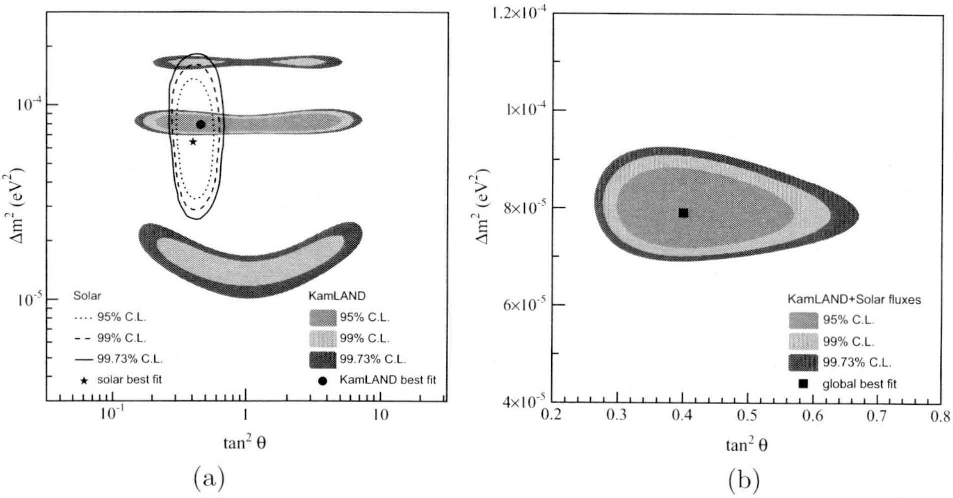

FIG. 12.10. (a) Allowed region in the $\tan^2\vartheta$-Δm^2 plane obtained in the KamLAND experiment (shaded regions) [103]. The lines show the region allowed by solar neutrino experiments [45]. (b) Result of a combined two-neutrino oscillation analysis of KamLAND and solar neutrino data.

large source–detector distance allowed the KamLAND experiment to measure the $\bar{\nu}_e$ disappearance due to the small solar Δm^2.

The ratio of measured to expected $\bar{\nu}_e$ events in KamLAND from March 2002 to January 2004 is [100]

$$R = 0.658 \pm 0.044 \pm 0.047\,, \tag{12.22}$$

which deviates from unity by about 5σ. Moreover, the observed deviations from the expected spectrum shown in Fig. 12.9 clearly indicate an oscillatory origin of the deficit. The best fit of the data is obtained for [103]

$$\Delta m^2 = 7.9^{+0.6}_{-0.5} \times 10^{-5}\,\mathrm{eV}^2\,. \tag{12.23}$$

The mixing angle has large uncertainty, as one can see from the allowed region in the $\tan^2\vartheta$-Δm^2 plane in Fig. 12.10a. Since the KamLAND allowed region overlaps with the LMA region of solar neutrino oscillations, also shown in Fig. 12.10a, KamLAND has confirmed the oscillation solution of the solar neutrino problem, excluding other mechanisms as the dominant effect. Figure 12.10b shows the allowed region in the $\tan^2\vartheta$-Δm^2 plane obtained with a combined fit of solar and KamLAND data, which gives [100]

$$\Delta m^2 = 7.9^{+0.6}_{-0.5} \times 10^{-5}\,\mathrm{eV}^2\,, \qquad \tan^2\vartheta = 0.40^{+0.10}_{-0.07}\,. \tag{12.24}$$

Comparing with eqn (12.23), one can see that the precision of the determination of Δm^2 is determined by the KamLAND data. This is due to the sensitivity of the spectral distortion in Fig. 12.9 to the value of Δm^2. On the other hand, the precision of the determination of $\tan^2\vartheta$ is determined by the solar data, as one can also understand from Fig. 12.10a.

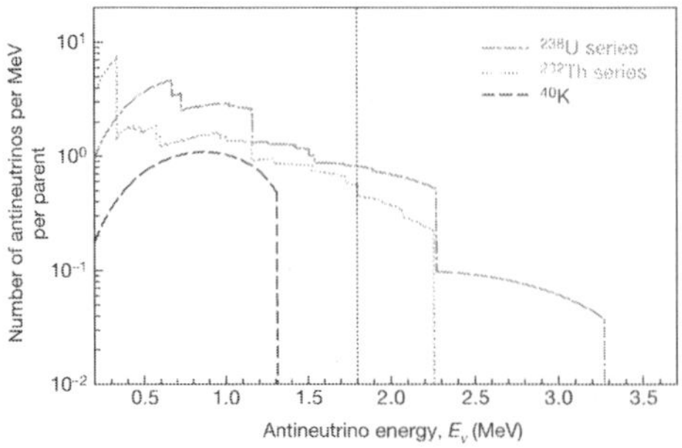

FIG. 12.11. The expected ^{238}U, ^{232}Th, and ^{40}K decay chain electron antineutrino energy distributions [102]. KamLAND can only detect electron antineutrinos to the right of the vertical dotted black line; hence it is insensitive to ^{40}K electron antineutrinos.

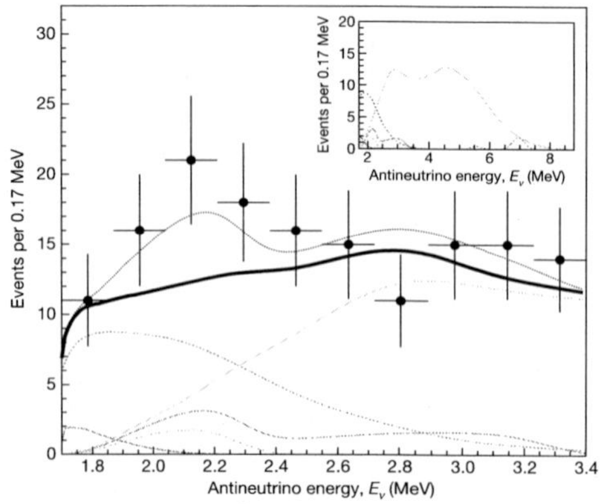

FIG. 12.12. $\bar{\nu}_e$ energy spectra in KamLAND [102]. Main panel: experimental points together with the total expectation (upper dotted line), the total expected spectrum excluding the geoneutrino signal (thick solid line), the expected signals from ^{238}U (dot-dashed line) and ^{232}Th (lower dotted line) geoneutrinos, and the backgrounds due to reactor $\bar{\nu}_e$ (upper dashed line), $^{13}\text{C}(\alpha, n)^{16}\text{O}$ reactions (middle dotted line), and random coincidences (lower dashed line). Inset: expected spectra extended to higher energy.

A future improvement of the determination of Δm^2 with a reactor experiment may be possible by adjusting the source–detector distance to the optimal distance corresponding to the first oscillation maximum, where the argument of the sine in eqn (7.74) is equal to $\pi/2$. In this case, the measurable disappearance of reactor $\bar{\nu}_e$'s is maximal, since the first oscillation maximum is that which suffers the smallest suppression due to the average over the L/E distribution, as shown in Fig. 7.2. In Fig. 12.3 one can see that the average source–detector distance in the KamLAND experiment is longer than the optimal distance and KamLAND is sensitive to averaged oscillations near the second maximum. For $\Delta m^2 \simeq 8 \times 10^{-5}\,\text{eV}^2$ and $E \simeq 4\,\text{MeV}$, the first oscillation maximum occurs at a source–detector distance of about 60 km.

Let us finally remark that KamLAND is the first experiment which was able to detect electron antineutrinos emitted in the decay chains of the radioactive isotopes ^{238}U and ^{232}Th in the Earth [102], which are called *geoneutrinos* [397, 779, 129, 703, 686, 888, 913, 440, 439, 830, 773, 443, 444, 452, 442]. Together with ^{40}K, the decay of these isotopes is one of the major sources of the heat generated in the Earth. However, KamLAND can observe only $\bar{\nu}_e$'s generated in the ^{238}U and ^{232}Th decay chains, since the energy spectrum of those generated in the ^{40}K decay chain is below the 1.8 MeV threshold, as shown in Fig. 12.11. In the KamLAND analysis, the geoneutrino spectra are calculated from a reference model of the Earth's composition [413], which assumes 16 TW radiogenic power from ^{238}U and ^{232}Th. The data and expected spectrum are shown in Fig. 12.12, together with the estimated backgrounds from various sources, among which the reactor neutrinos are dominant above about 2.3 MeV. One can see the excess due to geoneutrinos above the background, which is in agreement with the prediction of the geophysical model. The KamLAND measurement has opened a new field of research, which may lead to a future accurate determination of the composition of the Earth.

12.3 Accelerator experiments

As explained in section 7.5, accelerator experiments can be classified according to the method of production of the neutrino beam: pion decay in flight, muon decay at rest and beam dump.

In addition, the experiments with neutrino beams generated by pion decay in flight are further classified in three categories:

Wide band (WB) beam. These are experiments having a high-intensity neutrino beam with a wide energy spectrum which can span one or two orders of magnitude. This type of beam is convenient for investigating new oscillation signals in a wide range of values of Δm^2.

Narrow band (NB) beam. The neutrino beams of these experiments have a narrow energy spectrum, which is obtained by selecting the momenta of the parent pion and kaons. The resulting intensity of the neutrino flux of a NB beam is reduced comparing with a WB beam obtained from the same proton beam. NB beams are convenient for precise measurements of Δm^2.

Off-axis (OA). These are experiments which use a high-intensity WB beam with the detector shifted by a small angle from the axis of the beam, where the neutrino energy is almost monochromatic. The OA principle is a new concept [198], which will be implemented in future experiments (see section 12.3.3). We will explain it in subsection 12.3.3.

In the following two subsections we summarize the results of SBL accelerator experiments and the only LBL accelerator experiment K2K.

12.3.1 Short-baseline accelerator experiments

Many SBL accelerator experiments have been carried out in order to explore the different flavor transition channels, as one can see from the list in Table 12.1. All the experiments used neutrino beams from pion decay in flight, except for the $\bar{\nu}_\mu \to \bar{\nu}_e$ experiments and the beam dump experiments.

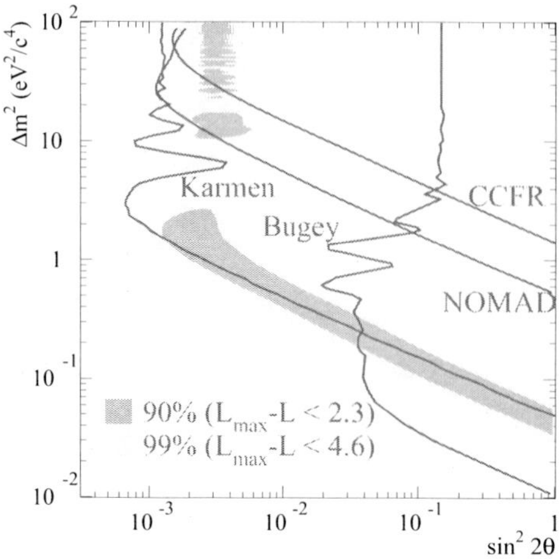

FIG. 12.13. Allowed (shadowed) region in the $\sin^2 2\vartheta$-Δm^2 plane for $\overset{(-)}{\nu}_\mu \to \overset{(-)}{\nu}_e$ oscillations obtained in the LSND experiment [37]. Also shown are the exclusion curves of the Bugey [363], CCFR [910], KARMEN [402], and NOMAD [128] experiments.

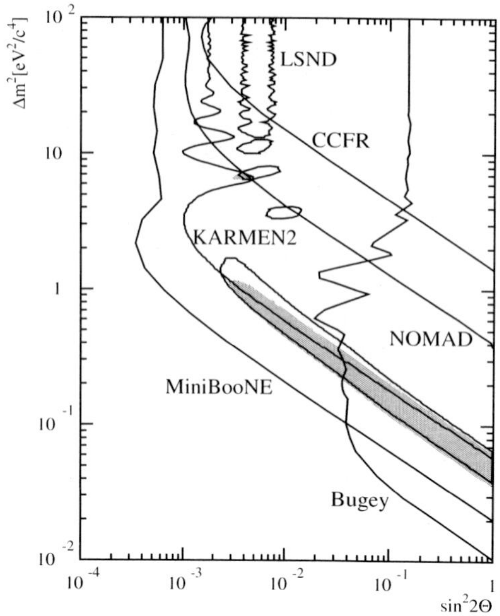

FIG. 12.14. 90% CL allowed region in the $\sin^2 2\vartheta$-Δm^2 plane for $\overset{(-)}{\nu}_\mu \to \overset{(-)}{\nu}_e$ oscillations obtained from a combined analysis of LSND and KARMEN data (shadowed region) [320]. Also shown are the allowed region of the LSND experiment [37], the exclusion curves of the Bugey [363], CCFR [910], KARMEN [402], and NOMAD [1032] experiments and the sensitivity curve of the MiniBooNE experiment [189].

All the SBL accelerator experiments did not find any indication of neutrino oscillations, except for the LSND experiment, which found a signal in the $\bar{\nu}_\mu \to \bar{\nu}_e$ channel [118, 120, 119, 37] and a weaker signal in the $\nu_\mu \to \nu_e$ channel [122, 121]. The allowed region in the $\sin^2 2\vartheta$-Δm^2 plane for $\overset{(-)}{\nu}_\mu \to \overset{(-)}{\nu}_e$ oscillations obtained in the LSND experiment [37] is shown in Fig. 12.13, together with the exclusion curves of other SBL experiments. One can see that part of the LSND-allowed region is not excluded by the results of other experiments. Figure 12.14 shows the result of combined statistical analysis of the data of the LSND and KARMEN experiments [320]. The resulting allowed region is in some parts not incompatible with the exclusion curves of other experiments. The LSND signal is under investigation in the MiniBooNE experiment at Fermilab [319, 189, 892]. This check is important, because a confirmation of the LSND signal could require an extension of the three-neutrino mixing scheme discussed in chapter 13 (see Refs. [231, 556]).

TABLE 12.1. SBL neutrino oscillation accelerator experiments.

Channel	Experiments
$\nu_\mu \to \nu_\mu$	CHARM [212]
$\overset{(-)}{\nu}_\mu \to \overset{(-)}{\nu}_\mu$	CDHSW [396], CCFR [983]
$\nu_\mu \to \nu_e$	BEBC [89], CHARM [212], LSND [121], NOMAD [116]
$\bar{\nu}_\mu \to \bar{\nu}_e$	LAMPF-0645 [460], LSND [37], KARMEN [105]
$\overset{(-)}{\nu}_\mu \to \overset{(-)}{\nu}_e$	BNL-E776 [265], CCFR [910], NuTeV [130]
$\nu_\mu \to \nu_\tau$	FNAL-E531 [1030], CHARM [212], CHORUS [422], NOMAD [115]
$\overset{(-)}{\nu}_\mu \to \overset{(-)}{\nu}_\tau$	CCFR [789]
$\nu_e \to \nu_\tau$	CHORUS [422], NOMAD [115]
$\overset{(-)}{\nu}_e \to \overset{(-)}{\nu}_\tau$	CCFR [823])
Beam dump	BEBC [465, 559], CHARM [388], CDHSW [208]

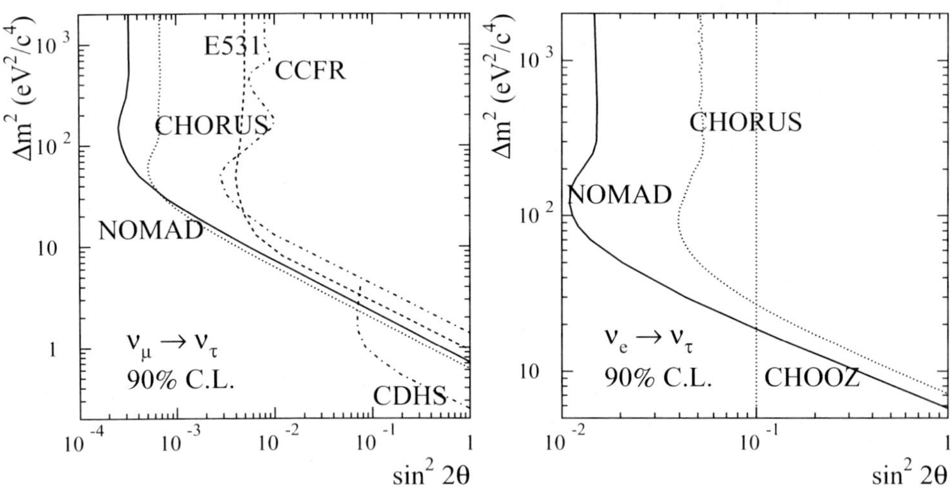

(a) (b)

FIG. 12.15. Exclusion curves in the $\sin^2 2\vartheta$-Δm^2 plane for $\overset{(-)}{\nu}_\mu \to \overset{(-)}{\nu}_\tau$ (a) and $\overset{(-)}{\nu}_e \to \overset{(-)}{\nu}_\tau$ (b) oscillations obtained in several experiments. Figure from Ref. [115].

Figure 12.15 shows the exclusion curves in the $\sin^2 2\vartheta$-Δm^2 plane for $\overset{(-)}{\nu}_\mu \to \overset{(-)}{\nu}_\tau$ and $\overset{(-)}{\nu}_e \to \overset{(-)}{\nu}_\tau$ oscillations obtained in several experiments. The exclusion curves of the CDHS $\equiv$ CDHSW [396] and CHOOZ [100] experiments, which are disappearance experiments, constrain, respectively, $\overset{(-)}{\nu}_\mu$ and $\overset{(-)}{\nu}_e$ disappearance into any other state, including $\overset{(-)}{\nu}_\tau$. The other curves correspond to appearance experiments listed in Table 12.1.

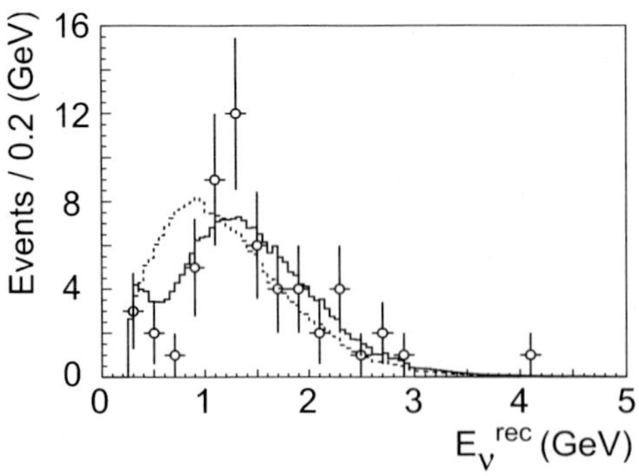

FIG. 12.16. Number of events in the K2K experiment as a function of the reconstructed neutrino energy [66]. Points with error bars are data. The solid line is the best-fit spectrum with the oscillation parameters in eqn (12.25). The dashed line is the expected spectrum without oscillation.

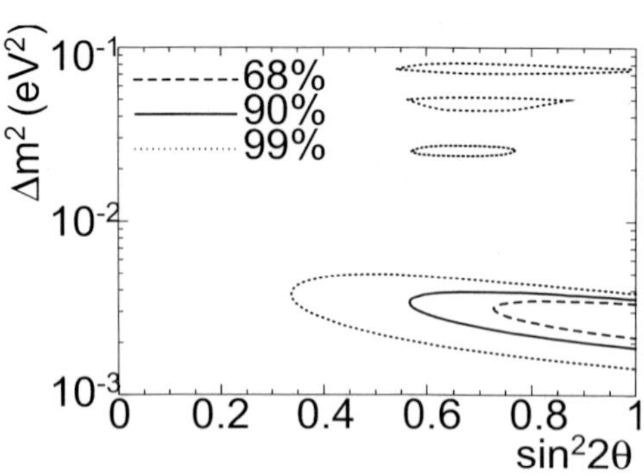

FIG. 12.17. Allowed region in the $\sin^2 2\vartheta$-Δm^2 plane for $\nu_\mu \to \nu_x$ oscillations obtained in the K2K experiment [66].

12.3.2 Long-baseline accelerator experiments: K2K

The K2K experiment is a Japanese accelerator LBL experiment with a distance of 250 km from the source in the KEK laboratory to the Super-Kamiokande detector in the Kamioka mine [48, 46, 47, 66]. The neutrino beam is a pulsed WB beam with a mean energy of 1.3 GeV. It is an almost pure ν_μ beam ($\sim 98\% \, \nu_\mu + \sim 1\% \, \bar{\nu}_\mu + \sim 1\% \, \nu_e$). Every 2.2 s about 6×10^{12} protons are extracted from the KEK proton-synchrotron (KEK-PS) in nine bunches lasting in total $1.1 \, \mu$s. The protons are focused on an aluminum target and the produced positive pions are focused towards a decay tunnel 200 m long, where they decay into antimuons and muon neutrinos. At the end of the decay tunnel there is an iron and concrete beam dump which stops all charged particles, except muons with energy greater than 5.5 GeV. After the beam dump there is a muon monitor and, after about 70 m of earth which eliminates all particles except neutrinos, there is a near neutrino detector system which is used to calibrate the neutrino beam (about 300 m from the production target). The near detector system consists of two detector sets: a 1 kton water Cherenkov detector and a fine-grained detector system. The water Cherenkov detector uses the same technology and analysis algorithms as the Super-Kamiokande far detector. It also has the same type and arrangement of PMTs.

The Super-Kamiokande has been described in section 10.6.2 in the context of solar neutrino detection. Here, it has to be added that the events for the K2K experiment are selected by synchronization with the proton extraction from the KEK-PS, which is done with high accuracy through the Global Positioning System (GPS). The K2K experiment has been divided, so far, in two phases. The first phase, called K2K-I, is a fraction of the Super-Kamiokande-I phase which lasted from June

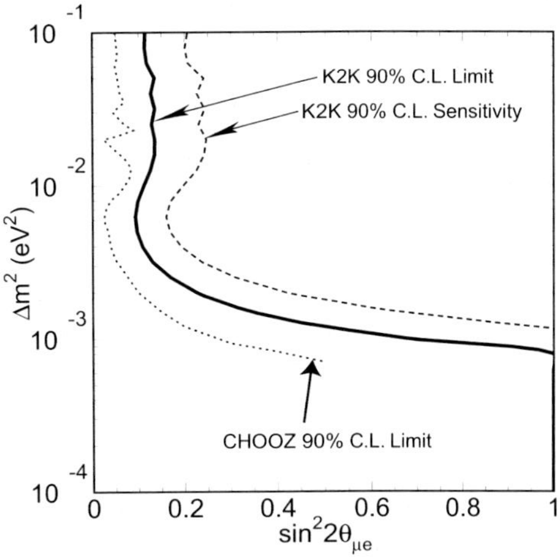

FIG. 12.18. Exclusion curve in the $\sin^2 2\vartheta$-Δm^2 plane for $\nu_\mu \to \nu_e$ oscillations obtained in the K2K experiment (February 2004) [47]. The dashed line indicates the sensitivity of the experiment. The dotted line shows the 90% CL exclusion curve of CHOOZ [100].

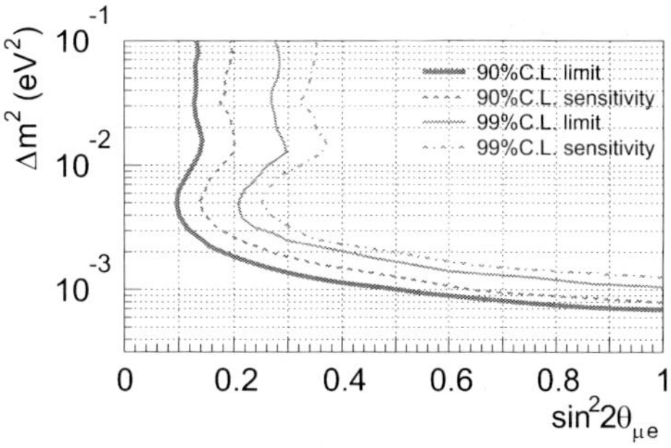

FIG. 12.19. The 90% and 99% CL upper bounds on the $\nu_\mu \to \nu_e$ oscillation parameters obtained in the K2K experiment (March 2006) [1078]. The sensitivities of the K2K experiment for each CL are indicated with dashed lines.

1999 to July 2001. On 12 November 2001 the Super-Kamiokande detector suffered the accidental loss of about half of the PMTs. The second phase, called K2K-II, coincides with the Super-Kamiokande-II phase from January 2003 to February 2004, in which 19% of the inner Super-Kamiokande detector has been covered with the surviving 5182 PMTs.

The K2K experiment has been built for a laboratory check of atmospheric neutrino oscillations. The optimal source–detector distance for measuring ν_μ disappearance with $E \simeq 1.3\,\text{GeV}$ and the atmospheric squared-mass difference $\Delta m^2 \simeq 2.5 \times 10^{-3}\,\text{eV}^2$ would be about 650 km, corresponding to the first oscillation maximum, where the argument of the sine in eqn (7.74) is equal to $\pi/2$. Such a source–detector distance would maximize the measurable disappearance of ν_μ's, since the first oscillation maximum is the only one which is almost unaffected by the average over the experimental L/E distribution, as illustrated in Fig. 7.2. The source–detector distance of 250 km in the K2K experiment has been determined by the pre-existing locations of the K2K laboratory and the Super-Kamiokande detector. Although it is not optimal, it has been sufficient for the measurement of ν_μ disappearance.

The K2K experiment observed 107 fully contained μ-like events in the 22.5 ktons fiducial volume of Super-Kamiokande, with an expected number without oscillations of 151^{+12}_{-10}. There is a discrepancy of about 3σ which indicates muon neutrino disappearance. Figure 12.16 shows the energy distribution of K2K events together with the spectrum expected without oscillations and the best-fit spectrum with oscillations, corresponding to the mixing parameters

$$\sin^2 2\vartheta = 1.0\,, \qquad \Delta m^2 = 2.8 \times 10^{-3}\,\text{eV}^2\,. \tag{12.25}$$

It is clear that the energy distribution of the data is in good agreement with the oscillation hypothesis.

Figure 12.17 shows the region in the $\sin^2 2\vartheta$-Δm^2 plane of neutrino oscillations allowed by the K2K data. This region is in good agreement with that in Fig. 11.30, which was obtained in the Super-Kamiokande atmospheric neutrino experiment. Therefore, the results of the K2K experiment provide a very important confirmation of the oscillations observed in atmospheric neutrino experiments, excluding other explanations of the atmospheric neutrino anomaly (see also the discussion in section 11.2.6).

The K2K collaboration has also investigated the possibility of $\nu_\mu \to \nu_e$ oscillations by looking at e-like events in the Super-Kamiokande detector synchronized with the proton extraction from the KEK-PS [47, 1078]. Since no events above the expected background were observed, K2K confirmed the CHOOZ and Palo Verde exclusion of $\nu_\mu \leftrightarrows \nu_e$ oscillations with a squared-mass difference larger than about $10^{-3}\,\text{eV}$. The 90% CL exclusion curve obtained by the K2K collaboration in February 2004 [47] is compared with the CHOOZ exclusion curve in Fig. 12.18. Figure 12.19 shows the 90% and 99% CL exclusion curves obtained by the K2K collaboration in March 2006 [1078]. The effective mixing angle $\vartheta_{\mu e}$ (which is equivalent to $\vartheta_{\mu e}^{\text{eff}}$ in eqn (13.16) in the case of three-neutrino mixing) is bounded by

$$\sin^2 2\vartheta_{\mu e} < 0.13 \quad (90\%\ \text{CL}) \quad \text{at} \quad \Delta m^2 = 2.8 \times 10^{-3}\,\text{eV}^2\,. \tag{12.26}$$

This limit can be interpreted as a bound on $|U_{e3}|$ in the case of three-neutrino mixing (see section 13.3.1).

12.3.3 Future LBL experiments: off-axis

There is intense activity aimed at the development of new LBL experiments, with the primary objective to discover $\overset{(-)}{\nu}_\mu \leftrightarrows \overset{(-)}{\nu}_e$ oscillations generated by the atmospheric Δm^2. Such a measurement would give information on the element U_{e3} of the neutrino mixing matrix in the case of three-neutrino mixing (see section 13.3.1), opening up the possibility to observe CP violation in neutrino oscillations and to distinguish the two three-neutrino mixing schemes in Fig. 13.1 through their different matter effects for LBL neutrino beams propagating in the Earth. The next accelerator LBL experiments are listed in Table 12.2. Other proposals for future LBL accelerator experiments are the so-called β-beams [1091, 796, 59, 576] and the Neutrino Factory [511, 58, 99, 59, 576]. β-beams are pure electron neutrino or antineutrino beams produced by the β-decays of accelerated radioactive isotopes with a short lifetime. A Neutrino Factory consists of a storage ring with a high current of muons whose decay produces an intense flux of electron and muons neutrinos ($\bar{\nu}_e + \nu_\mu$ in the case of μ^- and $\nu_e + \bar{\nu}_\mu$ in the case of μ^+).

Some of the future LBL experiments, such as the T2K and NOνA experiments in Table 12.2, will adopt the new off-axis (OA) configuration, which yields an almost

TABLE 12.2. The next LBL experiments with their oscillation channels, source–detector distance L, average neutrino energy E, and scheduled starting year.

Beam	Detector	Channels	L [km]	E [GeV]	Start
NuMI [694]	MINOS [375]	$\nu_\mu \to \nu_e$ $\nu_\mu \to \nu_\mu$	730	3.5	2005
CNGS [393]	OPERA [577, 691, 867] ICARUS [106, 101]	$\nu_\mu \to \nu_\tau$ $\nu_\mu \to \nu_e$	732	18	2007
T2K [633]	Super-Kamiokande	$\nu_\mu \to \nu_e$ $\nu_\mu \to \nu_\mu$	295	0.7	2009
NOνA [132, 80]	Low-Z Calorimeter [132]	$\nu_\mu \to \nu_e$ $\nu_\mu \to \nu_\mu$	~ 800	2	$\gtrsim 2010$

monochromatic neutrino flux in a detector shifted by a small angle from the axis of a high-intensity WB beam [198]. In the following, we explain the basic principles of OA experiments.

Let us consider a muon neutrino produced in the pion decay process $\pi^+ \to \mu^+ + \nu_\mu$. Since the final state has two particles, the conservation of energy and momentum implies that, in the rest frame of the pion which is the center-of-mass frame of the decay, the neutrino energy and momentum are given by

$$E_{\rm cm} = p_{\rm cm} = \frac{m_\pi}{2}\left(1 - \frac{m_\mu^2}{m_\pi^2}\right) \simeq 29.79\,\text{MeV}\,, \tag{12.27}$$

where we have neglected the neutrino mass. Let us now consider the laboratory frame, in which the high-energy pion moves along the z axis with velocity $v = p_\pi/E_\pi \simeq 1$ and a Lorentz boost factor $\gamma = \left(1 - v^2\right)^{-1/2} = E_\pi/m_\pi \gg 1$. The neutrino energy E and momentum component along the z axis p^z are given by

$$E = \gamma\left(E_{\rm cm} + v\,p^z_{\rm cm}\right)\,, \tag{12.28}$$

$$p^z = \gamma\left(v\,E_{\rm cm} + p^z_{\rm cm}\right)\,, \tag{12.29}$$

where $p^z_{\rm cm}$ is the momentum component along the z axis in the center-of-mass frame. We want to find the dependence of the neutrino energy E on the small off-axis angle θ of displacement of the detector with respect to the pion direction of flight along the z axis. Since $p^z = p\,\cos\theta$, eliminating $p^z_{\rm cm}$ in eqns (12.28) and (12.29) we obtain

$$E = \frac{E_{\rm cm}}{\gamma\left(1 - v\,\cos\theta\right)} \simeq \frac{\gamma\left(1+v\right)}{1 + \gamma^2\,\theta^2\,v\left(1+v\right)/2}\,E_{\rm cm} \simeq \frac{2\,\gamma}{1+\gamma^2\,\theta^2}\,E_{\rm cm}\,, \tag{12.30}$$

where the first approximation is due to $\cos\theta \simeq 1 - \theta^2/2$ and the second one to $v \simeq 1$. Using eqn (12.27) for E_{cm}, eqn (12.30) can be written as

$$E \simeq \left(1 - \frac{m_\mu^2}{m_\pi^2}\right) \frac{E_\pi}{1 + \gamma^2\,\theta^2} = \left(1 - \frac{m_\mu^2}{m_\pi^2}\right) \frac{E_\pi\, m_\pi^2}{m_\pi^2 + E_\pi^2\,\theta^2}\,. \tag{12.31}$$

This is the crucial equation which shows the monochromatic character of the neutrino energy off-axis.

For an on-axis detector, $\theta = 0$ and the neutrino energy is proportional to the pion energy, leading to a WB beam if the range of pion energies is wide. However, as soon as one moves slightly off-axis, both the numerator and denominator in eqn (12.31) increase with the pion energy, leading to a quenched dependence of E from E_π. This can be seen quantitatively by calculating the derivative of E with respect to E_π:

$$\frac{\mathrm{d}E}{\mathrm{d}E_\pi} \simeq \left(1 - \frac{m_\mu^2}{m_\pi^2}\right) \frac{1 - \gamma^2\,\theta^2}{\left(1 + \gamma^2\,\theta^2\right)^2}\,. \tag{12.32}$$

The derivative vanishes for $\theta = \gamma^{-1} = m_\pi/E_\pi$, which gives

$$E \simeq \left(1 - \frac{m_\mu^2}{m_\pi^2}\right) \frac{m_\pi}{2\,\theta} \simeq \frac{29.79\,\text{MeV}}{\theta}\,. \tag{12.33}$$

Thus, if $\langle E_\pi \rangle$ is the average energy of the pion beam, a detector at an off-axis angle $\theta \simeq m_\pi/\langle E_\pi \rangle$ receives an almost monochromatic neutrino beam with an average energy given by eqn (12.33).

The dependence of the neutrino energy E as a function of pion energy E_π as given by eqn (12.31) is illustrated in Fig. 12.20a. One can see that off-axis the neutrino energy E is almost constant in the range of pion energies shown in the figure. Another advantage of the OA configuration is that, as one can see from Fig. 12.20a, the neutrino energy is much smaller than the pion energy, leading to a shorter oscillation length and an enhanced sensitivity to small values of Δm^2.

The disadvantage of the OA configuration is that off-axis the neutrino flux is less intense that on-axis. However, the off-axis neutrino flux of a WB beam is much higher than that achievable with a traditional NB beam [847].

In order to calculate the intensity loss with respect to the on-axis configuration, let us first consider the rest frame of the pion, where the neutrino flux is isotropic. Along the z direction, with $z = 0$ and $z_{\text{cm}} = 0$ corresponding to the pion decay point, we have

$$\phi_{\text{cm}} = \frac{\mathrm{d}N}{\mathrm{d}\cos\theta_{\text{cm}}} \propto \frac{1}{z_{\text{cm}}^2}\,. \tag{12.34}$$

Since z is the instantaneous source–detector distance in the laboratory frame, we have $z_{\text{cm}} = \gamma\, z$. We need now to express $\cos\theta_{\text{cm}}$ in terms of $\cos\theta$. For this purpose, we note that in the direction orthogonal to the z axis we have $E\,\sin\theta = E_{\text{cm}}\,\sin\theta_{\text{cm}}$. The ratio of this equation to eqn (12.29) with $p^z = E\,\cos\theta$ and $p^z_{\text{cm}} = E_{\text{cm}}\,\cos\theta_{\text{cm}}$

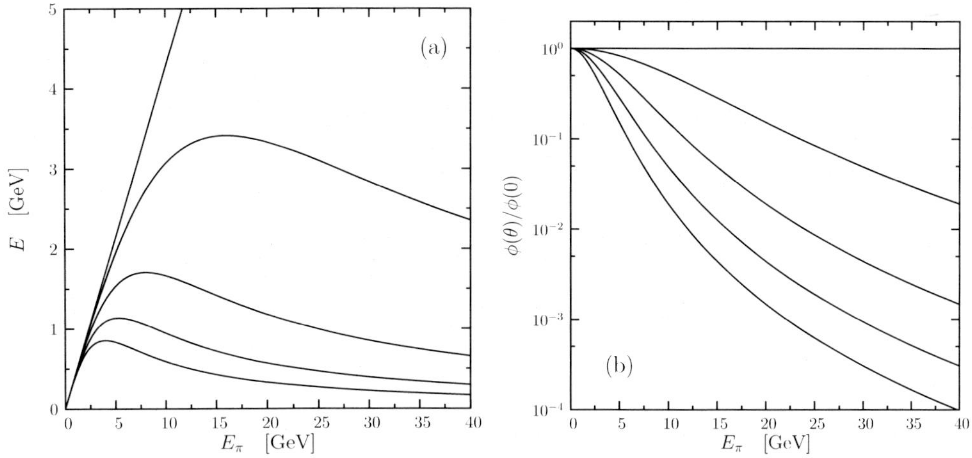

FIG. 12.20. Neutrino energy E as a function of pion energy E_π (a) and suppression of the off-axis neutrino flux with respect to the on-axis flux (b). The curves correspond to off-axis angles $\theta = 0.0°, 0.5°, 1.0°, 1.5°, 2.0°$.

yields

$$\tan\theta \simeq \frac{\sin\theta_{\text{cm}}}{\gamma\,(1+\cos\theta_{\text{cm}})} \,. \tag{12.35}$$

This is the relation between θ and θ_{cm}, from which one finds, for small off-axis angles,

$$\mathrm{d}\cos\theta_{\text{cm}} \simeq \left(\frac{2\,\gamma}{1+\gamma^2\,\theta^2}\right)^2 \mathrm{d}\cos\theta \,. \tag{12.36}$$

Thus, we finally obtain, in the laboratory frame,

$$\phi(\theta) = \frac{\mathrm{d}N}{\mathrm{d}\cos\theta} \propto \left(\frac{2}{1+\gamma^2\,\theta^2}\right)^2 \frac{1}{z^2} \,. \tag{12.37}$$

Figure 12.20b shows the suppression $\phi(\theta)/\phi(0)$ of the off-axis flux with respect to the on-axis flux for $\theta = 0.0°, 0.5°, 1.0°, 1.5°, 2.0°$. One can see that the suppression is rather severe, leading to the necessity of having a very intense beam (*superbeam*) and a very large detector.

13

PHENOMENOLOGY OF THREE-NEUTRINO MIXING

> Without experimentalists, theorists tend to drift. Without theorists, experimentalists tend to falter.
> **T.D. Lee** [725]

Solar and atmospheric neutrino experiments have shown that neutrinos oscillate with two different squared-mass differences, which we denote by Δm^2_{SOL} and Δm^2_{ATM}, respectively. This has been confirmed by independent measurements of the terrestrial KamLAND and K2K experiments. Therefore it is well established that there are at least two independent squared-mass differences. The minimal possibility of just two independent squared-mass differences is realized in three-neutrino mixing schemes, in which the left-handed components $\nu_{\alpha L}$ of the three flavor neutrinos ($\alpha = e, \mu, \tau$) are linear combinations of the left-handed components ν_{kL} of three neutrinos with masses m_k:

$$\nu_{\alpha L} = \sum_{k=1}^{3} U_{\alpha k}\, \nu_{kL}\,, \tag{13.1}$$

where U is the 3×3 unitary mixing matrix. The three light massive neutrinos could have a Dirac nature (see section 6.1.2) or a Majorana nature (see section 6.3) or they could be generated by a Dirac–Majorana mass term through the see-saw mechanism (see section 6.5).

Although in the case of three-neutrino mixing there are three squared-mass differences,

$$\Delta m^2_{21} \equiv m^2_2 - m^2_1\,, \quad \Delta m^2_{31} \equiv m^2_3 - m^2_1\,, \quad \Delta m^2_{32} \equiv m^2_3 - m^2_2\,, \tag{13.2}$$

only two of them are independent, since

$$\Delta m^2_{32} + \Delta m^2_{21} - \Delta m^2_{31} = 0\,. \tag{13.3}$$

Hence, the observed hierarchy

$$\Delta m^2_{\text{SOL}} \ll \Delta m^2_{\text{ATM}} \tag{13.4}$$

can be accommodated in the two types of three-neutrino mixing schemes shown schematically in Fig. 13.1. We choose the arbitrary labeling numbers of the massive

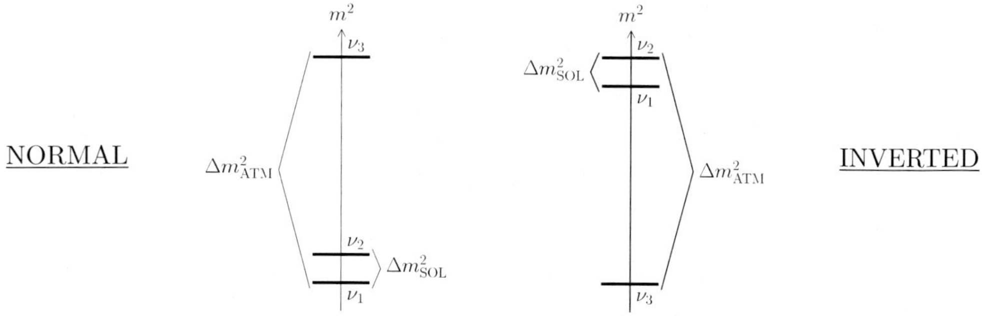

FIG. 13.1. The two three-neutrino schemes allowed by the hierarchy $\Delta m^2_{\rm SOL} \ll \Delta m^2_{\rm ATM}$.

neutrinos in order to have

$$\Delta m^2_{\rm SOL} = \Delta m^2_{21} \,, \qquad \Delta m^2_{\rm ATM} = |\Delta m^2_{31}| \,, \tag{13.5}$$

with

$$\Delta m^2_{21} \ll \Delta m^2_{31} \simeq \Delta m^2_{32} \,. \tag{13.6}$$

From the hierarchy in eqn 13.4, we have $\Delta m^2_{32} = \Delta m^2_{31} - \Delta m^2_{\rm SOL} \simeq \Delta m^2_{31}$. Therefore, considering $|\Delta m^2_{31}|$ or $|\Delta m^2_{32}|$ as the atmospheric squared-mass difference is the same in discussions of atmospheric and LBL experiments, which are not sensitive to $\Delta m^2_{\rm SOL}$. It is also the same in discussions of solar and VLBL experiments. These experiments are sensitive to $\Delta m^2_{\rm SOL}$, but can observe only the averaged transition probabilities generated by Δm^2_{32} and Δm^2_{31}, which do not depend on the exact values of these two squared-mass differences. A precise distinction between Δm^2_{32} and Δm^2_{31} is necessary only in discussions of LBL experiments which are sensitive to the subleading oscillations due to $\Delta m^2_{\rm SOL}$ (see Ref. [451]).

In this chapter, we discuss the main aspects of the phenomenology of three-neutrino mixing, with emphasis on the realistic schemes in Fig. 13.1. In sections 13.1 and 13.2 we discuss the general features of neutrino oscillations in vacuum and in matter, respectively. In section 13.3 we present the main results of the analysis of the oscillation data in the three-neutrino mixing framework.

13.1 Neutrino oscillations in vacuum

In the case of three-neutrino mixing, there are three flavor transition channels for neutrinos:

$$\nu_e \leftrightarrows \nu_\mu \,, \qquad \nu_e \leftrightarrows \nu_\tau \,, \qquad \nu_\mu \leftrightarrows \nu_\tau \,, \tag{13.7}$$

and the corresponding three channels for antineutrinos. The probabilities of neutrino oscillations depend on six independent parameters: two independent Δm^2 (often Δm^2_{21} and Δm^2_{31} are used), three mixing angles and one Dirac phase (ϑ_{12}, ϑ_{13}, ϑ_{23}, and δ_{13} in the parameterization given in eqn (6.191) of the Dirac part of the mixing matrix).

TABLE 13.1. Real parts of the quartic products of elements of the mixing matrix entering in the three-neutrino oscillation probabilities in eqns (13.9) and (13.10) with the parameterization in eqn (6.191) of the Dirac part of the mixing matrix. The last column gives the coefficients $s_{\alpha\beta;kj}$.

$\alpha\beta$	kj	$\mathfrak{Re}\left[U_{\alpha k}^{*}\, U_{\beta k}\, U_{\alpha j}\, U_{\beta j}^{*}\right]$	$s_{\alpha\beta;kj}$
	21	$-\frac{1}{4}\sin 2\vartheta_{12}\, c_{13}^2\, [\sin 2\vartheta_{12}\, (c_{23}^2 - s_{23}^2 s_{13}^2)$ $+\cos 2\vartheta_{12}\, \sin 2\vartheta_{23}\, s_{13}\cos\delta_{13}]$	$+1$
$e\mu$	32	$-s_{12}s_{23}c_{13}^2 s_{13}\,(s_{12}s_{23}s_{13} - c_{12}c_{23}\cos\delta_{13})$	$+1$
	31	$-c_{12}s_{23}c_{13}^2 s_{13}\,(c_{12}s_{23}s_{13} + s_{12}c_{23}\cos\delta_{13})$	-1
	21	$\frac{1}{16}\sin^2 2\vartheta_{12}\,\sin^2 2\vartheta_{23}\,(1+s_{13}^2)^2 - \frac{1}{4}\,(\sin^2 2\vartheta_{12} + \sin^2 2\vartheta_{23})\, s_{13}^2$ $-\frac{1}{16}\sin 4\vartheta_{12}\,\sin 4\vartheta_{23}\,(1+s_{13}^2)\, s_{13}\cos\delta_{13}$ $+\frac{1}{4}\sin^2 2\vartheta_{12}\,\sin^2 2\vartheta_{23}\, s_{13}^2\cos^2\delta_{13}$	$+1$
$\mu\tau$	32	$-\frac{1}{4}\sin 2\vartheta_{23}\, c_{13}^2\, [\sin 2\vartheta_{23}\, (c_{12}^2 - s_{12}^2 s_{13}^2)$ $+\sin 2\vartheta_{12}\, \cos 2\vartheta_{23}\, s_{13}\cos\delta_{13}]$	$+1$
	31	$\frac{1}{4}\sin 2\vartheta_{23}\, c_{13}^2\, [\sin 2\vartheta_{23}\, (c_{12}^2 s_{13}^2 - s_{12}^2)$ $+\sin 2\vartheta_{12}\, \cos 2\vartheta_{23}\, s_{13}\cos\delta_{13}]$	-1
	21	$\frac{1}{4}\sin 2\vartheta_{12}\, c_{13}^2\, [\sin 2\vartheta_{12}\, (c_{23}^2 s_{13}^2 - s_{23}^2)$ $+\cos 2\vartheta_{12}\, \sin 2\vartheta_{23}\, s_{13}\cos\delta_{13}]$	$+1$
τe	32	$-s_{12}c_{23}c_{13}^2 s_{13}\,(s_{12}c_{23}s_{13} + c_{12}s_{23}\cos\delta_{13})$	$+1$
	31	$-c_{12}c_{23}c_{13}^2 s_{13}\,(c_{12}c_{23}s_{13} - s_{12}s_{23}\cos\delta_{13})$	-1
	21	$\frac{1}{4}\sin^2 2\vartheta_{12}\, c_{13}^4$	0
ee	32	$\frac{1}{4}s_{12}^2\,\sin^2 2\vartheta_{13}$	0
	31	$\frac{1}{4}c_{12}^2\,\sin^2 2\vartheta_{13}$	0
	21	$\frac{1}{4}\sin^2 2\vartheta_{12}\,(c_{23}^4 + s_{23}^4 s_{13}^2) + \frac{1}{4}\,(1 - \frac{1}{2}\sin^2 2\vartheta_{12})\sin^2 2\vartheta_{23}\, s_{13}^2$ $+\frac{1}{4}\sin 4\vartheta_{12}\,\sin 2\vartheta_{23}\,(c_{23}^2 - s_{23}^2 s_{13}^2)\, s_{13}\cos\delta_{13}$ $-\frac{1}{4}\sin^2 2\vartheta_{12}\,\sin^2 2\vartheta_{23}\, s_{13}^2\cos^2\delta_{13}$	0
$\mu\mu$	32	$s_{23}^2 c_{13}^2\,(c_{12}^2 c_{23}^2 + s_{12}^2 s_{23}^2 s_{13}^2 - \frac{1}{2}\sin 2\vartheta_{12}\,\sin 2\vartheta_{23}\, s_{13}^2\cos^2\delta_{13})$	0
	31	$s_{23}^2 c_{13}^2\,(s_{12}^2 c_{23}^2 + c_{12}^2 s_{23}^2 s_{13}^2 + \frac{1}{2}\sin 2\vartheta_{12}\,\sin 2\vartheta_{23}\, s_{13}^2\cos^2\delta_{13})$	0
	21	$\frac{1}{4}\sin^2 2\vartheta_{12}\,(s_{23}^4 + c_{23}^4 s_{13}^2) + \frac{1}{4}\,(1 - \frac{1}{2}\sin^2 2\vartheta_{12})\sin^2 2\vartheta_{23}\, s_{13}^2$ $+\frac{1}{4}\sin 4\vartheta_{12}\,\sin 2\vartheta_{23}\,(s_{23}^2 - c_{23}^2 s_{13}^2)\, s_{13}\cos\delta_{13}$ $-\frac{1}{4}\sin^2 2\vartheta_{12}\,\sin^2 2\vartheta_{23}\, s_{13}^2\cos^2\delta_{13}$	0
$\tau\tau$	32	$c_{23}^2 c_{13}^2\,(c_{12}^2 s_{23}^2 + s_{12}^2 c_{23}^2 s_{13}^2 + \frac{1}{2}\sin 2\vartheta_{12}\,\sin 2\vartheta_{23}\, s_{13}^2\cos^2\delta_{13})$	0
	31	$c_{23}^2 c_{13}^2\,(s_{12}^2 s_{23}^2 + c_{12}^2 c_{23}^2 s_{13}^2 - \frac{1}{2}\sin 2\vartheta_{12}\,\sin 2\vartheta_{23}\, s_{13}^2\cos^2\delta_{13})$	0

As discussed in section 6.1, in the case of three-neutrino mixing all the imaginary parts of the rephasing invariant quartic products given in eqn (7.25) are equal up to a sign,

$$\mathfrak{Im}\left[U_{\alpha k}^{*}\,U_{\beta k}\,U_{\alpha j}\,U_{\beta j}^{*}\right] = s_{\alpha\beta;kj}\,J\,, \tag{13.8}$$

with the coefficients $s_{\alpha\beta;kj} = \pm 1$ as given in Table 13.1 and the Jarlskog invariant J given in eqn (6.26).

Using eqn (13.8) we can write the neutrino oscillation probability in eqn (7.38) as

$$P_{\nu_\alpha\to\nu_\beta}(L,E) = \delta_{\alpha\beta} - 4\sum_{k>j}\mathfrak{Re}\left[U_{\alpha k}^{*}\,U_{\beta k}\,U_{\alpha j}\,U_{\beta j}^{*}\right]\sin^2\left(\frac{\Delta m_{kj}^2 L}{4E}\right)$$
$$+\,2\,J\sum_{k>j} s_{\alpha\beta;kj}\,\sin\left(\frac{\Delta m_{kj}^2 L}{2E}\right)\,, \tag{13.9}$$

and the oscillation probability of antineutrinos as

$$P_{\bar\nu_\alpha\to\bar\nu_\beta}(L,E) = \delta_{\alpha\beta} - 4\sum_{k>j}\mathfrak{Re}\left[U_{\alpha k}^{*}\,U_{\beta k}\,U_{\alpha j}\,U_{\beta j}^{*}\right]\sin^2\left(\frac{\Delta m_{kj}^2 L}{4E}\right)$$
$$-\,2\,J\sum_{k>j} s_{\alpha\beta;kj}\,\sin\left(\frac{\Delta m_{kj}^2 L}{2E}\right)\,, \tag{13.10}$$

where we have changed the signs of the contributions of the imaginary parts of the quartic products in eqn (7.25) of elements of the mixing matrix, as explained in sections 7.3.2.

In Table 13.1 we have listed the real parts of the relevant quartic products of elements of the mixing matrix and the coefficients $s_{\alpha\beta;kj}$ for the transition channels $\overset{(-)}{\nu}_e \to \overset{(-)}{\nu}_\mu$, $\overset{(-)}{\nu}_\mu \to \overset{(-)}{\nu}_\tau$, $\overset{(-)}{\nu}_\tau \to \overset{(-)}{\nu}_e$ and for the survival channels $\overset{(-)}{\nu}_e \to \overset{(-)}{\nu}_e$, $\overset{(-)}{\nu}_\mu \to \overset{(-)}{\nu}_\mu$, $\overset{(-)}{\nu}_\tau \to \overset{(-)}{\nu}_\tau$. The oscillation probabilities for the T-conjugated transition channels $\overset{(-)}{\nu}_\mu \to \overset{(-)}{\nu}_e$, $\overset{(-)}{\nu}_\tau \to \overset{(-)}{\nu}_\mu$, $\overset{(-)}{\nu}_e \to \overset{(-)}{\nu}_\tau$ can be obtained by changing the sign of the corresponding coefficients $s_{\alpha\beta;kj}$.

13.1.1 CP and T violations

In the case of three-neutrino mixing, the CP asymmetries in eqn (7.61), which are equivalent to the T asymmetries in eqn (7.66), are given by

$$A_{\alpha\beta}^{\rm CP}(L,E) = 4\,J\sum_{k>j} s_{\alpha\beta;kj}\,\sin\left(\frac{\Delta m_{kj}^2 L}{2E}\right)\,. \tag{13.11}$$

From the values of the coefficients $s_{\alpha\beta;kj}$ listed in Table 13.1, it follows that all the CP asymmetries are equal up to a sign:

$$A_{e\mu}^{\rm CP} = A_{\mu\tau}^{\rm CP} = A_{\tau e}^{\rm CP} = -A_{\mu e}^{\rm CP} = -A_{\tau\mu}^{\rm CP} = -A_{e\tau}^{\rm CP}\,. \tag{13.12}$$

13.1.2 Hierarchy of Δm^2's

A simple case of three-neutrino mixing is realized when there is the hierarchy of squared-mass differences in eqn (13.6). This is the case which is of interest in practice, as explained in the introduction of this chapter, since we can identify the three-neutrino squared-mass differences with the solar and atmospheric squared-mass differences according to eqn (13.5). In the following we will consider the two types of three-neutrino mass spectra in Fig. 13.1, which can accommodate the observed hierarchy of squared-mass differences in eqn (13.4).

Atmospheric and LBL experiments, in which

$$\frac{|\Delta m^2_{31}|}{2} \left\langle \frac{L}{E} \right\rangle \sim \pi \,, \tag{13.13}$$

are sensitive to the oscillations due to Δm^2_{31}. In this case the hierarchy in eqn (13.6) implies that there is a dominance of the largest squared-mass difference Δm^2_{31}. Hence, the general discussion in section 7.7 applies, with the two groups of massive neutrinos $A = \{\nu_1 \,,\, \nu_2\}$ and $B = \{\nu_3\}$. The two-neutrino-like effective oscillation probability is given by eqn (7.125) for appearance experiments,

$$P^{\text{eff}}_{\nu_\alpha \to \nu_\beta}(L, E) = \sin^2 2\vartheta^{\text{eff}}_{\alpha\beta} \, \sin^2\left(\frac{\Delta m^2_{31} L}{4E}\right) \qquad (\alpha \neq \beta) \,, \tag{13.14}$$

and by eqn (7.133) for disappearance experiments,

$$P^{\text{eff}}_{\nu_\alpha \to \nu_\alpha}(L, E) = 1 - \sin^2 2\vartheta^{\text{eff}}_{\alpha\alpha} \, \sin^2\left(\frac{\Delta m^2_{31} L}{4E}\right) \,. \tag{13.15}$$

The effective mixing angles in appearance and disappearance experiments are given by eqn (7.136):

$$\sin^2 2\vartheta^{\text{eff}}_{\alpha\beta} = 4\,|U_{\alpha 3}|^2\,|U_{\beta 3}|^2 \quad (\alpha \neq \beta) \,, \qquad \sin^2 2\vartheta^{\text{eff}}_{\alpha\alpha} = 4\,|U_{\alpha 3}|^2 \left(1 - |U_{\alpha 3}|^2\right) \,. \tag{13.16}$$

Note the following important features of the oscillation probabilities in eqns (13.14) and (13.15):

1. All oscillation channels, $\overset{(-)}{\nu}_e \leftrightarrows \overset{(-)}{\nu}_\mu$, $\overset{(-)}{\nu}_\mu \leftrightarrows \overset{(-)}{\nu}_\tau$ and $\overset{(-)}{\nu}_\tau \leftrightarrows \overset{(-)}{\nu}_e$, are open and have the same oscillation length
$$L^{\text{osc}} = \frac{4\pi E}{\Delta m^2_{31}} \,. \tag{13.17}$$
2. As already remarked after eqn (7.136), the oscillation probabilities are independent of the phases in the mixing matrix and there is no possibility of revealing CP and T violations. In other words, $P^{\text{eff}}_{\nu_\alpha \to \nu_\beta} = P^{\text{eff}}_{\bar\nu_\alpha \to \bar\nu_\beta}$ and $P^{\text{eff}}_{\overset{(-)}{\nu}_\alpha \to \overset{(-)}{\nu}_\beta} = P^{\text{eff}}_{\overset{(-)}{\nu}_\beta \to \overset{(-)}{\nu}_\alpha}$.
3. The transition probabilities are determined by three parameters: Δm^2_{31}, $|U_{e3}|$, and $|U_{\mu 3}|$ (the unitarity of the mixing matrix implies that $|U_{\tau 3}| = 1 - |U_{e3}| - |U_{\mu 3}|$). The expressions of $|U_{e3}|$ and $|U_{\mu 3}|$ in terms of the mixing angles in the standard parameterization in eqn (6.191) are $|U_{e3}| = \sin^2\vartheta_{13}$ and $|U_{\mu 3}| =$

$\cos^2\vartheta_{13}\sin^2\vartheta_{23}$. Hence, the oscillation probabilities depend only on the two mixing angles ϑ_{13} and ϑ_{23}. They are independent of the mixing angle ϑ_{12}.

4. The two-neutrino form of the oscillation probabilities allows for a straightforward interpretation in terms of three-neutrino mixing of the analyses of experimental data which have been done using a two-neutrino mixing framework[70]. The two-neutrino squared-mass difference is identified with Δm^2_{31} and the two-neutrino mixing angle ϑ is identified with the appropriate effective mixing angle in eqn (13.16).

Let us now consider solar and VLBL experiments, in which

$$\frac{|\Delta m^2_{21}|}{2}\left\langle\frac{L}{E}\right\rangle \sim \pi\,. \tag{13.18}$$

Hence, these experiments are sensitive to the oscillations due to Δm^2_{21} and the discussion in section 7.8 applies: the small squared-mass difference Δm^2_{21} is active and oscillations due to the large squared-mass difference Δm^2_{31} are washed out. There is only one massive neutrino in each of the three groups considered in section 7.8: $\nu_1 \in A_1$, $\nu_2 \in A_2$, $\nu_3 \in B$. From eqns (7.148) and (7.149), the effective survival probabilities are given by

$$P^{\text{eff}}_{\overset{(-)}{\nu}_\alpha\to\overset{(-)}{\nu}_\alpha}(L,E) = \left(1-|U_{\alpha3}|^2\right)^2 P^{(1,2)}_{\overset{(-)}{\nu}_\alpha\to\overset{(-)}{\nu}_\alpha}(L,E) + |U_{\alpha3}|^4\,, \tag{13.19}$$

with the effective two-neutrino-like survival probability

$$P^{(1,2)}_{\overset{(-)}{\nu}_\alpha\to\overset{(-)}{\nu}_\alpha}(L,E) = 1-\sin^2 2\vartheta^{\text{eff}}_{\alpha\alpha}\,\sin^2\left(\frac{\Delta m^2_{21}L}{4E}\right)\,, \tag{13.20}$$

and the effective mixing angle given by

$$\sin^2 2\vartheta^{\text{eff}}_{\alpha\alpha} = 4\,\frac{|U_{\alpha1}|^2|U_{\alpha2}|^2}{\left(|U_{\alpha1}|^2+|U_{\alpha2}|^2\right)^2}\,. \tag{13.21}$$

The effective probability of $\nu_\alpha \to \nu_\beta$ transitions derived from eqn (7.156) is

$$\begin{aligned}P^{\text{eff}}_{\nu_\alpha\to\nu_\beta}(L,E) = {} & 4\,|U_{\alpha1}|^2\,|U_{\beta1}|^2\,\sin^2\left(\frac{\Delta m^2_{21}L}{4E}\right)\\ & + 4\,\mathfrak{Re}\left[U^*_{\alpha3}\,U_{\beta3}\,U_{\alpha1}\,U^*_{\beta1}\right]\,\sin^2\left(\frac{\Delta m^2_{21}L}{4E}\right)\\ & + 2\,s_{\alpha\beta;31}\,J\,\sin\left(\frac{\Delta m^2_{21}L}{2E}\right)\end{aligned}$$

[70] The important difference between a two-neutrino mixing scheme and the three-neutrino mixing scheme in the approximation under consideration is that a two-neutrino mixing scheme describes only transitions between two flavors, whereas in the three-neutrino mixing scheme simultaneous transitions among all three flavors are allowed, as remarked in item 1.

$$+ 2 |U_{\alpha 3}|^2 |U_{\beta 3}|^2 . \tag{13.22}$$

In practice, the effective survival probability in eqn (13.19) has been used in the analysis of experimental data of disappearance experiments (see chapter 13), but the effective transition probability in eqn (13.22) has never been used so far because there are no appearance experiments[71] which are sensitive to the oscillations generated by Δm^2_{21}. Note that the transition probabilities of neutrinos and antineutrinos are different if $J \neq 0$.

13.1.3 Trimaximal mixing

In the hypothetical but interesting case of trimaximal mixing presented in section 6.6.1, all the elements of the mixing matrix have the same absolute value equal to $1/\sqrt{3}$. This is the case when we have $\vartheta_{12} = \vartheta_{23} = \pi/4$, $s_{13} = 1/\sqrt{3}$, $\sin\delta_{13} = \pm 1$ in the parameterization given in eqn (6.191).

In the case of trimaximal mixing, all the survival probabilities are equal, as one can see immediately from eqn (7.39), which implies

$$P_{\nu_\alpha \to \nu_\alpha}(L, E) = 1 - \frac{4}{9} \sum_{k>j} \sin^2\left(\frac{\Delta m^2_{kj} L}{4E}\right) . \tag{13.23}$$

As an example, Fig. 13.2 shows the survival probability of ν_α as a function of $\langle L/E \rangle$ [km/GeV] Δm^2_{31} [eV^2] in the case of trimaximal mixing for $\Delta m^2_{21}/\Delta m^2_{31} = 10^{-3}$, averaged over a Gaussian L/E distribution with $\sigma_{L/E} = 0.2 \langle L/E \rangle$. The ratio $\Delta m^2_{21}/\Delta m^2_{31} = 10^{-3}$ implies the hierarchy of squared-mass differences in eqn (13.6). From Fig. 13.2 one can see that the survival probability starts to oscillate when $\langle L/E \rangle$ [km/GeV] Δm^2_{31} [eV^2] ~ 1. In this regime, Δm^2_{31} is the dominant squared mass difference and the effective survival probabilities are given by eqn (13.15) with the effective mixing angle $\vartheta^{\text{eff}}_{\alpha\alpha}$ given by

$$\sin^2 2\vartheta^{\text{eff}}_{\alpha\alpha} = \frac{8}{9} . \tag{13.24}$$

The oscillations due to Δm^2_{31} are damped for $\langle L/E \rangle$ [km/GeV] Δm^2_{31} [eV^2] $\gg 1$. In the interval $10 \lesssim \langle L/E \rangle$ [km/GeV] Δm^2_{31} [eV^2] $\lesssim 100$ the survival probability is averaged to

$$\langle P^{\text{eff}}_{\nu_\alpha \to \nu_\alpha} \rangle = 1 - \frac{1}{2} \sin^2 2\vartheta^{\text{eff}}_{\alpha\alpha} = \frac{5}{9} , \tag{13.25}$$

shown by the dashed horizontal line in Fig. 13.2.

When $\langle L/E \rangle$ is such that $\langle L/E \rangle$ [km/GeV] Δm^2_{31} [eV^2] $\sim 10^3$, the survival probability in Fig. 13.2 starts oscillations due to Δm^2_{21}, because

[71] Some solar neutrino experiment are sensitive to ν_μ's and ν_τ's generated by transitions from solar ν_e's. However, since they cannot distinguish ν_μ's from ν_τ's, they can measure only the inclusive transition probability $P_{\nu_e \to \nu_\mu} + P_{\nu_e \to \nu_\tau} = 1 - P_{\nu_e \to \nu_e}$, which is equivalent to a measurement of the survival probability of solar ν_e's.

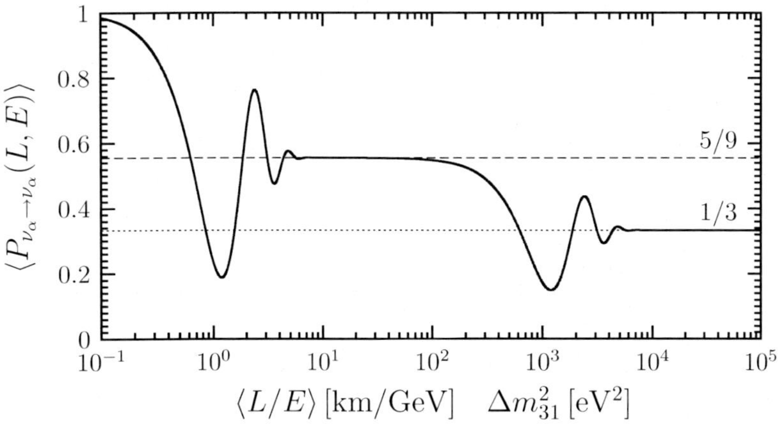

FIG. 13.2. Survival probability of ν_α as a function of $\langle L/E\rangle$ [km/GeV] Δm^2_{31} [eV^2] in the case of trimaximal mixing ($\vartheta_{12} = \vartheta_{23} = \pi/4$, $s_{13} = 1/\sqrt{3}$, $\sin\delta_{13} = \pm 1$; see section 6.6.1) for $\Delta m^2_{21}/\Delta m^2_{31} = 10^{-3}$, averaged over a Gaussian L/E distribution with $\sigma_{L/E} = 0.2\,\langle L/E\rangle$.

$\langle L/E\rangle$ [km/GeV] Δm^2_{21} [eV^2] ~ 1. In this regime, the small Δm^2_{21} is active and the effective survival probability is given by eqn (13.19) with $|U_{\alpha 3}|^2 = 1/3$,

$$P^{\text{eff}}_{\nu_\alpha\to\nu_\alpha}(L,E) = \frac{4}{9}\,P^{(1,2)}_{\nu_\alpha\to\nu_\alpha}(L,E) + \frac{1}{9}\,, \tag{13.26}$$

and the effective two-neutrino-like survival probability in eqn (13.20) with maximal effective mixing angle $\vartheta^{\text{eff}}_{\alpha\alpha}$,

$$\sin^2 2\vartheta^{\text{eff}}_{\alpha\alpha} = 1\,. \tag{13.27}$$

The oscillations due to Δm^2_{21} are damped for $\langle L/E\rangle$ [km/GeV] Δm^2_{31} [eV^2] $\gg 10^3$. The effective two-neutrino-like survival probability in eqn (13.20) averages to

$$\langle P^{(1,2)}_{\nu_\alpha\to\nu_\alpha}(L,E)\rangle = 1 - \frac{1}{2}\,\sin^2 2\vartheta^{\text{eff}}_{\alpha\alpha} = \frac{1}{2}\,, \tag{13.28}$$

implying that the averaged effective survival probability becomes

$$\langle P^{\text{eff}}_{\nu_\alpha\to\nu_\alpha}\rangle = \frac{1}{3}\,, \tag{13.29}$$

as shown by the dotted horizontal line in Fig. 13.2. This average value is the minimum possible average value of the survival probability for any mixing, as we have shown in eqn (7.44).

As discussed in section 6.6.1, a trimaximal Dirac mixing matrix is physically invariant under the cyclic permutations $e,\mu,\tau \leftrightarrows \mu,\tau,e \leftrightarrows \tau,e,\mu$ and $\nu_1,\nu_2,\nu_3 \leftrightarrows \nu_2,\nu_3,\nu_1 \leftrightarrows \nu_3,\nu_1,\nu_2$. Therefore, the transition probabilities are invariant under the same cyclic permutations. Taking also into account the CPT relation in eqn (7.57), we have the equalities

$$P_{\nu_e\to\nu_\mu} = P_{\nu_\mu\to\nu_\tau} = P_{\nu_\tau\to\nu_e} = P_{\bar\nu_\mu\to\bar\nu_e} = P_{\bar\nu_e\to\bar\nu_\tau} = P_{\bar\nu_\tau\to\bar\nu_\mu}\,, \tag{13.30}$$

$$P_{\nu_\mu \to \nu_e} = P_{\nu_e \to \nu_\tau} = P_{\nu_\tau \to \nu_\mu} = P_{\bar{\nu}_e \to \bar{\nu}_\mu} = P_{\bar{\nu}_\mu \to \bar{\nu}_\tau} = P_{\bar{\nu}_\tau \to \bar{\nu}_e} \,. \tag{13.31}$$

As we have shown in section 6.6.1, one of the exchanges $e \leftrightarrows \mu$ or $\mu \leftrightarrows \tau$ or $\tau \leftrightarrows e$ or $\nu_1 \leftrightarrows \nu_2$ or $\nu_2 \leftrightarrows \nu_3$ or $\nu_3 \leftrightarrows \nu_1$ transforms the two inequivalent trimaximal Dirac mixing matrices in eqn (6.373) with opposite Jarlskog invariant (see eqn (6.374)) one into the other. Therefore, under one of these exchanges, the contributions to the transition probabilities of the terms proportional to the Jarlskog invariant change sign. This means that the probabilities in eqns (13.30) and (13.31) are different and they are related by such a change of sign.

We can easily calculate one of the transition probabilities starting from the expression in eqn (7.118). Let us calculate $P_{\nu_e \to \nu_\mu}(L,E)$, which is one of the equal probabilities in eqn (13.30). Using the form in eqn (6.375) for the mixing matrix we obtain [529]

$$\begin{aligned}
P_{\nu_e \to \nu_\mu}(L,E) &= \left| \sum_k U_{ek}^* \, U_{\mu k} \, \exp\left(-i\,\frac{\Delta m^2_{k1} L}{2E}\right) \right|^2 \\
&= \frac{1}{9} \left| e^{\pm i2\pi/3} + e^{\mp i2\pi/3} \, \exp\left(-i\,\frac{\Delta m^2_{21} L}{2E}\right) + \exp\left(-i\,\frac{\Delta m^2_{31} L}{2E}\right) \right|^2 \\
&= \frac{1}{9} \left| 1 + \exp\left[-i\left(\frac{\Delta m^2_{21} L}{2E} \mp \frac{2\pi}{3}\right)\right] + \exp\left[-i\left(\frac{\Delta m^2_{31} L}{2E} \pm \frac{2\pi}{3}\right)\right] \right|^2 \\
&= 1 - \frac{4}{9} \left[\sin^2\left(\frac{\Delta m^2_{21} L}{4E} \mp \frac{\pi}{3}\right) + \sin^2\left(\frac{\Delta m^2_{31} L}{4E} \pm \frac{\pi}{3}\right) \right. \\
&\qquad \left. + \sin^2\left(\frac{\Delta m^2_{32} L}{4E} \mp \frac{\pi}{3}\right) \right] ,
\end{aligned} \tag{13.32}$$

where the upper and lower signs correspond, respectively, to a positive and a negative Jarlskog invariant (see eqn (6.374)). The probabilities in eqn (13.31) can be obtained by changing the signs of the phases $\frac{\pi}{3}$: for example

$$\begin{aligned}
P_{\bar{\nu}_e \to \bar{\nu}_\mu}(L,E) = 1 - \frac{4}{9} & \left[\sin^2\left(\frac{\Delta m^2_{21} L}{4E} \pm \frac{\pi}{3}\right) + \sin^2\left(\frac{\Delta m^2_{31} L}{4E} \mp \frac{\pi}{3}\right) \right. \\
& \left. + \sin^2\left(\frac{\Delta m^2_{32} L}{4E} \pm \frac{\pi}{3}\right) \right] .
\end{aligned} \tag{13.33}$$

We note that, in agreement with the discussion above, the probabilities in eqns (13.32) and (13.33) are invariant under the cyclic permutations $\nu_1, \nu_2, \nu_3 \leftrightarrows \nu_2, \nu_3, \nu_1 \leftrightarrows \nu_3, \nu_1, \nu_2$ and transform one into the other under one of the exchanges $\nu_1 \leftrightarrows \nu_2$ or $\nu_2 \leftrightarrows \nu_3$ or $\nu_3 \leftrightarrows \nu_1$.

Figure 13.3 shows the probabilities in eqns (13.32) and (13.33) averaged over a Gaussian L/E distribution with $\sigma_{L/E} = 0.2\,\langle L/E \rangle$, as functions of $\langle L/E \rangle\,[\text{km/GeV}]\,\Delta m^2_{31}\,[\text{eV}^2]$, for $\Delta m^2_{21}/\Delta m^2_{31} = 10^{-3}$ and $J = 1/6\sqrt{3}$, which implies that the upper signs in eqns (13.32) and (13.33) apply. One can see that the probabilities of $\nu_e \to \nu_\mu$ (solid line) and $\bar{\nu}_e \to \bar{\nu}_\mu$ (dashed line) transitions start to oscillate for $\langle L/E \rangle\,[\text{km/GeV}]\,\Delta m^2_{31}\,[\text{eV}^2] \sim 1$, where the effect of the small

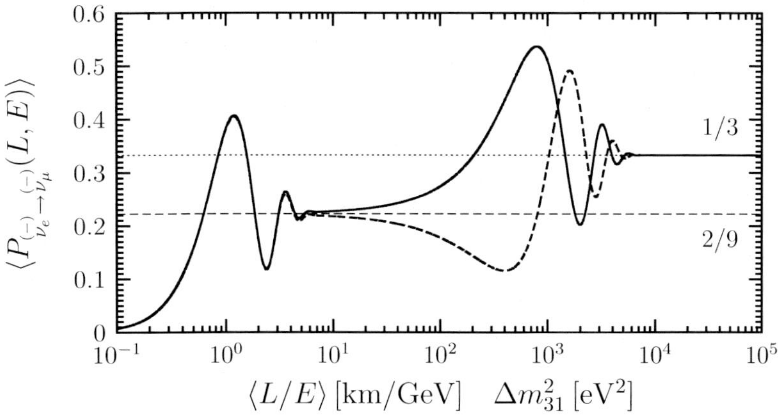

FIG. 13.3. Probabilities of $\nu_e \to \nu_\mu$ (solid line, eqn (13.32) with upper signs) and $\bar\nu_e \to \bar\nu_\mu$ (dashed line, eqn (13.33) with upper signs) transitions as functions of $\langle L/E\rangle$ [km/GeV] Δm^2_{31} [eV^2] in the case of trimaximal mixing with $J = 1/6\sqrt{3}$ ($\vartheta_{12} = \vartheta_{23} = \pi/4$, $s_{13} = 1/\sqrt{3}$, $\sin\delta_{13} = 1$; see section 6.6.1) for $\Delta m^2_{21}/\Delta m^2_{31} = 10^{-3}$, averaged over a Gaussian L/E distribution with $\sigma_{L/E} = 0.2\,\langle L/E\rangle$.

squared-mass difference Δm^2_{21} is negligible. In this case, since $\Delta m^2_{32} \simeq \Delta m^2_{31}$, the two probabilities are practically equal and are given by eqn (13.14) with effective mixing angle

$$\sin^2 2\vartheta^{\text{eff}}_{e\mu} = \frac{4}{9}\,. \tag{13.34}$$

For $\langle L/E\rangle$ [km/GeV] Δm^2_{31} [eV^2] $\gg 1$, the oscillations due to Δm^2_{31} are damped out and the two probabilities average to

$$\langle P_{\overset{(-)}{\nu}_e \to \overset{(-)}{\nu}_\mu}\rangle = \frac{2}{9}\,, \tag{13.35}$$

as shown by the dashed horizontal line in Fig. 13.3. For $\langle L/E\rangle$ [km/GeV] Δm^2_{31} [eV^2] $\gtrsim 10$, the effects of Δm^2_{21} start to be significant and the two probabilities deviate from each other. Since the sign of the phase $\pi/3$ is negative in the squared-sine depending on Δm^2_{21} of $P_{\nu_e\to\nu_\mu}(L,E)$, there is an initial cancellation between the contributions of $\Delta m^2_{21}L/4E$ and $\pi/3$. Since the squared-sine has negative sign, the probability of $\nu_e \to \nu_\mu$ transitions increases when the effect of Δm^2_{21} starts to become significant. The probability of $\bar\nu_e \to \bar\nu_\mu$ transitions suffers just the opposite effect, leading to an initial decrease. For $\langle L/E\rangle$ [km/GeV] Δm^2_{31} [eV^2] $\gg 10^3$, the oscillations due to Δm^2_{21} are damped out and both probabilities average to $1/3$.

The CP asymmetries can easily be calculated from eqns (13.30)–(13.33), yielding

$$A^{\text{CP}}_{e\mu}(L,E) = \pm\frac{2}{3\sqrt{3}}\left[\sin\left(\frac{\Delta m^2_{21}L}{2E}\right) - \sin\left(\frac{\Delta m^2_{31}L}{2E}\right) + \sin\left(\frac{\Delta m^2_{32}L}{2E}\right)\right]. \tag{13.36}$$

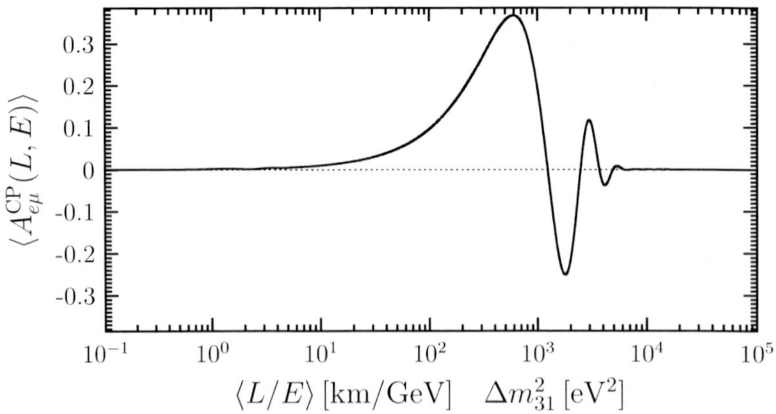

FIG. 13.4. Averaged CP asymmetry $\langle A^{\mathrm{CP}}_{e\mu}(L,E)\rangle$ as a function of $\langle L/E\rangle$ [km/GeV] Δm^2_{31} [eV^2] in the case of trimaximal mixing with $J = 1/6\sqrt{3}$ ($\vartheta_{12} = \vartheta_{23} = \pi/4$, $s_{13} = 1/\sqrt{3}$, $\sin\delta_{13} = 1$; see section 6.6.1) and $\Delta m^2_{21}/\Delta m^2_{31} = 10^{-3}$, averaged over a Gaussian L/E distribution with $\sigma_{L/E} = 0.2\,\langle L/E\rangle$. The vertical axis ranges from $-2/3\sqrt{3}$ to $2/3\sqrt{3}$.

The other CP asymmetries are equal to $A^{\mathrm{CP}}_{e\mu}(L,E)$ up to a sign, according to the general equalities in eqn (13.12) in three-neutrino mixing.

Figure 13.4 shows the CP asymmetry $A^{\mathrm{CP}}_{e\mu}(L,E)$ averaged over a Gaussian L/E distribution with $\sigma_{L/E} = 0.2\,\langle L/E\rangle$, as a function of $\langle L/E\rangle$ [km/GeV] Δm^2_{31} [eV^2] for trimaximal mixing with $J = 1/6\sqrt{3}$ and $\Delta m^2_{21}/\Delta m^2_{31} = 10^{-3}$. One can see that there is no CP asymmetry for $\langle L/E\rangle$ [km/GeV] Δm^2_{31} [eV^2] $\lesssim 1$, where Δm^2_{31} is the dominant squared mass difference, in agreement with the discussion in section 7.7. Indeed, since the phase due to Δm^2_{21} is negligible and $\Delta m^2_{32} \simeq \Delta m^2_{31}$, only the two sine functions in eqn (13.36) which depend on Δm^2_{31} and Δm^2_{32} contribute and cancel each other.

The CP asymmetry starts to deviate from zero at $\langle L/E\rangle$ [km/GeV] Δm^2_{31} [eV^2] $\sim 10^2$ and oscillates for $\langle L/E\rangle$ [km/GeV] Δm^2_{31} [eV^2] $\sim 10^3$, where $\langle L/E\rangle$ [km/GeV] Δm^2_{21} 1. This means that the small squared-mass difference Δm^2_{21} is active. In this regime, the two sine functions in eqn (13.36), which depend on Δm^2_{31} and Δm^2_{32}, are averaged to zero and the sine function which depends on Δm^2_{21} gives the nonvanishing oscillating contribution. Since the vertical axis in Fig. 13.4 ranges from $-2/3\sqrt{3}$ to $2/3\sqrt{3}$, one can see that in the first maximum the averaging procedure has very little effect. The averaging starts to damp the oscillations in a significant way at the first minimum and acts so quickly that the third maximum is barely visible.

For $\langle L/E\rangle$ [km/GeV] Δm^2_{31} [eV^2] $\gtrsim 10^4$, the CP asymmetry is damped to zero, because the averages of all the sine functions in eqn (13.36) vanish. This is a concrete example of the general property discussed at the end of section 7.3.2.

13.1.4 Bilarge mixing

As discussed in section 6.6.2, we have bilarge mixing if one of the elements of the mixing matrix is zero or much smaller than the others, so that it can be neglected in practice. In this case, with an appropriate parameterization, it is possible to express the mixing matrix in terms of only two mixing angles. Hence, the mixing matrix is real, the Jarlskog invariant is zero, and there are no CP or T violations in neutrino oscillations, yielding

$$P_{\nu_\alpha \to \nu_\beta} = P_{\bar{\nu}_\alpha \to \bar{\nu}_\beta} = P_{\bar{\nu}_\beta \to \bar{\nu}_\alpha} = P_{\nu_\beta \to \nu_\alpha} \,. \tag{13.37}$$

The oscillation probabilities in eqn (7.38) reduce to

$$P_{\nu_\alpha \to \nu_\beta}(L,E) = \delta_{\alpha\beta} - 4 \sum_{k>j} U_{\alpha k}\, U_{\beta k}\, U_{\alpha j}\, U_{\beta j} \, \sin^2\left(\frac{\Delta m^2_{kj} L}{4E}\right) . \tag{13.38}$$

As a useful example, let us consider the case in which $U_{e3} = 0$, which is a realistic approximation of the true mixing matrix, as we will see in section 13.3. In this case, the Dirac part of the mixing matrix can be parameterized as in eqn (6.378) in terms of the mixing angles ϑ_{12} and ϑ_{23},

$$U^{\mathrm{D}} = \begin{pmatrix} c_{12} & s_{12} & 0 \\ -s_{12}c_{23} & c_{12}c_{23} & s_{23} \\ s_{12}s_{23} & -c_{12}s_{23} & c_{23} \end{pmatrix} . \tag{13.39}$$

The electron neutrino is a mixture of ν_1 and ν_2 only and its survival and transition probabilities can be written in the two-neutrino-like form

$$P_{\nu_e \to \nu_\beta}(L,E) = \delta_{e\beta} + (1 - 2\delta_{e\beta}) \sin^2 2\vartheta^{\mathrm{eff}}_{e\beta} \sin^2\left(\frac{\Delta m^2_{21} L}{4E}\right) \qquad (\beta = e, \mu, \tau) \,, \tag{13.40}$$

with the effective mixing angles $\vartheta^{\mathrm{eff}}_{e\beta}$ given by

$$\vartheta^{\mathrm{eff}}_{ee} = \vartheta_{12} \,, \qquad \sin^2 2\vartheta^{\mathrm{eff}}_{e\mu} = \sin^2 2\vartheta_{12}\, c^2_{23} \,, \qquad \sin^2 2\vartheta^{\mathrm{eff}}_{e\tau} = \sin^2 2\vartheta_{12}\, s^2_{23} \,. \tag{13.41}$$

The survival probabilities of muon and tau neutrinos are given by

$$\begin{aligned} P_{\nu_\mu \to \nu_\mu}(L,E) = {} & 1 - \sin^2 2\vartheta_{23} \left[s^2_{12} \sin^2\left(\frac{\Delta m^2_{31} L}{4E}\right) + c^2_{12} \sin^2\left(\frac{\Delta m^2_{32} L}{4E}\right) \right] \\ & - \sin^2 2\vartheta_{12}\, c^4_{23} \sin^2\left(\frac{\Delta m^2_{21} L}{4E}\right) , \end{aligned} \tag{13.42}$$

$$\begin{aligned} P_{\nu_\tau \to \nu_\tau}(L,E) = {} & 1 - \sin^2 2\vartheta_{23} \left[s^2_{12} \sin^2\left(\frac{\Delta m^2_{31} L}{4E}\right) + c^2_{12} \sin^2\left(\frac{\Delta m^2_{32} L}{4E}\right) \right] \\ & - \sin^2 2\vartheta_{12}\, s^4_{23} \sin^2\left(\frac{\Delta m^2_{21} L}{4E}\right) , \end{aligned} \tag{13.43}$$

and the probability of $\nu_\mu \to \nu_\tau$ transitions is given by

$$P_{\nu_\mu \to \nu_\tau}(L,E) = \sin^2 2\vartheta_{23} \left[s^2_{12} \sin^2\left(\frac{\Delta m^2_{31} L}{4E}\right) + c^2_{12} \sin^2\left(\frac{\Delta m^2_{32} L}{4E}\right) \right]$$

$$- \frac{1}{4} \sin^2 2\vartheta_{12} \sin^2 2\vartheta_{23} \sin^2\left(\frac{\Delta m^2_{21} L}{4E}\right) . \tag{13.44}$$

Therefore, the data of neutrino oscillation experiments in which electron neutrinos or antineutrinos are produced or detected can be conveniently analyzed with a two-neutrino-like oscillation probability which depends only on two effective parameters. However, they can provide information only on Δm^2_{21} and ϑ_{12}. On the other hand, neutrino oscillation experiments that do not involve electron neutrinos or antineutrinos can yield information on all the five oscillation parameters, the two mixing angles ϑ_{12}, ϑ_{23} and the three squared-mass differences Δm^2_{21}, Δm^2_{31}, Δm^2_{32}, but they must be analyzed with more complicated formulas.

The oscillation probabilities in eqns (13.42)–(13.44) lead to simpler approximate expressions in the case of a hierarchy of squared-mass differences of the type in eqn (13.6), which, in fact, is of practical interest. Since $\Delta m^2_{31} \simeq \Delta m^2_{32}$, we have

$$P_{\nu_\mu \to \nu_\mu}(L, E) \simeq 1 - \sin^2 2\vartheta_{23} \sin^2\left(\frac{\Delta m^2_{31} L}{4E}\right) - \sin^2 2\vartheta_{12}\, c^4_{23} \sin^2\left(\frac{\Delta m^2_{21} L}{4E}\right) , \tag{13.45}$$

$$P_{\nu_\tau \to \nu_\tau}(L, E) \simeq 1 - \sin^2 2\vartheta_{23} \sin^2\left(\frac{\Delta m^2_{31} L}{4E}\right) - \sin^2 2\vartheta_{12}\, s^4_{23} \sin^2\left(\frac{\Delta m^2_{21} L}{4E}\right) , \tag{13.46}$$

$$P_{\nu_\mu \to \nu_\tau}(L, E) \simeq \sin^2 2\vartheta_{23} \sin^2\left(\frac{\Delta m^2_{31} L}{4E}\right) - \frac{1}{4} \sin^2 2\vartheta_{12} \sin^2 2\vartheta_{23} \sin^2\left(\frac{\Delta m^2_{21} L}{4E}\right) . \tag{13.47}$$

In experiments sensitive to the oscillations due to Δm^2_{31}, where eqn (13.13) is satisfied, the contributions of Δm^2_{21} are negligible and we have

$$P^{\text{eff}}_{\nu_\mu \to \nu_\mu}(L, E) = 1 - \sin^2 2\vartheta_{23} \sin^2\left(\frac{\Delta m^2_{31} L}{4E}\right) , \tag{13.48}$$

$$P^{\text{eff}}_{\nu_\tau \to \nu_\tau}(L, E) = 1 - \sin^2 2\vartheta_{23} \sin^2\left(\frac{\Delta m^2_{31} L}{4E}\right) , \tag{13.49}$$

$$P^{\text{eff}}_{\nu_\mu \to \nu_\tau}(L, E) = \sin^2 2\vartheta_{23} \sin^2\left(\frac{\Delta m^2_{31} L}{4E}\right) , \tag{13.50}$$

which agree with the general expressions in eqns (13.14) and (13.15). These experiments can give information on Δm^2_{31} and ϑ_{23}.

In experiments sensitive to the oscillations due to Δm^2_{21}, where the condition in eqn (13.18) holds, the oscillations due to the largest squared-mass difference Δm^2_{31} are washed out, leading to the effective probabilities

$$P^{\text{eff}}_{\nu_\mu \to \nu_\mu}(L, E) = 1 - \frac{1}{2} \sin^2 2\vartheta_{23} - \sin^2 2\vartheta_{12}\, c^4_{23} \sin^2\left(\frac{\Delta m^2_{21} L}{4E}\right) , \tag{13.51}$$

$$P^{\text{eff}}_{\nu_\tau \to \nu_\tau}(L, E) = 1 - \frac{1}{2} \sin^2 2\vartheta_{23} - \sin^2 2\vartheta_{12}\, s^4_{23} \sin^2\left(\frac{\Delta m^2_{21} L}{4E}\right) , \tag{13.52}$$

$$P^{\text{eff}}_{\nu_\mu \to \nu_\tau}(L, E) = \frac{1}{2} \sin^2 2\vartheta_{23} - \frac{1}{4} \sin^2 2\vartheta_{12} \sin^2 2\vartheta_{23} \sin^2\left(\frac{\Delta m^2_{21} L}{4E}\right), \qquad (13.53)$$

which are consistent with the general expressions in eqns (13.19) and (13.22). If the value of ϑ_{23} is not known independently from other experiments which are sensitive to the oscillations due to Δm^2_{31}, and thus measure the effective probabilities in eqns (13.48)–(13.50), the data from these experiments must be analyzed in terms of the three parameters Δm^2_{21}, ϑ_{12}, and ϑ_{23}.

13.2 Matter effects

In general, the discussion of oscillations in matter for three neutrino mixing is quite complicated, because of the possible interplay of the effects of the different Δm^2's (see Ref. [707]). Fortunately, in the case of the hierarchy of squared-mass differences in eqn (13.6), which is of interest in practice, the effects of the large and small squared-mass differences can be separated out, leading to a considerable simplification of the discussion and a clear understanding of the physical mechanisms at work in experiments. In section 13.1.2 we have already discussed the most important phenomenological aspects of oscillations in vacuum in this scenario. In matter, we will consider two regimes: the case of largest Δm^2 dominance in section 13.2.1 and the case of active small Δm^2 in section 13.2.2. However, before discussing these cases, we elaborate the evolution equation (9.54) in order to simplify irrelevant degrees of freedom.

As shown in eqn (4.77), the mixing matrix in the standard parameterization in eqn (6.31) can be written as

$$U = R^{23} W^{13} R^{12}, \qquad (13.54)$$

in the notation introduced in section 4.3.2. We do not consider additional Majorana phases in the case of Majorana neutrinos (see eqn (6.189)), since they are irrelevant for neutrino oscillations, as explained at the end of section 9.2. The parameterization in eqn (13.54) is convenient, since the rotation matrix R^{23} commutes with the matter potential matrix $\mathbb{A}$ in eqn (9.56):

$$R^{23\dagger} \mathbb{A} R^{23} = \mathbb{A}. \qquad (13.55)$$

Indeed, since a rotation in the 2-3 plane mixes ν_μ and ν_τ, which have the same matter potential, it is irrelevant for the matter effect. This property allows us to write the evolution equation (9.54) as

$$i \frac{\mathrm{d}}{\mathrm{d}x} \Psi_\alpha = \frac{1}{2E} R^{23} \left(W^{13} R^{12} \mathbb{M}^2 R^{12\dagger} W^{13\dagger} + \mathbb{A} \right) R^{23\dagger} \Psi_\alpha. \qquad (13.56)$$

Defining the new column matrix of amplitudes

$$\widetilde{\Psi}_\alpha = R^{23\dagger} \Psi_\alpha, \qquad (13.57)$$

we obtain the evolution equation

$$i \frac{\mathrm{d}}{\mathrm{d}x} \widetilde{\Psi}_\alpha = \frac{1}{2E} \left(W^{13} \, R^{12} \, \mathbb{M}^2 \, R^{12\dagger} \, W^{13\dagger} + \mathbb{A} \right) \widetilde{\Psi}_\alpha \, , \tag{13.58}$$

which is independent of ϑ_{23}. It is however to be noted that the amplitudes in the second and third entries of $\widetilde{\Psi}_\alpha$ do not have a definite flavor or mass character. Only the first entry, corresponding to $\nu_\alpha \to \nu_e$ transitions, is the same in Ψ_α and $\widetilde{\Psi}_\alpha$:

$$\begin{pmatrix} \psi_{\alpha e} \\ \psi_{\alpha\mu} \\ \psi_{\alpha\tau} \end{pmatrix} = \begin{pmatrix} \widetilde{\psi}_{\alpha 1} \\ c_{23} \, \widetilde{\psi}_{\alpha 2} + s_{23} \, \widetilde{\psi}_{\alpha 3} \\ -s_{23} \, \widetilde{\psi}_{\alpha 2} + c_{23} \, \widetilde{\psi}_{\alpha 3} \end{pmatrix} . \tag{13.59}$$

The flavor transition probabilities are given by

$$P_{\nu_\alpha \to \nu_e}(x) = |\psi_{\alpha e}(x)|^2 = |\widetilde{\psi}_{\alpha 1}(x)|^2 \, , \tag{13.60}$$

$$\begin{aligned} P_{\nu_\alpha \to \nu_\mu}(x) &= |\psi_{\alpha\mu}(x)|^2 \\ &= \cos^2\vartheta_{23} |\widetilde{\psi}_{\alpha 2}(x)|^2 + \sin^2\vartheta_{23} |\widetilde{\psi}_{\alpha 3}(x)|^2 + \sin 2\vartheta_{23} \mathfrak{Re}\left[\widetilde{\psi}^*_{\alpha 2}(x) \widetilde{\psi}_{\alpha 3}(x) \right] , \end{aligned} \tag{13.61}$$

$$\begin{aligned} P_{\nu_\alpha \to \nu_\tau}(x) &= |\psi_{\alpha\tau}(x)|^2 \\ &= \sin^2\vartheta_{23} |\widetilde{\psi}_{\alpha 2}(x)|^2 + \cos^2\vartheta_{23} |\widetilde{\psi}_{\alpha 3}(x)|^2 - \sin 2\vartheta_{23} \mathfrak{Re}\left[\widetilde{\psi}^*_{\alpha 2}(x) \widetilde{\psi}_{\alpha 3}(x) \right] . \end{aligned} \tag{13.62}$$

Moreover, the initial conditions for the evolution equation (13.58) depend on ϑ_{23}. For example, if the initial neutrino is a ν_μ, we have

$$\widetilde{\Psi}_\mu(0) = \begin{pmatrix} 1 & 0 & 0 \\ 0 & \cos\vartheta_{23} & \sin\vartheta_{23} \\ 0 & -\sin\vartheta_{23} & \cos\vartheta_{23} \end{pmatrix} \begin{pmatrix} 0 \\ 1 \\ 0 \end{pmatrix} = \begin{pmatrix} 0 \\ \cos\vartheta_{23} \\ -\sin\vartheta_{23} \end{pmatrix} . \tag{13.63}$$

Hence, the transition probabilities depend on ϑ_{23}, as they must, according to the discussion in section 7.3 of neutrino oscillations in vacuum. The meaning of the independence of the evolution equation (13.58) from ϑ_{23} is that this mixing angle is not modified in matter, whereas the mixing angles ϑ_{12} and ϑ_{13}, on which the evolution equation (13.58) depends, acquire effective values in matter, as we will see in the following.

Since for antineutrinos the matter potential changes sign (see section 9.1), in matter the oscillation probabilities of neutrinos and antineutrinos are different, even if the fundamental Lagrangian is CP invariant (i.e. even if the Jarlskog invariant of the mixing matrix is zero). In other words, there is a matter-induced CP violation in the oscillation probabilities due to the fact that the medium is not CP-invariant. Moreover, since the medium is also not CPT-invariant, there is a matter-induced CPT violation in the oscillation probabilities.

13.2.1 Δm^2_{31} dominance

In this subsection we consider atmospheric and LBL neutrino oscillation experiments in which eqn (13.13) holds. These experiments are sensitive to the largest squared-mass difference Δm^2_{31} and the small squared-mass difference Δm^2_{21} can be neglected in a study of the dominant effects due to Δm^2_{31}. Hence, in this case we have

$$\mathbb{M}^2 \simeq \text{diag}(0, 0, \Delta m^2_{31}) \,. \tag{13.64}$$

This approximation leads to a substantial simplification, since now R^{12} commutes with $\Delta\mathbb{M}$, leading to the evolution equation

$$i \frac{\mathrm{d}}{\mathrm{d}x} \widetilde{\Psi}_\alpha = \frac{1}{2E} \left(W^{13} \, \mathbb{M}^2 \, W^{13\dagger} + \mathbb{A} \right) \widetilde{\Psi}_\alpha \,. \tag{13.65}$$

Moreover, from eqns (4.62) and (4.77) we have $W^{13} = D^1(-\delta_{13}) R^{13} D^{1\dagger}(-\delta_{13})$ and, from the explicit expression of D^1 in eqn (4.56), one can immediately see that $D^{1\dagger}(-\delta_{13}) \mathbb{M}^2 D^1(-\delta_{13}) = \mathbb{M}^2$. Since we also have, from the explicit expression of $\mathbb{A}$ in eqn (9.56), $D^1(-\delta_{13}) \mathbb{A} D^{1\dagger}(-\delta_{13}) \mathbb{A}$, we can redefine

$$\widetilde{\Psi}_\alpha \to D^1(-\delta_{13}) \, \widetilde{\Psi}_\alpha \,, \tag{13.66}$$

in order to obtain the evolution equation

$$i \frac{\mathrm{d}}{\mathrm{d}x} \widetilde{\Psi}_\alpha = \frac{1}{2E} \left(R^{13} \, \mathbb{M}^2 \, R^{13\dagger} + \mathbb{A} \right) \widetilde{\Psi}_\alpha \,, \tag{13.67}$$

which depends only on the mixing angle ϑ_{13}. Since the redefinition in eqn (13.66) is just a phase shift of the first component of $\widetilde{\Psi}_\alpha$, it does not change the expressions of the flavor transition probabilities in eqns (13.60)–(13.62), which are, therefore, independent[72] of the CP-violating phase δ_{13}. Thus, CP violation is not observable in neutrino oscillations in matter, as well as in vacuum (see section 13.1.2), in experiments which are only sensitive to the oscillations generated by the largest squared-mass difference Δm^2_{31}. This does not mean, however, that the oscillation probabilities of neutrinos and antineutrinos are the same, because of the matter-induced CP-violation discussed above.

Explicitly, the evolution equation (13.67) reads

$$i \frac{\mathrm{d}}{\mathrm{d}x} \begin{pmatrix} \widetilde{\psi}_{\alpha 1} \\ \widetilde{\psi}_{\alpha 2} \\ \widetilde{\psi}_{\alpha 3} \end{pmatrix} = \frac{1}{2E} \begin{pmatrix} s^2_{13} \Delta m^2_{31} + A_{\text{CC}} & 0 & c_{13} s_{13} \Delta m^2_{31} \\ 0 & 0 & 0 \\ c_{13} s_{13} \Delta m^2_{31} & 0 & c^2_{13} \Delta m^2_{31} \end{pmatrix} \begin{pmatrix} \widetilde{\psi}_{\alpha 1} \\ \widetilde{\psi}_{\alpha 2} \\ \widetilde{\psi}_{\alpha 3} \end{pmatrix} . \tag{13.68}$$

Thus, the component $\widetilde{\psi}_{\alpha 2}$ is constant,

$$\widetilde{\psi}_{\alpha 2}(x) = \widetilde{\psi}_{\alpha 2}(0) \,, \tag{13.69}$$

and the evolution equation is reduced to an effective two-neutrino-like evolution equation for the flavor transition amplitudes $\widetilde{\psi}_{\alpha 1}$ and $\widetilde{\psi}_{\alpha 3}$. Neglecting an irrelevant

[72] One could have obtained the same result in a simpler way by using for the mixing matrix the parameterization $U = R^{23} \, R^{13} \, W^{12}$, instead of the standard one in eqn (13.54). Since W^{12} commutes with $\Delta\mathbb{M}$ in eqn (13.64), the evolution equation is obviously independent of the CP-violating phase in W^{12} [537].

common phase $\exp\left[-i\Delta m^2_{31}x/4E - \frac{i}{2}\int_0^x V_{\text{CC}}(x')\,\mathrm{d}x'\right]$, the evolution equation for these two components can be written in a form similar to that in eqn (9.59):

$$i\frac{\mathrm{d}}{\mathrm{d}x}\begin{pmatrix}\widetilde{\psi}_{\alpha 1}\\ \widetilde{\psi}_{\alpha 3}\end{pmatrix} = \frac{1}{4E}\begin{pmatrix}-\Delta m^2_{31}\cos 2\vartheta_{13} + A_{\text{CC}} & \Delta m^2_{31}\sin 2\vartheta_{13}\\ \Delta m^2_{31}\sin 2\vartheta_{13} & \Delta m^2_{31}\cos 2\vartheta_{13} - A_{\text{CC}}\end{pmatrix}\begin{pmatrix}\widetilde{\psi}_{\alpha 1}\\ \widetilde{\psi}_{\alpha 3}\end{pmatrix}. \tag{13.70}$$

Hence, the MSW effect discussed in section 9.3 occurs in the 1-3 sector. In particular, the effective mixing angle $\vartheta^{\text{M}}_{13}$ in matter is given by

$$\tan 2\vartheta^{\text{M}}_{13} = \frac{\tan 2\vartheta_{13}}{1 - \dfrac{A_{\text{CC}}}{\Delta m^2_{31}\cos 2\vartheta_{13}}}, \tag{13.71}$$

and there is a resonance when A_{CC} becomes equal to

$$A^{\text{R}}_{\text{CC}} = \Delta m^2_{31}\cos 2\vartheta_{13}. \tag{13.72}$$

The effective 3–1 squared-mass difference in matter is given by

$$\Delta m^2_{\text{M}31} = \sqrt{\left(\Delta m^2_{31}\cos 2\vartheta_{13} - A_{\text{CC}}\right)^2 + \left(\Delta m^2_{31}\sin 2\vartheta_{13}\right)^2}. \tag{13.73}$$

The discussion in section 9.3 for a medium with smooth density variations can be straightforwardly adapted to the present case. However, in practice we are interested in neutrino beams which propagate in the Earth, where the slab approximation discussed in section 9.4 is more appropriate. In this case, the flavor transition amplitudes are given by the analog of eqn (9.102):

$$\begin{aligned}\Psi_\alpha(x_n) = {} & \left[U_{\text{M}}\,\mathscr{U}_{\text{M}}(x_n - x_{n-1})\,U^\dagger_{\text{M}}\right]_{(n)}\left[U_{\text{M}}\,\mathscr{U}_{\text{M}}(x_{n-1} - x_{n-2})\,U^\dagger_{\text{M}}\right]_{(n-1)}\\ & \ldots\left[U_{\text{M}}\,\mathscr{U}_{\text{M}}(x_2 - x_1)\,U^\dagger_{\text{M}}\right]_{(2)}\left[U_{\text{M}}\,\mathscr{U}_{\text{M}}(x_1 - x_0)\,U^\dagger_{\text{M}}\right]_{(1)}\Psi_\alpha(x_0),\end{aligned} \tag{13.74}$$

where x_0 is the coordinate of the starting point, $x_1, x_2, \ldots, x_n$ are the boundaries of n slabs and the notation $[\ldots]_{(i)}$ indicates that all the matter-dependent quantities in the square brackets must be evaluated with the matter density in the ith slab, which extends from x_{i-1} to x_i. The unitary evolution operator in each slab diagonal basis is given by

$$\mathscr{U}_{\text{M}}(\Delta x) = \operatorname{diag}\left(\exp\left(-im^2_{\text{M}1}\Delta x/2E\right),\, 1,\, \exp\left(-im^2_{\text{M}3}\Delta x/2E\right)\right), \tag{13.75}$$

with the effective squared-masses

$$m^2_{\text{M}3,1} = \frac{1}{2}\left(\Delta m^2_{31} + A_{\text{CC}} \pm \Delta m^2_{\text{M}31}\right). \tag{13.76}$$

The second diagonal entry in $\mathscr{U}_{\text{M}}(\Delta x)$ is equal to unity because the corresponding effective squared-mass is equal to zero,

$$m^2_{\text{M}2} = 0, \tag{13.77}$$

as one can see immediately from eqn (13.68).

We determine the effective mixing matrix in matter $U_{\rm M}$ by noting that the evolution equation (13.68) is diagonalized by the orthogonal transformation

$$\widetilde{\Psi}_\alpha = R^{13}_{\rm M}\, \Phi_\alpha \,, \tag{13.78}$$

with

$$R^{13}_{\rm M} = \begin{pmatrix} \cos\vartheta^{\rm M}_{13} & 0 & \sin\vartheta^{\rm M}_{13} \\ 0 & 1 & 0 \\ -\sin\vartheta^{\rm M}_{13} & 0 & \cos\vartheta^{\rm M}_{13} \end{pmatrix} . \tag{13.79}$$

Thus, we obtain, from eqn (13.57),

$$\Psi_\alpha = U_{\rm M}\, \Phi_\alpha \,, \tag{13.80}$$

with the effective mixing matrix in matter

$$U_{\rm M} = R^{23}\, R^{13}_{\rm M} = \begin{pmatrix} \cos\vartheta^{\rm M}_{13} & 0 & \sin\vartheta^{\rm M}_{13} \\ -\sin\vartheta_{23}\, \sin\vartheta^{\rm M}_{13} & \cos\vartheta_{23} & \sin\vartheta_{23}\, \cos\vartheta^{\rm M}_{13} \\ -\cos\vartheta_{23}\, \sin\vartheta^{\rm M}_{13} & -\sin\vartheta_{23} & \cos\vartheta_{23}\, \cos\vartheta^{\rm M}_{13} \end{pmatrix} . \tag{13.81}$$

This effective mixing matrix does not depend on the mixing angle ϑ_{12} because we are working in the approximation $\Delta m^2_{21} = 0$, in which there is no mixing in the 1-2 sector. Note that the third column of the effective mixing matrix in matter has the same structure of the third column of the mixing matrix U (see eqn (6.191)), in terms of the mixing angle ϑ_{23} and the effective mixing angle in matter $\vartheta^{\rm M}_{13}$, which replaces the mixing angle ϑ_{13} (the CP-violating phase is irrelevant, as explained above). On the other hand, the first row of the effective mixing matrix in matter is simpler than that of the mixing matrix U, since the approximation $\vartheta_{12} \to 0$ implies that the electron neutrino is only affected by the mixing in the 1-3 sector. Physically, the special status of the electron neutrino is a consequence of its charged-current interactions with the electrons in the medium, represented by the potential $V_{\rm CC}$. The element $U_{{\rm M}e2}$ of the effective mixing matrix is equal to zero because the electron neutrino cannot mix with the second effective massive neutrino, whose squared-mass does not depend on $V_{\rm CC}$ (see eqn (13.77)).

Since the mixing angle ϑ_{23} does not change in matter, eqn (13.74) can be simplified to

$$\begin{aligned} \Psi_\alpha(x_n) = R^{23} & \left[R^{13}_{\rm M}\, \mathscr{U}_{\rm M}(x_n - x_{n-1})\, R^{13\dagger}_{\rm M} \right]_{(n)} \left[R^{13}_{\rm M}\, \mathscr{U}_{\rm M}(x_{n-1} - x_{n-2})\, R^{13\dagger}_{\rm M} \right]_{(n-1)} \\ & \ldots \left[R^{13}_{\rm M}\, \mathscr{U}_{\rm M}(x_2 - x_1)\, R^{13\dagger}_{\rm M} \right]_{(2)} \left[R^{13}_{\rm M}\, \mathscr{U}_{\rm M}(x_1 - x_0)\, R^{13\dagger}_{\rm M} \right]_{(1)} R^{23\dagger}\, \Psi_\alpha(x_0) \,, \end{aligned} \tag{13.82}$$

where one must take into account only the changes of the values of $\vartheta^{\rm M}_{13}$ and $\Delta m^2_{{\rm M}31}$ in the transitions between slabs with different matter densities.

In the simplest case of a constant matter density, the flavor transition probabilities at a distance L from the source can be calculated analytically from

$$\Psi_\alpha(L) = U_\mathrm{M}\, \mathscr{U}_\mathrm{M}(L)\, {U_\mathrm{M}}^\dagger\, \Psi_\alpha(0)\,. \tag{13.83}$$

Since $\psi_{\alpha\beta}(0) = \delta_{\alpha\beta}$, the effective probability $P^\mathrm{eff}_{\nu_\alpha\to\nu_\beta} = |\psi_{\alpha\beta}|^2$ of $\nu_\alpha \to \nu_\beta$ transitions is given by

$$P^\mathrm{eff}_{\nu_\alpha\to\nu_\beta} = \left| \sum_{k=1}^{3} U_{\mathrm{M}\alpha k}\, U_{\mathrm{M}\beta k}\, \exp\left(-i\frac{m^2_{\mathrm{M}k}\, L}{2\,E}\right) \right|^2 . \tag{13.84}$$

Since this expression has the same structure as that of the transition probability in vacuum, given in eqn (7.118), we can use the same method for the calculation of the transition probabilities. Using the unitarity relation

$$\sum_{k=1}^{3} U_{\mathrm{M}\alpha k}\, U_{\mathrm{M}\beta k} = \delta_{\alpha\beta}\,, \tag{13.85}$$

one obtains that the transition and survival probabilities are given by, respectively,

$$\begin{aligned} P^\mathrm{eff}_{\nu_\alpha\to\nu_\beta} &= 4\, U^2_{\mathrm{M}\alpha 3}\, U^2_{\mathrm{M}\beta 3}\, S^2_{31} - 4\, U_{\mathrm{M}\alpha 1}\, U_{\mathrm{M}\beta 1}\, U_{\mathrm{M}\alpha 2}\, U_{\mathrm{M}\beta 2}\, S^2_1 \\ &\quad - 4\, U_{\mathrm{M}\alpha 2}\, U_{\mathrm{M}\beta 2}\, U_{\mathrm{M}\alpha 3}\, U_{\mathrm{M}\beta 3} \left(S^2_3 - S^2_{31}\right) , \end{aligned} \tag{13.86}$$

$$\begin{aligned} P^\mathrm{eff}_{\nu_\alpha\to\nu_\alpha} &= 1 - 4\, U^2_{\mathrm{M}\alpha 3} \left(1 - U^2_{\mathrm{M}\alpha 3}\right) S^2_{31} - 4\, U^2_{\mathrm{M}\alpha 1}\, U^2_{\mathrm{M}\alpha 2}\, S^2_1 \\ &\quad - 4\, U^2_{\mathrm{M}\alpha 2}\, U^2_{\mathrm{M}\alpha 3} \left(S^2_3 - S^2_{31}\right) , \end{aligned} \tag{13.87}$$

with

$$S^2_1 \equiv \sin^2\left(\frac{m^2_{\mathrm{M}1}\, L}{4\,E}\right) , \qquad S^2_3 \equiv \sin^2\left(\frac{m^2_{\mathrm{M}3}\, L}{4\,E}\right) , \qquad S^2_{31} \equiv \sin^2\left(\frac{\Delta m^2_{\mathrm{M}31}\, L}{4\,E}\right) . \tag{13.88}$$

One can easily see that in vacuum these probabilities reduce to the two-neutrino-like forms in eqns (13.14) and (13.15). In general, however, in matter it is not possible to reduce the oscillation probabilities to the two-neutrino-like forms of the type in eqn (9.77), because in matter there are two independent squared-mass differences: $\Delta m^2_{\mathrm{M}31}$ and $\Delta m^2_{\mathrm{M}21} = -m^2_{\mathrm{M}1}$ (the third squared-mass difference is $m^2_{\mathrm{M}3} = \Delta m^2_{\mathrm{M}32} = \Delta m^2_{\mathrm{M}31} - \Delta m^2_{\mathrm{M}21}$). Such a reduction is only possible for the oscillation probabilities involving the electron neutrino, which has $U_{\mathrm{M}e2} = 0$:

$$P^\mathrm{eff}_{\nu_e\to\nu_e} = 1 - \sin^2 2\vartheta^\mathrm{eff}_{ee}\, S^2_{31}\,, \qquad P^\mathrm{eff}_{\nu_e\to\nu_\beta} = \sin^2 2\vartheta^\mathrm{eff}_{e\beta}\, S^2_{31} \qquad (\beta = \mu, \tau)\,, \tag{13.89}$$

with

$$\sin^2 2\vartheta^\mathrm{eff}_{ee} = \sin^2 2\vartheta^\mathrm{M}_{13}\,, \tag{13.90}$$

$$\sin^2 2\vartheta^\mathrm{eff}_{e\mu} = \sin^2\vartheta_{23}\, \sin^2 2\vartheta^\mathrm{M}_{13}\,, \tag{13.91}$$

$$\sin^2 2\vartheta^\mathrm{eff}_{e\tau} = \cos^2\vartheta_{23}\, \sin^2 2\vartheta^\mathrm{M}_{13}\,. \tag{13.92}$$

As remarked above, the oscillation probabilities in matter of neutrinos and antineutrinos are different, since the lack of CP-invariance of the medium induces, through the matter potential, a violation of the CP symmetry of the oscillation probabilities. The oscillation probabilities are also CPT asymmetric. However, for a constant medium the oscillation probabilities are T-invariant, as one can immediately see from the invariance of eqn (13.86) under the exchange $\alpha \leftrightarrows \beta$.

The resonance condition for neutrinos in eqn (13.72) can be satisfied for an appropriate matter density if the product $\Delta m^2_{31} \cos 2\vartheta_{13}$ is positive. In that case there cannot be a resonance for antineutrinos, which have a matter potential with opposite sign. Vice versa, if the product $\Delta m^2_{31} \cos 2\vartheta_{13}$ is negative, there can be a resonance for antineutrinos only.

Since we know that ϑ_{13} is small (see section 13.3.1), $\cos 2\vartheta_{13}$ is positive. Thus, it is possible to find the sign of Δm^2_{31} and distinguish the two schemes in Fig. 13.1 by measuring the matter effect in LBL experiments: if there is a resonance for neutrinos, $\Delta m^2_{31} > 0$ and we have a normal scheme; if there is a resonance for antineutrinos, $\Delta m^2_{31} < 0$ and we have an inverted scheme. Future LBL experiments (see section 12.3.3) are aimed at this goal.

13.2.2 Active Δm^2_{21}

We now consider solar and VLBL experiments, in which the validity of eqn (13.18) implies that the small squared-mass difference Δm^2_{21} is active and oscillations due to the large squared-mass difference Δm^2_{31} are averaged out. We also consider a matter potential

$$A_{\mathrm{CC}} \sim \Delta m^2_{21} \,, \tag{13.93}$$

which can affect the oscillations due to Δm^2_{21}.

For the study of this case, it is convenient to define the new column matrix of amplitudes

$$\widehat{\Psi}_\alpha = W^{13\dagger} \, R^{23\dagger} \, \Psi_\alpha \,, \tag{13.94}$$

which follows the evolution equation

$$i \, \frac{\mathrm{d}}{\mathrm{d}x} \, \widehat{\Psi}_\alpha = \frac{1}{2E} \, \widehat{\mathbb{M}}^2 \, \widehat{\Psi}_\alpha \,. \tag{13.95}$$

The effective squared-mass matrix

$$\widehat{\mathbb{M}}^2 = \left(R^{12} \, \mathbb{M}^2 \, R^{12\dagger} + W^{13\dagger} \, \mathbb{A} \, W^{13} \right) \tag{13.96}$$

is explicitly given by

$$\widehat{\mathbb{M}}^2 = \begin{pmatrix} s^2_{12} \Delta m^2_{21} + c^2_{13} A_{\mathrm{CC}} & c_{12} s_{12} \Delta m^2_{21} & -c_{13} s_{13} e^{-i\delta_{13}} A_{\mathrm{CC}} \\ c_{12} s_{12} \Delta m^2_{21} & c^2_{12} \Delta m^2_{21} & 0 \\ -c_{13} s_{13} e^{i\delta_{13}} A_{\mathrm{CC}} & 0 & \Delta m^2_{31} + s^2_{13} A_{\mathrm{CC}} \end{pmatrix} \,, \tag{13.97}$$

One must now note that the hierarchy of squared-mass differences in eqn (13.6) implies that the 3-3 entry in the effective squared-mass matrix $\widehat{\mathbb{M}}^2$ is much larger than all the other entries. Thus, to lowest order in the power expansion in the small quantities $\Delta m^2_{21}/\Delta m^2_{31} \sim A_{\text{CC}}/\Delta m^2_{31}$, the third eigenvalue of $\widehat{\mathbb{M}}^2$ is given by

$$\widehat{m}^2_{\text{M}3} = \Delta m^2_{31} \,. \tag{13.98}$$

In this approximation, the evolution of the amplitude $\widehat{\psi}_{\alpha 3}$ is decoupled from the evolution of the amplitudes $\widehat{\psi}_{\alpha 1}$ and $\widehat{\psi}_{\alpha 2}$, leading to

$$\widehat{\psi}_{\alpha 3}(x) = \exp\left(-i\frac{\Delta m^2_{31}\,x}{2\,E}\right)\widehat{\psi}_{\alpha 3}(0)\,. \tag{13.99}$$

Since, from eqns (9.45) and (13.94),

$$\widehat{\psi}_{\alpha k}(0) = \sum_{\rho=e,\mu,\tau} (W^{13\dagger}R^{23\dagger})_{k\rho}\,\psi_{\alpha\rho}(0) = (W^{13\dagger}R^{23\dagger})_{k\alpha} = (R^{23}W^{13})^*_{\alpha k}\,, \tag{13.100}$$

taking into account that $U_{\alpha 3} = (R^{23}W^{13})_{\alpha 3}$, we obtain

$$\psi_{\alpha\beta}(x) = \sum_{k=1,2} \left(R^{23}\,W^{13}\right)_{\beta k}\widehat{\psi}_{\alpha k}(x) + U^*_{\alpha 3}\,U_{\beta 3}\,\exp\left(-i\frac{\Delta m^2_{31}\,x}{2\,E}\right)\,. \tag{13.101}$$

Considering now the transition probability $P_{\nu_\alpha\to\nu_\beta}(x) = |\psi_{\alpha\beta}(x)|^2$, we note that eqns (13.6) and (13.18) imply that $\Delta m^2_{31}x/2E$ is very large. In this case, the interference between the two terms on the right-hand side of eqn (13.101) is washed out by the average over the experimental energy resolution, leading to the effective oscillation probability

$$P^{\text{eff}}_{\nu_\alpha\to\nu_\beta}(x) = \left|\sum_{k=1,2}(R^{23}W^{13})_{\beta k}\,\widehat{\psi}_{\alpha k}(x)\right|^2 + |U_{\alpha 3}|^2\,|U_{\beta 3}|^2\,. \tag{13.102}$$

This effective oscillation probability does not depend on Δm^2_{31}, but retains a dependence on the elements of the third column of the mixing matrix.

The amplitudes $\widehat{\psi}_{\alpha 1}$ and $\widehat{\psi}_{\alpha 2}$ remain to be determined. Their coupled evolution is governed by

$$i\,\frac{\mathrm{d}}{\mathrm{d}x}\begin{pmatrix}\widehat{\psi}_{\alpha 1}\\ \widehat{\psi}_{\alpha 2}\end{pmatrix} = \frac{1}{2E}\,\widehat{\mathbb{M}}^2_{2\times 2}\begin{pmatrix}\widehat{\psi}_{\alpha 1}\\ \widehat{\psi}_{\alpha 2}\end{pmatrix}, \tag{13.103}$$

with the effective 2×2 squared-mass matrix

$$\widehat{\mathbb{M}}^2_{2\times 2} = \frac{\Delta m^2_{21} + c^2_{13}A_{\text{CC}}}{2} + \frac{1}{2}\begin{pmatrix}-\cos 2\vartheta_{12}\Delta m^2_{21} + c^2_{13}A_{\text{CC}} & \sin 2\vartheta_{12}\Delta m^2_{21}\\ \sin 2\vartheta_{12}\Delta m^2_{21} & \cos 2\vartheta_{12}\Delta m^2_{21} - c^2_{13}A_{\text{CC}}\end{pmatrix}. \tag{13.104}$$

Apart from the irrelevant common term $(\Delta m^2_{21} + \cos^2\vartheta_{13}A_{\text{CC}})/2$, the evolution equation is the same as the two-generation evolution equation (9.59), with A_{CC}

replaced by $\cos^2\vartheta_{13}A_{CC}$. Therefore, the evolution equation (13.103) can be solved with the method discussed in section 9.3 through the analogy

$$\widehat{\psi}_{\alpha 1} \sim \psi_{ee}^{(1,2)}\,, \qquad \widehat{\psi}_{\alpha 2} \sim \psi_{e\mu}^{(1,2)}\,, \tag{13.105}$$

where $\psi_{ee}^{(1,2)}$ and $\psi_{e\mu}^{(1,2)}$ are the two-neutrino flavor amplitudes in the 1-2 sector. Obviously, this analogy becomes an exact equality in the limit of no mixing between the 1-2 sector and ν_3, i.e. $\vartheta_{23} = \vartheta_{13} = 0$.

The effective mixing angle in matter $\vartheta_{12}^{\mathrm{M}}$ is given by

$$\tan 2\vartheta_{12}^{\mathrm{M}} = \frac{\tan 2\vartheta_{12}}{1 - \dfrac{\cos^2\vartheta_{13}\,A_{CC}}{\cos 2\vartheta_{12}\,\Delta m^2}}\,. \tag{13.106}$$

There is a resonance for $A_{CC} = A_{CC}^{\mathrm{R}}$, with

$$\cos^2\vartheta_{13}\,A_{CC}^{\mathrm{R}} = \cos 2\vartheta_{12}\Delta m_{21}^2\,. \tag{13.107}$$

The effective squared-mass eigenvalues are

$$\widehat{m}_{\mathrm{M}2,1}^2 = \frac{1}{2}\left(\Delta m_{21}^2 + \cos^2\vartheta_{13}\,A_{CC} \pm \Delta m_{\mathrm{M}21}^2\right)\,, \tag{13.108}$$

with

$$\Delta m_{\mathrm{M}21}^2 = \sqrt{\left(\Delta m_{21}^2\,\cos 2\vartheta_{21} - \cos^2\vartheta_{13}\,A_{CC}\right)^2 + \left(\Delta m_{21}^2\,\sin 2\vartheta_{21}\right)^2}\,. \tag{13.109}$$

For the calculation of the transition probability in eqn (13.102), it is convenient to express the evolution of the amplitudes $\widehat{\psi}_{\alpha 1}$ and $\widehat{\psi}_{\alpha 2}$ through a unitary evolution matrix $\mathscr{U}(x)$ such that

$$\widehat{\psi}_{\alpha k}(x) = \sum_{j=1,2}\mathscr{U}_{kj}(x)\,\widehat{\psi}_{\alpha j}(0) = \sum_{j=1,2}\mathscr{U}_{kj}(x)\,(R^{23}W^{13})_{\alpha j}^{*} \qquad (k=1,2)\,. \tag{13.110}$$

Thus, we obtain

$$P_{\nu_\alpha\to\nu_\beta}^{\mathrm{eff}}(x) = \left|\sum_{k,j=1,2}(R^{23}W^{13})_{\alpha j}^{*}\,(R^{23}W^{13})_{\beta k}\,\mathscr{U}_{kj}(x)\right|^2 + |U_{\alpha 3}|^2\,|U_{\beta 3}|^2\,. \tag{13.111}$$

The evolution equation for $\mathscr{U}(x)$ is

$$i\,\frac{\mathrm{d}}{\mathrm{d}x}\,\mathscr{U}(x) = \frac{1}{2E}\,\widehat{\mathbb{M}}_{2\times 2}^2\,\mathscr{U}(x)\,, \tag{13.112}$$

with the initial condition

$$\mathscr{U}_{kj}(0) = \delta_{kj}\,. \tag{13.113}$$

In practice, it is useful only to calculate the survival probability of electron neutrinos, since the conditions in eqns (13.18) and (13.93) are satisfied only by

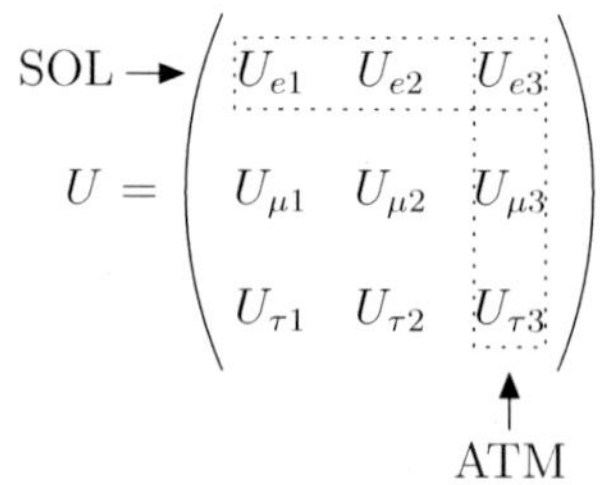

FIG. 13.5. Schematic description of the contributions of the elements of the mixing matrix to solar (SOL) and atmospheric (ATM) neutrino oscillations.

solar neutrinos. Since $(R^{23}W^{13})_{e1} = \cos\vartheta_{13}$, $(R^{23}W^{13})_{e2} = 0$ and $(R^{23}W^{13})_{e3} = \sin\vartheta_{13}\, e^{-i\delta_{13}}$, we obtain [957]

$$P^{\text{eff}}_{\nu_e\to\nu_e}(x) = \cos^4\vartheta_{13}\, |\mathscr{U}_{11}(x)|^2 + \sin^4\vartheta_{13}\,. \tag{13.114}$$

This simple result can also be written in terms of $|U_{e3}|$, which is parameterization-independent, as

$$P^{\text{eff}}_{\nu_e\to\nu_e}(x) = \left(1 - |U_{e3}|^2\right)^2 P^{(1,2)}_{\nu_e\to\nu_e}(x) + |U_{e3}|^4\,, \tag{13.115}$$

where $P^{(1,2)}_{\nu_e\to\nu_e}(x) = |\mathscr{U}_{11}(x)|^2$ is the effective two-neutrino-like survival probability of ν_e due to the mixing in the 1-2 sector. Since eqn (13.115) has the same structure as eqn (13.19), which gives the survival probability in vacuum, it can be considered as valid in general, with appropriate values of $P^{(1,2)}_{\nu_e\to\nu_e}$ in vacuum and in matter.

The expression in eqn (13.115) for the ν_e survival probability can be applied to the analysis of solar neutrinos in the three-neutrino mixing framework. In fact, it has been used in order to obtain information on the value of $|U_{e3}|$ from solar neutrino data [769, 150, 558, 451] (see section 13.3.1).

13.3 Analysis of oscillation data

In chapters 10 and 11 we discussed the results of solar and atmospheric neutrino experiments and their interpretation in terms of two-neutrino oscillations, which led to the establishment of the two squared-mass differences in eqn (13.4). Since these two Δm^2's are determined by the different scales of the oscillations in solar and atmospheric neutrino experiments, they cannot change when the data are analyzed in a three-neutrino mixing scheme. On the other hand, the information on the mixing of neutrino could drastically change when the separate two-neutrino analyses of the solar and atmospheric data is replaced by a combined three-neutrino analysis, because three-neutrino mixing, with its three mixing angles, has more degrees of freedom than a simple doubling of two neutrino mixing. Luckily, this is not the case in practice, since $|U_{e3}|$, which is the only element of the mixing matrix which enters in both solar and atmospheric neutrino oscillations, is small (see subsection 13.3.1).

In order to see that U_{e3} is the only element of the mixing matrix which enters in both solar and atmospheric neutrino oscillations, let us first consider solar neutrino experiments, which have observed the disappearance of electron neutrinos. In general, such disappearance depends only on the elements of the mixing matrix which connect ν_e with the three massive neutrinos. These are the three elements U_{e1}, U_{e2}, and U_{e3} in the first row of the mixing matrix, as shown schematically in Fig. 13.5. Considering now atmospheric neutrino experiments, the general evolution equation of flavor amplitudes discussed in section 13.2.1 depends only on the mixing angles ϑ_{23} and ϑ_{13}, which are determined by the elements in the third column of the mixing matrix through[73]

$$\sin\vartheta_{23} = \frac{|U_{\mu 3}|}{\sqrt{1-|U_{e3}|^2}}\,, \qquad \sin\vartheta_{13} = |U_{e3}|\,. \tag{13.116}$$

Therefore, atmospheric neutrino oscillations depend only on the third column of the elements of the mixing matrix, as shown schematically in Fig. 13.5, and U_{e3} is the only element of the mixing matrix which causes a correlation of solar and atmospheric neutrino oscillations.

The experimental bound on $|U_{e3}|$ is discussed in subsection 13.3.1. In subsection 13.3.2 we discuss the interpretation of the oscillation data in the approximate case of bilarge mixing (see sections 6.6.2 and 13.1.4) with $U_{e3} = 0$. However, since there is no reason to believe that U_{e3} is exactly zero, a correct analysis of the data must take into account a possible contribution of a small U_{e3}. In subsection 13.3.3 we discuss the results of such an analysis. Finally, in subsection 13.3.4 we discuss the implications for the absolute values of the neutrino masses.

13.3.1 Bound on $|U_{e3}|$

In section 13.1.2 it has been shown that the squared-mass hierarchy in eqn (13.6) implies that the survival probabilities of ν_α and $\bar{\nu}_\alpha$ in disappearance experiments sensitive to the oscillations due to $|\Delta m^2_{31}|$ are given by eqn (13.15) with the effective mixing angle in eqn (13.16), which depends only on $|U_{\alpha 3}|$. In particular, the CHOOZ and Palo Verde LBL reactor experiments (see section 12.2.2) were sensitive to $\bar{\nu}_e$ disappearance due to $|\Delta m^2_{31}|$, with an effective mixing angle given by

$$\sin^2 2\vartheta_{ee}^{\text{eff}} = 4\,|U_{e3}|^2\left(1-|U_{e3}|^2\right) = \sin^2 2\vartheta_{13}\,. \tag{13.117}$$

These experiments did not observe any disappearance of reactor ν_e's. Their exclusion curves in the Δm^2-$\sin^2 2\vartheta$ plane, shown in Figs. 12.6 and 12.7, can be interpreted in the framework of three-neutrino mixing simply by replacing $\sin^2 2\vartheta$ with $\sin^2 2\vartheta_{13}$ and Δm^2 with $|\Delta m^2_{31}|$. Thus, we have an upper bound $(\sin^2 2\vartheta_{13})_{\text{max}}$ on $\sin^2 2\vartheta_{ee}^{\text{eff}}$ as a function of $|\Delta m^2_{31}|$. In order to transform this limit into a bound

[73] Since the mixing angles are limited between 0 and $\pi/2$, the relations in eqn (13.116) determine ϑ_{23} and ϑ_{13} in a unique way.

on the value of $|U_{e3}|$, we invert eqn (13.117):

$$|U_{e3}|^2 = \frac{1}{2}\left(1 \pm \sqrt{1 - \sin^2 2\vartheta_{13}}\right) . \tag{13.118}$$

A small value of $\sin^2 2\vartheta_{13}$ corresponds either to a small value of $|U_{e3}|^2$ (minus sign) or to a value of $|U_{e3}|^2$ close to one (plus sign). However, the following argument shows that a value of $|U_{e3}|^2$ close to one is incompatible with solar and atmospheric neutrino oscillations.

Let us consider $|U_{e3}|^2 = 1 - \varepsilon$ with $\varepsilon \ll 1$. Then, since from the unitarity of the mixing matrix we have $|U_{e1}|^2 + |U_{e2}|^2 = \varepsilon$, the mixing of the electron neutrino with ν_1 and ν_2 is very small. In this case, the large observed disappearance of solar ν_e's cannot occur through Δm^2_{21}. With an analogous reasoning, since $|U_{\mu3}|^2+|U_{\tau3}|^2 = \varepsilon$, the mixing of the muon neutrino with ν_3 is very small and the large observed disappearance of atmospheric ν_μ's and $\bar{\nu}_\mu$'s cannot occur through Δm^2_{31}.

Therefore, the CHOOZ and Palo Verde bounds on $\sin^2 2\vartheta_{13}$ are compatible with solar and atmospheric oscillations only if $|U_{e3}|^2$ is small [228], i.e. for

$$|U_{e3}|^2 \leq \frac{1}{2}\left(1 - \sqrt{1 - (\sin^2 2\vartheta_{13})_{\max}}\right) . \tag{13.119}$$

From the 90% CL exclusion curve of the CHOOZ experiment in Fig. 12.6 one can obtain $(\sin^2 2\vartheta_{13})_{\max} \lesssim 0.18$ for $|\Delta m^2_{31}| \gtrsim 2 \times 10^{-3}\,\text{eV}^2$, which implies $|U_{e3}|^2 \lesssim 5 \times 10^{-2}$. A precise statistical analysis of the CHOOZ data, taking into account the constraints on $|\Delta m^2_{31}|$ given by the Super-Kamiokande atmospheric neutrino data (see section 11.2.3) yielded [448]

$$|U_{e3}|^2 < 5 \times 10^{-2} \qquad (99.73\%\ \text{CL}) . \tag{13.120}$$

The absence of a $\nu_\mu \to \nu_e$ oscillation signal in the LBL accelerator experiment K2K [47, 1078] (see section 12.3.2) confirms the CHOOZ and Palo Verde bounds on $|U_{e3}|^2$. In fact, the effective mixing angle for the $\nu_\mu \to \nu_e$ channel in the K2K experiment is, from eqn (13.16),

$$\sin^2 2\vartheta^{\text{eff}}_{\mu e} = 4\,|U_{\mu3}|^2\,|U_{e3}|^2 = \sin^2\vartheta_{23}\,\sin^2 2\vartheta_{13} , \tag{13.121}$$

where $\sin^2\vartheta_{23} \simeq 1/2$ from the atmospheric neutrino data (see eqns 11.74 and 13.131 and Table 13.2 below). The effective mixing angle $\vartheta^{\text{eff}}_{\mu e}$ coincides with $\vartheta_{\mu e}$ in Figs. 12.18 and 12.19, where we have also $\Delta m^2 = \Delta m^2_{31}$. These figures show that the value of $\vartheta^{\text{eff}}_{\mu e}$ is severely bounded by K2K data for $\Delta m^2_{31} \gtrsim 10^3\,\text{eV}^2$. Assuming $\sin^2\vartheta_{23} = 1/2$, we have $\sin^2 2\vartheta^{\text{eff}}_{\mu e} = \sin^2 2\vartheta_{13}/2 \simeq 2\sin^2\vartheta_{13} = 2|U_{e3}|^2$, for a small ϑ_{13}. Thus, we obtain, from the K2K bound in eqn (12.26), $|U_{e3}|^2 \lesssim 7 \times 10^{-2}$ (90% CL) at $\Delta m^2 = 2.8 \times 10^{-3}\,\text{eV}^2$.

Also the analysis of Super-Kamiokande atmospheric neutrino data in a three-neutrino mixing framework yield a bound on $\sin^2 2\vartheta_{13}$, since the measured electron neutrino flux does not show any anomaly with respect to the expectations without oscillations. The allowed Super-Kamiokande region in the $\sin^2 2\vartheta_{13}$-Δm^2_{31} plane is shown in Fig. 13.6. One can see that the Super-Kamiokande constraint on $\sin^2 2\vartheta_{13}$

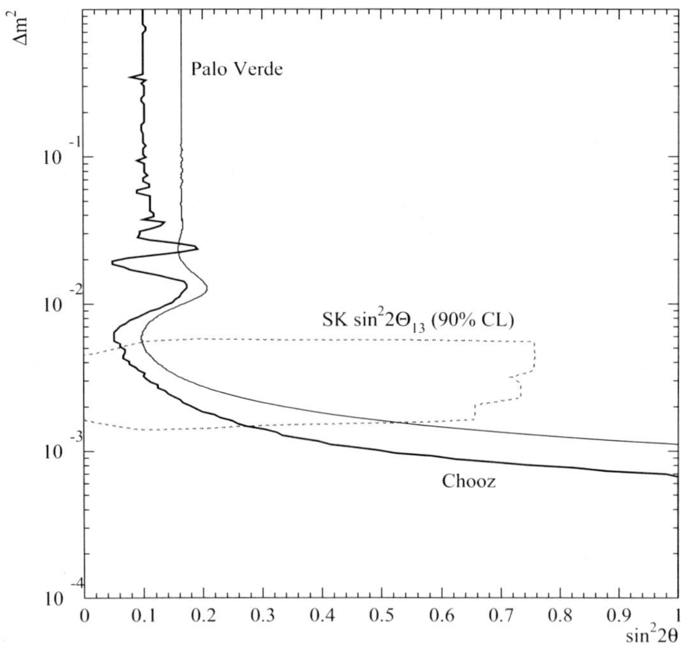

FIG. 13.6. Allowed Super-Kamiokande region in the plane of the oscillation parameters $\sin^2 2\vartheta = \sin^2 2\vartheta_{13}$ and $\Delta m^2 = \Delta m^2_{31}$. Also shown are the exclusion curves of the CHOOZ [100] and Palo Verde [255] experiments. Figure from Ref. [205].

is not as stringent as the CHOOZ and Palo Verde limits. However, the Super-Kamiokande result is important in order to restrict the allowed range of Δm^2_{31}, which is not constrained at all by CHOOZ and Palo Verde

Finally, also solar neutrino and KamLAND data constrain the value of $|U_{e3}|$. The survival probability of solar ν_e's depends on $|U_{e3}|$ through eqn (13.115), which takes into account the matter effects. The survival probability of the reactor $\bar{\nu}_e$'s observed in KamLAND, which propagate in vacuum, is given by eqn (13.19) with $\alpha = e$. Figure 13.7 shows the allowed region at 2σ in the parameter space $(\Delta m^2_{21}, \sin^2\vartheta_{12}, \sin^2\vartheta_{13})$ obtained in Ref. [451]. One can see that the solar+KamLAND bound on $|U_{e3}| = \sin^2\vartheta_{13}$ is comparable with the atmospheric+CHOOZ bound in eqn (13.120).

13.3.2 Bilarge mixing

The bound on $|U_{e3}|^2$ in eqn (13.120) implies that, in a first approximation, the data of solar and atmospheric neutrino experiments can be analyzed assuming bilarge mixing (see sections 6.6.2 and 13.1.4) with $U_{e3} = 0$. The mixing matrix, given in eqn (6.378), depends only on ϑ_{12} and ϑ_{23}. In this case, the oscillations of solar and atmospheric neutrinos are decoupled [228] and have two-neutrino expressions.

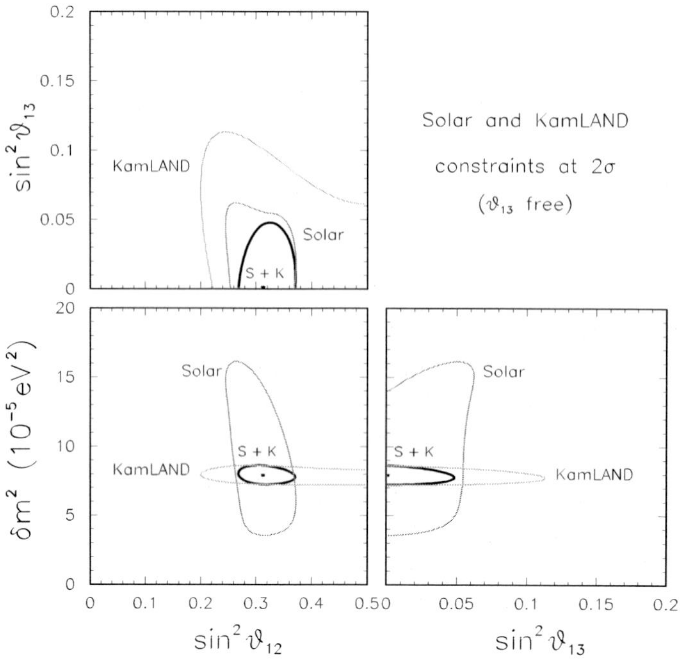

FIG. 13.7. Three-neutrino analysis of solar and KamLAND data (both separately and in combination) in the parameter space $(\delta m^2, \sin^2\vartheta_{12}, \sin^2\vartheta_{13})$, with $\delta m^2 \equiv \Delta m^2_{21}$ [451]. The contours represent projections of the region allowed at 2σ $(\Delta\chi^2 = 4)$.

Solar and atmospheric neutrino data give information, respectively, on ϑ_{12} and ϑ_{23}. In fact, for atmospheric neutrinos there are no matter effects (from eqn (13.71), $\vartheta_{13}^{\mathrm{M}} = \vartheta_{13} = 0$) and the oscillation probabilities are given by eqns (13.14) and (13.15). Since $U_{e3} = 0$, we have $\vartheta_{ee}^{\mathrm{eff}} = \vartheta_{e\mu}^{\mathrm{eff}} = \vartheta_{e\tau}^{\mathrm{eff}} = 0$ and $\vartheta_{\mu\tau}^{\mathrm{eff}} = \vartheta_{23}$. Hence, the atmospheric mixing angle coincides with ϑ_{23}. On the other hand, the survival probability of solar $\nu_e's$ in eqn (13.115) reduces to the effective two-neutrino-like survival probability due to the mixing in the 1-2 sector, which depends on ϑ_{12}.

In this approximation one can call

$$\vartheta_{\mathrm{SOL}} = \vartheta_{12}\,, \qquad \vartheta_{\mathrm{ATM}} = \vartheta_{23}\,, \tag{13.122}$$

and write the mixing matrix as

$$U = \begin{pmatrix} \cos\vartheta_{\mathrm{SOL}} & \sin\vartheta_{\mathrm{SOL}} & 0 \\ -\sin\vartheta_{\mathrm{SOL}}\cos\vartheta_{\mathrm{ATM}} & \cos\vartheta_{\mathrm{SOL}}\cos\vartheta_{\mathrm{ATM}} & \sin\vartheta_{\mathrm{ATM}} \\ \sin\vartheta_{\mathrm{SOL}}\sin\vartheta_{\mathrm{ATM}} & -\cos\vartheta_{\mathrm{SOL}}\sin\vartheta_{\mathrm{ATM}} & \cos\vartheta_{\mathrm{ATM}} \end{pmatrix}. \tag{13.123}$$

The massive neutrino composition of ν_e is

$$\nu_e = \cos\vartheta_{\mathrm{SOL}}\,\nu_1 + \sin\vartheta_{\mathrm{SOL}}\,\nu_2\,. \tag{13.124}$$

Solar electron neutrinos undergo transitions into the orthogonal state

$$\begin{aligned}\nu_{\text{SOL}} &= -\sin\vartheta_{\text{SOL}}\,\nu_1 + \cos\vartheta_{\text{SOL}}\,\nu_2 \\ &= \cos\vartheta_{\text{ATM}}\,\nu_\mu - \sin\vartheta_{\text{ATM}}\,\nu_\tau\,. \end{aligned} \tag{13.125}$$

This is a linear combination of ν_μ and ν_τ, whose relative amount is determined by the atmospheric mixing angle ϑ_{ATM}. Since atmospheric mixing is close to maximal, in a first approximation one can assume $\vartheta_{\text{ATM}} = \pi/4$. In this approximation, the mixing matrix is given by

$$U = \begin{pmatrix} \cos\vartheta_{\text{SOL}} & \sin\vartheta_{\text{SOL}} & 0 \\ -\sin\vartheta_{\text{SOL}}/\sqrt{2} & \cos\vartheta_{\text{SOL}}/\sqrt{2} & 1/\sqrt{2} \\ \sin\vartheta_{\text{SOL}}/\sqrt{2} & -\cos\vartheta_{\text{SOL}}/\sqrt{2} & 1/\sqrt{2} \end{pmatrix}. \tag{13.126}$$

Solar ν_e's undergo transitions into the state

$$\nu_{\text{SOL}} = \frac{1}{\sqrt{2}}\left(\nu_\mu - \nu_\tau\right), \tag{13.127}$$

which is an equal superposition of ν_μ and ν_τ. If one could detect separately the mu and tau neutrinos in the solar neutrino flux on the Earth, in this approximation, one would measure an equal amount of ν_μ's and ν_τ's. Furthermore, since the SNO charged-current to neutral-current ratio in eqn (10.54) is close to 1/3, the fluxes of solar ν_e, ν_μ, and ν_τ on the Earth are approximately the same and equal to 1/3 of the total flux.

As a final approximation, one can consider $\vartheta_{\text{SOL}} = \pi/6$, which gives $\tan^2\vartheta_{\text{SOL}} \simeq 0.33$. This value is within the 2σ measured range in eqn (10.79). In this approximation, the mixing matrix assumes the simple form

$$U = \begin{pmatrix} \sqrt{3}/2 & 1/2 & 0 \\ -1/2\sqrt{2} & \sqrt{3}/2\sqrt{2} & 1/\sqrt{2} \\ 1/2\sqrt{2} & -\sqrt{3}/2\sqrt{2} & 1/\sqrt{2} \end{pmatrix}. \tag{13.128}$$

Let us however emphasize that this expression for the mixing matrix is very approximate. In a realistic calculation it is better to consider the value of the mixing matrix obtained from the global analysis of oscillation data presented in the next section 13.3.3.

13.3.3 Global analysis of oscillation data

In this subsection we review the results obtained with a global analysis of all the oscillation data, with the exception of the LSND signal (see section 12.3.1), without any assumption on the value of U_{e3}. In such a global analysis, the five mixing parameters Δm^2_{21}, Δm^2_{31}, ϑ_{12}, ϑ_{23}, ϑ_{13} are considered as free parameters to be determined by the fit of the data. Since the current experiments are not sensitive to the CP-violating phase δ_{13}, the mixing matrix used in practice is

$$U = R^{23}\,R^{13}\,R^{12} = \begin{pmatrix} 1 & 0 & 0 \\ 0 & c_{23} & s_{23} \\ 0 & -s_{23} & c_{23} \end{pmatrix} \begin{pmatrix} c_{13} & 0 & s_{13} \\ 0 & 1 & 0 \\ -s_{13} & 0 & c_{13} \end{pmatrix} \begin{pmatrix} c_{12} & s_{12} & 0 \\ -s_{12} & c_{12} & 0 \\ 0 & 0 & 1 \end{pmatrix}$$

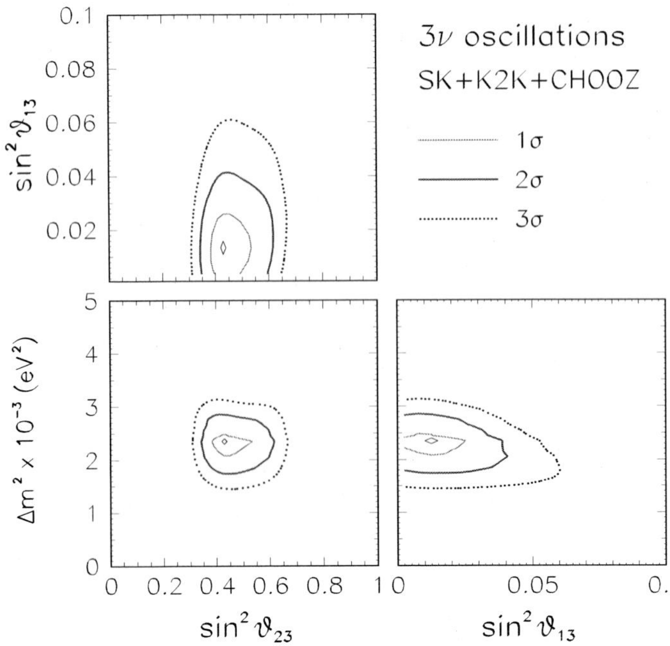

FIG. 13.8. Three-neutrino analysis of SK+K2K+CHOOZ data in the parameter space $(\Delta m^2, \sin^2\vartheta_{23}, \sin^2\vartheta_{13})$, with $\Delta m^2 \equiv \Delta m^2_{31}$ [451].

TABLE 13.2. Best-fit values, 2σ, 3σ, and 4σ intervals (one degree of freedom) for the three-neutrino oscillation parameters from global data [769].

Parameter	Best fit	2σ	3σ	4σ
Δm^2_{21} [10^{-5} eV^2]	8.1	7.5–8.7	7.2–9.1	7.0–9.4
Δm^2_{31} [10^{-3} eV^2]	2.2	1.7–2.9	1.4–3.3	1.1–3.7
$\sin^2\vartheta_{12}$	0.30	0.25–0.34	0.23–0.38	0.21–0.41
$\sin^2\vartheta_{23}$	0.50	0.38–0.64	0.34–0.68	0.30–0.72
$\sin^2\vartheta_{13}$	0.000	≤ 0.028	≤ 0.047	≤ 0.068

$$= \begin{pmatrix} c_{12}c_{13} & s_{12}c_{13} & s_{13} \\ -s_{12}c_{23} - c_{12}s_{23}s_{13} & c_{12}c_{23} - s_{12}s_{23}s_{13} & s_{23}c_{13} \\ s_{12}s_{23} - c_{12}c_{23}s_{13} & -c_{12}s_{23} - s_{12}c_{23}s_{13} & c_{23}c_{13} \end{pmatrix} . \tag{13.129}$$

From the discussion in the previous sections, it is clear that Δm^2_{21} and ϑ_{12} (R^{12}) are determined mainly by the solar and KamLAND data, Δm^2_{31} and ϑ_{23} (R^{23}) are determined mainly by the atmospheric and K2K data, and ϑ_{13} (R^{13}) is determined mainly by CHOOZ data.

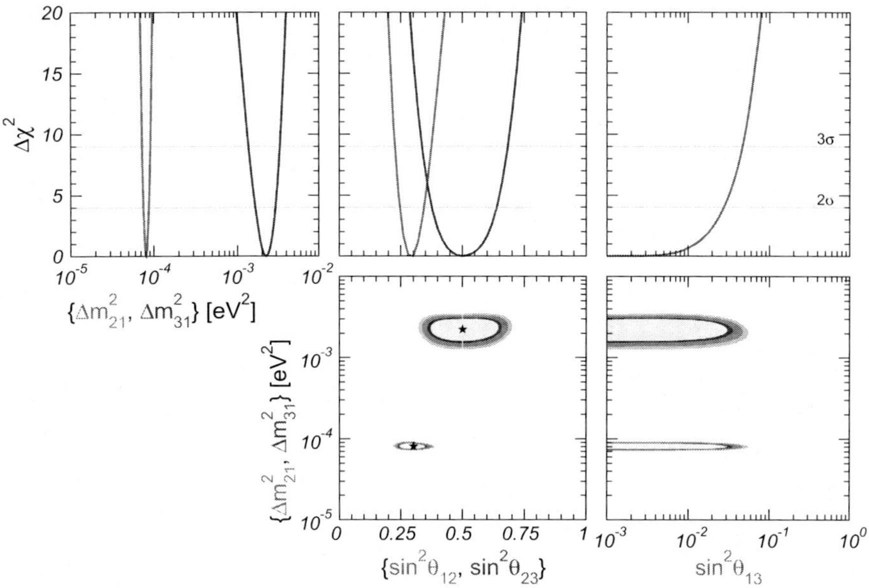

FIG. 13.9. Projections of the allowed regions from the global oscillation data at 90%, 95%, 99%, and 99.73% (3σ) CL for two degrees of freedom for various parameter combinations [769]. Also shown is $\Delta\chi^2$ as a function of the oscillation parameters $\sin^2\vartheta_{12}, \sin^2\vartheta_{23}, \sin^2\vartheta_{13}, \Delta m^2_{21}, \Delta m^2_{31}$, minimized with respect to all undisplayed parameters.

The global analysis of oscillation data presented in Ref. [451] yielded the values

$$\Delta m^2_{21} = 7.92\,(1 \pm 0.09) \times 10^{-5}\,\text{eV}^2 \qquad \sin^2\vartheta_{12} = 0.314\left(1^{+0.18}_{-0.15}\right)\,, \tag{13.130}$$

$$\Delta m^2_{31} = 2.4\left(1^{+0.21}_{-0.26}\right) \times 10^{-3}\,\text{eV}^2 \qquad \sin^2\vartheta_{23} = 0.44\left(1^{+0.41}_{-0.22}\right)\,, \tag{13.131}$$

$$\sin^2\vartheta_{13} = 0.9^{+2.3}_{-0.9} \times 10^{-2}\,, \tag{13.132}$$

where the intervals represent 2σ uncertainties (95% CL). Figure 13.7 and 13.8 show, respectively, the allowed regions in the parameter spaces $(\Delta m^2_{21}, \sin^2\vartheta_{12}, \sin^2\vartheta_{13})$ and $(\Delta m^2_{31}, \sin^2\vartheta_{23}, \sin^2\vartheta_{13})$. Other authors [555, 769, 554] obtained similar results. For example, the authors of Ref. [769] obtained the values of the mixing parameters listed in Table 13.2 and the allowed regions in Fig. 13.9.

From the best-fit values of the mixing angles in eqns (13.130)–(13.132), it is possible to infer the best-fit estimate of the absolute values of the elements of the mixing matrix in eqn (13.129):

$$|U|_{\text{bf}} \simeq \begin{pmatrix} 0.82 & 0.56 & 0.09 \\ 0.31 - 0.43 & 0.51 - 0.59 & 0.75 \\ 0.37 - 0.47 & 0.59 - 0.66 & 0.66 \end{pmatrix}. \tag{13.133}$$

The intervals in the μ, τ-2, 3 sector are due to the lack of any information on the value of the phase δ_{13}. Taking into account the 2σ uncertainties of the mixing angles

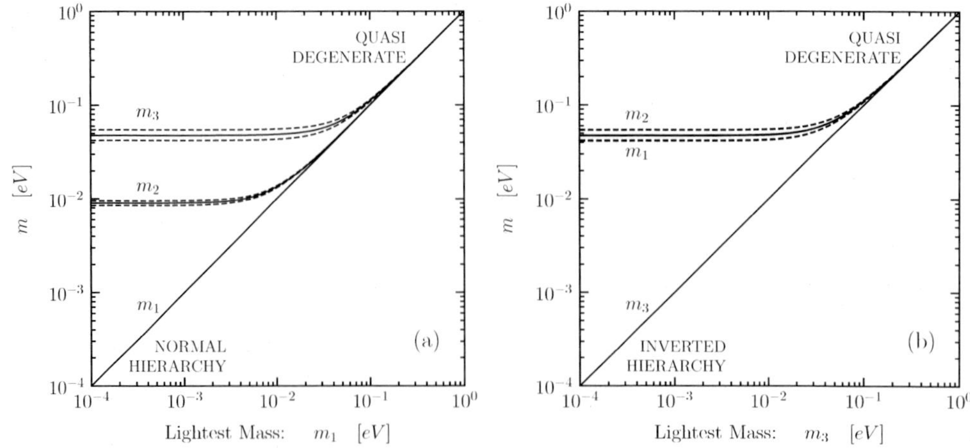

FIG. 13.10. Values of neutrino masses as functions of the lightest mass m_1 in the normal scheme and m_3 in the inverted scheme. Solid lines correspond to the best-fit values of the squared-mass differences in Table 13.2. Dashed lines enclose 3σ ranges.

in eqns (13.130)–(13.132), one can find the following intervals for the absolute values of the elements of the mixing matrix (neglecting the small correlations among the uncertainties of the mixing parameters):

$$|U|_{2\sigma} \simeq \begin{pmatrix} 0.78-0.86 & 0.51-0.61 & 0.00-0.18 \\ 0.19-0.57 & 0.39-0.73 & 0.61-0.80 \\ 0.20-0.57 & 0.40-0.74 & 0.59-0.79 \end{pmatrix}. \quad (13.134)$$

One should note that the neutrino mixing matrix is very different from the quark mixing matrix in eqn (4.2), which has large diagonal entries and small off-diagonal entries structured in the hierarchical way described by the Wolfenstein parameterization in eqn (4.83). All the entries in the neutrino mixing matrix are large, except for U_{e3}.

13.3.4 Absolute scale of neutrino masses

For the neutrino masses, although we do not know the absolute mass scale, we have the information in eqns (13.130) and (13.131) on the squared-mass differences. Thus, it is possible to express the neutrino masses as functions of only one unknown parameter representing the absolute mass scale. It is convenient to take this parameter as the value of the lightest mass, which is m_1 in the normal scheme and m_3 in the inverted scheme. In the normal scheme

$$m_2^2 = m_1^2 + \Delta m_{21}^2 = m_1^2 + \Delta m_{\text{SOL}}^2 , \quad m_3^2 = m_1^2 + \Delta m_{31}^2 = m_1^2 + \Delta m_{\text{ATM}}^2 , \quad (13.135)$$

and in the inverted scheme

$$m_1^2 = m_3^2 - \Delta m_{31}^2 = m_3^2 + \Delta m_{\text{ATM}}^2 \,, \quad m_2^2 = m_1^2 + \Delta m_{21}^2 = m_3^2 + \Delta m_{\text{ATM}}^2 + \Delta m_{\text{SOL}}^2 \,. \tag{13.136}$$

Figure 13.10 shows the values of the three neutrino masses as functions of the lightest mass in the two schemes. One can see that in both schemes there is quasidegeneracy of the three masses when

$$m_1 \simeq m_2 \simeq m_3 \simeq m_\nu \qquad \text{(quasidegenerate)} \,, \tag{13.137}$$

with

$$m_\nu \gg \sqrt{\Delta m_{\text{ATM}}^2} \simeq 5 \times 10^{-2}\,\text{eV} \,. \tag{13.138}$$

In this case, it is very difficult to distinguish the two schemes. On the other hand, the two schemes have very different features if the lightest mass is much smaller than $\sqrt{\Delta m_{\text{ATM}}^2}$. In this case, in the normal scheme there is a hierarchy of masses:

$$m_1 \ll m_2 \ll m_3 \qquad \text{(normal hierarchy)} \,. \tag{13.139}$$

In the inverted scheme there is a so-called *inverted hierarchy*

$$m_3 \ll m_1 \simeq m_2 \qquad \text{(inverted hierarchy)} \,, \tag{13.140}$$

in which m_1 and m_2 are quasidegenerate. In fact, in the inverted scheme m_1 and m_2 are always quasidegenerate, because their separation is due to the small solar squared-mass difference Δm_{SOL}^2.

Let us finally emphasize that, independently of the mass scheme, at least two neutrinos are massive, with masses larger than about $8 \times 10^{-3}\,\text{eV}$. This is clearly shown in Fig. 13.10.

14

DIRECT MEASUREMENTS OF NEUTRINO MASS

I don't see the logic of rejecting data just because they seem incredible.
Fred Hoyle

In this chapter we present the main aspects of the methods used for direct measurements of neutrino masses.

In the letter sent by Pauli on 4 December 1930 to the physics conference at Tubingen, in which he proposed the existence of the neutrino, he mentioned that the neutrino mass "should be of the same order of magnitude as the electron mass and in any event not larger than 0.01 proton masses" (see Ref. [855]). In 1933 Fermi [430, 432, 431] and Perrin [865] studied the effect of the neutrino mass in β-decays and concluded that the existing data were compatible with a massless neutrino. Further experiments showed that neutrinos could be massless and, in any case, the neutrino mass is much smaller than the electron mass. In 1952 a tritium decay experiment [717] yielded an upper limit of about 250 eV. After the formulation in 1957 of the two-component theory of the neutrino [711, 727, 919], stimulated by the discovery of parity violation in weak interactions [1073], it was widely thought that neutrinos are massless. This conviction led to the SM description of neutrinos as massless particles.

As we have seen in chapters 10–13, the results of neutrino oscillation experiments have recently proved that neutrinos are massive. Since these experiments give only information on the neutrino squared-mass differences, we currently know that there are at least two massive neutrinos, one with a mass larger than about $\sqrt{\Delta m^2_{21}} \simeq 9 \times 10^{-3}\,\text{eV}$ and another with a mass larger than about $\sqrt{\Delta m^2_{31}} \simeq 5 \times 10^{-2}\,\text{eV}$, as shown in Fig. 13.10. Further information about the absolute values of neutrino masses must be investigated with other methods.

In section 14.1 we discuss the effect of neutrino mass in β-decays, including the effects of neutrino mixing. We will see that the present upper bound for the neutrino masses in the case of three-neutrino mixing is about 2 eV. Hence, neutrinos are the lightest known massive particles, with a gap of about six orders of magnitude from the next lightest particle, the electron, as illustrated by the mass spectrum in Fig. 14.1.

In section 14.2 we discuss very briefly the effect of neutrino masses in pion and tau decays. The resulting experimental constraints on the neutrino masses are much less stringent than those obtained in β-decay experiments. However, they are historically interesting and can be used to constrain the mixing of ν_μ and ν_τ with heavy neutrinos beyond three-neutrino mixing.

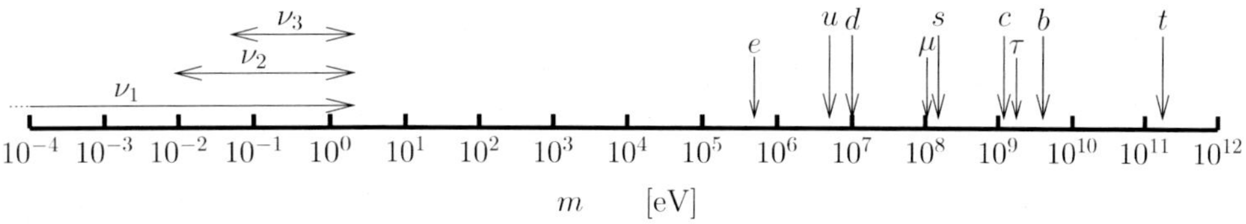

FIG. 14.1. Order of magnitude of the masses of leptons and quarks.

In section 14.3 we discuss neutrinoless double-β-decay, which is considered as the best way to investigate the Majorana nature of neutrinos. Let us emphasize the fundamental importance of determining the nature of neutrinos, which plays a crucial role in the theory of neutrino masses (see chapter 6). In particular, there are two known natural explanations of the smallness of neutrino masses with respect to those of the other elementary particles: the see-saw mechanism discussed in section 6.4.6 and the mechanism of effective generation of a Majorana mass term by physics beyond the SM discussed in section 6.2.7. In both cases massive neutrinos are Majorana particles, leading to the possibility of observing neutrinoless double-β-decay. In section 14.3 we will discuss in detail the contribution of massive neutrinos to the neutrinoless double-β-decay process, taking into account neutrino mixing in the case of three generations. Other important aspects of neutrinoless double-β-decay (the calculation of nuclear matrix elements and the inclusion of other mechanisms) are discussed in detail in the reviews in Refs. [605, 377, 1018, 997, 426, 1036].

14.1 Beta decay

The most sensitive known method to measure the electron neutrino mass is by observing the electron spectrum in nuclear β-decay

$$\mathcal{N}(A,Z) \to \mathcal{N}(A,Z+1) + e^- + \bar{\nu}_e\,, \tag{14.1}$$

where A and Z are, respectively, the mass and atomic numbers of the parent nucleus.

As we have seen in sections 6.1 and 6.2.1, the electron neutrino, in general, does not have a definite mass, but is a mixture of massive neutrinos. However, following the tradition, in this section we treat the electron neutrino as a mass eigenstate. We will discuss the effects of neutrino mixing in nuclear β-decay in section 14.1.1.

The differential decay rate in allowed[74] β-decays is given by

$$\frac{\mathrm{d}\Gamma}{\mathrm{d}E_e} = \frac{G_{\mathrm{F}}^2\, m_e^5}{2\,\pi^3} \cos^2\theta_{\mathrm{C}}\, |\mathcal{M}|^2\, F(Z,E_e)\, E_e\, p_e\, E_\nu\, p_\nu\,, \tag{14.2}$$

where θ_{C} is the Cabibbo angle, $\mathcal{M}$ is the nuclear matrix element, E_e (E_ν) and p_e (p_ν) are the electron (neutrino) energy and momentum, and $F(Z,E_e)$ is the Fermi

[74] Allowed β-decays are characterized by the fact that the matrix element $\mathcal{M}$ in eqn (14.2) is a constant, independent of the electron energy E_e.

function which describes the electromagnetic interaction of the produced electron with the final-state nucleus (taking into account also atomic corrections due to the surrounding electrons). The two factors $E_i\,p_i$ in eqn (14.2), with $i = e, \nu$, come from the phase-space factor of the final state: $\mathrm{d}^3 p_i = p_i^2\,\mathrm{d}p_i\,\mathrm{d}\cos\theta_i\,\mathrm{d}\phi_i = p_i\,E_i\,\mathrm{d}E_i\,\mathrm{d}\cos\theta_i\,\mathrm{d}\phi_i$, where θ_i and ϕ_i are the polar angular coordinates of $\vec{p}_i$.

Since the final nucleus $\mathcal{N}(A, Z+1)$ is much heavier than the final leptons, its kinetic energy is negligible and the total energy of the leptons is given by the mass difference of the initial and final nuclei. The neutrino energy is given by

$$E_\nu = Q_\beta - T\,, \tag{14.3}$$

where $T = E_e - m_e$ is the kinetic energy of the electron and Q_β is the so-called Q-value of the process, which is the total energy of the leptons minus the electron mass, given by

$$Q_\beta = M_i - M_f - m_e\,. \tag{14.4}$$

Here M_i and M_f are, respectively, the masses of the initial and final nuclei. The Q-value of the process is the maximal kinetic energy of the electron if the neutrino is massless. On the other hand, if the electron neutrino has a mass m_{ν_e}, the maximal kinetic energy of the electron is

$$T_{\mathrm{max}} = Q_\beta - m_{\nu_e}\,. \tag{14.5}$$

Since the neutrino momentum is given by

$$p_\nu = \sqrt{E_\nu^2 - m_{\nu_e}^2} = \sqrt{(Q_\beta - T)^2 - m_{\nu_e}^2}\,, \tag{14.6}$$

the differential decay rate in eqn (14.2) can be written, for $T \leq T_{\mathrm{max}}$, as[75]

$$\frac{\mathrm{d}\Gamma}{\mathrm{d}T} = \frac{G_{\mathrm{F}}^2\,m_e^5}{2\,\pi^3}\,\cos^2\theta_{\mathrm{C}}\,|\mathcal{M}|^2\,F(Z, E_e)\,E_e\,p_e\,(Q_\beta - T)\sqrt{(Q_\beta - T)^2 - m_{\nu_e}^2}\,, \tag{14.8}$$

where we used the equality $\mathrm{d}T = \mathrm{d}E_e$.

It is important to note that the effect of neutrino mass in β-decay is due to the kinematical relation in eqn (14.6), which is satisfied irrespective of the Dirac or Majorana nature of neutrinos. Thus, the direct measurement of neutrino mass

[75] For the measurement of a neutrino mass of the order of a few eV or smaller it is necessary to take into account also the excited energy levels of the electrons in the final atom (or molecule) through the differential decay rate

$$\frac{\mathrm{d}\Gamma}{\mathrm{d}T} = \frac{G_{\mathrm{F}}^2\,m_e^5}{2\,\pi^3}\,\cos^2\theta_{\mathrm{C}}\,|\mathcal{M}|^2\,F(Z, E_e)\,E_e\,p_e\sum_i W_i\left(Q_\beta^i - T\right)\sqrt{\left(Q_\beta^i - T\right)^2 - m_{\nu_e}^2}\,. \tag{14.7}$$

The sum runs over all the possible electronic final states i with branching ratios W_i and end-point energies $Q_\beta^i = Q_\beta - E_i^{\mathrm{ex}}$, where E_i^{ex} are the electron excitation energies measured from the ground state. In practice experimenters must take into account also the energy dissipation of the emitted electron in the source (which is minimized choosing a source as thin as possible) and in the measuring apparatus and effects due to the finite energy resolution of the apparatus. In our discussion we neglect these complications.

with β-decay experiments, as well as other kinematical methods, do not give any information on the Dirac or Majorana origin of the neutrino mass.

If the electron neutrino mass m_{ν_e} is small, its effect on the electron spectrum is maximal at the upper end of the spectrum, where T is close to $T_{\max}$ and $Q_\beta - T \sim m_{\nu_e}$. This is the so-called end-point of the spectrum. The problem in measuring the end-point of the electron spectrum is that very few events occur near the end-point. We can estimate the relative number of events occurring in an interval of energy ΔT below the end-point as follows. Below the end-point we have

$$T \simeq Q_\beta \qquad \Longrightarrow \qquad \begin{cases} E_e \simeq Q_\beta + m_e \\ p_e = \sqrt{E_e^2 - m_e^2} \simeq \sqrt{Q_\beta\,(Q_\beta + 2\,m_e)}\,. \end{cases} \tag{14.9}$$

Ignoring the neutrino mass and the Fermi function, we have

$$\left.\frac{\mathrm{d}\Gamma}{\mathrm{d}T}\right|_{\substack{T\simeq Q_\beta \\ m_{\nu_e}=0}} \propto (Q_\beta + m_e)\,\sqrt{Q_\beta\,(Q_\beta + 2\,m_e)}\,(Q_\beta - T)^2\,, \tag{14.10}$$

and

$$\int_{Q_\beta-\Delta T}^{Q_\beta} \frac{\mathrm{d}\Gamma}{\mathrm{d}T}\,\mathrm{d}T \propto (Q_\beta + m_e)\,\sqrt{Q_\beta\,(Q_\beta + 2\,m_e)}\,(\Delta T)^3\,. \tag{14.11}$$

The total number of events is proportional to

$$\int_0^{Q_\beta} \frac{\mathrm{d}\Gamma}{\mathrm{d}T}\,\mathrm{d}T \propto \int_0^{Q_\beta} (T + m_e)\,\sqrt{T\,(T + 2m_e)}\,(Q_\beta - T)^2\,\mathrm{d}T\,, \tag{14.12}$$

where we have neglected again the Fermi function and the neutrino mass. Since we are interested in an order-of-magnitude estimate, we consider $Q_\beta \gg m_e$, which leads to the approximation

$$\int_0^{Q_\beta} \frac{\mathrm{d}\Gamma}{\mathrm{d}T}\,\mathrm{d}T \propto Q_\beta^5\,. \tag{14.13}$$

Thus, the relative number of events occurring in an interval of energy ΔT below the end-point is given by

$$\frac{n(\Delta T)}{n} \propto \left(\frac{\Delta T}{Q_\beta}\right)^3\,. \tag{14.14}$$

One can obtain the same result considering $Q_\beta \ll m_e$. From eqn (14.14) it is clear that in order to maximize the fraction of decay events that occur near the end-point of the spectrum, it is desirable to have the Q-value as small as possible.

The most stringent information on the electron neutrino mass has been obtained with experiments measuring the end-point of the electron spectrum in the β-decay of tritium (see Ref. [622]),

$$^3\mathrm{H} \to {}^3\mathrm{He} + e^- + \bar{\nu}_e\,, \tag{14.15}$$

which has $Q_\beta = 18.574\,\mathrm{keV}$. The main reason why tritium β-decay experiments are the most sensitive to the electron neutrino mass is that tritium β-decay has one of the smallest Q-values among all known β-decays. Moreover, tritium β-decay is a superallowed transition between mirror nuclei[76] with a relatively short half-life (about 12.3 years), which implies an acceptable number of observed events during the experiment lifetime. Another advantage of tritium β-decay is that the atomic structure is less complicated than those of heavier atoms, leading to a more accurate calculation of atomic effects.

For the measurement of the electron neutrino mass it is convenient to define the Kurie function

$$\begin{aligned} K(T) &\equiv \sqrt{\frac{\mathrm{d}\Gamma/\mathrm{d}T}{\frac{G_{\mathrm{F}}^2 m_e^5}{2\pi^3} \cos^2\theta_{\mathrm{C}} \, |\mathcal{M}|^2 \, F(Z, E_e) \, E_e \, p_e}} \\ &= \left[(Q_\beta - T) \sqrt{(Q_\beta - T)^2 - m_{\nu_e}^2} \right]^{1/2} . \end{aligned} \tag{14.16}$$

If the electron neutrino mass is zero, the Kurie function is a linear function of the kinetic energy T of the electron:

$$K(T)|_{m_{\nu_e}=0} = Q_\beta - T \,. \tag{14.17}$$

Therefore, a nonzero electron neutrino mass implies a deviation of the Kurie function from the linear function in eqn (14.17). This is illustrated in Fig. 14.2, where the diagonal dotted line represents the linear function in eqn (14.17) and the dashed line is the Kurie function for $m_{\nu_e} = 5\,\mathrm{eV}$ (the solid line illustrates the case of neutrino mixing discussed in subsection 14.1.1). One can see that if the electron neutrino is massive, there is an early falloff of the electron spectrum near the end-point, which is shifted from $T_{\mathrm{max}} = Q_\beta$ to $T_{\mathrm{max}} = Q_\beta - m_{\nu_e}$. In principle, one could measure the neutrino mass by measuring this shift of the end-point of the electron spectrum, but in practice a precise measurement of the end-point is very difficult because the frequency of events goes to zero near the end point. Therefore, in practice, information on the value of the neutrino mass is obtained by looking for a distortion of the Kurie plot with respect to the linear function near the end-point. Using this technique, the Mainz [699] and Troitzk [751] tritium experiments obtained the most stringent upper bounds on the electron neutrino mass:

$$m_{\nu_e} < 2.3\,\mathrm{eV} \quad (95\%\,\mathrm{CL}) \quad [699] \,, \tag{14.18}$$

$$m_{\nu_e} < 2.5\,\mathrm{eV} \quad (95\%\,\mathrm{CL}) \quad [751] \,. \tag{14.19}$$

In these experiments the electron energy is measured with spectrometers using appropriate configurations of magnetic and electric fields that allow a high energy resolution. The Mainz and Troitzk collaborations have recently merged in a joint collaboration which is working at an improved experiment, called KATRIN [264], which will reach a sensitivity of about 0.2 eV.

[76] Superallowed transitions are allowed transitions between nuclei belonging to the same isospin multiplet. Mirror nuclei are pairs of nuclei which have equal numbers of protons and neutrons plus an extra proton in one case and an extra neutron in the other. In this case, the overlap of the initial and final nuclear wave functions is close to one, leading to a large nuclear matrix element.

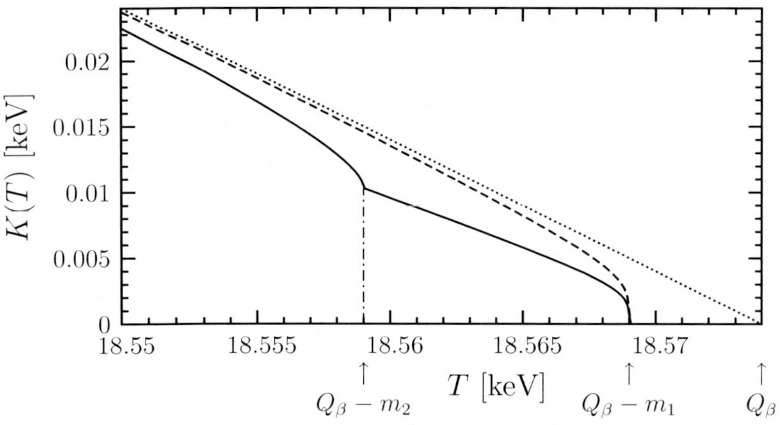

FIG. 14.2. Kurie plot for tritium β-decay. Dotted line: the linear Kurie function in eqn (14.17) for $m_{\nu_e} = 0$. Dashed line: Kurie function in eqn (14.16) for $m_{\nu_e} = 5\,\text{eV}$. Solid line: Kurie function in eqn (14.24) for two-neutrino mixing with $m_1 = 5\,\text{eV}$, $m_2 = 15\,\text{eV}$, and $\vartheta = \pi/4$.

An alternative method is to use cryogenic microcalorimeters, which measure the total energy released in β-decays. The main advantage of this method is the irrelevance of the atomic and molecular final state effects, which are difficult to calculate and are model-dependent. An experiment in Genoa using ^{187}Re, which has the smallest known Q-value ($Q_\beta = 2.47\,\text{eV}$ [501]), found

$$m_{\nu_e} < 26\,\text{eV} \quad (95\%\,\text{CL}) \quad [509]\,. \tag{14.20}$$

At the moment this limit is less restrictive than that obtained in tritium experiments, but considerable improvements are expected in the future.

14.1.1 Neutrino mixing

In the case of neutrino mixing, the electron neutrino field ν_e is a superposition of massive neutrino fields ν_k with weights given by the elements U_{ek} of the neutrino mixing matrix. The tritium β-decay in eqn (14.15) is, in fact, an incoherent sum of the separate decay modes

$$^3\text{H} \to {}^3\text{He} + e^- + \bar{\nu}_k\,. \tag{14.21}$$

The sum is incoherent because of the different final states. This means that the total decay rate is given by the sum of the decay rates relative to the processes in eqn (14.21). The probability of the decay mode in which the massive antineutrino $\bar{\nu}_k$ is produced is proportional to

$$\left| \langle e^-, \bar{\nu}_k | j^{\rho}_{W,L}{}^{\dagger} | 0 \rangle \, \langle {}^3\text{He} | j^{W,Q}_{\rho} | {}^3\text{H} \rangle \right|^2 = |U_{ek}|^2 \left| \overline{u_e} \gamma^\mu \left(1 - \gamma^5\right) v_{\nu_k} \, \langle {}^3\text{He} | j^{W,Q}_{\rho} | {}^3\text{H} \rangle \right|^2 , \tag{14.22}$$

where $j^{W,Q}_{\rho}$ is the hadronic weak charged current in eqn (3.176). The leptonic charged current $j^{\rho}_{W,L}$ has the same expression for Dirac and Majorana neutrinos (see eqns (6.12), (6.14), (6.184), and (6.187)):

$$j^{\rho}_{W,L} = 2 \sum_{\alpha=e,\mu,\tau} \overline{\nu_{eL}}\, \gamma^{\rho}\, \ell_{\alpha L} = 2 \sum_{\alpha=e,\mu,\tau} \sum_{k=1}^{3} \overline{\nu_{kL}}\, U^{*}_{\alpha k}\, \gamma^{\rho}\, \ell_{\alpha L} \,. \tag{14.23}$$

The mixing matrix U is different in the Dirac and Majorana cases, because in the Majorana case it contains two additional phases, as explained in section 6.3.1. However, the probability in eqn (14.22) depends only on the squared modulus $|U_{ek}|^2$, which is the same in the Dirac and Majorana cases. Therefore, also in the case of neutrino mixing the Dirac and Majorana character of neutrinos cannot be distinguished in tritium or other nuclear β-decays.

The Kurie function in tritium β-decay experiments is given by [958, 790, 687]

$$K(T) = \left[(Q_{\beta} - T) \sum_{k=1}^{3} |U_{ek}|^2 \sqrt{(Q_{\beta} - T)^2 - m_k^2} \right]^{1/2} . \tag{14.24}$$

This is a function of five parameters, the three neutrino masses and two mixing parameters (the unitarity of the mixing matrix implies that $\sum_{k=1}^{3} |U_{ek}|^2 = 1$). The main characteristics of the distortion of the Kurie function with respect to the linear function in eqn (14.17) are:

(a) A shift of the end-point of the spectrum from $T = Q$ to $T = Q - m_{\text{lht}}$, calling ν_{lht} the lightest massive neutrino component of ν_e (if $U_{e3} = 0$, $\nu_{\text{lht}} = \nu_1$ in both the normal and inverted schemes; otherwise, $\nu_{\text{lht}} = \nu_1$ in the normal scheme and $\nu_{\text{lht}} = \nu_3$ in the inverted scheme).
(b) Kinks at the electron kinetic energies $T_k = Q_{\beta} - m_k$, for $\nu_k \neq \nu_{\text{lht}}$, with corresponding strength determined by the value of $|U_{ek}|^2$.

This behavior of the Kurie function is illustrated by the solid line in Fig. 14.2, which describes the case of two-neutrino mixing ($U_{e3} = 0$) with $m_1 = 5\,\text{eV}$, $m_2 = 15\,\text{eV}$, and $\vartheta = \pi/4$ ($|U_{e1}|^2 = |U_{e2}|^2 = 1/2$).

If, in the future, effects of the neutrino masses are discovered in tritium or other β-decay experiments, a precise analysis of the data may reveal kinks of the Kurie function due to mixing of the electron neutrino with more than one massive neutrino[77]. In this case, the data will have to be analyzed using eqn (14.24).

However, so far tritium experiments did not find any effect of the neutrino masses and their data have been analyzed in terms of the one-generation Kurie function in eqn (14.16), leading to the upper bounds in eqns (14.18) and (14.19). How can this result be interpreted in the framework of three-neutrino mixing, in which eqn (14.24) holds? The exact expression of $K(T)$ in eqn (14.24) cannot be reduced to the one-generation Kurie function in eqn (14.16). In order to achieve such a reduction in an approximate way, one must note that, if an experiment does not

[77] In eqn (14.24) we considered the case of three-neutrino mixing, indicated by oscillation data (see chapter 13). However, tritium decay experiments (as well as other β-decay experiments) have the possibility of revealing a possible mixing of the electron neutrinos with more than three massive neutrinos, especially if the additional massive neutrinos are heavier than a few eV. In that case, the sum in eqn (14.24) must be extended to $k > 3$.

find any effect of the neutrino masses, its resolution for the measurement of $Q_\beta - T$ is much larger than the values of the neutrino masses. Considering $m_k \ll Q_\beta - T$, we have

$$\begin{aligned} K^2 &= (Q-T)^2 \sum_k |U_{ek}|^2 \sqrt{1 - \frac{m_k^2}{(Q-T)^2}} \simeq (Q-T)^2 \sum_k |U_{ek}|^2 \left[1 - \frac{1}{2}\frac{m_k^2}{(Q-T)^2}\right] \\ &= (Q-T)^2 \left[1 - \frac{1}{2}\frac{m_\beta^2}{(Q-T)^2}\right] \simeq (Q-T)^2 \sqrt{1 - \frac{m_\beta^2}{(Q-T)^2}} \\ &= (Q-T)\sqrt{(Q-T)^2 - m_\beta^2}\,, \end{aligned} \tag{14.25}$$

with m_β given by

$$m_\beta^2 = \sum_k |U_{ek}|^2 m_k^2\,. \tag{14.26}$$

The approximate expression of $K(T)$ in terms of m_β is the same as the expression in eqn (14.16) of the one-generation Kurie function in terms of m_{ν_e}. Therefore, m_β can be considered as the *effective electron neutrino mass in β-decay.* In the case of three-neutrino mixing, the upper bounds in eqns (14.18) and (14.19) must be interpreted as bounds on m_β:

$$m_\beta < 2.3\,\text{eV} \quad (95\%\,\text{CL})\,. \tag{14.27}$$

If future experiments do not find any effect of neutrino masses, they will provide more stringent bounds on the value of m_β.

In the standard parameterization of the mixing matrix in eqn (6.191), we have

$$m_\beta^2 = c_{12}^2\, c_{13}^2\, m_1^2 + s_{12}^2\, c_{13}^2\, m_2^2 + s_{13}^2\, m_3^2\,. \tag{14.28}$$

Although neutrino oscillation experiments do not give information on the absolute values of neutrino masses, they give information on the squared-mass differences Δm_{21}^2 and Δm_{31}^2 and on the mixing angles ϑ_{12} and ϑ_{13} (see Table 13.2). The values of the neutrino masses are determined according to eqns (13.135) and (13.136) in the normal and inverted schemes, respectively, as functions of the lightest mass (m_1 in the normal scheme and m_3 in the inverted scheme; see section 13). Therefore, also m_β can be considered as a function of the lightest mass, as shown in Fig. 14.3. The middle solid lines correspond to the best fit and the extreme solid lines delimit the 3σ allowed range. We have also shown with dashed lines the best-fit and 3σ ranges of the neutrino masses (same as in Fig. 13.10), which help us to understand their contribution to m_β.

From Fig. 14.3 one can see that, in the case of a normal mass hierarchy (normal scheme with $m_1 \ll m_2 \ll m_3$), the main contribution to m_β is due to m_2 or m_3 or both, because the upper limit for m_β is larger than the upper limit for m_2. The best-fit value of m_β is close to the lower limit, because the largest uncertainty is due to $\sin^2\vartheta_{13}$, which weights the contribution of the heaviest neutrino ν_3. Since the best-fit value of $\sin^2\vartheta_{13}$ is zero and coincides with its lower limit, the difference

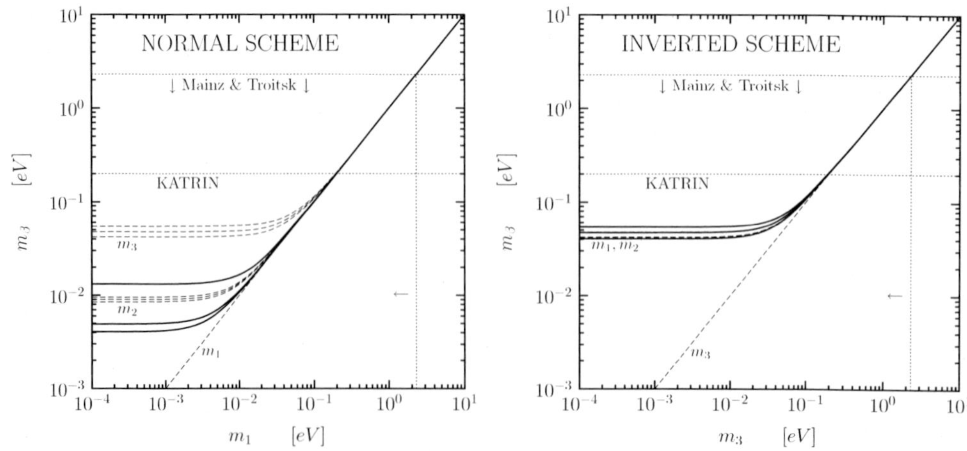

FIG. 14.3. Effective neutrino mass m_β in tritium β-decay experiments as a function of the lightest mass (m_1 in the normal scheme and m_3 in the inverted scheme; see Fig. 13.1). Middle solid lines correspond to the best-fit values in Table 13.2. Extreme solid lines enclose 3σ ranges. Dashed lines show the best-fit values and 3σ ranges of individual masses. In the inverted scheme, the best-fit values and 3σ ranges of m_1 and m_2 are practically the same and coincide with the best-fit value and 3σ range of m_β.

between the best-fit value of m_β and its lower limit is only due to the uncertainty of $\sin^2\vartheta_{12}\, m_2^2$, which is small.

In the case of an inverted mass hierarchy (inverted scheme with $m_3 \ll m_1 \lesssim m_2$), m_β has practically the same value as m_1 and m_2.

In the case of a quasidegenerate spectrum, m_β coincides with the approximately equal value of the three neutrino masses in both the normal and inverted schemes. In fact, in this case we have, from eqn (13.137),

$$m_\beta^2 \simeq m_\nu^2 \sum_{k=1}^{3} |U_{ek}|^2 = m_\nu^2\,, \tag{14.29}$$

and the uncertainties of the neutrino oscillation parameters are irrelevant.

Figure 14.3 shows that the Mainz and Troitsk experiments (eqns (14.18) and (14.19)) and the near-future KATRIN experiment [264] give information on the absolute values of neutrino masses in the quasidegenerate region in both normal and inverted schemes. From the Mainz upper bound in eqn (14.18), for the individual neutrino masses we obtain

$$m_k < 2.3\,\mathrm{eV} \quad (95\%\,\mathrm{CL})\,, \tag{14.30}$$

with $k = 1, 2, 3$.

One can note from Fig. 14.3 that the allowed ranges of m_β in the normal and inverted schemes in the case of a mass hierarchy are quite different and nonoverlapping: the lower limit for m_β in the inverted scheme is about $4 \times 10^{-2}\,\mathrm{eV}$, whereas the upper limit for m_β in the normal scheme is about 1.3×10^{-2} eV. If future experiments find an upper bound for m_β which is smaller than about 4×10^{-2} eV, the inverted scheme will be excluded, leaving the normal scheme as the only possibility.

Figure 14.3 shows also that β-decay experiments will not have to improve indefinitely for finding the effects of neutrino masses: the ultimate sensitivity is set at about 4×10^{-3} eV, which is the lower bound for m_β in the case of a normal mass hierarchy. Of course, if some β-decay experiment were to reveal the effects of neutrino masses, a more complicated analysis using the expression of $K(T)$ in eqn (14.24) will be needed. In that case, it may be possible to distinguish between the normal and inverted schemes even if both are allowed (i.e. $m_\beta \gtrsim 4 \times 10^{-2}$ eV).

Let us finally mention that β-decay measurements of various nuclei [937, 609, 155, 390, 502] allowed the derivation of constraints on the possible mixing of the electron neutrino with heavy neutrinos, beyond the three-generation light neutrinos.

14.2 Pion and tau decays

Accurate measurements of the kinematics in the decay of charged pions can give information on the neutrino masses. Let us consider the decay

$$\pi^+ \to \mu^+ + \nu_\mu \,, \tag{14.31}$$

which has been used in the most sensitive experiment performed at PSI [113]. Since the π^+ decay has a two-body final state, the mass of the neutrino can be determined by energy–momentum conservation if the momenta of the pion and muon are measured with sufficient accuracy. In the case of neutrino mixing, the muon neutrino is a superposition of different massive neutrinos. A measurement of the neutrino mass forces the superposition to collapse on the massive neutrino whose mass has been measured. Therefore, the decay rate must have peaks corresponding to the values of the neutrino masses, which are given by

$$m_k^2 = m_\pi^2 + m_\mu^2 - 2\, m_\pi \sqrt{m_\mu^2 + |\vec{p}_\mu|^2} \qquad (k = 1, 2, 3)\,, \tag{14.32}$$

for pions decaying at rest.

The value of the muon momentum measured in the PSI experiment [113] is

$$|\vec{p}_\mu| = 29.792\,00 \pm 0.000\,11\ \text{MeV}\,, \tag{14.33}$$

leading to

$$m_k < 0.17\,\text{MeV} \qquad (90\%\ \text{CL})\,, \tag{14.34}$$

for $k = 1, 2, 3$.

Also τ-decays have been used for the measurement of neutrino masses. The decays

$$\tau^- \to 2\,\pi^- + \pi^+ + \nu_\tau \,, \qquad \tau^- \to 3\,\pi^- + 2\,\pi^+ + \nu_\tau (+\pi^0) \tag{14.35}$$

have been studied in the ALEPH experiment [161], with the result

$$m_k < 18.2\,\text{MeV} \qquad (95\%\ \text{CL})\,, \tag{14.36}$$

for $k = 1, 2, 3$.

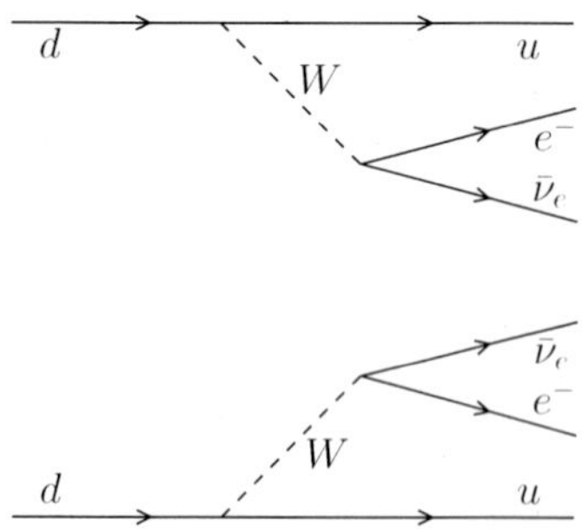

FIG. 14.4. Tree-level quark diagram of a $2\beta^-_{2\nu}$ process.

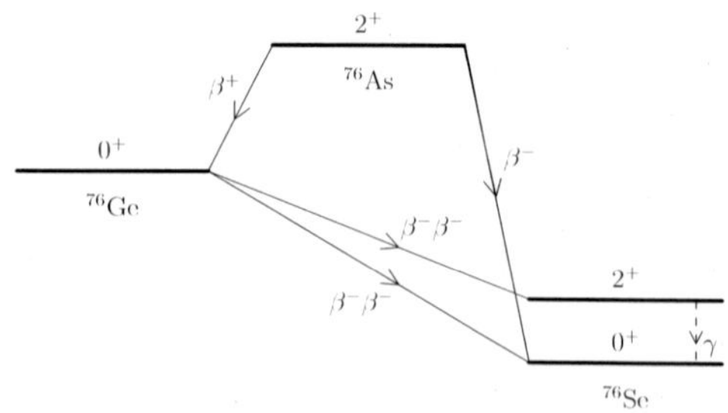

FIG. 14.5. Schematic illustration of the energy level structure of the $2\beta^-$-decay of ^{76}Ge into ^{76}Se.

The upper bounds on the neutrino masses obtained in pion and tau decay experiments are much less stringent than those obtained in tritium decay experiments (eqn (14.30)). Since it is unlikely that in the future the measurements of neutrino masses with pion and tau decays experiments may improve so much to reach a precision at the eV level, comparable with β-decay experiments, their interest lies mainly in the possibility of constraining the admixture of the muon and tau neutrinos with heavy neutrinos beyond three-neutrino mixing [114].

14.3 Neutrinoless double-beta decay

Neutrinoless double-β-decay experiments are considered as the most promising way to find if neutrinos are Majorana particles. In order to introduce neutrinoless double-β-decay, let us first consider the so-called two-neutrino double-β-decay processes ($2\beta_{2\nu}$) of the types

$$\mathcal{N}(A,Z) \to \mathcal{N}(A,Z+2) + 2\,e^- + 2\,\bar{\nu}_e \qquad (2\beta^-_{2\nu})\,, \tag{14.37}$$

$$\mathcal{N}(A,Z) \to \mathcal{N}(A,Z-2) + 2\,e^+ + 2\,\nu_e \qquad (2\beta^+_{2\nu})\,, \tag{14.38}$$

which were first suggested by M. Goeppert-Mayer in 1935 [546]. A $2\beta^-_{2\nu}$ ($2\beta^+_{2\nu}$) process consists of the simultaneous β^- (β^+) decay of two neutrons (protons) in the same nucleus. The $2\beta_{2\nu}$ processes are generated at second-order in the perturbative expansion of weak interactions in the SM. The tree-level quark diagram of $2\beta^-_{2\nu}$ decays is shown in Fig. 14.4. Figure 14.5 shows, as an example, the nuclear level structure of the $2\beta^-$-decay $^{76}\text{Ge} \to {}^{76}\text{Se}$.

The neutrinoless double-β-decay processes ($2\beta_{0\nu}$) of the types

$$\mathcal{N}(A,Z) \to \mathcal{N}(A,Z+2) + 2\,e^- \qquad (2\beta^-_{0\nu})\,, \tag{14.39}$$

$$\mathcal{N}(A,Z) \to \mathcal{N}(A,Z-2) + 2\,e^+ \qquad (2\beta^+_{0\nu})\,, \tag{14.40}$$

which have been proposed by W.H. Furry in 1939 [485], are forbidden in the SM. As explained in subsection 14.3.1, $2\beta_{0\nu}$-decays are possible if neutrinos are massive

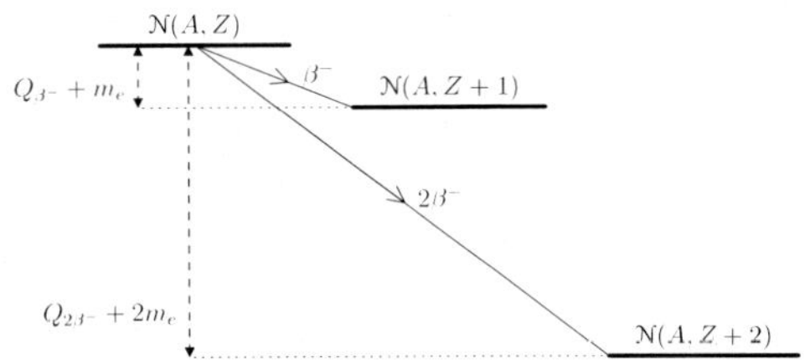

FIG. 14.6. Q-values of the $2\beta^-$-decay $\mathcal{N}(A,Z) \to \mathcal{N}(A,Z+2)$ ($Q_{2\beta^-}$) and of the β^--decay $\mathcal{N}(A,Z) \to \mathcal{N}(A,Z+1)$ (Q_{β^-}).

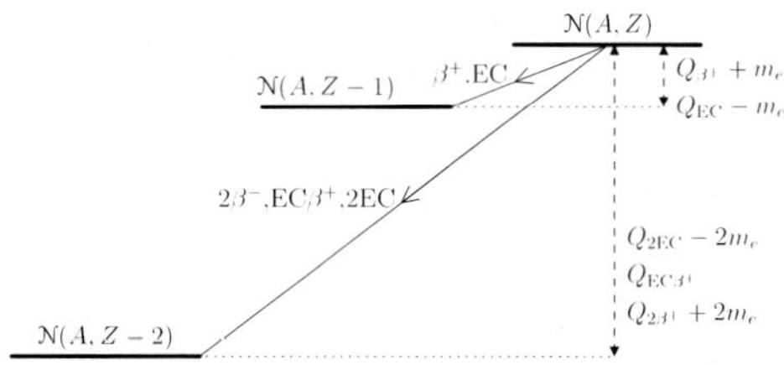

FIG. 14.7. Q-values of the $2\beta^+$ ($Q_{2\beta^+}$), ECβ^+ ($Q_{\text{EC}\beta^+}$) and 2EC ($Q_{2\text{EC}}$) decays $\mathcal{N}(A,Z) \to \mathcal{N}(A,Z-2)$ and of the β^+ (Q_{β^+}) and EC (Q_{EC}) decays $\mathcal{N}(A,Z) \to \mathcal{N}(A,Z-1)$.

Majorana particles. In this case, a nucleus which can decay through a $2\beta_{2\nu}$ process can also decay through a $2\beta_{0\nu}$ process, albeit with a different lifetime.

Since the $2\beta_{2\nu}$-decays have a four-body leptonic final state, the sum of the kinetic energies of the two decay electrons have a continuous spectrum from zero to the Q-value of the decay process (the recoil energy of the final nucleus is negligible), which is given by

$$Q_{2\beta} = M_i - M_f - 2\,m_e\,, \tag{14.41}$$

where M_i and M_f are, respectively, the masses of the initial and final nuclei (i.e. the energy levels of their ground states; if the transition occurs into an excited energy level of the final nucleus, M_f must be replaced with the appropriate energy). On the other hand, in the $2\beta_{0\nu}$-decays the sum of the kinetic energies of the two decay electrons is equal to the Q-value.

A nucleus $\mathcal{N}(A,Z)$ can decay through a $2\beta^{\mp}$ process if its ground state has an energy which is larger than the ground-state energy of the nucleus $\mathcal{N}(A,Z\pm2)$ plus twice the electron mass. Moreover, if a nucleus can decay through both the β and 2β processes, in practice the 2β-decay process is not observable, because its β-decay lifetime is much shorter than its 2β-decay lifetime. Therefore, in practice the 2β-decay of a nucleus is observable only if its β-decay is energetically forbidden or strongly suppressed because of a large change of spin. The β^--decay of a nucleus $\mathcal{N}(A,Z)$ is energetically forbidden if its ground-state energy is lower than the ground-state energy of the nucleus $\mathcal{N}(A,Z+1)$ plus the electron mass ($Q_{\beta^-} < 0$ in Fig. 14.6). Typically, in $2\beta^-$-decays the energy levels of the three nuclei $\mathcal{N}(A,Z)$, $\mathcal{N}(A,Z+1)$, and $\mathcal{N}(A,Z+2)$ are of the type depicted in Fig. 14.5, where the specific case of the ^{76}Ge, ^{76}As, and ^{76}Se nuclei is considered.

The 35 naturally occurring isotopes which can decay through the $2\beta^-$ process, with forbidden or suppressed β^--decay, are listed in Table 14.1. All of the initial and final nuclei in the $2\beta^-$ process listed in Table 14.1 are even-even, i.e. they have an even number of protons and neutrons. Their binding energy is larger than that of the intermediate odd-odd nuclei because of the pairing force acting between identical nucleons. For the same reason, all of the initial and final nuclei in the $2\beta^-$ process listed in Table 14.1 have a 0^+ ground state. Therefore, all ground-state to ground-state transitions are $0^+ \to 0^+$. However, ground-state transitions

TABLE 14.1. $2\beta^-$-decays of natural isotopes, with their Q-value and experimental measurements of the half-life of the 2ν and 0ν decay modes. The values are taken from Ref. [1024]. The percentage in parentheses is the confidence level. Results obtained in geochemical and radiochemical experiments are indicated, respectively, with (G) and (R).

$2\beta^-$-decay	$Q_{2\beta}$ [keV]	$T^{2\nu}_{1/2}$ [y]	$T^{0\nu}_{1/2}$ [y]
$^{46}Ca \to {}^{46}Ti$	990.4 ± 2.4		$> 1.0 \times 10^{17}$ (90%)
$^{48}Ca \to {}^{48}Ti$[a]	4272 ± 4	$4.2^{+3.3}_{-1.3} \times 10^{19}$	$> 1.5 \times 10^{21}$ (90%)
$^{70}Zn \to {}^{70}Ge$	1000.9 ± 3.4		$> 4.8 \times 10^{14}$
$^{76}Ge \to {}^{76}Se$	2039.006 ± 0.050	$(1.8 \pm 0.1) \times 10^{21}$	$> 1.9 \times 10^{25}$ (90%)
$^{80}Se \to {}^{80}Kr$	133.9 ± 3.7		
$^{82}Se \to {}^{82}Kr$	2995.1 ± 2.0	$(8.3 \pm 1.2) \times 10^{19}$	$> 2.7 \times 10^{22}$ (68%)
$^{86}Kr \to {}^{86}Sr$	1255.6 ± 2.4		
$^{94}Zr \to {}^{94}Mo$	1144.1 ± 2.0	$> 1.1 \times 10^{17}$ (90%)	$> 1.9 \times 10^{19}$ (90%)
$^{96}Zr \to {}^{96}Mo$[a]	3350.4 ± 2.9	$2.1^{+0.8}_{-0.4} \times 10^{19}$	$> 1.0 \times 10^{21}$ (90%)
$^{98}Mo \to {}^{98}Ru$	112 ± 6		$> 1.0 \times 10^{14}$
$^{100}Mo \to {}^{100}Ru$	3034 ± 6	$6.8^{+0.8}_{-0.9} \times 10^{18}$	$> 5.5 \times 10^{22}$ (90%)
$^{104}Ru \to {}^{104}Pd$	1300 ± 4		
$^{110}Pd \to {}^{110}Cd$	2000 ± 11	$> 6.0 \times 10^{16}$	$> 6.0 \times 10^{16}$
$^{114}Cd \to {}^{114}Sn$	536.8 ± 3.3	$> 9.2 \times 10^{16}$ (99%)	$> 2.0 \times 10^{20}$ (90%)
$^{116}Cd \to {}^{116}Sn$	2805.0 ± 3.8	$2.6^{+0.7}_{-0.4} \times 10^{19}$	$> 7.0 \times 10^{22}$ (90%)
$^{122}Sn \to {}^{122}Te$	366.2 ± 2.8		$> 5.8 \times 10^{13}$
$^{124}Sn \to {}^{124}Te$	2287.0 ± 1.5	$> 1.0 \times 10^{17}$	$> 2.4 \times 10^{17}$ (95%)
$^{128}Te \to {}^{128}Xe$	867.2 ± 1.0	$(2.2 \pm 0.3) \times 10^{24}$ (G)	$> 8.6 \times 10^{22}$ (90%)
$^{130}Te \to {}^{130}Xe$	2528.8 ± 1.3	$(7.9 \pm 1.0) \times 10^{20}$ (G)	$> 1.4 \times 10^{23}$ (90%)
$^{134}Xe \to {}^{134}Ba$	830.1 ± 3.0	$> 1.1 \times 10^{16}$	$> 8.2 \times 10^{19}$ (68%)
$^{136}Xe \to {}^{136}Ba$	2468 ± 7	$> 8.1 \times 10^{20}$ (90%)	$> 4.4 \times 10^{23}$ (90%)
$^{142}Ce \to {}^{142}Nd$	1416.9 ± 2.1	$> 1.6 \times 10^{17}$ (90%)	$> 1.5 \times 10^{19}$ (68%)
$^{146}Nd \to {}^{146}Sm$	70.2 ± 2.9		
$^{148}Nd \to {}^{148}Sm$	1928.8 ± 1.9	$> 3.0 \times 10^{18}$ (90%)	$> 3.0 \times 10^{18}$ (90%)
$^{150}Nd \to {}^{150}Sm$	3367.5 ± 2.2	$(6.8 \pm 0.8) \times 10^{18}$	$> 1.2 \times 10^{21}$ (90%)
$^{154}Sm \to {}^{154}Gd$	1251.0 ± 1.3	$> 2.3 \times 10^{18}$ (68%)	$> 2.3 \times 10^{18}$ (68%)
$^{160}Gd \to {}^{160}Dy$	1729.7 ± 1.3	$> 1.9 \times 10^{19}$ (90%)	$> 1.3 \times 10^{21}$ (90%)
$^{170}Er \to {}^{170}Yb$	653.6 ± 1.7	$> 3.2 \times 10^{17}$ (68%)	$> 3.2 \times 10^{17}$ (68%)
$^{176}Yb \to {}^{176}Hf$	1086.7 ± 1.9	$> 1.6 \times 10^{17}$ (68%)	$> 1.6 \times 10^{17}$ (68%)
$^{186}W \to {}^{186}Os$	488.0 ± 1.7	$> 5.9 \times 10^{17}$ (90%)	$> 2.7 \times 10^{20}$ (90%)
$^{192}Os \to {}^{192}Pt$	413.5 ± 3.0		$> 9.8 \times 10^{12}$
$^{198}Pt \to {}^{198}Hg$	1047 ± 3		$> 3.2 \times 10^{14}$
$^{204}Hg \to {}^{204}Pb$	416.3 ± 1.5		
$^{232}Th \to {}^{232}U$	842.2 ± 2.5		
$^{238}U \to {}^{238}Pu$	1145.0 ± 1.3	$(2.0 \pm 0.6) \times 10^{21}$ (R)	

[a] β^--decay energetically allowed but enormously suppressed.

to an excited state of the final nucleus may be energetically allowed, as in the case of the ^{76}Ge $\to$ ^{76}Se $2\beta^-$-decay in Fig. 14.5, in which there is an accessible 2^+ excited state of ^{76}Se. Since generally 1^+ states and states with negative parity have large excitation energies, in practice the only relevant ground-state to excited-state transitions are $0^+ \to 2^+$ for all the $2\beta^-$-decays in Table 14.1.

There are only six naturally occurring isotopes which can decay through the $2\beta^+$ process (see Ref. [605]). These isotopes have small Q-values and lifetimes which are much longer than the lifetimes of the $2\beta^-$ in Table 14.1. The reason for the rarity of $2\beta^+$-decaying isotopes and their small Q-values can be understood with the help of Fig. 14.7, which shows that the decay $\mathcal{N}(A,Z) \to \mathcal{N}(A,Z-1)$ can occur either through the β^+ process $\mathcal{N}(A,Z) \to \mathcal{N}(A,Z-1) + e^+ + \nu_e$ or through the electron-capture process $e^- + \mathcal{N}(A,Z) \to \mathcal{N}(A,Z-1) + \nu_e$. Since $Q_{\text{EC}} = Q_{\beta^+} + 2m_e$, the electron-capture process can occur even if the β^+ process is energetically forbidden ($Q_{\beta^+} < 0$). Thus, in order have an energetically forbidden $\mathcal{N}(A,Z) \to \mathcal{N}(A,Z-1)$ transitions, the ground-state energy of $\mathcal{N}(A,Z)$ must be smaller than the ground-state energy of the nucleus $\mathcal{N}(A,Z-1)$ minus the electron mass ($Q_{\text{EC}} < 0$ in Fig. 14.7). Considering as a reference energy the ground-state energy of the intermediate nucleus ($\mathcal{N}(A,Z+1)$ in $2\beta^-$-decays and $\mathcal{N}(A,Z-1)$ in $2\beta^+$-decays), the ground-state energy of the initial nucleus in a $2\beta^+$-decay must be at least $2m_e$ lower than in the case of a $2\beta^-$-decay. This implies that $2\beta^+$-decaying isotopes are more rare than $2\beta^-$-decaying isotopes. Moreover, for the same energy difference between the ground states of the intermediate and final nuclei, the energy difference between the ground states of the initial and final nucleus in a $2\beta^+$-decay is at least $2m_e$ lower than in the case of a $2\beta^-$-decay, leading to a correspondingly smaller Q-value. For these reasons, $2\beta^+$-decay has been less studied than $2\beta^-$-decay and in the following we will consider only $2\beta^-$-decays. In any case, the neutrino properties of $2\beta^+$-decays are the same as those of $2\beta^-$-decays. Let us only mention that, as shown in Fig. 14.7, $\mathcal{N}(A,Z) \to \mathcal{N}(A,Z-2)$ transitions can occur not only through $2\beta^+$ processes, but also through the $\text{EC}\beta^+_{2\nu}$ process $e^- + \mathcal{N}(A,Z) \to \mathcal{N}(A,Z-2) + e^+ + 2\,\nu_e$, the $2\text{EC}_{2\nu}$ process $2\,e^- + \mathcal{N}(A,Z) \to \mathcal{N}(A,Z-2) + 2\,\nu_e$, and their neutrinoless analogs.

There are three experimental methods for the observation of 2β-decay (see Ref. [605] for more details): geochemical, radiochemical and direct. In the geochemical method the 2β-decay half-life is measured by determining the abundance of the final nucleus in an old ore sample containing the initial nucleus. With this method the accumulation of the decay products over a very long time can be used to measure very long half-lives. In the radiochemical method a well-prepared artificial sample of the parent nucleus is used and the abundance of the final nucleus is measured after some years. The drawback of the geochemical and radiochemical methods is that the 2ν and 0ν decay modes cannot be distinguished. Hence, they are not interesting for our discussion on the search for $2\beta_{0\nu}$-decay. On the other hand, in direct experiments 2β-decay is measured in real time by measuring the energies of the decay electrons or their sum. The 2ν and 0ν decay modes are distinguished by the different distributions of the sum of the kinetic energies of the decay electrons: a continuous spectrum in the case of $2\beta_{2\nu}$-decay and a line at the Q-value in the case of $2\beta_{0\nu}$-decay.

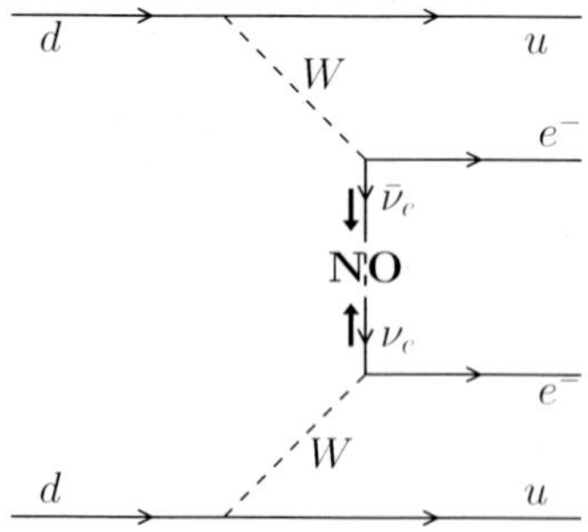

FIG. 14.8. Illustration of the two reasons why $2\beta^-_{0\nu}$ decay processes are forbidden in the SM: $\bar{\nu}_e \neq \nu_e$ and the helicity mismatch.

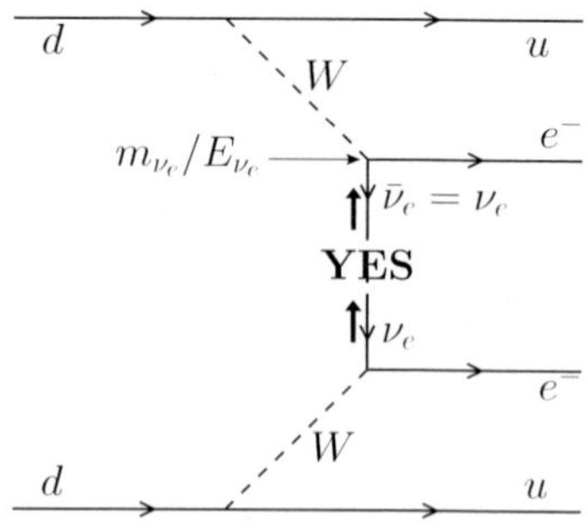

FIG. 14.9. Illustration of the particle–antiparticle and helicity matching conditions for $2\beta^-_{0\nu}$ decay.

Of the 35 $2\beta^-$ processes listed in Table 14.1, 28 processes have been studied experimentally in direct experiments and seven $2\beta^-_{2\nu}$ processes have been observed for the nuclei ^{48}Ca, ^{76}Ge, ^{82}Se, ^{96}Zr, ^{100}Mo, ^{116}Cd, and ^{150}Nd. Although the main experimental effort is directed towards the discovery of $2\beta^-_{0\nu}$, so far no experiment observed an unquestionable signal[78]. The lower bound on a $2\beta^-_{0\nu}$ half-life which is most sensitive to the Majorana neutrino mass is that obtained in the Heidelberg–Moscow ^{76}Ge experiment [680]:

$$T^{0\nu}_{1/2}(^{76}\text{Ge}) > 1.9 \times 10^{25}\,\text{y} \qquad (90\%\ \text{CL})\,. \tag{14.42}$$

The IGEX ^{76}Ge experiment [3] obtained the comparable limit $T^{0\nu}_{1/2}(^{76}\text{Ge}) > 1.57 \times 10^{25}$ y (90% CL).

For the future, many new $2\beta_{0\nu}$ experiments are planned and under preparation (see Refs. [405, 4]), since the quest for the Majorana nature of neutrinos is of fundamental importance.

14.3.1 Effective Majorana mass

In neutrinoless double-β-decay the conservation of the total lepton number is violated by two units:

$$\Delta L = \pm 2 \qquad (2\beta^{\mp}_{0\nu})\,. \tag{14.43}$$

Since in the SM the total lepton number is conserved (see section 3.5), the $2\beta_{0\nu}$-decay is forbidden in the SM. Figure 14.8 helps us to understand what is missing in the SM in order to make $2\beta^-_{0\nu}$ decay possible. Since no antineutrino is emitted, the two antineutrino lines in the $2\beta^-_{2\nu}$ tree-level quark diagram in Fig. 14.4 should be joined as indicated in Fig. 14.8 in order to form a virtual neutrino line which propagates between the two leptonic weak-interaction vertices. In the SM this is not possible for the two reasons illustrated in Fig. 14.8:

[78] There is a claim of a measurement of the $2\beta^-_{0\nu}$ of ^{76}Ge with $T^{0\nu}_{1/2} = 1.19^{+1.00}_{-0.17} \times 10^{25}$ y [679, 681]. However, this measurement is rather controversial [433, 2, 156].

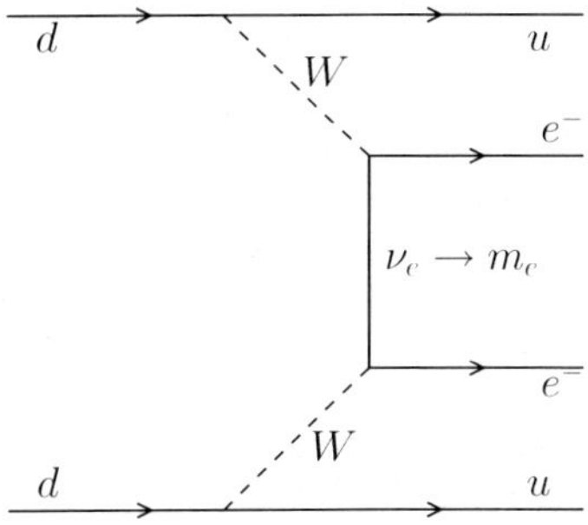

FIG. 14.10. Tree-level quark diagram of a $2\beta^-_{0\nu}$ process.

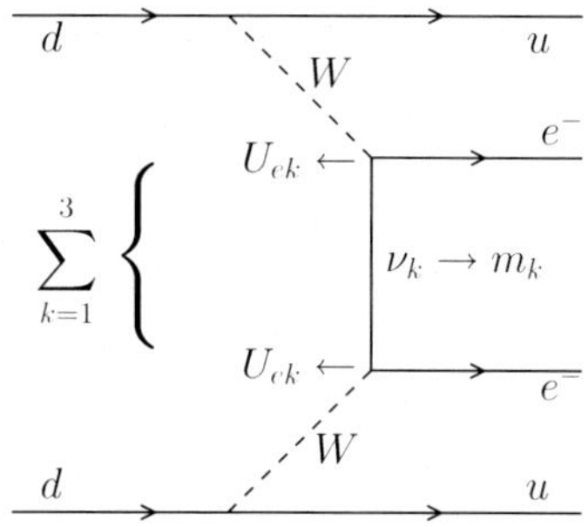

FIG. 14.11. Tree-level quark diagram of a $2\beta^-_{0\nu}$ process in the case of three--neutrino mixing.

The particle–antiparticle mismatch. A $\bar{\nu}_e$ emitted in the upper leptonic vertex cannot be absorbed in the lower leptonic vertex, which is capable only of absorbing a ν_e.

The helicity mismatch. The helicity of the neutral lepton emitted in the upper leptonic vertex is *positive* and the lower leptonic vertex can absorb only a neutral lepton with *negative* helicity.

Therefore, there are two necessary conditions for the occurrence of $2\beta_{0\nu}$-decay processes (for simplicity, for the moment we do not consider neutrino mixing, which will be taken into account later):

Particle–antiparticle matching: $\bar{\nu}_e = \nu_e$. The electron neutrino must be a Majorana particle. In this case, the total lepton number is not conserved (see section 6.2.4).

Helicity matching: $m_{\nu_e} \neq 0$. In this case, the upper leptonic vertex can emit a neutrino with negative helicity with relative amplitude m_{ν_e}/E_{ν_e} (see eqn (6.106)), which is absorbed by the lower leptonic vertex with relative amplitude equal to unity.

These two conditions are illustrated in Fig. 14.9. It is clear that, if these two conditions are satisfied, i.e. if the electron neutrino is a massive Majorana particle, the amplitude of a $2\beta_{0\nu}$-decay process is proportional to the Majorana mass m_{ν_e} of the electron neutrino (Fig. 14.10).

Let us now consider neutrino mixing with the SM weak leptonic current in eqn (14.23). Since, in this case, the electron neutrino field is a superposition of Majorana massive neutrino fields, the massive Majorana neutrinos propagate between the two leptonic weak interaction vertices, as illustrated in Fig. 14.11. For the propagation of the massive Majorana neutrino ν_k, each leptonic weak interaction vertex in Fig. 14.11 is described by $\overline{e_L}\,\gamma^\rho\,U_{ek}\,\nu_{kL}$, leading to a factor U_{ek}^2 in the amplitude of the process. In order to satisfy the helicity matching condition, there is also a factor m_k. Summing over the contributions of the three massive neutrinos, we obtain the *effective Majorana mass* in $2\beta_{0\nu}$-decay

$$m_{2\beta} = \sum_{k=1}^{3} U_{ek}^2 \, m_k \,. \tag{14.44}$$

The decay amplitude is proportional to $m_{2\beta}$. In fact, the propagator describing the internal neutrino line in Fig. 14.11 is given by

$$\begin{aligned} G(x_1 - x_2) &= \langle 0 | \mathrm{T} \left[\nu_{eL}(x_1) \nu_{eL}^T(x_2) \right] | 0 \rangle \\ &= \frac{1-\gamma^5}{2} \sum_{k=1}^{3} U_{ek}^2 \, \langle 0 | \mathrm{T} \left[\nu_k(x_1) \nu_k^T(x_2) \right] | 0 \rangle \left(\frac{1-\gamma^5}{2} \right)^T \\ &= - \, \frac{1-\gamma^5}{2} \sum_{k=1}^{3} U_{ek}^2 \, \langle 0 | \mathrm{T} [\nu_k(x_1) \overline{\nu_k}(x_2)] | 0 \rangle \, \mathcal{C} \, \frac{1-(\gamma^5)^T}{2} \\ &= - \, i \sum_{k=1}^{3} U_{ek}^2 \int \frac{\mathrm{d}^4 p}{(2\pi)^4} \frac{m_k}{p^2 - m_k^2} \, \mathrm{e}^{-ip \cdot (x_1 - x_2)} \, \frac{1-\gamma^5}{2} \, \mathcal{C} \,, \end{aligned} \tag{14.45}$$

where we have used the property[79] $\nu_k^T = -\overline{\nu_k}\,\mathcal{C}$, the property of $\mathcal{C}$ in eqn (2.347), the expression of the fermionic propagator in eqn (2.250), and the property $\left(1-\gamma^5\right) \not{p} \left(1-\gamma^5\right) = 0$. The squared neutrino mass in the denominator of the propagator is negligible in comparison with the average neutrino energy and momentum, $\langle p^0 \rangle \sim \langle |\vec{p}| \rangle \sim 1/R$ where R is the nuclear radius, which is approximated by the empirical formula $R = 1.2\, A^{1/3}\,\mathrm{fm}$. For $A \sim 100$, we have $\langle p^0 \rangle \sim \langle |\vec{p}| \rangle \sim 35\,\mathrm{MeV}$, which is much larger than the three light neutrino masses[80] (see eqn (14.30)). Hence, the neutrino propagator can be approximated by

$$G(x_1 - x_2) = -i \, m_{2\beta} \int \frac{\mathrm{d}^4 p}{(2\pi)^4} \frac{\mathrm{e}^{-ip \cdot (x_1 - x_2)}}{p^2 - m_k^2} \, \frac{1-\gamma^5}{2} \, \mathcal{C} \,, \tag{14.46}$$

which is proportional to the effective Majorana mass in eqn (14.44).

Note that the value of the effective Majorana mass $m_{2\beta}$ given by eqn (14.44) depends not only on the values of the Majorana neutrino masses, but also on the elements of the mixing matrix which connect the electron neutrino with the massive neutrinos (the first row of the mixing matrix). Moreover, each element, which is in general a complex number, appears in the sum as squared, leading to possible cancellations among the different mass contributions. In fact, in the standard parameterization in eqns (6.189), (6.191), and (6.192), we have, with $\lambda_1 = 0$,

$$m_{2\beta} = c_{12}^2 \, c_{13}^2 \, m_1 + e^{2i\lambda_2} \, s_{12}^2 \, c_{13}^2 \, m_2 + e^{2i(\lambda_3 - \delta_{13})} \, s_{13}^2 \, m_3$$

---

[79] From the Majorana condition $\nu_k = \nu_k^C = \mathcal{C}\,\overline{\nu_k}^T$, since $\mathcal{C}^T = -\mathcal{C}$ (eqn (2.346)), we obtain $\nu_k^T = \overline{\nu_k}\,\mathcal{C}^T = -\overline{\nu_k}\,\mathcal{C}$.

[80] We do not consider here the possibility of mixing of the electron neutrino with heavy massive neutrinos. In that case, if the heavy neutrino mass is of the order of $1/R$, the exact expression for the propagator in eqn (14.45) must be used. If the heavy neutrino mass is much larger, the squared mass in the denominator suppresses the contribution of the heavy neutrino.

$$= |U_{e1}|^2 \, m_1 + e^{i\alpha_2} \, |U_{e2}|^2 \, m_2 + e^{i\alpha_3} \, |U_{e3}|^2 \, m_3 \,, \tag{14.47}$$

with the convenient definitions

$$\alpha_2 \equiv 2\,\lambda_2 \,, \qquad \alpha_3 \equiv 2\,(\lambda_3 - \delta_{13}) \,. \tag{14.48}$$

Since the values of λ_2, λ_2, and δ_{13} are completely unknown, in the following we will consider $0 \leq \alpha_k < 2\pi$, for $k = 2,3$. The only special cases to be studied are those in which the CP symmetry is conserved. In section 6.3.2 we have seen that CP is conserved if $\delta_{13} = 0\,,\,\pi$ and $\lambda_{2,3} = 0\,,\,\pi/2\,,\,\pi\,,\,3\pi/2$. Hence, for the two phases α_2 and α_3 we have the physically relevant cases

$$\alpha_k = 0\,,\,\pi \quad \Rightarrow \quad e^{i\alpha_k} = \pm 1 \quad \text{(CP invariance)}\,, \tag{14.49}$$

for $k = 2,3$. Thus, if CP is conserved $m_{2\beta}$ is real and there are four possible cases:

$$(++) \quad \alpha_2 = \alpha_3 = 0 \qquad \Rightarrow \quad m_{2\beta} = |U_{e1}|^2 \, m_1 + |U_{e2}|^2 \, m_2 + |U_{e3}|^2 \, m_3 \,, \tag{14.50}$$

$$(--) \quad \alpha_2 = \alpha_3 = \pi \qquad \Rightarrow \quad m_{2\beta} = |U_{e1}|^2 \, m_1 - |U_{e2}|^2 \, m_2 - |U_{e3}|^2 \, m_3 \,, \tag{14.51}$$

$$(+-) \quad \alpha_2 = 0\,, \quad \alpha_3 = \pi \quad \Rightarrow \quad m_{2\beta} = |U_{e1}|^2 \, m_1 + |U_{e2}|^2 \, m_2 - |U_{e3}|^2 \, m_3 \,, \tag{14.52}$$

$$(-+) \quad \alpha_2 = \pi\,, \quad \alpha_3 = 0 \quad \Rightarrow \quad m_{2\beta} = |U_{e1}|^2 \, m_1 - |U_{e2}|^2 \, m_2 + |U_{e3}|^2 \, m_3 \,. \tag{14.53}$$

The $(++)$ case corresponds to the maximum possible value of $|m_{2\beta}|$, for given values of $|U_{ek}|^2 m_k$ $(k = 1,2,3)$, irrespective of CP invariance. In the other three cases, the possible cancellations among the three mass contributions due to the phases $e^{i\alpha_2}$ and $e^{i\alpha_3}$ are maximal. Which of the three cases gives the minimum value of $|m_{2\beta}|$ depends on the values of $|U_{ek}|^2 m_k$, for $k = 1,2,3$. Thus, the smallest and largest values of $|m_{2\beta}|$ for given values of U_{ek} and m_k $(k = 1,2,3)$ are obtained in the case of CP invariance.

14.3.2 Necessity of Majorana neutrinos

So far we have seen that $2\beta_{0\nu}$-decay occurs if neutrinos are massive Majorana particles (unless a fine-tuned cancellation occurs in the sum over the different mass contributions to $m_{2\beta}$). However, a $2\beta_{0\nu}$-decay process can occur through other mechanisms besides the exchange of Majorana massive neutrinos with SM interactions illustrated in Fig. 14.11. These other mechanisms involve new interactions and/or new particles beyond the SM (see Refs. [605, 377, 426, 1036]). In general, each of these mechanisms gives a contribution to the $2\beta_{0\nu}$-decay amplitude which is proportional to a small parameter which reflects the suppression of the corresponding new physics beyond the SM in low-energy phenomenology. For example, the helicity matching condition can be overcome if there is a $V + A$ interaction. In this case, the $2\beta_{0\nu}$-decay amplitude has a contribution proportional to a small effective $V + A$ coupling constant.

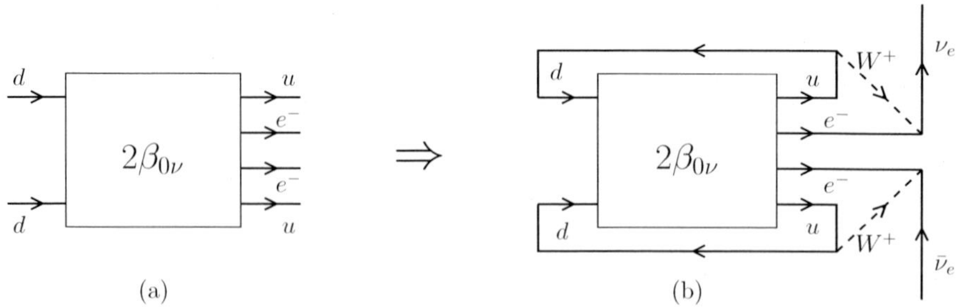

FIG. 14.12. (a) Black-box diagram of the elementary $2\beta_{0\nu}$ process $2\,d \to 2\,u + 2\,e^-$. (b) $\bar{\nu}_e \to \nu_e$ transition diagram through a $2\beta_{0\nu}$ black-box.

In view of these considerations, one may ask if the massive Majorana character of neutrinos is not only a sufficient, but also a necessary condition for $2\beta_{0\nu}$-decay. The answer is positive [935, 1005]. The reason is illustrated in Fig. 14.12. Figure 14.12a shows a schematic representation of $2\beta_{0\nu}$-decay in which the square box labeled $2\beta_{0\nu}$ includes all possible mechanisms at all orders in perturbation theory. Only the external lines of the quarks and the two final electrons are visible. Figure 14.12b shows that these external lines can be joined with SM weak interaction vertices to neutrino lines in order to obtain a diagram for a $\bar{\nu}_e \to \nu_e$ transition. This diagram generates a contribution to the Majorana mass of the electron neutrino through radiative corrections at some order of perturbation theory, even if there is no tree-level Majorana neutrino mass term. In order to have an electron neutrino Majorana mass it is necessary that the contribution generated by the diagram in Fig. 14.12b is not canceled by the contributions of other diagrams. Such a cancellation is unlikely, since it would require a fine-tuned arrangement of masses and mixing at all orders of perturbation theory, unless there is a global symmetry which forbids the electron neutrino Majorana mass term

$$\mathcal{L}^{\mathrm{M},\nu_e}_{\mathrm{mass}} = -\frac{1}{2}\, m_{ee} \left(-\nu_{eL}^T\, \mathcal{C}^\dagger\, \nu_{eL} + \overline{\nu_{eL}}\, \mathcal{C}\, \overline{\nu_{eL}}^T \right) . \tag{14.54}$$

However, the existence of such symmetry is incompatible with $2\beta_{0\nu}$-decay [1005]. The proof goes as follows. Since the other parts of the full Lagrangian must be invariant under the symmetry, it can only be a phase transformation of ν_{eL}:

$$\nu_{eL} \to e^{i\varphi_\nu}\, \nu_{eL} \, . \tag{14.55}$$

The symmetry can be either discrete or continuous. In the latter case it coincides with the global U(1) transformation associated with lepton number conservation, which is violated by $2\beta_{0\nu}$-decay. Hence, only the discrete case with

$$\varphi_\nu \neq 0 \tag{14.56}$$

remains. Introducing the associated phase transformations

$$e \to e^{i\varphi_e}\, e \,, \quad u \to e^{i\varphi_u}\, u \,, \quad d \to e^{i\varphi_d}\, d \,, \quad W^\rho \to e^{i\varphi_W}\, W^\rho \,, \tag{14.57}$$

from the invariance of charged-current weak interactions we have

$$\left.\begin{array}{lll}\overline{\nu_{eL}}\,\gamma^{\rho}\,e_L\,W_{\rho} & \Rightarrow & \varphi_\nu - \varphi_e - \varphi_W = 0 \\ \overline{u_L}\,\gamma^{\rho}\,d_L\,W_{\rho} & \Rightarrow & \varphi_u - \varphi_d - \varphi_W = 0\end{array}\right\} \quad \Rightarrow \quad \varphi_\nu = \varphi_d - \varphi_u + \varphi_e\,. \qquad (14.58)$$

On the other hand, the existence of the $2\beta_{0\nu}$-decay process $2\,d \to 2\,u + 2\,e^-$ requires that

$$\varphi_d - \varphi_u + \varphi_e = 0\,. \qquad (14.59)$$

Equations (14.58) and (14.59) imply $\varphi_\nu = 0$, which is incompatible with eqn (14.56). Therefore, if $2\beta_{0\nu}$-decay exists, there is no symmetry which can forbid the existence of a Majorana mass term for the electron neutrino, which is generated at some level of perturbation theory by the diagram in Fig. 14.12b.

For the sake of clarity, let us emphasize that if the main contribution to $2\beta_{0\nu}$-decay is due to a mechanism different from the exchange of Majorana massive neutrinos with SM interactions illustrated in Fig. 14.11, the induced Majorana mass for the electron neutrino could be very small. It is certainly much smaller than the value of $m_{2\beta}$ which would generate the same decay, since the amplitude of the process in Fig. 14.11 is suppressed by a factor G_F^2 with respect to the $2\beta_{0\nu}$-decay amplitude corresponding to the black-box.

It is also interesting to ask if the absence of $2\beta_{0\nu}$-decay in the case of Dirac massive neutrinos is consistent with the equivalence of a Dirac neutrino with two degenerate Majorana neutrinos discussed in section 6.4.4. Considering for simplicity only one generation, according to eqn (6.298) the electron neutrino field can be written as $\nu_e = (i\nu_1 + \nu_2)/\sqrt{2}$, where ν_1 and ν_2 are Majorana fields with the same mass $m_1 = m_2 = m_{\nu_e}^{\mathrm{D}}$, with $m_{\nu_e}^{\mathrm{D}}$ denoting the Dirac mass of ν_e. Then, the $2\beta_{0\nu}$-decay amplitude is described by the diagram in Fig. 14.11 with $U_{e1} = i$ and $U_{e2} = 1$. It follows that $m_{2\beta} = U_{e1}^2\,m_1 + U_{e2}^2\,m_2 = 0$, leading to the absence of $2\beta_{0\nu}$-decay in the Dirac case [1067]. Since the factors i and unity in front of ν_1 and ν_2 reflect the opposite CP parities of ν_1 and ν_2 (see section 6.4.4)), it is usually said that the contributions of the two Majorana neutrinos to $m_{2\beta}$ cancel because of their opposite CP parities.

14.3.3 Phenomenology

The $2\beta_{0\nu}$ half-life of a nucleus $\mathcal{N}$ is given by (see Refs. [605, 377, 997])

$$[T_{1/2}^{0\nu}(\mathcal{N})]^{-1} = G_{0\nu}^{\mathcal{N}}\,|\mathcal{M}_{0\nu}^{\mathcal{N}}|^2\,\frac{|m_{2\beta}|^2}{m_e^2}\,, \qquad (14.60)$$

where $\mathcal{M}_{0\nu}^{\mathcal{N}}$ and $G_{0\nu}^{\mathcal{N}}$ are, respectively, the nuclear matrix element and the phase space factor. The phase space factor can be calculated with small uncertainties (see, for example, Table 3.4 of Ref. [377] and Table 6 of Ref. [997]). For ^{76}Ge, we have [997]

$$G_{0\nu}^{^{76}\mathrm{Ge}} = 6.31 \times 10^{-15}\,\mathrm{y}^{-1}\,. \qquad (14.61)$$

On the other hand, the calculation of the nuclear matrix element $\mathcal{M}_{0\nu}^{\mathcal{N}}$ is very difficult, because it requires an accurate nuclear model (see Refs. [997, 405]). Table 14.2

TABLE 14.2. Realistic calculations of $|\mathcal{M}_{0\nu}^{^{76}\mathrm{Ge}}|$ (from Table 2 of Ref. [405]). The values of $|m_{2\beta}|$ correspond to $T_{1/2}^{0\nu}(^{76}\mathrm{Ge}) = 1.9 \times 10^{25}\,\mathrm{y}$, which is the current 90% CL lower bound (eqn (14.42)).

$\lvert\mathcal{M}_{0\nu}^{^{76}\mathrm{Ge}}\rvert$	$\lvert m_{2\beta}\rvert$ [eV]	Method	Reference
4.59	0.32	QRPA	[127]
4.32	0.34	QRPA	[1018]
4.21	0.35	QRPA	[820, 980]
4.04	0.37	RQRPA	[965]
3.64	0.41	QRPA	[321]
3.62	0.41	RQRPA	[426]
3.62	0.41	QRPA	[162]
3.45	0.43	Number-projected QRPA	[998]
3.41	0.43	QRPA	[846]
3.32	0.44	QRPA	[998]
3.21 – 3.82	0.39 – 0.46	Second QRPA	[987]
3.13	0.47	RQRPA with forbidden	[965]
2.95 – 3.16	0.47 – 0.50	RQRPA	[252]
2.68	0.55	QRPA with forbidden	[908]
2.40	0.62	RQRPA with forbidden	[908]
2.31 – 3.68	0.40 – 0.64	Full RQRPA	[987]
2.09	0.71	Full RQRPA	[966]
1.87 – 3.74	0.39 – 0.79	RQRPA	[987]
1.74	0.85	Large-scale shell model	[308]
1.71 – 4.45	0.33 – 0.86	QRPA	[987]
1.69 – 1.87	0.79 – 0.87	QRPA	[252]
1.50	0.98	QRPA with np pairing	[846]

shows a list of realistic calculations of the matrix element for the $^{76}\mathrm{Ge} \to {}^{76}\mathrm{Se}$ transition, $|\mathcal{M}_{0\nu}^{^{76}\mathrm{Ge}}|$, taken from Table 2 of Ref. [405]. One can see that the value of $|\mathcal{M}_{0\nu}^{^{76}\mathrm{Ge}}|$ obtained in these calculations lies in the range

$$1.5 \lesssim |\mathcal{M}_{0\nu}^{^{76}\mathrm{Ge}}| \lesssim 4.6\,. \tag{14.62}$$

Hence, there is a systematic uncertainty of about a factor of three for the extraction of the value of $|m_{2\beta}|$ from a measurement of $T_{1/2}^{0\nu}(^{76}\mathrm{Ge})$: from the bound in eqn (14.42), we obtain (see the $|m_{2\beta}|$ column in Table 14.2)

$$|m_{2\beta}| \lesssim 0.3 - 1.0\,\mathrm{eV}\,. \tag{14.63}$$

It is clear that in order to improve our knowledge of $|m_{2\beta}|$ an improvement in the reliability of the calculation of the nuclear matrix element would be very beneficial. On the other hand, let us emphasize that a positive signal of $2\beta_{0\nu}$-decay would prove the Majorana nature of neutrinos, regardless of the value of the nuclear matrix element. Measurements of $2\beta_{0\nu}$-decays of different nuclei could help to reduce the nuclear matrix element uncertainty [232, 240, 241].

As we have seen in section 13.3.3, the results of neutrino oscillation experiments gave us detailed information on the values of the mixing parameters in three-neutrino mixing. However, it is still unknown whether the neutrino mass scheme is normal or inverted (see Fig. 13.1) and which is the absolute value of the neutrino mass scale, except for the upper bound from tritium β-decay in eqn (14.30) (see also Fig. 14.3). In the following we discuss separately the implications of the results of neutrino oscillation experiments for $|m_{2\beta}|$ in the normal and inverted schemes.

14.3.3.1 Normal scheme In the normal scheme, since $\Delta m^2_{\text{SOL}} \ll \Delta m^2_{\text{ATM}}$, we have

$$m_{2\beta} \simeq |U_{e1}|^2 \, m_1 + e^{i\alpha_2} \, |U_{e2}|^2 \, \sqrt{m_1^2 + \Delta m^2_{\text{SOL}}} + e^{i\alpha_3} \, |U_{e3}|^2 \, \sqrt{m_1^2 + \Delta m^2_{\text{ATM}}} \, . \tag{14.64}$$

Figure 14.13 shows the behavior of $|m_{2\beta}|$ as a function of m_1 in the four CP-conserving cases in eqns (14.50)–(14.53).

In the quasidegenerate region ($m_1 \gg \sqrt{\Delta m^2_{\text{ATM}}} \simeq 5 \times 10^{-2}\,\text{eV}$), we have

$$m_{2\beta} \simeq m_1 \left(|U_{e1}|^2 + |U_{e2}|^2 \, e^{i\alpha_2} \right) , \tag{14.65}$$

since $|U_{e3}|^2 \ll |U_{e1}|^2, |U_{e2}|^2$. Hence, for the four CP-conserving cases we obtain:

$$(++)\,,\,(+-)\,: \quad m_{2\beta} \simeq m_1\,, \tag{14.66}$$

$$(--)\,,\,(-+)\,: \quad m_{2\beta} \simeq m_1 \left(|U_{e1}|^2 - |U_{e2}|^2 \right) \simeq m_1 \, \cos 2\vartheta_{12}\,. \tag{14.67}$$

This is in agreement with Fig. 14.13, which shows that in the quasidegenerate region $|m_{2\beta}|$ practically coincides with m_1 in the cases $(++)$ and $(+-)$ and is smaller than m_1 in the cases $(--)$ and $(-+)$, with an uncertainty due to the uncertainty of $\sin^2\vartheta_{12}$. Note that, for a given value of m_1 in the quasidegenerate region, $|m_{2\beta}|$ is bounded from below because the solar mixing angle is not maximal (see eqn (10.78)).

In the hierarchical region ($m_1 \ll m_2 \ll m_3$), we have

$$m_2 \simeq \sqrt{\Delta m^2_{\text{SOL}}}\,, \qquad m_3 \simeq \sqrt{\Delta m^2_{\text{ATM}}}\,. \tag{14.68}$$

In order to understand the behavior of $|m_{2\beta}|$ as a function of m_1, let us consider the best-fit values of the oscillation parameters in Table 13.2. Since the best-fit value of $|U_{e3}|$ is zero, we have

$$|m_{2\beta}|_{\text{bf}} \simeq \left| \cos^2\vartheta_{12}^{\text{bf}} \, m_1 + e^{i\alpha_2} \, \sin^2\vartheta_{12}^{\text{bf}} \, \sqrt{(\Delta m^2_{\text{SOL}})_{\text{bf}}} \right| \, . \tag{14.69}$$

Then, it is possible to have a vanishing $|m_{2\beta}|_{\text{bf}}$ in the CP-conserving cases $(--)$ and $(-+)$ if

$$m_1 \simeq \tan^2\vartheta_{12}^{\text{bf}} \, \sqrt{(\Delta m^2_{\text{SOL}})_{\text{bf}}} \simeq 4 \times 10^{-3}\,\text{eV}\,, \tag{14.70}$$

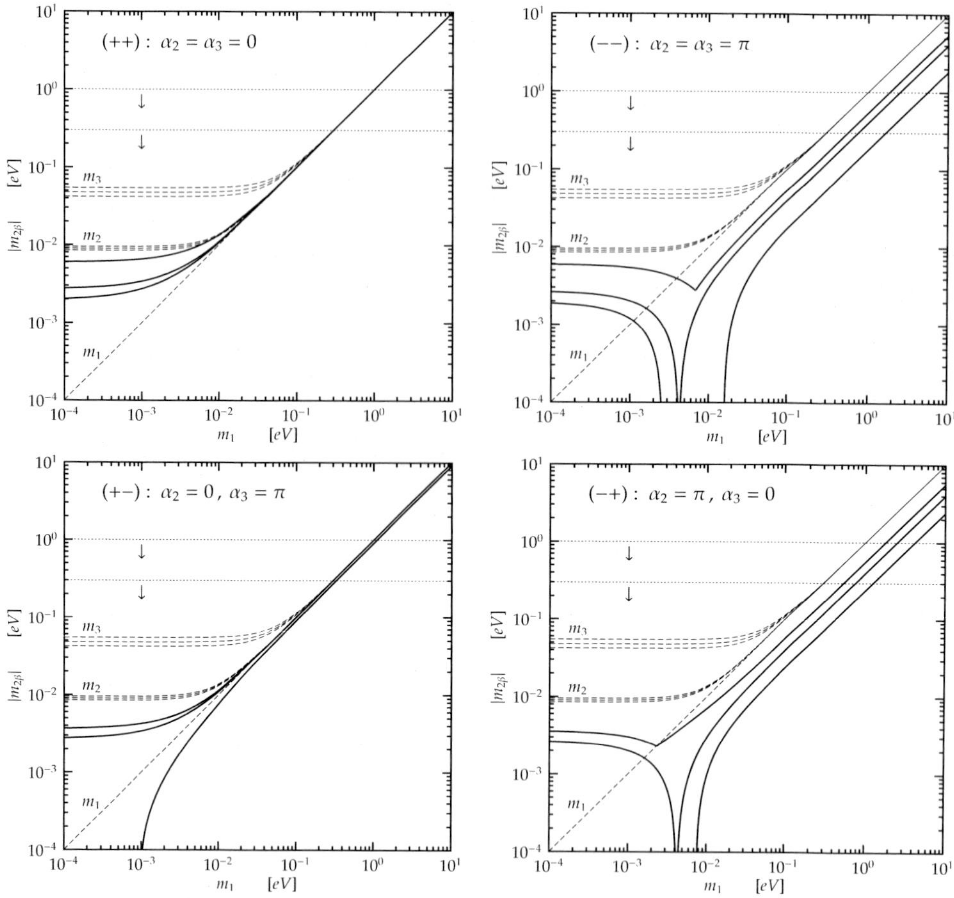

FIG. 14.13. Absolute value $|m_{2\beta}|$ of the effective Majorana neutrino mass in $2\beta_{0\nu}$-decay as a function of the lightest mass m_1 in the normal scheme in the four CP-conserving cases in eqns (14.50)–(14.53). Middle solid lines correspond to the best-fit values in Table 13.2. Extreme solid lines enclose 3σ ranges. Dashed lines show the best-fit values and 3σ ranges of individual masses. The two horizontal dotted lines correspond to the extremes of the upper bound range in eqn (14.63).

as shown in Fig. 14.13. For smaller values of m_1, we can approximate

$$|m_{2\beta}| \simeq \left| |U_{e2}|^2 \sqrt{\Delta m^2_{\rm SOL}} + e^{i(\alpha_3-\alpha_2)} \, |U_{e3}|^2 \sqrt{\Delta m^2_{\rm ATM}} \right| . \tag{14.71}$$

Hence, the value of $|m_{2\beta}|$ corresponding to the best-fit values of the oscillation parameters in Table 13.2 is

$$|m_{2\beta}|_{\rm bf} \simeq \sin^2 \vartheta_{12}^{\rm bf} \sqrt{(\Delta m^2_{\rm SOL})_{\rm bf}} \simeq 2.7 \times 10^{-3}\,{\rm eV} \tag{14.72}$$

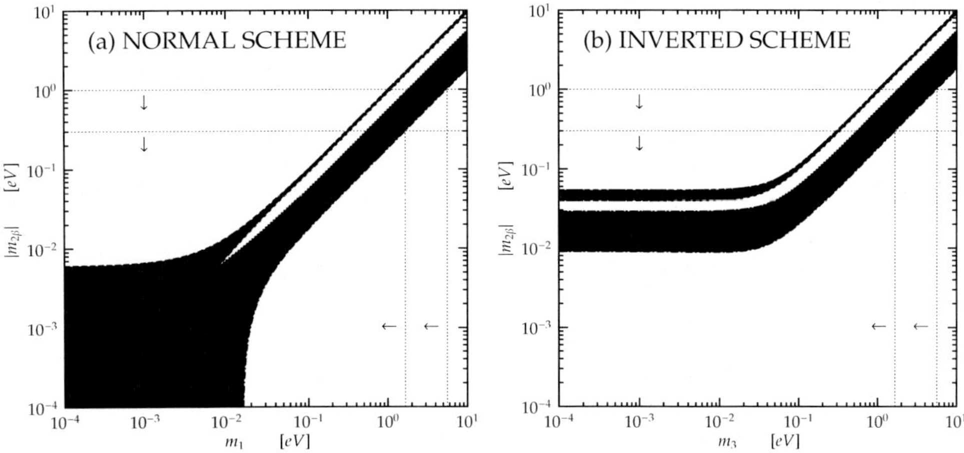

FIG. 14.14. Absolute value $|m_{2\beta}|$ of the effective Majorana neutrino mass in $2\beta_{0\nu}$-decay as a function of the lightest mass m_1 in the normal scheme (a) and m_3 in the inverted scheme (b). The shadowed region is the union of the 3σ regions in the four CP-conserving cases in eqns (14.50)–(14.53) (Fig. 14.13 in the normal scheme and Fig. 14.15 in the inverted scheme). The two horizontal dotted lines correspond to the extremes of the upper bound range in eqn (14.63). The two vertical dotted lines show the corresponding upper bounds for m_1 (a) and m_3 (b).

in all cases, irrespective of CP invariance. For the lower bound on $|m_{2\beta}|$, as one can see from Fig. 14.13, there can be a cancellation between the contributions of $\sqrt{\Delta m^2_{\rm SOL}}$ and $\sqrt{\Delta m^2_{\rm ATM}}$ in the CP-conserving cases $(+-)$ and $(-+)$, where $e^{i(\alpha_3-\alpha_2)} = -1$. This is due to the fact that there is an overlap of the allowed ranges of $|U_{e2}|^2 \sqrt{\Delta m^2_{\rm SOL}}$ and $|U_{e3}|^2 \sqrt{\Delta m^2_{\rm ATM}}$: from the 3σ ranges in Table 13.2 we have

$$2 \times 10^{-3} \lesssim |U_{e2}|^2 \sqrt{\Delta m^2_{\rm SOL}} \lesssim 4 \times 10^{-3}\,{\rm eV}\,, \qquad |U_{e3}|^2 \sqrt{\Delta m^2_{\rm ATM}} \lesssim 3 \times 10^{-3}\,{\rm eV}\,. \tag{14.73}$$

Note, however, that in the absence of a fine-tuned cancellation between the two contributions, the contribution due to $\sqrt{\Delta m^2_{\rm SOL}}$ may be dominant [521], with a lower bound $|m_{2\beta}| \gtrsim 2 \times 10^{-3}\,$eV.

The union of the allowed regions of $|m_{2\beta}|$ as a function of m_1 in the four CP-conserving cases is shown by the shadowed region in Fig. 14.14a. One can note that there is an unshaded strip within the shadowed band in Fig. 14.14a. Since for each value of m_1 the allowed interval for $|m_{2\beta}|$ is connected and the extreme values are obtained in the case of CP conservation, the band in Fig. 14.14a includes the possibility of CP violation and the unshaded strip corresponds to values of $|m_{2\beta}|$ which can be obtained only in the case of CP violation.

It is clear from Fig. 14.14a that in the hierarchical region, $|m_{2\beta}|$ is not bounded from below, but it is bounded from above by

$$|m_{2\beta}| \lesssim 6 \times 10^{-3}\,\text{eV} \quad \text{for} \quad m_1 \lesssim 10^{-3}\,\text{eV}\,. \tag{14.74}$$

Figure 14.14a shows also that the current upper bound range for $|m_{2\beta}|$ in eqn (14.63) probes the quasidegenerate region, with the corresponding upper bound range

$$m_k \lesssim 1.6 - 5.6\,\text{eV}\,, \tag{14.75}$$

for $k = 1, 2, 3$.

14.3.3.2 Inverted scheme In the inverted scheme, since $m_1 \simeq m_2 \simeq \sqrt{m_3^2 + \Delta m^2_{\text{ATM}}}$, we have

$$m_{2\beta} \simeq \left(|U_{e1}|^2 + e^{i\alpha_2}\,|U_{e2}|^2\right) \sqrt{m_3^2 + \Delta m^2_{\text{ATM}}} + e^{i\alpha_3}\,|U_{e3}|^2\, m_3\,. \tag{14.76}$$

It is important to note that there cannot be a complete cancellation between $|U_{e1}|^2$ and $|U_{e2}|^2$, even if $\alpha_2 = \pi$, because the solar mixing angle is not maximal (see eqn (10.78)) and there is no overlap between the allowed ranges of $|U_{e1}|^2$ and $|U_{e2}|^2$ at several σ's (see the 2σ ranges of $|U_{e1}|$ and $|U_{e2}|$ in eqn (13.134)).

Figure 14.15 shows the behavior of $|m_{2\beta}|$ as a function of m_3 in the four CP-conserving cases in eqns (14.50)–(14.53). One can see that, since $|U_{e3}|^2$ is small, $|m_{2\beta}|$ is practically the same in the cases $(++)$ and $(+-)$, as well as in the cases $(--)$ and $(-+)$.

In the quasidegenerate region ($m_3 \gg \sqrt{\Delta m^2_{\text{ATM}}} \simeq 5 \times 10^{-2}\,\text{eV}$), the behavior of $m_{2\beta}$ is similar to that in the normal scheme, given in eqns (14.65)–(14.67), with m_1 replaced by m_3.

In the inverted-hierarchy region ($m_3 \ll m_1 \simeq m_2$), since m_3 is very small, there cannot be a cancellation between $\left(|U_{e1}|^2 + e^{i\alpha_2}\,|U_{e2}|^2\right) \sqrt{m_3^2 + \Delta m^2_{\text{ATM}}}$ and $|U_{e3}|^2\, m_3$, even if $\alpha_3 = \pi$. Therefore, as one can see from Fig. 14.15, $|m_{2\beta}|$ is bounded from below. Neglecting m_3, we have

$$m_{2\beta} \simeq \left(|U_{e1}|^2 + e^{i\alpha_2}\,|U_{e2}|^2\right) \sqrt{\Delta m^2_{\text{ATM}}}\,. \tag{14.77}$$

Thus, in the four CP-conserving cases we can approximate

$$(++)\,,\,(+-)\,: \quad m_{2\beta} \simeq \sqrt{\Delta m^2_{\text{ATM}}}\,, \tag{14.78}$$

$$(--)\,,\,(-+)\,: \quad m_{2\beta} \simeq \sqrt{\Delta m^2_{\text{ATM}}}\left(|U_{e1}|^2 - |U_{e2}|^2\right) \simeq \sqrt{\Delta m^2_{\text{ATM}}}\,\cos 2\vartheta_{12}\,. \tag{14.79}$$

In fact, from Fig. 14.15 one can see that in the cases $(++)$ and $(+-)$ $m_{2\beta}$ practically coincides with $m_1 \simeq m_2$, and in the cases $(--)$ and $(-+)$ it is smaller, with an uncertainty due to the uncertainty of $\sin^2\vartheta_{12}$. In any case, $|m_{2\beta}|$ is bounded from below.

The shadowed region in Fig. 14.14b shows the union of the allowed regions of $|m_{2\beta}|$ as a function of m_3 in the four CP-conserving cases. As in the case of a

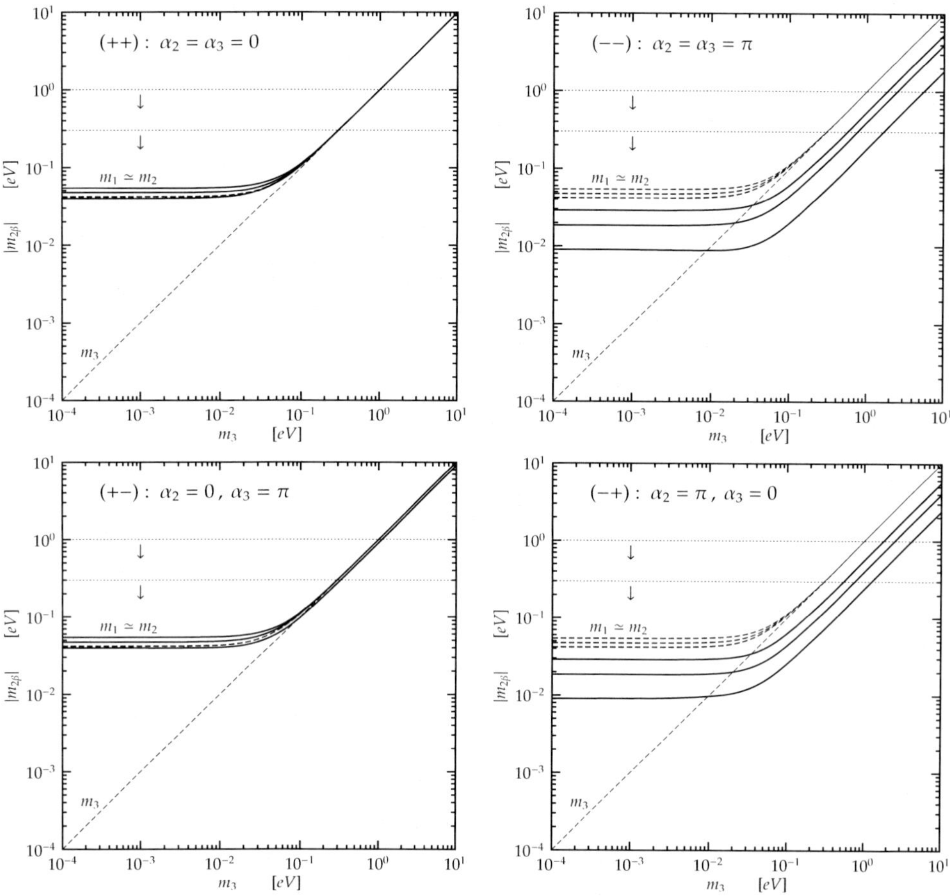

FIG. 14.15. Absolute value $|m_{2\beta}|$ of the effective Majorana neutrino mass in $2\beta_{0\nu}$-decay as a function of the lightest mass m_3 in the inverted scheme in the four CP-conserving cases in eqns (14.50)–(14.53). Middle solid lines correspond to the best-fit values in Table 13.2. Extreme solid lines enclose 3σ ranges. Dashed lines show the best-fit values and 3σ ranges of individual masses. The two horizontal dotted lines correspond to the extremes of the upper bound range in eqn (14.63). In the cases (++) and (+−), the best-fit values and 3σ ranges of m_1 and m_2 are practically the same and coincide with the best-fit value and 3σ range of $m_{2\beta}$.

normal hierarchy (Fig. 14.14a), the band in Fig. 14.14b includes the possibility of CP violation and the unshaded strip corresponds to values of $|m_{2\beta}|$ which can be obtained only in the case of CP violation.

Figure 14.14b shows that in the inverted-hierarchy region $|m_{2\beta}|$ is bounded from below and above:

$$9 \times 10^{-3} \lesssim |m_{2\beta}| \lesssim 5 \times 10^{-2}\,\text{eV} \quad \text{for} \quad m_3 \lesssim 10^{-2}\,\text{eV}\,. \tag{14.80}$$

Comparing this interval with the upper bound in eqn (14.74) in the case of a normal hierarchy, or comparing Figs. 14.14a and 14.14b, one can see that the allowed ranges for $|m_{2\beta}|$ in the cases of a normal or inverted hierarchy do not overlap. Thus, future experiments may be able to distinguish the two hierarchies. In particular, if future experiments do not observe $2\beta_{0\nu}$-decay with a sensitivity below $|m_{2\beta}| \simeq 9 \times 10^{-3}$, the inverted scheme will be excluded.

On the other hand, the normal and inverted schemes cannot be distinguished in the quasidegenerate region, which is probed by the current upper bound range for $|m_{2\beta}|$ in eqn (14.63). Comparing Figs. 14.14a and 14.14b, one can see that the resulting upper bound range for the neutrino masses in the inverted scheme is the same as that in the normal scheme, which is given in eqn (14.75).

15

SUPERNOVA NEUTRINOS

> There are two ways to slide easily through life; to believe everything or to doubt everything. Both ways save us from thinking.
> **A. Korzybski**

Supernovae (SNe) are extremely powerful explosions which terminate the life of some stars. Typically, some solar masses are ejected in the interstellar space with a kinetic energy of the order of 10^{51} erg. The ejecta contain heavy elements that are important for the chemical evolution of galaxies, stars, planets, and life. Some SNe produce a compact remnant, a neutron star or a black-hole, which may be observed.

Historically, after the realization in the 1920s that if the spiral nebulae are extragalactic star systems some of the novae observed in these spiral nebulae have been extremely bright (see Ref. [1025]), the study of SNe was initiated by W. Baade and F. Zwicky in the early 1930s [134]. They already suggested that the source of the enormous quantity of energy released in SNe is the gravitational collapse of a star to a neutron star and that SNe may be sources of cosmic rays. In the following years Zwicky and others started to organize systematic searches of SNe which have led to the present knowledge of thousands of SNe (see, for example, Ref. [842]). In the early 1960s Zwicky and collaborators suggested the currently adopted naming scheme of SNe: each supernova is designated with the SN prefix followed by the year of discovery, which is followed, in years with more than one supernova, by an upper-case letter A through Z for the first 26 SNe discovered in a given year or the lower-case letters aa, ab, ... for the following SNe discovered in the same year.

Some SNe that have exploded in our galaxy have been observed with the naked eye during the last 2000 years (see Ref. [561]). The most famous is the 1054 supernova, which produced the Crab nebula and the Crab pulsar (see Ref. [877]). The 1006 supernova is the brightest SN of all times. The last galactic SNe were observed with the naked eye in 1572 (Tycho Brahe) and 1604 (Joannes Kepler). In the last few centuries many SNe occurring in other galaxies have been observed with telescopes, since their light emission is comparable to that of an entire galaxy. Supernova SN1987A, which occurred on 23 February 1987 in the Large Magellanic Cloud, is the best studied of all SNe (see Refs. [996, 770, 850, 267] and references therein) and it is the only one which was detected also through its neutrino burst (see Refs. [1027, 695, 1058]). As we will see in the following, this first historical observation of neutrinos produced outside of the solar system (and even outside of our galaxy) is important not only for the study of SN dynamics, but also for the study of neutrino properties (in particular, neutrino masses).

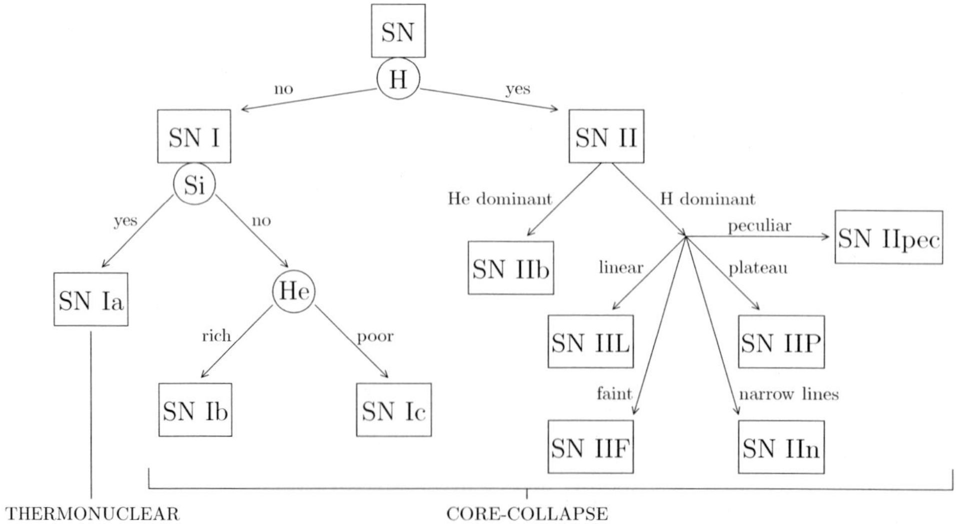

FIG. 15.1. The classification scheme of supernovae.

Introductions and reviews of the physics of SNe are given in Refs. [1025, 1026, 1071, 887, 1058, 638, 696, 1070]. Future perspectives for the detection of SN neutrinos have been reviewed in Refs. [309, 949]. This chapter is a revised and updated version of section 8 of Ref. [230].

15.1 Supernova types

For historical reasons, SNe are divided into different types characterized by their spectroscopic characteristics near maximum luminosity and by the properties of the light curve, which depend on the composition of the envelope of the SN progenitor star (see Fig. 15.1). The two wide categories called type I and type II are characterized by the absence or presence of hydrogen lines. However, the most important physical characteristic is the mechanism that generates the supernova, which distinguishes SNe of type Ia from SNe of type Ib, Ic and II, as shown in Fig. 15.1. This difference becomes noticeable in the light spectrum some months after maximum luminosity, when the ejecta become optically thin and the innermost regions become visible: the spectrum of SNe Ia is dominated by Fe emission lines, whereas SNe Ib, Ic, and II show O and C emission lines.

Typically, the optical emission of both type I and II SNe start with a rise in luminosity during the first week or two due to the expansion of the luminous surface. Type I SNe have typically a narrow luminosity peak, whereas type II have broad peaks, of the order of 100 days. After the peak, the luminosity decreases for about one year.

Type Ia SNe are thought to be generated by carbon–oxygen white dwarfs that have a close companion star from which the white dwarf can accrete mass. White dwarfs are the evolutionary products of stars which ended thermonuclear fuel burning. Weighting about one solar mass, they have a radius of about 5000 km and a density of the order of $10^6\,\mathrm{g\,cm^{-3}}$. The pressure of degenerate electrons support white dwarfs against the inward pull of gravity (see, for example, Ref. [952]). In 1931 Chandrasekhar discovered that white dwarfs have a maximum mass of about $1.4\,M_\odot$, above which the star collapses (see Ref. [311]).

In 1932 Landau [712] presented the following qualitative derivation[81] of the Chandrasekhar limit (see Ref. [952]), which can be applied also to the calculation of the limit of stability of neutron stars by replacing electrons with neutrons. Let us consider a white dwarf with N electrons and radius R. According to the Pauli exclusion principle, the volume per electron is $v \sim R^3/N$, which has a characteristic size $r \sim v^{1/3} \sim R/N^{1/3}$. Applying the Heisenberg uncertainty principle, the momentum of an electron is $p \sim r^{-1} \sim N^{1/3}/R$. The pressure that balances the inward pull of gravity is due to degenerate electrons. For small stellar masses these electrons are nonrelativistic and the star can find an equilibrium between pressure and gravity. For more massive stars, during contraction the central density becomes so high that the relativistic regime is reached. In this case, the electron mass can be neglected and the pressure energy is given by $E_P \simeq p \sim N^{1/3}/R$. Each electron is associated with a mass $m \simeq m_N/Y_e$, where m_N is the nucleon mass and Y_e is the electron fraction ($Y_e = N_p/(N_p + N_n)$, where N_p and N_n are, respectively, the numbers of protons and neutrons). Therefore, the gravitational energy per electron is $E_G \sim -G_\mathrm{N} Mm/R \sim -G_\mathrm{N} m_N^2 N Y_e^{-2}/R$, where $M \simeq mN \simeq m_N N/Y_e$ is the total mass. The equilibrium of the star is reached at a minimum of the total energy

$$E = E_P + E_G \sim \frac{N^{1/3}}{R} - \frac{G_\mathrm{N}\, m_N^2\, N\, Y_e^{-2}}{R}\,. \tag{15.1}$$

Since both terms scale as $1/R$, if the sign of E is negative, the total energy can be decreased without limit by decreasing R, leading to the collapse of the star. In order to have a stable configuration, the total energy must be nonnegative, yielding the upper bound

$$N < N_\mathrm{C} \sim \left(G_\mathrm{N}\, m_N^2\, Y_e^{-2}\right)^{-3/2} = \left(\frac{M_\mathrm{P}\, Y_e}{m_N}\right)^3 \simeq 2 \times 10^{57}\, Y_e^3\,. \tag{15.2}$$

The corresponding mass is the Chandrasekhar mass[82]:

$$M_\mathrm{C} \sim N_\mathrm{C}\, m_N\, Y_e^{-1} \sim 2 \times 10^{57}\, Y_e^2\, \mathrm{GeV} \simeq 2\, Y_e^2\, M_\odot\,. \tag{15.3}$$

A more accurate calculation yields (see Refs. [311, 952])

$$M_\mathrm{C} \simeq 5.83\, Y_e^2\, M_\odot\,. \tag{15.4}$$

For a typical value of $Y_e \simeq 0.5$, we have $M_\mathrm{C} \simeq 1.46 M_\odot$.

---

[81] We thank S.M. Bilenky for suggesting this instructive derivation.

[82] From the value of the solar mass in Table 10.1 and the appropriate conversion factor in Table A.1, we have $M_\odot \simeq 1.1 \times 10^{57}$ GeV.

When the mass of a white dwarf reaches the Chandrasekhar limit, the star becomes unstable, because the pressure of the degenerate electron gas can no longer sustain the gravitational weight. The white dwarf begins to collapse, triggering the fusion of carbon and oxygen to heavy nuclei, which liberates an enormous quantity of energy, causing the thermonuclear explosion of the star (see Ref. [1072]). This explosion disrupts the progenitor white dwarf and generates an expanding nebula without a central compact object.

The light emission of type Ia SNe is mainly due to the decay of ^{56}Ni, which is produced in great quantities during the collapse of the outer layer of the white dwarf core. It contains silicon absorption lines, because Si is a product of C and O fusion. The ^{56}Ni decays into ^{56}Co, which in turn decays into ^{56}Fe, whose emission lines dominate the spectrum after some months. Iron and other heavy elements are ejected in interstellar space by the explosion.

Since SNe of type Ia are all generated under similar physical circumstances, they have almost identical characteristics, the most important being the amount of released energy and the *light curve*, i.e. they release almost the same total light and the visible light decays at the same rate. This fact has been verified by the observation of many type Ia SNe in nearby galaxies of known distance.

An empirical relation between the duration of the peak phase of the light curve and the luminosity of type Ia SNe has been discovered by Phillips in 1993 [872] from the catalog of observed type Ia SNe in nearby galaxies with known distance. This width-luminosity relation (*broader is brighter*) allows the use of type Ia SNe as standard candles for the measurement of the distance of galaxies as far as 100 Mpc or more (see Ref. [964] and references therein).

The observation by the Hubble Space Telescope of SNe of type Ia in galaxies at cosmological distances has been used for the measurement of the Hubble constant and the deceleration parameter. Contrary to expectations, it has been found that the expansion of the Universe is accelerating [903, 863] (see section 16.2.3). This surprising behavior can be explained in the framework of the Standard Cosmological Model through the presence of a relatively large energy density of the vacuum, dubbed as dark energy (see section 16.3 and Ref. [841]).

For the point of view of neutrino physics, type Ib, Ic, and II SNe are much more interesting than type Ia SNe, simply because they produce a huge flux of neutrinos of all types. These SNe are generated by the collapse of the core of massive stars ($M \gtrsim 8\, M_\odot$), which leaves a compact remnant. During the few seconds following the collapse, the compact remnant is very hot and neutrinos of all types are copiously produced. Since the remnant and the surrounding envelope are optically thick, about 99% of the gravitational binding energy liberated by the collapse (about 3×10^{53} erg) is carried away by neutrinos. The average energy of the emitted neutrinos is of the order of 10 MeV, and their number is about 10^{58}, one order of magnitude larger than the lepton number of the collapsed core.

Type II SNe are thought to be generated by the core collapse of red (or blue as SN1987A) giant stars with a mass between about 8–9 and 40–60 solar masses. Since the size and mass of the hydrogen envelope can be very different from star to star, even if they have the same initial mass, the visible effects of the supernova explosion have a wide range of variability, leading to a further classification of type

II SNe into the subtypes (see Refs. [300, 816]): IIL if the decrease of the luminosity is approximately linear in time; IIP if the time evolution of the luminosity shows a plateau; IIF if the SN is faint; IIb if helium dominates over hydrogen; IIn if the spectrum shows narrow line emissions; IIpec if the SN has peculiar characteristics. Subtypes determined by spectral properties are denoted by lower-case letters and subtypes determined by properties of the light curve are denoted by upper-case letters. It is believed that the SN is of type Ib if the exploding star does not have a hydrogen envelope. If the helium shell is also missing, the SN is of type Ic. All these classes are not clear-cut and intermediate cases exist.

It is also believed that, in a typical SN of type IIP, the envelope mass is about $5-10\,M_\odot$ and the radius is of the order of 10^{15} cm. The shock wave produced by the collapse of the core (see section 15.3) rapidly ionizes the hydrogen in the envelope and causes its violent ejection. The energy from the recombination of hydrogen keeps the photosphere, from which we receive light, at an almost constant radius and temperature, leading to the luminosity plateau. The time duration of the plateau depends on the envelope mass. If the envelope mass is too small ($\lesssim 1 - 2\,M_\odot$), there is no plateau, the decrease of the luminosity is approximately linear in time, and the SN is classified as type IIL. Most SNe of type IIL are brighter than SNe of type IIP.

Supernova SN1987A was an extreme case of type IIP, since the luminosity increased for about 3 months after the collapse and the supernova was rather faint. Therefore, sometimes SN1987A is classified as IIP [300, 816], sometimes as IIF [842] and sometimes as IIpec [816]. Its faintness is very likely due to the compactness of the progenitor (a radius of about 10^{12} cm). In this case, much of the available energy is used in the expansion and the luminosity increases for some time because of the growing contribution of radioactive decay of heavy elements in inner shells, which become more visible as the envelope expands.

15.2 Supernova rates

A very important problem is how to estimate SN rates. Figure 15.2 shows the estimates of SN rates presented in Ref. [772]. One can see that the SN rates depend rather strongly on the galaxy type. No core-collapse SNe of type Ib/c and II have been observed in elliptical galaxies, which are very old and have little star formation that could produce short-lived massive stars ending their life with a core-collapse supernova explosion.

One of the most crucial questions for SN neutrino astronomy is the rate of core-collapse SNe in our galaxy, which could produce observable neutrino bursts with high statistics in neutrino telescopes. Estimates of the rate of core-collapse SNe in the Milky Way are summarized in Table 15.1. It is clear that the large uncertainty leaves the problem open to further study. The lack of observation of neutrinos from core-collapse SNe in our galaxy since the Baksan Underground Scintillator Telescope (see section 15.4.3) began observations in June 1980 imply an upper bound of 13 SNe per century in the Milky Way at 90% CL [62]. This measurement

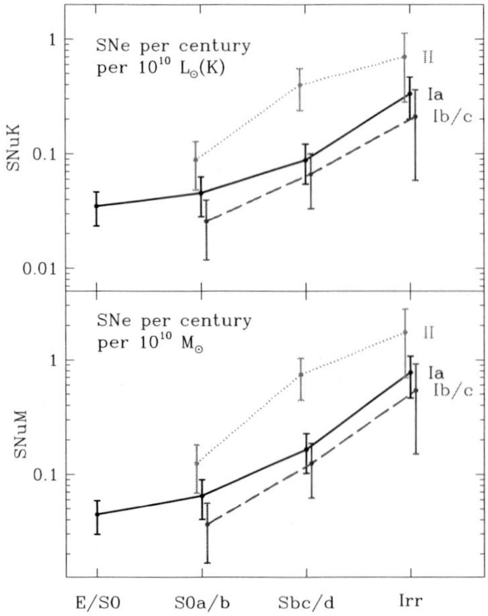

FIG. 15.2. SN rate as a function of the galactic morphological index [772]. Upper panel: SN rate expressed in SNuK (number of SNe per century per $10^{10} L_\odot$ of luminosity in the K band). Lower panel: SN rate expressed in SNuM (number of SNe per century per $10^{10}\, M_\odot$). The lines correspond to type Ia (solid), type II (dotted) and Ib/c (dashed), with 1σ error bars.

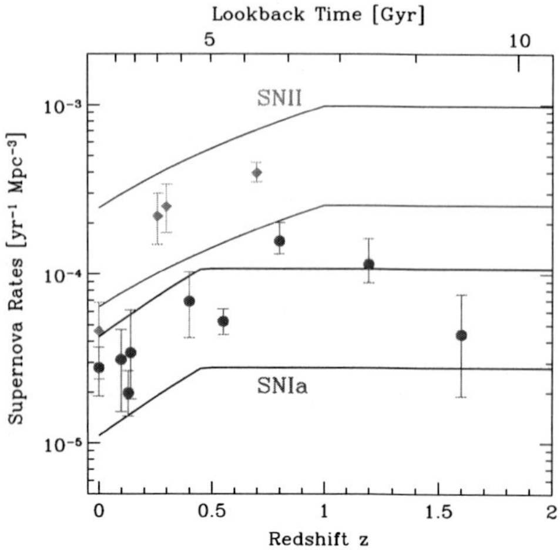

FIG. 15.3. The allowed bands for the rates of type Ia (SNIa) and core-collapse (SNII) supernovae as a functions of the redshift z [990]. The SNIa and SNII data are, respectively, from Refs. [347, 247] and Refs. [347, 299].

is consistent with the estimated rates and shows that the true rate cannot be much higher than the estimations.

The rate of core-collapse SNe is also important for the estimation of the *Relic Supernova Neutrino Background* (RSNB) (see Ref. [88]). Figure 15.3 shows the allowed bands found in Ref. [990] for the rates of type Ia and core-collapse supernovae as functions of the redshift z (defined in section 16.2.2). The Super-Kamiokande collaboration [768] searched for $\bar{\nu}_e$'s of the RSNB by means of the inverse neutron decay process in eqn (12.13), for neutrino energies above the end points of the ^{8}B and *hep* solar neutrino spectra (see Table 10.2). Since no signal was observed, the Super-Kamiokande data yielded the upper bound [768]

$$\Phi_{\bar{\nu}_e}(E > 19.3\,\text{MeV}) < 1.2\,\text{cm}^{-2}\,\text{s}^{-1} \qquad (90\%\ \text{CL})\,. \tag{15.5}$$

TABLE 15.1. Estimates of the rate of core-collapse SNe in the Milky Way.

Rate $[10^{-2}\,\mathrm{y}^{-1}]$	Reference
5.8 ± 2.4	Tammann (1982) [1006]
$1.2^{+1.7}_{-0.7}$	Ratnatunga & van den Bergh (1989) [890]
3^{+2}_{-1}	Strom (1990) [991]
4.0 ± 2.0	Muller et al. (1992) [818]
2.0 ± 1.1	Cappellaro et al. (1993) [297]
3.0 ± 1.5	van den Berg (1993) [1033]
$2.5^{+0.8}_{-0.5}$	Tammann et al. (1994) [1007]
5.7 ± 1.7	Strom (1994) [992]
1.3 ± 0.9	Cappellaro et al. (1997) [298]
3.4 ± 2.8	Timmes et al. (1997) [1016]
8.4 ± 2.8	Dragicevich et al. (1999) [389]
1.5 ± 1.0	Cappellaro & Turatto (2000) [300]
$1 - 2$	Reed (2005) [894]

15.3 Core-collapse supernova dynamics

Supernovae produced by the collapse of the core of massive stars produce large fluxes of neutrinos that could be detected on the Earth. Here we present a short description of the current standard theory of the dynamics of core-collapse SNe and the resulting neutrino flux (see Refs. [952, 289, 637, 754, 1071, 887, 1058, 638, 696, 1070] and references therein). As explained in section 15.1, core-collapse SNe are classified as of types II, Ib, or Ic depending on their spectroscopic characteristics at maximum luminosity. However, these characteristics depend only on the composition of the envelope, which plays no role in the collapse of the core and neutrino production. Hence, the following theory applies equally well to all types II, Ib, and Ic core-collapse SNe.

Core-collapse SNe are the final explosion of single stars with mass between about 8–9 and 40–60 solar masses. The explosion is due to the shock wave created when the core collapses to a proto-neutron star, which ejects the stellar envelope. Stars lighter than about 9 $M_\odot$ end their life as white dwarfs (but may explode as type Ia SNe if they belong to a multiple system). As illustrated in Figs. 15.4 and 15.5, stars heavier than about 40 $M_\odot$ can end their life in a supernova explosion if they have a sufficient initial metallicity [607], i.e. abundance of heavy elements (especially iron), which implies a larger photon opacity. During their life, these stars suffer significant mass losses through the stellar wind, leading to smaller envelopes which can explode when the core collapses to a proto-neutron star. The core of low and medium metallicity stars with masses between about 25 and 40 solar masses initially collapses to a proto-neutron star, generating a weak SN IIp, and later collapses to a black-hole because of the increase of mass of the proto-neutron star due to fallback of the envelope.

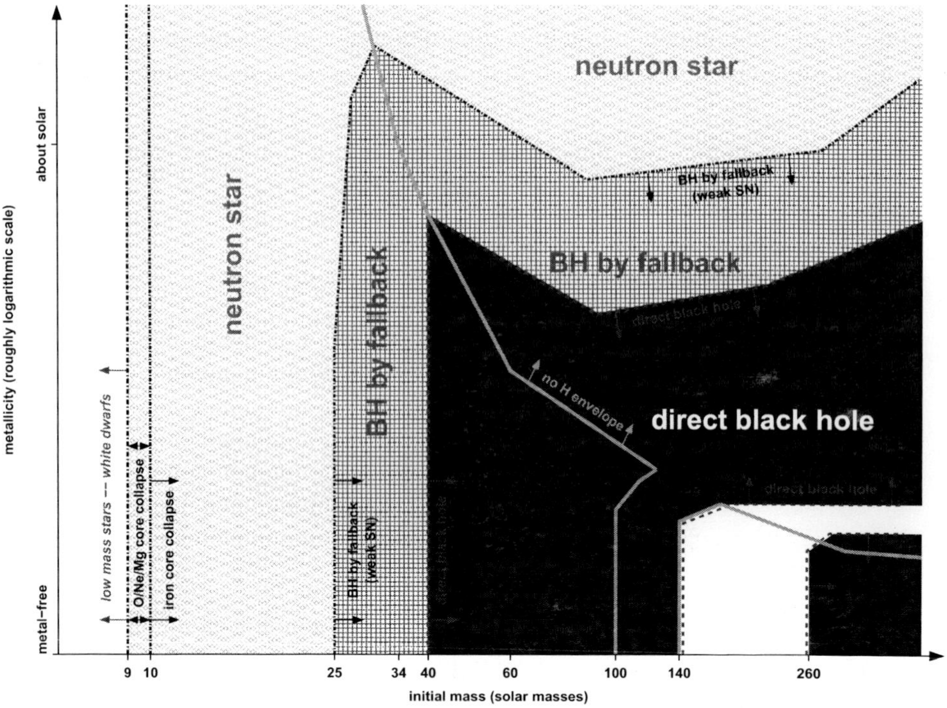

FIG. 15.4. Remnants of massive single stars as a function of initial mass (x axis) and initial metallicity (y axis) [607]. The line marked *no H envelope* separates the regimes where the stars keep their hydrogen envelope (left and lower right) from those where the hydrogen envelope is lost (upper right and small strip at the bottom between 100 and 140 $M_{\odot}$). The line marked *direct black-hole* indicates the border of the regime of direct black-hole formation (black). This domain is interrupted by a strip of pair-instability supernovae that leave no remnant (white). Outside the direct black-hole regime, at lower mass and higher metallicity, lies the regime of black-hole formation by fallback. Outside of this, lies the region corresponding to the formation of neutron stars. The lowest-mass neutron stars may be made by O/Ne/Mg core collapse instead of iron core collapse (vertical dash-dotted lines at the left). At even lower masses, the cores do not collapse and only white dwarfs are made (white strip at the very left).

Stars with mass in excess of about 10 solar masses are thought to undergo all the stages of nuclear fusion of hydrogen, helium, carbon, oxygen, neon, silicon (see Table 15.2), until the star has an onion-like structure shown in Fig. 15.6, with an iron core surrounded by shells composed of elements with decreasing atomic mass. At this point the iron core has a mass of about 1 solar mass, a radius of a few thousand km, a central density of about $10^{10}\,\mathrm{g\,cm^{-3}}$, a central temperature of about 1 MeV, and its weight is sustained by the pressure of degenerate relativistic

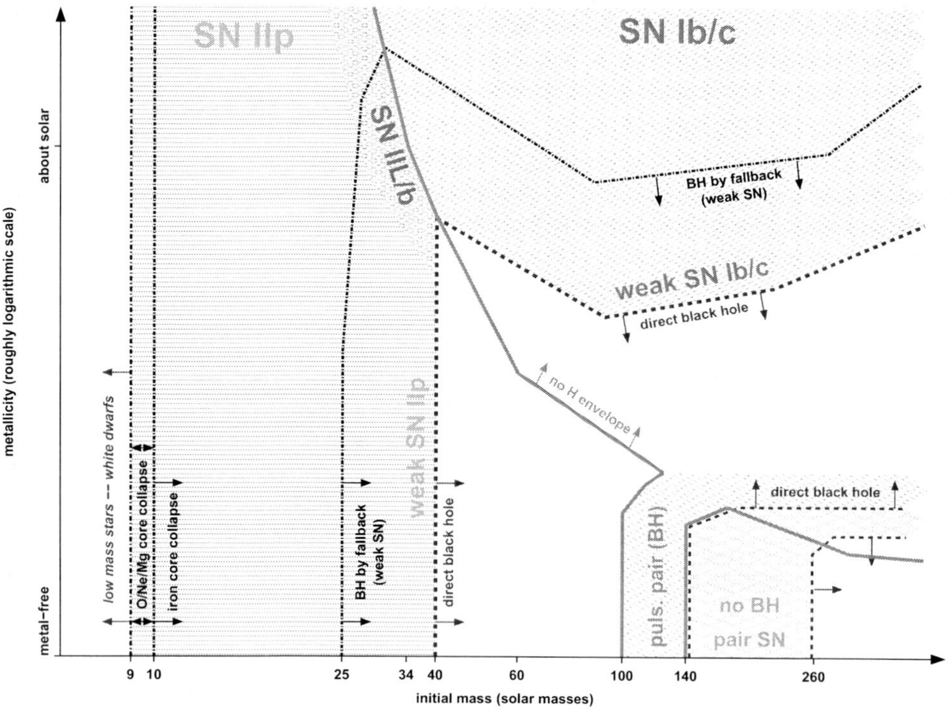

FIG. 15.5. Supernovae types of nonrotating massive single stars as a function of initial metallicity and initial mass [607]. The lines and axes have the same meaning as in Fig. 15.4.

electrons. Since iron is the most tightly bound nucleus, there remains no thermonuclear fuel to burn. The core contracts and the increased temperature causes photodissociation of iron through the process

$$\gamma + {}^{56}\mathrm{Fe} \to 13\,\alpha + 4\,n\,. \tag{15.6}$$

This reaction absorbs about 124 MeV of energy and reduces the kinetic energy and pressure of the electrons. Electron capture of nuclei,

$$e^- + \mathcal{N}(Z, A) \to \mathcal{N}(Z-1, A) + \nu_e\,, \tag{15.7}$$

and free protons,

$$e^- + p \to n + \nu_e\,, \tag{15.8}$$

favored by the high electron Fermi energy, additionally reduce the number and pressure of the electrons. At the onset of collapse, when the density of the iron core is not too high, the electron neutrinos produced by electron capture leave the core carrying away most of the kinetic energy of the captured electrons. The combined effect of iron photodissociation and electron capture lowers the electron pressure, decreasing the value of the Chandrasekhar mass, until the Chandrasekhar mass becomes

TABLE 15.2. Central temperature T_c, central density ρ_c, and time scale Δt of the evolutionary burning phases of Population I stars with initial masses $15\,M_\odot$ and $25\,M_\odot$ (values taken from Ref. [1071]). By comparison, the H-burning phase of a star with initial mass $1\,M_\odot$ lasts about 1.1×10^{10} y, with $T_c \simeq 1.57 \times 10^7$ K and $\rho_c \simeq 153\,\mathrm{g\,cm^{-2}}$. The thermonuclear life of such a star ends with the He-burning phase, which lasts about 1.1×10^8 y, with $T_c \simeq 1.25 \times 10^8$ K and $\rho_c \simeq 2.0 \times 10^4\,\mathrm{g\,cm^{-2}}$.

Phase	$15\,M_\odot$			$25\,M_\odot$		
	T_c [K]	ρ_c [g cm^{-3}]	Δt [yr]	T_c [K]	ρ_c [g cm^{-3}]	Δt [yr]
H	3.53×10^7	5.81	1.11×10^7	3.81×10^7	3.81	6.70×10^6
He	1.78×10^8	1.39×10^3	1.97×10^6	1.96×10^8	7.62×10^2	8.39×10^5
C	8.34×10^8	2.39×10^5	2.03×10^3	8.41×10^8	1.29×10^5	5.22×10^2
Ne	1.63×10^9	7.24×10^6	0.732	1.57×10^9	3.95×10^6	0.891
O	1.94×10^9	6.66×10^6	2.58	2.09×10^9	3.60×10^6	0.402
Si	3.34×10^9	4.26×10^7	5.01×10^{-2}	3.65×10^9	3.01×10^7	2.01×10^{-3}

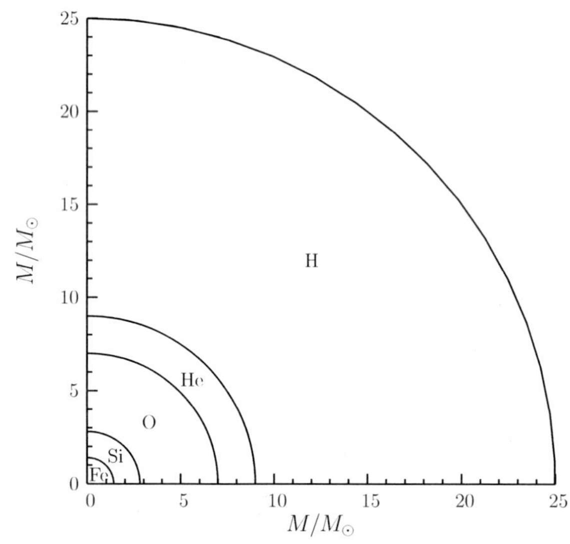

FIG. 15.6. Onion-like interior structure of a Population I star of $25\,M_\odot$ just before the onset of collapse (see Ref. [1025]). Fe represents assorted iron-peak elements: ^{48}Ca, ^{50}Ti, ^{54}Fe, ^{56}Fe, ^{58}Fe, ^{66}Ni. The Si shell contains less abundant amounts of S, O, Ar, Ca, the O shell contains less abundant amounts of Ne, C, Mg, Si, the He shell contains less abundant amounts of C, Ne, O, and the H shell contains less abundant amounts of He, Ne, O, N, C.

smaller than the core mass. At this moment the pressure of degenerate relativistic electrons can no longer sustain the weight of the core and the collapse commences. As the density and temperature increase, the processes in eqn (15.6)–(15.8) proceed faster, lowering further the electron pressure and favoring the collapse, which accelerates.

According to theory (see Figs. 15.4 and 15.5), stars with mass between about 9 and 10 solar masses burn hydrogen, helium, carbon, but the core does not get hot enough to burn oxygen. However, the core contains neon and magnesium at high density, which can undergo electron capture, reducing the electron pressure that sustains the core against gravity. As a result, the core collapses and, during the collapse, oxygen, neon and magnesium are converted to iron. Therefore, also in this case, the supernova explosion energy is produced by the collapse of an iron core.

The collapse of the core produces a neutron star and the huge liberated gravitational energy is released mainly as a flux of neutrinos, with a small fraction as electromagnetic radiation and kinetic energy of the ejecta, which form the visible explosion. The liberated gravitational energy is about $3 \times 10^{53}\,\mathrm{erg}$, of which only about 0.01% is transformed into electromagnetic radiation and about 1% is transformed into kinetic energy of the ejecta.

Let us examine in more detail the mechanism of formation of the neutron star, of neutrino production, and of supernova explosion.

The electron neutrinos produced by the electron capture processes in eqns (15.7) and (15.8) initially leave the core freely, carrying away energy and lepton number, since their mean free path is longer than the radius of the core. In this so-called *capture* phase electron neutrinos have a nonthermal spectrum and average energy that grows from about 12 to about 16 MeV (see Ref. [289]). The luminosity reaches about $10^{53}\,\mathrm{erg\,s^{-1}}$, but, in total, only about $10^{51}\,\mathrm{erg}$ are released before the core bounce, because the capture phase is too short (less than about 10 ms).

When the density of the inner part of the core (about $0.8\,M_{\odot}$) exceeds about $3 \times 10^{11}\,\mathrm{g\,cm^{-3}}$, neutrinos are trapped in the collapsing material leading to an adiabatic collapse with constant lepton number. During this stage, the inner part of the core collapses homologously, i.e. with subsonic velocity proportional to the radius. The outer part of the core collapses with supersonic free-fall velocity.

After about one second from the start of instability, the density of the inner core reaches the density of nuclear matter, about $10^{14}\,\mathrm{g\,cm^{-3}}$, and the pressure of degenerate nonrelativistic nucleons abruptly stops the collapse. The inner core settles into hydrostatic equilibrium, forming a proto-neutron star with a radius of about 10 km, while a supersonic shock wave caused by the halting and rebound of the inner core forms at its surface. The shock propagates outward through the outer iron core, which is still collapsing, with an initial velocity of the order of $100\,\mathrm{km\,msec^{-1}}$. The gas that is infalling at a velocity near free-fall is abruptly decelerated within the shock. Below the shock it falls much more slowly on the surface of the proto-neutron star, accreting it. Therefore, the proto-neutron star develops an unshocked core and a shocked mantle. The core has a radius of the order of 10 km with a density of the order of $10^{14}\,\mathrm{g\,cm^{-3}}$, as a nucleus. The mantle has a radius of about 100 km, with a density decreasing from the nuclear density of the core to about $10^9\,\mathrm{g\,cm^{-3}}$ at the surface of the proto-neutron star, where the density has a steep decrease of several orders of magnitude.

As the shock propagates through the infalling dense matter of the outer core, its energy is dissipated by the photodissociation of nuclei into protons and neutrons. Thus, the material behind the shock wave is mainly composed of free nucleons. Free protons have a high electron capture rate, leading to the transformation of most protons into neutrons, with huge production of electron neutrinos. These neutrinos pile up behind the shock, which is dense and opaque to them, until the shock reaches a zone with density about $10^{11}\,\mathrm{g\,cm^{-3}}$ (*shock breakout*) a few milliseconds after the bounce and the electron neutrinos behind the shock are released in a few milliseconds. This neutrino emission is usually called a *prompt electron neutrino burst* or *neutronization burst* or *breakout pulse*, to be distinguished from the thermal production of all neutrino flavors. The neutronization burst has a luminosity of about $6 \times 10^{51}\,\mathrm{erg\,s^{-1}}$ and carries away a few 10^{51} erg in a few milliseconds. In spite of his name, the neutronization burst is too short to carry away a significant part of the electron lepton number of the core, which remains trapped. Only the low-density periphery of the proto-neutron star is neutronized.

The energy loss due to photodissociation of nuclei and neutrino emission weakens the shock (about 1.5×10^{51} erg are dissipated for each 0.1 solar masses of photodissociated material). In the so-called *prompt* SN explosion scenario, the shock, although somewhat weakened, is able to expel the envelope of the star generating the SN explosion on a time scale of the order of 100 msec. If the star weighs more than about 10 solar masses, the shock is weakened and stalls about 100 ms after the bounce, at a radius of about 200–300 km, with insufficient energy to reach the outer layers of the star. Matter continues to fall through the stalled shock and be photodissociated. If too much matter lands on the proto-neutron star, the pressure of degenerate nucleons is not sufficient to maintain stability and the core collapses into a black-hole, presumably without a supernova explosion. The conditions that lead to a prompt supernova explosion, without a stalling shock, are controversial and are thought to depend on the mass of the progenitor star and on the equation of state of nuclear matter, which determines the energy transfered to the shock wave by the bounce.

If the shock stalls, a supernova explosion can be achieved only if the shock is revived by some mechanism that is able to renew its energy. The mechanism which is currently thought to be able to revive the shock is the energy deposition by the huge neutrino flux produced thermally in the proto-neutron star [326, 223]. It has recently been realized that the shock revival is helped by convection behind the shock, which can lead to an asymmetric explosion (see Ref. [1070]). Also acoustic power generated by oscillations of the accreting protoneutron star may help the shock revival [290]. If the shock is revived, a so-called *delayed* supernova explosion is produced on a time scale of the order of 0.5 s after the bounce.

Neutrinos of all flavors are produced in the hot core of the proto-neutron star (see Refs. [952]), which has a temperature of about 40 MeV, through electron–positron pair annihilation,

$$e^- + e^+ \to \nu + \bar{\nu}\,, \tag{15.9}$$

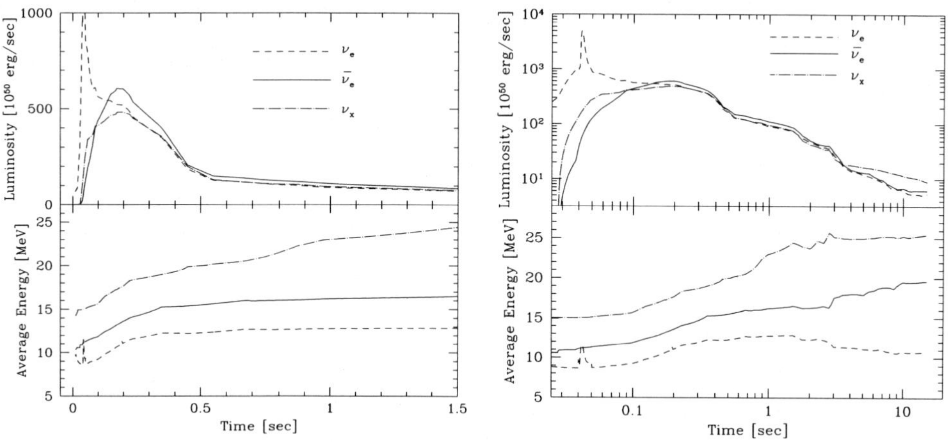

FIG. 15.7. Time evolution of neutrino luminosity and average energy obtained with the numerical SN model in Ref. [1021]. The time starts at the onset of collapse. The left panel shows the early phase in a linear time coordinate. The right panel shows the time evolution until about 10 s after the onset of collapse using a logarithmic time coordinate. The dashed lines are for ν_e, the solid lines for $\bar{\nu}_e$, and the dot-dashed lines for $\nu_\mu, \bar{\nu}_\mu, \nu_\tau, \bar{\nu}_\tau$, which are collectively denoted by ν_x. The neutronization burst is visible as a peak of luminosity and energy of electron neutrinos, which happens 40–50 msec after the onset of collapse.

electron–nucleon bremsstrahlung,

$$e^\pm + N \to e^\pm + N + \nu + \bar{\nu}\,, \tag{15.10}$$

nucleon–nucleon bremsstrahlung,

$$N + N \to N + N + \nu + \bar{\nu}\,, \tag{15.11}$$

plasmon decay

$$\gamma \to \nu + \bar{\nu}\,, \tag{15.12}$$

and photoannihilation

$$\gamma + e^\pm \to e^\pm + \nu + \bar{\nu}\,. \tag{15.13}$$

Electron neutrinos are also produced by the electron capture process in eqn (15.8), and electron antineutrinos are produced by positron capture on neutrons ($e^+ + n \to p + \bar{\nu}_e$). In spite of their weak interactions, these neutrinos are trapped in a SN core because of the very high matter density. Neutrinos can free-stream out of the mantle of a proto-neutron star only at a distance from the center where the matter density is low enough (of the order of $10^{11}\,\mathrm{g\,cm^{-3}}$) so that the neutrino mean free path is larger than radius of the core. The sphere from which neutrinos stream out freely is called the *neutrinosphere*, and it lies within the mantle of the proto-neutron star. Since neutrino interactions depend on flavor and energy, there are different energy-dependent neutrinospheres for different flavor neutrinos. More precisely, since the

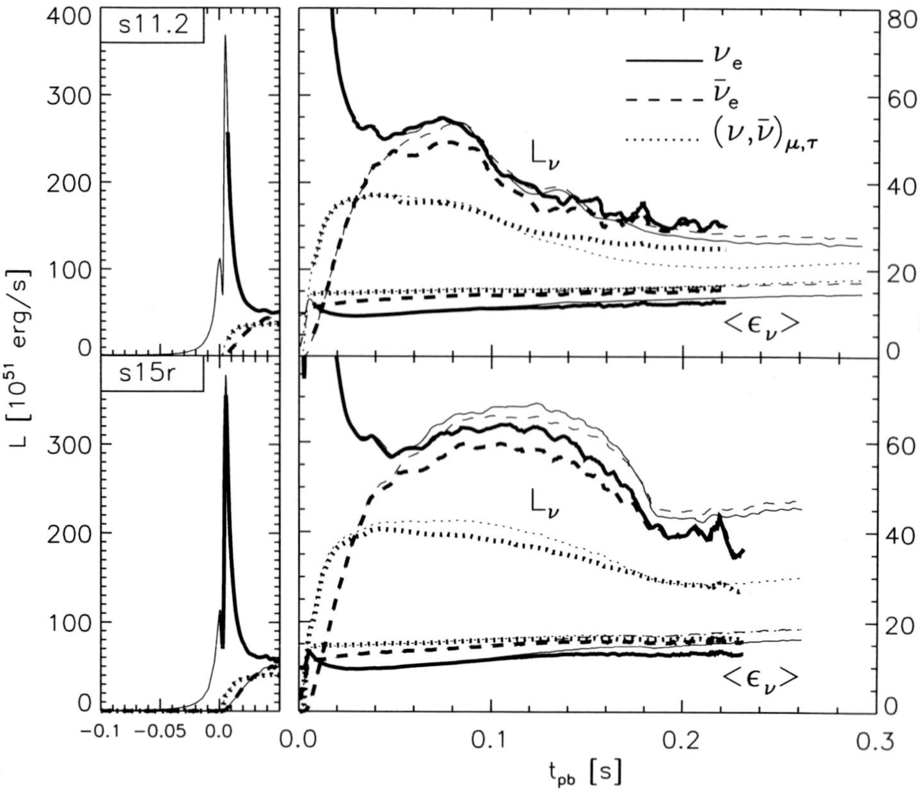

FIG. 15.8. Luminosities and mean energies (defined as ratio of energy flux to number flux) for ν_e, $\bar{\nu}_e$ and $\nu_{\mu,\tau}$, $\bar{\nu}_{\mu,\tau}$ as functions of time for a model with stellar mass $11.2\,M_\odot$ (s11.2) and a rotating model with stellar mass $15\,M_\odot$ (s15r) [279], measured by an observer comoving with the stellar medium at a radius of 500 km. The time origin is at the bounce. The left panels show the prompt ν_e burst. The right panels enlarge the post-bounce evolution. The thin lines represent results of 1D simulations for comparison.

medium is composed of protons, neutrons, and electrons, and the neutrino energy does not allow creation of muons and taus, the flavor neutrinos ν_e and $\bar{\nu}_e$ can interact with the medium through both charged-current and neutral-current weak processes, whereas the neutrinos ν_μ, $\bar{\nu}_\mu$, ν_τ, $\bar{\nu}_\tau$ can interact only through neutral-current weak processes, which are flavor-independent. Therefore, there are three energy-dependent neutrinospheres: one for ν_e, one for $\bar{\nu}_e$ and one for ν_μ, $\bar{\nu}_\mu$, ν_τ, $\bar{\nu}_\tau$. From now on, in this chapter, we will denote ν_μ, $\bar{\nu}_\mu$, ν_τ, $\bar{\nu}_\tau$ collectively as ν_x, as usually done in the literature. Each energy-dependent neutrinosphere emits a black-body thermal flux of neutrinos at the considered energy. The estimated radii of the neutrinospheres lie between about 50 and 100 km. As we have seen above, when the shock passes through the electron neutrino neutrinosphere (shock breakout) a few milliseconds after the bounce, a large flux of electron neutrinos is released in a few milliseconds in the neutronization burst. After the shock breakout, each neutrinosphere produces a thermal flux of the corresponding neutrino flavor.

The opacities of ν_e and $\bar{\nu}_e$ are dominated, respectively, by the charged-current weak interaction processes

$$\nu_e + p \to n + e^- \,, \tag{15.14}$$

$$\bar{\nu}_e + n \to p + e^+ \,. \tag{15.15}$$

These reactions allow exchanges of energy and lepton number between the neutrinos and the medium. For example, in the process in eqn (15.14) the neutrino energy is mainly transferred to the final electron[83] whose creation increases by one unit the lepton number of the medium.

Since the mantle of the proto-neutron star is neutron-rich, the opacity of ν_e of a given energy is larger than the opacity of $\bar{\nu}_e$ with the same energy, and the corresponding ν_e neutrinosphere has a larger radius than the $\bar{\nu}_e$ neutrinosphere. Therefore, for a fixed neutrino energy $\bar{\nu}_e$'s are emitted by a deeper and hotter layer than ν_e's, leading to a $\bar{\nu}_e$ mean energy larger than the ν_e mean energy. Moreover, the spectra do not have a perfect black-body shape (Fermi–Dirac distribution), but are *pinched*, i.e. both the low- and high-energy tails are suppressed with respect to the tails of a black-body thermal spectrum with the same mean energy. Figures 15.7 and 15.8 show the time evolution of neutrino luminosity and average energy obtained, respectively, with the numerical SN models in Ref. [1021] and [279]. Other similar estimates of the neutrino luminosity and average energy have been obtained with the numerical simulations in Refs. [735, 280]. Rough estimates of the time-integrated average energies are

$$\langle E_{\nu_e} \rangle \approx 10\,\text{MeV}\,, \quad \langle E_{\bar{\nu}_e} \rangle \approx 15\,\text{MeV}\,, \quad \langle E_{\nu_x} \rangle \approx 20\,\text{MeV}\,. \tag{15.16}$$

Figures 15.9 and 15.10 show, respectively, the energy spectra of ν_e's and $\bar{\nu}_e$'s in the numerical SN models of Ref. [1015] and [1021]. Note, however, that the model in Ref. [1015] did not lead to a successful SN explosion, which at present is difficult to reach in numerical simulations. The dashed curves in Fig. 15.10 correspond to Fermi–Dirac approximations with the number distribution

$$\frac{\mathrm{d}N}{\mathrm{d}E} = \frac{120\,L}{7\pi^4\,T^4}\,\frac{E^2}{e^{E/T}+1}\,, \tag{15.17}$$

where $T = 180\zeta(3)\langle E\rangle/7\pi^4 \simeq \langle E\rangle/3.1514$ is the effective temperature ($\zeta(3) \simeq 1.20206$ is the Riemann zeta function of 3; see eqns (A.130) and (A.131)) and $L = \int_0^\infty \mathrm{d}E\, E\, \mathrm{d}N/\mathrm{d}E$ is the total luminosity (released energy; eqn (A.128) is helpful for the calculation of the integral). In most numerical simulations the time-integrated

[83] The recoil kinetic energy of the final neutron is negligible. Indeed, momentum conservation implies that the momentum p_n of the final neutron is of the order of the momentum p_{ν_e} of the initial neutrino, which is practically equal to the neutrino energy, because of the smallness of neutrino masses. Since the neutrino energy is smaller than a few tens of MeV, the recoil kinetic energy of the neutron, $p_n^2/2m_n$, is suppressed by the large mass m_n of the neutron.

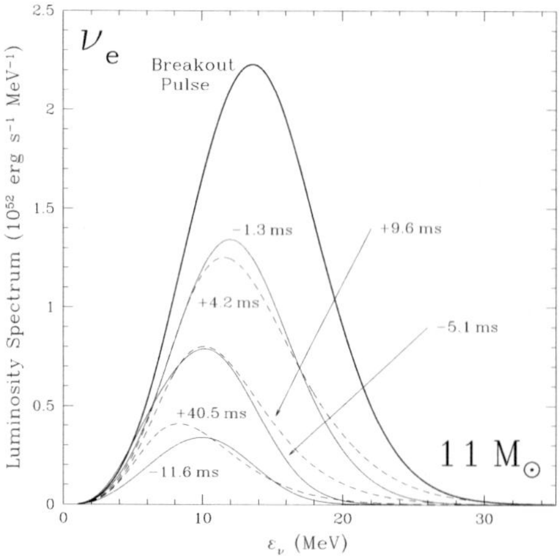

FIG. 15.9. Luminosity spectrum of ν_e's at infinity at various pre- (thin solid lines) and post-breakout (dashed lines) times for a SN with a progenitor of 11 solar masses [1015]. Time is measured relative to the peak breakout spectrum (thick solid line). The thin solid lines correspond to 11.6, 5.1, and 1.3 ms before the peak and the dashed lines to 4.2, 9.6, and 40.5 ms after the peak.

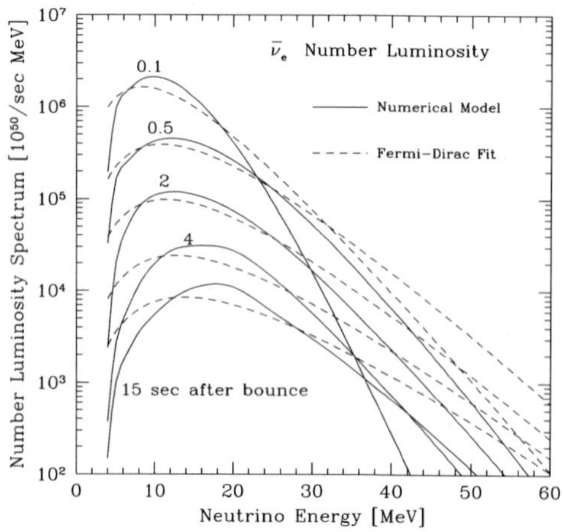

FIG. 15.10. Energy spectrum of $\bar{\nu}_e$'s in the numerical SN model of Ref. [1021]. The time (after the bounce) is indicated in the figure. The dashed lines are the Fermi–Dirac fits (with zero chemical potential) which have the same luminosity and average energy of the numerical curves.

luminosities of the different flavor neutrinos are approximately equal. A better fit of the spectra obtained in numerical SN simulations can be achieved adding a pinching parameter η which acts as an effective chemical potential:

$$\frac{\mathrm{d}N}{\mathrm{d}E} = \frac{L}{F(\eta)\,T^4}\,\frac{E^2}{e^{E/T-\eta}+1}\,, \tag{15.18}$$

with $F(\eta) = \int_0^\infty \mathrm{d}x\,x^3/(e^{x-\eta}+1)$. In this case, $\langle E\rangle/T \simeq 3.1514 + 0.1250\,\eta + 0.0429\eta^2 + \mathrm{O}(\eta^3)$. Typical values of η are $\eta_{\nu_e} \simeq 2$, $\eta_{\bar{\nu}_e} \simeq 3$, $\eta_{\nu_x} \simeq 1$ [664].

Another widely used parameterization of the neutrino spectrum is [664, 665]

$$\frac{\mathrm{d}N}{\mathrm{d}E} = \frac{(1+\beta)^{1+\beta}\,L}{\Gamma(1+\beta)\overline{E}^2}\left(\frac{E}{\overline{E}}\right)^{\beta} e^{-(1+\beta)E/\overline{E}}\,, \tag{15.19}$$

where L is the total energy released in neutrinos, $\overline{E}$ is the average neutrino energy, and β is a parameter. The fitting values of $\overline{E}$, β and L for the time-integrated $\bar{\nu}_e$ and ν_x spectra in the calculations of Refs.[1021, 1015] are listed in Table 15.3, which is reproduced from Ref. [88]. One can see that there are significant uncertainties on the value of the average neutrino energy.

TABLE 15.3. Fitting parameters of eqn (15.19) for the time-integrated $\bar{\nu}_e$ and ν_x (i.e $\nu_\mu, \bar{\nu}_\mu, \nu_\tau, \bar{\nu}_\tau$) spectra in three different calculations. Table reproduced from Ref. [88]. In Ref. [88] it is assumed that in the models of Ref. [1015] $L_{\bar{\nu}_e} = L_{\nu_x} = 5.0 \times 10^{52}$ erg, although such equipartition is not realized in the models, since the models of Ref. [1015] do not produce a SN explosion and the total energy released in neutrinos is unknown.

Ref.	Mass $[M_\odot]$	$\overline{E}_{\bar{\nu}_e}$ [MeV]	$\overline{E}_{\nu_x}$ [MeV]	$\beta_{\bar{\nu}_e}$	β_{ν_x}	$L_{\bar{\nu}_e}$ [erg]	L_{ν_x} [erg]
[1021]	20	15.4	21.6	3.8	1.8	4.9×10^{52}	5.0×10^{52}
	11	11.4	14.1	3.7	2.2	–	–
[1015]	15	11.4	14.1	3.7	2.2	–	–
	20	11.9	14.4	3.6	2.2	–	–

In the delayed supernova explosion scenario, the stalled shock lies at a radius of about 100–300 km, well outside of the neutrinosphere. The post-shock temperature is about 1.5 MeV and the density of the order of $10^8\,\mathrm{g\,cm^{-3}}$. The capture of a small fraction, about 5–10% [285], of the thermal flux of neutrinos emitted from the neutrinosphere could revive the shock, leading to the explosion. The largest energy deposition is due to electron neutrinos and antineutrinos, which have a charged-current cross-section on the free nucleons behind the shock that is larger than the neutral-current cross-section of all neutrino types.

If enough energy is deposited behind the shock, about a half second after the bounce the shock is revived and starts to sweep the outer layers of the star generating the explosion. However, most one-dimensional (i.e. spherically symmetric) computer simulations [889, 735, 286, 1015, 734] do not generate a successful explosion, which was recently obtained only by the Livermore group [1021] (they used the so-called *neutron finger convection* in the proto-neutron star to enhance the early neutrino luminosity, which leads to a large energy deposition behind the shock). In recent years several groups have performed multidimensional simulations which resulted in successful explosions [288, 469, 468, 290] and partial [281, 278] and complete [794, 279] failures. The multidimensionality of the simulations is important in order to take into account convection effects that enhance the efficiency of the neutrino energy deposition behind the shock. Intense computational efforts to perform accurate numerical multidimensional SN simulations are underway (see Refs. [302, 303, 305, 304]).

While the shock is stalled, matter continues to accrete on the proto-neutron star passing through the shock. During this so-called *accretion phase* the shocked hot material behind the shock, composed mainly of free nucleons, electrons and photons, is heated by the accretion and produces neutrinos and antineutrinos of all flavors through the processes in eqn (15.9)–(15.13). Since the stalled shock is out of the neutrinosphere, these neutrinos can free-stream out of the star and cause the so-called *hump* in the neutrino luminosity curve shown in Figs. 15.7 and 15.8.

The average neutrino energy is low during the hump because the dense matter in the shock is opaque to high-energy neutrinos. As shown in Fig. 15.7, as the shock gradually revives at about 0.5 s after the bounce, the matter density decreases and the average neutrino energy increases.

Summarizing, in the prompt explosion scenario there are two phases of the neutrino flux: first a brief and intense burst of prompt electron neutrinos from shock breakout, with a degenerate spectrum of high energy, which is, however, so brief that little energy and lepton number are carried away. Then there is a less intense thermal emission of neutrinos of all flavors, which last for a few seconds and carries away most of the binding energy of the neutron star. The total number of emitted neutrinos exceeds by an order of magnitude the original lepton number of the collapsed core.

In the delayed explosion scenario, in addition to the prompt electron neutrino burst and the thermal emission of neutrinos of all flavors one expects an accretion phase which prolongs the peak of the thermal neutrino luminosity over a time scale of about half a second (hump).

The delayed explosion scenario is a sort of standard model of core-collapse supernova explosion. However, the possibility of shock revival through neutrino heating is still under study (see Ref. [793, 285, 733, 795, 272]).

15.4 SN1987A

On 24 February 1987 a very bright SN of type II, SN1987A, was discovered in the Large Magellanic Cloud, which is a satellite galaxy of the Milky Way, at a distance of about 50 kpc from the solar system (see Refs. [1027, 1058]). At that time four large underground neutrino detectors potentially sensitive to SN neutrinos were in operation: Kamiokande-II [613, 615], IMB [242, 270, 1034], Baksan [64, 63, 318], and LSD [346]. These detectors observed an unusual number of events with energy of the order of 10 MeV within a time window of the order of 10 s in the hours before the optical discovery of SN1987A. The events observed in the Kamiokande-II, IMB, and Baksan happened at the same time (within uncertainties of the absolute time calibration of the detectors and the random occurrence of the events), whereas the LSD events have been recorded about five hours before those of the other detectors, at a time when the other detectors did not see any signal. Therefore, there is a controversy on the origin of the LSD events (see Refs. [361, 939]) and usually the LSD events are not included in the analysis of SN1987A data. In the following subsections we describe briefly the data of the Kamiokande-II, IMB and Baksan detectors, which are used to set limits on neutrino properties.

Supernova SN1987A is the best studied of all SNe not only because of the detection of its neutrinos but also because it was the first SN visible to the naked eye after the Kepler SN in 1604. Furthermore, it is the only SN for which the progenitor star is known: it was a blue supergiant star named Sanduleak $-69°202$ [969].

The evolution of the remnant of SN1987A have been extensively studied in all spectral bands: radio [770], infrared [266, 267], optical [799, 996, 560], ultraviolet

[678, 969], and x-rays [602, 291, 961, 848, 1088, 850, 849] (see also Refs. [1027, 1058]). Although no compact remnant has been identified with certainty so far, there is some indication of the presence of a 2.14 ms optical pulsar [799].

The observation of SN1987A neutrinos marked the beginning of *extrasolar system neutrino astronomy*[84]. It has been one of the great achievements of the Kamiokande detector, which was designed by Masatoshi Koshiba and earned him the 2002 Nobel Prize in Physics.

15.4.1 Kamiokande-II

The Kamiokande-II water Cherenkov detector has been described in section 10.6.1 in the context of solar neutrino detection. After the optical discovery of supernova SN1987A, the Kamiokande-II collaboration examined carefully their data looking for a significant number of events above background in a time interval of the order of 10 s and energy of the order of 10 MeV. They found such a collection of events at 7:35:35 UT of 23 February 1987. Unfortunately, before the discovery of supernova SN1987A the Kamiokande-II collaboration did not think that an accurate time measurement was necessary and the clock of the experiment was set by hand. As explained in Ref. [615], "it would be straightforward after SN1987A to have made an absolute calibration of the clock ..., but an abrupt power outage took place in the Kamioka mine on 25 February 1987, and precluded that alternative measure". Therefore, there is an uncertainty of about one minute in the Kamiokande-II determination of the time in which the SN1987A neutrino burst passed the Earth.

Electron antineutrinos with energy larger than 1.8 MeV can be detected with the *inverse neutron decay* reaction in eqn (12.13), with the cross-section in eqn (12.17). The produced positron can be observed in water Cherenkov detectors, as in Kamiokande-II. Since it is emitted almost isotropically, the information on the incoming neutrino direction is lost. On the other hand, the energy of the incident $\bar{\nu}_e$ can be measured through the relation in eqn (12.14). The Kamiokande-II detector could observe SN neutrinos also through the elastic scattering reaction in eqn (5.5), which was used for the solar neutrino detection. However, for SN neutrinos the elastic scattering cross-section is much smaller than the inverse neutron decay cross-section: neglecting m_e and $m_n - m_p$, from eqns (12.17) and (5.33), we have

$$\sigma_{\text{CC}}^{\bar{\nu}_e p} \simeq 9 \times 10^{-44} \, \frac{E_{\bar{\nu}_e}^2}{\text{MeV}^2} \, \text{cm}^2 \qquad \text{and} \qquad \sigma_{\text{ES}}^{\nu_e e} \simeq 9 \times 10^{-45} \, \frac{E_{\nu_e}}{\text{MeV}} \, \text{cm}^2 \, . \tag{15.20}$$

Table 15.4 shows the main characteristics of the 16 events measured in the Kamiokande-II detector during the supernova SN1987A neutrino burst. It is important to keep in mind that it is impossible to know with certainty which events have been really produced by neutrinos coming from SN1987A and which events are due to background. Therefore, in Table 15.4 we listed all the known events, taken from Refs. [615, 754], even those that are likely to be due to background and

[84] Solar neutrino astronomy was started by Raymond Davis Jr., co-winner of the 2002 Nobel Prize in Physics, with the Homestake chlorine experiment.

TABLE 15.4. Time t, energy E_e, and angle θ_e with respect to the direction opposite to SN1987A of the Kamiokande-II events [615, 754]. $B(E_e)$, P_B(prompt), and P_B(delayed) are, respectively, the event background rate and the probability that the event is due to background in the best-fit prompt and delayed supernova explosion models calculated in Ref. [754]. The event number 6 is reported in the original Kamiokande-II publication [615], but is excluded in their signal analysis because of a low number of hit photomultipliers. The events numbered 13–16 are not reported in the original Kamiokande-II publications, but have been used in the accurate analysis of Ref. [754] in which background effects were taken into account. The angles of these events have not been reported in Ref. [754].

			Kamiokande-II			
Event	Time t (s)	Energy E_e (MeV)	Angle θ_e (degrees)	$B(E_e)$ [754] (s^{-1})	P_B [754] (prompt)	P_B [754] (delayed)
1	0	20.0 ± 2.9	18 ± 18	1.6×10^{-5}	5.8×10^{-5}	2.4×10^{-5}
2	0.107	13.5 ± 3.2	40 ± 27	1.9×10^{-3}	6.3×10^{-3}	1.9×10^{-3}
3	0.303	7.5 ± 2.0	108 ± 32	2.9×10^{-2}	0.16	4.7×10^{-2}
4	0.324	9.2 ± 2.7	70 ± 30	1.2×10^{-2}	5.4×10^{-2}	1.7×10^{-2}
5	0.507	12.8 ± 2.9	135 ± 23	2.1×10^{-3}	7.6×10^{-3}	3.2×10^{-3}
6	0.686	6.3 ± 1.7	68 ± 77	3.7×10^{-2}	0.25	0.15
7	1.541	35.4 ± 8.0	32 ± 16	4.5×10^{-5}	1.2×10^{-3}	1.5×10^{-3}
8	1.728	21.0 ± 4.2	30 ± 18	8.2×10^{-5}	5.7×10^{-4}	1.0×10^{-3}
9	1.915	19.8 ± 3.2	38 ± 22	1.5×10^{-5}	9.9×10^{-5}	1.9×10^{-4}
10	9.219	8.6 ± 2.7	122 ± 30	1.5×10^{-2}	0.33	0.49
11	10.433	13.0 ± 2.6	49 ± 26	1.9×10^{-3}	0.11	0.12
12	12.439	8.9 ± 1.9	91 ± 39	1.6×10^{-2}	0.54	0.60
13	17.641	6.5 ± 1.6	?	3.8×10^{-2}	0.92	0.89
14	20.257	5.4 ± 1.4	?	2.9×10^{-2}	0.97	0.94
15	21.355	4.6 ± 1.3	?	2.8×10^{-2}	0.97	0.93
16	23.814	6.5 ± 1.6	?	3.8×10^{-2}	0.99	0.94

were excluded by all statistical analyses of the data, except for the very accurate and reliable analysis of Ref. [754] in which background effects were properly taken into account. For each event we list also the estimated background rate $B(E_e)$ calculated in Ref. [754], which depends on the event energy E_e, and the probabilities P_B(prompt) and P_B(delayed) that the event is due to background according to the best-fit prompt and delayed supernova explosion models (see section 15.3) calculated in Ref. [754].

The event number 6 is reported in the original Kamiokande-II publication [615], but is excluded in their signal analysis because of a low number of hit photomultipliers, which implies a significant probability that it is a background event. The events numbered 13–16 are not reported in the original Kamiokande-II publication [615] (although they can be seen in Fig. 9 of Ref. [615]), but have been used in the accurate analysis of Ref. [754]. From the last two columns one can see that, according to the calculation in Ref. [754], these events have a high probability to be due to background. However, the probability that at least one of them is a signal event is not negligible and it is correct to include them in the data analysis, as done in Ref. [754]. The angles of the events numbered 13–16 have not been reported in Ref. [754], since all the events have been assumed to be due either to background or to the inverse neutron decay reaction in eqn (12.13) in which the positron is emitted almost isotropically. This assumption is acceptable because of the dominance of the cross-section of the inverse neutron decay reaction (see eqn (15.20)). In fact, most authors agree that it is most likely that all Kamiokande events have been generated through the inverse neutron decay reaction [928].

Nevertheless, some authors [615, 912] have speculated on the fact that the first event point is almost in the opposite direction of the LMC[85], which could be an indication that it is due to an electron neutrino interacting in the detector through the elastic scattering process in eqn (5.5).

15.4.2 IMB

The IMB water Cherenkov detector has been described in section 11.2.2 in the context of atmospheric neutrino detection. On 23 February 1987 the IMB detector recorded eight neutrino-produced events with energies between 20 and 40 MeV in a time interval of 6 s starting from 7:35:41.37 UT (the clock had an absolute uncertainty of 50 msec and a relative uncertainty of 0.5 msec). The background rate is negligible, about two per day in the range 20–2000 MeV.

The important characteristics of the eight IMB events are listed in Table 15.5. Since these events are most likely due to the inverse neutron decay process in eqn (12.13), the neutrino energy is given by eqn (12.14).

Taking into account the trigger efficiency and about 13% dead time of the detector, the IMB collaboration estimated that 35 ± 15 neutrino events with energy above 20 MeV occurred in the detector [270].

15.4.3 Baksan

The Baksan Underground Scintillation Telescope [64, 63, 318] is located in the Baksan neutrino Observatory at a depth of 850 mwe in the Baksan Valley in North Caucasus, Russia. The telescope consists of 3150 parallelepipedal tanks filled with oil-based liquid scintillator viewed by a 15 cm photomultiplier. The energy threshold for SN neutrinos is about 10 MeV. The total target mass is about 330 ton. The background, mainly caused by cosmic ray muons and discharges in the photomultipliers, is relatively large. Therefore, only 1200 inner tanks with lower background and a mass of about 130 ton are used as signal triggers, and the inner tanks plus part of the external tanks are used as fiducial volume, with a mass of about 200 ton.

[85] In the first Kamiokande-II publication [613], the angle of the second event was reported to be 15 ± 27, pointing almost backward from the direction of the Large Magellanic Cloud. This angle was corrected in the second Kamiokande-II publication [615].

TABLE 15.5. IMB supernova SN1987A events from Ref. [270]. The time of each event is relative to the first one, which occurred at 7:35:41.37 UT of 23 February 1987. There is an additional systematic uncertainty in the energy scale estimated to be about 10%. The background rate is negligible (about two events per day).

Event	Time t (s)	IMB Energy E_e (MeV)	Angle θ_e (degrees)
1	0	38 ± 7	80 ± 10
2	0.412	37 ± 7	44 ± 15
3	0.650	28 ± 6	56 ± 20
4	1.141	39 ± 7	65 ± 20
5	1.562	36 ± 9	33 ± 15
6	2.684	36 ± 6	52 ± 10
7	5.010	19 ± 5	42 ± 20
8	5.582	22 ± 5	104 ± 20

As water Cherenkov detectors, the Baksan Underground Scintillation Telescope is mostly sensitive to electron antineutrinos which interact with protons through the inverse neutron decay reaction in eqn (12.13).

At the time of SN1987A, the Baksan Underground Scintillation Telescope had been in operation for about six years. During this period of time, including 23 February 1987, it never happened that more than seven events were observed in an interval of 20 s. The Baksan collaboration was expecting about 35 antineutrino events in the trigger mass and about 54 events in the fiducial mass for a SN at a distance of 10 kpc (i.e. within the Milky Way). In the period from 1 to 23 February 1987 the Baksan Underground Scintillation Telescope did not measure pulse clusters that differ significantly from the background. Therefore, the Baksan collaboration could not claim an independent observation of SN1987A neutrinos. However, when supplemented by the information of the Kamiokande-II and IMB observations, the Baksan collaboration identified five events in a 10 s interval that may overlap with the Kamiokande-II and IMB, taking into account an uncertainty of $^{+2}_{-54}$ s in the absolute Baksan clock measurement. The Baksan clock had a relative accuracy of about one millisecond and a nominal absolute accuracy of about 2 s, but on 11 March 1987 it was found that the clock had developed a forward shift in time of 54 s that could have happened in one step or gradually since 17 February 1987. Since the Baksan clock time of the five candidate SN events is about 30 s after the IMB events (which were measured with absolute time uncertainty of about 50 msec), the simultaneous occurrence of Baksan and IMB events is possible.

The main characteristics of the five Baksan events are listed in Table 15.6, where one can see that the background rate is rather high. For this reason, most authors did not include the Baksan data in the analysis of SN1987A neutrino events. However, the authors of Ref. [754] properly took into account the background rate and proved that the Baksan events are compatible with a SN signal. The probabilities

TABLE 15.6. Time t, energy E_e, and angle θ_e with respect to the direction opposite to SN1987A of the Baksan events [615, 754]. $B(E_e)$, P_B(prompt), and P_B(delayed) are, respectively, the event background rate and the probability that the event is due to background in the best-fit prompt and delayed supernova explosion models calculated in Ref. [754].

Event	Time t (s)	Energy E_e (MeV)	Baksan $B(E_e)$ [754] (s^{-1})	P_B [754] (prompt)	P_B [754] (delayed)
1	0	12.0 ± 2.4	8.4×10^{-4}	2.1×10^{-2}	4.9×10^{-3}
2	0.435	17.9 ± 3.6	1.3×10^{-3}	3.6×10^{-2}	1.9×10^{-2}
3	1.710	23.5 ± 4.7	1.2×10^{-3}	7.4×10^{-2}	0.12
4	7.687	17.6 ± 3.5	1.3×10^{-3}	0.30	0.35
5	9.099	20.3 ± 4.1	1.3×10^{-3}	0.55	0.52

P_B(prompt) and P_B(delayed) that an event is due to background according to the best-fit prompt and delayed supernova explosion models in Ref. [754] are listed in Table 15.6 and show that some of the Baksan events could be due to SN electron antineutrinos.

15.4.4 Comparison with supernova theory

The neutrino events have been compared with theoretical predictions in many papers [146, 971, 928, 929, 939, 701, 999, 754]. Although only about two dozens out of the estimated 10^{28} neutrinos that passed through the Earth were detected, these few events delivered us precious information about the physics of core-collapse SNe. Most authors agree that the detected neutrino events are compatible with the general features of the standard core-collapse SN scenario described in section 15.3.

In the accurate analysis in Ref. [754] it was found that models of supernova explosion with the delayed mechanism explained in section 15.3 are about 100 times more probable than prompt explosion models. The electron antineutrino average energy is about 15 MeV, as expected from the cooling of the proto-neutron star (see eqn (15.16)). The cooling time scale is about 4 s, and the time scale of the accretion component is about 0.7 s, in agreement with numerical calculations. The total inferred number of electron antineutrinos emitted is about 3×10^{57}, implying a binding energy of the neutron star of about 3×10^{53} erg, as expected from simple estimates (see section 15.3). Unfortunately, as explained in Ref. [754], the SN1987A neutrino data are too sparse to obtain more detailed information on the SN mechanism.

15.5 Neutrino mass

The basic idea of constraining neutrino masses from the observation of SN neutrinos was discussed first by Zatsepin in 1968 [1083] and later by Cabibbo [293] and Piran [874] in the early 1980s.

An extremely relativistic neutrino with mass $m \ll E$ propagates with a group velocity

$$v = \frac{p}{E} = \sqrt{1 - \frac{m^2}{2E^2}} \simeq 1 - \frac{m^2}{2E^2} \,. \tag{15.21}$$

If a neutrino flux is emitted by a source at a distance D, the time-of-flight delay of a massive neutrino with respect to a massless particle (as a photon or a graviton) emitted by the same source is

$$\Delta t = \frac{D}{v} - D \simeq \frac{m^2}{2E^2} D = 2.57 \left(\frac{m}{\text{eV}}\right)^2 \left(\frac{E}{\text{MeV}}\right)^{-2} \frac{D}{50\text{kpc}} \,\text{s} \,. \tag{15.22}$$

If neutrinos are emitted in a burst with intrinsic duration ΔT_0, the observation of events separated by a time interval larger than ΔT_0 would provide a direct measurement of the neutrino mass (assuming D known and E measurable). If the neutrino energy spectrum has mean value E and width ΔE, neutrinos produced at the same time with different energies would reach a detector at a distance D in the time interval

$$\Delta T \simeq \frac{m^2}{E^2} D \frac{\Delta E}{E} \,. \tag{15.23}$$

The model-independent sensitivity to the neutrino mass is found by requiring this time interval to be smaller than the intrinsic duration of the neutrino burst:

$$\Delta T < \Delta T_0 \leq \Delta T_{\text{obs}} \,, \tag{15.24}$$

where ΔT_{obs} is the observed time interval of the neutrino burst. The inequalities in eqn (15.24) imply the upper bound

$$m \lesssim E \sqrt{\frac{E}{\Delta E} \frac{\Delta T_{\text{obs}}}{D}} \simeq 14\,\text{eV} \left(\frac{E}{10\,\text{MeV}}\right) \sqrt{\frac{E}{\Delta E}} \left(\frac{\Delta T_{\text{obs}}}{10\,\text{s}}\right)^{1/2} \left(\frac{50\,\text{kpc}}{D}\right)^{1/2} \,. \tag{15.25}$$

It is clear that a large distance, a quick neutrino burst, a low neutrino energy, and a wide energy range are advantageous for the measurement of an effect due to the neutrino mass. Unfortunately, increasing the distance decreases the neutrino flux at the detector in proportion to D^{-2} and decreasing the energy decreases the detection event rate. In practice, the energy of neutrinos coming from a SN is of the order of 10 MeV and the existing detectors allow only the observation of neutrinos produced by SNe in our galaxy or in its satellites (the Small and Large Magellanic Clouds).

Supernova SN1987A occurred in the Large Magellanic Cloud, at a distance of about 50 kpc from the solar system. The measured neutrino burst had an average energy $E \simeq 15\,\text{MeV}$, a width $\Delta E \sim 15\,\text{MeV}$, and a time duration $\Delta T_{\text{obs}} \lesssim 12\,\text{s}$.

Assuming that the observed events are due to electron antineutrinos and neglecting neutrino mixing, eqn (15.25) implies the model-independent bound[86] [939]

$$m_{\nu_e} \lesssim 30\,\mathrm{eV}\,. \tag{15.26}$$

Many authors have calculated upper bounds on the electron antineutrino mass from the SN1987A neutrino data with specific model-dependent assumptions, often well-motivated, about the intrinsic spread of the neutrino burst, obtaining upper bounds for m_{ν_e} lying in the 5–30 eV range [143, 1004, 107, 928, 939, 701, 972, 13]. The accurate analysis in Ref. [754] yielded

$$m_{\nu_e} < 5.7\,\mathrm{eV} \qquad (95\%\ \mathrm{CL})\,. \tag{15.27}$$

15.6 Neutrino mixing

Since neutrino are mixed, an electron neutrino does not have a definite mass, but it is a superposition of different massive neutrinos. In this case, m_{ν_e} must be considered as an effective electron neutrino mass, which represents the masses of the massive neutrinos which have a substantial mixing with ν_e.

In the three-neutrino mixing scenario discussed in chapter 13, ν_e has large mixings only with ν_1 and ν_2. However, since the squared-mass differences are very small (see Table 13.2), the kinematical upper limit in eqns (15.26) and (15.27) apply to all three neutrino masses:

$$m_k \lesssim 30\,\mathrm{eV} \qquad (\text{model-independent})\,, \tag{15.28}$$

$$m_k < 5.7\,\mathrm{eV} \qquad (95\%\ \mathrm{CL}) \qquad [754]\,, \tag{15.29}$$

for $k = 1, 2, 3$.

It is, however, still possible that the electron antineutrino has a small mixing with one or more heavy massive neutrinos. The interaction probability of a heavy massive neutrino ν_h is proportional to $|U_{eh}|^2$, where U_{eh} is the element of the lepton mixing matrix connecting the electron neutrino to the heavy neutrino. Since $|U_{eh}|^2$ is small, the SN1987A Kamiokande-II and Baksan data do not place any constraint on m_h, because these detectors had a relatively high background rate and a weakly interacting heavy massive neutrino arriving on the Earth is indistinguishable from the background. On the other hand, the IMB detector had a negligible background and the IMB collaboration found no neutrino events in the 3.9 hours after the SN1987A neutrino burst. Since the IMB neutrino burst consisted of eight events, there is some probability that a heavy neutrino arriving much later interacts in the detector if $|U_{eh}|^2 \gtrsim 1/8$. From eqn (15.22), such a neutrino should be lighter than about 1.1 keV in order to arrive on the Earth within 3.9 hours. Therefore, using

[86] Note that the mass of a particle and its antiparticle are equal (see chapter 2).

the model-independent limit in eqn (15.26), we estimate

$$m_h \lesssim 30\,\mathrm{eV} \quad \text{or} \quad m_h \gtrsim 1.1\,\mathrm{keV} \quad \text{for} \quad |U_{eh}|^2 \gtrsim \frac{1}{8}\,. \tag{15.30}$$

For smaller values of $|U_{eh}|^2$, the SN1987A neutrino data do not put any constraint on m_h.

From the results of the CHOOZ long-baseline reactor $\bar{\nu}_e \to \bar{\nu}_e$ oscillation experiment (see section 12.2.2), we know that the mixing of ν_e with a heavy neutrino ν_h is small. In fact, the effective mixing angle for two-neutrino $\bar{\nu}_e \to \bar{\nu}_h$ oscillations is $\sin^2 2\vartheta_{eh}^{\text{eff}} = 4|U_{eh}|^2 \left(1 - |U_{eh}|^2\right)$ and the limit in eqn (12.19) implies that

$$|U_{eh}|^2 \lesssim 2.6 \times 10^{-2} \quad (90\%\ \mathrm{CL})\,. \tag{15.31}$$

In this case, the SN1987A neutrino data do not put any constraint on m_h. However, if the absolute value of the reactor $\bar{\nu}_e$ flux is considered to be uncertain, the spectral distribution of CHOOZ data do not put any limit on the mixing of ν_e with a heavy neutrino ν_h (see Fig. 12.6) and the SN1987A limit in eqn (15.30) becomes interesting.

Since the Kamiokande-II SN1987A events appear to be clustered in two groups separated by a time interval of about 10 s, some authors [335, 631] have claimed that there is evidence of two mass groupings at about 4 eV and 22 eV [335]. However, these authors had to assume that electron antineutrinos are emitted from the SN in a very short time, of the order of 0.1 s. This assumption is contrary to our understanding of the core-collapse SN mechanism, according to which electron antineutrinos are emitted during the cooling phase of the proto-neutron star on a time scale of about 10 s (see section 15.3). Moreover, the existence of neutrinos with masses of about 4 eV and 22 eV which have large mixing with the electron antineutrino is excluded by the tritium upper bound on the effective electron antineutrino mass (see section 14.1).

Other information on neutrino mixing was obtained from SN1987A data considering the effect of vacuum oscillations or MSW resonant transitions on the fluxes of different flavors [758, 806, 649, 648, 759, 332, 168]. Large $\bar{\nu}_x \leftrightarrows \bar{\nu}_e$ transitions are disfavored, because they would imply a spectrum of $\bar{\nu}_e$'s on the Earth which is more energetic than the one observed.

15.7 Other neutrino properties

For the sake of completeness, we briefly list some of the other neutrino properties that have been constrained by using the SN1987A neutrino data.

Since electron antineutrinos arrived at the Earth from a distance of about 50 kpc, their lifetime $\tau_{\bar{\nu}_e}$ is constrained by [615, 939]

$$\tau_{\bar{\nu}_e} \gtrsim 1.6 \times 10^5 \left(m_{\nu_e}/E_{\bar{\nu}_e}\right) \mathrm{yr}\,. \tag{15.32}$$

The total amount of emitted energy inferred from the measured $\bar{\nu}_e$ flux is compatible with the binding energy of a neutron star only if the number N_ν of neutrino flavors is limited by [409, 939, 701]

$$N_\nu \lesssim 6\,. \tag{15.33}$$

The cooling of the proto-neutron star constrains the Dirac masses of ν_μ and ν_τ by [886, 570, 506, 1029, 287]

$$\begin{aligned} &m_{\nu_\mu}^{\text{Dirac}} \lesssim 14\,\text{keV} \\ &m_{\nu_\tau}^{\text{Dirac}} \lesssim 14\,\text{keV} \quad \text{or} \quad m_{\nu_\mu}^{\text{Dirac}} \gtrsim 34\,\text{MeV}\,. \end{aligned} \tag{15.34}$$

The absence of γ emission accompanying the SN1987A neutrino burst implies a lower bound between about 10^6 and 10^{10} yr for the lifetime of a heavy massive neutrino with mass $2m_e < m_h \lesssim 100\,\text{MeV}$ which has a substantial mixing with the active light flavor neutrinos and decays via $\nu_h \to \nu_k + e^+ + e^-$ [140, 350, 1004].

The observed 10 s time scale of cooling of the proto-neutron star implies an upper bound [550, 719, 166, 165].

$$\mu_{\nu_e} \lesssim 10^{-12}\,\mu_B \tag{15.35}$$

for the electron neutrino magnetic moment which could flip neutrino helicity through scattering with electrons and nucleons or through interactions with the strong magnetic field, generating sterile right-handed neutrinos that escape freely, cooling the proto-neutron star in less than 1 s.

The absence of a similar cooling by right-handed neutrino emission constrains also the charge radius of right-handed neutrinos by [569]

$$\langle r^2 \rangle_R \lesssim 2 \times 10^{-33}\,\text{cm}^2\,. \tag{15.36}$$

The electric charge of the electron neutrino is bounded by [163]

$$q_{\nu_e} \lesssim 10^{-17}\,e\,, \tag{15.37}$$

otherwise the galactic magnetic field would have lengthened the neutrino path and neutrinos of different energy could not have arrived on the Earth within a few seconds.

15.8 Future

Several detectors sensitive to SN neutrinos are currently in operation (Super-Kamiokande [768], SNO [1040], LVD [35], KamLAND [995], AMANDA [49], Mini-BooNE [954]) or under preparation or study (see Refs. [309, 195, 791, 520, 949]). Many authors have studied future possibilities of SN neutrino detection and its potential sensitivity to neutrino masses (see Refs. [289, 441, 1022, 193, 190, 191, 192] and references therein). Current and future SN neutrino detectors are much larger than the detectors in operation during 1987. The order of magnitude of the number of events expected when the next galactic SN will explode is 10^4. Such impressive statistics will be precious in oder to test our understanding of SN physics and improve our knowledge of neutrino properties.

There is a general consensus among the experts in the field that future SN neutrino detections cannot be sensitive to an effective electron neutrino mass smaller than a few eV, because of the intrinsic spread in time of the neutrino burst. In Ref. [1022] it has shown that using the correlation between neutrino energy and arrival time implied by eqn (15.22), it is possible to reach a sensitivity of about 3 eV for the effective electron neutrino mass. The authors of Ref. [191, 192] have shown that an abrupt termination of the neutrino signal due to black-hole formation may allow the Super-Kamiokande detector to be sensitive to an electron neutrino mass as low as 1.8 eV.

However, since the current upper limit for the effective electron neutrino mass is already about 2 eV (see section 14.1) and the future KATRIN experiment [839] will be able to push the limit down to about 0.3 eV, a SN limit on m_{ν_e} of the order if 1 eV would not look extremely exciting. Therefore, several authors have concentrated on the possibility of constraining the effective masses of ν_μ and ν_τ [289, 441, 193, 190, 191, 192], whose laboratory limits are well above the eV scale (see section 14.1).

The flux of SN ν_μ, $\bar{\nu}_\mu$, ν_τ, and $\bar{\nu}_\tau$ is of the same order as that of ν_e and $\bar{\nu}_e$, but the problem is to distinguish them, because they can be observed only through neutral-current interactions, which are flavor blind (the energy is too low to produce μ or τ in charged-current reactions). Therefore, the ν_μ, $\bar{\nu}_\mu$, ν_τ, $\bar{\nu}_\tau$ signal can only be extracted on a statistical basis by subtracting the ν_e and $\bar{\nu}_e$ contributions from the measured neutral-current signal. The ν_e and $\bar{\nu}_e$ contributions are estimated from the ν_e and $\bar{\nu}_e$ charged-current signals.

Unfortunately, in the usual neutral-current neutrino interactions (as $\nu + d \to p + n + \nu$ used in SNO; see section 10.6.3), the energy of the neutrino is not determined. Therefore, it is not possible to use the correlation between neutrino energy and arrival time in eqn (15.22) for the measurement of neutrino masses, and the upper limit on the effective masses of ν_μ and ν_τ cannot be pushed below about 30 eV [289, 441, 193]. An interesting exception is the abrupt termination of the neutrino signal due to black-hole formation, which may allow a sensitivity to the effective masses of ν_μ and ν_τ as low as about 6 eV [191, 192]. Another promising technique [194] is the measurement of the recoil proton kinetic energy in neutral-current neutrino-proton elastic scattering,

$$\nu + p \to \nu + p\,. \tag{15.38}$$

For an incoming neutrino energy E_ν, the proton kinetic energy T_p ranges from zero to $T_p^{\max}$, given by

$$T_p^{\max} = \frac{2E_\nu^2}{m_p + 2E_\nu} \simeq \frac{2E_\nu^2}{m_p}\,, \tag{15.39}$$

which is obtained when the neutrino recoils backwards with momentum E_ν and the proton recoils forward with momentum $2E_\nu$. In this case, the neutrino energy can be directly obtained from the proton kinetic energy. Since the recoil protons have a kinetic energy of the order of 1 MeV, they are nonrelativistic and cannot be seen in water Cherenkov detectors. However, they can be observed in liquid scintillator detectors such as KamLAND [995] and Borexino [1008]. Unfortunately, the proton direction cannot be measured in scintillator detectors, denying the possibility of reconstructing the neutrino energy from simple kinematics on an event-by-event basis. However, the authors of Ref. [194] have shown that a fit of the proton kinetic energy distribution could allow a measurement of the neutrino temperature and the total neutrino energy with an accuracy of about 10%.

Obviously, a major problem in SN neutrino physics is the actual occurrence of a core-collapse SN at a galactic scale distance. As we have seen in section 15.2, the estimated rate of core-collapse SNe in the Milky Way is of the order of a few per century. Since the corresponding waiting time of a SN neutrino detector may be very long, there is active research to study the feasibility of huge detectors that could observe a few dozens of events produced by a SN in the local group of galaxies (see Ref. [309]).

16

COSMOLOGY

We must fall back upon the old axiom that when all other contingencies fail, whatever remains, however improbable, must be the truth.
A.C. Doyle, *The Adventure of the Bruce–Partington Plans*

16.1 Basic general relativity

The Standard Cosmological Model (see Refs. [1052, 832, 689, 858, 911, 856, 325, 213]) is based on the Einstein equation of gravity[87]

$$\mathfrak{R}^{\mu\nu} - \frac{1}{2}\,\mathfrak{R}\,g^{\mu\nu} - \Lambda\,g^{\mu\nu} = 8\pi\,G_{\rm N}\,T^{\mu\nu}\,. \tag{16.1}$$

Here, $g^{\mu\nu}$ is the symmetric metric tensor, which, in general relativity, describes the curvature of space-time. In an empty and flat space-time, $g^{\mu\nu}$ reduces to its Minkowski limit $\mathrm{diag}(1,-1,-1,-1)$, which is the metric tensor of special relativity (see appendix B). In eqn (16.1), Λ is the cosmological constant and $G_{\rm N}$ is Newton's constant of gravitation. The tensor $T^{\mu\nu}$ is the symmetric energy–momentum tensor of matter and radiation, which generates the gravitational field. The *Ricci tensor* $\mathfrak{R}^{\mu\nu}$ and *Ricci scalar* $\mathfrak{R}$ are defined by

$$\mathfrak{R}^{\mu\nu} = \mathfrak{R}^{\alpha\mu\nu}{}_{\alpha}\,, \qquad \mathfrak{R} = \mathfrak{R}^{\mu}{}_{\mu}\,, \tag{16.2}$$

where $\mathfrak{R}^{\mu}{}_{\nu\rho\sigma}$ is the *Riemann tensor*

$$\mathfrak{R}^{\mu}{}_{\nu\rho\sigma} = \frac{\partial\Gamma^{\mu}_{\nu\sigma}}{\partial x^{\rho}} - \frac{\partial\Gamma^{\mu}_{\nu\rho}}{\partial x^{\sigma}} + \Gamma^{\mu}_{\eta\rho}\,\Gamma^{\eta}_{\nu\sigma} - \Gamma^{\mu}_{\eta\sigma}\,\Gamma^{\eta}_{\nu\rho}\,. \tag{16.3}$$

The *Christoffel symbols*

$$\Gamma^{\mu}_{\alpha\beta} = \frac{1}{2}\,g^{\mu\rho}\left(\frac{\partial g_{\beta\rho}}{\partial x^{\alpha}} + \frac{\partial g_{\alpha\rho}}{\partial x^{\beta}} - \frac{\partial g_{\alpha\beta}}{\partial x^{\rho}}\right) \tag{16.4}$$

determine the motion of free-falling bodies through the *geodesic equation*

$$\frac{\mathrm{d}^2 x^{\mu}}{\mathrm{d}\tau^2} + \Gamma^{\mu}_{\alpha\beta}\,\frac{\mathrm{d}x^{\alpha}}{\mathrm{d}\tau}\,\frac{\mathrm{d}x^{\beta}}{\mathrm{d}\tau} = 0\,, \tag{16.5}$$

[87] We use the conventions of Ref. [689].

where $\mathrm{d}\tau$ is the infinitesimal proper-time interval given by

$$\mathrm{d}\tau^2 = g_{\alpha\beta}\,\mathrm{d}x^\alpha\,\mathrm{d}x^\beta\,, \tag{16.6}$$

with $g_{\alpha\beta}$ and $g^{\alpha\beta}$ related by

$$g^{\alpha\rho}\,g_{\rho\beta} = \delta^\alpha_\beta\,. \tag{16.7}$$

In general relativity, vectors and tensors have definite transformation properties under general coordinate transformations $x \to x'$. A contravariant four-vector V^μ transforms as $\mathrm{d}x^\mu$ and a covariant four-vector V_μ transforms as ∂_μ:

$$V'^\mu = \frac{\partial x'^\mu}{\partial x^\nu}\,V^\nu\,, \qquad W'_\mu = \frac{\partial x^\nu}{\partial x'^\mu}\,W_\nu\,. \tag{16.8}$$

Hence, the scalar product $V \cdot W = V^\mu W_\mu$ is invariant. The transformation rules in eqn (16.8) are easily generalized for tensors of higher rank with contravariant and covariant indices. For example, the rank-two covariant metric tensor $g_{\alpha\beta}$ transforms as

$$g'_{\alpha\beta} = \frac{\partial x^\rho}{\partial x'^\alpha}\,\frac{\partial x^\sigma}{\partial x'^\beta}\,g_{\rho\sigma}\,. \tag{16.9}$$

Hence, the infinitesimal proper-time interval in eqn (16.6) is invariant. Using eqn (16.7), it is possible to show that the rank-two contravariant metric tensor $g^{\alpha\beta}$ correctly transforms as

$$g'^{\alpha\beta} = \frac{\partial x'^\alpha}{\partial x^\rho}\,\frac{\partial x'^\beta}{\partial x^\sigma}\,g^{\rho\sigma}\,. \tag{16.10}$$

Note that the Christoffel symbols $\Gamma^\mu_{\alpha\beta}$ are *not* the components of a tensor, since

$$\Gamma'^\mu_{\alpha\beta} = \frac{\partial x'^\mu}{\partial x^\nu}\,\frac{\partial x^\rho}{\partial x'^\alpha}\,\frac{\partial x^\sigma}{\partial x'^\beta}\,\Gamma'^\nu_{\rho\sigma} + \frac{\partial x'^\mu}{\partial x^\nu}\,\frac{\partial^2 x^\nu}{\partial x'^\alpha \partial x'^\beta}\,. \tag{16.11}$$

The *Equivalence Principle*, based on the observed equality of inertial and gravitational masses, states that a free-falling observer does not experience any gravitational effect. This means that a free-falling observer can describe space-time with a metric which is locally flat, i.e. it reduces locally to the Minkowski metric and has locally vanishing Christoffel symbols. In this case, the geodesic equation (16.5) is reduced locally to the special relativity equation of motion for an inertial body, $\mathrm{d}^2x^\mu/\mathrm{d}\tau^2 = 0$.

Many macroscopic systems can be considered, with a good approximation, as *perfect fluids*, i.e. systems composed of weakly interacting objects in which an observer moving with the velocity of the fluid sees the fluid as isotropic. This happens if the mean free path between collisions is small with respect to the scale of lengths used by the observer, so that the observer sees an average distribution of objects. The energy–momentum tensor of a perfect fluid is

$$T^{\mu\nu} = (\rho + p)\, u^{\mu}\, u^{\nu} - p\, g^{\mu\nu}\,, \tag{16.12}$$

where ρ is the energy density, p is the pressure, and $u^{\mu} = \mathrm{d}x^{\mu}/\mathrm{d}\tau$ is the four-velocity of the fluid. In the rest frame of the fluid, where $u^{\mu} = (1, 0, 0, 0)$, we have

$$T^{\mu\nu} = \begin{pmatrix} \rho & 0 & 0 & 0 \\ 0 & p & 0 & 0 \\ 0 & 0 & p & 0 \\ 0 & 0 & 0 & p \end{pmatrix}. \tag{16.13}$$

Energy–momentum conservation implies that for any volume V

$$\mathrm{d}(\rho V) + p\,\mathrm{d}V = 0\,. \tag{16.14}$$

This is the first law of thermodynamics

$$đQ = \mathrm{d}U + đL\,, \tag{16.15}$$

where $đQ = 0$ is the heat given to the system, $\mathrm{d}U = \mathrm{d}(\rho V)$ is the variation of internal energy, and $đL = p\,\mathrm{d}V$ is the work done by the system. Here, $\mathrm{d}U$ is an exact differential and $U(T, V)$ is a function of state of the system (T is the temperature). On the other hand, $đQ$ and $đL$ are not exact differentials and cannot be integrated.

The cosmological constant Λ can be interpreted as the energy density of the vacuum, also called the *dark energy*, which is defined as the state of lowest attainable energy. With this interpretation, the contribution of the cosmological constant in the Einstein equation (16.1) can be moved from the left-hand side to the right-hand side as a contribution to the energy–momentum tensor given by

$$T^{\mu\nu}_{\Lambda} = \frac{\Lambda}{8\pi\, G_{\mathrm{N}}}\, g^{\mu\nu}\,. \tag{16.16}$$

Comparison with eqn (16.13) shows that in a locally inertial frame a cosmological constant is equivalent to a perfect fluid with constant energy density

$$\rho_{\Lambda} = \frac{\Lambda}{8\pi\, G_{\mathrm{N}}}\,, \tag{16.17}$$

and constant *negative* pressure

$$p_{\Lambda} = -\rho_{\Lambda}\,. \tag{16.18}$$

A negative pressure can be puzzling at first sight, but it is easily shown to be a consequence of the energy conservation equation (16.14), which can be written as

$$\mathrm{d}\rho = -\,(\rho + p)\,\mathrm{d}V\,. \tag{16.19}$$

For a normal fluid with positive pressure, an increase in volume implies a decrease of energy density. The energy density can remain constant only if $p = -\rho$. In other words, if we take a volume V with constant energy density ρ_Λ and increase it by $\mathrm{d}V$, the total energy increases by $\rho_\Lambda \mathrm{d}V$. In order to perform such an operation, energy must be supplied to the system by performing on it the work $-p_\Lambda\, \mathrm{d}V = \rho_\Lambda \mathrm{d}V$, which shows that $p_\Lambda = -\rho_\Lambda$.

Including in the energy–momentum tensor the vacuum contribution in eqn (16.16), we have

$$T^{\mu\nu} = T^{\mu\nu}_{\mathrm{M}} + T^{\mu\nu}_{\mathrm{R}} + T^{\mu\nu}_{\Lambda}\,, \tag{16.20}$$

where the subscripts M and R denote, respectively, matter and radiation. Einstein equation (16.1) can now be written in the simpler form

$$\mathfrak{R}^{\mu\nu} - \frac{1}{2}\,\mathfrak{R}\, g^{\mu\nu} = 8\pi\, G_{\mathrm{N}}\, T^{\mu\nu}\,. \tag{16.21}$$

Matter and radiation must be treated separately, because they have different equations of state. Matter implies nonrelativistic particles with vanishing pressure:

$$p_{\mathrm{M}} = 0\,, \tag{16.22}$$

whereas radiation implies relativistic particles with

$$p_{\mathrm{R}} = \frac{1}{3}\,\rho_{\mathrm{R}}\,. \tag{16.23}$$

It is convenient to write the general equation of state

$$p_i = w_i\, \rho_i\,, \tag{16.24}$$

with $i = \mathrm{M}, \mathrm{R}, \Lambda$ and

$$w_{\mathrm{M}} = 0 \qquad (\text{matter})\,, \tag{16.25}$$

$$w_{\mathrm{R}} = 1/3 \qquad (\text{radiation})\,, \tag{16.26}$$

$$w_{\Lambda} = -1 \qquad (\text{vacuum energy})\,. \tag{16.27}$$

In section 16.3, we will see that the different equations of state of matter, radiation, and vacuum energy imply that their energy densities evolve in very different ways in the history of the Universe.

16.2 Robertson–Walker metric

Since observations of the Universe have shown that it is spatially homogeneous and isotropic[88] on large scales ($\gtrsim 100\,\mathrm{Mpc}$), the Standard Cosmological Model (see Refs. [1052, 689, 858, 911, 856]) assumes that there is a frame in which the total

[88] Note that isotropy for two observers at different locations in space implies homogeneity. The assumption of spatial homogeneity and isotropy of the Universe is historically known as the *cosmological principle*. By contrast, the *perfect cosmological principle* assumes that the Universe does not change in time. This principle, which has generated the *steady state cosmological model*, has been refuted by observations.

matter and radiation of the Universe can be described on large scales by a perfect fluid with the energy momentum tensor in eqn (16.13). In such a *comoving* frame, the geometry of space-time is described by the Robertson–Walker metric

$$\mathrm{d}\tau^2 = \mathrm{d}t^2 - R^2(t)\left[\frac{\mathrm{d}r^2}{1-k\,r^2} + r^2\left(\mathrm{d}\theta^2 + \sin^2\theta\,\mathrm{d}\phi^2\right)\right]. \tag{16.28}$$

On average, galaxies are at rest in the comoving frame described by the Robertson–Walker metric, which is defined as the frame in which the cosmic microwave background radiation (CMBR) is isotropic.

The scale factor $R(t)$ in the Robertson–Walker metric has dimension of length and describes the expansion of the Universe as a function of time t. In eqn (16.28), k is the spatial curvature constant normalized to

$$k = \begin{cases} -1 & \text{open Universe}, \\ 0 & \text{flat Universe}, \\ +1 & \text{closed Universe}. \end{cases} \tag{16.29}$$

The comoving radial coordinate r is dimensionless and time-independent. If $k = 1$, the three-dimensional space is finite and positively curved. It can be considered as the three-dimensional surface of a hypersphere with radius $R(t)$ in the four-dimensional Euclidean space. In this case, the value of r ranges from 0 to 1. If $k = 0$, the three-dimensional space is infinite and flat, i.e. Euclidean. If $k = -1$, the three-dimensional space is infinite and negatively curved (negative curvature can be visualized as that of the two-dimensional surface of a saddle). For $k = 0$ and $k = -1$, the value of r ranges from 0 to infinity.

In order to illustrate the spatial geometry described by the Robertson–Walker metric, we can consider a circle of coordinate radius r in the $\phi = \text{const.}$ plane, whose circumference is $2\pi R(t) r$. The physical radius is given by $R(t) A_k(r)$, with

$$A_k(r) = \int_0^r \frac{\mathrm{d}r'}{\sqrt{1-k\,r'^2}} = \begin{cases} \arcsin r & \text{for} \quad k = +1, \\ r & \text{for} \quad k = 0, \\ \operatorname{arcsinh} r & \text{for} \quad k = -1. \end{cases} \tag{16.30}$$

Hence, the physical radius is equal to the Euclidean $R(t)\,r$ only for $k = 0$, i.e. if the spatial geometry is flat. If $k = +1$, the physical radius is larger than $R(t)\,r$. On the other hand, if $k = -1$, the physical radius is smaller than $R(t)\,r$. This is illustrated in Fig. 16.1, which shows $A_k(r)$ as a function of r in the three cases. Similarly, eqn (16.30) gives the radius of a two-dimensional sphere of coordinate radius r, having area $4\pi R^2(t) r^2$.

The physical distance, also called the *proper distance*, at the time t between two objects at coordinate radii $r_1 = 0$ and $r_2 = r$ is given by

$$d_{\mathrm{p}}(t) = R(t) \int_0^r \frac{\mathrm{d}r'}{\sqrt{1-k\,r'^2}}. \tag{16.31}$$

The physical distance in practice is not measurable, because its measurement would require a set of synchronized measurements of local distances all along the line

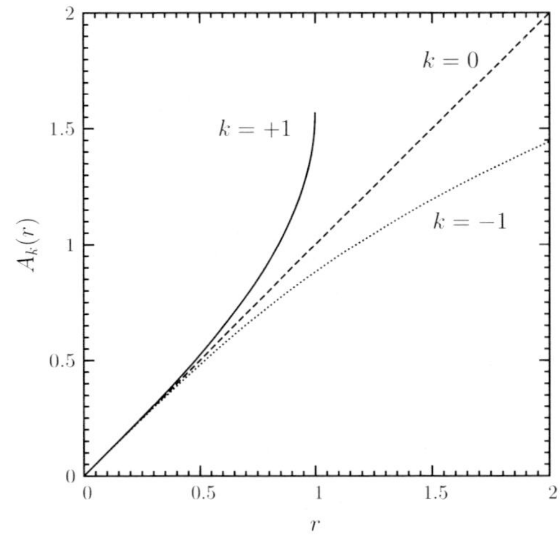

FIG. 16.1. The function $A_k(r)$ in eqn (16.30) for $k = +1$ (solid line), for $k = 0$ (dashed line), and for $k = -1$ (dotted line).

between the two objects. Since we are limited to observations from the Earth, we cannot measure the proper distance to comoving astrophysical objects. However, there are other distances which are measurable, and hence very useful for the cosmological interpretation of astrophysical data. The most interesting ones are the *luminosity distance* and the *angular diameter distance* presented, respectively, in sections 16.2.3 and 16.2.4.

16.2.1 Geodesic motion

It is useful to consider the geodesic equation (16.5) in order to find the motion of an object in free-fall in the Robertson–Walker metric. Let us write the geodesic equation as

$$\frac{\mathrm{d}u^\mu}{\mathrm{d}\tau} + \Gamma^\mu_{\alpha\beta}\, u^\alpha\, u^\beta = 0\,, \tag{16.32}$$

where $u^\mu = \mathrm{d}x^\mu/\mathrm{d}\tau$ is the four-velocity of an object. Let us consider the four-velocity[89] u^μ in the comoving frame described by the Robertson–Walker metric. Considering the $\mu = 0$ component and taking into account that the only non-vanishing components of $\Gamma^0_{\alpha\beta}$ in the comoving frame are $\Gamma^0_{kj} = -g_{kj}\dot{R}/R$, we obtain

$$\frac{\mathrm{d}u^0}{\mathrm{d}\tau} + \frac{\dot{R}}{R}\,|\vec{u}|^2 = 0\,, \tag{16.33}$$

[89] The four-velocity u^μ can be expressed in terms of the ordinary three-velocity $\vec{v} = \mathrm{d}\vec{x}/\mathrm{d}t$ as $u^\mu = (\gamma\,,\,\gamma\,\vec{v})$, where $\gamma = \left(1 - |\vec{v}|^2\right)^{-1/2}$, with $|\vec{v}|^2 = -g_{kj}\, v^k\, v^j$. The three-velocity $\vec{v}$ of an object in the comoving frame is usually called the *peculiar velocity*.

where $|\vec{u}|^2 = -g_{kj}u^k u^j$. Since[90] $u^2 = 1$, we have $(u^0)^2 - |\vec{u}|^2 = 1$ and $u^0\,\mathrm{d}u^0 = |\vec{u}|\,\mathrm{d}|\vec{u}|$, which allows us to write eqn (16.33) as

$$\frac{1}{u^0}\frac{\mathrm{d}|\vec{u}|}{\mathrm{d}\tau} + \frac{\dot{R}}{R}|\vec{u}| = 0\,. \tag{16.34}$$

Finally, since $u^0 = \mathrm{d}t/\mathrm{d}\tau$, we obtain

$$\frac{1}{|\vec{u}|}\frac{\mathrm{d}|\vec{u}|}{\mathrm{d}t} = -\frac{\dot{R}}{R} \qquad \Longrightarrow \qquad |\vec{u}| \propto R^{-1}\,. \tag{16.35}$$

Hence, as the Universe expands, the velocity $|\vec{u}|$ of an object in free-fall decreases as R^{-1}. This means that eventually the objects comes practically at rest in the comoving frame.

Since the four-momentum is $p^\mu = m\,u^\mu$, we obtain the important scaling relation of the three-momentum of a free-falling object as the Universe expands:

$$|\vec{p}| \propto R^{-1}\,. \tag{16.36}$$

16.2.2 Redshift

The variation with time of the scale factor $R(t)$ in the Robertson–Walker metric implies that the wavelength of freely propagating photons changes in proportion to $R(t)$. To see this, let us consider a wave propagating radially in the Robertson–Walker metric in eqn (16.28)[91]. Since $\mathrm{d}\tau = 0$ for a photon, the arrival time t_0 of a wave crest at $r = 0$ is related to its emission time t_e at the comoving radial coordinate $r = r_\mathrm{e}$ by the relation

$$\int_{t_\mathrm{e}}^{t_0}\frac{\mathrm{d}t}{R(t)} = \int_0^{r_\mathrm{e}}\frac{\mathrm{d}r}{\sqrt{1-k\,r^2}}\,. \tag{16.37}$$

If we now consider the following wave crest emitted at $t_\mathrm{e}+\delta t_\mathrm{e}$ at the same comoving radius r_e, its time of arrival $t_0+\delta t_0$ at $r = 0$ is given by

$$\int_{t_\mathrm{e}+\delta t_\mathrm{e}}^{t_0+\delta t_0}\frac{\mathrm{d}t}{R(t)} = \int_0^{r_\mathrm{e}}\frac{\mathrm{d}r}{\sqrt{1-k\,r^2}}\,. \tag{16.38}$$

Since the right-hand sides of eqns (16.37) and (16.38) are the same (the comoving distance is unchanged) and $R(t)$ is practically constant over the small intervals δt_e

[90] $u^2 = g_{\alpha\beta}\,u^\alpha\,u^\beta = g_{\alpha\beta}\,\frac{\mathrm{d}x^\alpha}{\mathrm{d}\tau}\,\frac{\mathrm{d}x^\beta}{\mathrm{d}\tau} = \frac{\mathrm{d}\tau^2}{\mathrm{d}\tau^2} = 1$.

[91] One can always choose the reference frame in such a way that an arbitrary free particle propagates radially.

and δt_0, by equating the left-hand sides, we obtain

$$\frac{\delta t_0}{R(t_0)} = \frac{\delta t_e}{R(t_e)} . \tag{16.39}$$

Since the wavelength λ of the radiation is equal to the speed of light times the time interval between the two crests ($c\,\delta t$), we obtain the relation[92]

$$\frac{\lambda_0}{\lambda_e} = \frac{R(t_0)}{R(t_e)} \equiv 1 + z , \tag{16.40}$$

where λ_e is the emitted wavelength, λ_0 is the observed wavelength, and we have defined the redshift

$$z \equiv \frac{\lambda_0 - \lambda_e}{\lambda_e} = \frac{\Delta\lambda}{\lambda} . \tag{16.41}$$

The redshift is a very useful quantity because the redshift of light coming from astronomical objects can be directly measured by observing the frequency shift of known emission or absorption lines in their spectra. Since astronomical observations show that distant galaxies have redshifted spectra (i.e. $z > 0$), the Universe is expanding.

Since the Universe is expanding, the scale factor $R(t)$ decreases going backward in time. This means that the present energy density in each comoving volume was confined to a smaller physical volume. In other words, the Universe was more dense than today and the density increases going backwards in time. If, in the past, the Universe was always expanding, there must have been an initial time $t = 0$ when $R(t = 0) = 0$ which is the start of the expansion. This is the so-called *Big Bang*. At $t = 0$ the energy density is infinite and our laws of physics break down. In fact, the laws of physics that we know already break down at the Plank time, when the energy density reaches the Plank density $M_P{}^4 \simeq 3 \times 10^{117}\,\mathrm{GeV\,cm^{-3}}$, where M_P is the Planck mass in eqn (A.167). At this enormous density, classical general relativity breaks down because of quantum effects. However, since we are interested in what happens after the Big Bang, we neglect the initial singularity approximating $t = 0$ with the Planck time.

16.2.3 Hubble's law

The Robertson–Walker metric implies an approximate linear relation between the distance of a comoving astronomical object and its redshift, if the redshift is small ($z \ll 1$). This is the famous *Hubble's law*, which is one of the pillars of the standard Big Bang model. In order to formulate Hubble's law, it is necessary to define a measurable distance of a comoving astronomical object. Since Hubble's law concerns

---

[92] The redshift equation (16.40) could also have been obtained from the scaling of momentum in eqn (16.36), using the quantum-mechanical relation $\lambda = h/|\vec{p}|$ between photon wavelength and momentum.

the light emitted by astronomical objects, we define the luminosity distance $d_{\rm L}$ by

$$d_{\rm L}{}^2 \equiv \frac{\mathscr{L}}{4\pi\,\mathscr{F}}\,, \tag{16.42}$$

where $\mathscr{L}$ is the absolute luminosity of the source, i.e. the energy produced per unit time by the source in its rest frame, and $\mathscr{F}$ is the measured flux, i.e. the energy per unit time per unit area measured on the Earth. The luminosity distance of a luminous astronomical object is measurable if its absolute luminosity is known. In order to connect the luminosity distance $d_{\rm L}$ to the redshift z, we note that the energy of photons coming from the source under consideration is decreased by a factor $(1+z)^{-1}$ due to the redshift relation in eqn (16.40). Moreover, the number of photons arriving at the detector per unit time is decreased by the same factor $(1+z)^{-1}$, which can be derived with the same reasoning that led to eqn (16.40). Hence, the measured energy flux is decreased with respect to that near the source by a factor $(1+z)^{-2}$. Assuming, without loss of generality, that the detector is located at $r = 0$, the measured flux at the time t_0 is given by

$$\mathscr{F} = \frac{\mathscr{L}}{4\pi\,R_0{}^2\,r_{\rm e}^2\,(1+z)^2}\,, \tag{16.43}$$

where $r_{\rm e}$ is the comoving radial coordinate of the source and $4\pi\,R_0{}^2\,r_{\rm e}^2$ is the area of the two-dimensional sphere centered on the source and passing through the detector, with $R_0 \equiv R(t_0)$. From eqns (16.42) and (16.43), we obtain

$$d_{\rm L} = R_0\,r_{\rm e}\,(1+z)\;. \tag{16.44}$$

In order to derive Hubble's law it is still necessary to express $r_{\rm e}$ in eqn (16.44) in terms of z. From eqn (16.37), we obtain

$$r_{\rm e} = S_k\left(\int_{t_{\rm e}}^{t_0}\frac{{\rm d}t}{R(t)}\right)\,, \tag{16.45}$$

where $t_{\rm e}$ is the emission time and

$$S_k(x) = \begin{cases} \sin x & \text{for} \quad k = +1\,, \\ x & \text{for} \quad k = 0\,, \\ \sinh x & \text{for} \quad k = -1\,. \end{cases} \tag{16.46}$$

For the luminosity distance, we obtain the expression

$$d_{\rm L} = (1+z)\,R_0\,S_k\left(\int_{t_{\rm e}}^{t_0}\frac{{\rm d}t}{R(t)}\right)\,. \tag{16.47}$$

Note that the luminosity distance of a comoving object is different from its proper distance, which, from eqns (16.31) and (16.37), is given by

$$d_{\rm P}(t_0) = R_0\int_0^r\frac{{\rm d}r'}{\sqrt{1-k\,r'^2}} = R_0\int_{t_{\rm e}}^{t_0}\frac{{\rm d}t'}{R(t')}\,. \tag{16.48}$$

The two distances coincide only in a flat and static Universe.

Although eqn (16.47) is the exact relation between $d_{\rm L}$ and z, it does not show the well-known linear dependence of $d_{\rm L}$ on z, which is the characteristic of Hubble's law. The reason is that the linear Hubble's law is valid as an approximation for small z. We can obtain such an approximation by expanding the S_k in eqn (16.47) in Taylor series about $z = 0$, which corresponds to the present time t_0. Expanding first $R(t)/R_0$ up to $(t - t_0)^2$, we obtain

$$(1+z)^{-1} = \frac{R(t)}{R_0} = 1 + H_0\,(t - t_0) - \frac{1}{2}\,q_0\,{H_0}^2\,(t - t_0)^2 + \mathcal{O}((t - t_0)^3)\,, \quad (16.49)$$

where $H_0 \equiv H(t_0)$ is the *Hubble constant* and $q_0 = q(t_0)$ is the *deceleration parameter*, with (a dot indicates a time derivative)

$$H(t) \equiv \frac{\dot{R}(t)}{R(t)}\,, \quad (16.50)$$

$$q(t) \equiv -\frac{\ddot{R}(t)}{H^2(t)\,R(t)}\,. \quad (16.51)$$

The function $H(t)$ is very important, because it represents the expansion rate of the Universe. The second-order term proportional to $(t - t_0)^2$ in the expansion in eqn (16.49) will give the first correction of order z^2 to the linear Hubble's law. Inverting eqn (16.49), we obtain $t(z)$:

$$t = t_0 - {H_0}^{-1}\left[z - \left(1 + \frac{q_0}{2}\right)z^2 + \mathcal{O}(z^3)\right]\,. \quad (16.52)$$

Using eqns (16.49) and (16.52), we can now calculate the integral in eqn (16.47):

$$\int_{t_e}^{t_0} \frac{\mathrm{d}t'}{R(t')} = {R_0}^{-1}\,{H_0}^{-1}\left[z - \frac{1}{2}\,(1 + q_0)\,z^2 + \mathcal{O}(z^3)\right]\,. \quad (16.53)$$

Finally, since $S_k(x) = x + \mathcal{O}(x^3)$, we obtain the desired relation:

$$H_0\,d_{\rm L} = z + \frac{1}{2}\,(1 - q_0)\,z^2 + \mathcal{O}(z^3)\,. \quad (16.54)$$

For small z ($z \ll 1$), we have the linear Hubble's law[93]

$$z \simeq H_0\,d_{\rm L}\,. \quad (16.55)$$

[93] The original version of Hubble's law was written as $v = H_0\,d_{\rm L}$, where v is the recession velocity of a comoving object, obtained by interpreting its redshift as a Doppler shift. In fact, for $v \ll 1$ the Doppler shift formula gives $1+z = \lambda_0/\lambda = \gamma\,(1+v) \simeq 1+v$. With this interpretation, eqn (16.55) is equivalent to the original version of Hubble's law. Although this interpretation is rather common, it is not correct. A Doppler shift is due to the fact that emitter and observers are at rest in different inertial frames. On the other hand, the cosmological redshift is measured by observers which are at rest in the comoving reference frame, detecting light emitted by sources at rest in the same comoving reference frame. The cosmological redshift is due to the cumulative expansion of space during photon propagation from source to observer.

Another apparent possibility arises from the time derivative of the proper distance in eqn (16.31): $\dot{d}_{\rm p}(t) = H(t)\,d_{\rm p}(t)$, where $\dot{d}_{\rm p}(t)$ could be interpreted as a velocity. This linear relation is exact, but useless. The reason is that, as remarked after eqn (16.31), the proper distances of astrophysical objects are not measurable.

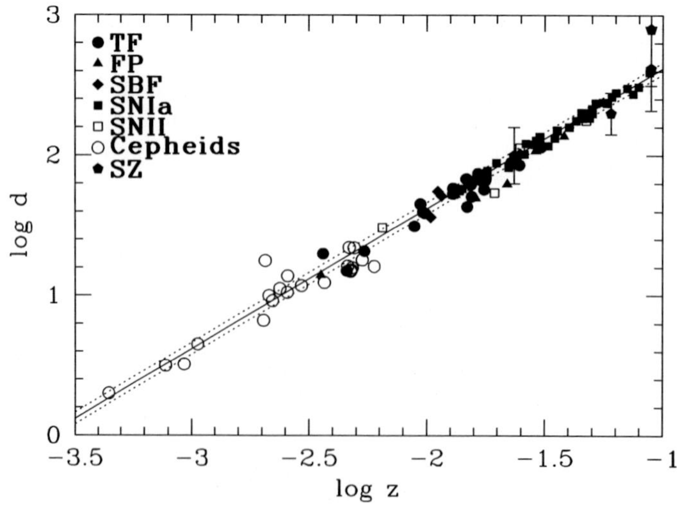

FIG. 16.2. Hubble diagram from Ref. [461]. d is the luminosity distance that we have denoted with d_L (see eqns (16.42)–(16.54)). The solid line corresponds to $H_0 = 72\,\mathrm{km\,s^{-1}\,Mpc^{-1}}$, and the dashed lines represent ±10% variations.

The Hubble constant is usually written as

$$H_0 = 100\,h\,\mathrm{km\,s^{-1}\,Mpc^{-1}}\,, \tag{16.56}$$

where h is the normalized dimensionless Hubble constant.

For many decades astronomers have struggled to measure the d_L and z of distant objects in order to infer the value of H_0 (see, for example, Ref. [914]), which is now known with about 10% precision from the Hubble Space Telescope Key Project: [461]:

$$h = 0.72 \pm 0.08\,. \tag{16.57}$$

Figure 16.2 shows the Hubble diagram from which this value was derived. The Hubble constant is perhaps the most important quantity in the Standard Cosmological Model, since it gives the present rate of expansion of the Universe, on which most of the other measurable quantities depend.

The inverse of the Hubble constant, called the *Hubble time*, gives the time scale of the age of the Universe:

$${H_0}^{-1} = 9.778\,h^{-1}\,\mathrm{Gyr} = 13.6 \pm 1.5\,\mathrm{Gyr}\,. \tag{16.58}$$

This can be understood by taking the first-order approximation in eqn (16.49): $R(t)/R_0 = 1 + H_0\,(t - t_0)$. Since $R(t = 0) = 0$, we have $t_0 = {H_0}^{-1}$. Beyond this crude approximation, the real age of the Universe depends on the type and amount of energy density, which can slow or accelerate the expansion rate, as we will see in the following section 16.3.

When expressed in distance units, the inverse of the Hubble constant is called the *Hubble distance*. It gives the distance covered by light in a Hubble time, i.e. the scale of the dimensions of the observable Universe:

$$H_0{}^{-1} = 2.998 \times 10^3\,h^{-1}\,\mathrm{Mpc} = 4.16 \pm 0.46\,\mathrm{Gpc}\,. \tag{16.59}$$

The deviations of the relationship between z and d_L from the linear Hubble's law at $z \gtrsim 1$ give information on the value of the deceleration parameter q_0. In 1998 remarkable observations of high-z supernovae of type Ia (SNIa; see section 15.1) showed that the deceleration parameter q_0 is *negative* [903, 863], implying that the expansion of the Universe is currently accelerating. This is contrary to the previous belief that the Universe is decelerating because of the attractive force of gravity. Thus the name *deceleration parameter*. As we will see in the following section 16.3, the accelerated expansion can be explained with a large contribution of vacuum energy to the average density of the Universe.

Figure 16.3[94] shows the Hubble diagram obtained with high-z SNIa whose measurements have been released in 1998 by the High-z SN Search Team [903] and the Supernova Cosmology Project [863]. On the vertical axis the luminosity distance is represented in terms of the so-called *distance modulus*, which is the difference between the apparent magnitude[95] m and the absolute magnitude M, related to the luminosity distance by $m - M = 5\log(d_\mathrm{L}/10\mathrm{pc})$. From the upper part of Fig. 16.3 one can see that at $z \sim 1$ the Hubble diagram deviates from the linear relation in eqn (16.55). The High-z SN Search Team found [903]

$$q_0 = -1.0 \pm 0.4\,. \tag{16.60}$$

The negative sign of q_0 is also illustrated in Fig. 16.4, obtained with more recent data [905], where one can see that $q_0 = 0$ is excluded with more than 99.7% confidence.

16.2.4 Angular diameter–redshift relation

Besides the luminosity distance–redshift relation there are other measurable relations which depend on the properties of the Universe. An important relationship is the angular diameter–redshift relation which connects the size of an object to its observed angular diameter. Let us consider an object at the comoving radius $r = r_\mathrm{e}$ which emits light at the cosmic time $t = t_\mathrm{e}$. Such light is now seen on the Earth, at $r = 0$, $t = t_0$. From the Robertson–Walker metric in eqn (16.28) the physical size $\mathrm{d}l$ of the object along the latitudinal coordinate θ is related to the observed angular diameter $\mathrm{d}\theta$ through

$$\mathrm{d}l = R(t_\mathrm{e})\,r_\mathrm{e}\,\mathrm{d}\theta\,. \tag{16.61}$$

The angular diameter distance d_A, defined as

$$d_\mathrm{A} \equiv \frac{\mathrm{d}l}{\mathrm{d}\theta}\,, \tag{16.62}$$

[94] The meanings of Ω_M and Ω_Λ are explained in eqn (16.79).

[95] The apparent magnitude m of an astrophysical object is given by $m = -2.5\log(\mathscr{F})+$ constant, where the constant depends on the units of the flux $\mathscr{F}$ and the energy band in which it is measured. The absolute magnitude M of an astrophysical object is equal to the apparent magnitude it would have if it were at a distance of 10 pc.

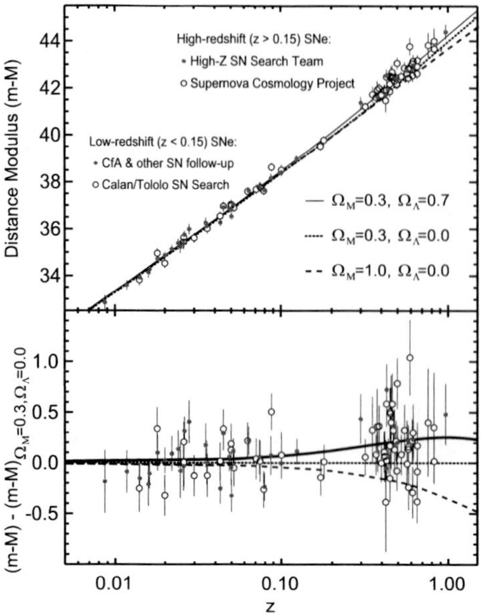

FIG. 16.3. Hubble diagram presented in Ref. [864]. The high-z SNIa were measured by the High-z SN Search Team [903] and the Supernova Cosmology Project [863].

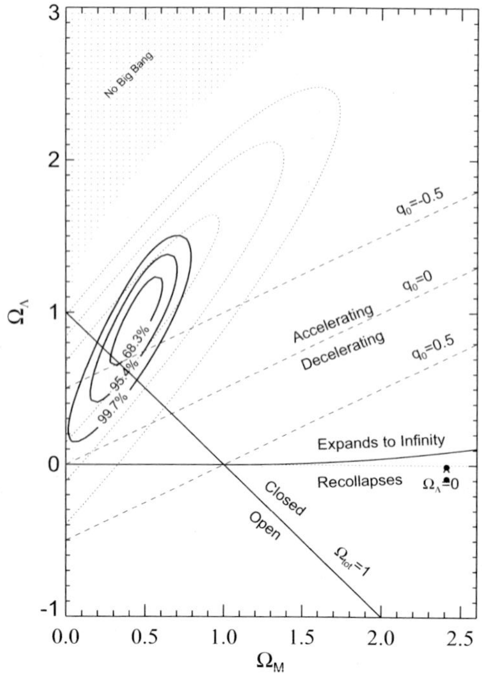

FIG. 16.4. Confidence intervals in the Ω_Λ-Ω_M plane [905]. The dotted contours are the previous results of Ref. [903]. Regions representing different cosmological scenarios are illustrated.

is given by

$$d_\mathrm{A} = R(t_\mathrm{e})\, r_\mathrm{e} = \frac{d_\mathrm{L}}{(1+z)^2}\,, \tag{16.63}$$

where we have used eqn (16.44) for the luminosity distance d_L. In a static and spatially flat Universe, $d_\mathrm{A} = d_\mathrm{L}$ is the time-independent physical distance of the object. Using the approximation in eqn (16.54) for small values of z we obtain

$$H_0\, d_\mathrm{A} = z - \frac{1}{2}\,(3 + q_0)\, z^2 + \mathcal{O}(z^3)\,. \tag{16.64}$$

Thus, one can get information on H_0 and q_0 also by measuring the angular diameter of far-away objects of a known size.

16.2.5 Particle horizon

Since the Universe had a beginning, it is possible that not all the Universe is in causal contact and for each observer there is a so-called *particle horizon*, or *causal horizon*, which limits the past light cone. In fact, considering, without loss of generality, an observer at $r = 0$, there may be a finite value $r_{\rm H}(t)$ of the comoving radial coordinate for which a light signal emitted towards the observer at the beginning of time at $r = r_{\rm H}(t)$ reaches the observer at a time t. In this case, $r_{\rm H}(t)$ is the radial coordinate of the particle horizon, since no signal emitted at $r > r_{\rm H}(t)$ can reach the observer before the time t.

Since the light signal is moving radially ($\mathrm{d}\theta = \mathrm{d}\phi = 0$) with $\mathrm{d}\tau = 0$, the Robertson–Walker metric implies that the physical distance to the particle horizon at the time t is given by

$$d_{\rm H}(t) = R(t) \int_0^{r_{\rm H}(t)} \frac{\mathrm{d}r'}{\sqrt{1 - k\,r'^2}} = R(t) \int_0^t \frac{\mathrm{d}t'}{R(t')} \,. \tag{16.65}$$

The behavior of $R(t)$ near the singularity at $t = 0$ is critical for the existence of the particle horizon. If $R(t) \propto t^\alpha$ for $t \to 0$, there is a particle horizon if $\alpha < 1$. In the time interval in which $R(t) \propto t^\alpha$ is approximately correct, we have the finite particle horizon $d_{\rm H}(t) = t/(1 - \alpha)$. Note that, in this case, we have $H(t) = \alpha/t$ and the deceleration parameter is constant, $q = (1 - \alpha)/\alpha$. Hence, there is a particle horizon if $q > 0$, i.e. if the expansion of the Universe decelerates. Although different comoving points are arbitrarily close for $t \to 0$, they cannot communicate, because the time variation of their proper distance diverges for $t \to 0$: from eqn (16.31) we have $\dot{d}_{\rm p}(t) = H(t)\, d_{\rm p}(t)$.

Often the integral over time in eqn (16.65) is called *conformal time* and denoted by

$$\eta \equiv \int_0^t \frac{\mathrm{d}t'}{R(t')} \,. \tag{16.66}$$

Conformal time allows one to write the Robertson–Walker metric as

$$\mathrm{d}\tau^2 = R^2(t) \left[\mathrm{d}\eta^2 - \frac{\mathrm{d}r^2}{1 - k\,r^2} - r^2 \left(\mathrm{d}\theta^2 + \sin^2\theta\, \mathrm{d}\phi^2 \right) \right] . \tag{16.67}$$

This expression shows explicitly that for $k = 0$ the Robertson–Walker metric is related to the Minkowski metric through the conformal transformation $g^{\mu\nu} \to R^2(t)\, g^{\mu\nu}$. The particle horizon at the conformal time η is simply written as

$$d_{\rm H}(\eta) = \eta\, R(\eta) \,. \tag{16.68}$$

16.3 Dynamics of expansion

In the Standard Cosmological Model, the large-scale structures of the Universe are described by a perfect fluid with the energy–momentum tensor in eqn (16.13) in the comoving frame with the Robertson–Walker metric in eqn (16.28). In this case (see Refs. [1052, 689, 858, 911, 856]), the Einstein equation (16.21) leads to the

Friedmann equation[96]

$$H^2 = \frac{8\pi\, G_{\rm N}}{3}\, \rho - \frac{k}{R^2}\,, \tag{16.70}$$

and the acceleration equation

$$\frac{\ddot{R}}{R} = -\frac{4\pi\, G_{\rm N}}{3}\, (\rho + 3\, p)\,. \tag{16.71}$$

By taking the derivative of the Friedmann equation (16.70) with respect to time and by using eqn (16.71), one can derive the covariant law of energy conservation

$$\dot{\rho} = -3\, H\, (\rho + p)\,, \tag{16.72}$$

which is equivalent to eqn (16.19) with $V \propto R^3$.

From the Friedmann equation (16.70) one can immediately see that $k = 0$, i.e. the Universe is spatially flat, if the density is equal to the *critical density*

$$\rho_{\rm c} \equiv \frac{3\, H^2}{8\pi\, G_{\rm N}}\,. \tag{16.73}$$

The present critical density is given by

$$\rho_{\rm c}^0 = \frac{3\, H_0{}^2}{8\pi\, G_{\rm N}} = (10.5369 \pm 0.0016)\, h^2\, {\rm keV\, cm^{-3}} = 5.5 \pm 1.2\, {\rm keV\, cm^{-3}}\,. \tag{16.74}$$

It is customary to express the energy density in terms of the dimensionless density

$$\Omega \equiv \frac{\rho}{\rho_{\rm c}}\,. \tag{16.75}$$

Rewriting the Friedmann equation (16.70) in the form[97]

$$\Omega - 1 = \frac{k}{H^2\, R^2} = \frac{k}{\dot{R}^2}\,, \tag{16.76}$$

one can see that Ω is constant only if it is equal to unity and the Universe is spatially flat ($k = 0$). If $\Omega > 1$, we have $k = 1$ and Ω increases when the expansion velocity $\dot{R}$ decreases. When $\dot{R} \to 0$ the dimensionless density Ω diverges because $\rho_{\rm c} \to 0$, but ρ stays finite, as one can see from the Friedmann equation (16.70) for $H = 0$.

---

[96] It is interesting to write the Friedmann equation (16.70) as

$$\frac{1}{2}\, \dot{R}^2 - G_{\rm N}\, \frac{\frac{4}{3}\, \pi\, \rho\, R^3}{R} = -\frac{1}{2}\, k\,, \tag{16.69}$$

which represents the energy-conservation equation for a comoving unit mass in the Newtonian gravitational potential due to the mass $\frac{4}{3}\, \pi\, \rho\, R^3$, with total energy $-k/2$. The value $-k/2$ is a consequence of Einstein equation which cannot be guessed by use of classical mechanics alone.

[97] Some authors use the definition $\Omega_k = 1 - \Omega = -k/R^2 H^2$ to represent curvature. We refrain from this practice, which could be confusing, since Ω_k does not have a definite sign and the associated $\rho_k = \Omega_k \rho_{\rm c}$ is not an energy density.

On the other hand, if $\Omega < 1$, we have $k = -1$ and Ω decreases with the expansion velocity $\dot{R}$. However, the expansion velocity cannot vanish, since $\Omega > 0$ implies that $\dot{R}^2 > 1$.

By considering the quantities in eqn (16.76) at the present time, we can express the curvature parameter k as

$$k = {H_0}^2 \, {R_0}^2 \left(\Omega^0 - 1 \right) , \tag{16.77}$$

where Ω^0 denotes the present value of Ω. Then, from eqn (16.76), the Friedmann equation can also be written as

$$H^2 = {H_0}^2 \left(1 + z\right)^2 \left(\frac{\Omega^0 - 1}{\Omega - 1} \right) . \tag{16.78}$$

It is useful to divide Ω into its matter, radiation, and vacuum contributions,

$$\Omega = \Omega_{\rm M} + \Omega_{\rm R} + \Omega_{\Lambda} , \tag{16.79}$$

with present values denoted, respectively, by Ω_Λ^0 $\Omega_{\rm M}^0$, and $\Omega_{\rm R}^0$. Since these three contributions have different equations of state which relate ρ and p, they evolve in different ways as functions of time. Using the general equation of state in eqn (16.24) for each component, the energy-conservation relation in eqn (16.72) becomes

$$\frac{\dot{\rho}_i}{\rho_i} = -3 \left(1 + w_i\right) \frac{\dot{R}}{R} , \tag{16.80}$$

with $i = {\rm M}, {\rm R}, \Lambda$. The solution of this equation is

$$\rho_i \propto R^{-3(1+w_i)} . \tag{16.81}$$

Therefore, the energy density decreases with the expansion of the Universe if $w_i > -1$. This is the case of matter and radiation, whereas the energy density of the vacuum remains constant:

$$w_{\rm M} = 0 \quad \Longrightarrow \quad \rho_{\rm M} \propto R^{-3} \propto \left(1 + z\right)^3 \qquad \text{(matter)} , \tag{16.82}$$

$$w_{\rm R} = 1/3 \quad \Longrightarrow \quad \rho_{\rm R} \propto R^{-4} \propto \left(1 + z\right)^4 \qquad \text{(radiation)} , \tag{16.83}$$

$$w_{\Lambda} = -1 \quad \Longrightarrow \quad \rho_{\Lambda} = \text{constant} \qquad \text{(vacuum energy)} . \tag{16.84}$$

The behavior of the energy density of matter is very easy to understand: since matter is composed of nonrelativistic particles, its energy density is given by the mass density. In this case, as the Universe expands, the energy density of matter in any comoving volume $V \propto R^3$ decreases proportionally to $V^{-1}(t)$. The energy density of radiation decreases as the scale factor $R(t)$ increases by one power of R faster than the decrease of the energy density of matter. This is the result of the redshift of radiation in eqn (16.40), due to the expansion, which implies that the energy of radiation decreases proportionally to $R^{-1}(t)$.

Since the energy density of the vacuum is constant, it dominates the dynamics of the expansion at large values of the time t (if the Universe expands forever). On the other hand, the energy density of radiation, which has the fastest increase when R decreases, dominates the dynamics of the expansion at small values of t. Matter may dominate at intermediate values of t.

Since the vacuum energy density is constant and the energy density of matter increases with redshift as given in eqn (16.82), we have

$$\frac{\rho_\Lambda}{\rho_M} = \frac{\rho_\Lambda^0}{\rho_M^0}\left(\frac{R}{R_0}\right)^3 = \frac{\rho_\Lambda^0}{\rho_M^0}(1+z)^{-3} . \tag{16.85}$$

Hence, there is a redshift z_{tr} given by

$$1 + z_{tr} = \left(\frac{\Omega_\Lambda^0}{\Omega_M^0}\right)^{1/3} , \tag{16.86}$$

at which there is a transition from a matter-dominated to a vacuum-dominated Universe. If $\Omega_\Lambda^0 > \Omega_M^0$, as indicated by current data (see section 17.8), we have $z_{tr} > 0$, which means that this transition has occurred in the past. For $z \gg z_{tr}$ the Universe was matter dominated, up to the redshift at which the energy density of radiation and matter were equal. Since, from eqns (16.82) and (16.83), we have

$$\frac{\rho_M}{\rho_R} = \frac{\rho_M^0}{\rho_R^0}\frac{R}{R_0} = \frac{\rho_M^0}{\rho_R^0}(1+z)^{-1} , \tag{16.87}$$

the redshift z_{eq} of matter–radiation equality is given by

$$1 + z_{eq} = \frac{\Omega_M^0}{\Omega_R^0} . \tag{16.88}$$

For $z \gg z_{eq}$ the Universe was radiation dominated.

Using eqns (16.82)–(16.84), the definition of ρ_c in eqn (16.73), and the Friedmann equation written in the form given in eqn (16.78), one can get the following expression for $\Omega - 1$ as a function of z:

$$\Omega - 1 = \frac{\Omega^0 - 1}{1 - \Omega^0 + \Omega_\Lambda^0 (1+z)^{-2} + \Omega_M^0 (1+z) + \Omega_R^0 (1+z)^2} . \tag{16.89}$$

This expression shows that if $|\Omega^0 - 1|$ is different from zero, $|\Omega - 1|$ increases with z only if the vacuum energy dominates. If matter or radiation dominates, $|\Omega - 1|$ decreases rapidly with z. In this case, a value of Ω^0 not far from unity requires a value of Ω at early times which is very close, but different from unity. Since Ω^0 is indeed close to unity and observations show that the Universe in the past was dominated first by radiation and then by matter (see section 17.8), in the Standard Cosmological Model there is a problem of fine-tuning of Ω at early times, which is called the *flatness problem*.

The behavior of $|\Omega - 1|$ as a function of $1 + z$ is illustrated in Fig. 16.5a for an open Universe with $\Omega_\Lambda^0 = 0.71$, $\Omega_M^0 = 0.25$, $\Omega_R^0 = 8 \times 10^{-5}$ (solid line) and a closed Universe with $\Omega_\Lambda^0 = 0.71$, $\Omega_M^0 = 0.27$, $\Omega_R^0 = 8 \times 10^{-5}$ (dashed line). As we will see in section 17.8, these values are quite realistic. In the figure it is difficult to distinguish

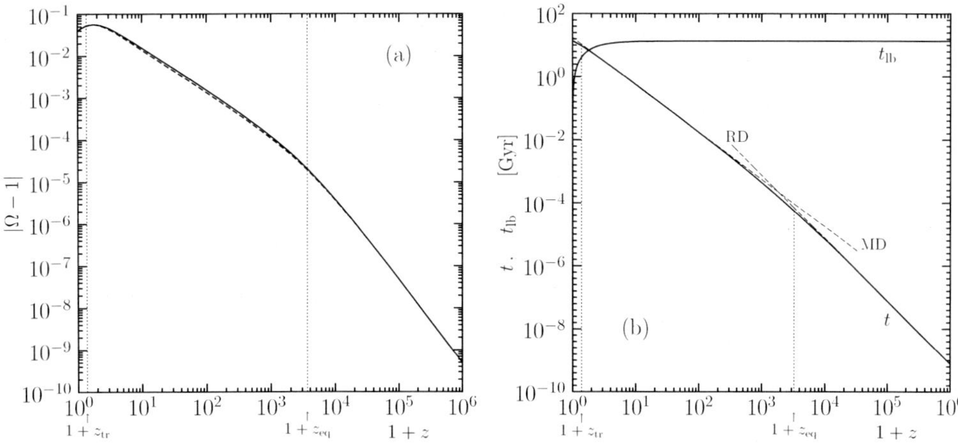

FIG. 16.5. (a) Value of $|\Omega - 1|$ as a function of $1+z$ according to eqn (16.89). Solid line: open Universe with $\Omega_\Lambda^0 = 0.71$, $\Omega_\mathrm{M}^0 = 0.27$, $\Omega_\mathrm{R}^0 = 8 \times 10^{-5}$. Dashed line: closed Universe with $\Omega_\Lambda^0 = 0.75$, $\Omega_\mathrm{M}^0 = 0.29$, $\Omega_\mathrm{R}^0 = 8 \times 10^{-5}$. The two vertical dotted lines show the redshifts of matter–vacuum equality and matter–radiation equality given by eqns (16.86) and (16.88), respectively. (b) Cosmic time t and lookback time t_lb as functions of $1+z$ for a flat Universe with $\Omega_\Lambda^0 = 0.73$, $\Omega_\mathrm{M}^0 = 0.27$, $\Omega_\mathrm{R}^0 = 8 \times 10^{-5}$. The dashed lines illustrate the behavior of t in a Universe which is dominated always by radiation (RD) or by matter (MD).

the two lines, which are slightly shifted only for $1 \lesssim z \lesssim 10^4$, which correspond to the matter-dominated era. For $z \lesssim 1$ the large value of Ω_Λ^0 causes an increase of $|\Omega - 1|$. For $z \gtrsim 1$ the vacuum energy is negligible and $|\Omega - 1|$ decreases rapidly with increasing z. For $1 \lesssim z \lesssim 10^4$, corresponding to the matter-dominated era, $|\Omega - 1| \propto (1+z)^{-1}$, and for $1 \lesssim z \lesssim 10^4$, corresponding to the radiation-dominated era, $|\Omega - 1|$ decreases faster and behaves as $\propto (1+z)^{-2}$. Note, however, that in the derivation of eqn (16.89) we have implicitly assumed that vacuum energy, matter, and radiation maintain their character as z increases. This assumption fails to represent real nature at large values of z, because, as we will see in section 16.8, in the hot early Universe, the particles which now contribute to the matter component (electrons, protons, and neutrons) were relativistic and contribute to the radiation component. Moreover, it is possible that in a very early era a new form of vacuum energy dominates the expansion of the Universe. This scenario, called *inflation*, could solve the flatness problem by driving $|\Omega - 1|$ to a very small value.

Using eqn (16.89) in the Friedmann equation (16.78), we obtain the expansion rate as a function of the redshift:

$$H(z) = H_0\,(1+z)\sqrt{1 - \Omega^0 + \Omega_\Lambda^0\,(1+z)^{-2} + \Omega_\mathrm{M}^0\,(1+z) + \Omega_\mathrm{R}^0\,(1+z)^2}\,. \quad (16.90)$$

This expression is convenient because it can be integrated to yield the time t as a function of z. Since $R = R_0 \, (1+z)^{-1}$, we have

$$\frac{dz}{dt} = -H\,(1+z)\,, \tag{16.91}$$

which yields, with $x = (1+z)^{-1}$,

$$t(z) = H_0{}^{-1} \int_0^{(1+z)^{-1}} \frac{dx}{\sqrt{1 - \Omega^0 + \Omega^0_\Lambda \, x^2 + \Omega^0_{\rm M} \, x^{-1} + \Omega^0_{\rm R} \, x^{-2}}}\,. \tag{16.92}$$

Taking $z = 0$, we obtain the current age of the Universe:

$$t_0 = H_0{}^{-1} \int_0^1 \frac{dx}{\sqrt{1 - \Omega^0 + \Omega^0_\Lambda \, x^2 + \Omega^0_{\rm M} \, x^{-1} + \Omega^0_{\rm R} \, x^{-2}}}\,. \tag{16.93}$$

As noted before eqn (16.58), the Hubble time $H_0{}^{-1}$ gives the scale of the age of the Universe. If the Universe is flat ($\Omega^0 = 1$), eqn (16.93) can be integrated analytically neglecting the radiation contribution (in this approximation, $\Omega^0_{\rm M} = 1 - \Omega^0_\Lambda$):

$$t_0 = \frac{2}{3}\, H_0{}^{-1} \, \frac{1}{\sqrt{\Omega^0_\Lambda}} \, \ln \frac{1 + \sqrt{\Omega^0_\Lambda}}{\sqrt{1 - \Omega^0_\Lambda}}\,. \tag{16.94}$$

In practice, this is a good approximation of the age of the Universe, since the radiation contribution is presently very small and dominates the density only at very early times (see section 17.8). The age of the Universe has to be longer than the age of the oldest stars in globular clusters, which is estimated to be 12.8 ± 1 Gyr [702]. From eqn (16.94) one can calculate that in order to have an age of the Universe which is longer than the Hubble time $H_0{}^{-1}$ it is necessary to have $\Omega^0_\Lambda \gtrsim 0.74$. This increase of the age of the Universe for large values of Ω^0_Λ is due to the fact that a vacuum-dominated Universe accelerates its expansion, as explained in section 16.7. In this case, the expansion was slower in the past, leading to a longer age of the Universe.

In eqns (16.92)–(16.94) we have neglected the above-mentioned break-down of eqn (16.89), and hence of eqn (16.90), for very large values of z, because in practice very large values of z give a negligible contribution to the age of the Universe (unless there is a previous exponential expansion driven by vacuum energy; in this case, we call *age of the Universe* the time elapsed after inflation). It is also useful to define the so-called *lookback time*, which does not suffer from this problem,

$$t_{\rm lb}(z) = t_0 - t(z) = H_0{}^{-1} \int_{(1+z)^{-1}}^1 \frac{dx}{\sqrt{1 - \Omega^0 + \Omega^0_\Lambda \, x^2 + \Omega^0_{\rm M} \, x^{-1} + \Omega^0_{\rm R} \, x^{-2}}}\,. \tag{16.95}$$

Figure 16.5b shows the cosmic time t and the lookback time $t_{\rm lb}$ as functions of $1+z$ for a flat Universe with $\Omega^0_\Lambda = 0.73$, $\Omega^0_{\rm M} = 0.27$, $\Omega^0_{\rm R} = 8 \times 10^{-5}$. The dashed lines represent the cosmic time t in the approximations of a radiation-dominated (RD) and matter-dominated (MD) Universe:

$$t(z) = \frac{1}{2}\, H_0{}^{-1} \, {\Omega^0_{\rm R}}^{-1/2} \, (1+z)^{-2} \qquad \text{(RD)}\,, \tag{16.96}$$

$$t(z) \simeq \frac{2}{3} H_0{}^{-1} \Omega_{\rm M}^0{}^{-1/2} (1+z)^{-3/2} \qquad \text{(MD)}\,. \tag{16.97}$$

One can see from the figure that the RD approximation is appropriate for $z \gtrsim z_{\rm eq} \simeq 3 \times 10^3$, corresponding to $t \lesssim t_{\rm eq} \simeq 6 \times 10^4\,{\rm yr}$. The MD approximation is successful for $z_{\rm tr} \lesssim z \lesssim z_{\rm eq}$, with $z_{\rm tr} \simeq 0.4$, corresponding to $t_{\rm tr} \simeq 10\,{\rm Gyr}$. For $z \lesssim z_{\rm tr}$ the Universe starts to be dominated by vacuum energy, which slows down the increase of t with decreasing z. From the curve of the lookback time $t_{\rm lb}$ it is clear that almost all the age of the Universe is covered by $z \lesssim 5$ and has been spent under matter domination and matter–vacuum domination.

From the expression of $H(z)$ in eqn (16.90), one can calculate the deceleration function q in eqn (16.51) as a function of z:

$$\begin{aligned} q(z) &= (1+z)\, H^{-1} \frac{{\rm d}H}{{\rm d}z} - 1 \\ &= (1+z) \frac{\frac{1}{2}\Omega_{\rm M}^0 - \Omega_\Lambda^0 (1+z)^{-3} + \Omega_{\rm R}^0 (1+z)}{1 - \Omega^0 + \Omega_\Lambda^0 (1+z)^{-2} + \Omega_{\rm M}^0 (1+z) + \Omega_{\rm R}^0 (1+z)^2}\,. \end{aligned} \tag{16.98}$$

Thus, the present deceleration parameter is related to the different energy densities by

$$q_0 = \frac{1}{2}\Omega_{\rm M}^0 - \Omega_\Lambda^0 + \Omega_{\rm R}^0\,. \tag{16.99}$$

It is clear that the only way to explain the negative value of q_0 in eqn (16.60), obtained with high-z SNIa, is by a large contribution of the vacuum energy, such that $\Omega_\Lambda^0 > \Omega_{\rm M}^0/2 + \Omega_{\rm R}^0$. This is illustrated in Fig. 16.4, where $q_0 = \Omega_{\rm M}^0/2 - \Omega_\Lambda^0$, because $\Omega_{\rm R}^0 \ll \Omega_{\rm M}^0$ in our Universe (see section 17.8). Note that a flat matter dominated Universe ($\Omega^0 \simeq \Omega_{\rm M}^0 \simeq 1$), which corresponds to $q_0 \simeq 0.5$, is practically excluded. Considering a flat Universe with $\Omega_\Lambda^0 = 1 - \Omega_{\rm M}^0$, the High-$z$ SN Search Team found [905]

$$\Omega_{\rm M}^0 = 0.29^{+0.05}_{-0.03}\,, \tag{16.100}$$

which is compatible with

$$\Omega_{\rm M}^0 = 0.25^{+0.08}_{-0.07}\,, \tag{16.101}$$

obtained by the Supernova Cosmology Project [684].

The expression for the expansion rate in eqn (16.90) allows one to calculate also the particle horizon in eqn (16.65) as a function of redshift. Using eqn (16.91), we obtain

$$d_{\rm H}(z) = H_0{}^{-1} (1+z)^{-1} \int_0^{(1+z)^{-1}} \frac{{\rm d}x}{x\sqrt{1 - \Omega^0 + \Omega_\Lambda^0 x^2 + \Omega_{\rm M}^0 x^{-1} + \Omega_{\rm R}^0 x^{-2}}}\,. \tag{16.102}$$

In fact, what is often interesting is not the value of the particle horizon at redshift z, but the corresponding value today, after it has been expanded by a factor $1+z$:

$$d_{\rm H}^0(z) = (1+z)\, d_{\rm H}(z) = H_0{}^{-1} \int_0^{(1+z)^{-1}} \frac{{\rm d}x}{x\sqrt{1 - \Omega^0 + \Omega_\Lambda^0 x^2 + \Omega_{\rm M}^0 x^{-1} + \Omega_{\rm R}^0 x^{-2}}}\,. \tag{16.103}$$

An important quantity for the formation of structures in the Universe, which will be discussed in section 17.7 is the amount of matter in the causal horizon as a function of redshift, which must be compared with the amount of matter in a typical galaxy such as our Milky Way, given by

$$M_{\rm gal} \sim 10^{12}\, M_\odot \,. \tag{16.104}$$

Since matter is comoving, we can calculate it from a measurement of the current energy density of matter using $d_{\rm H}^0(z)$: considering a sphere with radius $d_{\rm H}^0(z)/2$, we have

$$M_{\rm H}(z) = \frac{4\pi}{3}\left(\frac{d_{\rm H}^0(z)}{2}\right)^3 \rho_{\rm M}^0 = \frac{{H_0}^2}{16\, G_{\rm N}}\left(d_{\rm H}^0(z)\right)^3 \Omega_{\rm M}^0 \,. \tag{16.105}$$

It is useful to consider the case of dominance of one of the three components of Ω. Let us consider $\Omega_a \gg \Omega_b$ for $b \neq a$ at all times. If the Universe is spatially flat or the curvature term $-k/R^2$ is negligible, the Friedmann equation (16.70) can be integrated to yield

$$R \propto t^{2/[3(1+w_a)]} \,, \tag{16.106}$$

$$t = \frac{2}{3\,(1+w_a)}\,\frac{1}{H(t)} \,, \tag{16.107}$$

$$\rho_a \propto t^2 \,, \tag{16.108}$$

for $w_a > -1$. In this approximation the age of the Universe is shorter than the Hubble time in eqn (16.58) if $w_a > -1/3$. In the following sections 16.4 and 16.5 we discuss, respectively, the cases of dominance of matter and radiation, in which eqns (16.106)–(16.108) are applicable. In sections 16.6 and 16.7 we discuss the special cases of curvature and vacuum dominance.

16.4 Matter-dominated Universe

This is an important case, because the Universe is practically flat and was matter-dominated for a long time (for $3 \times 10^3 \gtrsim z \gtrsim 1$, corresponding to $6 \times 10^4\,{\rm yr} \lesssim t \lesssim 10\,{\rm Gyr}$).

If $\rho_{\rm M} \gg \rho_{\rm R}, \rho_\Lambda$ and the Universe is flat, or the curvature term $-k/R^2$ in the Friedmann equation (16.70) is negligible, eqns (16.106)–(16.108) are applicable with $a = {\rm M}$ and $w_{\rm M} = 0$, leading to

$$R \propto t^{2/3} \,, \tag{16.109}$$

$$t = \frac{2}{3}\,\frac{1}{H(t)} \,. \tag{16.110}$$

Hence, if the present Universe were dominated by matter[98], neglecting the early radiation-dominated era (see section 16.5), its age would be approximatively

$$t_0 \simeq \frac{2}{3}\,{H_0}^{-1} \simeq 9.1 \pm 1.0\,{\rm Gyr} \,, \tag{16.111}$$

[98] This was widely believed until 1998, when a large component of dark energy was discovered [903, 863] (see the end of section 16.2 and section 17.8).

in agreement with eqn (16.94) with $\Omega_\Lambda^0 = 0$. This age of the Universe is too short compared with the estimated age of globular clusters discussed after eqn (16.94).

In a flat matter-dominated Universe, the conformal time (see eqn (16.66)) is given by

$$\eta \propto t^{1/3} \propto R^{1/2}\,, \tag{16.112}$$

and the particle horizon (see eqn (16.65)) is given by

$$d_{\rm H}(t) = 3\,t = 2\,H^{-1}(t) \propto R^{3/2}\,. \tag{16.113}$$

One can see that the inverse of the expansion rate sets the scale of the particle horizon. Since

$$\frac{d_{\rm H}(t)}{R(t)} \propto R^{1/2} \propto t^{1/3}\,, \tag{16.114}$$

the particle horizon grows faster than the scale factor and with the passing of time the size of the comoving part of the Universe, which is in causal contact with each observer, increases.

If matter-domination were a good approximation for most of the history of the Universe, we would have a particle horizon

$$d_{\rm H}^0 = 3\,t_0 = 2\,{H_0}^{-1} = 27.2 \pm 3.0\,{\rm Gly} = 8.32 \pm 0.92\,{\rm Gpc}\,, \tag{16.115}$$

where we used the values of ${H_0}^{-1}$ in eqns (16.58) and (16.59) in order to express $d_{\rm H}^0$ in light-years and parsecs.

Let us now consider our Universe in the interval of redshift between matter–radiation equality in eqn (16.88) and matter–vacuum transition in eqn (16.86), when it was matter-dominated. Since the Universe is practically flat (see eqn (16.232)), for $z \gg 1$ the expansion rate as a function of the redshift in eqn (16.90) is approximately given by

$$H(z) \simeq H_0\,z^{3/2}\,\sqrt{\Omega_{\rm M}^0} \qquad (z_{\rm tr} \ll z \ll z_{\rm eq})\,. \tag{16.116}$$

In this case, the particle horizon as a function of redshift is given by

$$d_{\rm H}(z) \simeq 2\,{H_0}^{-1}\,z^{-3/2}\,(\Omega_{\rm M}^0)^{-1/2} \qquad (z_{\rm tr} \ll z \ll z_{\rm eq})\,. \tag{16.117}$$

From eqns (16.103) and (16.105), the corresponding amount of matter in the causal horizon is given by

$$M_{\rm H}(z) \simeq \frac{1}{2\,G_{\rm N}\,H_0\,z^{3/2}\,\sqrt{\Omega_{\rm M}^0}} \simeq 4 \times 10^{22}\,M_\odot\,z^{-3/2}\,(\Omega_{\rm M}^0)^{-1/2} \qquad (z_{\rm tr} \ll z \ll z_{\rm eq})\,. \tag{16.118}$$

Thus, a mass $M \gg M_{\rm H}(z_{\rm eq})$ entered the causal horizon during the matter-dominated era, at the redshift

$$z_{\rm H}(M) \sim 1 \times 10^7\,(\Omega_{\rm M}^0)^{-1/3}\left(\frac{M}{M_{\rm gal}}\right)^{-2/3}\,, \tag{16.119}$$

where $M_{\rm gal}$ is the galactic mass in eqn (16.104).

16.5 Radiation-dominated Universe

This is another important case, since the early Universe was dominated by radiation until the matter–radiation equality given by eqn (16.88).

Radiation dominates the Universe if $\rho_{\mathrm{R}} \gg \rho_{\mathrm{M}}, \rho_{\Lambda}$. If the Universe is flat or the curvature term $-k/R^2$ in the Friedmann equation (16.70) is negligible, from eqns (16.106)–(16.108) with $a = \mathrm{R}$ and $w_{\mathrm{R}} = 1/3$, we obtain

$$R \propto t^{1/2}\,, \tag{16.120}$$

$$t = \frac{1}{2}\,\frac{1}{H(t)}\,. \tag{16.121}$$

Hence, for a given value of H, the age of a radiation-dominated Universe is shorter than the age of a matter-dominated Universe, which is given by eqn (16.110). In the present Universe, the energy density of radiation has been much smaller than the energy density of matter for a long time. However, since the energy density of radiation decreases with the expansion of the Universe faster than the energy density of matter, radiation was dominating in the early Universe (see eqn (16.88)). In fact, the evolution of the early Universe is described by eqns (16.120) and (16.121).

In a radiation-dominated Universe we have

$$\eta \propto t^{1/2} \propto R\,, \tag{16.122}$$

and

$$d_{\mathrm{H}}(t) = 2\,t = H^{-1}(t) \propto R^2\,. \tag{16.123}$$

Hence, the particle horizon is equal to the inverse of the expansion rate. Since

$$\frac{d_{\mathrm{H}}(t)}{R(t)} \propto R \propto t^{1/2}\,, \tag{16.124}$$

the growth in time of the particle horizon with respect to the scale factor is faster than in the case of matter domination (eqn (16.114)). Also in this case there is an increase with time of the size of the comoving part of the Universe which is in causal contact with each observer.

Let us now consider our Universe before matter–radiation equality in eqn (16.88), when it was radiation-dominated. In this early radiation-dominated era the expansion rate as a function of the redshift in eqn (16.90) is approximately given by

$$H(z) \simeq H_0\,z^2\,\sqrt{\Omega_{\mathrm{R}}^0} \qquad (z \gg z_{\mathrm{eq}})\,, \tag{16.125}$$

with the particle horizon

$$d_{\mathrm{H}}(z) \simeq {H_0}^{-1}\,z^{-2}\,(\Omega_{\mathrm{R}}^0)^{-1/2} \qquad (z \gg z_{\mathrm{eq}})\,. \tag{16.126}$$

From eqns (16.103) and (16.105), the corresponding amount of matter in the causal horizon is given by

$$M_{\mathrm{H}}(z) \simeq \frac{\Omega_{\mathrm{M}}^0}{16\,G_{\mathrm{N}}\,H_0\,z^3\,(\Omega_{\mathrm{R}}^0)^{3/2}} \simeq 5 \times 10^{21}\,M_{\odot}\,z^{-3}\,z_{\mathrm{eq}}^{3/2}\,(\Omega_{\mathrm{M}}^0)^{-1/2} \qquad (z \gg z_{\mathrm{eq}})\,. \tag{16.127}$$

Note that this approximate expression for $M_{\rm H}(z)$ in the radiation-dominated era matches with that in eqn (16.118) for the matter-dominated era for $z = z_{\rm eq}/4$, with $z_{\rm eq}$ given by eqn (16.88).

From eqn (16.127), a mass $M \ll M_{\rm H}(z_{\rm eq})$ entered the causal horizon during the radiation-dominated era, at the redshift

$$z_{\rm H}(M) \sim 2 \times 10^3 \sqrt{z_{\rm eq}} \, (\Omega_{\rm M}^0)^{-1/6} \left(\frac{M}{M_{\rm gal}} \right)^{-1/3} , \tag{16.128}$$

where $M_{\rm gal}$ is the galactic mass in eqn (16.104).

16.6 Curvature-dominated Universe

In a curvature-dominated Universe the curvature term $-k/R^2$ in the Friedmann equation (16.70) dominates over the energy density contribution $8\pi G_{\rm N}\rho/3$. This is possible only if $\rho \ll \rho_{\rm c}$, which means that the Universe is open and $k = -1$. In this case, from the Friedmann equation we have

$$\dot{R} = 1 \qquad \Longrightarrow \qquad R = t\,. \tag{16.129}$$

Thus, a curvature-dominated Universe is open and always expanding, with the scale factor equal to time. Since the curvature term decreases only as R^{-2}, whereas $\rho_{\rm M} \propto R^{-3}$ and $\rho_{\rm R} \propto R^{-4}$, if $\rho_\Lambda = 0$, an open Universe eventually becomes curvature-dominated. On the other hand, if the Universe is curvature-dominated at one time, but $\rho_\Lambda \neq 0$, eventually the vacuum contribution in the Friedmann equation dominates.

16.7 Vacuum-dominated Universe

From the acceleration equation (16.71) one can see that if the vacuum energy with $p_\Lambda = -\rho_\Lambda$ dominates the Universe, the acceleration of the expansion rate is positive. Hence, the expansion accelerates instead of decelerating which is the case of matter or radiation domination. The rate of expansion for a vacuum-dominated Universe which is flat or with a negligible curvature term can easily be calculated from the Friedmann equation (16.70): since ρ_Λ is constant, H is also constant:

$$H = \sqrt{\frac{8\pi\, G_{\rm N}}{3}\,\rho_\Lambda}\,. \tag{16.130}$$

From the definition of H in eqn (16.50, we get the solution

$$R \propto e^{Ht}\,. \tag{16.131}$$

Hence, a vacuum-dominated Universe inflates exponentially if inflation lasts for a time much longer than H^{-1}.

It is believed that the Universe underwent an early era of vacuum dominance called *inflation* [579], which could explain some features of the observable Universe. Of course, the vacuum energy could not be exactly constant, otherwise inflation never ended. It is plausible that the vacuum energy was due to the potential energy of a scalar field, called *inflaton*, which was not constant. When the scalar field reached the minimum of the potential, inflation ended and the inflaton field decayed generating the hot plasma which formed the matter and radiation in the Universe. From then on, the Big Bang evolved according to the Standard Cosmological Model (see Ref. [689]).

One of the features of the Universe which is not explained if the Universe just started with the radiation-dominated era is its near flatness. As shown in Fig. 16.5a, in order to have Ω^0 equal to unity within a few percent, the value of Ω at early times must have been much closer to unity. Such a highly fine-tuned primordial value of Ω, called *flatness problem* could be explained with inflation. Let us denote with $\Omega_{\rm i}$, $H_{\rm i}$, and $R_{\rm i}$, respectively, the values of Ω, H, and R at the beginning of inflation. Since during inflation $H = H_{\rm i}$ is constant and $R = R_{\rm i} e^{Ht}$, we obtain, from eqn (16.76),

$$\frac{\Omega - 1}{\Omega_{\rm i} - 1} = \frac{H_{\rm i}^2 \, R_{\rm i}^2}{H^2 \, R^2} = e^{-2Ht} \,. \tag{16.132}$$

Thus, whatever the value of $\Omega_{\rm i} - 1$ during inflation, $\Omega - 1$ rapidly becomes very small, leading to a practically flat Universe.

Another important quantity during inflation is the horizon distance. From eqn (16.65) and the exponential growth of R in eqn (16.131), we get

$$d_{\rm H}(t) = d_{\rm H}(t_{\rm i}) + H^{-1} \left[e^{H(t-t_{\rm i})} - 1 \right] \sim H^{-1} \, e^{Ht} \propto R \,, \tag{16.133}$$

where $t_{\rm i}$ is the time of the beginning of inflation, which is assumed to be very small. Hence, during inflation the causal horizon grows exponentially, as the scale factor R. In fact, this behavior is what one could have expected, because regions of space which were in causal contact before inflation become separated by a physical distance proportional to the scale factor during the inflation of the Universe. This dramatic growth of the horizon distance during inflation could explain the so-called *horizon problem*: widely separated regions in the Universe which would be out of the respective causal horizons without inflation appear to have the same statistical properties. Thus, inflation explains the observed large-scale homogeneity of the Universe.

16.8 Thermodynamics of the early Universe

Since the scale factor $R(t)$ decreases going backwards in time, the density of matter and radiation in the early Universe was so large that there was thermal equilibrium due to rapid interactions of the particles. In this case, each particle species χ forms a dilute weakly-interacting gas whose number density n_χ, energy density ρ_χ and pressure density p_χ are given by

$$n_\chi = \frac{g_\chi}{(2\pi)^3} \int f_\chi(\vec{p})\, \mathrm{d}^3 p\,, \tag{16.134}$$

$$\rho_\chi = \frac{g_\chi}{(2\pi)^3} \int E_\chi(\vec{p})\, f_\chi(\vec{p})\, \mathrm{d}^3 p\,, \tag{16.135}$$

$$p_\chi = \frac{g_\chi}{(2\pi)^3} \int \frac{|\vec{p}|^2}{3E_\chi(\vec{p})}\, f_\chi(\vec{p})\, \mathrm{d}^3 p\,, \tag{16.136}$$

where g_χ is the number of internal degrees of freedom (spin) and $E_\chi(\vec{p}) = \sqrt{|\vec{p}|^2 + m_\chi^2}$ is the energy. The statistical distribution $f_\chi(\vec{p})$ depends on the energy E_χ, the *chemical potential* μ_χ, and the temperature[99] T_χ. It is given by

$$f_\chi(\vec{p}) = \frac{1}{e^{(E_\chi - \mu_\chi)/T_\chi} \pm 1}\,, \tag{16.137}$$

where the plus sign applies to fermions and the minus sign to bosons, respectively. If a particle species is in chemical equilibrium, its chemical potential is related to the chemical potentials of the other species with which it interacts. For example, the reaction $a+b \leftrightarrows c+d$ implies $\mu_a + \mu_b = \mu_c + \mu_d$. Since a particle and its antiparticle can annihilate into different numbers of photons (or other neutral channels), they have opposite chemical potentials. Photons and other self-conjugated particles have vanishing chemical potential. If a particle has a conserved charge, its chemical potential at equilibrium is different from zero if the number densities of particles and antiparticles are different. Since all the observed astrophysical objects appear to be neutral, the chemical potential associated with electric charge is zero or extremely small. On the other hand, since all the observed astronomical objects appear to be made of baryons and not antibaryons, the observed Universe has a finite baryon number. However, since the baryon-to-photon ratio is very small (see section 17.6), the chemical potential associated with baryon number can be neglected for most of the history of the Universe. Hence, in the following discussion we will often neglect chemical potentials.

The integrals in eqns (16.134)–(16.136) can be calculated analytically in the following useful approximations:

Nonrelativistic limit. For $m_\chi \gg T_\chi$ and $m_\chi \gg \mu_\chi$ we have, for both bosons and fermions,

$$f_\chi(\vec{p}) \simeq e^{(\mu_\chi - m_\chi)/T_\chi}\, e^{-|\vec{p}|^2/2mT_\chi}\,, \tag{16.138}$$

which leads to

$$n_\chi \simeq g_\chi \left(\frac{m_\chi T_\chi}{2\pi}\right)^{3/2} \exp\left(\frac{\mu_\chi - m_\chi}{T_\chi}\right)\,, \tag{16.139}$$

[99] We consider the general case in which different species may have thermal distributions with different temperatures. As we will see later, this is what happens to neutrinos, which have a different temperature than the photons, whose temperature was increased by e^+-e^- annihilations when electrons became nonrelativistic, after neutrino decoupling from the thermal bath.

$$\rho_\chi \simeq m_\chi n_\chi \left(1 + \frac{3}{2}\frac{T_\chi}{m_\chi}\right) \tag{16.140}$$

$$p_\chi \simeq n_\chi T_\chi \ll \rho_\chi \,. \tag{16.141}$$

The average energy is given by

$$\langle E_\chi \rangle \equiv \frac{\rho_\chi}{n_\chi} \simeq m_\chi + \frac{3}{2} T_\chi \,. \tag{16.142}$$

This is the case of nonrelativistic matter, with the equation of state in eqn (16.22).

Relativistic limit. For $T_\chi \gg m_\chi$ and $T_\chi \gg \mu_\chi$, we have

$$f_\chi(\vec{p}) \simeq \frac{1}{e^{|\vec{p}|/T_\chi} \pm 1} \,, \tag{16.143}$$

which leads to[100]

$$n_\chi \simeq \begin{cases} \dfrac{\zeta(3)}{\pi^2} g_\chi T_\chi^3 & (\chi = \text{boson})\,, \\ \dfrac{3}{4}\dfrac{\zeta(3)}{\pi^2} g_\chi T_\chi^3 & (\chi = \text{fermion})\,, \end{cases} \tag{16.144}$$

$$\rho_\chi \simeq \begin{cases} \dfrac{\pi^2}{30} g_\chi T_\chi^4 & (\chi = \text{boson})\,, \\ \dfrac{7}{8}\dfrac{\pi^2}{30} g_\chi T_\chi^4 & (\chi = \text{fermion})\,, \end{cases} \tag{16.145}$$

$$p_\chi \simeq \frac{1}{3}\rho_\chi \,, \tag{16.146}$$

where $\zeta(3) \simeq 1.20206$ is the Riemann zeta function of 3 (see eqns (A.130) and (A.131)). The last equation is the origin of the equation of state for radiation in eqn (16.23). The average energy is given by

$$\langle E_\chi \rangle \simeq \langle |\vec{p}_\chi| \rangle \simeq \begin{cases} \dfrac{\pi^4}{30\,\zeta(3)} T_\chi \simeq 2.701\, T_\chi & (\chi = \text{boson})\,, \\ \dfrac{7\pi^4}{180\,\zeta(3)} T_\chi \simeq 3.151\, T_\chi & (\chi = \text{fermion})\,. \end{cases} \tag{16.147}$$

A very important quantity is the number density of photons n_γ, given by the bosonic expression in eqn (16.144) with $g_\gamma = 2$. Since the present photon number density is known to a great accuracy from the measurements of the Cosmic Microwave Background Radiation (CMBR)(see eqn (16.210)), it is convenient to use it as a reference for the number densities of other particles.

For a nonrelativistic particle χ in equilibrium, the ratio of its number density and the photon number density is given by

$$\frac{n_\chi}{n_\gamma} = \frac{\pi^{1/2}}{2^{7/2}\,\zeta(3)} g_\chi \left(\frac{m_\chi}{T_\gamma}\right)^{3/2} \exp\left(\frac{\mu_\chi - m_\gamma}{T_\gamma}\right)$$

[100] The integrals can be evaluated using the formulas in eqns (A.127) and (A.128).

$$\simeq 0.13\, g_\chi \left(\frac{m_\chi}{T_\gamma}\right)^{3/2} \exp\left(\frac{\mu_\chi - m_\chi}{T_\gamma}\right), \tag{16.148}$$

with $T_\chi = T_\gamma$. Since $m_\chi \gg T_\gamma$, this ratio is exponentially suppressed.

The difference between the number densities of a fermion and its antifermion can be obtained from eqn (16.139) in the nonrelativistic limit. On the other hand, in the relativistic limit we have[101] [196]

$$n_\chi - n_{\bar{\chi}} = \frac{1}{6}\, g_\chi\, T_\chi^3 \left[\frac{\mu_\chi}{T_\chi} + \frac{1}{\pi^2}\left(\frac{\mu_\chi}{T_\chi}\right)^3\right] \qquad (T_\chi \gg m_\chi)\,. \tag{16.149}$$

This formula is useful, since it implies that in the relativistic limit

$$\eta_\chi \equiv \frac{n_\chi - n_{\bar{\chi}}}{n_\gamma} = \frac{\pi^2\, g_\chi}{12\,\zeta(3)}\left(\frac{T_\chi}{T_\gamma}\right)^3 \left[\frac{\mu_\chi}{T_\chi} + \frac{1}{\pi^2}\left(\frac{\mu_\chi}{T_\chi}\right)^3\right], \tag{16.150}$$

where we have used eqn (16.144) for the photon number density n_γ. Hence, if the asymmetry η_χ is small, it is approximately proportional to the chemical potential:

$$\eta_\chi \simeq \frac{\pi^2\, g_\chi}{12\,\zeta(3)}\left(\frac{T_\chi}{T_\gamma}\right)^3 \frac{\mu_\chi}{T_\chi} \simeq 0.68\, g_\chi \left(\frac{T_\chi}{T_\gamma}\right)^3 \frac{\mu_\chi}{T_\chi}\,. \tag{16.151}$$

The sum of the energy densities of a relativistic fermion and its antifermion are given by [196]

$$\rho_\chi + \rho_{\bar{\chi}} = \frac{7}{8}\,\frac{\pi^2}{15}\, g_\chi\, T_\chi^4 \left[1 + \frac{30}{7\,\pi^2}\left(\frac{\mu_\chi}{T_\chi}\right)^2 + \frac{15}{7\,\pi^4}\left(\frac{\mu_\chi}{T_\chi}\right)^4\right]. \tag{16.152}$$

A fermion with $\mu_\chi \gtrsim T_\chi$ is called *degenerate*, because its distribution has most of the lowest energy quantum levels filled and, if $g_\chi > 1$, there is a substantial probability to find more than one fermion in the same energy level. On the other hand, the antifermion distribution with chemical potential $\mu_{\bar{\chi}} = -\mu_\chi$ is strongly suppressed. In the limit $T_\chi \to 0$, all the fermion levels with energy smaller than μ_χ are occupied and all the levels with energy larger than μ_χ are empty. In this case, there are no antifermions and μ_χ is called the *Fermi energy*.

Since, in the early hot Universe, the energy density is dominated by relativistic particles, it is convenient to write it as (see eqn (16.145))

$$\rho = \frac{\pi^2}{30}\, g_\rho\, T_\gamma^4\,, \tag{16.153}$$

where T_γ is the photon temperature and the coefficient g_ρ is given by the sum over the contributions of all particles that populate the Universe:

$$g_\rho = \sum_x g_\rho^{(x)}\,. \tag{16.154}$$

[101] The integrals leading to eqns (16.149) and (16.152) can be obtained from the general formula in eqn (A.129).

From eqns (16.135) and (16.137), the contribution of the particle species χ to g_ρ is given by

$$g_\rho^{(\chi)} = g_\chi \frac{15}{\pi^4} \left(\frac{T_\chi}{T_\gamma} \right)^4 \int_{x_\chi}^{\infty} \mathrm{d}z \, \frac{z^2 \sqrt{z^2 - x_\chi^2}}{e^{z-\xi_\chi} \pm 1} \,, \tag{16.155}$$

with $x_\chi \equiv m_\chi / T_\chi$ and $\xi_\chi \equiv \mu_\chi / T_\chi$. The plus sign applies to fermions and the minus sign to bosons. In general, the integral in eqn (16.155) must be evaluated numerically. However, if the particle χ is relativistic ($T_\chi \gg m_\chi$ and $T_\chi \gg \mu_\chi$), the value of $g_\rho^{(\chi)}$ can be inferred from eqn (16.145) to be approximately given by

$$g_\rho^{(\chi)} \simeq \begin{cases} g_\chi \left(T_\chi / T_\gamma \right)^4 & (\chi = \text{boson}) \,, \\ (7/8) \, g_\chi \left(T_\chi / T_\gamma \right)^4 & (\chi = \text{fermion}) \,. \end{cases} \tag{16.156}$$

The contribution to g_ρ of relativistic particles is dominant in the early radiation-dominated era. If all the particles are either relativistic or nonrelativistic, eqn (16.154) is well approximated by

$$g_\rho = \sum_{\substack{\chi=\text{relativistic} \\ \text{bosons}}} g_\chi \left(\frac{T_\chi}{T_\gamma} \right)^4 + \frac{7}{8} \sum_{\substack{\chi=\text{relativistic} \\ \text{fermions}}} g_\chi \left(\frac{T_\chi}{T_\gamma} \right)^4 \,, \tag{16.157}$$

where we assumed $\mu_\chi \ll T_\chi$.

The value of g_ρ changes when the temperature drops below the mass of a particle in the plasma. This particle becomes nonrelativistic and its number, energy, and pressure densities are suppressed according to eqns (16.139)–(16.141). Then, this particle drops out of the sum in eqn (16.157). The transition between the two values of g_ρ must be calculated numerically using the exact expression in eqn (16.155) for the contribution of the particle which becomes nonrelativistic.

For $m_e/3 \simeq 0.2\,\text{MeV} \ll T_\gamma \ll 35\,\text{MeV} \simeq m_\mu/3$, the thermalized relativistic particles are the photons, electrons, positrons, and neutrinos, all with the same temperature, yielding

$$g_\rho = 2 + 2\,\frac{7}{8}\,2 + 6\,\frac{7}{8} = \frac{43}{4} = 10.75 \qquad (0.2\,\text{MeV} \ll T_\gamma \ll 35\,\text{MeV}) \,. \tag{16.158}$$

For $m_\mu/3 \simeq 35\,\text{MeV} \ll T_\gamma \ll 300\,\text{MeV} \simeq \Lambda_{\text{QCD}}$, where Λ_{QCD} is the energy scale of the quark–hadron phase transition, there are also muons and antimuons (contributing 7/2), leading to

$$g_\rho = \frac{57}{4} = 14.25 \qquad (35\,\text{MeV} \ll T_\gamma \ll 300\,\text{MeV}) \,. \tag{16.159}$$

For $T_\gamma \gg 300\,\text{MeV} \simeq \Lambda_{\text{QCD}}$, there are also three colors of u, d, and s quarks and antiquarks (contributing 63/2), eight gluons (contributing 16), c quarks and antiquarks (contributing 21/2 for $T_\gamma \gg m_c/3 \simeq 400\,\text{MeV}$), τ's and $\bar{\tau}$'s (contributing 7/2 for $T_\gamma \gg m_\tau/3 \simeq 600\,\text{MeV}$), b quarks and antiquarks (contributing 21/2 for $T_\gamma \gg m_b/3 \simeq 1.4\,\text{GeV}$), $W^\pm$'s and Z's (contributing 6 for $T_\gamma \gg m_Z/3 \simeq 30\,\text{GeV}$),

t quarks and antiquarks (contributing 21/2 for $T_\gamma \gg m_t/3 \simeq 60\,\mathrm{GeV}$), H (contributing 1 for $T_\gamma \gg m_H/3 \sim 100\,\mathrm{GeV}$), for a total

$$g_\rho = \frac{427}{4} = 106.75 \qquad (T_\gamma \gg 100\,\mathrm{GeV})\,. \tag{16.160}$$

We did not calculate the value of g_ρ for $T_\gamma \ll m_e/3 \simeq 0.2\,\mathrm{MeV}$, because at that temperature neutrinos are decoupled and have a temperature which is different from the photon temperature. The relation between the neutrino and photon temperatures and the value of g_ρ for $T_\gamma \ll m_e/3 \simeq 0.2\,\mathrm{MeV}$ will be calculated in section 17.2.

In the early radiation-dominated epoch, the curvature term, k/R^2, in the Friedmann equation (16.70) was negligible[102]. Neglecting also the contribution of nonrelativistic particles, and using eqn (16.153), the expansion rate is given by

$$H = \frac{2\,\pi^{3/2}}{3\sqrt{5}\,M_{\mathrm{P}}}\sqrt{g_\rho}\,T_\gamma^2 \simeq 0.2066\,\sqrt{g_\rho}\left(\frac{T_\gamma}{\mathrm{MeV}}\right)^2 \mathrm{s}^{-1}\,, \tag{16.161}$$

where M_{P} is the Planck mass in eqn (A.167). If g_ρ was constant (the case of a temperature-dependent g_ρ is discussed later), from eqn (16.121) we obtain the cosmic time as a function of temperature:

$$t(T_\gamma) = \frac{3\sqrt{5}\,M_{\mathrm{P}}}{4\,\pi^{3/2}}\,g_\rho^{-1/2}\,T_\gamma^{-2} \simeq 2.420\,g_\rho^{-1/2}\left(\frac{T_\gamma}{\mathrm{MeV}}\right)^{-2} \mathrm{s}\,. \tag{16.162}$$

Moreover, from eqn (16.120) it follows that the scale factor R and the temperature T_γ are inversely proportional:

$$T_\gamma \propto R^{-1}\,. \tag{16.163}$$

16.9 Entropy

Let us now consider the case of a temperature-dependent g_ρ. In order to find the relation between T_γ and R in this case, it is necessary to consider the second law of thermodynamics, which says that in eqn (16.15) $đQ = T\,\mathrm{d}S$, where $S(T,V)$ is a

[102] From eqn (16.77), the curvature term can be written as $k/R^2 = \left(\Omega^0 - 1\right) H_0{}^2\,(1+z)^2$. Since the photon density is proportional to R^{-4} (see eqn (16.83)), the photon contribution to the Friedmann equation is given by $8\pi\,G_{\mathrm{N}}\,\rho_\gamma/3 = \Omega_\gamma^0\,H_0{}^2\,(1+z)^4$. At early times, corresponding to large values of z, the photon contribution was much larger than the curvature term: even if $|\Omega^0 - 1|$ were as large as unity, the photon contribution would dominate over the curvature term for $1+z \gtrsim (\Omega_\gamma^0)^{-1/2} \simeq 140$, with $\Omega_\gamma^0 \simeq 5\times 10^{-5}$ (see eqn (16.212)). This value of the redshift is much smaller than the redshift of matter–radiation equality $z_{\mathrm{eq}} \simeq 3\times 10^3$ (see eqns (16.88) and (17.23)), beyond which radiation dominates.

function of state of the system called the *entropy*. From eqn (16.15) we have

$$T\,\mathrm{d}S(T,V) = \mathrm{d}[\rho(T)V] + p(T)\,\mathrm{d}V = V\,\frac{\mathrm{d}\rho(T)}{\mathrm{d}T}\,\mathrm{d}T + [\rho(T)+p(T)]\,\mathrm{d}V\,. \quad (16.164)$$

Hence,

$$\frac{\partial S(T,V)}{\partial T} = \frac{V}{T}\,\frac{\mathrm{d}\rho(T)}{\mathrm{d}T}\,, \qquad \frac{\partial S(T,V)}{\partial V} = \frac{\rho(T)+p(T)}{T}\,. \quad (16.165)$$

The integrability condition

$$\frac{\partial^2 S(T,V)}{\partial V \partial T} = \frac{\partial^2 S(T,V)}{\partial T \partial V} \quad (16.166)$$

implies that

$$\frac{\mathrm{d}p(T)}{\mathrm{d}T} = \frac{\rho(T)+p(T)}{T}\,. \quad (16.167)$$

Using this equality in eqn (16.15) with $đQ = T\,\mathrm{d}S$, we obtain

$$\mathrm{d}S = \mathrm{d}\left[\frac{\rho+p}{T}\,V\right]\,. \quad (16.168)$$

This means that, apart from an additive constant, the entropy $S(T,V)$ of the volume V at the temperature T is given by

$$S(T,V) = \frac{\rho+p}{T}\,V\,. \quad (16.169)$$

In thermal equilibrium the entropy $S(T,V)$ is conserved ($T\,\mathrm{d}S = đQ = 0$). It is useful to define the entropy density

$$s = \frac{S}{V} = \frac{\rho+p}{T}\,, \quad (16.170)$$

and the conservation of entropy in the expanding Universe implies that

$$s \propto R^{-3}\,. \quad (16.171)$$

It is convenient to write the entropy density of the plasma as a function of the photon temperature T_γ, which is the temperature of the plasma:

$$s = \frac{2\pi^2}{45}\,g_s\,T_\gamma^3\,, \quad (16.172)$$

where g_s is given by the sum over the effective coefficients of the interacting particles:

$$g_s = \sum_{\chi=\text{interacting}} g_s^{(\chi)}\,. \quad (16.173)$$

From eqns (16.135)–(16.137) and (16.170), the value of $g_s^{(\chi)}$ is given by

$$g_s^{(\chi)} = g_\chi\,\frac{15}{4\pi^4}\int_{x_\chi}^{\infty}\mathrm{d}z\,\frac{\left(4z^2 - x_\chi^2\right)\sqrt{z^2 - x_\chi^2}}{e^{z-\xi_\chi} \pm 1}\,, \quad (16.174)$$

where $x_\chi \equiv m_\chi/T_\gamma$ and $\xi_\chi \equiv \mu_\chi/T_\gamma$. The plus sign applies to fermions and the minus sign to bosons. The exact value of the integral in eqn (16.174) must be evaluated numerically. However, at temperatures at which the particle χ is relativistic

($T_\gamma \gg m_\chi$ and $T_\gamma \gg \mu_\chi$), the value of $g_s^{(\chi)}$ can be inferred from eqns (16.145) and (16.146):

$$g_s^{(\chi)} \simeq \begin{cases} g_\chi & (\chi = \text{boson}), \\ (7/8)\, g_\chi & (\chi = \text{fermion}). \end{cases} \tag{16.175}$$

On the other hand, for a nonrelativistic particle ($m_\chi \gg T_\gamma$ and $m_\chi \gg \mu_\chi$), from eqns (16.139)–(16.141) we obtain

$$g_s^{(\chi)} \simeq g_\chi \frac{45}{2^{5/2}\, \pi^{7/2}} \left(\frac{m_\chi}{T_\gamma} \right)^{5/2} \exp\left(\frac{\mu_\chi - m_\chi}{T_\gamma} \right), \tag{16.176}$$

for both bosons and fermions. It is clear that the contribution to the entropy of a nonrelativistic particle is strongly suppressed by the exponential factor with respect to the contribution of a relativistic particle.

Since the contribution to the entropy of relativistic particles is dominant, it is common to write

$$g_s = \sum_{\substack{\chi=\text{interacting}\\ \text{relativistic}\\ \text{bosons}}} g_\chi + \frac{7}{8} \sum_{\substack{\chi=\text{interacting}\\ \text{relativistic}\\ \text{fermions}}} g_\chi \,. \tag{16.177}$$

Equation (16.177) gives the value of g_s when the interacting particles are either relativistic or nonrelativistic. When a relativistic interacting particle becomes non-relativistic as a consequence of the temperature decrease of the expanding Universe, its entropy is transferred to the other particles and its contribution to g_s must be calculated solving numerically the exact relevant expression in eqn (16.174).

Note that when the approximation in eqn (16.177) is valid, the entropy density s is proportional to the number density of relativistic interacting particles. Therefore, conservation of entropy implies that the number of relativistic interacting particles in a comoving volume is constant when g_s is constant.

From eqns (16.171) and (16.172), we obtain

$$T_\gamma \propto g_s^{-1/3}\, R^{-1} \,. \tag{16.178}$$

This equation is the generalization of eqn (16.163) including possible variations of the number of relativistic interacting particles. During the cooling of the expanding Universe, when a particle species becomes nonrelativistic, its entropy is transferred to the remaining interacting relativistic particles through a change of the factor $g_s^{-1/3}$. Since in this case the value of g_s decreases, the cooling of the plasma is slowed down with respect to the cooling for constant g_s, given by $T_\gamma \propto R^{-1}$.

Equation (16.178) allows us to calculate the behavior of T_γ as a function of $1+z$:

$$T_\gamma = T_\gamma^0 \left(\frac{g_s^0}{g_s} \right)^{1/3} \frac{R_0}{R} = T_\gamma^0 \left(\frac{2}{g_s} \right)^{1/3} (1+z) \,, \tag{16.179}$$

where T_γ^0 is the present temperature of the CMBR (see eqn (16.209)) and

$$g_s^0 = g_s^{(\gamma)} = g_\gamma = 2 \tag{16.180}$$

is the present value of g_s. A useful quantity is the photon number density n_γ, which is related to its present value n_γ^0 by

$$n_\gamma = n_\gamma^0 \, \frac{2}{g_s} \, (1+z)^3 \, . \tag{16.181}$$

Figure 16.6a shows the behavior of T_γ as a function of $1+z$ up to $z \simeq 10^{12}$. One can see that T_γ is proportional to $1+z$ everywhere, except at $T_\gamma \sim m_e/3 \sim 0.2\,\text{MeV}$. At that temperature the electrons and positrons became nonrelativistic and their entropy was transferred to the photons through e^+-e^- annihilations. The decrease of temperature is slowed down because of the change of value of g_s from

$$g_s = g_s^{(\gamma)} + g_s^{(e^\pm)} = 2 + \frac{7}{8}\,4 = \frac{11}{2} \qquad \text{for} \qquad \frac{m_e}{3} \ll T_\gamma \ll \frac{m_\mu}{3} \, , \tag{16.182}$$

to the present value in eqn (16.180) (at $T_\gamma \sim m_\mu/3 \sim 35\,\text{MeV}$ there is another change of g_s due to μ^+-μ^- annihilations). From eqn (16.179) we obtain

$$T_\gamma = T_\gamma^0 \left(\frac{4}{11}\right)^{1/3} (1+z) \qquad \text{for} \qquad \frac{m_e}{3} \ll T_\gamma \ll \frac{m_\mu}{3} \, . \tag{16.183}$$

In Fig. 16.6b we have plotted T_γ as a function of cosmic time t. We used eqn (16.92) for $T_\gamma < 10^4\,\text{eV}$. For $T_\gamma > 10^4\,\text{eV}$ eqn (16.92) is not accurate, because it has been derived under the implicit assumption that the numbers of relativistic and nonrelativistic particles are constant. This assumption breaks down at the time of electron–positron annihilation. Since from eqn (16.178) we have

$$H = \frac{\dot{R}}{R} = -\frac{\dot{T}_\gamma}{T_\gamma} \left(1 + \frac{1}{3} \frac{\mathrm{d}\ln g_s}{\mathrm{d}\ln T_\gamma}\right) , \tag{16.184}$$

in the early radiation-dominated Universe, from eqn (16.161) we obtain

$$t(T_\gamma) = \frac{3\sqrt{5}\,M_\mathrm{P}}{2\,\pi^{3/2}} \int_{T_\gamma}^{\infty} \frac{\mathrm{d}T'_\gamma}{\sqrt{g_\rho}\,{T'_\gamma}^3} \left(1 + \frac{1}{3} \frac{\mathrm{d}\ln g_s}{\mathrm{d}\ln T'_\gamma}\right) . \tag{16.185}$$

One can easily check that if g_ρ and g_s were constant, $t(T_\gamma)$ would be given by eqn (16.162). Since g_s increases with T_γ, when g_s changes the variation of T_γ is slower than for constant g_s. In other words, when g_s decreases during the expansion of the Universe, the cooling of the plasma is slowed down with respect to the cooling $T_\gamma \propto t^{-1/2}$ for constant g_s.

An important application of the relation in eqn (16.183) is the calculation of the temperature of relic neutrinos, which is shown by the dash-dotted lines in Figs. 16.6a and b. This application will be discussed in section 17.2.

16.10 Decoupling

In general, a particle decouples from the primordial plasma when its interaction rate Γ becomes smaller than the rate of change of the temperature, $|\dot{T}_\gamma|/T_\gamma$ or the

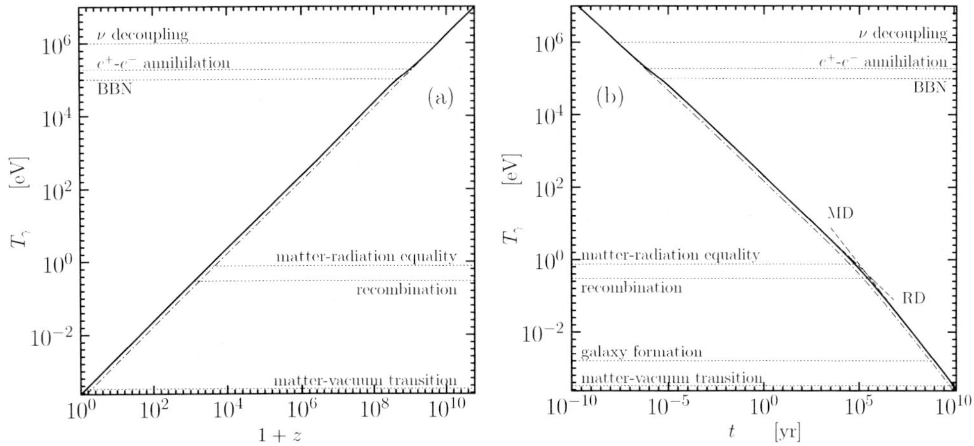

FIG. 16.6. (a) Photon temperature T_γ (solid curve) as a function of $1+z$ from eqn (16.179). The dash-dotted line shows the behavior of the neutrino temperature $T_\nu = T_\nu^0 (1+z)$, with T_ν^0 given in eqn (17.17). (b) Photon temperature T_γ (solid curve) as a function of cosmic time t. The dash-dotted line shows the neutrino temperature T_ν as a function of cosmic time t. The dashed lines illustrate the behavior of T_γ in a Universe which is dominated always by radiation (RD) or by matter (MD).

expansion rate H. When this happens, the interactions are not rapid enough to adjust the momentum distribution function to the change in temperature and the particle drops out of equilibrium, i.e. decouples. Neglecting changes in g_ρ for an order of magnitude estimate, since $T_\gamma \propto R^{-1}$, we have $\dot{T}_\gamma / T_\gamma = -H$, which leads to the decoupling relation

$$\Gamma \sim H \,. \tag{16.186}$$

Another way to understand this decoupling relation is to consider the mean free path of the particle given by Γ^{-1}. Since the particle horizon in a radiation-dominated Universe is given by H^{-1} (see eqn (16.123)), when $\Gamma \lesssim H$ the mean free path becomes larger than the particle horizon, leading to decoupling.

After decoupling, the number of decoupled particles in a comoving volume remains constant, and the number density scales as R^{-3}:

$$n_\chi = n_\chi^{\chi\text{-dec}} \left(\frac{R}{R_{\chi\text{-dec}}} \right)^{-3} . \tag{16.187}$$

From eqn (16.181) the ratio between the present number density of a decoupled particle χ and the present photon number density is given by

$$\frac{n_\chi^0}{n_\gamma^0} = \frac{2}{g_s^{\chi\text{-dec}}} \frac{n_\chi^{\chi\text{-dec}}}{n_\gamma^{\chi\text{-dec}}} . \tag{16.188}$$

This quantity is useful, because the present photon number density is known with great accuracy from measurements of the CMBR (see eqn (16.210)).

Furthermore, since the momentum scales as R^{-1} (see eqn (16.36)),

$$|\vec{p}| = |\vec{p}|_{\chi\text{-dec}} \left(\frac{R}{R_{\chi\text{-dec}}}\right)^{-1} , \tag{16.189}$$

for the momentum distribution function we have

$$f_\chi(\vec{p}) = \frac{\mathrm{d}^3 n_\chi}{\mathrm{d}^3 p} = \left.\frac{\mathrm{d}^3 n_\chi}{\mathrm{d}^3 p}\right|_{\chi\text{-dec}} = f_\chi(\vec{p})|_{\chi\text{-dec}} \,. \tag{16.190}$$

Since we have T $T_\chi^{\chi\text{-dec}} = T_\gamma^{\chi\text{-dec}}$ at the decoupling, the momentum distribution function after decoupling is given by

$$f_\chi(\vec{p}) = \left[\exp\left(\frac{\sqrt{|\vec{p}|^2\,(R/R_{\chi\text{-dec}})^2 + m_\chi^2} - \mu_\chi^{\chi\text{-dec}}}{T_\gamma^{\chi\text{-dec}}}\right) \pm 1\right]^{-1} . \tag{16.191}$$

In general, it is not possible to define an effective temperature and an effective chemical potential (which must be momentum-independent) which describe the evolution of the momentum distribution function after the decoupling. Such a simple description is possible only in the relativistic and nonrelativistic limits which we are going to discuss. Note, however, that these limits are of practical interest: most particles decouple when they are either relativistic or nonrelativistic, since the coincidence of the values of their mass and decoupling temperature is unlikely.

If the particle χ decouples when it is relativistic ($T_\gamma^{\chi\text{-dec}} \gg m_\chi$ and $T_\gamma^{\chi\text{-dec}} \gg \mu_\chi^{\chi\text{-dec}}$), it is called a *hot relic*. From eqn (16.191), after decoupling its momentum distribution function is approximately given by

$$f_\chi(\vec{p}) \simeq \left[\exp\left(\frac{|\vec{p}|\,(R/R_{\chi\text{-dec}})}{T_\gamma^{\chi\text{-dec}}}\right) \pm 1\right]^{-1} = \frac{1}{e^{|\vec{p}|/T_\chi} \pm 1} , \tag{16.192}$$

with the effective temperature

$$T_\chi = T_\gamma^{\chi\text{-dec}} \left(\frac{R}{R_{\chi\text{-dec}}}\right)^{-1} . \tag{16.193}$$

Hence, after decoupling a hot relic maintains a relativistic momentum distribution function (compare eqn (16.192) with eqn (16.143)), with an effective temperature scaling as R^{-1}. Note that this behavior is independent of the mass of the particle (as long as $m_\chi \ll T_\gamma^{\chi\text{-dec}}$): even if the particle χ becomes nonrelativistic after decoupling, its momentum distribution function has the relativistic form in eqn (16.192).

Since we have, from eqn (16.178),

$$T_\gamma = T_\gamma^{\chi\text{-dec}} \left(\frac{g_s}{g_s^{\chi\text{-dec}}} \right)^{-1/3} \left(\frac{R}{R_{\chi\text{-dec}}} \right)^{-1} , \tag{16.194}$$

the relation between T_χ and T_γ after decoupling is

$$T_\chi = \left(\frac{g_s}{g_s^{\chi\text{-dec}}} \right)^{1/3} T_\gamma \, . \tag{16.195}$$

This relation is very useful, because it allows us to calculate the present temperature of χ from the well-known present temperature of the CMBR (given in eqn (16.209)). After decoupling, we also have the useful relation

$$n_\chi = B \left(\frac{g_s}{g_s^{\chi\text{-dec}}} \right) n_\gamma \, , \tag{16.196}$$

between the number density of χ and the photon number density. The factor B is equal to 1 or 3/4 if χ is a boson or a fermion, respectively.

Consider now a *cold relic*, i.e. a particle χ which decouples when it is nonrelativistic. From eqn (16.191), after decoupling its momentum distribution function is approximately given by

$$f_\chi(\vec{p}) \simeq \exp\left(\frac{\mu_\chi^{\chi\text{-dec}} - m_\chi - |\vec{p}|^2 \left(R/R_{\chi\text{-dec}}\right)^2 / 2m_\chi}{T_\gamma^{\chi\text{-dec}}} \right) = e^{(\mu_\chi - m_\chi)/T_\chi} \, e^{|\vec{p}|^2/2m_\chi T_\chi} \, , \tag{16.197}$$

with the effective temperature

$$T_\chi = T_\gamma^{\chi\text{-dec}} \left(\frac{R}{R_{\chi\text{-dec}}} \right)^{-2} , \tag{16.198}$$

and the effective chemical potential

$$\mu_\chi = m_\chi + (\mu_\chi^{\chi\text{-dec}} - m_\chi) \left(\frac{T}{T_{\chi\text{-dec}}} \right) = m_\chi + (\mu_\chi^{\chi\text{-dec}} - m_\chi) \left(\frac{R}{R_{\chi\text{-dec}}} \right)^{-2} . \tag{16.199}$$

Hence, after decoupling a cold relic maintains a nonrelativistic momentum distribution function (compare eqn (16.197) with eqn (16.138)), with an effective temperature which scales as R^{-2}, decreasing with the expansion of the Universe faster than in the relativistic case. Note the behavior of the chemical potential, which is just a parameter in the distribution function, because the decoupled particle is noninteracting. Its behavior has been chosen in order to keep $(\mu_\chi - m_\chi)/T_\chi$ equal to its decoupling value $(\mu_\chi^{\chi\text{-dec}} - m_\chi)/T_\gamma^{\chi\text{-dec}}$. In particular, even if $\mu_\chi^{\chi\text{-dec}} = 0$, the chemical potential after decoupling is different from zero. Moreover, after decoupling, we have $\mu_\chi \neq -\mu_{\bar{\chi}}$: from eqn (16.199) and $\mu_\chi^{\chi\text{-dec}} = -\mu_{\bar{\chi}}^{\chi\text{-dec}}$, we have $\mu_\chi + \mu_{\bar{\chi}} = 2m_\chi \left[1 - \left(R/R_{\chi\text{-dec}}\right)^{-2} \right]$.

It is useful to calculate the present ratio of the number density of a cold relic χ and the photon number density. Equation (16.148) gives the value of this ratio at χ decoupling:

$$\frac{n_{\chi}^{\chi\text{-dec}}}{n_{\gamma}^{\chi\text{-dec}}} \simeq 0.13\, g_{\chi} \left(\frac{m_{\chi}}{T_{\gamma}^{\chi\text{-dec}}} \right)^{3/2} \exp\left(\frac{\mu_{\chi}^{\chi\text{-dec}} - m_{\chi}}{T_{\gamma}^{\chi\text{-dec}}} \right) . \tag{16.200}$$

Then, from eqn (16.188), the present ratio is given by

$$\frac{n_{\chi}^{0}}{n_{\gamma}^{0}} \simeq 0.26\, \frac{g_{\chi}}{g_{s}^{\chi\text{-dec}}} \left(\frac{m_{\chi}}{T_{\gamma}^{\chi\text{-dec}}} \right)^{3/2} \exp\left(\frac{\mu_{\chi}^{\chi\text{-dec}} - m_{\chi}}{T_{\gamma}^{\chi\text{-dec}}} \right) . \tag{16.201}$$

This rather simple equation allows an approximate calculation of the present number density of a cold relic once its decoupling temperature is known.

A more accurate calculation of the number density n_{χ} of a relic particle χ can be performed by solving the Boltzmann equation (see Refs.[689, 856, 379])

$$\frac{\mathrm{d}n_{\chi}}{\mathrm{d}t} = -3\, H\, n_{\chi} - \langle \sigma v \rangle \left[n_{\chi}^{2} - (n_{\chi}^{\mathrm{eq}})^{2} \right] , \tag{16.202}$$

where n_{χ}^{eq} is the number density that the particle species χ would have if it were in thermal equilibrium, which is given by eqn (16.134) with $T_{\chi} = T_{\gamma}$.

The first term on the right-hand side of eqn (16.202) represents the decrease of the number density n_{χ} due to the expansion of the Universe: in the absence of the interaction term on the right-hand side, the solution of eqn (16.202) would be $n_{\chi} \propto R^{-3}$, which is a simple geometrical decrease of the number density.

The cross-section σ in eqn (16.202) refers to all the possible processes of χ-$\bar{\chi}$ annihilation, which can change the number of χ's and $\bar{\chi}$'s in a comoving volume. The angular brackets represent thermal averaging and v is the particle velocity. The quantity $\langle \sigma v \rangle\, n_{\chi}^{2}$ represents annihilation of χ-$\bar{\chi}$ pairs and the quantity $\langle \sigma v \rangle\, (n_{\chi}^{\mathrm{eq}})^{2}$ represent the inverse processes of creation of χ-$\bar{\chi}$ pairs from other particles in thermal equilibrium. The rate of these inverse processes is related to the equilibrium number density n_{χ}^{eq} by the principle of detailed balance. Indeed, in a static Universe ($H = 0$), there is equilibrium if the number density remains constant, which implies $n_{\chi} = n_{\chi}^{\mathrm{eq}}$.

In order to scale out the effect of the expansion of the Universe, it is convenient to consider the number of particles in a comoving volume R^{3},

$$N_{\chi} = n_{\chi}\, R^{3} . \tag{16.203}$$

The Boltzmann equation (16.202) assumes the simple form (see Ref. [856])

$$\frac{\mathrm{d}\ln N_{\chi}}{\mathrm{d}\ln R} = -\frac{\Gamma}{H} \left[1 - \left(\frac{N_{\chi}^{\mathrm{eq}}}{N_{\chi}} \right)^{2} \right] , \tag{16.204}$$

where

$$\Gamma = \langle \sigma v \rangle\, n_{\chi} \tag{16.205}$$

is the annihilation rate. The ratio Γ/H regulates the rate of change of the number of particles in a comoving volume as the Universe expands. The solutions of eqn (16.204) when either Γ or H dominate are easily seen to be

$$N_\chi = N_\chi^{\mathrm{eq}} \qquad \text{for} \qquad \Gamma \gg H\,, \tag{16.206}$$

$$N_\chi = \text{constant} \qquad \text{for} \qquad \Gamma \ll H\,. \tag{16.207}$$

In general, the transition between the two regimes, which gives the value of the constant relic N_χ for $\Gamma \ll H$, must be calculated by solving the differential equation numerically. However, it can be solved approximately in the relativistic and nonrelativistic limits.

If the particle χ decouples when it is relativistic, before decoupling we have $N_\chi = N_\chi^{\mathrm{eq}} \propto T_\gamma^3 R^3$. If g_ρ is constant during decoupling, N_χ^{eq} remains constant before, during, and after decoupling, because T_γ scales as R^{-1} (see eqn (16.163)). Therefore, for a hot relic $N_\chi = N_\chi^{\mathrm{eq}}$ is a solution of the Boltzmann equation (16.204) during and after decoupling. Since $N_\chi \propto T_\chi^3 R^3$ remains constant after decoupling, we have $T_\chi \propto R^{-1}$, in agreement with eqn (16.193), obtained in a different way.

If the particle χ decouples when it is nonrelativistic, from eqn (16.139) we have $N_\chi^{\mathrm{eq}} \propto T_\gamma^{3/2} \exp[(\mu_\chi - m_\chi)/T_\gamma]\, R^3$. The Boltzmann factor e^{-m_χ/T_γ} implies that N_χ^{eq} decreases rapidly as the Universe expands and cools. On the other hand, according to eqn (16.207) N_χ becomes constant after decoupling. Hence, there is a deviation of N_χ from N_χ^{eq} at decoupling. The relic number density in eqn (16.201) was derived in the approximation in which the constant relic N_χ is given by the value of N_χ^{eq} at the decoupling temperature $T_\gamma^{\chi\text{-dec}}$ defined by $\Gamma = H$. In such an approximation, the small deviation of N_χ from N_χ^{eq} for $T_\gamma > T_\gamma^{\chi\text{-dec}}$ and the small evolution of N_χ for $T_\gamma < T_\gamma^{\chi\text{-dec}}$ are neglected.

16.11 Cosmic microwave background radiation

The cosmic microwave background radiation (CMBR), discovered in 1965 by Penzias and Wilson [860, 368], is one of the main sources of information on the physics and history of the Universe. It was generated after the temperature of the expanding Universe became low enough to allow electrons and nuclei to form neutral atoms. This process, called the *recombination*, occurred at (see Refs. [689, 858, 856])

$$T_\gamma^{\mathrm{rec}} \simeq 0.3\,\mathrm{eV}\,, \qquad z_{\mathrm{rec}} \simeq 1.1 \times 10^3\,, \qquad t_{\mathrm{rec}} = 3.8 \times 10^5\,\mathrm{yr}\,. \tag{16.208}$$

Before the recombination, matter and photons were in thermal equilibrium through rapid scattering of photons on free electrons. At the recombination, the number of free electrons dropped dramatically and the Universe became transparent to photons, leading to *photon decoupling*. After the primordial photons have decoupled from matter, they traveled undisturbed until the present time, when we see them as the CMBR. The only change that they have suffered is the redshift of their wavelength by the factor $z_{\mathrm{rec}} \simeq 1100$, which implies a temperature $T_\gamma^0 \simeq 3 \times 10^{-4}\,\mathrm{eV} \simeq 3\,\mathrm{K}$. In fact, the observed CMBR has the perfect blackbody spectrum

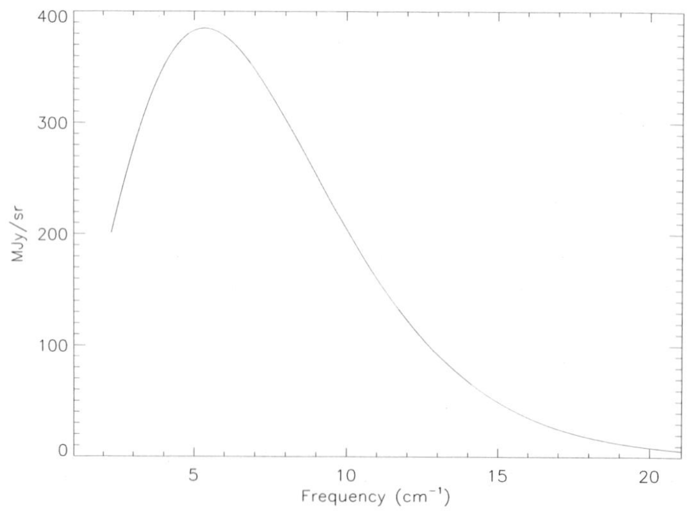

FIG. 16.7. CMBR spectrum and its blackbody fit (indistinguishable) from Ref. [446]. Experimental uncertainties are a small fraction of the line thickness.

shown in Fig. 16.7, with temperature [944]

$$T_\gamma^0 = 2.725 \pm 0.001\,\mathrm{K} = (2.348 \pm 0.001) \times 10^{-4}\,\mathrm{eV}\,, \tag{16.209}$$

which was measured with the FIRAS detector on the COBE satellite [782, 446, 783].

From eqns (16.144) and (16.145) with $g_\gamma = 2$, the CMBR temperature corresponds to the number and energy densities

$$n_\gamma^0 = \frac{2\,\zeta(3)}{\pi^2}\,(T_\gamma^0)^3 = 410.5 \pm 0.5\,\mathrm{cm}^{-3}\,, \tag{16.210}$$

$$\rho_\gamma^0 = \frac{\pi^2}{15}\,(T_\gamma^0)^4 = 0.2604 \pm 0.0004\,\mathrm{eV\,cm}^{-3}\,. \tag{16.211}$$

The contribution to the energy density of the Universe is

$$\Omega_\gamma^0 = (2.471 \pm 0.004) \times 10^{-5}\,h^{-2} = (4.7 \pm 1.0) \times 10^{-5}\,. \tag{16.212}$$

In the last equality we used the value of h in eqn (16.57). Since $\Omega^0 \simeq 1$ (see the following eqn (16.232)), the current contribution of CMBR to the energy density of the Universe is practically negligible.

In 1992 the DMR detector on the COBE satellite measured small variations of the temperature of the CMBR in different directions [967]. As we will see in the following, these anisotropies give important cosmological information.

The anisotropies are fluctuations $\Delta T_\gamma(\theta,\phi) \equiv T_\gamma(\theta,\phi) - T_\gamma$ of the temperature $T_\gamma(\theta,\phi)$ with respect to the average temperature T_γ in eqn (16.209), which depend on the polar angles θ and ϕ. It is convenient to expand the anisotropies in spherical harmonics (given in eqn (A.138)):

$$\frac{\Delta T_\gamma}{T_\gamma}(\theta,\phi) = \sum_{\ell=0}^{\infty}\sum_{m=-\ell}^{\ell} a_{\ell m}\, Y_\ell^m(\theta,\phi)\,, \tag{16.213}$$

where $a_{\ell m}$ are the multipole moments. Since the spherical harmonic Y_ℓ^m corresponds to angular variations $\Delta\theta \sim \pi/\ell$, low values of ℓ correspond to anisotropies on large scales, whereas large values of ℓ correspond to anisotropies on small scales. From the orthonormality relation of spherical harmonics in eqn (A.139), the multipole moments are given by $a_{\ell m} = \int \frac{\Delta T_\gamma}{T_\gamma}(\theta,\phi)\, Y_\ell^{m*}(\theta,\phi)\, \mathrm{d}\cos\theta\, \mathrm{d}\phi$. Note that the monopole moment a_{00} of the fluctuations is equal to zero, because T_γ is the monopole moment of the temperature: $T_\gamma = \int T_\gamma(\theta,\phi)\, Y_\ell^{m*}(\theta,\phi)\, \mathrm{d}\cos\theta\, \mathrm{d}\phi$. The first nontrivial multipole moments of the temperature anisotropies are the dipole moments with $\ell = 1$. The COsmic Background Explorer (COBE) [967] and Wilkinson Microwave Anisotropy Probe (WMAP) [206] experiments measured a dipole amplitude which is interpreted as the result of a Doppler shift due to the motion of the solar system with respect to the comoving frame. The measured velocity of the barycenter of the solar system is $368 \pm 2\,\mathrm{km\,s^{-1}}$, which implies a velocity of the local group of galaxies of $627 \pm 22\,\mathrm{km\,s^{-1}}$. Since the dipole does not carry cosmological information, it is subtracted from the data for the study of cosmological CMBR anisotropies. The multipoles with $\ell \geq 2$ are interpreted as due to perturbations in the energy density of the Universe at the time of last scattering of the photons ($z_{\mathrm{rec}} \simeq 1100$).

In order to understand the treatment of perturbations of the cosmic energy density, it is important to realize from the beginning that given a cosmological model in which perturbations are generated, it is not possible to calculate the exact amount of perturbations in the observable Universe, because we do not know the initial conditions. However, a proper cosmological model must be able to predict the statistical properties of the perturbations, extracted from a statistical ensemble of possible universes. The real Universe is assumed to be a member of such an ensemble and the observed perturbations are a realization of the statistical distribution of possible perturbations. Comparison of a cosmological model with observations is done by comparing the predicted and measured statistical properties of the perturbations. Of course, we do not have experimental access to an ensemble of universes (which does not necessarily exist). Thus, from the observational point of view the statistical properties of perturbations are obtained by averaging over large volumes or different directions in the sky. According to the *ergodic* assumption, the perturbations in widely separated parts of the Universe are independent and averaging over a sufficiently large volume is equivalent to the average over an ensemble of universes.

It is convenient to work with measurable quantities which depend on the statistical distribution of the temperature fluctuations. A widely used quantity is the two-point correlation function (see Refs. [1060, 628, 275, 180, 510, 626, 206])

$$C(\theta) \equiv \left\langle \frac{\Delta T_\gamma}{T_\gamma}(\theta_1,\phi_1)\, \frac{\Delta T_\gamma}{T_\gamma}(\theta_2,\phi_2) \right\rangle , \tag{16.214}$$

where θ is the angle, given in eqn (A.141), between the directions (θ_1, ϕ_1) and (θ_2, ϕ_2). The average is done over the model-dependent statistical distribution of temperature fluctuations. Since the fluctuations are assumed to be statistically isotropic, i.e. with statistical properties which are independent on the direction of view, the two-point correlation function is the same for all directions (θ_1, ϕ_1) and (θ_2, ϕ_2) which subtend the same angle θ. Hence, $C(\theta)$ depends only on θ.

Since $\Delta T_\gamma / T_\gamma$ is real, we can replace the second one in eqn (16.214) with its complex conjugate and obtain, from eqn (16.213),

$$C(\theta) = \sum_{\ell_1, m_1} \sum_{\ell_2, m_2} \langle a_{\ell_1 m_1} \, a^*_{\ell_2 m_2} \rangle \, Y^{m_1}_{\ell_1}(\theta_1, \phi_1) \, Y^{m_2 *}_{\ell_2}(\theta_2, \phi_2) \, . \tag{16.215}$$

The average over the statistical distribution of temperature fluctuations is equivalent to an average over the statistical distribution of the multipoles. If the multipoles are independent random variables, the matrix of covariances $\langle a_{\ell m} \, a^*_{\ell' m'} \rangle$ is diagonal, i.e. $\langle a_{\ell m} \, a^*_{\ell' m'} \rangle \propto \delta_{\ell \ell'} \delta_{m m'}$. Moreover, if the temperature fluctuations are statistically isotropic, the variances of the multipoles do not depend on m, leading to

$$\langle a_{\ell m} \, a^*_{\ell' m'} \rangle = C_\ell \, \delta_{\ell \ell'} \, \delta_{m m'} \, , \tag{16.216}$$

with

$$C_\ell = \frac{1}{2\ell + 1} \sum_{m=-\ell}^{\ell} \langle |a_{\ell m}|^2 \rangle \, . \tag{16.217}$$

The set of C_ℓ's forms the so-called *angular power spectrum*. Using eqn (16.216) in eqn (16.215), and the addition theorem of spherical harmonics in eqn (A.140), we obtain

$$C(\theta) = \sum_{\ell} \frac{2\ell + 1}{4\pi} \, C_\ell \, P_\ell(\cos\theta) \, . \tag{16.218}$$

Therefore, the two-point correlation function gives information on the angular power spectrum: from the relation of orthogonality of Legendre polynomials in eqn (A.134) we have

$$C_\ell = \int_{-1}^{1} C(\theta) \, P_\ell(\cos\theta) \, \mathrm{d}\cos\theta \, . \tag{16.219}$$

If the temperature fluctuations have a Gaussian distribution, as predicted by simple inflationary models (see Refs. [762, 182]), the statistical distribution of each multipole has zero mean ($\langle a_{\ell m} \rangle = 0$) and is fully characterized by only one parameter, the variance C_ℓ. In this case, the two-point correlation function $C(\theta)$ completely characterizes the temperature fluctuations.

The angular power spectrum C_ℓ is a theoretical quantity that can be evaluated in each cosmological model. In order to test the cosmological models it is necessary to compare the theoretical angular power spectrum C_ℓ with the experimental angular

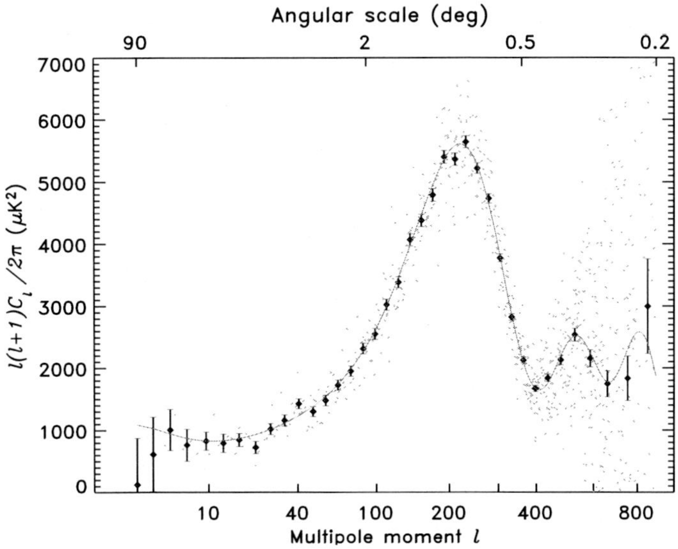

FIG. 16.8. Angular power spectrum of CMBR temperature anisotropies measured by WMAP [970] (black dots with error bar). The gray dots are unbinned data. The solid line represent the angular power spectrum in the best-fit flat ΛCDM model, in which $h = 0.72 \pm 0.05$, $\Omega_B^0 h^2 = 0.024 \pm 0.001$, $\Omega_M^0 h^2 = 0.14 \pm 0.02$.

power spectrum $\widehat{C}_\ell$, obtained from the data using eqn (16.217):

$$\widehat{C}_\ell = \frac{1}{2\ell + 1} \sum_{m=-\ell}^{\ell} |\widehat{a}_{\ell m}|^2 \,, \tag{16.220}$$

where $\widehat{a}_{\ell m}$ are the measured multipoles. Figure 16.8 shows the angular power spectrum of CMBR temperature anisotropies measured by WMAP [970], compared with the theoretical angular power spectrum obtained in the best-fit flat ΛCDM model (see section 17.8), in which $h = 0.72 \pm 0.05$, $\Omega_B^0 h^2 = 0.024 \pm 0.001$, $\Omega_M^0 h^2 = 0.14 \pm 0.02$. One can note that the error bars of the experimental data points are large for small and large values of ℓ. Since large values of ℓ correspond to fluctuations with small size, i.e. small angle of view, the largeness of the error bars is due to the finite resolution of the detector. Hence, in that case the uncertainty can be reduced by experiments with a better angular resolution. On the other hand, the largeness of the error bars at small values of ℓ is due to an intrinsic uncertainty called *cosmic variance*. It is due to the fact that for small values of ℓ the number $2\ell + 1$ of measurable multipoles is small and the estimate $\widehat{C}_\ell$ of C_ℓ has a large uncertainty. If the multipoles are random Gaussian variables, $\sum_m |\widehat{a}_{\ell m}|^2 / C_\ell$ has a χ^2 distribution with $2\ell + 1$ degrees of freedom (see, for example, Ref. [400]), and its variance is $2(2\ell+1)$. Taking into account that $\mathrm{Var}(ax) = a^2 \,\mathrm{Var}(x)$, the uncertainty of the measure of C_ℓ is given by

$$\sqrt{\mathrm{Var}(\widehat{C}_\ell)} = \sqrt{\frac{2}{2\ell+1}}\,\widehat{C}_\ell\,. \tag{16.221}$$

Therefore, even in a perfect experiment it is impossible to know the value of C_ℓ with a fractional uncertainty better than $(\ell+1/2)^{-1/2}$.

The peaks in the angular power spectrum in Fig. 16.8 are called *acoustic peaks*, because they are due to gravity-driven density oscillations in the primordial plasma. It is assumed that some mechanism, such as inflation, has generated an initial spectrum of density perturbations which could evolve dynamically only after matter–radiation equality, at $z_{\mathrm{eq}} \simeq 3\times 10^3$ and $t_{\mathrm{eq}} \simeq 6\times 10^4\,\mathrm{yr}$ (see eqn (17.23) and Fig. 16.5b), because before matter–radiation equality their evolution was damped by the dominating photon pressure. Since the initial perturbations are small, it is possible to study their evolution using a linear theory, i.e. neglecting all the effects which are quadratic or higher-order in the perturbations. In this case, it is convenient to Fourier transform the relative density fluctuations,

$$\delta(t,\vec{x}) \equiv \frac{\rho(t,\vec{x}) - \langle\rho(t)\rangle}{\langle\rho(t)\rangle} = \int \frac{\mathrm{d}^3k}{(2\pi)^3}\,\delta(t,\vec{k})\,e^{i\vec{k}\cdot\vec{x}}\,. \tag{16.222}$$

This trick allows one to transform differential equations into algebraic ones. In the linear theory, the algebraic equations for the amplitude of each fluctuation mode with wavenumber $\vec{k}$ are independent. In other words, the amplitude $\delta(t,\vec{k})$ of each fluctuation mode evolves in time independently of the others and can be conveniently studied separately. If $\vec{x}$ is a physical distance, $\vec{k}$ is a physical wavenumber (contrasted with the often used comoving wavenumber given by $\vec{k}R$), and

$$\lambda(t) = \frac{2\pi}{|\vec{k}|} \tag{16.223}$$

is the corresponding physical wavelength. In the linear approximation, the wavelength of each Fourier mode evolves independently as the Universe expands.

If the physical wavelength of a Fourier mode is larger than the particle horizon, its amplitude cannot evolve dynamically. Since its comoving wavelength is fixed, its physical wavelength grows with the scale factor. As we have seen in eqns (16.114) and (16.124), the particle horizon grows faster than the scale factor both in a matter-dominated and in a radiation-dominated Universe. Hence, if there is enough time, a Fourier mode may enter the causal horizon before recombination. After entering the causal horizon, gravity causes an increase of the density of overdense regions and a decrease of the density of underdense regions, until the photon pressure reverts the process, leading to a series of density oscillations called *cosmic acoustic oscillations*. These oscillations stop at recombination, when photons decouple. The CMBR carries a picture of the last oscillations just before photon decoupling, which manifest themselves in the acoustic peaks in Fig. 16.8. The angular scale $\theta \sim \pi/\ell$ shown on the top axis corresponds to the physical wavelength

$\lambda(t_{\rm rec})$ through eqn (16.61):

$$\theta = \frac{\lambda(t_{\rm rec})}{d_{\rm A}{}^{\rm rec}} = \frac{\lambda(t_{\rm rec})}{R_{\rm rec}\, r_{\rm rec}} , \tag{16.224}$$

where $d_{\rm A}{}^{\rm rec} = R_{\rm rec}\, r_{\rm rec}$ is the angular diameter distance of the last-scattering surface, which we see now at comoving coordinate $r_{\rm rec}$. From eqn (16.45) we have

$$r_{\rm rec} = S_k\left(\int_{t_{\rm rec}}^{t_0} \frac{{\rm d}t'}{R(t')}\right) = S_k(\eta_0 - \eta_{\rm rec}) \simeq S_k(\eta_0) , \tag{16.225}$$

where η_0 and $\eta_{\rm rec}$ are, respectively, the present conformal time and the much smaller conformal time at recombination. Thus, we have

$$d_{\rm A}{}^{\rm rec} \simeq R_{\rm rec}\, S_k(\eta_0) , \tag{16.226}$$

which shows that the angular diameter distance of the last-scattering surface depends on the geometry of the Universe. This important fact allows us to probe the geometry of the Universe by measuring the locations of the acoustic peaks in the angular power spectrum, if the wavelength corresponding to the peaks is known. In fact, it is possible to calculate rather precisely the wavelength of the peaks (see Refs. [628, 275, 626]), which depends on the rapidity of the acoustic oscillations, governed by the speed of sound

$$v_{\rm s} \equiv \sqrt{\frac{\partial p}{\partial \rho}} . \tag{16.227}$$

In the primordial plasma, pressure was dominated by photons, leading to $v_{\rm s} \simeq 1/\sqrt{3}$. Since the first peak corresponds to the wavelength which has undergone the first maximal compression, it is given by the distance that sound has traveled from the Big Bang to the recombination time, which is the so-called *sound horizon* given by an expression analogous to that for the particle horizon in eqn (16.68):

$$\lambda_1(t_{\rm rec}) = v_{\rm s}\, \eta_{\rm rec}\, R_{\rm rec} \simeq \frac{1}{\sqrt{3}}\, \eta_{\rm rec}\, R_{\rm rec} . \tag{16.228}$$

From eqns (16.224)–(16.228), the first peak corresponds to the angle

$$\theta_1 \simeq \frac{1}{\sqrt{3}}\, \frac{\eta_{\rm rec}}{S_k(\eta_0)} . \tag{16.229}$$

For a flat matter-dominated Universe, from eqn (16.112) we have $\eta_{\rm rec}/\eta_0 = (1 + z_{\rm rec})^{-1/2}$. For $z_{\rm rec} \simeq 1100$, we obtain

$$\theta_1 \simeq 0.02 \simeq 1^\circ \qquad (\text{flat MD}) , \tag{16.230}$$

corresponding to the angular momentum

$$\ell_1 \simeq \frac{\pi}{\theta_1} \simeq 200 . \tag{16.231}$$

From Fig. 16.8 one can see that the first peak is indeed close to $\ell = 200$, which is evidence for an almost flat Universe. A detailed calculation allowed the WMAP

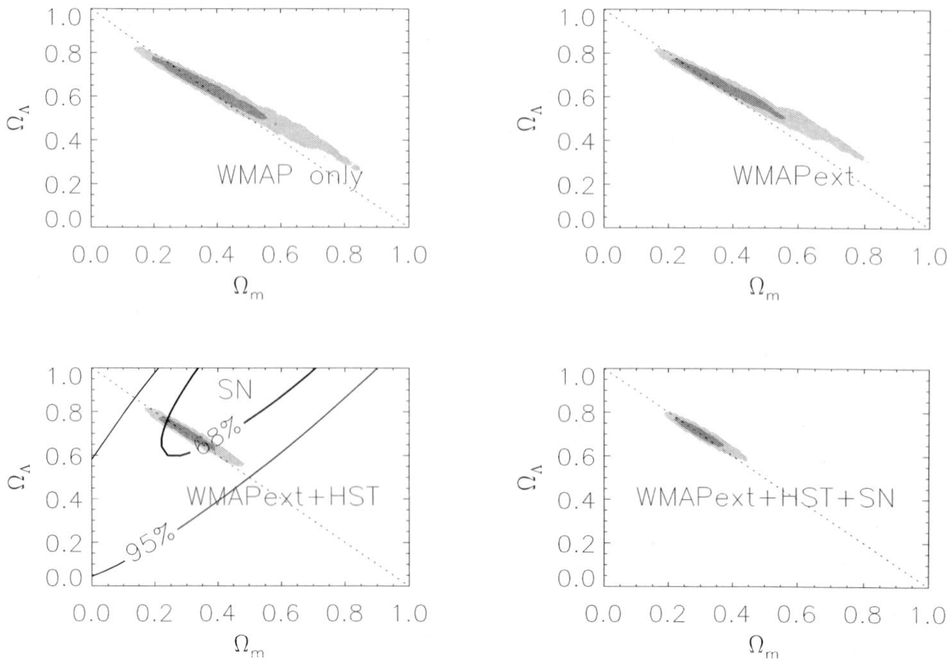

FIG. 16.9. Constraints on the geometry of the Universe in the $\Omega^0_{\rm M}$-Ω^0_Λ plane [970]. The dark and light areas are, respectively, 68% and 95% likelihood regions for four combinations of data: WMAP only, WMAPext (WMAP + other CMBR experiments), WMAP + Hubble Space Telescope Key Project, WMAP + Hubble Space Telescope Key Project + high-z supernovae data.

collaboration [970] to obtain from their data the constraints on the geometry of the Universe shown in Fig. 16.9. The most stringent constraint on the value of the total energy density of the Universe has been obtained by combining WMAP data with the Hubble Space Telescope Key Project determination of h in eqn (16.57) [461] and high-z supernovae data [903, 904]:

$$\Omega^0 = 1.02 \pm 0.02 \,. \tag{16.232}$$

Thus, the Universe is flat or almost flat, with very stringent limits on the possible deviation from flatness.

The other peaks in Fig. 16.8 correspond to multiples of ℓ_1. Their amplitudes are progressively damped by the viscosity of the primordial plasma and by the average over the finite interval of time of photon decoupling. These mechanisms imply that the amplitudes of the peaks give information on the amount of matter. In fact, the WMAP collaboration could measure with remarkable accuracy the baryon and matter densities as given by

$$\Omega^0_{\rm B}\, h^2 = 0.0224 \pm 0.0009 \,, \tag{16.233}$$

$$\Omega^0_{\rm M}\, h^2 = 0.135^{+0.008}_{-0.009} \,. \tag{16.234}$$

The clear discrepancy between the two numbers is evidence that most of the matter in the Universe is nonbaryonic. This so-far unknown form of matter is usually called *dark matter* and its density is written as $\Omega_{\rm DM}$. Strictly speaking one should distinguish between baryonic and nonbaryonic dark matter, because only about 10% of the baryonic matter is luminous [484]:

$$\Omega^0_{\rm lum}\, h^2 = 0.0024^{+0.0012}_{-0.0009}\,. \tag{16.235}$$

However, from the astrophysical point of view there is no problem in the existence of a large amount of dark baryonic matter in the form of dead stars (black dwarfs, neutron stars, black-holes), failed stars (brown dwarfs) and giant planets (see Ref. [1028]). Hence, unless otherwise specified, the term *dark matter* usually refers to nonbaryonic dark matter.

The baryon density in eqn (16.233) implies that the baryon-to-photon ratio is very small. From eqns (16.74), (16.210), and (A.169), this quantity, usually denoted by η (easily distinguished from the conformal time in eqn (16.66) from the context), is given by

$$\eta \equiv \frac{n_{\rm B}}{n_\gamma} \simeq \frac{\rho_{\rm B}}{m_p\, n_\gamma} = \frac{\rho_{\rm c}\,\Omega_{\rm B}}{m_p\, n_\gamma} \simeq 2.7\times 10^{-8}\,\Omega^0_{\rm B}\, h^2 = (6.1\pm 0.3)\times 10^{-10}\,. \tag{16.236}$$

Since the Universe at present is composed of baryons, but not antibaryons, η is equal to the baryon asymmetry $\eta_{\rm B} \equiv (n_{\rm B} - n_{\bar{\rm B}})/n_\gamma$. The ratio η has remained constant after electron–positron annihilation at $T_\gamma \sim 0.2\,{\rm MeV}$, because both $n_{\rm B}$ and n_γ have decreased proportionally to R^{-3}. From eqn (16.183) it follows that before electron–positron annihilation the value of η was larger by a factor 11/4, keeping that value up to the quark–hadron phase transition at $T_\gamma \sim 200\,{\rm MeV}$.

17

RELIC NEUTRINOS

> The most incomprehensible thing about the world is that it is comprehensible.
> **Albert Einstein**

Relic neutrinos are an important product of the standard hot Big Bang. Neutrinos were in thermal equilibrium in the hot plasma which filled the early Universe through weak interactions with the other particles. As the Universe expanded and cooled, the rates of weak interaction processes decreased and neutrinos decoupled when these rates became smaller than the expansion rate (decoupling is discussed in section 16.10). Since for the three known light neutrinos with masses smaller than about 1 eV the decoupling occurred when they were relativistic, these neutrinos are hot relics.

Relic neutrinos pervade space, but their temperature T_ν^0 is extremely small, being of the order of $10^{-4}\,\text{eV}$ (see eqn (17.17)). This implies that their weak interaction cross-section with matter is extremely small, of the order of

$$\sigma \sim G_F^2 (T_\nu^0)^2 \sim 10^{-64}\,\text{cm}^2\,, \tag{17.1}$$

if neutrinos are massless. If neutrinos are massive and nonrelativistic, the cross-section is larger, but still very small, being of the order of

$$\sigma \sim G_F^2 m_\nu^2 \sim 10^{-56} \left(\frac{m_\nu}{1\,\text{eV}}\right)^2 \text{cm}^2\,. \tag{17.2}$$

Therefore, the direct detection of relic neutrinos is an extremely difficult task with present experimental techniques [838, 984, 732, 294, 715, 907, 515, 906].

However, as we will see in this chapter, relic neutrinos are an essential ingredient of the Standard Cosmological Model. The ability of the Standard Cosmological Model to explain the astrophysical data is indirect evidence of the existence of the relic neutrino background. Moreover, the astrophysical observations provide us of information on the neutrino properties, in particular on their mass.

In the following sections 17.1, 17.2, 17.3, and 17.4, we discuss, respectively, the decoupling of light neutrinos at $T_\gamma \sim 1\,\text{MeV}$, the effect of electron–positron annihilation at $T_\gamma \sim m_e/3 \sim 0.2\,\text{MeV}$, the present neutrino temperature and the energy density of light massive neutrinos. In section 17.5, we consider hypothetical heavy active flavor neutrinos. In sections 17.6 and 17.7, we review, respectively, the main properties of Big-Bang nucleosynthesis and large-scale structure formation, which are cornerstones of the Standard Cosmological Model and depend on relic neutrinos and their properties. In section 17.8 we present the results of recent global fits of cosmological data, with an emphasis on the resulting limits on neutrino masses. Finally, in sections 17.9 and 17.10 we review, respectively, the cosmological constraints on the number of neutrino species and on the neutrino asymmetries.

17.1 Neutrino decoupling

In the early Universe, when $1\,\text{MeV} \lesssim T_\gamma \lesssim 100\,\text{MeV}$, neutrinos were kept in equilibrium with the primordial plasma by the weak interactions

$$\nu + \bar{\nu} \leftrightarrows e^+ + e^- \,, \tag{17.3}$$

$$\overset{(-)}{\nu} + e^\pm \leftrightarrows \overset{(-)}{\nu} + e^\pm \,. \tag{17.4}$$

The reactions with nucleons were negligible, because the number density of the nonrelativistic nucleons was much smaller than the density of relativistic electrons and positrons. The interaction rate for each neutrino was given by

$$\Gamma = n \, \langle \sigma v \rangle \,, \tag{17.5}$$

where n is the number density of target particles, σ is the cross-section and $v \simeq 1$ is the neutrino velocity. The angle brackets denote thermal averaging. For the weak interaction processes in eqns (17.3) and (17.4), we have

$$\langle \sigma v \rangle \sim G_{\text{F}}^2 \, T_\gamma^2 \,, \tag{17.6}$$

where the temperature T_γ of the thermal bath gives the order of magnitude of the energies of the relativistic particles participating in the reactions. Since the number density of relativistic particles was $n \sim T_\gamma^3$, the interaction rate for each neutrino was

$$\Gamma \sim G_{\text{F}}^2 \, T_\gamma^5 \,. \tag{17.7}$$

One can see that the neutrino interaction rate decreases rapidly with the decrease of temperature due to the expansion of the Universe.

In order to find the neutrino decoupling temperature, we use the relation in eqn (16.186),

$$\Gamma \sim H \,. \tag{17.8}$$

Using the expression of H as a function of T_γ for a radiation-dominated Universe in eqn (16.161),

$$H \sim \frac{T_\gamma^2}{M_{\text{P}}} \,, \tag{17.9}$$

we obtain the decoupling temperature

$$T_\gamma^{\nu\text{-dec}} \sim \left(M_{\text{P}} \, G_{\text{F}}^2 \right)^{-1/3} \sim 1\,\text{MeV} \,. \tag{17.10}$$

More precise calculations, which take into account the different interactions of $\overset{(-)}{\nu}_e$ and $\overset{(-)}{\nu}_{\mu,\tau}$, yield (see Ref. [379])

$$T_\gamma^{\overset{(-)}{\nu}_e\text{-dec}} \simeq 1.3\,\text{MeV} \,, \qquad T_\gamma^{\overset{(-)}{\nu}_{\mu,\tau}\text{-dec}} \simeq 1.5\,\text{MeV} \,. \tag{17.11}$$

The decoupling temperature of $\overset{(-)}{\nu}_e$ is smaller than that of $\overset{(-)}{\nu}_{\mu,\tau}$ because $\overset{(-)}{\nu}_e$ interact with electrons and positrons more strongly than $\overset{(-)}{\nu}_{\mu,\tau}$ (see section 5.1.1).

17.2 Electron-positron annihilation

Since neutrinos were relativistic at decoupling, their temperature evolves after the decoupling according to eqn (16.193):

$$T_\nu \propto R^{-1} \qquad \text{for} \qquad T_\gamma < T_\gamma^{\nu\text{-dec}} \sim 1\,\text{MeV}\,. \tag{17.12}$$

Since before the decoupling when the temperature was equal to the photon temperature, T_γ scaled also as R^{-1} for $T_\gamma \ll m_\mu/3 \sim 35\,\text{MeV}$, we have

$$T_\nu = T_\nu^0\,(1+z) \qquad \text{for} \qquad T_\gamma \ll \frac{m_\mu}{3}\,. \tag{17.13}$$

On the other hand, as we have explained in section 16.9, the photon temperature does not evolve as R^{-1}, because at a temperature of about $m_e/3 \simeq 0.2\,\text{MeV}$, after neutrino decoupling, electrons and positrons became nonrelativistic and their entropy was transferred to the photons, which are the only remaining relativistic particles in equilibrium. This was done through e^+-e^- annihilations into photons.

Comparing eqn (16.179) and eqn (17.13) for $T_\gamma \gg T_\gamma^{\nu\text{-dec}}$, when $T_\nu = T_\gamma$, we easily obtain the following relation between the present neutrino and photon temperatures:

$$T_\nu^0 = \left(\frac{4}{11}\right)^{1/3} T_\gamma^0\,. \tag{17.14}$$

Unfortunately, this very precise theoretical prediction has not been checked experimentally so far, because of the enormous difficulty of detecting relic neutrinos due to their extremely weak interactions.

Since both the neutrino and photon temperatures scaled as R^{-1} after e^+-e^- annihilation, they have maintained the same ratio since that time:

$$T_\nu = \left(\frac{4}{11}\right)^{1/3} T_\gamma \simeq 0.7138\,T_\gamma \qquad (T_\gamma \ll 0.2\,\text{MeV})\,. \tag{17.15}$$

This relation can also be obtained from eqn (16.195) with the present value of g_s in eqn (16.180) and the value of g_s at neutrino decoupling given in eqn (16.182).

Using eqn (17.15), we can calculate the value of g_ρ for $T_\gamma \ll 0.2\,\text{MeV}$:

$$g_\rho = g_\gamma + 6\,\frac{7}{8}\left(\frac{T_\nu}{T_\gamma}\right)^4 = 2 + \frac{21}{4}\left(\frac{4}{11}\right)^{4/3} \simeq 3.363 \qquad (T_\gamma \ll 0.2\,\text{MeV})\,, \tag{17.16}$$

where we have used the fact that there are three left-handed neutrinos and three right-handed antineutrinos.

17.3 Neutrino temperature

From the measured temperature of the CMBR in eqn (16.209) we can infer the present neutrino temperature with remarkable accuracy:

$$T_\nu^0 = \left(\frac{4}{11}\right)^{1/3} T_\gamma^0 = 1.945 \pm 0.001\,\mathrm{K} = (1.676 \pm 0.001) \times 10^{-4}\,\mathrm{eV}\,. \qquad (17.17)$$

Let us compare this temperature with the constraints on the values of neutrino masses obtained in neutrino oscillations experiments (see section 13.3.4). Since at least two neutrino species have masses larger than about $8\times10^{-2}\,\mathrm{eV}$ (see Fig. 13.10), at least the neutrinos belonging to these two species are nonrelativistic in the relic neutrino background. From eqn (17.13), the massive neutrino ν_j with mass m_j became nonrelativistic at redshift

$$z_{\nu_j\text{-nr}} \sim \frac{m_j}{3\,T_\nu^0} \simeq 2.0 \times 10^3 \left(\frac{m_j}{\mathrm{eV}}\right)\,. \qquad (17.18)$$

However, as explained in section 16.10, since neutrinos were relativistic at the decoupling, the number density of relic neutrinos is given by the relativistic expression in eqn (16.144) independently from the values of their masses. In other words, light neutrinos are hot relics and contribute to the hot dark matter in the Universe. From eqns (16.179) and (16.193), for any value of the light neutrino masses, (which are much smaller than $T_\nu^{\nu\text{-dec}} \sim 1\,\mathrm{MeV}$), the present neutrino temperature is given by

$$T_\nu^0 = T_\nu^{\nu\text{-dec}} \left(1 + z_{\nu\text{-dec}}\right)^{-1} = \left(\frac{4}{11}\right)^{1/3} T_\gamma^0\,, \qquad (17.19)$$

in agreement with eqn (17.17).

For any value of the light neutrino masses, the present number density of each neutrino generation is given by eqn (16.144) with the neutrino temperature in eqn (17.17) (see also eqn (16.196)):

$$n_\nu^0 + n_{\bar\nu}^0 = \frac{3}{2}\frac{\zeta(3)}{\pi^2}\, g_\nu\, (T_\nu^0)^3 = \frac{6}{11}\frac{\zeta(3)}{\pi^2}\, (T_\gamma^0)^3 = \frac{3}{11}\, n_\gamma^0 = 111.9 \pm 0.1\,\mathrm{cm}^{-3}\,, \qquad (17.20)$$

where $g_\nu = 1$, because one has to take into account only left-handed neutrinos and right-handed antineutrinos which were in thermal equilibrium in the primordial plasma for $T_\gamma \gtrsim T_\gamma^{\nu-\mathrm{dec}}$.

An important quantity is the present contribution of neutrinos to the energy density of the Universe. Massless neutrinos and neutrinos with masses much smaller than the effective neutrino temperature in eqn (17.17) are still relativistic. For each of them, the contribution to the energy density of the Universe is given by the photon contribution in eqn (16.212) times $(4/11)^{4/3}$:

$$\Omega^0_{\nu\text{-relativistic}} = \left(\frac{4}{11}\right)^{4/3} \Omega_\gamma^0 = (0.6414 \pm 0.001) \times 10^{-5}\, h^{-2} = (1.2 \pm 0.3) \times 10^{-5}\,. \qquad (17.21)$$

The total radiation density in the case of three relativistic neutrinos is given by

$$\Omega_{\mathrm{R}}^{0} = \left[1 + 3\left(\frac{4}{11}\right)^{4/3}\right]\Omega_{\gamma}^{0} = (4.395 \pm 0.007) \times 10^{-5}\, h^{-2} = (8.4 \pm 1.8) \times 10^{-5}\,, \tag{17.22}$$

which is very small, practically negligible in comparison with the matter and vacuum contributions (see section 17.8). This is the case of the SM, where neutrinos are massless.

Assuming that all the three light neutrinos were relativistic before matter–radiation equality, we can calculate the redshift z_{eq} of matter–radiation equality from eqns (16.88) and (17.22):

$$z_{\mathrm{eq}} \simeq 2.3 \times 10^{4}\, \Omega_{\mathrm{M}}^{0} \simeq 3.1 \times 10^{3}\,, \tag{17.23}$$

where we have used the values of $\Omega_{\mathrm{M}}^{0}\, h^{2}$ in eqn (16.234) and h in eqn (16.57). From eqn (17.18), we have $z_{\nu_j\text{-nr}} < z_{\mathrm{eq}}$ for $m_j \lesssim 1.5\,\mathrm{eV}$. Therefore, the estimate of z_{eq} in eqn (17.23) is valid if all the light neutrino masses are lighter than about $1.5\,\mathrm{eV}$. Since the laboratory upper bound on the light neutrino masses is of the order of 2 eV (see section (14.1.1)), the estimate of z_{eq} in eqn (17.23) is rather plausible. The corresponding photon temperature is

$$T_{\gamma}^{\mathrm{eq}} \simeq 0.73\,\mathrm{eV}\,. \tag{17.24}$$

An important consequence of the value of z_{eq} in eqn (17.23) is that the mass that could form a galaxy, given in eqn (16.104), entered the horizon shortly before the time of matter–radiation equality, as one can see from eqn (16.128) (the corresponding eqn (16.119) in a matter-dominated Universe is not applicable, because it would give a redshift of the order of 10^{7}, which is out of the range corresponding to a matter-dominated Universe).

17.4 Energy density of light massive neutrinos

Nonrelativistic massive neutrinos may give a significant contribution to the energy density of the Universe. From eqn (16.140), for each massive neutrino species ν_j, the energy density of left-handed ν_j's and right-handed $\bar{\nu}_j$'s is given by the mass m_j times the number density in eqn (17.20):

$$\rho_{\nu_j}^{0} + \rho_{\bar{\nu}_j}^{0} = m_j\left(n_{\nu}^{0} + n_{\bar{\nu}}^{0}\right)\,, \tag{17.25}$$

which corresponds to

$$\Omega_{\nu_j}^{0}\, h^{2} = \frac{m_j}{94.14\,\mathrm{eV}}\,. \tag{17.26}$$

It is common to express the total contribution of neutrinos to the energy density of the Universe as

$$\Omega_{\nu}^{0}\, h^{2} = \frac{\sum_j m_j}{94.14\,\mathrm{eV}}\,, \tag{17.27}$$

where we have omitted the negligible contribution of massless neutrinos, given by eqn (17.21). Thus, the neutrino energy density is proportional to the sum of neutrino masses. This value is relevant for the present energy balance if there are neutrinos with masses of the order of 1 eV or more.

Equation (17.27) is very important from a historical point of view, because one can obtain, in a straight forward way, a powerful bound on the sum of neutrino masses from the well-known fact that the Universe is not over-closed by matter. From the WMAP measurement of the matter content of the Universe in eqn (16.234), we obtain

$$\Omega_\nu^0 \, h^2 < \Omega_{\rm M}^0 \, h^2 \simeq 0.14 \qquad \Longrightarrow \qquad \sum_j m_j \lesssim 13 \, {\rm eV} \, . \tag{17.28}$$

This type of limit on the masses of light neutrinos (depending on the constraint imposed on $\Omega_\nu^0 \, h^2$) is often called the *Gershtein–Zeldovich limit* [518] or *Cowsik–McClelland limit* [336].

The bound in eqn (17.28) is less stringent than the limit on neutrino masses obtained in tritium β-decay experiments (see section 14.1.1). However, we will see in section 17.7 that massive neutrinos have a significant effect on the formation of structures in the early Universe. Using this effect, rich experimental surveys of structures at high redshift lead to a constraint on the sum of neutrino masses which is more powerful than that in eqn (17.28) and competitive with the bound obtained in tritium β-decay (see section 17.8).

17.5 Energy density of heavy neutrinos

The existence of more than three active flavor neutrinos with mass smaller than $m_Z/2 \simeq 45 \, {\rm GeV}$ has been excluded in 1990 by LEP measurements of the invisible width of the Z boson (see section 5.1.3). Before this very important result, it was common to consider the contribution to the energy density of the Universe of heavy active flavor neutrinos ν_h, which were nonrelativistic at the decoupling. Thus, such a ν_h is called a cold relic. We present, in the following, an estimate of such an energy density for historical completeness and also because its method can be generalized to the calculation of the energy density of other particles with different interactions.

First, we must estimate the decoupling temperature, which is different from that of light flavor neutrinos because of the different interaction rate. Since, for a nonrelativistic particle, the average energy is practically equal to its mass (see eqn (16.142)), the ν_h-$\bar{\nu}_h$ annihilation cross-section of heavy active flavor neutrinos with mass $m_{\nu_h} \ll m_Z/2 \simeq 45 \, {\rm GeV}$ is of order

$$\langle \sigma v \rangle \sim G_{\rm F}^2 \, m_{\nu_h}^2 \, . \tag{17.29}$$

The case $m_{\nu_h} \gtrsim m_Z/2 \simeq 45 \, {\rm GeV}$, in which one must take into account the Z-boson propagator, will be discussed at the end of this section.

Considering the nonrelativistic number density in eqn (16.139) in the nondegenerate case, the ν_h-$\bar{\nu}_h$ annihilation rate, which changes the number of ν_h's and

$\bar{\nu}_h$'s in a comoving volume, is given by

$$\Gamma \sim 0.1\, G_{\mathrm{F}}^2\, m_{\nu_h}^{7/2}\, T_\gamma^{3/2}\, e^{-m_{\nu_h}/T_\gamma} \,. \tag{17.30}$$

From eqn (16.161), the expansion rate of the Universe is given by

$$H \sim \frac{\sqrt{g_\rho}}{M_{\mathrm{P}}}\, T_\gamma^2 \,. \tag{17.31}$$

The decoupling relation $\Gamma \sim H$ implies that the decoupling temperature $T_\gamma^{\nu_h\text{-dec}}$ is given by

$$\frac{m_{\nu_h}}{T_\gamma^{\nu_h\text{-dec}}} - \ln \frac{m_{\nu_h}}{T_\gamma^{\nu_h\text{-dec}}} \sim \ln \frac{G_{\mathrm{F}}^2\, M_{\mathrm{P}}}{10\, \sqrt{g_\rho}} + 3\, \ln m_{\nu_h} \,. \tag{17.32}$$

Considering $m_{\nu_h} \gtrsim 1\,\mathrm{GeV}$, we have $g_\rho \sim 10^2$ and $\ln(G_{\mathrm{F}}^2\, M_{\mathrm{P}}/10\,\sqrt{g_\rho}) \sim 15$ for m_{ν_h} in units of GeV. Since the logarithm on the left-hand side of eqn (17.32) can be neglected, the decoupling temperature is approximately given by (see also Refs. [689, 379])

$$\frac{m_{\nu_h}}{T_\gamma^{\nu_h\text{-dec}}} \sim 15 + 3\, \ln\left(\frac{m_{\nu_h}}{\mathrm{GeV}}\right) \,. \tag{17.33}$$

From eqn (16.201), we can calculate the present ratio of the number density of $\nu_h + \bar{\nu}_h$'s and the photon number density: for a nondegenerate ($\mu_{\nu_h}^{\nu_h\text{-dec}} = 0$) Dirac neutrino ($g_{\nu_h} = 2$) and $g_s^{\nu_h\text{-dec}} = g_\rho^{\nu_h\text{-dec}} \sim 10^2$ we have

$$\frac{n_{\nu_h}^0 + n_{\bar{\nu}_h}^0}{n_\gamma^0} \sim 10^{-2} \left(\frac{m_{\nu_h}}{T_\gamma^{\nu_h\text{-dec}}}\right)^{3/2} e^{-m_{\nu_h}/T_\gamma^{\nu_h\text{-dec}}} \,. \tag{17.34}$$

Assuming that m_{ν_h} is roughly of the order of 1 GeV, we obtain, from eqns (17.33) and (17.34),

$$\frac{n_{\nu_h}^0 + n_{\bar{\nu}_h}^0}{n_\gamma^0} \sim 2 \times 10^{-7} \left(\frac{m_{\nu_h}}{\mathrm{GeV}}\right)^{-3} \,. \tag{17.35}$$

From $\rho_{\mathrm{c}}^0/h^2 \simeq 8 \times 10^{-47}\,\mathrm{GeV}^4$ from eqn (16.74) and $n_\gamma^0 \simeq 3 \times 10^{-39}\,\mathrm{GeV}^3$ from eqn (16.210), the contribution of ν_h to the energy density of the Universe is given by

$$\Omega_{\nu_h}^0\, h^2 = \frac{m_{\nu_h}\, (n_{\nu_h}^0 + n_{\bar{\nu}_h}^0)}{\rho_{\mathrm{c}}^0/h^2} \sim 10 \left(\frac{m_{\nu_h}}{\mathrm{GeV}}\right)^{-2} \,. \tag{17.36}$$

The dependence of $\Omega_{\nu_h}^0\, h^2$ on $m_{\nu_h}^{-2}$ implies that an upper bound on $\Omega_{\nu_h}^0\, h^2$ implies a lower bound for m_{ν_h}. This behavior is opposite to that which was obtained in section (17.4) for light neutrinos.

From the dark matter measurement in eqn (16.234) we obtain

$$\Omega_{\nu_h}^0\, h^2 < \Omega_{\mathrm{M}}^0\, h^2 \simeq 0.14 \qquad \Longrightarrow \qquad m_{\nu_h} \gtrsim 8\,\mathrm{GeV} \,. \tag{17.37}$$

Requiring instead that $\Omega_{\nu_h}^0\, h^2 < 1$, we would have obtained $m_{\nu_h} \gtrsim 3\,\mathrm{GeV}$, which is close to the so-called *Lee–Weinberg bound*, $m_{\nu_h} \gtrsim 2\,\mathrm{GeV}$, obtained through a more precise calculation in Ref. [723] and, independently, in Refs. [629, 1044, 927] (details are discussed in Refs. [700, 690, 973]).

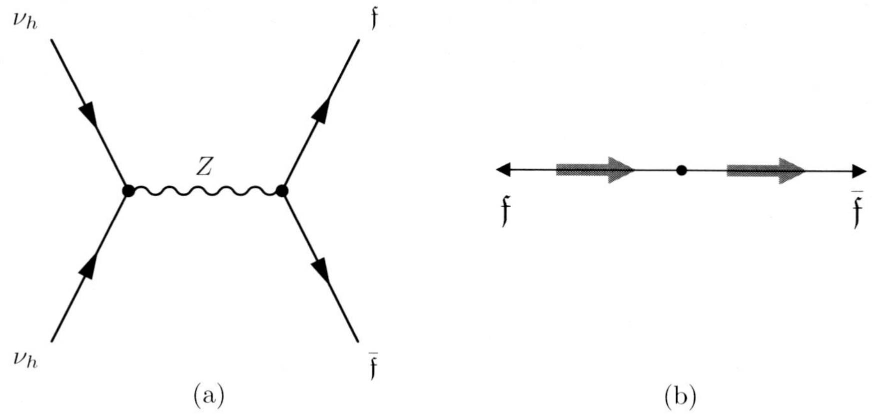

FIG. 17.1. (a) Diagram of the annihilation channel in eqn (17.38) for a heavy Majorana neutrino ν_h. (b) Configuration of the final state of the annihilation channel in eqn (17.38) generated by the term $\overline{\mathfrak{f}_L}\,\gamma^\mu\left(g_V^{\mathfrak{f}} - g_A^{\mathfrak{f}}\,\gamma^5\right)\mathfrak{f}_L$ in eqn (17.40). The thin and thick arrows represent, respectively, momentum and spin.

For heavy Majorana neutrinos, the annihilation cross-section is suppressed with respect to that in eqn (17.29) [700] because Fermi statistics requires the antisymmetric state of annihilating Majorana fermions to be in a p-wave [547]. In order to understand this statement, let us consider the annihilation channel of two heavy Majorana neutrinos into $\mathfrak{f}$-$\bar{\mathfrak{f}}$,

$$\nu_h + \nu_h \to \mathfrak{f} + \bar{\mathfrak{f}}\,, \tag{17.38}$$

where $\mathfrak{f}$ is a fermion with $m_{\mathfrak{f}} < m_{\nu_h}$, through the diagram in Fig. 17.1a. The creation of the $\mathfrak{f}$-$\bar{\mathfrak{f}}$ pair occurs through the weak neutral current

$$j^\mu_{Z,\mathfrak{f}} = \bar{\mathfrak{f}}\,\gamma^\mu\left(g_V^{\mathfrak{f}} - g_A^{\mathfrak{f}}\,\gamma^5\right)\mathfrak{f}\,, \tag{17.39}$$

with the vector and axial couplings given in Table 3.6 (page 78). Decomposing $\mathfrak{f}$ into its left-handed and right-handed chiral components, we have

$$j^\mu_{Z,\mathfrak{f}} = \overline{\mathfrak{f}_L}\,\gamma^\mu\left(g_V^{\mathfrak{f}} - g_A^{\mathfrak{f}}\,\gamma^5\right)\mathfrak{f}_L + \overline{\mathfrak{f}_R}\,\gamma^\mu\left(g_V^{\mathfrak{f}} - g_A^{\mathfrak{f}}\,\gamma^5\right)\mathfrak{f}_R\,. \tag{17.40}$$

Since typically $m_{\mathfrak{f}} \ll m_{\nu_h}$, we can consider $\mathfrak{f}$ as massless in a first approximation. In this case, $\mathfrak{f}_L$ describes particles with negative helicity and antiparticles with positive helicity, whereas $\mathfrak{f}_R$ describes particles with positive helicity and antiparticles with negative helicity (see section 2.9.2). This means that the term $\overline{\mathfrak{f}_L}\,\gamma^\mu\left(g_V^{\mathfrak{f}} - g_A^{\mathfrak{f}}\,\gamma^5\right)\mathfrak{f}_L$ can create a negative helicity $\mathfrak{f}$ and a positive helicity $\bar{\mathfrak{f}}$, as illustrated in Fig. 17.1b. In the center-of-mass frame the total angular momentum will be $J = 1$ (assuming the favored s-wave orbital configuration). The same holds for the other term. Thus, the total angular momentum of the initial state must also be $J = 1$. The problem arises from the fact that, if ν_h is a Majorana particle, the initial state must be antisymmetric, because the two ν_h's in the initial state are identical. This can be obtained either with a spin-0 and s-wave configuration or a spin-1 and p-wave configuration. In fact, a spin-0 state

$$|0,0\rangle = \frac{1}{\sqrt{2}} \left(|\uparrow\downarrow\rangle - |\downarrow\uparrow\rangle \right) . \tag{17.41}$$

is antisymmetric and an s-wave wavefunction is symmetric, since the exchange of the two particles is equivalent to a parity transformation in the center-of-mass frame (see eqn (A.142)). On the other hand, a p-wave wavefunction is antisymmetric (see again eqn (A.142)) and the spin-1 states

$$|1,1\rangle = |\uparrow\uparrow\rangle \, , \qquad |1,0\rangle = \frac{1}{\sqrt{2}} \left(|\uparrow\downarrow\rangle + |\downarrow\uparrow\rangle \right) , \qquad |1,-1\rangle = |\downarrow\downarrow\rangle \tag{17.42}$$

are symmetric. Between the two configurations, the total angular momentum $J = 1$ can be obtained only with the spin-1 and p-wave configuration. The p-wave results in a dependence of the annihilation cross-section on the squared momentum $|\vec{p}|^2 \simeq m_h^2 \, v^2$, instead of m_h^2 in eqn (17.29). Therefore, the annihilation cross-section is suppressed with respect to that in eqn (17.29) by an additional factor v^2,

$$\langle \sigma v \rangle \sim G_{\mathrm{F}} \, m_{\nu_h}^2 \, v^2 \, , \tag{17.43}$$

which is small at the decoupling temperature, when ν_h was nonrelativistic. A smaller annihilation cross-section implies a larger decoupling temperature and, through eqn (17.34), a larger relic number density, which is only partially compensated by the halving of the number of degrees of freedom. A calculation of these effects leads to a lower bound for the mass of a heavy Majorana neutrino which is about three times larger than the corresponding one in the Dirac case [700].

The contribution of a stable Dirac or Majorana neutrino with mass m to the cosmological energy density as a function of m is shown in Fig. 17.2 for $1\text{eV} < m \lesssim 10\,\text{GeV}$. The left part of the figure shows the increase of the contribution proportional to m for a light neutrino, according to eqn (17.26). In the right part of the figure, for $m \gtrsim 1\text{MeV}$, the neutrino is nonrelativistic at decoupling and its contribution decreases as m^{-2}, according to eqn (17.36).

Figure 17.3 shows the relic density $\Omega_\nu \, h^2$ of a stable Dirac or Majorana neutrino up to a mass of 10^3 TeV. As mentioned at the beginning of this section, since there are only three active flavor neutrinos with mass smaller than $m_Z/2 \simeq 45\,\text{GeV}$, the lower bound in eqn (17.37) is interesting only for historical and pedagogical reasons. Instead, the part of Fig. 17.3 above 45 GeV is more interesting. In order to understand the qualitative behavior of $\Omega_{\nu_h} \, h^2$ for $m_{\nu_h} \gtrsim 45\,\text{GeV}$, it is useful to find the dependence of $\Omega_{\nu_h} \, h^2$ on the thermally averaged annihilation cross-section $\langle \sigma v \rangle$. For this purpose, we write the annihilation rate Γ in eqn (17.30) in terms of $\langle \sigma v \rangle$ as

$$\Gamma \sim 0.1 \, \langle \sigma v \rangle \, m_{\nu_h}^{3/2} \, T_\gamma^{3/2} \, e^{-m_{\nu_h}/T_\gamma} \, . \tag{17.44}$$

From $\Gamma \sim H$, we obtain, with the expansion rate in eqn (17.31),

$$\frac{m_{\nu_h}}{T_\gamma^{\nu_h\text{-dec}}} \sim \ln \frac{M_{\mathrm{P}} \, \langle \sigma v \rangle \, m_{\nu_h}}{10 \, \sqrt{g_\rho}} \, . \tag{17.45}$$

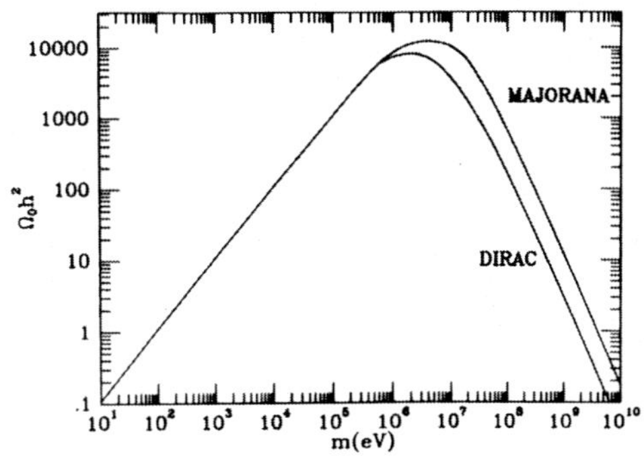

FIG. 17.2. Contribution of a stable Dirac or Majorana neutrino with mass m to the cosmological energy density as a function of m for $1\text{eV} < m \lesssim 10\,\text{GeV}$ [689].

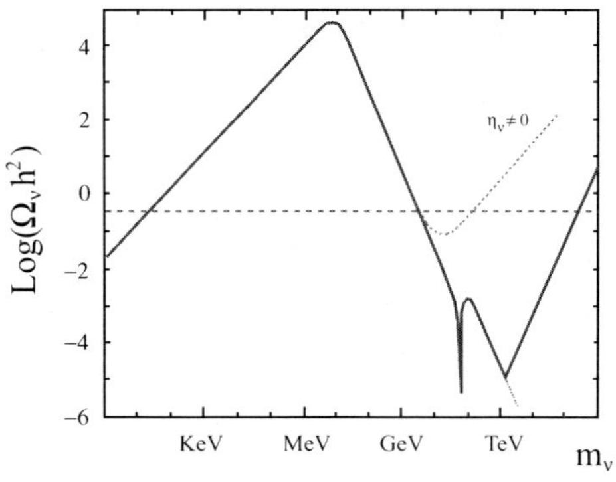

FIG. 17.3. Relic density $\Omega_\nu h^2$ of a stable Dirac neutrino as a function of its mass m_ν, including a possible asymmetry $\eta_\nu = 5\times 10^{-11}$ [650, 834].

Substituting this value in eqn (17.34), we have

$$\frac{n^0_{\nu_h} + n^0_{\bar{\nu}_h}}{n^0_\gamma} \sim 10^{-2} \left(\ln \frac{M_\text{P} \langle \sigma v \rangle m_{\nu_h}}{10\sqrt{g_\rho}} \right)^{3/2} \frac{10\sqrt{g_\rho}}{M_\text{P} \langle \sigma v \rangle m_{\nu_h}} . \tag{17.46}$$

Neglecting the logarithm in eqn (17.46), we finally obtain the approximate relation

$$\Omega^0_{\nu_h} h^2 \propto \langle \sigma v \rangle^{-1} , \tag{17.47}$$

which is consistent with eqn (17.36) when $\langle \sigma v \rangle$ is given by eqn (17.29).

From Fig. 17.3 one can see that at $m_{\nu_h} = m_Z/2 \simeq 45\,\text{GeV}$ there is a deep drop of $\Omega_{\nu_h} h^2 = \Omega_\nu h^2$ due to the Z-boson pole in the ν_h-$\bar{\nu}_h$ annihilation process, which enhances $\langle \sigma v \rangle$ dramatically.

Consider next the case of $m_Z/2 < m_{\nu_h} \lesssim m_W$, i.e. $45\,\text{GeV} < m_{\nu_h} \lesssim 80\,\text{GeV}$. In this regime, the Fermi constant in eqn (17.29) is replaced by the Z-boson propagator $\left(p^2 + m_Z^2\right)^{-1} \sim \left(m^2_{\nu_h} + m_Z^2\right)^{-1} \sim m^{-2}_{\nu_h}$, leading to $\langle \sigma v \rangle \propto m^{-2}_{\nu_h}$. Equation (17.47) shows that, in this case, $\Omega^0_{\nu_h} h^2$ increases approximately as $m^2_{\nu_h}$.

For $m_{\nu_h} \gtrsim m_W \simeq 80\,\text{GeV}$, the annihilation cross-section is enhanced by the $\nu_h + \bar{\nu}_h \to W^+ + W^-$ channel [415], which leads to $\langle \sigma v \rangle \propto m^2_{\nu_h}$. Hence, $\Omega^0_{\nu_h} h^2$ decreases approximately as $m^{-2}_{\nu_h}$, up to a mass of a few TeV. At these very high masses, the tree-level annihilation cross-section violates the unitarity bound [568, 414]. This means that the annihilation process is nonperturbative. The value of $\Omega^0_{\nu_h} h^2$ in Fig. 17.3 for $m_{\nu_h} \gtrsim 1\,\text{TeV}$ can be estimated using the unitarity limit, leading to $\langle \sigma v \rangle \propto m^{-2}_{\nu_h}$ and $\Omega^0_{\nu_h} h^2 \propto m^2_{\nu_h}$. One can see from the figure that this increase of $\Omega^0_{\nu_h} h^2$ implies an upper bound for $m^2_{\nu_h}$ of about 300 TeV, above which ν_h over-closes the Universe.

In summary, a heavy stable weakly interacting neutrino may have a mass between a few GeV and a few hundred TeV. However, if the heavy neutrino has an asymmetry $\eta_{\nu_h} \equiv (n_{\nu_h} - n_{\bar{\nu}_h})/n_\gamma$, its present energy density could be determined by the asymmetry [630], instead of the annihilation cross-section. The present energy density due to the asymmetry, $m_{\nu_h}|\eta_{\nu_h}|n_\gamma^0$, implies

$$\Omega^0_{\nu_h} h^2 \simeq 4 \times 10^7 \, |\eta_{\nu_h}| \left(\frac{m_{\nu_h}}{\text{GeV}}\right) . \tag{17.48}$$

If this value of $\Omega^0_{\nu_h} h^2$ is greater than that in eqn (17.36), the asymmetry dominates. The behavior of $\Omega^0_{\nu_h} h^2$ for $\eta_\nu = 5 \times 10^{-11}$ is shown by the dotted line in Fig. 17.3 for $m_{\nu_h} \gtrsim 10\,\text{GeV}$, where the asymmetry is dominant. It is clear that, with such large values of the asymmetry, the allowed window for m_{ν_h} is strongly reduced or disappears entirely.

17.6 Big-Bang nucleosynthesis

Big-Bang nucleosynthesis (BBN) is one of the cornerstones of the Big Bang model (see Refs. [1052, 217, 689, 423, 924, 940, 837, 683, 438, 834]). Already in the late 1940s Alpher, Bethe, Gamow and Herman [504, 70, 71, 72] realized that the conditions in the hot early Universe was suitable for the formation of light nuclei. Their insight led to the prediction of the existence of the CMBR, which was discovered many years later by Penzias and Wilson [860, 368].

The following studies led to the understanding that the BBN is responsible for the formation of the four light isotopes ^{2}He, ^{3}He, ^{4}He, ^{7}Li, whose primordial abundances can be inferred from astrophysical measurements. It is not possible to form heavier nuclei, because there are no stable nuclei with atomic mass 5 and 8.

The BBN is very powerful, in the sense that it gives us information about the Universe at a temperature of about 0.1 MeV, corresponding to a time of about 1 s, much earlier than the time of photon decoupling probed ($\sim 10^5\,\text{yr}$) by our observations of the CMBR.

Before the neutrino decoupling, around $T_\gamma \simeq 1\,\text{MeV}$, the weak interaction processes

$$n + e^+ \leftrightarrows p + \bar{\nu}_e \, , \tag{17.49}$$

$$p + e^- \leftrightarrows n + \nu_e \tag{17.50}$$

$$n \leftrightarrows p + e^- + \bar{\nu}_e \tag{17.51}$$

were in equilibrium. The amount of nucleosynthesis depends crucially on the number of neutrons present in the thermal plasma, because neutrons are necessary to make nuclei (at the time of BBN, the temperature and density were high enough to form nuclei only from the pre-existing neutrons and protons). When the weak interaction processes in eqns (17.49)–(17.51) were in equilibrium, the number densities of protons and neutrons were given by eqn (16.139), leading to the ratio

$$\frac{n_n}{n_p} \simeq \exp\left[-\frac{m_n - m_p - (\mu_n - \mu_p)}{T_\gamma}\right] , \tag{17.52}$$

with $m_n - m_p \simeq 1.293\,\text{MeV}$. We have neglected the small difference between the proton and neutron masses in the factor in front of the exponential. The thermal equilibrium of the weak interaction processes in eqns (17.49)–(17.51) implies that the chemical potentials of protons, neutrons, electrons, and neutrinos are related by

$$\mu_p + \mu_e = \mu_n + \mu_{\nu_e} \,. \tag{17.53}$$

The chemical potentials of protons and neutrons were different from zero, because the Universe is made of baryons[103]. The chemical potential of electrons is small, because from electrical charge neutrality of the Universe and the value of the baryon-to-photon ratio η in eqn (16.236), we have $n_e/n_\gamma = n_p/n_\gamma \sim n_{\mathrm{B}}/n_\gamma = \eta \sim 10^{-9}$. Since η was almost constant up to $T_\gamma \sim 200\,\mathrm{MeV}$, from eqn (16.151) we obtain $\mu_e/T_\gamma \sim 10^{-9}$. On the other hand, the electron neutrino chemical potential μ_{ν_e} is not constrained by other measurements. In the standard BBN, it is assumed that μ_{ν_e} is also negligible (the effect of a nonnegligible μ_{ν_e} is discussed in section 17.10). In this case, from eqn (17.53), we get $(\mu_n - \mu_p)/T_\gamma \sim 10^{-9}$, which is negligible with respect to $(m_n - m_p)/T_\gamma \sim 1$ for $T_\gamma \sim 1\,\mathrm{MeV}$. Thus, the neutron-to-proton ratio in thermal equilibrium before BBN was given by

$$\frac{n_n}{n_p} \simeq \exp\left[-\frac{m_n - m_p}{T_\gamma}\right] \,. \tag{17.54}$$

The weak interaction processes in eqns (17.49)–(17.51) go out of equilibrium at the freeze-out temperature (see Ref. [379])

$$T_\gamma^{\mathrm{fr}} \simeq 0.7 \left(\frac{g_\rho}{10.75}\right)^{1/6} \mathrm{MeV} \,. \tag{17.55}$$

The dependence on $g_\rho^{1/6}$ comes from the equality at freeze-out of the interaction rate $\Gamma \propto G_{\mathrm{F}}^2 T_\gamma^5$ (analogous to eqn (17.7)) and the expansion rate $H \propto g_\rho^{1/2} T_\gamma^5$ (according to eqn (16.161)).

In the standard BBN, we have $g_\rho = 10.75$ for $T_\gamma \gtrsim 0.2\,\mathrm{eV}$ (see eqn (16.158)), leading to the freeze-out temperature

$$T_\gamma^{\mathrm{fr}} \simeq 0.7\,\mathrm{MeV} \,. \tag{17.56}$$

From eqn (16.162), the freeze-out time is

$$t_{\mathrm{fr}} \simeq 1.5\,\mathrm{s} \,. \tag{17.57}$$

At freeze-out, the neutron-to-proton ratio is given by

$$\left(\frac{n_n}{n_p}\right)_{\mathrm{fr}} \simeq \exp\left[-\frac{m_n - m_p}{T_\gamma^{\mathrm{fr}}}\right] \simeq \frac{1}{6} \,. \tag{17.58}$$

After freeze-out, the neutron-to-proton ratio is affected only by the neutron decay process

$$n \to p + e^- + \bar{\nu}_e \,. \tag{17.59}$$

Although the nuclear binding energies of light nuclei are larger than 0.7 MeV, as one can see from Table 10.5, the nucleosynthesis effectively occurs between $T_\gamma \simeq$

[103] If the chemical potentials of baryons were equal to zero, the number of baryons and antibaryons in the hot early Universe would have been the same, leading to total annihilation during the cooling due to the expansion.

0.3 MeV and $T_\gamma \simeq 0.1$ MeV. This is due to the large number of photons in the high-energy tail of the distribution, which break up nuclei as soon as they are formed. In other words, the high entropy of the Universe implies that the isotope $\mathcal{I}$ with atomic number Z and atomic mass A becomes thermodynamically favored at the temperature $T_\mathcal{I}$, approximately given by (see Refs. [689, 940])

$$T_\mathcal{I}\,[1.5\,\ln(m_p/T_\mathcal{I}) - \ln\eta] \simeq \frac{B(A,Z)}{A-1}\,, \tag{17.60}$$

where $B(A,Z)$ is the binding energy. For $\eta \simeq 6\times 10^{-10}$ (from eqn (16.236)), we have $T_{^2\mathrm{H}} \simeq 0.07\,\mathrm{MeV}$, $T_{^3\mathrm{He}} \simeq 0.11\,\mathrm{MeV}$, $T_{^4\mathrm{He}} \simeq 0.28\,\mathrm{MeV}$, and $T_{^7\mathrm{Li}} \simeq 0.07\,\mathrm{MeV}$. Although ^{4}He becomes thermodynamically favored before ^{2}H, it cannot be produced in large quantities before $T_{^2\mathrm{H}}$ is reached because its production occurs through the chains of reactions

$$p\,(n,\gamma)\,{}^2\mathrm{H}\,({}^2\mathrm{H},p)\,{}^3\mathrm{H}\,({}^2\mathrm{H},n)\,{}^4\mathrm{He}\,, \tag{17.61}$$

and

$$p\,(n,\gamma)\,{}^2\mathrm{H}\,({}^2\mathrm{H},n)\,{}^3\mathrm{He}\,({}^2\mathrm{H},p)\,{}^4\mathrm{He}\,, \tag{17.62}$$

which require a substantial previous formation of ^{2}H. This is the so-called *deuterium bottleneck*, which leads to the effective BBN temperature

$$T_{\mathrm{BBN}} \simeq 0.07\,\mathrm{MeV}\,, \tag{17.63}$$

which in turn corresponds, from eqn (16.162), to the BBN time

$$t_{\mathrm{BBN}} \simeq 150\,\mathrm{s}\,. \tag{17.64}$$

Since the neutron lifetime is about 886 s (see eqn (A.158)), some neutrons decay in the time interval from freeze-out to nucleosynthesis, lowering the neutron-to-proton ratio at the nucleosynthesis value to

$$\left(\frac{n_n}{n_p}\right)_{\mathrm{BBN}} \simeq \frac{1}{7}\,. \tag{17.65}$$

Since almost all the neutrons end up in forming ^{4}He, which is the most tightly bound stable light nucleus (see Table 10.5), the resulting mass fraction of ^{4}He, usually denoted by Y_p, is given by

$$Y_\mathrm{p} \simeq \left(\frac{2\,n_n}{n_p+n_n}\right)_{\mathrm{BBN}} = \frac{2\,(n_n/n_p)_{\mathrm{BBN}}}{1+(n_n/n_p)_{\mathrm{BBN}}} \simeq 0.25\,. \tag{17.66}$$

The value of Y_p inferred from astrophysical data requires the subtraction of ^{4}He produced in stars and other astrophysical corrections. Recent estimates of Y_p are $Y_\mathrm{p} = 0.2391 \pm 0.0020$ [760], $Y_\mathrm{p} = 0.2421 \pm 0.0021$ [635], $Y_\mathrm{p} = 0.249 \pm 0.009$ [836], with a conservative allowed range $0.232 \leq Y_\mathrm{p} \leq 0.258$ [836]. These values are in

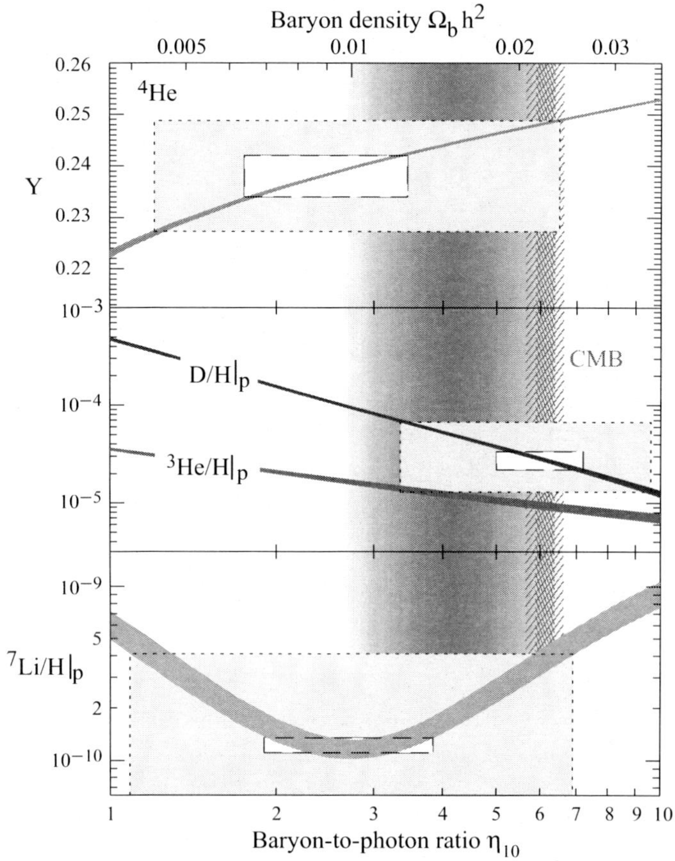

FIG. 17.4. Theoretical predictions of the BBN abundances of light nuclei as functions of $\eta_{10} \equiv 10^{10}\eta$ [438]. $Y \equiv Y_{\mathrm{p}}$ is the mass fraction of ^{4}He. The abundances of $D \equiv {}^2$H, ^{3}He, and ^{7}Li are given as number densities relative to the number density of $p \equiv {}^1\mathrm{H} \equiv \mathrm{H}$. The boxes indicate the observed abundances: the small boxes indicate 2σ statistical uncertainties, whereas the large boxes indicate 2σ statistical plus systematic uncertainties added in quadrature. The narrow vertical band corresponds to the WMAP measurement of η in eqn (16.236).

remarkable agreement with the rough estimation in eqn (17.66), confirming the basic formulation of the standard BBN. In the following, we will use the value

$$Y_{\mathrm{p}} = 0.249 \pm 0.009 \tag{17.67}$$

from Ref. [836], in which some likely sources of systematic uncertainties in the determination of Y_{p} were taken into account.

Figure 17.4 shows the results of a precise calculation of the BBN abundances of light nuclei as functions of the baryon-to-photon ratio $\eta_{10} \equiv 10^{10}\eta$. One can see that the sensitivity of the ^{4}He abundance Y_{p} to the value of η is weak. The slow increase of $Y_{\rm p}$ with η is due to the increase of the BBN temperature (i.e. the ^{2}H formation temperature from eqn (17.60)), which anticipate the time at which BBN occurs, leaving less time for neutrons to decay. With $3 \lesssim \eta_{10} \lesssim 10$ and a neutron lifetime of 886 s (see eqn (A.158)), the amount of ^{4}He is approximately given by [256, 217, 1048, 424]

$$Y_{\rm p} \simeq 0.227 + 0.010 \, \ln \eta_{10} \,, \tag{17.68}$$

which gives

$$Y_{\rm p} \simeq 0.245 \qquad \text{for} \qquad \eta_{10} \simeq 6.1 \,. \tag{17.69}$$

From a more accurate numerical calculation, the authors of Ref. [344] obtained

$$Y_{\rm p} = 0.2484^{+0.0004}_{-0.0005} \qquad \text{for} \qquad \eta_{10} = 6.14 \pm 0.25 \,. \tag{17.70}$$

On the other hand, the primordial deuterium abundance is very sensitive to the value of η. From Fig. 17.4, one can see that the deuterium abundance decreases almost exponentially as η increases. This rapid depletion of deuterium is caused by the process $^2\text{H}(p,\gamma)^3\text{He}$, whose rate is proportional to the proton number density. Since deuterium can only be depleted in stars and other astrophysical processes, the present abundance of deuterium is an upper limit to the primordial value and a measurement of the deuterium abundance implies an upper limit on the baryon density [896, 420]. For these reasons, deuterium is considered as the best *cosmic baryometer*.

Unfortunately, the astrophysical evolution of ^{3}He is not sufficiently well known to infer a useful estimate of the primordial abundance of ^{3}He from astrophysical observations. The estimate of the primordial abundance of ^{7}Li from the data suffers also from large systematic uncertainties, as one can see from Fig. 17.4.

Figure 17.4 shows that there is agreement among the predicted and measured abundances for [438]

$$3.4 \leq \eta_{10} \leq 6.9 \qquad (95\% \text{ CL}) \,. \tag{17.71}$$

Taking into account the difficulty of measurements and the huge extrapolation back in time, this is an impressive feat indeed, which makes the BBN a pillar of the Standard Cosmological Model.

The narrow vertical band in Fig. 17.4 shows the WMAP measurement of η in eqn (16.236), which is in good agreement with the deuterium abundance. Since this determination of η is very precise, it is possible to assume it as an input in the BBN calculation. In this case, the standard BBN becomes a theory without undetermined parameters. This fact is very useful, since it allows one to use the measured light element abundances to constrain possible physics beyond the SM, which can affect the BBN. The BBN constraints on the number of neutrino generations and on neutrino asymmetries are to be discussed, respectively, in sections 17.9 and section 17.10.2.

17.7 Large-scale structure formation

The formation of large-scale structures (LSS) is one of the main topics in modern cosmology. It allows one to test a cosmological model against the observations, which have reached an impressive wealth of information with the 2dF Galaxy Redshift Survey (2dFGRS) [861, 327, 324], the Sloan Digital Sky Survey (SDSS) [9, 28], and other surveys of astrophysical objects. As we will see in the following, the LSS formation gives information on the values of the masses of light neutrinos, which contribute to the hot dark matter in the Universe, since they decoupled when they were relativistic (see section 17.1).

The comparison of the measured distribution of large objects in the Universe with theoretical predictions is usually done through the so-called *power spectrum* of density fluctuations. In order to derive the power spectrum, let us consider the two-point correlation function

$$\xi(t, y) \equiv \langle \delta(t, \vec{x})\, \delta(t, \vec{x} + \vec{y}) \rangle\,, \tag{17.72}$$

where $\delta(t, \vec{x})$ is the field of relative density fluctuations in eqn (16.222). The angle brackets indicate an average over a model-dependent ensemble of universes (see the discussion before eqn (16.214)). Since it is assumed that the fluctuations are statistically homogeneous in space, the two-point correlation function does not depend on $\vec{x}$. Moreover, since it is assumed that the fluctuations are statistically isotropic, the two-point correlation function depends only on $y \equiv |\vec{y}|$.

Consider now the correlation function in Fourier space $\langle \delta(t, \vec{k})\, \delta(t, \vec{k}^{\prime}) \rangle$. Since from eqn (16.222) we have

$$\delta(t, \vec{k}) = \int \mathrm{d}^3x\, \delta(t, \vec{x})\, e^{-i\vec{k}\cdot\vec{x}}\,, \tag{17.73}$$

we obtain the correlation function in Fourier space

$$\langle \delta(t, \vec{k})\, \delta(t, \vec{k}^{\prime}) \rangle = (2\pi)^3\, \delta^3(\vec{k} + \vec{k}^{\prime})\, P(k, t)\,, \tag{17.74}$$

with the *power spectrum* $P(k, t)$ given by the Fourier transform of the two-point correlation function:

$$P(k, t) = \int \mathrm{d}^3y\, \xi(t, y)\, e^{-i\vec{k}\cdot\vec{y}} = 4\pi \int_0^\infty \xi(t, y)\, \frac{\sin(ky)}{ky}\, y^2\, \mathrm{d}y\,. \tag{17.75}$$

It is often convenient to consider a finite volume in which the wavenumber $\vec{k}$ is quantized by periodic boundary conditions. In this case, we have $(2\pi)^3\, \delta^3(\vec{k} + \vec{k}^{\prime}) \to \delta_{\vec{k}, -\vec{k}^{\prime}}$. Taking into account that $\delta(t, -\vec{k}) = \delta^*(t, \vec{k})$, which follows from the reality of $\delta(t, \vec{x})$ in eqn (16.222), the power spectrum can be written as

$$P(k, t) = \langle |\delta(t, \vec{k})|^2 \rangle\,. \tag{17.76}$$

Hence, the power spectrum is the variance of the distribution of fluctuations in Fourier space. If the fluctuations have a Gaussian distribution, as predicted by simple inflationary models (see Refs. [762, 182]), their distribution depends only on one parameter, the variance (obviously, the mean is equal to zero). Thus, Gaussian fluctuations are completely characterized by the power spectrum.

According to the standard picture (see Refs. [689, 856, 325]), the Universe started in an almost perfectly homogeneous state, with tiny *primordial* density inhomogeneities which could have been generated by the amplification of quantum fluctuations in the inflationary era (see Refs. [762, 182]). In many models the primordial power spectrum $P_\star(k)$ have, at least approximately, a power-law form

$$P_\star(k) = A\,k^{n_s}\,, \tag{17.77}$$

where the power n_s is called the *spectral index*. The simplest primordial power spectrum, predicted by simple inflationary models (see Refs. [762, 182]), is the so-called *scale-invariant* or *Harrison–Zeldovich* primordial power spectrum with $n_s = 1$ [598, 859, 1086]. The reason why it is called *scale-invariant* is that the fluctuations of the gravitational potential have an amplitude independent of scale (see Refs. [689, 325]). If $n_s \neq 1$, the power spectrum is called *tilted*.

In the linear theory, the density fluctuations $\delta(t,\vec{k})$ at some cosmological time t can be written as $\delta(t,\vec{k}) = T(t,k)\,\delta_\star(\vec{k})$, where $\delta_\star(\vec{k})$ are the primordial density fluctuations and $T(t,k)$ is the *transfer function*. Since $P_\star(k) = \langle|\delta_\star(\vec{k})|^2\rangle$, we have, for the power spectrum,

$$P(k,t) = T^2(t,k)\,P_\star(k)\,. \tag{17.78}$$

When the time is not specified, the present time is implicitly assumed:

$$P(k) \equiv P(k,t_0) = T^2(t_0,k)\,P_\star(k) \equiv T^2(k)\,P_\star(k)\,. \tag{17.79}$$

The calculable effects of the evolution of the Universe after the primordial era (inflation or whatever) on the observed power spectrum $P(k)$ are encoded in the transfer function $T^2(k)$. In the following, we review the fundamental aspects of the evolution of the density perturbations in inflationary scenarios and the resulting effects on the power spectrum.

During inflation the particle horizon grows exponentially, as the scale factor R (see eqn (16.133)). For the sake of clarity, we will call the *inflated horizon*, denoted by d_{IH}, the particle horizon obtained from eqn (16.65) by assuming that $t = 0$ is the beginning of inflation. It corresponds to the size of the regions in the Universe which have been in causal contact since the beginning of inflation. On the other hand, following a common convention, we will simply call the *horizon*, denoted by d_{H}, the particle horizon obtained from eqn (16.65) by assuming that $t = 0$ is the end of inflation. Hence, the horizon corresponds to the size of the regions in the Universe which have been in causal contact since the end of inflation. In other words, the horizon coincides with the particle horizon in the Standard Cosmological Model without inflation. As illustrated in Fig. 17.5, the physical wavelengths of primordial fluctuations were stretched beyond the horizon by the exponential expansion of the Universe in the inflationary era. However, primordial fluctuations were always within the inflated horizon, as required by causality: any physical process which generated the fluctuations could have acted only within the causal horizon. After the end of inflation, the horizon grew faster than the scale factor $R(t)$, both in the radiation-dominated era ($d_{\mathrm{H}} \propto R^2$; see eqn (16.123)) and in the matter-dominated era ($d_{\mathrm{H}} \propto R^{3/2}$; see eqn (16.113)). Thus, gradually, the physical wavelengths of the density perturbations entered the horizon. Let us emphasize that for the physical processes which led to structure formation, the horizon is relevant, not the inflated

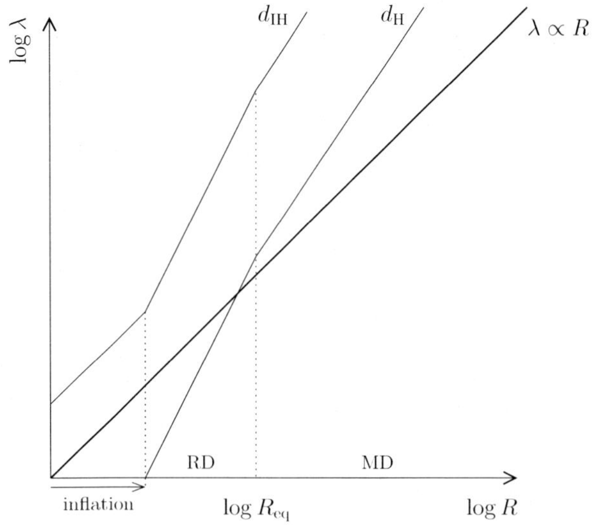

FIG. 17.5. Schematic description of the behavior of physical wavelengths $\lambda \propto R$ of primordial fluctuations with respect to the horizon $d_{\rm H}$ and the inflated horizon $d_{\rm IH}$.

horizon, because all the particles were generated by the decay of the inflation field at the end of inflation. In this sense, the end of inflation can be considered as the origin of time for the standard hot Big Bang.

Let us first determine the behavior of the fluctuations with wavelengths larger than the horizon, called *superhorizon perturbations*. We will then discuss the behavior of the perturbations after they entered the horizon.

Since superhorizon perturbations are not affected by causal processes, their behavior is governed only by the expansion of the Universe and can be estimated with the following approximate reasoning. Consider a flat Universe, in agreement with inflation, in which the average energy density $\langle\rho\rangle$ obeys the Friedmann equation

$$H^2 = \frac{8\pi\, G_{\rm N}}{3}\, \langle\rho\rangle\,. \qquad (17.80)$$

Now consider a Universe with the same expansion rate H, but slightly different energy density ρ. In this case, the Friedmann equation must contain the curvature term:

$$H^2 = \frac{8\pi\, G_{\rm N}}{3}\, \rho - \frac{k}{R^2}\,, \qquad (17.81)$$

where $k = \pm 1$, defined in eqn (16.29) (this k should not be confused with the wavenumber). For the relative density difference, we get

$$\delta = \frac{\rho - \langle\rho\rangle}{\langle\rho\rangle} = \frac{k}{H^2\, R^2} \propto \begin{cases} R^2 \propto t & \text{(RD)}\,, \\ R \propto t^{2/3} & \text{(MD)}\,. \end{cases} \qquad (17.82)$$

The behaviors in the radiation dominated (RD) and matter-dominated (MD) eras have been obtained from eqns (16.120), (16.121) and eqns (16.109), (16.110), respectively. Thus, we see that the amplitude of superhorizon fluctuations grows with the expansion of the Universe, independently of the wavenumber.

When a perturbation enters the horizon, more complicated things happen. Luckily, however, their study is simplified by the fact that within the horizon an approximate Newtonian analysis is sufficient for most purposes. In general, if there are several components of the energy density, $\rho = \sum_i \rho_i$, the evolution of the density perturbations $\delta_i(t, \vec{k})$ of the ith component is given by the differential equation (see Ref. [689])

$$\ddot{\delta}_i(t, \vec{k}) + 2\,H\,\dot{\delta}_i(t, \vec{k}) + (v_{\mathrm{s}}^i)^2\,k^2\,\delta_i(t, \vec{k}) - 4\pi\,G_{\mathrm{N}}\sum_j \rho_j\,\delta_j(t, \vec{k}) = 0\,, \tag{17.83}$$

where $k \equiv |\vec{k}|$ and v_{s}^i is the sound velocity of the ith component, given by the generalization of eqn (16.227). In the following subsections we will discuss separately the evolution of the density perturbations of baryonic matter (section 17.7.1), cold dark matter (section 17.7.2) and hot dark matter (section 17.7.3).

By definition hot dark matter (HDM) and cold dark matter (CDM) are composed of weakly interacting massive particles (WIMPs) which participated in structure formation in the form of a collisionless fluid. The distinction between HDM and CDM lies in their average velocities:

Hot dark matter is composed of hot relic WIMPs with a number density of the same order of magnitude as the photon density in the CMBR. This means that HDM particles must have decoupled when they were relativistic, at a temperature below the quark–hadron phase transition temperature of about 300 MeV. This was the time when the cooling of the CMBR with the expansion of the Universe was strongly suppressed by the transfer of entropy from the annihilations of nonrelativistic hadrons and antihadrons. The temperature of WIMPs which decoupled before the quark–hadron phase transition would be much smaller than the CMBR temperature, leading to a negligible number density.

Neutrinos with masses of the order of 1 eV are natural candidates for HDM. Note that, although HDM particles are nonrelativistic today, they still have a relatively large average velocity. In fact, if χ is a HDM particle, from eqn (16.147) we have $\langle|\vec{p}_\chi|\rangle \simeq 3T_\chi$. Since today $|\vec{p}_\chi| = m_\chi|\vec{v}_\chi|$, the average velocity is given by

$$|\vec{v}_\chi^{\,0}| \simeq 150\left(\frac{T_\chi^0}{T_\nu^0}\right)\left(\frac{m_\chi}{1\,\mathrm{eV}}\right)^{-1}\mathrm{km\,s^{-1}}\,, \tag{17.84}$$

where we have normalized the temperature with respect to the present neutrino temperature in eqn (17.17).

When HDM particles were relativistic, the HDM perturbations within the horizon were erased by the so-called *free-streaming* discussed in section 17.7.3. This is due to the random particle velocities close to the velocity of light, which allowed the HDM particle to escape the overdense regions, leading to the equalization of the HDM density within the horizon.

Cold dark matter is composed of cold relic WIMPs. Since they decoupled while they were nonrelativistic, their number density may be very small and the mass very high, with negligible thermal velocity. Heavy massive neutrinos of the type discussed in section 17.5 are possible candidates. Other possible candidates are sterile heavy neutrinos and supersymmetric particles (neutralinos, sneutrinos, gravitinos, axinos). It is also possible to have CDM which was produced non-thermally, for example superheavy particles (wimpzillas), axions, and primordial black-holes (see Refs. [646, 392, 219]).

17.7.1 Baryonic matter

The evolution of the density perturbations of baryonic matter (BM) are of primary interest, because baryons form the observable objects in the Universe. The formation of these objects must have occurred through the collapse of large structures due to gravitational instability. As we will see, this could happen after the recombination. Before the recombination, the growth of baryon density perturbations was hindered by the pressure of photons with which baryons interacted.

The theory of gravitational instability of a collisional fluid, as BM, was studied by Jeans in 1902 in a Newtonian framework. He showed that the evolution of density fluctuations of a homogeneous background depends on the balance between self-gravitation, which induces the collapse of overdense regions, and pressure, which tends to restore hydrostatic equilibrium. If self-gravitation dominates over pressure, an overdense region can accumulate material from its surroundings, increasing its density and eventually collapsing into a gravitationally bound object.

One can derive approximately the Jeans criterion for fluctuation growing by comparing the characteristic time scales of gravitational collapse and pressure response. If the fluctuation has a scale λ and a mass M, the gravitational acceleration at its surface is $g = G_{\text{N}}M/\lambda^2 \sim G_{\text{N}}\rho\lambda$, with a time scale for gravitational collapse given by $\tau_g \sim \sqrt{\lambda/g} \sim (G_{\text{N}}\rho)^{-1/2}$. On the other hand, the time scales of pressure response is given by $\tau_p \sim \lambda/v_{\text{s}}$, where v_{s} is the sound velocity in eqn (16.227). A fluctuation can grow if gravitational collapse occurs before pressure forces can respond to restore hydrostatic equilibrium: $\tau_g \lesssim \tau_p$, which implies $\lambda \gtrsim \lambda_{\text{J}}$, with the *Jeans length* $\lambda_{\text{J}} \sim v_{\text{s}}/\sqrt{G_{\text{N}}\rho}$.

From eqn (17.83), the evolution of baryonic perturbations $\delta_{\text{B}}(t,\vec{k})$ in a baryonic matter-dominated Universe ($\rho \simeq \rho_{\text{B}}$) is given by (see Refs. [689, 856, 325])

$$\ddot{\delta}_{\text{B}}(t,\vec{k}) + 2\,H\,\dot{\delta}_{\text{B}}(t,\vec{k}) + \left[v_{\text{s}}^2\,k^2 - 4\pi\,G_{\text{N}}\,\rho\right]\delta_{\text{B}}(t,\vec{k}) = 0\,. \tag{17.85}$$

Since the qualitative behavior of the solution depends on the sign of the quantity in square brackets[104], the physical Jeans wavenumber k_{J} is given by

$$k_{\text{J}}^2 = \frac{4\pi\,G_{\text{N}}\,\rho}{v_{\text{s}}^2}\,, \tag{17.86}$$

[104] In a static Universe ($H = 0$) the solution of eqn (17.85) has the form $\delta_{\text{B}}(t,\vec{k}) \propto e^{\pm i\omega t}$, with $\omega = \sqrt{v_{\text{s}}^2\,k^2 - 4\pi\,G_{\text{N}}\,\rho}$. In this case, it is clear that $\delta_{\text{B}}(t,\vec{k})$ oscillates for $k > k_{\text{J}}$, whereas for $k < k_{\text{J}}$ there is an exponentially growing solution.

with the corresponding physical Jeans length

$$\lambda_{\rm J} = \frac{2\pi}{k_{\rm J}} = v_{\rm s}\sqrt{\frac{\pi}{G_{\rm N}\rho}}\,, \tag{17.87}$$

in agreement with the estimate above.

The solution of eqn (17.85) for $k > k_{\rm J}$ implies that fluctuations with a wavelength smaller than the Jeans length oscillate as acoustic waves. On the other hand, for $k < k_{\rm J}$ there are unstable solutions in which the perturbations grow with time. To see this in a simple way, consider $k \ll k_{\rm J}$ and a flat matter-dominated Universe. In this case, using eqn (16.110) and $4\pi G_{\rm N}\rho = 3H^2/2$ from the Friedmann equation, eqn (17.85) reduces to

$$\ddot{\delta}_{\rm B}(t,\vec{k}) + \frac{4}{3\,t}\,\dot{\delta}_{\rm B}(t,\vec{k}) - \frac{2}{3\,t^2}\,\delta_{\rm B}(t,\vec{k}) = 0\,. \tag{17.88}$$

One can immediately check that there is a growing solution

$$\delta_{\rm B}(t,\vec{k}) \propto t^{2/3}\,. \tag{17.89}$$

These growing modes can generate structures after the recombination.

The reason why baryonic structures can be generated only after the recombination is that before the recombination the Jeans length is very large, because of the pressure of photons, even in the matter dominated era. This depends on the large sound speed $v_{\rm s}$. In the baryon–photon plasma before recombination, the energy density is $\rho = \rho_{\rm B} + \rho_\gamma$ and the pressure is provided by photons: $p = p_\gamma = \rho_\gamma/3$. In order to calculate $v_{\rm s}$ through eqn (16.227), one must take into account that we are considering adiabatic perturbations. Since both entropy and baryon number are conserved, the entropy per baryon,

$$s_{\rm B} = \frac{s}{n_{\rm B}}\,, \tag{17.90}$$

is conserved. Taking into account that $s \propto T_\gamma^3$, $\rho_\gamma \propto T_\gamma^4$, and $\rho_{\rm B} = m_{\rm B} n_{\rm B}$, we obtain, from $\delta s_{\rm B} = 0$,

$$\frac{\delta\rho_{\rm B}}{\rho_{\rm B}} = 3\,\frac{\delta T_\gamma}{T_\gamma} = \frac{3}{4}\,\frac{\delta\rho_\gamma}{\rho_\gamma} \qquad \Longrightarrow \qquad \frac{\partial\rho_{\rm B}}{\partial\rho_\gamma} = \frac{3}{4}\,\frac{\rho_{\rm B}}{\rho_\gamma}\,. \tag{17.91}$$

Hence, the sound speed is given by

$$v_{\rm s}^2 = \frac{1}{3}\left(1 + \frac{\partial\rho_{\rm B}}{\partial\rho_\gamma}\right)^{-1} = \frac{1}{3}\left(1 + \frac{3}{4}\,\frac{\rho_{\rm B}}{\rho_\gamma}\right)^{-1} = \frac{1}{3}\left(1 + \frac{3}{4}\,\frac{\Omega_{\rm B}^0}{\Omega_\gamma^0}\,(1+z)^{-1}\right)^{-1}\,, \tag{17.92}$$

for $z > z_{\rm rec} \simeq 1.1\times 10^3$. Using the values of Ω_γ^0 and $\Omega_{\rm B}^0$ in eqns (16.212) and (16.233), one can find that before recombination $1/6 \lesssim v_{\rm s}^2 \leq 1/3$. From eqn (17.87), considering a flat Universe, the baryon Jeans length before recombination is bounded by

$$\lambda_{\rm JB} \gtrsim \sqrt{\frac{\pi}{6\,G_{\rm N}\rho}} = \frac{2\pi}{3\,H}\,. \tag{17.93}$$

Comparing with eqn (16.113), one can see that the baryon Jeans length before recombination was larger than the horizon. This means that before recombination only the baryon perturbations with superhorizon wavelength can grow according to eqn (17.82). When the perturbations enter the horizon, their growth stops, because the perturbation amplitude starts to oscillate in response to the photon distribution. This scenario is in agreement with the discussion in section 16.11 of the generation of CMBR anisotropies.

After recombination, the baryon Jeans length drops dramatically, because the photon pressure disappears. The relevant energy density for the evaluation of the baryon sound speed is the baryon energy density ρ_{B}. From eqns (16.140) and (16.141), the baryon pressure is $p_{\text{B}} = \rho_{\text{B}} T_{\text{B}} / m_{\text{B}}$. The resulting baryon sound speed is given by

$$v_{\text{s}}^2 = \frac{\partial p_{\text{B}}}{\partial \rho_{\text{B}}} = \frac{T_{\text{B}}}{m_{\text{B}}} + \frac{\rho_{\text{B}}}{m_{\text{B}}} \frac{\partial T_{\text{B}}}{\partial \rho_{\text{B}}} \simeq \frac{5}{3} \frac{T_{\text{B}}}{m_{\text{B}}} \,. \tag{17.94}$$

Since baryons are nonrelativistic, $T_{\text{B}} \ll m_{\text{B}}$, the baryon sound speed after recombination is very small. In this case, the Jeans length is much smaller than the horizon. Furthermore, since, from eqn (16.198), after recombination T_{B} decreases proportionally to R^{-2}, the baryonic Jeans length decreases proportionally to $R^{-1} \propto 1+z$. Thus, after recombination subhorizon baryonic perturbations larger than the Jeans length can grow according to eqn (17.89) until they reach $\delta_{\text{B}}(t, \vec{k}) \sim 1$. At that point, the linear theory no longer applies and the perturbations rapidly collapse to form structures.

It is very plausible that the matter in the Universe is composed not only of baryons, but there is also a large amount of cold dark matter. As we will see in section 17.7.2, subhorizon cold dark matter perturbations start to grow at matter–radiation equality, which occurs earlier than recombination. Thus, at recombination, the CDM perturbations are much larger than the baryon perturbations. It can be shown that in this case the growth of baryon perturbations is enhanced by the presence of the CDM perturbations. As a result, the baryon perturbations rapidly reach the same amplitude as the CDM perturbations (see Refs. [689, 856, 325]). In this case, the transfer functions of the baryon and CDM density fluctuations are practically equal.

17.7.2 Cold dark matter

Cold dark matter is composed of cold relic WIMPs. This means that during the structure formation CDM is a gas of nonrelativistic collisionless particles. Since the velocity dispersion of such a gas is very small, its free-streaming is practically negligible. The evolution of subhorizon CDM density perturbations $\delta_{\text{CDM}}(t, \vec{k})$ in a smooth radiation background is given by eqn (17.83) with $i = \text{CDM}$, $v_{\text{s}}^{\text{CDM}} \simeq 0$ and negligible $\delta_j(t, \vec{k})$ for $j \neq \text{CDM}$ (see Refs. [689, 856, 325]):

$$\ddot{\delta}_{\text{CDM}}(t, \vec{k}) + 2\, H\, \dot{\delta}_{\text{CDM}}(t, \vec{k}) - 4\pi\, G_{\text{N}}\, \rho_{\text{CDM}}\, \delta_{\text{CDM}}(t, \vec{k}) = 0 \,. \tag{17.95}$$

This equation is similar to eqn (17.85) for baryonic perturbations, with the difference that the sound velocity of CDM is negligible and that we have considered a model in which ρ_{CDM} may be different from the total energy density ρ. This happens in the radiation-dominated era, when $\rho \simeq \rho_{\text{R}}$. In order to solve eqn (17.95), it is convenient to define

$$y \equiv \frac{\rho_{\rm CDM}}{\rho_{\rm R}} = \frac{R}{R_{\rm eq}} \,. \tag{17.96}$$

Then, from the Friedmann equation for a flat Universe, we have $H = \sqrt{8\pi G_{\rm N}\rho_{\rm R}(1+y)/3}$ and, hence, eqn (17.95) can be written as

$$\delta''_{\rm CDM} + \frac{2+3y}{2y\,(1+y)}\,\delta'_{\rm CDM} - \frac{3}{2y\,(1+y)}\,\delta_{\rm CDM} = 0 \,, \tag{17.97}$$

where a prime denotes the derivative with respect to y. One can immediately check that there is a growing solution,

$$\delta_{\rm CDM} \propto y + \frac{2}{3} \,, \tag{17.98}$$

which has $\delta''_{\rm CDM} = 0$.

Before the matter–radiation equality, when the Universe was radiation-dominated, y was very small. During this era, the CDM perturbations did not grow. From eqn (17.98), one can see that they were practically frozen at a constant value:

$$\delta_{\rm CDM} \simeq \text{constant} \qquad \text{for} \qquad t \ll t_{\rm eq} \,. \tag{17.99}$$

Only after matter–radiation equality, y became large and the CDM perturbations could grow in the usual way (see eqn (17.89)):

$$\delta_{\rm CDM} \propto y \propto R \propto t^{2/3} \qquad \text{for} \qquad t \gg t_{\rm eq} \,. \tag{17.100}$$

The stagnation of CDM perturbation during the radiation-dominated era is called the *Meszaros effect* [792]. Physically, it is due to the fact that the CDM gravitational self-interaction $4\pi\, G_{\rm N}\, \rho_{\rm CDM}$ is too slow in comparison with the large expansion rate $H \simeq \sqrt{8\pi G_{\rm N}\rho_{\rm R}/3}$.

In summary, for scales greater than the horizon, CDM perturbations grow as other perturbations according to eqn (17.82). The growth is gradually stopped when a CDM perturbation enters the horizon during the radiation-dominated era and can smoothly increase to that in eqn (17.100) after the matter–radiation equality.

Let us emphasize that the above treatment is valid as long as the perturbations are small, $\delta_{\rm CDM} \ll 1$. When $\delta_{\rm CDM}$ grows to the order of unity, the linear approximation breaks down and the evolution of perturbations with different wavelength is no longer independent. Numerical simulations show that the nonlinear behavior of the perturbations lead to collapse and formation of structures.

The structures observed in our Universe have statistical properties compatible with those obtained in numerical simulations of a Universe in which matter is mainly composed of CDM. The biggest problem of CDM models is that so far CDM has never been directly detected and the nature of CDM is completely unknown (although there are many plausible candidates; see Refs. [646, 392, 219]).

17.7.3 Hot dark matter

Hot dark matter is composed of hot relic WIMPs with a number density of the same order of magnitude as the CMBR density. Relic neutrinos with masses of the order of 1 eV are HDM, since they decoupled at a temperature of the order of 1 MeV while they were relativistic and their number density is a fraction 3/11 of the CMBR (see section 17.3).

Since WIMPs decoupled much earlier than the time of matter–radiation equality, when galactic-size masses entered the horizon, they were a collisionless gas during LSS formation. As long as the HDM gas is relativistic, the HDM perturbations within the horizon are erased by *free streaming*: the random particle velocities close to the velocity of light disperse all HDM overdensities with sizes smaller than the horizon. Free-streaming ceases when the HDM gas becomes nonrelativistic, at redshift $z_{\rm nr}$. Thus, only the HDM perturbations with wavelength larger than the horizon distance at $z_{\rm nr}$ survive and can partake in the generation of structures in the Universe. Since the horizon distance at $z_{\rm nr}$ is typically much larger than the volume corresponding to galactic-size masses, in a Universe dominated by HDM, the formation of structures must proceed according to a top-down mechanism, in which very large-scale structures, of supercluster size, form first. Later, these structures fragment and generate galaxies. It turns out that the observed Universe has statistical properties which are different from those of a Universe generated in this way. Hence, it is currently believed that our Universe is not dominated by HDM. The observations favor a CDM-dominated Universe in which the formation of structures proceed according to a bottom-up mechanism: small-scale structures of galactic size form first, and later merge to form clusters and superclusters. The LSS observations are used in order to constrain the amount of HDM, which suppresses the formation of small-scale structures through free-streaming.

Considering specifically a light neutrino ν_j with mass m_j, let us estimate the current physical free-streaming scale $\lambda^0_{\nu_j\text{-fs}}$ as the present physical distance corresponding to the horizon distance at the redshift $z_{\nu_j\text{-nr}}$, when ν_j became nonrelativistic. Assuming a neutrino mass smaller than about 1.5 eV, it became non-relativistic after the matter–radiation equality (see the discussion after eqn (17.23)). In this case, in a first approximation, we can use the expression for the horizon distance in a matter-dominated Universe, given in eqn (16.117). Using the value of $z_{\nu_j\text{-nr}}$ in eqn (17.18), the current physical free-streaming scale is given by

$$\begin{aligned} \lambda^0_{\nu_j\text{-fs}} &\simeq z_{\nu_j\text{-nr}}\, d_{\rm H}(z_{\nu_j\text{-nr}}) \simeq 2\, H_0{}^{-1}\, z_{\nu_j\text{-nr}}^{-1/2}\, (\Omega^0_{\rm M})^{-1/2} \\ &\simeq 1.3\times 10^2 \left(\frac{m_j}{\rm eV}\right)^{-1/2} (\Omega^0_{\rm M})^{-1/2}\, h^{-1}\,{\rm Mpc}\,. \end{aligned} \tag{17.101}$$

Neutrino free-streaming suppresses the power spectrum of density fluctuations for physical wavenumbers larger than the current physical free-streaming wavenumber

$$k^0_{\nu_j\text{-fs}} \simeq \frac{2\pi}{\lambda^0_{\nu_j\text{-fs}}} \simeq 4.7\times 10^{-2} \left(\frac{m_j}{\rm eV}\right)^{1/2} \sqrt{\Omega^0_{\rm M}}\, h\,{\rm Mpc}^{-1}\,. \tag{17.102}$$

The suppression of the power spectrum of density fluctuations in the case of massive neutrinos is illustrated in Fig. 17.6, where we have plotted the ratio of the CDM transfer functions for massive and massless neutrinos (as explained at the end of section 17.7.1, the baryon and CDM transfer functions are practically the same). One can see that massive neutrinos suppress the transfer function for

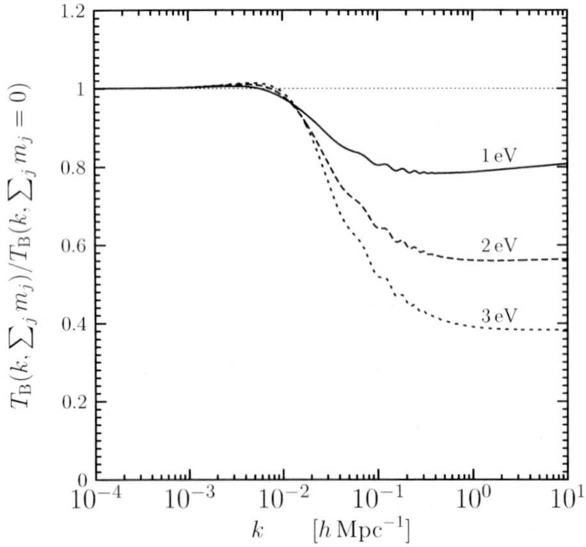

FIG. 17.6. Ratio of the CDM transfer functions for massive and massless neutrinos, obtained with the CMBFAST code [946] in a flat ΛCDM model with $h = 0.72$, $\Omega_B^0 = 0.05$, $\Omega_\Lambda^0 = 0.70$ and three quasidegenerate massive neutrinos. Solid line: $\sum_j m_j = 1\,\text{eV}$ ($\Omega_\nu^0 \simeq 0.02$). Long-dashed line: $\sum_j m_j = 2\,\text{eV}$ ($\Omega_\nu^0 \simeq 0.04$). Short-dashed line: $\sum_j m_j = 3\,\text{eV}$ ($\Omega_\nu^0 \simeq 0.06$). The CDM density is $\Omega_{\text{CDM}}^0 = 1 - \Omega_\Lambda^0 - \Omega_B^0 - \Omega_\nu^0$.

values of k approximately larger than $k_{\nu_j\text{-fs}}^0$, within an order of magnitude. The partial suppression for values of k just below $k_{\nu_j\text{-fs}}^0$ is due to the high-energy tail of the neutrino distribution and residual free-streaming after neutrinos have become nonrelativistic but still have high velocities.

The amount of the suppression of the power spectrum can be evaluated approximately as follows [1012]. Let us consider the matter-dominated era, when the subhorizon CDM perturbations could grow, as explained in section 17.7.2. However, in section 17.7.2 we did not take into account the possible presence of a HDM component. Next, let us consider a HDM component made of massive neutrinos with masses of the order of 1 eV, which became nonrelativistic shortly after the matter–radiation equality. After the neutrinos became nonrelativistic, they contributed to the matter-dominated energy density of the Universe, but their perturbations are suppressed below the free-streaming scale. In this case, below the free-streaming scale only the CDM density perturbations $\delta_{\text{CDM}}(t, \vec{k})$ are nonnegligible (baryonic perturbations contribute only after the recombination, as explained in section 17.7.1; in this approximate derivation we neglect the baryon contribution, which is small since $\Omega_B^0/\Omega_M^0 \simeq 0.17$ from eqns (16.233) and (16.234)). Hence, from the general evolution equation (17.83), we obtain again, for the CDM density perturbations, the evolution equation (17.95). However, in this case we are considering the matter-dominated era, with $\rho \simeq \rho_{\text{CDM}} + \rho_\nu$. From the Friedmann equation for a flat Universe, it follows that

$$4\pi\, G_{\rm N}\, \rho_{\rm CDM} = \frac{3}{2}\, H^2\, \Omega_{\rm CDM}\,, \tag{17.103}$$

where $\Omega_{\rm CDM} = \rho_{\rm CDM}/\rho$. Since, in a matter dominated Universe, $H = 2/3t$ (see eqn (16.110)), the evolution equation (17.95) in this case can be written as

$$\ddot{\delta}_{\rm CDM}(t,\vec{k}) + \frac{4}{3\,t}\,\dot{\delta}_{\rm CDM}(t,\vec{k}) - \frac{2}{3\,t^2}\,\Omega_{\rm CDM}\,\delta_{\rm CDM}(t,\vec{k}) = 0\,. \tag{17.104}$$

This equation can easily be solved by assuming

$$\delta_{\rm CDM}(t,\vec{k}) \propto t^p\,. \tag{17.105}$$

One finds the two solutions for p [262]:

$$p_\pm = \pm\frac{\sqrt{1+24\,\Omega_{\rm CDM}} \mp 1}{6}\,. \tag{17.106}$$

The exponent p_+ gives a growing solution. Note that for $\Omega_{\rm CDM} = 1$, which corresponds to $\Omega_\nu = 0$, i.e. massless neutrinos, we have $p_+ = 2/3$ and one recovers the growth of CDM perturbations in a CDM-dominated Universe given in eqn (17.100). On the other hand, if there is a neutrino component, we have $p_+ < 2/3$ and the evolution of CDM perturbations is suppressed. For $\Omega_\nu \ll 1$, we have

$$p_+ \simeq \frac{2}{3}\left(1 - \frac{3}{5}\,\Omega_\nu\right)\,. \tag{17.107}$$

In this case, the growth of CDM perturbations is approximately given by (the contribution of the solution corresponding to p_- rapidly disappears)

$$\delta_{\rm CDM}(t,\vec{k}) \propto t^{\frac{2}{3}\left(1-\frac{3}{5}\,\Omega_\nu\right)} \propto R^{1-\frac{3}{5}\,\Omega_\nu}\,. \tag{17.108}$$

CDM perturbations grew in this way from the matter–radiation equality to the matter–vacuum transition, when their growth was slowed down by an effect similar to the Meszaros effect discussed in section 17.7.2. Hence, the growing factor of CDM perturbations is

$$\frac{\delta_{\rm CDM}(t_{\rm tr},\vec{k})}{\delta_{\rm CDM}(t_{\rm eq},\vec{k})} \sim \left(\frac{R_{\rm tr}}{R_{\rm eq}}\right)^{1-\frac{3}{5}\,\Omega_\nu} = \left(\frac{1+z_{\rm eq}}{1+z_{\rm tr}}\right)^{1-\frac{3}{5}\,\Omega_\nu} = \left(\frac{(\Omega^0_{\rm M})^{4/3}}{\Omega^0_{\rm R}\,(\Omega^0_\Lambda)^{1/3}}\right)^{1-\frac{3}{5}\,\frac{\Omega^0_\nu}{\Omega^0_{\rm M}}}\,, \tag{17.109}$$

where we have written Ω_ν in the matter-dominated era as the present fraction $\Omega^0_\nu/\Omega^0_{\rm M}$, and we have used the eqns (16.86) and (16.88) for $1+z_{\rm tr}$ and $1+z_{\rm eq}$, respectively. From eqn (17.76), the ratio of the power spectra for massive and massless neutrinos is given by

$$\frac{P(k,\Omega_\nu^0)}{P(k,\Omega_\nu^0=0)} \sim \left(\frac{(\Omega_{\rm M}^0)^{4/3}}{\Omega_{\rm R}^0\,(\Omega_\Lambda^0)^{1/3}}\right)^{-\frac{6}{5}\frac{\Omega_\nu^0}{\Omega_{\rm M}^0}} = \exp\left[-\frac{6}{5}\frac{\Omega_\nu^0}{\Omega_{\rm M}^0}\ln\left(\frac{(\Omega_{\rm M}^0)^{4/3}}{\Omega_{\rm R}^0\,(\Omega_\Lambda^0)^{1/3}}\right)\right]. \tag{17.110}$$

Thus, for $\Omega_\nu^0 \ll \Omega_{\rm M}^0$, the suppression of the power spectrum below the neutrino free-streaming scale estimated in eqn (17.101) is approximately given by

$$\frac{\Delta P(k)}{P(k)} \sim -\frac{6}{5}\frac{\Omega_\nu^0}{\Omega_{\rm M}^0}\ln\left(\frac{(\Omega_{\rm M}^0)^{4/3}}{\Omega_{\rm R}^0\,(\Omega_\Lambda^0)^{1/3}}\right). \tag{17.111}$$

Since the neutrino perturbations are erased below the free-streaming scale, the suppression of the power spectrum is proportional to the ratio $\Omega_\nu^0/\Omega_{\rm M}^0$, which gives the relative neutrino energy density during the matter-dominated era (Ω_ν in eqn (17.108)).

Considering a flat Universe with $\Omega_{\rm M}^0 \simeq 0.3$ and $\Omega_{\rm R}^0$ given by eqn (17.22), we obtain (see also Ref. [627])

$$\frac{\Delta P(k)}{P(k)} \sim -10\,\frac{\Omega_\nu^0}{\Omega_{\rm M}^0} \sim -\left(\frac{\sum_j m_j}{1\text{eV}}\right)\left(\frac{0.1}{\Omega_{\rm M}^0\,h^2}\right). \tag{17.112}$$

From this suppression, one can constrain the amount of HDM and, as a consequence, the sum of the masses of light neutrinos from measurements of the power spectrum.

17.8 Global fits of cosmological data

The current astronomical observations indicate that the Universe is described on large scales by a spatially flat Robertson–Walker metric, with an energy density which consists of about 5% of baryons, about 25% of CDM, and about 70% of dark energy[105]. The statistical distribution of structures is in good agreement with that predicted by CDM models with a scale-invariant primordial power spectrum. This cosmological model is called the *flat* Λ*CDM model* or *concordance model*, since it is in good agreement with all data.

Figures 17.7 and 17.8 show, respectively, summaries of the experimental data on the angular power spectrum of CMBR temperature anisotropies [1010] and the current matter power spectrum [1012], which are used in global cosmological fits. The angular power spectrum of CMBR temperature anisotropies in Fig. 17.7 extends the WMAP data points in Fig. 16.8 with other measurements at large values of ℓ. The current matter power spectrum in Fig. 17.8 is obtained mainly from CMBR data at small wavenumbers (large wavelengths), LSS data at intermediate wavenumbers, and *Lyman-*α *forest* data at large wavenumbers (small wavelengths). It is assumed

[105] Considering a general dark energy equation of state $p_{\rm DE} = w_{\rm DE}\,\rho_{\rm DE}$, the first year data of the Supernova Legacy Survey [117] combined with SDSS data [401] imply $w_{\rm DE} = -1.023 \pm 0.090 \pm 0.054$ in a spatially flat ΛCDM model [117], in agreement with the value $w_{\rm DE} = -1$ for a cosmological constant (see eqn (16.84)).

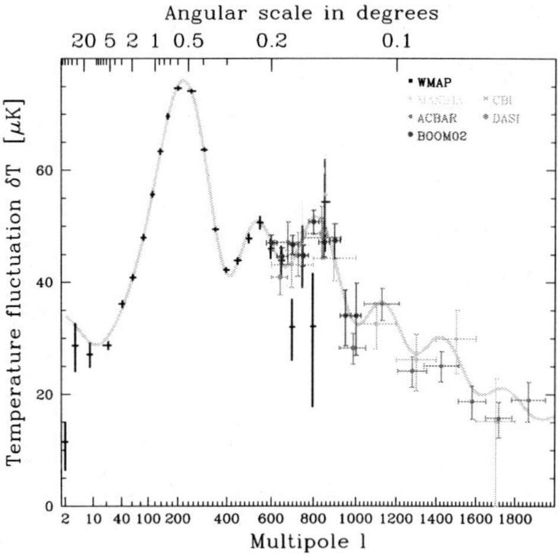

FIG. 17.7. Angular power spectrum of CMBR temperature anisotropies from Ref. [1010]. Besides the WMAP data [206], there are data from Boomerang [916], DASI [582], MAXIMA [722], VSA [1009, 369], CBI [962, 893], and ACBAR [706].

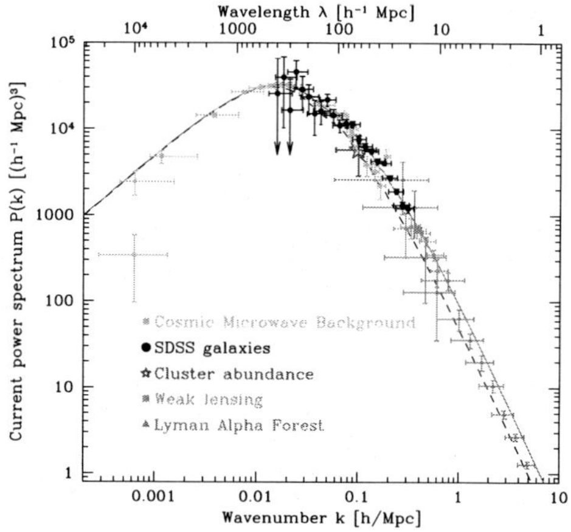

FIG. 17.8. Current matter power spectrum $P(k)$ from Ref. [1012] (see also Ref. [1011]). The solid curve shows the theoretical prediction for a flat ΛCDM model with a scale-invariant primordial power spectrum, matter density $\Omega_{\mathrm{M}}^{0} = 0.28$, Hubble constant $h = 0.72$, and baryon fraction $\Omega_{\mathrm{B}}^{0}/\Omega_{\mathrm{M}}^{0} = 0.16$. The dashed curve shows a suppression of small-scale power by about a factor of two due to a replacement of 7% of the CDM with massive neutrinos having $\sum_j m_j = 1\,\mathrm{eV}$.

that galaxies trace dark matter up to a constant factor b, called the *bias*, which implies that the galaxy power spectrum $P_{\mathrm{g}}(k)$ and the dark matter power spectrum $P_{\mathrm{DM}}(k)$ are related by

$$P_{\mathrm{g}}(k) = b^2\, P_{\mathrm{DM}}(k)\,. \tag{17.113}$$

The value of the bias factor is usually extracted from the fits of the data in the framework of a specific cosmological model.

The Lyman-α forest data are absorption lines in the spectrum of distant quasars due to intermediate hydrogen clouds which absorb the Lyman-α line at wavelength $\lambda_\alpha = 1215.67\,\text{Å}$, corresponding to an electron transition from the ground state to the first excited state. If the quasar is at redshift z_{q}, the emission Lyman-α line of the quasar has observed wavelength $\lambda_\alpha^{\mathrm{q}} = (1 + z_{\mathrm{q}})\,\lambda_\alpha$. Since the absorption clouds are at different redshifts, smaller than the quasar redshift, it is common to observe a *forest* of absorption Lyman-α lines below the emission Lyman-α line of the quasar. At still smaller wavelengths, the intermediate clouds can also absorb the Lyman-β line at $\lambda_\beta = 1025.72\,\text{Å}$, corresponding to an electron transition from

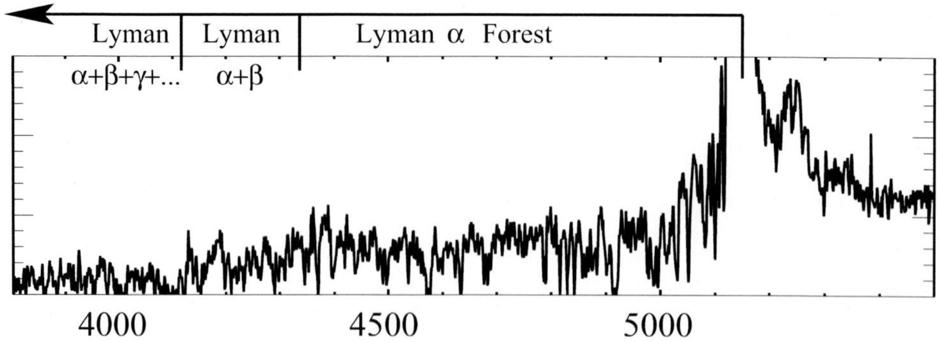

FIG. 17.9. Lyman-α forest in the spectrum of quasar Q2139-4434, at redshift $z_\mathrm{q} \simeq 3.23$ [374]. x axis: measured wavelength in Angstrom. y axis: flux intensity.

the ground state to the second excited state. Therefore, the Lyman-α forest lies in the region between $\lambda_\beta^\mathrm{q} = (1 + z_\mathrm{q})\,\lambda_\beta$ and $\lambda_\alpha^\mathrm{q}$, where only Lyman-$\alpha$ photons can be absorbed. An example is shown in Fig. 17.9, for a quasar at redshift $z_\mathrm{q} \simeq 3.23$, with $\lambda_\alpha^\mathrm{q} \simeq 5144$ Å and $\lambda_\beta^\mathrm{q} \simeq 4339$ Å. One can see the indicated Lyman-α forest between these two observed wavelengths. From the characteristics of the absorption lines, it is possible to infer the power spectrum of density fluctuations for relatively small wavelengths $\lambda \sim 2 - 20\,h^{-1}\,\mathrm{Mpc}$, as shown in Fig. 17.8 (see Refs. [338, 891, 337, 788]). Unfortunately, since the extraction of the power spectrum from Lyman-α data is very difficult and model dependent, the systematic uncertainty may be large [948]. For this reason several authors do not consider Lyman-α data.

An important feature of the matter power spectrum in Fig. 17.8 is the peak at $\lambda \sim 300\,h^{-1}\,\mathrm{Mpc}$. This peak corresponds, approximately, to the horizon distance at the matter–radiation equality. The reason is that wavelengths larger than the horizon distance at matter–radiation equality, on the left of the peak in Fig. 17.8, entered the horizon during the matter-dominated era and the corresponding CDM fluctuations could always grow, independently of λ, before the horizon-entering according to eqn (17.82) and after the horizon-entering according to eqn (17.100). Hence, the power spectrum on the left of the peak is proportional to the primordial power spectrum $P_\star(k)$, which is consistent with being scale-invariant ($P(k) \propto P_\star(k) \propto k$ on the left of the peak). On the other hand, wavelengths smaller than the horizon distance at the matter–radiation equality, on the right of the peak in Fig. 17.8, entered the horizon during the radiation-dominated, and the Meszaros effect, discussed in section 17.7.2, suppressed the growth of the corresponding CDM fluctuations from horizon-entering to the matter–radiation equality. As one can see from Fig. 17.8, the suppression was larger for smaller wavelengths, which passed a longer time within the horizon during the radiation-dominated era. In order to check this interpretation, we can get an order-of-magnitude estimate of the current length corresponding to the horizon distance at the matter–radiation equality using the approximation for the radiation-dominated era in eqn (16.126):

$$d_\mathrm{H}^0(z_\mathrm{eq}) = z_\mathrm{eq}\,d_\mathrm{H}(z_\mathrm{eq}) \sim {H_0}^{-1}\,z_\mathrm{eq}^{-1}\,(\Omega_\mathrm{R}^0)^{-1/2} = {H_0}^{-1}\,z_\mathrm{eq}^{-1/2}\,(\Omega_\mathrm{M}^0)^{-1/2} \simeq 100\,h^{-1}\,\mathrm{Mpc}\,. \tag{17.114}$$

Within a factor of three, this rough approximation coincides with the peak of the power spectrum in Fig. 17.8. Obviously, the exact location of the peak can be obtained only with a detailed calculation of the evolution of the density fluctuations through the transition between radiation and matter domination, which spans a large redshift interval, as one can see, for example, in Fig. 16.5b. The solid curve in Fig. 17.8 is the result of such a detailed calculation in a flat ΛCDM model which fits the power spectrum.

The dashed curve in Fig. 17.8 shows the change in the power spectrum due to a replacement of 7% of the CDM with HDM composed of massive neutrinos having $\sum_j m_j = 1\,\text{eV}$. One can see that the power at small scales is suppressed, according to the discussion in section 17.7.3. The lack of such suppression in the measured power spectrum implies an upper limit for the sum of the masses of light neutrinos.

In Table 17.1, we have summarized the information on the main cosmological parameters in flat ΛCDM models obtained with global fits of astronomical data by the WMAP collaboration [970] and the SDSS collaboration [1010]. The WMAP fit in the second column allowed a *running spectral index* with constant slope. However, one can see that the spectral index is compatible with scale-invariance ($n_s = 1$ within $2.3\,\sigma$ and $\mathrm{d}n_s/\mathrm{d}\ln k = 0$ within $1.9\,\sigma$). The SDSS fits in columns 3–6 assume a constant spectral index. In the fit in column 3, the spectral index is assumed to be equal to unity, whereas in the other column it is obtained as a result of the fit. One can see that in any case it is compatible with scale invariance within less than $1.4\,\sigma$. The other parameters have compatible values, within the uncertainties.

From the point of view of neutrino physics the most important results are in the last two rows of Table 17.1, which give the upper bound for the neutrino density and the corresponding upper bound for the sum of the masses of the three light neutrinos (from eqn (17.27)). These bounds follow from the small-scale suppression of the power spectrum approximated by eqn (17.112) and illustrated by the dashed curve in Fig. 17.8. One can see that the bounds are rather stringent, especially those obtained in Refs. [970, 947] which included Lyman-α information on small-scale structures, whose power spectrum is very sensitive to the presence of neutrino HDM.

As shown in the summary in Table 17.2 (see also the reviews in Refs. [595, 403, 596]), other authors obtained similar results analyzing sets of cosmological data which do not overlap completely. In particular, the authors of Ref. [449] performed two fits, without and with SDSS Lyman-α data [788], obtaining, respectively, the upper bounds 1.4 eV and 0.47 eV for the sum of light neutrino masses. These results illustrate the power of Lyman-α data. The rather stringent result obtained by the SDSS collaboration in Ref. [945] is due to an extraction from their data of the bias factor b (see eqn (17.113)), which is usually a free parameter in the fit.

All the upper limits on the sum of neutrino masses, which depends on the type of analysis and on the data considered, can be summarized by the approximate upper bound range

$$\sum_j m_j \lesssim 0.5 - 1\,\text{eV}\,. \tag{17.115}$$

TABLE 17.1. Parameters of flat ΛCDM models obtained with global fits of astronomical data by the WMAP collaboration [970] and the SDSS collaboration [1010]. The values without uncertainty are assumed. The missing values are not reported in the cited papers (but can be derived). The value of the spectral index n_s reported by the WMAP collaboration [970] refers to the physical wavenumber $k = 0.05\,\text{Mpc}^{-1}$. The upper limits in the last two rows have 95% confidence.

Parameter	WMAP [970]	SDSS [1010]	SDSS [1010]	SDSS [1010]	SDSS [947]
n_s	0.93 ± 0.03	1	$0.966^{+0.025}_{-0.020}$	$0.972^{+0.041}_{-0.027}$	$0.989^{+0.026}_{-0.023}$
$\dfrac{dn_s}{d\ln k}$	$-0.031^{+0.016}_{-0.018}$	0	0	0	0
h	$0.71^{+0.04}_{-0.03}$	$0.708^{+0.023}_{-0.024}$	$0.685^{+0.027}_{-0.026}$	$0.645^{+0.048}_{-0.040}$	$0.710^{+0.023}_{-0.022}$
$\Omega^0_B\,h^2$	0.0224 ± 0.0009	$0.0238^{+0.0006}_{-0.0006}$	$0.0228^{+0.0010}_{-0.0008}$	$0.0234^{+0.0014}_{-0.0011}$	$0.0236^{+0.0009}_{-0.0009}$
Ω^0_B	0.044 ± 0.004				
$\Omega^0_M\,h^2$	$0.135^{+0.008}_{-0.009}$	$0.1471^{+0.0090}_{-0.0080}$	$0.1459^{+0.0077}_{-0.0071}$	$0.158^{+0.015}_{-0.012}$	
Ω^0_M	0.27 ± 0.04	$0.293^{+0.039}_{-0.031}$	$0.309^{+0.040}_{-0.032}$	$0.380^{+0.087}_{-0.074}$	$0.284^{+0.025}_{-0.023}$
Ω^0_Λ	0.73 ± 0.04	$0.707^{+0.031}_{-0.039}$	$0.691^{+0.032}_{-0.040}$	$0.620^{+0.074}_{-0.087}$	
t_0 [Gyr]	13.7 ± 0.2	$13.40^{+0.13}_{-0.12}$	$13.62^{+0.20}_{-0.20}$	$13.65^{+0.25}_{-0.28}$	
$\Omega^0_\nu\,h^2$	< 0.0072	0	0	< 0.018	
$\sum_j m_j$ [eV]	< 0.68	0	0	< 1.74	< 0.42

This upper bound range is depicted in Fig. 17.10, which shows the sum of neutrino masses in the two schemes of three-neutrino mixing indicated by neutrino oscillation data (see chapter 13), as a function of the lightest mass. In the quasidegenerate region ($m_1 \simeq m_2 \simeq m_3 \gg \sqrt{\Delta m^2_{31}} \simeq 5 \times 10^{-2}\,\text{eV}$), the sum of neutrino masses is the same in both normal and inverted schemes. On the other hand, in the hierarchical region the sum of neutrino masses in the inverted scheme is about twice as large as in the normal scheme, because it gets the contributions of the two almost equal heavy masses with value $m_1 \simeq m_2 \simeq \sqrt{\Delta m^2_{31}}$. One can see from the figure that the approximate cosmological upper bound range in eqn (17.115) limits the sum of neutrino masses in the quasidegenerate region, corresponding to an upper limit of one-third of the sum for each mass in both schemes:

$$m_j \lesssim 0.17 - 0.33\,\text{eV} \qquad (j = 1, 2, 3)\,. \tag{17.116}$$

Although this limit does not allow one to distinguish the normal and inverted schemes, it is very stringent in comparison with the direct kinematical measurement of the effective mass in tritium β-decay (see eqn (14.30)).

TABLE 17.2. Upper limits on the sum of light neutrino masses obtained by different authors from global fits of cosmological data using WMAP [206] CMBR data and the other CMBR data listed in the third column, LSS data listed in the fourth column, and other data listed in the fifth column.

Ref.	Max $\sum_j m_j$ (95% CL) [eV]	CMBR	LSS	Other data
Spergel et al. [970]	0.68	CBI [962] ACBAR [706]	2dFGRS [861]	SNIa [903, 904] H_0 [461] bias [1035] Lyα [337, 545]
Hannestad [592]	1.01	WTJZ [1050]	2dFGRS [861]	SNIa [863] H_0 [461]
Tegmark et al. [1010]	1.74		SDSS [9]	
Barger et al. [170]	0.75	WTJZ [1050]	2dFGRS [861] SDSS [1011]	H_0 [461]
Crotty et al. [340]	0.6	ACBAR [706]	2dFGRS [861] SDSS [1011]	SNIa [1019] H_0 [461]
Seljak et al. [945]	0.54		SDSS [7]	bias [945]
Seljak et al. [947]	0.42		SDSS [1011]	bias [945] Lyα [786, 787, 788] SNIa [905]
Fogli et al. [449]	1.4	Boomerang [916] DASI [582] MAXIMA [722] CBI [893] VSA [369]	2dFGRS [861] SDSS [1011]	SNIa [905] H_0 [461]
Fogli et al. [449]	0.47	Boomerang [916] DASI [582] MAXIMA [722] CBI [893] VSA [369]	2dFGRS [861] SDSS [1011]	SNIa [905] H_0 [461] Lyα [788]
Hannestad [596]	0.65		SDSS [1011]	SNIa [905] H_0 [461] Lyα [337, 545]
MacTavish et al. [764]	1.20	Boomerang [780] DASI [582] MAXIMA [722] ACBAR [706] CBI [893] VSA [369]	2dFGRS [861] SDSS [1011]	SNIa [905] H_0 [461]
Sanchez et al. [921]	1.16	ACBAR [706] CBI [893] VSA [369]	2dFGRS [324]	

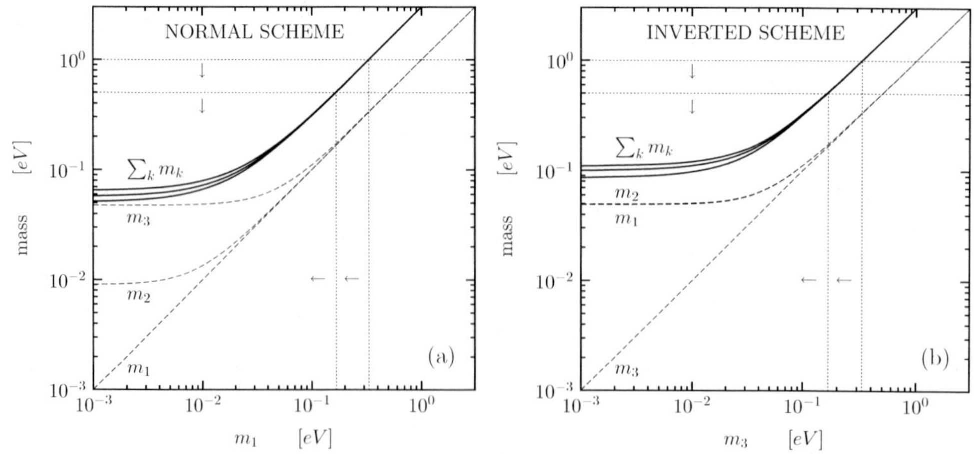

FIG. 17.10. Sum of neutrino masses in the two schemes of three-neutrino mixing indicated by neutrino oscillation data, as a function of the lightest mass (m_1 in the normal scheme in (a) and m_3 in the inverted scheme in (b)). The three solid lines represent the best-fit and 2σ uncertainty band obtained from the squared–mass differences in eqns (13.130) and (13.131). The two horizontal dotted lines represent the approximate cosmological upper bound range in eqn (17.115), and the two vertical dotted lines give the corresponding upper bound range for the lightest mass. The dashed curves show the three individual masses.

In the future, high-precision astrophysical data on the CMBR and large galaxy surveys may allow us to explore the more interesting hierarchical region, reaching a sensitivity to neutrino masses of the order of a few 10^{-2} eV [591, 658, 730, 1049], where the normal and inverted schemes may be distinguished. One should note from Fig. 17.10 that a future limit of about 8×10^{-2} eV on the sum of neutrino masses would imply the exclusion of the inverted scheme.

17.9 Number of neutrinos

It is well known that there are only three light active neutrino flavors (see section 5.1.3). However, it is possible that the effective number of neutrinos N_ν in the Universe is larger than three if there are light sterile neutrinos (see chapter 6) or other neutral relativistic nonstandard particles which have been thermalized during the early expansion of the Universe. In this case, the energy density of radiation is increased by a change of g_ρ given by

$$\Delta g_\rho = \frac{7}{4} \Delta N_\nu \,, \tag{17.117}$$

where $\Delta N_\nu \equiv N_\nu - 3$ is the effective number of additional neutrinos. In general, ΔN_ν is given by

$$\Delta N_\nu = \sum_{\chi=\text{fermions}} \frac{g_\chi}{2} \left(\frac{T_\chi}{T_\nu} \right)^4 + \frac{8}{7} \sum_{\chi=\text{bosons}} \frac{g_\chi}{2} \left(\frac{T_\chi}{T_\nu} \right)^4 , \tag{17.118}$$

where the sums run over relativistic nonstandard particles (sterile neutrinos and others) which have been thermalized in the early Universe. If such particles exist, they are very weakly interacting (otherwise they would have been already detected) and must have decoupled at a temperature much higher than the BBN temperature. After their decoupling their temperature scaled as R^{-1}, without being affected by subsequent entropy releases due to the annihilations of massive heavy particles when they became nonrelativistic. This implies that the temperature at BBN was much lower than the neutrino temperature, suppressing their contribution to ΔN_ν. This argument can be used to constrain the interactions of these nonstandard particles, which must be weak enough to give a sufficiently high decoupling temperature.

Among others (see Refs. [924, 379, 438, 834]), BBN data have been used to constrain: the magnetic moments of Dirac neutrinos, which would imply a production of right-handed neutrinos through scattering [817] or through spin rotation in the primordial magnetic field [953, 761]; the lifetime of heavy neutrinos [927, 370], whose decay products would increase the entropy of the Universe; and the mixing with sterile neutrinos, which can be produced through oscillations [378, 416, 651, 417, 164, 418, 956, 833, 238, 675, 676, 364, 674, 10, 677, 365, 386].

Note that even in the SM with three neutrinos, there are two effects which give contributions to ΔN_ν [371]. The first one is the correction to the neutrino distribution functions due to the fact that neutrinos are not completely decoupled during electron–positron annihilation, which gives a contribution $\Delta N_\nu \simeq 0.03$ [382, 383, 425, 379]. The second one is due to finite-temperature electromagnetic corrections to the energy density of the relativistic photon–electron–positron plasma, leading to $\Delta N_\nu \simeq 0.01$ [752, 771]. Hence,

$$\Delta N_\nu^{\text{SM}} \simeq 0.04 . \tag{17.119}$$

In the following subsections 17.9.1–17.9.3, we discuss the constraints on ΔN_ν which have been obtained, respectively, from CMBR, CMBR + LSS, and BBN measurements.

17.9.1 Limit from CMBR

The effective number of additional neutrinos ΔN_ν can be constrained by the measurements of the anisotropies of the CMBR [593, 594, 339, 873, 592]. Since the Universe is not completely matter-dominated at the recombination, the addition of radiation density due to ΔN_ν increases the expansion rate, leading to an enhancement of the first acoustic peak. Moreover, the peaks are shifted to larger values of ℓ, because the increased radiation density implies a smaller conformal time of recombination η_{rec} and a smaller sound horizon in eqn (16.228) [873]. A fit of WMAP

[206] data leads to [592]

$$\Delta N_\nu = -0.9^{+6.7}_{-2.2} \qquad (95\% \text{ CL}) . \tag{17.120}$$

17.9.2 Limit from CMBR + LSS

Also the formation of large-scale structures is sensitive to ΔN_ν, since an increase of the energy density of radiation delays the time of matter–radiation equality. As explained in section 17.7.2, the growth of CDM perturbations within the horizon is practically frozen during the radiation-dominated era. This implies that a delay of matter–radiation equality suppresses the power spectrum on the right of the peak in Fig. 17.8, which correspond to short fluctuation wavelengths which have been within the horizon during the radiation-dominated era for a longer time. Since the peak corresponds to the horizon distance at matter–radiation equality, a delay of matter–radiation equality shifts the peak to a larger wavelength, i.e. to a lower wavenumber.

Several authors have derived bounds on ΔN_ν from a combined analysis of CMBR and LSS data [339, 873, 592, 340]. A fit of WMAP [206] CMBR data and 2dFGRS [861, 784] LSS data yielded [592]

$$\Delta N_\nu = 0.1^{+3.9}_{-2.8} \qquad (95\% \text{ CL}) . \tag{17.121}$$

Using also the Hubble Space Telescope Key Project measurement of h in eqn (16.57), the Supernova Cosmology Project constraint $\Omega_{\text{M}}^0 = 0.28 \pm 0.14$ [863], and other CMBR measurements [1050], the author of Ref. [592] obtained

$$\Delta N_\nu = 1.0^{+3.0}_{-2.1} \qquad (95\% \text{ CL}) . \tag{17.122}$$

The improvement with respect to the range in eqn (17.120), obtained with CMBR data alone, is apparent.

17.9.3 Limit from BBN

The BBN limits on physics beyond the SM are obtained mainly from the ^{4}He abundance Y_{p}, which is rather sensitive to the neutron-to-proton ratio (see eqn (17.66)). The neutron-to-proton ratio at BBN depends on the value of the freeze-out temperature, which is given by $\Gamma \sim H$, where Γ is the interaction rate of the processes in eqns (17.49)–(17.51). During BBN the Universe was radiation-dominated and the expansion rate H was given by eqn (16.161), which depends on the amount of relativistic particles through g_ρ. Since the freeze-out temperature was above the electron–positron annihilation temperature, the value of g_ρ can be written as

$$g_\rho = 10.75 + \frac{7}{4} \Delta N_\nu , \tag{17.123}$$

where 10.75 is the SM value (see eqn (16.158)). If $N_\nu > 0$, g_ρ and H are larger than in the SM, leading to a higher freeze-out temperature. In this case, the neutron-to-proton ratio is larger than in the SM, yielding a larger amount of ^{4}He [857, 981].

From eqn (17.66), the change $\Delta Y_{\rm p}$ of the ^{4}He abundance due to a small change $\Delta(n_n/n_p)_{\rm BBN}$ of the neutron-to-proton ratio at BBN is given by

$$\Delta Y_{\rm p} \simeq Y_{\rm p}\left(1-\frac{Y_{\rm p}}{2}\right)\frac{\Delta(n_n/n_p)_{\rm BBN}}{(n_n/n_p)_{\rm BBN}}. \tag{17.124}$$

From eqns (17.58) and (17.55), we have

$$\frac{\Delta(n_n/n_p)_{\rm BBN}}{(n_n/n_p)_{\rm BBN}} \simeq \frac{m_n-m_p}{T_\gamma^{\rm fr}}\frac{\Delta T_\gamma^{\rm fr}}{T_\gamma^{\rm fr}} \simeq \frac{1}{6}\frac{m_n-m_p}{T_\gamma^{\rm fr}}\frac{\Delta g_\rho}{g_\rho}. \tag{17.125}$$

Taking the standard values $Y_{\rm p} \simeq 0.25$, $T_\gamma^{\rm fr} \simeq 0.7\,{\rm MeV}$, $g_\rho = 10.75$, and $\Delta g_\rho = (7/4)\,\Delta N_\nu$ from eqn (17.8), we obtain (see also [256, 217, 1048, 424, 367])

$$\Delta Y_{\rm p} \simeq \frac{7\,Y_{\rm p}}{24\,g_\rho}\left(1-\frac{Y_{\rm p}}{2}\right)\frac{m_n-m_p}{T_\gamma^{\rm fr}}\,\Delta N_\nu \simeq 0.011\,\Delta N_\nu\,. \tag{17.126}$$

Using this approximation, the experimental value of $Y_{\rm p}$ in eqn (17.67) and the standard theoretical prediction in eqn (17.70) imply the bound

$$|\Delta Y_{\rm p}| \lesssim 0.009 \qquad \Longrightarrow \qquad |\Delta N_\nu| \lesssim 0.82\,. \tag{17.127}$$

The precisely calculated abundances of ^{4}He, ^{2}He, ^{3}He, ^{7}Li are shown in Fig. 17.11 as functions of η for N_ν from 2 to 7. Taking into account the experimental uncertainties, it is clear that $Y_{\rm p}$ is the abundance most sensitive to N_ν. A fit of the ^{4}He and ^{2}He data and the use the WMAP value of η in eqn (16.236) yield [345]

$$\eta = 6.10^{+0.24}_{-0.22}\times 10^{-10}\,, \qquad N_\nu = 3.24^{+0.61}_{-0.57}\,, \qquad \Delta N_\nu < 1.44\,(95\%\ {\rm CL})\,. \tag{17.128}$$

Hence, the BBN abundances are in good agreement with the SM value $N_\nu = 3$ and additional degrees of freedom are constrained, albeit not very severely.

17.10 Neutrino asymmetry

In the Standard Cosmological Model it is assumed that neutrinos are very far from degenerate, with asymmetries

$$\eta_{\nu_\alpha} \equiv \frac{n_{\nu_\alpha}-n_{\bar{\nu}_\alpha}}{n_\gamma} = \frac{\pi^2}{12\,\zeta(3)}\left(\frac{T_\nu}{T_\gamma}\right)^3\left[\xi_{\nu_\alpha}+\frac{\xi_{\nu_\alpha}^3}{\pi^2}\right] \tag{17.129}$$

which are zero or negligibly small. In eqn (17.129) we have rewritten eqn (16.150) with $g_{\nu_\alpha}=1$ and the usual notation $\xi_{\nu_\alpha} \equiv \mu_{\nu_\alpha}/T_\nu$.

It is natural to expect that the neutrino asymmetries are of the same order of magnitude as the charged lepton and baryon asymmetries, leading to negligible cosmological effects. However, since there is no constraint on the values of the cosmological neutrino asymmetries from simple considerations (such as electrical neutrality of astrophysical objects, which constraints the electron asymmetry),

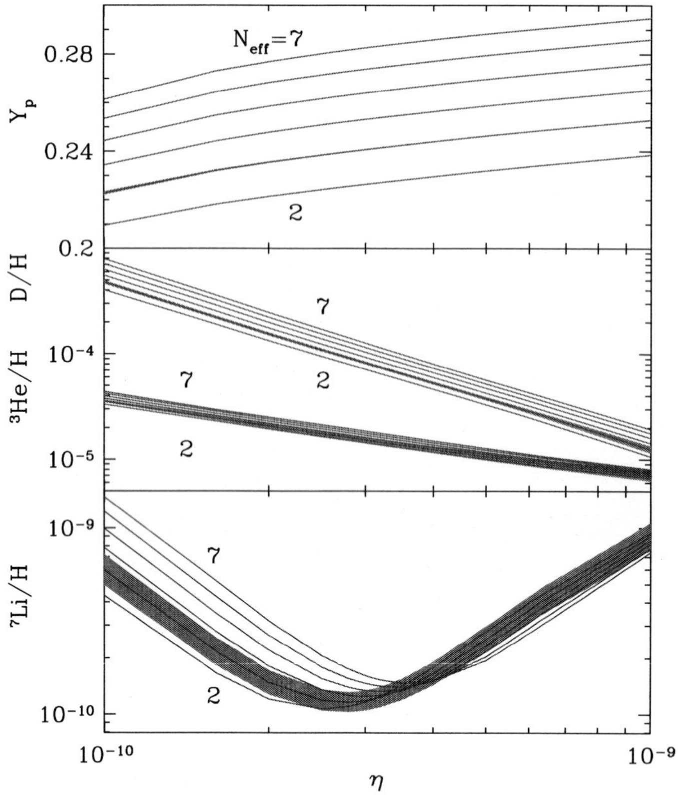

FIG. 17.11. Theoretical predictions of the BBN abundances of light nuclei as functions of η for $N_{\text{eff}} \equiv N_\nu$ from 2 to 7, as indicated by the numbers in the figure [343]. The bands for $N_\nu = 3$ represent the 1σ uncertainties.

it is interesting to consider the cosmological effects of relatively large neutrino asymmetries.

From eqn (16.152), the energy density of neutrinos and antineutrinos of flavor α is given by

$$\rho_{\nu_\alpha} + \rho_{\bar{\nu}_\alpha} = \frac{7}{8}\frac{\pi^2}{15}T_\nu^4\left[1 + \frac{30}{7}\left(\frac{\xi_{\nu_\alpha}}{\pi}\right)^2 + \frac{15}{7}\left(\frac{\xi_{\nu_\alpha}}{\pi}\right)^4\right]. \tag{17.130}$$

If ξ_{ν_α} is large, the neutrino energy density can be much larger than in the nondegenerate case.

Comparing eqn (17.130) and eqn (16.153), one can see that the contribution of degenerate neutrinos to g_ρ is given by the quantity in square brackets in eqn (17.130) times 7/4. Then, from eqn (17.123), the contribution of neutrino chemical potentials to the expansion rate is equivalent to

$$\Delta N_\nu^\xi = \sum_{\alpha=e,\mu,\tau} \frac{30}{7} \left[\left(\frac{\xi_{\nu_\alpha}}{\pi} \right)^2 + \frac{1}{2} \left(\frac{\xi_{\nu_\alpha}}{\pi} \right)^4 \right] . \tag{17.131}$$

Neutrino asymmetries are always equivalent to a positive effective number of additional neutrinos.

Since light neutrinos were relativistic at decoupling, from eqn (16.193) we have $T_\nu \propto R^{-1}$ after decoupling. Moreover, after decoupling the number densities of neutrinos and antineutrinos decreased as R^{-3}, since they were noninteracting. Hence, from eqn (16.149) one can see that after neutrino decoupling the value of $\xi_{\nu_\alpha} \equiv \mu_{\nu_\alpha}/T_\nu$ remained constant during the expansion of the Universe, i.e.

$$\mu_{\nu_\alpha} \propto T_\nu . \tag{17.132}$$

For light degenerate neutrinos, the neutrino chemical potentials in eqn (17.130) cannot be too large, otherwise the Universe would be over-closed. In analogy with the bound on the sum of neutrino masses in eqn (17.28), taking $\mu_{\nu_\alpha} \gg T_\nu$ in eqn (17.130) and requiring that $\sum_\alpha \rho^0_{\nu_\alpha} + \rho^0_{\bar{\nu}_\alpha} \lesssim \rho^0_{\mathrm{M}}$, we obtain

$$\sum_{\alpha=e,\mu,\tau} (\mu^0_{\nu_\alpha})^4 \lesssim 8\,\pi^2 \left(\Omega^0_{\mathrm{M}}\, h^2\right) \frac{\rho^0_{\mathrm{c}}}{h^2} \simeq \left(5.5 \times 10^{-3}\,\mathrm{eV}\right)^4 , \tag{17.133}$$

where we have used the value of $\Omega^0_{\mathrm{M}}\, h^2$ in eqn (16.234) and the value of ρ^0_{c}/h^2 in eqn (16.74). This bound on the neutrino chemical potentials is not very restrictive. In fact, from the neutrino temperature in eqn (17.17), we have

$$\sum_{\alpha=e,\mu,\tau} (\xi_{\nu_\alpha})^4 \lesssim (33)^4 . \tag{17.134}$$

In principle, the asymmetries of the three flavor neutrinos could be very different. However, it has been shown in Refs. [930, 757, 380, 1069, 11] that the three-neutrino oscillation parameters obtained from experimental data (see section 13.3.3) imply that oscillations equilibrate the distribution functions of the three flavors before neutrino decoupling. This implies that the three flavor asymmetries ξ_{ν_e}, ξ_{ν_μ}, and ξ_{ν_τ} are practically equal after decoupling. In this case, the bound in eqn (17.134) leads to

$$-\,25 \lesssim \xi_{\nu_e} \simeq \xi_{\nu_\mu} \simeq \xi_{\nu_\tau} \lesssim 25 . \tag{17.135}$$

17.10.1 Limit from CMBR and LSS

The equivalence in eqn (17.131) of a lepton asymmetry with a positive effective number of additional neutrinos allows us to derive from constraints on ΔN_ν corresponding constraints on the lepton asymmetries [593].

Taking into account the equality of the asymmetries of the three flavor neutrinos, from eqn (17.131), we have

$$\xi_{\nu_e} \simeq \xi_{\nu_\mu} \simeq \xi_{\nu_\tau} \simeq \pi \left[\sqrt{1 + \frac{7}{45} \Delta N_\nu^\xi} - 1 \right]^{1/2}, \tag{17.136}$$

for $\Delta N_\nu^\xi \geq 0$. Using the upper bound for ΔN_ν in eqn (17.122), $\Delta N_\nu^\xi \leq \Delta N_\nu \lesssim 4$, we obtain

$$-1.6 \lesssim \xi_{\nu_e} \simeq \xi_{\nu_\mu} \simeq \xi_{\nu_\tau} \lesssim 1.6, \tag{17.137}$$

which is much more stringent than the allowed range in eqn (17.135).

17.10.2 Limit from BBN

The good agreement of the standard BBN with astrophysical observations implies constraints on the neutrino asymmetries, which must be small enough not to modify too much the standard BBN predictions of the light element abundances [1046, 895, 196, 1077, 197, 1013, 1014, 835, 656, 688, 424, 597, 342].

Large neutrino asymmetries have two effects on BBN (see Refs. [385, 767, 924, 379]):

1. An electron neutrino asymmetry modifies the freeze-out neutron-to-proton ratio in eqn (17.58) into

$$\left(\frac{n_n}{n_p}\right)_{\text{fr}} \simeq \exp\left[-\frac{m_n - m_p}{T_\gamma^{\text{fr}}} - \xi_{\nu_e}\right], \tag{17.138}$$

which follows from eqns (17.52) and (17.53), with $\xi_{\nu_e} \equiv \mu_{\nu_e}^{\text{fr}}/T_{\nu_\alpha}^{\text{fr}}$. A positive chemical potential of electron neutrinos decreases the neutron-to-proton ratio at freeze-out, leading to an abundance of ^{4}He which is smaller than in the standard BBN. The variation ΔY_{p} of the value of Y_{p} with respect to the standard one can be calculated straightforwardly from eqn (17.138) for a small ξ_{ν_e} [367]: since

$$\Delta \left(\frac{n_n}{n_p}\right)_{\text{BBN}} \simeq -\xi_{\nu_e} \left(\frac{n_n}{n_p}\right)_{\text{BBN}}, \tag{17.139}$$

we obtain, from eqn (17.124) (see also [217, 367, 380]),

$$\Delta Y_{\text{p}} \simeq -\xi_{\nu_e} Y_{\text{p}} \left(1 - \frac{Y_{\text{p}}}{2}\right) \simeq -0.22\, \xi_{\nu_e}. \tag{17.140}$$

Comparing this variation with that in eqn (17.126) one can see that for the production of primordial ^{4}He a small ξ_{ν_e} is equivalent to

$$\Delta N_\nu^{\xi_{\nu_e}}(Y_{\text{p}}) \simeq -20\, \xi_{\nu_e}. \tag{17.141}$$

2. Large neutrino chemical potentials increase the energy density of the Universe in comparison with the standard BBN scenario, leading to a faster expansion.

Since neutrons have less time to decay between freeze-out and the onset of nucleosynthesis, the neutron-to-proton ratio during nucleosynthesis is larger than the standard BBN, leading to a larger abundance of ^{4}He.

Since these two effects work in opposite directions if ξ_{ν_e} is positive, it is difficult to exclude a conspiracy between the values of ξ_{ν_e} and ξ_{ν_μ}, ξ_{ν_τ} which could lead to cancellations. In general, for the primordial ^{4}He production, we have

$$\Delta N_\nu(Y_\mathrm{p}) = \Delta N_\nu + \Delta N_\nu^\xi + \Delta N_\nu^{\xi_{\nu_e}}(Y_\mathrm{p}) \,, \qquad (17.142)$$

where ΔN_ν, given by eqn (17.118), represents the contribution to the energy density of relativistic nonstandard particles (sterile neutrinos and others), ΔN_ν^ξ, given by eqn (17.131), represents the contribution to the energy density of neutrino chemical potentials, and $\Delta N_\nu^{\xi_{\nu_e}}(Y_\mathrm{p})$, given by eqn (17.141), represents the effect on Y_p of ξ_{ν_e} due to the variation of the neutron-to-proton ratio at freeze-out. Since $\Delta N_\nu^{\xi_{\nu_e}}(Y_\mathrm{p})$ is negative for a positive ξ_{ν_e}, it can compensate the effects of ΔN_ν and ΔN_ν^ξ, leading to a small [424, 342] or even negative total value of $\Delta N_\nu(Y_\mathrm{p})$. Note, however, that a cancellation requires some fine tuning among the values of the neutrino chemical potentials and the number of nonstandard particles thermalized before BBN. For simplicity, in the following, we will discuss the effects of ΔN_ν^ξ and $\Delta N_\nu^{\xi_{\nu_e}}(Y_\mathrm{p})$ separately.

From the contribution of neutrino chemical potentials to the expansion rate in eqn (17.131) and the bound on ΔN_ν in eqn (17.127) we obtain

$$\sum_{\alpha=e,\mu,\tau} \xi_{\nu_\alpha}^4 \lesssim (2.5)^4 \qquad (95\%\ \mathrm{CL}) \,, \qquad (17.143)$$

which is much more stringent that the bound in eqn (17.134).

From eqn (17.141) and the bound on ΔN_ν in eqn (17.127), obtained from Y_p, we find

$$|\xi_{\nu_e}| \lesssim 0.04 \,. \qquad (17.144)$$

Hence, ξ_{ν_e} is much more tightly constrained than ξ_{ν_μ} and ξ_{ν_τ}.

So far, we did not take into account the equilibration of the values of the three flavor neutrino asymmetries through neutrino oscillations before neutrino decoupling and BBN [930, 757, 380, 1069, 11]. This implies that the three flavor asymmetries ξ_{ν_e}, ξ_{ν_μ}, and ξ_{ν_τ} are practically equal at BBN and the bound in eqn (17.144) applies to all of them:

$$-0.04 \lesssim \xi_{\nu_e} \simeq \xi_{\nu_\mu} \simeq \xi_{\nu_\tau} \lesssim 0.04 \,. \qquad (17.145)$$

This is the most stringent bound on the asymmetries of the three flavor neutrinos which can be obtained from the present experimental data.

The cosmological scenario becomes much more complicated if there are light sterile neutrinos which mix with the active ones. It has been shown in Ref. [455] that active–sterile oscillations can generate large neutrino asymmetries. This mechanism has been studied in detail in Refs. [456, 202, 955, 454, 381, 366, 367, 757, 419, 364, 365, 169, 8].

APPENDIX A

CONVENTIONS, USEFUL FORMULAS, AND PHYSICAL CONSTANTS

A.1 Conventions

In this book, we use natural units in which

$$c = \hbar = k_{\mathrm{B}} = 1\,, \tag{A.1}$$

where c is the velocity of light (eqn (A.154)), $\hbar$ is the reduced Planck constant (eqn (A.166)), and k_{B} is the Boltzmann constant (eqn (A.145)). The natural units imply the conversion factors in Table A.1.

Other conventions that we use in the book are:

– Space-time contravariant four-vector:

$$x^{\mu} = (x^0, x^1, x^2, x^3) = (t, \vec{x})\,. \tag{A.2}$$

– Gradient covariant four-vector:

$$\partial_{\mu} = \frac{\partial}{\partial x^{\mu}} = \left(\frac{\partial}{\partial x^0}, \frac{\partial}{\partial \vec{x}}\right) = \left(\partial_0, \vec{\nabla}\right)\,. \tag{A.3}$$

– Energy–momentum contravariant four-vector:

$$p^{\mu} = (p^0, \vec{p})\,. \tag{A.4}$$

– Contravariant and covariant representations of a four-vector V:

$$V^{\mu} = (V^0, \vec{V}) = g^{\mu\nu}\, V_{\nu}\,, \qquad V_{\mu} = (V^0, -\vec{V}) = g_{\mu\nu}\, V^{\nu}\,. \tag{A.5}$$

– Metric tensor:

$$g_{\mu\nu} = g^{\mu\nu} = \mathrm{diag}(1, -1, -1, -1)\,. \tag{A.6}$$

– Scalar product between two four-vectors A and B:

$$A \cdot B = A^{\mu} B_{\mu} = g_{\mu\nu} A^{\mu} B^{\nu} = g^{\mu\nu} A_{\mu} B_{\nu} = A^0 B^0 - \vec{A} \cdot \vec{B}\,. \tag{A.7}$$

– Norm of a four-vector V:

$$|V|^2 = V \cdot V = (V^0)^2 - \vec{V}^2\,. \tag{A.8}$$

Four-vectors are divided in three groups according to the sign of their norm:

$$|V|^2 > 0 \qquad \text{time-like} \tag{A.9}$$

TABLE A.1. Conversion factors for natural units (suggested by Ref. [379]). It is also useful to know that $1\,\text{in} \equiv 2.54\,\text{cm}$, $1\,\text{Å} \equiv 10^{-8}\,\text{cm}$, $1\,\text{fm} \equiv 10^{-13}\,\text{cm}$, $1\,\text{J} \equiv 10^{7}\,\text{erg}$, $1\,\text{W} \equiv 1\,\text{J}\,\text{s}^{-1}$.

	s^{-1}	cm^{-1}	K	eV	amu	erg	g
s^{-1}	1	3.336×10^{-11}	7.638×10^{-12}	6.582×10^{-16}	7.066×10^{-25}	1.055×10^{-27}	1.173×10^{-48}
cm^{-1}	2.998×10^{10}	1	2.290×10^{-1}	1.973×10^{-5}	2.118×10^{-14}	3.162×10^{-17}	3.518×10^{-38}
K	1.309×10^{11}	4.367	1	8.617×10^{-5}	9.251×10^{-14}	1.381×10^{-16}	1.536×10^{-37}
eV	1.519×10^{15}	5.068×10^{4}	1.160×10^{4}	1	1.074×10^{-9}	1.602×10^{-12}	1.783×10^{-33}
amu	1.415×10^{24}	4.721×10^{13}	1.081×10^{13}	9.315×10^{8}	1	1.492×10^{-3}	1.661×10^{-24}
erg	9.483×10^{26}	3.163×10^{16}	7.243×10^{15}	6.242×10^{11}	6.701×10^{2}	1	1.113×10^{-21}
g	8.522×10^{47}	2.843×10^{37}	6.510×10^{36}	5.610×10^{32}	6.022×10^{23}	8.988×10^{20}	1

$$|V|^2 = 0 \qquad \text{light-like} \tag{A.10}$$

$$|V|^2 < 0 \qquad \text{space-like} \tag{A.11}$$

Light-like four-vectors are orthogonal to themselves. A four-vector that is orthogonal to a time-like four-vector is space-like, and vice versa. For a particle with mass m

$$p^2 = m^2 \qquad \Longrightarrow \qquad p^0 = \sqrt{\vec{p}^{\,2} + m^2} \equiv E\,. \tag{A.12}$$

– Kronecker delta:

$$\delta_{kj} = \delta^{kj} = \delta^k_j\,. \tag{A.13}$$

– Totally antisymmetric tensors of rank 2:

$$\epsilon^{kj} = \epsilon_{kj}\,, \quad \text{with} \quad \epsilon^{12} = 1\,, \tag{A.14}$$

$$\epsilon_{ij}\,\epsilon_{k\ell} = \begin{vmatrix} \delta_{ik} & \delta_{i\ell} \\ \delta_{jk} & \delta_{j\ell} \end{vmatrix}\,, \tag{A.15}$$

$$\sum_i \epsilon_{ij}\,\epsilon_{i\ell} = \delta_{j\ell}\,, \tag{A.16}$$

$$\sum_{i,j} \epsilon_{ij}\,\epsilon_{ij} = 2\,. \tag{A.17}$$

– Totally antisymmetric tensors of rank 3:

$$\epsilon^{kjl} = \epsilon_{kjl}\,, \quad \text{with} \quad \epsilon^{123} = 1\,, \tag{A.18}$$

$$\epsilon_{ijk}\,\epsilon_{\ell mn} = \begin{vmatrix} \delta_{i\ell} & \delta_{im} & \delta_{in} \\ \delta_{j\ell} & \delta_{jm} & \delta_{in} \\ \delta_{k\ell} & \delta_{km} & \delta_{kn} \end{vmatrix}\,, \tag{A.19}$$

$$\sum_i \epsilon_{ijk}\,\epsilon_{imn} = \begin{vmatrix} \delta_{jm} & \delta_{jn} \\ \delta_{km} & \delta_{kn} \end{vmatrix}\,, \tag{A.20}$$

$$\sum_{i,j} \epsilon_{ijk}\,\epsilon_{ijn} = 2\,\delta_{kn}\,, \tag{A.21}$$

$$\sum_{i,j,k} \epsilon_{ijk}\,\epsilon_{ijk} = 6\,. \tag{A.22}$$

– Totally antisymmetric tensors of rank 4:

$$\epsilon^{\mu\nu\rho\sigma} = -\epsilon_{\mu\nu\rho\sigma}\,, \quad \text{with} \quad \epsilon^{0123} = 1\,, \tag{A.23}$$

$$\epsilon^{\mu\nu\rho\sigma}\,\epsilon^{\alpha\beta\gamma\delta} = -\begin{vmatrix} g^{\mu\alpha} & g^{\mu\beta} & g^{\mu\gamma} & g^{\mu\delta} \\ g^{\nu\alpha} & g^{\nu\beta} & g^{\nu\gamma} & g^{\nu\delta} \\ g^{\rho\alpha} & g^{\rho\beta} & g^{\rho\gamma} & g^{\rho\delta} \\ g^{\sigma\alpha} & g^{\sigma\beta} & g^{\sigma\gamma} & g^{\sigma\delta} \end{vmatrix}\,, \tag{A.24}$$

$$\sum_{\sigma} \epsilon^{\mu\nu\rho\sigma}\,\epsilon^{\alpha\beta\gamma}{}_{\sigma} = -\begin{vmatrix} g^{\mu\alpha} & g^{\mu\beta} & g^{\mu\gamma} \\ g^{\nu\alpha} & g^{\nu\beta} & g^{\nu\gamma} \\ g^{\rho\alpha} & g^{\rho\beta} & g^{\rho\gamma} \end{vmatrix}\,, \tag{A.25}$$

$$\sum_{\rho,\sigma} \epsilon^{\mu\nu\rho\sigma}\,\epsilon^{\alpha\beta}{}_{\rho\sigma} = -2\begin{vmatrix} g^{\mu\alpha} & g^{\mu\beta} \\ g^{\nu\alpha} & g^{\nu\beta} \end{vmatrix}\,, \tag{A.26}$$

$$\sum_{\nu,\rho,\sigma} \epsilon^{\mu\nu\rho\sigma}\,\epsilon^{\alpha}{}_{\nu\rho\sigma} = -6\,g^{\mu\alpha}\,, \tag{A.27}$$

$$\sum_{\mu,\nu,\rho,\sigma} \epsilon^{\mu\nu\rho\sigma}\,\epsilon_{\mu\nu\rho\sigma} = -24\,. \tag{A.28}$$

A.2 Pauli matrices

– Definition:

$$\sigma^1 = \sigma_1 = \begin{pmatrix} 0 & 1 \\ 1 & 0 \end{pmatrix} \qquad \sigma^2 = \sigma_2 = \begin{pmatrix} 0 & -i \\ i & 0 \end{pmatrix} \qquad \sigma^3 = \sigma_3 = \begin{pmatrix} 1 & 0 \\ 0 & -1 \end{pmatrix}\,. \tag{A.29}$$

– Properties:

$$\left(\sigma^k\right)^2 = 1\,, \tag{A.30}$$

$$\sigma^k = \left(\sigma^k\right)^\dagger\,, \tag{A.31}$$

$$\left(\sigma^k\right)^T = \left(\sigma^k\right)^*\,, \tag{A.32}$$

$$\sigma^2\sigma^k\sigma^2 = -\left(\sigma^k\right)^T\,. \tag{A.33}$$

– Commutation and anticommutation:

$$\left[\sigma^k, \sigma^j\right] = 2i\sum_{\ell} \epsilon^{kj\ell}\,\sigma^\ell\,, \tag{A.34}$$

$$\left\{\sigma^k, \sigma^j\right\} = 2\delta^{kj}\mathbf{1}\,. \tag{A.35}$$

– Products:

$$\sigma^i\sigma^j = \delta^{ij} + i\sum_k \epsilon^{ijk}\sigma^k\,, \tag{A.36}$$

$$\sigma^i\sigma^j\sigma^k = i\epsilon^{ijk} + \delta^{ij}\sigma^k - \delta^{ik}\sigma^j + \delta^{jk}\sigma^i\,. \tag{A.37}$$

– Traces:

$$\mathrm{Tr}\big[\sigma^i\big] = 0\,, \tag{A.38}$$

$$\mathrm{Tr}\big[\sigma^i\sigma^j\big] = 2\delta^{ij}\,, \tag{A.39}$$

$$\mathrm{Tr}\big[\sigma^i\sigma^j\sigma^k\big] = 2i\epsilon^{ijk}\,, \tag{A.40}$$

$$\mathrm{Tr}\big[\sigma^i\sigma^j\sigma^k\sigma^\ell\big] = 2\left[\delta^{ij}\delta^{k\ell} - \delta^{ik}\delta^{j\ell} + \delta^{i\ell}\delta^{jk}\right]. \tag{A.41}$$

– Every 2×2 matrix X can be written as

$$X = \frac{1}{2}\left(\mathrm{Tr}[X] + \sum_k \mathrm{Tr}\big[X\sigma^k\big]\,\sigma^k\right). \tag{A.42}$$

A.3 Dirac matrices

– Defining relations:

$$\{\gamma^\mu,\gamma^\nu\} \equiv \gamma^\mu\gamma^\nu + \gamma^\nu\gamma^\mu = 2\,g^{\mu\nu}\,, \tag{A.43}$$

$$\gamma^0\,\gamma^{\mu\dagger}\,\gamma^0 = \gamma^\mu\,. \tag{A.44}$$

– Properties ($k = 1, 2, 3$):

$$(\gamma^0)^\dagger = \gamma^0\,, \qquad (\gamma^k)^\dagger = -\gamma^k \qquad \Longleftrightarrow \qquad (\gamma^\mu)^\dagger = \gamma_\mu \equiv g_{\mu\nu}\,\gamma^\nu\,. \tag{A.45}$$

$$(\gamma^0)^2 = \mathbf{1}\,, \qquad (\gamma^k)^2 = -\mathbf{1} \qquad \Longrightarrow \qquad \gamma^\mu\gamma_\mu = 4\,, \tag{A.46}$$

– Definition and properties of γ^5:

$$\gamma^5 \equiv \gamma_5 \equiv i\,\gamma^0\,\gamma^1\,\gamma^2\,\gamma^3 = -\frac{i}{4!}\,\epsilon_{\mu\nu\rho\sigma}\,\gamma^\mu\,\gamma^\nu\,\gamma^\rho\,\gamma^\sigma\,, \tag{A.47}$$

$$\big\{\gamma^5,\gamma^\mu\big\} = 0\,, \tag{A.48}$$

$$\big(\gamma^5\big)^2 = \mathbf{1}\,, \tag{A.49}$$

$$\big(\gamma^5\big)^\dagger = \gamma^5\,, \tag{A.50}$$

$$\gamma^\mu\,\gamma^5 = \frac{i}{6}\,\epsilon^{\mu\nu\rho\sigma}\,\gamma_\nu\,\gamma_\rho\,\gamma_\sigma\,. \tag{A.51}$$

– Definition and properties of $\sigma^{\mu\nu}$:

$$\sigma^{\mu\nu} \equiv \frac{i}{2}\,[\gamma^\mu,\gamma^\nu] = i\,\gamma^\mu\,\gamma^\nu - i\,g^{\mu\nu} = i\,g^{\mu\nu} - i\,\gamma^\nu\,\gamma^\mu\,, \tag{A.52}$$

$$\big[\gamma^5\,,\,\sigma^{\mu\nu}\big] = 0\,, \tag{A.53}$$

$$\gamma^0\,(\sigma^{\mu\nu})^\dagger\,\gamma^0 = \sigma^{\mu\nu}\,. \tag{A.54}$$

$$\sigma^{\mu\nu}\,\gamma^5 = \frac{i}{2}\,\epsilon^{\mu\nu\rho\sigma}\,\sigma_{\rho\sigma}\,. \tag{A.55}$$

– Definition and properties of $\vec{\Sigma}$:

$$\Sigma^k \equiv \frac{1}{2}\sum_{j,l} \epsilon^{kjl}\,\sigma^{jl} = \frac{i}{2}\sum_{j,l} \epsilon^{kjl}\,\gamma^j\,\gamma^l = \gamma^0\,\gamma^k\,\gamma^5\,, \tag{A.56}$$

$$\vec{\Sigma} = (i\gamma^2\gamma^3, i\gamma^3\gamma^1, i\gamma^1\gamma^2) = (\sigma^{23}, \sigma^{31}, \sigma^{12})\,, \tag{A.57}$$

$$\sigma^{jk} = \sum_\ell \epsilon^{jk\ell}\,\Sigma^\ell\,. \tag{A.58}$$

$$[\Sigma^k, \Sigma^j] = 2\,i\sum_l \epsilon^{kjl}\,\Sigma^l\,, \tag{A.59}$$

$$\{\Sigma^k, \Sigma^j\} = 2\,\delta^{kj}\,, \tag{A.60}$$

$$(\Sigma^k)^\dagger = \Sigma^k\,, \tag{A.61}$$

$$[\Sigma^k, \gamma^0] = [\Sigma^k, \gamma^5] = 0\,, \tag{A.62}$$

$$[\Sigma^k, \gamma^j] = 2i\sum_\ell \epsilon^{kj\ell}\,\gamma^\ell\,. \tag{A.63}$$

– Products of γ matrices:

$$\gamma^\mu\,\gamma^\nu = g^{\mu\nu} - i\,\sigma^{\mu\nu}\,, \tag{A.64}$$

$$\gamma^\mu\,\gamma^\nu\,\gamma^5 = g^{\mu\nu}\,\gamma^5 + \frac{1}{2}\,\epsilon^{\mu\nu\rho\sigma}\,\sigma_{\rho\sigma}\,, \tag{A.65}$$

$$\gamma^\mu\,\gamma^\nu\,\gamma^\rho = g^{\mu\nu}\,\gamma^\rho - g^{\mu\rho}\,\gamma^\nu + g^{\nu\rho}\,\gamma^\mu + i\,\epsilon^{\mu\nu\rho\sigma}\,\gamma_\sigma\,\gamma_5\,, \tag{A.66}$$

$$\gamma^\mu\,\gamma^\alpha\,\gamma_\mu = -2\,\gamma^\alpha\,, \tag{A.67}$$

$$\gamma^\mu\,\gamma^\alpha\,\gamma^\beta\,\gamma_\mu = 4\,g^{\alpha\beta}\,, \tag{A.68}$$

$$\gamma^\mu\,\gamma^\alpha\,\gamma^\beta\,\gamma^\rho\,\gamma_\mu = -2\,\gamma^\rho\,\gamma^\beta\,\gamma^\alpha\,. \tag{A.69}$$

– Traces:

$$\mathrm{Tr}[\gamma^{\mu_1}] = 0\,, \quad \mathrm{Tr}[\gamma^{\mu_1}\,\gamma^{\mu_2}\,\gamma^{\mu_3}] = 0\,, \quad \ldots\,, \quad \mathrm{Tr}[\gamma^{\mu_1}\,\gamma^{\mu_2}\,\ldots\,\gamma^{\mu_{2n+1}}] = 0\,, \tag{A.70}$$

$$\mathrm{Tr}[\gamma^{\mu_1}\,\gamma^{\mu_2}\,\ldots\,\gamma^{\mu_{n-1}}\,\gamma^{\mu_n}] = \mathrm{Tr}[\gamma^{\mu_n}\,\gamma^{\mu_{n-1}}\,\ldots\,\gamma^{\mu_2}\,\gamma^{\mu_1}]\,, \tag{A.71}$$

$$\mathrm{Tr}[\gamma^{\mu_1}\,\gamma^{\mu_2}] = 4\,g^{\mu_1\mu_2}\,, \tag{A.72}$$

$$\mathrm{Tr}[\gamma^{\mu_1}\,\gamma^{\mu_2}\,\gamma^{\mu_3}\,\gamma^{\mu_4}] = 4\,(g^{\mu_1\mu_2}\,g^{\mu_3\mu_4} - g^{\mu_1\mu_3}\,g^{\mu_2\mu_4} + g^{\mu_1\mu_4}\,g^{\mu_2\mu_3})\,, \tag{A.73}$$

$$\begin{aligned}\mathrm{Tr}[\gamma^{\mu_1}\gamma^{\mu_2}\gamma^{\mu_3}\gamma^{\mu_4}\cdots\gamma^{\mu_{2n-1}}\gamma^{\mu_{2n}}] &= g^{\mu_1\mu_2}\mathrm{Tr}\left[\gamma^{\mu_3}\gamma^{\mu_4}\cdots\gamma^{\mu_{2n-1}}\gamma^{\mu_{2n}}\right]\\ &\quad - g^{\mu_1\mu_3}\mathrm{Tr}\left[\gamma^{\mu_2}\gamma^{\mu_4}\cdots\gamma^{\mu_{2n-1}}\gamma^{\mu_{2n}}\right]\\ &\quad + \cdots\\ &\quad + g^{\mu_1\mu_{2n}}\mathrm{Tr}\left[\gamma^{\mu_2}\gamma^{\mu_3}\gamma^{\mu_4}\cdots\gamma^{\mu_{2n-1}}\right]\,.\end{aligned} \tag{A.74}$$

$$\mathrm{Tr}\left[\gamma^5\right] = 0\,, \quad \mathrm{Tr}\left[\gamma^{\mu_1}\,\gamma^5\right] = 0\,, \quad \mathrm{Tr}\left[\gamma^{\mu_1}\,\gamma^{\mu_2}\,\gamma^5\right] = 0\,, \quad \mathrm{Tr}\left[\gamma^{\mu_1}\,\gamma^{\mu_2}\,\gamma^{\mu_3}\,\gamma^5\right] = 0\,, \tag{A.75}$$

$$\mathrm{Tr}\left[\gamma^{\mu_1}\,\gamma^{\mu_2}\,\gamma^{\mu_3}\,\gamma^{\mu_4}\,\gamma^5\right] = -4\,i\,\epsilon^{\mu_1\mu_2\mu_3\mu_4}\,. \tag{A.76}$$

– Left-handed and right-handed projectors:

$$P_L = \frac{1-\gamma^5}{2}\,, \qquad P_R = \frac{1+\gamma^5}{2}\,, \tag{A.77}$$

$$P_L^2 = P_L\,, \qquad P_R^2 = P_R\,, \qquad P_L\,P_R = P_R\,P_L = 0\,. \tag{A.78}$$

– Defining equations of four-momentum and helicity eigenstate spinors:

$$(\not{p} - m)\,u^{(h)}(p) = 0\,, \tag{A.79}$$

$$(\not{p} + m)\,v^{(h)}(p) = 0\,. \tag{A.80}$$

$$\frac{\vec{p}\cdot\vec{\Sigma}}{|\vec{p}|}\,u^{(h)}(p) = h\,u^{(h)}(p)\,, \tag{A.81}$$

$$\frac{\vec{p}\cdot\vec{\Sigma}}{|\vec{p}|}\,v^{(h)}(p) = -h\,v^{(h)}(p)\,. \tag{A.82}$$

– Normalization:

$$\overline{u^{(h)}}(p)\,u^{(h')}(p) = 2\,m\,\delta_{hh'}\,, \tag{A.83}$$

$$\overline{v^{(h)}}(p)\,v^{(h')}(p) = -2\,m\,\delta_{hh'}\,. \tag{A.84}$$

– Useful relations:

$$\overline{u^{(h)}}(p)\,\gamma^\mu\,u^{(h')}(p) = \overline{v^{(h)}}(p)\,\gamma^\mu\,v^{(h')}(p) = 2\,p^\mu\,\delta_{hh'}\,, \tag{A.85}$$

$$\overline{u^{(h)}}(p)\,\gamma^5\,u^{(h')}(p) = \overline{v^{(h)}}(p)\,\gamma^5\,v^{(h')}(p) = 0\,, \tag{A.86}$$

$$u^{(h)\dagger}(p^0,\vec{p})\,v^{(h')}(p^0,-\vec{p}) = 0\,, \tag{A.87}$$

$$\left.\begin{array}{l}\gamma^0\,u^{(-h)}(p^0,-\vec{p}) = \eta(\vec{p},h)\,u^{(h)}(p^0,\vec{p})\,,\\ \gamma^0\,v^{(-h)}(p^0,-\vec{p}) = -\eta^*(\vec{p},h)\,v^{(h)}(p^0,\vec{p})\,,\end{array}\right\}\ \text{with}\ \left\{\begin{array}{l}|\eta(\vec{p},h)|^2 = 1\,,\\ \eta(-\vec{p},-h) = \eta^*(\vec{p},h)\,,\end{array}\right. \tag{A.88}$$

$$\left.\begin{array}{l}\gamma^5\,v^{(-h)}(p) = \zeta(h)\,u^{(h)}(p)\,,\\ \gamma^5\,u^{(-h)}(p) = -\zeta^*(h)\,v^{(h)}(p)\,,\end{array}\right\}\quad \text{with}\quad \left\{\begin{array}{l}|\zeta(h)|^2 = 1\,,\\ \zeta(-h) = -\zeta(h)\,.\end{array}\right. . \tag{A.89}$$

– Gordon identities:

$$\overline{u^{(h_2)}}(p_2)\,\gamma^\mu\,u^{(h_1)}(p_1) = \frac{1}{2\,m}\,\overline{u^{(h_2)}}(p_2)\left[(p_2+p_1)^\mu + i\,\sigma^{\mu\nu}\,(p_2-p_1)_\nu\right]u^{(h_1)}(p_1)\,, \tag{A.90}$$

$$\overline{v^{(h_2)}}(p_2)\,\gamma^\mu\,v^{(h_1)}(p_1) = -\frac{1}{2\,m}\,\overline{v^{(h_2)}}(p_2)\left[(p_2+p_1)^\mu + i\,\sigma^{\mu\nu}\,(p_2-p_1)_\nu\right]v^{(h_1)}(p_1)\,. \tag{A.91}$$

– Projectors on components with positive and negative energy:

$$\Lambda_+(p) = \frac{m+\not{p}}{2m} = \sum_{h=\pm 1}\frac{u^{(h)}(p)\,\overline{u^{(h)}}(p)}{2m}\,, \tag{A.92}$$

$$\Lambda_-(p) = \frac{m - \not{p}}{2m} = -\sum_{h=\pm 1} \frac{v^{(h)}(p)\, \overline{v^{(h)}}(p)}{2m} \,. \tag{A.93}$$

– Projectors on components with definite energy and helicity:

$$\Lambda_+^h(p) = \left(\frac{m + \not{p}}{2m}\right)\left(\frac{1 + \gamma^5 \not{s}_h}{2}\right) = \frac{u^{(h)}(p)\, \overline{u^{(h)}}(p)}{2m} \,, \tag{A.94}$$

$$\Lambda_-^h(p) = \left(\frac{m - \not{p}}{2m}\right)\left(\frac{1 + \gamma^5 \not{s}_h}{2}\right) = -\frac{v^{(h)}(p)\, \overline{v^{(h)}}(p)}{2m} \,, \tag{A.95}$$

with the polarization four-vector

$$s_h^\mu = h\left(\frac{|\vec{p}|}{m}, \frac{E}{m}\frac{\vec{p}}{|\vec{p}|}\right), \qquad s_h^2 = -1 \,, \qquad s_h \cdot p = 0 \,, \tag{A.96}$$

– Charge conjugation matrix:

$$\mathcal{C}\, \gamma_\mu^T\, \mathcal{C}^{-1} = -\gamma_\mu \,, \tag{A.97}$$

$$\mathcal{C}^\dagger = \mathcal{C}^{-1} \,, \tag{A.98}$$

$$\mathcal{C}^T = -\mathcal{C} \,. \tag{A.99}$$

$$\mathcal{C}\, (\gamma^5)^T\, \mathcal{C}^{-1} = \gamma^5 \,, \tag{A.100}$$

$$\mathcal{C}\, (\sigma^{\mu\nu})^T\, \mathcal{C}^{-1} = -\sigma^{\mu\nu} \,. \tag{A.101}$$

$$u^{(h)}(p) = \mathcal{C}\, \overline{v^{(h)}}^T(p) \,, \qquad v^{(h)}(p) = \mathcal{C}\, \overline{u^{(h)}}^T(p) \,, \tag{A.102}$$

– Dirac representation:

$$\gamma_{\mathrm{D}}^0 = \begin{pmatrix} \mathbf{1} & 0 \\ 0 & -\mathbf{1} \end{pmatrix}, \qquad \vec{\gamma}_{\mathrm{D}} = \begin{pmatrix} 0 & \vec{\sigma} \\ -\vec{\sigma} & 0 \end{pmatrix}, \qquad \gamma_{\mathrm{D}}^5 = \begin{pmatrix} 0 & \mathbf{1} \\ \mathbf{1} & 0 \end{pmatrix}, \tag{A.103}$$

$$\vec{\Sigma}_{\mathrm{D}} = \begin{pmatrix} \vec{\sigma} & 0 \\ 0 & \vec{\sigma} \end{pmatrix}, \qquad \mathcal{C}_{\mathrm{D}} = i\, \gamma_{\mathrm{D}}^2\, \gamma_{\mathrm{D}}^0 = -i \begin{pmatrix} 0 & \sigma^2 \\ \sigma^2 & 0 \end{pmatrix}, \tag{A.104}$$

$$\sigma_{\mathrm{D}}^{0k} = i\, \alpha_{\mathrm{D}}^k = i \begin{pmatrix} \sigma^k & 0 \\ 0 & -\sigma^k \end{pmatrix}, \qquad \sigma_{\mathrm{D}}^{kj} = \sum_\ell \epsilon^{kj\ell}\, \Sigma_{\mathrm{D}}^\ell = \sum_\ell \epsilon^{kj\ell} \begin{pmatrix} \sigma^\ell & 0 \\ 0 & \sigma^\ell \end{pmatrix}, \tag{A.105}$$

$$u_{\mathrm{D}}^{(h)}(p) = \begin{pmatrix} \sqrt{E+m}\, \chi^{(h)}(\vec{p}) \\ h\, \sqrt{E-m}\, \chi^{(h)}(\vec{p}) \end{pmatrix}, \tag{A.106}$$

$$v_{\mathrm{D}}^{(h)}(p) = \begin{pmatrix} -\sqrt{E-m}\, \chi^{(-h)}(\vec{p}) \\ h\, \sqrt{E+m}\, \chi^{(-h)}(\vec{p}) \end{pmatrix}. \tag{A.107}$$

$$\zeta(h) = -h \,. \tag{A.108}$$

– Chiral representation

$$\gamma^0_{\rm C} = \begin{pmatrix} 0 & -\mathbf{1} \\ -\mathbf{1} & 0 \end{pmatrix}, \qquad \vec{\gamma}_{\rm C} = \begin{pmatrix} 0 & \vec{\sigma} \\ -\vec{\sigma} & 0 \end{pmatrix}, \qquad \gamma^5_{\rm C} = \begin{pmatrix} \mathbf{1} & 0 \\ 0 & -\mathbf{1} \end{pmatrix}, \tag{A.109}$$

$$\gamma^\mu_{\rm C} = \begin{pmatrix} 0 & \bar{\sigma}^\mu \\ -\sigma^\mu & 0 \end{pmatrix}, \qquad \text{with} \qquad \sigma^\mu = (\mathbf{1}\,, \vec{\sigma})\,, \qquad \bar{\sigma}^\mu = (-\mathbf{1}\,, \vec{\sigma})\,, \tag{A.110}$$

$$\vec{\Sigma}_{\rm C} = \begin{pmatrix} \vec{\sigma} & 0 \\ 0 & \vec{\sigma} \end{pmatrix}, \qquad \mathcal{C}_{\rm C} = i\,\gamma^2_{\rm C}\,\gamma^0_{\rm C} = -i \begin{pmatrix} \sigma^2 & 0 \\ 0 & -\sigma^2 \end{pmatrix}. \tag{A.111}$$

$$\sigma^{0k}_{\rm C} = i\,\alpha^k_{\rm C} = i \begin{pmatrix} \sigma^k & 0 \\ 0 & -\sigma^k \end{pmatrix}, \qquad \sigma^{kj}_{\rm C} = \sum_\ell \epsilon^{kj\ell}\,\Sigma^\ell_{\rm C} = \sum_\ell \epsilon^{kj\ell} \begin{pmatrix} \sigma^\ell & 0 \\ 0 & \sigma^\ell \end{pmatrix}, \tag{A.112}$$

$$u^{(h)}_{\rm C}(p) = \begin{pmatrix} -\sqrt{E + h\,|\vec{p}|}\,\chi^{(h)}(\vec{p}) \\ \sqrt{E - h\,|\vec{p}|}\,\chi^{(h)}(\vec{p}) \end{pmatrix}, \tag{A.113}$$

$$v^{(h)}_{\rm C}(p) = -\,h \begin{pmatrix} \sqrt{E - h\,|\vec{p}|}\,\chi^{(-h)}(\vec{p}) \\ \sqrt{E + h\,|\vec{p}|}\,\chi^{(-h)}(\vec{p}) \end{pmatrix}, \tag{A.114}$$

$$\zeta(h) = -h\,. \tag{A.115}$$

– Two-component helicity eigenstate spinors:

$$\frac{\vec{p}\cdot\vec{\sigma}}{|\vec{p}|}\,\chi^{(h)}(\vec{p}) = h\,\chi^{(h)}(\vec{p})\,, \tag{A.116}$$

$$\left(\chi^{(h)}(\vec{p})\right)^\dagger \chi^{(h')}(\vec{p}) = \delta_{hh'}\,, \tag{A.117}$$

$$i\sigma^2 \left(\chi^{(h)}(\vec{p})\right)^* = -h\,\chi^{(-h)}(\vec{p})\,, \tag{A.118}$$

$$\chi^{(-h)}(-\vec{p}) = \eta(\vec{p},h)\,\chi^{(h)}(\vec{p}) \quad \text{with} \quad \begin{cases} |\eta(\vec{p},h)|^2 = 1\,, \\ \eta(-\vec{p},-h) = \eta^*(\vec{p},h)\,, \\ \eta(\vec{p},-h) = -\eta^*(\vec{p},h)\,, \end{cases} \tag{A.119}$$

$$\left(\chi^{(h)}(\vec{p})\right)^\dagger \sigma^k\,\chi^{(h)}(\vec{p}) = \frac{p^k}{h\,|\vec{p}|}\,. \tag{A.120}$$

For $\vec{p} = |\vec{p}|\,(\sin\theta\cos\phi, \sin\theta\sin\phi, \cos\theta)$,

$$\chi^{(+)}(\vec{p}) = \begin{pmatrix} \cos\frac{\theta}{2} \\ \sin\frac{\theta}{2}\,e^{i\phi} \end{pmatrix}, \qquad \chi^{(-)}(\vec{p}) = \begin{pmatrix} -\sin\frac{\theta}{2}\,e^{-i\phi} \\ \cos\frac{\theta}{2} \end{pmatrix}, \tag{A.121}$$

$$\eta(\vec{p},h) = h\,e^{-ih\phi}\,. \tag{A.122}$$

A.4 Mathematical formulas

– Dirac δ-function:

$$\int f(x)\,\delta(x)\,\mathrm{d}x = f(0)\,, \tag{A.123}$$

$$\delta(g(x)) = \sum_a \frac{\delta(x-a)}{|g'(a)|}\,, \quad \text{for} \quad g(a)=0\,, \quad g'(a)\neq 0\,. \tag{A.124}$$

– For a matrix $\boldsymbol{M}$[106]

$$\mathrm{Det}\left[e^{\boldsymbol{M}}\right] = e^{\mathrm{Tr}\boldsymbol{M}}\,. \tag{A.125}$$

– Useful series:

$$\sum_{k=0}^{\infty} x^k = (1-x)^{-1} \qquad (x<1)\,. \tag{A.126}$$

– Useful integrals[107]:

$$\int_0^\infty \frac{x^n}{e^x-1}\,\mathrm{d}x = n!\,\zeta(n+1)\,, \tag{A.127}$$

$$\int_0^\infty \frac{x^n}{e^x+1}\,\mathrm{d}x = \left(1-2^{-n}\right) n!\,\zeta(n+1)\,, \tag{A.128}$$

$$\sum_{s=\pm1}\int_0^\infty \frac{s^{n+1}\,x^n\,\mathrm{d}x}{e^{x-s\xi}+1} = \frac{\xi^{n+1}}{n+1} + \sum_{k=0}^{n-1}\left[1+(-1)^{n+k+1}\right]\left(1-2^{k-n}\right)\frac{n!}{k!}\,\zeta(n-k+1)\,\xi^k\,. \tag{A.129}$$

– Riemann zeta function:

$$\zeta(n) \equiv \sum_{k=1}^{\infty} k^{-n} \qquad (k>1)\,, \tag{A.130}$$

$$\zeta(2) = \frac{\pi^2}{6}\,, \qquad \zeta(3) \simeq 1.2020569032\,, \qquad \zeta(4) = \frac{\pi^4}{90}\,, \tag{A.131}$$

$$\sum_{k=1}^{\infty}(-1)^{k-1}\,k^{-n} = \left(1-2^{1-n}\right)\zeta(n)\,. \tag{A.132}$$

---

106 The proof is easy: $\mathrm{Det}\left[e^{\boldsymbol{M}}\right] = \mathrm{Det}\left[\lim_{n\to\infty}\left(1+\frac{\boldsymbol{M}}{n}\right)^n\right] = \lim_{n\to\infty}\left(\mathrm{Det}\left[1+\frac{\boldsymbol{M}}{n}\right]\right)^n = \lim_{n\to\infty}\left(1+\frac{\mathrm{Tr}\boldsymbol{M}}{n}\right)^n = e^{\mathrm{Tr}\boldsymbol{M}}$.

107 The integrals in eqns (A.127) and (A.128) can be evaluated by using the series in eqn (A.126), which implies $(e^x \pm 1)^{-1} = e^{-x}(1\pm e^{-x})^{-1} = e^{-x}\sum_{k=0}^{\infty}\left(\mp e^{-x}\right) = \sum_{k=1}^{\infty}(\mp 1)^{k-1}\,e^{-kx}$. The calculation is done through integration by parts, taking into account the definition of the Riemann zeta function in eqn (A.130) and the property in eqn (A.132).

The integral in eqn (A.129) can be calculated with the change of variable $y = x - s\xi$, using the identity $(e^y+1)^{-1} + \left(e^{-y}+1\right)^{-1} = 1$ [410].

– Legendre polynomials ($\ell \geq 0$):

$$P_\ell(x) = \frac{1}{2^\ell\,\ell!}\,\frac{\mathrm{d}^\ell}{\mathrm{d}x^\ell}\,(x^2-1)^\ell \qquad \text{(Rodrigues' formula)}, \tag{A.133}$$

$$\int_{-1}^{+1} P_k(x)\,P_\ell(x)\,\mathrm{d}x = \frac{2}{2\ell+1}\,\delta_{k\ell}\,. \tag{A.134}$$

– Associated Legendre polynomials ($|m| \leq \ell$):

$$P_\ell^m(x) = \left(1-x^2\right)^{m/2}\frac{\mathrm{d}^m}{\mathrm{d}x^m}P_\ell(x) = \frac{1}{2^\ell\,\ell!}\left(1-x^2\right)^{m/2}\frac{\mathrm{d}^{\ell+m}}{\mathrm{d}x^{\ell+m}}\left(x^2-1\right)^\ell\,, \tag{A.135}$$

$$P_\ell^{-m}(x) = (-1)^m\,\frac{(\ell-m)!}{(\ell+m)!}\,P_\ell^m(x)\,, \qquad P_\ell^m(-x) = (-1)^{\ell+m}\,P_\ell^m(x)\,, \tag{A.136}$$

$$\int_{-1}^{+1} P_k^m(x)\,P_\ell^m(x)\,\mathrm{d}x = \frac{2}{2\ell+1}\,\frac{(\ell+m)!}{(\ell-m)!}\,\delta_{k\ell}\,. \tag{A.137}$$

– Spherical harmonics:

$$Y_\ell^m(\theta,\phi) = (-1)^m\left[\frac{2\ell+1}{4\pi}\,\frac{(\ell-m)!}{(\ell+m)!}\right]^{1/2} P_\ell^m(\cos\theta)\,e^{im\phi}\,, \tag{A.138}$$

$$\int_{-1}^{+1}\int_0^{2\pi} Y_\ell^m(\theta,\phi)\,Y_{\ell'}^{m'\,*}(\theta,\phi)\,\mathrm{d}\cos\theta\,\mathrm{d}\phi = \delta_{\ell\ell'}\,\delta_{mm'}\,. \tag{A.139}$$

– Addition theorem:

$$\sum_{m=-\ell}^{\ell} Y_\ell^m(\theta_1,\phi_1)\,Y_\ell^{m*}(\theta_2,\phi_2) = \frac{2\ell+1}{4\pi}\,P_\ell(\cos\theta)\,, \tag{A.140}$$

where θ is the angle between the directions (θ_1,ϕ_1) and (θ_2,ϕ_2), given by

$$\cos\theta = \cos\theta_1\,\cos\theta_2 + \sin\theta_1\,\sin\theta_2\,\cos(\phi_1-\phi_2)\,. \tag{A.141}$$

– Parity:

$$Y_\ell^m(\theta,\phi) \xrightarrow{\mathrm{P}} Y_\ell^m(\pi-\theta,\pi+\phi) = (-1)^\ell\,Y_\ell^m(\theta,\phi)\,. \tag{A.142}$$

A.5 Physical constants

In this section we give a list of values of physical constants used in the book. The values without reference are taken from the *Review of Particle Physics* in Ref. [400]. The one-σ uncertainties in the last digits are given in parentheses.

Atomic Mass Unit $\qquad \mathrm{amu} \equiv M(^{12}\mathrm{C})/12 = 931.494\,043\,(80)\,\mathrm{MeV}\,,$ (A.143)

Avogadro Number	$N_\mathrm{A} = 6.022\,1415\,(10) \times 10^{23}\,\mathrm{mol}^{-1}\,,$	(A.144)
Boltzmann Constant	$k_\mathrm{B} = 8.617\,343\,(15) \times 10^{-5}\,\mathrm{eV\,K}^{-1}\,,$	(A.145)
CMBR Temperature	$T_\gamma = 2.725\,(1)\,\mathrm{K}\,,$	(A.146)
Conversion Constant	$\hbar\, c = 1.973\,269\,68\,(17) \times 10^{-5}\,\mathrm{eV\,cm}\,,$	(A.147)
Earth Mass	$M_\oplus = 5.972\,3\,(9) \times 10^{27}\,\mathrm{g}\,,$	(A.148)
Earth Mean Equatorial Radius	$R_\oplus \simeq 6.378140 \times 10^{8}\,\mathrm{cm}\,,$	(A.149)
Electron Mass	$m_e = 0.510\,998\,918\,(44)\,\mathrm{MeV}\,,$	(A.150)
Fermi Constant	$G_\mathrm{F} = 1.166\,37\,(1) \times 10^{-5}\,\mathrm{GeV}^{-2}\,,$	(A.151)
Hubble Constant	$H_0 = 72\,(8)\,\mathrm{km\,s^{-1}\,Mpc^{-1}}$ [461],	(A.152)
Fine-Structure Constant	$\alpha^{-1} \equiv (e^2/4\pi\hbar c)^{-1} = 137.035\,999\,11\,(46)\,,$	(A.153)
Light's Velocity	$c \equiv 299\,792\,458\,\mathrm{m\,s}^{-1}\,,$	(A.154)
Muon Lifetime	$\tau_\mu = 2.197\,03\,(4) \times 10^{-6}\,\mathrm{s}\,,$	(A.155)
Muon Mass	$m_\mu = 105.658\,369\,(9)\,\mathrm{MeV}\,,$	(A.156)
Newton Constant	$G_\mathrm{N} = 6.6742\,(10) \times 10^{-11}\,\mathrm{m^3\,kg^{-1}\,s^{-2}}\,,$	(A.157)
Neutron Lifetime	$\tau_n = 885.7\,(8)\,\mathrm{s}\,,$	(A.158)
Neutron Magnetic Moment	$\mu_n = -1.913\,042\,73\,(45)\,\mu_\mathrm{N}\,,$	(A.159)
Neutron Mass	$m_n = 939.565\,36\,(8)\,\mathrm{MeV}\,,$	(A.160)
Parsec	$\mathrm{pc} = 3.085\,677\,580\,7\,(4) \times 10^{16}\,\mathrm{m} \simeq 3.262\,\mathrm{ly}\,,$	(A.161)
Pion $\pi^\pm$ Lifetime	$\tau_{\pi^\pm} = 2.603\,3\,(5) \times 10^{-8}\,\mathrm{s}\,,$	(A.162)
Pion $\pi^\pm$ Mass	$m_{\pi^\pm} = 139.570\,18\,(35)\,\mathrm{MeV}\,,$	(A.163)
Pion π^0 Lifetime	$\tau_{\pi^0} = 8.4\,(6) \times 10^{-17}\,\mathrm{s}\,,$	(A.164)
Pion π^0 Mass	$m_{\pi^0} = 134.976\,6\,(6)\,\mathrm{MeV}\,,$	(A.165)
Planck Constant Reduced	$\hbar = 6.582\,119\,15\,(56) \times 10^{-16}\,\mathrm{eV\,s}\,,$	(A.166)
Planck Mass	$M_\mathrm{P} \equiv \sqrt{\hbar c/G_\mathrm{N}} = 1.220\,90\,(9) \times 10^{19}\,\mathrm{GeV}\,,$	(A.167)
Proton Magnetic Moment	$\mu_p = 2.792\,847\,351\,(28)\,\mu_\mathrm{N}\,,$	(A.168)
Proton Mass	$m_p = 938.272\,029\,(80)\,\mathrm{MeV}\,,$	(A.169)
Tau Mass	$m_\tau = 1776.99^{+0.29}_{-0.26}\,\mathrm{MeV}\,,$	(A.170)
Weak Mixing Angle	$\sin^2\vartheta_\mathrm{W} = 0.231\,49\,(15)\,,$	(A.171)

APPENDIX B

SPECIAL RELATIVITY

Physical laws are expressed in terms of functions of the spatial coordinates $\vec{x}$ and the temporal coordinate t. However, physical laws describe events which occur independently of the system of coordinates, also called the *reference frame*, which is adopted by an observer. Therefore, the transformations between different reference frames and the quantities which are independent of the reference frame play a crucial role in the formulation of physical theories.

According to the *principle of relativity*, the physical laws are equally valid in all reference frames. In other words, there is no privileged frame of reference. Therefore, the structure of the equations representing physical laws must be independent of the adopted reference frame. This is achieved by expressing physical laws through *covariant* equations, i.e. equations written in tensorial form. This is called the *principle of covariance.*

The theory of special relativity treats the relations between different inertial reference frames, which are systems of coordinates in which free particles propagate with uniform motion, in the absence of gravity. The relative velocity of different inertial reference frames is constant. They are related by Lorentz transformations, which are a consequence of the observed invariance of the velocity of light (see Refs. [183, 951]).

B.1 The Lorentz group

A Lorentz transformation is a linear homogeneous coordinate transformation

$$x^\mu \to x'^\mu = \Lambda^\mu{}_\nu \, x^\nu \,, \tag{B.1}$$

from a reference frame S with coordinates $x^\mu = (x^0, x^1, x^2, x^3)$ to a reference frame S' with coordinates $x'^\mu = (x'^0, x'^1, x'^2, x'^3)$. The constant 4×4 matrix $\Lambda^\mu{}_\nu$ must satisfy the constraints in eqn (B.14). In each reference frame $x^0 = t$ is the time coordinate and $\vec{x} = (x^1, x^2, x^3)$ is the three-vector of spatial coordinates. Hence, the coordinates of x^μ are often indicated as $x^\mu = (x^0, \vec{x})$ or $x^\mu = (t, \vec{x})$. The indices μ and ν in eqn (B.1) are called *Lorentz indices.* In eqn (B.1) and in the following equations, Lorentz indices run from 0 to 3 and repeated indices, one low and one high, are summed from 0 to 3.

Four-vectors are geometrical objects which form a vector space. They are represented in each reference frame by four real numbers called *components.* The components of a four-vector transform as the coordinates under Lorentz transformations. We denote four-vectors through their four components in a given reference frame: for example, $V^\mu = (V^0, V^1, V^2, V^3) = (V^0, \vec{V})$, with $\vec{V} = (V^1, V^2, V^3)$. Then, under the Lorentz transformation in eqn (B.1) a four-vector V^μ transforms as

$$V^\mu \to V'^\mu = \Lambda^\mu{}_\nu \, V^\nu \,. \tag{B.2}$$

In particular, x^μ is the coordinate four-vector. Note that the index of a four-vector is always placed in the upper position.

Covariant four-vectors[108], denoted with a low index, e.g. $V_\mu = (V_0, V_1, V_2, V_3)$, are geometrical objects analogous to four-vectors, whose components transform as the components of the four-gradient $\partial_\mu \equiv \partial/\partial x^\mu = (\partial_0, \partial_1, \partial_2, \partial_3)$. Since $\partial_\mu x^\nu = \delta^\nu_\mu = \partial'_\mu x'^\nu$, the four-gradient transforms as

$$\partial_\mu \to \partial'_\mu = (\Lambda^{-1})^\nu{}_\mu \, \partial_\nu \,. \tag{B.3}$$

High indices are called *contravariant* and low indices are called *covariant*. Thus, sometimes four-vectors are also called *contravariant four-vectors* in order to distinguish them from the covariant four-vectors.

Tensors are geometrical objects whose components have one or more indices and transform as products of contravariant and/or covariant four-vectors. The number of indices is the *rank* of the tensor. Contravariant and covariant four-vectors are tensors of rank one, or, more precisely, of rank $\binom{1}{0}$ and $\binom{0}{1}$, respectively. For example, $T^\mu{}_\nu$ is a tensor of rank two or, more precisely, of rank $\binom{1}{1}$, which transforms as

$$T^\mu{}_\nu \to T'^\mu{}_\nu = \Lambda^\mu{}_\alpha \, (\Lambda^{-1})^\beta{}_\nu \, T^\alpha{}_\beta \,. \tag{B.4}$$

The scalar product of two four-vectors V^μ and W^μ is defined by

$$V \cdot W \equiv g_{\mu\nu} \, V^\mu \, W^\mu \,, \tag{B.5}$$

where $g_{\mu\nu}$ is the *metric tensor*. The norm of a four-vector V^μ is given by

$$|V|^2 \equiv V \cdot V \,. \tag{B.6}$$

The metric tensor allows the lowering of the indices of vectors and tensors. For example, for each four-vector V^μ there is a corresponding covariant four-vector V_μ given by

$$V_\mu = g_{\mu\nu} \, V^\nu \,. \tag{B.7}$$

A tensor can be represented in contravariant (e.g. $T^{\mu\nu}$), covariant (e.g. $T_{\mu\nu} = g_{\mu\rho} g_{\nu\sigma} T^{\rho\sigma}$), or mixed form (e.g. $T^\mu{}_\nu = g_{\nu\rho} T^{\mu\rho}$). The contravariant metric tensor

[108] In differential geometry, covariant four-vectors are called *dual four-vectors* or *one-forms* (see Refs. [808, 1047, 941]). They form the dual vector space of linear functions of four-vectors into the real numbers.

$g^{\mu\nu}$, which allows the raising of indices, is defined by

$$g_{\mu\rho}\, g^{\rho\nu} = \delta_{\mu}^{\nu} \,. \tag{B.8}$$

For example,

$$V^{\mu} = g^{\mu\nu}\, V_{\nu} \,, \qquad T^{\mu\nu} = g^{\mu\rho}\, T_{\rho}{}^{\nu} \,. \tag{B.9}$$

Using contravariant and covariant indices, the scalar product in eqn (B.5) can be written in the alternative forms

$$V \cdot W = V^{\mu}\, W_{\mu} = V_{\mu}\, W^{\mu} = g^{\mu\nu}\, V_{\mu}\, W_{\nu} \,. \tag{B.10}$$

Let us introduce the *infinitesimal space-time interval* $\mathrm{d}\tau$, also called the *infinitesimal proper-time interval*[109], which is given by

$$\mathrm{d}\tau^2 = g_{\mu\nu}\, \mathrm{d}x^{\mu}\, \mathrm{d}x^{\nu} = (\mathrm{d}x^0)^2 - (\mathrm{d}x^1)^2 - (\mathrm{d}x^2)^2 - (\mathrm{d}x^3)^2 = (\mathrm{d}t)^2 - (\mathrm{d}\vec{x})^2 \,, \tag{B.11}$$

with the metric tensor

$$g_{\mu\nu} = \mathrm{diag}(1, -1, -1, -1) \,. \tag{B.12}$$

Space-time with this metric is called *Minkowski space-time.* Equation (B.8) implies that the contravariant and covariant metric tensors are equal:

$$g^{\mu\nu} = \mathrm{diag}(1, -1, -1, -1) = g_{\mu\nu} \,. \tag{B.13}$$

The invariance of the velocity of light implies that $\mathrm{d}\tau^2$ is Lorentz-invariant (for light rays $\mathrm{d}\tau^2 = 0$ in all reference frames). Hence, the matrix $\Lambda^{\mu}{}_{\nu}$ is subject to the condition

$$g_{\mu\nu}\, \Lambda^{\mu}{}_{\rho}\, \Lambda^{\nu}{}_{\sigma} = g_{\rho\sigma} \,, \tag{B.14}$$

which is the fundamental relation defining the Lorentz transformations. This condition implies the Lorentz-invariance of the metric tensor and of the scalar product. It can also be written as

$$(\Lambda^T)_{\rho}{}^{\mu}\, g_{\mu\nu}\, \Lambda^{\nu}{}_{\sigma} = g_{\rho\sigma} \,, \tag{B.15}$$

or in the compact matrix form

$$\Lambda^T\, g\, \Lambda = g \,. \tag{B.16}$$

Equation (B.8) allows one to write eqn (B.15) as

$$g^{\alpha\rho}\, (\Lambda^T)_{\rho}{}^{\mu}\, g_{\mu\nu}\, \Lambda^{\rho}{}_{\nu} = \delta_{\nu}^{\alpha} \,, \tag{B.17}$$

which shows that

$$(\Lambda^{-1})^{\alpha}{}_{\nu} = g^{\alpha\rho}\, (\Lambda^T)_{\rho}{}^{\mu}\, g_{\mu\nu} \,, \quad \text{i.e.} \quad \Lambda^{-1} = g\, \Lambda^T\, g \,. \tag{B.18}$$

[109] In the rest frame, where $\mathrm{d}x^1 = \mathrm{d}x^2 = \mathrm{d}x^3 = 0$, we have $\mathrm{d}\tau = \mathrm{d}x^0$.

The Lorentz transformations form the Lie group[110] $\mathscr{L}$. Taking the determinant of eqn (B.16) and taking into account that $\mathrm{Det}M^T = \mathrm{Det}M$ for any matrix M, one finds

$$(\mathrm{Det}\Lambda)^2 = 1 \qquad \Longrightarrow \qquad \mathrm{Det}\Lambda = \pm 1\,. \tag{B.19}$$

The Lorentz transformations with $\mathrm{Det}\Lambda = +1$ are called *proper* and form the subgroup $\mathscr{L}_+$ of the Lorentz group. The Lorentz transformations with $\mathrm{Det}\Lambda = -1$ are called *improper* and form the set $\mathscr{L}_-$, which is not a group.

Taking $\rho = \sigma = 0$ in eqn (B.14), we obtain

$$\left(\Lambda^0{}_0\right)^2 = 1 + \sum_{i=1}^{3} \left(\Lambda^i{}_0\right)^2\,, \tag{B.20}$$

which implies that

$$\left(\Lambda^0{}_0\right)^2 \geq 1 \qquad \Longrightarrow \qquad \Lambda^0{}_0 \geq +1 \quad \text{or} \quad \Lambda^0{}_0 \leq -1\,. \tag{B.21}$$

The subgroup $\mathscr{L}^\uparrow$ of Lorentz transformations with $\Lambda^0{}_0 \geq +1$ is called *orthochronous*. The set $\mathscr{L}^\downarrow$ of antichronous Lorentz transformations with $\Lambda^0{}_0 \leq -1$ do not form a group.

Therefore, the set of all Lorentz transformations is divided into four subsets according to the signs of $\mathrm{Det}\Lambda$ and $\Lambda^0{}_0$. The subgroup $\mathscr{L}_+^\uparrow$ of proper orthochronous Lorentz transformations is called the *restricted Lorentz group*. The restricted Lorentz group is a six-parameter continuous group. This can be seen by considering an infinitesimal Lorentz transformation

$$\Lambda^\mu{}_\nu = \delta^\mu_\nu + \varepsilon\,\omega^\mu{}_\nu\,, \tag{B.22}$$

with infinitesimal ε. The condition in eqn (B.14), which defines Lorentz transformations, implies that ω is an antisymmetric matrix:

$$\omega_{\mu\nu} = -\omega_{\nu\mu}\,. \tag{B.23}$$

Since ω is a 4×4 matrix, it contains six independent components[111]. A finite restricted Lorentz transformation matrix Λ can be written as an infinite product of infinitesimal transformations, leading to the exponential form

$$\Lambda^\mu{}_\nu = \lim_{N\to\infty} \left(\delta^\mu_{\rho_1} + \frac{\omega^\mu{}_{\rho_1}}{N}\right)\left(\delta^{\rho_1}_{\rho_2} + \frac{\omega^{\rho_1}{}_{\rho_2}}{N}\right)\cdots\left(\delta^{\rho_{N-1}}_\nu + \frac{\omega^{\rho_{N-1}}{}_\nu}{N}\right)$$

[110] A group $\mathcal{G}$ is a set of objects, called *elements of* $\mathcal{G}$, which can be combined to form an operation called the *product*, denoted by ab, for $a, b \in \mathcal{G}$. The product must satisfy the following four conditions:

1. $ab \in \mathcal{G}$, if $a, b \in \mathcal{G}$.
2. $(ab)c = a(bc)$, if $a, b, c \in \mathcal{G}$ (associativity).
3. There is a unit element $I \in \mathcal{G}$ such that $Ia = aI = a$ for all $a \in \mathcal{G}$.
4. Each $a \in \mathcal{G}$ has an inverse $a^{-1} \in \mathcal{G}$ such that $aa^{-1} = a^{-1}a = I$.

A Lie group is a group whose elements can be parameterized and the parameters of a product are analytic functions of the parameters of the factors.

[111] The number of independent components of a $N \times N$ antisymmetric matrix is $N(N-1)/2$.

$$= \lim_{N\to\infty} \left[\left(\mathbf{1} + \frac{\omega}{N} \right)^N \right]^\mu{}_\nu = [e^\omega]^\mu{}_\nu \, . \tag{B.24}$$

Any Lorentz transformation can be written as a product of a restricted Lorentz transformation times one of the following three discrete transformations:

Space inversion: $P = \mathrm{diag}\,(1, -1, -1, -1) \in \mathscr{L}_-^\uparrow$,
Time inversion: $T = \mathrm{diag}\,(-1, 1, 1, 1) \in \mathscr{L}_-^\downarrow$,
Total inversion: $PT = \mathrm{diag}\,(-1, -1, -1, -1) \in \mathscr{L}_+^\downarrow$.

Spatial rotations have

$$\Lambda^0{}_0 = 1 \, , \quad \Lambda^0{}_k = \Lambda^k{}_0 = 0 \, , \quad \Lambda^k{}_j = R^k{}_j \, , \tag{B.25}$$

where R is a unimodular ($\mathrm{Det}R = +1$) orthogonal ($R^T = R^{-1}$) matrix. Spatial rotations form a subgroup of the restricted Lorentz group. For a rotation by an angle θ ($0 \leq \theta \leq \pi$) around the unit vector $\vec{n}$ we have

$$R^k{}_j = \cos\theta \, \delta^k_j + (1 - \cos\theta) \, n^k \, n^j + \sin\theta \, \epsilon^{kjl} \, n^l \, . \tag{B.26}$$

Given a unimodular orthogonal matrix R, the angle θ and the unit vector $\vec{n}$ are given by

$$\cos\theta = \frac{\mathrm{Tr}R - 1}{2} \, , \qquad n^k = \frac{\epsilon^{kjl} \, R^j{}_l}{2 \sin\theta} \, . \tag{B.27}$$

A Lorentz transformation which connects two coordinate systems with different velocities is called a *boost*. The elements of the matrix $\Lambda^\mu{}_\nu$ of a Lorentz boost that connects a coordinate system x with a coordinate system x' moving with velocity $\vec{v}$ relative to x are

$$\Lambda^0{}_0 = \gamma \, , \qquad \Lambda^0{}_k = \Lambda^k{}_0 = -\gamma \, v^k \, , \qquad \Lambda^k{}_j = \delta^k_j - (1 - \gamma) \, \frac{v^k \, v^j}{v^2} \, , \tag{B.28}$$

with

$$v \equiv |\vec{v}|^2 \qquad \text{and} \qquad \gamma \equiv \left(1 - v^2\right)^{-1/2} \, . \tag{B.29}$$

For a boost with velocity v in the direction of the x^k axis we have

$$\Lambda^0{}_0 = \Lambda^k{}_k = \gamma = \cosh\varphi \, , \qquad \Lambda^0{}_k = \Lambda^k{}_0 = -\gamma \, v = -\sinh\varphi \, , \tag{B.30}$$

with the so-called *rapidity* φ, given by

$$\tanh\varphi = v \qquad \Longleftrightarrow \qquad \varphi = \ln[\gamma \, (1 + v)] = \frac{1}{2} \ln \frac{1 + v}{1 - v} \, . \tag{B.31}$$

Rapidities are quite useful because they add in successive boosts along the same axis. Indeed, two successive boosts along the same axis with velocities and rapidities $\vec{v}_1$, φ_1 and $\vec{v}_2$, φ_2 are equivalent to a boost along the same axis with velocity $\vec{v}$ and rapidity φ given by

$$\vec{v} = \frac{\vec{v}_1 + \vec{v}_2}{1 + \vec{v}_1 \cdot \vec{v}_2} \, , \qquad \varphi = \varphi_1 + \varphi_2 \, . \tag{B.32}$$

In general a orthochronous Lorentz transformation can always be decomposed into a rotation times a boost (see Ref. [951]). To show this, let us write the transformation matrix $\Lambda^{\mu}{}_{\nu}$ as

$$\Lambda^{\mu}{}_{\nu} = \begin{pmatrix} \gamma & -\boldsymbol{a}^T \\ -\boldsymbol{b} & \boldsymbol{M} \end{pmatrix} , \tag{B.33}$$

where $\boldsymbol{a}$ and $\boldsymbol{b}$ are 3×1 column matrices and $\boldsymbol{M}$ is a 3×3 matrix, and $\gamma \geq 0$. From eqn (B.18), we get

$$(\Lambda^{-1})^{\mu}{}_{\nu} = \begin{pmatrix} \gamma & \boldsymbol{b}^T \\ \boldsymbol{a} & \boldsymbol{M}^T \end{pmatrix} . \tag{B.34}$$

The relations $\Lambda^{-1}\Lambda = \mathbf{1}$ and $\Lambda\Lambda^{-1} = \mathbf{1}$ lead, respectively, to the constraints

$$\gamma^2 - \boldsymbol{b}^2 = 1 , \qquad \boldsymbol{M}^T \boldsymbol{b} = \gamma \, \boldsymbol{a} , \qquad \boldsymbol{M}^T \boldsymbol{M} = \boldsymbol{a} \, \boldsymbol{a}^T + \mathbf{1} , \tag{B.35}$$

$$\gamma^2 - \boldsymbol{a}^2 = 1 , \qquad \boldsymbol{M} \, \boldsymbol{a} = \gamma \, \boldsymbol{b} , \qquad \boldsymbol{M} \, \boldsymbol{M}^T = \boldsymbol{b} \, \boldsymbol{b}^T + \mathbf{1} . \tag{B.36}$$

The origin of the reference frame S', with coordinates $x'^{\mu} = (x'^0, 0, 0, 0)$, in the reference frame S has coordinates given by $x = \Lambda^{-1}x'$: $x^0 = \gamma x'^0$, $\vec{x} = \boldsymbol{a} x'^0$. Hence, in the reference frame S the origin of the reference frame S' moves with constant velocity

$$\vec{v} = \frac{\Delta \vec{x}}{\Delta x^0} = \frac{\boldsymbol{a}}{\gamma} . \tag{B.37}$$

From eqn (B.28) we can write the corresponding boost as

$$(\Lambda_{\vec{v}})^{\mu}{}_{\nu} = \begin{pmatrix} \gamma & -\gamma \, \vec{v}^T \\ -\gamma \, \vec{v} & \mathbf{1} + \dfrac{\gamma^2}{1+\gamma} \, \vec{v} \, \vec{v}^T \end{pmatrix} = \begin{pmatrix} \gamma & -\boldsymbol{a}^T \\ -\boldsymbol{a} & \mathbf{1} + \dfrac{\boldsymbol{a} \, \boldsymbol{a}^T}{1+\gamma} \end{pmatrix} , \tag{B.38}$$

Let us now consider $\Lambda\Lambda_{\vec{v}}^{-1}$, which should correspond to a rotation. Indeed, since $\Lambda_{\vec{v}}^{-1} = \Lambda_{-\vec{v}}$, we obtain

$$[\Lambda \, \Lambda_{\vec{v}}^{-1}]^{\mu}{}_{\nu} = \begin{pmatrix} 1 & \mathbf{0}^T \\ \mathbf{0} & \boldsymbol{R} \end{pmatrix} = (\Lambda_{\boldsymbol{R}})^{\mu}{}_{\nu} , \qquad \text{with} \qquad \boldsymbol{R} = \boldsymbol{M} - \frac{\boldsymbol{b} \, \boldsymbol{a}^T}{1+\gamma} , \tag{B.39}$$

and $\mathbf{0}$ is the zero 3×1 column matrix. $\Lambda_{\boldsymbol{R}}$ represents a rotation, because the relations in eqns (B.35) and (B.36) imply that $\boldsymbol{R}^T = \boldsymbol{R}^{-1}$. Hence, the general orthochronous Lorentz transformation Λ in eqn (B.33) can be written as the product of the rotation $\Lambda_{\boldsymbol{R}}$ and the boost $\Lambda_{\vec{v}}$:

$$\Lambda = \Lambda_{\boldsymbol{R}} \, \Lambda_{\vec{v}} . \tag{B.40}$$

Using the relations in eqns (B.35) and (B.36) one can easily show that Λ can also be written as the product of the boost $\Lambda_{\boldsymbol{R}\vec{v}}$ and the rotation $\Lambda_{\boldsymbol{R}}$:

$$\Lambda = \Lambda_{\boldsymbol{R}\vec{v}} \, \Lambda_{\boldsymbol{R}} . \tag{B.41}$$

B.2 Representations of the Lorentz group

The Lorentz transformation of four-vectors in eqns (B.1) and (B.2) is the defining representation of the Lorentz group. A general representation of the Lorentz group is made by objects Ψ that transform as

$$\Psi \to \Psi' = \mathcal{D}(\Lambda)\,\Psi\,, \tag{B.42}$$

with operators $\mathcal{D}(\Lambda)$ that satisfy the multiplication rule of the Lorentz group: for two Lorentz transformations Λ_1 and Λ_2

$$\mathcal{D}(\Lambda_2)\,\mathcal{D}(\Lambda_1) = \mathcal{D}(\Lambda_2\Lambda_1)\,. \tag{B.43}$$

For an infinitesimal Lorentz transformation in eqn (B.22) the operator $\mathcal{D}(\Lambda)$ must be infinitesimally close to the identity and can be written as

$$\mathcal{D}(\mathbf{1} + \varepsilon\omega) = \mathbf{1} + \frac{i}{2}\,\varepsilon\,\omega_{\mu\nu}\,J^{\mu\nu}\,, \tag{B.44}$$

where $J^{\mu\nu} = -J^{\nu\mu}$ is a set of six operators which are the generators of the Lorentz group in the representation under consideration. If the operators $J^{\mu\nu}$ are Hermitian, the representation is unitary ($\mathcal{D}^\dagger = \mathcal{D}^{-1}$).

The condition in eqn (B.43) implies that the generators $J^{\mu\nu}$ must satisfy the commutation relations

$$[J^{\mu\nu}, J^{\rho\sigma}] = i\,(J^{\mu\rho}\,g^{\nu\sigma} - J^{\nu\rho}\,g^{\mu\sigma} - J^{\mu\sigma}\,g^{\nu\rho} + J^{\nu\sigma}\,g^{\mu\rho})\,. \tag{B.45}$$

The problem of finding all the representations of the restricted Lorentz group is reduced to finding all the sets of operators that satisfy the commutation relations in eqn (B.45).

The defining four-vector representation is a matrix representation of the Lorentz group. Comparing eqns (B.22) and (B.44), we obtain the generators

$$[J^{\rho\sigma}]^{\mu}{}_{\nu} = -i\,(g^{\rho\mu}\,\delta^{\sigma}_{\nu} - g^{\sigma\mu}\,\delta^{\rho}_{\nu})\,. \tag{B.46}$$

These matrices are Hermitian and generate a unitary representation.

In general, the expression of $\mathcal{D}(e^{\omega})$ for a finite restricted Lorentz transformation $\Lambda = e^{\omega}$ in eqn (B.24) can be written as an infinite product of infinitesimal transformations by using the multiplication property in eqn (B.43), leading to the exponential form

$$\begin{aligned}\mathcal{D}(e^{\omega}) &= \mathcal{D}\bigg(\lim_{N\to\infty}\left[\mathbf{1} + \frac{\omega}{N}\right]^{N}\bigg) = \lim_{N\to\infty}\left[\mathcal{D}\Big(\mathbf{1} + \frac{\omega}{N}\Big)\right]^{N} \\ &= \lim_{N\to\infty}\left[\mathbf{1} + \frac{i}{2}\,\frac{1}{N}\,\omega_{\mu\nu}\,J^{\mu\nu}\right]^{N} = \exp\left(\frac{i}{2}\,\omega_{\mu\nu}\,J^{\mu\nu}\right).\end{aligned} \tag{B.47}$$

The six generators $J^{\mu\nu}$ can be divided into the angular momentum three-vector operators[112]

$$J^k = -\frac{1}{2}\sum_{j,l} \epsilon^{kjl}\, J^{jl} \quad \Longrightarrow \quad \vec{J} = (-J^{23}, -J^{31}, -J^{12}) \tag{B.49}$$

and the boost three-vector operator

$$K^k = J^{0k} \quad \Longrightarrow \quad \vec{K} = (J^{01}, J^{02}, J^{03})\,, \tag{B.50}$$

that satisfy the commutation relations

$$[J^k, J^j] = i\sum_l \epsilon^{kjl}\, J^l\,, \tag{B.51}$$

$$[J^k, K^j] = i\sum_l \epsilon^{kjl}\, K^l\,, \tag{B.52}$$

$$[K^k, K^j] = -i\sum_l \epsilon^{kjl}\, J^l\,. \tag{B.53}$$

The commutation relations in eqn (B.51) are those of angular momentum operators, which generate three-dimensional rotations. Since the angular momentum operators satisfy a closed algebra, rotations form a group. On the other hand, eqn (B.53) shows that Lorentz boosts do not form a group.

The operator $\mathcal{D}(\mathbf{1} + \varepsilon\omega)$ in eqn (B.44) can be written as

$$\mathcal{D}(\Lambda) = \mathbf{1} + i\,\varepsilon \sum_k \lambda_k\, K^k + i\,\varepsilon \sum_k \eta_k\, J^k\,, \tag{B.54}$$

with

$$\lambda_k = \omega_{0k}\,, \tag{B.55}$$

$$\eta_k = -\frac{1}{2}\sum_{j,l} \epsilon_{kjl}\, \omega_{jl}\,. \tag{B.56}$$

For a boost with rapidity φ in the direction of the x^k axis, we have

$$\omega^0{}_k = \omega^k{}_0 = -\varphi \quad \Longrightarrow \quad \lambda_k = -\varphi \quad \Longrightarrow \quad \mathcal{D}^k_{\text{boost}}(\varepsilon\varphi) = \mathbf{1} - i\,\varepsilon\,\varphi\, K^k\,. \tag{B.57}$$

In the case of a boost with finite rapidity φ in the direction of the x^k axis we have

$$\mathcal{D}^k_{\text{boost}}(\varphi) = e^{-i\,\varphi\, K^k}\,. \tag{B.58}$$

For a rotation by an angle θ around the x^k axis we have

$$\omega^j{}_l = \epsilon^j{}_{lk}\,\theta \quad \Longrightarrow \quad \eta_k = \theta \quad \Longrightarrow \quad \mathcal{D}^k_{\text{rot}}(\varepsilon\theta) = \mathbf{1} + i\,\varepsilon\,\theta\, J^k\,. \tag{B.59}$$

For a rotation by a finite angle θ around the x^k axis we have

$$\mathcal{D}^k_{\text{rot}}(\theta) = e^{i\,\theta\, J^k}\,. \tag{B.60}$$

[112] The inverse relation is

$$J^{kj} = -\sum_l \epsilon^{kjl}\, J^l\,. \tag{B.48}$$

B.2.1 Fields

Under a Lorentz transformation in eqn (B.1), a multicomponent field

$$\psi(x) \equiv \begin{pmatrix} \psi_1(x) \\ \vdots \\ \psi_n(x) \end{pmatrix} \tag{B.61}$$

transforms as

$$\psi(x) \to \psi'(x') = \mathcal{S}(\Lambda)\,\psi(x)\,, \tag{B.62}$$

where $\mathcal{S}(\Lambda)$ is a $n \times n$ matrix that depends on the spin of the particle. For example, for a scalar field $n = 1$ and $\mathcal{S}(\Lambda) = 1$ and for a vector field $n = 4$ and $\mathcal{S}(\Lambda) = \Lambda$.

For an infinitesimal Lorentz transformation in eqn (B.22) let us write $\mathcal{S}(\mathbf{1} + \varepsilon\omega)$ as

$$\mathcal{S}(\mathbf{1} + \varepsilon\omega) = \mathbf{1} + \frac{i}{2}\,\varepsilon\,\omega_{\mu\nu}\,S^{\mu\nu}\,, \tag{B.63}$$

where $S^{\mu\nu}$ is the spin part of the generators of the Lorentz group in the representation under consideration.

Taking into account the transformation of coordinates, the full transformation of $\psi(x)$ under a Lorentz transformation is given by

$$\psi(x) \to \psi'(x') = \mathcal{S}(\Lambda)\,\psi(x) = \mathcal{S}(\Lambda)\,\psi(\Lambda^{-1}x') = \mathcal{D}(\Lambda)\,\psi(x')\,. \tag{B.64}$$

Using eqn (B.63) and taking into account the antisymmetric character of $\omega_{\mu\nu}$ (see eqn (B.23)), for the infinitesimal Lorentz transformation in eqn (B.22), we have

$$\begin{aligned} \psi'(x) &= \left(\mathbf{1} + \frac{i}{2}\,\varepsilon\,\omega_{\mu\nu}\,S^{\mu\nu}\right)\psi(x - \varepsilon\omega x) \\ &= \left\{\mathbf{1} + \frac{i}{2}\,\varepsilon\,\omega_{\mu\nu}\Big[-i\,(x^\mu\,\partial^\nu - x^\nu\,\partial^\mu) + S^{\mu\nu}\Big]\right\}\psi(x)\,. \end{aligned} \tag{B.65}$$

Therefore, the operator $\mathcal{D}(\mathbf{1} + \varepsilon\omega)$ that implements the infinitesimal Lorentz transformation in eqn (B.22) on $\psi(x)$ is given by

$$\mathcal{D}(\mathbf{1} + \varepsilon\omega) = \mathbf{1} + \frac{i}{2}\,\varepsilon\,\omega_{\mu\nu}\Big[-i\,(x^\mu\,\partial^\nu - x^\nu\,\partial^\mu) + S^{\mu\nu}\Big]\,. \tag{B.66}$$

Comparing this expression with eqn (B.44), we obtain the generators of the Lorentz group:

$$J^{\mu\nu} = -i\,(x^\mu\,\partial^\nu - x^\nu\,\partial^\mu) + S^{\mu\nu}\,. \tag{B.67}$$

These generators can be written as

$$J^{\mu\nu} = L^{\mu\nu} + S^{\mu\nu}\,, \tag{B.68}$$

with the space-time part

$$L^{\mu\nu} = -i\,(x^\mu\,\partial^\nu - x^\nu\,\partial^\mu) = -\,(x^\mu\,P^\nu - x^\nu\,P^\mu)\,, \tag{B.69}$$

where $P^\mu = i\partial^\mu$ (see eqn (B.86)), which is the same for all fields. Since $L^{\mu\nu}$ satisfy the commutation relations in eqn (B.45) and commute with $S^{\mu\nu}$, in general,

$S^{\mu\nu}$ must satisfy the commutation relations in eqn (B.45) of the Lorentz group generators.

The space part of the angular momentum operator is

$$L^k = -\frac{1}{2} \sum_{j,l} \epsilon^{kjl} L^{jl} = \sum_{j,l} \epsilon^{kjl} x^j P^l = \left(\vec{x} \times \vec{P} \right)^k , \tag{B.70}$$

implying that $\vec{L}$ is the usual spatial angular momentum operator $\vec{L} = \vec{x} \times \vec{P}$. Denoting with S^k the spin part of the angular momentum operator,

$$S^k = -\frac{1}{2} \sum_{j,l} \epsilon^{kjl} S^{jl} , \tag{B.71}$$

the total angular momentum operator $\vec{J}$ can be written in general as the sum of its spatial and spin parts:

$$\vec{J} = \vec{L} + \vec{S} . \tag{B.72}$$

B.3 The Poincaré group and its representations

The Poincaré group (also known as the *inhomogeneous Lorentz group*) is the Lorentz group augmented with space-time translations. A Poincaré transformation is a coordinate transformation

$$x^\mu \to x'^\mu = \Lambda^\mu{}_\nu \, x^\nu + a^\mu , \tag{B.73}$$

with four arbitrary real constants a^μ representing space-time translations.

A general representation of the Poincaré group is constituted by objects Ψ which transform as

$$\Psi \to \Psi' = \mathcal{D}(\Lambda, a) \, \Psi , \tag{B.74}$$

with operators $\mathcal{D}(\Lambda, a)$ which satisfy the multiplication rule of the Poincaré group:

$$\mathcal{D}(\Lambda_2, a_2) \, \mathcal{D}(\Lambda_1, a_1) = \mathcal{D}(\Lambda_2 \Lambda_1, \Lambda_2 a_1 + a_2) . \tag{B.75}$$

For an infinitesimal Poincaré transformation

$$\Lambda^\mu{}_\nu = \delta^\mu_\nu + \varepsilon \, \omega^\mu{}_\nu , \qquad a^\mu = \varepsilon \, b^\mu , \tag{B.76}$$

with infinitesimal ε, the operator $\mathcal{D}(\mathbf{1} + \varepsilon\omega, \varepsilon b)$ can be written as

$$\mathcal{D}(\mathbf{1} + \varepsilon\omega, \varepsilon b) = \mathbf{1} + \frac{i}{2} \, \varepsilon \, \omega_{\mu\nu} \, J^{\mu\nu} - i \, \varepsilon \, b_\mu \, P^\mu , \tag{B.77}$$

where $J^{\mu\nu} = -J^{\nu\mu}$ are the six generators of the Lorentz group and P^μ are the four generators of space-time translations.

The condition in eqn (B.75) leads to the following Lie algebra of the Poincaré group generators:

$$[J^{\mu\nu}, J^{\rho\sigma}] = i \left(J^{\mu\rho} \, g^{\nu\sigma} - J^{\nu\rho} \, g^{\mu\sigma} - J^{\mu\sigma} \, g^{\nu\rho} + J^{\nu\sigma} \, g^{\mu\rho} \right) , \tag{B.78}$$

$$[P^\mu, J^{\rho\sigma}] = i\,(P^\rho\, g^{\mu\sigma} - P^\sigma\, g^{\mu\rho})\,, \tag{B.79}$$

$$[P^\mu, P^\rho] = 0\,. \tag{B.80}$$

Furthermore, the angular momentum operators in eqn (B.49) and boost operators in eqn (B.50) satisfy the following commutation relations with the energy and momentum operators:

$$[J^k, P^j] = i\,\epsilon^{kjl}\, P^l\,, \tag{B.81}$$

$$[J^k, P^0] = 0\,, \tag{B.82}$$

$$[K^k, P^j] = i\, P^0\, \delta^{kj}\,, \tag{B.83}$$

$$[K^k, P^0] = i\, P^k\,. \tag{B.84}$$

Analogously to eqn (B.64), the transformation of a field $\psi(x)$ under a Poincaré transformation in eqn (B.73) is given by

$$\psi(x) \to \psi'(x') = \mathcal{S}(\Lambda)\,\psi(x) = \mathcal{S}(\Lambda)\,\psi(\Lambda^{-1}(x'-a)) = \mathcal{D}(\Lambda)\,\psi(x')\,. \tag{B.85}$$

In this case, the generators $J^{\mu\nu}$ of the Lorentz group are given by eqn (B.67) and the generators P^μ of space-time translations are given by

$$P^\mu = i\,\partial^\mu\,. \tag{B.86}$$

Covariance of physical laws under Poincaré transformations imply that all quantities defined in Minkowski space-time must belong to a representation of the Poincaré group. By definition, the states that describe elementary particles belong to irreducible representations of the Poincaré group. These representations can be classified by the eigenvalues of the Casimir operators, which are the functions of the generators that commute with all the generators. This property implies that the eigenvalues of the Casimir operators remain invariant under group transformations.

There are two Casimir operators of the Poincaré group. The first one is

$$P^2 = P_\mu\, P^\mu\,. \tag{B.87}$$

The corresponding eigenvalues

$$p^2 = m^2 \tag{B.88}$$

are the squared masses of particles. In the real world we observe only time-like or light-like four-momenta, i.e. particles with positive or zero mass. Furthermore, the temporal components p^0 of the four-momenta of physical particles, which correspond to energy and whose sign is a Lorentz invariant, are always positive.

The second Casimir operator of the Poincaré group is

$$W^2 = W_\mu\, W^\mu\,, \tag{B.89}$$

where W_μ is the Pauli–Lubanski four-vector

$$W_\mu = \frac{1}{2}\,\epsilon_{\mu\nu\rho\sigma}\, J^{\nu\rho}\, P^\sigma\,, \tag{B.90}$$

such that

$$[W^\mu, P^\rho] = 0\,. \tag{B.91}$$

In three-vector notation, taking into account that $\vec{L}\cdot\vec{P} = 0$, we have

$$W^0 = \vec{J}\cdot\vec{P} = \vec{S}\cdot\vec{P}\,, \qquad \vec{W} = \vec{J}\,P^0 - \vec{K}\times\vec{P}\,. \tag{B.92}$$

Hence, W^0 depends only on $\vec{P}$ and the spin operator $\vec{S}$.

We can evaluate the Lorentz-invariant W^2 in the rest frame of a particle with mass m, where

$$W^0 = 0\,, \qquad \vec{W} = m\,\vec{S}\,, \tag{B.93}$$

which lead to

$$W^2 = -m^2\,\vec{S}^2\,. \tag{B.94}$$

Since from the nonrelativistic quantum theory of angular momentum we know that the eigenvalue of $\vec{S}^2$ is $s(s+1)$, where s is the spin of the particle (half-integer or integer), we see that the eigenvalue of the relativistic invariant W^2,

$$w^2 = -m^2\,s\,(s+1)\;, \tag{B.95}$$

gives the spin of the particle.

A complete set of commuting observables is composed of P^2, the three components of $\vec{P}$, W^2 and one of the four components of W^μ. The eigenvalues of P^2 and W^2 or, equivalently, the mass m and the spin s, distinguish (possibly together with other quantum numbers) different particles. Given a particle with mass m and spin s, the eigenvalues of $\vec{P}$ and one of the four components of W^μ distinguish different states of the same particle. It is convenient to use the helicity operator $\widehat{h}$ defined by

$$\widehat{h} = \frac{W^0}{s\,|\vec{P}|} = \frac{\vec{S}\cdot\vec{P}}{s\,|\vec{P}|}\,, \tag{B.96}$$

which is proportional to W^0. Therefore the states of a particle with mass m and spin s can be distinguished by the eigenvalues of $\vec{P}$ and $\widehat{h}$.

APPENDIX C

LAGRANGIAN THEORY

C.1 Variational principle and field equations

Let us consider a set of n real fields $\psi_r(x)$, with $r = 1, \ldots, n$ (for example a set of n real scalar fields, or the electromagnetic field $A^\mu(x)$ with four components, $\mu = 0, 1, 2, 3$) and a real *Lagrangian*[113]

$$\mathscr{L}(x) = \mathscr{L}(\psi_r(x), \partial_\mu \psi_r(x)) \,. \tag{C.1}$$

The Lagrangian formalism is more suitable for the study of relativistic field theories than the Hamiltonian formalism, because the Lagrangian is a Lorentz scalar, whereas the Hamiltonian represents the energy of the fields and transforms as the time component of the energy–momentum four-vector.

Let us define the *action* as

$$I(\Omega) \equiv \int_\Omega \mathrm{d}^4x \, \mathscr{L}(\psi_r(x), \partial_\mu \psi_r(x)) \,, \tag{C.2}$$

where Ω is an arbitrary space-time region. According to the *variational principle*, the fields must be such that the action is stationary,

$$\delta_\mathrm{v} I(\Omega) = 0 \,, \tag{C.3}$$

for infinitesimal variations of the fields that vanish on the hypersurface S surrounding the space-time region Ω. These are variations of the type

$$\psi_r(x) \to \psi_r(x) + \delta_\mathrm{v} \psi_r(x) \,, \tag{C.4}$$

with

$$\delta_\mathrm{v} \psi_r(x)\big|_S = 0 \,. \tag{C.5}$$

The variation of the action under the transformation in eqns (C.4) and (C.5) is

$$\delta_\mathrm{v} I(\Omega) = \int_\Omega \mathrm{d}^4x \sum_r \left[\frac{\partial \mathscr{L}}{\partial \psi_r} \, \delta_\mathrm{v} \psi_r + \frac{\partial \mathscr{L}}{\partial(\partial_\mu \psi_r)} \, \underbrace{\delta_\mathrm{v}(\partial_\mu \psi_r)}_{\partial_\mu(\delta_\mathrm{v} \psi_r)} \right]$$

[113] Strictly speaking, $\mathscr{L}(x)$ is a Lagrangian density, which is a function of time and space. The corresponding Lagrangian is $L(t) = \int \mathrm{d}^3x \, \mathscr{L}(x)$, which is a function of time only. However, in quantum field theory, the Lagrangian density is the main quantity of interest and in this book we omit, for simplicity, the word *density*.

$$= \int_\Omega \mathrm{d}^4x \sum_r \left[\frac{\partial \mathscr{L}}{\partial \psi_r} \delta_\mathrm{V}\psi_r + \partial_\mu \left(\frac{\partial \mathscr{L}}{\partial(\partial_\mu \psi_r)} \delta_\mathrm{V}\psi_r \right) - \partial_\mu \left(\frac{\partial \mathscr{L}}{\partial(\partial_\mu \psi_r)} \right) \delta_\mathrm{V}\psi_r \right] . \tag{C.6}$$

By using the Gauss theorem and the constraint in eqn (C.5), one obtains

$$\int_\Omega \mathrm{d}^4x \, \partial_\mu \left(\frac{\partial \mathscr{L}}{\partial(\partial_\mu \psi_r)} \delta_\mathrm{V}\psi_r \right) = \int_S \mathrm{d}S_\mu \, \frac{\partial \mathscr{L}}{\partial(\partial_\mu \psi_r)} \delta_\mathrm{V}\psi_r = 0 \, , \tag{C.7}$$

where $\mathrm{d}S_\mu(x)$ is the surface element (see eqn (2.292)). Then, the variational principle in eqn (C.3) leads to

$$0 = \delta_\mathrm{V} I(\Omega) = \int_\Omega \mathrm{d}^4x \sum_r \left[\frac{\partial \mathscr{L}}{\partial \psi_r} - \partial_\mu \frac{\partial \mathscr{L}}{\partial(\partial_\mu \psi_r)} \right] \delta_\mathrm{V}\psi_r \, . \tag{C.8}$$

Since the variations $\delta_\mathrm{V}\psi_r(x)$ are arbitrary and independent for different r, the fields must satisfy the *Euler–Lagrange equations*

$$\partial_\mu \frac{\partial \mathscr{L}}{\partial(\partial_\mu \psi_r)} - \frac{\partial \mathscr{L}}{\partial \psi_r} = 0 \qquad (r = 1, \ldots, n) \, . \tag{C.9}$$

It is clear that the scalar character of the Lagrangian is crucial in order to guarantee the Lorentz covariance of the field equations in eqn (C.9).

C.2 Canonical quantization

In quantum field theory, the canonical quantization of fields is implemented by imposing on the fields ψ_r and their canonically conjugated momenta

$$\pi_r = \frac{\partial \mathscr{L}}{\partial(\partial_0 \psi_r)} \tag{C.10}$$

the equal-time relations

$$[\psi_r(t, \vec{x} \, , \, \pi_s(t, \vec{y}]_\pm = i \, \delta_{rs} \, \delta^3(\vec{x} - \vec{y}) \, , \tag{C.11}$$

$$[\psi_r(t, \vec{x} \, , \, \psi_s(t, \vec{y}]_\pm = [\pi_r(t, \vec{x} \, , \, \pi_s(t, \vec{y}]_\pm = 0 \, , \tag{C.12}$$

where the plus and minus subscripts denote, respectively, anticommutators for fermion fields and commutators for boson fields.

C.3 Noether's theorem

Noether's theorem establishes a connection between symmetries under continuous transformations and conservation laws. Let us consider an infinitesimal transformation of the n fields ψ_r,

$$\psi_r(x) \to \psi_r(x) + \varepsilon \, \delta\psi_r(x) \, , \tag{C.13}$$

where ε is infinitesimal and the variation $\delta\psi_r(x)$ is not subject to any constraint, but the fields are required to satisfy the field equations in eqn (C.9). The transformation in eqn (C.13) is a symmetry if the field equations in eqn (C.9) remain invariant, i.e. if the action in eqn (C.2) is invariant up to a surface term (the constraint in eqn (C.5) implies that a surface term does not contribute to the field equations). Therefore, the transformation in eqn (C.13) is a symmetry if the Lagrangian remains invariant up to a four-divergence:

$$\mathscr{L}(x) \to \mathscr{L}(x) + \varepsilon\, \partial_\mu \mathcal{I}^\mu(x) \qquad \Longleftrightarrow \qquad \delta\mathscr{L}(x) = \varepsilon\, \partial_\mu \mathcal{I}^\mu(x)\,, \tag{C.14}$$

for any $\mathcal{I}^\mu(x)$. From eqn (C.6) with $\delta_{\mathrm{V}}\psi_r$ replaced by $\varepsilon\delta\psi_r$, by using the field equations in eqn (C.9), one can see that the variation of the Lagrangian under the transformation in eqn (C.13) is

$$\delta\mathscr{L} = \varepsilon\, \partial_\mu \left(\sum_r \frac{\partial \mathscr{L}}{\partial(\partial_\mu \psi_r)}\, \delta\psi_r \right) . \tag{C.15}$$

Thus, the transformation in eqn (C.13) is a symmetry if

$$\partial_\mu \left(\sum_r \frac{\partial \mathscr{L}}{\partial(\partial_\mu \psi_r)}\, \delta\psi_r \right) = \partial_\mu \mathcal{I}^\mu \qquad \Longleftrightarrow \qquad \partial_\mu \left(\sum_r \frac{\partial \mathscr{L}}{\partial(\partial_\mu \psi_r)}\, \delta\psi_r - \mathcal{I}^\mu \right) = 0\,. \tag{C.16}$$

This is the conservation equation

$$\partial_\mu\, j^\mu = 0\,, \tag{C.17}$$

for the current

$$j^\mu = \sum_r \frac{\partial \mathscr{L}}{\partial(\partial_\mu \psi_r)}\, \delta\psi_r - \mathcal{I}^\mu\,. \tag{C.18}$$

The conservation equation (C.17) implies the existence of a charge

$$\mathsf{Q} = \int \mathrm{d}^3x\, j^0(x) = \int \mathrm{d}^3x \left(\sum_r \frac{\partial \mathscr{L}}{\partial(\partial_0 \psi_r)}\, \delta\psi_r - \mathcal{I}^0 \right) , \tag{C.19}$$

which is conserved in time:

$$\frac{\partial \mathsf{Q}}{\partial t} = 0\,. \tag{C.20}$$

Since all the quantities in the definition of Q in eqn (C.19) are real, in quantum field theory the conserved charge is an Hermitian operator ($\mathsf{Q}^\dagger = \mathsf{Q}$). Hence, it is a measurable quantity and its constancy in time allows the use of its eigenvalues for the classification of states.

The canonical quantization relations in eqns (C.11) and (C.12) imply that the charge Q is the generator of the symmetry transformation in eqn (C.13) if $\int \mathrm{d}^3y \left[\mathcal{I}^0(t,\vec{y})\,,\, \psi_r(t,\vec{x})\right] = 0$. In this case, one can derive the commutation relations

$$[\mathsf{Q}\,,\, \psi_r(x)] = \pm i\, \delta\psi_r(x)\,, \tag{C.21}$$

where the upper and lower signs apply, respectively, to fermion and boson fields. The infinitesimal symmetry transformation in eqn (C.13) is generated by

$$e^{\mp i\varepsilon\mathsf{Q}}\, \psi_r(x)\, e^{\pm i\varepsilon\mathsf{Q}} = \psi_r(x) \mp i\,\varepsilon\, [\mathsf{Q}\,,\, \psi_r(x)] = \psi_r(x) + \varepsilon\, \delta\psi_r(x)\,. \tag{C.22}$$

C.4 Space-time translations

Using Noether's theorem, one can show that the symmetry under infinitesimal space-time translations

$$x^\mu \to x'^\mu = x^\mu + \varepsilon\, \delta x^\mu \tag{C.23}$$

implies the conservation of the energy–momentum four-vector. Since, the variations of the fields $\psi_r(x)$ and the Lagrangian $\mathscr{L}(x)$ are given by

$$\delta\psi_r(x) = -\varepsilon\, \delta x^\mu\, \partial_\mu \psi_r(x)\,, \qquad \delta\mathscr{L}(x) = -\varepsilon\, \delta x^\mu\, \partial_\mu \mathscr{L}(x)\,, \tag{C.24}$$

in this case, we have

$$\mathcal{I}^\mu = -\delta x^\mu\, \mathscr{L}(x)\,. \tag{C.25}$$

Thus, eqn (C.16) leads to

$$\partial_\mu \left(\sum_r \frac{\partial \mathscr{L}}{\partial(\partial_\mu \psi_r)}\, \delta x^\nu\, \partial_\nu \psi_r - \delta x^\mu\, \mathscr{L} \right) = 0\,. \tag{C.26}$$

Since the variation δx^μ is arbitrary, there are four conservation laws:

$$\partial_\mu \mathcal{T}^{\mu\nu} = 0 \qquad (\nu = 0,1,2,3)\,, \qquad \text{with} \qquad \mathcal{T}^{\mu\nu} = \sum_r \frac{\partial \mathscr{L}}{\partial(\partial_\mu \psi_r)}\, \partial^\nu \psi_r - g^{\mu\nu}\, \mathscr{L}\,. \tag{C.27}$$

Here $\mathcal{T}^{\mu\nu}$ is the *energy–momentum tensor*, whose columns are the four conserved currents $\mathcal{T}^{\mu 0}$, $\mathcal{T}^{\mu 1}$, $\mathcal{T}^{\mu 2}$, $\mathcal{T}^{\mu 3}$. Hence, there are four constant quantities ($\partial_0\, \mathsf{P}^\nu = 0$), which form the *energy–momentum four-vector*:

$$\mathsf{P}^\nu = \int \mathrm{d}^3x\, \mathcal{T}^{0\nu}(x) = \int \mathrm{d}^3x \left(\sum_r \frac{\partial \mathscr{L}}{\partial(\partial_0 \psi_r)}\, \partial^\nu \psi_r - g^{0\nu}\, \mathscr{L} \right) \qquad (\nu = 0,1,2,3)\,. \tag{C.28}$$

If $\int \mathrm{d}^3y \left[\mathscr{L}^0(t,\vec{y})\,,\, \psi_r(t,\vec{x})\right] = 0$, eqn (C.21) implies that P^ν are the generators of space-time translations of quantized fields through the commutator

$$[\mathsf{P}^\nu\,,\, \psi_r(x)] = \mp i\, \partial^\nu \psi_r(x)\,, \tag{C.29}$$

where the upper and lower signs apply, respectively, to fermion and boson fields. Under a finite space-time translation

$$x^\mu \to x'^\mu = x^\mu + \delta x^\mu\,, \tag{C.30}$$

the fields $\psi_r(x)$ transform according to

$$\psi_r(x) \to e^{\mp i \delta x \cdot \mathsf{P}}\, \psi_r(x)\, e^{\pm i \delta x \cdot \mathsf{P}} = \psi_r(x - \delta x)\,. \tag{C.31}$$

Choosing $\delta x = x$, we obtain the general space-time behavior of the fields with respect to their value in the origin:

$$\psi_r(x) = e^{\pm i x \cdot \mathsf{P}}\, \psi_r(0)\, e^{\mp i x \cdot \mathsf{P}}\,. \tag{C.32}$$

Obviously, any product of the fields behaves in the same way.

C.5 Lorentz transformations

Under an infinitesimal Lorentz transformation in eqn (B.22) we have

$$\delta x^\mu = \varepsilon\, \omega^\mu{}_\nu\, x^\nu\,. \tag{C.33}$$

From eqns (B.65) and (B.67), the variation of the fields is

$$\delta\psi(x) = \frac{i}{2}\,\varepsilon\, \omega_{\mu\nu}\, J^{\mu\nu}\, \psi(x)\,, \tag{C.34}$$

where $\psi(x)$ is the column matrix of the field components in eqn (B.61). Since the variation of the Lagrangian is given by

$$\delta\mathscr{L}(x) = -\delta x^\mu\, \partial_\mu \mathscr{L}(x) = -\varepsilon\, \omega^\mu{}_\nu\, x^\nu\, \partial_\mu \mathscr{L}(x) = \partial_\mu \left(-\varepsilon\, \omega^\mu{}_\nu\, x^\nu\, \mathscr{L}(x)\right)\,, \tag{C.35}$$

we have, in this case,

$$\mathcal{I}^\mu = -\varepsilon\, \omega^\mu{}_\nu\, x^\nu\, \mathscr{L}(x)\,. \tag{C.36}$$

Taking into account the antisymmetry of $\omega_{\alpha\beta}$, eqn (C.16) leads to

$$\frac{1}{2}\,\varepsilon\, \omega_{\alpha\beta}\, \partial_\mu \left[i\, \frac{\partial\mathscr{L}}{\partial(\partial_\mu\psi)}\, J^{\alpha\beta}\, \psi(x) + \left(g^{\mu\alpha}\, x^\beta - g^{\mu\beta}\, x^\alpha\right) \mathscr{L}(x)\right] = 0\,. \tag{C.37}$$

Since the six independent components of $\omega_{\alpha\beta}$ are arbitrary, there are six conservation laws,

$$\partial_\mu\, \mathcal{J}^{\mu\alpha\beta} = 0\,, \quad \text{with} \quad \mathcal{J}^{\mu\alpha\beta}(x) = i\, \frac{\partial\mathscr{L}}{\partial(\partial_\mu\psi)}\, J^{\alpha\beta}\, \psi(x) + \left(g^{\mu\alpha}\, x^\beta - g^{\mu\beta}\, x^\alpha\right) \mathscr{L}(x)\,, \tag{C.38}$$

one for each of the six independent components of $\mathcal{J}^{\mu\alpha\beta}$ for fixed μ. Indeed, $\mathcal{J}^{\mu\alpha\beta}$ is antisymmetric in the indices α and β. From eqns (B.67) and (C.27), $\mathcal{J}^{\mu\alpha\beta}$ can be written as

$$\mathcal{J}^{\mu\alpha\beta}(x) = x^\alpha\, \mathcal{T}^{\mu\beta}(x) - x^\beta\, \mathcal{T}^{\mu\alpha}(x) + i\, \frac{\partial\mathscr{L}}{\partial(\partial_\mu\psi)}\, S^{\alpha\beta}\, \psi(x)\,. \tag{C.39}$$

There are six conserved quantities ($\partial_0\, \mathsf{J}^{\alpha\beta} = 0$) given by the six independent components of the antisymmetric angular momentum tensor

$$\mathsf{J}^{\alpha\beta} = \int \mathrm{d}^3x\, \mathcal{J}^{0\alpha\beta}(x) = \int \mathrm{d}^3x \left(x^\alpha\, \mathcal{T}^{0\beta}(x) - x^\beta\, \mathcal{T}^{0\alpha}(x) + i\, \frac{\partial\mathscr{L}}{\partial(\partial_0\psi)}\, S^{\alpha\beta}\, \psi(x)\right)\,. \tag{C.40}$$

C.6 Complex fields

If the fields ψ_r are complex (non-Hermitian in the case of quantized fields), each field has two degrees of freedom that can be represented by the real and imaginary parts or ψ_r and ψ_r^* ($\psi_r^\dagger$ in the case of quantized fields). In this case, the expression in eqn (C.18) for the conserved current must be modified to

$$j^\mu = \sum_r \left(\frac{\partial \mathscr{L}}{\partial(\partial_\mu \psi_r)} \, \delta\psi_r + \delta\psi_r^* \, \frac{\partial \mathscr{L}}{\partial(\partial_\mu \psi_r^*)} \right) - \mathcal{I}^\mu \, . \tag{C.41}$$

The equations following eqn (C.18) must be modified in a similar way, by adding the contribution obtained from the variation of ψ_r^*. In particular, we have

$$\mathsf{Q} = \int d^3x \left[\sum_r \left(\frac{\partial \mathscr{L}}{\partial(\partial_0 \psi_r)} \, \delta\psi_r + \delta\psi_r^* \, \frac{\partial \mathscr{L}}{\partial(\partial_0 \psi_r^*)} \right) - \mathcal{I}^0 \right] , \tag{C.42}$$

$$\mathcal{T}^{\mu\nu} = \sum_r \left(\frac{\partial \mathscr{L}}{\partial(\partial_\mu \psi_r)} \, (\partial^\nu \psi_r) + (\partial^\nu \psi_r^*) \, \frac{\partial \mathscr{L}}{\partial(\partial_\mu \psi_r^*)} \right) - g^{\mu\nu} \, \mathscr{L} \, , \tag{C.43}$$

$$\mathsf{P}^\mu = \int d^3x \left[\sum_r \left(\frac{\partial \mathscr{L}}{\partial(\partial_0 \psi_r)} \, (\partial^\mu \psi_r) + (\partial^\mu \psi_r^*) \, \frac{\partial \mathscr{L}}{\partial(\partial_0 \psi_r^*)} \right) - g^{0\mu} \, \mathscr{L} \right] , \tag{C.44}$$

$$\mathcal{J}^{\mu\alpha\beta} = x^\alpha \, \mathcal{T}^{\mu\beta} - x^\beta \, \mathcal{T}^{\mu\alpha} + i \left(\frac{\partial \mathscr{L}}{\partial(\partial_\mu \psi)} \, S^{\alpha\beta} \, \psi - \psi^\dagger \, S^{\alpha\beta\dagger} \, \frac{\partial \mathscr{L}}{\partial(\partial_\mu \psi^\dagger)} \right) , \tag{C.45}$$

$$\mathsf{J}^{\alpha\beta} = \int d^3x \left[x^\alpha \, \mathcal{T}^{0\beta} - x^\beta \, \mathcal{T}^{0\alpha} + i \left(\frac{\partial \mathscr{L}}{\partial(\partial_0 \psi)} \, S^{\alpha\beta} \, \psi - \psi^\dagger \, S^{\alpha\beta\dagger} \, \frac{\partial \mathscr{L}}{\partial(\partial_0 \psi^\dagger)} \right) \right] . \tag{C.46}$$

C.7 Global gauge symmetry

According to Noether's theorem, charge conservation is a consequence of invariance of the Lagrangian ($\delta\mathscr{L} = 0$) under phase transformations of complex fields,

$$\psi_r(x) \to e^{i\theta} \, \psi_r(x) \qquad (r = 1, \ldots, n) \, , \tag{C.47}$$

where θ is an arbitrary parameter. The most well-known conserved charges are the electric charge and the baryon and lepton numbers.

The transformations in eqn (C.47) are called *global gauge transformations*, where the adjective *global* indicates that the parameter θ does not depend on space-time. Since the variations of $\psi_r(x)$ and $\psi_r^*(x)$ for an infinitesimal $\delta\theta$ are

$$\delta\psi_r = i \, \psi_r \, \delta\theta \, , \qquad \delta\psi_r^* = -i \, \psi_r^* \, \delta\theta \qquad (r = 1, \ldots, n) \, , \tag{C.48}$$

and $\partial_\mu \mathcal{I}^\mu = 0$ (because $\delta\mathscr{L} = 0$), the conserved current ($\partial_\mu j^\mu = 0$) is

$$j^\mu = i \sum_r \left(\frac{\partial \mathscr{L}}{\partial(\partial_\mu \psi_r)} \, \psi_r - \psi_r^* \, \frac{\partial \mathscr{L}}{\partial(\partial_\mu \psi_r^*)} \right) , \tag{C.49}$$

and the conserved charge ($\partial_0 \mathsf{Q} = 0$) is

$$\mathsf{Q} = \int d^3x \, j^0(x) = i \int d^3x \sum_r \left(\frac{\partial \mathscr{L}}{\partial(\partial_0 \psi_r)} \, \psi_r - \psi_r^* \, \frac{\partial \mathscr{L}}{\partial(\partial_0 \psi_r^*)} \right) . \tag{C.50}$$

The global gauge transformation in eqn (C.47) involves a common variation of the phases of all the n fields $\psi_r(x)$. Such a transformation belongs to the abelian group U(1) of continuous phase transformations. Let us now consider a nonabelian Lie group of continuous transformations $\mathcal{G}$ of order N (for example SU(N)[114]), whose transformations depend on N real parameters θ_a ($a = 1, \ldots, N$).

In general, the transformations belonging to a nonabelian group $\mathcal{G}$ mix the n complex fields $\psi_r(x)$. We assume that a set of fields $\psi_r(x)$ is a n-dimensional irreducible representation of the group $\mathcal{G}$. Writing the n fields ψ_r in the matrix form

$$\Psi \equiv \begin{pmatrix} \psi_1 \\ \vdots \\ \psi_n \end{pmatrix} , \tag{C.51}$$

their variation under an infinitesimal transformation parameterized by $\delta\theta_a$ ($a = 1, \ldots, N$) is given by

$$\delta\Psi = i\, L_a\, \Psi\, \delta\theta_a\,, \qquad \delta\Psi^\dagger = -i\, \Psi^\dagger\, L_a\, \delta\theta_a\,, \tag{C.52}$$

with an implicit summation of the index a from 1 to N. Here L_a ($a = 1, \ldots, N$) are the $n \times n$ Hermitian matrices which form the n-dimensional representation of the generators of the group and satisfy the commutation relations

$$[L_a\,,\, L_b] = i \sum_{c=1}^{N} f_{abc}\, L_c \qquad (a, b = 1, \ldots, N)\,, \tag{C.53}$$

where the real numbers f_{abc} are the structure constants of the group[115]. Since the infinitesimal parameters $\delta\theta_a$ are arbitrary, Noether's theorem implies the existence of a conserved current ($\partial_\mu j_a^\mu = 0$) for each generator of the group:

$$j_a^\mu = i \left(\frac{\partial \mathscr{L}}{\partial(\partial_\mu \Psi)}\, L_a\, \Psi - \Psi^\dagger\, L_a\, \frac{\partial \mathscr{L}}{\partial(\partial_\mu \Psi^\dagger)} \right) \qquad (a = 1, \ldots, N)\,. \tag{C.54}$$

The corresponding N conserved charges ($\partial_0 \mathsf{Q}_a = 0$) are

$$\mathsf{Q}_a = \int \mathrm{d}^3x\, j_a^0(x) = i \int \mathrm{d}^3x \left(\frac{\partial \mathscr{L}}{\partial(\partial_0 \Psi)}\, L_a\, \Psi - \Psi^\dagger\, L_a\, \frac{\partial \mathscr{L}}{\partial(\partial_0 \Psi^\dagger)} \right) \quad (a = 1, \ldots, N)\,. \tag{C.55}$$

For quantized fields, these conserved charges generate the group transformations in eqn (C.52) through

$$[\mathsf{Q}_a\,,\, \Psi(x)] = \mp L_a\, \Psi(x)\,, \tag{C.56}$$

where the upper and lower signs apply, respectively, to fermion and boson fields. Indeed, using the canonical quantization relations in eqns (C.11) and (C.12) one

---

114 U(N) is the group of unitary transformations of dimension N and can be decomposed into the direct product U(N) = SU(N)×U(1) where SU(N) is the group of unitary unimodular transformations of dimension N.

115 For example, the pion triplet π^+, π^0, π^- forms an irreducible representation of the isospin group SU(2)$_{\mathrm{I}}$, whose structure constants are $f_{abc} = \varepsilon_{abc}$. The three-dimensional representation of the generators is $(L_a)_{jk} = -i\varepsilon_{ajk}$.

can derive the commutation relations in eqn (C.53) for the conserved charges:

$$[\mathsf{Q}_a\,,\,\mathsf{Q}_b] = i\,f_{abc}\,\mathsf{Q}_c\,. \tag{C.57}$$

Thus, the conserved charges form a representation of the generators of the symmetry group.

APPENDIX D

GAUGE THEORIES

In this appendix we present the salient features of gauge theories (section D.1). These are theories based on a local gauge invariance of the Lagrangian. It was known for a long time that the Maxwell equations and quantum electrodynamics (QED) are locally gauge invariant. However, the idea of a generalization of the local gauge principle to nonabelian groups and its application to the creation of new field theories were discovered only in 1954 by Yang and Mills [1080]. The Standard Model of electroweak interactions, discussed in chapter 3, is a gauge theory, as well as the theory of quantum chromodynamics (QCD), which is sketched in section D.2.

D.1 General formulation of gauge theories

Let us consider a Lie group $\mathcal{G}$ of order N. The N generators T_a $(a = 1, \ldots, N)$ satisfy the commutation relations (Lie algebra)

$$[T_a, T_b] = i f_{abc} T_c \qquad (a, b = 1, \ldots, N), \tag{D.1}$$

where the real numbers f_{abc} are the structure constants of the group. The generators define the infinitesimal elements of the group, i.e. those which are infinitesimally close to the identity:

$$g(\underline{\epsilon}) = 1 + i\,\underline{\epsilon} \cdot \underline{T}, \tag{D.2}$$

where $\underline{T} \equiv (T_1, \ldots, T_N)$ and $\underline{\epsilon} \equiv (\epsilon_1, \ldots, \epsilon_N)$ is a set of N real infinitesimal parameters. The finite elements of the group, $g(\underline{\theta})$ with $\underline{\theta} \equiv (\theta_1, \ldots, \theta_N)$, are obtained by integration of the infinitesimal elements.

In a field theory, a n-dimensional unitary representation of $\mathcal{G}$ is formed by

(a) A set of n fields ψ_r $(r = 1, \ldots, n)$, which can be written in matrix form as in eqn (C.51). They will be called *matter fields*, since they usually describe spin 1/2 fermions or scalar Higgs bosons.
(b) A set $\underline{L} = (L_1, L_2, \ldots, L_N)$ of N Hermitian $n \times n$ matrices which form an n-dimensional representation of the generators, i.e. that satisfy the commutation relations in eqn (C.53). The matrices L_a $(a = 1, \ldots, N)$ are independent and are conveniently chosen to satisfy the orthonormality condition

$$\mathrm{Tr}(L_a L_b) = \frac{1}{2}\,\delta_{ab}\,. \tag{D.3}$$

For each element of the group, there is a mapping $g(\underline{\theta}) \mapsto U(\underline{\theta})$, where $U(\underline{\theta})$ is the unitary $n \times n$ matrix

$$U(\underline{\theta}) = e^{i\underline{\theta} \cdot \underline{L}} \equiv \exp\left(i \sum_{a=1}^{N} \theta_a \, L_a \right) , \tag{D.4}$$

which generates the transformation of the matter fields:

$$\Psi(x) \to \Psi'(x) = U(\underline{\theta}) \, \Psi(x) \, . \tag{D.5}$$

If the n-dimensional matrix representation of the generators is known, the values of the structure constants can be determined through

$$f_{abc} = -2 \, i \, \mathrm{Tr}([L_a, L_b] L_c) \, , \tag{D.6}$$

which is a consequence of eqns (C.53) and (D.3).

We have already considered the symmetry of the Lagrangian under global gauge transformations in section C.7, where we have seen that Noether's theorem implies the conservation of the currents in eqn (C.54), one for each generator of the group. Let us now consider *local gauge transformations*, i.e. transformations of the type in eqn (D.5) with parameters θ_a ($a = 1, \ldots, N$) which depend on the space-time coordinates x: $\theta_a = \theta_a(x)$.

Since global gauge transformations are a subset of the corresponding local gauge transformations, we consider a Lagrangian of the type in eqn (C.1) which is invariant under global gauge transformations of the type in eqn (D.5). However, this Lagrangian cannot be invariant under the corresponding local gauge transformations, because the field derivative $\partial_\mu \Psi(x)$ does not transform as $\Psi(x)$ and its transformation is inhomogeneous:

$$\partial_\mu \Psi(x) \to \partial_\mu \Psi'(x) = \partial_\mu \big[U(\underline{\theta}(x)) \, \Psi(x) \big] = U(\underline{\theta}(x)) \big[\partial_\mu \Psi(x) \big] + \big[\partial_\mu U(\underline{\theta}(x)) \big] \Psi(x) \, . \tag{D.7}$$

For example, if the matter fields ψ_r describe spin 1/2 fermions, the kinetic term in the Dirac Lagrangian (see eqn (2.1)) transforms as

$$\overline{\Psi}(x) \overleftrightarrow{\partial\!\!\!/} \Psi(x) \to \overline{\Psi'}(x) \overleftrightarrow{\partial\!\!\!/} \Psi'(x) = \overline{\Psi}(x) \overleftrightarrow{\partial\!\!\!/} \Psi(x) + \overline{\Psi}(x) \big[U^\dagger(\underline{\theta}(x)) \overleftrightarrow{\partial\!\!\!/} U(\underline{\theta}(x)) \big] \Psi(x) \, . \tag{D.8}$$

The Lagrangian can be made invariant under local gauge transformations by replacing the ordinary derivative ∂_μ with a *covariant derivative* D_μ such that the transformation of $D_\mu \Psi$ is homogeneous and equal to that of Ψ:

$$D_\mu \Psi \to D'_\mu \Psi' = U(\underline{\theta}) \, D_\mu \Psi \, . \tag{D.9}$$

Note that in a local gauge transformation there is a transformation $D_\mu \to D'_\mu$ of the covariant derivative. Since $\Psi(x) = U^{-1}(\underline{\theta}) \, \Psi'$, the transformation of D_μ which satisfies eqn (D.9) is

$$D_\mu \to D'_\mu = U(\underline{\theta}) \, D_\mu \, U^{-1}(\underline{\theta}) \, . \tag{D.10}$$

This transformation can be achieved by defining the covariant derivative as

$$D_\mu \equiv \partial_\mu + i \, g \, \underline{A}_\mu \cdot \underline{L} \, , \tag{D.11}$$

where g is a real constant and $\underline{A}_\mu(x) \equiv (A_1^\mu(x), \ldots, A_N^\mu(x))$ is a set of N real vectorial *gauge fields*. Their transformation under a local gauge transformation is defined in order to fulfill eqn (D.10): with the definition in eqn (D.11), the condition in eqn (D.10) becomes

$$\begin{aligned}\partial_\mu + i\,g\,\underline{A}'_\mu \cdot \underline{L} &= U(\underline{\theta})\big[\partial_\mu + i\,g\,\underline{A}_\mu \cdot \underline{L}\big]U^{-1}(\underline{\theta}) \\ &= U(\underline{\theta})\left\{U^{-1}(\underline{\theta})\,\partial_\mu + \left[\partial_\mu U^{-1}(\underline{\theta})\right]\right\} + i\,g\,U(\underline{\theta})\,\underline{A}_\mu \cdot \underline{L}\,U^{-1}(\underline{\theta}) \\ &= \partial_\mu + U(\underline{\theta})\left[\partial_\mu U^{-1}(\underline{\theta})\right] + i\,g\,U(\underline{\theta})\,\underline{A}_\mu \cdot \underline{L}\,U^{-1}(\underline{\theta})\,. \end{aligned} \tag{D.12}$$

Thus, the transformation of $\underline{A}_\mu \cdot \underline{L}$ is given by

$$\underline{A}_\mu \cdot \underline{L} \to \underline{A}'_\mu \cdot \underline{L} = U(\underline{\theta})\left(\underline{A}_\mu \cdot \underline{L} - \frac{i}{g}\,\partial_\mu\right)U^{-1}(\underline{\theta})\,. \tag{D.13}$$

The explicit transformation of the gauge fields A_a^μ can be found by considering the infinitesimal transformation

$$U(\underline{\epsilon}(x)) = 1 + i\,\underline{\epsilon}(x)\cdot\underline{L}\,. \tag{D.14}$$

We obtain

$$\delta\underline{A}_\mu \cdot \underline{L} = i\left[\underline{\epsilon}\cdot\underline{L}\,,\,\underline{A}_\mu\cdot\underline{L}\right] - \frac{1}{g}\,\partial_\mu\underline{\epsilon}\cdot\underline{L}\,, \tag{D.15}$$

which can be written as

$$\sum_{a=1}^{N}\delta A_a^\mu\,L_a = -\sum_{a,b,c=1}^{N} f_{abc}\,\epsilon_a\,A_b^\mu\,L_c - \frac{1}{g}\sum_{a=1}^{N}\partial_\mu\epsilon_a\,L_a\,. \tag{D.16}$$

Using the orthonormality relations in eqn (D.3), one can finally find that the variation of A_a^μ is given by

$$\delta A_a^\mu = \sum_{b,c=1}^{N} f_{abc}\,A_b^\mu\,\epsilon_c - \frac{1}{g}\,\partial_\mu\epsilon_a \qquad (a = 1,\ldots,N)\,. \tag{D.17}$$

This expression is often written as

$$\delta\underline{A}_\mu = \underline{A}_\mu \times \underline{\epsilon} - \frac{1}{g}\,\partial_\mu\underline{\epsilon}\,, \tag{D.18}$$

with[116]

$$\left(\underline{A}^\mu \times \underline{\epsilon}\right)_a \equiv \sum_{b,c=1}^{N} f_{abc}\,A_b^\mu\,\epsilon_c\,. \tag{D.19}$$

The expression in eqn (D.17) shows that the transformation of the gauge fields is independent of the representation $\underline{L}$ of the group generators. One should also note

[116] If the group is SU(2), we have $N = 3$ and $f_{abc} = \varepsilon_{abc}$. In this case, the notation in eqn (D.19) coincides with the usual notation for the vectorial product of two three-vectors.

that the variation in eqn (D.17) of the gauge fields A_a^μ is a generalization to the nonabelian case of the well-known variation of the electromagnetic field,

$$\delta A_\mu = -\frac{1}{e}\,\partial_\mu \theta \tag{D.20}$$

under the local $\mathrm{U}(1)_Q$ symmetry. The additional term $\sum_{b,c} f_{abc}\, A_b^\mu\, \epsilon_c$ in eqn (D.17) is characteristic of nonabelian groups.

The substitution in the Lagrangian of the ordinary derivative ∂_μ with the covariant derivative D_μ implies the introduction of terms which couple the gauge fields A_a^μ to the matter fields ψ_r. For example, if the fields ψ_r describe fermions with spin 1/2, from the kinetic term of the Dirac Lagrangian we obtain

$$i\overline{\Psi}\not{D}\Psi = i\overline{\Psi}\not{\partial}\Psi - g\,\overline{\Psi}\,\underline{\not{A}}\cdot\underline{L}\,\Psi\,. \tag{D.21}$$

Therefore, a local gauge symmetry of the Lagrangian implies:

(A) The existence of a set of vectorial gauge fields.
(B) The existence of interactions of the gauge fields with the matter fields with strength given by the *coupling constant* g.

It now still remains to determine the Lagrangian kinetic term for the gauge fields A_a^μ. For this purpose, we search for an appropriate generalization to the nonabelian case of the electromagnetic tensor

$$F^{\mu\nu} = \partial^\mu A^\nu - \partial^\nu A^\mu\,, \tag{D.22}$$

and of the electromagnetic kinetic term

$$\mathcal{L}_{(\gamma)} = -\frac{1}{4}\,F^{\mu\nu}\,F_{\mu\nu}\,. \tag{D.23}$$

The generalization of the kinetic term to the nonabelian case is straightforward:

$$\mathcal{L}_{(A)} = -\frac{1}{4}\,\underline{A}^{\mu\nu}\cdot\underline{A}_{\mu\nu} = -\frac{1}{2}\mathrm{Tr}\big[\,(\underline{A}^{\mu\nu}\cdot\underline{L})\,(\underline{A}_{\mu\nu}\cdot\underline{L})\,\big]\,, \tag{D.24}$$

with the set of tensors $\underline{A}^{\mu\nu} \equiv (A_1^{\mu\nu},\ldots,A_N^{\mu\nu})$. On the other hand, a naive generalization of eqn (D.22) to $A_a^{\mu\nu} = \partial^\mu A_a^\nu - \partial^\nu A_a^\mu$ fails, because the corresponding kinetic term is not invariant under the gauge transformation in eqn (D.13). In order to find the correct expression of $\underline{A}^{\mu\nu}$ let us consider the commutator $[D^\mu, D^\nu]$:

$$\begin{aligned}
[D^\mu, D^\nu] &= [\partial^\mu + i\,g\,\underline{A}^\mu\cdot\underline{L}\,,\ \partial^\nu + i\,g\,\underline{A}^\nu\cdot\underline{L}] \\
&= \underbrace{[\partial^\mu,\partial^\nu]}_{0} + i\,g\underbrace{[\partial^\mu\,,\ \underline{A}^\nu\cdot\underline{L}]}_{\partial^\mu\underline{A}^\nu\cdot\underline{L}} + i\,g\underbrace{[\underline{A}^\mu\cdot\underline{L}\,,\ \partial^\nu]}_{-\partial^\nu\underline{A}^\mu\cdot\underline{L}} - g^2\,[\underline{A}^\mu\cdot\underline{L}\,,\ \underline{A}^\nu\cdot\underline{L}] \\
&= i\,g\,\big(\partial^\mu\underline{A}^\nu\cdot\underline{L} - \partial^\nu\underline{A}^\mu\cdot\underline{L} + i\,g\,[\underline{A}^\mu\cdot\underline{L}\,,\ \underline{A}^\nu\cdot\underline{L}]\big) \\
&= i\,g\,\big(\partial^\mu\underline{A}^\nu - \partial^\nu\underline{A}^\mu - g\,\underline{A}^\mu\times\underline{A}^\nu\big)\cdot\underline{L}\,.
\end{aligned} \tag{D.25}$$

We see that in the commutator the derivatives which do not act on the gauge fields disappear. Moreover, from the transformation rule of the covariant derivative in eqn (D.10), the transformation of $[D^\mu, D^\nu]$ is given by

$$[D^\mu, D^\nu] \to [D'^\mu, D'^\nu] = U(\underline{\theta})\,[D^\mu, D^\nu]\,U^{-1}(\underline{\theta})\,. \tag{D.26}$$

Therefore, $\mathrm{Tr}\Big\{[D^\mu, D^\nu]\,[D_\mu, D_\nu]\Big\}$ is invariant under local gauge transformations and the correct expression for $\underline{A}^{\mu\nu}\cdot\underline{L}$ in eqn (D.24) is

$$\underline{A}^{\mu\nu}\cdot\underline{L} = -\frac{i}{g}\,[D^\mu, D^\nu] = (\partial^\mu\underline{A}^\nu - \partial^\nu\underline{A}^\mu - g\,\underline{A}^\mu\times\underline{A}^\nu)\cdot\underline{L}\,, \tag{D.27}$$

leading to

$$A_a^{\mu\nu} = \partial^\mu A_a^\nu - \partial^\nu A_a^\mu - g\sum_{b,c=1}^{N} f_{abc}\,A_b^\mu\,A_c^\nu \qquad (a = 1, \ldots, N)\,. \tag{D.28}$$

Note that the generalization to the nonabelian case of the electromagnetic tensor in eqn (D.22) requires the addition of a new term which is quadratic in the gauge fields. Therefore, the Lagrangian in eqn (D.24) contains not only the kinetic term of the gauge fields, but also self-interaction terms of the gauge fields. In fact, we have explicitly

$$\begin{aligned}\mathcal{L}_{(A)} = &-\frac{1}{4}(\partial^\mu\underline{A}^\nu - \partial^\nu\underline{A}^\mu)\cdot(\partial_\mu\underline{A}_\nu - \partial_\nu\underline{A}_\mu)\\ &+\frac{1}{2}\,g\,(\partial^\mu A_a^\nu - \partial^\nu A_a^\mu)\,f_{abc}\,A_{b\mu}\,A_{c\nu}\\ &-\frac{1}{4}\,g^2 f_{abc}\,f_{ade}\,A_b^\mu\,A_c^\nu\,A_{d\mu}\,A_{e\nu}\,.\end{aligned} \tag{D.29}$$

The first line is the kinetic term, the second line corresponds to trilinear interaction vertices and the third line represents quadrilinear interaction vertices. This means that a nonabelian gauge theory is quite different from an abelian one, in which the gauge fields are not self-interacting (the photon has no electric charge): in the nonabelian case the gauge fields can carry a charge and interact among themselves.

The coupling constant g plays a fundamental role, since it determines the strength of the self-interaction of the gauge fields, as well as the strength of their interactions with the matter fields. A very important characteristic of nonabelian gauge theories is that the matter fields which belong to multiplets of the same type have the same charge g. In fact, the interaction of the matter fields with the gauge fields is determined by the expression in eqn (D.11) of the covariant derivative and the matrix representation of the generators for each type of multiplet (singlets, doublets, triplets, etc.) is uniquely determined by the commutation relations in eqn (C.53). For example, the commutation relations in eqn (C.53) do not allow a rescaling of the matrices L_a. On the other hand, in an abelian gauge theory the generators are represented by scalars which can have arbitrary values in different representations.

The two cases are exemplified by the two symmetries $\mathrm{SU}(2)_L$ and $\mathrm{U}(1)_Y$ of the $\mathrm{SU}(2)_L \times \mathrm{U}(1)_Y$ electroweak group in the SM discussed in section 3.1. The matrix representation of the generators I_a of the $\mathrm{SU}(2)_L$ group in the doublet and singlet representations are fixed uniquely according to eqns (3.8) and (3.17), respectively. Therefore, the action of the generators I_a on all left-handed doublets (right-handed singlets) is the same. On the other hand, the action generator of the $\mathrm{U}(1)_Y$ group can assume different values for different multiplets of the same type, as one can see from eqns (3.9) and (3.18).

Let us finally remark that a mass term for the gauge fields,

$$\frac{1}{2} m^2 \underline{A}^\mu \cdot \underline{A}_\mu \tag{D.30}$$

is forbidden by the local symmetry. Therefore, if the symmetry is unbroken, the gauge fields are massless. In the SM, the gauge bosons W and Z acquire mass through a spontaneous symmetry breaking generated by the Higgs mechanism (see section 3.4).

D.2 Quantum chromodynamics

Quantum chromodynamics (QCD) is the gauge theory of strong interactions, based on the symmetry group $\mathrm{SU}(3)_\mathrm{C}$ of order $N = 8$, where the subscript C indicates a new degree of freedom called *color*. It is assumed that each quark field $q(x)$ ($q = u, d, c, s, t, b$) has three color degrees of freedom:

$$q(x) = \begin{pmatrix} q_1(x) \\ q_2(x) \\ q_3(x) \end{pmatrix} . \tag{D.31}$$

The transformation of the quark fields under a local $\mathrm{SU}(3)_\mathrm{C}$ gauge transformation is given by

$$q(x) \to q'(x) = U(\underline{\theta}(x))\, q(x) , \tag{D.32}$$

with

$$U(\underline{\theta}(x)) = \exp\left(i \sum_{a=1}^{8} \theta_a(x) L_a \right) , \tag{D.33}$$

where $L_a = \lambda_a/2$ are the eight generators of $\mathrm{SU}(3)_\mathrm{C}$ in the three-dimensional representation. λ_a are the eight 3×3 Hermitian traceless Gell-Mann matrices

$$\lambda_1 = \begin{pmatrix} 0 & 1 & 0 \\ 1 & 0 & 0 \\ 0 & 0 & 0 \end{pmatrix} , \qquad \lambda_2 = \begin{pmatrix} 0 & -i & 0 \\ i & 0 & 0 \\ 0 & 0 & 0 \end{pmatrix} , \qquad \lambda_3 = \begin{pmatrix} 1 & 0 & 0 \\ 0 & -1 & 0 \\ 0 & 0 & 0 \end{pmatrix} ,$$

$$\lambda_4 = \begin{pmatrix} 0 & 0 & 1 \\ 0 & 0 & 0 \\ 1 & 0 & 0 \end{pmatrix} , \qquad \lambda_5 = \begin{pmatrix} 0 & 0 & -i \\ 0 & 0 & 0 \\ i & 0 & 0 \end{pmatrix} ,$$

$$\lambda_6 = \begin{pmatrix} 0 & 0 & 0 \\ 0 & 0 & 1 \\ 0 & 1 & 0 \end{pmatrix}, \qquad \lambda_7 = \begin{pmatrix} 0 & 0 & 0 \\ 0 & 0 & -i \\ 0 & i & 0 \end{pmatrix}, \qquad \lambda_8 = \frac{1}{\sqrt{3}} \begin{pmatrix} 1 & 0 & 0 \\ 0 & 1 & 0 \\ 0 & 0 & -2 \end{pmatrix}. \tag{D.34}$$

The generators $L_a = \lambda_a/2$ satisfy the orthonormality relations in eqn (D.3). The structure constants f_{abc} of the group can be found through eqn (D.6):

$$f_{abc} = -\frac{i}{4}\,\mathrm{Tr}([\lambda_a, \lambda_b]\lambda_c)\,. \tag{D.35}$$

The QCD gauge fields are the eight vectorial gluon fields G_a^μ ($a = 1, \ldots, 8$), whose variation under a local gauge transformation is of the type in eqn (D.17):

$$\delta G_a^\mu = \sum_{b,c=1}^{8} f_{abc}\, G_b^\mu\, \epsilon_c - \frac{1}{g_s}\,\partial_\mu \epsilon_a \qquad (a = 1, \ldots, 8)\,. \tag{D.36}$$

where g_s is the QCD coupling constant.

The QCD Lagrangian is

$$\mathcal{L}_{\mathrm{QCD}} = \sum_q \overline{q}\,(i\,\not{D} - m_q)\,q - \frac{1}{4}\,\underline{G}^{\mu\nu}\cdot\underline{G}_{\mu\nu}\,, \tag{D.37}$$

where D_μ is the covariant derivative

$$D_\mu = \partial_\mu + i\,g_s\,\underline{G}_\mu\cdot\underline{L}\,, \tag{D.38}$$

and $\underline{G}^{\mu\nu} \equiv (G_1^{\mu\nu}, \ldots, G_8^{\mu\nu})$ is the set of eight gluonic tensors

$$G_a^{\mu\nu} = \partial^\mu G_a^\nu - \partial^\nu G_a^\mu - g_s\, f_{abc}\, G_b^\mu\, G_c^\nu \qquad (a = 1, \ldots, 8)\,. \tag{D.39}$$

The gluons carry color charges and self-interact with trilinear and quadrilinear couplings generated by $\underline{G}^{\mu\nu}\cdot\underline{G}_{\mu\nu}$ in analogy with eqn (D.29). Since the $\mathrm{SU}(3)_\mathrm{C}$ is unbroken, the gluons are massless.

There are some similarities between QCD and QED, since both theories are based on unbroken local gauge symmetries, with massless gauge bosons. However, the nonabelian character of the color symmetry implies that gluons are self-interacting and the QCD phenomenology is much richer and more difficult to study than the QED phenomenology. Moreover, since the QCD coupling constant g_s is not small, the theory cannot be solved in a perturbative way in the low-energy domain. At high energies, however, the renormalized coupling constant becomes small (for example, $\alpha_s(m_Z) = 0.1187 \pm 0.0020$ [400], with $\alpha_s \equiv g_s^2/4\pi$), allowing a perturbative treatment (see, for example, [821, 721]).

APPENDIX E

FEYNMAN RULES OF THE STANDARD ELECTROWEAK MODEL

The Feynman rules allow one to calculate the transition amplitude $\mathcal{A}$ of a given process. In the following, we summarize the Feynman rules of the Standard Electroweak Model in the unitary gauge, for the calculation of tree diagrams. This is sufficient for the topics discussed in the book. The calculation of higher order diagrams with loops is much more complicated (see, for example, Refs. [25, 634, 314, 917, 720, 721]). In section E.4 we give general formulas suitable for the calculation of the cross-section or decay rate of a process from its amplitude.

The Feynman rules for the calculation of the tree-level amplitude $\mathcal{A}$ of a given process are as follows:

1. Draw all connected tree diagrams which contribute to the process under consideration by using the external lines in section E.1, the internal lines in section E.2 and the vertices in section E.2.
2. For each external line, write down the corresponding quantity in section E.1.
3. For each internal line, write down the corresponding propagator in section E.2.
4. For each vertex, write down the corresponding quantity in section E.3.
5. Enforce energy–momentum conservation in each vertex.
6. Assign a relative factor -1 to diagrams which differ only by an interchange of two external lines (the overall sign is irrelevant).

E.1 External lines

p is the particle momentum, r is the spin index for fermions and α is the spin index for spin-one bosons.

Incoming fermion $\mathfrak{f}$:

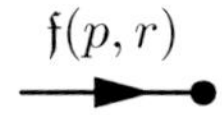

$$\Longrightarrow \quad u_{\mathfrak{f}}^{(r)}(p) \tag{E.1}$$

Outgoing fermion $\mathfrak{f}$:

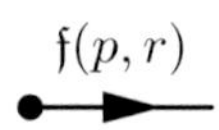

$$\Longrightarrow \quad \overline{u_{\mathfrak{f}}^{(r)}}(p) \tag{E.2}$$

Incoming antifermion $\bar{f}$:

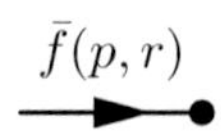

$$\Longrightarrow \quad \overline{v_{\mathfrak{f}}^{(r)}}(p) \tag{E.3}$$

Outgoing antifermion $\bar{f}$: $\bar{f}(p,r)$  $\Longrightarrow \quad v_{\mathfrak{f}}^{(r)}(p)$ (E.4)

Incoming photon: $A(p,\alpha)$  $\Longrightarrow \quad \varepsilon^{(\alpha)}(p)$ (E.5)

Outgoing photon: $A(p,\alpha)$  $\Longrightarrow \quad \varepsilon^{(\alpha)*}(p)$ (E.6)

Incoming Z: $Z(p,\alpha)$  $\Longrightarrow \quad \varepsilon^{(\alpha)}(p)$ (E.7)

Outgoing Z: $Z(p,\alpha)$  $\Longrightarrow \quad \varepsilon^{(\alpha)*}(p)$ (E.8)

Incoming W^+: $W^+(p,\alpha)$  $\Longrightarrow \quad \varepsilon^{(\alpha)}(p)$ (E.9)

Outgoing W^+: $W^+(p,\alpha)$  $\Longrightarrow \quad \varepsilon^{(\alpha)*}(p)$ (E.10)

Incoming W^-: $W^-(p,\alpha)$ 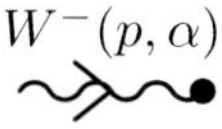 $\Longrightarrow \quad \varepsilon^{(\alpha)*}(p)$ (E.11)

Outgoing W^-: $W^-(p,\alpha)$ 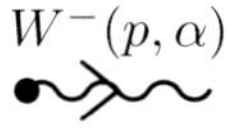 $\Longrightarrow \quad \varepsilon^{(\alpha)}(p)$ (E.12)

Incoming or outgoing Higgs: $H(p)$ 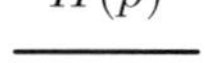 $\Longrightarrow \quad 1$ (E.13)

E.2 Internal lines

Fermion propagator: $\mathfrak{f}(p)$ 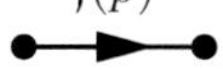

$$\Longrightarrow \quad G^{(\mathfrak{f})}(p) = i\,\frac{\not{p} + m}{p^2 - m_{\mathfrak{f}}^2 + i\epsilon} \tag{E.14}$$

Photon propagator: $A(p)$ μ  ν

$$\Longrightarrow \quad G_{\mu\nu}^{(A)}(p) = i\,\frac{-g_{\mu\nu}}{p^2 + i\epsilon} \tag{E.15}$$

W propagator:

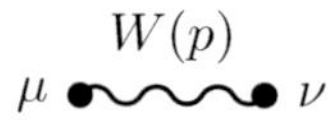

$$\Longrightarrow \quad G^{(W)}_{\mu\nu}(p) = i\,\frac{-g_{\mu\nu} + \dfrac{p_\mu p_\nu}{m_W^2}}{p^2 - m_W^2 + i\epsilon} \tag{E.16}$$

Z propagator:

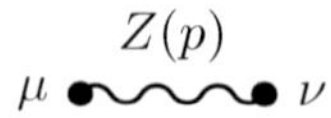

$$\Longrightarrow \quad G^{(Z)}_{\mu\nu}(p) = i\,\frac{-g_{\mu\nu} + \dfrac{p_\mu p_\nu}{m_Z^2}}{p^2 - m_Z^2 + i\epsilon} \tag{E.17}$$

Higgs propagator:

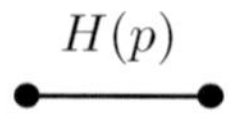

$$\Longrightarrow \quad G^{(H)}_{\mu\nu}(p) = \frac{i}{p^2 - m_H^2 + i\epsilon} \tag{E.18}$$

E.3 Vertices

The coefficient $q_{\mathfrak{f}}$ is the charge of the fermion $\mathfrak{f}$ in units of the elementary charge e: $q_\nu = 0$, $q_e = q_\mu = q_\tau = -1$, $q_u = q_c = q_t = 2/3$, $q_d = q_s = q_b = -1/3$. The coefficients g_V^f and g_A^f are given in Table 3.6 (page 78). From eqn (3.42) we have $g \sin\vartheta_{\mathrm{W}} = e$.

EM vertices:

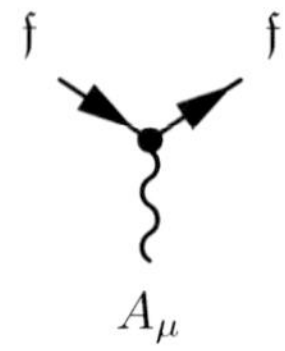

$$\Longrightarrow \quad -\,i\,e\,q_{\mathfrak{f}}\,\gamma_\mu \tag{E.19}$$

NC vertices:

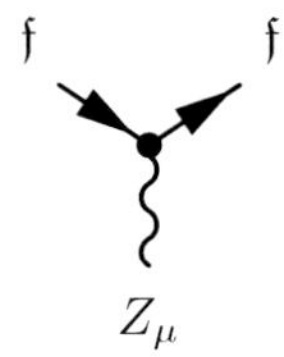

$$\Longrightarrow \quad -\,i\,\frac{g}{2\cos\vartheta_{\mathrm{W}}}\,\gamma_\mu\left(g_V^f - g_A^f\gamma^5\right) \tag{E.20}$$

CC lepton vertices:

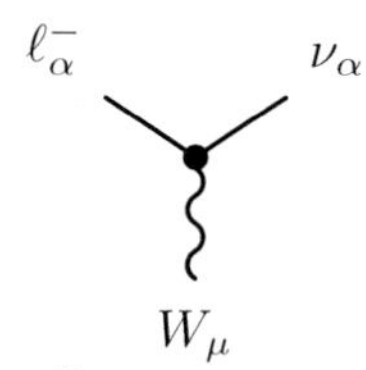

$$\Longrightarrow \quad -\,i\,\frac{g}{2\sqrt{2}}\,\gamma_\mu\left(1-\gamma^5\right) \tag{E.21}$$

CC quark vertices:

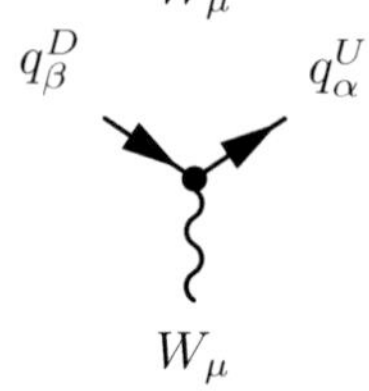

$$\Longrightarrow \quad -\,i\,\frac{g}{2\sqrt{2}}\,\gamma_\mu\left(1-\gamma^5\right)\,V_{\alpha\beta} \tag{E.22}$$

CC quark vertices:

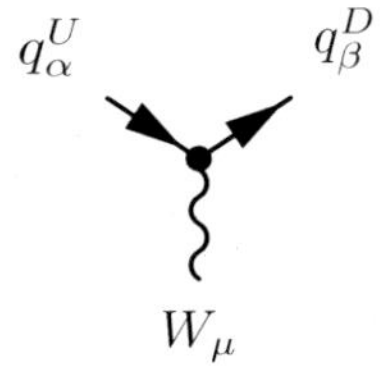

$$\Longrightarrow \quad -i\,\frac{g}{2\sqrt{2}}\,\gamma_\mu\left(1-\gamma^5\right)V^*_{\alpha\beta} \tag{E.23}$$

Trilinear W-γ vertex:

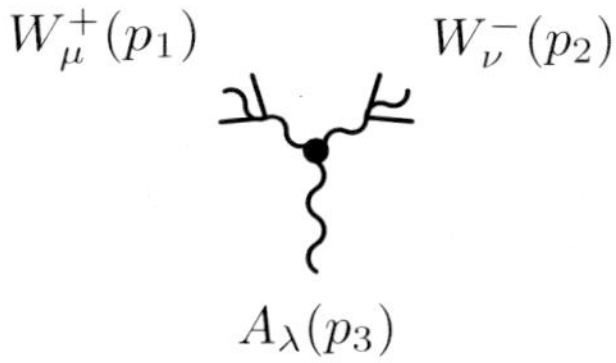

$$\Longrightarrow \quad \begin{aligned} ie\,[&(p_1-p_2)_\lambda\, g_{\mu\nu} \\ &+(p_2-p_3)_\mu\, g_{\nu\lambda} \\ &+(p_3-p_1)_\nu\, g_{\lambda\mu}] \end{aligned} \tag{E.24}$$

Trilinear W-Z vertex:

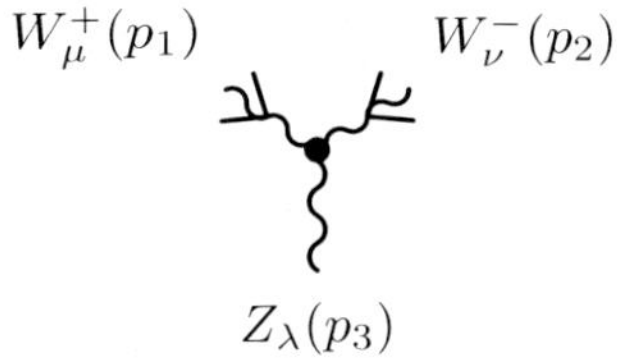

$$\Longrightarrow \quad \begin{aligned} ig\cos\vartheta_{\rm W}[&(p_1-p_2)_\lambda\, g_{\mu\nu} \\ &+(p_2-p_3)_\mu\, g_{\nu\lambda} \\ &+(p_3-p_1)_\nu\, g_{\lambda\mu}] \end{aligned} \tag{E.25}$$

Quadrilinear W-γ vertex:

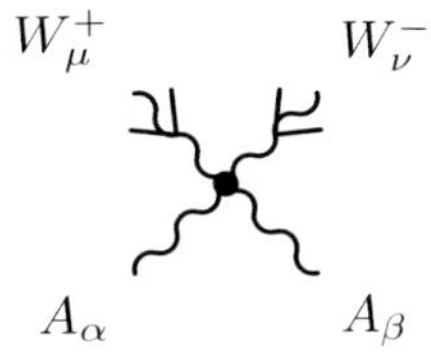

$$\Longrightarrow \quad \begin{aligned} -ie^2\,[&2g_{\mu\nu}g_{\alpha\beta} \\ &-g_{\mu\alpha}g_{\nu\beta}-g_{\mu\beta}g_{\nu\alpha}] \end{aligned} \tag{E.26}$$

Quadrilinear W-Z vertex:

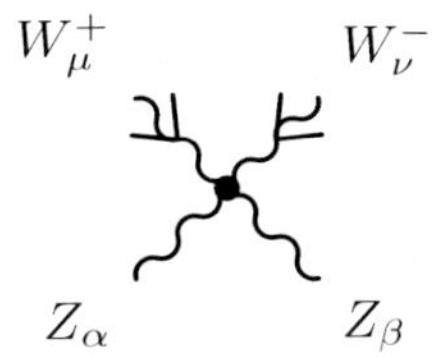

$$\Longrightarrow \quad \begin{aligned} -ig^2\,\cos^2\vartheta_{\rm W}\,[&2g_{\mu\nu}g_{\alpha\beta} \\ &-g_{\mu\alpha}g_{\nu\beta}-g_{\mu\beta}g_{\nu\alpha}] \end{aligned} \tag{E.27}$$

Quadrilinear W-Z-A vertex:

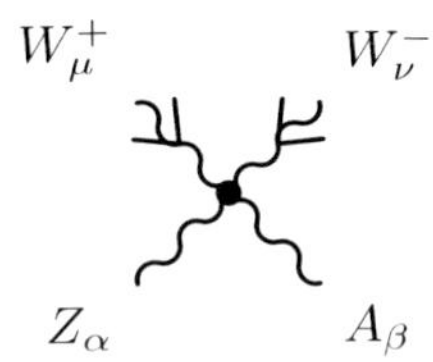

$$\Longrightarrow \quad \begin{aligned} -ieg\,\cos\vartheta_{\rm W}\,[&2g_{\mu\nu}g_{\alpha\beta} \\ &-g_{\mu\alpha}g_{\nu\beta}-g_{\mu\beta}g_{\nu\alpha}] \end{aligned} \tag{E.28}$$

Quadrilinear W vertex:

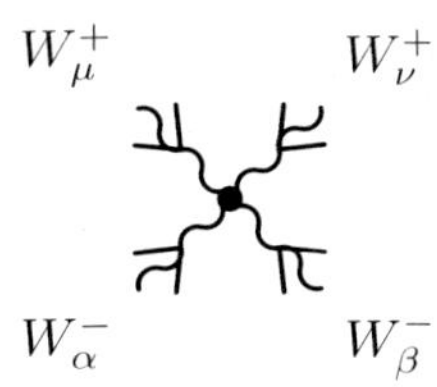

$$\Longrightarrow \quad \begin{aligned} ig^2\,[&2g_{\mu\nu}g_{\alpha\beta} \\ &-g_{\mu\alpha}g_{\nu\beta}-g_{\mu\beta}g_{\nu\alpha}] \end{aligned} \tag{E.29}$$

Trilinear fermion–H vertex:

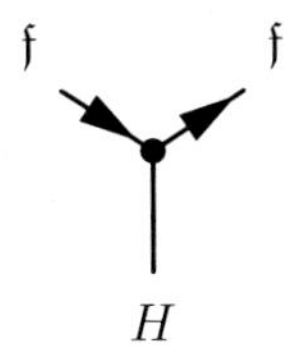

$$\Longrightarrow \quad -i\,\frac{gm_{\mathfrak{f}}}{2m_W}g_{\mu\nu} \tag{E.30}$$

Trilinear W-H vertex:

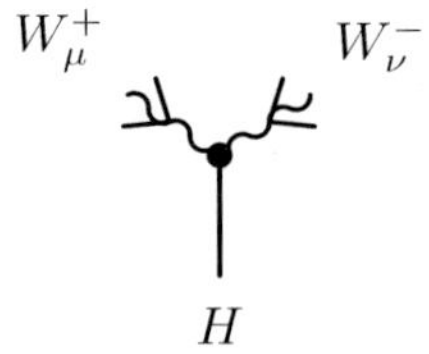

$$\Longrightarrow \quad igm_W g_{\mu\nu} \tag{E.31}$$

Trilinear Z-H vertex:

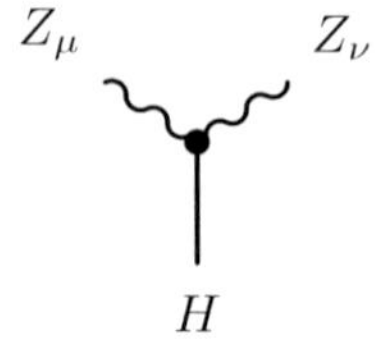

$$\Longrightarrow \quad i\frac{gm_Z}{\cos\vartheta_{\mathrm{W}}} g_{\mu\nu} \tag{E.32}$$

Trilinear Higgs vertex:

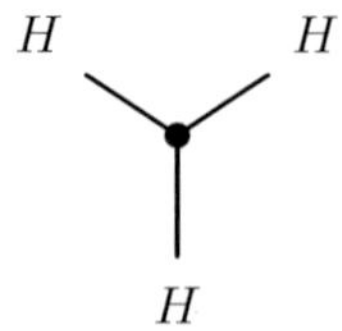

$$\Longrightarrow \quad -i\frac{3gm_H^2}{2m_W} g_{\mu\nu} \tag{E.33}$$

Quadrilinear W-H vertex:

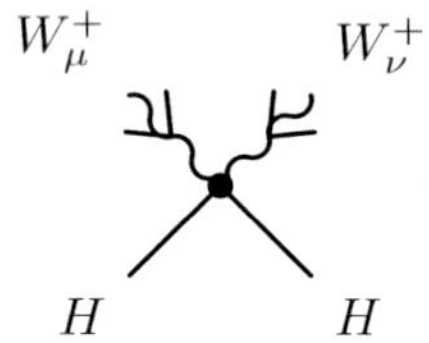

$$\Longrightarrow \quad i\frac{g^2}{2} g_{\mu\nu} \tag{E.34}$$

Quadrilinear Z-H vertex:

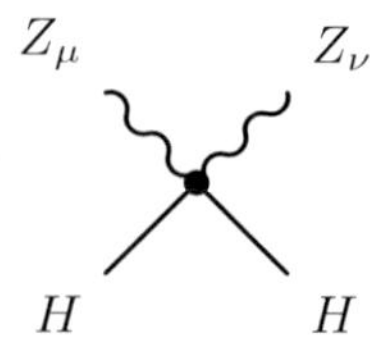

$$\Longrightarrow \quad i\frac{g^2}{\cos\vartheta_{\mathrm{W}}} g_{\mu\nu} \tag{E.35}$$

Quadrilinear Higgs vertex:

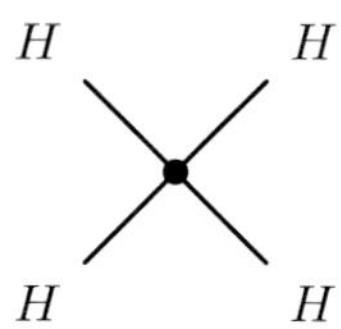

$$\Longrightarrow \quad i\frac{3g^2m_H^2}{4m_W^2} \tag{E.36}$$

E.4 Cross-sections and decay rates

The differential cross-section of a process with two particles a, b in the initial state and N_{f} particles in the final state is given by

$$\mathrm{d}\sigma = S\overline{\sum_{\mathrm{spin}}} \frac{(2\pi)^4\delta^4(P_{\mathrm{i}} - P_{\mathrm{f}})\,|\mathcal{A}|^2}{4\sqrt{(p_a\cdot p_b)^2 - m_a^2\,m_b^2}} \prod_{\mathrm{f}=1}^{N_{\mathrm{f}}} \frac{\mathrm{d}^3 p_{\mathrm{f}}}{(2\pi)^3\,2E_{\mathrm{f}}}\,, \tag{E.37}$$

where p_a, p_b are the four-momenta of the two initial particles, m_a, m_b are their masses, $p_{\mathrm{f}} = (E_{\mathrm{f}}, \vec{p}_{\mathrm{f}})$ is the four-momentum of the fth final particle, $P_{\mathrm{i}} = p_a + p_b$ and

$P_\mathrm{f} = \sum_{\mathrm{f}=1}^{N_\mathrm{f}} p_\mathrm{f}$ are, respectively, the total four-momenta of the initial and final states, $\mathcal{A}$ is the total amplitude of the process (the sum of the amplitudes of the diagrams contributing to the process), and the symbol $\overline{\sum}_\mathrm{spin}$ indicates an average over the unobserved spin states of the initial particles and a sum over the unobserved spin states of the final particles. S is a statistical factor given by the product of a factor $1/n!$ for each set of n identical particles.

The cross-section of a process of the type

$$a + b \to c + d\,, \tag{E.38}$$

with unpolarized particles, depends on the four-momenta of the particles only through the three Lorentz-invariant Mandelstam variables

$$s = (p_a + p_b)^2 = (p_c + p_d)^2 = (p_a + p_b) \cdot (p_c + p_d)\,, \tag{E.39}$$

$$t = (p_a - p_c)^2 = (p_d - p_b)^2 = (p_a - p_c) \cdot (p_d - p_b)\,, \tag{E.40}$$

$$u = (p_a - p_d)^2 = (p_c - p_b)^2 = (p_a - p_d) \cdot (p_c - p_b)\,, \tag{E.41}$$

where we used the energy–momentum conservation

$$p_a + p_b = p_c + p_d\,. \tag{E.42}$$

Only two Mandelstam variables are independent, since[117]

$$s + t + u = m_a^2 + m_b^2 + m_c^2 + m_d^2\,. \tag{E.43}$$

The decay width of a particle with mass m in a final state with N_f particles is given by

$$\mathrm{d}\Gamma = S \overline{\sum_\mathrm{spin}} \frac{(2\pi)^4 \delta^4(p - P_\mathrm{f})\, |\mathcal{A}|^2}{2m} \prod_{\mathrm{f}=1}^{N_\mathrm{f}} \frac{\mathrm{d}^3 p_\mathrm{f}}{(2\pi)^3\, 2E_\mathrm{f}}\,, \tag{E.44}$$

where p is the four-momentum of the initial fermion.

[117] This relation can be straightforwardly obtained by summing all the expressions for the three Mandelstam variables in eqns (E.39)–(E.41)).

BIBLIOGRAPHY

[1] National Nuclear Data Center, NuDat: `http://www.nndc.bnl.gov/nudat`.

[2] C. E. Aalseth et al., *Mod. Phys. Lett.*, A17, 1475–1478, 2002, `hep-ex/0202018`.

[3] C. E. Aalseth et al., *Phys. Rev.*, D65, 092007, 2002, `hep-ex/0202026`.

[4] C. Aalseth et al., 2004, `hep-ph/0412300`.

[5] P. A. Aarnio et al., *Phys. Lett.*, B231, 539, 1989.

[6] S. Abachi et al., *Phys. Rev. Lett.*, 74, 2632–2637, 1995, `hep-ex/9503003`.

[7] K. Abazajian et al., *Astron. J.*, 128, 502, 2004, `astro-ph/0403325`.

[8] K. Abazajian, N. F. Bell, G. M. Fuller, and Y. Y. Y. Wong, *Phys. Rev.*, D72, 063004, 2005, `astro-ph/0410175`.

[9] K. Abazajian et al., *Astron. J.*, 126, 2081, 2003, `astro-ph/0305492`.

[10] K. N. Abazajian, *Astropart. Phys.*, 19, 303, 2003, `astro-ph/0205238`.

[11] K. N. Abazajian, J. F. Beacom, and N. F. Bell, *Phys. Rev.*, D66, 013008, 2002, `astro-ph/0203442`.

[12] A. I. Abazov et al., *Phys. Rev. Lett.*, 67, 3332–3335, 1991.

[13] L. F. Abbott, A. De Rujula, and T. P. Walker, *Nucl. Phys.*, B299, 734, 1988.

[14] J. N. Abdurashitov et al., *Phys. Lett.*, B328, 234–248, 1994.

[15] J. N. Abdurashitov et al., *Phys. Rev. Lett.*, 77, 4708–4711, 1996.

[16] J. N. Abdurashitov et al., *Phys. Rev.*, C59, 2246–2263, 1999, `hep-ph/9803418`.

[17] J. N. Abdurashitov et al., *Phys. Rev. Lett.*, 83, 4686–4689, 1999, `astro-ph/9907131`.

[18] J. N. Abdurashitov et al., *Phys. Rev.*, C60, 055801, 1999, `astro-ph/9907113`.

[19] J. N. Abdurashitov et al., *J. Exp. Theor. Phys.*, 95, 181–193, 2002, `astro-ph/0204245`.

[20] J. N. Abdurashitov et al., *Phys. Rev.*, C73, 045805, 2006, `nucl-ex/0512041`.

[21] J. N. Abdurashitov et al., *Astropart. Phys.*, 25, 349–354, 2006, `nucl-ex/0509031`.

[22] F. Abe et al., *Phys. Rev. Lett.*, 74, 2626–2631, 1995, `hep-ex/9503002`.

[23] K. Abe et al., *Phys. Rev. Lett.*, 87, 091802, 2001, `hep-ex/0107061`.

[24] H. Abele, *Nucl. Instrum. Meth.*, A440, 499–510, 2000, Workshop on Particle Physics with Slow Neutrons, Grenoble, France, 22–24 Oct 1998.

[25] E. S. Abers and B. W. Lee, *Phys. Rep.*, 9, 1–141, 1973.

[26] C. V. Achar et al., *Phys. Lett.*, 18, 196–199, 1965.

[27] C. V. Achar et al., *Phys. Lett.*, 19, 78–80, 1965.

[28] J. Adelman-McCarthy et al., *Astrophys. J. Suppl.*, 162, 38, 2006, `astro-ph/0507711`.

[29] B. Adeva et al., *Phys. Lett.*, B231, 509, 1989.

[30] S. L. Adler, *Phys. Rev.*, 143, 1144–1155, 1966.

[31] S. L. Adler, *Phys. Rev.*, 177, 2426–2438, 1969.
[32] A. I. Afonin et al., *JETP Lett.*, 45, 247–251, 1987.
[33] M. Aglietta et al., *Europhys. Lett.*, 8, 611–614, 1989.
[34] M. Aglietta et al., *Europhys. Lett.*, 15, 559–564, 1991.
[35] M. Aglietta et al., *Nucl. Phys. Proc. Suppl.*, 110, 410–413, 2002, astro-ph/0112312.
[36] V. Agrawal, T. K. Gaisser, P. Lipari, and T. Stanev, *Phys. Rev.*, D53, 1314–1323, 1996, hep-ph/9509423.
[37] A. Aguilar et al., *Phys. Rev.*, D64, 112007, 2001, hep-ex/0104049.
[38] B. Aharmim et al., *Phys. Rev.*, D70, 093014, 2004, hep-ex/0407029.
[39] B. Aharmim et al., *Phys. Rev.*, D72, 052010, 2005, hep-ex/0507079.
[40] B. Aharmim et al., *Phys. Rev.*, C72, 055502, 2005, nucl-ex/0502021.
[41] S. P. Ahlen et al., *Phys. Lett.*, B357, 481–486, 1995.
[42] Q. R. Ahmad et al., *Phys. Rev. Lett.*, 87, 071301, 2001, nucl-ex/0106015.
[43] Q. R. Ahmad et al., *Phys. Rev. Lett.*, 89, 011301, 2002, nucl-ex/0204008.
[44] Q. R. Ahmad et al., *Phys. Rev. Lett.*, 89, 011302, 2002, nucl-ex/0204009.
[45] S. Ahmed et al., *Phys. Rev. Lett.*, 92, 181301, 2004, nucl-ex/0309004.
[46] M. H. Ahn et al., *Phys. Rev. Lett.*, 90, 041801, 2003, hep-ex/0212007.
[47] M. H. Ahn et al., *Phys. Rev. Lett.*, 93, 051801, 2004, hep-ex/0402017.
[48] S. H. Ahn et al., *Phys. Lett.*, B511, 178–184, 2001, hep-ex/0103001.
[49] J. Ahrens et al., *Astropart. Phys.*, 16, 345–359, 2002, astro-ph/0105460.
[50] E. K. Akhmedov, *Nucl. Phys.*, B538, 25–51, 1999, hep-ph/9805272.
[51] E. K. Akhmedov, *Sov. J. Nucl. Phys.*, 47, 301–302, 1988.
[52] E. K. Akhmedov, 1997, hep-ph/9705451, Fourth International Solar Neutrino Conference, Heidelberg, Germany, 8–11 Apr 1997.
[53] M. Z. Akrawy et al., *Phys. Lett.*, B231, 530, 1989.
[54] W. M. Alberico and S. M. Bilenky, *Phys. Part. Nucl.*, 35, 297, 2004, hep-ph/0306239.
[55] W. M. Alberico, S. M. Bilenky, C. Giunti, and C. Maieron, *Z. Phys.*, C70, 463–472, 1996, hep-ph/9508277.
[56] W. M. Alberico, S. M. Bilenky, and C. Maieron, *Phys. Rep.*, 358, 227–308, 2002, hep-ph/0102269.
[57] W. M. Alberico et al., *Nucl. Phys.*, A651, 277–286, 1999, hep-ph/9812388.
[58] C. Albright et al., 2000, hep-ex/0008064.
[59] C. Albright et al., 2004, physics/0411123.
[60] J. Alcaraz et al., *Phys. Lett.*, B490, 27–35, 2000.
[61] J. Alcaraz et al., *Phys. Lett.*, B494, 193–202, 2000.
[62] E. N. Alekseev and L. N. Alekseeva, *J. Exp. Theor. Phys.*, 95, 5–10, 2002, astro-ph/0212499.
[63] E. N. Alekseev, L. N. Alekseeva, I. V. Krivosheina, and V. I. Volchenko, *Phys. Lett.*, B205, 209–214, 1988.
[64] E. N. Alekseev, L. N. Alekseeva, V. I. Volchenko, and I. V. Krivosheina, *JETP Lett.*, 45, 589–592, 1987.
[65] E. N. Alekseev et al., *Phys. Part. Nucl.*, 29, 254–256, 1998.
[66] E. Aliu et al., *Phys. Rev. Lett.*, 94, 081802, 2005, hep-ex/0411038.
[67] W. W. M. Allison et al., *Phys. Lett.*, B391, 491–500, 1997, hep-ex/9611007.

[68] W. W. M. Allison et al., *Phys. Lett.*, B449, 137–144, 1999, hep-ex/9901024.
[69] W. W. M. Allison et al., *Phys. Rev.*, D72, 052005, 2005, hep-ex/0507068.
[70] R. A. Alpher, H. Bethe, and G. Gamow, *Phys. Rev.*, 73, 803–804, 1948.
[71] R. A. Alpher and R. C. Herman, *Phys. Rev.*, 74, 1737–1742, 1948.
[72] R. A. Alpher and R. C. Herman, *Phys. Rev.*, 75, 1089–1095, 1949.
[73] G. Altarelli, *Phys. Rep.*, 81, 1, 1982.
[74] G. Altarelli and F. Feruglio, *Phys. Rep.*, 320, 295–318, 1999, hep-ph/9905536.
[75] G. Altarelli and F. Feruglio, *New J. Phys.*, 6, 106, 2004, hep-ph/0405048.
[76] M. Altmann et al., *Phys. Lett.*, B490, 16–26, 2000, hep-ex/0006034.
[77] M. Altmann et al., *Phys. Lett.*, B616, 174, 2005, hep-ex/0504037.
[78] L. W. Alvarez, 1949, University of California Radiation Laboratory Report UCRL 328.
[79] U. Amaldi, W. de Boer, and H. Furstenau, *Phys. Lett.*, B260, 447–455, 1991.
[80] I. Ambats et al., 2005, hep-ex/0503053.
[81] M. Ambrosio et al., *Phys. Lett.*, B434, 451–457, 1998, hep-ex/9807005.
[82] M. Ambrosio et al., *Phys. Lett.*, B478, 5–13, 2000.
[83] M. Ambrosio et al., *Phys. Lett.*, B517, 59–66, 2001, hep-ex/0106049.
[84] M. Ambrosio et al., *Phys. Lett.*, B566, 35, 2003, hep-ex/0304037.
[85] L. Anchordoqui, T. Paul, S. Reucroft, and J. Swain, *Int. J. Mod. Phys.*, A18, 2229, 2003, hep-ph/0206072.
[86] D. Anderson, *Theory of the Earth*, Blackwell Scientific Publications, 1989.
[87] K. Anderson et al., 2004, hep-ex/0402041, White Paper Report.
[88] S. Ando and K. Sato, *New J. Phys.*, 6, 170, 2004, astro-ph/0410061.
[89] C. Angelini et al., *Phys. Lett.*, B179, 307, 1986.
[90] P. Anselmann et al., *Phys. Lett.*, B285, 390–397, 1992.
[91] P. Anselmann et al., *Phys. Lett.*, B285, 376–389, 1992.
[92] P. Anselmann et al., *Phys. Lett.*, B314, 445–458, 1993.
[93] P. Anselmann et al., *Phys. Lett.*, B327, 377–385, 1994.
[94] P. Anselmann et al., *Phys. Lett.*, B342, 440–450, 1995.
[95] P. Anselmann et al., *Phys. Lett.*, B357, 237–247, 1995.
[96] H. M. Antia and S. Basu, *Astrophys. J.*, 620, L129, 2005, astro-ph/0501129.
[97] M. Apollonio et al., *Phys. Lett.*, B420, 397–404, 1998, hep-ex/9711002.
[98] M. Apollonio et al., *Phys. Lett.*, B466, 415–430, 1999, hep-ex/9907037.
[99] M. Apollonio et al., 2002, hep-ph/0210192.
[100] M. Apollonio et al., *Eur. Phys. J.*, C27, 331, 2003, hep-ex/0301017.
[101] P. Aprili et al., 2002, CERN-SPSC-2002-027.
[102] T. Araki et al., *Nature*, 436, 499–503, 2005.
[103] T. Araki et al., *Phys. Rev. Lett.*, 94, 081801, 2005, hep-ex/0406035.
[104] F. Ardellier et al., 2004, hep-ex/0405032.
[105] B. Armbruster et al., *Phys. Rev.*, D65, 112001, 2002, hep-ex/0203021.
[106] F. Arneodo et al., 2001, hep-ex/0103008.
[107] W. D. Arnett and J. L. Rosner, *Phys. Rev. Lett.*, 58, 1906, 1987.
[108] G. Arnison et al., *Phys. Lett.*, B122, 103–116, 1983.
[109] G. Arnison et al., *Phys. Lett.*, B126, 398–410, 1983.

[110] Y. Ashie et al., *Phys. Rev. Lett.*, 93, 101801, 2004, hep-ex/0404034.
[111] Y. Ashie et al., *Phys. Rev.*, D71, 112005, 2005, hep-ex/0501064.
[112] M. Asplund, N. Grevesse, and J. Sauval, 2004, astro-ph/0410214, International Symposium on Cosmic Abundances as Records of Stellar Evolution and Nucleosynthesis in Honor of Professor David Lambert, Austin, Texas, 17–19 June 2004.
[113] K. Assamagan et al., *Phys. Rev.*, D53, 6065–6077, 1996.
[114] K. Assamagan et al., *Phys. Lett.*, B434, 158–162, 1998.
[115] P. Astier et al., *Nucl. Phys.*, B611, 3–39, 2001, hep-ex/0106102.
[116] P. Astier et al., *Phys. Lett.*, B570, 19–31, 2003, hep-ex/0306037.
[117] P. Astier et al., *Astron. Astrophys.*, 447, 31, 2006, astro-ph/0510447.
[118] C. Athanassopoulos et al., *Phys. Rev. Lett.*, 75, 2650–2653, 1995, nucl-ex/9504002.
[119] C. Athanassopoulos et al., *Phys. Rev. Lett.*, 77, 3082–3085, 1996, nucl-ex/9605003.
[120] C. Athanassopoulos et al., *Phys. Rev.*, C54, 2685–2708, 1996, nucl-ex/9605001.
[121] C. Athanassopoulos et al., *Phys. Rev. Lett.*, 81, 1774–1777, 1998, nucl-ex/9709006.
[122] C. Athanassopoulos et al., *Phys. Rev.*, C58, 2489–2511, 1998, nucl-ex/9706006.
[123] B. Aubert et al., *Phys. Rev. Lett.*, 87, 091801, 2001, hep-ex/0107013.
[124] J. J. Aubert et al., *Phys. Rev. Lett.*, 33, 1404–1406, 1974.
[125] G. Audi et al., *Nucl. Phys.*, A729, 3–128, 2003.
[126] J. E. Augustin et al., *Phys. Rev. Lett.*, 33, 1406–1408, 1974.
[127] M. Aunola and J. Suhonen, *Nucl. Phys.*, A643, 207–221, 1998.
[128] D. Autiero, 1998, 29th International Conference on High-Energy Physics (ICHEP 98), Vancouver, British Columbia, Canada, 23–29 Jul 1998.
[129] C. Avilez, G. Marx, and B. Fuentes, *Phys. Rev.*, D23, 1116, 1981.
[130] S. Avvakumov et al., *Phys. Rev. Lett.*, 89, 011804, 2002, hep-ex/0203018.
[131] D. Ayres et al., MINOS Proposal, NuMI-L-63.
[132] D. Ayres et al., 2002, hep-ex/0210005.
[133] D. S. Ayres et al., *Phys. Rev.*, D29, 902, 1984.
[134] W. Baade and F. Zwicky, *Phys. Rev.*, 46, 76–77, 1934.
[135] P. Bagnaia et al., *Phys. Lett.*, B129, 130–140, 1983.
[136] J. N. Bahcall, *Phys. Rev. Lett.*, 12, 300–302, 1964.
[137] J. N. Bahcall, *Neutrino Astrophysics*, Cambridge University Press, 1989.
[138] J. N. Bahcall, *Phys. Rev.*, D49, 3923–3945, 1994.
[139] J. N. Bahcall, *Phys. Rev.*, C65, 025801, 2002, hep-ph/0108148.
[140] J. N. Bahcall, A. Dar, and T. Piran, *Nature*, 326, 135, 1987.
[141] J. N. Bahcall et al., *Phys. Rev.*, C54, 411–422, 1996, nucl-th/9601044.
[142] J. N. Bahcall, W. A. Fowler, I. Iben, and R. L. Sears, *Astrophys. J.*, 137, 344, 1963.
[143] J. N. Bahcall and S. L. Glashow, *Nature*, 326, 476, 1987.
[144] J. N. Bahcall and P. I. Krastev, *Phys. Rev.*, D53, 4211–4225, 1996, hep-ph/9512378.

[145] J. N. Bahcall, M. H. Pinsonneault, and S. Basu, *Astrophys. J.*, 555, 990–1012, 2001, astro-ph/0010346.

[146] J. N. Bahcall, T. Piran, W. H. Press, and D. N. Spergel, *Nature*, 327, 682–685, 1987.

[147] J. N. Bahcall, *Phys. Rev.*, C56, 3391, 1997, hep-ph/9710491.

[148] J. N. Bahcall, *SLAC Beam Line*, 31N1, 2–12, 2001, astro-ph/0009259.

[149] J. N. Bahcall, N. A. Bahcall, and G. Shaviv, *Phys. Rev. Lett.*, 20, 1209–1212, 1968.

[150] J. N. Bahcall, M. C. Gonzalez-Garcia, and C. Pena-Garay, *JHEP*, 08, 016, 2004, hep-ph/0406294.

[151] J. N. Bahcall and C. Pena-Garay, *New J. Phys.*, 6, 63, 2004, hep-ph/0404061.

[152] J. N. Bahcall and M. H. Pinsonneault, *Phys. Rev. Lett.*, 92, 121301, 2004, astro-ph/0402114.

[153] J. N. Bahcall, A. M. Serenelli, and S. Basu, *Astrophys. J.*, 621, L85–L88, 2005, astro-ph/0412440.

[154] J. N. Bahcall, A. M. Serenelli, and S. Basu, *Astrophys. J. Supp. Ser.*, 165, 400–431, 2006, astro-ph/0511337.

[155] M. Y. Bahran and G. R. Kalbfleisch, *Phys. Lett.*, B354, 481–485, 1995.

[156] A. Bakalyarov et al., *Phys. Part. Nucl. Lett.*, 2, 77, 2005, hep-ex/0309016.

[157] A. B. Balantekin, *Phys. Rev.*, D58, 013001, 1998, hep-ph/9712304.

[158] A. Balantekin, 2006, hep-ph/0601113, Origin of Matter and the Evolution of Galaxies (OMEG05), Tokyo, Japan.

[159] A. J. Baltz and J. Weneser, *Phys. Rev.*, D35, 528, 1987.

[160] M. Banner et al., *Phys. Lett.*, B122, 476–485, 1983.

[161] R. Barate et al., *Eur. Phys. J.*, C2, 395–406, 1998.

[162] C. Barbero, F. Krmpotic, A. Mariano, and D. Tadic, *Nucl. Phys.*, A650, 485–497, 1999, nucl-th/9902040.

[163] G. B. Barbiallini and G. Cocconi, *Nature*, 329, 21, 1987.

[164] R. Barbieri and A. Dolgov, *Nucl. Phys.*, B349, 743–753, 1991.

[165] R. Barbieri, R. N. Mohapatra, and T. Yanagida, *Phys. Lett.*, B213, 69, 1988.

[166] R. Barbieri and R. N. Mohapatra, *Phys. Rev. Lett.*, 61, 27, 1988.

[167] D. Y. Bardin and G. Passarino, *The standard model in the making: Precision study of the electroweak interactions*, Oxford University Press, 1999, The International Series of Monographs on Physics, 104.

[168] V. Barger, P. Huber, and D. Marfatia, *Phys. Lett.*, B617, 167, 2005, hep-ph/0501184.

[169] V. Barger, J. P. Kneller, P. Langacker, D. Marfatia, and G. Steigman, *Phys. Lett.*, B569, 123, 2003, hep-ph/0306061.

[170] V. Barger, D. Marfatia, and A. Tregre, *Phys. Lett.*, B595, 55, 2004, hep-ph/0312065.

[171] V. D. Barger, K. Whisnant, S. Pakvasa, and R. J. N. Phillips, *Phys. Rev.*, D22, 2718, 1980.

[172] V. D. Barger et al., *Phys. Lett.*, B462, 109–114, 1999, hep-ph/9907421.

[173] V. D. Barger, J. G. Learned, S. Pakvasa, and T. J. Weiler, *Phys. Rev. Lett.*, 82, 2640–2643, 1999, astro-ph/9810121.

[174] V. D. Barger, K. Whisnant, D. Cline, and R. J. N. Phillips, *Phys. Lett.*, B93, 194, 1980.
[175] V. D. Barger, K. Whisnant, and R. J. N. Phillips, *Phys. Rev.*, D22, 1636, 1980.
[176] V. Barone and E. Predazzi, *High-energy particle diffraction*, Springer, 2002.
[177] G. Barr, T. K. Gaisser, and T. Stanev, *Phys. Rev.*, D39, 3532–3534, 1989.
[178] G. Barr et al., *Phys. Rev.*, D70, 023006, 2004, astro-ph/0403630.
[179] S. M. Barr, T. K. Gaisser, P. Lipari, and S. Tilav, *Phys. Lett.*, B214, 147, 1988.
[180] R. B. Barreiro, *New Astron. Rev.*, 44, 179–204, 2000, astro-ph/9907094.
[181] P. H. Barrett, L. M. Bollinger, G. Cocconi, Y. Eisenberg, and K. Greisen, *Rev. Mod. Phys.*, 24, 133–178, 1952.
[182] N. Bartolo, E. Komatsu, S. Matarrese, and A. Riotto, *Phys. Rep.*, 402, 103, 2004, astro-ph/0406398.
[183] A. Barut, *Electrodynamics and Classical Theory of Fields and Particles*, Dover, 1980.
[184] M. Battaglia et al., 2003, hep-ph/0304132, First Workshop on the CKM Unitarity Triangle, CERN 13–16 February 2002.
[185] G. Battistoni et al., *Phys. Lett.*, B118, 461, 1982.
[186] G. Battistoni et al., *Phys. Lett.*, B133, 454, 1983.
[187] G. Battistoni et al., *Astropart. Phys.*, 12, 315–333, 2000, hep-ph/9907408.
[188] G. Battistoni, A. Ferrari, T. Montaruli, and P. R. Sala, *Astropart. Phys.*, 19, 269, 2003, hep-ph/0207035.
[189] A. Bazarko, *Nucl. Phys. Proc. Suppl.*, 91, 210–215, 2000, hep-ex/0009056.
[190] J. F. Beacom, 1999, hep-ph/9901300, 22nd Symposium on Nuclear Physics, Oaxtepec, Morelos, Mexico, 5–8 Jan 1999.
[191] J. F. Beacom, R. N. Boyd, and A. Mezzacappa, *Phys. Rev. Lett.*, 85, 3568, 2000, hep-ph/0006015.
[192] J. F. Beacom, R. N. Boyd, and A. Mezzacappa, *Phys. Rev.*, D63, 073011, 2001, astro-ph/0010398.
[193] J. F. Beacom and P. Vogel, *Phys. Rev.*, D58, 093012, 1998, hep-ph/9806311.
[194] J. F. Beacom, W. M. Farr, and P. Vogel, *Phys. Rev.*, D66, 033001, 2002, hep-ph/0205220.
[195] J. F. Beacom and M. R. Vagins, *Phys. Rev. Lett.*, 93, 171101, 2004, hep-ph/0309300.
[196] G. Beaudet and P. Goret, *Astron. Astrophys.*, 49, 415–419, 1976.
[197] G. Beaudet and A. Yahil, *Astrophys. J.*, 218, 253–262, 1977.
[198] D. Beavis et al., 1995, BNL-52459.
[199] R. Becker-Szendy et al., *Phys. Rev. Lett.*, 69, 1010–1013, 1992.
[200] R. Becker-Szendy et al., *Phys. Rev.*, D46, 3720–3724, 1992.
[201] J. S. Bell and R. Jackiw, *Nuovo Cim.*, A60, 47–61, 1969.
[202] N. F. Bell, R. Foot, and R. R. Volkas, *Phys. Rev.*, D58, 105010, 1998, hep-ph/9805259.
[203] R. Bellotti et al., *Phys. Rev.*, D53, 35–43, 1996.
[204] R. Bellotti et al., *Phys. Rev.*, D60, 052002, 1999, hep-ex/9905012.

[205] C. Bemporad, G. Gratta, and P. Vogel, *Rev. Mod. Phys.*, 74, 297, 2002, hep-ph/0107277.
[206] C. L. Bennett et al., *Astrophys. J. Supp. Ser.*, 148, 1–27, 2003, astro-ph/0302207.
[207] A. C. Benvenuti et al., *Phys. Rev. Lett.*, 32, 800–803, 1974.
[208] P. Berge et al., *Z. Phys.*, C56, 175–180, 1992.
[209] C. Berger et al., *Phys. Lett.*, B227, 489, 1989.
[210] C. Berger et al., *Phys. Lett.*, B245, 305–310, 1990.
[211] H. E. Bergeson, G. L. Cassiday, and M. B. Hendricks, *Phys. Rev. Lett.*, 31, 66–70, 1973.
[212] F. Bergsma et al., *Z. Phys.*, C40, 171, 1988.
[213] L. Bergstrom and A. Goobar, *Cosmology and Particle Astrophysics*, Springer, 2004.
[214] J. Bernabeu, G. C. Branco, and M. Gronau, *Phys. Lett.*, B169, 243–247, 1986.
[215] V. Bernard, L. Elouadrhiri, and U. G. Meissner, *J. Phys.*, G28, R1–R35, 2002, hep-ph/0107088.
[216] H. J. Bernstein and F. E. Low, *Phys. Rev. Lett.*, 59, 951, 1987.
[217] J. Bernstein, L. S. Brown, and G. Feinberg, *Rev. Mod. Phys.*, 61, 25, 1989.
[218] S. Berridge et al., 2004, hep-ex/0410081.
[219] G. Bertone, D. Hooper, and J. Silk, *Phys. Rep.*, 405, 279, 2005, hep-ph/0404175.
[220] H. Bethe and R. Peierls, *Nature*, 133, 532, 1934.
[221] H. A. Bethe, *Phys. Rev.*, 55, 434–456, 1939.
[222] H. A. Bethe, *Phys. Rev. Lett.*, 56, 1305, 1986.
[223] H. A. Bethe and R. Wilson, James, *Astrophys. J.*, 295, 14–23, 1985.
[224] M. Beuthe, *Phys. Rev.*, D66, 013003, 2002, hep-ph/0202068.
[225] M. Beuthe, *Phys. Rep.*, 375, 105–218, 2003, hep-ph/0109119.
[226] K. Bhattacharya and P. B. Pal, *Proc. Ind. Natl. Sci. Acad.*, 70, 145–161, 2004, hep-ph/0212118.
[227] S. M. Bilenky, *Introduction to Feynman diagrams and electroweak interactions physics*, Editions Frontieres, 1994.
[228] S. M. Bilenky and C. Giunti, *Phys. Lett.*, B444, 379, 1998, hep-ph/9802201.
[229] S. M. Bilenky and C. Giunti, *Int. J. Mod. Phys.*, A16, 3931–3949, 2001, hep-ph/0102320.
[230] S. M. Bilenky, C. Giunti, J. A. Grifols, and E. Masso, *Phys. Rep.*, 379, 69–148, 2003, hep-ph/0211462.
[231] S. M. Bilenky, C. Giunti, and W. Grimus, *Prog. Part. Nucl. Phys.*, 43, 1, 1999, hep-ph/9812360.
[232] S. M. Bilenky and J. A. Grifols, *Phys. Lett.*, B550, 154–159, 2002, hep-ph/0211101.
[233] S. M. Bilenky and J. Hosek, *Phys. Rep.*, 90, 73, 1982.
[234] S. M. Bilenky, J. Hosek, and S. T. Petcov, *Phys. Lett.*, B94, 495, 1980.
[235] S. M. Bilenky and S. T. Petcov, *Rev. Mod. Phys.*, 59, 671, 1987.
[236] S. M. Bilenky and B. Pontecorvo, *Sov. J. Nucl. Phys.*, 24, 316–319, 1976.
[237] S. M. Bilenky and B. Pontecorvo, *Phys. Rep.*, 41, 225, 1978.

[238] S. M. Bilenky, C. Giunti, W. Grimus, and T. Schwetz, *Astropart. Phys.*, 11, 413–428, 1999, hep-ph/9804421.

[239] S. M. Bilenky and B. Pontecorvo, *Nuovo Cim. Lett.*, 17, 569, 1976.

[240] S. Bilenky, A. Faessler, and F. Simkovic, *Phys. Rev.*, D70, 033003, 2004, hep-ph/0402250.

[241] S. Bilenky and S. Petcov, 2004, hep-ph/0405237.

[242] R. M. Bionta et al., *Phys. Rev. Lett.*, 58, 1494, 1987.

[243] R. M. Bionta et al., *Phys. Rev.*, D38, 768–775, 1988.

[244] J. D. Bjorken, *Phys. Rev.*, 163, 1767–1769, 1967.

[245] J. D. Bjorken, *Phys. Rev.*, 179, 1547–1553, 1969.

[246] J. D. Bjorken and I. Dunietz, *Phys. Rev.*, D36, 2109, 1987.

[247] G. Blanc et al., *Astron. Astrophys.*, 423, 881, 2004, astro-ph/0405211.

[248] J. Blatt and V. Weisskopf, *Theoretical Nuclear Physics*, Springer-Verlag, 1979.

[249] A. Blondel and D. Denegri, *Cambridge Monogr. Part. Phys. Nucl. Phys. Cosmol.*, 14, 165–186, 2000.

[250] S. A. Bludman, D. C. Kennedy, and P. G. Langacker, *Nucl. Phys.*, B374, 373–391, 1992.

[251] S. A. Bludman, D. C. Kennedy, and P. G. Langacker, *Phys. Rev.*, D45, 1810–1813, 1992.

[252] A. Bobyk, W. A. Kaminski, and F. Simkovic, *Phys. Rev.*, C63, 051301, 2001, nucl-th/0012010.

[253] F. Boehm et al., *Phys. Rev.*, D62, 072002, 2000, hep-ex/0003022.

[254] F. Boehm et al., *Phys. Rev. Lett.*, 84, 3764–3767, 2000, hep-ex/9912050.

[255] F. Boehm et al., *Phys. Rev.*, D64, 112001, 2001, hep-ex/0107009.

[256] A. M. Boesgaard and G. Steigman, *Ann. Rev. Astron. Astrophys.*, 23, 319, 1985.

[257] M. Boezio et al., *Phys. Rev.*, D62, 032007, 2000, hep-ex/0004014.

[258] J. Boger et al., *Nucl. Instrum. Meth.*, A449, 172–207, 2000, nucl-ex/9910016.

[259] D. Bohm, *Quantum Theory*, Prentice Hall, 1959.

[260] M. M. Boliev et al., *Yad. Fiz.*, 34, 1418–1421, 1981, In Russian.

[261] M. M. Boliev et al., *Bull. Russ. Acad. Sci. Phys.*, 55, NO.4126, 1991.

[262] J. R. Bond, G. Efstathiou, and J. Silk, *Phys. Rev. Lett.*, 45, 1980–1984, 1980.

[263] M. Born and E. Wolf, *Principles of Optics*, Pergamon Press, 1959.

[264] L. Bornschein et al., *eConf*, C030626, FRAP14, 2003, hep-ex/0309007, XIII Physics in Collision Conference(PIC03), Zeuthen, Germany, June 2003.

[265] L. Borodovsky et al., *Phys. Rev. Lett.*, 68, 274–277, 1992.

[266] P. Bouchet et al., *Astrophys. J.*, 611, 394, 2004, astro-ph/0312240.

[267] P. Bouchet et al., *Astrophys. J.*, 650, 212–227, 2006, astro-ph/0601495.

[268] C. Bouchiat, J. Iliopoulos, and P. Meyer, *Phys. Lett.*, B38, 519–523, 1972.

[269] G. C. Branco and L. Lavoura, *Phys. Lett.*, B208, 123, 1988.

[270] C. B. Bratton et al., *Phys. Rev.*, D37, 3361, 1988.

[271] R. J. Breene Jr., *Rev. Mod. Phys.*, 29, 94, 1957.

[272] S. W. Bruenn, 2005, astro-ph/0506313, Workshop on Open Issues in Understanding Core Collapse Supernovae, Seattle, Washington, 22–24 June 2004.
[273] H. Budd, A. Bodek, and J. Arrington, 2003, hep-ex/0308005, NuInt02, Dec. 2002, Irvine, CA, USA.
[274] E. V. Bugaev and V. A. Naumov, *Phys. Lett.*, B232, 391–397, 1989.
[275] E. F. Bunn, 1996, astro-ph/9607088, 1996 NATO Advanced Study Institute on "The Cosmic Background Radiation".
[276] A. J. Buras, *Rev. Mod. Phys.*, 52, 199, 1980.
[277] A. J. Buras, M. E. Lautenbacher, and G. Ostermaier, *Phys. Rev.*, D50, 3433–3446, 1994, hep-ph/9403384.
[278] R. Buras, H.-T. Janka, M. Rampp, and K. Kifonidis, *Astron. Astrophys.*, 457, 281–308, 2006, astro-ph/0512189.
[279] R. Buras, M. Rampp, H.-T. Janka, and K. Kifonidis, *Phys. Rev. Lett.*, 90, 241101, 2003, astro-ph/0303171.
[280] R. Buras, H.-T. Janka, M. T. Keil, G. G. Raffelt, and M. Rampp, *Astrophys. J.*, 587, 320, 2003, astro-ph/0205006.
[281] R. Buras, M. Rampp, H. T. Janka, and K. Kifonidis, *Astron. Astrophys.*, 447, 1049–1092, 2006, astro-ph/0507135.
[282] E. M. Burbidge, G. R. Burbidge, W. A. Fowler, and F. Hoyle, *Rev. Mod. Phys.*, 29, 547, 1957.
[283] H. Burkhardt, J. Lowe, G. J. Stephenson, and T. Goldman, *Phys. Rev.*, D59, 054018, 1999, hep-ph/9803365.
[284] H. Burkhardt, J. Lowe, G. J. Stephenson, and T. Goldman, *Phys. Lett.*, B566, 137, 2003, hep-ph/0302084.
[285] A. Burrows and T. A. Thompson, 2002, astro-ph/0210212, ESO/MPA/MPE Workshop (an ESO Astrophysics Symposium) "From Twilight to Highlight: The Physics of Supernovae", Garching bei Munchen, Germany, July 29-31, 2002.
[286] A. Burrows, T. Young, P. Pinto, R. Eastman, and T. A. Thompson, *Astrophys. J.*, 539, 865–887, 2000.
[287] A. Burrows, R. Gandhi, and M. S. Turner, *Phys. Rev. Lett.*, 68, 3834–3837, 1992.
[288] A. Burrows, J. Hayes, and B. A. Fryxell, *Astrophys. J.*, 450, 830, 1995, astro-ph/9506061.
[289] A. Burrows, D. Klein, and R. Gandhi, *Phys. Rev.*, D45, 3361–3385, 1992.
[290] A. Burrows, E. Livne, L. Dessart, C. Ott, and J. Murphy, *Astrophys. J.*, 640, 878–890, 2006, astro-ph/0510687.
[291] D. N. Burrows et al., *Astrophys. J.*, 543, L149–L152, 2000, astro-ph/0009265.
[292] N. Cabibbo, *Phys. Rev. Lett.*, 10, 531–532, 1963.
[293] N. Cabibbo, 1980, Proc. of "Astrophysics and Elementary Particles, Common Problems", Rome, Italy, 21–23 February 1980, p. 209.
[294] N. Cabibbo and L. Maiani, *Phys. Lett.*, B114, 115, 1982.
[295] J. Callan, Curtis G. and D. J. Gross, *Phys. Rev. Lett.*, 22, 156–159, 1969.
[296] A. G. W. Cameron, *Pub. Astron. Soc. Pac.*, 69, 201, 1957.

[297] E. Cappellaro et al., *Astron. Astrophys.*, 273, 383, 1993, astro-ph/9302017.

[298] E. Cappellaro et al., *Astron. Astrophys.*, 322, 431–441, 1997, astro-ph/9611191.

[299] E. Cappellaro, R. Evans, and M. Turatto, *Astron. Astrophys.*, 351, 459–466, 1999, astro-ph/9904225.

[300] E. Cappellaro and M. Turatto, 2000, astro-ph/0012455, The Influence of Binaries on Stellar Population Studies, Brussels 21–25 Aug. 2000.

[301] C. Y. Cardall, *Phys. Rev.*, D61, 073006, 2000, hep-ph/9909332.

[302] C. Y. Cardall, 2002, astro-ph/0212438, 4th International Workshop on the Identification of Dark Matter (IDM2002), York, England 2–6 September 2002.

[303] C. Y. Cardall, *Nucl. Phys. Proc. Suppl.*, 145, 295, 2005, astro-ph/0502232, NOW2004, Conca Specchiulla (Otranto, Italy), September 11-17, 2004.

[304] C. Y. Cardall et al., *J. Phys. Conf. Ser.*, 16, 390, 2005, astro-ph/0510706, SciDAC 2005, Scientific Discovery through Advanced Computing, San Francisco, CA, 26–30 June 2005.

[305] C. Y. Cardall, A. O. Razoumov, E. Endeve, and A. Mezzacappa, 2005, astro-ph/0510704, Open Issues in Understanding Core Collapse Supernovae, National Institute for Nuclear Theory, University of Washington, 22–24 June 2004.

[306] D. Casper et al., *Phys. Rev. Lett.*, 66, 2561–2564, 1991.

[307] V. Castellani, S. Degl'Innocenti, G. Fiorentini, M. Lissia, and B. Ricci, *Phys. Rep.*, 281, 309–398, 1997, astro-ph/9606180.

[308] E. Caurier, F. Nowacki, A. Poves, and J. Retamosa, *Phys. Rev. Lett.*, 77, 1954, 1996.

[309] F. Cei, *Int. J. Mod. Phys.*, A17, 1765–1776, 2002, hep-ex/0202043, Second International Workshop on Matter, Anti-Matter and Dark Matter, Trento (Italy), 29–30 October 2001.

[310] J. Chadwick, *Nature*, 129, 312, 1932.

[311] S. Chandrasekhar, *An Introduction to the Study of Stellar Structure*, University of Chicago Press, 1938.

[312] L.-L. Chau and W.-Y. Keung, *Phys. Rev. Lett.*, 53, 1802, 1984.

[313] H. H. Chen, *Phys. Rev. Lett.*, 55, 1534–1536, 1985.

[314] T. P. Cheng and L. F. Li, *Gauge theory of elementary particle physics*, Oxford University Press, 1984.

[315] M. V. Chizhov and S. T. Petcov, *Phys. Rev. Lett.*, 83, 1096–1099, 1999, hep-ph/9903399.

[316] M. V. Chizhov and S. T. Petcov, *Phys. Rev.*, D63, 073003, 2001, hep-ph/9903424.

[317] J. H. Christenson, J. W. Cronin, V. L. Fitch, and R. Turlay, *Phys. Rev. Lett.*, 13, 138–140, 1964.

[318] A. E. Chudakov, Y. S. Elensky, and S. P. Mikheev, *JETP Lett.*, 46, 373–377, 1987.

[319] E. Church et al., 1997, nucl-ex/9706011.

[320] E. D. Church, K. Eitel, G. B. Mills, and M. Steidl, *Phys. Rev.*, D66, 013001, 2002, hep-ex/0203023.

[321] O. Civitarese and J. Suhonen, *Nucl. Phys.*, A729, 867–883, 2003, nucl-th/0208005.
[322] R. Clark et al., *Phys. Rev. Lett.*, 79, 345–348, 1997.
[323] B. T. Cleveland et al., *Astrophys. J.*, 496, 505–526, 1998.
[324] S. Cole et al., *Mon. Not. Roy. Astron. Soc.*, 362, 505, 2005, astro-ph/0501174.
[325] P. Coles and F. Lucchin, *Cosmology: The Origin and Evolution of Cosmic Structure*, John Wiley, 2002.
[326] S. A. Colgate and R. H. White, *Astrophys. J.*, 143, 626–681, 1966.
[327] M. Colless, 2003, astro-ph/0305051.
[328] E. D. Commins and P. H. Bucksbaum, *Weak Interactions of Leptons and Quarks*, Cambridge University Press, 1983.
[329] G. Comsa, *Phys. Rev. Lett.*, 51, 1105, 1983.
[330] E. U. Condon and R. W. Gurney, *Phys. Rev.*, 33, 127, 1929.
[331] G. Costa and F. Zwirner, *Riv. Nuovo Cim.*, 9N3, 1–134, 1986.
[332] M. L. Costantini, A. Ianni, and F. Vissani, *Phys. Rev.*, D70, 043006, 2004, astro-ph/0403436.
[333] S. Coutu et al., *Phys. Rev.*, D62, 032001, 2000, hep-ex/0004010.
[334] C. L. Cowan, F. Reines, F. B. Harrison, E. C. Anderson, and F. N. Hayes, *Phys. Rev.*, 90, 493–494, 1953.
[335] R. Cowsik, *Phys. Rev.*, D37, 1685, 1988.
[336] R. Cowsik and J. McClelland, *Phys. Rev. Lett.*, 29, 669–670, 1972.
[337] R. A. C. Croft et al., *Astrophys. J.*, 581, 20–52, 2002, astro-ph/0012324.
[338] R. A. C. Croft, D. H. Weinberg, N. Katz, and L. Hernquist, *Astron. J.*, 495, 44, 1998, astro-ph/9708018.
[339] P. Crotty, J. Lesgourgues, and S. Pastor, *Phys. Rev.*, D67, 123005, 2003, astro-ph/0302337.
[340] P. Crotty, J. Lesgourgues, and S. Pastor, *Phys. Rev.*, D69, 123007, 2004, hep-ph/0402049.
[341] M. F. Crouch et al., *Phys. Rev.*, D18, 2239–2252, 1978.
[342] A. Cuoco et al., *Int. J. Mod. Phys.*, A19, 4431, 2004, astro-ph/0307213.
[343] R. H. Cyburt, B. D. Fields, and K. A. Olive, *Astropart. Phys.*, 17, 87–100, 2002, astro-ph/0105397.
[344] R. H. Cyburt, B. D. Fields, and K. A. Olive, *Phys. Lett.*, B567, 227, 2003, astro-ph/0302431.
[345] R. H. Cyburt, B. D. Fields, K. A. Olive, and E. Skillman, *Astropart. Phys.*, 23, 313–323, 2005, astro-ph/0408033.
[346] V. L. Dadykin et al., *JETP Lett.*, 45, 593–595, 1987.
[347] T. Dahlen et al., *Astrophys. J.*, 613, 189–199, 2004, astro-ph/0406547.
[348] G. Danby et al., *Phys. Rev. Lett.*, 9, 36–44, 1962.
[349] A. Dar, *Phys. Rev. Lett.*, 51, 227, 1983.
[350] A. Dar and S. Dado, *Phys. Rev. Lett.*, 59, 2368, 1987.
[351] G. V. Dass and K. V. L. Sarma, *Phys. Rev.*, D30, 80, 1984.
[352] K. Daum, *Z. Phys.*, C66, 417–428, 1995.
[353] R. Davis, *Phys. Rev.*, 97, 766–769, 1955.
[354] R. Davis, *Phys. Rev. Lett.*, 12, 303–305, 1964.

[355] R. Davis, *Prog. Part. Nucl. Phys.*, 32, 13–32, 1994.

[356] R. Davis, D. S. Harmer, and K. C. Hoffman, *Phys. Rev. Lett.*, 20, 1205–1209, 1968.

[357] R. d'E. Atkinson and F. G. Houtermans, *Z. Phys.*, 54, 656, 1929.

[358] A. de Gouvea, A. Friedland, and H. Murayama, *Phys. Rev.*, D60, 093011, 1999, hep-ph/9904399.

[359] A. de Gouvea, A. Friedland, and H. Murayama, *Phys. Lett.*, B490, 125–130, 2000, hep-ph/0002064.

[360] S. De Leo, C. C. Nishi, and P. P. Rotelli, 2002, hep-ph/0208086.

[361] A. De Rujula, *Phys. Lett.*, B193, 514, 1987.

[362] D. Decamp et al., *Phys. Lett.*, B231, 519, 1989.

[363] Y. Declais et al., *Nucl. Phys.*, B434, 503–534, 1995.

[364] P. Di Bari, *Phys. Rev.*, D65, 043509, 2002, hep-ph/0108182.

[365] P. Di Bari, *Phys. Rev.*, D67, 127301, 2003, astro-ph/0302433.

[366] P. Di Bari and R. Foot, *Phys. Rev.*, D61, 105012, 2000, hep-ph/9912215.

[367] P. Di Bari and R. Foot, *Phys. Rev.*, D63, 043008, 2001, hep-ph/0008258.

[368] R. H. Dicke, P. J. E. Peebles, P. G. Roll, and D. T. Wilkinson, *Astrophys. J.*, 142, 414–419, 1965.

[369] C. Dickinson et al., *Mon. Not. Roy. Astron. Soc.*, 353, 732, 2004, astro-ph/0402498.

[370] D. A. Dicus, E. W. Kolb, V. L. Teplitz, and R. V. Wagoner, *Phys. Rev.*, D17, 1529–1538, 1978.

[371] D. A. Dicus et al., *Phys. Rev.*, D26, 2694, 1982.

[372] M. Diemoz, F. Ferroni, E. Longo, and G. Martinelli, *Cambridge Monogr. Part. Phys. Nucl. Phys. Cosmol.*, 14, 417–498, 2000.

[373] A. S. Dighe, Q. Y. Liu, and A. Y. Smirnov, 1999, hep-ph/9903329.

[374] M. Dijkstra, A. Lidz, and L. Hui, *Astrophys. J.*, 605, 7, 2004, astro-ph/0305498.

[375] M. V. Diwan, *eConf*, C0209101, TH08, 2002, hep-ex/0211026, Seventh International Workshop on Tau Lepton Physics (TAU02), Santa Cruz, Ca, USA, Sept 2002.

[376] M. Doi, T. Kotani, H. Nishiura, K. Okuda, and E. Takasugi, *Phys. Lett.*, B102, 323, 1981.

[377] M. Doi, T. Kotani, and E. Takasugi, *Prog. Theor. Phys. Suppl.*, 83, 1, 1985.

[378] A. D. Dolgov, *Sov. J. Nucl. Phys.*, 33, 700–706, 1981.

[379] A. D. Dolgov, *Phys. Rep.*, 370, 333–535, 2002, hep-ph/0202122.

[380] A. D. Dolgov et al., *Nucl. Phys.*, B632, 363–382, 2002, hep-ph/0201287.

[381] A. D. Dolgov, S. H. Hansen, S. Pastor, and D. V. Semikoz, *Astropart. Phys.*, 14, 79–90, 2000, hep-ph/9910444.

[382] A. D. Dolgov, S. H. Hansen, and D. V. Semikoz, *Nucl. Phys.*, B503, 426–444, 1997, hep-ph/9703315.

[383] A. D. Dolgov, S. H. Hansen, and D. V. Semikoz, *Nucl. Phys.*, B543, 269–274, 1999, hep-ph/9805467.

[384] A. D. Dolgov, A. Y. Morozov, L. B. Okun, and M. G. Shchepkin, *Nucl. Phys.*, B502, 3, 1997, hep-ph/9703241.

[385] A. D. Dolgov and Y. B. Zeldovich, *Rev. Mod. Phys.*, 53, 1–41, 1981.

[386] A. Dolgov and F. Villante, *Nucl. Phys.*, B679, 261, 2004, hep-ph/0308083.
[387] J. C. D'Olivo, J. F. Nieves, and M. Torres, *Phys. Rev.*, D46, 1172–1179, 1992.
[388] J. Dorenbosch et al., *Z. Phys.*, C40, 497, 1988.
[389] P. M. Dragicevich, D. G. Blair, and R. R. Burman, *Mon. Not. Roy. Astron. Soc.*, 302, 693–699, 1999.
[390] O. Dragoun et al., *J. Phys.*, G25, 1839–1858, 1999.
[391] D. Drakoulakos et al., 2004, hep-ex/0405002.
[392] M. Drees and G. Gerbier, *Phys. Lett.*, B592, 216, 2004, The Review of Particle Properties 2004.
[393] D. Duchesneau, *eConf*, C0209101, TH09, 2002, hep-ex/0209082, Seventh International Workshop on Tau Lepton Physics (TAU02), Santa Cruz, CA, USA, Sept 2002.
[394] I. Dunietz, *Ann. Phys.*, 184, 350, 1988.
[395] I. Dunietz, O. W. Greenberg, and D. di Wu, *Phys. Rev. Lett.*, 55, 2935, 1985.
[396] F. Dydak et al., *Phys. Lett.*, B134, 281, 1984.
[397] G. Eder, *Nucl. Phys.*, 78, 657–662, 1966.
[398] K. Eguchi et al., *Phys. Rev. Lett.*, 90, 021802, 2003, hep-ex/0212021.
[399] K. Eguchi et al., *Phys. Rev. Lett.*, 92, 071301, 2004, hep-ex/0310047.
[400] S. Eidelman et al., *Phys. Lett.*, B592, 1, 2004.
[401] D. J. Eisenstein et al., *Astrophys. J.*, 633, 560, 2005, astro-ph/0501171.
[402] K. Eitel, *Nucl. Phys. Proc. Suppl.*, 91, 191–197, 2000, hep-ex/0008002, 19th International Conference on Neutrino Physics and Astrophysics – Neutrino 2000, Sudbury, Ontario, Canada, 16–21 June 2000.
[403] O. Elgaroy and O. Lahav, *New J. Phys.*, 7, 61, 2005, hep-ph/0412075.
[404] S. Eliezer and A. R. Swift, *Nucl. Phys.*, B105, 45, 1976.
[405] S. R. Elliott and J. Engel, *J. Phys.*, G30, R183, 2004, hep-ph/0405078.
[406] J. R. Ellis, *Nucl. Phys.*, A684, 53–70, 2001, hep-ph/0005322, 16th International Conference on Few-Body Problems in Physics (FB 16), Taipei, Taiwan, China, 6–10 Mar 2000.
[407] J. R. Ellis and M. Karliner, *Phys. Lett.*, B213, 73, 1988.
[408] J. R. Ellis, S. Kelley, and D. V. Nanopoulos, *Phys. Lett.*, B260, 131–137, 1991.
[409] J. R. Ellis and K. A. Olive, *Phys. Lett.*, B193, 525, 1987.
[410] H. T. Elze, W. Greiner, and J. Rafelski, *J. Phys.*, G6, L149–L153, 1980.
[411] J. J. Engelmann, P. Ferrando, A. Soutoul, P. Goret, and E. Juliusson, *Astron. Astrophys.*, 233, 96–111, 1990.
[412] F. Englert and R. Brout, *Phys. Rev. Lett.*, 13, 321–322, 1964.
[413] S. Enomoto, E. Ohtani, K. Inoue, and A. Suzuki, 2005, hep-ph/0508049.
[414] K. Enqvist and K. Kainulainen, *Phys. Lett.*, B264, 367–372, 1991.
[415] K. Enqvist, K. Kainulainen, and J. Maalampi, *Nucl. Phys.*, B317, 647–664, 1989.
[416] K. Enqvist, K. Kainulainen, and J. Maalampi, *Phys. Lett.*, B244, 186–190, 1990.

[417] K. Enqvist, K. Kainulainen, and J. Maalampi, *Phys. Lett.*, B249, 531–534, 1990.

[418] K. Enqvist, K. Kainulainen, and J. Maalampi, *Nucl. Phys.*, B349, 754–790, 1991.

[419] K. Enqvist, K. Kainulainen, and A. Sorri, *JHEP*, 04, 012, 2001, hep-ph/0012291.

[420] R. I. Epstein, J. M. Lattimer, and D. N. Schramm, *Nature*, 263, 198–202, 1976.

[421] J. Erler and P. Langacker, *Phys. Lett.*, B592, 2004, hep-ph/0407097, The Review of Particle Properties 2004.

[422] E. Eskut et al., *Phys. Lett.*, B497, 8–22, 2001.

[423] R. Esmailzadeh, G. D. Starkman, and S. Dimopoulos, *Astrophys. J.*, 378, 504–518, 1991.

[424] S. Esposito, G. Mangano, G. Miele, and O. Pisanti, *JHEP*, 09, 038, 2000, astro-ph/0005571.

[425] S. Esposito, G. Miele, S. Pastor, M. Peloso, and O. Pisanti, *Nucl. Phys.*, B590, 539–561, 2000, astro-ph/0005573.

[426] A. Faessler and F. Simkovic, *J. Phys.*, G24, 2139–2178, 1998, hep-ph/9901215.

[427] B. Faid, G. L. Fogli, E. Lisi, and D. Montanino, *Astropart. Phys.*, 10, 93–105, 1999, hep-ph/9805293.

[428] J. Favier, R. Kossakowski, and J. P. Vialle, *Phys. Rev.*, D68, 093006, 2003, astro-ph/0305460.

[429] S. A. Fayans, *Sov. J. Nucl. Phys.*, 42, 590, 1985.

[430] E. Fermi, *Ricerca Scientifica*, 2, 12, 1933.

[431] E. Fermi, *Z. Phys.*, 88, 161–177, 1934.

[432] E. Fermi, *Nuovo Cim.*, 11, 1–19, 1934.

[433] F. Feruglio, A. Strumia, and F. Vissani, *Nucl. Phys.*, B637, 345–377, 2002, hep-ph/0201291.

[434] R. P. Feynman and M. Gell-Mann, *Phys. Rev.*, 109, 193–198, 1958.

[435] R. P. Feynman, *Phys. Rev. Lett.*, 23, 1415–1417, 1969.

[436] R. P. Feynman, *Photon-Hadron Interactions*, W. Benjamin, 1972.

[437] J. H. Field, *Eur. Phys. J.*, C30, 305–325, 2003, hep-ph/0211199.

[438] B. Fields and S. Sarkar, *Phys. Lett.*, B592, 202, 2004, astro-ph/0406663, The Review of Particle Properties 2004.

[439] G. Fiorentini et al., *Phys. Lett.*, B558, 15, 2003, hep-ph/0301042.

[440] G. Fiorentini, F. Mantovani, and B. Ricci, *Phys. Lett.*, B557, 139, 2003, nucl-ex/0212008.

[441] G. Fiorentini and C. Acerbi, *Astropart. Phys.*, 7, 245–262, 1997, astro-ph/9701232.

[442] G. Fiorentini, M. Lissia, F. Mantovani, and B. Ricci, *Phys. Lett.*, B629, 77, 2005, hep-ph/0508048.

[443] G. Fiorentini, M. Lissia, F. Mantovani, and R. Vannucci, *Nucl. Phys. Proc. Suppl.*, 143, 53, 2005, hep-ph/0409152.

[444] G. Fiorentini, M. Lissia, F. Mantovani, and R. Vannucci, *Phys. Rev.*, D72, 033017, 2005, hep-ph/0501111.

[445] G. Fiorentini, V. A. Naumov, and F. L. Villante, *Phys. Lett.*, B510, 173–188, 2001, hep-ph/0103322.
[446] D. J. Fixsen et al., *Astrophys. J.*, 473, 576, 1996, astro-ph/9605054.
[447] J. W. Flanagan, J. G. Learned, and S. Pakvasa, *Phys. Rev.*, D57, 2649–2652, 1998, hep-ph/9709438.
[448] G. L. Fogli et al., *Phys. Rev.*, D66, 093008, 2002, hep-ph/0208026.
[449] G. L. Fogli et al., *Phys. Rev.*, D70, 113003, 2004, hep-ph/0408045.
[450] G. L. Fogli, E. Lisi, A. Marrone, D. Montanino, and A. Palazzo, *Phys. Rev.*, D66, 053010, 2002, hep-ph/0206162.
[451] G. L. Fogli, E. Lisi, A. Marrone, and A. Palazzo, *Prog. Part. Nucl. Phys.*, 57, 742–795, 2006, hep-ph/0506083.
[452] G. L. Fogli, E. Lisi, A. Palazzo, and A. M. Rotunno, *Phys. Lett.*, B623, 80, 2005, hep-ph/0505081.
[453] G. L. Fogli, E. Lisi, D. Montanino, and A. Palazzo, *Phys. Rev.*, D62, 113004, 2000, hep-ph/0005261.
[454] R. Foot, *Phys. Rev.*, D61, 023516, 2000, hep-ph/9906311.
[455] R. Foot, M. J. Thomson, and R. R. Volkas, *Phys. Rev.*, D53, 5349–5353, 1996, hep-ph/9509327.
[456] R. Foot and R. R. Volkas, *Phys. Rev.*, D55, 5147–5176, 1997, hep-ph/9610229.
[457] W. A. Fowler, *Rev. Mod. Phys.*, 56, 149–179, 1984.
[458] P. H. Frampton and S. L. Glashow, *Phys. Rev.*, D25, 1982, 1982.
[459] P. H. Frampton and R. N. Mohapatra, *Phys. Rev.*, D50, 3569–3571, 1994, hep-ph/9312230.
[460] S. J. Freedman et al., *Phys. Rev.*, D47, 811–829, 1993.
[461] W. L. Freedman et al., *Astrophys. J.*, 553, 47–72, 2001, astro-ph/0012376.
[462] A. Friedland, *Phys. Rev. Lett.*, 85, 936–939, 2000, hep-ph/0002063.
[463] A. Friedland, *Phys. Rev.*, D64, 013008, 2001, hep-ph/0010231.
[464] A. Friedland and C. Lunardini, *Phys. Rev.*, D68, 013007, 2003, hep-ph/0304055.
[465] P. Fritze et al., *Phys. Lett.*, B96, 427, 1980.
[466] H. Fritzsch and P. Minkowski, *Phys. Lett.*, B62, 72, 1976.
[467] H. Fritzsch and Z.-z. Xing, *Prog. Part. Nucl. Phys.*, 45, 1–81, 2000, hep-ph/9912358.
[468] C. L. Fryer and M. S. Warren, *Astrophys. J.*, 601, 391–404, 2004, astro-ph/0309539.
[469] C. Fryer and M. Warren, *Astrophys. J.*, 574, L65–L68, 2002, astro-ph/0206017.
[470] S. Fukuda et al., *Phys. Rev. Lett.*, 85, 3999–4003, 2000, hep-ex/0009001.
[471] S. Fukuda et al., *Phys. Rev. Lett.*, 86, 5656–5660, 2001, hep-ex/0103033.
[472] S. Fukuda et al., *Phys. Rev. Lett.*, 86, 5651–5655, 2001, hep-ex/0103032.
[473] S. Fukuda et al., *Phys. Lett.*, B539, 179–187, 2002, hep-ex/0205075.
[474] Y. Fukuda et al., *Phys. Lett.*, B335, 237–245, 1994.
[475] Y. Fukuda et al., *Phys. Rev. Lett.*, 77, 1683–1686, 1996.
[476] Y. Fukuda et al., *Phys. Rev. Lett.*, 81, 1562–1567, 1998, hep-ex/9807003.
[477] Y. Fukuda et al., *Phys. Lett.*, B433, 9–18, 1998, hep-ex/9803006.

[478] Y. Fukuda et al., *Phys. Rev. Lett.*, 81, 1158–1162, 1998, hep-ex/9805021.
[479] Y. Fukuda et al., *Phys. Lett.*, B436, 33–41, 1998, hep-ex/9805006.
[480] Y. Fukuda et al., *Phys. Rev. Lett.*, 82, 1810–1814, 1999, hep-ex/9812009.
[481] Y. Fukuda et al., *Phys. Rev. Lett.*, 82, 2644–2648, 1999, hep-ex/9812014.
[482] Y. Fukuda et al., *Phys. Rev. Lett.*, 82, 2430–2434, 1999, hep-ex/9812011.
[483] Y. Fukuda et al., *Phys. Lett.*, B467, 185–193, 1999, hep-ex/9908049.
[484] M. Fukugita, C. J. Hogan, and P. J. E. Peebles, *Astrophys. J.*, 503, 518, 1998, astro-ph/9712020.
[485] W. Furry, *Phys. Rev.*, 56, 1184–1193, 1939.
[486] T. Futagami et al., *Phys. Rev. Lett.*, 82, 5194–5197, 1999, astro-ph/9901139.
[487] T. K. Gaisser, *Cosmic Rays and Particle Physics*, Cambridge University Press, 1990.
[488] T. K. Gaisser et al., *Phys. Rev.*, D54, 5578–5584, 1996, hep-ph/9608253.
[489] T. K. Gaisser and M. Honda, *Ann. Rev. Nucl. Part. Sci.*, 52, 153, 2002, hep-ph/0203272.
[490] T. K. Gaisser and J. S. O'Connell, *Phys. Rev.*, D34, 822–825, 1986.
[491] T. K. Gaisser and T. Stanev, *Phys. Lett.*, B592, 228–234, 2004, The Review of Particle Properties 2004.
[492] T. K. Gaisser, T. Stanev, M. Honda, and P. Lipari, 2001, 27th International Cosmic Ray Conference (ICRC 2001), Hamburg, Germany, 7–15 Aug 2001.
[493] T. K. Gaisser and T. Stanev, *Phys. Rev.*, D30, 985, 1984.
[494] T. K. Gaisser and T. Stanev, *Phys. Rev.*, D57, 1977–1982, 1998, astro-ph/9708146.
[495] T. K. Gaisser, T. Stanev, S. A. Bludman, and H.-s. Lee, *Phys. Rev. Lett.*, 51, 223–226, 1983.
[496] T. K. Gaisser, *Astropart. Phys.*, 16, 285–294, 2002, astro-ph/0104327.
[497] T. K. Gaisser, *Phys. Scripta*, T121, 51, 2005, astro-ph/0502380, Nobel Symposium 129 Neutrino Physics.
[498] T. K. Gaisser, 2005, astro-ph/0501195, 14th International School of Cosmic Ray Astrophysics: "Neutrinos and Explosive Events in the Universe", July, 2004.
[499] T. K. Gaisser, F. Halzen, and T. Stanev, *Phys. Rep.*, 258, 173–236, 1995, hep-ph/9410384.
[500] T. K. Gaisser and T. Stanev, *Nucl. Phys.*, A777, 98–110, 2006, astro-ph/0510321.
[501] M. Galeazzi, F. Fontanelli, F. Gatti, and S. Vitale, *Phys. Rev.*, C63, 014302, 2001.
[502] M. Galeazzi, F. Fontanelli, F. Gatti, and S. Vitale, *Phys. Rev. Lett.*, 86, 1978–1981, 2001.
[503] G. Gamow, *Z. Phys.*, 51, 204, 1928.
[504] G. Gamow, *Phys. Rev.*, 70, 572–573, 1946.
[505] G. Gamow and E. Teller, *Phys. Rev.*, 49, 895–899, 1936.
[506] R. Gandhi and A. Burrows, *Phys. Lett.*, B246, 149–155, 1990.
[507] Y. Gando et al., *Phys. Rev. Lett.*, 90, 171302, 2003, hep-ex/0212067.

[508] G. T. Garvey, W. C. Louis, and D. H. White, *Phys. Rev.*, C48, 761–765, 1993.

[509] F. Gatti, *Nucl. Phys. Proc. Suppl.*, 91, 293–296, 2001, 19th International Conference on Neutrino Physics and Astrophysics – Neutrino 2000, Sudbury, Ontario, Canada, 16–21 June 2000.

[510] E. Gawiser and J. Silk, *Phys. Rep.*, 333, 245–267, 2000, astro-ph/0002044.

[511] S. Geer, *Phys. Rev.*, D57, 6989–6997, 1998, hep-ph/9712290.

[512] M. Gell-Mann, P. Ramond, and R. Slansky, 1979, In "Supergravity", p. 315, edited by F. van Nieuwenhuizen and D. Freedman, North Holland, Amsterdam.

[513] M. Gell-Mann and M. Levy, *Nuovo Cim.*, 16, 705, 1960.

[514] M. Gell-Mann and A. Pais, *Phys. Rev.*, 97, 1387–1389, 1955.

[515] G. B. Gelmini, *Phys. Scripta*, T121, 131, 2005, hep-ph/0412305, Nobel Symposium on Neutrino Physics, Enkoping, Sweden, Augus 19-24, 2004.

[516] C. Q. Geng and R. E. Marshak, *Phys. Rev.*, D39, 693, 1989.

[517] H. Georgi and S. L. Glashow, *Phys. Rev.*, D6, 429, 1972.

[518] S. S. Gershtein and Y. B. Zeldovich, *JETP Lett.*, 4, 120–122, 1966.

[519] S. Gerstein and Y. Zeldovich, *Sov. Phys. JETP*, 2, 576, 1956.

[520] Y. Giomataris and J. Vergados, *Phys. Lett.*, B634, 23, 2006, hep-ex/0503029.

[521] C. Giunti, *Phys. Rev.*, D61, 036002, 2000, hep-ph/9906275.

[522] C. Giunti, *Mod. Phys. Lett.*, A16, 2363, 2001, hep-ph/0104148.

[523] C. Giunti, *JHEP*, 11, 017, 2002, hep-ph/0205014.

[524] C. Giunti, 2003, hep-ph/0302045.

[525] C. Giunti, *Physica Scripta*, 67, 29–33, 2003, hep-ph/0202063.

[526] C. Giunti, *Found. Phys. Lett.*, 17, 103–124, 2004, hep-ph/0302026.

[527] C. Giunti, *Am. J. Phys.*, 72, 699, 2004, physics/0305122.

[528] C. Giunti and C. W. Kim, *Phys. Rev.*, D58, 017301, 1998, hep-ph/9711363.

[529] C. Giunti, C. W. Kim, and J. D. Kim, *Phys. Lett.*, B352, 357–364, 1995, hep-ph/9411219.

[530] C. Giunti, C. W. Kim, and W. P. Lam, *Phys. Rev.*, D43, 164–169, 1991.

[531] C. Giunti, C. W. Kim, J. A. Lee, and U. W. Lee, *Phys. Rev.*, D48, 4310–4317, 1993, hep-ph/9305276.

[532] C. Giunti, C. W. Kim, and U. W. Lee, *Mod. Phys. Lett.*, A6, 1745–1755, 1991.

[533] C. Giunti, C. W. Kim, and U. W. Lee, *Phys. Rev.*, D44, 3635–3640, 1991.

[534] C. Giunti, C. W. Kim, and U. W. Lee, *Phys. Lett.*, B274, 87–94, 1992.

[535] C. Giunti, C. W. Kim, and U. W. Lee, *Phys. Rev.*, D45, 2414–2420, 1992.

[536] C. Giunti, C. W. Kim, and U. W. Lee, *Phys. Lett.*, B421, 237–244, 1998, hep-ph/9709494.

[537] C. Giunti, C. W. Kim, and M. Monteno, *Nucl. Phys.*, B521, 3–36, 1998, hep-ph/9709439.

[538] C. Giunti, 2004, hep-ph/0402217.

[539] C. Giunti, *J. Phys. G: Nucl. Part. Phys.*, 34, R93–R109, 2007, hep-ph/0608070.

[540] C. Giunti and C. W. Kim, *Found. Phys. Lett.*, 14, 213–229, 2001, hep-ph/0011074.
[541] C. Giunti and M. Laveder, *Mod. Phys. Lett.*, 12, 1155–1168, 2001, hep-ex/0002020.
[542] C. Giunti and M. Tanimoto, *Phys. Rev.*, D66, 113006, 2002, hep-ph/0209169.
[543] S. L. Glashow, *Nucl. Phys.*, 22, 579–588, 1961.
[544] S. L. Glashow, J. Iliopoulos, and L. Maiani, *Phys. Rev.*, D2, 1285–1292, 1970.
[545] N. Y. Gnedin and A. J. S. Hamilton, *Mon. Not. Roy. Astron. Soc.*, 334, 107–116, 2002, astro-ph/0111194.
[546] M. Goeppert-Mayer, *Phys. Rev.*, 48, 512–516, 1935.
[547] H. Goldberg, *Phys. Rev. Lett.*, 50, 1419, 1983.
[548] M. L. Goldberger and S. B. Treiman, *Phys. Rev.*, 111, 354–361, 1958.
[549] M. Goldhaber, L. Grodzins, and A. W. Sunyar, *Phys. Rev.*, 109, 1015–1017, 1958.
[550] I. Goldman, Y. Aharonov, G. Alexander, and S. Nussinov, *Phys. Rev. Lett.*, 60, 1789, 1988.
[551] E. Golowich and P. B. Pal, *Phys. Rev.*, D41, 3537–3540, 1990.
[552] J. J. Gomez-Cadenas et al., 2001, hep-ph/0105297, 9th International Symposium on Neutrino Telescopes, Venice, Italy, 6–9 Mar 2001.
[553] P. Gondolo, G. Ingelman, and M. Thunman, *Astropart. Phys.*, 5, 309–332, 1996, hep-ph/9505417.
[554] M. C. Gonzalez-Garcia, *Phys. Scripta*, T121, 72, 2005, hep-ph/0410030, Nobel Symposium on Neutrino Physics, Haga Slott, Enkoping, Sweden.
[555] M. C. Gonzalez-Garcia and C. Pena-Garay, *Phys. Rev.*, D68, 093003, 2003, hep-ph/0306001.
[556] M. Gonzalez-Garcia and Y. Nir, *Rev. Mod. Phys.*, 75, 345–402, 2003, hep-ph/0202058.
[557] M. Goodman, *Nucl. Phys. Proc. Suppl.*, 145, 186, 2005, hep-ph/0501206, 2004 Neutrino Oscillation Workshop, Otranto Italy.
[558] S. Goswami and A. Y. Smirnov, *Phys. Rev.*, D72, 053011, 2005, hep-ph/0411359.
[559] H. Grassler et al., *Nucl. Phys.*, B273, 253, 1986.
[560] G. J. M. Graves et al., *Astrophys. J.*, 629, 944, 2005, astro-ph/0505066.
[561] D. A. Green and F. R. Stephenson, 2003, astro-ph/0301603.
[562] O. W. Greenberg, *Phys. Rev.*, D32, 1841, 1985.
[563] Z. D. Greenwood et al., *Phys. Rev.*, D53, 6054–6064, 1996.
[564] K. Greisen, *Ann. Rev. Nucl. Part. Sci.*, 10, 63–108, 1960.
[565] K. Greisen, *Phys. Rev. Lett.*, 16, 748–750, 1966.
[566] N. Grevesse and A. J. Sauval, *Space Sci. Rev.*, 85, 161–174, 1998.
[567] V. N. Gribov and B. Pontecorvo, *Phys. Lett.*, B28, 493, 1969.
[568] K. Griest and M. Kamionkowski, *Phys. Rev. Lett.*, 64, 615, 1990.
[569] J. A. Grifols and E. Masso, *Phys. Rev.*, D40, 3819, 1989.
[570] J. A. Grifols and E. Masso, *Phys. Lett.*, B242, 77, 1990.
[571] M. Gronau, A. Kfir, and R. Loewy, *Phys. Rev. Lett.*, 56, 1538, 1986.

[572] D. J. Gross and R. Jackiw, *Phys. Rev.*, D6, 477–493, 1972.
[573] D. J. Gross and C. H. Llewellyn Smith, *Nucl. Phys.*, B14, 337–347, 1969.
[574] Y. Grossman and M. P. Worah, 1998, hep-ph/9807511.
[575] Y. Grossman and H. J. Lipkin, *Phys. Rev.*, D55, 2760–2767, 1997, hep-ph/9607201.
[576] A. Guglielmi, M. Mezzetto, P. Migliozzi, and F. Terranova, 2005, hep-ph/0508034.
[577] M. Guler et al., 2000, CERN-SPSC-2000-028.
[578] G. S. Guralnik, C. R. Hagen, and T. W. B. Kibble, *Phys. Rev. Lett.*, 13, 585–587, 1964.
[579] A. H. Guth, *Phys. Rev.*, D23, 347–356, 1981.
[580] M. M. Guzzo, A. Masiero, and S. T. Petcov, *Phys. Lett.*, B260, 154–160, 1991.
[581] T. J. Haines et al., *Phys. Rev. Lett.*, 57, 1986–1989, 1986.
[582] N. W. Halverson et al., *Astrophys. J.*, 568, 38–45, 2002, astro-ph/0104489.
[583] F. Halzen, *Eur. Phys. J.*, C46, 669–687, 2006, astro-ph/0602132.
[584] F. Halzen and D. Hooper, *Rept. Prog. Phys.*, 65, 1025–1078, 2002, astro-ph/0204527.
[585] W. Hampel et al., *Phys. Lett.*, B388, 384–396, 1996.
[586] W. Hampel et al., *Phys. Lett.*, B420, 114–126, 1998.
[587] W. Hampel et al., *Phys. Lett.*, B436, 158–173, 1998.
[588] W. Hampel et al., *Phys. Lett.*, B447, 127–133, 1999.
[589] C. Hamzaoui, *Phys. Rev. Lett.*, 61, 35, 1988.
[590] T. Han and D. Hooper, *New J. Phys.*, 6, 150, 2004, hep-ph/0408348.
[591] S. Hannestad, *Phys. Rev.*, D67, 085017, 2003, astro-ph/0211106.
[592] S. Hannestad, *JCAP*, 0305, 004, 2003, astro-ph/0303076.
[593] S. Hannestad, *Phys. Rev. Lett.*, 85, 4203–4206, 2000, astro-ph/0005018.
[594] S. Hannestad, *Phys. Rev.*, D64, 083002, 2001, astro-ph/0105220.
[595] S. Hannestad, *New J. Phys.*, 6, 108, 2004, hep-ph/0404239.
[596] S. Hannestad, *Nucl. Phys. Proc. Suppl.*, 145, 313, 2005, hep-ph/0412181, NOW2004 workshop, Conca Specchiulla, Italy, September 11-17, 2004.
[597] S. H. Hansen, G. Mangano, A. Melchiorri, G. Miele, and O. Pisanti, *Phys. Rev.*, D65, 023511, 2002, astro-ph/0105385.
[598] E. R. Harrison, *Phys. Rev.*, D1, 2726–2730, 1970.
[599] F. J. Hasert et al., *Phys. Lett.*, B46, 138–140, 1973.
[600] F. J. Hasert et al., *Phys. Lett.*, B46, 121–124, 1973.
[601] F. J. Hasert et al., *Nucl. Phys.*, B73, 1, 1974.
[602] G. Hasinger, B. Aschenbach, and J. Trumper, *Astron. Astrophys.*, 312, L9–L12, 1996, astro-ph/9606149.
[603] S. Hatakeyama et al., *Phys. Rev. Lett.*, 81, 2016–2019, 1998, hep-ex/9806038.
[604] W. C. Haxton, *Phys. Rev.*, D35, 2352, 1987.
[605] W. C. Haxton and G. J. Stephenson, *Prog. Part. Nucl. Phys.*, 12, 409–479, 1984.
[606] X. G. He, G. C. Joshi, and R. R. Volkas, *Phys. Rev.*, D41, 278–280, 1990.

[607] A. Heger, C. L. Fryer, S. E. Woosley, N. Langer, and D. H. Hartmann, *Astrophys. J.*, 591, 288–300, 2003, astro-ph/0212469.
[608] S. W. Herb et al., *Phys. Rev. Lett.*, 39, 252–255, 1977.
[609] K. H. Hiddemann, H. Daniel, and O. Schwentker, *J. Phys.*, G21, 639–650, 1995.
[610] P. W. Higgs, *Phys. Rev. Lett.*, 13, 508–509, 1964.
[611] P. W. Higgs, *Phys. Lett.*, 12, 132–133, 1964.
[612] P. W. Higgs, *Phys. Rev.*, 145, 1156–1163, 1966.
[613] K. Hirata et al., *Phys. Rev. Lett.*, 58, 1490–1493, 1987.
[614] K. S. Hirata et al., *Phys. Lett.*, B205, 416, 1988.
[615] K. S. Hirata et al., *Phys. Rev.*, D38, 448–458, 1988.
[616] K. S. Hirata et al., *Phys. Rev. Lett.*, 63, 16, 1989.
[617] K. S. Hirata et al., *Phys. Rev. Lett.*, 65, 1301–1304, 1990.
[618] K. S. Hirata et al., *Phys. Rev. Lett.*, 65, 1297–1300, 1990.
[619] K. S. Hirata et al., *Phys. Rev.*, D44, 2241–2260, 1991.
[620] K. S. Hirata et al., *Phys. Rev. Lett.*, 66, 9–12, 1991.
[621] K. S. Hirata et al., *Phys. Lett.*, B280, 146–152, 1992.
[622] E. Holzschuh, *Rept. Prog. Phys.*, 55, 1035–1091, 1992.
[623] M. Honda, T. Kajita, K. Kasahara, and S. Midorikawa, *Phys. Rev.*, D52, 4985–5005, 1995, hep-ph/9503439.
[624] M. Honda, T. Kajita, K. Kasahara, and S. Midorikawa, *Phys. Rev.*, D64, 053011, 2001, hep-ph/0103328.
[625] J. Hosaka et al., *Phys. Rev.*, D73, 112001, 2006, hep-ex/0508053.
[626] W. Hu and S. Dodelson, *Ann. Rev. Astron. Astrophys.*, 40, 171, 2002, astro-ph/0110414.
[627] W. Hu, D. J. Eisenstein, and M. Tegmark, *Phys. Rev. Lett.*, 80, 5255–5258, 1998, astro-ph/9712057.
[628] W. T. Hu, 1995, astro-ph/9508126.
[629] P. Hut, *Phys. Lett.*, B69, 85, 1977.
[630] P. Hut and K. A. Olive, *Phys. Lett.*, B87, 144–146, 1979.
[631] H. Huzita, *Mod. Phys. Lett.*, A2, 905–911, 1987.
[632] M. Ito, T. Kaneko, and M. Nakagawa, *Prog. Theor. Phys.*, 79, 13, 1988.
[633] Y. Itow et al., 2001, hep-ex/0106019, Third Workshop on Neutrino Oscillations and Their Origin (NOON 2001), Kashiwa, Japan, 5–8 Dec 2001.
[634] C. Itzykson and J. B. Zuber, *Quantum Field Theory*, McGraw-Hill, 1980, International Series In Pure and Applied Physics.
[635] Y. I. Izotov and T. X. Thuan, *Astrophys. J.*, 602, 200–230, 2004, astro-ph/0310421.
[636] J. D. Jackson, *Classical Electrodynamics*, John Wiley, 1975, Second Edition.
[637] H.-T. Janka, *Astron. Astrophys.*, 368, 527–560, 2001, astro-ph/0008432.
[638] H.-T. Janka et al., 2002, astro-ph/0212314.
[639] C. Jarlskog, *Z. Phys.*, C29, 491–497, 1985.
[640] C. Jarlskog, *Phys. Rev. Lett.*, 55, 1039, 1985.
[641] C. Jarlskog, *Phys. Rev.*, D36, 2128, 1987.
[642] C. Jarlskog, *Phys. Rev.*, D35, 1685, 1987.
[643] C. Jarlskog and R. Stora, *Phys. Lett.*, B208, 268, 1988.

[644] F. A. Jenkins and H. E. White, *Fundamentals of Optics*, McGraw-Hill, 1981.
[645] C. K. Jung, C. McGrew, T. Kajita, and T. Mann, *Ann. Rev. Nucl. Part. Sci.*, 51, 451–488, 2001.
[646] G. Jungman, M. Kamionkowski, and K. Griest, *Phys. Rep.*, 267, 195–373, 1996, hep-ph/9506380.
[647] M. Kachelriess, *Comptes Rendus Physique*, 5, 441, 2004, hep-ph/0406174.
[648] M. Kachelriess, A. Strumia, R. Tomas, and J. W. F. Valle, *Phys. Rev.*, D65, 073016, 2002, hep-ph/0108100.
[649] M. Kachelriess, R. Tomas, and J. W. F. Valle, *JHEP*, 01, 030, 2001, hep-ph/0012134.
[650] K. Kainulainen and K. A. Olive, 2002, hep-ph/0206163.
[651] K. Kainulainen and A. Sorri, *JHEP*, 02, 020, 2002, hep-ph/0112158.
[652] H. Kaiser, S. A. Werner, and E. A. George, *Phys. Rev. Lett.*, 51, 1106, 1983.
[653] T. Kajita and Y. Totsuka, *Rev. Mod. Phys.*, 73, 85–118, 2001.
[654] K. Kanaya, *Prog. Theor. Phys.*, 64, 2278, 1980.
[655] T. Kaneko, *Prog. Theor. Phys.*, 78, 532, 1987.
[656] H.-S. Kang and G. Steigman, *Nucl. Phys.*, B372, 494–520, 1992.
[657] D. B. Kaplan and A. Manohar, *Nucl. Phys.*, B310, 527, 1988.
[658] M. Kaplinghat, L. Knox, and Y.-S. Song, *Phys. Rev. Lett.*, 91, 241301, 2003, astro-ph/0303344.
[659] B. Kayser and L. Stodolsky, *Phys. Lett.*, B359, 343–350, 1995.
[660] B. Kayser, *Phys. Rev.*, D24, 110, 1981.
[661] B. Kayser, 1997, hep-ph/9702327, 28th International Conference on High-energy Physics (ICHEP 96), Warsaw, Poland, 25–31 Jul 1996.
[662] B. Kayser, 1997, SLAC-PUB-7123.
[663] E. Kearns, *Nucl. Phys. Proc. Suppl.*, 70, 315–323, 1999, hep-ex/9803007.
[664] M. T. Keil et al., *Astrophys. J.*, 590, 971, 2003, astro-ph/0208035.
[665] M. T. Keil, 2003, astro-ph/0308228.
[666] T. W. B. Kibble, *Phys. Rev.*, 155, 1554–1561, 1967.
[667] K. Kiers, S. Nussinov, and N. Weiss, *Phys. Rev.*, D53, 537–547, 1996, hep-ph/9506271.
[668] K. Kiers and N. Weiss, *Phys. Rev.*, D57, 3091–3105, 1998, hep-ph/9710289.
[669] C. W. Kim, J. Kim, and W. K. Sze, *Phys. Rev.*, D37, 1072, 1988.
[670] C. W. Kim and A. Pevsner, *Neutrinos in Physics and Astrophysics*, Harwood Academic Press, 1993, Contemporary Concepts in Physics, Vol. 8.
[671] C. W. Kim, W. K. Sze, and S. Nussinov, *Phys. Rev.*, D35, 4014, 1987.
[672] J. E. Kim, P. Langacker, M. Levine, and H. H. Williams, *Rev. Mod. Phys.*, 53, 211, 1981.
[673] S. F. King, *Rept. Prog. Phys.*, 67, 107, 2004, hep-ph/0310204.
[674] D. P. Kirilova, *Astropart. Phys.*, 19, 409–417, 2003, astro-ph/0109105.
[675] D. P. Kirilova and M. V. Chizhov, *Phys. Rev.*, D58, 073004, 1998, hep-ph/9707282.
[676] D. P. Kirilova and M. V. Chizhov, *Nucl. Phys.*, B591, 457–468, 2000, hep-ph/9909408.
[677] D. Kirilova, *Int. J. Mod. Phys.*, D13, 831, 2004, hep-ph/0209104.

[678] R. P. Kirshner, G. Sonneborn, D. M. Crenshaw, and G. E. Nassiopoulos, *Astrophys. J.*, 320, 602–608, 1987.

[679] H. V. Klapdor-Kleingrothaus, A. Dietz, H. L. Harney, and I. V. Krivosheina, *Mod. Phys. Lett.*, A16, 2409–2420, 2001, `hep-ph/0201231`.

[680] H. V. Klapdor-Kleingrothaus et al., *Eur. Phys. J.*, A12, 147–154, 2001.

[681] H. Klapdor-Kleingrothaus, I. Krivosheina, A. Dietz, and O. Chkvorets, *Phys. Lett.*, B586, 198–212, 2004, `hep-ph/0404088`.

[682] O. Klein, *Nature*, 161, 897, 1948.

[683] J. P. Kneller and G. Steigman, *New J. Phys.*, 6, 117, 2004, `astro-ph/0406320`.

[684] R. A. Knop et al., *Astrophys. J.*, 598, 102, 2003, `astro-ph/0309368`.

[685] M. Kobayashi and T. Maskawa, *Prog. Theor. Phys.*, 49, 652–657, 1973.

[686] M. Kobayashi and Y. Fukao, *Geophys. Res. Lett.*, 18, 633–636, 1991.

[687] I. Y. Kobzarev, B. V. Martemyanov, L. B. Okun, and M. G. Shchepkin, *Sov. J. Nucl. Phys.*, 32, 823, 1980.

[688] K. Kohri, M. Kawasaki, and K. Sato, *Astrophys. J.*, 490, 72–75, 1997, `astro-ph/9612237`.

[689] E. W. Kolb and M. S. Turner, *The Early Universe*, Addison-Wesley, 1990, Frontiers in Physics, 69.

[690] E. W. Kolb and K. A. Olive, *Phys. Rev.*, D33, 1202, 1986.

[691] M. Komatsu, P. Migliozzi, and F. Terranova, *J. Phys.*, G29, 443, 2003, `hep-ph/0210043`.

[692] O. C. W. Kong, *Phys. Rev.*, D55, 383–396, 1997, `hep-ph/9608246`.

[693] E. J. Konopinski, *Rev. Mod. Phys.*, 27, 254–257, 1955.

[694] S. E. Kopp, 2005, `physics/0508001`, U.S. Particle Accelerator Conference, Knoxville, TN (May 2005).

[695] M. Koshiba, *Phys. Rep.*, 220, 229–381, 1992.

[696] K. Kotake, K. Sato, and K. Takahashi, *Rept. Prog. Phys.*, 69, 971, 2006, `astro-ph/0509456`.

[697] P. I. Krastev and S. T. Petcov, *Phys. Lett.*, B207, 64, 1988, [Erratum-ibid. B214, 661 (1988)].

[698] P. I. Krastev and A. Y. Smirnov, *Phys. Lett.*, B226, 341–346, 1989.

[699] C. Kraus et al., *Eur. Phys. J.*, 40, 447, 2005, `hep-ex/0412056`.

[700] L. M. Krauss, *Phys. Lett.*, B128, 37, 1983.

[701] L. M. Krauss, *Nature*, 329, 689–694, 1987.

[702] L. M. Krauss, *Phys. Rep.*, 333, 33–45, 2000, `astro-ph/9907308`.

[703] L. M. Krauss, S. L. Glashow, and D. N. Schramm, *Nature*, 310, 191–198, 1984.

[704] M. R. Krishnaswamy et al., *Proc. R. Soc. London*, A323, 489, 1971.

[705] M. R. Krishnaswamy et al., *Phys. Lett.*, B57, 105, 1975.

[706] C. Kuo et al., *Astrophys. J.*, 600, 32, 2004, `astro-ph/0212289`.

[707] T. K. Kuo and J. Pantaleone, *Rev. Mod. Phys.*, 61, 937, 1989.

[708] T. K. Kuo and J. Pantaleone, *Phys. Rev.*, D39, 1930, 1989.

[709] V. A. Kuzmin, *Sov. Phys. JETP*, 22, 1051, 1966.

[710] H. Kwon et al., *Phys. Rev.*, D24, 1097–1111, 1981.

[711] L. Landau, *Nucl. Phys.*, 3, 127, 1957.

[712] L. D. Landau, *Phys. Z. Sowjetunion*, 1, 285, 1932.
[713] P. Langacker, S. T. Petcov, G. Steigman, and S. Toshev, *Nucl. Phys.*, B282, 589, 1987.
[714] P. Langacker, *Phys. Rep.*, 72, 185, 1981.
[715] P. Langacker, J. P. Leveille, and J. Sheiman, *Phys. Rev.*, D27, 1228, 1983.
[716] P. Langacker and J. Liu, *Phys. Rev.*, D46, 4140–4160, 1992, hep-ph/9206209.
[717] L. M. Langer and R. J. D. Moffat, *Phys. Rev.*, 88, 689–694, 1952.
[718] T. Lasserre, *Nucl. Phys. Proc. Suppl.*, 149, 29, 2005, hep-ex/0411083, Sixth International Workshop on Neutrino Factories & Superbeams, July 26-Aug 1, 2004, Osaka, Japan.
[719] J. M. Lattimer and J. Cooperstein, *Phys. Rev. Lett.*, 61, 23–26, 1988.
[720] E. Leader and E. Predazzi, *Camb. Monogr. Part. Phys. Nucl. Phys. Cosmol.*, 3, 1, 1996.
[721] E. Leader and E. Predazzi, *Camb. Monogr. Part. Phys. Nucl. Phys. Cosmol.*, 4, 1–431, 1996.
[722] A. T. Lee et al., *Astrophys. J.*, 561, L1–L6, 2001, astro-ph/0104459.
[723] B. W. Lee and S. Weinberg, *Phys. Rev. Lett.*, 39, 165–168, 1977.
[724] F.-F. Lee and G.-L. Lin, *Astropart. Phys.*, 25, 64, 2006, hep-ph/0412383.
[725] T. D. Lee, *Symmetries, Asymmetries, and the World of Particles*, University of Washington Press, 1988.
[726] T. D. Lee, M. Rosenbluth, and C.-N. Yang, *Phys. Rev.*, 75, 905, 1949.
[727] T. D. Lee and C. N. Yang, *Phys. Rev.*, 105, 1671, 1957.
[728] T. D. Lee and C.-N. Yang, *Phys. Rev.*, 104, 254–258, 1956.
[729] S. D. Leo, G. Ducati, and P. Rotelli, *Mod. Phys. Lett.*, A15, 2057–2068, 2000, hep-ph/9906460.
[730] J. Lesgourgues, S. Pastor, and L. Perotto, *Phys. Rev.*, D70, 045016, 2004, hep-ph/0403296.
[731] R. R. Lewis, *Phys. Rev.*, D21, 663, 1980.
[732] R. R. Lewis, *Phys. Rev.*, D21, 663, 1980.
[733] M. Liebendoerfer, 2004, astro-ph/0405029, 12th Workshop on "Nuclear Astrophysics", Ringberg Castle, March 22-27, 2004.
[734] M. Liebendoerfer, M. Rampp, H.-T. Janka, and A. Mezzacappa, *Astrophys. J.*, 620, 840, 2005, astro-ph/0310662.
[735] M. Liebendorfer et al., *Phys. Rev.*, D63, 103004, 2001, astro-ph/0006418.
[736] P. Lipari, *Astropart. Phys.*, 1, 195–227, 1993.
[737] P. Lipari, *Astropart. Phys.*, 14, 171–188, 2000, hep-ph/0003013.
[738] P. Lipari, *Astropart. Phys.*, 14, 153–170, 2000, hep-ph/0002282.
[739] H. J. Lipkin, *Phys. Lett.*, B477, 195, 2000.
[740] H. J. Lipkin, *Phys. Lett.*, B348, 604–608, 1995, hep-ph/9501269.
[741] H. J. Lipkin, 1999, hep-ph/9901399.
[742] H. J. Lipkin, *Phys. Lett.*, B579, 355–360, 2004, hep-ph/0304187.
[743] E. Lisi, A. Marrone, and D. Montanino, *Phys. Rev. Lett.*, 85, 1166–1169, 2000, hep-ph/0002053.
[744] E. Lisi, A. Marrone, D. Montanino, A. Palazzo, and S. T. Petcov, *Phys. Rev.*, D63, 093002, 2001, hep-ph/0011306.

[745] P. J. Litchfield, *Nucl. Instrum. Meth.*, A451, 187–191, 2000.
[746] D. W. Liu et al., *Phys. Rev. Lett.*, 93, 021802, 2004, hep-ex/0402015.
[747] J. Liu, *Phys. Rev.*, D45, 1428–1431, 1992.
[748] Q. Y. Liu, S. P. Mikheyev, and A. Y. Smirnov, *Phys. Lett.*, B440, 319–326, 1998, hep-ph/9803415.
[749] Y. Liu, L. Derome, and M. Buenerd, *Phys. Rev.*, D67, 073022, 2003, astro-ph/0211632.
[750] C. H. Llewellyn Smith, *Phys. Rep.*, 3, 261, 1972.
[751] V. M. Lobashev et al., *Phys. Lett.*, B460, 227–235, 1999.
[752] R. E. Lopez, S. Dodelson, A. Heckler, and M. S. Turner, *Phys. Rev. Lett.*, 82, 3952–3955, 1999, astro-ph/9803095.
[753] Lord Rayleigh, *Phil. Mag.*, 27, 298, 1889.
[754] T. J. Loredo and D. Q. Lamb, *Phys. Rev.*, D65, 063002, 2002, astro-ph/0107260.
[755] R. Loudon, *The Quantum Theory of Light*, Oxford University Press, 1992.
[756] J. Lowe et al., *Phys. Lett.*, B384, 288–292, 1996, hep-ph/9605234.
[757] C. Lunardini and A. Y. Smirnov, *Phys. Rev.*, D64, 073006, 2001, hep-ph/0012056.
[758] C. Lunardini and A. Y. Smirnov, *Phys. Rev.*, D63, 073009, 2001, hep-ph/0009356.
[759] C. Lunardini and A. Y. Smirnov, *Astropart. Phys.*, 21, 703, 2004, hep-ph/0402128.
[760] V. Luridiana, A. Peimbert, M. Peimbert, and M. Cervino, *Astrophys. J.*, 592, 846, 2003, astro-ph/0304152.
[761] B. W. Lynn, *Phys. Rev.*, D23, 2151, 1981.
[762] D. H. Lyth and A. Riotto, *Phys. Rep.*, 314, 1–146, 1999, hep-ph/9807278.
[763] E. Ma, *New J. Phys.*, 6, 104, 2004, hep-ph/0405152.
[764] C. J. MacTavish et al., *Astrophys. J.*, 647, 799, 2006, astro-ph/0507503.
[765] E. Majorana, *Nuovo Cim.*, 14, 171–184, 1937.
[766] Z. Maki, M. Nakagawa, and S. Sakata, *Prog. Theor. Phys.*, 28, 870, 1962.
[767] R. A. Malaney and G. J. Mathews, *Phys. Rep.*, 229, 145–219, 1993.
[768] M. Malek et al., *Phys. Rev. Lett.*, 90, 061101, 2003, hep-ex/0209028.
[769] M. Maltoni, T. Schwetz, M. Tortola, and J. Valle, *New J. Phys.*, 6, 122, 2004, hep-ph/0405172.
[770] R. N. Manchester et al., *Astrophys. J.*, 628, L131, 2005, astro-ph/0506475.
[771] G. Mangano, G. Miele, S. Pastor, and M. Peloso, *Phys. Lett.*, B534, 8–16, 2002, astro-ph/0111408.
[772] F. Mannucci et al., *Astron. Astrophys.*, 433, 807–814, 2005, astro-ph/0411450.
[773] F. Mantovani, L. Carmignani, G. Fiorentini, and M. Lissia, *Phys. Rev.*, D69, 013001, 2004, hep-ph/0309013.
[774] W. J. Marciano and Z. Parsa, *J. Phys.*, G29, 2629, 2003, hep-ph/0403168.
[775] W. J. Marciano and A. Sirlin, *Phys. Rev. Lett.*, 71, 3629–3632, 1993.
[776] M. Maris and S. T. Petcov, *Phys. Lett.*, B457, 319, 1999, hep-ph/9903303.
[777] M. A. Markov and I. M. Zheleznykh, *Nucl. Phys.*, 27, 385–394, 1961.

[778] R. E. Marshak, Riazuddin, and C. P. Ryan, *Theory of Weak Interactions in Particle Physics*, Wiley-Interscience, 1969.

[779] G. Marx, *Czech. J. Phys.*, B19, 1471–1479, 1969.

[780] S. Masi et al., 2005, `astro-ph/0507509`.

[781] I. Masina, *Int. J. Mod. Phys.*, A16, 5101–5200, 2001, `hep-ph/0107220`.

[782] J. C. Mather et al., *Astrophys. J.*, 354, L37–L40, 1990.

[783] J. C. Mather, D. J. Fixsen, R. A. Shafer, C. Mosier, and D. T. Wilkinson, *Astrophys. J.*, 512, 511–520, 1999, `astro-ph/9810373`.

[784] Y. X. Max Tegmark, Andrew J. S. Hamilton, *Mon. Not. Roy. Astron. Soc.*, 335, 887, 2002, `astro-ph/0111575`.

[785] A. B. McDonald et al., *Rev. Sci. Instrum.*, 75, 293, 2004, `astro-ph/0311343`.

[786] P. McDonald et al., *Astrophys. J.*, 543, 1–23, 2000, `astro-ph/9911196`.

[787] P. McDonald et al., *Astrophys. J.*, 562, 52–75, 2001, `astro-ph/0005553`.

[788] P. McDonald et al., *Astrophys. J. Suppl.*, 163, 80, 2006, `astro-ph/0405013`.

[789] K. S. McFarland et al., *Phys. Rev. Lett.*, 75, 3993–3996, 1995, `hep-ex/9506007`.

[790] B. H. J. McKellar, *Phys. Lett.*, B97, 93, 1980.

[791] D. N. McKinsey and K. J. Coakley, *Astropart. Phys.*, 22, 355, 2005, `astro-ph/0402007`.

[792] P. Meszaros, *Astron. Astrophys.*, 37, 225–228, 1974.

[793] A. Mezzacappa, 2000, `astro-ph/0010580`, Nuclei in the Cosmos 2000, University of Aarhus, Aarhus, Denmark, June 27-July 1, 2000.

[794] A. Mezzacappa et al., 1996, `astro-ph/9612107`.

[795] A. Mezzacappa, 2004, `astro-ph/0410085`, Supernovae as Cosmological Lighthouses, Padua, Italy, June 16-19, 2004.

[796] M. Mezzetto, *Nucl. Phys. Proc. Suppl.*, 143, 309, 2005, `hep-ex/0410083`, 21st International Conference on Neutrino Physics and Astrophysics (Neutrino 2004), Paris, France, 14–19 June 2004.

[797] M.Honda, T.Kajita, K.Kasahara, and S.Midorikawa, *Phys. Rev.*, D70, 043008, 2004, `astro-ph/0404457`.

[798] A. A. Michelson, *Astrophys. J.*, 2, 251, 1895.

[799] J. Middleditch et al., 2000, `astro-ph/0010044`, 5th CTIO/ESO Workshop and 1st CTIO/ESO/LCO Workshop: SN 1987A: Ten Years After, La Serena, Chile, 22-28 Feb 1997.

[800] L. A. Mikaelyan, *Phys. Atom. Nucl.*, 65, 1173, 2002, `hep-ph/0210047`.

[801] S. P. Mikheev and A. Y. Smirnov, *Sov. J. Nucl. Phys.*, 42, 913–917, 1985.

[802] S. P. Mikheev and A. Y. Smirnov, *Nuovo Cim.*, C9, 17–26, 1986.

[803] S. P. Mikheev and A. Y. Smirnov, *Sov. Phys. JETP*, 65, 230–236, 1987.

[804] S. P. Mikheev and A. Y. Smirnov, *Sov. Phys. Usp.*, 30, 759–790, 1987.

[805] J. A. Minahan, P. Ramond, and R. C. Warner, *Phys. Rev.*, D41, 715, 1990.

[806] H. Minakata and H. Nunokawa, *Phys. Lett.*, B504, 301–308, 2001, `hep-ph/0010240`.

[807] P. Minkowski, *Phys. Lett.*, B67, 421, 1977.

[808] C. Misner, K. Thorne, and J. Wheeler, *Gravitation*, W.H. Freeman and Company, 1973.

[809] E. Mocchiutti et al., 2003, 28th International Cosmic Ray Conferences (ICRC 2003), Tsukuba, Japan, 31 Jul – 7 Aug 2003.

[810] R. N. Mohapatra, *New J. Phys.*, 6, 82, 2004.

[811] R. N. Mohapatra, 2004, hep-ph/0412379, SEESAW25, Paris, June, 2004.

[812] R. N. Mohapatra and P. B. Pal, *Massive Neutrinos in Physics and Astrophysics*, World Scientific, 1998, Lecture Notes in Physics, Vol. 60.

[813] R. N. Mohapatra and A. Y. Smirnov, *J. Phys. Conf. Ser.*, 53, 44–82, 2006, hep-ph/0603118.

[814] R. N. Mohapatra and G. Senjanovic, *Phys. Rev. Lett.*, 44, 912, 1980.

[815] R. Mohapatra et al., 2004, hep-ph/0412099.

[816] M. J. Montes, Supernova Taxonomy: http://rsd-www.nrl.navy.mil/7212/montes/snetax.html.

[817] J. A. Morgan, *Phys. Lett.*, B102, 247–250, 1981.

[818] R. A. Muller et al., *Astrophys. J.*, 384, L9–L13, 1992.

[819] F. D. Murnaghan, *The Unitary and Rotation Groups*, Spartan Articles, 1962.

[820] K. Muto, E. Bender, and H. V. Klapdor, *Z. Phys.*, A334, 177–186, 1989.

[821] O. Nachtmann, *ELementary Particle Physics: Concepts and Phenomena*, Springer-Verlag, 1990.

[822] M. Nakahata et al., *J. Phys. Soc. Jap.*, 55, 3786, 1986.

[823] D. Naples et al., *Phys. Rev.*, D59, 031101, 1999, hep-ex/9809023.

[824] M. Nauenberg, *Phys. Lett.*, B447, 23–30, 1999, hep-ph/9812441.

[825] S. H. Neddermeyer and C. D. Anderson, *Phys. Rev.*, 51, 884–886, 1937.

[826] M. M. Nieto, *Hyperfine Interact.*, 100, 193, 1996, hep-ph/9509370.

[827] D. Notzold, *Phys. Rev.*, D36, 1625, 1987.

[828] D. Notzold and G. Raffelt, *Nucl. Phys.*, B307, 924, 1988.

[829] S. F. Novaes, 2000, hep-ph/0001283, 10th Jorge Andre Swieca Summer School: Particle and Fields, Sao Paulo, Brazil, 31 Jan – 12 Feb 1999.

[830] H. Nunokawa, W. J. C. Teves, and R. Z. Funchal, *JHEP*, 0311, 020, 2003, hep-ph/0308175.

[831] S. Nussinov, *Phys. Lett.*, B63, 201–203, 1976.

[832] H. Ohanian, *Gravitation and Spacetime*, W.W. Norton & Company, 1976.

[833] N. Okada and O. Yasuda, *Int. J. Mod. Phys.*, A12, 3669–3694, 1997, hep-ph/9606411.

[834] K. A. Olive, 2005, astro-ph/0503065, TASI 2004.

[835] K. A. Olive, D. N. Schramm, D. Thomas, and T. P. Walker, *Phys. Lett.*, B265, 239–244, 1991.

[836] K. A. Olive and E. D. Skillman, *Astrophys. J.*, 617, 29, 2004, astro-ph/0405588.

[837] K. A. Olive, G. Steigman, and T. P. Walker, *Phys. Rep.*, 333, 389–407, 2000, astro-ph/9905320.

[838] R. Opher, *Astron. Astrophys.*, 37, 135–137, 1974.

[839] A. Osipowicz et al., 2001, hep-ex/0109033.

[840] Y. Oyama et al., *Phys. Rev.*, D39, 1481, 1989.

[841] B. R. P. J. E. Peebles, *Rev. Mod. Phys.*, 75, 599, 2003, astro-ph/0207347.

[842] Padova – Asiago Supernova Group, Asiago Supernova Catalogue: `http://web.pd.astro.it/supern/snean.txt`.
[843] S. Pakvasa, *Nuovo Cim. Lett.*, 31, 497, 1981.
[844] P. B. Pal and T. N. Pham, *Phys. Rev.*, D40, 259, 1989.
[845] J. Pantaleone, *Phys. Rev.*, D46, 510–523, 1992.
[846] G. Pantis, F. Simkovic, J. D. Vergados, and A. Faessler, *Phys. Rev.*, C53, 695–707, 1996, `nucl-th/9612036`.
[847] A. Para and M. Szleper, 2001, `hep-ex/0110032`.
[848] S. Park et al., *Adv. Space Res.*, 35, 991–995, 2005, `astro-ph/0501561`.
[849] S. Park et al., 2006, `astro-ph/0604201`.
[850] S. Park, S. A. Zhekov, D. N. Burrows, and R. McCray, *Astrophys. J.*, 634, L73, 2005, `astro-ph/0510442`.
[851] S. J. Parke, *Phys. Rev. Lett.*, 57, 1275–1278, 1986.
[852] E. A. Paschos and L. Wolfenstein, *Phys. Rev.*, D7, 91–95, 1973.
[853] L. Pasquali, M. H. Reno, and I. Sarcevic, *Phys. Rev.*, D59, 034020, 1999, `hep-ph/9806428`.
[854] W. Pauli, *General Principles of Quantum Mechanics*, Springer-Verlag, 1980, English translation of W. Pauli, "Allgemeine prinzipien der wellenmechanik", volume XXIV, Part 1 of Handbuch der Physik, 1933, Springer, Berlin, Germany.
[855] W. Pauli, *Cambridge Monogr. Part. Phys. Nucl. Phys. Cosmol.*, 14, 1–22, 2000.
[856] J. A. Peacock, *Cosmological Physics*, Cambridge University Press, 1999.
[857] P. J. E. Peebles, *Phys. Rev. Lett.*, 16, 410–413, 1966.
[858] P. J. E. Peebles, *Principles of Physical Cosmology*, Princeton University Press, 1993.
[859] P. J. E. Peebles and J. T. Yu, *Astrophys. J.*, 162, 815–836, 1970.
[860] A. A. Penzias and R. W. Wilson, *Astrophys. J.*, 142, 419–421, 1965.
[861] W. J. Percival et al., *Mon. Not. Roy. Astron. Soc.*, 327, 1297, 2001, `astro-ph/0105252`.
[862] M. L. Perl et al., *Phys. Rev. Lett.*, 35, 1489–1492, 1975.
[863] S. Perlmutter et al., *Astrophys. J.*, 517, 565–586, 1999, `astro-ph/9812133`.
[864] S. Perlmutter and B. P. Schmidt, 2003, `astro-ph/0303428`.
[865] F. Perrin, *Comptes Rendues*, 197, 1625, 1933.
[866] M. E. Peskin, 1994, `hep-ph/9405255`.
[867] H. Pessard et al., 2005, `hep-ex/0504033`, 32nd International Conference on High-Energy Physics (ICHEP 04), Beijing, China, 16–22 Aug 2004.
[868] S. T. Petcov, *Phys. Lett.*, B191, 299, 1987.
[869] S. T. Petcov, *Phys. Lett.*, B200, 373–379, 1988.
[870] S. T. Petcov, *Phys. Lett.*, B214, 139, 1988.
[871] S. T. Petcov, *Phys. Lett.*, B434, 321–332, 1998, `hep-ph/9805262`.
[872] M. M. Phillips, *Astrophys. J.*, 413, L105–L108, 1993.
[873] E. Pierpaoli, *Mon. Not. Roy. Astron. Soc.*, 342, L63, 2003, `astro-ph/0302465`.
[874] T. Piran, *Phys. Lett.*, B102, 299–302, 1981.
[875] P. Pizzochero, *Phys. Rev.*, D36, 2293–2296, 1987.

[876] V. Plyaskin, *Phys. Lett.*, B516, 213–235, 2001, hep-ph/0103286.

[877] F. Polcaro and A. Martocchia, 2005, astro-ph/0511187, IAU Symposium no.230, "Populations of High Energy Sources in Galaxies", Dublin (Ireland), August 15-19, 2005.

[878] B. Pontecorvo, 1946, Chalk River Report PD 205.

[879] B. Pontecorvo, *Phys. Rev.*, 72, 246, 1947.

[880] B. Pontecorvo, *Sov. Phys. JETP*, 6, 429, 1957.

[881] B. Pontecorvo, *Sov. Phys. JETP*, 7, 172–173, 1958.

[882] B. Pontecorvo, *Sov. Phys. JETP*, 10, 1236–1240, 1960.

[883] B. Pontecorvo, *Sov. Phys. JETP*, 26, 984–988, 1968.

[884] J. Pulido, *Phys. Rep.*, 211, 167–199, 1992.

[885] G. Puppi, *Lett. Nuovo Cim.*, 5, 587–588, 1948.

[886] G. Raffelt and D. Seckel, *Phys. Rev. Lett.*, 60, 1793, 1988.

[887] G. G. Raffelt, *Nucl. Phys. Proc. Suppl.*, 110, 254–267, 2002, hep-ph/0201099, TAUP 2001: Topics in Astroparticle and Underground Physics, Assergi, Italy, 8–12 Sep 2001.

[888] R. S. Raghavan et al., *Phys. Rev. Lett.*, 80, 635–638, 1998.

[889] M. Rampp and H. T. Janka, *Astrophys. J.*, 539, L33–L36, 2000, astro-ph/0005438.

[890] K. U. Ratnatunga and S. van den Bergh, *Astrophys. J.*, 343, 713–717, 1989.

[891] M. Rauch, *Ann. Rev. Astron. Astrophys.*, 36, 267–31, 1998, astro-ph/9806286.

[892] H. Ray et al., *Int. J. Mod. Phys.*, A20, 3062, 2005, hep-ex/0411022, Meeting of the Division of Particles and Fields (DPF2004).

[893] A. C. S. Readhead et al., *Astrophys. J.*, 609, 498–512, 2004, astro-ph/0402359.

[894] B. C. Reed, *Astronomical J.*, 130, 1652–1657, 2005, astro-ph/0506708.

[895] H. Reeves, *Phys. Rev.*, D6, 3363–3368, 1972.

[896] H. Reeves, J. Audouze, W. A. Fowler, and D. N. Schramm, *Astrophys. J.*, 179, 909–930, 1973.

[897] F. Reines, *Rev. Mod. Phys.*, 68, 317–327, 1996.

[898] F. Reines and C. L. Cowan, *Phys. Rev.*, 90, 492–493, 1953.

[899] F. Reines, C. L. Cowan, F. B. Harrison, A. D. McGuire, and H. W. Kruse, *Phys. Rev.*, 117, 159–173, 1960.

[900] F. Reines et al., *Phys. Rev. Lett.*, 15, 429–433, 1965.

[901] F. Reines et al., *Phys. Rev.*, D4, 80–98, 1971.

[902] P. Renton, *Electroweak Interactions: an Introduction to the Physics of Quarks and Leptons*, Cambridge University Press, 1990.

[903] A. G. Riess et al., *Astron. J.*, 116, 1009–1038, 1998, astro-ph/9805201.

[904] A. G. Riess et al., *Astrophys. J.*, 560, 49–71, 2001, astro-ph/0104455.

[905] A. G. Riess et al., *Astrophys. J.*, 607, 665, 2004, astro-ph/0402512.

[906] A. Ringwald, 2005, hep-ph/0505024, XI International Workshop on "Neutrino Telescopes", Feb 22-25, 2005, Venice, Italy.

[907] A. Ringwald and Y. Y. Y. Wong, *JCAP*, 0412, 005, 2004, hep-ph/0408241.

[908] V. A. Rodin, A. Faessler, F. Simkovic, and P. Vogel, *Phys. Rev.*, C68, 044302, 2003, nucl-th/0305005.

[909] C. E. Rolfs and W. S. Rodney, *Cauldrons in the Cosmos*, The University of Chicago Press, 1988.

[910] A. Romosan et al., *Phys. Rev. Lett.*, 78, 2912–2915, 1997, hep-ex/9611013.

[911] M. Roos, *Introduction to Cosmology*, John Wiley, 1994.

[912] S. P. Rosen, *Phys. Rev.*, D37, 1682, 1988.

[913] C. G. Rothschild, M. C. Chen, and F. P. Calaprice, *Geophys. Res. Lett.*, 25, 1083, 1998, nucl-ex/9710001.

[914] M. Rowan-Robinson, *The Cosmological Distance Ladder: Distance and Time in the Universe*, W.H. Freeman & Company, 1985.

[915] S. Rudaz, *Phys. Rev.*, D41, 2619–2621, 1990.

[916] J. E. Ruhl et al., *Astrophys. J.*, 599, 786, 2003, astro-ph/0212229.

[917] L. H. Ryder, *Quantum Field Theory*, Cambridge University Press, 1985.

[918] J. J. Sakurai, *Nuovo Cimento*, 7, 649, 1958.

[919] A. Salam, *Nuovo Cim.*, 5, 299, 1957.

[920] A. Salam, 1969, Proc. of the 8th Nobel Symposium on 'Elementary Particle Theory, Relativistic Groups and Analyticity', Stockholm, Sweden, 1968, edited by N. Svartholm, p. 367-377.

[921] A. G. Sanchez et al., *Mon. Not. Roy. Astron. Soc.*, 366, 189, 2006, astro-ph/0507583.

[922] M. Sanchez et al., *Phys. Rev.*, D68, 113004, 2003, hep-ex/0307069.

[923] T. Sanuki et al., *Astrophys. J.*, 545, 1135, 2000, astro-ph/0002481.

[924] S. Sarkar, *Rept. Prog. Phys.*, 59, 1493–1610, 1996, hep-ph/9602260.

[925] K. Sasaki, *Z. Phys.*, C32, 149–152, 1986.

[926] E. Sassaroli, Y. N. Srivastava, and A. Widom, 1995, hep-ph/9509261.

[927] K. Sato and M. Kobayashi, *Prog. Theor. Phys.*, 58, 1775, 1977.

[928] K. Sato and H. Suzuki, *Phys. Rev. Lett.*, 58, 2722, 1987.

[929] K. Sato and H. Suzuki, *Phys. Lett.*, B196, 267, 1987.

[930] M. J. Savage, R. A. Malaney, and G. M. Fuller, *Astrophys. J.*, 368, 1–11, 1991.

[931] S. Schael et al., *Phys. Rept.*, 427, 257, 2006, hep-ex/0509008.

[932] J. Schechter and J. W. F. Valle, *Phys. Rev.*, D21, 309, 1980.

[933] J. Schechter and J. W. F. Valle, *Phys. Rev.*, D22, 2227, 1980.

[934] J. Schechter and J. W. F. Valle, *Phys. Rev.*, D25, 774, 1982.

[935] J. Schechter and J. W. F. Valle, *Phys. Rev.*, D25, 2951, 1982.

[936] L. I. Schiff, *Quantum Mechanics*, McGraw-Hill, 1955.

[937] S. Schonert et al., *Nucl. Phys. Proc. Suppl.*, 48, 201–203, 1996.

[938] S. Schonert, *Nucl. Phys. Proc. Suppl.*, 110, 277–287, 2002, hep-ex/0202021, TAUP2001 – Topics in Astroparticle and Underground Physics, LNGS, Italy (September 8-12, 2001).

[939] D. N. Schramm, *Comments Nucl. Part. Phys.*, 17, 239, 1987.

[940] D. N. Schramm and M. S. Turner, *Rev. Mod. Phys.*, 70, 303–318, 1998, astro-ph/9706069.

[941] B. F. Schutz, *A First Course in General Relativity*, Cambridge University Press, 1985.

[942] M. Schwartz, *Phys. Rev. Lett.*, 4, 306–307, 1960.

[943] S. S. Schweber, *An Introduction to Relativistic Quantum Field Theory*, Harper & Row, 1962.

[944] D. Scott and G. Smoot, *Phys. Lett.*, B592, 221, 2004, astro-ph/0406567, The Review of Particle Properties 2004.

[945] U. Seljak et al., *Phys. Rev.*, D71, 043511, 2005, astro-ph/0406594.

[946] U. Seljak and M. Zaldarriaga, CMBFAST: http://www.cmbfast.org/.

[947] U. Seljak et al., *Phys. Rev.*, D71, 103515, 2005, astro-ph/0407372.

[948] U. Seljak, P. McDonald, and A. Makarov, *Mon. Not. Roy. Astron. Soc.*, 342, L79, 2003, astro-ph/0302571.

[949] M. Selvi, *Nucl. Phys. Proc. Suppl.*, 145, 339–342, 2005.

[950] G. Senjanovic, 2005, hep-ph/0501244, SEESAW25: International Conference on the Seesaw Mechanism and the Neutrino Mass, Paris, France, 10–11 June 2004.

[951] R. Sexl and H. Urbantke, *Relativity, Groups, Particles: Special Relativity and Relativistic Symmetry in Field and Particle Physics*, Springer, 2001.

[952] S. L. Shapiro and S. A. Teukolsky, *Black Holes, White Dwarfs, and Neutron Stars: the Physics of Compact Objects*, John Wiley, 1983.

[953] S. Shapiro and I. Wasserman, *Nature*, 289, 657, 1981.

[954] M. K. Sharp, J. F. Beacom, and J. A. Formaggio, *Phys. Rev.*, D66, 013012, 2002, hep-ph/0205035.

[955] X. Shi, G. M. Fuller, and K. Abazajian, *Phys. Rev.*, D60, 063002, 1999, astro-ph/9905259.

[956] X. Shi, D. N. Schramm, and B. D. Fields, *Phys. Rev.*, D48, 2563–2572, 1993, astro-ph/9307027.

[957] X. Shi and D. N. Schramm, *Phys. Lett.*, B283, 305–312, 1992.

[958] R. E. Shrock, *Phys. Lett.*, B96, 159, 1980.

[959] R. E. Shrock, *Phys. Rev.*, D24, 1232, 1981.

[960] R. E. Shrock, *Phys. Rev.*, D24, 1275, 1981.

[961] P. Shtykovskiy, A. Lutovinov, M. Gilfanov, and R. Sunyaev, *Astron. Lett.*, 31, 258, 2005, astro-ph/0411731.

[962] J. L. Sievers et al., *Astrophys. J.*, 591, 599, 2003, astro-ph/0205387.

[963] G. Sigl, *Annals Phys.*, 301, 53–64, 2002, hep-ph/0205051, WHEPP-7 Conference, Allahabad, India (January, 2002).

[964] M. Signore and D. Puy, *New Astron. Rev.*, 45, 409, 2001, astro-ph/0010634.

[965] F. Simkovic, G. Pantis, J. D. Vergados, and A. Faessler, *Phys. Rev.*, C60, 055502, 1999, hep-ph/9905509.

[966] F. Simkovic, J. Schwieger, M. Veselsky, G. Pantis, and A. Faessler, *Phys. Lett.*, B393, 267–273, 1997, nucl-th/9612037.

[967] G. F. Smoot et al., *Astrophys. J.*, 396, L1–L5, 1992.

[968] M. Smy et al., *Phys. Rev.*, D69, 011104, 2004, hep-ex/0309011.

[969] G. Sonneborn, B. Altner, and R. P. Kirshner, *Astrophys. J.*, 323, L35–L39, 1987.

[970] D. N. Spergel et al., *Astrophys. J. Supp. Ser.*, 148, 175–194, 2003, astro-ph/0302209.

[971] D. N. Spergel, T. Piran, A. Loeb, J. Goodman, and J. N. Bahcall, *Science*, 237, 1471, 1987.

[972] D. N. Spergel and J. N. Bahcall, *Phys. Lett.*, B200, 366, 1988.

[973] M. Srednicki, R. Watkins, and K. A. Olive, *Nucl. Phys.*, B310, 693, 1988.

[974] Y. N. Srivastava, S. Palit, A. Widom, and E. Sassaroli, 1998, `hep-ph/9807543`.

[975] Y. N. Srivastava and A. Widom, 1997, `hep-ph/9707268`.

[976] Y. N. Srivastava, A. Widom, and E. Sassaroli, *Phys. Lett.*, B344, 436–440, 1995.

[977] Y. N. Srivastava, A. Widom, and E. Sassaroli, *Z. Phys.*, C66, 601–605, 1995.

[978] Y. N. Srivastava, A. Widom, and E. Sassaroli, *Eur. Phys. J.*, C2, 769, 1998.

[979] A. M. Stasto, *Int. J. Mod. Phys.*, A19, 317, 2004, `astro-ph/0310636`.

[980] A. Staudt, K. Muto, and H. V. Klapdor-Kleingrothaus, *Europhys. Lett.*, 13, 31–36, 1990.

[981] G. Steigman, D. N. Schramm, and J. E. Gunn, *Phys. Lett.*, B66, 202–204, 1977.

[982] R. I. Steinberg, 1993, `hep-ph/9306282`.

[983] I. E. Stockdale et al., *Z. Phys.*, C27, 53, 1985.

[984] L. Stodolsky, *Phys. Rev. Lett.*, 34, 110, 1975.

[985] L. Stodolsky, *Phys. Rev.*, D36, 2273, 1987.

[986] L. Stodolsky, *Phys. Rev.*, D58, 036006, 1998, `hep-ph/9802387`.

[987] S. Stoica and H. V. Klapdor-Kleingrothaus, *Nucl. Phys.*, A694, 269–294, 2001.

[988] R. F. Streater and A. S. Wightman, *PCT, spin and statistics, and all that*, Addison-Wesley, 1989, Advanced Book Classics.

[989] J. C. Street and E. C. Stevenson, *Phys. Rev.*, 52, 1003–1004, 1937.

[990] L. E. Strigari, J. F. Beacom, T. P. Walker, and P. Zhang, *JCAP*, 0504, 017, 2005, `astro-ph/0502150`.

[991] R. G. Strom, 1990, NATO Advanced Study Institute on Neutron Stars: Their Birth, Evolution, Radiation and Winds, Erice, Sicily, Italy, September 5-17, 1988, p. 253.

[992] R. G. Strom, *Astron. Astrophys.*, 288, L1–L4, 1994.

[993] E. C. G. Sudarshan and R. E. Marshak, *Phys. Rev.*, 109, 1860–1862, 1958.

[994] F. Suekane, 2004, `hep-ex/0407016`, 5th Workshop on Neutrino Oscillations and their Origin (NOON04), Tokyo, Japan, Feb., 2004.

[995] F. Suekane et al., 2004, `physics/0404071`, "KEK-RCNP International School and Mini-Workshop for Scintillating Crystals and their Applications in Particle and Nuclear Physics". (Nov. 17-18, 2003, KEK, Japan).

[996] B. E. K. Sugerman et al., *Astrophys. J.*, 627, 888, 2005, `astro-ph/0502268`.

[997] J. Suhonen and O. Civitarese, *Phys. Rep.*, 300, 123–214, 1998.

[998] J. Suhonen, O. Civitarese, and A. Faessler, *Nucl. Phys.*, A543, 645–660, 1992.

[999] H. Suzuki and K. Sato, *Prog. Theor. Phys.*, 79, 725, 1988.

[1000] M. Suzuki, *Phys. Lett.*, B592, 495–496, 2004, The Review of Particle Properties 2004.

[1001] G. 't Hooft, *Nucl. Phys.*, B35, 167–188, 1971.

[1002] G. 't Hooft, *Nucl. Phys.*, B33, 173–199, 1971.
[1003] G. 't Hooft and M. J. G. Veltman, *Nucl. Phys.*, B44, 189–213, 1972.
[1004] M. Takahara and K. Sato, *Mod. Phys. Lett.*, A2, 293, 1987.
[1005] E. Takasugi, *Phys. Lett.*, B149, 372, 1984.
[1006] G. A. Tammann, 1982, Supernovae: a Survey of Current Research, Cambridge, England, June 29-July 10, 1981, p. 371-403.
[1007] G. A. Tammann, W. Loeffler, and A. Schroder, *Astrophys. J. Suppl.*, 92, 487–493, 1994.
[1008] R. Tartaglia, *Nucl. Instrum. Meth.*, A461, 327–328, 2001.
[1009] A. C. Taylor et al., *Mon. Not. Roy. Astron. Soc.*, 341, 1066, 2003, astro-ph/0205381.
[1010] M. Tegmark et al., *Phys. Rev.*, D69, 103501, 2004, astro-ph/0310723.
[1011] M. Tegmark et al., *Astrophys. J.*, 606, 702, 2004, astro-ph/0310725.
[1012] M. Tegmark, *Phys. Scripta*, T121, 153, 2005, hep-ph/0503257, "Neutrino Physics", Proceedings of Nobel Symposium 129.
[1013] N. Terasawa and K. Sato, *Prog. Theor. Phys.*, 72, 1262–1265, 1984.
[1014] N. Terasawa and K. Sato, *Prog. Theor. Phys.*, 80, 468, 1988.
[1015] T. A. Thompson, A. Burrows, and P. A. Pinto, *Astrophys. J.*, 592, 434, 2003, astro-ph/0211194.
[1016] F. X. Timmes, R. Diehl, and D. H. Hartmann, 1997, astro-ph/9701242.
[1017] J. Tiomno and J. A. Wheeler, *Rev. Mod. Phys.*, 21, 153, 1949.
[1018] T. Tomoda, *Rept. Prog. Phys.*, 54, 53–126, 1991.
[1019] J. L. Tonry et al., *Astrophys. J.*, 594, 1, 2003, astro-ph/0305008.
[1020] S. Toshev, *Phys. Lett.*, B196, 170, 1987.
[1021] T. Totani, K. Sato, H. E. Dalhed, and J. R. Wilson, *Astrophys. J.*, 496, 216–225, 1998, astro-ph/9710203.
[1022] T. Totani, *Phys. Rev. Lett.*, 80, 2039–2042, 1998, astro-ph/9801104.
[1023] Y. Totsuka, *Rept. Prog. Phys.*, 55, 377–430, 1992.
[1024] V. Tretyak and Y. Zdesenko, *Atomic Data and Nuclear Data Tables*, 80, 83, 2002.
[1025] V. Trimble, *Rev. Mod. Phys.*, 54, 1183–1224, 1982.
[1026] V. Trimble, *Rev. Mod. Phys.*, 55, 511–563, 1983.
[1027] V. Trimble, *Rev. Mod. Phys.*, 60, 859–871, 1988.
[1028] V. Trimble, *Ann. Rev. Astron. Astrophys.*, 25, 425–472, 1987.
[1029] M. S. Turner, *Phys. Rev.*, D45, 1066–1075, 1992.
[1030] N. Ushida et al., *Phys. Rev. Lett.*, 57, 2897–2900, 1986.
[1031] J. W. F. Valle, *Phys. Lett.*, B199, 432, 1987.
[1032] V. Valuev, 1997, International Europhysics Conference on High-Energy Physics (HEP 97), Jerusalem, Israel, 19–26 Aug 1997.
[1033] S. van den Bergh, *Comments on Astrophys.*, 17, 125–130, 1993.
[1034] J. C. Van Der Velde et al., *Nucl. Instrum. Meth.*, A264, 28–31, 1988.
[1035] L. Verde et al., *Mon. Not. Roy. Astron. Soc.*, 335, 432–440, 2002, astro-ph/0112161.
[1036] J. D. Vergados, *Phys. Rep.*, 361, 1–56, 2002, hep-ph/0209347.
[1037] G. S. Vidyakin et al., *JETP Lett.*, 59, 390–393, 1994.
[1038] P. Vilain et al., *Phys. Lett.*, B335, 246–252, 1994.

[1039] P. Vilain et al., *Phys. Lett.*, B364, 121–126, 1995.
[1040] C. J. Virtue, *Nucl. Phys. Proc. Suppl.*, 100, 326–331, 2001, astro-ph/0103324.
[1041] P. Vogel, *Phys. Rev.*, D29, 1918, 1984.
[1042] P. Vogel and J. F. Beacom, *Phys. Rev.*, D60, 053003, 1999, hep-ph/9903554.
[1043] L. V. Volkova and G. T. Zatsepin, *Sov. J. Nucl. Phys.*, 14, 117, 1971.
[1044] M. I. Vysotsky, A. D. Dolgov, and Y. B. Zeldovich, *JETP Lett.*, 26, 188–190, 1977.
[1045] D. J. Wagner and T. J. Weiler, *Phys. Rev.*, D59, 113007, 1999, hep-ph/9801327.
[1046] R. V. Wagoner, W. A. Fowler, and F. Hoyle, *Astrophys. J.*, 148, 3–49, 1967.
[1047] R. M. Wald, *General Relativity*, The University of Chicago Press, 1984.
[1048] T. P. Walker, G. Steigman, D. N. Schramm, K. A. Olive, and H.-S. Kang, *Astrophys. J.*, 376, 51–69, 1991.
[1049] S. Wang et al., *Phys. Rev. Lett.*, 95, 011302, 2005, astro-ph/0505390.
[1050] X. Wang, M. Tegmark, B. Jain, and M. Zaldarriaga, *Phys. Rev.*, D68, 123001, 2003, astro-ph/0212417.
[1051] S. Weinberg, *Phys. Rev. Lett.*, 19, 1264–1266, 1967.
[1052] S. Weinberg, *Gravitation and Cosmology*, John Wiley, 1972.
[1053] S. Weinberg, *Phys. Rev. Lett.*, 43, 1566–1570, 1979.
[1054] S. Weinberg, *Phys. Rev.*, 112, 1375–1379, 1958.
[1055] S. Weinberg, *Phys. Rev.*, D22, 1694, 1980.
[1056] H. A. Weldon and A. Zee, *Nucl. Phys.*, B173, 269, 1980.
[1057] J. Wentz et al., *Phys. Rev.*, D67, 073020, 2003, hep-ph/0301199.
[1058] J. C. Wheeler, *Am. J. Phys.*, 71, 11, 2003, astro-ph/0209514.
[1059] H. E. White, *Introduction to Atomic Spectra*, McGraw-Hill Kogakusha, ltd., 1934.
[1060] M. J. White, D. Scott, and J. Silk, *Ann. Rev. Astron. Astrophys.*, 32, 319–370, 1994.
[1061] Widom, A., Srivastava, and Y. N., 1996, hep-ph/9605399.
[1062] F. Wilczek and A. Zee, *Phys. Rev. Lett.*, 43, 1571–1573, 1979.
[1063] D. H. Wilkinson, *Nucl. Phys.*, A377, 474–504, 1982.
[1064] R. G. Winter, *Lett. Nuovo Cim.*, 30, 101–104, 1981.
[1065] L. Wolfenstein, *Phys. Rev.*, D17, 2369, 1978.
[1066] L. Wolfenstein, *Phys. Rev.*, D18, 958–960, 1978.
[1067] L. Wolfenstein, *Phys. Lett.*, B107, 77, 1981.
[1068] L. Wolfenstein, *Phys. Rev. Lett.*, 51, 1945, 1983.
[1069] Y. Y. Y. Wong, *Phys. Rev.*, D66, 025015, 2002, hep-ph/0203180.
[1070] S. Woosley and H.-T. Janka, *Nature Phys.*, 1, 147–154, 2005, astro-ph/0601261.
[1071] S. E. Woosley, A. Heger, and T. A. Weaver, *Rev. Mod. Phys.*, 74, 1015–1071, 2002.
[1072] S. E. Woosley and T. A. Weaver, *Ann. Rev. Astron. Astrophys.*, 24, 205, 1986.
[1073] C. S. Wu, E. Ambler, R. W. Hayward, D. D. Hoppes, and R. P. Hudson, *Phys. Rev.*, 105, 1413–1414, 1957.

[1074] D.-d. Wu, 1985, LBL-19982.
[1075] D.-d. Wu, *Phys. Rev.*, D33, 860, 1986.
[1076] Z.-Z. Xing, *Int. J. Mod. Phys.*, A19, 1, 2004, hep-ph/0307359.
[1077] A. Yahil and G. Beaudet, *Astrophys. J.*, 206, 26–29, 1976.
[1078] S. Yamamoto et al., *Phys. Rev. Lett.*, 96, 181801, 2006, hep-ex/0603004.
[1079] T. Yanagida, 1979, Workshop on the Baryon Number of the Universe and Unified Theories, Tsukuba, Japan, 13–14 Feb 1979.
[1080] C.-N. Yang and R. L. Mills, *Phys. Rev.*, 96, 191–195, 1954.
[1081] J. Yoo et al., *Phys. Rev.*, D68, 092002, 2003, hep-ex/0307070.
[1082] G. Zacek et al., *Phys. Rev.*, D34, 2621–2636, 1986.
[1083] G. T. Zatsepin, *JETP Lett.*, 8, 205, 1968.
[1084] G. T. Zatsepin and V. A. Kuzmin, *Sov. Phys. JETP*, 14, 1294, 1962.
[1085] G. T. Zatsepin and V. A. Kuzmin, *JETP Lett.*, 4, 78–80, 1966.
[1086] Y. B. Zeldovich, *Mon. Not. Roy. Astron. Soc.*, 160, 1–3, 1972.
[1087] G. P. Zeller et al., *Phys. Rev. Lett.*, 88, 091802, 2002, hep-ex/0110059.
[1088] S. Zhekov et al., *Astrophys. J.*, 628, L127, 2005, astro-ph/0506443.
[1089] M. V. Zobeck, *Handbook of Space Astronomy and Astrophysics*, Cambridge University Press, 1990, 2nd Edition.
[1090] M. Zralek, *Acta Phys. Polon.*, B29, 3925, 1998, hep-ph/9810543.
[1091] P. Zucchelli, *Phys. Lett.*, B532, 166–172, 2002.

INDEX